Soybeans

Chemistry, Production, Processing, and Utilization

Soybeans

Chemistry, Production, Processing, and Utilization

Editors

Lawrence A. Johnson

Pamela J. White

Richard Galloway

UNITED SOYBEAN BOARD
Making Your Checkoff Pay Off.

AOCS PRESS
Urbana, Illinois

AOCS Mission Statement

To be a global forum to promote the exchange of ideas, information, and experience, to enhance personal excellence, and to provide high standards of quality among those with a professional interest in the science and technology of fats, oils, surfactants, and related materials.

AOCS Press, Urbana, IL 61802

ISBN 978-1-893997-64-6

Library of Congress Cataloging-in-Publication Data

Soybeans : chemistry, production, processing, and utilization / editors, Lawrence A. Johnson, Pamela J. White, Richard Galloway.

p. cm.
Includes bibliographical references and index.
ISBN 978-1-893997-64-6 (alk. paper)
1. Soybean I. Johnson, Lawrence Alan, 1947- II. White, Pamela J. III. Galloway, Richard.

SB205.S7S557 2008
633.3'4--dc22

2008005938

The paper used in this book is acid-free and falls within the guidelines established to ensure permanence and durability.

Contents

Preface

We are pleased to offer to students, researchers, industry practitioners, and all who are interested in the world's most versatile crop, the most complete and authoritarian book on soybeans: *Soybeans: Chemistry, Production, Processing and Utilization.* This is one of several books comprising the AOCS Monograph Series on Oilseeds published by AOCS Press of the American Oil Chemists' Society, which provides the latest and most comprehensive information on plant sources of fats, oils and protein meals of vital importance in feeding the world and providing the many biobased products we consume every day.

Ancient Chinese literature provides ample evidence that soybean was one of the first plants to be domesticated and cultivated for food. Today, soybeans are recognized to provide sources of functional foods and food ingredients with potential health benefits, possibly playing roles in preventing cardiovascular disease and cancer protection qualities. Soybeans are grown for both oil and protein. Indeed, no other widely grown crop is more versatile in providing food, feed, fuel, and biobased products. With energy prices again on the rise, soybeans will become even more important in providing the fuels and industrial products so important to maintaining our lifestyles. With soybeans, modern agriculture can indeed deliver both food and fuel. The advent of renewable fuels has radically altered how we use soybeans from just five years ago and the present book will bring the reader up to date with these and other major changes. Probably no other crop has been studied as much as soybeans and this book attempts to summarize that knowledge base. Soybeans are often referred to as the *miracle crop*, and if you doubt this notion, we think you will become convinced once reading this book.

We strove to make this book as complete as possible, with ample references to assist the reader in finding additional information on a particular topic. No other book focuses on all aspects of the soybean. We modeled some chapters after those included in *The Practical Handbook of Soybeans*, edited by David R. Erickson and published in 1995 by the AOCS Press. The present book was intended to be the *one-stop* reference on soybeans, providing information with broad appeal, yet with sufficient depth to meet the needs of both experts in the subject matter as well as individuals with cursory knowledge of the topic.

As we considered who should contribute to this book, we chose the most internationally recognized authorities on each chapter topic. Much to our surprise and relief, all our "first choices" for chapter authors enthusiastically agreed to assist with this project, for which we are very grateful. All chapters underwent multiple reviews. We gratefully acknowledge these authorities, noted on following pages, for their timely and rigorous reviews that made this book better. All our chapter contributors and reviewers aim to provide the most accurate and complete information to our readers.

Lastly, we are grateful to AOCS staff, especially Jodey Schonfeld and Brock Peoples, who guided the authors through the process, kept the editors "on track" and worked very hard to make this book a success.

Lawrence A. Johnson
Pamela J. White
Richard Galloway

1

The History of the Soybean

Theodore Hymowitz
Department of Crop Sciences, University of Illinois, Urbana-Champaign,IL 61801

Introduction

The soybean [*Glycine* max (L.) Merr.], together with wheat [*Triticum aestivum* L.], maize [*Zea mays* L.], rice [*Oryza sativa* L.], barley [*Hordeum vulgare* L.], sugarcane [*Saccharum officinarum* L.], sorghum [*sorghum bicolor* (L.) Moench], potato [*Solanum tuberosum* L.], oats [*Avena sativa* L.], cassava [*Manihot esculenta* Crantz], sweet potato [*Ipomoea batatas* (L.) Lam.], and sugar beet [*Beta vulgaris* L.], are the principal food plants for humans (Harlan, 1992; Kasmakoglu, 2004).

Of the food plants, the soybean is unique in that the traditional foods in Asia made from the soybean (e.g., tofu, miso, and soy sauce) bear no semblance to or association with the crop growing in the field. The word *soy* comes from the Japanese word *shoyu* and first appeared in a Japanese dictionary published in 1597 (Shurtleff & Aoyagi, 1983). The popularity of tofu (bean curd) in China took place during the latter half of the Song Dynasty (960–1279 CE) (Shinoda, 1971). Miso is fermented soybean paste that originated in China around the first century BCE. Today, Westerners refer to it by its Japanese name (Shurtleff & Aoyagi, 1983). The Chinese word for soy sauce is *jiang-you.* Supposedly, it originated prior to the Zhou Dynasty (before 211 BCE) (Shurtleff & Aoyagi, 1983).

In the West, the two main products of the soybean are seed oil and the protein-containing meal. Soybean seeds contain 18–23% oil and 38–44% protein on a moisture-free basis. The oil is converted to margarine, mayonnaise, shortening, salad oils, and salad dressings. The meal is used primarily as a source of high-protein feeds for the production of pork, poultry, eggs, fish, beef, and milk. The soybean protein also is used in the form of protein concentrates and isolates, and texturized protein for human consumption (Hymowitz & Newell, 1981). Today, soy is taken for granted without appreciable forethought as to by whom, when, where, and how the soybean was domesticated in China for human use; by whom and when the soybean was disseminated throughout the world; and where the wild relatives of the soybean are and can they be exploited for the development of improved cultivars (Hymowitz, 2004).

Unfortunately, the popular literature concerned with the historical development of the soybean is fraught with errors and misconceptions that keep recycling from one publication or Web site to another without proper documentation (Hymowitz & Shurtleff, 2005).

In the past, studies on the domestication of the soybean were extremely difficult for two main reasons: i) the soybean is autochthonous to the Orient, where Western scientists were at a linguistic disadvantage with respect to historical records. However, in the past 40 years, classical Chinese works were translated into English; establishment of international soybean symposia (e.g., the World Soybean Congress) enables Chinese and Western academicians interested in soybean history to meet and discuss common issues on a regular basis; and lastly, molecular studies on soybean germplasm resources are beginning to answer questions that were not asked previously; and ii) many libraries were loathe to permit research scholars to handle fragile pages of archived manuscripts, books, and newspapers. However, today commercial companies scan and digitize many key documents and place them on the Internet, and these documents are available on commercial and public Web sites, especially at large research institutions.

This chapter attempts to combine information from many disciplines to establish a solid foundation for understanding the history of the soybean.

The Genus *Glycine* and its Immediate Allies

The genus *Glycine* Willd. is a member of the family Fabaceae/Leguminosae, subfamily Papilionoideae, and tribe Phaseoleae. The Phaseoleae is the most economically important tribe. It contains members that have considerable importance as sources of food and feed, for example, *Glycine max*—soybean; *Cajanus cajan* (L.) Millsp.—pigeon pea; *Lablab purpureus* (L.) Sweet—hyacinth bean; *Phaseolus* spp.—common bean, lima bean, tepary bean; *Psophocarpus tetragonolobus* (L.) DC.—winged bean; and *Vigna* spp.—azuki bean, cow pea, and Bambarra groundnut (Hymowitz & Singh, 1987).

Within the tribe Phaseoleae, Lackey (1977a) recognized 16 genera of the subtribe Glycininae, which he subdivided into two groups, Glycine and Shutaria, based upon morphological alliances. The Glycine group is distributed in the Old World with the exception of *Teramnus*, which has a pantropical distribution. The Shuteria group represents all of the other Glycininae. Polhill (1994) transferred *Calopogonium* and *Pachyrhizus* from the subtribe Diocleinae *sensu* Lackey (1977a) to Glycininae and reorganized 18 genera within Glycininae (see Table 1.1.).

Lee and Hymowitz (2001) studied the phylogenetic relationships among 13 genera of the subtribe Glycininae inferred from chloroplast DNA *rps*16 intron sequence variation. Phylogenies estimated using parsimony and neighbor-joining methods revealed that: (a) the genera *Teramnus* and *Amphicarpea* are closely related to *Glycine* and (b) the genus *Pueraria* regarded as closely related to the genus *Glycine* is not

Table 1.1. Genera, Number of Species, 2*n* Number, and Geographical Distribution in the Sub-tribe Glycininae[a]

Genus	No. of Species	2*n*	Geographic Distribution
Amphicarpaea	4	20, 22, 40	Asia, Africa and North America
Calopogonum	9	36	South and Central America
Cologonia	9	44	Central and S. America, Mexico
Dumasia	10	-	Asia, Africa
Diphyllarium	1	20	Indochina
Eminia	4	22	Tropical Africa
Glycine	25	38, 40, 78, 80	Asia, Australia
Mastersia	2	22, 44	Indo - Malaya
Neonotonia	2	22	Africa, Asia
Nogra	4	22	Asia
Pachyrhizus	5	22	Neotropics
Pseudeminia	4	22	Tropical Africa
Pseudovigna	2	22	Tropical Africa
Pueraria	18	22	Asia
Shuteria	4	22	Indo-Malaya
Sinodolichos	2	-	Asia
Teramnus	9	28	Pantropical
Teyleria	3	44	Asia

[a]Adapted from Lackey (1977a) and Polhill (1994).

monophyletic and should be divided into at least four genera, an idea previously supported by Lackey (1977a).

Pueraria montana var. lobata (Willd.) Maesen and A.M. Almeida (ILDIS, 2006) commonly is known as kudzu. These days it thrives as a weed throughout the southeastern part of the United States. Kudzu also acts as an alternate host for the economically important pathogen *Phakopsora pachyrhizi* Syd. The fungal pathogen known as *soybean rust* over winters on kudzu in frost-free environments along the U.S. Gulf Coast. It was first identified in the continental United States in 2004. Soybeans are very susceptible to soybean rust and, if infected and left untreated, the plants quickly defoliate and die. How much damage will occur to the soybean crop in the future by the pathogen is uncertain.

The Taxonomic History of the Genus *Glycine*

Glycine has a confused taxonomic history, which dates back to the time of its first inception. The name *Glycine* was originally introduced by Linnaeus in the first edition of his *Genera Plantarum* (Linnaeus, 1737), and is based on *Apios* of Boerhaave (Linnaeus, 1754). *Glycine* is derived from the Greek *glykys* (sweet) and probably refers to the sweetness of the edible tubers produced by *G. apios* L. (Henderson, 1881), now *Apios americana* Medik. In the *Species Plantarum* of 1753, Linnaeus listed eight *Glycine* spp. (Table 1.2.). All of these were subsequently moved to other genera, although *G. javanica* remained as the lectotype for the genus until 1966 (Hitchcock and Green, 1947). Thus, when *G. apios* became *A. americana*, the original justification for the name *Glycine* was removed from the genus. Therefore, the Greek *glykys* does not refer to any of the current *Glycine* species (Hymowitz & Singh, 1987).

The cultivated soybean was described by Linnaeus in 1753 as both *Phaseolus max*, based on specimens that he saw, and *Dolichos soja*, which he compiled from the descriptions of other writers. Later this gave rise to a great deal of confusion concerning the correct nomenclature of the soybean. Linnaeus apparently had the soybean in mind when he described *D. soja*, but, although *P. max* was based on actual specimens of the soybean, Linnaeus apparently intended the name to apply to the mung bean of India (Piper, 1914; Piper & Morse, 1923). It was not until several years later that he obtained seed of *D. soja* and grew the plants at Uppsala, Sweden. Only then was he able to see that *P. max* and *D. soja* were the same plant and that the mung bean was still without a name. Thus, in *Mantissa Plantarum* published in 1767, Linnaeus described the mung bean for the first time under *P. mungo* (Hymowitz & Newell, 1981).

Table 1.2. The Species of *Glycine* According to Linnaeus (1753) and Their Subsequent Classification[a]

Glycine Species	Currently
Apios	Apios
Frutescens	Wisteria
Abrus	Abrus
Tomentosa	Rhynchosia
Comosa	Amphicarpeae
Javanica	Neonotonia
Bracteata	Amphicarpeae
Bituminosa	Fagelia

[a]Adapted from Hymowitz and Singh (1987) and Lackey (1977b).

Since then, the correct nomenclature for the soybean has been the subject of much debate (Lawrence, 1949; Paclt, 1949; Piper, 1914; Piper & Morse, 1923; Ricker & Morse, 1948). Currently the combination *G. max* proposed by Merrill in 1917 is widely accepted as the valid designation for the soybean.

According to Bentham, by the time of De Candolle's *Prodromus* in 1825, "the genera *Glycine* and *Dolichos* had become the receptacle for all the Phaseoleae, which had no striking character to distinguish them" (Bentham, 1865). This led to an enormous proliferation of species attributed to *Glycine*, such that 286 species were eventually listed in *Index Kewensis*, with additional subspecies and taxonomic varieties bringing the total to 323 (Hermann, 1962). Bentham arranged the genus into three sections containing 11 species (Bentham, 1864, 1865): *Leptolobium* that comprised six species of Australian origin; *Johnia* that included *G. javanica*, the sole remaining Linnaean species of African and Asian origin; and *Soja* that included the cultivated soybean.

Hermann (1962) published a revision of the genus *Glycine* and its allies. He brought together the pertinent literature on *Glycine* nomenclature and listed those species that were published as *Glycine* in the past but later were excluded from the genus. According to his classification, *Glycine* consists of three subgenera: (i) *Leptocyamus* (Benth.) F.J. Herm., which includes six primarily Australian species; (ii) *Glycine;* and (iii) *Soja* (Moench) F.J. Herm., composed of the soybean and its wild annual counterpart described as *G. ussuriensis* by Regel and Maack (1861). In addition, Hermann found that name changes had to be made because of earlier homonyms. Thus, *G. sericea* became *G. canescens, G. tomentosa* became *G. tomentella,* and variety *latifolia* of *G. tabacina* was no longer considered distinct (Hymowitz & Singh, 1987).

Further revision became necessary when Verdcourt (1966) chanced to examine Linnaeus's specimen of *G. javanica* during the preparation of *Flora of Tropical East Africa.* He discovered that the type specimen was not *G. javanica* but rather a *Pueraria* with an abnormal inflorescence. To avoid major alterations in nomenclature of economically important legume genera, Verdcourt proposed that the name *Glycine* be conserved from a later author, Willdenow (1802), and that *G. clandestina* should become the type for the genus. Thus, the original type specimen became a synonym of *Pueraria montana* (Lour.) Merr. However, all those plants previously regarded as *G. javanica* L. were thus without a name, and for these Verdcourt adopted the name *G. wightii* (R. Grah. Ex Wight and Arn.) Verdc. Verdcourt also altered the subgeneric names to reflect the change in type. In addition, Verdcourt apparently overlooked the possibility that *Soja* Moench (1794) had priority over Willdenow (1802). Therefore, Lackey (1977b) proposed to conserve the generic name *Glycine* Willdenow over *Soja* Moench.

In 1970, Verdcourt proposed that *G. soja* is the valid designation of the wild annual relative of the soybean since Siebold and Zuccarini described it in 1846 as a new species and not based on *D. soja* L. Therefore, *G. soja* predates *G. ussuriensis* Regal and Maack of 1861 (Verdcourt, 1970).

In 1977, Lackey proposed the removal of *G. wightii* from the genus and suggested a new designation *Neonotonia wightii* (R. Grah. Ex Wight and Arn.) Lackey (Lackey, 1977a, 1977b). Thus, the last Linnaean *Glycine* was removed from the genus.

Since 1976, plant taxonomists have described 17 additional perennial *Glycine* species. This was due primarily to extensive plant exploration activities undertaken by U.S. and Australian scientists (e.g., Anonymous, 1988; Brown et al., 2002; Brown et al. 1985; Hymowitz, 1982, 1989, 1998; Hymowitz & Newell, 1981; Newell, 1981; Pfeil & Craven, 2002; Pfeil et al., 2001, 2006; Tindale, 1984, 1986a, 1986b; Tindale & Craven, 1988, 1993).

The genus *Glycine* Willd., as currently delimited, is divided into two subgenera *Glycine* and *Soja* (Moench) F.J. Herm. (Table 1.3.). The subgenus *Glycine* comprises 23 wild perennial species. The subgenus *Soja* includes the cultigen *G. max* (L.) Merr. and its annual wild counterpart, *G. soja* Sieb and Zucc. Hymowitz (2004) and Hymowitz and Singh (1987) presented tables showing the evolution of *Glycine* nomenclature.

Various breeding programs (Stalker, 1980) have effectively employed wild relatives of crop plants as sources of genetic diversity. From a taxonomic point of view the 23 perennial members of the subgenus *Glycine* are candidates for gene exchange with the soybean and therefore potentially useful for broadening the germplasm base of the crop (Hymowitz, 1998). For example, investigations show that the wild perennial *Glycine* species carry resistance to diseases such as soybean rust (*Phakopsora pachyrhizi* Sydow) (Schoen et al., 1992), soybean brown spot (*Septoria glycines* Hemmi.) (Lim & Hymowitz, 1987), powdery mildew (*Microsphaera diffusa* Cke. and Pk.) (Mignucci & Chamberlain, 1978), phytophthora root rot (*Phytophthora sojae* H.J. Kaufmann and J.W. Gerdemann (Kenworthy, 1989), white mold (*Sclerotinia sclerotiorum* L.b. De Bary) (Hartman et al. 2000), sudden death syndrome [(*Fusarium solani* (Mart.) Sacc.)] (Hartman et al., 2000), tobacco ringspot (Orellana, 1981), yellow mosaic

Table 1.3. The Genus Glycine, 3-Letter Code, 2*n* Number, Genome, and Distribution[a]

Subgenus Glycine	Code	2n	Genome	Geographic Distribution
1. G. *albicans* Tind. And Craven	ALB	40	I	Australia
2. G. *aphyonota* B. Pfeil	APH	40	?	Australia
3. G. *arenarea* Tind.	ARE	40	H	Australia
4. G. *argyria* Tind.	ARG	40	A	Australia
5. G. *canescens* F.J. Herman	CAN	40	A	Australia
6. G. *clandestina* Wendl.	CLA	40	A	Australia
7. G. *curvata* Tind.	CUR	40	C	Australia

Table 1.3., cont. The Genus Glycine, 3-Letter Code, 2*n* Number, Genome, and Distribution[a]

Subgenus Glycine	Code	2n	Genome	Geographic Distribution
8. *G. cyrtoloba* Tind.	CYR	40	C	Australia
9. *G. falcata* Benth.	FAL	40	F	Australia
10. *G. gracei* B.E. Pfeil and Craven	GRA	40	A	Australia
11. *G. hirticaulis* Tind. And Craven	HIR	40 80	H ?	Australia Australia
12. *G. lactovirens* Tind. And Craven	LAC	40	I	Australia
13. *G. latifolia* (Benth.) Newell and Hymowitz	LAT	40	B	Australia
14. *G. latrobeana* (Meissn.) Benth.	LTR	40	A	Australia
15. *G. microphylla* (Benth.) Tind.	MIC	40	B	Australia
16. *G. montis-douglas* B.E. Pfeil and Craven	MON	40	?	Australia
17. *G. peratosa* B. Pfeil and Tind.	PER	40	A	Australia
18. *G. pindanica* Tind. And Craven	PIN	40	H	Australia
19. *G. rubiginosa* Tind. and B. Pfeil	RUB	40	A	Australia
20. *G. stenophita* B. Pfeil and Tind.	STE	40	B	Australia
21. *G. syndetika* B.E. Pfeil and Craven	SYN	40	A	Australia
22. *G. tabacina* (Labill.) Benth.	TAB	40 80	B Complex	Australia Australia, W.C. and S. Pacific Islands
23. *G. tomentella* Hayata	TOM	38	E	Australia
		40	D	Australia, PNG
		78	Complex	Australia, PNG
		80	Complex	Australia, PNG, Indonesia, Philippines, Taiwan
Subgenus *Soja* (Moench) F.J. Herm.				
24. *G. soja* Sieb. and Zucc.	SOJ	40	G	China, Japan, Korea, Russia, Taiwan (Wild Soybean)
25. *G. max* (L.) Merr.	MAX	40	G	Cultigen (Soybean)

[a]Adapted from Hymowitz (2004) and Pfeil et al. (2006).

virus (Singh et al., 1974), alfalfa mosaic virus (Horlock et al., 1997), and soybean cyst nematode (*Heterodera glycines* Ichinohe) (Riggs et al., 1998). The wild perennial *Glycine* species are tolerant to certain herbicides (Loux et al., 1987; Hart et al., 1988), salt-tolerant (Hymowitz et al., 1987) and lacking the Bowman-Birk Inhibitor (Domagalski et al., 1992) the p34 allergen (Joseph et al., 2006), and lectin (Mettu et al., 1995). The Bowman-Birk Inhibitor, the p34 allergen, and lectin are biologically active components of seed within the *Glycine* species.

Thus far, only Singh et al. (1990, 1993) have reported successful backcrossed-derived fertile progeny from the soybean and a wild perennial relative, *G. tomentella*.

Geographical Origin of the Genus *Glycine*

"The base number for Phaseoleae is almost certainly $x = 11$, which is also probably basic in all tribes" (Goldblatt, 1981). Goldblatt also pointed out that aneuploid reduction ($x = 10$) is prevalent throughout the Papilionoideae. Previously, Darlington and Wylie (1955) proposed that an $x = 10$ basic chromosome number for the cultivated soybean. Based upon the above views and on recent taxonomic, cytological and molecular systematics research on the genus *Glycine* and allied genera, a putative ancestor of the genus *Glycine* with $2n = 20$ arose in Southeast Asia (Kumar & Hymowitz, 1989; Lee & Hymowitz, 2001; Singh & Hymowitz, 1999; Singh et al., 2001). However, such a progenitor is either extinct or yet to be collected and identified in Southeast Asia (Fig. 1.1).

Singh et al. (2001) assume that the path of migration northward (Fig. 1.1) from the ancestral region to China from a common progenitor is: wild perennial ($2n = 4x = 40$, unknown or extinct) → wild annual ($2n = 4x = 40$; *G. soja)* → soybean ($2n = 4x = 40$; *G. max*, cultigen). All of the *Glycine* species studied by Singh and Hymowitz (1985a) exhibited diploid-like meiosis, are primarily inbreeders, and produce cleistogamous seed.

Allopolyploidization (interspecific hybridization followed by chromosome doubling) via unreduced gametes probably played a major role in the speciation of the genus *Glycine*. This assumption infers that the 40-chromosome *Glycine* species and the 80-chromosome *G. tabacina, G. tomentella,* and *G. hirticaulis* are tetraploid and octoploid, respectively. The expression of four rDNA loci in *G. curvata* and *G. cyrtoloba* (Singh et al., 2001) strongly supports a hypothesis of allotetraploid origin that was originally proposed on the basis of cytogenetic evidence (Singh & Hymowitz, 1985a, 1985b; Xu et al., 2000) and molecular studies (Lee & Verma, 1984; Shoemaker et al., 1996).

Hymowitz et al. (1990), based upon cytogenetic studies, hypothesized that the disjunct allopolyploid distribution of *G. tabacina* and *G. tomentella* between Australia and the islands of the west-central Pacific region was due to long-distance dispersal by migrating shore birds. That hypothesis was verified by Doyle et al. (1990a, 1990b) who examined chloroplast DNA and histone H3-D polymorphism patterns within the *G. tabacina* polyploidy complex.

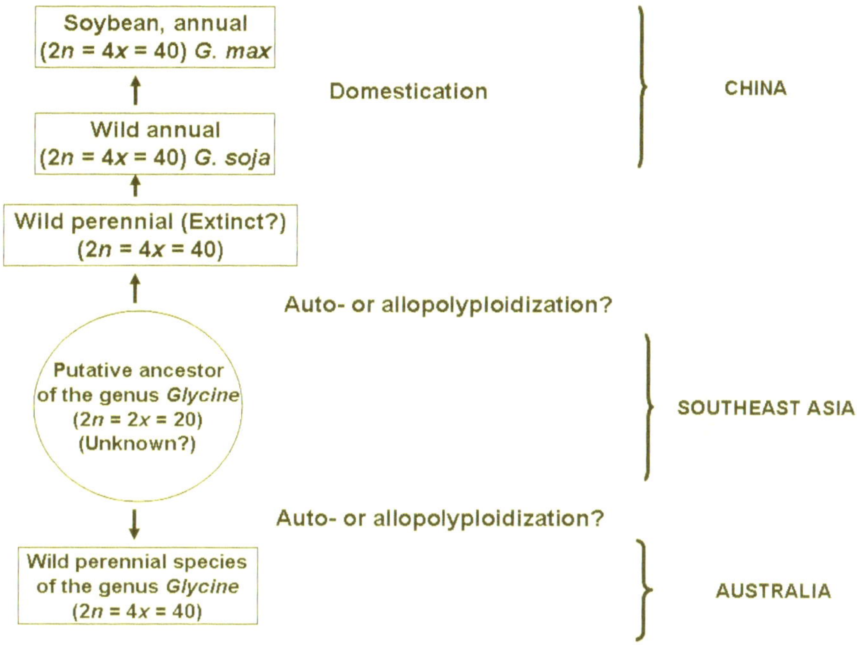

Fig. 1.1. Geographical origin of the genus *Glycine*. Adapted from Hymowitz, 2004.

Domestication of the Soybean

The farmers of China domesticated the soybean. Linguistic, geographical, and historical evidence suggest that the soybean emerged as a domesticate during the Zhou Dynasty (ca. 1125 to 256 BCE) in the eastern half of north China. Domestication is a process of trial and error and not a time-datable event. In the case of the soybean, this process probably took place during the Shang Dynasty (ca. 1766 to ca. 1125 BCE) (Bray, 1984; Ho 1969, 1975; Hymowitz, 1970; Hymowitz & Newell, 1980). The movement of the soybean land races within China is associated with the development and consolidation of territories and the degeneration of Chinese dynasties (Ho, 1969). In addition, the new dynasties arose either in the north or northwest China. Thus, the movement of people and cultivated plants in China primarily was from the north to the south.

Unfortunately, soybean historical literature and soybean-associated Internet Web sites are replete with factual errors. The misinformation keeps recycling from one publication or Web site to another without documentation. Attempts to correct these errors are met with stiff resistance (Hymowitz & Shurtleff, 2005). Apparently, myths and legends make better stories than the truth; for example, Morse (1950) reported that the first written record of the soybean is in the book *Pen Ts'ao Kong Mu,* which is a description of plants of China by Emperor Shennong in 2838 BCE. According to Chinese mythology, Emperor Shennong was the Father of Agriculture, the God of Wind, and the Patron of Pharmacists. Supposedly, Shennong taught his subjects how to use the plow and sow grain, and he kept people healthy by prescribing for their ailments natural herbs that had medicinal value. He is often portrayed having the head of an ox and the body of a man. No fewer than six different years (i.e., 2838, 2828, 2737, 2700, 2448, and 2383 BCE) are calculated as the publication date for Shennong's book (Hymowitz, 1970).

We must dispel the enchanting myths about Emperor Shennong because they appear to be fabrications of ethnocentric Han historians (Western Han Dynasty: 206 BCE–24 CE; Eastern Han Dynasty 25 BCE–220 CE), as is the emperor himself. For example, none of Professor Ho's carefully documented works mentions Shennong (Ho, 1969, 1975). In discussing the antiquity of the soybean, Ho comments that the beginnings of the domestication of the soybean may never be exactly known. We know only that the plant was probably first domesticated successfully in the eastern half of North China, probably not too much earlier than the eleventh century B.C.

Hymowitz and Shurtleff (2005) traced the origin of the Emperor Shennong soybean myth in the English language. The earliest citation seen was by Wells (1861). He referred to Shennong as the fabled farmer of agriculture. However, he did not link Shennong to the soybean. Rein (1889) noted that Shennong spread the practice of agriculture about the year 2700 BCE. This is the earliest English document suggesting that the soybean was one of the five major crops of China. The connection between the soybean and Shennong traced back to the 1893 publication of Breschneider's classical book on Chinese botany. Within the past 110 years, a great deal of archaeological, historical, and ethnobotanical research has debunked the authenticity of the Emperor Shennong, the date of his reign, and his relationship to the soybean. Amazingly, the myth of Emperor Shennong is erroneously cited in the soybean literature as a fact. In addition, statements such as "the soybean is one of the oldest cultivated crops" or "it has been cultivated for over 5000 years" are incorrect (Hymowitz & Shurtleff, 2005).

Dissemination of the Soybean

The history of the dissemination of the soybean is, of course, only partially known. We must recognize that it is not uncommon for traders, travelers, emissaries, and government officials to leave few or no records. Then again, "it is foolish to believe that a certain plant can be introduced into a new area only once and then only by a certain route" (Ho, 1955).

From about the first century A.D. to the Age of Discovery (fifteenth–seventeenth century A.D.), soybeans were introduced into many Asian countries with land races eventually developing in Indonesia, Japan, Malaysia, Myanmar, Nepal, North India,

Philippines, Thailand, and Vietnam. These regions compose a secondary gene center. The movement of the soybean throughout the period was due to the establishment of sea and land trade routes, for example, the Silk Road (Boulnois, 1966); the migration of certain tribes from China, for example, the Thais (Prince Dhaninavat, 1961); and the rapid acceptance of the seeds as a staple food by other cultures, for example, the Indonesians (Hymowitz, 1990; Hymowitz & Newell, 1980).

Soybean seed protein extracts from over 2,000 accessions obtained from 16 Asian countries or regions were analyzed by polyacrylamide gel electrophoresis (Hymowitz and Kaizuma, 1979, 1981) to determine the allelic distribution of the Kunitz trypsin inhibitor and β-amylase. By combining the frequency of the alleles in various populations with available historical, agronomic and biogeographical literature, they developed hypotheses concerning the dissemination of the soybean from China (the primary gene center) to other countries or regions in Asia (the secondary gene centers). The dissemination concept was based partly upon the pioneer studies of Nagata (1960), who used primarily physiological and morphological data to point out possible paths of dissemination of the soybean from China to the rest of Asia.

The suggested paths of dissemination of the soybean from the eastern half of north China to other regions in Asia are shown in Figure 1.2 and summarized below:

1. The soybeans grown in the former U.S.S.R. (Asia) came from Northeast China.

2. The soybeans grown in Korea are derived from two or three possible sources—Northeast China, North China, and the introduction of soybeans from Japan especially in the southern part of Korea.

3. The soybeans grown in Japan were derived from the intermingling of two possible sources of germplasm—Korea and Central China. The first points of contact were probably in Kyushu, and from there the soybean moved slowly northward to Hokkaido. In addition, the soybean moved southward from Kyushu to the Ryukyu Islands, where they came in contact with the soybeans moving northward from Taiwan. The earliest Japanese reference to the soybean is in KoJiKi or "Records of Ancient Matters," which was published in 712 CE (Chamberlain, 1906).

4. The soybeans originally grown in Taiwan came from Coastal China.

5. The germplasm source for the soybeans grown in Southeast Asia is Central and South China.

6. The soybeans grown in the northern half of the Indo-Pakistan subcontinent came from Central China.

Fig. 1.2. Paths of migration of the soybean from China. Adapted from Hymowitz and Kaizuma (1979; 1981).

7. The soybeans grown in Central India were introduced from Japan, South China, and Southeast Asia.

Early Western Knowledge of the Soybean

Pre-Marco Polo

According to Harlan (1992), "On the whole, Far Eastern agriculture may be characterized as introverted with very little dispersal until well into modern historical times, and many crops did not move out until the arrival of European shipping in the late fifteenth century and early sixteenth century A.D." However, some exceptions existed; for example, Greek Theophrastus (370 to ca 295 BCE) described rice (Hort, 1919). According to Laufer (1919), silk dealers may have transmitted the peach and the apricot, first to Iran (in the second or first century BCE) and then to Greece and Rome (in the first century CE).

In the first two centuries of the Common Era exploration by land in Asia was very slow and in one direction. Chinese goods, such as silk, reached the West but in limited quantities. The declining Roman Empire and the early Byzantine Empire saw very little exploration take place except for the opening up of the Silk Road north of the Caspian Sea. The Silk Road was not a single road. Rather, it was an interconnected series of ancient trade routes through the Asian continent linking Xi'an, China, with Asia minor (Turkey) (Boulnois, 1966). The rise of Islam in the seventh century made travel from Europe to Asia via land routes very dangerous. Thus, this period of time (eighth to fifteenth centuries) is described as a period of scientific stagnation in Europe (Cary & Warmington, 1929; Wright, 1925).

Marco Polo Era (Thirteenth to Fifteenth Centuries CE)

We must consider Marco Polo (Sept. 15, 1254–Jan. 8, 1324 CE), a Venetian merchant, the first botanical explorer of the modern era. For 17 years, Kublai Khan employed him. Although Polo was a keen observer of Chinese traditions and described many plants and animals utilized in China, he made obvious omissions such as tea, fishing with cormorants, footbinding, chop sticks, and soy. Polo probably ate soy products but was unable to associate the food products made from soy with the crop growing in the fields (Penzer, 1929; Olschki, 1960; Rugoff, 1961).

At least five European contemporaries of Marco Polo visited China. They were John of Pian de Capine [1246 CE], William of Rubruck [1254 CE], John of Monte Corvino [1305 CE], Odoric of Pordenone [1323 ce], and John de Marginolli [1342 CE]. All were Franciscans. Their mission in China was to try to convert the royal family and save souls. Except for William, the Franciscan missionaries hardly mentioned plants in the course of their travels (Bretschneider, 1962; Komroff, 1928; Olschki, 1943; Yule 1866; 2002). William of Rubruck was an exception. A keen observer of Chinese culture and foods consumed, he was the first Westerner to suggest the soybean or soy foods. In 1254 he wrote, "The monk said he only ate on Sunday, when this lady sent him a meal of cooked dough with vinegar to drink." Rockhill (1900), the translator, noted that the dish called *mien* by the Chinese is the most common article of diet in northern China and Mongolia. "The vinegar or soy is used to season the water in which the paste has been cooked and is drunk as a soup." Rubruck never mentioned soy. However, the dish called *mien* as noted by Rockhill is often flavored with soy sauce. Thus, this is an indirect mention of the use of a soy product.

Another contemporary of Marco Polo who traveled to China in 1325 was Abu Abdullah Muhammad Ibn Battuta (Ibn Battuta), a Moroccan Islamic scholar. Unfortunately, he makes no mention of soybean in the accounts of his journeys (Yule, 1866).

In 1589, when John Huyghen Van Linschoten came across a banana, he called it an Indian fig (Burnell & Tiele, 1855). In other words, he described a new plant using terminology available to a European. Likewise, a Westerner seeing the soybean in the field might describe it as peas or beans. However, the products of the soybean, such as tofu, soy sauce, or soy milk, would remain unknown.

In Yule (1866) three sentences appear circa 1330 that appear to describe soybean products. "In the empire of Boussaye aforesaid growth a certain manner of trees which from their sap are of great help to the folk of the country. For there be some of them which from their bark give forth a white liquor like milk, sweet, savory, and abundant (*soy milk* [italics added by author]), and the people of the country make drink and food of it as if it were goat's milk (*tofu* [italics added by author]) and that right gladly. And when they cut those trees anywhere, whether it be in the branches or elsewhere, they give fourth where they were cut a manner of juice in great plenty which juice hath the colour and savour of wine (*soy sauce* [italics added by author])." The above suggests a garbled mistranslation of the soybean probably from Chinese to Latin to French to English. This perhaps is the second oldest citation found in Western literature about the soybean and/or soybean products. However, use caution in citing the above as it is a speculative guess by the author.

The Age of Discovery for the Soybean

In 1509, the Portuguese navigator Diego Lopez de Sequeira stepped upon the shores of Malacca (on the southwestern coast of the Malay Peninsula). That act established the possibility of trade by European countries with Asian countries, in particular, the French, English, and Dutch with Asian countries bypassing the slow overland routes. Over time, European countries established trading colonies or factories from India to China, and Japan to Indonesia. Employed at these colonies were well-trained individuals such as medical doctors and botanists, as well as the ever-present travelers and missionaries. These individuals published their logs, diaries and even books about their experiences and observations in the colonies. This resulted in the accumulation of knowledge about the use of the peculiar bean used to produce various food products (Boxer, 1953, 1967, 1968, 1979, 1988; Burnell & Tiele, 1885; Dulles, 1931; Eames, 1974; Ray, 1999; Wills, 1974). Note several examples given below.

Valignano (1954) was an Italian Jesuit priest who focused his attention on the need for European missionaries in Japan to learn Japanese. In 1583, among the foods he purchased for his provisions were rice, dried fish, and *miso.*

Francesco Carletti, the Florentine, who visited Nagasaki, Japan in 1597, wrote in his memoirs that the Japanese flavor fish dishes with a certain sauce called *misol* (*miso*) made from a bean that is grown in various localities (Carletti, 1964).

In 1613, John Saris was the captain of the *Clove* on the first English voyage to

Japan. In his log he wrote the following about the food habits of the Japanese: "Of cheese they have plenty. Butter they make none, neither will they eat any milk..." Almost certainly, he mistook tofu for cheese.

Boxer (1967) provides an account of a Yedo (Tokyo) jail by Spanish Franciscan Diego de San Francisco in 1615. "The official ration was a handful of rice daily. On the other hand, the guards could sometimes be bribed to allow prisoners' friends to smuggle a little rice, soy, or fish by way of supplementing the starvation diet."

John Nieuhoff noted that in 1656 the Dutch East India ambassadors, Peter de Goyer and Jacob de Keyzer, to the Emperor of China received daily as a part of their rations 5 tael (1 tael = ca. 40 g) of *mison* (*miso*). Their secretaries received daily as a part of their rations one measure of *taufoe* (*tofu*) and 4 tael of *mison* (*miso*) (Pinkerton, 1811).

In 1665, Friar Domingo Navarette described tofu as a common and cheap food of China. "They drew the milk out of the Kidney-Beans and turning it, make great Cakes of it like Cheeses... All the Mass is as white as the very Snow... Alone it is insipid, but very good dress'd as I say and excellent fry'd in Butter. It is incredible what vast quantities of it are consum'd in China, and very hard to conceive there should be such abundance of Kidney-Beans. That Chinese who has *Teu Fu* (*tofu*) herbs and rice, need no other sustenance to work..." (Cummins, 1962).

The Dutch East India Company exported from Japan soy sauce as early as 1673–1674 (Boxer, 1988). In 1673, the ship *In Laeren* carried 12 tubs of soy sauce as cargo, and in 1674 the ship *In Hasenburg* had an unlisted amount of soy. The Dutch had a trading monopoly with the Japanese from 1641 until 1853. The trade took place at Deshima, an artificial island in Nagasaki Bay. Due to the consequence of war with England and France in 1672, the Dutch ships from Japan proceeded to Batavia, and from there goods were shipped via the Malaccas to the British colonies in Bengal, Surat, etcetera. From the British colonies soy was shipped to London. Thus, the products of commerce such as soy sauce reached Europe before soybean seed. And the Dutch were primarily responsible for making soy sauce known to the Europeans (Burkill, 1935). Indeed, when soy sauce became an export to Europe, it became an immediate success. For example, King (1679) noted that when eating in London: "Mango and Saio are two sorts of sauces brought from the East Indies." (King, 1972). "Saio" almost certainly refers to soy sauce.

In 1689, interest in soy sauce extended to the English factory in Surat (then part of the Mughal Empire). Ovington (1929) spoke concerning English, Portuguese, and Indian styles of cooking, "Bambou and mangoe achan (pickle) and souy the choices of all sauces, are always ready to whet the appetite."

In 1688, Capt. William Dampier while visiting the Kingdom of Tonquin (Vietnam) made the following observation in his diary "...Nuke-mum (fish sauce)... a good Sauce for Fowls, not only by the Natives, but also by the Europeans, who esteem it equal with Soy. I have been told that Soy is made partly with a fishy Composition, and it seems most likely by the Taste: tho' a Gentleman of my Acquaintance, who was very intimate with one that sailed often from Tonquin to Japan, from whence the true Soy comes, told me, that it is made only with Wheat, and a sort of Beans mixt with Water and Salt."

By 1705, European pharmacologists were familiar with the soybean from Japan and its culinary value (Dale, 1705). Lockyer (1711) wrote that "Soy comes in Tubs from Jappan and the best Ketchup from Tonqueen…both are made and sold very cheap in China."

However, it was not until 1712, when Engelbert Kaempfer, who lived in Japan from 1690 to 1692 as a medical officer of the Dutch East India Company, published his book *Amoenitatum Exoticarum* that the Western world fully understood the connection between the cultivation of soybeans and its utilization as a food plant. Kaempfer's drawing of the soybean is accurate, and his detailed description of how to make soy sauce is correct.

By the 1750s, soy sauce was common in England. Cookbooks (Glasse, 1983) mentioned it, newspapers advertised it for sale (Watkinson, 1750), and by 1760 silversmiths handcrafted soy cruets (Hughes, 1955).

The earliest report seen in Western literature for the use of soybean seed for animal feed was by Le Comte (1697). "All the Northern and Western Provinces (in China) bear wheat, barley, several kinds of millet, and tobacco, with black and yellow pease, with which they feed horses as we do with oats." Bretschneider (1898) concurs that black and yellow peas to which Le Comte refers are varieties of the soybean.

Modern Dissemination of Soybean Seed

Dr. William Roxburgh, employed as the Director of the Honorable East India Company Botanic Garden near Calcutta, in his *Flora Indica* noted that in 1798 soybean seed received from the Moluccas (Indonesia) were planted in the garden.

For five years (1672–1677) Paul Hermann, an employee of the Dutch East India Company, collected plants on Ceylon (Sri Lanka). When he returned home, he became Professor of Medicine and Botany at Leiden. His *Musaeum Zelanicum,* published in 1717, contains the earliest documentation seen for soybean in Sri Lanka.

The first record by a European of soybeans in Indonesia is by George Everhard Rumphius (1628–1702), an employee of the Dutch East India Company (Merrill, 1917). His book, *Herbarium Amboinense,* published in 1747, 45 years after his death, was based on observations made by him between 1653 and 1670. Rumphius noted that the soybean was used both for food (tofu) and as a green manure.

The soybean reached Europe quite late. It must have reached the Netherlands before 1737 as Linnaeus described the soybean in *Hortus Cliffortianus,* which was based on plants cultivated in the garden at Hartecamp. In 1740, soybean seeds sent by missionaries in China were planted in the Jardin des Plantes, Paris, France. In 1790, soybeans were planted at the Royal Botanic Garden at Kew, England. In 1804, they were planted near Dubrovnik, Croatia, and prior to 1817 in the Vojvodina Region, Serbia. In the Netherlands, France, and England, the soybeans were grown for taxonomic or display purposes. However, the soybeans grown in Croatia and Serbia were harvested, cooked, mixed with cereal grain, and fed to chickens for increased egg production (Aiton, 1812; Buconjie n.d.; Djukic, 1975; Linnaeus, 1737; Paillieux, 1880).

The earliest seen report for soybean distribution in Canada was by T.V.P. (see Peticolas) of Mount Carmel, Ohio, in 1855. He reported that seeds were distributed from Texas to Canada. However, nothing is known as to who planted the seed, or where, and no results were reported. Thus, the first practical introduction of soybeans into Canada was by Zavitz in 1893 (Beversdorf, 1995). The Ontario Agricultural College employed him, and for 30 years he evaluated and selected soybean introductions for both fodder and grain production.

In 1882 D'Utra published the earliest confirmed report seen on the introduction of the soybean into South America. The Bahia School of Agriculture cultivated the soybean. Another early citation for the soybean was by Dafert (1893). The Agronomic Institute in Campinas, the State of Sao Paulo, Brazil, evaluated soybeans as a forage crop.

As in Europe, soy sauce reached the English Colonies in the New World prior to the introduction of soybean seed (New York Gazette, 1750). In 1765, Samuel Bowen introduced Chinese vetches (soybean) into the Colony of Georgia. He obtained the soybean seed in China (Hymowitz & Harlan, 1983). Henry Yonge, the Surveyor General of Georgia, planted soybeans on his farm at the request of Samuel Bowen in 1765 (Yonge, 1767). From 1766 on, Mr. Bowen planted soybeans on his property, "Greenwich," located in Thunderbolt, Georgia, a few kilometers east of Savannah (Hymowitz & Harlan, 1983). Today, the property is used as a city cemetery. The soybeans grown by Bowen were used to manufacture soy sauce and vermicelli (soy sprouts). In addition, he manufactured a sago powder substitute from sweet potatoes. The products were exported to England and sold in major cites along the Atlantic coast (Dunlap's Pennsylvania Packet, 1774; Newport Mercury, 1771; New York Gazette, 1777).

On July 1, 1767, Samuel Bowen received a patent, number 878, for his "new invented method of preparing and making sago, vermicelli, and soy from plants growing in America, to be equal in goodness to those made in the East Indies" (Woodcraft, 1854). Samuel Bowen was awarded a gold medal from the Society of Arts, Manufacturers, and Commerce and received a present of 200 guineas from King George III. In addition, Bowen sent soy sauce and soybeans to the American Philosophical Society in Philadelphia and was elected to membership of the society (Lesley, 1884). Unfortunately, when Sam Bowen died in London on December 30, 1777, his soybean enterprise in Georgia ended.

The second earliest document seen for the introduction of the soybean to North America was by Benjamin Franklin. In 1770, Franklin sent soybean seeds to his friend

John Bartram in Philadelphia (Smyth, 1907). John Bartram probably planted the soybean seed sent to him by Franklin and/or Bowen in his garden, which was situated on the west bank of the Schuylkill River below Philadelphia (Fox, 1919; Bartram, 1807; Bartram, 2004).

Dr. James Mease (1804) apparently is the first person in American literature to use the word *soybean*. Most probably he coined the word to refer to the bean from which soy sauce was produced.

For many years, Mease's 1804 soybean report was considered the earliest citation in American literature (Piper & Morse, 1916). However, the 1983 publication by Hymowitz and Harlan clearly demonstrated that the introduction of the soybean into the Colony of Georgia by Samuel Bowen in 1765 was 39 years earlier than the Mease publication. Yet, Web sites and soybean commodity literature continue to cite Mease's publication as the earliest introduction (Hymowitz & Shurtleff, 2005).

In 1851, the soybean was introduced to Illinois and subsequently throughout the U.S. Corn Belt. The introduction came about through a series of very unusual circumstances. In December 1850, the barque *Auckland* left Hong Kong for San Francisco carrying sugar and other general merchandise. About 500 miles off the coast of Japan, the ship came across a Japaneese junk foundering on the sea. The Japanese crew was removed from the junk and placed aboard the *Auckland*, which continued on to San Francisco. In San Francisco, the Japanese fishermen were not permitted to go ashore because of the possibility of spreading diseases. By coincidence, waiting for a passenger ship to take him back to Alton, Illinois, via the Panama overland route was Dr. Benjamin Franklin Edwards. Dr. Edwards examined the Japanese fishermen, declared them free of any contagious diseases, and received as a gift a packet of soybeans that he carried back to Alton. Mr. John H. Lea, an Alton horticulturist, planted the soybeans in his garden in the summer of 1851. In 1852, the multiplied soybeans were grown in Davenport, Iowa, by Mr. J.J. Jackson and in Cincinnati, Ohio, by Mr. A.H. Ernst. In 1853, Mr. Ernst distributed soybean seeds to the New York State Agricultural Society, the Massachusetts Horticultural Society, and the Commissioner of Patents. The two societies and the Commissioner of Patents sent soybean seeds to dozens of farmers throughout the United States (Hymowitz, 1986). Thus, by the end of 1854 the soybean seeds brought by Dr. Benjamin Franklin Edwards in 1851 from San Francisco to Alton, Illinois, were grown, disseminated, and evaluated by farmers in several states.

Amazingly, one of the Japanese fishermen rescued by the crew of the *Auckland* remained in the United States. As a 14-year-old he took the name Joseph Heco (ne Hizozaemon), learned to read and write English, and became a U.S. citizen. He wrote a book in 1895 that confirmed the *Auckland* incident from the Japanese point of view.

In 1854, when Commodore Matthew Perry's Expedition opened Japan to Western trade, the expedition's surgeon, Dr. Daniel Green, observed that the Japanese grew a peculiar kind of bean called the Japan pea (i.e., soybean) (Perry, 1856). In mid-1854, the expedition's agriculturist, Dr. James Morrow, obtained soybean seed and sent them to the Commissioner of Patents; subsequently the seeds were distributed to farmers (Browne, 1855; Cole, 1947). Thus, from 1855 onward, to distinguish between soybean seed sources in farmers' reports is difficult. Did their soybean seeds originate from the Illinois accession or the Perry Expedition? Perhaps they grew soybeans from both sources. Graff (1949) cited post-1854 soybean evaluation reports from Connecticut, Delaware, Indiana, Kentucky, Maryland, Missouri, New York, North Carolina, Ohio, Pennsylvania, and Virginia.

Because the Perry Expedition (1852–1854) is so well-documented, the soybeans sent from Japan to the United States received an enormous amount of publicity. On the other hand, time obscures the specific details concerning the earlier introduction of soybeans into Illinois by Dr. Benjamin Franklin Edwards in 1851.

In 1878, while in Europe, Dr. George H. Cook and James Nielson of the New Jersey Agricultural Experiment Station obtained soybean seed at the Bavarian Agricultural Experiment Station and at the Vienna Exposition. The seeds were planted at the College Farm in May 1879, and harvested in October. The results were encouraging. This is the first report of soybeans tested at a Land Grant institution in the United States (Cook, 1879). Within a short time, soybean seeds were introduced from Japan and China and grown by McBryde (Tennessee), Sturtevant (Cornell University), Brooks (Hatch, Massachusetts), and Georgeson (Kansas). During the last two decades of the nineteenth century, soybeans were grown at almost every agricultural station in the country. The crop was tested for use in pastures as hay, silage, and soiling, alone or in combinations with other crops. Feeding experiments were conducted with horses, poultry, sheep, cattle, and milk cows. All parts of the plant were chemically analyzed. Some experimenters lauded the value of the soybean while others considered it worthless (Brooks, 1890; Georgeson et al., 1890; McBryde, 1882; Sturtevant, 1883).

In 1888, in Germany, Hellriegal and Wilfarth demonstrated that legumes fix nitrogen when nodulated by a microorganism present in soil extracts. In 1893, W.P. Brooks then conducted what is a classic experiment. He placed never-before- cropped soil into pots and planted seed from three soybean cultivars originally from Japan. In one series of pots he added a pinch of dust collected from the floor where soybeans had been thrashed, and the other series of pots were his control. The results were striking. In the pots receiving a pinch of dust, the plants were greener, more vigorous, and the seed yields much larger than the controls. In addition, nodules were found on the roots of the plants that received the pinch of dust. Soil from Brook's experiment was sent to New Jersey and Kansas stations, and his results were confirmed. Commercial soybean inoculum was made available by 1905. This was the first major technological advance in the successful establishment of the soybean in North America.

In 1898, the Office of Foreign Seed and Plant Introduction was established within the USDA to centralize introduction activities. Introduced plants were assigned permanent numbers under the Plant Introduction (P.I.) designation system. The first soybean listed in the P.I. system was P.I. 480 from South Ussurie, Siberia. The seeds were received from Professor N.E. Hansen, of the South Dakota Agricultural College in March 1898 (Hymowitz, 1990).

Two major technological advances occurred during the first quarter of the twentieth century. In 1917, Osborne and Mendel demonstrated that unheated soybean meal is inferior in nutritional quality to properly heated soybean meal. Thus, the value of soybean seed meal as a feed and the potential for the development of a soybean processing industry were established.

In 1920, Garner and Allard recognized the significance of length of day in the flowering behavior of soybeans and termed the response *photoperiodism*. An understanding of the photoperiod in relation to cultivar adaptation is of extreme importance to the plant breeder. Today, in North America, soybeans are classified into 13 maturity groups (MG) based upon the effects of day length on timing of the appearance of first flowers. In Canada and northern parts of the United States, most cultivars are indeterminate and have relatively short crop durations; they are classified as MG 000, 00, and 0. In the central states, cultivars from MG II, III, IV, and V are grown. Those adapted to the subtropical and tropical zones are often determinate, have relatively long crop durations, and are classified in MG IX and X.

William J. Morse joined the United States Department of Agriculture (USDA) in 1907. With great singleness of purpose and dedication, his entire career focused on encouraging soybean production and rooting the soybean industry in the United States (Hymowitz, 1984; Shurtleff, 1981).

No single factor contributes more to the increase in production of the soybean in the United States than the development of new cultivars by public and private soybean breeders through the introduction of germplasm from China, Japan, and Korea. USDA scientists undertook two major soybean exploration trips. From August 1924 through December 1926, P.H. Dorsett collected soybean germplasm in Northeast China. He sent back to the United States about 1500 soybean accessions. From March 1929 to February 1931, P.H. Dorsett and W.J. Morse collected soybean germplasm in Japan, Korea, and China (Hymowitz, 1984). They sent back to the United States about 4500 soybean accessions. Unfortunately, during the first five decades of this century, the USDA was not much concerned with the preservation of soybean germplasm. Hence, many of the accessions Dorsett and Morse introduced were either discarded, or seed viability was lost due to lack of preservation facilities.

When William Morse retired in 1949, Martin G. Weiss replaced him. Weiss with Jackson L. Cartter of the U.S. Regional Soybean Laboratory at Urbana, Illinois, initiated the development of a comprehensive soybean germplasm collection. In 1951, Edgar E. Hartwig was appointed curator of the southern collection at Stoneville, Mississippi. In 1954, Richard L. Bernard became the curator of the northern collection located at Urbana.

Today, both the Southern and Northern soybean collections are merged. The collection contains over 20,000 strains of soybeans, wild soybeans, and wild perennial *Glycine* species (Table 1.4). Dr. R.L. Nelson of USDA/Urbana is the curator of the collection. His e-mail address is rlnelson@uiuc.edu. The seed are distributed free of charge to U.S. as well as non–U.S. institutions. For example, in 2006, Dr. Nelson (personal communication) distributed 19,737 seed lots representing 8731 accessions. Seed were sent to 36 states as well as to 15 countries. Thus, unlike the past, today testing soybean accessions for various traits including adaptability to specific regions is relatively easy.

Conclusion

The closest genera to the genus *Glycine* are *Teramnus, Amphicarpeae,* and *Pueraria.* All evidence points to Laos–Cambodia–Vietnam as the region where the genus originated. From this region the genus moved north and south. In the north, the farmers of the eastern half of North China domesticated the soybean *Glycine max* from its wild annual counterpart, *G. soja* Sieb. and Zucc. The domestication process took place ca. the eleventh century BCE. In the south, ca. two dozen wild perennial *Glycine* species evolved and are indigenous to Australia. These wild perennial *Glycine* species are potential candidates for providing genes to improve soybean cultivars.

The dissemination of the soybean out from its heartland to other countries was a slow process and initially localized to China's neighbors. The soybean and/or soy products moved rapidly from China to Europe during the Age of Discovery. The association of the soybean growing in the field with its main traditional products such as tofu, soy sauce, and miso was a mystery to the West. However, Kaempfer's book, published in 1712, provided the recipes to make traditional products from soybeans.

In 1765, Samuel Bowen introduced the soybean into the Colony of Georgia. He

Table 1.4. USDA Soybean Germplasm Collection and Number of Strains in Each Group as of December 31, 2006[a]

Collection	No. of Strains
Public Cultivars	718
FC and PI Strains	16,791
Genetic Types (T-Lines)	196
Genetic Isolines	641
Wild Annual Soybean (*G. soja*)	1,116
Wild Perennial *Glycine* species	919
	20,381

[a] Information provided by R. Nelson, USDA/ARS. Urbana, Illinois.

obtained seed in China while employed by the Honorable East India Company. Mr. Bowen planted soybeans on his property "Greenwich," located in a suburb of Savannah. He received patent number 878 for making soy sauce from plants grown in the Colony of Georgia. Dr. James Mease was the first person to use the word "soybean," in English.

After World War II, the USDA developed a national soybean germplasm collection. Currently, the collection contains over 20,000 strains. It is the primary source for new genetic traits for the improvement of soybean cultivars as well as for basic genetic studies.

Acknowledgments

I wish to thank Dr. Christine DuBois, Dr. Jules Janick, and Mr. Bill Shurtleff for reviewing the manuscript. Malcolm Obourn and Matthew Houlihan, undergraduate laboratory assistants, were of great help in manuscript preparation. The author takes sole responsibility for the correctness of the text and any typos. The text was written without any outside funding.

References

Aiton, W.T. *Hortus Kewensis*; Longman, Hurst, Rees, Orne, and Brown: London, 1812; Vol. 4.

Anonymous. Final report on the Australian *Glycine* and *Gossypium* exploration mission in 1987. *Plant Genetic Resources Newsl. 73/74* **1988,** *Rep. 88/9*, p.55.

Bartram, J. *A Catalogue of Trees, Shrubs, and Herbaceous Plants...*; Bartram and Reynolds: Philadelphia, 1807.

Bartram, W. *William Bartram's "Common Place Book"*; The John Bartram Association: Philadelphia, 2004.

Bentham, G. *Flora Australiensis;* L. Reeve: London, 1864; Vol. 2.

Bentham, G. On the genera *Sweetia* Sprengel and *Glycine* Linn., simultaneously published under the name of *Leptolobium. J. Linn. Soc. Bot.* **1865,** *8*, 259–267.

Beversdorf, W.D.; R.I. Buzzell; G.R. Ablett; H.D. Voldeng. Soybean. In *Harvest of Gold: The History of Field Crop Breeding in Canada;* (A.E. Slinkard, D.R. Knott, Eds.) University Extension Press: Saskatoon, Canada, 1995.

Boulnois, L. *The Silk Road* (translation by D. Chamberlain); Dutton: New York, 1966.

Boxer, C.R. *South China in the Sixteenth Century;* Printed for Hakluyt Soc.: London, 1953.

Boxer, C.R. *The Christian Century in Japan, 1549–1650.* University of California Press: Berkeley, 1967; p. 348.

Boxer, C.R. *Fidalgos in the Far East 1550–1770;* Second revised edition reprinted with corrections; Oxford: Asia (historical reprints); Oxford University Press: London, 1968.

Boxer, C.R. Jan compagnie in war and peace, 1602–1799; *A Short History of the Dutch East India Company*; Asia Studies Series; Heinemann Asia: Hong Kong, 1979.

Boxer, C.R. *Dutch Merchants and Mariners in Asia, 1602–1795*; Variorum Reprints: London, 1988.

Bray, F. Joseph Needham, science and civilization in China. *Biology and Biological Technology. Part II: Agriculture;* Cambridge University Press: Cambridge, England, 1984; Vol. 6.

Bretschneider, E.V. *Botanicum Sinicum. Notes on Chinese Botany from Native and Western ources.* II. J.R. Asiatic Soc.: North China Branch, Ser 2.25, **1893**; 1–468.

Bretschneider, E.V. *History of European Botanical Discoveries in China*; Zentral-Antiquariat Der Deutschen Demokratischen Republik: Leipzig, 1898 (Reprint 1962).

Brooks, W.P. Results of a trial of a few Japanese crops. IV. Varieties of soja bean. *Massachusetts (Hatch) agric. Exp. Sta. Bull.* **1890,** *7,* 14–17.

Brooks, W.P. Soya Beans. The Agriculture Division: Massachusetts (Hatch) Exp. Sta. 5th Ann. Rep.1893; p. 152.

Brown, A.H.D.; J.L. Doyle; J.P. Grace; J.J. Doyle. Molecular phylogenetic relationships within and among diploid races of *Glycine tomentella* (Leguminosae). *Aust. Syst. Bot.* **2002**, *15*, 37–47.

Brown, A.H.D.; J.E.Grant; J.J. Burdon; J.P. Grace; R. Pullen. In *Collection and Utilization of Wild Perennial* Glycine; R. Shibles, Ed.; World Soybean Research Conference, Iowa State University, Ames, Iowa, Aug. 12–17, 1984. Westview Press: Boulder, CO, 1985; pp. 345–352.

Browne, D.J. Preliminary remarks. *Report of the Commissioner of Patents for the year 1854*, 1855.

Buconjie, P.A. n.d. Monograph of Dubrovnik, 1800–1810. Unpublished. Church of St. Luka Marunic: Dubrovnik. (In Serbo-Croatian, translation by Bogdan Belic).

Burkill, I.H. *A Dictionary of the Economic Products of the Malay Peninsula;* Published on behalf of the Governments of the Straits Settlements and Federated Malay States by the crown agents for the colonies: London, 1935.

Burnell, A.C.; P.A. Tiele. *The Voyage of John Huyghen Van Linschoten to the East Indies.* (From the old English translation of 1598). Hakluyt Society: London, 1885.

Carletti, F. *My Voyage Around the World.* (translation by H. Weinstock). Pantheon Books: New York, 1964; p. 110.

Cary, N.; H. Warmington. *The Ancient Explorers;* Methuen and Co. Ltd, London, 1929.

Chamberlain, B.H. Ko-Ji-Ki, records of ancient matters. *Trans. Asiat. Soc. Japan.* (translation) **1906,** *10*, Supplement.

Cole, A.B. *A Scientist with Perry in Japan. The Journal of Dr. James Morrow;* University of North Carolina Press: Chapel Hill, NC, 1947.

Cummins, J.S. *The Travels and Controversies of Friar Domingo Navarrete 1618–1686*; Hakluyt Soc.: Cambridge, England, 1962.

Dafert, F.W. In *Relatorio annual do Instituto Agronomico do Estado de Sar Paulo (Brazil) em Campinas 1892.* (In *Portuguese*), 1893.

Dale, S. Samuelis Dale pharmacologiae.... Sam. Smith and Benj. Walford, Londini (In *Latin*), 1705.

Dampier, W. *Voyages and Descriptions,* Third edition; In three parts, viz 1. A supplement of the voyage round the world...2. Two voyages to Campeachy...3. A discourse of trade-winds, breezes, storms...Third edition. Printed for James Knapton: London, 1705; Vol. 2, p.78.

Domagalski, J.M.; K.O. Kollipara; A.H. Bates; D.L. Brandon; M. Friedman; T. Hymowitz. Nulls for the major Bowman-Birk protease inhibitor in the genus. *Glycine. Crop Sci.* **1992,** *32,* 1502–1505.

Darlington, C.D.; and A.P. Wylie. *Chromosome Atlas of Flowering Plants.* Allen and Unwin, London, 1955

Doyle, J.J.; J.L. Doyle; A.H.D. Brown. Analysis of polyploidy complex in *Glycine* with chloroplast and nuclear DNA. *Aust. Syst. Bot.* **1990a,** *3,* 125–136.

Doyle, J.J.; J.L. Doyle; A.H.D. Brown. A chloroplast-DNA phytogeny of the wild perennial relatives of soybean (*Glycine* subgenus *Glycine*): Congruence with morphological and crossing groups. *Evolution* **1990b,** *44,* 371–389.

Dulles, F.R. *Eastward Ho! The First English Adventures to the Orient,* R. Chancellor, A. Jenkinson, J. Lancaster, W. Adame, Sir T. Roe, Eds.; Houghton Mifflin Co.: Boston, MA, 1931.

Dunlap's Pennsylvania Packet. Advertisement. Bowen's sago…*Dunlap's Pennsylvania Packet* or, the *General Advertiser.* March 7, 1774.

D'Utra, G. Soja. *Jornal do Agricultor (Brazil)* **1882,** *4,* 185–188 (In *Portuguese*).

Eames, J.B. *The English in China. Being an Account of the Intercourse and Relations Between England and China from the Year 1600 to the Year 1843 and Summary of Later Developments;* Curzon Press: London, 1974 (originally published in 1909).

Ernst, A.H. Japan peas; Report of the Commissioner of Patents for the year 1852. Part II Agriculture; 1853; p. 448.

Fox, R.H. *Dr. John Fothergill and His Friends;* Macmillan: London, 1919.

Garner, W.W.; H.A. Allard. Effect of the relative length of day and night and other factors of the environment in growth and reproduction in plants. *J. Agric. Res* **1920,** *.18,* 553–606.

Georgeson, C.C.; H.M. Cottrell; W. Shelton. Experiments with forage plants. Soybeans. *Kansas Agric. Exp. Sta. Bull.* **1890,** *18,* 175.

Gibb, H.A.R. *Travels of Ibn Battuta, A.D. 1325–1354*; translated with revisions and notes from the Arabic text (C. Defremery, B.R. Sanquinetti, Eds.); R.M. McBride and Co.: New York, 1958..

Glasse, H. *The Art of Cookery Made Plain and Easy by a Lady;* Prospect Books: London, 1983 (Report of 1747).

Goldblatt, P. Cytology and the phylogeny of Legumminosae. *Advances in Legume Systematics. Part 2;* R.M. Polhill, P.H. Raven, Eds.; Royal Botanic Garden: Kew, England, 1981; pp. 427–463.

Graff, H.F. The early impact of Japan upon American agriculture. *Agric. Hist.* **1949,** *23,* 110–116.

Harlan, J.R. *Crops and Man,* Second ed.; American Society of Agronomy: Madison, WI, 1992.

Hart, S.E.; D.S. Glenn; W.J. Kenworthy. *Herbicide Tolerance of Wild Glycine Species;* Proceedings 42nd Annual Meeting of the Northeastern Weed Science Society, 1988, p. 54.

Hartman, G.L.; M.E. Gardner; T. Hymowitz; G.C. Naidoo. Evaluation of perennial *Glycine* species for resistance to soybean fungal pathogens that cause Sclerotenia stem rot and sudden death syndrome. *Crop Sci.* **2000,** *4,* 545–549.

Heco, J. *The Narrative of a Japanese; What He Has Seen and the People He Has Met in the Course of the Last Forty Years.* Maruzen; Tokyo, 1895.

Hellriegel, H.; H. Wilfarth. *Studies on Nitrogen Utilization in Grasses and Legumes.* Beilage heft zu der Zeitschrift des Vereins fur Rubenzucker–Industrie Deutchen Reichs, 1888, p. 234. (In *German*).

Henderson, P. *Henderson's Handbook of Plants;* Henderson and Co.: New York, 1881.

Herman, F.J. A revision of the genus *Glycine* and its immediate allies. U.S.D.A. *Tech. Bull.* **1962,** *1268,* 1–79.

Hermann, P. Musaeum zeylanicum, sive, catalogus plantarum in Zeylana sponte nascentium. Apud Isaacum Severinum, Lugdoni Batavorum (In *Latin*); 1717.

Hitchcock, A.S.; M.L. Green. Species lectotypical generum Linnaei. *Brittonia* **1947,** *6,* 114–118.

Ho, P.-T. The introduction of American food plants in China. *Am. Anthrop.* **1955,** *57,* 191–201.

Ho, P.-T. The loess and the origin of Chinese agriculture. *Am. Hist. Rev.* **1969,** *75,* 1–36.

Ho, P.-T. *The Cradle of the Eastern University of Hong Kong.* Hong Kong, 1975.

Horlock, C.M.; D.S. Teakle; R.M. Jones. Natural infection of the native pasture legume, *Glycine latfolia* by alfalfa mosaic virus in Queensland. *Australasian Plant Path.* **1997,** *26,* 115–116.

Hort, A. Theophastus. *Enquiry into Plans and Minor Works on Odours and Weather Signs;* English translation by Sir Arthur Hort; The Loeb classical library, Harvard University Press: Cambridge, MA, 1961 (reprint of 1916).

Hughes, G.B. *Old English Cruets. Country Life (London)*; 1955; Jan. 20, pp. 178–180.

Hymowitz, T. On the domestication of the soybean. *Econ. Bot.* **1970,** *24,* 408–421.

Hymowitz, T. Exploration for a wild relative of the soybean on Vanuatu. *Naika* **1982,** *7,* 1–4.

Hymowitz, T. Dorsett-Morse soybean collection trip to East Asia: 50 year retrospective. *Econ Bot.* **1984,** *38,* 378–388.

Hymowitz, T. Introduction of the soybean to Illinois. *Econ. Bot.* **1986,** *41,* 28–32.

Hymowitz, T. Exploration for wild perennial *Glycine* species on Taiwan. *Soybean Genet. Newsl.* **1989,** *16,* 92–93.

Hymowitz, T. Soybeans: the success story. *Advances in New Crops* J. Janick, J. E. Simon, Eds. Timber Press: Portland, Oregon, 1990; pp. 159–163.

Hymowtiz, T. Plant exploration trip to Taiwan and islet of Kinmen (Quemoy). *Soybean Genet. Newsl.* **1998,** *Vol, 25,* 127.

Hymowitz, T. Speciation and cytogenetics. *Soybeans: Improvement Production and Uses;* H.R. Boerma, J.E. Specht, Eds. 3rd ed. Agron. Monogr. 16, ASA, CSSA and SSSA: Madison, WI, 2004; pp. 97–131.

Hymowitz, T.; J.R. Harlan. Introduction of soybean to North America by Samuel Bowen in 1765. *Econ. Bot.* **1983,** *37,* 371–379.

Hymowitz, T.; N. Kaizuma. Dissemination of soybeans (*Glycine max*): Seed protein electrophoresis profiles among Japanese cultivars. *Econ. Bot.* **1979,** *33,* 311–319.

Hymowitz, T.; N. Kaizuma. Soybean seed protein electroporesis profiles from 15 Asian countries on regions. Hypotheses on paths of dissemination of soybeans from China. *Econ. Bot.* 1981, *35,* 10–23.

Hymowitz, T.; C.A. Newell. Taxonomy, speciation, domestication, dissemination, germplasm resources and variation in the genus *Glycine. Advances in Legume Science;* R.J. Summerfield, A.H. Bunting, Eds.; Royal Botanic Gardens: Kew, England, 1980; pp.251–264.

Hymowitz, T.; C.A. Newell. Taxonomy of the genus *Glycine*, domestication and uses of soybeans. *Econ. Bot.* **1981,** *35,* 272–288.

Hymowitz, T.; W.R. Shurtleff. Debunking soybean myths and legends in the historical and popular literature. *Crop Sci.* **2005,** *45,* 473–476.

Hymowitz, T.; R.J. Singh. Taxonomy and speciation. *Soybeans: Improvement, Production and Uses,* Second ed.; J.R. Wilcox, Ed.; Agron. Mono. 16, ASA, CSSA, and SSSA: Madison, WI, 1987; pp. 23–48.

Hymowitz, T.; R.J. Singh; K.P. Kollipara. The genomes of the *Glycine. Plant Breeding Rev.* **1998,** *16,* 289–317.

Hymowitz, T.; R.J. Singh; P.P. Larkin. Long distance dispersal. The case for the allopolyploids *Glycine tabacina* Benth. and *G. tomentella* Hayata in the west-central pacific. *Micronesica* **1990,** *23,* 5–13.

Hymowitz, T.; J.T. Woolley; D.B. Peters. Preliminary investigations on the salt tolerance of wild perennial *Glycine* species. *Soybean Genet. Newsl.* **1987,** *14,* 271–272.

International Legume Database and Information Service (ILDIS). http://www.ILDIS.org. School of Plant Sciences: The University of Reading, England, 2006.

Joseph, L.M.; T. Hymowitz; M.A. Schmidt; E.M. Herman. Evaluation of *Glycine* germplasm for nulls of the immunodominant allergen P34/GlymBd30k. *Crop Sci.* **2006,** *46,* 1755–1763.

Kaempfer, E. *Amoenitatum Exoticarum…H.W. Meyeri, Lemgoviae.* (In *Latin*), 1712.

Kasmakoglu, H. *FAO Production Yearbook—2003;* Rome, Italy, 2004; Vol. 57.

Kenworthy, W.J. Potential genetic contributions of wild relatives to soybean improvement. *World Soybean Research Conference IV. Proc. Buenos Aires, March 5–9, 1989, Argentina;* A.J. Pascale, Ed.; Association Soja: Buenos Aires, Argentina, 1989; pp. 883–888.

King, P.K. *Life and Letters of John Locke, with Extracts from His Journals and Common-place Books. With a General Index.* B. Franklin: New York, 1972 (Reprint of 1884).

Komroff, M. *Contemporaries of Marco Polo.* Jonathan Cape: London, 1928.

Kumar, P.S.; T. Hymowitz. Where are the diploid ($2n = 2x = 20$) genome donors of *Glycine* Wild. (Leguminosae, Papilionoideae)? *Euphytica* **1989,** *40,* 221–226.

Lackey, J.A. A Synopsis of the Phaseoleae (Leguminosae, Papilionoideae). Ph.D. dissertation, Iowa State University: Ames, IA, 1977a.

Lackey, J.A. *Neonotonia*, a new generic name to include *Glycine wightii* (Arnott) Verdcourt (Leguminosae, Papilionoideae). *Phytologia* **1977b**, *37,* 209–212.

Lackey, J.A. Tribe 19. Phaseoleae DC. (1825). *Advances in Legume Systematics;* R.M. Polhill, P.H. Raven, Eds.; Royal Botanic Gardens: Kew, England, 1981; pp. 301–327.

Laufer, B. Sino-Iranica. *Chinese Contributions to the History of Civilization in Ancient Iran;* Ch'eng-wen Publishing Company: Taipei, 1919. (Reprint of 1967).

Lawrence, G.H.M. Name of the soybean. *Science* **1949,** *110,* 566–567.

LeComte, L. Nouveaux memoires sur l'etat de la chine. J. Anison: Paris. (In *French*), 1696.

Lee, J.; T. Hymowitz. A molecular phylogenetic study of the subtribe Glycinnae (Leguminosae) derived from the chloroplast DNA rps16 intron sequence. *Am. J. Bot.* **2001,** *88,* 2064–2073.

Lee, J.S.; D.P.S. Verma. Structure and chromosomal arrangement of leghemoglobin genes in kidney bean suggest divergence in soybean leghemoglobin gene loci following tetrapoidization. *EMBO J.* **1984,** *3,* 2745–2752.

Lesley, J.P. (Ed.) *Early Proceedings of the American Philosophical Society for the Promotion of Useful Knowledge: Compiled from the Manuscript Minutes of Its Meeting 1749 to 1838.* McCalla and Stavely: Philadelphia, PA, 1884; Vol. 1.

Lim, S.M.; T. Hymowitz. Reactions of perennial wild species of genus *Glycine* to *Septoria glycines. Plant Dis.* **1987,** *17,* 891–893.

Linnaeus, C. *Hortus Cliffortianus. Historiae Naturalis Classica.* J. Cramer, H.K. Swann, Eds.; Stechert-Hafner: New York.. (In *Latin*), 1737, Reprint 1968; Vol. 63.

Linnaeus, C. *Species Plantarum,* First ed.; Conrad Wishoff: Leiden, The Netherlands. (In *Latin*), 1753.

Linnaeus, C. *Genera Plantarum,* Fifth ed.; Lars Salvius: Stockholm. (In *Latin*), 1754.

Linnaeus, C. Mantissa Plantarum. *Historiae Naturalis Classica;* J. Cramer, H.K. Swann, Eds. Vol. 7. Lars Salvius: Stockholm, 1767; Vol. 7; Reprint 1961---Hafner: New York. (In *Latin*).

Lockyer, C. *An Account of the Trade in India, Containing Rules for Good Government in Trade*...Printed for the author and sold by Samuel Crouch: London, 1711.

Loux, M.M.; R.A. Lieble; T. Hymowitz. Examination of wild perennial *Glycine* species for glyphosate tolerance. *Soybean Genet. Newsl.* **1987,** *14,* 268–271.

McBryde, J.M. *Experiments with Other Crops: Soybean*; School of Agriculture, Horticulture and Botany: University of Tennessee. 3rd Annual Rep. 1882; p. 127.

Mease, J. (Ed.) *Willich's Domestic Encyclopedia.* 1st American edition, 1804; Vol. 5:13,

Merrill, Ed. *An Interpretation of Rumphius's Herbarium Amboinense.* Bureau of Printing: Manila, Philippines, 1917.

Mettu, S.L.; O. Vodkin; and T. Hymowitz. Investigation of lectin nulls in perennial *Glycine* species. *Soyean Genet. Newsl.* **1995,** *22,* 212–217.

Mignucci, J.S.; and D.W. Chamberlain. Interaction of *Microsphaera diffusa* with soybeans and other legumes. *Phyopath* **1978,** *68,* 169–173.

Nagata, T. *Studies on the Differentiation of Soybeans in Japan and the World.* Mem. Hyogo University of Agriculture 3 (Agron. Series No. 4); 1960; pp. 63–102.

Newell, C.A. Distribution of *Glycine tabacina* (Labill.) Benth. In the West-Central Pacific. *Micronesica* **1981,** *17,* 59–65.

Newell, C.A.; T. Hymowitz. A reappraisal of the subgenus *Glycine. Am. J. Bot.* **1978,** 65, 168–179.

Newport Mercury. Advertisement. Bowen's Sago Powder and Soy. The Newport Mercury. August 12, 1771.

New York Gazette. Advertisement. Just imported by the last ships from London... Rochell and

Sarp...Indian soy. New York Gazette, or Weekly Post-boy. Dec. 17, 1750. Issue 413, p. 3.

New York Gazette. Advertisement. John Amiel Jr. Has just opened for sale...The New York Gazette; and the Weekly Mercury. January 20, 1777.

Olschki, L. Marco Polo's precursors. Johns Hopkins Press: Baltimore, MD, 1943.

Olschki, L. *Marco Polo's Asia.* University of California Press: Berkeley, CA, 1960.

Orellana, R.G. Resistance to bud blight in introductions from the germplasm of wild soybean. *Plant Disease.* **1981,** *65,* 594–595.

Osborne, T.B.; L.B. Mendel. The use of soy bean as food. *J. Biol. Chem.* **1917,** *32,* 369–387.

Ovington, J. A voyage to Suratt in the year 1689. Oxford University Press: London, 1929. (Reprint of 1696).

Paclt. Nomenclature of the soybean. Science **1949,** *109,* 339.

Paillieux, M.A. Le soya: sa composition chimique, ses varietes', sa culture et ses usages. *Bulletin de la Societe D'Acclimation* **1880,** 3rd seris, Tome VII. *27,* 414–471, 538–596 (In *French*).

Penzer, N.M. *The Most Notable and Famous Travels of Marco Polo Together with the Travels of Nicolo de' Conti,* edited from the Elizabethan translation of Jolin Frampton, with introduction notes and appendixes by N.M. Penzer. The Argonaut Press: London, 1929.

Perry, M.C. *Narrative of the Expedition of an American Squadron to the China Seas and Japan.* 3 vol. F.L. Hawks, Ed.; House of Representatives Document 97, 33 Congress, 2 Session, Washington, DC, 1856.

Peticolas, T.V. (TVP) Japan pea. *Country Gentleman* **1855,** *5,* 232.

Pfeil, B.E.; L.A. Craven. New taxa in *Glycine* (Fabaceae, Phaseolae) from north-western Australia. *Austr. Syst. Bot.* **2002,** *15,* 565–573.

Pfeil, B.E.; L.A. Craven; A.H.D. Brown; B.G. Murray; J.J. Doyle. Three new species of northern Australian *Glycine* (Fabaceae, Phaseolae), *G. gracei, G. monti-douglas* and *G. syndetika. Aust. Syst. Bot.* **2006,** *19,* 245–258.

Pfeil, B.E.; M.D. Tindale; and L.A. Craven. A review of *Glycine clandestina* species complex (Fabaceae, Phaseolae) reveals to species. *Austr. Syst. Bot.* **2001,** *14,* 891–900.

Pinkerton, J. *A General Collection of the Best and Most Interesting Voyages and Travels in All Parts of the World.* Longman, Hurst, Rees, Orme and Brown: London, 1811; Vol. 7.

Piper, C.V. The name of the soybean: a chapter in its botanical history. *J. Am. Soc. Agron.* **1914,** *6,* 75–84.

Piper, C.V.; W.J. Morse. The soybean, with special reference to its utilization for oil, cake, and other. USDA Bull. 439, 1916.

Piper, C.V.; W.J. Morse. *The Soybean;* McGraw-Hill: New York, 1923.

Polhill, R.M. 1994. Classification of the Leguminosae. XXXV-XLVII. *In* F.A. Bisby, J. Buckingham, and J.B. Harborne (eds.), Phytochemical dictionary of the Leguminosae. Chapman and Hall, New York.

Prince Dhaninavat. 1961. Thai migrations. *In* Symposium on historical archeological and linguistic studies on Southern China, South-East Asia and the Hong Kong region, F.S. Drake (general ed.). University of Hong Kong Press, Hong Kong.

Ray, I. 1999. The *French* East India Company and the trade of the Indian Ocean: a collection of essays; edited by Lakshmi Subramanian. Munshiram Manoharial Publishers, New Delhi.

Regel, E.A. von (Ed.); R. Maack. Tentamen florae ussuriensis, Se. 7. *Mem. Acad. Imp. Sci.* **1861,** *4 no. 4,* St. Petersbourg. (In *Latin*).

Rein, J.J. *The Industries of Japan.* (Together with an account of its agriculture, forestry, arts and commerce. From travels and researches undertaken at the cost of the Prussian government). Hodder and Stoughton: London, 1889.

Ricker, P.L.; W.J. Morse. The correct botanical name for the soybean. *J. Am. Soc. Agron.* **1948,** *40,* 190–191.

Riggs, R.D.; S. Wang; R.J. Singh; T. Hymowitz. Possible transfer of resistance to *Heterodera glycines* from *Glycine tomentella* (2*n* = 78) to soybean. Supplement to the *J. of Nematology* **1998,** *30,* 547–552.

Rockhill, W.W. *The Journey of William Rubruck to the Eastern Parts of the World 1253–55;* Printed for the Hakluyt Society: London, 1900; p. 206.

Roxburgh, W. Flora Indica, or Descriptions of Indian Plants; Today and Tomorrow's Printers and Publishers: New Delhi, 1971 (reprinted literatim from Carey's edition of 1832).

Rugoff, M. *The Travels of Marco Polo. A Signet Classic;* The New American Library of World Literature, Inc.: New York, 1961.

Rumphius, G.E. *Herbaruim Amboinense. Amstelaedami.* (In *Latin*), 1747.

Satow, E.M. *The Voyage of Captain John Saris to Japan, 1613.* Printed for the Hakluyt Society: London, 1900.

Schoen, D.J.; J.J. Burdon; A.H.D. Brown.. Resistance of *Glycine tomentella* to soybean leaft rust *Phakopsora pachyrhizi* in relation to ploidy level and geographical distribution. *Theor. Appl. Genet.* **1992,** *83,* 827–832.

Shinoda, O. A treatise on tofu (bean curd). *The Continent Magazine* **1971,** *42,*172–178 (In *Chinese,* translation by E. Wolff).

Shoemaker, R.C.; K. Polzun; J. Labate; J. Specht; E.C. Brummer; T. Olsen; N. Young; V. Concibido; J. Wilcox; J.P. Tamulonis et al. Genome duplication in soybean (*Glycine* subgenus *soja*). *Genetics* **1996,** *144,* 329–338.

Shurtleff, W. *William Morse: The Father of Soybeans in America.* Soyfoods No 5. pp.56–60, Summer; 1981.

Shurtleff, W.; A. Aoyagi. *The Book of Miso.* Second Ed.; Ten Speed Press: Berkeley, California, 1983.

Siebold, P.F. de,; J.G. Zuccarini. Florae Japonicae. Abh. Bayer Akad. *Wiss. Math-Phys.* **1846,** *4,* 109–204. (In *German*).

Singh, B.B.; S.C. Gupta; B.D. Singh. Sources of field resistance to rust and yellow mosaic diseases of soybean. *Indian J. Genet. Plant Breed.* **1974,** *34,* 400–404.

Singh, R.J.; T. Hymowitz. Diploid-like meiotic behavior in synthesized amphiploids of the genus *Glycine* Willd. subgenus *Glycine. Can. J. Genet. Cytol.* **1985a,** *27,* 655–660.

Singh, R.J., and T. Hymowitz. The genomic relationships among six wild perennial species of the genus *Glycine* subgenus *Glycine* Willd. *Theor. Appl. Genet.* **1985b,** *71,* 221–230.

Singh, R.J.; T. Hymowitz. Soybean genetic resources and crop improvement. *Genome* **1999,** *42,* 605–616.

Singh. R.J.; H.H. Kim; T. Hymowitz. Distribution of rDNA loci in the genus *Glycine* Willd. *Theor. Appl. Genet.* **2001,** *103,* 212–218.

Singh, R.J.; K.P. Kollipara; T. Hymowitz, Backcross-derived progeny from soybean and *Glycine tomentella* Hayata intersubgeneric hybrids. *Crop Sci.* **1990,** *30,* 871–874.

Singh, R.J.; K.P. Kollipara; T. Hymowitz. Backcross(BC_2–BC_4)-derived fertile plants from *Glycine max* and *G. tomentella* intersubgeneric hybrids. *Crop Sci.* **1993,** *33,* 1002–1007.

Smyth, A.H. *Writings of Benjamin Franklin;* Macmillan: New York, 1907; Vol. 5.

Stalker, H.T. Utilization of wild species for crop improvement. *Adv. Agron.* **1980,** *33,* 112–147.

Sturtevant, E.L. *Soja Bean;* New York Agricultural Experiment Station; 1st Ann. Rep. for the year 1882; 1883; pp. 24, 74.

Tindale, M.D. Two new eastern Australian species of *Glycine* willd. (Fabaceae). *Brunonia* **1984,** *7,* 207–213.

Tindale, M.D. A new north Queensland species of *Glycine* willd. (Fabaceae). *Brunonia* **1986a,** *9,* 99–103.

Tindale, M.D. Taxonomic notes on three Australian and Norfolk Island species of *Glycine* Willd. (Fabaceae: Phaseolae) including the choice of a neotype for *G. clandestina* Wendl. *Brunonia* **1986b,** *9,* 179–191.

Tindale, M.D.; L.D. Craven. Three new species of *Glycine* (Fabaceae: Phaseolae) from North-Western Australia with notes on amphicarpy in the genus. *Aust. Syst. Bot.* **1988,** *1,* 399–410.

Tindale, M.D.; L.A. Craven. *Glycine pindanica* (Fabaceae: Phaseolae), a new species from West Kimberley, Western Australia. *Aust. Syst. Bot.* **1993,** *6,* 371–376.

Valignano, A. Sumario de las cosas de Japan (1583). Adiciones del sumario de Japan (1592); J.L. Alvarez-Taladniz, Ed.; Sophia University: Tokyo (In *Spanish*), 1954; p. 93.

Verdcourt, B. A proposal concerning *Glycine* L. *Taxon* **1966,** *15,* 34–36.

Verdcourt, B. Studies in the Leguminosae-Papilionoideae for the Flora of Tropical East Africa. *II Kew Bulletin* **1970,** *24,* 235–307.

Watkinson, J. John Watkinson , grocer, tea dealer, and company. Next door to the coffee-house in Newbrough, Scarbrough, sells the following articles at the lowest prices. York? England. Based on information from English Short Title Catalogue. Eighteenth Century Collections Online, Gale Group. Document from Bodleian Library: Oxford, England, 1750.

Wells, W.S. Notices of Chinese agriculture and its principal products.. *Report of the Commissioner of Patents, Agriculture. For the year 1860.* Washington, DC, 1861; . pp. 467–476.

Willdenow, K.L. *Species Plantarum.* Third ed.; Berlin. (In *Latin*), 1802.

Wills, J.E. Jr. *Pepper, Guns, and Parleys. The Dutch East India Company and China, 1622–1681.* Harvard University Press: Cambridge, MA, 1974.

Woodcraft, B. *Titles of Patents of Invention, Chronologically Arranged from March 2, 1617 to October 1, 1852;* 158 Queens Printing Office: London, 1854; Vol. 1.

Wright, J.K. *The Geographical Lore of the Time of the Crusades.* American Geographical Society of New York. Dover Publications Inc.: New York, 1925 (reprint of 1965).

Xu, S.J.; R.J. Singh; T. Hymowitz. Monosomics in soybean; origin, identification, cytology, and breeding behavior. *Crop Sci.* **2000,** *40,* 985–989.

Yonge, H. Gentleman's Mag. **1767,** *37,* 253.

Yule, H. *Cathay and the Thithers; Being a Collection of Medieval Notices of China.* Printed for the Hakluyt Society: London, 1866; Vol. 2, p.244.

Yule, H. (translated by). *The Travels of Friar Odoric. A 14th Century Journal of the Blessed Odoric of Pordenone;* William B. Eerdmans Publishing Co.: Grand Rapids, MI, 2002.

2

Breeding, Genetics, and Production of Soybeans

James H. Orf
Department of Agronomy and Plant Genetics, University of Minnesota, St. Paul, MN 55108 USA

Soybeans are one of the major crops in the United States and the world. Soybean breeding, genetics, and production have undergone some dramatic changes in the last few decades. Undoubtedly, this rapid change will continue as general scientific discoveries and innovation are applied to soybean. This chapter briefly discusses the current situation in soybean breeding, genetics, and production.

Soybean Production

Soybean is the oilseed with the greatest production on a worldwide basis. Production of soybeans is increasing and is expected to continue to increase as demand for soybean oil for human consumption and biodiesel and demand for high-protein meal for animal feed grows in both developed and developing countries. Currently the United States is the largest producer of soybeans followed by Brazil, Argentina, and China (USDA–FAS, 2007). Table 2.1 shows the most recent data on world supply.

Asia has grown soybeans for centuries. China is considered to be the area where soybean was domesticated (Hymowitz, 1970), from which it then spread to other countries. Probst and Judd (1973) presented a review of the origin and early history of soybean. They also reviewed the early history of the use of soybean in the United States. The soybean was first introduced into the United States in 1765 by Samuel Bowen (Hymowitz & Harlan, 1983). Additional information on soybean production and use, especially in the United States through 2002, was presented by Smith and Huyser (1987) and Wilcox (2004). Somewhere in the middle part of the twentieth century the United States became the largest soybean producer and remains so currently. In 2006/07 the United States produced about 37% of the world's soybean supply, Brazil about 25%, Argentina about 19% and China about 7% (Table 2.1).

Soybean hectarage, production, and yield in the United States from 1924 through 2006 are shown in Table 2.2. During that time, harvested area ranged from 181,000 hectares in 1925 to 30,214,000 hectares in 2006, yields from 0.74 tons per hectare in 1924 to 2.89 tons per hectare in 2005, and total production from 132,000 metric

Table 2.1. Soybeans: World Supply and Distribution (thousand metric tons)[1]

		2002/03	2003/04	2004/05	2005/06	Apr 2006/07
Production						
	United States	75,010	66,778	85,013	83,368	86,770
	Brazil	52,000	51,000	53,000	57,000	58,800
	Argentina	35,500	33,000	39,000	40,500	45,500
	China, People's Republic	16,510	15,394	17,400	16,350	16,200
	India	4,000	6,800	5,850	6,300	7,300
	Paraguay	4,500	3,911	4,050	4,000	5,500
	Canada	2,336	2,263	3,042	3,161	3,500
	Other	6,918	7,366	8,387	9,254	9,925
	Total	196,774	186,512	215,742	219,933	233,495
Imports						
	China, People's Republic	21,417	16,933	25,802	28,317	31,000
	EU-25	16,872	14,638	14,544	13,934	14,338
	Japan	5,087	4,688	4,295	3,957	4,100
	Mexico	4,230	3,797	3,640	3,667	3,775
	Taiwan	2,351	2,217	2,256	2,498	2,550
	Thailand	1,779	1,407	1,517	1,473	1,500
	Korea, Republic of	1,516	1,368	1,240	1,190	1,275
	Indonesia	1,238	1,059	1,112	1,187	1,270
	Iran	533	883	976	1,084	1,200
	Turkey	756	612	1,046	863	1,050
	Other	7,330	6,457	7,278	5,826	6,643
	Total	63,109	54,059	63,706	63,996	68,701
Exports						
	United States	28,423	24,128	29,860	25,778	29,393
	Brazil	19,629	20,417	20,137	25,911	26,100
	Argentina	8,714	6,741	9,568	7,249	7,550
	Paraguay	2,806	2,776	2,888	2,400	3,300
	Canada	726	897	1,093	1,326	1,550
	Other	776	1,029	1,089	1,404	1,571
	Total	61,074	55,988	64,635	64,068	69,464

[1]Source: USDA/FAS, April 2007.

tons in 1925 to 86,848,000 metric tons in 2006. Over this time period, the trends for area, yield, and production were upward.

In 2006, soybean area and production were reported from 31 states, all in the eastern half of the United States (Table 2.3). The leading states in terms of production are Iowa (16% of total), Illinois (15%), Minnesota (10%), Indiana (9%), Nebraska (8%), Ohio (7%), Missouri (6%), South Dakota (4%), North Dakota (4%), and Arkansas (3%). In the last 10 years, production shifted from the southern and eastern parts of the soybean growing area to the northern and western areas, as noted by the following production shifts. In 1969, the North Central states of Iowa, Illinois, Minnesota, Indiana, Nebraska, Ohio, Missouri, South Dakota, North Dakota, Michigan, and Wisconsin produced 69% of the total U.S. soybeans while the Southern states of Arkansas, Mississippi, Louisiana, South Carolina, Georgia, and Alabama produced 19%. In 2006, the production in the North Central States was 84% and in the South only 6% of the total U.S. production. A number of reasons could account for this shift including greater yield potential (and thus greater breeding efforts) in the North Central area, more diseases, insects, and other challenges in the South and more available area to shift to soybeans in the North Central states.

Brazil is the second-largest soybean producer in the world (Table 2.1). Soybean production has increased slightly in the last five years. Reports from within and outside Brazil indicate that large areas in the Cerrados ecological zone, especially in the states of Mato Grosso, Mato Grosso do Sul, Goias and Bahia, and perhaps even in the tropical rainforest zone, are available for expansion of soybean production. A number of challenges exist for soybean production and export in Brazil including poor transportation infrastructure, diseases and insects, and higher input costs. Nevertheless, Brazil will surpass the United States as the largest soybean producer in the world in the not-too-distant future. Brazil has well-developed research organizations and is able to consistently produce high yields with the rapid adoption of new technologies (Wilcox, 2004).

Argentina is the third-largest soybean producer in the world and second-largest in South America behind Brazil (Table 2.1). Some expansion of soybean production in Argentina in the last five years has occurred, but not nearly as much opportunity exists there for additional expansion of soybean production as in Brazil. One can attribute the expansion of soybean production in Argentina at least in part to more favorable economic policies by the government, the use of minimum and no-tillage production systems, the adoption of double-cropping soybean after wheat, and improvements in storage and transportation infrastructure (Wilcox, 2004).

China continues to be a major producer of soybeans. The production of soybeans remains about the same since 2002/03 (Table 2.1). The provinces of Heilongjiang, Liaoning, and Inner Mongolia produce about 45% of the total in China (Wilcox, 2004). Most of the soybeans in these areas are seeded in the spring. About 30% of the production is double-cropped behind wheat, especially in Henan, Shandong, Hebei,

Table 2.2. Soybeans: Hectare, Yield, and Production, United States 1924 to 2006[1]

Year	Hectares Harvested (000)	Yield per Harvested Hectare T/Ha	Production (000) MT
1924	181	0.74	134
1925	168	0.79	132
1926	189	0.75	142
1927	230	0.82	189
1928	235	0.91	214
1929	287	0.89	257
1930	435	0.87	379
1931	462	1.01	470
1932	405	1.01	412
1933	423	0.87	368
1934	630	1.00	630
1935	1,181	1.13	1332
1936	955	0.96	918
1937	1,047	1.20	1257
1938	1,229	1.37	1686
1939	1,748	1.40	2455
1940	1,947	1.09	2125
1941	2,385	1.22	2920
1942	4,007	1.28	5108
1943	4,211	1.23	5179
1944	4,149	1.26	5252
1945	4,350	1.21	5261
1946	4,022	1.34	5340
1947	4,621	1.10	5078
1948	4,326	1.43	6189
1949	4,245	1.50	6379
1950	5,592	1.46	8151
1951	5,514	1.40	7730
1952	5,846	1.39	8140
1953	6,006	1.22	7332
1954	6,904	1.34	9290
1955	7,541	1.35	10179

Table 2.2., cont. Soybeans: Hectare, Yield, and Production, United States 1924 to 2006[1]

Year	Hectares Harvested (000)	Yield per Harvested Hectare T/Ha	Production (000) MT
1956	8,351	1.46	12237
1957	8,447	1.56	13168
1958	9,717	1.63	15806
1959	9,166	1.58	14516
1960	9,580	1.58	15120
1961	10,936	1.69	18484
1962	11,181	1.63	18229
1963	11,589	1.64	18228
1964	12,471	1.53	19093
1965	13,952	1.65	23034
1966	14,801	1.71	25291
1967	16,121	1.65	26598
1968	16,763	1.79	30154
1969	16,741	1.84	30866
1970	17,111	1.79	30702
1971	17,296	1.85	32037
1972	18,502	1.87	34611
1973	22,545	1.87	42155
1974	20,793	1.59	33132
1975	21,715	1.94	42177
1976	20,007	1.75	35102
1977	23,421	2.06	48140
1978	25,784	1.98	50905
1979	28,489	2.16	61581
1980	27,464	1.78	48965
1981	26,796	2.02	54183
1982	28,124	2.12	59664
1983	25,323	1.76	44558
1984	26,776	1.89	50690
1985	24,948	2.29	57178
1986	23,616	2.24	52915
1987	23,155	2.28	52784

Cont. on p. 38.

Table 2.2., cont. Soybeans: Hectare, Yield, and Production, United States 1924 to 2006[1]

Year	Hectares Harvested (000)	Yield per Harvested Hectare T/Ha	Production (000) MT
1988	23,236	1.81	42190
1989	24,113	2.17	52401
1990	22,887	2.29	52463
1991	23,494	2.30	54173
1992	23,584	2.53	59665
1993	23,209	2.19	50931
1994	24,628	2.78	68505
1995	24,925	2.37	59227
1996	25,656	2.53	64839
1997	27,990	2.61	73242
1998	28,529	2.61	74665
1999	29,341	2.46	72288
2000	29,325	2.56	75123
2001	29,555	2.66	78742
2002	29,361	2.55	75077
2003	29,353	2.28	66838
2004	29,953	2.84	85089
2005	28,857	2.89	83419
2006	30,214	2.87	86848

[1]Source: USDA-NASS, April 2007

and Anhui provinces. The remaining production is in the south and frequently follows rice. Although China was the world's largest soybean producer in the past, some soybean production areas are planted with other crops. In the last few years, soybean use increased dramatically in China as the standard of living increased, and a greater demand arose for soybean oil and meat products from animals that consume soybean meal. Recently, China became the largest importer of soybeans (Table 2.1).

India ranks fifth in soybean production on a worldwide basis (Table 2.1). India recently expanded its soybean production (Wilcox, 2004). Major soybean-growing states include Madhya, Pradesh, Maharashtra, and Rajasthan. As population and living standards increase in India, soybean production is expected to increase there.

Paraguay has become a significant producer of soybean because of good land for growing soybeans and favorable transportation for export (Table 2.1). Cultivars developed for Argentina or Brazil can be planted in Paraguay.

Table 2.3. Soybean Production by State, 2006 (million metric tons)[1]

Alabama	0.08	Nebraska	6.82
Arkansas	2.92	New Jersey	0.08
Delaware	0.15	New York	0.25
Florida	0.01	North Carolina	1.18
Georgia	0.10	North Dakota	3.27
Illinois	13.13	Ohio	5.91
Indiana	7.73	Oklahoma	0.10
Iowa	13.88	Pennsylvania	0.46
Kansas	2.68	South Carolina	0.31
Kentucky	1.64	South Dakota	3.56
Louisiana	0.80	Tennessee	1.20
Maryland	0.43	Texas	0.10
Michigan	2.44	Virginia	0.43
Minnesota	8.68	West Virginia	0.02
Mississippi	1.17	Wisconsin	1.96
Missouri	5.29		

[1]Source: USDA/NASS, April 2007.

Canada has a limited area where they can grow soybeans. Recently, Canada ranked eighth in soybean production in the world. Cultivars and production practices are similar to those in the Midwest United States.

Soybean imports for the last five marketing years appear in Table 2.1. The People's Republic of China has become the largest importer of soybeans. Until about 30 years ago China was an exporter of soybeans (Wilcox, 2004). The European Union continues to be a major importer of soybean since a large demand exists for the protein in soybean meal, and most countries do not have large areas that are favorable for soybean production. Japan also has been a major soybean importer for many years. Japan imports soybeans for human food use as well as for crushing. Other important importing countries include Mexico, Taiwan, The Republic of Korea, Indonesia, Iran, and Turkey (Table 2.1).

About 29% of the world's production of soybeans was exported in 2006/07 (Table 2.1). Although Brazil has increased exports in recent years, the United States remains the number one soybean exporter. Argentina is also a major exporter of soybeans. Paraguay exports over half of the soybeans they produce. Many of the soybeans exported from Canada are used for human food.

Production practices vary considerably around the world. Sizes of fields vary from a few square meters to thousands of hectares. The work of growing soybeans may be mainly done by hand or almost totally mechanized. While most soybeans are grown under rain-fed conditions, irrigation is used for at least some production in many countries. Inputs also vary from no to large amounts of fertilizer and pesticides throughout the growing season. Although production situations do vary, space limitations in this chapter do not allow a complete discussion of all situations; therefore the following material highlights the major management practices used in production of soybeans.

Since no one management system is best for all situations, the soybean grower needs to consider all production variables to meet the unique characteristics of each individual field. The production of a successful crop includes everything from land preparation, to planting, to harvest and storage, to sale of the grain.

Soybean production generally begins with tillage that is designed to prepare a seedbed, to incorporate fertilizer and herbicides, and to control weeds. Tillage systems include full tillage (also called clean tillage; little if any residue is left on the soil surface), reduced tillage (up to 30% residue is left on the surface), conservation or minimum tillage (greater than 30% residue remains on the surface), and no-till (no tillage is done prior to or during the growing season) (Hoeft et al., 2000; Heatherly & Elmore, 2004). Secondary tillage before planting may involve the use of a disc or field cultivator to further prepare the seedbed or to control weeds that may have germinated. Post planting tillage is done to control weeds. Rotary hoeing is usually done shortly after soybean emergence while inter-row cultivation with a row crop cultivator may be done one to three times before the canopy closes.

Since soybean is a leguminous crop, it fixes its own nitrogen in association with *Bradyrhizobium japonicum*. If soybeans have not been grown on the field or it has been many years since soybeans were raised, an inoculant should be applied at planting to establish the bacteria in the soil (Hoeft et al., 2000). The specific amount of other major and minor nutrients to apply to the soil depends on the results of soil tests and the yield level anticipated. Heatherly and Elmore (2004) discuss lime and fertility needs for soybeans in greater detail.

Cultivar selection is a very important step in achieving maximum soybean production. Improved cultivars are available for all soybean-producing areas. Selecting a cultivar should be done on an individual field basis. Important aspects of a cultivar include yield potential in its area of adaptation, resistance to diseases, nematodes and insects, tolerance to various abiotic stresses (including soil pH, drought, and salt), levels of protein and oil, and tolerance to herbicides. In the United States, a relative maturity system is used to indicate where cultivars are considered full season. Other parts of the world use a number of other systems to classify when and where cultivars should be planted. In most soybean-growing areas of the world, many cultivars (both publicly and/or privately developed) are available with different characteristics that are suitable for almost any given environmental situation. In most cases, public or private organizations conduct cultivar tests to aid growers in selecting the best cultivar(s) for their fields. Use of high-quality seed helps assure good results.

Planting date varies considerably around the world but is mainly determined by temperature and/or rainfall (i.e., water availability to produce a crop). The length of the growing season certainly influences planting date in temperate regions. Row width and plant density (spacing within the row) also vary markedly in different countries. However, the general recommendation is to plant 275,000 to 350,000 seeds per hectare. Optimum planting depth for most cultivars and soils is 2.5 to 4 cm. It is important to make sure good seed-soil contact occurs at planting for uniform emergence.

Soybeans are grown as a single crop or as a double crop planted after harvest of a previous crop or even as an intercrop in some situations (Johnson, 1987; Heatherly & Elmore, 2004). Adjustments of cultivars, planting dates, row spacing, and plant characteristics may be needed for different situations.

Weed control is one of the primary management factors that leads to high yields. Most large-scale production systems manage weeds with the use of herbicides. Depending on the compound or compounds used, these chemicals may be applied preplant, pre-emerge or post-emerge. The appropriate weed management regime requires careful consideration of the situation in each field (Buhler & Hartzler, 2004). With the introduction of the glyphosate tolerance gene into soybeans in the last decade, most soybean producers in the United States and Argentina (and many in Brazil) plant cultivars with the glyphosate-tolerant gene and have gone to mainly using glyphosate post-emergence for weed management.

Lack of adequate water at critical growth stages limits soybean production and yield improvements (Specht et al., 1999). Thus, soybeans are irrigated in some environments to optimize yields. Determining the need and timing of irrigation to make maximum use of the applied water requires careful management on a daily basis (Heatherly & Elmore, 2004).

Diseases, nematodes, and insects can cause major yield losses if they are not controlled. In most cases, chemicals are available to help control the problem, but the chemicals and their application can be expensive. Soybean breeders continue to work hard to develop resistant cultivars in areas where specific diseases or nematodes or insects are problems on a regular basis. Space precludes a detailed discussion of management of these pests. Recent summaries of management practices for diseases, nematodes, and insects are found in Grau et al. (2004), Tolin and Lacey (2004), Niblack et al. (2004), and Boethel et al. (2004).

Soybeans are harvested after the plant has matured (the leaves drop and the pods turn brown). In some instances, frost may prematurely kill plants in northern growing regions, or a dessicant may hasten dry down (Heatherly & Elmore, 2004). Soybeans are usually harvested after the seed moisture drops below 14%. If they are harvested at a higher moisture, they may need drying before they can be safely stored. If soybeans become too dry, cracked seed coats or split seeds may be caused by harvesting or grain transfer equipment. Adjustments to machines may be necessary to limit damage.

If the harvested soybeans are not sold directly from the field, they are stored. Market-grade soybeans can be stored up to about three years at 12% moisture (Johnson, 1987). Monitor stored soybeans on a regular basis for insects, moisture damage, or other problems that can occur during storage, especially in warmer climates.

Soybean Genetics

Many topics are available for discussion under soybean genetics. These include qualitative genetics, quantitative genetics, cytogenetics, molecular genetics, and the rapidly developing field of soybean genomics. The sequencing and subsequent annotation of the soybean genome and development of the soybean genetic map impact all of the areas of soybean genetics.

Qualitative Genetics

The first report of the inheritance of a qualitative trait of soybean in the literature (soybean pubescence color) occurred about a century ago (Piper & Morse, 1910). Subsequently, geneticists reported loci controlling a number of different types of traits including diseases, nematodes, insects, herbicide reaction, nodulation response, rooting response/reaction, growth and morphology, fertility/sterility, physiology, pigmentation, isozymes, seed storage proteins, and seed fatty acids (Palmer et al., 2004).

Breeders and pathologists have studied loci controlling reaction to many soybean diseases. Diseases that have had alleles for resistance published in the literature include: bacterial blight, caused by *Pseudomonas syringae* pv. *glycinea* (Coerper) Young, Dye, and Wilkie; bacterial pustule, caused by *Xanthomonas campestris* pv. *glycines* (Nakano) Dye; brown stem rot, caused by *Phialophora gregata* (Allington and Chamberlain) W. Gams; frogeye leaf spot, caused by *Cercospora sojina* Hara; downy mildew, caused by *Peronospora manshurica* (Naum.) Syd ex Guam; powdery mildew, caused by *Microsphaera diffusa* CKe. & Pk.; phytophthora root rot, caused by *Phytophthora sojae* (Kaufmann and Gerdemann); stem canker, caused by *Diaporthe phaseolorum* (CKe. & Ell.) Sacc var *caulivora* Athow and Caldwell and *D. phaseolorum* f. sp. meridionalis (Morgan-Jones); sudden death syndrome, caused by *Fusarium solani* (Mart.) Appel and Wollenweb. emend. W.C. Snyder & H.N. Hans. (also designated *F. solani* f. sp. *glycines*); soybean rust, caused by *Phakospora pachyrhizi* Syd. & Syd.); soybean mosaic virus; peanut mottle virus; cowpea chlorotic mottle virus; and beanpod mottle virus. Palmer et al.(2004) discusses a summary of the loci and alleles for disease reaction. The development of commercial cultivars with qualitative resistance to one or more diseases significantly reduced economic losses to diseases in soybeans. Although many diseases can cause occasional widespread losses, currently the disease receiving much attention in North and South America is soybean rust, which has the potential to almost completely destroy soybean yields over wide areas unless it is controlled (mainly by chemical sprays currently).

Many species of plant-parasitic nematodes feed on soybeans (Caviness & Riggs, 1976). Qualitative genes were reported on three species. Five loci were reported for soybean cyst nematode, *Heterodera glycines* Ichinoe (Caldwell et al., 1960; Matson & Williams, 1965; Rao-Arelli et al., 1992; Rao-Arelli, 1994). Allelic differences from resistant parents at the *Rhg*1 locus were observed (Brucker et al., 2005). Resistance to root knot nematode (*Melodogyne incognita* (Kofoid & White) Chitwood) and the reniform nematode (*Rotylenchulus reniformis* Linford & Oliveira) were reported (Williams et al., 1981; Luzzi et al., 1994). Many commercial cultivars have resistance to nematodes, especially soybean cyst nematode.

Some soybean genotypes show differences in the degree of sensitivity or tolerance to herbicides. Reports of sensitivity to bentazon, metribuzin, and chlorimuron and tolerance to sulfonylurea herbicides are summarized by Palmer et al. (2004). The most well-known herbicide-tolerant line (and the source of all glyphosate-tolerant soybeans) is soybean line 40-3-2 (Padget et al., 1995). The glyphosate-tolerant gene was inserted using transformation and is reported to behave as a single dominant gene. The majority of soybeans planted in the United States, Brazil, and Argentina have the glyphosate-tolerant transgene.

The control of nodulation of soybean with nitrogen-fixing microsymbionts occurs at several loci. These include non-nodulating, ineffective, and hypernodulating reactions (Palmer et al., 2004). These nodulation reactions may be in association with the slow-growing bacteria *Bradyrhizobium japonicum*, *B. elkanii*, and *B. liaoningense*, the intermediate growing *Mesorhizobium tianshanense* and the fast-growing bacterium *Sinerhizobium fredii* (Kyukendall, 2005; Kyukendall et al., 2000). Other root characters affect root fluorescence and necrotic root mutants (Palmer et al., 2004).

Flowering and maturity of soybeans resulting in cultivars that belong to the various maturity groups are controlled by at least seven loci and a long juvenile trait. One of the seven loci is a response to fluorescent light (Buzzell, 1971). The long juvenile trait is actually delayed flowering under short-day conditions (Ray et al., 1995). Palmer et al. (2004) provides a more detailed discussion of qualitative loci affecting flowering and maturity.

Two loci control stem termination in soybean.. The most common locus Dt_1, when it is homozygous dominant, gives an indeterminate plant found in almost all northern cultivars (maturity groups 000–IV), whereas, the homozygous recessive genotype gives a determinate plant found in almost all southern cultivars (maturity groups V–X) (Bernard, 1972). The second locus Dt_2 produces a semideterminate type (Bernard, 1972). Palmer et al. (2004) summarize other loci affecting the growth of the stem, petioles, and influorescence.

Dwarfness in soybean can be produced from at least twelve different qualitative loci. Another twelve loci are involved in producing different leaf forms, which include five and seven foliolate leaf types (Palmer et al., 2004). These authors also summarize the different pubescence types that occur in soybean.

Several authors reported qualitative loci controlling fertility-sterility in soybean At some loci more than two alleles are present. The different fertility-sterility types are maintained in the Soybean Genetic Type Collection that is part of the USDA Soybean Germplasm Collection (Carter et al., 2004). The Soybean Genetic Type Collection is currently managed by Dr. R.L. Nelson, USDA-ARS, University of Illinois,

Department of Crop Sciences, National Soybean Research Center, 110 West Peabody Drive, Urbana, IL 61801, USA.

Genes affecting reaction to nutritional factors including phosphorus, iron, and chloride and the presence/absence of a constitutive nitrate reductase enzyme occur in soybean. Palmer et al. (2004) discuss these as well as leaf flavonol glucosides.

Chlorophyll deficiency or retention, caused by qualitative genes, occurs in most plant species. In soybean the phenotypes can be caused by nuclear genes, cytoplasmic genes, or a nuclear–cytoplasmic interaction. These types also are maintained in the Soybean Genetic Type Collection.

The color of soybean flowers, pubescence, pods, hilum, or seed coat is a qualitative trait frequently used to describe soybean cultivars and genotypes. The most common flower colors are purple and white. The most common pubescence types are gray and tawny (brown). Seed coat and hilum colors that occur often are yellow, black, brown, buff, and imperfect black. Pod color is usually brown or tan. Qualitative genes also control the distribution of color on the hilum or seed coats. Palmer et al. (2004) summarized the genes controlling pigmentation in flowers, pubescence, pods, and seeds.

The first simply inherited molecular markers used were isozymes. Isozymes are detected as either mobility variants or null variants and are observed using starch, polyacrylamide, or acrylamide/starch gel electrophoresis. Most of the isozyme reports occurred before the mid-1990s, and Palmer et al. (2004) summarized them. Since the mid 1980s, DNA marker polymorphisms have been used extensively in soybean. These markers are usually considered as simply inherited and include restriction fragment length polymorphisms (RFLP), microsatellite or simple sequence repeat markers (SSR), random amplified polymorphic DNA (RAPD), amplification fragment length polymorphism markers (AFLP), and single nucleotide polymorphisms (SNP). Shoemaker et al. (2004) discuss the various marker systems and their use in soybean genomics. Molecular markers associated with qualitative loci are found at URL: http://soybase.ncgr.org.

Cho et al. (1989) studied the inheritance of the glycinin subunits of soybean seed storage protein. They identified five genes, one of which had a third allele. Palmer et al. (2004) discuss these loci and loci associated with protease inhibitors and peroxidase.

Qualitative genes affecting the levels of all five of the major seed fatty acids were identified, which has impacted the use of soybean oil for human and industrial purposes. The fatty acid content of normal commodity soybeans is approximately 11% palmitate, 4% stearate, 24% oleate, 54% linoleate, and 7% linolenate (Fehr, 1991). Palmer et al. (2004) discuss all qualitative genes affecting fatty acid composition.

Because of worldwide interest in the health concerns of consuming *trans* fats, and subsequent food-labeling requirements for *trans*-fat content in the United Stastes and elsewhere, great interest in low-linolenate soybean oil (less than 3% linolenate) has arisen (Chapters: *Lipids*, *Food Uses for Soybean Oil and Alternatives to* Trans *Fatty Acids in Foods*, and *Human Nutrition Value of Soybean Oil and Soy Protein*; Fehr & Hammond, 1996). The low linolenate trait is simply inherited so most efforts have been to combine low linolenate with good agronomic performance. Commercial cultivars with low linolenate are available from a number of soybean seed enterprises.

Quantitative Genetics

In soybean, many economically important plant characteristics, including yield and the most important seed traits, are quantitatively inherited, meaning they are measured as a continuous range in phenotype. Since the environment can have a relatively large effect on quantitative traits, information on photoperiod (latitude and planting date) and temperature (especially during seed filling and maturation) are needed to make meaningful comparisons of experiments. Traits such as yield, plant height, length of the growing period, length of the seed filling period, and seed composition are quantitative traits that are significantly influenced by the environment.

The genetic components of quantitative traits have traditionally been ascertained by determining the heritability of that trait (Burton, 1987; 1997). Then, heritability estimates are used to predict gain from selection for quantitative traits in breeding populations (Brim, 1973). Since the early 1990s, molecular markers have generally been used to find quantitative trait loci (QTL) for important quantitative traits, which can be accomplished through an analysis of the association of a phenotypic trait and genetic marker data from a population of lines segregating for loci (genes) that influence that trait. The analysis allows the quantitative trait to be partitioned into a set of discrete QTLs (Tanksley et al., 1989). Detailed summarizations of QTL mapping in soybean are found in SoyBase (http://soybase.ncgr.org).

Summaries of studies on heritability of quantitative traits in soybean were published by Brim and Stuber (1973) and Burton (1987) and more recently on QTL's by Orf et al. (2004). Among the agronomic traits studied are yield, plant height, lodging, days to flower, days to maturity, reproductive period, seed weight, stem diameter, and iron chlorosis score. Quantitative genetic studies on seed composition include protein content, oil content, sugar content, and various fatty acids, amino acids, and simple sugars. Physiological traits studied are characteristics such as water use efficiency, photosynthetic rate, flooding tolerance, aluminum tolerance, salt tolerance, specific leaf weight, leaf length, leaf width, leaf area, and early plant vigor. Other traits include hypocotyl length, soybean aphid resistance (*Aphis. glycines* Matsumaxa), sudden death syndrome resistance, Sclerotinia stem rot resistance (caused by *Sclerotinia sclerotionum* (Lib) deBary), and corn earworm resistance (*Helicoverpa zea*, Boddie).

Breeders and geneticists are beginning to consider using marker-assisted selection for some quantitative traits. Data collected from breeding programs and the soybean germplasm collection suggest a large range in values for most traits exists. As more information about the complete soybean genome becomes available, possibly greater gains from selection for quantitative traits may result.

Cytogenetics

The cultivated soybean belongs to the genus *Glycine*. Two subgenera are within the genus: *Glycine* (perennials) and *Soja* (Moench) F.J. Herm. (annuals). The subgenus *Glycine* contains 22 perennial species while the subgenus *Soja* contains the cultivated soybean *Glycine max* (L.) Merr. and its annual wild progenitor *Glycine soja* Sieb & Zucc. Details of the species in the genus *Glycine* and genomic relationships are found in Hymowitz (2004). Crosses between *Glycine max* and *Glycine soja* are generally successful but between *Glycine max* and any of the perennial species extremely difficult (only using embryo rescue techniques) or has not been successful to date.

Glycine max has 20 chromosomes so its diploid number is 2n = 40, although many cytogeneticists consider it an autopolyploid or allopolyploid with diploid-like meiosis (Hymowitz, 2004). Techniques for counting soybean mitotic chromosomes are detailed in Xu et al. (1998) and for meiotic chromosomes by Singh and Hymowitz (1985). The ideogram of the pachytene chromosomes and the genomic relationship of *G. max* and *G. soja* was published (Singh & Hymowitz, 1988). Soybean has chromosomes that are smaller than most other crop plants; thus fewer cytogenetic and cytological studies are reported in the literature. Hymowitz (2004) reported on the identification of primary trisomics, tetrasomics, monosomics, translocations, inversions, and monosomic alien addition lines in soybean. The assignment of molecular linkage groups to many individual soybean chromosomes using primary trisomics was reported (Zea et al. 2003; 2006). Walling et al. (2006) also have added to the understanding of soybean chromosome structure. The materials and research should be useful for future genetic and cytogenetic studies and potentially for making further improvements through breeding.

Molecular Genetics and Genomics

The first molecular genetic map of the soybean genome (based on RFLP markers) was published by Keim et al. (1990). The concept is based on the idea that DNA polymorphisms could be used to develop molecular genetic maps (Botstein et al., 1980). As new technologies were developed, especially the use of the polymerase chain reaction (Mullis et al., 1986) as a tool to detect polymorphism in genomes, additional classes of DNA markers were used in soybeans. The latest molecular genetic linkage map is at http://soybase.ncgr.org. Details of DNA markers used in soybeans and the development of molecular genetic maps are given by Shoemaker et al. (2004).

Some consider soybean to be a model crop system, especially since its genome is sequenced in the united States, China, and Japan. Other aspects that make it a potential model system include its densely saturated genetic map (see SoyBase), a genetic transformation system useful for research and breeding (Parrott et al., 2004), and genetic and cytogenetic materials. By all estimates, the soybean genome has about 1.1 Mbp/C (Arumuganathan & Earle, 1991), and the total length of the genome is currently over 3000 cM (SoyBase).

The fields of molecular genetics and genomics of soybean are changing very rapidly as new technologies from other projects, especially the human genome project, are applied to soybean. As the field of genomics develops and changes, new terms such as functional genomics, proteomics, and metabolomics are discussed. Genomics refers to the study of all genes, regulatory sequences, and structure and function of the genome. Proteomics refers to analyzing hundreds or thousands of proteins at a time, while metabolomics refers to profiling the metabolites of cells and/or tissues. Shoemaker et al. (2004) state, "In the broadest view, functional genomics is defined as the process of generating, integrating and using information from genomics (sequencing), gene expression profiling (microarrays and chips), proteomics, metabolic profiling and large-scale genotyping and trait analysis to understand the function of genes." Shoemaker et al. (2004) summarized soybean genomics to the date of the publication; however, the most current publicly available information for soybean is found in SoyBase (http://soybase.ncgr.org).

Soybean Breeding

As noted earlier in this chapter, China domesticated soybean. The selection of cultivars with different characteristics occurred over the centuries. Exactly when and where deliberate and continuous breeding and selection of soybean occurred are not known. The first cultivars grown in the United States were direct introductions from Asian countries—mainly China and Japan. No actual breeding was done: just selection for uniformity of phenotypic characters if the introduction was a mixture of genotypes. Although a few people were interested in soybeans for use as food, the main consideration in the late nineteenth and early twentieth centuries was the suitability of cultivars for forage production. Breeding efforts for improved soybean cultivars began in the first part of the twentieth century. During that time, processors began to extract oil from soybean seed. It was not until 1941 that the hectarage grown for grain exceeded that grown for forage (Hartwig, 1973). Much of the early breeding efforts were aimed at increased yield, higher oil content, and the elimination of shattering. Soybean breeding and cultivar improvement made a major step forward with the establishment of the U.S. Regional Soybean Industrial Products Laboratory in Urbana, Illinois, in 1936. This laboratory worked in cooperation with the experiment stations from 12 North Central states. The program also did research and measurements on protein and oil in soybean cultivars. The cooperative work continued and expanded to 12 southern states and eventually to all agricultural experiment stations interested in soybean research. The increased efforts by the United States Department of Agriculture (USDA) and state agricultural stations has resulted in hundreds of publicly developed soybean cultivars since the mid 1940s. These cultivars ranged from maturity group 00 to VIII. The passage of the Plant Variety Protection Act in 1970 (revised in 1994) provided incentives and legal protection for soybean cultivar development. Publicly developed soybean cultivars dominated until the mid 1980s.

The use of privately developed cultivars by soybean producers increased significantly during the 1980s. After the introduction of glyphosate-tolerant cultivars in 1996, the proportion of the total hectarage planted to private varieties has risen to over 90% (Parrott & Clemente, 2004; USDA–NASS, 2007).

Conventional Breeding Methods

Conventional breeding methods have been very successful in improving the productivity, hazard resistance, and quality of soybean. Breeding for direct improvement of yield remains the trait of greatest emphasis by breeders as it is the trait that is of greatest interest by producers. Breeding to improve or protect yield through hazard resistance or breeding for enhanced quality is also an important part of all breeding programs and may require additional or special breeding methodologies. Progress in breeding has been made for many traits including yield; resistance to pathogens, insects, and nematodes; tolerance to herbicides and production hazards; and improvement in seed protein, oil, and other quality traits as well as other agronomic characteristics such as standability and adaptability.

In general, each breeding method that leads to genetic improvement begins with the breeder making choices as to the parents or starting material to be used to create segregating populations. Those populations are then advanced toward homozygosity, without selection or with selection that may involve various techniques, to produce relatively homozygous lines that are then subject to yield and other trait evaluations. The breeding method (or cycle) is complete when the best line(s) are released as improved pure-line cultivars or improved germplasm. The pure-line cultivar is what is grown by the farmer.

Soybean breeders and geneticists use many different breeding methods for cultivar and/or germplasm development. Most, if not all, methods have a number of aspects in common. They include the objectives of the breeding/genetics program, selection of parents, type of populations and selection, and inbred line development.

Objectives

The objectives of a breeding program sometimes dictate which breeding method(s) might be best used. Even though this section discusses conventional breeding methods, the identification of objectives is equally important (and the same) for molecular-based breeding efforts. Although the objectives for a specific cross or program may be limited and highly dependent on the individual situation, many breeding/genetics programs at least have some of the following traits or characteristics as consideration for selection.

In almost all cases, yield or productivity is the character of greatest importance or at least among the characters of greatest importance. Since yield is a quantitative trait, it is the most challenging trait to breed for in a genetic improvement program. Over the decades, considerable progress has occurred in improving yield. Undoubtedly, further yield increases will occur in the future.

The yield potential of a cultivar or germplasm line will not be realized if it is injured by diseases, insects, or nematodes. In almost all breeding programs, resistance to some pest or pests is part of the objectives. The amount of emphasis placed on pest resistance depends on the regularity and severity with which the particular pest problem(s) occur in the target breeding area and the level of economic loss that can occur from the pest. Specific resistance, general resistance, or tolerance can provide protection against economic loss. Specific resistance is usually conferred by one or a few major genes and can be easily transferred to susceptible cultivars. The disadvantage of specific resistance is the fact that it may not provide protection to new races of a pest. General or field resistance (or sometimes called field tolerance) is mainly responsible for reduced levels of infection but does not confer immunity like specific resistance. This type of resistance is generally quantitative (conferred by many genes). It provides protection against multiple races of pests, but is much more difficult to transfer from a breeding standpoint. Tolerance to a pest is usually defined as a cultivar or germplasm line that suffers less loss in productivity than a nontolerant line even though both lines have similar levels of the pest present. Tolerance is a result of even more complex genetics and interactions than general resistance and, thus, is not frequently used in breeding programs.

Maturity is an important trait for a breeding program in a particular area. Breeders generally work with lines adapted to their target environment; however, if parents of unadapted maturities are used, modified techniques may be needed. Also, for crosses of parents of widely differing maturities, the number of adapted segregating progenies in populations may be limited; thus, larger populations are required in order to obtain a given number of progeny of the desired maturity. Although maturity is generally considered a quantitative character, several major genes for maturity were reported (Palmer & Kilen, 1987; Palmer et al., 2004).

Lodging resistance, plant height, and stem termination are traits important in cultivar development. Major genes control stem termination; however, the final plant height and lodging resistance of adapted cultivars is considered quantitative and must be selected for using field trials in the target environment.

Shattering resistance is generally present in most improved cultivars; however, many plant introductions or germplasm lines may shatter especially under warm and/or dry conditions. Although major genes exist for shattering resistance, several minor or modifying genes make shattering resistance challenging to select for since the climatic conditions that induce shattering can vary from year to year.

Seed size may be an important characteristic especially for special purpose or food-type cultivars. One can select especially large or small seed size for using methods like mass selection or bulk breeding. The inheritance for seed size is quantitative, but selecting for extremes in populations can result in many lines with the targeted seed size.

Seed quality, that is the appearance of the seed, is a trait measured in some cultivar development programs. Unfavorable weather conditions and/or certain diseases may cause undesirable seed quality. Selection for disease resistance can improve seed quality. Germinability is also sometimes part of seed quality. Poor germination tends to be a greater problem in low latitudes. Since seed quality, including germinability, is a complex trait, lines need to be evaluated from field plots (many times with delayed harvest) over several years.

Seed composition is a very important trait in soybean. Since soybean is used for both oil and protein, breeders generally try to aim for 40% protein and 20% oil (on a dry matter basis). In most cases, to date, soybeans are not marketed on composition; however, protein and oil content has been considered in the special purpose/food soybean market for many years. Recently, limited markets are available for commodity soybeans with specified oil and/or protein levels. Although both oil and protein levels are quantitative traits, breeding can readily alter them. As technology for rapidly, accurately, and inexpensively measuring oil and protein content in soybeans becomes available, seed composition will need to be a trait specifically selected for breeders.

Other seed composition traits, besides protein and oil content, have been explored, and to a limited extent, incorporated into commercial cultivars. Among the traits commercialized to date are low-linolenic acid, low or no lipoxygenase, low-saturated fatty acids, reduced trypsin inhibitor (no Kunitz trypsin inhibitor), and higher levels of sulfur-containing amino acids. Several other traits including mid-oleic acid (50–60%), combinations of altered fatty acids, higher levels of other essential amino acids, higher sucrose content, lower oligiosaccharides, higher isoflavones and other desirable traits, for special purpose and/or food soybeans are being selected for in some breeding programs and may become of greater importance in the future. Many, but not all, of these traits are controlled by a few major genes, but most also have modifiers. Thus, breeders will need to assess the levels of the traits, in the lines, in their breeding programs.

Resistance or tolerance to several different production hazards also may be important traits in some breeding programs. Among the traits that have received attention are tolerance to iron-deficiency chlorosis (high pH), acidity tolerance (low pH), drought tolerance, manganese tolerance, salt tolerance, flooding tolerance, and high nitrate tolerance, to name a few. Most of these traits are quantitative in nature and require special field and/or greenhouse or laboratory conditions to assess the breeding line or cultivars' response to the particular hazard.

In recent years, resistance or tolerance to herbicide injury has become a very important objective for cultivar development. Currently, tolerance to glyphosate is present in most cultivars in the United States, Argentina, and Brazil. This trait was introduced via transformation and is simply inherited. Resistance or tolerance to other herbicides including metribuzin, dicamba, glufosinate, and 2,4D have been reported and in some cases tolerant versions of cultivars released. The resistance or tolerance is generally simply inherited and possibly was introduced into soybean via transformation.

As biotechnology and transformation techniques continue to improve other traits or characters will likely be introduced into soybeans. In the near term, most of those traits will be simply inherited making the incorporation of the traits relatively easy regardless of the breeding method used.

Selection of Parents

Selection of parents is an extremely important part of any soybean breeding method. The selection of parents sets in motion the whole cascade of events in succeeding generations of all breeding methods. The parents used to create segregating populations can be from many different sources such as existing cultivars, adapted elite breeding lines, unadapted germplasm with special desired traits, or even exotic germplasm. Generally, elite parents of diverse origin are more likely to produce progeny that are superior to either parent (and superior to existing cultivars) than parents that are closely related (Burton, 1987). The way parents are selected depends on many factors, including the trait(s) of interest, the purpose of the cross, the relative importance of characters other than yield, the ancestry of the lines, and the resources and time available. Parents may be selected on the basis of comparative evaluation *per se*, by test-cross evaluation or other methods that may identify germplasm with good combining ability. In many cases, *per se* evaluation data are readily available in the form of breeder-directed or fee-based yield performance tests or from government-required tests. If the objective is to identify parental germplasm with favorable alleles not presented in existing cultivars, test-cross evaluations may be a better approach. Kenworthy (1980) suggested a method for soybean. Another test-cross method developed by St. Martin et al. (1996) outlines a procedure for identifying germplasm lines with the potential to contribute favorable alleles for improving pure-line cultivars of soybean. Another method for improving yield suggested by Henderson (1975), and more fully explored by Panter and Allen (1995), is the use of the best linear unbiased predictions (the use of a mixed linear model).

Selection of parents will continue to be a very challenging but extremely important aspect that determines the success of all breeding procedures for genetic improvement. In many programs where resources are quite limited, use of existing comparative data and/or the best linear unbiased predictions appears to be very useful. If more time and resources are available, one can use test-cross evaluations to identify parental germplasm with favorable alleles not present in current cultivars or breeding lines. With the increased availability of molecular and genomic data on individual cultivars and genotypes, this data will become more valuable in assisting breeders/geneticists in the selection of parents (Orf et al., 2004).

Once the parents are selected, a cross or crosses are made to initiate populations. Populations can be developed with different numbers of parents and varying percentages of each parent before inbreeding and selection are begun. The majority of soybean cultivars were selected from populations that resulted from two- or three-parent hybridizations involving existing cultivars, breeding lines, or other germplasm (Fehr, 1987b). Multiple parent populations (more than two parents) are less common; however, three, four, or as many as eight parents are used to develop breeding populations. Backcross populations involve the use of a nonrecurrent parent and the repeated use of the recurrent parent in crossing. Backcross populations were generally developed to transfer genes for pest resistance or other simply inherited traits from an agronomically unacceptable parent into an elite cultivar or breeding line. Although some researchers refer to recurrent selection populations, it is probably more appropriate to refer to a population of a specific cycle of recurrent selection. In recurrent selection, many (sometimes dozens) of parents are present.

Inbreeding, Selection, and Line Evaluation

After a cross or crosses are made, the populations are then advanced through several generations of selfed inbreeding. A number of factors need to be considered during inbreeding. Among them are the method of inbreeding (including possible selection) and the number of generations of self-pollination to allow before lines are derived for potential cultivar evaluation. Lines can be derived from a population in the F_2 or in any of the more advanced generations of inbreeding. Selection can be practiced among plants during early generations of inbreeding, before yield tests are initiated or later among lines during yield testing. The amount and effectiveness of selection depend on the heritability of the trait or character and the environment where the population or lines are grown. Visual or easily determined selection is mainly carried out during early generations of inbreeding, while selection based on data from unreplicated or replicated plots is carried out in later generations. At some point in the inbreeding process, nearly homozygous lines are created from individually harvested inbred plants. These lines are then extensively evaluated to identify those that are superior in performance to existing cultivars. The methods of inbred line development include pedigree, bulk, single seed descent, mass selection, and early generation testing. Other methods or procedures used in soybean cultivar development include backcrossing and population improvement using recurrent selection that may involve a genetic male sterility system.

Pure Line Method

Although selection for desired plant types has been going on for centuries and led to new and better cultivars, it was not until the early twentieth century that scientists developed theories and methods that were routinely applied to genetic improvement of soybeans. The first technique or method employed by breeders/geneticists was the pure line method of breeding. In this method no artificial hybridization occurs but rather the breeder selects individual plants from an already existing "mixed" cultivar, meaning a cultivar that has several different phenotypes and, thus, different genotypes. Since soybean is a self-pollinating species, it is assumed each phenotype is essentially homozygous and true breeding. By selecting individual plants that are then planted out in progeny rows, meaning a row of plants that are from the seed of an individual plant, the breeder can observe and select those progeny rows with the desired trait(s) or characteristics for an improved or new cultivar. Selected progeny rows, which are generally 2–3 meters long, are harvested and the seed from those rows generally used to plant multi-row, multi-location, replicated yield trials in the area where the breeder works. The yield test includes standard or check cultivars and/or genotypes. The performance of the experimental lines (selected progeny rows) in a test is compared to the performance of the checks, and only those that are superior to the checks are saved for further testing in succeeding years. Besides yield, other traits such as maturity, lodging, seed composition, and hazard resistance may, in most cases, also need to be compared with the check genotypes and used as criteria for deciding which experimental lines will be further evaluated. The first-year yield trials are mainly used to eliminate the unpromising lines that are inferior to the checks rather than trying to identify the best experimental line or lines. The breeder may evaluate the selected lines from the first-year yield trials a second year in local yield trials. Again, yield performance is the trait of greatest interest, but each experimental line in the second-year's tests is also rated for other traits, as noted in the first-year evaluations, and compared to the check genotypes. The data from the two years are combined, and only those experimental lines that are superior to the check cultivars or genotypes are saved for further testing. Thus the emphasis, using all the data collected, shifts to identifying and selecting those lines that are truly superior to the checks.

After the best lines are identified by local testing, the lines are then entered into regional testing (multi-state or multi-country testing). The regional testing involves experimental lines from several breeders as well as check cultivars or lines. In many cases, the regional tests are government-sponsored or official government trials. The procedure for regional trials varies but may require two to three years of testing before a line is considered for release as a cultivar. The final decision on release is generally made by the institution or company that employs the breeder and is usually made by a cultivar release committee or administrative group.

As the experimental line is being evaluated, usually during the regional testing, the line undergoes a purification and seed multiplication process so significant quantities of seed are available upon release. The purification process involves selecting a single plant or a limited number of plants (20–100) to be grown out in a progeny row(s) for observation of phenotypic and/or molecular characteristics. Seed from the progeny row or a bulk of the uniform rows is increased to form the initial breeder's seed. The breeder's seed is then further multiplied over several generations, with or without the use of winter seed increases, to provide sufficient seed of the new cultivar at time of release for sale to soybean growers.

The process of testing, seed multiplication, and release is as briefly described above. But most methods are carried out with some modifications by breeders, depending on the program and the trait(s) that are being selected.

Pedigree

The pedigree method has been used since the rediscovery of Mendel's laws and has led to many successful soybean cultivars. In the pedigree method, the ancestral lineage of each line tracing back to individual F_2 plants is recorded. Thus, care must be taken to keep accurate records so each selection can be traced back to the original hybridization. The size of the F_2 population is a subject of considerable discussion among breeders and depends on the resources of the program including personnel, equipment, land area, laboratory space, as well as the philosophy of the breeder: for example, does the breeder favor more crosses with few plants selected from each cross or fewer crosses with more plants selected from each cross? The literature suggests 2,000–5,000 F_2 plants, with 5–10% of those plants selected. The F_2 population should be space planted so the phenotype of each F_2 can be observed. Keep in mind that factors exist such as competitiveness, G x E interactions, genetic components (remaining heterozygosity, epistasis, dominance, etc.), and the interplay of these factors that can influence the phenotype observed in the F_2. The pedigree method allows selected plants (and their progeny) to be observed in additional generations (if selected) so that those plants that are truly desired from a phenotypic standpoint are continued to the yield testing phase (Fehr, 1987a). Also some debate persists as to whether the F_3 progeny row is space planted or planted at "normal" densities. Since individual F_3 plants need to be selected from the F_3 progeny rows, most breeders use a density lower than for commercial production of soybeans (in the range of 50 to 75%). With each successive generation of inbreeding, the additive genetic variability within lines is reduced and additive genetic variability between lines increases. Thus, in the pedigree method, the idea is to retain the maximum number of lines that trace back to different F_2 plants. In practice, this means that the number of plants selected within lines decreases with each generation of inbreeding: for example, four plants are selected from an $F_{2:3}$ line while only two plants are selected from an $F_{4:5}$ line. The number of generations that pedigree selection is practiced also depends on the resources of the breeding program or when a line appears uniform for the trait(s) being selected. Breeders generally harvest lines in bulk to begin replicated yield testing after the F_4, F_5 or F_6 generation and discard any heterogeneous lines for the trait(s) under selection.

The pedigree method has been used to develop many soybean cultivars. This method allows the breeder to discard inferior phenotype material early in the inbreeding process, allows the breeder to minimize the relationship among retained lines, and provides phenotypic observations over several generations in different environments. However, this method requires considerable land, labor, and other resources as well as extensive record keeping, usually requires an experienced breeder to make selections, and is not effective in environments where genetic variability for trait(s) is not expressed. This last point means that the pedigree method is not well-suited for inbreeding in greenhouses or winter nurseries and, thus, only one generation per year can be completed; therefore the pedigree method takes a longer time before a line is released. Thus, the pedigree method is no longer widely used in soybean breeding especially in production areas where current cultivars are rapidly replaced by newer cultivars.

Bulk

In the bulk method, plants in segregating population(s) are harvested together each cycle of inbreeding, and a sample of the harvested seed is used to plant the next generation. When the desired level of inbreeding is reached, single plants are harvested and grown as progeny rows and then selected lines advanced to yield evaluations. In the classical application of the bulk method, the main force acting on the population is natural selection. Natural selection favors those traits or characteristics that increase a plant's competitiveness (that is, it allows a plant to produce more seed than another) and may include such traits as tall height, late maturity, and resistance to natural hazards (diseases or insects). Generally, bulk populations are planted at "normal" densities; however, some breeders may use a reduced density to encourage less competitive genotypes.

The beginning population(s) for the bulk method can vary considerably from two-parent to multiple-parent crosses, to crosses of unadapted by adapted parents, to a mixture of many F_2 populations such that one or only a few bulks contain many crosses. The number of generations of bulk harvest can also vary considerably from two or three to as many as eight to ten. The greater number of generations gives natural selection more time to work. Although natural selection is the "classical" way to conduct bulk selection, many breeders will use some combination(s) of natural and "breeder imposed" (mass) selection for their populations. The bulk method can produce lines with the desired traits if the environment where the method is carried out favors those traits. The bulk method is an easy way to maintain populations during inbreeding and should increase the frequency of the desired genotypes in the population(s). Since bulk selection depends on natural selection, the trait or traits that are "selected for" may change from year to year depending on the natural environment. Also, since the breeder depends on natural selection, only one generation per year is accomplished, thus extending the time it takes to develop a cultivar (Sleper & Poehlman, 2006).

Mass Selection

Mass selection in a heterogenous population that is undergoing inbreeding (self-pollination) is one of the oldest methods of breeding. Some breeders consider mass selection a variation of the bulk method; only in the case of mass selection, the breeder does the selection rather than "nature." Mass selection may be done from either a positive or negative approach. Positive selection is when the desired phenotypes are selected and rebulked. Negative selection is when the undesirable phenotypes are culled or removed from the population. Generally, mass selection results in a population that is selected and therefore improved for one or more traits or characteristics. Some examples of traits in soybean that are subject to mass selection are maturity, seed size, and seed composition. If mass selection is practiced in the classical sense, the resulting cultivar is heterogeneous for some traits since a relatively large number of plants is bulked to form the cultivar. However, if many traits are mass selected, the resulting cultivar (at least phenotypically) may appear uniform. Mass selection may also be used to maintain the purity of cultivars by the rouging of "off-type" plants and/or seed (negative selection). Mass selection is only used in environments or situations where the trait(s) or character(s) is expressed, and its effectiveness depends on the heritability of the character on a plant or seed basis (it does not work well for characters with low heritability). In many soybean breeding programs, some form of mass selection is used at some point in the cultivar development process.

Single-seed Descent (SSD)

Single-seed descent is a method to rapidly inbreed a population before beginning evaluation. Goulden (1941) initially proposed it, and Brim (1966) more fully described it.

Single-seed descent (SSD) is referred to as a modified pedigree method by Brim (1966). In the strict sense, SSD refers to harvesting and planting only one seed from each plant from the F_2 generation on until plant selection is done. This means, due to failure of germination or a plant to reach maturity or to produce one seed, that not all F_2 plants are represented when generation advance is completed. In practice, most breeders harvest one (or two) pod(s), thresh the pods, and take a sample that is approximately the same number of seeds as the previous generation. This maintains the population size and also provides a remnant. Thus, not every F_2 plant is represented only once. Some are not represented, whereas others are represented more than once. This situation often is referred to as a modified SSD, or as a modified bulk. The general idea of SSD is to advance the population to a desired level of homozygosity via inbreeding as rapidly as possible, with the use of greenhouses, growth chambers, or off-season nurseries, and then begin evaluation of progeny rows.

Single-seed descent is currently the most widely used breeding procedure for inbreeding soybean populations. It is popular because a breeder can obtain two, or three, or almost four generations per year with the use of greenhouses, growth chambers, or tropical nurseries. Thus, a cross is made and a population initially evaluated in about two years, compared to six years for the pedigree or bulk method. This savings of three or four years means, for example, that a cultivar is released in six years instead of ten years with the same amount of yield testing. The use of SSD, however, does not allow for the observation of plants or progeny in early selfing generations and does not allow natural or artificial selection. SSD has contributed to the more rapid turnover of soybean cultivars in the last decade and will continue to do so in the future. Single-seed descent also requires fewer resources and time; thus, it will continue to be the method of choice for soybean breeding programs where cultivar development is the primary objective.

Early Generation Testing (EGT)

Early generation testing is designed to identify bulk hybrid populations that have the greatest potential to produce superior lines. Two methods are used: i) testing of bulk populations or ii) testing of F_2-derived lines that represent the population. The first method may yield test the population, with replication and locations depending on seed supply, in the F_2, F_3 and F_4 generations and only select plants from those populations that are superior to check cultivars, or a certain percentage of all populations tested.

The second method is for using F_2-derived lines for yield testing. Immer in 1941 described the concept of EGT. Some variations of EGT suggest selecting lines at different stages (generations) within populations, among populations or among individuals within populations. Other variations include using EGT to identify the superior populations from which to select lines. Recall all populations were advanced in the pedigree, bulk, or SSD methods. If lines, rather than the whole population, are yield tested in EGT, the main limitation is the amount of seed from F_2-derived lines for the first yield tests. Some breeders grow F_2-derived progeny rows and delay yield testing for one generation. As with the other methods, except SSD, only one generation per year can be grown since yield tests need to be in the target environment. As yield tests are conducted, evaluations for other traits or characters are usually carried out. This method, in addition to being time-consuming, is also expensive due to the land and labor costs of yield testing. Several successful soybean cultivars have been developed using EGT testing.

Backcross

The backcross method is used to add a highly heritable characteristic (allele) to a cultivar or line (recurrent parent) for which it is deficient. The term *backcrossing*, as originally described by Harlan and Pope (1922), refers to the repeated crossing of hybrid progeny back to the recurrent parent. The simplest type of backcrossing scheme is a dominant allele that can be evaluated on a single-plant basis before flowering.

Modifications are needed if the trait cannot be evaluated before flowering. Extra crosses need to be made or, if the allele is recessive, selfing or progeny tests are needed and/or blind backcrosses can be made. Many descriptions suggest five backcrosses; however, some breeders may use fewer or more backcrosses. If molecular markers are used, F_1 or BCn F_1 plants with a larger percentage than average of the recurrent parent alleles in the genome can be used for the crosses so the recovery of the recurrent parent proceeds more quickly, and fewer backcrosses are needed to recover the recurrent parent to a given level. Genes closely linked to the allele being transferred may not be eliminated during backcrossing and thus may make the complete recovery of the recurrent parent phenotype difficult. Details on the aspects noted above are shown in many plant breeding textbooks (Fehr, 1987; Sleper & Poehlman, 2006).

Backcrossing has been used for many years in soybean breeding for various simply inherited traits like disease resistance, leaflet shape, and more recently the Roundup Ready gene. It is possible to backcross more than one trait at a time, but a larger number of plants are needed and extra evaluations are needed to maintain all the desired traits during backcrossing. Because the number of plants used in backcrossing is relatively small and in many cases phenotypic evaluations for the trait are relatively easy, breeders use the greenhouse, growth chamber, and/or off-season nurseries to get several generations per year, as discussed in the SSD section. Thus, backcrossing can be accomplished in about two years and evaluations in another two or three years; so the "new" backcross version of a cultivar can be released relatively quickly. Despite the shorter development time, it is still very important to choose a recurrent parent that is outstanding in almost all traits except the trait to be incorporated by backcrossing.

Recurrent Selection

Recurrent selection is a cyclic method of population improvement but does not directly lead to release of cultivars. The basic steps in a cycle of recurrent selection are intermating, evaluation, and selection. The main challenge in using recurrent selection in soybean is the difficulty of the intermating step. Since recurrent selection is designed to improve the frequency of favorable alleles (for the trait undergoing selection) in a population for quantitative traits, further breeding efforts are needed to release a cultivar from a recurrent selection population. Recent summaries of recurrent selection studies in soybean by Lewers and Palmer (1997) and Orf et al. (2004) discuss traits or characters investigated, selection methods, and intermating methods as well as marker-assisted recurrent selection techniques.

Use of Male Sterility in Soybean Breeding

The use of genetic male sterility to facilitate crossing especially in recurrent selection schemes has been used to some extent since the 1970s (Brim & Stuber, 1973; Lewers et al., 1996). Specht and Graaf (1990) described a breeding method called male-sterile-facilitated cyclic breeding (MSFCB) for cultivar development. The authors suggest this method combines the best aspects of conventional breeding and diallele selective mating as described by Jensen (1970). Briefly, the MSFCB method involves placing annually chosen elite parents in a checkboard row pattern in an isolation nursery containing rows of male sterile parents. Insects transfer the pollen from the elite parents to the male sterile plants. At least one F_1 seed is harvested from each male sterile plant. It is suggested the F_1 plants be grown in a winter nursery and plants threshed in bulk to provide F_2 seed. The majority of the F_2 seed is advanced for cultivar development using single seed descent with male fertile plants. The cyclic part of the method is continued by using a small portion of the F_2 seed for the next year's male sterile plants in the isolation nursery. The male fertile plants are rogued at flowering. A number of high-yielding cultivars were released using the MSFCB scheme.

Mutation Breeding

Mutation breeding can help to develop improved cultivars of soybean. However, most breeders agree that mutation breeding is best used when a desired trait or character is not found in the germplasm that can be used for crossing. Since the frequency of the desired mutation (genetic change) is usually very low, the breeder needs to screen a large number of plants (10,000s to 100,000s) using a rapid, inexpensive procedure or technique. There are many aspects in a mutation breeding program to consider. Beginning with the trait or character, as well as the screening procedure, the choice of line(s) to subject to mutagenesis, the mutagenic agent to use, the type of plant material to treat, and the details of the treatment (dose, condition of the plant material, treatment conditions, etc.). Seeds are the most common plant material treated. The treated seed is usually planted in isolation (to prevent inadvertent crossing). The plants that grow from the treated seed are considered the M1 generation and produce M2 seed. The breeding methods used during the selfing generations after mutagenesis are the same as those used for populations developed from crosses (pedigree, bulk, SSD, EGT) with slight modifications for screening for the desired trait or character. Much greater detail on all aspects of mutation breeding is found in Fehr (1987).

Transformation

Soybeans have had a number of traits added to the genome via transformation. The only trait commercialized to date is glyphosate tolerance. The other transformation events are in various stages of testing and/or worldwide regulatory approval. From a breeding perspective, once a plant is stably transformed, i.e., has undergone several generations of selfing, and is shown to pass the trait on to its progeny, the trait is treated as qualitatively inherited and used as such in any breeding method.

Use of Genetic Markers in Soybean Breeding

The use of genetic markers for cultivar development in soybean breeding programs is just beginning. A need exists to develop methods for the widespread application of marker-based techniques in breeding programs. Currently, genetic markers used in soybean breeding programs complement rather than replace traditional breeding methods. Some current and potential uses of genetic markers in cultivar development programs include the selection of parents, their use in backcrossing to speed up the recovery of the recurrent parent by reducing the number of backcross generations, elimination of undesirable linked loci, or aiding in genotyping for the trait being incorporated, and for marker-assisted selection (Orf et al., 2004). Uses of marker-assisted selection to date include the development of H7242RR (Orf et al., 2004), assisting in the incorporation of soybean aphid resistance (Li et al., 2007), and use for several traits by private sector companies. As more molecular markers and "breeder friendly" markers, such as single nucleotide polymorphisms, become available, marker assisted selection will be used more extensively to complement or supplement the more traditional proven breeding methods for cultivar development.

Hybrid Soybean Cultivars

Some research suggests that the use of F_1 hybrid seed might improve the productivity or other characteristics in soybean. The growing of F_1 hybrids on a commercial scale is not possible currently due to the difficulty of producing large quantities of hybrid seed economically. Progress toward commercial use of F_1 hybrids appears to be occurring, but no reports appear of widespread large- scale testing of hybrids, let alone commercial soybean grain production from hybrids. Palmer et al. (2001) listed five components that are critical for developing hybrid soybean on a commercial scale. They are i) parental combinations that produce heterosis levels superior to the best pure line cultivars, ii) a stable male-sterile, female fertile sterility system, iii) a selection system to obtain 100% female (pod parent) plants that set seed normally and are harvested mechanically, iv) an efficient pollen transfer mechanism from pollen parent to pod parent, and v) an economical level of seed increase for the seedsmen and growers that ultimately benefits the consumer. Progress is being made with regard to some of the components, but it is not likely F_1 hybrids will be commercialized in the near future.

Conclusion

The soybean is an important source of oil and protein for the world. The United States, Brazil, Argentina, and China are major soybean-producing countries. China and the European Union are the largest importers of soybeans. Soybean production practices vary widely around the world. However, the trend in many countries is toward large scale, high input, soybean production practices. Soybean genetics, cytogenetics, molecular genetics, and genomics have made important contributions to soybean breeding and soybean production. As the sequencing of the soybean genome is completed, new opportunities will arise in each of these areas that will impact soybean breeding and ultimately soybean production. Soybean breeding methods for development of commercial cultivars continue to evolve. Although soybean breeders are successful in producing cultivars using traditional conventional breeding methods, the most widely used methods have shifted in recent decades to wide use of single seed descent, and with the advent of transformation, renewed use of backcrossing. Many breeders are now using molecular-based plant breeding methods and techniques, such as marker-assisted selection, as part of their cultivar development program. The challenge is to introgress the new or modified conventional and molecular technologies into existing cultivar development programs so that progress in soybean cultivar development continues in the future. As more is discovered about the soybean genome and soybean genetics and as future trends in soybean production change, both classical and molecular breeding techniques will contribute to developing soybean cultivars for food, feed, and fuel.

References

Arumuganathan, K.; E.D. Earle. Nuclear DNA content of some important plant species. *Plant Mol. Biol. Rep.* **1991,** *9,* 208–219.

Bernard, R.L. Two genes affecting stem termination in soybeans. *Crop Sci.* **1972,** *11,* 242–244.

Boethel, D.J. Integrated management of soybean insects. *Soybeans: Improvement, Production and Uses;* H.R. Boerman, J.E. Specht, Eds.; Madison, WI: American Society of Agronomy, 2004; pp. 853–882.

Boerman, H.R.; J.E. Specht, Eds.; American Society of Agronomy: Madison, WI, 2004; pp. 97–136.

Botstein, D.; R.L. White,; M. Skolnick; R.W. Davis. Construction of genetic linkage map in man using restriction fragment length polymorphisms. *Am. J. Hum. Genet.* **1980,** *37,* 314–331, 1980.

Brim, C.A. Quantative genetics and breeding. *Soybeans: Improvement, Production and Uses;* B.E. Caldwell, Ed.; American Society of Agronomy: Madison, WI, 1973; pp 155–186.

Brim, C.A. A modified pedigree method of selection in soybeans. *Crop Sci.* **1966,** *6,* 220.

Brim, C.A.; C.W. Stuber. Application of genetic male sterility to recurrent selection schemes in soybeans. *Crop Sci.* **1973,** *13,* 528–530.

Brucker, E.; S. Carlson; E. Wright; T. Niblack; B. Diers. *Rhg1* allelles from soybean PI437654 and PI88788 respond differentially to isolates of *Heterodera glycines* in the greenhouse. *Theor. Appl. Genet.* **2005,** *111,* 44–49.

Buhler, D.D.; R.G. Hartzler. Weed biology and management. *Soybeans: Improvement, Production and Uses.* H.R. Boerman; J.E. Specht; Eds. American Society of Agronomy: Madison, WI , 2004; pp 883–918.

Burton, J.W. Soyabean (*Glycine max* (L.) Merr.). *Field Crops Res.* **1997,** 53, 171–186.

Burton, J.W. Quantitative genetics: results relevant to soybean breeding. *Soybeans: Improvement, Production and Uses,* Second ed.; J.R. Wilcox, Ed.; Agron. Monogr. 16. ASA, CSSA, and SSSA: Madison, WI, 1987; pp. 211–247.

Buzzell, R.I. Inheritance of a soybean flowering response to fluorescent-day length conditions. *Can.*

J. Genet. Cytol. **1971,** *13,* 703–707.

Caldwell, B.E.; C.A. Brim; J.P. Ross. Inheritance of resistance of soybeans to cyst nematode *Heterodera glycines. Agron. J.* **1960,** *52,* 635–636.

Carter, T.E. Jr.; R.L. Nelson; C.H. Sneller; Z. Cai. Genetic diversity in soybean. *Soybeans: Improvement, Production and Uses.* H.R. Boerma, J.E. Specht, Eds. American Society of Agronomy: Madison, WI , 2004; pp. 303–416.

Caviness, C.E.; R.D. Riggs. Breeding for nematode resistance. *Proceedings of the World Soybean Research Conference I, Champaign, IL, 3–8 Aug., 1975.* L.D. Hill, Ed. The Interstate Printers and Publishers: Danville, IL, 1976; pp. 594–601.

Cho, T.J.; C.S. Davies; N.C. Nelson. Inheritance and organization of glycinin genes in soybean. *Plant Cell* **1989,** *1,* 329–337.

Fehr, W.R. A plant breeder's response to changing needs in soybean oil quality. *Proceedings XXI Soybean Research Conference, Chicago, IL, Dec. 10–11, 1991.* D. Wilkinson, Ed. American Seed Trade Association: Washington, D.C., 1991; pp. 79–87.

Fehr, W.R. Breeding methods for cultivar development. *Soybeans: Improvement, Production and Uses.* Second ed.; J.R. Wilcox, Ed., Agron. Monogr. 16. ASA, CSSA, and SSSA: Madison, WI, 1987a; pp. 249–293.

Fehr, W.R. *Principles of Cultivar Development. I. Theory and Technique.* Macmillan Publishing Co.: New York, NY, 1987b.

Goulden, C.H. Problems in plant selection. *Proceedings of the 7th International Genetical Congress,* Edinburgh, Scotland, 1941, pp. 132–133.

Grau, C.R.; A.E. Dorrance; J. Bend; J.S. Russin. Fungal diseases. *Soybeans: Improvement, Production and Uses.* H.R. Boerman, J.E. Specht, Eds. American Society of Agronomy: Madison, WI, 2004; pp. 679–764.

Harlan, H.V; M.N. Pope. The use and value of backcrosses in small grain breeding. *J. Hered.* **1922,** *13,* 319–322.

Hartwig, E.E. Varietal development. *Soybeans: Improvement Production and Uses.* B.T. Caldwell, Ed. American Society of Agronomy: Madison, WI, 1973; pp. 187–210.

Heatherly, L.G.; R.W. Elmore, Managing inputs for peak production. *Soybeans: Improvement, Production and Uses.* H.R. Boerman, J.E. Specht, Eds. American Society of Agronomy: Madison, WI, 2004; pp. 451–536.

Henderson, C.R. Best linear unbiased estimation and prediction under a selection model. *Biometics* **1975,** *31,* 423–477.

Hoeft, R.G.; E.D. Nafziger; R.R. Johnson; S.R. Aldrich. Modern Corn and Soybean Production. MCSP Publ.: Champaign, IL, 2000.

Hymowitz, T. Speciation and cytogenetics. *Soybeans: Improvement, Production and Uses.* Amer Society of Agronomy: Madison, WI, 1987.

Hymowitz, T.; J.R. Harlan. The introduction of soybean to North America by Samuel Bowen in 1765. *Econ. Bot.* **1983,** *37,* 371–379.

Hymowitz, T. On the domestication of the soybean. *Econ. Bot.* **1970,** *24,* 408–421.

Immer, F.R. Relation between yielding ability and homozygosis in barley crosses. *Agron. J.* **1941,** *33,* 200–206.

Jensen, N.F. A diallele selective mating system for cereal breeding. *Crop Sci.* **1970,** *10,* 629–635..

Johnson, R.R. Crop management. *Soybeans: Improvement, Production and Uses.* J.R. Wilcox, Ed. American Society of Agronomy: Madison, WI, 1987; pp 355–390.

Keim, P.; B.W. Diers; T.C. Olsen; R.C. Shoemaker. RELP mapping in soybean: Association between marker loci and variation in quantitative traits. *Genetics* **1990,** 735–742, 1990.

Kenworthy, W.J. Strategies for introgressing exotic germplasm in breeding programs. *Proc. World Soybean Res. Conf. II, Raleigh, NC.* F.T. Corbin, Ed. Westview Press: Boulder, CO, 1980; pp. 217–233.

Kyukendall, L.D. Genus *Bradyrhizobius,* family *Bradyrhizobiaceae. The Alpha-, Beta-, Delta- and Epsilonproteobacteria, The Proteobacteria, Part C, Bergey's Manual of Systematic Bacteriology.* Second ed.; D.J. Brenner, N.R. Krieg, J.T. Staley, G.M. Garrity, Eds.; Springer Verlag: New York, 2005; *2,* pp. 438–443, 505–506.

Kuykendall, L.D.; Hashem, R.B.; Dadson, R.B.; Elkan, G.H. Nitrogen fixation. In: J. Clederberg; M. Alexander; Ed.; *Encyclopedia of Bacteriology, I,* New York: Academic Press, 2000, pp 492-505.

Lewers, K.S.; R.G. Palmer. Recurrent selection in soybean. *Plant Breed. Rev.* **1997,** *15,* 275–313.

Lewers, K.S.; S.K. St. Martin; B.R. Hedges; M.P. Widrlechner; R.G. Palmer. Hybrid soybean seed production: Comparison of three methods. *Crop Sci.* **1996,** *36,* 1560–1567.

Li, Y.; C. Hill; S. Carlson; B. Diers; G. Hartman. Soybean aphid resistance genes in the soybean cultivars. Dowling and Jackson map to linkage group M. *Mol. Breed.* **2007,** *19,* 25–34, 2007.

Luzzi, B.M.; H.R. Boerma; R.S. Hussey. A gene for resistance to the southern root-knot nematode in soybean. *J. Hered.* **1994,** *85,* 484–486.

Matson, A.G.; L.F. Williams. Evidence of a fourth gene for resistance to the soybean cyst nematode. *Crop Sci.* **1965,** *5,* 477.

Mullis, K.; F. Faleona,; S. Scharf; R. Saiki; G. Horn; H. Erlich. Specific enzymatic amplification of DNA in vitro: The polymerase chain reaction. Cold Spring Harbor Symp. *Quant. Biol.* **1986,** *51,* 263–273.

Niblack, T.L.; G.L. Tylka; R.D. Riggs. Nematode pathogens of soybean. *Soybeans: Improvement, Production and Uses.* H.R. Boerman, J.E. Specht, Eds.; American Society of Agronomy: Madison, WI, 2004; pp. 821–852.

Orf, J.H.; B.W. Diers; H.R. Boerma. Genetic improvement: Conventional and molecular-based strategies. *Soybeans: Improvement, Production, and Uses, Third ed.;* H.R. Boerma; J.E. Specht, Eds; Agron. Monogr. 16. ASA, CSSA, and SSSA: Madison, WI, 2004; pp. 417–450.

Padgett, S.R.; K.H. Kolacz; X. Delannay; D.B. Re; B.J. LaValle; C.N. Tinius; W.K. Rhodes; Y.I. Otero; G.F. Barry; D.A. Eichholtz; et al. Development, identification, and characterization of a glyphosate-tolerant soybean line. *Crop Sci.* **1995,** *35,* 1451–1461.

Palmer, R.G.; J. Gai; H. Sun; J.W. Burton. Production and evaluation of hybrid soybean. *Plant Breed. Rev.* **2001,** *21,* 263–307.

Palmer, R.G.; T.C. Kilen. Quantitative genetics and cytogenetics. *Soybeans: Improvement, Production and Uses, Second ed.;* J.R. Wilcox, Ed.. Agron. Monogr. 16. ASA, CSSA, and SSSA: Madison,

WI, 1987; pp. 135–209.

Palmer, R.G.; T.W. Pfeiffer; G.R. Buss; T.C. Kilen. Qualitative genetics. *Soybeans: Improvement, production and uses,* Third ed.; H.R. Boerma, J.E. Specht, Eds. Agron. Monogr. 16. ASA, CSSA, and SSSA: Madison, WI, 2004: pp. 137–233.

Panter, D.M.; F.L. Allen. Using best linear unbiased predictions to enhance breeding for yield in soybean. I. Choosing parents. *Crop Sci.* **1995,** *35,* 397–405.

Parrott, W.P.; T.E. Clemente, Transgenic soybean. *Soybeans: Improvement, Production and Uses.* H.R. Boerma; J.E. Specht, Eds.; American Society of Agronomy: Madison, WI, 2004; pp. 265–302.

Piper, C.G.; W.J. Morse. The soybean: History, varieties and field studies. *USDA Bureau of Plant Industry Bulletin 197.* U.S. Gov. Print. Office: Washington, D.C., 1910.

Probst, A.H.; R.W. Judd. Origin, U.S. history and development and world distribution. *Soybeans: Improvement, Production and Uses.* B.E. Caldwell, Ed. American Society of Agronomy: Madison, WI, 1973; pp. 1–15.

Rao-Arelli, A.P. Inheritance of resistance to *Heterodera glycines* race 3 in soybean accessions. *Plant Dis.* **1994,** *78,* 898–900.

Rao-Arelli, A.P.; S.C. Anand.; J.A. Wrather. Soybean resistance to soybean cyst nematode rate 3 is conditioned by an additional dominant gene. *Crop Sci.* **1992,** *32,* 862–864.

Ray, J.D.; K. Hinson; J.E. Mankono; M.F. Brand-Malo. Genetic control of a long-juvenile trait in soybean. *Crop Sci* **1995,** *35,* 1001–1006.

Shoemaker, R.C.; P.R. Cregan; L.O. Vodkin. Soybean genomics. *Soybeans: Improvement, Production and Uses;* H.R. Boerma, J.E. Specht, Eds.; American Society of Agronomy: Madison, WI, 2004; pp. 235–263.

Singh, R.J.; T. Hymowitz. The genomic relationships between *Glycine max* (L.) Merr. and *G. soja* Sieb. and Zucc. as revealed by pachytene chromosome analysis. *Theor. Appl. Genet.* **1988,** *76,* 705–711.

Singh, R.J.; T. Hymowitz. Diploid-like meiotic behavior in synthesized amiphiploids of the genus *Glycine* Willd. subgenus *Glycine. Can. J. Genet. Cytol.* 1985, ***27,*** 655–660.

Sleper, D.A.; J.M. Poehlman. *Breeding Field Crops.* Blackwell Publishing: Ames, IA, 2006.

Specht, J.E.; G.L. Graef. Breeding methodologies for chickpea: New avenues to greater productivity. *Chickpea in the Nineties.* B.J. Walby, S.D. Hall, Eds. Proc. 2nd Int. Workshop on Chickpea Improvement ICRISAT Center, India; 1990; pp. 217–223.

Specht, J.E.; D.E. Hume; S.V. Kumundini. Soybean yield potential—A genetic and physiological perspective. *Crop Sci.* **1999,** *39,* 1560–1570.

St. Martin, S.K.; K.S. Lewers; R.G. Palmer; B.R. Hedges. A testcross procedure for selecting exotic strains to improve pure-line cultivars in predominately self-fertilizing species. *Theor. Appl. Genet.* **1996,** *92,* 78–82.

Tanksley, S.D.; N.D. Young; A.H. Paterson; M.W. Bonierbale. RFLP mapping in plant breeding: New tools for an old science. *Biotechnology* **1989,** *7,* 257–264.

Tolin, S.A.; G.H. Lacy. Viral, bacterial and phytoplasmal diseases of soybean. *Soybeans: Improvement, Production and Uses* In: H.R. Boerman, J.E. Specht, Eds.; American Society of Agronomy: Madison, WI, 2004; pp. 765–820.

USDA-Foreign Agricultural Service. April 2007.

USDA-National Agricultural Statistics Service. April 2007.

Walling, J.G.; R. Shoemaker; N. Young; J. Mudge; S. Jackson. Chromosome-level homeology in Paleopolyploid soybean (*Glycine max.*) revealed through integration of genetic and chromosome maps. *Genetics* **2006,** *172,* 1893–1900.

Wilcox, J.R. World distribution and trade of soybean. *Soybeans: Improvement, Production and Uses.* H.R. Boerman, J.E. Specht, Eds.; American Society of Agronomy: Madison, WI, 2004; pp. 1–14.

Williams, C.; D.F. Gillman; D.S. Fontenot; W. Birchfield. Inheritance of reaction to the reniform nematode in soybean. *Crop Sci.* **1981,** *21,* 93–94.

Xu, S.J.; R.J. Singh; T. Hymowitz. A modified producer for mitotic chromosome counts in soybean. *Soybean Genet. Newsl.* **1998,** *25,* 107.

Zou, J.J.; R.J. Singh; J. Lee; S.J. Xu; P.B. Cregan; T. Hymowitz. Assignment of molecular linkage groups to soybean chromosomes by primary trisomics. *Theor. Appl. Genet.* **2003,** *107,* 745–750.

Zou, J.J.; R.J. Singh; J. Lee; S.J. Xu; T. Hymowitz. SSR markers exhibit trisomic segregation distortion in soybean. *Crop Sci.* **2006,** *46,* 1456–1461.

3

Harvesting, Storing, and Post-Harvest Management of Soybeans

Carl J. Bern[1], H. Mark Hanna[1], and William F. Wilcke[2]
[1]Department of Agricultural and Biosystems Engineering, Iowa State University, Ames, IA; [2]Department of Bioproducts and Biosystems Engineering; University of Minnesota, St. Paul, MN

Introduction

This chapter describes systems and procedures for harvesting, drying, handling, and storing soybeans. These operations are interdependent and all must be carried out correctly so that the quantity and quality of the soybeans are as high as is practicable when the crop is marketed and processed. Recommendations for harvesting, drying, handling, and storing soybeans are presented along with some history, recent developments, and trends for the future.

Moisture Content

Arguably, the most important property of soybeans associated with their harvesting, drying, handling, and storing is moisture content. Soybean moisture content is defined assuming soybeans consist of two components: water and dry matter. Then,

$$Mwb = \frac{water\ weight}{water\ weight + dry\ matter\ weight}(100) \qquad [1]$$

where, Mwb = percentage moisture content, wet basis, and

$$Mdb = \frac{water\ weight}{dry\ matter\ weight}(100) \qquad [2]$$

where, Mdb = percentage moisture content, dry basis.

Wet-basis moisture contents are used in the grain trade and in most other instances. Dry-basis moisture contents are sometimes used in research, especially related to the drying process. Unless otherwise noted, all moisture contents stated in this chapter are on wet basis.

Moisture Measurement

To make good management decisions about soybean harvesting, drying, and storing, the moisture content of the soybeans should be known to within one percentage point of the true moisture content. Moisture can be measured by direct methods, which actually separate the water and dry matter in a sample, or by indirect methods, which measure some other parameter that is correlated with moisture content. The hot-air oven procedure is the most prominent direct soybean moisture determination method. This procedure assumes that all water can be removed by heating the soybeans in a hot-air oven over a prescribed time. ASABE Standard 352.2 for hot-air determination of moisture in whole soybeans is 103°C for 72 h, using a 15 g sample (ASABE, 2006d). Percentage wet-basis moisture is calculated as the percentage weight loss to the sample during 72 h in the oven. This procedure yields reliable results, but the 72 h heating time makes the oven method very inconvenient. AOCS Official Method Ac-2-41 specifies 130°C for 3 h as the procedure for whole soybeans (AOCS, 2007).

Electronic Moisture Meters

Electronic moisture meters are available to quickly and accurately determine the moisture content of most grains and oilseeds, including soybeans. Most modern electronic moisture meters utilize the capacitance principle in which moisture is calculated from electrical measurements on a capacitor having a sample of seeds packed between its plates, or a combination of capacitance and conductance, which additionally takes into account the electrical resistance of the seeds between the plates. This approach to grain moisture measurement is indirect and, thus, these meters must be calibrated using experimental data for each grain from some direct method, such as the hot-air oven method. Figure 3.1 shows two electronic moisture meters. Advertised repeatabilities (precision) are 0.5 percentage point and 0.1 percentage point of moisture for the portable and commercial models, respectively.

Fig. 3.1. Capacitance-type electronic moisture testers (courtesy of Dickey-john Corporation, Auburn, IL).

Equilibrium Moisture Content

When a sample of soybeans or other hygroscopic material is placed in air at constant temperature and relative humidity, moisture will exchange between the sample and the air until the sample reaches a constant moisture level known as its *equilibrium moisture content* for that air condition. This happens regardless of whether the sample is initially wet or dry. Table 3.1 shows equilibrium moistures for soybeans over a range of relative humidities and temperatures. Some examples will illustrate use of the table. Soybeans stored in a bin, which is not being aerated by a fan, will bring the air within the soybean bulk into equilibrium with the soybeans because the mass of the soybeans is much greater than the mass of the air. Example, what is the relative humidity of air within a bin of soybeans at 13% moisture and 16°C (60°F)? Answer: the interstitial air will be at 70% relative humidity (Table 3.1). Soybeans being aerated with outside air will eventually come to equilibrium with the outside air because the mass of the outside air is far greater than the mass of the soybeans. Example: soybeans at 16% moisture are being aerated with outside air at 10°C (50°F) and 50% relative humidity. To what moisture content will the air eventually dry the soybeans? Answer: 9.5%.

Soybean Harvesting

As with many other grain and oilseed crops, harvesting is delayed until the crop is mature and the moisture content of soybeans is at levels for acceptable storage (i.e., about 13%). Because weather conditions can adversely affect harvesting, many growers desire to complete harvesting within about 10 days of fieldwork. Hot, dry winds can rapidly lower moisture content, so soybeans more easily shatter from pods or storms can lodge bean stalks. Timing of initial harvesting is also affected by the total acreage to be harvested (often including a corn crop requiring harvesting during the same period) or "green" weed areas that may be difficult for the combine to handle until a killing frost occurs. The harvesting objective is to collect the soybeans from plants in the field at or below the safe storage moisture content with minimal damage.

Table 3.1. Equilibrium Moisture Contents (percentage wet basis) for Soybeans

Temperature		Relative Humidity (%)				
(°C)	(°F)	50	60	70	80	90
0.0	32	10.0	11.8	13.7	16.2	19.8
4.4	40	9.8	11.5	13.5	16.0	19.6
10.0	50	9.5	11.2	13.2	15.7	19.4
16.0	60	9.2	11.0	13.0	15.4	19.1
21.0	70	8.9	10.7	12.7	15.2	18.9
27.0	80	8.6	10.4	12.5	15.0	18.7

Damage is often considered in terms of USDA grading standards specifying maximum amounts of split soybeans [split soybeans pass through a 4.0 mm (10/64 in.) x 19.0 mm (¾ in.) slotted sieve but are held on a 3.2 mm (8/64 in.) round-hole sieve], damaged soybeans (soybeans with cracked seed coat visible to naked eye), and foreign material [smaller pieces passing through a 3.2 mm (8/64 in.) round-hole sieve]. Although USDA grain-damage standards are relatively lenient (USDA no. 1 grade: 10% split soybeans, 2% damaged soybeans, and 1% foreign material), requirements may vary with different customers. Some end-users desire high bean purity levels and specify little or no commingling of other soybean varieties or grains.

Self-propelled combines used in North American for harvesting soybeans combine cutting the plants near the base, threshing the beans from the pods, and separating the beans from other materials. Major functional areas of the combine (Fig. 3.2) include gathering the crop by the gathering head, threshing by the rotor/cylinder and concave, separating by the straw walkers or the rear section of the rotor/concave, cleaning by the cleaning shoe, and storing soybeans in the grain tank. Various conveyors (augers, elevators, feederhouse chain) transfer the soybeans and other plant biomaterials between processing steps in the combine. Details presented here focus primarily on operational procedures and recommendations for settings during soybean harvesting. A more detailed description of the combine internal mechanisms can be found in Hanna and Quick (2007).

Operation and adjustment of combine settings are always trade-offs between balancing soybean quality (i.e., damage) and maintaining machine losses in the field at or below an acceptable level for a given crop throughput. The American Society of Agricultural and Biological Engineers (ASABE) standard S343.3 defines combine soybean throughput capacity as throughput achieved with no more than 1% loss occurring in the threshing and separating system (ASABE, 2005).

Functional Areas and Settings

Gathering Head

A grain platform or soybean head is used to gather the crop into the combine. Important components include the cutterbar for cutting plants near the base, the reel for guiding plants onto the gathering head, and the auger for pulling plants from across the head into a central feederhouse (Fig. 3.3). About 90% of the losses due to machine operation may occur at the head. Losses can take several forms and can give clues to corrective adjustments. Stubble loss occurs when plants are cut too high and leave pods still attached in the field at the plant base. Lodging loss occurs if pods are present on uncut stems bent below the cutterbar from prior plant lodging. Loose stem losses are pods still attached to cut stems and often occur as stems are carried back out of the head by the reel or bouncing off the other parts of the head. Loose bean and pod losses (soybeans and pods separated from stems) may occur from shat-

Fig. 3.2. Major combine areas.

tering of pods or beans off the plants at the head, from unthreshed pods inside the combine, or from soybeans not separated from materials other than grain (MOG) in the separating or cleaning sections of the combine. Dry soybeans [e.g., 9–10% moisture content wet basis (m.c.)] are significantly more prone to shattering during field harvesting than wetter soybeans (e.g., 12–13% m.c.). Because field moisture content can vary diurnally, harvesting is occasionally limited to morning or evening hours to avoid shatter losses.

The cutterbar sections should be in good condition (i.e., serrated and sharp) and held in close tolerance with guards by hold-down clips and wear plates (if present) for good shear-cutting the plants between the knife plates and guards. The cutterbar should be in register (i.e., knives oscillating from directly under one guard to directly under an adjacent guard). Bats on the reel, often equipped with pick-up fingers, are used to guide plants across the cutterbar and on to the gathering head. Adjustments include reel speed and position. Speed is measured by the reel index, which is the ratio of peripheral reel speed to combine travel speed. A reel index of 1.0 is where the reel speed just matches the combine travel speed and minimizes movement of plants relative to the head. The reel index typically varies from 1.1 (crop standing well, good harvest conditions) to 1.5 or greater (lodged crop). Sufficient speed should be used to

Fig. 3.3. Grain platform used as gathering head for soybean harvest.

aid and control crop feeding, but not so fast as to promote shatter loss or carry plants back out of the head. The cross-auger position may be adjustable to allow smoother feeding transition from the cutterbar into the feederhouse area.

Head losses can be caused by different component areas. Research has suggested most losses occur from the cutterbar, with lesser losses from the cross-auger, and lesser still from the reel (Quick, 1973). Losses due to the combine (i.e., machine losses as opposed to pre-harvest loss already on the ground prior to combine operation) can be 70 kg/ha (1 bu/A) or greater for each 2.5 cm (1 in.) of uncut stubble. Using a floating rather than a fixed cutterbar is common as is an automatic height-control adjustment for the head, which senses the ground surface. After-market attachments include crop-lifters on the cutterbar to help raise lodged crop and pneumatic "air" reels to blow plant material into the head.

Feederhouse

A chain with cross-slats transports material up the bottom of the enclosed feederhouse and then into the threshing area. A front drum guides the chain onto plant material. Most combines have a position adjustment for the front drum (upper position for corn, lower position for small grains/other crops). Because many combines used for soybean harvesting frequently alternate with corn harvesting, the position of the front drum should be set in the lower position unless crop volume is unusually large.

Rotor/Cylinder and Concave

Two predominant thresher types, rotor and cylinder, are used. In both cases, crop material is fed into the opening between a rotating cylinder and the surrounding open-mesh wire or grid concave. Rasp bars on the cylinder or on the front portion of the rotor help thresh soybeans from the pods. Crop flow typically enters a rotor thresher parallel to the axis of the rotor. Rear sections of the rotor (Fig. 3.4) are used to aid further separation of soybeans from pods, stems, and leaves by using centrifugal force to throw smaller, heavier soybeans through openings in the concave while larger, lighter pods and stems exit the rear of the rotor. Crop flow enters a cylinder-type thresher tangentially, and oscillating straw walker sieves are used for further separation by gravity of smaller, heavier soybeans from MOG. Although subsequent separating further divides soybeans from MOG, most separation (70–90%) occurs in the threshing area.

Threshing action should be aggressive enough to detach soybeans from the pods, but not so aggressive as to cause excessive damage to the soybean seed coats. Suggested rotor or cylinder speed range and concave clearance for soybeans in the combine operator's manual can be used as starting points but will need to be evaluated and adjusted in the field. Rotor and cylinder speeds are expressed in rev/min (RPM). Typical peripheral speed of the rasp bars at the edge of the cylinder for soybean harvest is in the range of 15 m/s (3000 ft/min). Rotor and cylinder diameters vary, so suggested rotary speeds will vary with combine model according to the relationship peripheral speed = 3.14 x diameter x rev/min. As an example, rasp bars turning at 380 rev/min on a 0.76 m (30 in.) diameter rotor have a peripheral speed of 15 m/s (3000 ft/min). Typical concave clearance is around 2 cm (¾ in.).

Fig. 3.4. Concave sections covering rotor of an axial-flow rotor thresher and separator.

Rotor or cylinder speed should be decreased to just below the point where damage is observed, but still maintain acceptably low threshing losses (<1%) of soybeans left in pods. Excessive cylinder speed or improperly loading the combine at extremely low travel speed can greatly increase soybean damage (Fig. 3.5). Concave clearance between the rotor or cylinder and the concave should be only as narrow as required for adequate threshing. A key concept to minimize damage is to use as much "crop-on-crop" threshing (rubbing with biomaterial) as possible rather than striking pods with metal rasp bars.

Combine operators should make changes in rotor/cylinder speed and concave clearance as field conditions dictate. A round-wire concave may be gentler on damage than a rectangular-wire concave. If soybeans are threshing easily, using a concave with wider spacing between wires allows quicker removal from the thresher and less potential for damage. Conversely, some operators have used filler plates between the bars near the front of the concave to hold the crop in the threshing zone ("crescent") longer if pods are green, tough, and difficult to thresh. Harvesting may also be limited to drier portions of the day (late morning, afternoon) to improve threshing if plant material is green.

Cleaning Shoe

The cleaning shoe (Fig. 3.6) receives the grain and smaller pieces of biomaterial from the threshing and separating areas. It consists of two or more oscillating sieves or screens and a fan. The fan supplies an air blast to fluidize the mat of biomaterial, allowing easier separation of heavier whole soybeans from lighter pieces of pods, stems,

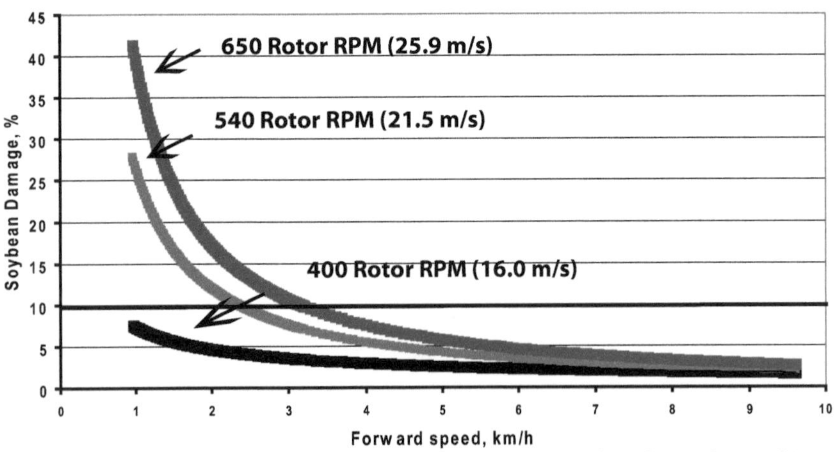

Fig. 3.5. Soybean damage increases with excessive rotor speed or if travel speed is too low to load the combine (Quick, Personal communication; John Deere 9650 STS).

Fig. 3.6. Top sieve (chaffer) of cleaning shoe as viewed from the rear of the combine.

and leaves. Air flow may be directed by a wind board or divided into two streams so that greater air flow is directed toward initial separation near the front of the oscillating sieves. The top sieve is often referred to as the *chaffer*.

Fan speed and sieve openings are adjustable. Sufficient air flow should be used to fluidize the biomaterial mat for maximum separation efficiency. Begin with a fan speed near the upper end of the manufacturer's suggested operating range and reduce it to just below the point where soybeans are blown from the combine out the rear of the shoe. Sieve openings should be wide enough so that whole soybeans fall through without being carried to the rear tailings return auger while unthreshed pods stay on top before falling into the tailings auger. Adjustments in the cleaning shoe help clean soybeans in the grain tank or limit tailings return flow or loss out the rear of the combine. Adjustments should normally be first to air flow, then chaffer openings, and finally lower sieve openings (perhaps slightly narrower than the chaffer opening).

Other Combine Areas

Sharp edges anywhere in the combine can damage soybean seed coats and affect quality. Worn auger edges, sharp edges on feederhouse chain slats, or chipped surfaces on chrome-plated rasp bars can damage soybeans. Chrome-alloy rasp bars don't chip like chrome-plated rasp bars and can be desirable after an initial break-in period.

Although MOG (stems, pods, leaves) can improve soybean quality (as long as it is properly separated) by cushioning soybean flow through the combine, large amounts of MOG exiting the combine rear can be detrimental to subsequent planting unless it is uniformly spread on the soil surface. Most combine operators chop soybean straw exiting the rear of the combine and use a straw spreader to distribute non-soybean biomaterial back on the ground. A second spreader of one or two spinning discs may be used at the rear of the cleaning shoe to spread chaff.

A yield monitor and global positioning system (GPS) are commonly used to record yield information. The yield monitor uses an impact-plate sensor at the top of the clean-grain elevator. Yield is often adjusted by a moisture content sensor mounted along the clean-grain elevator. Auto-steering may be used with GPS to reduce operator fatigue or allow easier use of a full cutting swath width during nighttime operation when soybeans may be slightly damper and less prone to shatter loss.

Safety During Field Harvesting

Some of the most traumatic injuries with potential for fatality involve becoming caught in moving parts around the gathering head or the head falling on a worker during failure of the hydraulic support system. Always disengage power, stop the engine, and remove the key before doing any work on the combine. Make sure the gathering head is mechanically locked and blocked before getting underneath it. Replace all guards, shields, and access doors before operating equipment. Although acutely traumatic injuries may be well known, less severe but more frequent injuries result from falling off equipment. The combine operator's station and engine compartment are 2.5–3 m (8–10 ft) above the ground. Use handrails and shoes with non-skid soles. Keep steps, platforms, and other areas free of crop residue, mud, and ice or snow. Cleaning flammable crop residue from the combine is also important for fire safety. Two class ABC fire extinguishers should be carried on the combine, a smaller 10 lb extinguisher in the operator's cab and a larger 20 lb extinguisher at the ground level.

Loss Measurements and Summary of Combine Adjustments for Quality

Losses should be evaluated numerically by counting soybeans on the ground across the width of the gathering head after the combine passes. Every 43 soybeans/m^2 (4 soybeans/ft^2) equals approximately 67 kg/ha (1 bu/A) loss. Because losses at the gathering head (e.g., due to cutterbar, reel, or cross-auger operation) may be the major portion of machine field losses, if losses are excessive, it is advisable to check head loss by stopping the combine in the middle of the field, backing it up about 5 m (15 ft), and observing losses in front of the combine traversed by the gathering head, but not the rear of the combine. If losses are high, check for pre-harvest field losses by examining an unharvested area and subtract this value from total losses behind the combine or at the gathering head to determine losses due to the machine. The difference between total machine loss (evaluated behind the combine less any pre-harvest loss) and head loss (evaluated in a harvested area in front of the combine less any pre-harvest loss) equals losses due to threshing and separating processes inside the machine. Total machine losses should be less than 67 kg/ha (1 bu/A) if the crop is standing well and harvesting conditions are reasonable (Gliem et al., 1990).

Rotor or cylinder speed should be reduced to the point where either no soybean damage is visible or minor damage is acceptable while threshing loss is limited. If harvesting conditions are reasonable, a goal is to observe no soybean damage in the grain tank with threshing loss less than 1%. Concave clearance should be widened to just below the point where threshing of soybeans from pods is compromised. Make adjustments one-at-a-time to evaluate changes. Rotor/cylinder speed is usually initially adjusted before concave clearance (if clearance is judged to be wide enough to limit crop damage).

If the tailings return is heavy, the grain tank sample is dirty, or soybeans are exiting the cleaning shoe, make adjustments in the cleaning shoe area. Adjust airflow to just below the point where whole soybeans are blown out the rear, then check that the chaffer and sieve openings are allowing whole soybeans to fall through to the clean grain auger while unthreshed pods are carried across the sieves to the tailings return. Excessive unthreshed pods require more aggressive threshing. Excessive whole soybeans in the tailings return indicate too small sieve openings or perhaps excessive air flow.

Good combine operation results from an attentive operator observing soybean quality in the grain tank and evaluating machine field losses on the ground. If soybeans are threshing easily in good conditions, gentler threshing action will supply excellent quality soybeans with few machine losses. If soybeans are more difficult to thresh, a careful balance should be evaluated between soybean damage and acceptable field losses.

Clean-out for Identity-preserved Soybean Customers

As markets evolve, some end-users desire limited or no commingling of soybean varieties for subsequent processing and use. Combines can hold 30–70 kg (60–150 lb) of grain and other biomaterial after the grain tank has been emptied by the unloading auger. Cleaning out the combine between crops and between harvestings of different soybean varieties becomes important to lower commingling of crops to an acceptable value. Depending on the size of combine, degree of familiarity with cleaning, and level of commingling desired, clean-out may take several hours. Accessible areas of the combine are opened and compressed- and vacuum-air along with picks (e.g., flathead screwdrivers) are used for cleaning. The greatest amounts of residual crop material are often contained in the grain tank, rock trap, and head (Hanna et al., 2006).

Soybean Drying

In most years, weather conditions at harvest allow soybeans to dry to 13% moisture content (wet basis) or less in the field. But in some years, weather conditions prevent soybeans from drying to 13% moisture, and sometimes, growers harvest at moisture contents >13% to avoid the harvesting losses that can occur at lower moisture contents (soybeans can be harvested without too much mechanical damage[1] up to about 18% moisture). If soybeans are harvested at moisture contents much above 13%, artificial drying is necessary (Siemens & Hirning, 1974; McKenzie, 1973).

There is not much published research on soybean drying. Most drying recommendations are based on limited experience or are extrapolated from corn-drying recommendations. In most cases, dryers that were designed for corn can be adapted for use with soybeans.

Natural-air Drying

Using unheated air to dry soybeans usually works well, but it is a slow process (2–6 weeks, depending on initial moisture, airflow, and weather). Bins used for natural-air drying should have full-perforated floors and fairly large drying fans. Fan power requirements depend on desired airflow and depth of beans. For example, delivery of 1.1 m^3/(min-t) [1.0 cfm/bu (ft^3 of air per min per bu of beans in the bin)] through a 5.5 m [18 ft] depth of soybeans would require about 0.02 kW/t [0.6 hp (horsepower) per 1000 bu] of beans in the bin, while delivery of 1.7 m^3/(min-t) [1.5 cfm/bu] through 5.5 m (18 ft) of beans would take about 0.044 kW/t [1.6 hp/1000 bu] (ASABE, 2006a,c).

Management of natural-air soybean dryers is similar to that for natural-air corn dryers, except that soybean moisture values need to be about two percentage points lower than those recommended for corn (Wilcke & Morey, 1995). In the central United States, a natural-air drying airflow of 1.1 m^3/(min-t) [1 cfm/bu] is suggested for drying 17–18% moisture beans, 0.83 m^3/(min-t) [0.75 cfm/bu] for 15–17% moisture beans, and 0.6 m^3/(min-t) [0.5 cfm/bu] for 13–15% moisture beans. In the northern United States, higher airflow is needed since fewer days are available for drying in the fall. In northern areas, use 1.1 m^3/(min-t) [1.0 cfm/bu] to dry soybeans that are 16% moisture or less, 1.38 m^3/(min-t) [1.25 cfm/bu] for 17% moisture beans, and 1.65 m^3/(min-t) [1.5 cfm/bu] for 18% moisture beans. See Wilcke and Morey (1995) and Hellevang (1983) for information on equipping and managing natural-air dryers.

Because natural-air drying is a slow process, it will be difficult to use one bin to dry both beans and corn in the same year. It is unlikely that the beans will be dry before corn harvesting unless the soybeans are only slightly wetter than 13%, or unless a shallow drying depth is used.

Low-temperature Drying

Early in the fall, especially in years with warm, dry weather, it is possible to dry soybeans to <13% moisture with no supplemental heat (see the previous section on natural-air drying). However, late in the fall, or in years with cool, damp weather,

[1]In this document, mechanical damage refers to chipped, broken, or split soybean seeds.

soybeans might not dry to 13% and it can be helpful to add a small amount of supplemental heat to the air in natural-air dryers. The temperature of the drying air should not be increased more than 1.7–2.8°C (3–5°F) though, or soybeans will over dry (to <13% moisture) and increased splitting of soybean seeds is likely. Research has shown that exposing soybeans to relative humidities of <40% can cause excessive splitting. For every 11.1°C (20°F) that air is heated, its relative humidity is cut approximately in one-half, so it doesn't take very much heat to produce relative humidities <40% (ASABE, 2006b).

Some alternatives to adding supplemental heat to natural-air drying bins include:

- Turning off the fan when weather gets cold in the fall, keeping the beans cold during winter, and resuming drying when average temperatures climb above freezing in the spring.

- Installing bigger fans so that drying is finished earlier in the fall when weather is better.

- Using manual or automatic control to turn off the fan during periods of high humidity. Fan control will increase the amount of time required for drying, but it will result in drier beans.

High-temperature Drying

Many kinds of gas-fired corn dryers can be used to dry soybeans, but heated-air corn dryers should be used carefully with soybeans. Soybeans split easily if they are dried too fast or are roughly handled. The drying air temperature should be set lower than for corn and dryers that recirculate the crop during drying should be avoided. Column-type dryers can often be operated at 48.9–60.0°C (120–140°F) without causing too much mechanical soybean damage (especially splits), although some trial-and-error might be required to set the dryers properly. Soybeans leaving the dryer should be examined carefully and the drying air temperature should be reduced if there are too many splits. If the soybeans will be saved for seed, drying temperatures should be kept under 43.3°C (110°F) to avoid killing the embryo.

Crops dried in gas-fired dryers must be cooled within a day or so to remove dryer heat. This can be done in the dryer or in aerated storage bins. Stored beans should be aerated again later in the fall to cool them to within about 8.3°C (15°F) of the average winter temperature.

Reconditioning Over-dry Soybeans

In years with exceptionally warm, dry falls, soybeans are sometimes harvested at moisture contents well under 13% moisture. Although it is illegal to add liquid water to increase soybean moisture content, it is possible, given enough time and a high enough airflow per bushel, to increase the moisture content of soybeans by aerating them with humid air. But here are some practical concerns and limitations (Morey et al., 2003) if a storage manager decides to attempt soybean reconditioning:

- The process is quite slow—even with the high airflow per bushel (0.83–1.1 m^3/(min-t) [0.75–1.0 cfm/bu] available on bins equipped for drying. It would be difficult to accomplish significant reconditioning using the low airflow aeration systems common on storage bins.

- Fan control is tricky and some beans could end up too wet for safe storage.

- There will be layers of wet beans and dry beans unless there is some way to mix the layers in the bin or during unloading of the bin.

- Swelling that accompanies rewetting will increase stress on bin walls.

Table 3.1 shows the moisture content that soybeans would come to if exposed to different combinations of temperature and relative humidity for long periods of time. Continuously aerated beans would tend to lose moisture during periods of low humidity and tend to gain moisture during periods of high humidity. To recondition soybeans to 13% moisture during normal fall temperatures in the northern United States of -1.1–16°C (30–60°F), the fan should be controlled so that it operates during weather that has an average relative humidity of 65–70%. Table 3.1 indicates that bean moisture increases sharply as relative humidity increases, which means that it is quite easy to rewet a layer of soybeans to a moisture content that is too high for safe storage.

During reconditioning, the moisture of the whole bin doesn't change at once. A rewetting zone develops and moves slowly through the bin in the direction that the airflow is moving. This is similar to the way a drying zone moves through a drying bin. In most cases, there are not enough high-humidity hours available in the fall to move a rewetting zone all the way through the bin. And in many cases, depending on how the fan is controlled, the parts of the bin that have been rewet will be too wet for safe storage. It would be best to mix the wet layers with the dry layers to reduce spoilage risk and to avoid drying charges for the wet layers when the beans are sold. Mixing can be accomplished to a limited extent by emptying the bin and moving the beans through a grain handling system. The most effective way to mix the beans, though, would be to use an in-bin stirring system. In fact, bin dryers equipped with stirring augers are a good choice for reconditioning soybeans. Vertical stirring augers have been shown to cause little mechanical damage to shelled corn and it is expected that stirrers would cause little mechanical damage to soybeans.

If the initial moisture content of the beans is <10%, controlling the fan so that it only runs when relative humidity of the air reaching the beans is >55% should result in rewetting. If a single humidistat is used to turn the fan on anytime humidity is greater than 55%, average humidity during the hours the fan operates should be well above 55% and the beans are likely to rewet to at least 13%. Since humidity is almost always higher at night than it is during the day, an alternative to a humidistat would be a timer set to run the fan only during nighttime hours.

If beans can't be mixed after reconditioning, it is important to avoid rewetting them to moisture levels that are too high for safe storage. Approaches to prevent excessive rewetting include:

- Reducing the humidity setting on the humidistat that controls the fan so that the fan runs during drier conditions.

- Adding a second humidistat that stops the fan when relative humidity reaches very high levels.

- Installing a sophisticated microprocessor-based controller that monitors both temperature and humidity and only runs the fan when air conditions will bring the crop to the desired moisture content (for either drying or rewetting).

The disadvantage of the last two approaches is that the fan doesn't run as many hours as it would with a single-humidistat control and less total moisture would be added. Running the fan at high humidities and then mixing the wet and dry beans would result in greater average moisture content.

Reconditioning time depends primarily on airflow per bushel and weather conditions. It is fastest when airflow per bushel is high and air is warm and humid. Reconditioning will be most successful in a bin equipped as a drying bin—one that has a full perforated floor and a fan that can deliver at least 0.83 $m^3/(min\text{-}t)$ [0.75 cfm/bu]. Even with this airflow, it would probably take at least a month of fan operation to move a rewetting front all the way through the bin. Keep in mind that the fan shouldn't be run continuously because in a typical fall in the northern United States, continuous fan operation would result in drying rather than rewetting. Attempts to use storage bins equipped with low-airflow aeration systems to recondition crops are usually not very successful—mainly because it just takes too long to move the rewetting front very far into the bin.

Soybeans swell when they absorb moisture, and experience during floods indicates that soaking the bottom few feet of beans in a bin can result in enough pressure to rupture bin walls. Not enough research on reconditioning soybeans through the use of airflow has been done to know whether this procedure can damage bins, but based on work with other crops (Kebeli et al., 1998), the process is likely to increase stress on bin walls. Using a vertical-stirring auger to mix layers of dry and wet beans might be one way to reduce outward pressure generated during rewetting.

To increase chances of success in using airflow to recondition soybeans:

- Use a bin equipped with a full perforated floor and a fan that can deliver at least 0.83 m^3/(min-Mg) [0.75 cfm/bu].

- If available, use a bin equipped with stirring equipment. If stirring equipment is not available, consider transferring the beans to another bin to mix the wet and dry layers.

- Use timers, humidistats, programmable controllers, or some other type of automatic control to limit fan operation to weather conditions that will cause rewetting.

- Keep reconditioned beans cool during storage to reduce chances of spoilage.

- Watch carefully for signs of moldy beans and for excessive stress on the bin.

Soybean Handling

Soybean-handling systems are designed to move soybeans from one location to another at a desired flow rate, while minimizing bean damage, dust generation, energy use, and cost. As handling system design evolves, performance improvements are made in each of these areas. Current trends are moving toward larger and larger handling capacities and increased automation of equipment. There is also increasing emphasis on effective cleanout so as to minimize cross contamination between lots of soybeans.

Hauling Soybeans

Soybeans are hauled over the road in tractor-pulled wagons, single-axle trucks, tandem-axle trucks, and semi-trailers (semis). Costs favor semis and there is a trend toward greater use of these large carriers. Their use has more than doubled in the U.S. Midwest over the last decade. Compared to semis, hauling costs are about 150, 200, and 400% higher for double-axle trucks, single-axle trucks and tractor-pulled wagons, respectively (Peterson, 2006). Semis are also safer, and make marketing and delivery at long distances more and more feasible.

Conveying Soybeans

All of the conventional grain conveyors are used with soybeans (Table 3.2). Each conveyor type has characteristics which make it a good choice in certain applications. Energy-efficient conveyors tend to have the highest purchase price, and low-cost conveyors tend to be the least energy efficient. Considering both ownership costs

Table 3.2. Conveyors for Soybeans

Type of Conveyor	Power Requirement per Unit Capacity	Advantages	Disadvantages
Auger	High	• Simple, widely available in many sizes. • Low cost. • Available for horizontal, inclined, vertical applications; portable, wheeled, or fixed. • Adapted to most grain and feed materials. • Useful as mixer, flow meter, force feeder, or agitator.	• High torque and power required. • Medium to heavy wear. • Noisy, if not bearing-supported and/or operated full. • High grain damage if not operated full. • Rotating screw dangerous where exposed. • Power requirement for wet grain is much higher.
Bucket	Low	• Efficient, compact. • Low maintenance. • Quiet. • High capacity for vertical lift. • Reliable and adaptable to automation. • Easily cleaned.	• Difficult to erect. • Expensive. • Grain damage high for large drop heights. • Elevator head service is difficult.
Belt	Low	• Good for long distances. • Low power requirement. • Quiet. • Least handling damage. • Capacity only affected by grain weight. • Self cleaning.	• Limited in angle of elevation. • Expensive. • Belt maintenance.
Bulk or mass flow	Medium	• Good for long distances. • Quiet. • Low maintenance. • Little grain damage.	• Expensive
Pneumatic	High	• Flexible installation. • Easily cleaned. • Convenient grain delivery to many locations. • Reduces dust at intake.	• High power requirement. • Creates dust at discharge, usually requires separation equipment.

Adapted from MWPS (1987)

and energy costs, low-cost, high-energy conveyors are best for applications where the conveyor is used only a few days per year, and high-cost, energy-efficient conveyors are most applicable where they are used day-in and day-out (this generalization may not hold for pneumatic conveyors).

Auger Conveyors

An auger conveyor (or screw conveyor) consists of a helicoid screw rotating within an enclosed tube or open U-tube (Fig. 3.7). Its invention is credited to Archimedes (287-212 BC) who used it to pump water from the holds of ships (CEMA, 2003). With only one moving part in contact with grain, low-cost augers are simple and find extensive use especially in intermittent use conveying systems. Augers are able to convey grain at any angle from horizontal to vertical, can be permanently installed or portable, and are available in a wide variety of sizes and configurations. Their high power requirements and high susceptibilities to abrasion wear from grain limit their applicabilities for high use applications. Internal pinch points in auger conveyors can damage soybeans, particularly when the auger is operating at partial capacity.

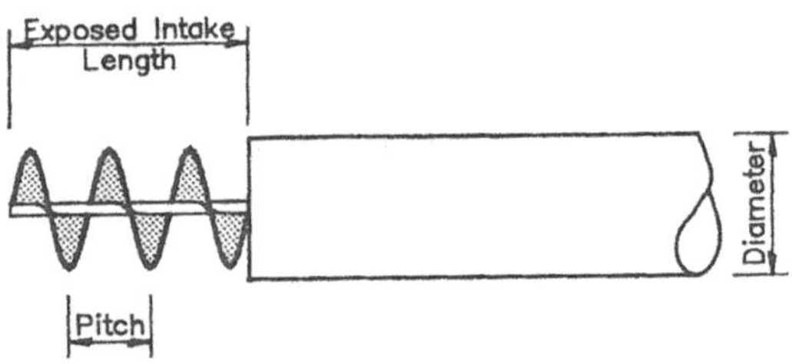

Fig. 3.7. Auger conveyor (MWPS, 1987).

Auger conveyors can be dangerous and it is important to follow good safety practices when they are used. Because of their length and portability, special care must be taken to keep conveyors away from overhead power lines. Contact with a power line can cause death. Also, be careful to keep hands, feet, hair, and clothing away from rotating shafts and the intake of the conveyor. Serious injury or death can result from entanglement.

Table 3.3 can be used to estimate capacities and power requirements for augers carrying dry soybeans. An example will illustrate its use: A 254 mm (10 in) diameter auger 10 m (32.8 ft) long turning at 360 rev/min is set at a 35 degree angle with the horizontal and is conveying dry soybeans. Estimate the flow rate and the power requirement. From Table 3.3, the flow rate is about 57.9 Mg/h (2120 bu/h) and the power requirement for a 10 m (32.8 ft) auger is about 8.6 kW (11.5 hp).

Table 3.3. Estimated Auger Capacity and Power for Dry (≤14% Moisture) Soybeans

Nominal auger dia., mm	Auger dia., in.	Auger speed, rpm	Incline Angle from Horizontal 0°				Incline Angle from Horizontal 25°			
			Mg/h	Bu/h	W/m	hp/10'	Mg/h	Bu/h	W/m	hp/10'
100	4	900	11.5	420	160	0.66	10.2	375	240	1
152	6	600	30.7	1130	270	1.1	27.6	1010	400	1.7
203	8	450	45.2	1660	380	1.5	40.7	1490	590	2.4
254	10	360	67.5	2480	540	2.2	60.8	2230	830	3.4
305	12	300	92.5	3390	670	2.8	83.3	3050	1000	4.3
356	14	260	127	4670	920	3.7	115	4210	1400	5.8
406	16	225	164	6030	1200	4.8	148	5430	1800	7.5

Nominal auger dia., mm	Auger dia., in.	Auger speed, rpm	Incline Angle from Horizontal 35°				Incline Angle from Horizontal 45°			
			Mg/h	Bu/h	W/m	hp/10'	Mg/h	Bu/h	W/m	hp/10'
100	4	900	9.82	360	240	1	9.2	338	270	1.1
152	6	600	26.4	968	430	1.8	24.3	893	430	1.8
203	8	450	38.7	1420	590	2.4	36.0	1320	620	2.5
254	10	360	57.9	2120	860	3.5	53.6	1970	860	3.5
305	12	300	79.2	2900	1100	4.4	73.4	2690	1100	4.4
356	14	260	109	4000	1500	5.9	101	3710	1500	6.1
406	16	225	141	5150	1900	7.7	131	4790	1900	7.8

Nominal auger dia., mm	Auger dia., in.	Auger speed, rpm	Incline Angle from Horizontal 90°			
			Mg/h	Bu/h	W/m	hp/10'
100	4	900	5.52	202	215	0.9
152	6	600	14.5	533	350	1.4
203	8	450	21.5	788	480	2
254	10	360	32.1	1180	670	2.8
305	12	300	44	1610	860	3.5
356	14	260	60.5	2220	1200	4.7
406	16	225	78.1	2860	1500	6.2

Adapted from MWPS (1987)

Bucket Elevators

Centrifugal-discharge bucket elevators, called legs, convey grain vertically and are the heart of most handling-storage systems. After being lifted in the bucket elevator, grain can flow by gravity into storage bins, dryers, or carriers (i.e., trucks or rail cars). Since grain is carried in buckets, which are fastened to a belt supported on anti-friction bearings, the power required is usually only 10–20% more than what is required to lift the grain. Capacity is simply the product of belt speed, bucket capacity, and spacing. Any other type of vertical conveyor would be chosen only for its lower initial cost.

Belt Conveyors

A belt conveyor consists of a rubber and fabric belt, often troughed for grain, supported by idlers and extending between two pulleys. The belt is driven by the pulley at the discharge end. Where there is a need to move soybeans horizontally or up an incline no more than 12–16° at a very high rate, a belt conveyor is the method of choice. Since grain is carried on a belt supported on anti-friction bearings, grain damage is almost zero and the power requirement is low. High initial cost limits their use to day-in and day-out applications. In spite of their high initial cost, belt conveyors often find use for conveying soybean seed because of their very low damage production.

En-masse Conveyors

En-masse conveyors (also called bulk or mass flow conveyors) use chains and flights to move grain in "slug flow" (en-masse) inside a box enclosure (Fig. 3.8). Load is carried on the floor and flights return on the top. Grain depth can be much greater than flight height. The enclosure retains dust, protects grain from weather, and provides

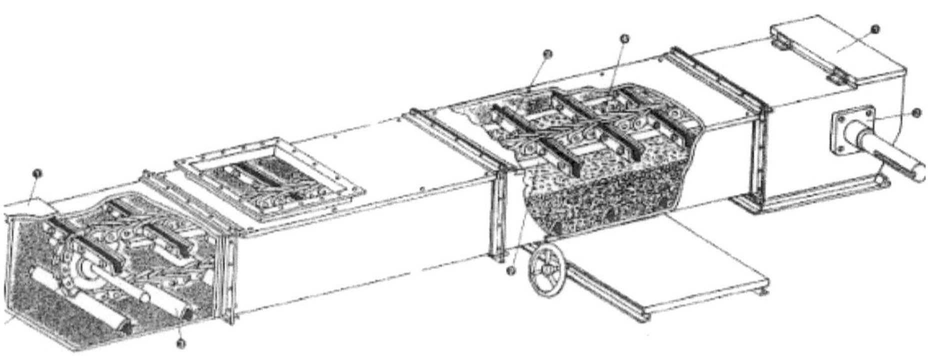

Fig. 3.8. En-masse conveyor (courtesy of Union Iron Works, Decatur, IL).

structural support. Metal-to-metal sliding contact is avoided by using low-friction UHMW-polyethylene flight inserts. However, since the grain slides along the conveyor surface, power requirements are higher than for belt conveyors doing the same job. During loading, grain falls through the return drive line. Discharge can be under the drive pulley, or at any intermediate point. With appropriate flights, the en-masse conveyor can operate at any incline, including straight up. It can also be built with part of its length horizontal and part inclined for an application like grain removal from a dump pit. The horizontal section extends over the length of the pit, and the inclined section carries grain to a level where it can load a bucket conveyor.

Pneumatic Conveyors

Pneumatic grain conveyors move grain in a conduit by means of an air stream moving at speeds up to about 25 m/s (5000 ft/min). Energy for conveying is input through a blower or air pump driven by a dedicated engine, a tractor power-take-off, or an electric motor. Except for low-flow-rate systems, like what might be used to remove dry grain from a continuous dryer, power requirements are so high that electric motors may not be feasible. The airstream pushes grain in a positive pressure system and pulls grain like a vacuum cleaner in negative pressure systems. Grain does not pass through the blower. Some pneumatic-conveyor systems use both positive- and negative-pressure conveyors, and place the engine/blower unit between the conveyor sections.

The extremely high power requirement and high purchase cost for a pneumatic conveyor can be justified in applications where the pneumatic unit offers a big convenience increase. One example is conveying soybean seed from a gravity wagon or truck to planter seed boxes using a positive pressure engine-driven pneumatic conveyor. One person can easily move the flexible output spout from box to box and quickly fill the entire planter, with no spillage of seed.

Soybean Storage

Following harvesting (and artificial drying if necessary), soybeans are placed in bulk storage until processed. As are all seeds and grains, soybeans are subject to invasion by fungi and insects while they are in storage. The most important factor for successful storage is soybean moisture content. Table 3.4 shows the maximum recommended moisture contents for storing soybeans at normal bin storage temperatures. At the moisture contents specified, the relative humidity in the microenvironment around each soybean is low enough so that fungal growth is too slow to be objectionable over the specified storage period. It is important to note that fungi respond to individual soybean moisture levels, and not to the average moisture of a soybean lot. These moisture contents are lower than those recommended for corn at the same conditions (15, 14, and 13%, respectively). These differences are due to the much higher oil content of the soybeans (about 22% on dry weight basis) compared to corn (about 4% on dry weight basis). Since oil is immiscible with water, the non-oil portion of the soybean, which is hydrophilic and susceptible to fungi, has higher moisture content. As a result, the moisture content of the entire soybean must be lower than for corn for comparable periods of safe storage.

Table 3.4. Maximum Soybean Moisture Contents for Safe Storage

Storage Period	Maximum Safe Storage Moisture Content, % Wet Basis
Stored up to 6 months	13
Stored 6–12 months	12
Stored more than 12 months	11

From Wilcke et al. (2004).

In spite of this situation, soybeans generally present fewer storage problems than corn since the soybean crop usually dries to a lower moisture level in the field prior to harvest. Fungi likely to be present on soybeans stored at 25°C at various relative humidities and equilibrium moisture contents are shown in Table 3.5.

Table 3.5. Fungi Likely to Be Present on Soybeans Stored at 25°C and Various Equilibrium Moisture Contents

Soybean Moisture Content, %	Interstitial Air Relative Humidity, %	Fungi
11–12	65–70	*Aspergillis halophilicus*
12–14	70–75	*A. restrictus, A. glaucus, Wallemia sebi*
14–16	75–80	*A. candidus, A. ochraceus,* plus the above
16–19	80–85	*A. flavus, Penicillium* spp., plus the above
19–23	85–90	Any of the above

Adapted from Sauer et al., (1992).

Moisture Migration

When soybeans (and other grains) are placed in storage after harvest, grain temperatures can be quite warm, like the outside air. As weather becomes cooler in the fall, bin walls and grain near the walls cools down. Since bulk grain is a good insulator, grain farther in from the wall remains warmer and, as a result, downward convection currents near the wall cause air to rise through grain in the center of the bin. As this air rises through grain in the bin center, it can pick up moisture from the grain. When this air moves into cooler grain layers near the top, it can saturate and deposit moisture in the grain. This phenomenon is called *moisture migration* (Fig. 3.9). As warmer weather and solar heat raise the temperature in the bin headspace, this wetted grain becomes a suitable environment for storage fungi which can form a moldy crust in the soybeans near the top (Fig. 3.10). A crust may also form near the cold north wall of the bin. Periodic aeration, which cools the grain in the fall and warms the grain in the spring, can eliminate moisture migration problems. See Wilcke et al. (2004) for aeration guidelines.

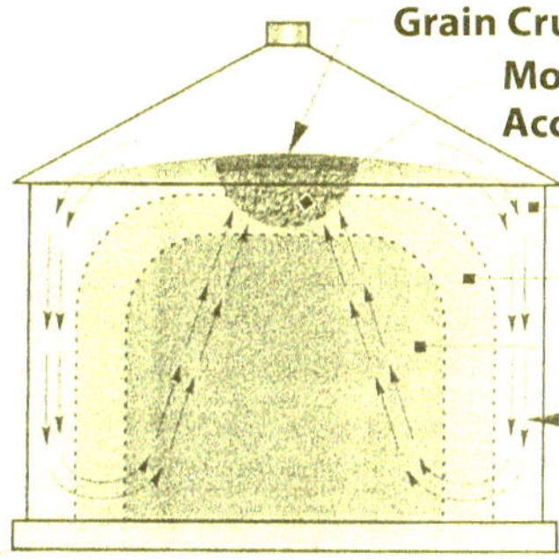

Fig. 3.9. Moisture migration in warm grain stored into cold weather (Source: Wilcke et al., 2004).

Fig. 3.10. Crust formed at top of a bin of soybeans due to moisture migration (Source: photo courtesy of Vince McFadden).

Free Fatty Acid

The common systems for soybean processing separate soybeans into oil and meal. Meal quality is not particularly sensitive to soybean quality (other than protein content), but soybean oil quality is sensitive to soybean quality. The free fatty acid (FFA) level of the oil in stored soybeans is a commonly used quality indicator. The FFA level can increase over time in storage, and the rate of formation increases with storage temperature and moisture content, and also with various kinds of soybean damage. FFA level can be used as a predictor of refining losses and costs as the crude oil is refined. At harvest, FFA level is typically in the range of 0.05–0.7% (Hiromi et al., 1992). Derocher et al. (2005) tracked FFA levels in Asgro 2601 soybeans stored for 18 months at 9 and 14% moisture, and at 27 and 10°C (80 and 50°F). At harvest, FFA averaged 0.5%. Soybeans stored at 9% moisture did not increase in FFA over 18 months at either 27 or 10°C. The FFA of 14% moisture soybeans increased, on average, 0.07 percentage points per month of storage if stored at 27°C, so the FFA increased from 0.5 to 1.8% after 18 months. The increase was only 0.007 percentage points per month if stored at 10°C, so the FFA increased from 0.5 to 0.7% after 18 months. Dry, cool, undamaged soybeans will have a low rate of FFA formation in the oil.

Mycotoxins

Mycotoxins are compounds produced by fungi under certain conditions and most are carcinogenic and/or toxic. Fungi strains which produce mycotoxins can, and usually do, grow without producing mycotoxins. Hawk (2004) lists aflatoxin, deoxynivalenol (DON or vomitoxin), fumonisn, zearalenone, ochratoxin A, and T2 as the six mycotoxins of most concern in the grain industry. Although soybeans can support many of the fungi strains capable of producing mycotoxins, they are rarely produced on soybeans. Occurrences on corn are much more frequent. Jacobsen et al. (1995) analyzed and processed 24 soybean samples from elevators in Illinois, Iowa, and Michigan in 1986. The soybean harvest was delayed in this area during 1986 due to abnormally warm, humid weather, and soybeans at or near maturity during this time showed extensive damage from field molds at harvest. Samples were drawn from soybean lots either refused or heavily discounted by elevators. Four of Hawk's six important mycotoxins were present. Zearalenone, DON, and T2 were detected in whole soybeans, hulls, meal, and oil, while aflatoxin B(1) was found in soybean hulls. Surprisingly, FFA levels at 0.36 to 0.57% were within the normal range.

Insect Problems with Stored Soybeans

Stored soybeans are susceptible to attack by stored grain insects; however, insect problems in soybeans are much less common than in grains like corn or wheat. Insect problems in stored soybeans can be reduced or eliminated by following good management practices. Clean, dry soybeans should be placed in clean, dry bins. New crop soybeans should never be placed on top of old crop soybeans in a bin. Storage of soybeans for over a year should be avoided. Management of soybean temperature and moisture by aeration can be used to control insect problems since insects require certain minimum temperature and moisture levels to feed and reproduce. Most visible insect action stops when the grain temperature falls below 10°C (50°F).

References

AOCS. Official Method Ac-2-41 Moisture and Volatile Matter. American Oil Chemists' Society: Champaign, IL, 2007.

ASABE. D241.4, Density, Specific Gravity, and Mass-Moisture Relationships of Grain for Storage. American Society of Agricultural and Biological Engineers: St. Joseph, MI, 2006a.

ASABE. D271.2, Psychrometric Data. American Society of Agricultural and Biological Engineers: St. Joseph, MI, 2006b.

ASABE. D272.3, Resistance to Airflow of Grains, Seeds, Other Agricultural Products, and Perforated Metal Sheets. American Society of Agricultural and Biological Engineers: St. Joseph, MI, 2006c.

ASABE. S343.3. Terminology for Combines and Grain Harvesting. In *ASABE Standards,* American Society of Agricultural and Biological Engineers: St. Joseph, MI, 2005.

ASABE. S352.2. Moisture Measurement—Unground Grain and Seeds. In *ASABE Standards 2006;* American Society of Agricultural and Biological Engineers: St Joseph, MI, 2006d.

CEMA. Screw Conveyors. *CEMA Book 350, 3rd edn.*; Conveyor Equipment Manufacturer's Association: Naples, FL, 2003.

DeRocher, B.D.; C.J. Bern; C.R. Hurburgh. Effects of Long-Term Storage on Soybean Quality. *Final Report, Independent Study Project;* Agricultural and Biosystems Engineering Department: IA State University, Ames, IA, 2006.

Dickey-John. GAC 2100 Agri. http://www.dickey-john.com/products/ag/product_detail.php?products_id=; 1999.

Gleim, J.A.; R.G. Holmes; R.K. Wood. Corn and Soybean Harvesting Losses. Paper no. 90-1563. American Society of Agricultural and Biological Engineers: St. Joseph, MI, 1990.

Hanna, H.M.; D.H. Jarboe; G.R. Quick. Grain Residuals and Time Requirements for Combine Cleaning. Paper no. 066082. American Society of Agricultural Engineers: St. Joseph, MI, 2006.

Hanna, H.M.; G.R. Quick. Grain Harvesting Machinery Design. In *Food Machinery Handbook,* M. Kutz, Ed.; William Andrew Publishing: Norwich, New York, 2007; pp. 93–111.

Hawk, A. Mycotoxins 101, Part 1. *Grain J.* **2004** *32*:10–11.

Hellevang, K.J. *Natural Air/Low Temperature Crop Drying, EB-35.* North Dakota State University Extension Service: Fargo, ND, 1983.

Jacobsen, B.J.; K.S. Harlin; P. Swanson; R.J. Lambert; J.B. Sinclair. Occurrence of fungi and mycotoxins associated with field mold damaged soybeans in the Midwest. *Plant Disease* **1995,** *79,* 86–89.

Kebeli, K.V.; R.A. Bucklin; D.S. Ellifritt; K.V. Chau. The Effects of Changes in Grain Moisture

Content on the Loads in Grain Bins. Paper no. 984018. American Society of Agricultural and Biological Engineers: St. Joseph, MI, 1998.

McKenzie, B.A. Drying Soybeans with Heated and Unheated Air, AE-84. W. Lafayette, Indiana. Agricultural Engineering Department, Purdue University Cooperative Extension Service: West Lafayette, IN, 1973.

Morey, R.V.; W.F. Wilcke; D.J. Hansen. Aeration strategies for reconditioning dry soybeans. *Applied Eng. Agri.* **2003,** *19,* 433–446.

MWPS. *Grain Drying, Handling, and Storage Handbook, 2nd Edn.* Midwest Plan Service: IA State University, Ames, IA, 1987.

Peterson, G. Semis Favored Way to Market. *Successful Farming* **2006,** *104*(11A), 36–39.

Quick, G.R. Laboratory Analysis of the Combine Header. *Trans ASAE* **1973,** *16,* 5–12.

Sauer, D.B.; R.A. Meronuck; C.M. Christensen. Microflora. In *Storage of Cereal Grains and Their Products, 4th Edn.* D.B. Sauer, D.B. Ed.; American Association of Cereal Chemists: St Paul, MN, 1992.

Siemens, J.C.; H.J. Hirning. Harvesting and Drying Soybeans, Circular 1094. University of Illinois Cooperative Extension Service: Champaign, IL. 1974.

Wilcke, W.F.; K.J. Hellevang; J.P. Harner; D.E. Maier; W.W. Casady. Managing Dry Grain. In *Storage, 2nd edn.;* AED 20, Midwest Plan Service: IA State University, Ames, IA, 2004.

Wilcke, W.F.; R.V. Morey. Natural-Air Corn Drying in the Upper Midwest, BU-6577. University of Minnesota Extension: St. Paul, MN, 1995.

Wilcke, W.; K. Hellevang; V. Morey. Soybean Drying, Handling, and Storage. In The Minnesota Soybean Field Book, MI-7290; J.M. Bennett J.M., Ed.; University of Minnesota Extension: St. Paul, MN. 1999.

4

Effect of Pests and Diseases on Soybean Quality

John Rupe[1] and Randall G. Luttrell[2]
[1]Department of Plant Pathology, [2]Department of Entomology, University of Arkansas, Fayetteville, Arkansas

Diseases and insect pests are important factors limiting soybean production. In 1994, diseases resulted in nearly 11% yield loss in the top-ten soybean-producing countries in the world (Wrather et al., 2001) while in 2006 in the southern United States, disease losses averaged nearly 9% (Koenning, 2007). Insect damage causes losses of similar magnitude. These losses come from a reduction in yield and grain quality and, for soybeans grown for seed, reductions in seed quality. In this chapter, grain quality refers to the physical and chemical properties of the soybean seed that are important in its utilization, such as the types and quantities of oil and protein found in the seed. Seed quality refers to the ability of the seed to produce a vigorous seedling. Besides grain and seed quality issues, the international shipment of soybean grain and seed raises phytosanitary issues involving the movement of exotic pathogens between countries.

Research concentrates on the effect of diseases and insects on seed quality. Grain quality is not an important factor in selling soybeans, but that may be changing. The United Soybean Board is sponsoring the Better Bean Initiative (BBI) that includes goals to identify oil and protein traits that will improve the value of soybean and to incorporate these traits into commercial germplasm (Durham, 2003; Wilson, 2004). The current goals of BBI are to reduce the saturated fatty acids in the oils, specifically palmitic and stearic fatty acids, to reduce linolenic and increase oleic fatty acids to increase protein, and to lower phytate-phosphorus. Only limited research has occurred on the effect of diseases and insects on these traits, but this will change as markets develop for these traits, and rapid, accurate assessment procedures for these traits are developed. In this chapter, we present the major diseases and insects affecting soybean seed and grain quality, and review the basic control measures. Most of these diseases and insects directly affect the soybean seed by growing on or in the seed or by feeding on the seed. Many other diseases and insect pests directly affect soybean seed and grain quality by attacking other parts of the soybean plant, limiting nutrients to the developing seed and shortening the life of the plant. We discuss a few of these diseases and insects.

Diseases

Pathogens that directly affect soybean seeds fall into three main groups: fungi, viruses, and bacteria. Fungi are the largest group of pathogens affecting seed, but viruses and bacteria cause important soybean seed diseases.

Fungi

Phomopsis Seed Decay

Phomopsis seed decay is the most important and the most studied seed disease of soybean. Fungi in the Diaporthe/Phomopsis complex cause this disease, also called pod and stem blight (Kulik & Sinclair, 1999; Kulik & Yaklich, 1982). This complex contains fungi with a sexual stage, the *Diaporthe* species, and one fungus with only the asexual reproductive stage, *Phomopsis longicolla*. Of these fungi, *P. longicolla* is more frequently isolated from seed, followed by *Diaporthe phaseolorum* var. *sojae*. These two fungi are the causal agents of pod and stem blight. Two other closely related fungi cause stem canker diseases: *D. phaseolorum* var. *caulivora*, northern stem canker, and *D. phaseolorum* var. *meridionales*, southern stem canker. Both of these fungi cause cankers that kill the plant, but also can infect seed to some extent. All four fungi are spread by splashing rain and primarily infect the lower part of the plant.

The pod and stem blight fungi infect the plant at any time, but colonization beyond the point of infection only happens as the plant dies either through senescence or injury. Then the fungi form black sporulating structures called pycnidia on stems, petioles, and pods (Fig. 4.1). Seed infection does not begin until the plant reaches physiological maturity, and the level of seed infection depends on how quickly the pods dry to harvest maturity. Wet weather or seed maturation under hot, moist conditions increases seed infection by *P. longicolla* (Kulik & Sinclair, 1999). These infections usually occur in the seed coat and are symptomless, but can significantly reduce seed germination. If wet weather delays the harvest, then more extensive colonization of the seed can occur resulting in shrunken, wrinkled seed covered with white mycelium (Fig. 4.2). Not only does this severe infection reduce yield, it also changes the chemical composition of the seed. We tend to associate severe pod and stem blight with increases in seed oil and protein content (Bradley et al., 2002; Hepperly & Sinclair, 1978). Oil from heavily infected seed may appear discolored and have a rancid odor, and the flour may have an off taste (Hepperly & Sinclair, 1978). Severely infected seeds have increased levels of linoleic and linolenic acids and have decreased levels of palmitic and oleic acids compared to symptomless seeds (Wrather et al., 2003). Degradation of cotyledondary proteins such as lipoxygenase, conglycinins, and glycinins is associated with *P. longicolla* as is degradation of β-amylase in the seed coat and the cotyledons (Velicheti et al., 1992).

Environment, the cultivar, infection by other pathogens, and foliar fungicides determine seed infection by *P. longicolla*. Warm, wet weather favors infection of the pod

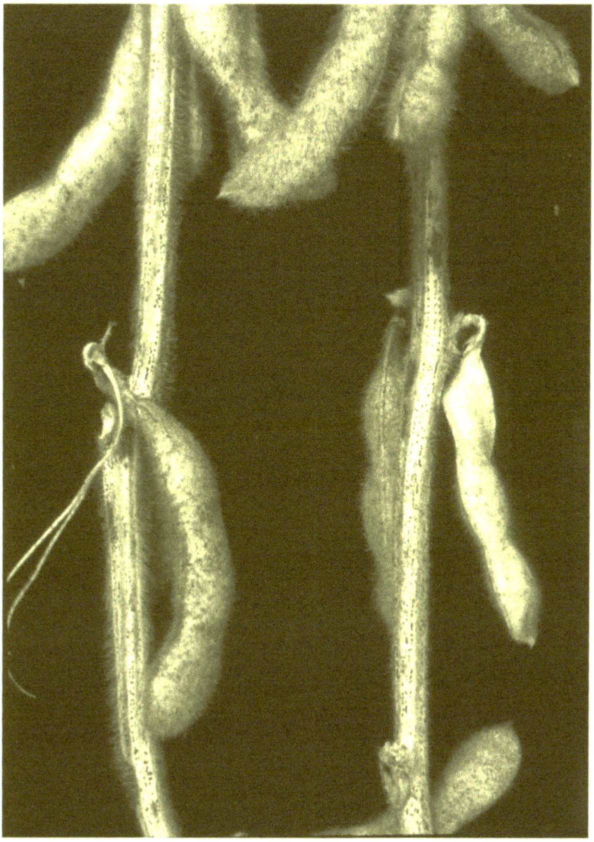

Fig. 4.1. Pod and stem blight on senescent soybean stems showing rows of black pycnidia. (Photo courtesy of J. Walters)

and then, after physiological maturity, the seed. As a result, early planting of early-season cultivars usually results in increased seed infection by *P. longicolla* and reduced germination (Mayhew & Caviness, 1994). Delaying planting and/or planting later-maturing cultivars can greatly reduce seed infection and increase germination. Cultivars vary in their reaction to *P. longicolla*, and several resistance genes can dramatically reduce seed infection levels (Jackson et al., 2005; Minor et al., 1993; Zimmerman & Minor, 1993). Other pathogens can affect seed infection. Plants infected by the soybean mosaic virus (Hepperly et al., 1979; Ross, 1977) or the bean pod mottle virus often have higher levels of *P. longicolla* (Gergerich, 1999). Seed levels of *P. longicolla* negatively correlate to seed levels of *Cercospora kikuchii* (the purple seed stain fungus) (McGee et al., 1980; Pathan et al., 1989). Applying foliar fungicides during reproductive development of the soybean crop can protect seed against infection by these fungi and increase seed germination (Padgett et al., 2003; Tenne & Sinclair, 1978).

Fig. 4.2. Mycelium of *Phomopsis* spp. growing on the surface of soybean seeds. (Photo courtesy of D. Mueller)

Purple Seed Stain

Purple seed stain, caused by *C. kikuchii*, is a common soybean seed disease resulting in a deep-purple discoloration of the seed coat, although symptomless infections are common (Fig. 5.3). The purple stain is due to a toxin, cercosporin, produced by the fungus and is an important component in disease development in the plant (Velicheti & Sinclair, 1994). Cercosporin is a perylenequinone that is active only in light where it converts into an excited triplet state that reacts with oxygen to produce activated oxygen species (Daub & Chung, 2007; Upchurch, 1995). These activated oxygen species result in peroxidation of membrane lipids, killing the plant cells and allowing the fungus to colonize the seed. Infections by *C. kikuchii* usually occur in the seed coat, but heavily infected seed have lower germination, increased free fatty acid content (oleic acid) and protein, and lower oil content (Pathan et al., 1989). They also may increase lipids and lower saccharide levels (Katsube, 1980) and may affect cooking quality (Taira et al., 1990).

Cercospora kikuchii degrades lipoxygenase in the seed coat but not in the cotyledons (Velicheti et al., 1992). Seed infection by C. *kikuchii* may (Pathan et al., 1989;

Tyler & Overton, 1981; Wilcox & Abney, 1973) or may not affect seed germination (Galli et al., 2005). *Cercospora kikuchii* causes both purple seed stain and Cercospora leaf blight, but the two diseases do not appear to be linked (McGee et al., 1980; Orth & Schuh, 1994). Different subgroups within the pathogen species may cause these diseases (Cai & Schneider, 2005). Most seed infection occurs late in seed development (Ortiz et al., 1988) and increases when wet weather delays harvest (Fujita, 1990). Foliar fungicides can reduce the incidence of seed infection (Padgett et al., 2003; Tenne & Sinclair, 1978). Some soybean cultivars resist purple seed stain (Orth & Schuh, 1994), and the description of at least one dominant resistance gene does exist (Jackson et al., 2006).

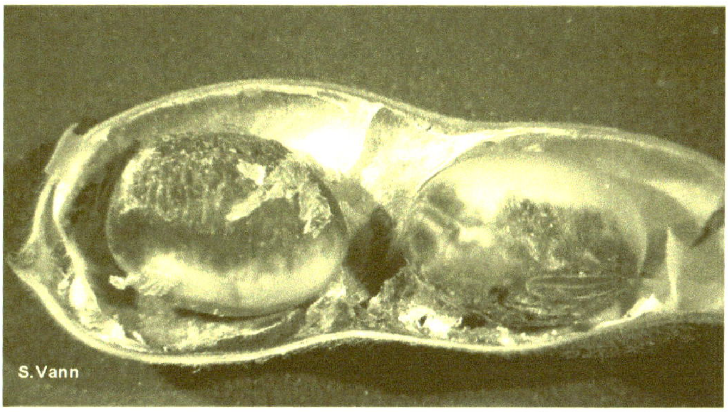

Fig. 4.3. Soybean seed with purple seed stain caused by *Cercospora kikuchii*. (Photo courtesy of S. Vann)

Sclerotinia Stem Rot

Sclerotinia stem rot, caused by *Sclerotinia sclerotiorum*, is an important disease in the U.S. Midwest and in Argentina and Brazil, causing reduced yield and seed size, infecting seed, and contaminating seed with fungal sclerotia (Fig. 4.4) (Grau & Hartman, 1999). The fungus survives as small, black sclerotia (compact masses of fungal hyphae 2–22 mm dia). Under cool, wet conditions, the sclerotia germinate to form a fruiting structure, an apothecium, which produces ascospores. These ascospores infect the plant through the flowers, which provide the nutrients necessary for the spores to infect the plant. The fungus then grows on the inside and outside of the stem, petioles, and pods as a white, cottony mycelium. Sclerotia form in and on all diseased tissue. Diseased plants wilt, possibly severely reducing yields. Seed may become infected and may become covered in fungal mycelium, or the infections may be symptomless. Sometimes sclerotia of the fungus replace the seed. Since some of the sclerotia are the same size and density as soybean seed, sclerotia often mix with the seed.

Fig. 4.4. Black sclerotia and soybean seed covered with mycelium of Sclerotinia sclerotiorum. (Photo courtesy of D. Mueller)

Disease incidence negatively correlated to seed germination, oil content, and seed weight and positively correlated to visual seed symptoms and to the number of sclerotia mixed in the seed (Hoffman et al., 1998). These sclerotia and the infected seed are important means of spreading the pathogens to new fields (Yang et al., 1998). Practices utilized to maximize soybean yields (narrow row spacing, high plant populations, high soil fertility, early planting, irrigation, and narrow rows) and no tillage (which leaves the sclerotia on the soil surface) favor disease (Grau & Hartman, 1999). Some cultivars are moderately resistant, and planting should include disease-free seed uncontaminated by sclerotia.

Yeast Spot

Yeast spot, caused by *Nematospora coryli* and *N. lycopersici*, can discolor soybean seed and lower oil content (Sinclair, 1999d). The mouth parts of several species of stink bug transmit these pathogens that infect the seed when the insect feeds on the embryo (Sinclair, 1999c, Russin & Boethel, 1994). Early infection during pod development may lead to pod abortion while later feeding can lead to light- or cream-colored sunken spots on seeds (Sinclair, 1999c). Cotelydons may become white and have a cheesy texture that results in shrunken, wrinkled seeds at maturity.

Other Fungi

Other fungi that primarily impact plant foliage, steams, or roots may infect seed, such as *Cercospora sojina* (causal agent of frogeye leaf spot) (Phillips, 1999a), *Macrophomina phaseolina* (causal agent of charcoal rot), *Peronospora manschurica* (the downy mildew pathogen) (Phillips, 1999b), and *Colletotrichum truncatum* (causal agent of anthracnose) (Manandhar & Hartman, 1999). Reduced seed germination can occur with *C. truncatum* (Chandrasekaran & Rajappan, 2002), but with most of the other fungi the effects on seed germination or other aspects of seed or grain quality are not well documented. Since these fungi cause other diseases, seed infected by them may be subject to quarantine restrictions.

Many diseases and insects through premature senescence or plant death can indirectly affect soybean seed quality. Soybean rust, caused by *Phakopsora pachyrhizi*, attacks soybean leaves primarily, but not seed, leading to defoliation and early senescence. Yield losses are due to a reduction in seed number and seed size. Seed size losses of 12 to 20% can occur (Arias et al., 2005; Dupleich et al., 2005).

Likewise, drought damage can shorten the seed maturation period, often resulting in smaller seed and an increase in hard seed, although seed germination is usually not affected (Vieira et al., 1992). If the seed maturation period is shortened too much, seed may be smaller, wrinkled, and green (Franca-Neto et al., 2005). While these studies did not assess seed composition, presumably since the types of oils, proteins, and sugars deposited in the seed change as the seeds mature (Wilson, 2004), anything affecting seed maturation also will affect seed composition.

Fungi that do not appear to cause disease, but are found in the seed, can infect soybean. These fungi include *Alternaria alternata*, *Aspergillus* spp., *Cephlasporium* spp., *Chaetomium* spp., *Cladosporum* spp., *Curvularia lunata*, *Cylindrocarpon* spp., *Fusarium* spp., *Rhizopus* spp., *Nematospora coryli*, *Penicillium* spp., *Phoma* spp., *Trichothecium roseum*, *Ulocladium botryritus*, and others (Bringel et al., 2001; Hepperly, 1985; Ibraheem et al., 1987; Jordan et al., 1986; Kilpatrick & Hartwig, 1955; Lee, 1984; Padule et al., 2004; Solanke et al., 1997; Vishwadhar & Sarbhoy, 1994; Zad, 1987). In general, pod wounding (Lori & Sarandon, 1989), pod feeding damage by the bean leaf beetle (Coeoptera: Chrysomelidae) (Obopile & Hammond, 2001; Shortt et al., 1982), or freezing (Osorio & McGee, 1992) reportedly increase seed infection from a number of fungi.

Mycotoxins

Some of the seed-infecting fungi are known to produce mycotoxins. *Fusarium graminearum*, once thought to be a saprophyte on soybean seed and now reported to cause stem and pod diseases in soybean (Pioli et al., 2004), produces the mycotoxin, nivaleonol (Martinelli et al., 2004). Isolates of *F. oxysporum*, *F. verticillioides*, *F. semitectum*, *F. solani*, *F. graminearum*, and *F. lateritium* from soybean seed collected in Korea produce apicidin (Park et al., 1999). Fusarin C is a product of *F. verticillioides* on several crop seeds including soybean, but soybean is the poorest substrate of the crops tested (Bacon et al., 1989). In fact, mycotoxins are usually not a significant problem in soybean (Nesheim & Garnett, 1995; Wilson et al., 1995). Lipoxygenases in soybean seed reportedly inhibit the production of aflatoxin by *Aspergillus* spp. (Burow et al., 1997), but later research did not find a correlation between lipoxygenases and aflatoxin production (Mellon & Cotty, 2002).

In the United States and Canada, mycotoxins are found in soybean seed, but only in severely damaged seed, because of delayed harvest due to severe weather. *Aspergillus* spp. were isolated from the seed, and aflatoxins were found in seed harvested in half of the Maryland fields sampled after a very wet August, September, and October (Bean et al., 1972). Similar weather in Canada resulted in badly discolored seed (Clear et al., 1989). *Alternaria* spp. were isolated from the dark discolored seed, and these seeds had detectable levels of HT-2 toxin. *Fusarium graminearum* and *F. sporotrichioides* infected the reddish discolored seed that had both deoxynivalenol and HT-2 toxin, primarily in the seed coat. In the Midwest, wet weather at the end of the season resulted in seed infection by *Alternaria alternata*, *F. graminearum*, and *Aspergillus* spp. and detectable levels of zearalenone, zeralenol, diacetoxyscirpenol, and addexoynivalenol (Jacobsen et al., 1995). Although mycotoxins are produced in soybean, they do not appear to be an important problem in most temperate climates. More problems may arise with mycotoxins in tropical climates where the soybeans mature and are in storage under warm temperatures. In Egypt, over 70 species of fungi including several *Aspergillus* spp. appeared in soybean seed lots (el-Kady & Youssef, 1993). Aflatoxin was the only mycotoxin recovered and was found in 35% of the seed samples. In addition, storage conditions of soybean grain can affect the level of infection by these mycotoxin-producing fungi (Narayanaswamy et al., 2000; Pessu et al., 2005).

Viruses

Another major pathogen group that infects seed is viruses. Being composed of a protein coat and a nucleic acid core, viruses are much simpler pathogens than the fungi. The nucleic acid is usually a single-stranded, positive-sense RNA, but some plant pathogenic viruses have double-stranded RNA or DNA, and may exist in a single protein particle or in multiple particles. Unlike fungi that attack by absorbing plant nutrients from the cells, viruses take over the cell's metabolic processes, forcing the cells to make new viruses. As a result, many symptoms of viral infection appear as growth abnormalities such as stunting, malformed leaves and fruit, and mosaic leaf patterns. Most of the viruses that are important economic pathogens of a crop become systemic in the infected plant and are often found in the seed. Seed infection levels may be low, but these seed infections can be important for moving the virus to new fields. While many viruses are pathogens on soybean, three are particularly important: soybean mosaic virus, bean pod mottle virus, and tobacco ringspot virus.

Soybean Mosaic Virus (SMV)

Soybean mosaic virus (SMV) is a member of the potyvirus group (Hill, 1999). The virus particle is a long, flexuous rod containing a single strand of RNA. Over 30 species of aphid including *Aphis glycines*, the soybean aphid, transmit the virus particle. Field incidences of SMV correlate with flights of the soybean aphid (Burrows et al., 2005). With any of the aphid vectors, the stylet transmits SMV which means that the virus is acquired and transmitted rapidly, and after a few feedings by the aphid (Hill, 1999). SMV also infects the seed. Seed infection rates and yield losses are higher if the plant is infected before rather than after flowering (Nakano et al., 1988; Ren et al., 1997b). Infected seed can result in infected plants that often serve as the initial points of inoculum in a field with later infections resulting from aphid feeding. Seed infection can be as high as 75% depending on the soybean cultivar and the strain of the virus, but is usually less than 5% (Hill, 1999). Infection with SMV may result in smaller, less vigorous seed with lower oil, but increased protein and amino acid content (El-Amrety et al., 1999; Suteri, 1980). Hilum bleeding (Fig. 4.5), where the color of the hilum spreads out into the seed coat, is often associated with SMV-in-

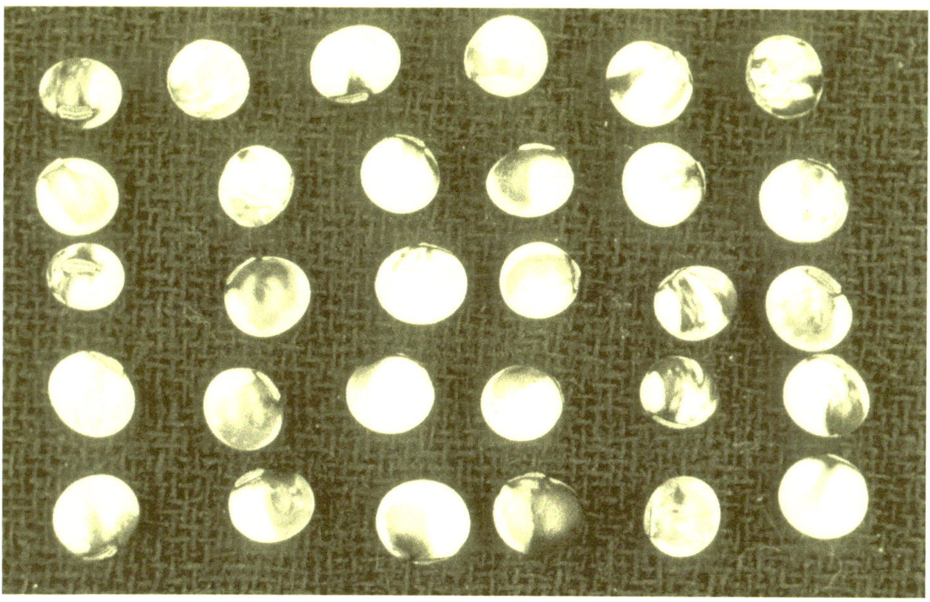

Fig. 4.5. Soybean seed with hilum bleeding due to infection by soybean mosaic virus. (Photo courtesy of J. Walters)

fected seed; however, other stresses can cause this symptom, and SMV infections can be symptomless.

Yield loss due to SMV, as with most viruses, depends on when infection occurs, but is usually not more than 8 to 25% unless the plant is co-infected with bean pod mottle virus. Then, yield loss can be as much as 66 to 86% (Hill, 1999; Ross, 1968). A great deal of variability exists between strains of SMV in the level of symptoms caused and in which cultivars they can infect (Hill, 1999). At least eight single genes for resistance are located at three alleles. Fifteen strains of SMV are identified using a set of eight differential cultivars. Managing SMV depends on the type of resistance the plant has and the strain of the virus (Bowers & Goodman, 1991; Silva et al., 2003). Resistance is more important in late- than early-planted soybean, because the insect vectors increase population densities during the season with SMV infection typically higher in late-planted soybeans (Ren et al., 1997a). Since SMV infection predisposes the soybean to seed infection by *Phomopsis longicolla*, the use of SMV-resistant cultivars can reduce the incidence of seed infection by this fungus as well (Koning et al., 2002; Ross, 1977). The most effective management strategy is to plant virus-free seed.

Bean Pod Mottle Virus (BPMV)

Bean pod mottle virus (BPMV) is a comovirus that contains its single-stranded RNA in two isometric particles, and both particles are necessary for disease (Gergerich, 1999). The bean leaf beetle, *Cerotoma trifurcata*, and several other beetles transmit BPMV. The beetle acquires and transmits the virus quickly, but can retain the virus for several days. Infected plants produce fewer and smaller seeds with losses as high as 60%, but usually less than 10%. Co-inoculation of BPMV with SMV results in a synergistic increase in disease loss (Ross, 1977). Infection by BPMV also predisposes the plant to increased seed infection by *P. longicolla* (Gergerich, 1999). Stems on infected plants may remain green, even after the pods are mature, hindering harvest. BPMV has a wide host range, including noncrop host plants that serve as an inoculum source for the field. Reducing weed pressure at the edge of the field and planting virus-free seed may reduce the incidence of BPMV. No plant resistance to BPMV is available in commercial soybean cultivars. Planting date does not have a consistent effect on BPMV infection (Krell et al., 2005); however, controlling the first generation of beetles reduces infection by the virus (Krell et al., 2005b).

Tobacco Ringspot Virus (TRSV)

Tobacco ringspot virus (TRSV) causes a disease in soybean called bud blight and may be a contributor to greenbean syndrome (Demski & Kuhn, revised by Hartman, 1999). TRSV is a nepovirus that the nematode *Xiphimena americanum can transmit*, but it is also seedborne. The RNA genome of the virus has two parts, each contained in a polyhedral particle. Both parts of RNA genome are necessary for infection. While *X. americanum* can transmit TRSV, seed infection is a more important means of dispersal, and the virus seems to spread from weed hosts on the edge of the field by an unknown vector. The terminal bud of infected plants dies and crooks. Later, other buds die and pods abort. The dead areas become brittle. If infection occurs early enough, few, if any, seed form, and the plant remains green until frost. Later infections may result in infected seed. The main controls are to use virus-free seed and avoid growing soybeans in fields with a history of this disease.

Bacteria

Bacteria can contaminate or infect soybean seed that can serve as an important means of spreading these pathogens from field to field. Bacterial blight, caused by *Pseudomonas savastanoi* pv. *gylcinea* (Sinclair, 1999c), and bacterial pustule, *Xanthomnas axonopodis* pv. *glycines* (Sinclair, 1999b), are both found in and on soybean seed. Seed infection by *P. savastanoi* pv. *gylcinea* correlates with plant surface populations of the bacterium and the intensity of foliar symptoms (Stefani et al., 1998).

Bacillus Seed Decay

Bacillus seed decay, caused by *Bacillus subtilis* (Sinclair, 1999a), damages the seed directly. Under moist, hot (over 30°C) conditions, *B. subtilis* can cause a seed decay in the soil or in storage, producing a slimy soft rot of the seed. This appears to be primarily a problem in tropical climates (Scortichini et al., 1989), but also can be a problem in temperate climates and is related to reduced seed germination (Schiller et al., 1977).

Insect Pests

Although more than 700 species of plant-feeding insects are found in soybean (Way, 1994), most of the damage is attributed to as few as eight species in the United States. These include the velvetbean caterpillar, *Anticarsia gemmatalis*; soybean looper, *Pseudoplusia includens*; green cloverworm, *Plathypena scabra*; Mexican bean beetle, *Epilachna varivestis*; bean leaf beetle, *Cerotoma trifurcata*; southern green stink bug, *Nezara viridula*; green stink bug, *Acrosternum hilare*; and corn earworm, *Helicoverpa zea*. Other pest species of occasional importance include the three-cornered alfalfa hopper, *Spissistilus festinus*; lesser cornstalk borer, *Elasmopalpus lignosellus*; tobacco budworm, *Heliothis virescens*; beet armyworm, *Spodoptera exigua*; twospotted spider mite, *Tetranychus urticae*; and several grasshoppers, *Melanoplus* spp. Insect plant injury falls into four categories: plant injury caused by root-feeding, stem-feeding, leaf-feeding, and pod-feeding (Higley, 1994). Within each of these broad categories, different feeding behaviors among pests may differentially impact the plant and resulting plant injury. Insect feeding may cause stand reductions, leaf mass removals, leaf photosynthetic rate reductions, leaf senescence, light reductions, assimilate removals, architecture modifications, and phenological disruptions. Plant feeding by insects directly affects plant carbon gain by removing photosynthetic leaf tissue and indirectly by inducing costly defensive compounds and disrupting the water retention capabilities of the plant (Aldea et al., 2005). All of these physiological impacts may impact seed size and seed quality. Water loss is a critical aspect of soybean yield reduction, and insect injury often results in more severe yield impacts under water-stress conditions.

Direct effects on soybean seed and grain quality are typically associated with pod or seed feeders, which include various stink bugs, corn earworm, bean leaf beetle, grasshoppers, several armyworms, and occasionally high population densities of soybean looper and velvetbean caterpillar (Steffey et al., 1994; Turnipseed, 1976; Hammond et al., 1991; Zeiss & Klubertanz, 1994).

Soybean has a tremendous capacity to compensate for insect damage. As a general rule, soybean can tolerate as much as 40% defoliation before bloom and 25% after bloom with no reduction in yield (Lorenz et al., 2000). Pod-feeding insects typically reduce yields when damage occurs during early- to mid-reproductive growth; however, soybean can compensate completely for pod losses as high as 80% if the damage occurs at the early stages of pod set (Smith & Bass, 1972). While damage early in reproductive stages often causes pod abortion or individual seed loss, the plant can compensate for this loss via increased retention of other pods or increased seed size (Hammond et al., 1991). Compensation by the soybean plant is limited when plants are in mid-reproductive development (R5–R6). At or after this point, pod damage can result in significant yield losses (Kincade et al., 1971; Thomas et al., 1974). In late reproductive stages (R6–R7), pod and seed feeding, especially by stink bugs that feed directly on mature seed, seldom results in yield loss but can cause significant seed damage. We often overlook this damage and its association with disease organisms, because yields are not affected. This may change with an increased awareness of seed quality and price discounts for damaged seed.

Insects reduce soybean seed quality through seed-coat mottling, shrinking, and alteration in oil and protein content and reduced germination (Hammond et al., 1991). Palmitic, stearic, and oleic acids increase and linoleic and linolenic acids decrease (Todd et al., 1973). These reductions in soybean grain quality usually have had little effect on the price paid to farmers unless yields are reduced. However, stink bug damage to soybean seed can discolor the seed and result in discounted prices for export soybeans (Miner, 1966). If this damage is excessive enough, the elevator could reject the crop. Certified soybean seed or edible soybean grain is especially sensitive to insect damage, because of higher per unit value, and may be rejected if damage levels are too high.

Stink Bugs

Stink bugs are the most important insect pests affecting soybean seed quality. Common species in U.S. soybean include southern green stink bug, green stink bug, and brown stink bug, *Euschistus servus*. Several other species, including *Euchistus tristigmus*, *Piezodorus guildinii* and *Euchistus variolarius*, may be pests in localized areas (McPherson et al., 1994). All of these species have similar impacts on soybean seed quality. Stink bug populations appear to be increasing as a pest problem in the southern United States, and growers in some regions expect to apply one to two insecticide applications annually to control these pests. Increased production of early-season, indeterminate soybean cultivars appears to be an associated factor with this rise in pest status (Baur et al., 2000; Gore et al., 2006; McPherson et al., 1993; Smith, 2006).

Stink bugs may feed on plant stems, foliage, blooms and seed, but prefer young tender reproductive tissues and developing seed (McPherson et al., 1994). They feed by puncturing the plant tissue with their piercing and sucking mouthparts. Damage results from the loss of plant fluid, injection of digestive enzymes, abortion of fruit, and damaged seed. Stink bug feeding often predisposes the plant to infection by pathogenic and decay organisms. Various fungi and bacteria can infect the pod at the feeding site resulting in localized or general pod decay (Fig. 4.6). Seed damaged

Fig. 4.6. Stink bug damage to soybean pods. Pod discoloration due to fungi and bacteria introduced at the feeding site. Pod in the center is undamaged. (Photo courtesy of R. Luttrell)

by stink bug feeding may stop growing resulting in small wrinkled seed that may remain green (Fig. 4.7). These seed usually do not germinate. Stink bug damage may result in severe direct and indirect losses. A strong linear relationship exists between stink bug density, seed damage, and yield loss (Daugherty et al., 1964; McPherson et al., 1974; Yeargan, 1977). Stink bug feeding causes delayed maturity (Boethel et al., 2000), reduced seed germination (Jensen & Newsom, 1972; Yeargan, 1977), increased incidence of seedborne *Fusarium* spp., reduced seed viability, and delayed harvest up to six weeks (Russin et al., 1988). Feeding of most stink bug species (Russin & Boethel, 1994) transmits yeast spot, *Nematospora coryli*. The southern green stink bug also transmits 13 genera of bacteria to soybean including several species that are seedborne (*Bacillus, Pseudomonas, and Xanthomonas*) and is associated with increased frequencies of *Fusarium* spp. and *Alternaria tenuissima* seed infection (Russin & Boethel, 1994). Heavy seed damage reduces oil content, increases protein content, increases percentage of small seed, and increases discolored or moldy seed in storage (Daugherty et al., 1964; Miner, 1961; Miner, 1966; Todd et al., 1973; Yeargan, 1977). Stored grain typically deteriorates with increased storage moisture and increased breakage, splits and damaged seed coats, but heavy damage from stink bug feeding can result in more severe storage problems over time (Miner & Dumas, 1980). Even in the absence of yield loss, stink bug damaged seed can result in significant dockage when sold as grain (Evans et al., 1997).

Fig. 4.7. Stink bug damage to soybean seed. Damaged seed on the left, healthy seed on the right.

Other Pod-feeding Insects

Other pod-feeding insects, including bean leaf beetle, corn earworm, and several species of grasshoppers, also make soybean seed more susceptible to decay and disease (Hammond, 1996; Smelser & Pedigo, 1992; Turnipseed, 1976). These pests have chewing mouthparts, and they mechanically damage pods and seed by removing large chunks of plant tissue while feeding. They seldom completely consume an entire pod, but the damaged pod and seed are open to infection and water loss. This results in additional split and cracked beans in storage.

Soybean Aphid

Recently, the soybean aphid, *Aphis glycine*, was introduced into the North Central United States from Asia. This insect is a serious concern because of its population-growth potential and its ability to transmit a wide range of plant diseases. Yield losses as great as 30% were reported in some provinces of China when high densities of the aphid infested soybean (Wu et al., 2004). The soybean aphid was first found in Minnesota in 2000 and has consistently expanded its range into the U.S. heartland of soybean production since then (Ragesdale et al., 2004). The insect utilizes *Rhamnus* spp. (buckthorn) as an alternative and early-season host, moving to soybean in mid to late summer. This is the only known aphid that reproduces and has colonies on U.S. soybean. Ostlie (2004) and DiFonzo and Hines (2002) measured yield losses in insecticide field trials in the United States as high as those observed in China. Heavy infestations also reduce photosynthesis (Macedo et al., 2003). This aphid not only causes direct damage by feeding, but it also is an effective vector of soybean mosaic virus and alfalfa mosaic virus (Hill et al., 2001). At present, the main control for soybean aphid is the use of insecticides, and it may become a key pest that will require routine insecticide applications. These applications in turn may trigger subsequent pest problems because of the disruption of natural controls. Insecticide treatments are typically triggered at >40 aphids per leaf with 84% of the plants infested. This is the equivalent of ~250 aphids/plant, and densities of thousands of aphids per plant are not uncommon (Hodgson et al., 2004).

Insect Management

To a large extent, soybean insect pests are successfully managed across the United States. by implementation of diversified IPM programs (Hammond et al., 1991; Todd et al., 1994) based on timely sampling of individual fields and preservation of natural controls. Insecticide treatments typically are reserved for emergency situations, and U.S. soybean has not historically had a key pest species that warranted routine treatment. The introduction of the soybean aphid and expanded concern for higher stink bug populations in the United States may be challenging this historical trend as insecticide use seems to be increasing in some production regions. Increased pest abundance, especially the emergence of key pest species that routinely require corrective action, could trigger additional pest problems and further accelerate pest damage and control costs.

Many of the more important insect pests of soybean are polyphagous species building populations on a number of cultivated and wild host plants. This is especially true for the pod-feeding stink bugs and corn earworm, as well as the recently introduced soybean aphid. Effective soybean insect management has long relied on a fundamental understanding of insect preference for different crops and different stages of soybean crop development (Hammond et al., 1991). Producers can avoid many soybean pest problems if the crop is not suitable for colonization when these polyphagous species are abundant (Todd et al., 1994). As soybean seed quality becomes a more important aspect of crop value and production goals, changing IPM programs will focus on specific management strategies to optimize both yield and seed quality. Seed-quality issues that are impacted by insects and disease organisms are a good example of the need to examine multiple pest problems and optimize total production goals.

Grading Damaged Seed

Visual grading of soybean seeds can identify damage caused by a number of diseases and pests including Phomopsis seed decay, purple seed stain, and stink bug damage (Kulik & Yaklich, 1982; Sinclair, 1992). A model was developed using visual-light-identified seed damage from a number of diseases with an overall accuracy of 88% (Ahmad et al., 1999). More recently, near-infrared spectroscopy was able to identify healthy seed and seed damaged by *Phomopsis*, *Cercospora kikuchii*, soybean mosaic virus, and downy mildew with accuracies of 100, 99, 84, 94, and 96%, respectively (Wang et al., 2004). Shatadal and Tan (2003) used an image analysis system to accurately classify heat-damaged and green-frost-damaged seeds, but the method did not work well for stink bug-damaged seed. In the future, these rating systems and other techniques will relate seed damage by disease and insects to seed and grain quality.

Conclusion

More research is expected on the effects of diseases and insects on soybean seed quality in the future. Increasingly, soybean is marketed based on oil and protein content. With the United Soybean Board's Better Bean Initiative, specific quality traits are being developed, such as low phytate or linolenic acid soybean to meet industry needs (Durham, 2003; Wilson, 2004). Soybeans with these traits will command higher prices, so the effect of diseases and insects on the seed quality will be more important. Seed certification is becoming more important as soybean seed companies become global in not only developing cultivars but also in producing seed for planting, thereby increasing the danger of introducing foreign diseases and insects into areas with no resistant cultivars. Finally, as more and more direct human consumption of soybean products occurs, the quality of the oils and proteins and the risks of mycotoxins become increasingly important.

References

Ahmad, I.; J. Reid; M. Paulsen; J. Sinclair. Color classifier for symptomatic soybean seeds using image processing. *Plant Dis.* **1999**, 83, 320–327.

Aldea, M.; J.G. Hamilton; J.P. Resti; A.R. Zangerl; M.R. Berenbaum; E.H. DeLucia. Indirect effects of insect herbivory on leaf gas exchange in soybean. *Plant, Cell, Environ.* **2005,** *28*, 402–411.

Arias, N.; J. Hegglin; L. Fredman; C. Villon. Evaluacion del efecto de la roya asiatica sobre los rendimientosa del cultivos de soja. *Actualizacion Tecnica Soja EEA INTA Parana. Serie Extension* **2005**, no. 3456–58.

Bacon, C.W.; D.R. Marijanovic; W.P. Norred; D.M. Hinton. Production of fusarin C on cereal and soybean by *Fusarium moniliforme. Appl. Environ. Microbiol.* **1989**, *55*, 2745–2748.

Baur, M.E.; D.L. Goethel; M.L. Boyd; G.R. Bowers; M.O. Way; L.G. Heatherly; J. Rabb; L. Ashlock. Arthropod populations in early soybean production systems in the mid-south. *Environ. Entomol.* **2000**, *29*, 312–328.

Bean, G.A.; J.A. Schillinger; W.L. Klarman. Occurrence aflatoxins and aflatoxin-producing strains of *Aspergillus* sp. in soybeans. *Appl. Microbiol.* **1972**, *24*, 437–439.

Boethel, D.J.; J.S. Russin; A.T. Weir; M.B. Layton; J.S. Mink; M.L. Boyd. Delayed maturity associated with southern green stink bug (Heteroptera: Pentatomidae) injury at various stink bug phonological stages. *J. Economic Entomol.* **2000**, *93*, 707–712.

Bowers, G.R., Jr.; R.M. Goodman.; Strain specificity of soybean mosaic virus seed transmission in soybean. *Crop Sci.* **1991**, *31*, 1171–1174.

Bradley, C.A.; G.L. Hartman; L.M. Wax; W.L. Pedersen. Quality of harvested seed associated with soybean cultivars and herbicides under weed-free conditions. *Plant Dis.* **2002**, *86*, 1036–1042.

Bringel, J.M.M.; M.H.D. de Moraes; J.O.M. Menten; I.P. Bedendo. Health and physiological quality of soybean seeds produced in the south region of Maranhao/Brazil. *Summa Phytopathologica* **2001**, *27*, 438–441.

Burow, G.B.; T.C. Nesbitt; J. Dunlap; N.P. Keller. Seed lipxygenase products modulate Aspergillus mycotoxin biosynthesis. *MPMI* **1997,** *10*, 380–387.

Burrows, M.E.L.; C.M. Boerboom; J.M. Gaska; C.R. Grau. The relationship between *Aphis glycines* and soybean mosaic virus incidence in different pest management systems. *Plant Dis.* **2005,** *89*, 926–934.

Cai, G.; R. Schneider. Vegetative compatibility groups in *Cercospora kikuchii*, the causal agent of Cercospora leaf blight and purple seed stain in soybean. *Phytopathol.* **2005**, *95*, 257–261.

Chandrasekaran, A.; K. Rajappan. Effect of *Colletotrichum truncatum* seed infection on seed weight and seedling vigor of soybean. *Plant Dis. Res.* **2002**, *17*, 72–73.

Clear, R.M.; T.W. Nowicki; J.K. Daun. Soybean seed discoloration by *Alternaria* spp. and *Fusarium* spp., effects on quality and production of fusariotoxins. *Canadian J. Plant Pathol.* **1989**,

11, 308–312.

Daub, M.E.; K.-R. Chung. Cercosporin: A photoactivated toxin in plant disease. **2007**, www.apsnet.org/online/feature/Cercosporin.

Daugherty, D.M.; M.H. Neustradt; C.W. Gehrker; L.E. Cavannah; L.F. Williams; D.E. Green. An evaluation of damage to stink bugs by brown and green stink bugs. *J. Economic Entomol.* **1964**, *57*, 719–722.

Demski, J.W.; C.W. Kuhn, revised by G.L. Hartman. Tobacco ringspot. *Compendium of Soybean Diseases*. Fourth ed.; G.L. Hartman. J.B. Sinclair. J.C. Rupe Eds.; APS Press: St. Paul, MN, **1999;** pp. 66–68.

DiFonzo, C.; R. Hines. Soybean aphid in Michigan: Update from the 2001 season. Michigan State University, East Lansing, MI. **2002**, MSU Extension Bulletin E-2748.

Dupleich, J.; A.N. Formento; N. Wourterlood; V. Rolon; C. Pons; Barolin; J. Velazquez; I. Vicentin. Efecto de la roya asiatica (*Phakopsora pachyrhizi*) sobre el rendimiento del cultivo de soja (*Glycine max*). **2005**, *Actualizacion Tecnica Soja. EEA INTA Parana. Serie Extension no. 3454–55.*

Durham, D. The United Soybean Board's better bean initiative: Building United States soybean competitiveness from the inside out. *AgBioForum* **2003**, *6*(1&2), 23–26.

El-Amrety, A.S.; H.M. El-Said; D.E. Salem. Effect of soybean mosaic virus infection on quality of soybean seeds. *Agric. Res. Rev.* **1985**, *63*, 155–164.

el-Kady, I.A.; M.S. Youssef. Survey of mycoflora and mycotoxins in Egyptian soybean seeds. *J. Basic Microbiol.* **1993**, *33*, 371–378.

Evans, M.G.; E.P. Christmas; C.B. Southard. USDA grading standards and moisture conversion table for soybean. *Agronomy Guide, Purdue University Cooperative Extension Service* 1997.

Franca-Neto, J.B.; G.P. Padua; M.L.M. de Carvalho; O. Costa; P.S.R. Brumatti; F.C. Krzyzanowski; N.P. de Costa; A.A. Henning; D.P. Sanches. Semente esverdeada de soja e sua qualidade fisiologica. Circular Tecnica 38 EMBRAPA, **2005**.

Fujita, Y. Ecology and control of purple seed stain of soybean caused by *Cercospora kikuchii. Bulletin of the Tohoku National Agricultural Experiment Station* **1990**, 51–109.

Galli, J.A.; R.D.C. Panizzi; S.A. Fessel; F. de Simoni; M.F. Ito. Effect of *Colletotrichum dematium* var. *truncata* and *Cercospora kikuchii* on soybean seed germination. *Revista Brasileira De Sementes* **2005**, *27*, 182–187.

Gergerich, R.C. Bean pod mottle. *Soybean Disease Compendium*, Fourth ed.; G.L. Hartman; J.B. Sinclair; J.C. Rupe, Eds.; APS Press: St. Paul, MN, 1999; pp. 61–62.

Gore, J.; C.A. Abel; J.J. Adamczyk; G. Snodgrass. Influence of soybean planting date and maturity group on stink bug (Heteroptera: Pentatomidae) populations. *Environ. Entomol.* **2006**, *35*, 531–536.

Grau, C.R.; G.L. Hartman. Sclerotinia stem rot. *Compendium of Soybean Diseases*. Fourth ed.; G.L. Hartman, J.B. Sinclair, J.C. Rupe, Eds.; APS Press: St. Paul, MN, 1999, pp. 47–49.

Hammond, R.B. Soybean Insect IPM. *Radcliffe's IPM World Textbook*. University of Minnesota, 1996, http:ipmworld.umn.edu/ipmsite.htm.

Hammond, R.B.; R.A. Higgins; T.P. Mack; L.P. Pedigo; E.J. Bechinski. Soybean pest management. *CRC Handbook of Pest Management in Agriculture*, Second ed.; D. Pimentel, Ed.; CRC Press Inc.:

Boca Raton, FL, 1991; Vol. 3, pp. 341–472.

Hepperly, P.R. *Fusarium* species and their association with soybean seed under humid tropical conditions in Puerto Rico. *J. Agric. University of Puerto Rico* **1985**, *69*, 25–33.

Hepperly, P.R.; G.R. Bowers Jr.; J.B. Sinclair; R.M. Goodman. Predisposition to seed infection by *Phomopsis sojae* in soybean plants infected by soybean mosaic virus. *Phytopathology* **1979,** *69*, 846–848.

Hepperly, P.R.; J.B. Sinclair. Quality losses in Phomopsis-infected soybean seeds. *Phytopathology* **1978**, *68*, 1684–1687.

Higley, L.G. Insect injury to soybean. *Handbook of Soybean Insect Pests*; L.G. Higley; D.J. Boethel, Eds.; Entomology Society of America: Landham, MD, 1994, pp. 11–13.

Hill, J.H. Soybean mosaic. *Compendium of Soybean Diseases*; Fourth ed.; G.L. Hartman; J.B. Sinclair; J.C. Rupe, Eds.; APS Press: St. Paul, MN, 1999; pp. 70–71.

Hill, J.H.; H.R. Alleman; D.B. Hogg; C.R. Grau. First report of transmission of soybean mosaic virus and aflafa mosaic virus by *Aphis glycines* in the new world. *Plant Dis.* **2001**, *85*, 561.

Hodgson, E.W.; E.C. Burkness; W.D. Hutchison; D.W. Ragsdale. Enumerative and binomial sequential sampling plans for soybean aphid (Homoptera: Aphididae) in soybean. *J. Economic Entomol.* **2004**, *97*, 2127–2136.

Hoffman, D.D.; G.L. Hartman; D.S. Mueller; R.A. Leitz; C.D. Nickell; W.L. Pedersen. Yield and seed quality of soybean cultivars infected with *Sclerotinia sclerotiorum. Plant Dis.* **1998**, *82*, 826–829.

Ibraheem, S.A.; A.M. Okesha; K.T. Mhathem. Interrelationship between protein and oil content of soybean seed with some associated fungi. *J. Agric. Water Resources Res., Plant Production* **1987**, *6*, 53–66.

Jackson, E.W.; P. Fenn; P.Y. Chen. Inheritance of resistance to Phomopsis seed decay in soybean PI 80837 and MO/PSD-0259 (PI 562694). *Crop Sci.* **2005**, *45*, 2400–2404.

Jackson, E.W.; P. Fenn; P.Y. Chen. Inheritance of resistance to purple seed stain caused by *Cercospora kikuchii* in PI 80837 soybean. *Crop Sci.* **2006**, *46*, 1462–1466.

Jacobsen, B.J.; K.S. Harlin; S.P. Swanson; R.J. Lambert; V.R. Beasley; J.B. Sinclair; L.S. Wei. Occurrence of fungi and mycotoxins associated with field mold damaged soybean in the Midwest. *Plant Dis.* **1995**, *79*, 86–88.

Jensen, R.L.; L.D. Newsom. Effect of stink bug-damaged soybean seed on germination, emergence and yield. *J. Econ. Entomol.* **1972,** *65*, 261–264.

Jordan, E.G.; J.B. Manandhar; P.N. Thapliyal; J.B. Sinclair. Factors affecting soybean seed quality in Illinois. *Plant Dis.* **1986,** *70*, 246–248.

Katsube, T. The effect of soybean purple blotch on growth, yield and some chemical components of seeds (in Japanese*). Society of Plant Protection North Japanese Annual Report* **1980**, *31*, 64–66.

Kilpatrick, R.A.; E.E. Hartwig. Fungus infection of soybean seed as influenced by stink bug injury. *Plant Dis. Reporter* **1955,** *39*, 177–180.

Kincade, R.T.; M.L. Laster; E.E. Hartwig. Simulated pod injury to soybeans. *J. Economic Entomol.* **1971**, *64*, 984–985.

Koenning, S.R. Southern United States soybean disease loss estimates for 2006. 2007: http://www.cipm.ncsu.edu/ent/SSDW/SSDWloss2006.pdf.

Koning, G.; D. TeKrony; S. Ghabrial; T. Pfeiffer. Soybean mosaic virus (SMV) and the SMV resistance gene (Rsv1): Influence on *Phomopsis* spp. seed infection in an aphid free environment. *Crop Sci.* **2002**, *42*, 178–185.

Krell, R.K.; L.P. Pedigo; M.E. Rice; M.E. Westgate; J.H. Hill. Using planting date to manage bean pod mottle virus in soybean. *Crop Protection* **2005**, *24*, 909–914.

Kulik, M.M.; J.B. Sinclair. Phomopsis seed decay. *Compendium of Soybean Diseases*, Fourth ed.; G.L. Hartman, J.B. Sinclair, J.C. Rupe, Eds.; APS Press: St. Paul, MN, 1999; pp. 31–32.

Kulik, M.M.; R.W. Yaklich. Relationship of the appearance of soybean seeds to seed-borne infection by *Diaporthe phaseolorum* var. *sojae* and other aspects of seed quality. *Seed Sci. Technol.* **1982**, *10*, 335–342.

Lee, D.H. Fungi associated with soybean seed, their pathogenicity and seed treatment. *Korean J. Mycol.* **1984,** *12*, 27–33.

Lorenz, G.; D. Johnson; G. Studebaker; C. Alle; S. Young. Insect pest management in soybean. *Arkansas Soybean Handbook*; Arkansas Soybean Promotion Board and Arkansas Cooperative Extension Service: Little Rock, AR, 2000; pp. 84–94.

Lori, G.A.; S.J. Sarandon. Pathogenicity of *Fusarium* spp. incidence on soybean seed quality. *Agronomie* **1989**, *9*, 77–82.

Macedo, T.B.; C.S. Bastos; L.G. Higley; K.R. Ostlie; S. Madhavan. Photosynthetic responses of soybean to soybean aphid (Homoptera: Aphididae) injury. *J. Econ. Entomol.* **2003**, *96*, 188–193.

Manandhar, J.B.; G.L. Hartman. Anthracnose. *Compendium of Soybean Diseases*, Fourth ed.; G.L. Hartman, J.B. Sinclair, J.C. Rupe, Eds.; APS Press: St. Paul, MN, 1999; pp. 13–14.

Martinelli, J.A.; C.A.C. Bocchese; W. Xie; K. O'Donnell; H.C. Kistler. Soybean pod blight and root rot caused by lineages of *Fusarium graminearum* and the production of mycotoxins. *Fitopatologia Brasileira* **2004**, *29*, 492–498.

Mayhew, W.L.; C.E. Caviness. Seed quality and yield of early-planted, short-season soybean genotypes. *Agronomy J.* **1994**, *86*, 16–19.

McGee, D.C.; C.L. Brandt; J.S. Burris. Seed mycoflora of soybeans relative to fungal interactions, seedling emergence, and carry over of pathogens to subsequent crops. *Phytopathology* **1980,** *70*, 615–617.

McPherson, R.M.; G.K. Douce; R.D. Hudson. Annual variation in stink bug (Hereroptera: Pentatomidae) seasonal abundance and species composition in Georgia soybean and its impact on yield and quality. *J. Economic Entomol.* **1993**, *28*, 61–72.

McPherson, R.M.; L.D. Newsom; B.F. Farthing. Evaluation of four stink bug species from three genera affecting soybean yield and quality in Louisiana. *J. Economic Entomol.* **1979**, *72*, 188–194.

McPherson, R.M.; J.W. Todd; K.V. Yeargan. Stink bugs. *Handbook of Soybean Insect Pests*, L.G. Higley, D.G. Boethel, Eds.; Enomology Society of America: Landham, MD, 1994; pp. 87–90.

Mellon, J.E.; P.J. Cotty. No effect of soybean lypoxygenase on aflatoxin production in *Aspergillus flavus*-inoculated seeds. *J. Food Protection* **2002**, *65*, 1984–1987.

Miner, F.D. Biology and control of stink bugs on soybean. *Arkansas Agricultural Experiment Station Bulletin 70840*, 1966.

Miner, F.D. Sink bug damage to soybeans. *Arkansas Agric. Exp. Station Farm Res.* **1961**, *10*, 12.

Miner, F.D.; B.A. Dumas. Effect of green stink bug damage on soybean seed quality before and after storage. *Arkansas Agric. Exp. Station Bulletin 84419*, 1980.

Minor, H.C.; E.A. Brown; B. Doupnik, Jr.; R.W. Elmore; M.S. Zimmerman. Registration of Phomopsis seed decay resistant soybean germplasm MO/PSD-0259. *Crop Sci.* **1993**, *33*, 1105.

Nakano, M.; T. Usugi; A. Shinkai. Effect of inoculation time of soybean mosaic virus on yield and seed quality of soybean. *Proceedings of the Association for Plant Protection of Kyushu* **1988,** *34*, 13–16.

Narayanaswamy, S.; M. Ravikumar; R. Sreerama. Storability of soybean [*Glycine max* (L.) Merrill] seed as influenced by packaging. *Mysore J. Agric. Sci.* **2000**, *34*, 227–232.

Nesheim, S.; W.E. Garnett. Regulatory aspects of mycotoxins in soybean and soybean products. *J. Am. Oil Chem. Soc.* **1995**, *72*, 1421–1423.

Obopile, M.; R.B. Hammond. Effects of delayed harvest on soybean seed quality following bean leaf beetle (Coleoptera: Chrysomelidae) pod injury. *J.Kansas Entomol. Soc.* **2001,** *74*, 40–48.

Orth, C.E.; W. Schuh. Resistance of 17 soybean cultivars to foliar, latent, and seed infection by *Cercospora kikuchii. Plant Dis.* **1994**, *78*, 661–664.

Ortiz, C.; S.R. de Cianzio; P.R. Hepperly. Fungi and insect damage to soybean seeds

harvested at immature stages in tropical environments. *J. Agric. Univ Puerto Rico* **1988**, *72*, 73–79.

Osorio, J.A.; D.C. McGee. Effects of freeze damage on soybean seed mycoflora and germination. *Plant Dis.* **1992**, *76*, 879–882.

Ostlie, K. Soybean aphid reduces yield: Harvest results from insecticide strip trials. 2004, http://www.soybean.umn.edu/crop/insects/aphid/studyresults/htm.

Padgett, B.; R. Schneider; K. Whitam. Foliar-applied fungicides in soybean disease management. *Louisiana Agric.* **2003**, *46*, 7–9.

Padule, D.N.; S.S. Chaudhary; S.B. Gawade. Effect of different grades of seeds of soybean and threshing methods on association of seed borne fungi, seed germination and seedling vigor index *Jawaharlal Nehru Krishi Vishwa Vidalaya Res. J.* **2004**, *38*, 34–37.

Park, J.-S.; K.-R. Lee; J.-C. Kim; S.-H. Lim; J.-A. Seo; Y.-W. Lee. A hemorrhagic factor (apicidin) produced by toxic *Fusarium* isolates from soybean seeds. *Appl. Environ. Microbiol.* **1999**, *65*, 126–130.

Pathan, M.A.; J.B. Sinclair; R.D. McClary. Effects of *Cercospora kikuchii* on soybean seed germination and quality. *Plant Dis.* **1989,** *73*, 720–723.

Pessu, P.O.; M.N. Adindu; O.C. Umeozor. Effects of long-term storage on the quality of soybean, *Glycine max* (L.) Merrill, in different containers in southern Nigeria. *Global J. Pure Appl. Sci.* **2005**, *11*, 165–168.

Phillips, D.V. Frogeye leaf spot. *Compendium of Soybean Diseases*, Fourth ed.; G.L. Hartman, J.B. Sinclair, J.C. Rupe, Eds.; APS Press: St. Paul, MN, 1999a; pp. 20–21.

Phillips, D.V. Downy mildew. *Compendium of Soybean Diseases*, Fourth ed.; G.L. Hartman, J.B. Sinclair, J.C. Rupe, Eds.; APS Press: St. Paul, MN, 1999b; pp. 18–19.

Pioli, R.N.; L. Mozzoni; E.N. Morandi. First report of pathogenic association between *Fusarium graminearum* and soybean. *Plant Dis.* **2004**, *72*, 88–88.

Ragesdale, D.W.; D.J. Voegtlin; R.J. O'Neil. Soybean aphid biology in North America. *Annals Entomol. Soc. Am.* **2004**, *97*, 204–218.

Ren, Q.; T.W. Pfeiffer; S.A. Ghabrial. Soybean mosaic virus resistance improves productivity of double-cropped soybean. *Crop Sci.* **1997a,** *37*, 1712–1718.

Ren, Q.; T.W. Pfeiffer; S.A. Ghabrial. Soybean mosaic virus incidence level and infection time: Interaction effects on soybean. *Crop Sci.* **1997b,** *37*, 1706–1711.

Ross, J.P. Effect of aphid-transmitted soybean mosaic virus on yields of closely related resistant and susceptible soybean lines. *Crop Sci.* **1977,** *17*, 869–872.

Russin, J.S.; D.J. Boethel. Soybean pest interactions. *Handbook of Soybean Insect Pests*, L.G. Higley; D.J. Boethel, Eds.; Entomological Society of America: Landham, MD, 1994; pp. 120–121.

Russin, J.S.; D.B. Orr; M.B. Layton; D.J. Boethel. Incidence of microorganisms in soybean seeds damaged by stink bug feeding. *Phytopathology* **1988**,*78*, 306–310.

Schiller, C.T.; M.A. Ellis; F.D. Tenne; J.B. Sinclair. Effect of *Bacillus subtilis* on soybean seed decay, germination, and stand inhibition. *Plant Dis. Reporter* **1977**, *61*, 213–217.

Scortichini, M.; M.P. Rossi; B. Ricci; B. Ndzoumba. Soybean (*Glycine max* [L.] Merr.) seed decay associated with *Bacillus subtilis* (Ehrenberg) Cohn, in Gabon. *FAO Plant Protection Bull.* **1989**, *37*, 87–91.

Shatadal, P.; J. Tan. Identifying damaged soybeans by color image analysis. *Appl. Engin. Agric.* **2003,** *19*, 65–69.

Shortt, B.J.; J.B. Sinclair; M.R. Helms; M.R. Jeffords; M. Kogan. Soybean seed quality losses associated with bean leaf beetles and *Alternaria tenuissima. Phytopathology* **1982,** *72*, 615.

Silva, M.F.; A.M.R. Almeida; C.A. Arias. Evaluation of losses caused by two strains of soybean mosaic virus in two soybean cultivars. *Fitopatologia Brasileira* **2003**, *28*, 597–601.

Sinclair, J.B. Bacillus seed decay. *Compendium of Soybean Diseases*, Fourth ed.; G.L. Hartman, J.B. Sinclair, J.C. Rupe, Eds.; APS Press: St. Paul, MN, 1999a; pp. 8–9.

Sinclair, J.B. Bacterial pustule. *Compendium of Soybean Diseases*, Fourth ed.; G.L. Hartman, J.B. Sinclair, J.C. Rupe, Eds.; APS Press: St. Paul, MN, 1999b; pp. 6–7.

Sinclair, J.B. Bacterial blight. *Compendium of Soybean Diseases*, Fourth ed.; G.L. Hartman, J.B. Sinclair, J.C. Rupe, Eds.; APS Press: St. Paul, MN, 1999c; pp. 5–6.

Sinclair, J.B. Discoloration of soybean seeds - an indicator of quality. *Plant Dis.* **1992**, *76*, 1087–1091.

Sinclair, J.B. Yeast spot. *Compendium of Soybean Diseases*, Fourth ed.; G.L. Hartman, J.B. Sinclair, J.C. Rupe, Eds.; APS Press: St. Paul, 1999d; pp. 75–76.

Smelser, R.B.; L.P. Pedigo. Soybean seed yield and quality reduction by bean leaf beetle (Coleoptera: Chrysomelidae) pod injury. *J. Econ. Entomol.* **1992**, *85*, 2399–2403.

Smith, J.F. Refining the use of early soybean as a trap crop for stink bugs in Arkansas' changing soybean production systems. University of Arkansas: Fayetteville, AR, 2006.

Smith, R.H.; M.H. Bass. Relationship of artificial pod removal to soybean yields. *J. Econ. Entomol.* **1972**, *63*, 606–609.

Solanke, R.B.; S.S. Kore; S.M. Sudewad. Detection of soybean seed borne pathogens and effect of fungicides. *J. Maharashtra Agric. Univer.* **1997**, *22*, 168–170.

Stefani, E.; D. Caffier; N. Fiore. The economic impact of the bacterial blight of soybean under European agroclimatic conditions. *J. Plant Pathol.* **1998**, *80*, 211–218.

Steffey, K.L.; M.E. Gray; L.G. Higley. Introduction to identification and diagnosis of injury. *Handbook of Soybean Insect Pests*; L.G. Higley, D.J. Boethel, Eds.; Entomology Society of America: Landham, MD, 1994; pp. 17–34.

Suteri, B.D. Oil content of soybean seeds infected with two strains of soybean mosaic virus. *Indian Phytopathol.* **1980**, *33*, 139–140.

Taira, H.; H. Taira; Y. Kokubu; S. Otake. Suitability of food processing on purple specked soybean seeds by *Cercospora kikuchii* (in Japanese). *National Food Res. Inst. Annual Report* **1990**, *36*, 14–20.

Tenne, F.D.; J.B. Sinclair. Control of internally seedborne micro-organisms of soybean with foliar fungicides in Puerto Rico. *Plant Dis. Reporter* **1978**, *62*, 459–463.

Thomas, G.D.; C.M. Ignoffo; C.E. Morgan; W.A. Dickerson. Southern green stink bug: Influence on yield and quality of soybean. *J. Econ. Entomol.* **1974**, *67*, 501–503.

Todd, J.W.; M.D. Jellum; D.B. Leuck. Effects of southern green stink bug damage on fatty acid composition of soybean oil. *Environ. Entomol.* **1973**, *2*, 685–689.

Todd, J.W.; R.M. McPherson; D.J. Boethel. Management tactics for soybean insects. *Handbook of Soybean Insect Pests*; L.G. Higley, D.J. Boethel, Eds.; Entomological Society of Amereica, Landham, MD, 1994; pp. 115–116.

Turnipseed, S.G. Soybean entomology. *Annl Rev. Entomol.* **1976**, *21*, 247–282.

Tyler, D.D.; J.R. Overton. No-tillage and conventional tillage effects on soybean seed quality. *1981 Conference Abstracts*; American Society of Agronomy, Madison, WI.

Upchurch, R. Genetic regulation of cercosporin production in *Cercospora kikuchii*. *J. Am. Oil Chem. Soc.* **1995**, *72*, 1435–1438.

Velicheti, R.; K. Kollipara; J. Sinclair; T. Hymowitz. Selective degradation of proteins by *Cercospora kikuchii* and *Phomopsis longicolla* in soybean seed coats and cotyledons. *Plant Dis.* **1992,** *76*, 779–782.

Velicheti, R.K.; J.B. Sinclair. Production of cercosporin and colonization of soybean seed coats by *Cercospora kikuchii*. *Plant Dis.* **1994**, *78*, 342–346.

Vieira, R.D.; D.M. TeKrony; D.B. Egli. Effect of drought and defoliation stress in field on soybean seed germination and vigor. *Crop Sci.* **1992**, *32*, 471–475.

Vishwadhar; A.K. Sarbhoy. Soybean seed health hazard in different agro-climatic region of India. *Indian J. Hill Farming* **1994**, *7*, 192–202.

Wang, D.; F. Dowell; M. Ram; W. Schapaugh. Classification of fungal-damaged soybean seeds using near-infrared spectroscopy. *Int. J. Food Properties* **2004**, *7*, 75–82.

Way, M.O. Status of soybean insect pests in the United States. *Handbook of Soybean Insect Pests*; L.G. Higley, D.J. Boethel, Eds.; Entomological Society of America: Landham, MD, 1994; pp. 15–16.

Wilcox, J.; T. Abney. Effects of *Cercospora kikuchii* on soybeans. *Phytopathology* **1973**, *63*, 796–797.

Wilson, R.F. Seed composition. *Soybeans: Improvement, Production, and Uses*; Third Ed.; H.R. Boerma, J.E. Specht, Eds.; American Society of Agronomy, Crop Science Society of America, Soil Science Society of America: Madison, WI, 2004; pp. 621–677.

Wilson, R.F.; W.P. Novitzky; G.P. Fenner. Effect of fungal damage on seed composition and quality of soybeans. *J. Am. Oil Chem. Soc.* **1995,** *72*, 1425–1429.

Wrather, J.; T. Anderson; D. Arsyad; Y. Tan; L. Ploper; A. Porta Puglia; H. Ram; J. Yorinori. Soybean disease loss estimates for the top ten soybean-producing countries in 1998. *Can. J. Plant Pathol., Revue Canadienne De Phytopathologie.* **2001,** *23*, 115–121.

Wrather, J.A.; D.A. Sleper; W.E. Stevens; J.G. Shannon; R.F. Wilson. Planting date and cultivar effects on soybean yield, seed quality, and *Phomopsis* sp. seed infection. *Plant Dis.* **2003**, *87*, 529–532.

Wu, Z.S.; D. Schenk Hamlin; W.Y. Zhan; D.W. Ragsdale; G.E. Heimpel. The soybean aphid in china: A historical review. *Ann. Entomol. Soc. Am.* **2004**, *97*, 209–218.

Yang, X.B.; F. Workneh; P. Lundeen. First report of sclerotium production by *Sclerotinia sclerotiorum* in soil on infected soybean seeds. *Plant Dis.* **1998**, *82*, 264.

Yeargan, K.V. Effects of green stink bug damage on yield and quality of soybeans. *J. Econ. Entomol.* **1977**, *70*, 619–622.

Zad, S.J. Soybean seed-borne diseases. Mededelingen Van De Faculteit Landbouwwetenschappen, *Rijksuniversiteit Gent* **1987,** *52*, 825–829.

Zeiss, M.R.; T.H. Klubertanz. Sampling programs for soybean arthropods. *CRC Handbook of Sampling Methods for Arthropods in Agriculture*; L.P. Pedigo, G.D. Buntin, Eds.; CRC Press Inc.: Boca Raton, FL, 1994; pp. 539–601.

Zimmerman, M.S. Minor HC; Inheritance of Phomopsis seed decay resistance in soybean PI 417479. *Crop Sci.* **1993,** *33*, 96–100.

5

Economics of Soybean Production, Marketing, and Utilization

Peter D. Goldsmith
Soybean Industry Endowed Associate Professor in Agricultural Strategy, Executive Director, the National Soybean Research Laboratory. University of Illinois, Urbana-Champaign, IL, 61801.

Introduction

Soybean production now occupies close to 6% of the world's arable land. Soybean expansion is occurring much faster than with other major grains or oilseeds. Since 1993, soybean hectares grew two times the overall global economy (Fig. 5.1). Soybeans increasingly are being employed as the modern input of choice for buyers. They are mainly used as intermediate food, feed, and industrial inputs, not final consumer products, therefore remaining somewhat invisible in the economy.

Only 2% of soybean protein is consumed directly by humans in the form of soy food products such as tofu, soy hamburger, or soy milk analogs. All but a very small percentage of the other 98% is processed into soybean meal and fed to livestock, such as poultry and pigs.

In this way, soybean demand is essentially a derived demand for meat. Soybean has risen to become a leading crop because the income elasticity of meat is high. Consumers shift their consumption from grains, such as rice and wheat, to meat and other animal products as personal incomes rise around the world.

In 1961 the average annual per capita consumption globally of poultry and pork was 17.69 pounds; by 2003 it almost doubled to 34.39 pounds (FAO, 2005). The change in the world's GDP over the period was almost three times as large, making the income elasticity of poultry 0.32 (1961–2003). For each 1% of per capita income increase, poultry consumption increased .32% worldwide. For many individual countries, the income elasticities exceeded 1.00 over the time period. A majority of the transition from grains to meat is occurring in developing countries where consumption patterns are shifting much more quickly. For example, India and China, with an income elasticity of poultry of 1.07, increased their poultry consumption more than 15% per year during the last 10 years and now consume 1.54 million tons and 14.7 million tons per year, respectively (FAO, 2007). Similar changes occurred with pork. The Philippines and Vietnam are leading new consumers of pork with an income elasticity of pork consumption of 1.38, over four times the world level. They

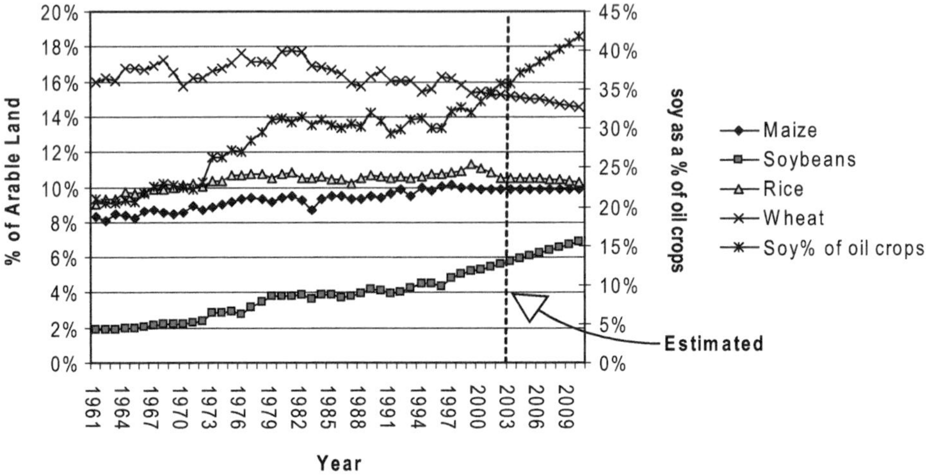

Fig. 5.1. Land use of major crops. Source: FAO, 2005 and author's calculations.

have increased their consumption by more than 10% per year during the last 10 years and now consume over 1.1 million tons of pork annually.

Soybean demand in Europe is also increasing because a strong demand for vegetarian animal feeds to replace bone meal exists, following the outbreak of BSE in the United Kingdom in the mid 1990s. Finally, fish feed derived from wild fish stocks is in decline, creating exciting new markets in the high-growth aquaculture-producing regions of Asia and South America.

The role of soybean oil completes another part of the story of the rapid rise of the global soybean complex over the last 40 years. An important complementarity exists between the meal and the oil since processors have broad and active markets because of soybean's value proposition: meal for animal feed; and oil for food and bioenergy (Fig. 5.2).

Though the primary business of soybean processing is to produce animal protein feed, 17 to 20% of the soybean is an oil coproduct. In the 1960s, human health concerns about cholesterol caused an increase in demand for soybean oil. Food manufacturers shifted away from animal by-products as a source of fat or oil. Then in the 1970s, concerns arose about saturated fats from tropical oils. This caused another spike in demand as U.S. food manufacturers switched from palm oil to less saturated oils, such as soy. These events created significant opportunities for soybean oil to become the preferred oil for food manufacturers. Now evidence links the presence of trans-fatty acids, found in processed soybean oil, with heart disease. For some products, soybean oil is partially hydrogenated to improve products' appearance, stability,

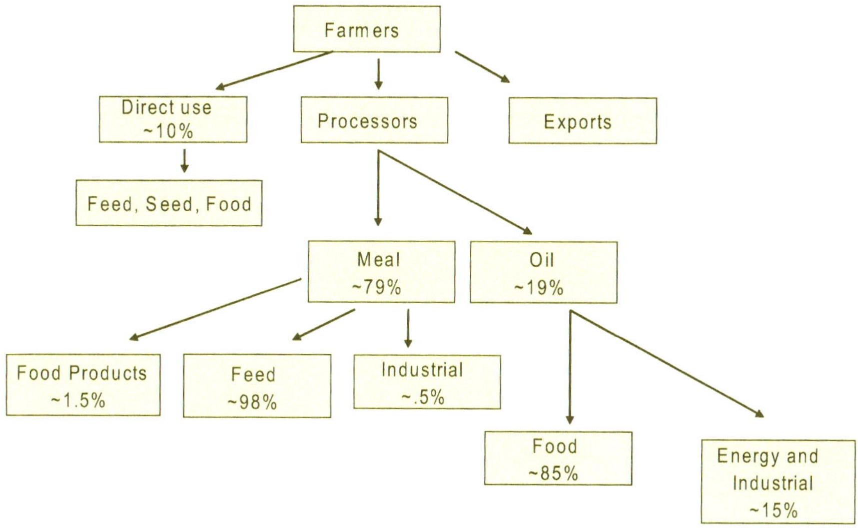

Fig. 5.2. Soybean complex: Structure of the industry. Source: U.S. International Trade Commission, 2003 and author's calculations.

shelf-life, and mouthfeel. Some shift away from soy oil resulted as food processors reformulated recipes away from partially hydrogenated oils. Plant breeders have now developed soybean varieties with low levels of linolenic acid. These "Low-Lin" varieties allow food processors and consumers to benefit from the functionality of soybean oil without adversely affecting health.

The increase in the consumption of processed foods in Western diets increased the demand for low-cost and highly functional oils. In the United States, Brazil, and China, the oil added will likely come from soybeans. Today's supermarkets are full of processed foods with vegetable oil on the ingredient list. Oil is added for taste, nutrition, and cooking performance. In the last ten years the consumption of soybean oil in Brazil and China increased 15% and 40% per year, respectively. Brazil consumes 30 kilograms of soybean oil per capita, while China, who has only begun to integrate soybean oil into their food system, consumes only 4 kilograms per capita. The United States consumes about 27 kilograms per capita, and has seen its consumption of soybean oil decline 21% over the last ten years. Recently though, biodiesel production created a new and significant market for soybean oil, and now accounts for 15% of U.S. soybean oil demand (USDA, 2007a).

Recent Trends in Soybean Production

Historically the United States has been, and continues to be, the leading soybean-producing and -exporting country in the world (Fig. 5.3). South America, led by Brazil, Argentina, Paraguay, and Bolivia, as a region, recently surpassed the United States' output and now produces 48% of the world's needs (Fig. 5.4). The United States still exports 40% of the world's soybeans, with the leading global importers being China (43%), The Netherlands (7%), and Japan (6%) (FAO, 2007).

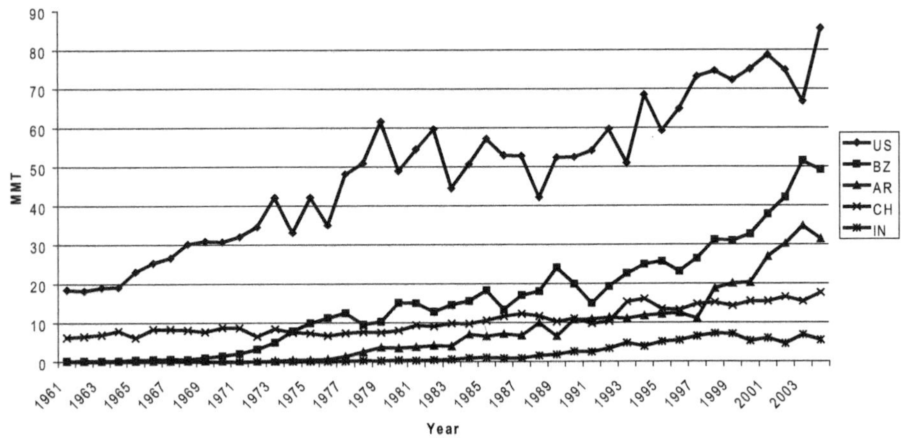

Fig. 5.3. World soybean production. Source: FAO, 2005 and author's calculations.

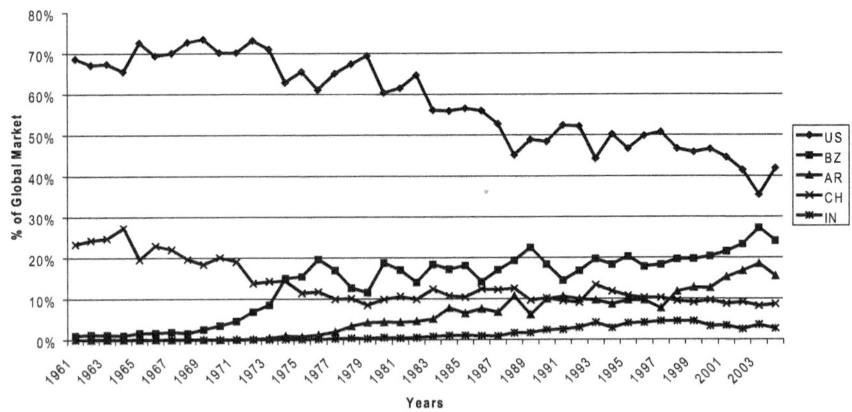

Fig. 5.4. Global soybean market shares. Source: FAO, 2005 and author's calculations.

A fundamental shift in soybean processing investment occurred, away from the United States and Europe toward China, Argentina, and Brazil. Capital for soybean processing is increasingly invested outside the United States because of superior procurement economies, lower costs of plant operation, and close proximity to high-growth livestock industries (Goldsmith et al., 2004). Demand growth is most active outside the United States so foreign crush facilities many times are better able to supply these new customers. For example, two of the fastest growing poultry and pork sectors are in Brazil and China, which are able to utilize their domestically produced meal (Fig. 5.5).

Argentina and Brazil are today's leading exporters of soybean meal, together capturing 64% of world exports; while France, The Netherlands, and Italy lead soybean meal imports with 23%. The United States still holds the most soybean-processing capacity, followed by China and Brazil (Fig. 5.6).

Since the mid-1990s, China dedicated itself to increasing its processing capacity. They shifted domestic policy to favor soybean meal for livestock feed, and soybean oil for human consumption. This policy causes China to import large quantities of soybeans, mostly from Brazil and the United States, to fuel its growing processing industry. China's demand combined with Brazil's relatively small-animal industry resulted in Brazil exporting 73% of the soybeans it produces (production + a small amount of imports), 48% in the form of meal and 52% as raw soybeans.

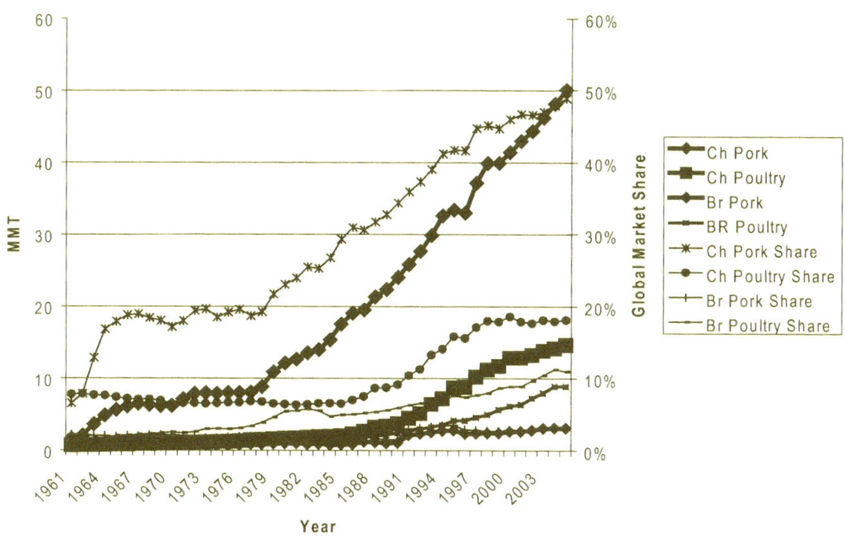

Fig. 5.5. Pork and poultry production in China and Brazil. Source: FAO, 2005 and author's calculations.

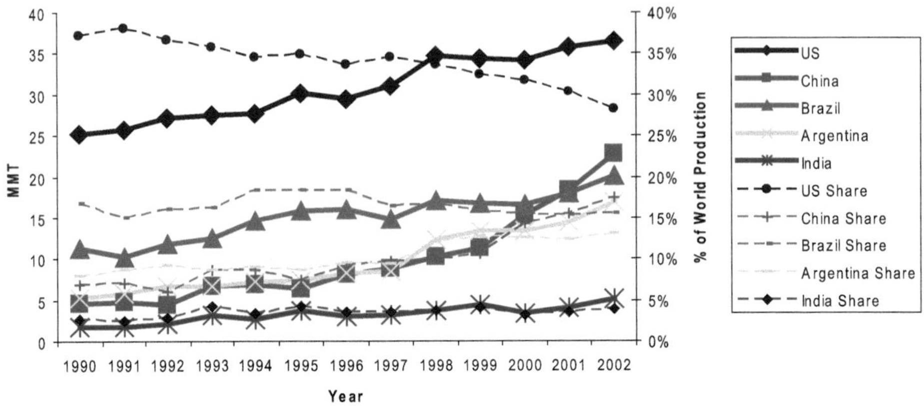

Fig. 5.6. Leading soybean meal producers. Source: FAO, 2005 and author's calculations.

Argentina also is a major exporter with superior logistics due to geography. The main soybean-growing region lies within 500 kilometers of the deep water port at Rosario. Argentina maintains tax polices favoring processing over direct grain exportation. Argentina exports 97% of its soybeans, 74% in the form of meal and only 26% as raw soybeans.

Alternatively, the United States is primarily a domestic crusher of its soybeans and soybean meal, producing six times the soybean meal it exports (Fig. 5.7). This, in part, is due to its large domestic agro–industrial complex that increasingly focuses on domestic demand, not exports. The percentage of U.S. soybeans that were exported decreased 30% since 2000, while at the same time the percentage of U.S.-produced meal exported declined 44%. Correspondingly, Argentina and Brazil, with much smaller agro–industrial/livestock complexes, export most of what they produce to meet the world's growing demand.

Soybean Industry in the United States

The United States produced 3.2 billion bushels of soybeans on 74.6 million acres in 2006 (Fig. 5.8). Since 1986, the size of the U.S. soybean crop increased 3.05% per year. The leading states producing soybeans are found in the Midwest: Iowa, Illinois, Minnesota, Indiana, and Missouri (Fig. 5.9). These five states produced about 51% of national production in 2006.

In 2007 the number of acres planted with soybeans in the United States fell 15% to about 63 million acres. Most of those acres lost involved a switch by Midwest farmers to more corn in response to demand from corn-based ethanol processors. Historically, corn and soybeans are complementary crops in the Midwest planted in a 50:50

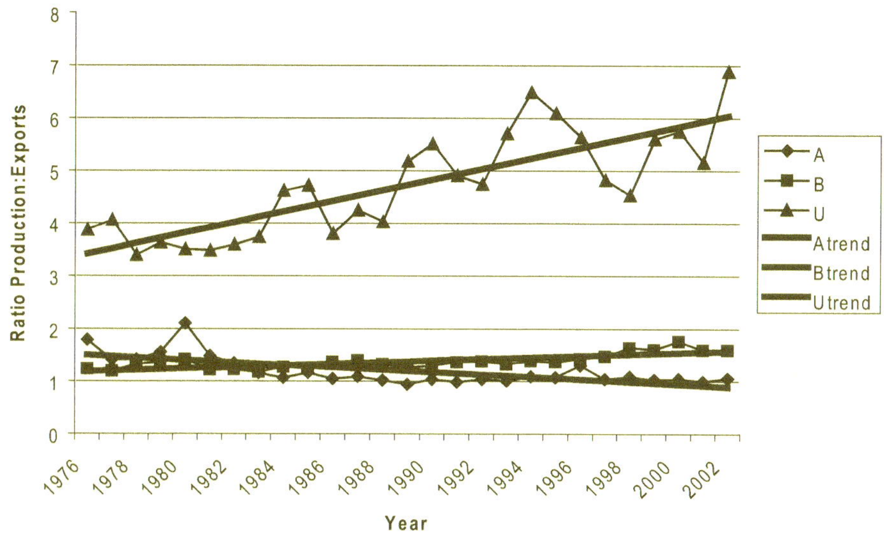

Fig. 5.7. Soybean meal domestic use ratios. Source: FAO, 2005 and author's calculations.

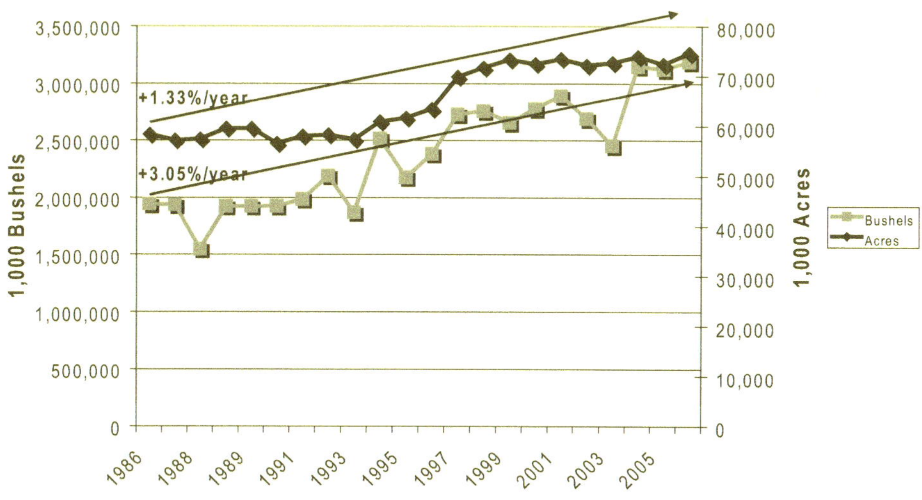

Fig. 5.8. United States soybean acres harvested and production (1986–2006). Source: USDA, NASS, 2007b and author's calculations.

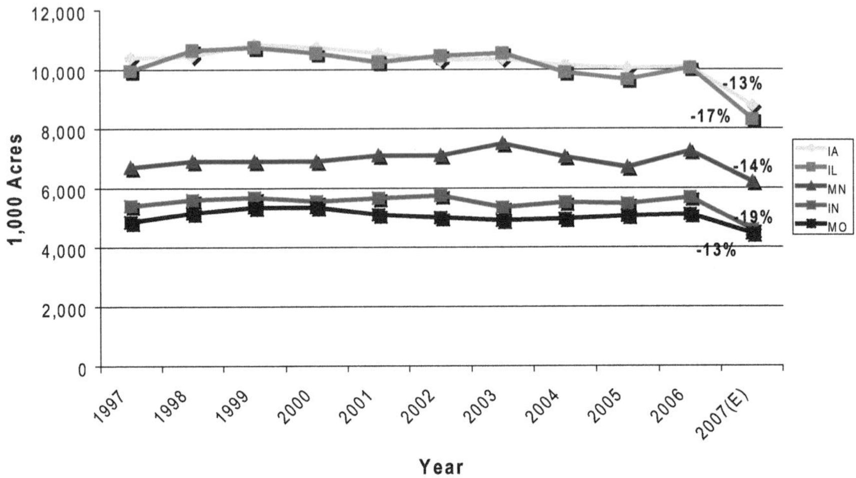

Fig. 5.9. Soybean harvest in the United States: Leading states (1997–2007). Source: USDA, NASS, 2007b and author's calculations.

annual rotation. Fertilizer costs, disease management, and risk were optimized when soybeans and corn were annually alternated. The higher prices for corn due to ethanol demand, combined with new technologies such as stacked biotechnology traits, may make continuous corn production more viable. The historical complementary relationship between corn and soybeans would end if continuous corn production was technically, economically, and environmentally viable. The two crops would then become substitutes and begin to compete for the same acres.

The value of the national soybean crop approaches $20 billion, an increase of over 35% since 1995 (Fig. 5.10). Increases in the value of the national crop since 1995 were not due to rising nominal prices for soybeans as the annual average price fluctuated within a range of $4.38 to $7.34 per bushel. The expansion of soybean acres (+21%) and yields (+21%) over the last ten years explains the industry's growth.

Soybean prices are seasonal. According to daily cash prices from the Chicago Board of Trade, the highest average daily prices occur in May at $6.11 per bushel (Fig. 6.11). The maximum average daily prices though are found in March at $10.41. The lowest average daily prices at $5.39 occur at harvest in September. The minimum average daily price of $3.88 though occurs in July. Average minimum prices show less seasonality than maximum prices because government-support programs provide an effective safety net against price declines. Seasonal soybean prices vary most in the spring as measured by the coefficient of variation. Prices vary in March and April almost twice as much as in September.

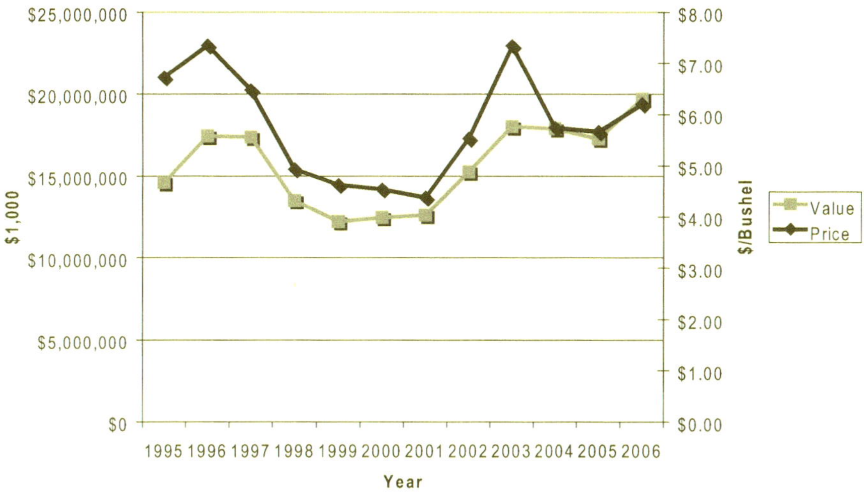

Fig. 5.10. United States average prices received and national crop value for soybeans (1995–2006). Source: USDA, NASS, 2007b.

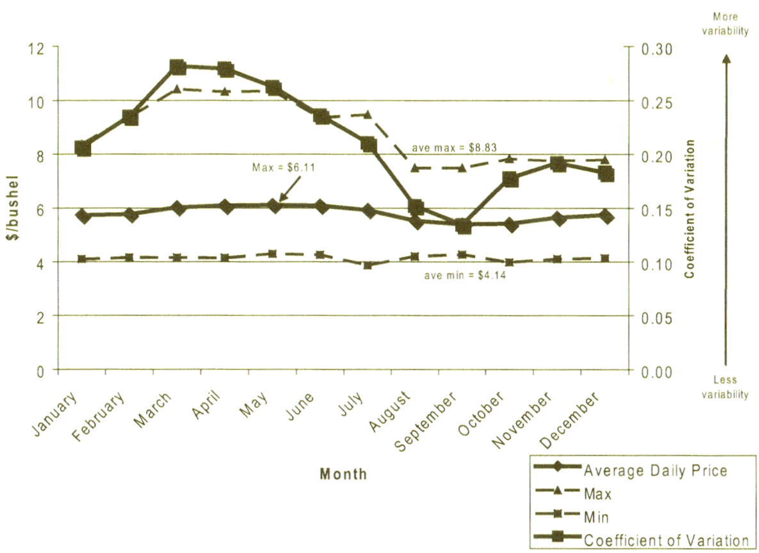

Fig. 5.11. Average monthly soybean prices. Source: Barchart.com, 2007 and author's calculations.

Soybean Yield, Price, and Revenue Trends

Soybean yields increased about 0.40 bushels or 2.55% per year since 1940 (Fig. 5.12). Over the last ten years the increases were more moderate at about 1.5% per year.

Soybean revenue per acre, though flat in recent years, increased over 8% per year or $4.30 per acre since 1940 due to both increases in yield and nominal prices received for soybeans. Profitability conditions though over the last ten years were not favorable. The flattening of the yield curve, little growth in nominal prices, and increasing costs of production made soybeans profitable in only two of the nine years between 1997 and 2005.

The lack of profitability comes at a time of increasing demand for corn. Thus, the opportunity cost of planting soybeans is rising as alternatives, such as corn, become more attractive. Yields improvements have not kept pace with corn, adding to the challenge for producers to stick with soybean production (Fig. 5.13). Alternatively, the high opportunity costs of producing soybeans are not present in the lower latitude-high growth soybean regions of the world such as South America. This presents South America with a comparative advantage in soybean production, while the Midwest in the United States maintains a comparative advantage in corn production.

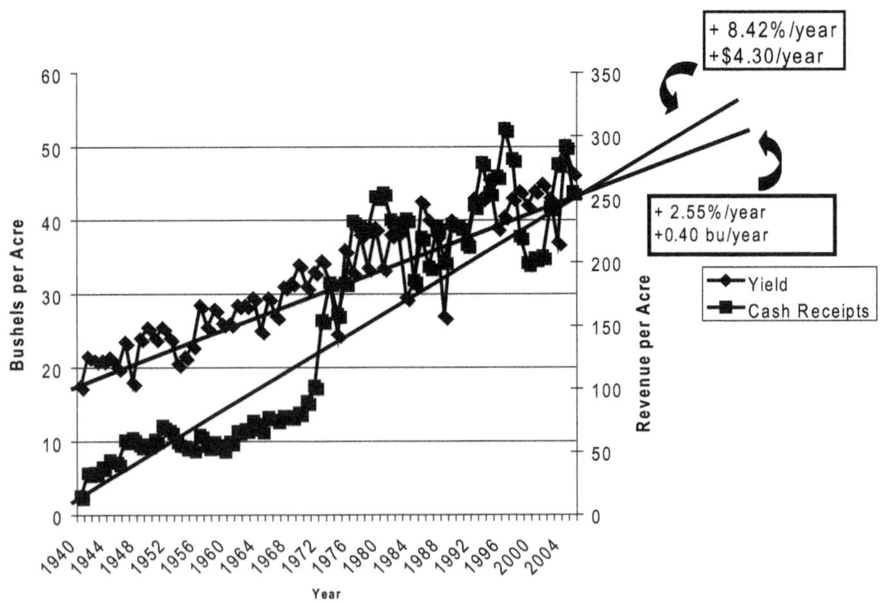

Fig. 5.12. U.S. soybean yield and revenue trends: 1940–2005. Source: USDA, NASS, 2007b and author's calculations.

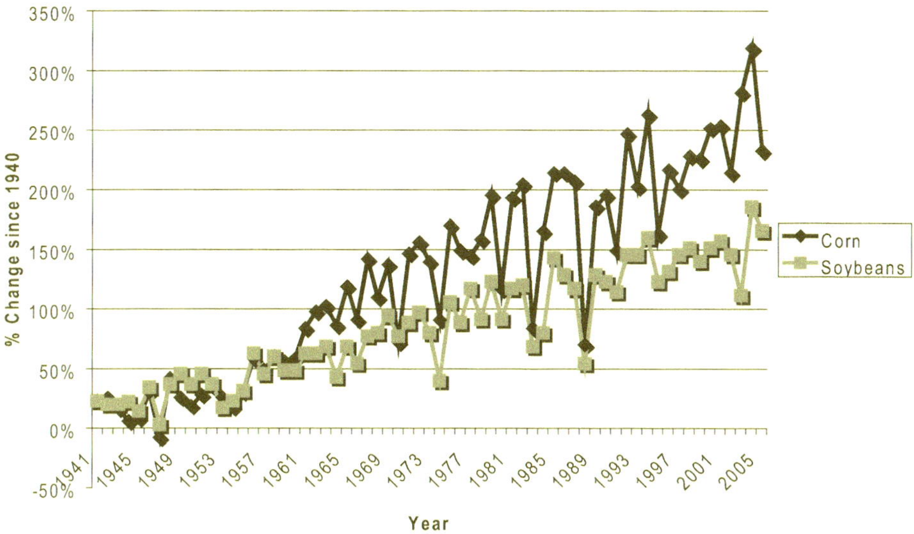

Fig. 5.13. U.S. soybean and corn yield indices: 1940–2005. Source: USDA, NASS, 2007b and author's calculations.

Incentives for seed companies to invest in soybean research dampened in recent years with the maturing of the U.S. soybean industry. Research and development has lagged because weak intellectual property rights in the new high-growth regions make protecting self-pollinated seed technologies, such as soybeans, difficult (Goldsmith et al., 2006). The soybean seed market in the United States and other countries with strong intellectual property protection environments has not been growing nearly as rapidly as the markets in South America and Asia. The recent expansion of the corn–ethanol demand may even portend a decline in soybean acres in the United States, thus actually reducing the size of the global soybean seed market. Fewer acres in soybeans in the United States would create greater incentives to shift research and development dollars to other crops where the intellectual property is protected and the markets are expanding, such as corn.

Pesticide Usage

In 2005, 80 million pounds, about one pound per acre, of pesticides were applied to soybean crops in the United States. The most common pesticide (79%) was glyphosate isopropylamine salt (Fig. 5.14). The popularity of glyphosate is due to the fact that 91% of the U.S. soybean crop utilizes transgenic seed specifically resistant to the herbicide. This allows farmers the ability to spray their fields once for weeds without killing the soybeans. It is estimated that 77 million pounds of herbicide were ap-

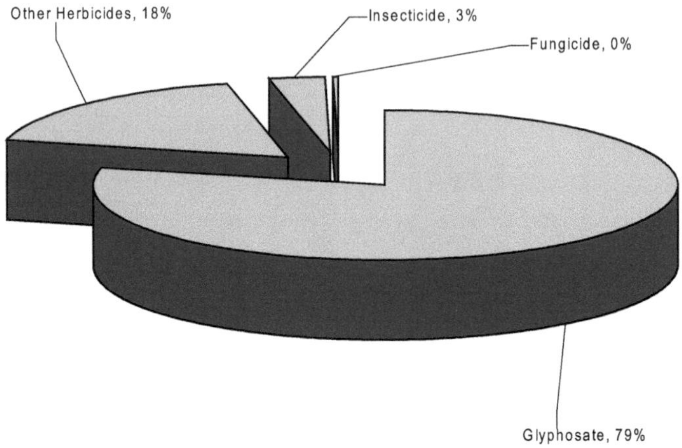

Fig. 5.14. Pesticide usage on soybeans in the United States (2005). Source: USDA, NASS, 2007a and author's calculations.

plied to the U.S. soybean crop, of which 63 million pounds were glyphosate (USDA, NASS 2007a). Only minor amounts of insecticides and fungicides are applied to soybeans in the United States.

Recently Asian rust became an increasing concern for U.S. farmers. Significant preparations were implemented even though a major outbreak of the disease has yet to occur. Farmers began scouting for the fungus, learning about treatment procedures, and actively monitoring spore movements throughout the country.

In 2005, 2% of the soybean acres received applications of fungicide, mostly occurring in three Southern states: Louisiana, Arkansas, and Tennessee (USDA, NASS, 2007a). To date, rust's impact is minimal. Most outbreaks were in the South, where relatively few soybeans are produced and the disease showed up late in the growing season. The real risk is that an early spring storm might bring rust spores up into the heart of the soybean belt from warm over-wintering locations in the Deep South. Infection during critical crop-growth stages would occur under such conditions, and treatment would be necessary to minimize losses.

Rust presents two basic challenges for producers: first, a devastating fungal disease, eliminating 100% of the crop if left untreated; second, knowing when, where, and how much to spray. (Organic soybean producers are extremely vulnerable because there is no known organic treatment for rust.) The uncertainty associated with managing rust is exacerbated because U.S. producers have very little experience with the disease, and an outbreak in any given year is a probabilistic event, more likely not occurring than occurring. Brazil has much more experience managing the disease because significant annual widespread outbreaks have occurred since 2003. Farmers there have experience with the disease, already make many passes over the fields treating other pests, and know with certainty that rust is present in some form in any given year.

The most important tactic for producers when operating under the threat of rust is scouting and being aware of infestation levels nationally, regionally, and locally. Numerous Web sources, sentinel plots, and weather forecasting models were assembled to help soybean producers manage the risks associated with rust.

It is unlikely that an infestation is limited to individual farms so producers need to assure themselves of access to product and spraying services. Costs per acre are about $15 per application (Goldsmith & Schnitkey, 2005). Anywhere from one to three applications may be necessary. The environmental implications to ecosystems of widespread fungicide spraying in soybean fields are not well understood. Fungicide usage historically is not widely practiced on U.S. soybeans, so experience and analyses are limited. Concern exists that important good fungi are killed at the same time farmers are trying to deal with rust's devastating effects.

Cost of Production

The United States has long held a competitive advantage in the production of grains and oilseeds. That is, it was the most efficient producer of soybeans in the world. As a result, the United States produced over 73% of the world's soybeans up until 1973 (FAO, 2007).

Since that time, its competitive and comparative advantages eroded, to the point that the United States now produces less than 40% of the world's soybeans. Three factors are at play. First, farmers' competitive advantage in the production of soybeans declined as the costs of production, especially those corresponding to fixed costs, rose sharply compared to other soybean-growing regions of the world. The sharp increase in fixed costs is primarily due to increased land values in the United States. Second, other countries, such as Brazil, through their own research and development efforts, improved their soybean production efficiency over the last 30 years. Third, U.S. farmers do not have a comparative advantage producing soybeans because Brazilian opportunity costs are lower, especially with the advent of widespread Midwest U.S. corn-based ethanol production. Soybean production is relatively less efficient compared to other activities, most notably corn production. As a result, resources, research, and investment at the margin gravitated away from soybeans and toward corn production.

A positive feedback effect is a critical element of comparative advantage. For example, yields and productivity of soybeans over time will decline as research is diverted from soybeans to corn because of the higher returns to corn and their direct substitutability. Market correction may not occur if soybean research in the Southern Hemisphere can replace the lost research in the United States. This, in fact, has not occurred completely. Weak intellectual property rights in the rapidly expanding soybean regions of the world reduced soybean research and development incentives (Goldsmith et al., 2006).

The cost of production for soybeans averages about $250 per acre: with about 35% being operating costs and 65% being nonoperating costs (Fig. 5.15). A producer with U.S. average yields in 2006 of 42.7 bushels per acre needs a price of $5.85 per bushel to break even. Between 1997 and 2005, the overall cost of production in the United States soybean belt increased 13.3% (2% per year) per year, while yields increased only 10.5%.

Operating Costs

Seed costs rose 65%, or 8% per year, since 1995 and now comprise 35% of total operating costs, and are the largest operating cost category (Fig. 5.16). The chemical component of producing soybeans though fell by half since 1995 as producers adopted transgenic soybeans (Fig. 6. 17). Interestingly, seed costs increased $12.90 per acre since 1997 while during the same period chemical costs fell $12.78. This is due to the substitution of seed technology for chemicals in fighting crop pests. The next smallest

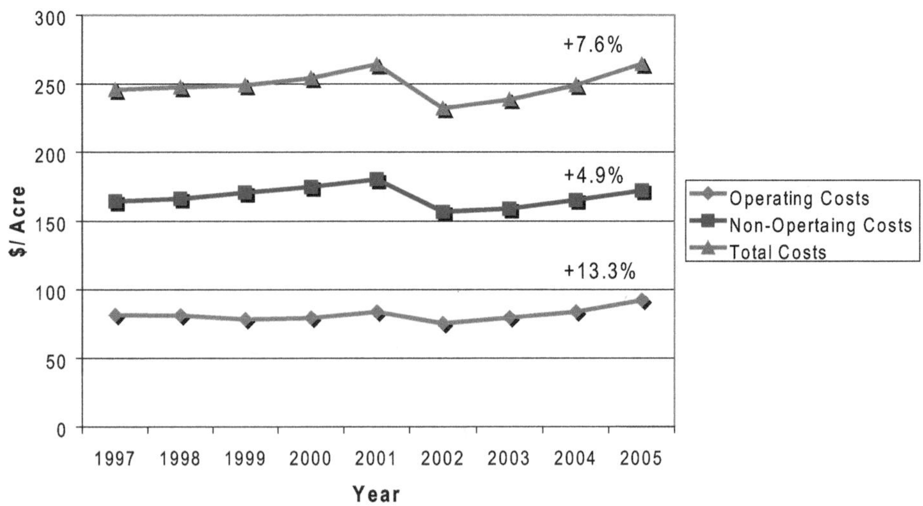

Fig. 5.15. Soybean cost trends: Overview (1997–2005). Source: Illinois Farm Business and Farm Management, 2006.

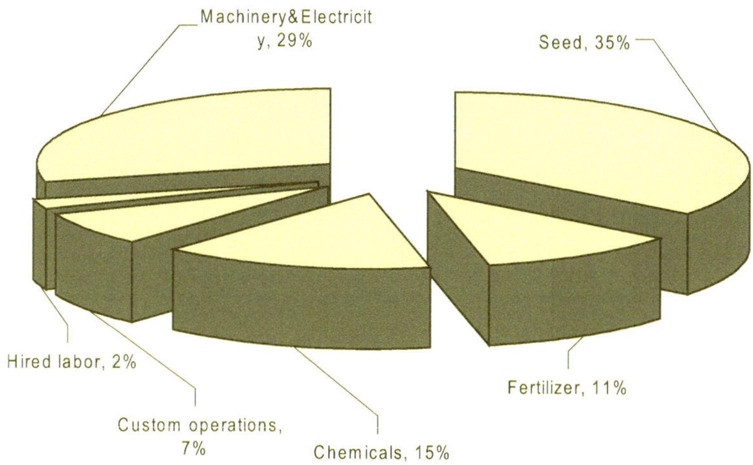

Fig. 5.16. Soybean operating cost breakdown (2005). Source: USDA, Economic Research Service, 2007b and author's calculations.

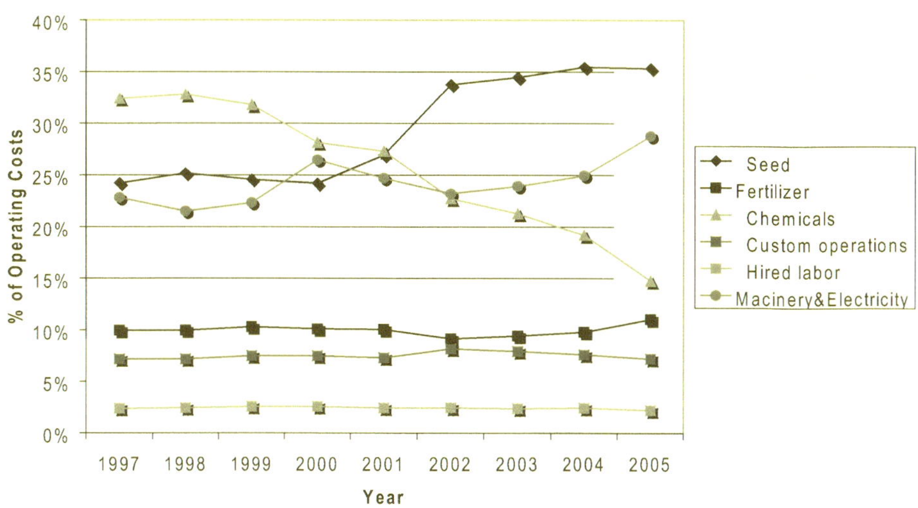

Fig. 5.17. Selected items as a percentage of operating costs (1997–2005). Source: USDA, Economic Research Service, 2007b and author's calculations.

component is custom application hiring charges. U.S. farmers tend to own their own equipment and produce their own crops. Machinery-related items and electricity, the final cost category, comprise about 25% of operating costs and rose about 5% per year since 1995. Custom charges amount to only 7% of operating costs and rose about 2% per year. The third-largest cost area is chemicals, comprising 15% of operating costs. Fertilizer is the fourth-largest cost area for soybean producers. Soybean fertilizer costs rose 26%, or 3% per year, since 1995 and now comprise about 11% of total operating costs. Hired labor is the smallest component of operating costs, comprising 2% of the total. Labor costs increased 5% since 1997.

Nonoperating Costs

Slightly less than two-thirds of the cost of producing soybeans involves nonoperating (fixed) costs. These costs rose much more slowly, less than 1% per year, than operating costs (Fig. 5.18). The most significant nonoperating cost is land. It comprises 50% of nonoperating costs, and 33% of all costs to produce soybeans in the United States (Fig. 5.19). Land costs are also the fastest growing nonoperating cost, increasing at a rate of 1.62% per year since 1997. The second-largest nonoperating cost at 29% involves machinery and equipment recovery (depreciation). These costs increased only slightly over the time period. All of the remaining fixed costs—opportunity cost of unpaid labor, taxes and insurance, and general farm overhead—comprise about 20% of nonoperating costs and fell about 1% per year since 1997.

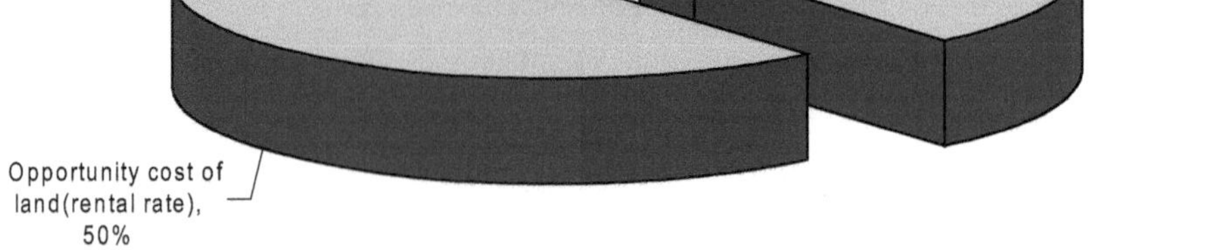

Fig. 5.18. Soybean nonoperating cost breakdown (2005). Source: USDA, Economic Research Service, 2007b and author's calculations.

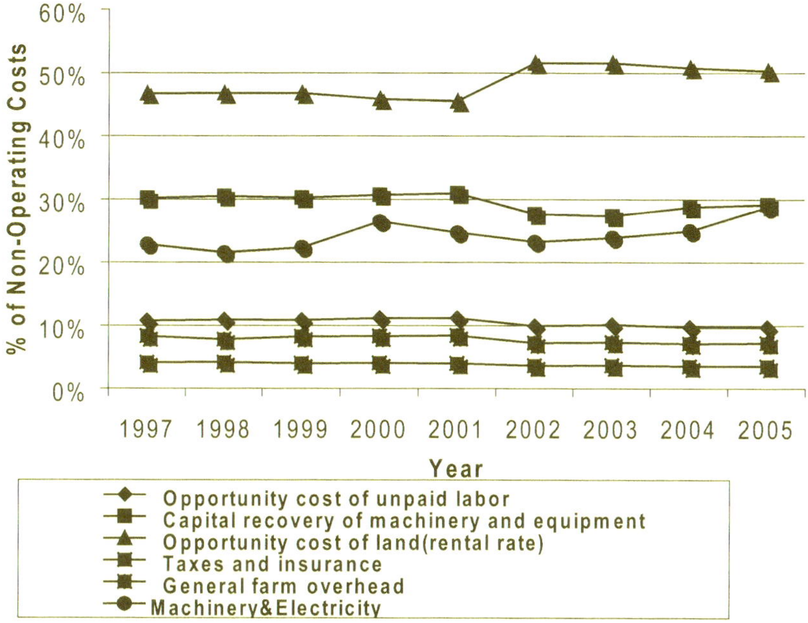

Fig. 5.19. Selected items as a percentage of nonoperating costs (1997–2005). Source: USDA, Economic Research Service, 2007b and author's calculations.

Net Revenue

Rising costs, limited yield growth, and flat nominal prices for soybeans resulted in net losses six out of the nine years from 1997 to 2005 (Fig. 5.20). The worst annual losses over the period occurred in 2001 when the average producer lost $85 per acre or 32% of total costs.

Government Payments

Three sources of government funding for soybean producers are available: Marketing Loan Agreements/Loan Deficiency Payments (MLA/LDP), Direct Payments (DP), and Counter Cyclical Payments (CCP) (USDA Economic Research Service, 2002, 2007d, 2007e). The MLA program loans money at a set price per bushel, and farmers may receive payment of the loan in the form of cash or soybeans. The current (2007) loan rate is $5.00/bushel (USDA, 2007e). When prices fall below that level,

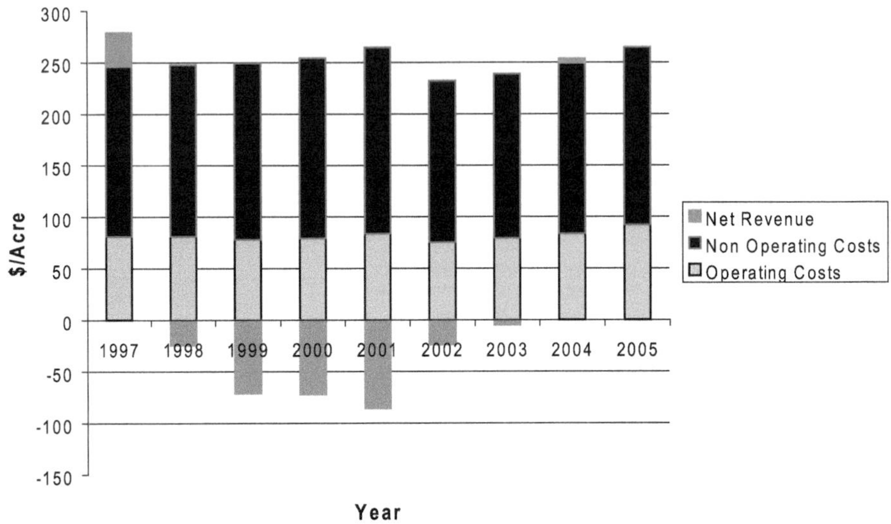

Fig. 5.20. Average soybean revenue and costs per acre (1997–2005). Source: Illinois Farm Business and Farm Management, 2006.

the producer pays back the loan as soybeans and nets the difference. The need for an exchange of grain is not necessary though. The LDP option allows farmers to simply receive the difference as cash without a loan being taken.

Direct payments are paid to soybean farmers at a rate of $0.44/bushel on 85% of their base acres at the level of their program yields. Finally, the countercyclical payments provide soybean farmers payments when prices fall below $5.80. The maximum a farmer may receive is: $75,000 from the MLA/LDP program; $40,000 in direct payments; and $65,000 from countercyclical payments. A total cap is $360,000 per individual for all forms of government payment.

Government payments averaged 6–11% of gross soybean revenue since 1989, with payments per acre averaging about $20–30 per acre (Edwards, 2000; Schnitkey, 2004). Total government payments to U.S. soybean producers would then be about $2 billion annually (this assumes 1999 soybean acreage of 73.7 million acres and Iowa's average payment level from 1989 to 1999 of $26/acre) assuming such levels of payments. The impact on producer net revenue is much greater, where government payments ranged from 0% to 180% of net revenue between 1989 and 1999 in Iowa (Fig. 5.21). Loss years occurred in 1991, 1993, and 1998; government payments were sufficient in 1991 and 1998 to ensure net positive returns, while in 1993 farmers still incurred net losses. As noted above, when land costs are included in the farmer's

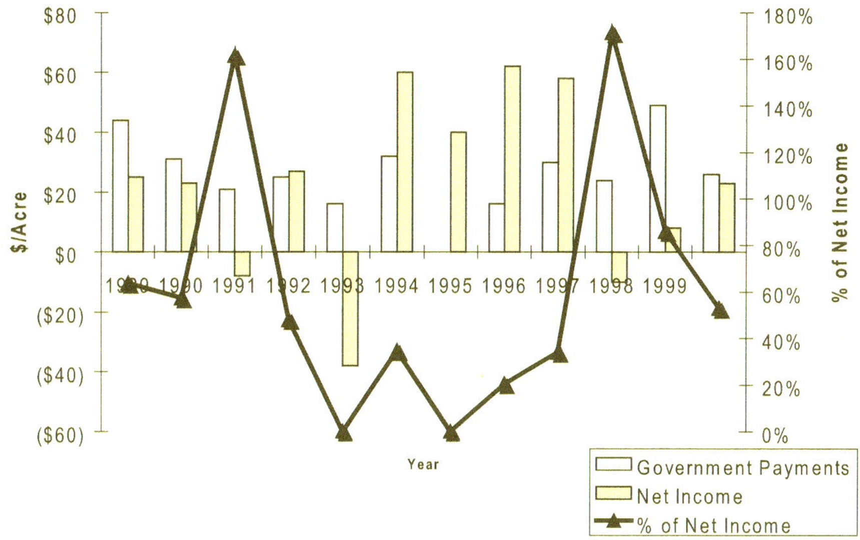

Fig. 5.21. Government payments to soybean acres: Iowa 1989–1999. Source: Edwards, 2000 and author's calculations.

budget, margins become very tight and government payments can become a very important source of net income for soybean farmers.

Soybean Processing

Soybean Meal

Though numerous high-profile closings occurred, plant numbers and the number of employees actually rose since 1977. Currently, 117 soybean processing plants exist in the United States representing 61 different firms (U.S. Census Bureau, 2002). Most operations (64%) employ 20 or more workers in their facilities. Illinois is the leading state in terms of number of establishments, 22, and processing capacity, 15%.

Soybean meal production in the United States servicing the growing domestic livestock industry increased about 2.6% per year since 1987 (Fig. 5.22). Production of soybeans though increased one-third faster at over 4% per year. At the same time, soybean meal exports only increased at a rate of 0.76% per year. As a result, whole soybean exports of the excess supply began increasing but generate less value to the economy. The shift to whole soybean exports indicates that while a market still remains, even though declining, for U.S. soybeans, the production of soybean meal and oil is more efficiently conducted overseas. Correspondingly, domestic utilization of

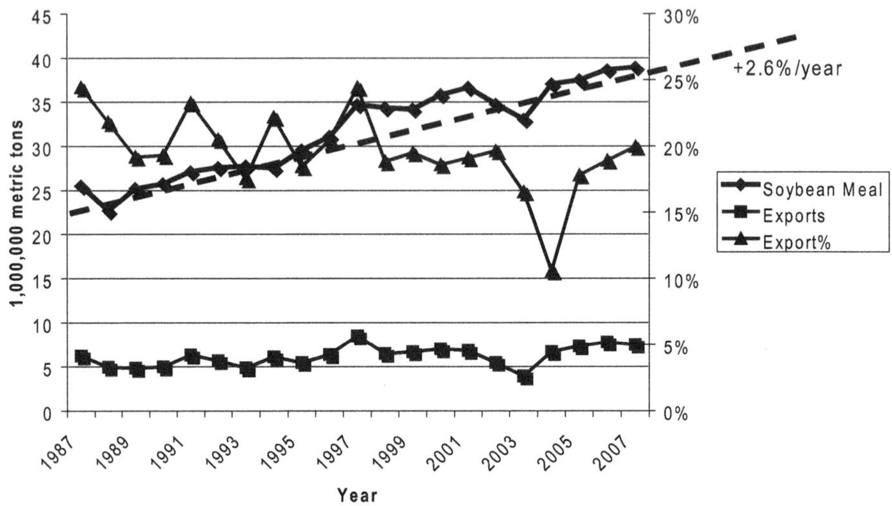

Fig. 5.22. U.S. soybean meal production: 1989–2003. Source: U.S. Department of Commerce, 2007.

domestically produced soybeans (in the form of meal) was as low as 50% as recently as 1995, but now has risen to 63%.

New, larger, more modern soybean processing plants were built offshore to serve the growing livestock demand in Asia or to leverage large supplies of easily originated soybeans in South America (Goldsmith et al., 2004). The United States' share of the global soybean meal market fell 27% since 1995. The trend of increasing domestic utilization, slowing growth of domestic crush, and increasing exports of whole soybeans is consistent with the recent trend of global investment in soybean processing in China, Argentina, and, to a lesser degree, Brazil. Not withstanding, the United States is still the leading soybean meal processor in the world, with 26% of capacity (Fig. 5.23).

Crush Margins

About 98% of all soybean meal is used for livestock feed (Fig. 5.24). Feed utilization levels dip slightly during the summer when crush margins are at their worst and processors market soy protein to the higher valued soy food sector.

Soybean meal prices averaged about $209 per metric ton with little trend in nominal terms over the last ten years (Fig. 5.25). The coefficient of variation is 23% with the range in prices being $144 to $305, at two standard deviations.

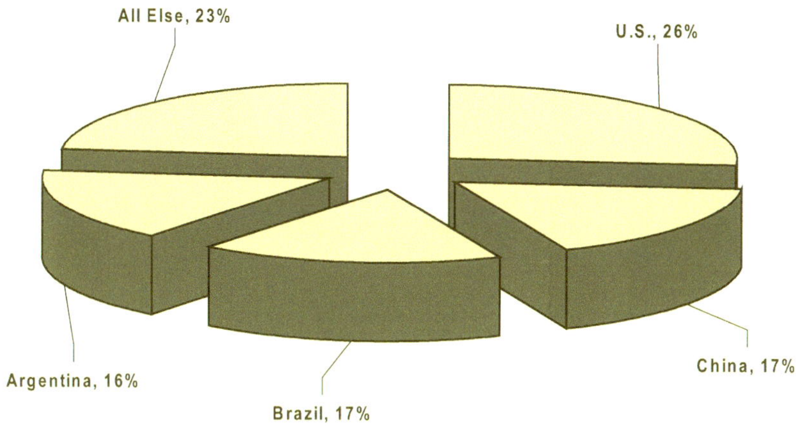

Fig. 5.23. World soybean crushing capacity shares: 2005. Source: FAO, 2007 and author's calculations.

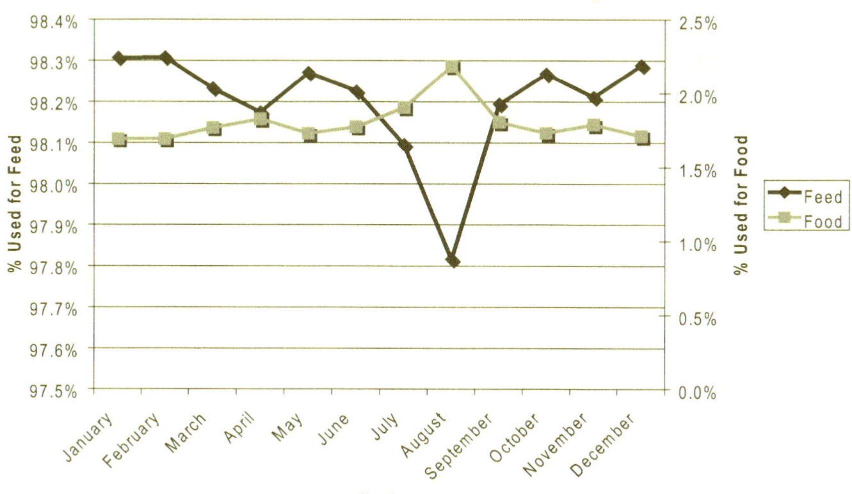

Fig. 5.24. Monthly soybean meal usage levels. Source: USDA, NASS, 2007b and author's calculations.

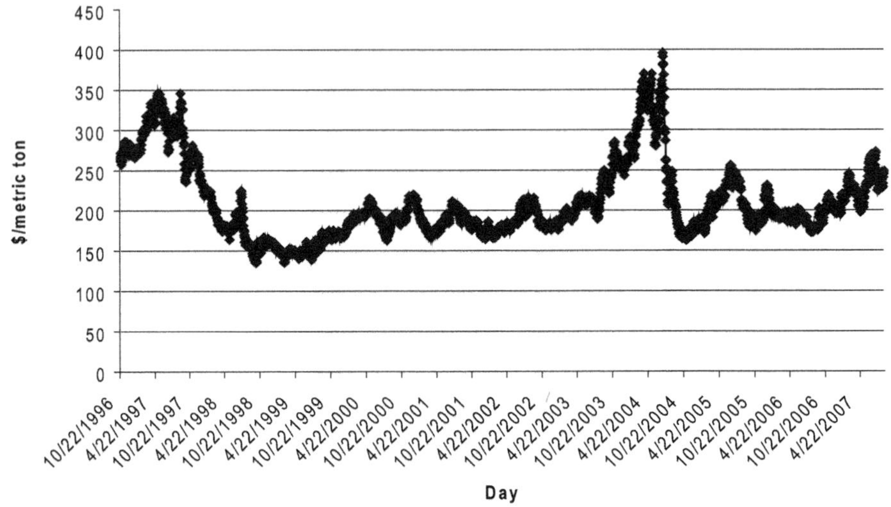

Fig. 5.25. Daily U.S. soybean meal prices: 1996–2007. Source: Barchart.com, 2007 and author's calculations.

During the period of 1996–2007, average monthly crush margins averaged $40 per ton or 19% of the cost of the soybeans (Fig. 5.26). The margin provides the funds to operate the plant and provides profits to the firm. These relatively small margins make input and output marketing, risk management, and economies of scale very important for crusher profitability. Not only are soybean crushing margins thin, they can be volatile. The coefficient of variation of the margin is 27%, with the margin's range at two standard deviations falling between $18 and $61 per metric ton. The highest margins are found during and immediately following the North American harvest, while the tightest margins occur prior to the South American harvest (Fig. 5.27).

Soybean Oil

The United States produced 2.7 million gallons of soybean oil in 2005 and could produce over 4.7 million gallons if the entire U.S. soybean crop was crushed (Fig. 6. 28). About 90% of that U.S.-produced soybean oil is used domestically. As a result, the United States provides only 5% of the world's exports of soybean oil (Fig. 5.29). The increasing global demand for soybean oil is met more competitively by offshore crushers.

Soybean oil prices show a gradual rise from the lows of 2000 due to increasing

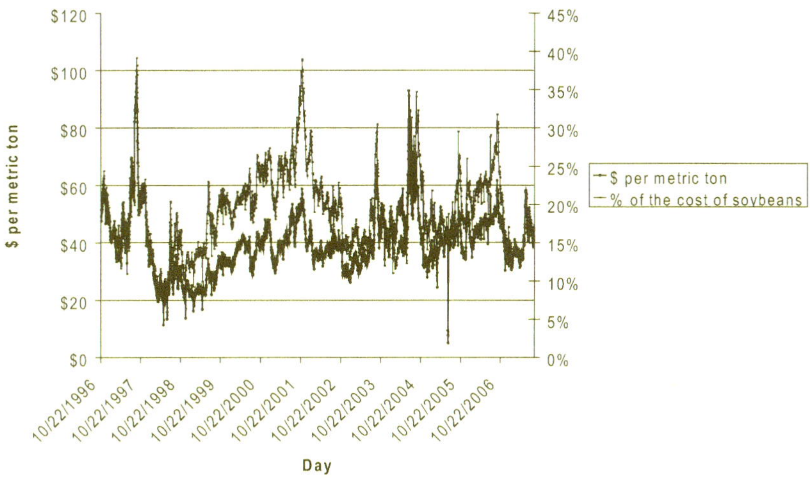

Fig. 5.26. U.S. daily crush margin: 1996–2007. Source: Barchart.com, 2007 and author's calculations.

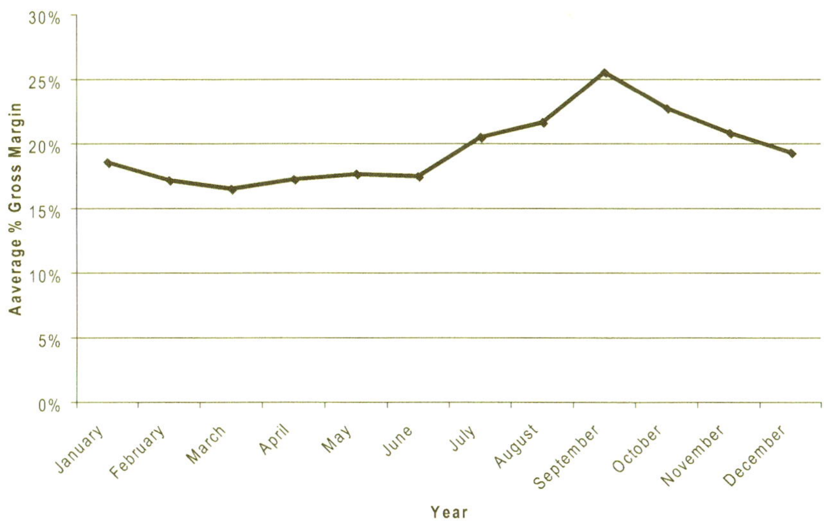

Fig. 5.27. Seasonality of crush margin in the United States: 1996–2005. Source: Barchart.com, 2007 and author's calculations.

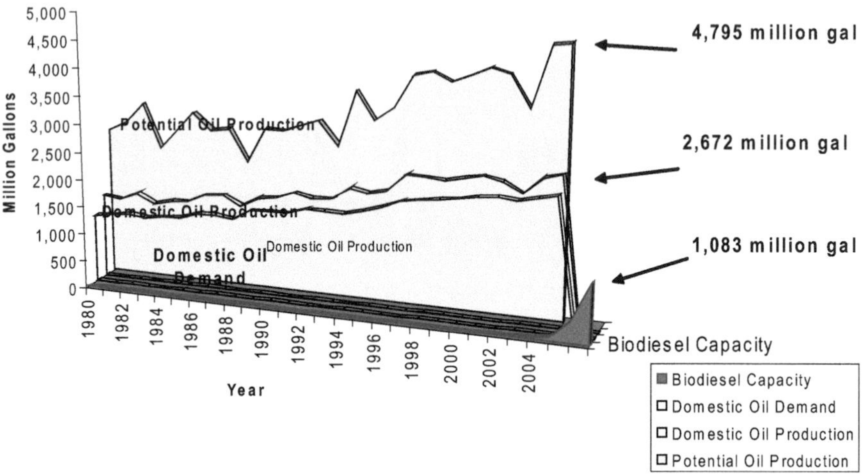

Fig. 5.28. Domestic oil supply and demand with biodiesel capacity. Source: USDA, NASS, 2005 and author's calculations.

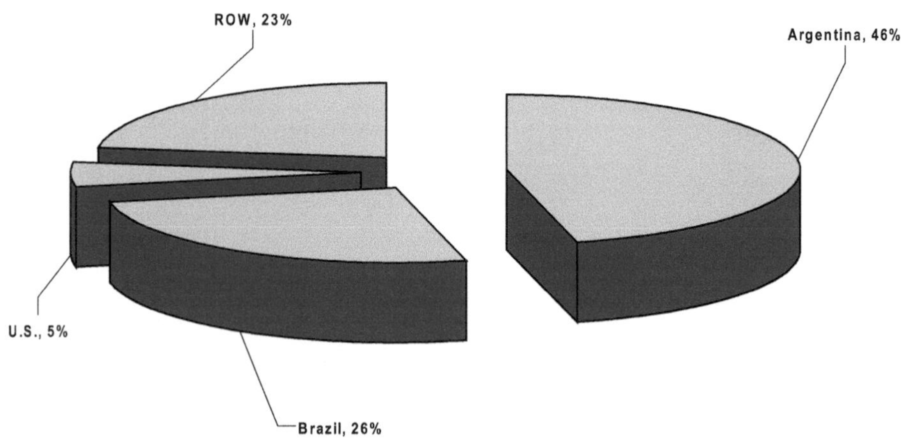

Fig. 5.29. Leading soybean oil exporters : 1995–2005. Source: FAO, 2007 and author's calculations.

food demand in Asia, and more recently the shift to plant-based oils for bioenergy purposes. These two positive effects offset the dampening of demand for soybean oil due to *trans*-fat health concerns and labeling requirement. Prices in 2007 are up 60% compared to 2006 (Fig. 5.30). Soybean oil prices' volatility is moderate: ± about 25%, as measured by the coefficient of variation (Fig. 5.31). Soybean oil prices range between $365 and $605 per metric ton when using the ten-year average price of $485 per ton and a the standard deviation of $120. The sensitivity of food oil markets to supply and demand shocks has significant implications for biodiesel producers who already are operating with thin margins.

Biodiesel

In recent years, soybean oil was increasingly used to make biodiesel. Up until 2005, domestic demand was stable, causing little impact on prices. In 2005, only 34 thousand metric tons, or 2.8%, of the 9.1 million tons of soybean oil produced in the United States went for biodiesel usage (USDA, 2007c). Recent expansion, though, in bioenergy, and continued economic growth in Asia caused oil prices to rise. Demand in the United States is still small relative to current potential supplies of domestically produced soybean oil.

Soybean oil-based biodiesel in the United States is not profitable without government support. Wholesale diesel prices have to break $2.00/gallon just to provide

Fig. 5.30. U.S. soybean oil prices: 1996–2007. Source: Barchart.com, 2007 and author's calculations.

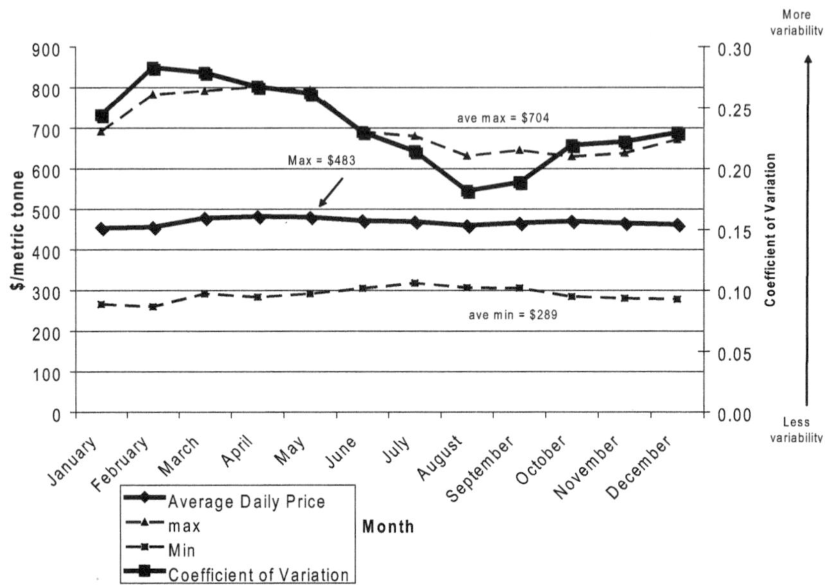

Fig. 5.31. Average daily soybean oil prices: 1997–2006. Source: Barchart.com, 2007 and author's calculations.

any working margin (Fig. 5.32). Nevertheless, the federal government support of \$1.00/gallon (~\$293/mt of oil), usage mandates, and state and local incentives make biodiesel production economically viable in the United States in the near term.

Currently, 58 biodiesel plants operate in the United States with a total capacity to produce 1.3 million tons of fuel and an average capacity per plant of 22 thousand metric tons (Fig. 5.33). An additional 53 plants are planned for construction with a total capacity of 2.4 million tons and an average size of 45 thousand tons. The industry is young and highly fragmented as 80% of the production is produced by 33% of the firms (Fig. 5.34). Soybeans would feed about 47% of the planned biodiesel production. The 940 thousand metric tons of additional oil demand would still only equal 10% of the 2005 U.S. oil supply, up from the current 3%. So the direct price effects from U.S. biodiesel expansion would be moderate. Most of the price effects on food oils will come from much more rapid expansion of biodiesel utilization in Europe and Asia.

The production of biodiesel can occur using a variety of feedstocks. Alternative sources can be other oilseeds, tropical oils, waste oils, and animal fat. Soybean oil is plentiful and relatively inexpensive because it is a by-product of soybean meal production. In this way, it makes sense for the early stages of bioenergy development in the United States to use readily available soybean oil supplies and the associated

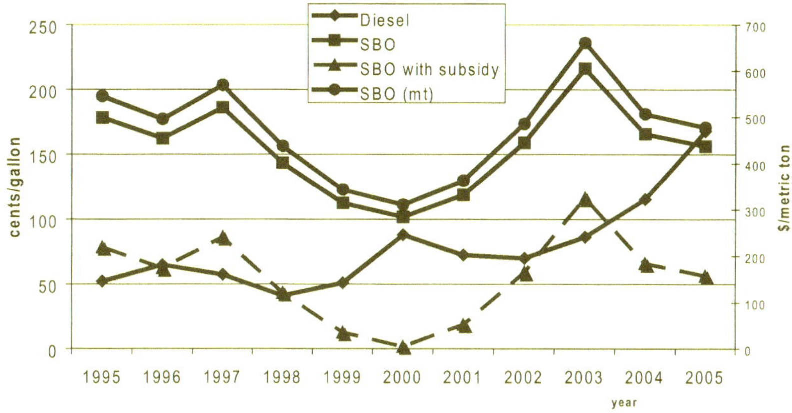

Fig. 5.32. Diesel fuel and soybean oil wholesale price relationship (1995–2005). Source: Barchart.com, 2007 and author's calculations.

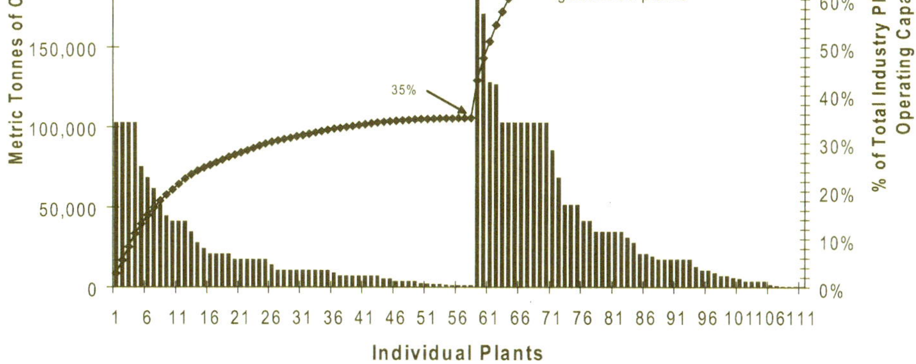

Fig. 5.33. Distribution of biodiesel capacity in the United States (2006). Source: Biodiesel Board, 2007 and author's calculations.

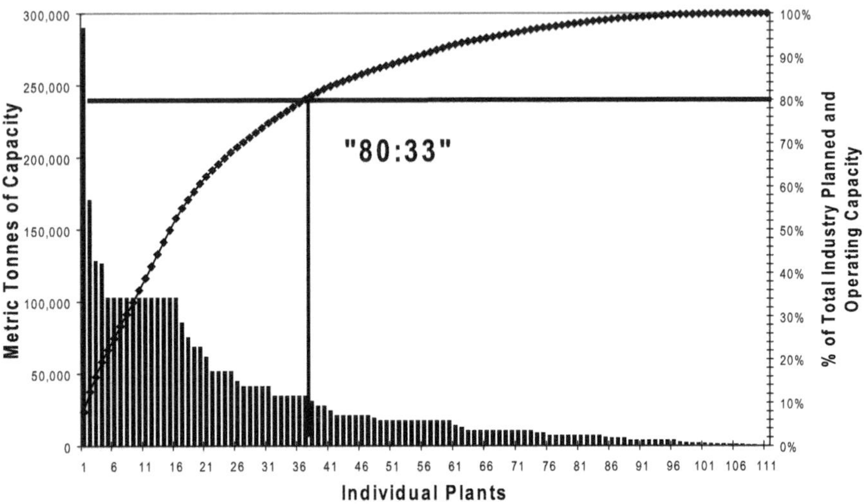

Fig. 5.34. Distribution of biodiesel capacity in the United States (planned and operating) (2006). Source: Biodiesel Board, 2007 and author's calculations.

infrastructure. Other regions of the world, such as Southeast Asia, do not have a soybean complex on which to build a biodiesel industry. Their oil of choice is palm, from which oil is the main product, and not a by-product. The economics of palm oil-based diesel are comparable to soybean oil-based models when factoring in soybean meal production and the proximity of the large agro–industrial complex in the United States.

The largest planned biodiesel plant, at close to 300,000 metric tons, will use camelina oil, not soybean oil. Of the 111 planned or operating plants, 62% of the feedstock will be soybean oil (Fig. 5.35). Most of the plant capacity (55%) will be located in the Midwest and 26% in the South (Fig. 5.36). Biodiesel production in the United States does not need to be co-located with the main soybean-producing regions in the Midwest because soybean oil is readily transportable, and a variety exists of alternative oils, animal fats, and waste oils, both domestic and imported, that can feed a biodiesel plant.

Research and Development

A third factor placing the U.S. soybean industry at a crossroads is the weakened state of global research and development because soybean intellectual property is vulnerable to pirating (Goldsmith et al., 2006). Corn is marketed as hybrids while soybeans are

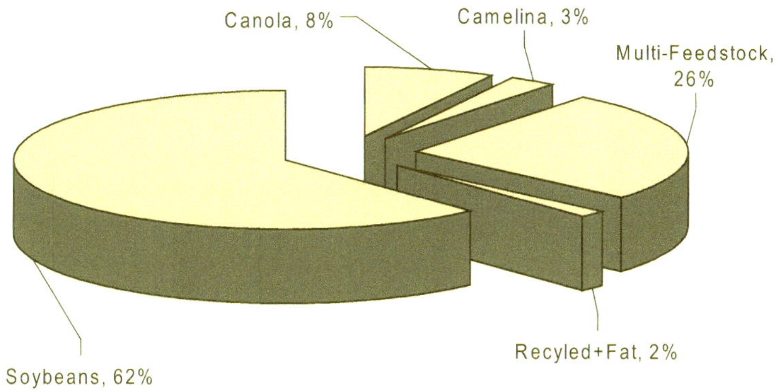

Fig. 5.35. Distribution of biodiesel feedstock types (planned and operating) (Spring 2006). Source: Biodiesel Board, 2007 and author's calculations.

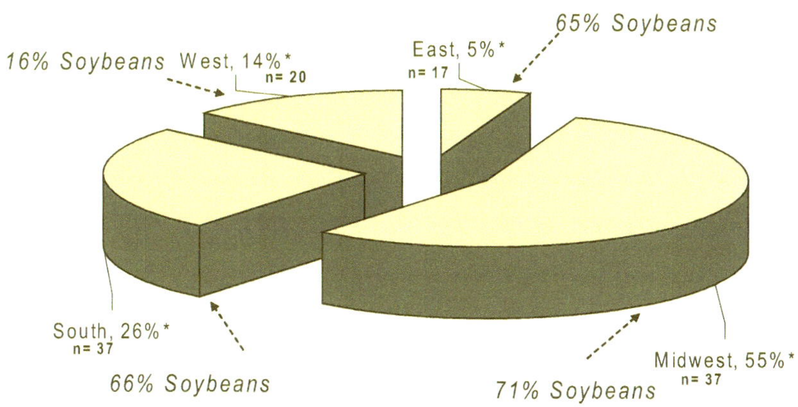

Fig. 5.36. Location of biodiesel plants (planned and operating, Spring 2006). * of capacity. Source: Biodiesel Board, 2007 and author's calculations.

self-pollinating. Low germination rates occur when replanting seed from hybridized crops like corn. Alternatively, very high germination rates and plant productivity result when replanting saved or pirated seed from self-pollinated crops, such as soybeans or wheat. As a result, farmers have to purchase seed to produce hybridized crops, but can replant seed from self-pollinated crops collected at the previous year's harvest. In the United States, as well as most developed countries, soybean seed companies employ patent protection or contracts to ensure that farmers purchase seed every year and no intellectual property theft occurs. Many developing countries, especially the high-growth soybean countries in South America, have weak intellectual property protection environments. Saving seed and reselling seed are widespread.

Soybean farmers can lower the cost of production by saving seed from the harvest and not needing to pay for seed inputs. The savings would be significant as seed costs are an expensive input.

But not purchasing seed denies seed development companies the revenue associated with selling self-pollinated seed products. Additionally, independent seed companies, called *brown baggers*, can multiply and sell large quantities of self-pollinated seed and compete directly with the formal seed company, often with their own product. Brown bagging (the commercial multiplication and sale of pirated seed), like farmer-saved seed, decreases the incentives for seed developers to invest in research and development because seed sales are reduced.

Though a lot of soybeans are grown in the United States, the market is mature. Private research and development companies are increasingly less interested in investing in soybean research because most of the new opportunities lie in the countries where intellectual property rights are weak and profitability is unlikely. Unlike in the United States, it is often legal for farmers to save seed from newly released varieties. Also governments are often unable to stop the practice of brown bagging.

The weak incentives to invest in soybean research and development in countries where soybeans are expanding rapidly reduce the level worldwide of soybean research activity compared to other crops more easily protected. Protecting the intellectual property associated with self-pollinated crops is critical to maintaining the proper incentives for active research and development investment. Nevertheless, U.S. producers continue to invest close to $100M annually in research and development through their checkoff programs. These investments are an important complement to declining private and government soybean research dollars focused on the U.S. market.

Conclusion

The future of soybean production and soybean utilization is bright because of the growing demand for protein. The United States continues to be the world's largest soybean producer with some of the world's lowest operating and logistics costs. New opportunities emerged with biodiesel that portend a significant new market for soybean oil.

Long term though, the U.S. soybean industry is at a crossroads as new opportunities arise for corn-based bioenergy and farmers shift acres from soybeans to corn. It is still unclear how well the continuous corn model will fair. Only recently have farmers begun experimenting with an unbalanced rotation between corn and soybeans.

U.S. producers no longer have a comparative advantage in soybean production, as corn demand opportunities expand. As a result, increases in the worldwide soybean market are being met by competitors overseas. New corn technologies and management practices allow U.S. producers to better combat disease and insects in continuous corn. The new ability to continually grow corn combined with strong demand may break the historic complementary relationship between corn and soybeans in the United States. Increasing soybean yields, disease resistance, and overall profitability would help soybeans better compete with corn. As well, the soybean industry may look to reinvigorate the complementary relationship with corn by pursuing research that reduces pest management and fertilizer costs, and improves soil quality and environmental performance when soybeans are rotated with corn.

At the same time, the theory of comparative advantages teaches that the shift to corn in the United States also is a function of the low opportunity costs and competitiveness of soybean producers outside the United States. For example, in the Cerrado region of Brazil, soybeans are well-adapted, grow well, and have yields equaling those in the United States. The opportunity costs of planting soybeans are low in the region because few comparable alternatives exist, like those found in the Midwest in the United States. Even though U.S. soybean producers are globally competitive, the opportunity costs of foregoing corn production are increasingly high.

Critical to the vitality of the U.S. soybean complex is the continued expansion of the livestock sector. For example, over the last ten years, poultry production in the United States grew only two-thirds of the world rate, and exports grew only one-fourth of the global rate. This indicates that poultry production is not only growing faster outside the United States, but new demand is being met by offshore producers, such as Brazil and China. China would source some of their soybeans from the United States, while Brazil will not. The challenge for the U.S. soybean industry is how to make U.S. soybeans and soybean meal a preferred input for the rapidly expanding livestock and aquaculture industries overseas.

The domestic livestock sector is the soybean industry's best customer. But recently, environmental management, animal welfare, and ample labor supply issues weighed heavily on the profitability and expansion of the industry (Armbruster, 2006). At the same time, continued urbanization pressures reduce the number of siting locations for new ventures and significant labor demands that make nonfarm and crop agriculture opportunities appealing to livestock operators. Therefore, it is in the interest of the soybean complex to keep the livestock industry viable, serving not only the domestic but the fast-growing overseas markets.

Another strategic issue facing the U.S. soybean complex is the increasing supply of dried distiller grains and solubles (DDGS). One-third of the output of a corn-based ethanol plant is a mid-level protein product that can be used for livestock feed. DDGS are already aggressively being integrated into ruminant feed rations. The high-fiber levels currently limit the feeding of DDGS to nonruminants (pork, poultry, and aquaculture). Significant ongoing research though is being conducted to help improve the nonruminant feeding quality of DDGS. This would ensure a broader market for this by-product and in turn improve the profitability of ethanol plants. The large and new supply of DDGS increases the supply of protein in the market and has the potential to drive down the price of protein and the relative value of soybean meal. The industry needs to keep improving and marketing the value of soybean meal so that it is the protein source of choice for livestock and aquaculture feeders.

An impact to the soybean complex is that the shift to a bioenergy economy may shift the relative contribution to crush margin of the protein and oil content of soybeans. DDGS production will tend to lower the price of soybean meal while demand for biodiesel will tend to elevate the price of soybean oil. The weakness of crush margins going forward provides significant incentive for soybean crushers to integrate downstream and produce higher-valued biodiesel in order to take advantage of current public policy favoring bioenergy.

The challenge is that soybeans are not an oil crop, as only 19% of the soybean is oil. Raising those levels is difficult as a negative correlation exists with both protein and yield. The fall in the value of soybean meal does more harm to margin than a rise in oil prices. So simply shifting the industry's focus to oil is not economically viable. For example, to maintain a 20% crush margin, assuming soybeans are $220 per metric ton, soybean oil would have to exceed $700 per metric ton if soybean meal prices fell below $200 per metric ton (Table 5.1). Unfortunately, the number of days over the last ten years that soybean meal was below $200 was far greater than the number of days soybean oil was above $700. Notably, if oil demand was so strong as to cause prices to dramatically rise, a greater interest would arise among processors and farmers to switch from soybeans. Higher oil crops are much more efficient at producing oil than a protein crop such as soybeans.

Finally, the U.S. soybean industry needs research and development more than ever to counter the effects of declining private and government funding. Historically, much of the world's soybean R&D occurred, or had its origins, in the United States. Much of Brazil's work with low-latitude soybeans had its origins in the Southern United States in the 1960s. Weak intellectual property rights created environments of underinvestment in research and development in South America, especially by private seed companies. As a result, R&D globally is less intense due to the maturity of the U.S. market and the lack of comparative advantage in the United States, combined with weak private incentives for soybean R&D outside the United States. Such effects of reduced investment do not show up at the margin, but over time yield growth rates will slow, disease resistance will lessen, innovations will occur less often, and farmer profitability will suffer.

Table 5.1. Crush Margin Percentages Involving Soybean Meal and Soybean Oila[a]

Soybean Meal ($/mt)	Soybean Oil ($/mt)									
	250	300	350	400	450	500	550	600	650	700
100	-75%	-63%	-53%	-44%	-36%	-29%	-22%	-16%	-11%	-6%
125	-51%	-42%	-34%	-27%	-21%	-15%	-10%	-5%	-1%	3%
150	-33%	-26%	-20%	-14%	-9%	-4%	0%	4%	8%	11%
175	-18%	-13%	-8%	-3%	1%	5%	8%	12%	15%	18%
200	-7%	-2%	2%	6%	9%	12%	15%	18%	21%	23%
225	3%	6%	10%	13%	16%	19%	21%	24%	26%	28%
250	11%	14%	17%	19%	22%	24%	27%	29%	31%	33%
275	17%	20%	23%	25%	27%	29%	31%	33%	35%	37%
300	23%	25%	28%	30%	32%	34%	35%	37%	39%	40%
325	28%	30%	32%	34%	36%	37%	39%	40%	42%	43%

[a]Assume cost of soybeans = $220/mt.

References

Armbruster, W. (ed.) The future of animal agriculture in North America. Farm Foundation; 2006; pp. 161.

Barchart.com. 2007. http://barchart.com/.

Biodiesel Board, 2007. http://www.biodiesel.org/.

Edwards, W. Comparing Cash Rental Rates to Net Returns to Land. Iowa State University. April, 2000. http://www.extension.iastate.edu/AgDM/articles/edwards/EdwApr00.htm.

FAO. FAOSTATS; Food and Agricultural Organization; 2005.

http://faostat.fao.org/site/336/default.aspx.

FAO. FAOSTATS, Food and Agricultural Organization; 2007.

http://faostat.fao.org/site/336/default.aspx.

Farm Business and Farm Management. Survey of Illinois Farmers; 2006.

Goldsmith, P.D.; B. Li; J. Fruin; R. Hirsch. Global shifts in agro-industrial capital and the case of soybean crushing: Implications for managers and policy makers. *Int. Food Agribusiness Management Rev.* **2004**, *7*, 87–115.

Goldsmith, P.D.; G. Ramos; C. Steiger. Intellectual property piracy in a north-south context: Empirical evidence. *Agric. Econ.* **2006**, *35*, 335–349.

Goldsmith, P.D.; G. Schnitkey. Soybean rust scenario model: Crop year 2005 decision-making requires planning. *Feedstuffs* **2005**, *77*, March 7.

Schnitkey, G. The economics of adding more corn to corn-soybean rotations. FarmDoc; 2004. http://www.farmdoc.uiuc.edu/manage/newsletters/fefo04_20/fefo04_20.html.

United States International Trade Commission. Industry and Trade Summary: Oilseeds. USITC Publication 3576. February, 2003; pp. 33.

U.S. Census Bureau. 2002. http://www.census.gov/cir/www/311/m311j.html.

U.S. Department of Commerce. Monthly Reports on Oilseed Crushing. 2007.

USDA, Economic Research Service. 2002. http://www.ers.usda.gov/Briefing/FarmPolicy/2002malp.htm.

USDA, Economic Research Service. Soybeans and Oil Crops: Market Outlook. 2007a. http://www.ers.usda.gov/Briefing/SoybeansOilcrops/2007baseline.htm.

USDA, Economic Research Service. 2007b. http://www.ers.usda.gov/data/CostsandReturns/data/current/C-Soyb.xls.

USDA, Economic Research Service. 2007c. http://usda.mannlib.cornell.edu/usda/current/OCS/OCS-08-13-2007.pdf.

USDA, Economic Research Service 2007d. http://www.ers.usda.gov/Briefing/FarmPolicy/Direct-Payments.htm.

USDA, Economic Research Service 2007e. http://www.ers.usda.gov/Briefing/FarmPolicy/Counter-CyclicalPay.htm.

USDA, NASS. 2007a. http://www.pestmanagement.info/nass/app_usage.cfm.

USDA, NASS. 2007b. http://www.usda.gov/nass/pubs/agr02/02_ch3.pdf.

6

Measurement and Maintenance of Soybean Quality

Marvin R. Paulsen
Department of Agricultural & Biological Engineering, University of Illinois, Champaign-Urbana, IL, 61801

Introduction

Soybeans [*Glycine max* (L.) Merr.] is a native crop of China and one of the oldest oilseed crops in the world. Soybeans are an important source of dietary protein and oil for humans and animals and can aid in reducing chronic diseases. They are also used for soyfoods, and the oil has use as a renewable biofuel. Whether soybeans are crushed for soybean meal and crude oil or used directly for food, industrial chemicals, or fuel, the quality of soybeans will never be higher than that at harvest. From harvest onward, quality at best can only be maintained. In fact, major effort is made solely to slow the normal rate of deterioration so that soybeans can provide an ample year's supply with some carryover and sufficient seed for future propagation.

Since quality deterioration can only be slowed, knowledge of methods for identifying and measuring quality is very important. Good measurement methods provide the first step in assessing quality and, subsequently, determining how various harvesting, handling, drying, and storing practices affect the rate of quality change.

One of the first approaches to measuring and assuring quality of U.S. soybeans is the testing and reporting of factors, which are included in the U.S. Soybean Grading Standards (*U.S. Standards for Soybeans*, 1994). While the U.S. Soybean Grading Standards provide a good measure of quality, particularly for commodity soybeans, they do not provide all of the information that soybean buyers and consumers may need. Therefore, other quality factors, in addition to the U.S. Soybean Grading Standards, are also discussed. Moisture is an important factor that affects marketability and storability. Germination and seed vigor test methodologies determine the quality as to whether the soybeans are suitable for seed. One of the measurement techniques that is seeing more widespread use because of its speed and ability to nondestructively test samples is near-infrared (NIR) spectroscopy. Near-infrared absorbances occur in response to electromagnetic radiation impinging upon molecules with chemical bonds containing hydrogen, usually -OH, -CH, and -NH bonds. With enhanced chemometric software and high-speed computers interfaced to spectrometers, improvements in NIR spectroscopic techniques occurred in recent years, which have enabled this technology to gain increased acceptance at elevators and processing plants throughout the world. Therefore, recent advances in NIR spectroscopy relative to soybean quality measurements are included in the discussion of many of the chemical quality factors.

Quality Factors in the U.S. Soybean Grading Standards

The quality factors for the *U.S. Standards for Soybeans* (1994) include: heat-damaged kernels, total-damaged kernels, foreign material, splits, and soybeans of other colors (Table 6.1). All are reported to the nearest 0.1%. The standards traditionally included test weight, but as of September 1, 2007, test weight was changed from a grade-determining factor to an informational factor only. Hill (1991) summarized the ideal soybeans standards and recommended in 1991 that test weight in soybeans be included as a nongrade determining factor. This change made test weight become an informational factor only, and test weight is reported to the nearest 0.1 lb/bu as opposed to the previous method of rounding and reporting to the nearest 0.5 lb/bu. Test weight is the weight of soybeans contained in a Winchester bushel (defined as 35.239 L [2,150.42 in^3]) as is determined by funneling soybeans from a drop height of 5.1 cm (2.0 in) into a quart kettle and weighing the mass contained.

USDA defines soybeans as an oilseed "that consists of 50 percent or more of whole or broken soybeans (*Glycine max* (L.) Merr.) that will not pass through an 3.2-mm (8/64-in) round-hole sieve and not more than 10.0 percent of other grains for which standards have been established under the United States Grain Standards Act" FGIS (2004).

Splits

Splits in soybeans are defined as those soybeans that have more than one-fourth of the bean broken and removed but otherwise not damaged (FGIS, 2007). The percentage of splits is determined by sieving about 125 g of soybeans on a 4.0 × 19.0 mm (10/64 × ¾ in) slotted sieve, a 3.2 × 19.0 mm, or 3.6 × 19.0 mm oblong-hole sieve. U.S. No. 1 soybeans are allowed up to 10% splits, and most lots of U.S. soybeans are able to make No. 1 grade on splits. U.S. grade No. 2 soybeans have between 10 and 20% splits.

Foreign Material

Foreign material is defined as material passing through a 3.2 mm (8/64 in) round-hole sieve plus all material, other than soybeans, that remains on top of that sieve. U.S. No. 1 grade soybeans are allowed 1% foreign material, a value that is usually difficult to meet without cleaning.

Table 6.1. Soybeans Grades and Grade Requirements

Grading Factors	Grades U.S. No.			
	1	2	3	4
Maximum percent limits of:				
Damaged kernels: Heat (part of total)	0.2	0.5	1.0	3.0
Total	2.0	3.0	5.0	8.0
Foreign material	1.0	2.0	3.0	5.0
Splits	10.0	20.0	30.0	40.0
Soybeans of other colors [1]	1.0	2.0	5.0	10.0
Maximum count limits of:				
Other materials: Animal filth	9	9	9	9
Castor beans	1	1	1	1
Crotalaria seeds	2	2	2	2
Glass	0	0	0	0
Stones [2]	3	3	3	3
Unknown foreign substance	3	3	3	3
Total [3]	10	10	10	10

U.S. Sample grade are soybeans that:
(a) Do not meet the requirements for U.S. Nos. 1, 2, 3, or 4; or
(b) Have a musty, sour, or commercially objectionable foreign odor (except for garlic odor); or
(c) Are heating or otherwise of distinctly low quality.

[1] Disregard for mixed soybeans; [2] In addition to the maximum count limit, stones must exceed 0.1 % of the sample weight; [3] Includes any combination of animal filth, castor beans, crotalaria seeds, glass, stones, and unknown foreign substances. The weight of stones is not applicable for total other material. (Effective September 2007)
Source: Official United States Standards for Grain, USDA-GIPSA-FGIS, 2007) from: http://archive.gipsa.usda.gov/reference-library/handbooks/grain-insp/grbook2/soybean.pdf

Total-damaged Soybeans

Damaged soybeans are soybeans or pieces of soybeans that are heat-damaged, badly ground-damaged, badly weathered-damaged, diseased, frost-damaged, germ-damaged, insect-bored, mold-damaged, sprout-damaged, stinkbug-stung, or otherwise materially damaged. Frost-damaged kernels are kernels that have green coloring in a cross-sectional cut of the kernel. Stinkbug-stung kernels are considered damaged at the rate of one-fourth of the actual percentage of the stung kernel. Examples of these types of damage are shown in interpretive line slides used by the Federal Grain

Inspection System (Fig. 6.1) (Friedrich, 2007). The percentage of damaged soybeans is determined after the foreign material is first removed. The term *total damage* refers to damaged soybeans from all sources, including heat damage.

Heat-damaged Soybeans

Heat-damaged soybeans are soybeans that are materially discolored due to heat or mold damage. Heat damage is a subset of total damage.

Soybeans of Other Colors

Classes of green, black, and brown soybeans were changed to "Soybeans of Other Colors" in September 1985. Soybeans are divided broadly into two classes based on color: yellow soybeans and mixed soybeans. Yellow soybeans are those with yellow or green seed coats; their cross sections are yellow or have a yellow tinge, but they do not contain more than 10.0% of soybeans of other colors. Mixed soybeans are soybeans that do not meet the color requirement for yellow soybeans. Soybeans of other colors

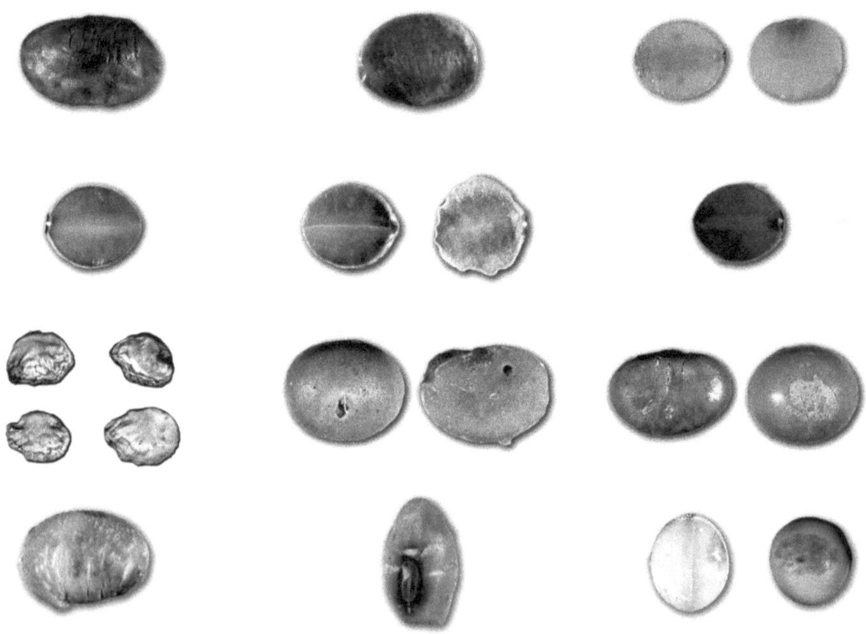

Fig. 6.1. Federal Grain Inspection Service interpretive line slides of various types of damage in soybeans (Source: http://www.gipsa.usda.gov/GIPSA/documents/GIPSA_Documents/soydamage.pdf).

have green, black, brown, or bicolored seedcoats. Bicolored soybeans have seed coats of two colors; one of the colors is either brown or black and must cover 50% of the seed-coat area. The hilum of the seed coat is not taken into consideration for determining bicolored soybeans (*U.S. Standards for Soybeans*, 1994).

Special Grades

Three special grades were established by FGIS in 1994 and are used to identify unusual conditions. The "special grade" designation is listed on the certificate of inspection but does not affect the "numerical grade" or the "sample grade" designation. The special grades are: (i) garlicky soybeans, (ii) infested soybeans, and (iii) purple mottled or stained soybeans.

Garlicky soybeans are soybeans that contain five or more green garlic bulblets or an equivalent quantity of dry or partly dry bulblets in a 1.0-kg portion. Infested soybeans are soybeans that are infested with live weevils or other insects injurious to stored grain. Purple mottled or stained soybeans are soybeans that are discolored with pink or purple seed coats, dirt or a dirt-like substance, or pokeberry stains, as determined on a portion of 400 g with the use of an FGIS Interpretive Line Slide or Print (FGIS, 2004).

FGIS Grading Steps for Soybeans (FGIS, 2004)

In the official determination of grades for soybeans, a certain step sequence must be followed to assure sample randomness and adequate sample size. The first step is to examine for heating, odor, animal filth, castor beans, crotalaria seeds, glass, stones, garlic, insect infestation, purple mottled or stained soybeans, unknown foreign substances, and other unusual conditions. Step two is to determine the moisture content. The third step is to measure test weight per bushel, if done for informational purposes. For step four the sample is divided into representative portions, and the soybean class, damaged kernels, heat-damaged kernels, foreign material, soybeans of other colors, and splits are determined (all of the tests in step four are determined after foreign material is removed). Step five is optional but, if requested, percentages of oil and protein on a portion with foreign material removed may be determined.

Summary of FGIS Inspections for Soybeans (FGIS, 2004)

A summary of FGIS GIPSA inspections for soybeans over the 1994 to 2004 calendar years for soybean crop years of 1993 to 2003 is shown in Table 6.2 (Brumm & Hurburgh, 2003; Brumm et al., 2005). On the average, about 91% of the U.S. exported soybeans is grade U.S. No. 2. Moisture contents average about 12.1%, which is 0.9 percentage points below the typical market moisture of 13.0%. Damaged kernels averaged 1.1% which was well below the 2.0% allowed for U.S. No. 1 grade. Foreign

Table 6.2. Summary of GIPSA Grain Inspection Data for Soybeans

	GIPSA Export Inspection Data						Iowa State Survey	
Year*	No. 2 YSB, %	Moisture, %	Foreign Material, %	Damaged Kernels, %	Protein, %	Oil, %	Protein, %	Oil, %
1994	90.3	12.6	1.7	1.1	35.5	18.4	35.5	18.1
1995	92.3	12.2	1.7	1.0	35.2	18.5	35.4	18.2
1996	92.2	12.1	1.7	1.1	35.1	18.5	35.5	18.0
1997	90.9	12.6	1.6	0.8	35.3	18.4	35.0	18.2
1998	90.0	12.2	1.6	1.0	35.5	18.8	35.3	18.1
1999	89.4	12.0	1.6	0.9	35.3	18.8	35.3	18.9
2000	90.0	11.4	1.7	1.0	35.0	18.5	35.4	18.6
2001	89.5	11.5	1.7	1.3	35.8	18.5	35.6	18.8
2002	93.1	12.1	1.5	1.4	35.2	18.8	35.5	19.0
2003	92.6	12.2	1.4	1.5	35.3	18.9	35.5	19.0
2004	94.1	12.1	1.5	1.4	35.2	18.8	35.2	18.7
2005							34.9	19.5
Averages	91.3	12.1	1.6	1.1	35.3	18.6	35.3	18.6

Source: USDA Grain Inspection Packers and Stockyards Administration and Iowa State University. Protein and oil content based on 13.0% moisture content.*Crop was grown in the previous year.
Sources: Brumm and Hurburgh, 2003; Brumm et al., 2005.

material contents averaged 1.6%, which was below the 2% allowed for U.S. No. 2 yellow soybeans. Foreign material is usually the grade factor that causes soybeans to shift to a lower grade. In 2002–2004 when foreign material was at its lowest average of 1.4 to 1.5%, the highest percentage of soybeans made No. 2 grade. While protein and oil are factors outside the grading standards, they averaged 35.3 and 18.6%, respectively, and were in exact agreement with survey results conducted by Brumm et al. (2005) at Iowa State University.

Moisture

Moisture Measurement Methods

Moisture is not a grade-determining factor, but moisture is determined on all soybean samples. Moisture content determines how much dry matter is bought or sold. It is also very important for assessing storability, and it affects end-use value. The official moisture meter for testing soybeans is the GAC 2100 as of August 1, 1998 (DICKEY-john Corp, Auburn, IL). This moisture meter uses approximately 400–450 g of sample. This instrument is usually updated annually on August 1 with a new set of K1–K9 constants provided from the Web site at: http://www.dickey-john.com/_dl/2-0563B_6.pdf.

Electronic moisture meters usually operate on a dielectric principle and/or kernel surface conductance with compensation for sample temperature and density. Thus, electronic moisture meters measure electrical properties that are calibrated to oven moisture measurements. The typical air-oven reference methods used for whole soybeans are: the AOCS Method Ac 2-41, ASABE Standard S352.2, and AACC Method 44-15a.

In the AOCS Method Ac 2-41 (1987), the soybean sample is first cleaned on a 3.2 mm (8/64-in) round-hole sieve. Then 10 g of whole soybeans are placed in a tared moisture dish and placed in a forced-draft air oven at 130 ± 3°C for 3 h.

ASABE Standard S352.2 (2006) provides an unground method of measuring moisture for soybeans. Here 15.0 g of soybeans are weighed into tared moisture dishes and are dried in an air oven at 103 ± 1°C for 72 h. Dried samples are placed into a desiccator to cool before weighing. Replicate determinations should check within 0.2% moisture.

Place samples in heavy-gauge aluminum moisture dishes approximately 55 mm diameter and 15 mm height. The moisture dishes should have slip covers and should be dried for 1 h at 103 or 130°C before using and then cooled in a desiccator. Place hot samples removed from the oven first in an airtight desiccator containing activated alumina or suitable desiccant. Make weighings to the nearest 0.1 mg after samples reach room temperature (usually 30–40 min).

The AACC Method 44-15a (1999) is applicable to soybeans but has an air oven method that uses 15 g of unground soybeans in a tared moisture dish. Then they are heated at 103 ± 1°C for 72 h. Replicate determinations should check within 0.2% moisture. This whole soybean method is the same as the ASABE Standard S352.2 (2006).

AACC Method 44-15a (1999) is the same as the ASABE Standard S352.2 (2006), but also describes a two-stage method for soybeans above 10% moisture and a one-stage method for soybeans <10% moisture. The AACC 44-15a method is a two-stage method that is used by the Federal Grain Inspection System for measuring soybean moisture content (Pierce, 2007). A tared moisture dish is filled with unground soybeans and weighed. The sample is allowed to sit above a heated oven with the dish cover off for 14–16 h to partially dry. The sample is weighed, and the weight loss is determined. When the sample is below or equal to 10% moisture content, it is placed into the one-stage process for drying. In the one-stage process, soybeans below or equal to 10% moisture are ground (30–40 g), and then a 2 to 3 g portion is placed into a tared moisture dish. The sample is placed in an air oven at 103 ± 1°C for 60 min (time starts after the oven regains temperature from being loaded; oven should regain temperature 15–20 min after loading). Replicate determinations should check within 0.2% moisture.

Hartwig and Hurburgh (2007) compared 20 samples of 1987 crop-year soybeans with three air-oven methods. They found the AOCS Ac 2-41 method and the USDA method had an average difference of 0.04 percentage points and a standard deviation of 0.18 percentage points and can be used interchangeably with the USDA method for calibration of moisture devices. On the average, the AACC Method 44-18 was 0.15 and 0.19 percentage points higher than the AOCS and USDA methods, respectively. The AACC Method 44-18 they used was a modified two-stage air-oven method that used ground samples below 13% moisture and an air oven at 135 ± 2°C for 2 h. This method was eliminated by AACC in 1995.

In addition to the air-oven methods for moisture measurement, there is also a Karl Fischer titration method. In this method, moisture is considered to be the moisture that is extracted by methyl alcohol in a sealed container with ground soybeans for a specified time period (Paulsen, 1991). This water–alcohol extract is placed in a reaction chamber where the water reacts with iodine. The amount of iodine used in titration is a measure of the amount of water in the soybeans. Soybean samples tested with the Karl Fischer method usually show about 0.2 percentage points lower moisture than those tested with the USDA air-oven method (Paulsen, 1991). The standard deviations of differences from duplicate samples were lower (0.07 percentage points) for the USDA air-oven method than for the Karl Fischer method (0.11 percentage points) (Paulsen, 1991).

Desired Moisture Contents

The desired moisture content for soybeans depends on how long they are stored before using, as well as their geographical location and climate conditions. For soybeans in the Midwest, the Midwest Plan Service (1988) recommends 14% as maximum moisture content if soybeans are to be sold by spring. If stored up to one year, they should be at 12% moisture, and if stored beyond 1 yr they should be at 11% moisture content. All of the moisture contents should be further reduced by 1 percentage point if the soybeans are not clean or of good quality.

One of the problems with these moisture recommendations is that market moisture content for soybeans is considered to be 13.0%. If soybeans are sold to an elevator or soybean processor at moistures >13.0%, the seller is charged for drying and for shrinkage until the soybeans are at exactly 13.0% moisture. If the soybeans are sold at any moisture content <13%, the producer is not paid for the extra dry matter delivered. Hence, the market signals that soybeans are wanted at precisely 13.0% moisture, and if a producer has to aerate or dry soybeans to 11 or 12% in order to store safely through the summer, they are faced with either losing the opportunity to sell full-moisture weight at soybean prices or having their soybeans become molded in the storage bins due to too high moisture content during storage.

The basic formula for fungal metabolism of starch converted to sugars from soybeans is $C_6H_{12}O_6 + 6\ O_2 = 6\ H_2O + 6\ CO_2 + 677.2$ Kcal of heat. Once conditions become favorable for fungal metabolism to occur, the by-products of the reaction are additional moisture and heat. This produced moisture further deteriorates storage conditions, and enables accelerated degradation with other fungal species, which can raise moistures and temperatures even higher. This fungal metabolism process causes mold damage and dry matter loss of starches/sugars. The metabolism process produces 14.7 g of CO_2 per kg of dry matter consumed, and, by all estimates, a loss of only 0.5% dry matter reduces grade (due to mold damage) by one numerical level.

Effects of Moisture

Frankel et al. (1987) stored 800 g of soybean seeds at rewetted moistures of 13, 16, and 20% for 19–50 days at 41°C. Soybeans at 13 and 16% initial moisture reached 41–48°C while the 20% moisture soybeans reached 47–49°C. They found free fatty acids (FFA) climbed from 0.20% to about 1.25% for the 13% moisture soybeans in 49 days. For the 16% moisture soybeans, FFA elevated from 0.5% to about 2.0% in about 27 days. For the 20% moisture soybeans, FFA went from 0.6 to 2.3% in about 28 days. Phosphorus in crude oil before degumming from the 13% moisture-stored soybeans dropped from 1044 ppm to about 400 ppm during the 50-day period. For the 16% moisture beans, it dropped faster from 850 ppm to 0 in 27 days; and for the 20% moisture beans, it dropped from 500 ppm to 0 in about 20 days. The reduction in phosphorus was explained by formation of nonhydratable phosphatides, which are made up of Ca and Mg salts of phosphatides (Frankel et al., 1987). They also scanned samples from 1100–2500 nm by using a computerized spectrophotometer, and when second derivatives were taken, they found absorption bands at 2260 nm (R = 0.86) and 1810 nm (R = –0.72) that were correlated to FFA.

Germination and Seed Vigor

The ability of soybeans to germinate and vigorously grow is an important quality trait for seed production but often for food and commercial use as well. When quality of a seed declines, germination is one of the first attributes that is lost. Factors that can reduce germinability and viability usually start in the field. Prior to harvest, disease and fungal invasion are leading contributors to reduced germination. Even imbibing moisture during rain periods close to harvest leading to swelling and later drying reduces germination. Frost damage and insect damage also contribute to reduced germination. Other contributors are mechanical damage due to impact in dropping and conveying equipment, which increase splits, seed-coat cracks, and possible impact damage to the embryonic region of the soybean.

Germination Tests

Germination is the ability of a seed to produce an embryonic root and stem within a specified period of time. The most common test is the warm germination test, which typically involves placing 100 whole seeds on moist paper tissue that is maintained in a chamber at air conditions of 25°C (77°F) and 95–100% relative humidity for 5–8 days. At the end of the standard warm germination, seedlings are inspected to have normal development of all essential seedling structures. This is the germination percentage reported as seed-labeling information required for all seed offered for sale (Beuerlein, 2005). For seed use, >90% warm germination is desired.

The cold germination test is also a very common test and more closely predicts the ability of seeds to germinate under adverse early-season field conditions. The cold test typically involves planting 100 whole seeds in an unsterilized soil–sand mixture maintained at 10°C (50°F) for 7 days, followed by conditions of the warm germination test for 4 days (Nave & Paulsen, 1979). For seed use, >80% cold germination is desired. Actual field germination normally is between the cold test and the warm-germination percentage.

Vigor Tests

Vigor relates to the ability of a seed to produce a healthy seedling quickly while under less than ideal conditions (Nafziger, 2002b). Soybeans may maintain high germinability for a long period of time, but seedling vigor starts to drop more quickly (Fig. 6.2). Hence, it is important to conduct vigor tests for seeds that were in storage for a period of time.

A commonly used vigor test is the accelerated aging test, where 100 whole kernels are exposed to 40°C (104°F) for 32 h, then to 10°C (50°F) for 7 days, followed by

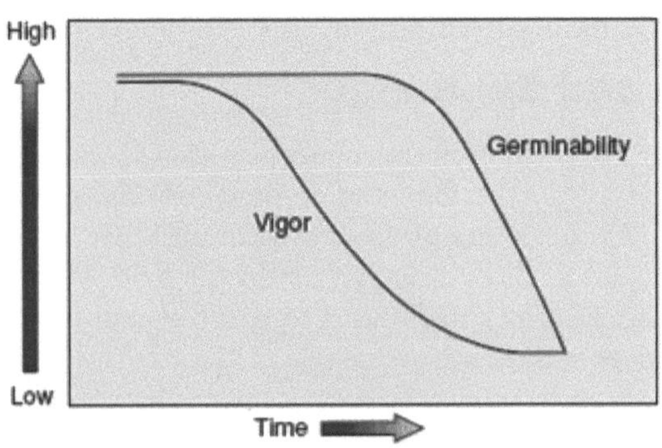

Fig. 6.2. Typical changes in soybean seed vigor relative to germinability during storage (Source: Nafziger, 2002b).

conditions of the warm-germination test for 6 days (Nave & Paulsen, 1979). This test accelerates the deterioration of seed vigor and helps to identify seed that is quickly declining in quality even though warm-germination percentages may still be holding up.

The tetrazolium test is another one for determining viability or vigor index. It consists of pre-moistening 100 whole seeds for 12 h in paper towels, and then soaking the seeds for 5 to 7 h in a 1.0% tetrazolium solution. In this test, healthy living tissue stains a pink color; damaged tissue stains a dark purple; and nonliving tissue remains white (AOSA, 2000; Delouche et al., 1962). The location and the relative amount of the nonliving tissue allow a seed analyst to sort the 100 seeds into various categories that would be nongerminable, weakly germinable, and highly germinable and vigorous. In particular, if tissues near the hypocotyl are nonliving, the seed is nongerminable.

The electrical conductivity is another test that is often used to detect loss in soybean vigor. Basically, the conductivity of a solution of water depends on the concentration of any dissolved salts or other chemicals that ionize the solution. The purer the water, the lower the conductivity: for example, electrical conductivity decreases if a sample changes from sea water, to drinking water, to deionized water. If a given mass of soybean seeds is placed in water, a certain amount of electrolytes will leach out through the seed coat into the solution. As more electrolytes leach into the water, conductivity increases, and it indicates a low-vigor seed lot. Since electrical conductivity changes with temperature, to maintain the solution at a constant temperature is important. Electrical conductivity is expressed in µmhos/g, where mhos are units of conductivity = to 1/ resistance in ohms. Yaja et al. (2005) reported that soybeans initially at 74 µmhos/g increased after 120 days of storage in Thailand. Soybeans were stored at 12% moisture at temperatures of 15, 20, 25, and 30°C, and at ambient room temperature, had electrical conductivities of 141, 143, 146, and 149 µmhos/g of seeds, respectively. Hence, storage time alone almost doubled the electrical conductivity, and it became progressively higher as storage temperatures increased. Warm germination for these samples started at 93% and after 120 days lessened to 76, 78, 75, 72, and 71% for soybeans stored at 12% moisture at temperatures of 15, 20, 25 and 30°C, and at ambient room conditions, respectively (Yaja et al., 2005). Thus, as electrical conductivity increases, germination decreases. Many seed producers have developed correlations of electrical conductivity tests for expected field emergence or vigor of soybean seeds based on their experiences.

Effects of Impact Damage and Flooding on Seed Vigor

A study by Wuebker et al. (2001) investigated the effect of impact damage on seed vigor with a rotating horizontal cylinder. Each revolution of the cylinder lifted 250 g of seeds with a horizontal paddle and allowed the seeds to drop within the cylinder. Impact damage was assessed visibly by an increase in seed-coat cracks and splits and by loss in germination. Rotation causing 1000 small impacts reduced germination of one variety from 85 to 72%. For the same variety, 2500 impacts reduced germination to 68%. Table 6.3 shows that damaged seed had a lower germination percentage than undamaged soybeans. It also shows the germination reduction was greater at 15°C than at 25°C. Evidently, conditions are more favorable for germination at 25°C, so damage at that temperature made less difference (Wuebker et al., 2001). However, for both temperatures, seedling dry weight was reduced for the damaged soybeans, and higher seedling dry weights resulted when germinating at 25°C rather than at 15°C.

One of the reasons soybean seeds may not germinate after planting is excess moisture after the initial imbibition has begun. Wuebker et al. (2001) looked at the effect of flooding for 1–4 days on damaged and undamaged soybeans at 15 and 25°C. In general, flooded seeds germinated better under the 25°C conditions than at 15°C for the first 3 days (Table 6.4). Again, damaged seeds had significantly lower germination than undamaged seeds.

Other Soybean Quality Factors

Other soybean quality factors broadly include physical factors, chemical factors, processing characteristics, and some environmental factors. Environmental factors were included because of interest in climate change and its likely affect on soybean production and quality factors.

Table 6.3. Effect of Impact Damage and Germination Temperature on Germination Percentages and Total Seedling Dry Weight Yield of Soybeans

Damage, Germination Temperature	Germination, %	Total Seedling Dry Weight Yield, g
Not damaged, 15 °C	45a*	0.24a
Not damaged, 25 °C	59b	0.52b
Damaged, 15 °C	37c	0.19c
Damaged, 25 °C	43a	0.38d

* Means with same letter within a column were not significantly different at $P = 0.05$ level. N = 324.
Source: Wuebker et al., 2001.

Table 6.4. Effect of Flooding Soybean Seeds After Imbibition and Impact Damage on Mean Germination Percentage After 1 to 4 Days

Damage	Control Not Flooded	Flooded
Not damaged	71a*	48a
Damaged	54b	37b

* Means with same letter within a column were not significantly different at P = 0.05 level. N = 324.
Source: Wuebker et al., 2001.

Physical Factors

Whole Soybeans and Seed Weight

Whole soybeans are soybean seeds that are not missing any parts of the seed and with seed coats fully intact. This can be contrasted with soybean splits: those soybeans that have more than one-fourth of the bean broken and removed but otherwise not damaged (FGIS, 2007). If a soybean was seven-eighths present, it would not be a split soybean, but it also could not be considered to be a whole soybean. Whole soybeans are important for seed germination determinations, for increased resistance to fungal invasion, for improved storage life, and for less oxidation of fats and other components.

Soybean seed weight is often expressed as mg per seed, g per 100 seeds, or seeds per 453.59 g (1 lb); however, the seeds being weighed should be based on whole soybeans. Thus, seed weight provides a very simple test for indicating relative size of soybeans assuming density is fairly constant.

Nafziger (2002a) stated that seed size is reduced due to high stress during the seed-fill period but that seed size has little effect on soybean emergence in the field or on final yield of soybeans. Even differences in plant height (less than 4 in) are too small to have significant effects on yield (Nafiziger, 2002a). He further adds that the size of seed produced from the same variety has the same genetic potential and is not affected by whether the parent seed was large or small. However, soybean planters must be calibrated to compensate for small seed sizes because if soybeans are planted excessively thick, the soybeans will be more prone to lodging in the field. Soybean varieties can be determinate (meaning vegetative growth on the main plant stem terminates when flowering starts) or indeterminate. The advantage of determinate varieties is to have good resistance to lodging in a high-yield environment. However, high-yield environments are not easily assured, and most soybeans in Illinois are now planted to indeterminate varieties (Nafiziger, 2002a).

Identity-preserved soybeans for food use are one of the primary markets where soybean size is important. Generally, tofu and soymilk processors prefer large soybeans (4,400–5,500 seeds/kg); natto producers prefer small soybeans (10,000–11,000 seeds/kg); and processors of miso prefer medium-sized soybeans that are uniform in size (Brumm, 2004). In soy products where minimal fiber is wanted, large soybeans have less seed coat (fiber) relative to total seed weight.

Mian et al. (1996) stated that in South Korea the average weight of black soybeans for food use with rice was 300–350 mg/seed (2857–3333 seeds/kg) while for sprouts it was smaller, about 120 mg/seed (8333 seeds/kg). They found between seven and nine independent marker loci were responsible for explaining 73 and 74% of the seed weight variation in two different populations of soybeans, respectively. The marker loci were very consistent across environments and across years. For the two populations, seed weight of high and low progeny ranged from 116–227 mg/seed (4405–8621 seeds/kg).

Matsue et al. (2005) examined growth and productivity of soybean plants produced from seeds of different seed weights that were stored for extended periods of time but had still maintained 80% or higher warm germination. This study provided valuable information for the practice of storing seed beans for very long periods of time at 5°C and 40% relative humidity. Soybean seeds were from crop years 1992, 1995, 2000, and 2002, and had been in storage for 10.6, 7.6, 2.6, and 0.6 years, respectively. These seeds had 100-seed weights of 30.1, 30.0, 29.7, and 36.0 g and warm germinations of 60.0, 97.0, 98.7, and 100.0%, respectively. Plants produced from these stored seeds had shortened main stems (29.0 cm) for the 1992 soybeans and 36.9, 37.2, and 42.9 cm for the newer seeds, respectively. Pods produced per plant were 51.1 for the 1992 soybeans and 89.3, 87.7, and 93.6 for the newer seeds, respectively. Seed yield for the 1992 soybeans was 179 g/m^2 of test plot and 262, 265, and 299 g/m^2 for the newer seeds, respectively. The protein content for soybeans produced from all years of seeds was 44.8 to 45.5% with no significant difference found between years. The 100-seed weight of soybeans produced was 28.9–29.5 g with no significant difference found between years. Their study concluded that soybeans stored up to 7.6 yr having higher than 80% germination produced plants having normal growth and yield productivity, but those stored 10.6 yr had poorer plant growth and seed yields.

Li and Burton (2002) stated that seed density is a part of grain yield that usually correlates positively with protein concentration (R = 0.06–0.71), and it usually correlates weakly with seed yield (R = –0.20 to +0.41). They found seed density correlated positively with protein for three different soybean populations, but the correlation was statistically significant for only two of the three populations. When they correlated seed weight to yield, they obtained significant positive correlations in two of three populations (R = 0.81, 0.96, and 0.19). Seed-weight correlations with protein were mixed and weak (R = –0.17, 0.36, and 0.16). Seed-weight correlations with oil were positive (R = 0.46, 0.03, and 0.13) but significant in only one of the three populations. They concluded that selection for a seed density and seed-weight index may be an inexpensive way to select for increased protein.

Seed-coat Cracks

The seed coat of a soybean is an often overlooked component. Yet the intact seed coat has vital functions. It can affect rate of moisture uptake or imbibition during germination, the vigor of a seedling, the rate of electrolyte leakage during storage, and general ability to store with some initial protection from insects and fungal invasion. Seed-coat cracks are simply fissures or openings in the protective covering that surrounds soybean cotyledons. Seed-coat cracks in soybeans are undesirable because they can also lead to increased chance for splitting and breakage during handling. Seed-coat cracks are also greatly affected by heated-air drying; whenever drying air falls below 40% relative humidity, expect increases in seed-coat cracks.

Several methods exist for determining seed-coat crack percentages, including a soak test where soybeans are immersed in a 0.1% solution of sodium hypochlorite for 5 min. Soybeans with seed-coat cracks or openings quickly absorb the sodium hypochlorite solution and swell to two to three times their original size (Rodda et al., 1973).

Paulsen and Nave (1979) developed a seed-coat crack detection procedure using indoxyl acetate. A 0.1% indoxyl acetate solution was prepared. Then 100 whole soybeans were placed in a wire-mesh basket and immersed for 10 sec in the indoxyl acetate solution. Soybeans were removed and immediately sprayed with a 20% household ammonia-distilled-water solution for 10 sec using an atomizer or very fine mist. Next, soybeans were air-dried with an air blower with no heat added. Areas of soybeans with seed-coat cracks, scratches, abrasions, or imperfections in the seed coat stain a blue-green color. Samples were divided into three categories: those with seed-coat abrasions, those with small holes in the seed coat, and those with seed-coat cracks. Paulsen and Nave (1979) found that the indoxyl acetate test detected minute seed-coat damage, which was not easily seen by visual inspection and was effective in detecting higher levels of seed-coat damage than the soak test. Also, the indoxyl acetate test does not have any detrimental effect on warm germination percentages.

Color and Morphological Properties

Color and morphological properties of soybean seeds are affected by fungi, fungal damage, and immaturity. Frost damage may affect immaturity. Hobbs et al. (2003) studied infections of soybean plants by soybean mosaic virus and bean-pod mottle virus. They stated the color of mottling is determined by genes that determine the color of the hilum and that in food-grade soybeans a dark pigment on the seed coat reduces consumer acceptance, particularly for natto soybeans where a uniform yellow color is desired. Casady et al. (1992) and Ahmad et al. (1999) used machine vision for detection of soybeans with fungal damage due to *Alternaria* spp., *Cercospora* spp., *Fusarium* spp., *Phomopsis* spp., and soybean mosaic virus. Ahmad et al. (1999) reported that soybean seeds affected by *Alternaria* spp. are usually small, shriveled, with light- to dark-brown coloring on parts of the seeds. Those with *Cercospora* spp. had colors ranging from light to dark brown, purple, or gray. *Fusarium* spp. caused a pink to red coloration; while *Phomopsis* spp.-affected seeds are shriveled and elongated with cracked or chalky seed-coat appearance. Soybean mosaic virus and bean-pod mottle virus cause yellow soybeans to have a mottled brown or black coloring that varies with hilum color. The mottling causes an appearance of bleeding of the hilum color over the seed coat, yet the soybean seed coat stays very smooth and round about the soybean. Immature soybeans vary in color, from whitish green to dark green, and often occur as a result of frost on younger soybean pods. The Casady et al. (1992) study used chromaticity coordinates, and they found that seeds infected by *Cercospora kikuchii* were found with 97% accuracy, asymptomatic or healthy seeds were identified with 94% accuracy, and the collective group of *Alternaria* spp., *Fusarium* spp., and *Phomopsis* spp. combined was identified with 85% accuracy. The Ahmad et al. (1999) study used a red, green, blue (RGB) color feature multivariate decision model to distinguish between asymptomatic and symptomatic seeds. A linear discriminant analysis was used and obtained 88% classification accuracy between asymptomatic and symptomatic soybeans. Individual classification accuracies were: *Alternaria* spp. 30%; *Cercospora* spp. 83%; *Fusarium* spp. 62%; *Phomopsis* spp. 45%; soybean mosaic virus (black) 81%; soybean mosaic virus (brown) 87%; green seeds 91%; and asymptomatic seeds 97%.

One of the factors affecting color and immaturity is frost damage. Soybeans that have not reached maturity and are frost-damaged will be green or elongated yellow color and will shrink considerably when dried (Hurburgh et al., 2007). Pritchard (1983) reported that frost damage affects soybean oil content, color, and refining loss of the oil. The oil from frost-damaged soybeans is hard to extract, of low quality, and often <16% of the bean (Hurburgh et al., 2007). Frost damage and also damage due to poor storage conditions cause degradation of hydratable phosphatides which leads to high oil-refining losses (Pritchard, 1983). Elevated levels of FFA occur when soybeans are stored above 13.0% moisture, and storage fungi produce hydrolytic lipases. Similarly, prolonged field exposure to wet conditions also activates lipase enzyme in the seed and leads to higher oil-refining losses.

Urbanski et al. (1980) froze immature soybean pods at –5.5°C for 6 h. Freeze-damaged soybeans had similar oil and protein contents and trypsin inhibitor activities as the unfrozen soybeans. Frozen soybeans had greener color and lower lipoxygenase activity. In addition, oil from freeze-damaged soybeans contained 0.26% FFA, which rose to 1.63% after 14 mo of storage. Oil from undamaged soybeans contained 0.14% FFA, which increased to only 0.48% after storage. Sternberg et al. (1990) reported that freeze damage occurs at about –3.9°C (25°F), causing plant maturation to stop and eventually leading to plant death. Frost damage occurs at –2.2 to –1.7°C (28–29°F) when only the leaves are killed and seeds can actually continue to mature. These soybean seeds tend to lose most green color except for a slight tinge of green on some seed coats. Mounts et al. (1990) reported that color was an important factor for the crude soybean oil and that a predominant green color in the crude oil is undesirable. Using Lab values on a Hunter colorimeter utilizing AOCS Method Cc 13b-45 (Wang & Johnson, 2001) can measure the color of soybean oil.

Hilum Color

Hilum color is also an important consideration for many buyers of soybeans to be used for food uses. Many buyers prefer a light-colored (yellow or clear) hilum as opposed to a buff, brown, or dark hilum. Clear hilums are usually desired for soymilk and tofu production. Soybeans that are small with clear hilums and thin seed coats are preferred for natto beans. Hilum color is determined by the genetics of the soybean variety and determined by visual inspection. For the hilum test, usually 125 g of whole soybeans with foreign material removed are inspected visually.

Acoustical Properties

Acoustical properties of soybeans can be used to help distinguish between healthy and diseased soybeans. Misra et al. (1990) measured acoustic properties of soybeans by transmitting sound waves through soybeans using acoustic transmission and by an impact force method. In the impact force method, a seed is dropped on an acoustic transducer creating an impulse wave. The acoustic transmission method was slow but was able to predict the mass of individual soybeans. The impact force method showed that diseased soybeans had a narrower bandwidth than healthy soybeans. Soybeans with wrinkled surfaces and diseased and damaged soybeans were detected from healthy soybeans based on wide variations at low frequencies.

Chemical Factors

This section on chemical factors related to quality in soybeans is divided into subparts on protein and oil, fatty acids, amino acids, tests for protein, carbohydrates and sugars, and other factors that are often discussed, particularly as soybeans are enhanced for more specific end uses. The other factors include tests for fiber, phosphorus, tocopherols, and isoflavones. The intent is to discuss the importance of these factors, to provide background, and, because of increased use of near-infrared spectroscopy as a measurement method for whole and ground soybeans, to include that technology in the discussion of test measurements. While many primary and other methods for measuring these chemical factors are available, this chapter does not intend to cover those methods.

Protein and Oil Contents

Naeve and Orf (2006) conducted a soybean measurement survey of 1593 samples for the American Soybean Association and found that in 2006 average protein of soybeans grown in the United States was 34.51% (13% moisture basis) and average oil was 19.17% (13% moisture basis) (Table 6.5).

As a general rule, southern regions of the United States tend to produce soybeans with higher oil contents than more northern regions. This is usually offset by northern regions producing slightly higher protein contents than southern regions. However, in years where yields are particularly high, protein (units of nitrogen) must be divided over more tonnes of soybeans per ha, causing percentages to be proportionately lower. Protein content in 2006 was 0.83 percentage points below the average for the past 20 years; while oil in 2006 was 0.5 percentage points above the average for the past 20 yr (Table 6.5). As a general rule, if protein content increases by one percentage point,

Table 6.5. Protein and Oil Measurements by Nir of 2006 Crop Year Soybean Samples Regions of the United States

U.S. Region and States	Number of Samples	Average Protein, %*	Range of Protein, %	Average Oil, %*	Range of Oil, %
Western–IA, KS, MN, MO, NE, ND, SD	990	34.27	27.8–41.4	19.12	16.0–22.5
Eastern–IL, IN, MI, OH, WI	452	34.68	28.1–38.8	19.00	16.0–22.2
Midsouth–AK, KY, LA, MS, OK, TN, TX	109	34.85	30.4–39.3	19.90	17.5–22.7
Southeast–AL, FL, GA, NC, SC	14	35.46	32.3–37.8	20.12	18.2–21.1
East coast–DE, MD, NJ, NY, PA, VA	28	34.89	31.6–39.1	19.08	16.0–22.7
All samples above for U.S.	1593	34.26	27.9–41.4	19.23	16.0–22.7
All U.S. using weighted crop averages for 2006		34.51		19.17	
U.S. averages for 1986–2006		35.34		18.67	

* Protein and oil percentages are expressed at 13% moisture content based on whole kernel Perten DA 7200 diode array near-infrared measurements.
Source: Naeve and Orf, 2006.

oil content will decrease by one-half percentage point; but exceptions occur, and in those cases likely carbohydrates also changed. High yields in 2006 may partially explain lower protein content in 2006. Average yields were about 2.89 MT/ha, which is equal to the 2005 yield and at that time was the highest soybean yield in U.S. history (Naeve & Orf, 2006).

Brumm and Hurburgh (2006) stated that many processors are beginning to offer premiums for soybeans with higher levels of protein and oil contents to the extent that they can be identified based on geographical growing region. Since the sum of the total constituents can be no more than 100%, this statement implies that to be higher in both protein and oil contents, the soybeans have to be lower in another constituent, presumably carbohydrates. To be of high enough value for premiums to be offered also implies that the mix of desirable amino acids and fatty acids was not altered detrimentally by soybean plants that yield more oil and more protein.

Brumm and Hurburgh (2006) discussed three major soybean production changes that occurred over the past 10 yr. Those changes are: (i) more of the western Corn Belt states are producing more soybeans; (ii) RoundUp Ready® soybeans have become widely available; and (iii) average soybean yields have steadily gone up at an average annual rate of about 0.4 bu/acre. Their study looked at whole kernel NIR data for protein and oil composition in soybeans (adjusted to 13.0% moisture) for the years of 1986 through 2004. Based on 10,240 samples from 1986 to 2003, they found average protein content of 35.5% and average oil content of 18.6%; and for 16,890 samples from 1994 to 2004 they found average protein content of 35.4% and average oil content of 18.6%, very close to U.S. averages for 1986–2006 listed in Table 6.5. Thus, as yields rose, protein content decreased by 0.1 percentage point, and oil content was unchanged. Generally, as yields increase, available nitrogen is divided over more units, and therefore, protein content decreases. Brumm and Hurburgh (2006) concluded that increased planting of RoundUp Ready® soybeans was not causing the change in protein content. They concluded that protein content in the western Corn Belt states tended to be about 1 percentage point lower than that in other growing regions. Brumm and Hurburgh (2006) also found higher oil contents (+0.3 percentage points) in soybeans grown in Southern states as compared to more northern states.

Fatty Acids

Soybean oil is made up of five predominant fatty acids: palmitic, stearic, oleic, linoleic, and linolenic acids. Fatty acids in oil are characterized by number of carbon atoms (usually 16 or 18) and the number of double bonds between carbon atoms (ranging from 0, 1, 2, or 3) (Table 6.6) (Brumm, 2004). A fatty acid with no double bonds is called saturated. A fatty acid with only one double bond is monounsaturated, and fatty acids with two or more double bonds are polyunsaturated. Fatty acids are expressed as a percentage of the oil, and the sum of the fatty acids percentages should equal 100.

The saturated fatty acids are palmitic and stearic. Soybeans with lower palmitic acid may have cardiovascular benefits to consumers (Cherrak et al., 2003). Linolenic acid is oxidatively unstable, and soybean oil is often partially hydrogenated to reduce linolenic levels and to reduce rancidity (Naeve & Orf, 2006). Thus, soybeans with low linolenic acid do not require hydrogenation, which produces *trans* isomers of the fatty acids. Since January 2006, the U.S. government has required manufacturers of packaged food to label *trans*-fat content because of its adverse effect on human health. Several soybean seed companies have developed and marketed soybean varieties that are naturally low in linolenic fatty acid, about 1–3% instead of the normal 8%. Oil from these soybeans can be used without partial hydrogenation; therefore, it can be used to make products that are labeled "*trans*-fat free" (Naeve & Orf, 2006). Food servings labeled as partially hydrogenated can contain up to 0.5 g of *trans* fat and still be considered *trans*-fat free (Haynes, 2007). In 2006, about 300,000 ha of low-linolenic soybeans were produced in the United States, which, although less than 1% of production, is a growing trend (Naeve & Orf, 2006). A typical level of linolenic acid in crude oil from commodity soybeans is about 7.4%, while that from low-linolenic soybeans is about 3.1% (Table 6.6).

Oleic acid is a monounsaturated fatty acid, and for foods with high levels of oleic acid, the need is reduced for hydrogenation, resulting in foods with less *trans* fat. Major efforts are underway by seed companies to develop soybean lines higher in oleic acid. A typical level of oleic acid in crude oil from commodity soybeans is about 25%, while that from high-oleic soybeans is about 79% (Table 6.6).

Wang and Johnson (2001) processed normal soybeans and several types of genetically enhanced soybeans with four extraction methods. The genetically enhanced soybeans were high-oleic with 79.2% oleic acid, low saturated fat with 8.4% total saturated fatty acids, low-linolenic with 3.1% linolenic acid, lipoxygenase-free, and an experimental line of high-cysteine soybeans. The fatty acid compositions of some of these soybeans are compared to oil from commodity soybeans (Table 6.6). Comparisons of fatty acids of soybeans to corn oil, canola, and lard are also given in Table 6.6 (Liu, 1999).

Protein, oil, and fiber percentages for oil from high-oleic, low-saturated fat, low-linolenic, lipoxygenase-free, high-cysteine soybeans, and commodity soybeans were determined by Wang and Johnson (2001). Table 6.7 shows that low-linolenic soybeans have much higher oil content (20.2 %) than other soybean types. Fiber percentages were lower for low-saturated fatty acid soybeans than for the other soybean types.

Kovalenko et al. (2006b) tested 1400 whole soybean samples from eight crop

Table 6.6. Fatty Acid Compositions in Percent for Oil From Commodity, Four Types of Genetically Modified Soybeans, and Other Oil Sources

Oil	Palmitic acid 16:0	Stearic Acid 18:0	Oleic Acid 18:1	Linoleic Acid 18:2	Linolenic Acid 18:3
Commodity soybean oil	10.82	4.89	25.21	51.61	7.47
Lipoxygenase-free oil	10.15	4.60	33.14	45.42	6.68
High-oleic	6.72	3.80	79.22	7.15	3.12
Low-saturated fatty acid	4.61	3.82	22.43	62.02	7.12
Low-linolenic acid	10.74	4.55	25.03	56.60	3.07
Canola oil	3.9	1.9	64.1	18.7	9.2
Corn oil	12.2	2.2	27.5	57.0	0.9
Lard	24.8	12.3	45.1	9.9	0.1

Sources: Wang and Johnson, 2001; Liu, 1999.

Table 6.7. Soybean Seed Composition in Percentage at 13.0% Moisture Content as Measured by Near-Infrared Spectroscopy* for Oil from Commodity and Six Types of Genetically Modified Soybeans

Oil	Protein, %	Oil, %	Fiber, %
Commodity soybean oil	34.7	18.1	5.3
Lipoxygenase-free oil	39.8	17.2	5.0
High-oleic	37.8	18.5	5.5
Low-saturated fat	36.7	17.8	4.7
Low-linolenic acid	35.5	20.2	5.0
Low-stachyose	38.0	16.8	5.2
High-cystine	38.4	17.1	5.1

* Denotes Foss Infratec 1229 whole kernel spectrometer was used.
Source: Wang and Johnson, 2001.

years between 1991 and 2003 with Foss Infratec 1225, 1229, and 1241 spectrometers for fatty acids. They thinned samples to more uniformly distribute samples over individual fatty acid ranges; samples with abnormally high or low variations in second derivatives were eliminated as outliers. Sample numbers then ranged from 616 to 976. One-fourth of these samples was randomly placed in a validation data set. They used partial least squares (PLS), artificial neural networks (ANN), and support vector machines (SVM) software for calibrations. Their calibration for the sum of palmitic and stearic acids, called total saturates, obtained highest R^2 values (0.91, 0.92, and 0.94) for the three chemometric methods, respectively. Palmitic acid (ratio of standard deviation of lab values to standard error of prediction, RPD of 2.4) was predicted with less error than stearic acid (RPD = 1.8). Oleic acid was predicted with R^2 of 0.76–0.81 and with RPDs of 2.1–2.3. Linoleic acid was predicted with R^2 of 0.73–0.76 and RPDs of 1.9–2.0. Linolenic acid was predicted with R^2 of 0.67–0.74 and RPDs of 1.7– 2.0. Since RPDs close to 1.0 are considered to provide negligible information and RPDs of 3 to 5 would be more ideal, those above 2 may be useful, and those above 1.8 may still provide some useful screening information.

Nimaiyar et al. (2004) also found stearic acid to be the most difficult fatty acid to predict using Fourier-Transform near-infrared spectroscopy over the 833 to 2500 nm range. Stearic acid had the lowest R value and lowest RPD value of all five fatty acids tested.

Amino Acids

Much work has been done on the measurement of crude protein in soybeans and in breeding efforts to raise protein levels, yet the balance of amino acids, which make up the protein, has the greatest effect on feed and food value for the products made from soybeans. About 18 amino acids exist, which are usually reported when laboratory analysis for amino acids is performed. The official laboratory method that is commonly used for determining amino acids in soybeans is AOAC 982.30 E (a,b,c) Ch. 45.3.05 (AOAC, 1990). Of primary importance are the limiting amino acids, which in soybeans are usually considered to be lysine, methionine, threonine, tryptophan, and cystine (Williams et al., 1984). Liu (1999) reported that soybeans like other leguminous crops are low in sulfur-containing amino acids. Methionine is the most limiting sulfur-containing amino acid, followed by cysteine and threonine. But soy protein is high in lysine, which is low in most cereal crops, so a mixture of soy protein with cereal grains helps to balance out lysine and methionine.

One observation is that many individual amino acids increase as crude protein content increases. As is apparent from Table 6.8, as the protein content of soybean meal goes up (in this case from 44 to 48%), in all cases the percentage of each amino acid contained goes up, but not necessarily in the exact ratio of the protein increase.

Kovalenko et al. (2006a) also found amino acids to be highly correlated to the crude protein percentage. Zarkadas et al. (1993) and Liu (1999) reported that: (i) glutamic acid is the most highly abundant amino acid in soybeans; (ii) two acidic amino acids, glutamic acid and aspartic acid, make up about one-fourth of the amino acids present; (iii) the basic amino acids, lysine, arginine, and histidine, make up about one-fifth of the amino acids; (iv) five amino acids with hydrophobic side chains, glycine, alanine, valine, leucine, and isoleucine, comprise about 19– 20% of the total protein; (v) the aromatic amino acids, phenylalanine, tyrosine, and tryptophan, are about 9–10% of the total protein; and (vi) proline comprises about 5% of the total protein.

Table 6.8. Essential Amino Acids and Typical Amino Acid Content (in Percent) for 44 and 48% Soybean Meal

Amino Acid	Soybean Meal 44% Protein	Soybean Meal 48% Protein
Arginine	3.4	3.6
Histidine	1.1	1.3
Isoleucine	2.5	2.6
Leucine	3.4	3.8
Lysine	2.9	3.0
Methionine	0.65	0.70
Cystine	0.67	0.71
Phenylalanine	2.2	2.7
Threonine	1.7	2.0
Tryptophan	0.6	0.7

Source: Dale and Batal, 2003.

Nimaiyar (2004) used a Perkin-Elmer Spectrum One Fourier-Transform NIR (FT-NIR) spectrometer to scan ground soybean samples in reflectance mode. She had a calibration set of 74 samples and a validation set of 24 samples. Of 15 amino acids, she was able to best predict (% of dry weight) aspartic acid, serine, glutamic acid, and alanine. Threonine and tyrosine could not be predicted.

As seen in the Bajjalieh (2006) data set of Table 6.9, threonine and tyrosine were also difficult to predict with low RPDs of 1.54 and 1.24, respectively. Table 6.9 shows a summary of the minimum, maximum, and means for the Bajjalieh (2006) data set, which is compared to the Bluebook (2005) means for mature raw soybean seeds in g/100 g. The Bajjalieh (2006) data set consisted of multiple soybean crop-years with about 564 samples. The columns under NIR calibration were derived from the calibration file of the Bajjalieh (2006) data set. The calibration was not optimized, and it likely under- estimates how well the calibration can really do. However, the RPDs ranged from 1.24 to 6.79. Ideally, RPDs above 2 and higher are desirable. Crude protein is a very solid calibration with an RPD of 6.79. Technically, the RPD should be determined from the Standard Error of Prediction (SEP) and not from the SEC, since no validation is set, no SEP exists. Experience usually shows SEP values from a validation set are normally slightly higher than the SEC values. Thus, the RPDs shown in Table 6.9 are probably slightly better than they would be if obtained from a validation set.

Some of the early work on amino acid NIR calibrations was performed by Pazdernik et al. (1997) (Table 6.10). They developed NIR equations to measure 17 amino acids in ground and whole soybean samples on a NIRSystems 6500. They had 90 samples in their calibration set and 26 in the validation set derived from 408 soybean lines. Samples were grown in Minnesota in 1994 and 1995. Samples were analyzed first as whole beans and then ground in a Fred Stein mill for 1 min. Results are reported in grams per kilograms of crude protein.

Table 6.10 indicates that in all cases the ground soybeans were predicted with higher RPDs than the whole soybeans. However, the RPDs ranged from 1 to 1.79 for whole seeds and from 1.09 to 1.9 for ground soybeans. RPDs were calculated from the R2 value using the formula by Funk (2004):

$$RPD = 1/\ (1-R^2)^{0.5}$$

Table 6.11 shows updated work on amino acid calibrations at the University of Minnesota, noting that RPDs are higher than those derived from the Pazdernik (1997) publication. In the University of Minnesota work, amino acids were expressed as a percentage of total protein, whereas, in the other studies, amino acids were expressed as a percentage of total soybean weight, dry basis.

Kovalenko et al. (2006a) determined amino acid composition (% of total weight on a dry basis) in whole soybeans using a FOSS Infratec 1241 (FOSS North America,

Table 6.9. Summary of Amino Acids With Initial Nir Calibration Data Set Results[1]

Amino Acid	NIR Calculation (from Bajjalieh data set)					NIR Calibration			Bluebook Means, g/100g dry wt.
	Min	Max	Mean	Std Dev	N	R^2	SEC	RPD*	
Aspartic Acid	3.47	6.48	4.82	0.458	564	0.87	0.167	2.73	4.59
Threonine	1.26	1.96	1.58	0.096	564	0.58	0.057	1.54	1.59
Serine	1.38	2.90	2.08	0.235	564	0.87	0.073	2.71	2.12
Glutamic Acid	5.08	9.97	7.50	0.734	564	0.88	0.257	2.83	7.07
Alanine	1.41	2.57	1.77	0.132	564	0.78	0.057	2.12	1.72
Glycine	1.34	2.22	1.77	0.136	564	0.80	0.060	2.23	1.69
Cystine	0.43	1.13	0.75	0.155	141	0.94	0.036	4.22	0.59
Valine	1.37	2.50	1.96	0.195	564	0.71	0.105	1.85	1.82
Methionine	0.42	1.13	0.66	0.122	141	0.89	0.030	2.97	0.49
Isoleucine	1.39	2.37	1.86	0.173	564	0.69	0.095	1.81	1.77
Leucine	0.30	4.10	3.16	0.306	564	0.81	0.111	2.28	2.97
Tryptophan	1.10	1.86	1.38	0.109	564	0.70	0.058	1.82	0.53
Phenylalanine	1.54	2.78	2.07	0.178	564	0.81	0.076	2.27	1.91
Histidine	0.84	1.47	1.09	0.101	564	0.80	0.044	2.26	0.98
Lysine	2.07	3.28	2.62	0.186	564	0.80	0.081	2.25	2.43
Arginine	2.18	5.03	3.08	0.418	564	0.86	0.145	2.64	2.83
Proline	1.53	2.89	2.08	0.184	564	0.82	0.075	2.36	2.14
Tyrosine	0.37	0.65	0.52	0.056	70	0.35	0.045	1.24	1.38
Crude Protein	30.70	54.79	41.85	3.447	564	0.98	0.492	6.79	36.50

* Calculated RPD from $RPD = 1/(1-R^2)^{0.5}$ (Source: Funk, 2004)

[1] Project funding for this investigation was provided by the United Soybean Board (USB). Dr. Nick Bajjalieh was the Principle Investigator with contributions to calibration development and data assembly, analysis and release made by Caltest, LLC, Ballston Lake, NY.

Source: Bajjalieh, 2006.

Eden Prairie, MN), DICKEY-john OmegAnalyzerG (DICKEY-john Corp, Auburn, IL), Perten DA 7200 (Perten Instruments Inc., Springfield, IL), Bruker Optics/ Cognis QTA (Brucker Optics Inc., Billerica, MA), and an ASD LabSpec Pro (Analytical Spectral Devices Inc., Boulder, CO) for 18 amino acids. Partial least squares (PLS) and support vector machines (SVM) regression models performed significantly better than artificial neural networks (ANN). They used a calibration data set of 526 samples

Table 6.10. Amino Acids for 116 Soybean Samples (G/ Kg Crude Protein at 13.0% Moisture Content); Calibrations for 26 Ground and Whole Soybean Samples Using a Nirsystems 6500 Spectrophotometer

Amino Acid	HPLC Laboratory Data				Ground Soybeans			Whole Soybeans		
	Mean	Min.	Max.	Std. Dev.	SEPC	R^2	RPD*	SEPC	R^2	RPD*
Asparagine	98.6	76.4	117.1	11.7	7.1	0.65	1.32	10.50	0.54	1.19
Glutamine	193.4	162.1	217.8	17.8	10.2	0.72	1.44	17.80	0.49	1.15
Serine	54.5	44.1	69.4	6.1	2.9	0.61	1.26	4.60	0.39	1.09
Threonine	42.0	34.5	46.7	3.1	2.6	0.70	1.40	3.50	0.53	1.18
Glycine	44.1	26.4	59.7	7.5	2.1	0.63	1.29	3.10	0.52	1.17
Alanine	43.2	35.2	48.6	5.5	2.8	0.58	1.23	4.90	0.16	1.01
Proline	53.6	43.3	69.1	0.2	3.1	0.64	1.30	4.40	0.46	1.13
Valine	47.7	40.2	52.2	3.6	2.1	0.78	1.60	2.90	0.50	1.15
Arginine	81.3	70.4	96.7	6.7	2.7	0.85	1.90	3.70	0.67	1.35
Methionine	13.0	10.5	20.8	4.4	1.4	0.59	1.24	2.70	0.48	1.14
Isoleucine	49.4	41.9	58.2	5.8	1.6	0.84	1.84	2.60	0.83	1.79
Leucine	71.7	61.3	85.7	6.3	2.6	0.80	1.67	5.10	0.46	1.13
Phenylalanine	53.2	48.8	62.6	4.6	2.5	0.64	1.30	4.50	0.43	1.11
Cystine	19.8	11.2	25.7	3.8	1.7	0.61	1.26	3.20	0.49	1.15
Lysine	62.0	55.1	70.4	4.7	1.8	0.68	1.36	3.10	0.52	1.17
Histidine	33.8	29.8	39.1	2.1	4.1	0.40	1.09	5.90	0.06	1.00
Tyrosine	38.5	31.6	51.2	4.9	1.8	0.68	1.36	4.90	0.42	1.10

* Calculated RPD from RPD = $1/(1-R^2)^{0.5}$ (from Funk, 2004). Source: Pazdernik, 1997.

from 1997 to 2001 crop years and a validation set of 147 samples from 2002 crop year. They used multiplicative scatter correction and differentiation (either first or second derivatives depending on machine). They stated that their calibrations for cysteine and tryptophan had particularly low RPDs and that cysteine and tryptophan did not correlate with spectral information. One of their key findings was that 16 of 18 amino acids tested (excludes cysteine and tryptophan) had R^2 correlations with crude protein ranging from 0.53 to 0.91. Cysteine and tryptophan were 0.37 and 0.20, respectively. Crude protein alone provides a very strong absorbance for infrared energy; hence, NIR can predict crude protein very well (RPD = 6.79 in Table 6.9). This fact means that amino acids having a very high correlation to crude protein may in fact just be measuring a response to crude protein. The applicability of NIR for measuring amino acids as a percent of the protein in soybeans appears to have usefulness for some amino acids as evidenced by relatively high R^2 values or RPD values of 2 or higher (Table 6.11).

Table 6.11. Prediction of Amino Acids (% of protein) at University of Minnesota Using NIRSystems 6500.

Constituent	N	Mean	Min	Max	SEC	R^2	RPD*
Aspartic Acid	554	11.53	5.88	17.17	0.55	0.92	2.49
Serine	562	5.86	1.99	9.73	0.37	0.92	2.49
Threonine	560	4.38	2.38	6.39	0.26	0.85	1.92
Proline	564	5.58	3.70	7.46	0.30	0.76	1.55
Arginine	538	7.78	6.34	9.22	0.30	0.62	1.27
Methionine	334	1.70	0.00	3.79	0.24	0.88	2.09
Isoleucine	571	4.86	2.66	7.06	0.29	0.84	1.86
Histidine	538	3.04	1.58	4.50	0.28	0.68	1.36
Tryptophan	67	1.02	0.56	1.48	0.15	0.10	1.00
Cystine	308	1.26	0.00	3.97	0.20	0.95	3.30
Glutamic Acid	551	16.84	4.16	29.52	0.90	0.95	3.35
Glycine	545	5.58	0.00	11.90	0.49	0.95	3.11
Alanine	539	4.80	2.56	7.04	0.20	0.93	2.64
Valine	553	5.52	2.37	8.67	0.26	0.94	2.86
Leucine	557	7.24	2.95	11.54	0.41	0.92	2.52
Phenylalanine	531	5.13	3.32	6.94	0.33	0.70	1.39
Lysine	548	6.04	2.57	9.51	0.39	0.89	2.15
Tyrosine	548	3.90	1.72	6.07	0.29	0.85	1.87
Protein	312	41.44	31.74	51.14	0.52	0.97	4.47

* Calculated RPD from RPD = $1/(1-R^2)^{0.5}$ from Funk (2004). (Source: FSMIP, 2005)

Tests for Protein

2S, 7S, and 11S Proteins

Classification of soy proteins are rather complex. Two of the proteins from soybeans are commonly known as glycinin and β-conglycinin, which are believed to have originally been derived from the genus name of the soybean plant, *Glycine* (Liu, 1999). Other proteins have an enzymatic function that includes hemagglutinin, trypsin inhibitors, and lipoxygenases (Liu, 1999). A method of identifying proteins is based on sedimentation coefficients when ultracentrifugation is used to separate seed proteins. With the proper buffer conditions, soy proteins fall into four fractions after ultracentrifugation, known as 2*S*, 7*S*, 11*S*, and 15*S* (Liu, 1999). The letter *S* is named after a Svedburg unit and is determined by:

$$S = (dc/dt)\ w^2 c$$

where S = Svedburg units, with a unit of 10^{-13} sec (can range from 1 to 200); *c* = distance to the center of the centrifuge; *t* = time in sec; *w* = angular velocity.

Liu (1999) stated that 11*S* and 15*S* are pure proteins. The 11*S* fraction is glycinin and accounts for about 33% of the extractable protein. The 15*S* fraction is believed to be a polymer of glycinin, which accounts for about 10% of the extractable protein (Liu, 1999). The 2*S* fraction contains the Kunitz and Bowman-Birk trypsin inhibitors and cyctochrome C; while the 7*S* fraction contains conglycinin, α-amylase, lipoxygenase, and hemagglutinin (Liu, 1999). Soy food processors are interested mainly in the 7*S* fraction that is mostly β-conglycinin and the 11*S* fraction that is mostly glycinin. Glycinin is believed to be high quality compared to β-conglycinin because it contains a large number of sulfur-containing amino acids (methionine and cysteine) (Brumm, 2004).

Lipoxygenase is an enzyme found in the 7*S* fraction of soy protein. It is responsible for off-flavors in soyfoods. This enzyme catalyzes the oxygenation or breakdown of lipids contained in the soyfood. Soybean varieties were developed that do not have one, two, or all three of the lipoxygenase enzymes. If all three lipoxygenase enzymes are missing, soyfoods made from those soybeans tend to be mild and lacking a distinctive bitter or beany flavor.

Trypsin is an enzyme in mammals that helps degrade and digest protein. One of the 2*S* fractions contains a trypsin inhibitor that inhibits or slows the trypsin enzyme, thus, reducing the nutritional content of products made from soybeans. To overcome this trypsin inhibitor, a heat treatment is used in commercial processing during the desolventizing-toasting of soybean meal. The heat treatment usually consists of heating the soybean meal to about 100°C for at least 15 min to inactivate this enzyme. Under NOPA trading rules, there is a urease activity test to perform on soybean meal to check if heating was sufficient to denature or deactivate the urease enzyme (and therefore also the more heat-labile trypsin inhibitor). The urease test involves mixing soybean meal with urea and water. Urease, present naturally in raw soybeans, releases ammonia from the urea. Since ammonia is alkaline, a pH meter can measure it. Urease activity of properly heated soybean meal should be in the range of 0.02 to 0.30 pH units of rise. A larger increase in pH indicates that a high amount of urease is still present, meaning that the heat treatment was not adequate for inactivating the urease enzyme; hence, the trypsin inhibitor also would not have been sufficiently inactivated.

Protein Solubility Test (KOH)

The protein solubility test (KOH) or nitrogen solubility index (NSI) is an indicator of solubility of protein in an alkaline solution of potassium hydroxide (KOH). It indicates the percentage of total nitrogen (protein) that is soluble. Protein solubility is important for tofu and soymilk production and for estimating the level of toasting or damage in soybean meal. Soybean meal should have a protein solubility index when using 0.2% KOH of 73 to 85% or higher if urease is within its specification (NOPA, 2007).

Protein Dispersibility Index (PDI)
The protein dispersibility index (PDI) determines the amount of water-dispersible protein (nitrogen x 6.25) as a percentage of total protein that is in soybean products. Protein dispersibility is very important for tofu and soymilk and is considered to be a better measure than NSI for determining amino acid digestibility of heat-treated soybean meal (Brumm, 2004). The PDI helps to distinguish quality of soybean meals that are already considered high quality based on the urease and the KOH protein solubility test. A PDI range of 15–40% is the acceptable range based on NOPA trading rules (NOPA, 2007).

Carbohydrates and Sugars

Besides protein and oil contents, several other compositional quality factors are of interest for soybeans. While soybeans contain about 40% protein and 21% oil on a dry basis, they also contain about 35% carbohydrates (Liu, 1999). The soybean contains about 11% soluble carbohydrates on a dry basis (Maughan et al., 2000). The primary carbohydrate in soybeans is sucrose (40–68%) of the total sugars followed by stachyose and raffinose (Hymowitz & Collins, 1974). Two other sugars commonly found are glucose and fructose, but they are typically low, only 0.1 to 1.0% of the total sugars. One of the primary interests in sugars is for food-grade soybeans. Food-grade soybeans are used largely for production of tofu, natto, and miso. Sucrose content was reported to be positively correlated with oil content and negatively correlated with protein content in soybeans (Hymowitz et al., 1972). While this trend was not shown by the Naeve and Orf (2006) study, clearly, if both protein and oil were desired to be increased, the implied unstated result is that it must be desirable to reduce carbohydrates.

Naeve and Orf (2006) conducted a study on carbohydrate analysis of 22 soybean samples from northern states including: Illinois, Indiana, Iowa, Minnesota, Nebraska, North Dakota, Ohio, and South Dakota; and 18 samples from southern states including: Arkansas, Kansas, Louisiana, Mississippi, Missouri, and Texas (Table 6.12).

Naeve and Orf (2006) found higher levels of total sugars in northern-grown soybeans than in those grown in southern states. For this data set, total soluble sugars ranged from 22.1 to 141.0 mg/g (include sucrose, raffinose, and stachyose).

Geater et al. (2000) evaluated soybean seed traits desirable for soybeans used for Natto. Natto is a product of fermentation that is consumed in Japan and is made from very small soybeans, usually less than 80 mg/seed. Natto soybeans are usually high in carbohydrate content, have clear hila color, and thin seed coats that allow high water absorption. Geater et al. (2000) measured water absorption, water loss, hardness of steamed seeds, and darkness of color of seeds for natto traits. Water absorption is

Table 6.12. Carbohydrate Analysis of Soybean Samples from Northern and Southern U.S. States*.

States	Protein, %	Oil, %	Fiber, %	Soluble carbohydrates			Total
				Sucrose, mg/g	Raffinose, mg/g	Stachyose, mg/g	sugars, mg/g
Northern	36.2	20.8	4.78	57.8	7.63	43.4	108.8
Southern	38.3	21.0	5.05	23.1	7.94	30.3	61.4
Minimum				2.36	4.27	15.5	22.1
Maximum				83.8	9.51	55.9	141.0

* Chemical analyses were performed by University of Missouri Experiment Station Chemical Lab, Columbia, MO.
Source: Naeve and Orf, 2006.

determined by changes in seed weight after a soaking interval in water. High water absorption is desired for natto, as well as some other soybean uses, and it is enhanced by having soft seeds that are relatively low in calcium. Chen (2004) reported calcium among soybean lines ranges from 0.17 to 0.52%.

Geater et al. (2000) measured total sugar, free sugar, sucrose, raffinose, stachyose, protein, oil, fiber, protein + oil, protein + oil + fiber, and seed size as seed traits. Seed size ranged from 59 to 79 mg/seed, and within this range, seed size did not correlate with any of the natto or seed traits. All of the seed traits except oil, stachyose, and seed size were significantly correlated to natto quality. Only the combination of protein + oil was significantly correlated with natto quality and with the seed traits. They concluded that protein + oil would be useful criteria for the selection of breeding cultivars for natto use.

Stombaugh et al. (2004) stated that a significant part of the dry matter in soybeans is made up of polysaccharides in the seed cell wall. They reported cell-wall polysaccharides average about 165 g/kg of dry matter (range 160.0 to 198.7) in whole soybean seeds and are significantly negatively correlated with oil ($R = -0.26$), the sum of protein plus oil ($R = -0.26$), and seed weight ($R = -0.33$). Most cell-wall polysaccharides are found in the cotyledons of soybean seeds and are composed of approximately 76% pectin and 24% cellulose plus hemicellulose. Cell-wall polysaccharides had significant positive correlations to rhamnose (0.69), fucose (0.33), arabinose (0.42), xylose (0.80), mannose (0.25), galactose (0.70), glucose (0.75), uronic acids (0.71), and pectin (0.87). Stombaugh et al. (2004) concluded that decreasing cell-wall polysaccharides would improve seed quality.

Based on Li and Burton (2002) work that an increase in an index of seed weight and seed density could predict an increase in protein, it follows from Stombaugh et al. (2004) that if seed weight were to increase, it is likely that cell-wall polysaccharides would decrease, leading to an expected increase in protein.

Fiber

Cellulose, lignin, and hemicellulose make up most of the structural carbohydrates in soybeans. The seed coat is about 8% of the weight of a soybean, and it makes up a large portion of the crude fiber in soybeans (Sessa & Wolf, 2001). Crude fiber can range from 4 to 8% but averages about 5% for most soybeans on a dry weight basis (Brumm, 2004). Large-seeded soybeans tend to have lower fiber percentages than small soybeans. Mullin and Xu (2001) reported that the major constituents of soybean seed coats were cellulose (14–25 g/100 g), hemicellulose (14–20), pectin (10–12), protein (9–12), uronic acid (7–11), ash (4–5), and lignin (3–40), all in g/100 g (Table 6.13).

The seed coats of six varieties of soybeans averaged about 4.17, 11.17, 16.62, and 17.73 g/100 g dry weight of lignin, pectin, hemicellulose, and cellulose, respectively (Table 6.13). Table 6.13 also shows some of the variability in fiber constituents among several varieties of soybeans. Table 6.14 shows calcium, phosphorous, and magnesium percentages and iron and zinc in ppm for six varieties of soybean seed coats and for cotyledons (Mullin & Xu, 2001). Calcium in seed coats ranged from 0.11 to 0.29% for different varieties. Based on previous discussion, low calcium in seed coats is usually associated with a softer seed coat with high water absorption capability. Phosphorous contents in seed coats averaged 0.85%, which was considerably higher than that for cotyledons with only 0.09%. Magnesium content was about the same in seed coats as in cotyledons. Iron content was much higher in cotyledons, 382 ppm, versus 82 ppm in seedcoats. Zinc content was slightly higher in cotyledons than in seed coats (Table 6.14).

In addition to fiber, soybean seed coats are a source of a potential anti-carcinogenic agent called the Bowman-Birk-inhibitor (Sessa & Wolf, 2001).

Table 6.13. Constituents of Soybean Seedcoats, g/100 g Dry Weight*

Variety	Ash	Uronic Acid	Lignin	Pectin	Hemi-cellulose	Protein	Cellulose
Ox 951	3.83 c	6.63 e	2.45 c	9.7 d	20.1 a	10.25 c	14.14 d
Harosoy	4.67 a	8.24 d	5.85 a	12.0 a	17.9 b	11.94 a	19.88 b
Bobcat	4.60 a	9.15 c	4.44 ab	10.9 c	18.1 b	19.12 e	14.28 d
Harovinton	4.20 b	8.53 d	3.98 b	12.0 a	14.8 c	8.75 e	15.86 cd
AC Onrei	3.90 c	10.46 a	4.24 ab	11.6 b	15.0 c	9.38 d	24.5 a
Tachana-gaha	4.60 a	9.84 b	4.04 b	10.8 c	13.8 d	11.50 b	17.72 bc
Mean	4.30	8.81	4.17	11.17	16.62	11.82	17.73

* The same letter within columns indicates means are not significantly different at the 0.01 level.
Source: Mullin and Xu, 2001.

Table 6.14. Contents of Soybean Seedcoats and Means (standard error) for Cotyledons on a Dry Weight Basis

Variety	Ca, %	P, %	Mg, %	Fe, ppm	Zn, ppm
Ox 951	0.27	0.90	0.30	86.7	63.0
Harosoy	0.21	0.90	0.32	80.3	58.1
Bobcat	0.21	0.74	0.26	80.3	46.0
Harovinton	0.29	0.88	0.33	81.1	65.3
AC Onrei	0.21	0.87	0.30	83.8	62.9
Tachanagaha	0.11	0.82	0.24	76.9	57.4
Seedcoat mean	0.22 (0.03)	0.85 (0.02)	0.29 (0.01)	81.5 (1.4)	58.8 (2.8)
Cotyledon mean	0.56 (0.03)	0.09 (0.01)	0.26 (0.26)	381.8 (14.4)	68.2 (8.2)

Source: Mullin and Xu, 2001.

Phosphorus

The nutritive value of soybeans used for soybean meal is affected by its phytate level. Basically, total phosphorus content is the sum of phytate P, inorganic P, and other P. Generally if phytate P is reduced, the inorganic P increases, which is a nutritionally desirable effect. Erdman (1979) reported that phytate P in soybean meal is mostly unavailable to swine, poultry, and nonruminant animals. This is because nonruminants have very little phytase enzymes in their digestive system to enable phytate P to become available. Phytate P also binds metals that are nutritionally needed, such as zinc, calcium and magnesium, so they become less available. Hence, if through breeding programs, phytate P can be decreased causing inorganic P to be increased, less supplemental inorganic P would need to be added to rations and less undigested phosphorus would be in fecal material (Ertl et al., 1998). Deak and Johnson (2007) reported that phytic acid is at about 1–2% concentration in soybeans and that high-sucrose/low-stachyose soybeans typically contain significantly lower amounts of phytate than commodity soybeans.

Oltmans et al. (2004) reported that conventional soybeans have about 4.3 g/kg phytate P and about 0.7 g/kg of inorganic P. Oltmans et al. (2005) stated that a mutant line of soybeans with reduced phytate P and higher inorganic P was developed by Wilcox et al. (2000). Oltmans et al. (2005) developed low phytate (LP) lines that had 24.9% phytate P, 37.9% inorganic P, and 37.2% other P while their normal phosphate (NP) lines averaged 71.4% phytate P, 3.9% inorganic P, and 24.8% other P forms.

Delwiche et al. (2006) used a NIRSystems 6500 spectrophotometer to collect reflectance data for inorganic P from ground soybeans. Their samples ranged from 334 to 3370 mg/kg based on 150 samples. By using 14 factors, they achieved an R^2 of 0.87 and square root of mean square of residuals (RMSD) of 263 mg/kg, and on a validation set of 41 samples, they obtained an R^2 of 0.86, RMSD of 248, and RPD (ratio of standard deviation of lab data to RMSD) of 2.7. They also did predictions on whole-bean samples in transmission mode, which did not perform as well as the ground beans (RPD = 1.5). Bajjalieh (2006) scanned 88 samples for phytate P on a NIRSystems 6500 and had a range of 0.12 to 0.53 g/kg. A calibration made from his calibration file obtained an R^2 of 0.89, standard error of calibration (SEC) of 0.029, and an RPD of 2.97, based on the SEC value. In both studies, the RPD values were above 2.5, indicating NIR appeared to be doing a reasonable job of measuring phytate P or inorganic P.

Tocopherol and Isoflavones

Soybean oil provides a major source of naturally-derived tocopherols that are antioxidant molecules that help prevent lipid peroxidation (Rani et al., 2007). Chu and Lin (1993) reported that tocopherol content of soybean oil is reduced by poor storage conditions, high damage percentages, and in general low-quality soybeans. Tocopherol content can also be reduced by soybean oil refining, bleaching, and deodorizing; but deodorization is primarily responsible for decreasing tocopherol contents in soybean oil products (Chu & Lin, 1993). Whole soybeans at a moisture content of about 12.0% were exported to Taiwan from the United States. The soybeans were remoistened to obtain 13 moisture levels between 12 and 18%. The soybeans were equilibrated for one week, and part of the soybeans were stored for three more weeks before extracting oil and determining tocopherol contents (Chu & Lin, 1993). Figure 6.3 shows that tocopherol content decreases with increases in moisture content and that with only three weeks of storage tocopherols decrease markedly at moisture contents >15%.

Tocopherol content in oil from cracked soybeans at 12% moisture declined by about 10% over a 30-day period, as shown in Fig. 6.4. However, a huge reduction in tocopherol content occurs for 15 and 18% moisture cracked soybeans when compared to the 12% moisture cracked soybeans.

A comparison of Figures 6.3 and 6.4 shows that oil from cracked soybeans reduced tocopherol content, but moisture contents >12% have a large effect on reducing tocopherol contents. Table 6.15 shows the effect of cracked soybeans at 12, 15, and 18% moisture contents on tocopherol contents of oil. Moisture had a significant effect on reducing tocopherol content as moistures increase. The level of cracked soybeans had no significant effect on tocopherol for the 12% soybeans. The level of cracked soybeans had a significant effect on tocopherol for the 18% moisture soybeans for both the initial and for the two-week storage period.

Means in the same vertical column with different letters (c–g) are significantly different at the 0.05 level. Means in the same row with different letters (h–j) are significantly different at the 0.05 level.

Rani et al. (2007) reported that four isomers of tocopherols, α, β, γ, and Δ, are found in soybean seeds. In looking at 66 soybean genotypes grown in India, they

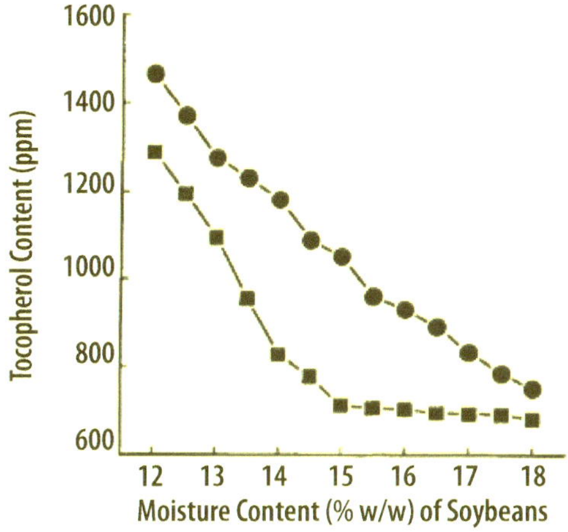

Fig. 6.3. Total tocopherol content of crude soybean oil extracted from intact soybean as a function of moisture content. Soybeans were remoistened to moistures and equilibrated for one week (circles) and equilibrated for one week and stored for three more weeks (squares). Values are averages of duplicate analyses of two replicates (Source: Chu and Lin, 1993).

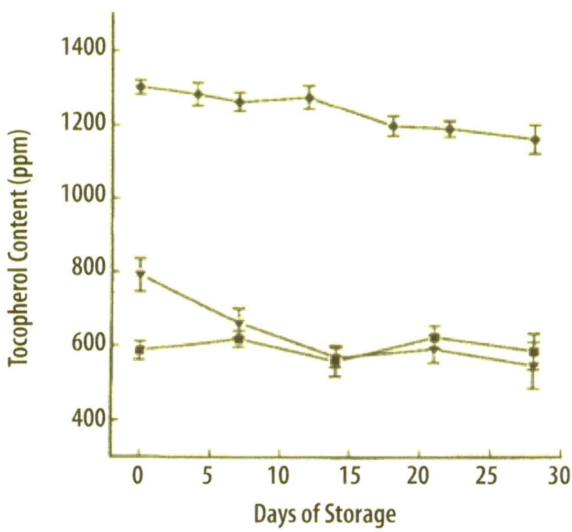

Fig. 6.4. Effects of moisture content and storage time of cracked soybeans on total tocopherol contents of crude soybean oils (Source: Chu and Lin, 1993). Cracked soybeans at moistures of 12% (diamonds), 15% (triangles), and 18% (squares) were equilibrated for one week before storage. Each data point is the mean of three samples with the standard deviation shown by error bars.

Table 6.15. Effect of Soybean Moisture and Cracking Level on Tocopherol Content of Crude Oil in ppm[a].

Moisture, %	Cracked Soybeans, %	Initial Period	After 2-week Storage
12	0	1476 (35)c,h	1322 (57) c,h
	10	1444 (67) c,h	1302 (70) c,h
	15	1421 (35) c,h	1284 (56) c,i
	20	1400 (62) c,h	1255 (61) c,i
15	0	991 (44) d,h	786 (32) d,i
	10	984 (37) d,h	776 (22) d,i
	15	951 (27) d,h	717 (31) d,e,j
	20	937 (23) d,h	694 (28) e,f,j
18	0	706 (35) e,h	675 (27) e,f,h
	10	687 (47) e,f,h	641 (21) g,h
	15	668 (33) e,f,h	616 (21) g,h
	20	633 (25) f,h	601 (22) g,h

[a] Values represent average of duplicate analyses on 3 replications, mean (standard deviation).
Source: Chu and Lin, 1993.

found average contents of 269, 40, 855, 241, and 1405 μg/g of oil for tocopherols isomers of α, β, γ, Δ, and total, respectively.

Soybeans also contain isoflavones, a nonnutritive but physiologically active component. The total isoflavone content of soybeans is about 0.25% (Wolf, 1976). The isoflavone content in the hypocotyl axis is about five to six times higher than in the cotyledons (Kudou et al., 1991). Isoflavones in soy protein reduce the risk factors for breast cancer in premenopausal women (Cassidy et al., 1994). Herman et al. (1995) and Anthony et al. (1996) showed that isoflavones in soy protein also decreased LDL and increased HDL cholesterol concentrations in male and female monkeys that decrease the risk for cardiovascular diseases. The two principal isoflavones are genistein and daidzein; a third isoflavone called glycitein is present in smaller quantities (Naim et al., 1974). They also stated that total isoflavones consist of about 64% genistin, 23% daidzin, and 13% glycetin and that all three compounds exist in the glycoside form in food. The glycoside form binds to a glucose residue as genistin, daidzin, or glycitin. Once the glycoside form is consumed, bacteria in the intestinal tract remove the glucose to produce the aglycone forms called genistein, daidzein, and glycitein. Table 6.16 shows HPLC reference values for the three combined forms of isoflavones (Nimaiyar, 2004).

The samples in Table 6.16 were used by Nimaiyar (2004) for developing calibrations using a Fourier-Transform NIR spectrometer for measuring the three isoflavone types and total isoflavones for ground samples of soybeans.

Table 6.16. HPLC laboratory Reference (ppm) Values for the Total of the Three Forms of Three Types of Isoflavones and Total Isoflavones.

Isoflavones	N	Min.	Max.	Mean	Std. Dev.
Daidzein, daidzin, and 6"-o-malonyl daidzin	160	118	2505	754	480
Genistein, genistin, and 6"-o-malonyl genistin	179	120	2177	862	501
Glycitein, glycitin, and 6"-o-malonyl glycitin	173	36	547	276	115
Total isoflavones	186	201	5924	1945	1210

Source: Nimaiyar, 2004.

Soybean Fact Sheet (2003) reported that isoflavone levels in soybeans are related to temperature during pod filling and possibly to potassium fertility. High air temperatures during pod fill and low levels of potassium in the soil tend to cause lower isoflavone concentrations. This fact was given as a competitive advantage for producing soybeans in the Midwest versus producing in Brazil. A niche market developed for non-GMO soybeans of specific varieties with high protein/isoflavone content. Soybeans were processed at Bloomington, Illinois, and sent to Protein Technologies (St. Louis, MO). In 2003, about 12,000 acres of high isoflavone soybeans were contracted in central Illinois. The contracts typically do not state minimal levels of isoflavones, but the varieties planted were previously identified for containing high levels of isoflavones. Premiums have been up to $0.55/bu with an additional premium of $0.20/bu if high isoflavone levels are tested present at delivery (*Soybean Fact Sheet,* 2003).

Mycotoxins

Literature exists indicating that aflatoxin and mycotoxins were found in soybeans, but this risk was quite low (Meronuck, 1991). Infections of *Fusarium* in the field can lead to Deoxynivalenol and Zearalenone. In a mycotoxin study of 1,046 samples of soybeans, only two were found to contain any mycotoxin, and those were both sample grade and contained aflatoxin (Meronuck, 1991). Wilson (1995) also stated that mycotoxins are not considered a significant problem in soybeans. Mycotoxins produced by fungi may directly or indirectly affect chemical quality of soybean seed. In highly infected seed, pathogenic agents can form nonhydratable phospholipids; can increase fatty acids; can oxidize lipids, which can create undesirable color in processed oils; can reduce lecithin; and can reduce soybean meal quality. Wilson (1995) also reported that the pathogens metabolize the carbohydrate fraction, leaving concentrations of oil and protein to become higher due to the loss of carbohydrates.

Processability Factors

Processability of soybeans is affected by moisture, by physical factors, and by many chemical factors. Ultimately, the processor wants soybeans that are low in free fatty acid (FFA), generally less than 0.75% in crude degummed soybean oil. Mounts et al. (1990) stated that FFA provides a measure of enzymatic hydrolysis of triglycerides in the oil and that FFA increases as splits and other damage to seeds increase. Some Brazilian soybeans had higher oil contents than U.S. soybeans but had about 1.0% FFA, indicating higher refining losses during processing for Brazilian beans than for U.S. soybeans. Increases in FFA often coincide with elevated iron levels above the normal 1–3 ppm range (Pritchard, 1983).

Pritchard (1983) reported that damage to soybeans causes hydratable phosphatides to degrade to nonhydratable phosphatides. As nonhydratable phosphatides increase, higher refining losses occur. Low nonhydratable phospholipids (NHP) in degummed soybean oil (<0.3%) are desirable (Mounts et al., 1990). Otherwise, as NHP increases, neutral oil loss in refining increases. They reported an enzymatic action in soybean seed causes formation of NHP and suggested that crude soybean oil containing <400 ppm phosphorus is damaged and will have increased NHP. Equivalent phospholipid content in percent is equal to phosphorus in percent multiplied by 30 (Mounts et al., 1990).

Another processability factor is TOTOX value that is related to the degree of oil deterioration due to lipoxygenase action on polyunsaturated fatty acids in the oil (Mounts et al., 1990). Products of oxidation degrade flavor and stability of edible oils. A low TOTOX value <3.0 meq/kg is desirable. TOTOX was calculated as two times the peroxide value as determined by AOCS Method Ca9f-57 plus the anisidine value (AOCS, 1987).

Wang and Johnson (2001) reported on test measurement methods that were major indicators of soybean oil quality. These tests included peroxide value, anisidine value, FFA content, phospholipid content, total tocopherol content, oxidative stability index, color, and moisture content. For soybean meal, they reported on urease activity, protein dispersibility index (PDI), rumen bypass or rumen undegradable protein, trypsin inhibitor activity, moisture content, residual oil content, protein content, fiber content, color, amino acid profiles, and protein solubility under alkaline (KOH) conditions.

Wang and Johnson (2001) processed soybeans using extrusion-expelling and solvent extraction. They found that commodity soybeans with extruded-expelled crude oil had significantly less tocopherols (918 versus 1324 ppm), nonsignificantly higher peroxides (1.50 meq/kg versus 1.14 meq/kg), and nonsignificantly higher oxidation (anisidine value 1.45 versus 1.12) than crude oil from soybeans that were solvent-extracted. High-oleic soybean oil was hydrolyzed to a greater degree during solvent extraction processing than during extrusion-expelling processing. Oils with more unsaturated fatty acid content had more oxidative degradation if extracted by extrusion–expelling than if by solvent extraction.

Peroxide value (PV) is an indication of primary lipid oxidation products in oil or the level of oxidation in soybean oil. Yildiz et al. (2002) measured oxidation with the standard method (AOCS Method Cd 8-53—AOCS, 1987) and used NIRSystem 6500 spectrometer to detect PV values in soybean and corn oils. Peroxide values ranged from 0.30 to 20.85 meq/kg. They obtained a very good calibration with R = 0.99 and SEP = 0.52 meq/kg using a first derivative math treatment and 13 PLS factors.

Environmental Factors

Changes in soybean quality due to environmental factors have not been widely considered, but environmental changes occurring due to increases in carbon dioxide may be a future consideration. Some of the early work in this area was done by Bernacchi et al. (2007) where soybeans were grown in plots with normal levels of CO_2 (~375 ppm) and elevated levels (about 550 ppm). They measured leaf stomatal conductance and evapotranspiration and found a 10% reduction in stomatal conductance caused an 8.6% decrease in evapotranspiration. They also found that the soybeans grown under the elevated CO_2 conditions had a large increase in photosynthesis and seed yield and evapotranspiration decreased by 9–16%. This indicates that with increased CO_2 the stomata close, resulting in less transpiration and apparently more water going into streams and rivers. Evidently, in some plants, leaves grown under high CO_2 levels have increased sugar that makes such leaves more desirable to certain insects.

Sinclair et al. (1987) investigated soybean seed growth under different partial pressures of oxygen. They looked at partial pressures of 0.10, 0.21 (ambient), 0.42, and 0.62 by either adding O_2 gas to ambient air or adding N_2 gas to ambient air, with no attempt to control CO_2 level. They found mean dry soybean seed weight to be 129, 167, 199, and 194 mg/seed for the partial pressures of 0.10, 0.21, 0.42, and 0.62, respectively. Thus, at low oxygen levels, seed weight was significantly lower. However, by increasing oxygen from 0.42 to 0.62, seed weight did not change significantly. In an earlier experiment with partial pressure of 0.05 oxygen, the plants produced little or no seed.

Conclusion

Soybean quality factors are evolving with changes in breeding, genetic modification, climate, environment, newer technologies for measurement, and with our increased understanding of the chemistry and functionality of soybeans. Other factors driving change are consumers becoming more cognizant of health benefits from soybean-derived products and the need for reduced-*trans* fat foods. The move toward production of biofuels also impacts soybean oil usage. As stated earlier, quality can be no higher than it was at harvest; but with increased knowledge of methods for measuring quality and improved technology for identifying high quality, it is highly possible for producers, managers of stored soybeans, and processors to have highly acceptable soybean quality for food, processing, marketing, and industrial uses.

References

AACC. Method 44-15a; Moisture—Air-oven methods. American Society of Cereal Chemists: St. Paul, MN, 1999.

Ahmad, I.S.; J.F. Reid; M.R. Paulsen; J.B. Sinclair. Color classifier for symptomatic soybean seeds using image processing. *Plant Dis.* **1999,** *83,* 320–327.

Anthony, M.S.; T.B. Clarkson; C.L. Hughes; T.M. Morgan; G.L. Burke. Soybean isoflavones improve cardiovascular risk factors without affecting the reproductive system of perpubertal rhesus monkeys. *J. Nutr.* **1996,** *126,* 43.

AOAC. Method 982.30e. *Official Methods of the Association of Official Analytical Chemists,* Fifteenth ed.; AOAC International: Arlington, VA, 1990.

AOCS. Method Ac 2-41. Moisture and volatile matter. Society, Fifth ed.; AOCS Press: Champaign, IL, 1987. Method Cd 8-53; Method Ca9f-57.

AOSA. *Tetrazolium Testing Handbook Contribution No. 29.* Association of Official Seed Analysts: , 2000.

ASABE. Standard S352.2. Moisture measurement-unground grain and seeds. American Society of Agricultural and Biological Engineers: St. Joseph, MI, 2006.

Bajjalieh, N.; Calibration data set of calibration file for soybean NIR testing on NIRSsystems 6500 from Caltest, LLC, Ballston Lake, NY made possible by United Soybean Board funding. Personal Communication, 2006.

Bernacchi, C.J.; B.A. Kimball; D.R. Quarles; S.P. Long; D.R. Ort. Decreases in stomatal conductance of soybean under open-air elevation of CO_2 are closely coupled with decreases in ecosystem evapotranspiration. *Plant Physiol.* **2007,** *143,* 134–144.

Beuerlein, J. Variety evaluation, seed quality, and the seed law. Chapter 11. *Ohio Agronomy Guide,* Fourteenth ed.; Ohio State University Extension. 2005. http://ohioline.osu.edu/b472/index.html accessed 5-15-2007.

Bluebook. The annual directory of the world oilseed industry. *Soya & Oilseed Bluebook*; Soyatech Publication. ISBN 0-9723216-2-4. Soyatech, Inc.: Bar Harbor, ME, 2005.

Brumm, T.J. Quality and grading factors of IP soybeans. *Managing Grain after Harvest;* C.J. Bern, T.J. Bern, Eds.; Department of Agricultural and Biosystems Engineering, Iowa State University: Ames, IA, 2004.

Brumm, T.J.; C.H. Hurburgh, Jr. Quality of the 2003 soybean crop from the United States. American Soybean Association Quality Mission to Japan, U.S. Soybean Export Council. Agricultural & Biosystems Engineering: Iowa State University: Ames, IA. Nov. 2003. 16 pp.

Brumm, T.J.; C.H. Hurburgh, Jr. Changes in long-term soybean compositional patterns. Letter to the editor. *J. Am. Oil Chem. Soc.* **2006,** *83,* 981–983.

Brumm, T.J.; C.H. Hurburgh, Jr.; S. Naeve. Quality of the 2005 soybean crop from the United States. American Soybean Association, U.S. Soybean Export Council, 2005. http://www.asajapan.org/english/food/quality/05.html

Casady, W.W.; M.R. Paulsen; J.F. Reid; J.B. Sinclair. A trainable algorithm for inspection of soybean seed quality. *Trans. ASAE* **1992,** *35,* 2027–2034.

Cassidy, A.; S. Bingham; K.D.R. Setchell. Biological effects of a diet of soy protein rich in isoflavones on the menstrual cycle of premenopausal women. *Am. J. Clin. Nutr.* **1994,** *60,* 333–339.

Chen, P. Developing high quality identity-preserved soybeans for the specialty soyfood market. *Production and Marketing of Identity-Preserved Soybeans;* American Soybean Association, St. Louis, MO, 2004; pp. 23–31.

Cherrak, C.M.; V.R. Pantalone; E.J. Meyer; D.L. Ellis; S.L. Melton; D.R. West; J.R. Mount, Low-palmitic, low-linolenic soybean development. *J. Am. Oil Chem. Soc.* **2003,** *80,* 539–543.

Chu, Y.; J. Lin. Factors affecting the content of tocopherol in soybean oil. *J. Am. Oil Chem. Soc.* **1993,** *70,* 1263–1268.

Dale, N.; A. Batal. Ingredient analysis table. *Feedstuffs* **2003,** *75(38),* 16 Sep 17, 2003.

Deak, N.A.; L.A. Johnson. Fate of phytic acid in producing soy protein ingredients. *J. Am. Oil Chem. Soc.* **2007,** *84,* 369–376.

Delouche, J.C.; T.W. Still; M. Raspet; M. Lienhard. The tetrazolium test for seed viability. *Tech. Bull. 51.* Mississippi State University. Agricultural Experiment Station: State College, MS, 1962.

Delwiche, S.R.; L.O. Pordesimo; A.M. Scaboo; V.R. Pantalone. Measurement of inorganic phosphorus in soybeans with near-infrared spectroscopy. *J. Agric. Food Chem.* **2006,** *54,* 6951–6956.

DICKEY-john. 2007. http://www.dickey-john.com/_dl/2-0563B_6.pdf.

Erdman, J.W., Jr. Oilseed phytates: Nutritional implications. *J. Am. Oil Chem. Soc.* **1979,** *56,* 736–741.

Ertl, D.S.; K.A. Young; V. Raboy. Plant genetic approaches to phosphorus management in agricultural production. *J. Environ. Qual.* **1998,** *27,* 299–304.

FGIS. 2004 (revised Sept. 1, 2007). http://archive.gipsa.usda.gov/reference-library/handbooks/grain-insp/grbook2/soybean.pdf

Friedrich, R.L. Images of soybean damage. USDA-GIPSA, Digital Media Group, Kansas City, MO. Personal Communication. 2007.

FSMIP. Market improvement for northern soybeans through direct marketing. 12-25-g-0367. Minnesota Dept. of Agriculture, Agricultural Marketing Services Division: St. Paul, MN, 2005.

Frankel, E.N.; A.M. Nash; J.M. Snyder. A methodology study to evaluate quality of soybeans stored at different moisture levels. *J. Am. Oil Chem. Soc.* **1987,** *64,* 987–992.

Funk, D. *Principles of NIR Spectroscopy for Applied Research Workshop.* American Society of Agricultural and Biological Engineering Meeting Aug. 2004, Ottawa, Canada.

Geater, C.W., W.R. Fehr; L.A. Wilson. Association of soybean seed traits with physical properties of Natto. *Crop Sci.* **2000,** *40,* 1529–1534.

Hartwig, R.A.; C.R. Hurburgh, Jr. A justification for using a whole-grain oven method to determine soybean moisture content. Iowa State University. University Extension: Ames, IA, 2007. http://www.extension.iastate.edu/grain/info/moisturerelations.htm

Haynes, F. *Low Fat Cooking;* 2007. http://lowfatcooking.about.com/od/faqs/f/hydrogenated.htm

Herman, C.; T. Adiercreutz; B.R. Goldin; S.L. Gorbach; K.A.V. Hbckerstedt. Soybean phytoestrogen intake and cancer risk. *J. Nutr.* **1995,** *125,* 757S.

Hill, L.D. Uniformity by 2000: the summary of an international workshop on maize and soybean quality. *Uniformity by 2000: An International Workshop n Maize and Soybean Quality;* L.D. Hill, Ed.; Scherer Communications: Urbana, IL, 1991.

Hobbs, H.A.; G.L. Hartman; Y. Wang; C.B. Hill; R.L. Bernard; W.L. Pedersen; L.L. Domier. Occurence of seedcoat mottling in soybean plants inoculated with *Bean pod mottle virus* and *Soybean mosaic virus. Plant Dis.* **2003,** *87,* 1333–1336.

Hurburgh, C.R.; R. Elmore; P. Pedersen. Frost damage to corn and soybeans. PM 1635. Iowa State University, University Extension. http://www.extension.iastate.edu/NR/rdonlyres/1ECB8840-4333-43AA-B3E4-250062F51BBC/63865/FrostDamageCornSoybeans.pdf

Hymowitz, T.; F.I. Collins. Variability of sugar content in seed of *Glycine max* (L.) Merrill and *G. soja* Sieb and Zucc. *Agron. J.* **1974,** *66,* 239–240.

Hymowitz, T; F.I. Collins; J. Panczer; W.M. Walker. Relationship between the content of oil, protein, and sugar in soybean seed. *Agron. J.* **1972,** *66,* 613–616.

Kovalenko, I.V.; G.R. Rippke; C.R. Hurburgh. Determination of amino acid composition of soybeans (*Glycine max*) by near-infrared spectroscopy. *J. Agric. Food Chem.* **2006a,** *54,* 3485–3491.

Kovalenko, I.V.; G.R. Rippke; C.R. Hurburgh. Measurement of soybean fatty acids by near-infrared spectroscopy: linear and nonlinear calibration methods. *J. Am. Oil Chem. Soc.* **2006b**, *83,* 421–427.

Kudou, S.; Y. Fleury; D. Welti; D. Magnaolato; T. Uchida;, K. Kitamura; K. Okubo. Malonyl isoflavones glucosides in soybean seeds (*Glycine max* Merrill). *Agric. Biol. Chem.* **1991,** *55,* 2227–2233.

Li, H.; J.W. Burton. Selecting increased seed density to increase indirectly soybean seed protein concentration. *Crop Sci.* **2002,** *42,* 393–398.

Liu, K. Chemistry and nutritional value of soybean components. *Soybeans—Chemistry, Technology, and Utilization;* A Chapman & Hall Food Science Book. Aspen Publishers, Inc.: Gaithersburg, MD, 1991.

Matsue, Y.; O. Uchikawa; H. Sato; K. Tanaka. Productivity of soybean seeds stored for various periods. *Plant Prod. Sci.* **2005,** *8,* 393–396.

Maughan, P.J.; M.A. Saghai Maroof; G.R. Buss. Identification of quantitative trait loci controlling sucrose content in soybean (*Glycine max*). *Mol. Breeding* **2000,** *6,* 105–111.

Meronuck, R.A. The effect of fungi on soybean quality. *Uniformity by 2000: Highlights of an International Workshop on Maize and Soybean Quality;* L.D. Hill, Ed.; Scherer Communications: Urbana, IL, 1991; pp. 275–282.

Mian, M.A.R.; M.A. Bailey; J.P. Tamulonis; E.R. Shipe; T.E. Carter, Jr.; W.A. Parrott; D.A. Ashley; R.S. Hussey; H.R. Boerma. Molecular markers associated with seed weight in two soybean populations. *Theor. Appl. Genet.* **1996,** *93,* 1011–1016.

Midwest Plan Service. *Grain Drying, Handling and Storage Handbook;* MWPS-13. Second ed.; Midwest Plan Service. Iowa State University: Ames, IA, 1988.

Misra, M.K.; B. Koerner; A. Pate; C.P. Burger. Acoustic properties of soybeans. *Trans. ASAE* **1990,** *33.* 671–677.

Mounts, T.L.; J.M. Snyder; R.T. Hinsch; A.J. Bongers; A.R. Class. Quality of soybeans in export. *J. Am. Oil Chem. Soc.* **1990,** *67,* 743–746.

Mullin, W.J.; W. Xu. Study of soybean seedcoat components and their relationship to water absorption. *J. Agric. Food Chem.* **2001,** *49,* 5331–5335.

Naeve, S.L.; J.H. Orf. Quality of the U.S. soybean crop: 2006. *Report to the American Soybean Association and U.S. Soybean Export Council Quality Mission to Asia,* Nov. 13–21, 2006.

Nafziger, E.D. Soybeans. *Illinois Agronomy Handbook,* Twenty-third ed.; University of Illinois: Urbana-Champaign, IL, 2002.

Nafziger, E.D. Seed. *Illinois Agronomy Handbook,* Twenty-third ed.; University of Illinois: Urbana-Champaign, IL, 2002.

Naim, M.; B. Gestetner; S. Zilkah; Y. Birk; A. Bondi. Soybean isoflavones: Characterization, determination, and antifungal activity. *J. Agr. Food Chem.* **1974,** *22,* 806–810.

Nave, W.R.; M.R. Paulsen. Soybean seed quality as affected by planter meters. *Trans. ASAE.* **1979,** *22,* 739–745.

Nimaiyar, S. Measurement of Isoflavones, Fatty Acids, and Amino Acids in Soybeans Using FT-NIR Spectroscopy. M.S. Thesis. University of Illinois, Urbana- Champaign, IL, 2004.

Nimaiyar, S.; M.R. Paulsen; R.L. Nelson. Rapid analysis of fatty acids in soybeans using FT-NIR. *ASAE Paper No. 04-6118. 2004;* American Society of Agricultural and Biological Engineering: St. Joseph, MI, 2004.

NOPA. National Oilseeds Processors Association. *Specifications for Soybean Meal;* Washington, DC, 2007.

Oltmans, S.E.; W.R. Fehr; G.A. Welke; S.R. Cianzio. Inheritance of low-phytate phosphorus in soybean. *Crop Sci.* **2004,** *44,* 433–435.

Oltmans, S.E.; W.R. Fehr; G.A. Welke; V. Raboy; K.L. Peterson. Agronomic and seed traits of soybean lines with low-phytate phosphorus. *Crop Sci.* **2005,** *45,* 593–598.

Paulsen, M.R. Analyzing moisture measurement—Direct moisture measurement reference methods. *Uniformity by 2000: Highlights of an International Workshop on Maize and Soybean Quality;* L.D. Hill, Ed.; Scherer Communications: Urbana, IL, 1991; pp. 391–399.

Paulsen, M.R.; W.R. Nave. Improved indoxyl acetate test for detecting soybean seedcoat damage. *Trans. ASAE* **1979,** *22,* 1475–1479.

Pazdernik, D.L.; A.S. Killam; J.H. Orf. Analysis of amino and fatty acid composition in soybean seed, using near infrared reflectance spectroscopy. *Agron. J.* **1997,** *89,* 679–685.

Pierce, R.O. FGIS-GIPSA-USDA. Personal communication by e-mail. March 22, 2007.

Pritchard, J.R. Oilseed quality requirements for processing. *J. Am. Oil Chem. Soc.* **1983,** *80,* 322–332.

Rani, A.; V. Kumar; S.K. Verma; A.K. Shakya; G.S. Chauhan. Tocopherol content and profile of soybean: Genotypic variability and correlation studies. *J. Am. Oil Chem. Soc.* **2007**, *84,* 377–383.

Rodda, E.D.; M.P. Steinberg; L.S.Wei. Soybean damage detection and evaluation for food use. *Trans. ASAE* **1973,** *16,* 365–366.

Sessa, D.J.; W.J. Wolf. Bowman-Birk inhibitors in soybean seed coats. *Ind. Crop Prod.* **2001,** *14,* 73–83.

Sinclair, T.R.; J.P. Ward; C.A. Randall. Soybean seed growth in response to long-term exposures to differing oxygen partial pressures. *Plant Physiol.* **1987,** *83,* 467–468.

Soybean Fact Sheet. Illinois Specialty Farm Products, High isoflavones soybeans update 2003. http://web.aces.uiuc.edu/value/factsheets/soy/fact-isoflavone-soy.htm Accessed 10-15-2007.

Sternberg, K.J.; L.S. Wei; A.I. Nelson. Effect of simulated freeze damage on soybean seed composition and functionality. *J. Am. Oil Chem. Soc.* **1990,** *67,* 308–314.

Stombaugh, S.K.; J.H. Orf; H.G. Jung; K. Chase; K.G. Lark; D.A. Somers. Quantitative trait loci associated with cell wall polysaccharides in soybean seed. *Crop Sci.* **2004,** *44,* 2101–2106.

Urbanski, G.E.; L.S. Wei; A.I. Nelson. Effect of freeze damage on soybean quality and storage stability. *J. Food Sci.* **1980,** *45,* 208–212.

U.S. Standards for Soybeans. Subpart J. 1994. http://archive.gipsa.usda.gov/reference-library/standards/810soybean.pdf Accessed 3-2-2007."

Wang, T.; L.A. Johnson. Survey of soybean oil and meal qualities produced by different processes. *J. Am. Oil Chem. Soc.* **2001,** *78,* 311–318.

Wilcox, J.R.; G.S. Premachandra; K.A. Young; V. Raboy. Isolation of high seed inorganic P low-phytate soybean mutant. *Crop Sci.* **2000,** *40,* 601–1605.

Williams, P.C.; K.R. Preson; K. Norris; P.M. Starkey. Determination of amino acids in wheat and barley by near-infrared reflectance spectroscopy. *J. Food Sci.* **1984,** *49,* 17–20.

Wilson, R.R. Dealing with the problems of fungal damage in soybean and other oilseeds. *J. Am. Oil Chem. Soc.* **1995,** *72,* 1413–1414.

Wolf, W.J. Chemistry and technology of soybean. *Adv. Cereal Sci. Technol.* **1976,** 1, 325.

Wuebker, E.F.; R.E. Mullen; K. Koehler. Flooding and temperature effects on soybean germination. *Crop Sci.* **2001,** *41,* 1857–1861.

Yaja, J.; E. Pawelzik; S. Vearasilp. Prediction of soybean seed quality in relation to seed moisture content and storage temperature. Tropentag 2005 Conference on International Agricultural Research for Development. Stuttgart-Hohenheim, Germany, 2005.

Yildiz, G.; R.L. Wehling; S.L. Cuppett. Monitoring PV in corn and soybean oils by NIR spectroscopy. *J. Am. Oil Chem. Soc.* **2002,** *79,* 1085–1089.

Zarkadas, C.G.; Z. Yu; H.D. Voldeng; A. Minero-Amador. Assessment of the protein quality of a new high-protein soybean cultivar by amino acid analysis. *J. Agric. Food Chem.* **1993,** *41,* 616–623.

7

Lipids

Jose A. Gerde and Pamela J. White
Department of Food Science & Human Nutrition, Iowa State University, Ames, IA 50011

Introduction

Lipids represent one of the most important classes of components in soybeans. Economically, soybean oil comprises approximately 29% of the world's fat and oil production (Golbitz, 2007). Soy lipids are primarily in the soybean cotyledon and comprise about 20% of its weight. Physiologically, soybean lipids have a broad spectrum of functions, including being a part of membranes, acting as an energy reserve, and serving as the solvent medium for many lipid-soluble substances. In general, neutral lipids are soluble in organic solvents and are not soluble in water. Some lipid compounds, however, contain polar groups which, along with the hydrophobic part, impart an amphiphilic character to the molecule, thus favoring the formation of micelles from these compounds.

The predominant fatty acids (FA) present in soybeans are palmitic (16:0), stearic (18:0), oleic (18:1), linoleic (18:2), and linolenic (18:3). Table 7.1 shows the common name, the systematic name, structure, and abbreviation of these FA. Other FA, such as arachidic (20:0) and behenic (22:0) acids, are present in minor quantities (<1%). The FA consist of a hydrocarbon chain with a carboxylic group at position 1 and typically contains an even number of carbon atoms. This molecule also can include points of unsaturation (double bonds). Palmitic acid, with 16 carbons, and stearic acid with 18 carbons, are the typical saturated (no double bonds) FA in soybean oil. Oleic, linoleic, and linolenic acids have 18 carbon atoms, with one, two, and three double bonds, respectively. In all the cases, the geometric configuration of the double bonds in the native oil is *cis*, and when there is more than one double bond, they are methylene interrupted (Fig. 7.1).

Normally, the FA are linked to a glycerol molecule through an ester bond to form glycerides. Depending on the number of FA, the molecule will be a mono-, di-, or triacylglyceride (TAG) (Fig. 7.2). In mature undamaged seeds, <1% of the FA occur as free FA (not linked to any other molecule). In phospholipids, the glycerol moiety also contains a phosphate ester in place of one FA. This phosphate group can

Table 7.1. Names of Fatty Acids Found in Soybean Oils and Their Structures

Systematic Name	Common Name	Abbreviation	Structure
Hexadecanoic Acid	Palmitic	16:0	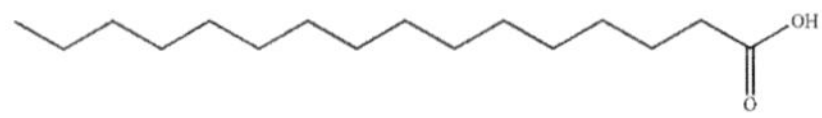
Octadecanoic Acid	Stearic	18:0	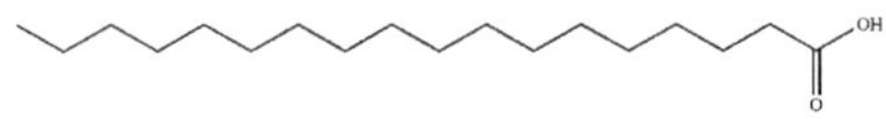
9c-Octadec-enoic Acid	Oleic	18:1 Δ^{9}	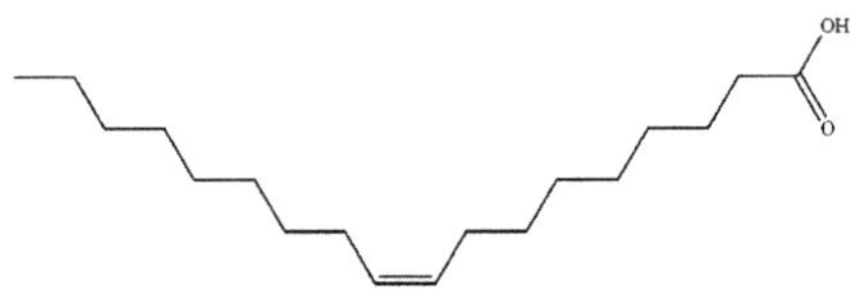
9c,12c-Octa-decadienoic Acid	Linoleic	18:2 $\Delta^{9,12}$	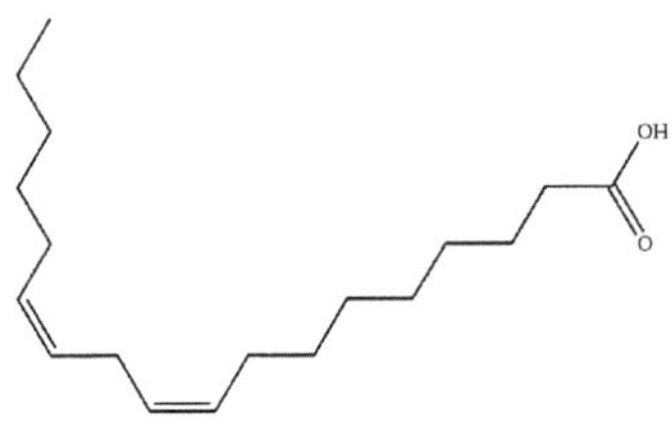
9c,12c,15c-Octadecatri-enoic Acid	Linolenic	18:3 $\Delta^{9,12,15}$	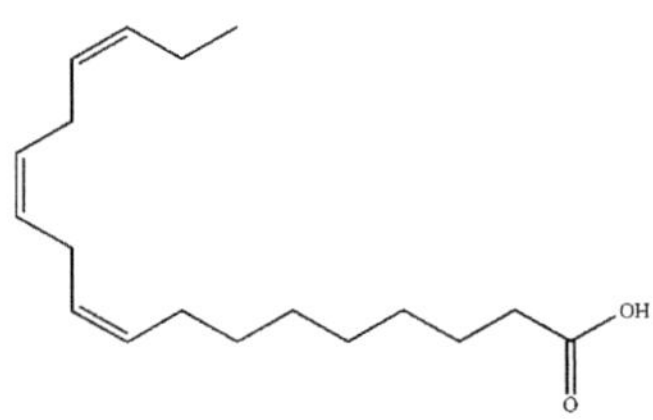
Eicosanoic Acid	Arachidic	20:0	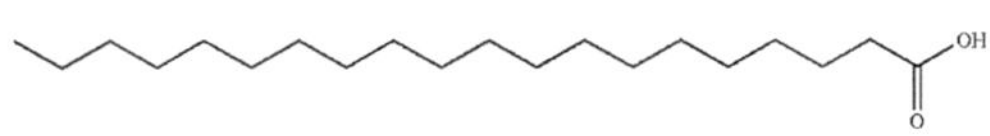
Docosanoic Acid	Behenic	22:0	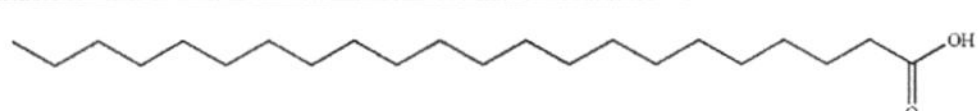
Tetracosanoic Acid	Lignoceric	24:0	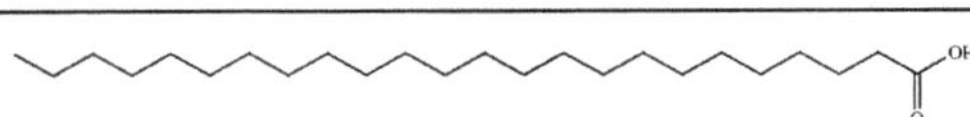

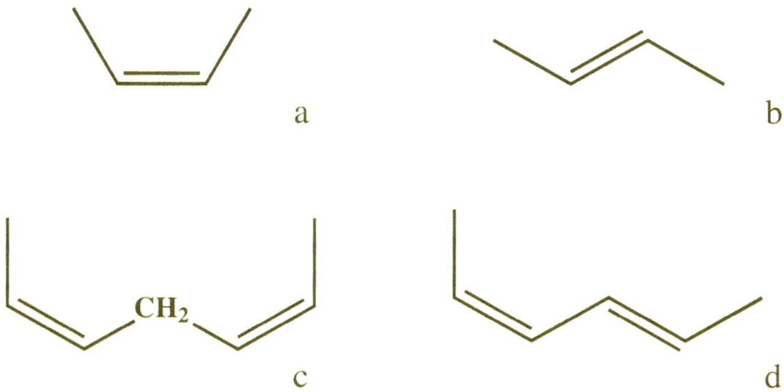

Fig. 7.1. Double bond configurations in fatty acids associated with natural or processed soybean oils: (a) *cis* double bond; (b) *trans* double bond; (c) methylene interrupted double bonds; (d) conjugated double bonds (conjugated dienes).

be linked to other groups, frequently containing nitrogen (Fig. 7.3). The FA distribution in soybean TAG and phospholipid molecules is shown in Fig. 7.4. Sphingolipids are the result of the amide linkage between a FA and sphingosine, which can also be linked to a sugar or a phosphate group. Other lipid compounds present in very low amounts include sterols (steroid alcohols), tocopherols, and pigments (chlorophyll).

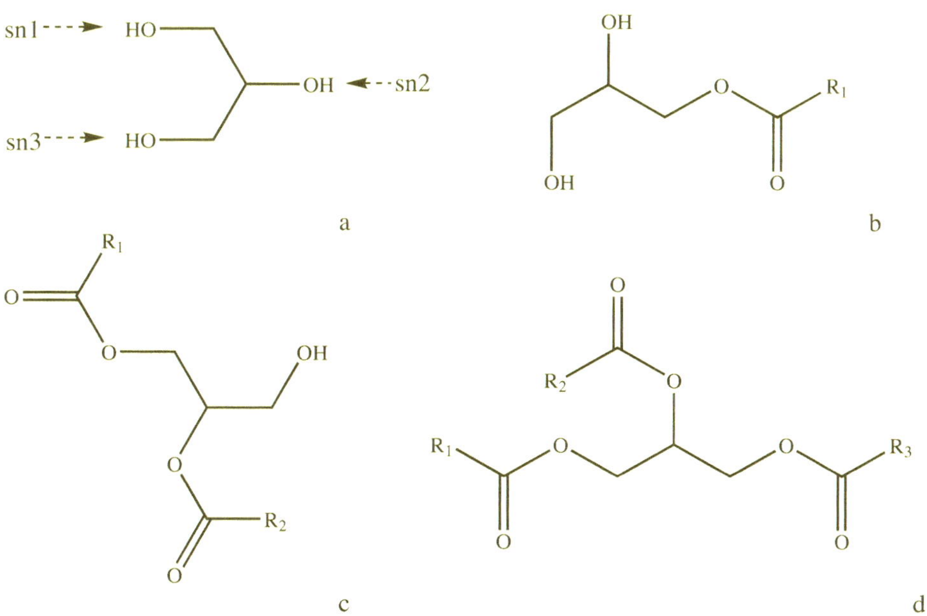

Fig. 7.2. General structures for stereospecific naming of (a) glycerol molecules, (b) monoacylglyceride, (c) diacylglyceride, (d) triacylglyceride. R_1, R_2, and R_3: hydrocarbon chains.

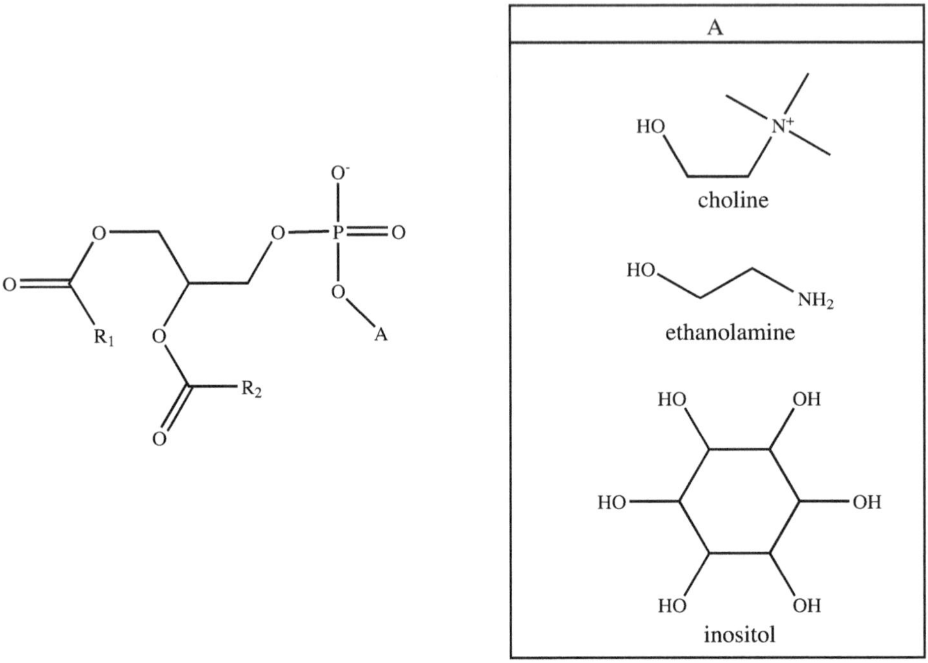

Fig. 7.3. General structure of phospholipids. R_1, R_2: hydrocarbon chains. Point 'A' is likely composed of structures noted in the box.

Glyceride Biosynthesis

Fatty Acid Biosynthesis

Initiation

In plants, FA are synthesized in the stroma of the plastids (Lynen, 1961; Ohlrogge & Browse, 1995; Voelker & Kinney, 2001; Sasaki & Nagano, 2004), where 2-carbon units from malonyl-acyl carrier protein (ACP) are incorporated into the growing acyl-ACP chain. Malonyl-CoA is synthesized from acetyl-CoA and CO_2 by acetyl-CoA carboxylase. Malonyl-CoA:ACP acyltransacylase produces malonyl-ACP from ACP and malonyl-CoA (Ohlrogge & Browse, 1995) (Fig. 7.5).

Elongation and Monounsaturation

The FA elongation stage is catalyzed by β-ketoacyl synthase III (KAS III), a single polypeptide enzyme with several catalytic sites, which allows the enzyme to promote different reactions during this stage (Voelker & Kinney, 2001). The condensation of malonyl-ACP and acetyl-CoA to yield 3-ketobutyryl-ACP and CO_2 is the first reaction. The second reaction is the reduction of 3-ketobutyl-ACP to 3-hydroxylacyl-

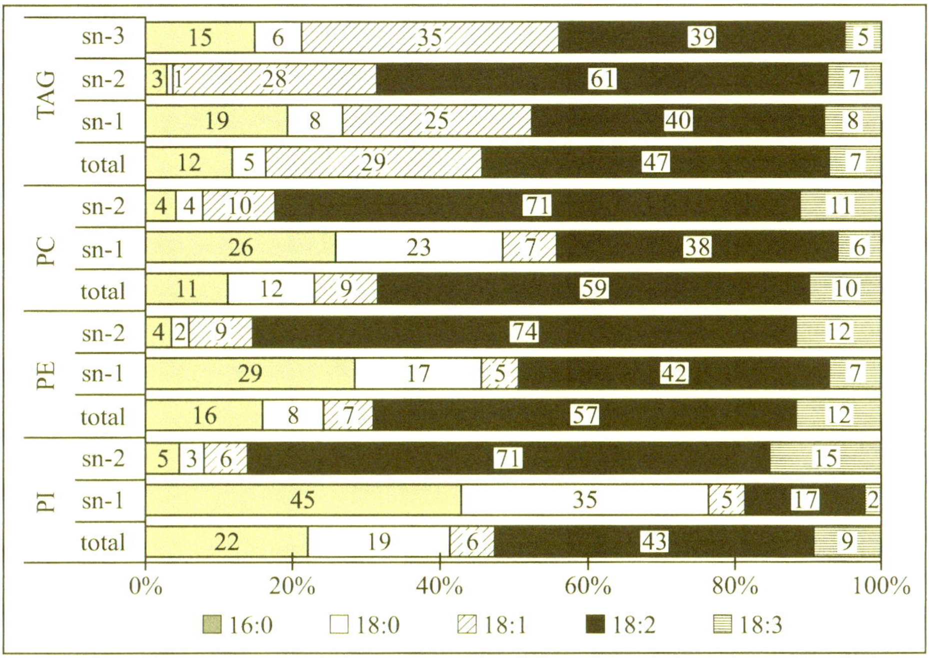

Fig. 7.4. Stereospecific distribution of fatty acyl groups in commodity soybean triacylglycerides (TAG), phospahtidylcholine (PC), phosphatidyletanolamine (PE), and phosphatidylinositol (PI). 16:0, palmitic acid; 18:0, stearic acid; 18:1, oleic acid; 18:2, linoleic acid; 18:3, linolenic acid (Harp & Hammond, 1998; Wang et al., 1997; Hammond et al., 2005).

ACP, with subsequent dehydration to enoyl-ACP, which is further reduced to butyryl-ACP (Voelker & Kinney, 2001). Later steps of 2-carbon unit additions are catalyzed by KAS I (Shimakata & Stumpf, 1983). Once palmitoyl-ACP is formed, KAS II may add an extra 2-carbon unit to form stearoyl-ACP (Shimakata & Stumpf, 1982). The formation of oleoyl-ACP is catalyzed by Δ^9 stearoyl-ACP desaturase, an enzyme that promotes the formation of double bonds in the Δ^9 position (Shanklin & Cahoon, 1998). This enzyme also desaturates palmitoyl-ACP in position Δ^9. However, the apparent specificity factor for stearoyl-ACP is much greater than for palmitoyl-ACP (Gibson, 1993). The presence of two Δ^9 stearoyl-ACP desaturases coded by two different genes has been reported in soybean (Byfield et al., 2006).

The three main products formed during the FA synthesis in the stroma of the plastid are hydrolyzed by two possible acyl-ACP thioesterases (FAT): FAT A and FAT B. FAT A has preference for oleoyl-ACP but also hydrolyzes stearoyl- and palmitoyl-ACP. FAT B, which has preference for saturated FA-ACP and especially palmitoyl-ACP, can also hydrolyze oleoyl-ACP (Voelker & Kinney, 2001).

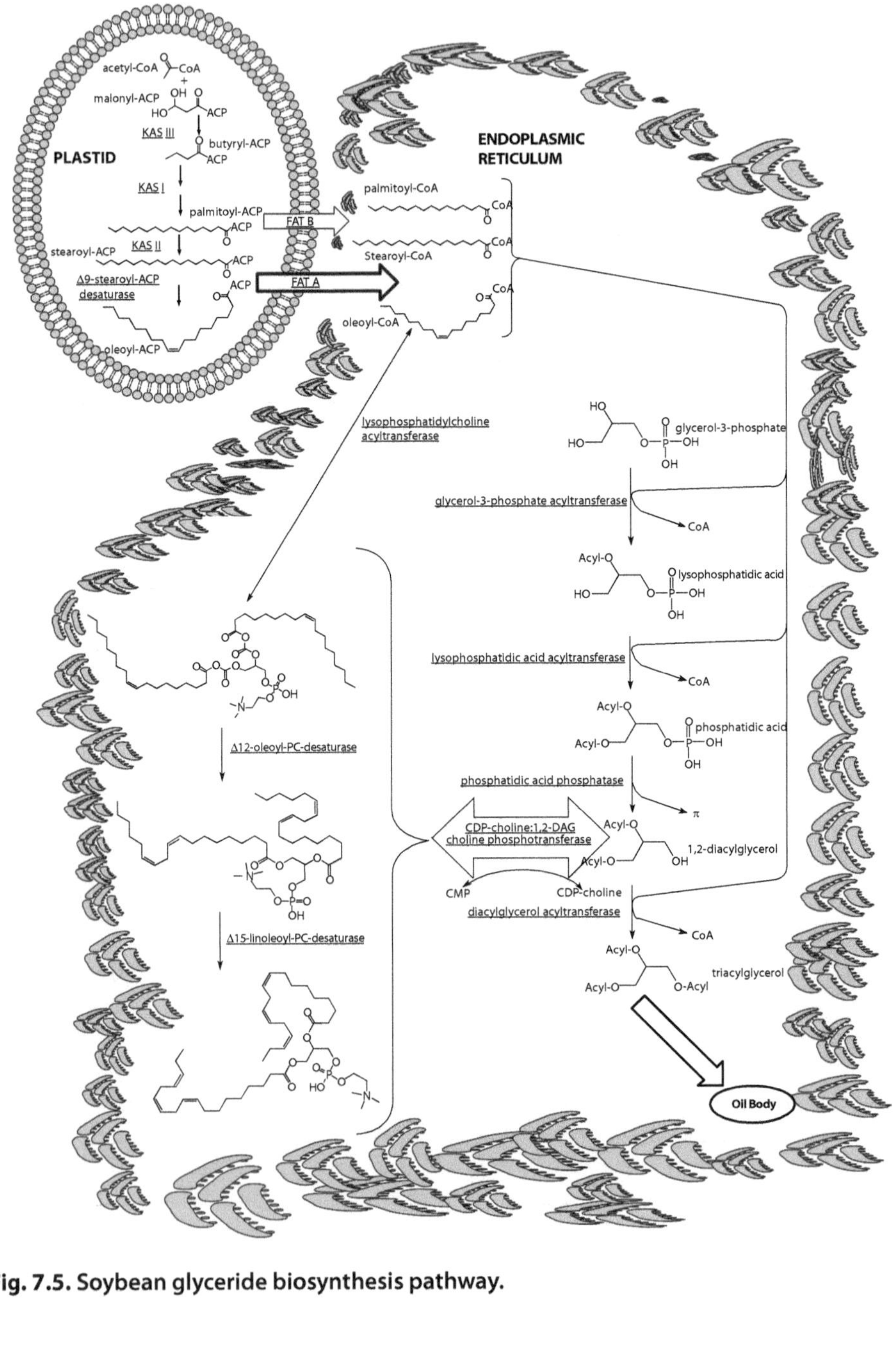

Fig. 7.5. Soybean glyceride biosynthesis pathway.

Triacylglyceride and Phospholipid Biosynthesis

Free fatty acids (FFA) are exported to the cytosol and esterified to CoA to form acyl-CoA. The export mechanism is not fully understood. Koo et al. (2004) proposed three possible mechanisms for the FA export in pea chloroplasts: intervening transfer proteins, facilitated diffusion, or acyl-CoA synthesis in the inner membrane of the chloroplast. Once in the cytosol, acyl-CoA is transesterified to glycerol-3-phosphate by glycerol-3-phosphate acyltransferase, to form lysophosphatidic acid with the FA ester in the *sn*-1 position. The esterification of another FA from acyl-CoA to lysophosphatidic acid is catalyzed by lysophosphatidic acid acyltransferase to form phosphatidic acid (PA) (Frentzen, 1998; Voelker & Kinney, 2001). This process occurs principally in the endoplasmic reticulum (ER). Phosphatidic acid can be hydrolyzed to yield diacylglycerol (DAG), which can be converted into TAG via esterification with acyl-CoA by DAG acyltransferase (Wilson, 2004; Lung & Weselake, 2006). In soybeans this enzyme has preference for linoleoyl-CoA over palmitoyl- and stearoyl-CoA (Cao & Huang, 1986). DAG acyltransferase also can catalyze the interesterification between two DAG molecules to yield TAG and MAG (Lung & Weselake, 2006). DAG also can be directed to phospholipid synthesis by reacting with cytidine 5'-phosphate (CDP)-choline or CDP-ethanolamine to form phosphatidylcholine (PC) or phosphatidylethanolamine (PE). If PA is converted to DAG-CDP, it can react with myo-inositol to yield phosphatidylinositol (Wilson, 2004; Ohlrogge & Browse, 1995). The formation of PC via CDP-choline:1,2-DAG choline phosphotransferase is reversible (Goode & Dewey, 1999).

Polyunsaturated Fatty Acid Biosynthesis

The FA can become polyunsaturated by further desaturation of oleic acid (18:1) esters present in phospholipids (Wilson et al., 1980). In soybean seeds, the formation of linoleic acid (18:2) is catalyzed by a microsomal enzyme, ω6-desaturase (FAD2) (Heppard et al., 1996). Two isoforms of this enzyme exist: FAD2-1 (seed specific and a key element in the control of 18:2 production) and FAD2-2 (found in vegetative tissues and seed) (Heppard et al., 1996). Linolenic acid is produced by desaturation of 18:2 esterified to PC; however, the possibility of desaturation of 18:2 occurring when esterified to something other than PC has not been ruled out. This reaction in oilseeds also occurs in cotyledon microsomes (Yadav et al., 1993). A family of three ω3 microsomal desaturases (FAD3A, FAD3B, and FAD3C) is responsible for this reaction (Bilyeu et al., 2003).

The FA from PC may be transesterified to DAG by phospholipid:diacylglycerol acyltransferase (PDAT) to form TAG (Wilson, 2004; Lung & Weselake, 2006). Lysophosphatidylcholine acyltransferase (LPCAT) can catalyze the FA exchange with acyl-CoA, contributing in this way with polyunsaturated acyl-CoA to the TAG synthesis (Lung & Weselake, 2006). Finally, the formation of CDP-choline and DAG via CDP-choline:1,2-DAG choline phosphotransferase also supplies DAG containing polyunsaturated FA to the TAG synthesis (Lung & Weselake, 2006).

Oil Bodies

Reserve lipids are stored during seed development in cotyledon oil bodies, also known as spherosomes, oleosomes, or lipid bodies (Huang, 1992). The origin of these organelles is still a matter of debate, however the most accepted possible starting point is the ER (Murphy, 2001). In soybean seeds, the oleosome diameter ranges between 0.2 and 0.5 μ (Liu, 1997). Their surface is covered by a phospholipid monolayer with proteins embedded: oleosins, caleosins (Murphy, 2004), and steroleosins (Lin & Tzen, 2004). Oleosins are the major protein in the oil body, consisting of 24 kD proteins (Herman, 1987; Herman et al., 1990), which contribute to the stability of the organelles, and prevent their coalescence (Huang, 1996).

Fatty Acid Composition of Typical and Modified Soybean Oils

Typical commodity soybean oil is rich in polyunsaturated FA (linoleic, 18:2 and linolenic acid, 18:3), with some contributions from saturated and monounsaturated FA (Table 7.2). The oil is suitable for many industrial and food uses. For industrial use, soybean oils with the typical FA composition noted are successfully used in the production of inks, lubricants, fuels, etcetera (United Soybean Board, 2005). They can be epoxidized to make plasticizers and coatings or interesterified to produce methyl soyate, which is used in solvent production. Soybean oil or methyl soyate also can be derivatized to produce soap, polyurethanes, composites, and paints, among other products (United Soybean Board, 2005).

Considerations of Fatty Acids in Oils for Food Use

Altering the typical FA make-up of soybean and other oils for food use is popular to improve oil nutritional and functional quality. One of the saturated FA, palmitic (16:0) acid, is thought to increase low-density lipoprotein (LDL)-cholesterol, which can lead to cardiovascular disease, when compared to lauric acid (12:0) (Denke & Grundy, 1992; Kris-Etherton & Yu, 1997), stearic acid (18:0), and oleic acid (18:1) (Bonanome & Grundy, 1988; Kris-Etherton & Yu, 1997). There is increasing evidence that stearic acid is neutral in its effects on plasma cholesterol concentration, and was reported to have the same effect as 18:1 (Bonanome & Grundy, 1988; Thijssen & Mensink, 2005). The U.S. nutritional labeling considers 16:0 and 18:0 to be saturated FA and their amounts are combined to equal the total saturated FA listed in the food. Oleic acid, the only major monounsaturated FA in soybean oil, has neutral to positive health benefits, because it lowers total and LDL-cholesterol (Bonanome &

Table 7.2. Typical Fatty Acid Composition of Commodity Soybean Oil

Fatty Acid	Content
Dodecanoic (lauric)	0-0.1
Tetradecanoic (myristic)	0-0.2
Hexadecanoic (palmitic)	8.0-13.5
Hexadecenoic (palmitoleic)[a]	0-0.2
Heptadecanoic (margaric)	0-0.1
Heptadecenoic[a]	0-0.1
Octadecanoic (stearic)	2.0-5.4
Octadecenoic (oleic)[a]	17.0-30.0
Octadecadienoic (linoleic)[a]	48.0-59.0
Octadecatrienoic (linolenic)[a]	4.5-11.0
Eicosanoic (arachidic)	0.1-0.6
Eicosenoic[a]	0-0.5
Eicosadienoic[a]	0-0.1
Docosanoic (behenic)	0-0.7
Docosenoic[a]	0-0.3
Tetracosanoic (lignoceric)	0-0.5

[a] The source does not specify position or geometric isomers. The authors consider them to be all *cis*.Source: Codex, (2005)

Grundy, 1988). Linoleic acid is an ω-6 FA, which is beneficial in the right amounts, but pro-inflammatory if too much is consumed, as described later. Linolenic acid, an ω-3 FA, is beneficial for heart and cardiovascular function (de-Lorgeril et al., 1994). It has recently received much attention in the United States by nutritionists who are worried about a decreasing ratio of dietary ω-3 to ω-6 FA. The recommended ratio of these FA types is 1:2.3 (Kris-Etherton et al., 2000). The relationship of FA ingestion to nutrition and health is covered in great detail in the Chapter: *Human Nutrition Value of Soybean Oil and Protein.*

Trans Fats

Although polyunsaturated FA generally are nutritionally desirable, their polyunsaturation also contributes to oxidative instability and degradation during frying. Until the late 1980s, soybean oils sold in retail stores in the United States as salad/cooking oils were routinely "brush" hydrogenated to bring the 18:3 concentration <3%, a level shown to significantly improve the oxidative and frying stabilities of the oil (Frankel, 1998). Concerns over reduction of the nutritionally desirable 18:3, and more particularly the ratio of ω-3 to ω-6 FA in U.S. diets, prompted oil manufacturers to simply eliminate the "brush" hydrogenation and to market the soybean oils as is, with the typical FA composition shown in Table 7.2.

Until very recently, "partial hydrogenation" was used widely, especially in the preparation of commercial frying fats, margarines, and shortenings. Frying fats are more stable to high-temperature abuse after hydrogenation, which increases the saturated FA and reduces the unsaturated FA, whereas the semi-solid functional characteristics of margarines and shortenings are provided by a greater percentage of saturated (and *trans*) FA. Margarines and shortenings also are stabilized against oxidative rancidity by increasing saturated FA and reducing unsaturated FA. A consequence of hydrogenation is an increase in *trans* FA isomers (Frankel, 1998). The consumption of fats with a high content of *trans* FA has been associated with negative health effects, including increased blood serum low-density lipoprotein cholesterol levels and decreased blood serum high-density lipoprotein levels (Mensink & Katan, 1990). Therefore, the U.S. Food and Drug Administration has required a labeling declaration of *trans* fats in foods containing 0.5 g or more *trans* fats per serving (FDA, 2003) since January 1, 2006. Canada initiated a similar labeling law on December 12, 2005 (Canadian Food Inspection Agency, 2003). Other countries, and private businesses, also have *trans*-fat labeling or nutritional guidelines aimed to limit *trans* fat intake. For example, on Nov. 24, 2005, Marks and Spencer retailer, a major company in the United Kingdom, banned the sales of hydrogenated fats coming from processed foods (Marks & Spencer, 2003). Also, the American Heart Association in the United States currently recommends *trans*-fat intakes of <1% of total calories consumed (Liechtenstein et al., 2006).

Fatty Acid Modifications to Increase Soybean Oil Stability, Enhance Oil/Fat Function, and/or Improve Nutrition

Soybean cultivars with altered FA composition have been developed in an effort to avoid hydrogenation and the subsequent *trans* fats formation. Breeders have targeted several traits to increase the oxidative stability of the oils. As stated previously, the most influential characteristic regarding oil stability is the 18:3 content. When soybean oils from low-linolenic-acid-containing varieties (A16 and A87–191039 with 1.8 and 1.9% 18:3, respectively) were tested during frying of bread cubes, the cubes showed superior oxidative and flavor stability than commodity-type soybean and canola oils containing 5.9–6.8 and 10.3% 18:3, respectively (Liu & White, 1992a). Similarly, potatoes fried in soybean oil containing 1.9 and 2.9% 18:3 (N85–2176 and N89–2009) had equal or better flavor scores than potatoes fried in commodity and partially hydrogenated soybean oils (Mounts et al., 1994). High-stearic (A6, 20.2% 18:0 and 5.9% 18:3), low-linolenic (A5, 3.5% 18:3), and commodity soybean oils (Hardin and BSR–101, with 6.9 and 8.1% 18:3, respectively) were compared during intermittent frying of bread cubes. The flavor stability of the bread cubes fried in those oils heated for 40 h was A5 > A6 > Hardin = BSR 101 (Miller & White, 1988b), demonstrating that the reduction in the 18:3 content had more influence on the flavor than the increase in 18:0 content during frying. During storage at 28 and 60 °C, oils from A5 and A6 cultivars had better performance (lower peroxide values and lower conjugated dienoic acid contents) than commodity soybean oil when compared by using chemical tests, but sensory evaluation did not show differences (Miller & White, 1988a).

Soybean oil with ~1% 18:3 is commonly referred to as ultra-low 18:3 oil. The beans are being grown and marketed by three companies, Asoyia LLC, Zeeland Farm Services, Inc., and Innovative Growers, LLC (Iowa State University, 2007). The oil containing <3% 18:3 is referred to as low-18:3 soybean oil. Companies providing low-18:3 soybean oils with ~3% 18:3 include TREUS™, a brand created from the alliance of Bunge and DuPont/Pioneer Hybrid companies, and Vistive™, an oil brand created from a network of several companies, including Monsanto, Cargill, Zeeland, and Agriculture Grain Processing (AGP).

The soybean industry continues to address the sometimes conflicting issues of soybean oil functionality and nutritional quality, and has been actively producing new cultivars containing oils with new and improved nutritional benefits, yet desirable functional FA profiles. In addition to the low-18:3 soybean oils developed to enhance oxidative and frying stabilities, high-oleic acid soybean oils, targeted for development by the "Better Bean Initiative" (Burton, 2006), also enhance oxidative stability. High-oleic and decreased linolenic acid soybean oil containing 79% 18:1 and 3.8% 18:3 had improved oxidative and frying stabilities when compared to commodity soybean oil (Su & White, 2004a; 2004b). However, oils containing high levels of 18:1 (around 75 to 80%) produced fried foods with less desirable flavors than more polyunsaturated oils (Warner, 2004). The high 18:1 oils are so stable during frying that they do not create the desirable flavorful breakdown products of typical or even mid-oleic oils, whose 18:2 levels provide substrate for breakdown to form some desirable (and some undesirable) volatile compounds we associate with good fried-food flavor (Warner & Gupta, 2005). Oil containing low-18:3 and increased 16:0 concentrations (A17) was more stable than commodity oils during frying and room-temperature storage (Liu & White, 1992a, 1992b).

Soybean oils with a high content of saturated FA (18:0 and 16:0) also have been tested in food applications, especially as feedstock for margarine production. Kok et al. (1999) developed a *trans*-free margarine using interesterified soybean oil with a high content of 16:0 (23%) and 18:0 (20%) and commodity soybean oil. The sensory and instrumental evaluations showed that the product was harder and more difficult to spread (Kok et al., 1999). High 18:0 soybean oils from three different cultivars (A6, Iowa State University, Ames, IA; HS-1, Jacob Hartz Seed Co., Stuttgart, AR; and A90–143073, Pioneer Hi-Bred International, Inc., Waterloo, IA) were interesterified.

Solid fat index (SFI) and drop point were determined for the interesterified and the natural products. The drop point was higher in the interesterified products, whereas the SFI was lower. The amount of solids present in the interesterified products over a range of temperatures (10–40°C) was high enough to meet the requirements of soft-tub margarine (List et al., 1997).

Several other new soybean oils currently being developed include those with saturated FA levels <7%. This type of oil, coupled with a screw press process and physical refining, is being marketed by Innovative Growers, LLC and Zeeland Farm Services, Inc. for health-food markets, with consumers using this oil as an all-purpose, home-use salad/cooking oil. The corresponding changes in the FA composition of the soybean oil means that the oil is particularly high in PUFA while maintaining low levels of saturated FA, which is nutritionally desirable. Nutritional labeling laws in the United States (FDA, 1993) require the amount of saturated FA to be declared if there is >0.5 g per serving. A more recent type of soybean oil combines either a low or ultra-low 18:3 composition with a mid-oleic acid level. For example, a mid-oleic soybean oil with 55–60% 18:1 and <3% 18:3 is being developed by Monsanto, with commercial availability aimed for 2008. Iowa State University should have a commercially available version of soybean oil with a mid-oleic acid concentration of about 52% and an ultra-low linolenic acid concentration of about 1.1% by the Fall of 2007, available through Asoyia LLC (Iowa State University, 2007). The mid-oleic seed for the development of the Iowa State University seed originated from Japan, with Professor Yutaki Takagi, of Saga University, who licensed the seed to Iowa State University (W. Fehr, personal communication). The intended use for soybean oil with mid-oleic, and low- or ultra-low 18:3 levels is for frying and baking.

An oil with a mid-oleic acid level of 55–60%, a low-linolenic acid level of <3% or an ultra-low linolenic acid level of 1%, and a low saturated FA composition of less than 7% would be highly desirable as a multi-purpose oil. Monsanto anticipates having soybean oil with 3% 18:3 after the year 2011. A soybean with a mid-oleic acid oil composition of 55–60%, an ultra-low-18:3 composition of 1%, and a low saturated FA composition of <7% is being developed at Iowa State University.

Screw-pressed soybean oils with 1.5 and 2.6% 18:3 (IA2064 and IA3018, respectively), and physically refined, together with a commodity-type soybean oil processed in the same way, were examined during commercial-like frying of French fries. Both 18:3-reduced oils performed better than the commodity oil (Gerde et al., 2007). Other food applications for soybean oils are further explained in the Chapter: *Food Uses for Soybean Oil and Alternatives to* Trans *Fatty Acids in Foods.*

Breeding to Obtain Soybean Oil for Non-edible Products

The introduction of novel FA through plant genetic modifications has allowed the production of oils with unique characteristics valuable in industrial processes. An example is the introduction of calendic acid (8, 10, 12 all *cis* – 18:3) in a proportion of approximately 20% (Cahoon et al., 2001; Cahoon, 2003). Calendic acid, originally found in marigold seed (*Calendula oficinalis*) is a conjugated polyunsaturated FA, which has drying properties that make the oil useful in coating products (Cahoon, 2003). The production of the 20-carbon monounsaturated FA, (5-eicosenoic acid), also has been achieved by inserting genes from meadowfoam (*Limnantes alba*) into soybeans (Cahoon et al., 2000; Cahoon, 2003). Oils rich in these FA could be used as industrial lubricants. The challenge is to produce soybeans containing these novel FA while maintaining the agronomic performance typical of commodity-type soybeans (Cahoon, 2003).

Biodiesel made with commodity soybean oil tends to crystallize when used at low temperature (-2°C). Biodiesel made from soybean oil with a low content of 16:0, together with branched chain alcohols, has a decreased crystallization temperature (-7°C) (Lee et al., 1995). In the same way, when methyl soyate made from low-16:0 soybean oil was winterized to decrease its saturation, and thus its crystallization temperature, the yield of the process was better than that from winterization of methyl soyate from commodity soybean oil (Lee et al., 1996).

Environmental Effects on Oil Composition

Temperature and climate affect the FA composition in many plant species. Typically, there is an inverse relationship between temperature and unsaturation level (Neidleman, 1987). The percentages of 16:0 and 18:0 in soybeans are not affected by different growing temperatures. However, polyunsaturated FA (18:2 and 18:3) decrease in soybeans grown at high temperatures. In contrast, the proportion of 18:1 (monounsaturated) increases with increased growing temperature (Wolf et al., 1982; Rennie & Tanner, 1989; Dornbos & Mullen, 1992; Primono et al., 2002). The reason for these FA changes could be due to reduced desaturase activity at increased temperature (Cheesbrough, 1989). In soybean cell suspension cultures, the level of 18:1 was higher in soybean cells grown at 25 than at 15 °C. Under the same comparative conditions, the level of 18:3 decreased at the higher temperature (MacCarthy & Stumpf, 1980).

Soybean oil FA composition seems to be affected by differences between day and night growing temperatures. Soybeans grown under moderate daytime temperatures and increased night temperatures produced a higher content of 18:2. The opposite effect was observed in beans grown at relatively high day temperatures and increased night temperatures, resulting in a lesser amount of 18:1 being desaturated to 18:2 (Gibson & Mullen, 1996).

The impact of changes in FA composition of the storage oil from soybeans can be extremely important, especially given the recent efforts by plant breeders to alter FA composition for specific food and industrial markets. If FA compositions are not stable and change substantially based on environment, oil ends up with neither desired nor predicted FA compositions. This lack of stability would produce huge economic losses for oil companies. Thus, finding genes imparting changes in FA composition that are stable to environmental factors is very important. For example, soybean genotypes with increased content of 18:1 (N 97-3363-4, N 98-4445A) showed very low FA content stability when grown in different environments (Columbia, MO; Portageville, MO; Sandhills, NC; Stoneville, MS) (Oliva et al., 2006). Soybeans with high 18:0 profiles (RG6, RG7, RG8) also were affected by the growing conditions (Primono et al., 2002) and growing years (A6) (Schnebly & Fehr, 1993). Genotypes with reduced 18:3 (RG10, C1640, IA 3017, IA 3018, S01-9370) were stable under changing growing conditions (Oliva et al., 2006; Primono et al., 2002).

Non-glyceride Lipid Components

Sphingolipids

Sphingolipid Structure

The topic of soybean sphingolipids has recently stimulated interest in soy foods from a human health perspective, because sphingolipid consumption has been associated with decreased colon and skin cancer in rats (Schmelz, 2000) and decreased plasma cholesterol levels (Kobayashi et al., 1997). Sphingolipids are cell membrane constituents involved in regulating cell metabolism (Lynch & Dunn, 2004). They have three characteristic components: 1) a FA, 2) a sphingosine or one of its derivatives, and 3) a polar head group. The FA is combined via an amide linkage at the amino group of the sphingosine (Fig. 7.6). Thus, the molecule has two non-polar tails and a polar head group. The resulting compound, the parent structure of all sphingolipids, is called ceramide (Fig. 7.7).

In higher plants, including soybeans, the sphingosine derivative is sphinganine, which is dihydrosphingosine (Sullards et al., 2000; Sperling & Heinz, 2003). Sphinganine can also be hydroxylated and unsaturated to yield 4,8-sphingadienine (Sullards et al., 2000) (Fig. 7.6). In soybeans, the most abundant sphingolipid is glucosylceramide (GlcCer). Most of the soybean GlcCer (>95%) contains 4,8-sphingadienine and α-hydroxypalmitic acid (Sullards et al., 2000). When the ceramide is linked through its primary hydroxyl group with a sugar molecule, the resulting compound is known as a cerebroside (Gutierrez & Wang, 2004). The glucose contributes to its polar character. Ceramide (Cer), without glucose, also is an important soybean sphingolipid, although it is found in much lower amounts than GlcCer (Wang et al., 2006b).

Presence and Content of Sphingolipids

The level of GlcCer in soybeans, measured as ceramide monohexoside using high-performance liquid-chromatography (HPLC) with evaporative light scattering detection, was 8.0 mg/100 g dry weight basis (dwb) (Sugawara & Miyazawa, 1999). Gutierrez et al. (2004) investigated several methods to determine GlcCer content in

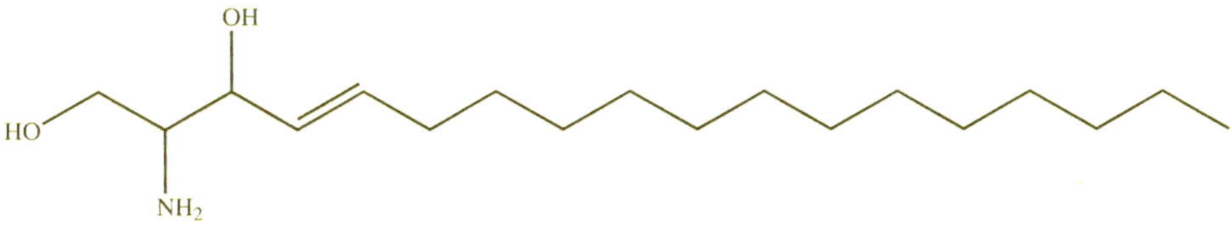

sphingosine

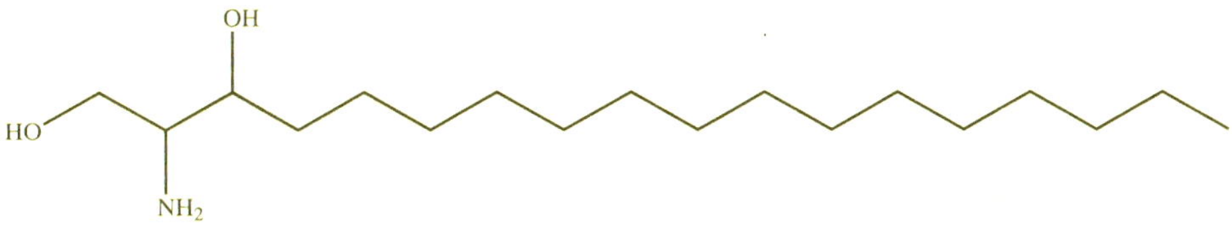

dihydrosphingosine

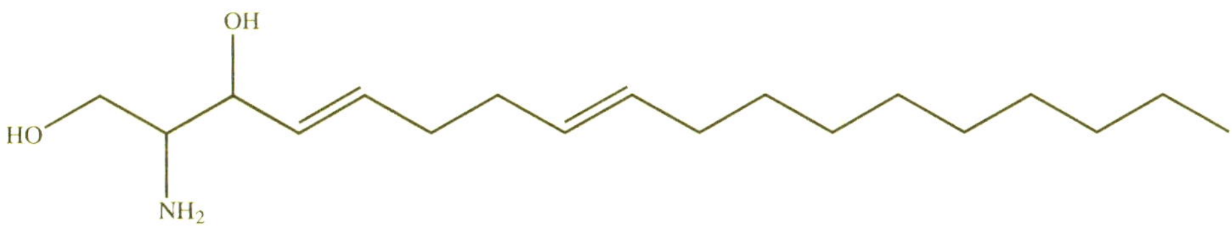

4,8-sphingadienine

Fig. 7.6. Structures of sphingosine and its derivatives present in soybeans.

soybeans, concluding that solvent partition extraction followed by preparative silica chromatography and HPLC quantification was the most effective procedure. The GlcCer content in soybean seeds from ten genotypes with a broad range of FA profiles, ranged from 142 to 493 nmol/g, dwb. There was no environmental effect on the GlcCer amount, but there was a tendency for immature seeds to contain higher levels of GlcCer than mature seeds (Gutierrez et al., 2004). Indeed, Wang et al. (2006a) later confirmed that the content of both GlcCer and Cer significantly decreased during seed development, with GluCer dropping from 522.8 nmol/g at 28 days after flowering (DAF) to 135.8 nmol/g at 68 DAF, and Cer dropping from 51.4 nmol/g at 28 DAF to 22.2 nmol/g at 68 DAF, all on a dwb.

Ranges of 83.4 to 397.6 nmol/g for GlcCer and of 8.4 to 20.7 nmol/g for Cer

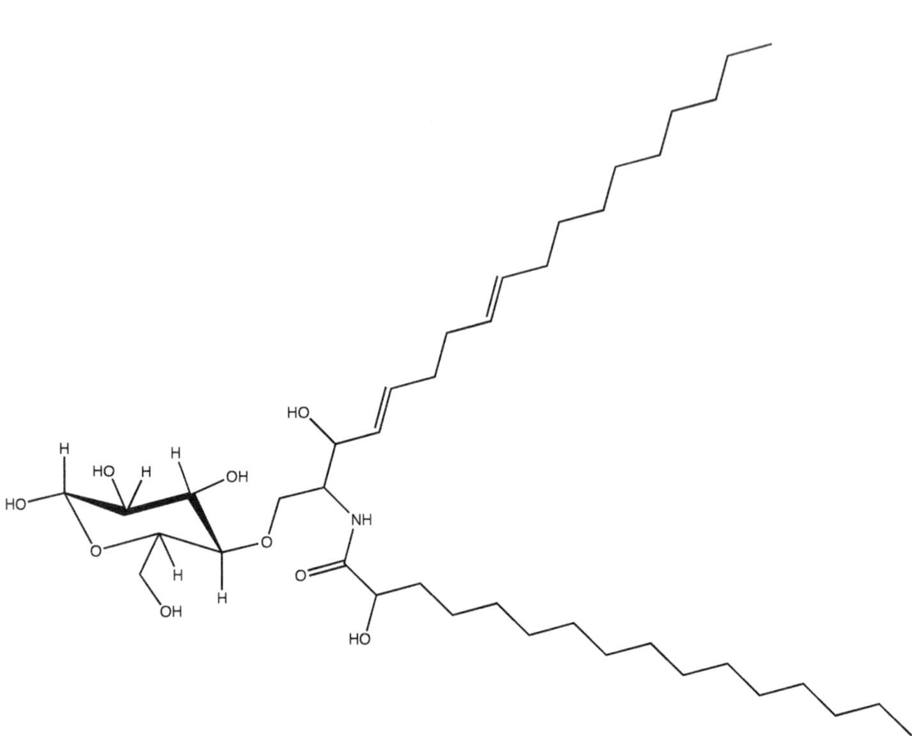

Fig. 7.7. Sphingolipid structure: glucosylceramide (GlcCer).

were found (on a dwb) in soybean seeds containing storage oil ranging between 3.7 and 40.7% 16:0 by using HPLC fitted with an evaporative light-scattering detector (Wang et al., 2006b). There was a positive correlation between Cer and GlcCer concentrations. Previous work (Merrill et al., 1988) had shown a positive relationship between free 16:0 in a cell culture and long-chain sphingoid base biosynthesis. Thus, Wang et al. (2006b) hypothesized that soybeans whose storage oil had high percentages of 16:0 might also have high concentrations of sphingolipids; however, the 16:0 concentration did not correlate with the sphingolipid concentration.

During soybean oil extraction, GlcCer remained mostly in the defatted soy flakes (Gutierrez & Wang, 2004). However, further processing to obtain soy protein isolates and concentrates resulted in GlcCer losses of between 48 and 74% (Gutierrez & Wang, 2004).

Tocopherols and Tocotrienols in Soybeans and Soybean Oil

Tocopherol and Tocotrienol Structures

Tocopherols, also known as tocols, are compounds derived from 2-methyl-2-(4,8,12-trimethyltridecyl)chroman-6-ol, and tocotrienols are compounds derived from

2-methyl-2-(4,8,12-trimethyltrideca-3,7,11-trienyl) chroman-6-ol (IUPAC, 1981). Tocopherols and tocotrienols differ in that the terpenic side chain of the tocopherols is saturated, whereas the side chain of the tocotrienols contains three double bonds. The α-, β-, γ-, and δ-tocopherol analogs differ in the number and position of methyl substituents they contain (Fig. 7.8).

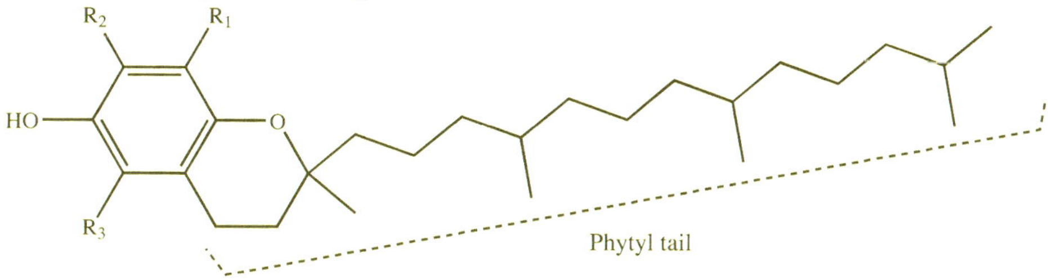

α-tocopherol	R_1= -CH_3	R_2= -CH_3	R_3= -CH_3
β-tocopherol	R_1= -CH_3	R_2= -H	R_3= -CH_3
γ-tocopherol	R_1= -CH_3	R_2= -CH_3	R_3= -H
δ-tocopherol	R_1= -CH_3	R_2= -H	R_3= -H

Fig. 7.8. Structure of tocopherols and different analogs present in soybean.

Presence and Contents of Tocopherols and Tocotrienols

Tocopherols exhibit antioxidant properties and contribute significantly to the oxidative stability of oils (Sherwin, 1976). Typical tocopherol concentrations for crude soybean oils, representing 14 lines of soybeans exhibiting conventional FA compositions grown in the Midwest, are reported in Table 7.3 (Dolde et al., 1999). Others have reported ratios of approximately 1:13:5 for α-, γ-, and δ-tocopherols in soybean oils (Jung et al., 1989; Evans et al., 2002). In general, tocotrienols were not detected in soybeans (Ko et al., 2003).

Table 7.3. Tocopherol Concentrations (ppm) in Crude Soybean Oil from Soybeans Grown in the Midwest

Tocopherol	Mean[a]	Range
α-Tocopherol	96	44-158
β-Tocopherol	11	2-29
γ-Tocopherol	1048	926-1559
δ-Tocopherol	372	254-477
Total	1527	1363-2195

[a] n = 14 lines of soybeans.
Source: Dolde et al., (1999)

Typical refining, bleaching, and deodorization of soybean oil decrease the total tocopherol concentration in the refined oil to 800-1100 ppm. However, the relative proportions of the tocopherol analogs were similar before and after processing (Jung et al., 1989). Chemical refining promoted greater tocopherol loss than physical refining and the loss of α-tocopherol was greater than that of the other analogs (Verleyen et al., 2002b). The greater the temperature and the longer the deodorization and physical refining time the greater the tocopherol loss (Jawad et al., 1984).

Antioxidant and Vitamin Properties of Tocopherols

Tocopherols, located in the plastids and thylakoid membranes of plants, protect the cell against highly oxidizing oxygen molecules produced during photosynthesis (Sattler et al., 2003). The tocopherol analogs vary in their antioxidant activities, and the relative effectiveness varies with the conditions. For example, α-tocopherol had the highest relative in vivo antioxidant activity, followed in order by β-, γ-, and δ-analogs, whereas under in vitro conditions the results were variable (Kamal-Eldin & Appelqvist, 1996). The in vivo vitamin E activity, as measured by Leth and Sondergaard (1977) (rat resorption-gestation test) paralleled the in vivo antioxidant activity (Kamal-Eldin & Appelqvist, 1996). When each one of the four analogs was tested separately in solution under 760 torr of oxygen at 30 °C, the order did not differ from that obtained from in vivo conditions (Burton & Ingold, 1981). When tested in in vitro systems of 18:2 and 18:2-methyl ester at 37 and 47 °C, γ-tocopherol was more stable than α-tocopherol (Gottstein & Grosch, 1990). Tocopherol stability and antioxidant activity were tested in corn oil heated at 70 °C and aerated at 100 mL/min. The order of the antioxidant activity was γ->δ->β->α-tocopherol (Chow & Draper, 1974). When tested in menhaden oil at 37 and 50 °C, both γ- and δ-analogs had greater antioxidant capacity than α-tocopherol (Olcott & Van Der Veen, 1968). In general, when tested in oils, fats, and lipoproteins, the order of the antioxidant activity was in the opposite direction from that obtained with in vivo studies, with the antioxidant activity being: δ- > γ-> β- and > α-tocopherol analogs (Kamal-Eldin & Appelqvist, 1996). Temperature, light, presence, and concentration of other pro- and anti-oxidants all impact the antioxidant effectiveness of the tocopherol analogs.

Effect of Tocopherol Levels on Soybean Oil Stability

Optimal tocopherol concentrations to maximize oxidative stability in soybean oil were 100, 250, and 500 ppm for α-, γ-, and δ-tocopherol, respectively, when tested individually in the dark at 55°C (Jung & Min, 1990). Similarly, at temperatures ranging from 40 to 60°C in the dark, optimal concentrations for α- and γ-tocopherols were ~100 and ~300 ppm, respectively; however, δ-tocopherol did not exhibit an optimum concentration under these conditions (Evans et al., 2002). Tocopherol concentrations are critical, because the compounds can act as pro-oxidants when in excess in the presence of other oxidation-promoting compounds, such as peroxides or metals (Kamal-Eldin & Appelqvist, 1996). Indeed, at greater than optimal concentrations, individual tocopherols and tocopherol mixtures were pro-oxidant, a behavior enhanced by increasing oil temperature from 40 to 60°C (Evans et al., 2002).

Warner (2005) tested tocopherol-stripped soybean and sunflower oils, to which pure tocopherols had been replaced in proportions typically found in these oils. At 60°C under darkness, soybean and sunflower oils with typical soybean tocopherol composition (low α- and high γ- and δ-) had better oxidative stabilities than did those with the typical sunflower tocopherol composition (high α- and low γ- and δ-). In contrast, when tested under light conditions at 30°C, oils with high α-tocopherol (sunflower composition) were more stable than oils with high γ- and δ- (soybean composition) (Warner, 2005), likely a result of the higher capacity of the α-analog to prevent singlet oxygen oxidation (Frankel, 1998; Warner, 2005).

Accumulation and Distribution of Tocopherols in the Seed

The rate of tocopherol accumulation in soybeans was maximum between 30 and 45 DAF, coinciding with the period when the oil accumulation rate also was maximum (Almonor et al., 1998). Positive correlations were noted in oils of mature soybeans between γ-tocopherol and 18:3 concentrations and between γ-tocopherol and growing temperature (Almonor et al., 1998).

Tocopherols are unevenly distributed in the seed, with the embryonic axis containing the greatest concentration followed by the cotyledon and then the seed coat (Yoshida et al., 1998, 2006a; Ko et al., 2003) (Table 7.4). In the embryonic axis, γ-tocopherol was present in the greatest amounts (74.2-80.0%) followed by δ-tocopherol (12.4–15.1%), α-tocopherol (6.8–12.0%) and β-tocopherol (0.4–0.7%) (Yoshida et al., 2006b). The high percentage of 18:3 in the axis may be a reason for the high γ-tocopherol concentration in the axis (Yoshida et al., 2006b), as noted in the previous paragraph.

Table 7.4. Tocopherol Concentrations (ppm) in the Oil Extracted from Various Soybean Seed Sections

Seed Section	α-Tocopherol	β-Tocopherol	γ-Tocopherol	δ-Tocopherol
Cotyledon	25[a]-160[b]	33[a]-40[b]	640[b]-933[a]	320[b]-900[a]
Axis	150[a]-1000[b]	33[a]-100[b]	900[b]-2150[b]	100[b]-400[a]
Seed coat	33[a]-183[b]	33[a]-50[b]	133[b]-288[b]	75[b]-225[b]

[a]Approximate concentrations derived from graphical data. Yoshida et al. (1998). $n = 3$ soybean cultivars.

[b]Approximate concentrations derived from graphical data. Yoshida et al. (2006a). $n = 4$ soybean lines.

Phytosterols in Soybeans

Phytosterol Structures

Phytosterols, triterpenes originating from squalene, are steroid compounds naturally present in plants. Their core structures are cyclopenta[*a*]phenanthrene with a hydroxyl group at C3 (IUPAC, 1989), possible methyl substituents at C10 and C13, and an alkyl side chain at C17 (Fig. 7.9). In plants, phytosterols may exist as free sterols, as steryl esters, steryl glycosides, or acylated steryl glycosides. Free sterols and part of the steryl conjugates are incorporated in the cell membranes (Wojciechowski, 1991) where they play a role in preserving the functionality of the membrane (Piironen et al., 2000). Steryl esters are mostly found within the cell as a reserve (Piironen et al., 2000). An early study demonstrated that the proportion of free and conjugated sterols during soybean seed maturation remained relatively constant (50–70% free sterols, 5–20% steryl esters, 5–20% steryl glycosides, and 10–30% acylated steryl glycosides) (Katayama & Katoh, 1973).

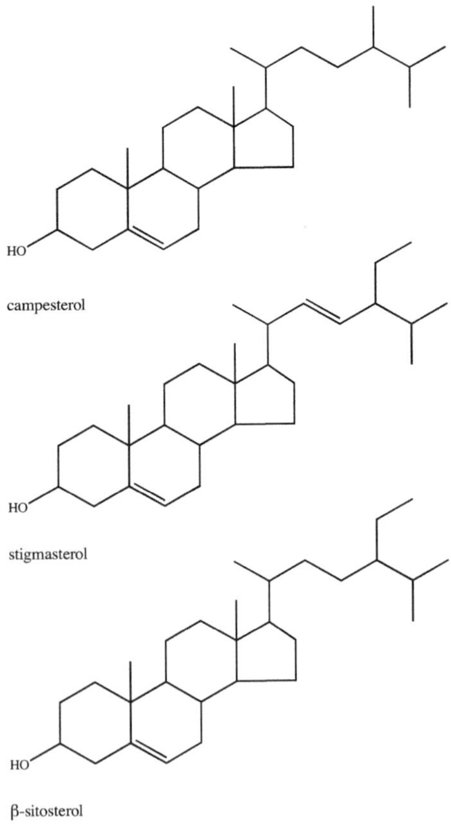

Fig. 7.9. Structures of the most abundant sterols in soybeans.

Presence and Contents of Phytosterols

There are several types of phytosterols present in soybean seeds, with campesterol, stigmasterol, and β-sitosterol present in the greatest amounts. Others present in much smaller quantities include Δ5-avenasterol, Δ7-avenasterol, and Δ5-stigmastenol. Table 7.5 shows typical ranges of these sterols in mature soybeans. The composition of the free fraction differs substantially from the esterified one (Verleyen et al., 2002a) as noted in Table 7.6. Campesterol, sitosterol, and stigmasterol decrease the membrane permeability and regulate its fluidity by limiting the movement of fatty acyl chains (Piironen et al., 2000). When included in soybean phosphatidylcholine bilayers, sitosterol was very effective in decreasing the water permeability (Schuler et al., 1991).

The FA composition of soy steryl esters differs from that of soy oil triacylglycerols. The proportion of 18:2 in the steryl esters was much lower than in the triacylglycerols (Ferrari et al., 1997). In contrast, the percentages of 18:3, 20:0, 22:0, and 24:0 were greater in the steryl esters than in the oil (Ferrari et al., 1997).

Table 7.5. Ranges of Phytosterol Concentrations in Oil from Mature Soybeans

Phytosterol	Range (ppm)
β-Sitosterol	650[a]-2360[b]
Campesterol	248[a]-1310[b]
Stigmasterol	219[a]-770[b]
Δ5-Avenasterol	27[a]-135[b]
Δ5-Stigmastenol	28[c]-150[e]
Δ7-Avenasterol	20[d]-40[e]
Total	1210[a]-4050[b]

[a] Ferrari et al. (1997)
[b] Vlahakis & Hazebroek (2000)
[c] Jawad et al. (1984)
[d] Weihrauch & Gardner (1978)
[e] Dutta & Appelqvist (1996)

Table 7.6. Distribution (%) of Phytosterols in Free and Esterified Fractions from Soybean Oil[a]

Sterol	Esterified	Free
Campesterol	10.4	22.6
Stigmasterol	7.8	21.7
β-Sitosterol	67.3	55.7
Δ5-Avenasterol	14.5	-

[a] Estimated from Verleyen et al. (2002a)

Processing Effects on Sterol Content and Composition

During oil refining, both the free and esterified sterols were reduced with total sterol losses ranging between 18% (Ferrari et al., 1996) and 34% (Verleyen et al., 2002b). During the neutralization stage of chemical refining, 20% of the sterols, mostly the free sterols, were lost in the soapstock (Verleyen et al., 2002b). Sterols also can be removed from oil during deodorization. Free sterols are volatile under deodorization conditions, whereas steryl esters are not (De Greyt & Kellens, 2005). When chemically neutralized, the oil arrives at the deodorizer with a very low content of free FA so there is little to no esterification with free sterols. In contrast, the content of free FA is higher in physically refined oils; thus, under deodorization conditions, the free FA can react with free phytosterols to yield non-volatile steryl esters, which increase the esterified phytosterol fraction (Verleyen et al., 2002b). The free phytosterol fraction and the total phytosterol composition differed substantially in composition (Verleyen et al., 2002a; Ferrari et al., 1997). Figure 7.10 shows the relative proportions of the phytosterols in each fraction obtained during refining. Although there were some sterol losses during oil refining, as previously mentioned, the proportions of the various sterols in the fractions did not seem to be altered (Ferrari et al., 1997).

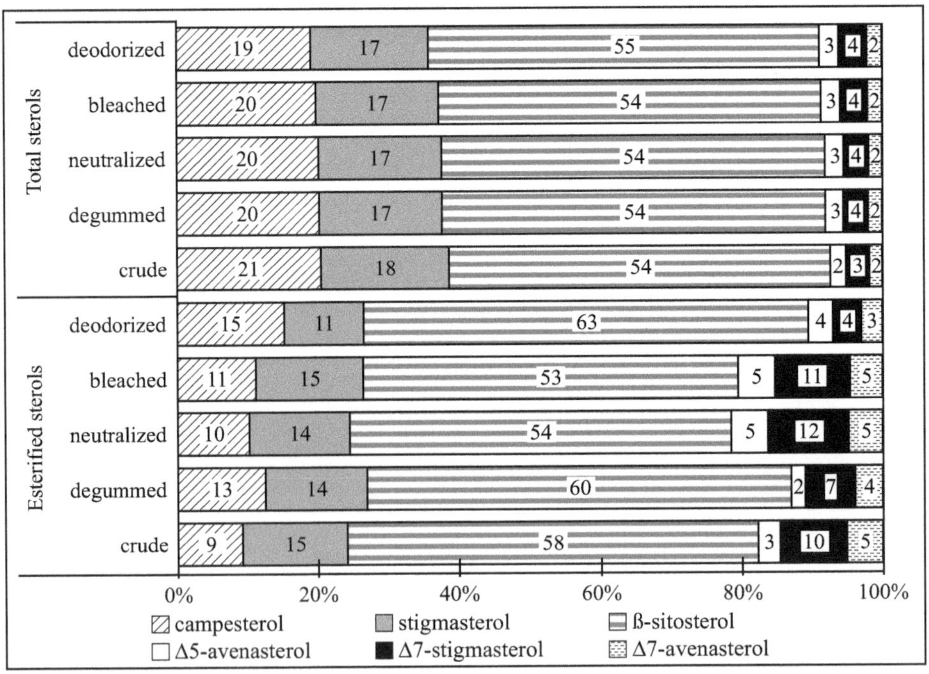

Fig. 7.10. Distribution (%) of soybean oil phytosterols after various stages of processing (Ferrari et al., 1997).

Significant amounts of sterols are lost during oil heating at frying temperatures (Ghavami & Morton, 1984). However, the antioxidant properties of Δ7-avenasterol, a minor sterol in soybean oil, were reported at frying temperatures (Sims et al., 1972; Yan & White, 1990).

Environmental Effects on Sterol Content and Composition

The phytosterol content is affected by many factors. For example, in soybeans genetically modified to alter FA composition, the total sterol content increased when the growing temperature increased (Vlahakis & Hazebroek, 2000). In addition, the composition changed, with a greater proportion of campesterol and a lesser proportion of both stigmasterol and β-sitosterol occurring at warmer growing temperatures. For commodity soybeans with typical FA compositions planted in the Midwest, the planting locations (Johnston, IA; LaSalle, IL; Jasper, MI; Napoleon, OH; Pocahontas, IA) and the genotypes (2396, 2506, 2835 by Asgrow, Urbandale, IA; 2990 by Agripro, Ames, IA; JACK; 9255, 9281 by Pioneer, Des Moines, IA; 1990, 2918 by Novartis, Minneapolis, MN; 2621 by Stine, Adel, IA; 2660; YB30M) significantly affected the total phytosterol content but no single parameter was responsible for the changes. Also, no significant correlations occurred between either the sterol and tocopherol contents, or the sterol and FA unsaturation levels (Vlahakis & Hazebroek, 2000). In 510 cultivars (263 Japanese and 247 non-Japanese) harvested between 1981 and 1996, the total sterol content was influenced by the variety; however, the sterol composition was not affected (Yamaya et al., 2007). In two cultivars (Kuromame and Shirodaizu) tested in different planting locations within Japan there was a tendency for the sterol concentration to be greater in seeds harvested in warmer areas, but the sterol composition was not affected (Yamaya et al., 2007). Also, in agreement with a previous study (Vlahakis & Hazebroek, 2000), the correlation between concentrations of tocopherols and phytosterols was not significant (Yamaya et al., 2007).

Effects of Sterols on Health

The health effects of phytosterols have been studied in great detail in recent years, with evidence suggesting that phytosterol consumption decreases blood cholesterol levels (Kritchevsky & Chen, 2005). The intake of 2 g/d of phytosterols or phytostanols may reduce the low-density lipoprotein (LDL) cholesterol level in blood serum by 10% (Katan et al., 2003). Dressings (Italian and ranch) fortified with soybean sterol esters successfully reduced the LDL-cholesterol level by 17% in mildly hypercholesterolemic adults when compared to their level at the beginning of the study, three weeks earlier (Judd et al., 2002). The intake of more than 3.2 g/d had no additional cholesterol lowering effect (Clifton et al., 2004). However, a regular intake of 6.6 g/d also may produce a reduction in the level of plasma carotenoids, which is not a desirable nutritional outcome. This change was accompanied by an increase in the plasma phytosterol level (Clifton et al., 2004).

Lipoxygenase

Lipoxygenase Enzymes in Soybean Seed

Lipoxygenases (linoleate:oxygen oxidoreductase, EC 1.13.11.12, LOX) are enzymes belonging to a group of non-heme-iron-containing proteins (Brash, 1999). They catalyze the oxidation of the FA containing a *cis, cis*-1,4-pentadiene group (Song et al., 1990; Siedow, 1991). The typical substrates of plant lipoxygenases are linoleic (18:2) and α-linolenic acid (18:3). Lipoxygenases can be classified as 9-LOX and 13-LOX depending on the type of hydroperoxide formed, which yield 9- and 13- hydroperoxides, respectively (Feussner & Wasternack, 2002) (Fig. 7.11).

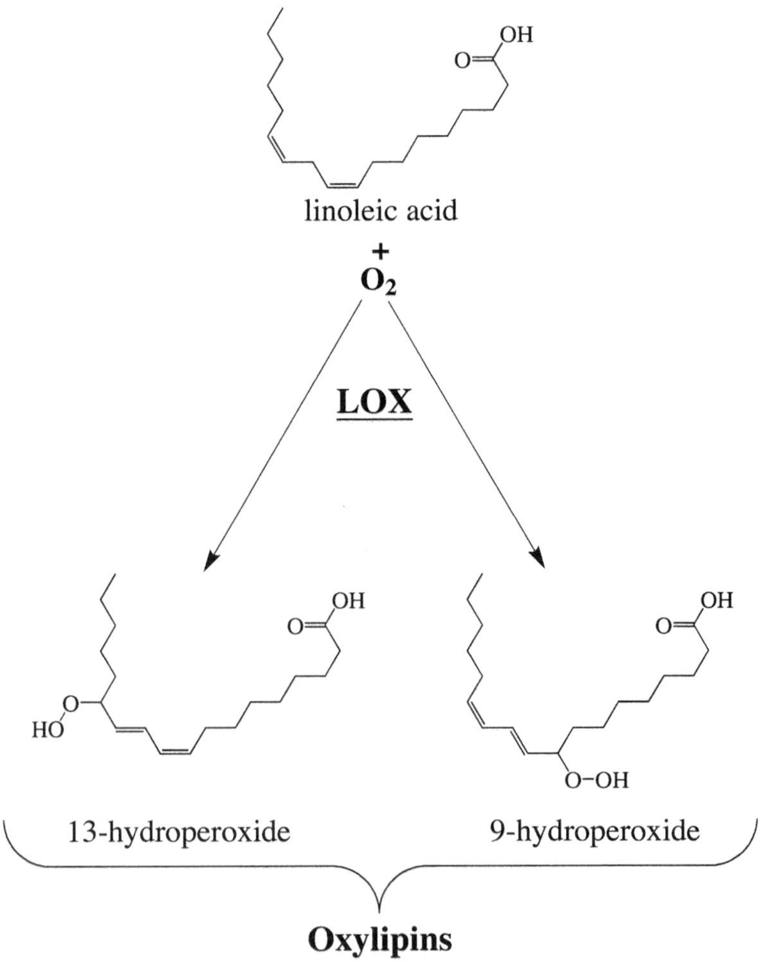

Fig. 7.11. General illustration of soybean lipoxygenase-catalyzed reactions.

In soybeans, eight lipoxygenase isozymes have been identified (Brash, 1999). However, mature seed cotyledons contain primarily three lipoxygenase isozymes: LOX-1, LOX-2, and LOX-3 (Axelrod et al., 1981). The 13-hydroperoxide is the only product produced in LOX-1-promoted oxidation of 18:2, whereas LOX-2 produced roughly equal amounts of 13- and 9-hydroperoxide from 18:2 (Axelrod et al., 1981). These authors also reported that LOX-3 produced 65% of the 13-isomer and 35% of the 9- isomer; however, Christopher et al. (1972) noted that the proportion of the products of LOX-3 was highly dependent on the reaction conditions. The presence of calcium ions increased the LOX-2 activity without favoring any specific isomer at pH 7-9. However, when calcium was not present in the system, the ratio of 13- to 9-isomer produced by LOX-2 changed from 38:62 at pH 7 to 60:40 at pH 9 (Christopher et al., 1972). Another study showed that under in vitro conditions (linoleic acid as substrate and aeration at 20°C), LOX-2 produced 13- and 9-hydroperoxy linoleic acids in a ratio of 4:1, and LOX-3 in a ratio of 1:2 (Fukushige et al., 2005). The resulting hydroperoxides can be further derivatized by LOX and other enzymes to produce substances with sensory properties that may affect quality in food products. Hexanal, which has been associated with undesirable beany flavor, is one example of a secondary product produced by lipoxygenase (Fujimaki et al. 1965; Wilson, 1996).

The activities of LOX-1 and LOX-2 are highest at pH values 9 and 6.5, respectively, for 18:2 substrates (Axelrod et al., 1981). LOX-3 is active in a wide range of pHs, centering at pH 7 (Axelrod et al., 1981). LOX-2 and -3 were more susceptible to heat treatment than LOX-1. After 20 min at 70 °C, LOX-2 and LOX-3 were inactive, whereas LOX-1 required 120 min at the same temperature to be totally inactivated (Hildebrand & Kito, 1984). LOX-2 and -3 had a marked preference for 18:3 over 18:2 as the substrate, whereas LOX-1 showed a higher relative activity with 18:2 (Kato et al., 1992).

The three isozymes are localized in the cytoplasm of the cotyledon cells (Song et al., 1990; Wang et al., 1999) and in their protein-storage vacuoles. Their physiological roles are not yet fully understood. Some studies propose that their function might be the oxygenation of FA to facilitate their transport to the glyoxisomes (Song et al., 1990; Vernooy-Gerritsen et al., 1984). Later studies suggested they are not involved in the lipid mobilization during germination, but rather work as storage proteins (Siedow, 1991; Wang et al., 1999). No harmful consequences have been observed in mutants lacking lipoxygenases (Siedow, 1991). The lipoxygenases may play a role in the defense of the seed during its development by acting to form jasmonic acid and other oxylipins (oxygenated polyunsaturated FA derivatives) (Wang et al., 1999; Blée, 1996).

During germination, LOX-1, -2, and -3 activities decrease (Song et al., 1990; Kato et al., 1992; Wang et al., 1999) and new isozymes appear: LOX-4, -5, and -6. LOX-4 produced 13- and 9- hydroperoxy linolenic acid in a ratio of 46:54 and both LOX-5 and -6 produced these hydroperoxides radicals in a ratio of 85:15 (Kato et al.,

1992). Their maximum activities were achieved at pH 6.5, with a preference for 18:3 over 18:2 as the substrate (Kato et al., 1992). The LOX-4, -5 and -6 also are cytoplasmic enzymes, which are not directly associated with triacylglyceride mobilization during seed germination and, in the same way as LOX-1, -2, and -3, they may be part of a defense mechanism (Wang et al., 1999). Another study showed the presence of a membrane-bound lipoxygenase in germinating soybean cotyledons (Fornaroli et al., 1999). This membrane-bound enzyme had some similarities to LOX-1, such as its optimum pH, size, and preference to produce 13-hydroperoxy linoleic acid. Soluble cytosolic lipoxygenases were proposed to be translocated to the cell membrane (Fornaroli et al., 1999).

Effect of Lipoxygenases on Food Quality

The effect of lipoxygenases on food quality also has been studied extensively. The enzymes are involved in the synthesis of substances with grassy or beany sensory properties undesirable in food products (Wilson, 1996), a factor leading to the development of mutant soybean lines lacking lipoxygenase isozymes (Davies & Nielsen, 1986, 1987; Narvel et al., 1998). Lipoxygenase-null genotypes had yields, seed weights, and protein contents the same as those of normal lines (Narvel et al., 1998). Genotypes combining the lipoxygenase-null and low 18:3 content traits had lower total oil amounts, but greater protein contents and seed weights than typical cultivars (Reinprecht et al., 2006). When mutant genotype seeds lacking LOX-1 and LOX-3, LOX-2 and LOX-3, or LOX-3, and a line containing the three isozymes were tested for longevity during storage no differences were noted (Trawatha et al., 1995).

High-temperature stabilities of oils from soybeans lacking LOX-2 and -3 or LOX-2 (with low and normal 18:3 concentrations) were tested. The lack of LOX did not improve these oils (Shen et al., 1997). Also, the sensory quality of crude oil from LOX-1 null soybeans was no better than that from commodity varieties (Engeseth et al., 1987). Similarly, oils from LOX-free beans were not significantly improved in flavor or oxidative stability (King et al., 1998). The flavor of bread, meat patties, and beverage products was not improved when LOX-free soybeans were used in place of normal soybean products (King et al., 2001). In contrast, soymilk and tofu made from LOX-free soybeans had less cooked beany aroma and flavor compared to products made from commodity beans (Torres-Penaranda et al., 1998). Likely, the very bland flavors of soymilk and tofu enhanced the impact of LOX in the products made from commodity beans, making them more apparent. As a result of these and other studies, LOX-null soybeans are not being developed extensively by seed companies and universities.

Conclusion

Much work has helped characterize the lipids in soybeans and the understanding of their biosynthesis. In addition, there has been great progress in breeding soybeans to contain a variety of fatty acid profiles so that soybeans can be grown to produce specific fatty acid arrangements. Although there are many reports of the minor constituents in soybeans, we are just beginning to fully understand the interrelations of these minor constituents and their impact on lipid and oil quality, as well as their nutritional contributions. Future research on the lipids in soybeans should focus on these factors further to improve our overall understanding of oil quality, stability, and nutritional value so that plant breeders can incorporate these minor desired traits into soybeans, just as they have done with modified fatty acid profiles. The use of soybean, as well as other vegetable oils, in biodiesel production also will be enhanced by these types of studies.

References

Almonor, G.O.; G.P. Fenner; R.F. Wilson. Temperature effects on the tocopherol composition in soybeans with genetically modified improved oil quality. *J. Am. Oil Chem. Soc.* **1998,** *75,* 591–596.

Axelrod, B.; T.M. Cheesbrough; S. Laakso. Lipoxygenase from soybeans. *Methods Enzymol.* **1981,** *71,* 441–451.

Bilyeu, K.D.; L. Palavalli; D.A. Sleper; P.R. Beuselink. Three microsomal omega-3 fatty-acid desaturase genes contribute to soybean linolenic acid levels. *Crop Sci.* **2003,** *43,* 1833–1838.

Blée, E. Phytooxylipins: The Peroxygenase Pathway. In *Lipoxygenase and Lipoxygenase Pathway Enzymes.* Piazza, G.J. Ed.; AOCS Press: Champaign, IL, 1996; pp. 138–161.

Bonanome, A.; S.M. Grundy. Effect of dietary stearic acid on plasma cholesterol and lipoprotein levels. *N. Engl. J. Med.* **1988,** *318,* 1244–1248.

Brash, A.R. Lipoxygenases: occurrence, functions, catalysis, and acquisition of substrate. *J. Biol. Chem.* **1999,** *274,* 23679–23682.

Burton, G.W.; K.U. Ingold. Autoxidation of biological molecules. 1. The antioxidant activity of vitamin E and related chain-breaking phenolic antioxidants in vitro. *J. Am. Chem. Soc.* **1981,** *103,* 6472–6477.

Burton, J. Improving Soybean Oil Quality Through the Better Bean Initiative. *97th AOCS Annual Meeting and Expo Book of Abstracts.* Saint Louis, MO, 2006. AOCS: Champaign, IL,

Byfield, G.E.; H. Xue.; R.G. Upchurch. Two genes from soybean encoding soluble Δ9-stearoyl-ACP desaturases. *Crop Sci.* **2006,** *46,* 840–846.

Cahoon, E.B. Genetic enhancement of soybean oil for industrial uses: prospects and challenges. *AgBioForum* **2003,** *6,* 11–13, 2003.

Cahoon, E.B.; E.F. Marrillia; K.L. Stecca; S.E. Hall; D.C. Taylor; A.J. Kinney. Production of fatty acid components of meadowfoam oil in somatic soybean embryos. *Plant Physiol.* **2000,** *124,* 243–251.

Cahoon, E.B.; K.G. Ripp; S.E. Hall; A.J. Kinney. Formation of conjugated delta 8, delta 10 double bonds by delta 12-oleic acid desaturase related enzymes: biosynthetic origin of calendic acid. *J. Biol. Chem.* **2001,** *276,* 2637–2643.

Canadian Food Inspection Agency. Guide to Food Labeling and Advertising, 2003. http://www.inspection.gc.ca/english/fssa/labeti/guide/toce.shtml (accessed October 2006).

Cao, Y.Z.; A.H.C. Huang. Diacylglycerol acyltransferase in maturing oil seeds of maize and other species. *Plant Physiol.* **1986,** *82,* 813–820.

Cheesbrough, T.M. Changes in the enzymes for fatty acid synthesis and desaturation during acclimation of developing soybean seeds to altered growth temperature. *Plant Physiol.* **1989,** *90,* 760–764.

Chow, C.K.; H.H. Draper. Oxidative stability and antioxidant activity of the tocopherols in corn and soybean oils. *Internat. J. Vit. Nutr. Res.* **1974,** *44,* 396–403.

Christopher, J.P.; E.K. Pistorius; F.E. Regnier; B. Axelrod. Factors influencing the positional specificity of soybean lipoxygenase. *Biochim. Biophys. Acta* **1972,** *289,* 82–87.

Clifton, P.M.; M. Noakes; D. Ross; A. Fassoulakis; M. Cehun; P. Nestel. High dietary intake of phytosterol esters decreases carotenoids and increases plasma plant sterol levels with no additional cholesterol lowering. *J. Lipid Res.* **2004,** *45,* 1493–1499.

Codex Alimentarius. Codex Standard for Named Vegetable Oils CODEX-STAN 210 (Ammended 2005). http://www.codexalimentarius.net/web/standard_list.do?lang=en (accessed June 2007).

Davies, C.S.; N.C. Nielsen. Genetic analysis of a null-allele for lipoxygenase-2 in soybean. *Crop Sci.* **1986,** *26,* 460–463.

Davies, C.S.; N.C. Nielsen. Registration of soybean germplasm that lacks lipoxygenase isozymes. *Crop Sci.* **1987,** *27,* 370–371.

De Greyt, W.; M. Kellens. Deodorization. In *Bailey's Industrial Oil and Fat Products,* 6th Edn.; Shahidi, F. Ed.; John Wiley and Sons, Inc.: Hoboken, NJ, 2005; Vol. 5, pp. 341–383.

de-Lorgeril, M.; S. Renaud; N. Mamelle; P. Salen; J.L. Martin; I. Monjaud; J. Guidollet; P. Touboul; J. Delaye. Mediterranean alpha-linolenic acid-rich diet in secondary prevention of coronary heart disease. *Lancet* **1994,** *343,* 1454–1459.

Denke, M.A.; S.M. Grundy. Comparison of effects of lauric acid and palmitic acid on plasma lipids and lipoproteins. *Am. J. Clin. Nutr.* **1992,** *56,* 895–898.

Dolde, D.; C. Vlahakis; J. Hazebroek. Tocopherols in breeding lines and effects of planting location, fatty acid composition, and temperature during development. *J. Am. Oil Chem. Soc.* **1999,** *76,* 349–355.

Dornbos, D.L.; R.E. Mullen. Soybean seed protein and oil contents and fatty acid composition adjustments by drought and temperature. *J. Am. Oil Chem. Soc.* **1992,** *69,* 228–231.

Dutta, P.C.; L.A. Appelqvist. Saturated sterols (stanols) in unhydrogenated and hydrogenated edible vegetable oils and in cereal lipids. *J. Sci. Food Agric.* **1996,** *71,* 383–391.

Engeseth, N.J.; B.P. Klein; K. Warner. Lipoxygenase isoenzymes in soybeans: effect on crude oil quality. *J. Food. Sci.* **1987,** *53,* 1015–1019.

Evans, J.C.; D.R. Kodali; P.B. Addis. Optimal tocopherol concentrations to inhibit soybean oil oxidation. *J. Am. Oil Chem. Soc.* **2002,** *79,* 47–51.

Ferrari R.Ap.; W. Esteves; K.D. Mukherjee; E. Schulte. Alteration of sterols and steryl esters in vegetable oils during industrial refining. *J. Agric. Food Chem.* **1997,** *45,* 4753–4757.

Ferrari R.Ap.; E. Schulte; W. Esteves; L. Brühl; K.D. Mukherjee. Minor constituents of vegetable oils during industrial processing. *J. Am. Oil Chem. Soc.* **1996,** *73,* 587–592.

Feussner, I.; C. Wasternack. The lipoxygenase pathway. *Annu. Rev. Plant Physiol Plant Mol. Biol.* **2002,** *53,* 275–297.

Food and Drug Administration, 21 CFR Part 101 Food Labeling: *trans* Fatty Acids in Nutrition Labeling; Consumer Research to Consider Nutrient Content and Health Claims and Possible Footnote or Disclosure Statements; Final Rule and Proposed Rule, *Federal Register* Vol. 68 133:41433–41506, 2003. http://www.a257.g.akamaitech.net/7/257/2422/14mar20010800/edocket.a ccess.gpo.gov/2003/pdf/03-17525.pdf (accessed June 2006).

Food and Drug Administration, 21 CFR Part 101 Food Labeling: Mandatory Status of Nutrition Labeling and Nutrient Content Revision, Format for Nutrition Label; Final Rule, *Federal Register* Vol. 58 3:2079–2205, 1993. http://www.foodrisk.org/FedRegister/FedRegisterVol58PartIV-2079-2205.pdf (accessed June 2007).

Fornaroli, S.; S. Petrussa; E. Braidot; A. Vianello; F. Macri. Purification of a plasma membrane-bound lipoxygenase from soybean cotyledons. *Plant Sci.* **1999,** *145,* 1–10.

Frankel, E.N. *Lipid Oxidation.* The Oily Press: Dundee, Scotland, 1998.

Frentzen, M. Acyltransferases from Basic Science to Modified Seed Oils. *Fett/Lipid* **1998,** *100,* 161–166.

Fujimaki, M., S. Arai; N. Kirigaya; Y. Sakurai. Studies of flavour compounds in soybean. Part 1. Aliphatic carbonyl compounds. *Agric. Biol. Chem.* **1965,** *29,* 855–858.

Fukushige, H.; C. Wang; T.D. Simpson; H.W. Gardner; D.F. Hildebrand. Purification and identification of linoleic acid hydroperoxides generated by soybean seed lipoxygenases 2 and 3. *J. Agric. Food Chem.* **2005,** *53,* 5691–5694.

Ghavami, M.; I.D. Morton. Effect of heating at deep-fat frying temperature on the sterol content of soya bean oil. *J. Sci. Food Agric.* **1984,** *35,* 569–572.

Gerde, J.; C. Hardy; W. Fehr; P.J. White. Frying performance of no-*trans,* low-linolenic acid soybean oils. *J. Am. Oil Chem. Soc.* **2007,** *84,* 557–563.

Gibson, K.J. Palmitoleate formation by soybean stearoyl-acyl carrier protein desaturase. *Biochim. Biophys. Acta* **1993,** *1169,* 231–235.

Gibson, L.R.; R.E. Mullen. Soybean seed composition under high day and night growth temperatures. *J. Am. Oil Chem. Soc.* **1996,** *73,* 733–737.

Golbitz, P. Statistics. *Soya and Oilseed Bluebook.* Soyatech LLC: Bar Harbor, ME, 2007; pp. 345–390.

Goode, J.H.; R.E. Dewey. Characterization of aminoalcoholphosphotransferases from *Arabidopsis thaliana* and soybean. *Plant Physiol. Biochem.* **1999,** *37,* 445–457.

Gottstein, T.; W. Grosch. Model study of different antioxidant properties of α- and γ-tocopherol in fats. *Fat Sci. Technol.* **1990,** *92,* 139–144.

Gutierrez, E.; T. Wang. Effect of processing on sphingolipid content in soybean products. *J. Am. Oil Chem. Soc.* **2004,** *81,* 971–977.

Gutierrez, E.; T. Wang; W.R. Fehr. Quantification of sphingolipids in soybeans. *J. Am. Oil Chem. Soc.* **2004,** *81,* 737–742.

Hammond, E.G.; L.A. Johnson; C. Su; T. Wang; P.J. White. Soybean Oil. In *Bailey's Industrial Oil and Fat Products*, 6th Edn.; Shahidi, F. Ed.; John Wiley and Sons, Inc.: Hoboken, NJ, 2005; vol. 2, pp. 577–653.

Harp, T.K.; E.G. Hammond. Stereospecific analysis of soybean triacylglycerols. *Lipids* **1998,** *33,* 209–216.

Heppard, E.P.; A.J. Kinney; K.L. Stecca; G.H. Miao. Developmental and growth temperature regulation of two different microsomal ω6-desaturase genes in soybeans. *Plant Physiol.* **1996,** *110,* 311–319.

Herman, E.M. Inmunogold-localization and synthesis of an oil-body membrane protein in developing soybean seeds. *Planta* **1987,** *172,* 336–345.

Herman, E.M.; D.L. Melroy; T.J. Buckhout. Apparent processing of soybean oil body protein accompanies the onset of oil mobilization. *Plant Physiol.* **1990,** *94,* 341–349.

Hildebrand, D.F.; M. Kito. Role of lipoxygenases in soybean seed protein quality. *J. Agric. Food Chem.* **1984,** *32,* 815–819.

Huang, A.H.C. Oil bodies and oleosins in seeds. *Annu. Rev. Plant Physiol. Plant Mol. Biol.* **1992,** *43,* 177–200.

Huang, A.H.C. Oleosins and oil bodies in seeds and other organs. *Plant Physiol.* **1996,** *110,* 1055–1061.

Iowa State University. Healthier Soybean Oils, 2007. http://www.notrans.iastate.edu/purchasing.html (accessed June 2007).

IUPAC. The Nomenclature of Steroids. 1989. http://www.chem.qmul.ac.uk/iupac/steroid/3S01.html#3S10 (accessed June 2007).

IUPAC. Nomenclature of Tocopherols and Related Compounds, 1981. http://www.chem.qmul.ac.uk/iupac/misc/toc.html (accessed June 2007).

Jawad, I.M.; S.P. Kochhar; B.J.F. Hudson. The physical refining of edible oil 2. Effect on unsaponifiable components. *Lebensm. Wiss. Technol.* **1984,** *17,* 155–150.

Judd, J.T.; D.J. Baer; S.C. Chen, B.A. Clevidence; R.A. Muesing; M. Kramer; G.W. Meijer. Plant sterol esters lower plasma lipids and most carotenoids in mildly hypercholesterolemic adults. *Lipids* **2002,** *37,* 33–42.

Jung, M.Y.; D.B. Min. Effects of α-, γ-, and δ-tocopherols on oxidative stability of soybean oil. *J. Food Sci.* **1990,** *55,* 1464–1465.

Jung, M.Y.; S.H. Yoon; D.B. Min. Effects of processing steps on the contents of minor compounds and oxidation of soybean oil. *J. Am. Oil Chem. Soc.* **1989,** *66,* 118–120.

Kamal-Eldin, A.; L.A. Appelqvist. The chemistry and antioxidant properties of tocopherols and tocotrienols. *Lipids* **1996,** *31,* 671:701.

Katan, M.B.; S.M. Grundy; P. Jones; M. Law; T. Miettinen; R. Paoletti. Efficacy and safety of plant stanols and sterols in the management of blood cholesterol levels. *Mayo Clin Proc.* **2003,** *78,* 965–978.

Katayama, M.; M. Katoh. Soybean sterols during maturation of seeds i. Accumulation of sterols in the free form, fatty acid esters, acylated glucosides and non-acylated glucosides. *Plant Cell Physiol.* **1973,** *14,* 681–688.

Kato, T.; H. Ohta; K. Tanaka; D. Shibata. Appearance of new lipoxygenases in soybean cotyledons after germination and evidence for expression of a major lipoxygenase gene. *Plant Physiol.* **1992,** *98,* 324:330.

King, J.M.; S.M. Chin; L.K. Svendsen; C.A. Reitmeier; L.A. Johnson; W.R. Fehr. Processing of lipoxygenase-free soybeans and evaluation in foods. *J. Am. Oil Chem. Soc.* **2001,** *78,* 353–360.

King, J.M.; L.K. Svendsen; W.R. Fehr; J.M. Narvel; P.J. White. Oxidative and flavor stability of oil from lipoxygenase-free soybeans. *J. Am. Oil Chem. Soc.* **1998,** *75,* 1121–1126.

Ko, S.N.; C.L. Kim; H. Kim; C.T. Kim; S.H. Chung; B.S. Tae; I.H. Kim. Tocol levels in milling fractions of some cereal grain and soybean. *J. Am. Oil Chem. Soc.* **2003,** *80,* 585–589.

Kobayashi, T.; T. Shimizugawa; T. Osakabe; S. Watanabe; H. Okuyama. A long-term feeding of sphingolipids affected the levels of plasma cholesterol and hepatic triacylglycerol but not tissue phospholipids and sphingolipids. *Nutr. Res.* ***1997,*** *17,* 111–114.

Kok, L.L.; W.R. Fehr; E.G. Hammond; P.J. White. *trans*-free margarine from highly saturated soybean oil. *J. Am. Oil Chem. Soc.* **1999,** *76,* 1175–1181.

Koo, A.J.K.; J.B. Ohlrogge; M. Pollard. On the export of fatty acids from the chloroplast. *J. Biol. Chem.* **2004,** *279,* 16101–16110.

Kris-Etherton, P.; S. Yu. Individual fatty acid effects on plasma lipids and lipoproteins: human studies. *Am. J. Clin. Nutr.* **1997,** *65,* 1628S–1644S.

Kris-Etherton, P.M.; D.S. Taylor; S. Yu-Poth; P. Huth; K. Moriarty; V. Fishell; R.L. Hargrove; G. Zhao; T.D. Etherton. Polyunsaturated fatty acids in the food chain in the united states. *Am. J. Clin. Nutr.* **2000,** *71,* 179S–188S.

Kritchevsky, D.; S.C. Chen. Phytosterols-health benefits and potential concerns: a review. *Nutr. Res.* **2005,** *25,* 413–428.

Lee, I.; L.A. Johnson; E.G. Hammond. Reducing the crystallization temperature of biodiesel by winterizing methyl soyate. *J. Am. Oil Chem. Soc.* **1996,** *73,* 631–636.

Lee, I.; L.A. Johnson; E.G. Hammond. Use of branched-chain esters to reduce the crystallization temperature of biodiesel. *J. Am. Oil Chem. Soc.* **1995,** *72,* 1155–1160.

Leth, T.; H. Sondergaard. Biological activity of vitamin e compounds and natural materials by the resorption-gestation test, and chemical determination of vitamin E activity in foods and feeds. *J. Nutr.* **1977,** *107,* 2236–2243.

Liechtenstein, A.H.; L.J. Appel; M. Brands; M. Carnethon; S. Daniels; H.A. Franch; B. Franklin; P. Kris-Etherton; W.S. Harris; B. Howard; et al. Diet and lifestyle recommendations revision 2006–a scientific statement from the american heart association nutrition committee. *Circulation* **2006,** *114,* 82–96.

Lin, L.J.; J.T.C. Tzen. Two distinct steroleosins are present in seed oil bodies. *Plant Physiol, Bioch.* **2004,** *42,* 601–608.

List, G.R.; T.L. Mounts; F. Orthoefer; W.E. Neff. Effect of interesterification on the structure and physical properties of high-stearic acid soybean oils. *J. Am. Oil Chem. Soc.* **1997,** *74,* 327–329.

Liu, H.R.; P.J. White. High-temperature stability of soybean oils with altered fatty acid compositions. *J. Am. Oil Chem. Soc.* **1992a,** *69,* 533–537.

Liu, H.R.; P.J. White. Oxidative stability of soybean oils with altered fatty acid compositions. *J. Am.*

Oil Chem. Soc. **1992b,** *69,* 528–532.

Liu, K. *Soybeans: Chemistry, Technology, and Utilization.* Chapman and Hall: New York, NY, 1997.

Lung, S.C.; R.J. Weselake. Diacylglycerol acyltransferase: a key mediator of plant triacylglycerol synthesis. *Lipids* **2006,** *41,* 1073–1088.

Lynch, D.V.; T. Dunn. An introduction to plant sphingolipids and a review of recent advances in understanding their metabolism and function. *New Phytol.* **2004,** *161,* 677–702.

Lynen, F. Biosynthesis of saturated fatty acids. *Fed. Proc.* **1961,** *20,* 941–951.

MacCarthy, J.J.; P.K. Stumpf. Fatty-acid composition and biosynthesis in cell suspension cultures of *Glycine max* (L.) Merr., *Catharanthus roseus* G. Don and *Nicotiana tabacum* L. *Planta* **1980,** *147,* 384–388.

Marks Spencer. Fats, 2003. http://www.marksandspencer.com/gp/node/n/43580031/026-6356728-6306864?ie=UTF8andmnSBrand=core (accessed June 2007).

Mensink, R.P.; M.B. Katan. Effects of dietary *trans* fatty acids on high-density and low-density lipoprotein cholesterol levels in healthy subjects. *N. Engl. J. Med.* **1990,** *323,* 439–445.

Merrill, A.H.; E. Wang; R.E. Mullins. Kinetics of long-chain sphingoid base biosynthesis in intact lm cells: effects of varying the extracellular concentrations of serine and fatty acid precursors of this pathway. *Biochem.* **1988,** *27,* 340–345.

Miller, L.A.; P.J. White. Oxidative stabilities of low-linolenate, high-stearate and common soybean oils. *J. Am. Oil Chem. Soc.* **1988a,** *65,* 1334–1338.

Miller, L.A.; P.J. White. High-temperature stabilities of low-linolenate, high-stearate and common soybean oils. *J. Am. Oil Chem. Soc.* **1988b,** *65,* 1324–1327.

Mounts, T.L.; K. Warner; G.R. List; W.E. Neff; R.F. Wilson. Low-linolenic acid soybean oils—alternatives to frying oils. *J. Am. Oil Chem. Soc.* **1994,** *71,* 495–499.

Murphy, D.J. The biogenesis and functions of lipid bodies in animals, plants and microorganisms. *Prog. Lipid Res.* **2001,** *40,* 325–438.

Murphy, D.J. The roles of lipid bodies and lipid-body proteins in the assembly and trafficking of lipids in plant cells. *16th International Plant Lipid Symposium.* Budapest, Hungary, 2004, pp. 55–62.

Narvel, J.M.; W.R. Fehr; G.A. Welke. Agronomic and seed traits of soybean lines lacking seed lipoxygenases. *Crop Sci.* **1998,** *38,* 926–928.

Neidleman, S.L. Effects of temperature on lipid unsaturation. *Biotechnol. Genet. Eng. Rev.* **1987,** *5,* 245–268.

Ohlrogge, J.; J. Browse. Lipid biosynthesis. *The Plant Cell* **1995,** *7,* 957–970.

Olcott, H.S.; J. Van Der Veen. Comparison of antioxidant activities of tocol and its methyl derivatives. *Lipids* **1968,** *3,* 331–334.

Oliva, M.L.; J.G. Shannon; D.A. Sleper; M.R. Ellersieck; A.J. Cardinal; R.L. Paris; J.D. Lee. Stability of fatty acid profile in soybean genotypes with modified seed oil composition. *Crop Sci.* **2006,** *46,* 2069–2075.

Piironen, V.; D.G. Lindsay; T.A. Miettinen; J. Toivo; A.M. Lampi. Plant sterols: biosynthesis, biological function, and importance to human nutrition. *J. Sci. Food Agric.* **2000,** *80,* 939–966.

Primono, V.S; D.E. Falk; G.R. Ablett; J.W. Tanner; I. Rajcan. Genotype x environment interactions, stability, and agronomic performance of soybean with altered fatty acid profiles. *Crop Sci.* **2002,** *42,* 37–44.

Reinprecht, Y.; V.W. Poysa; I. Rajcan; G.R. Ablett; K.P. Pauls. Agronomic performance of soybean with seed lipoxygenase nulls and low linolenic acid content. *Can. J. Plant Sci.* **2006,** *86,* 379–387.

Rennie, B.D.; J.W. Tanner. Fatty acid composition of oil from soybean seeds grown at extreme temperatures. *J. Am. Oil Chem. Soc.* **1989,** *66,* 1622–1624.

Sasaki, Y.; Y. Nagano. Plant acetyl-coa carboxylase: structure, biosynthesis, regulation, and gene manipulation for plant breeding. *Biosci. Biotechnol. Biochem.* **2004,** *68,* 1175–1184.

Sattler, S.E.; E.B. Cahoon; S.J. Coughlan; D. DellaPenna. Characterization of tocopherol cyclases from higher plants and cyanobacteria. Evolutionary implications for tocopherol synthesis and function. *Plant Physiol.* **2003,** *132,* 2184–2195.

Schmelz, E.M. Dietary sphingomyelin and other sphingolipids in health disease. *Nutr. Bull.* **2000,** *25,* 135–139.

Schnebly, S.R.; W.R. Fehr. Effect of years and planting dates on fatty acid composition of soybean genotypes. *Crop Sci.* **1993,** *33,* 716–719.

Schuler, I.; A. Milon; Y. Nakatani; G. Ourisson; A.M. Albrecht; P. Benveniste; M.A. Hartmann. Differential effects of plant sterols on water permeability and on acyl chain ordering of soybean phosphatidylcholine bilayers. *Proc. Natl. Acad. Sci. USA,* **1991,** *88,* 6926–6930.

Shanklin, J.; E.B. Cahoon. Desaturation and related modifications of fatty acids. *Annu. Rev. Plant Physiol. Plant Mol. Biol.* **1998,** *49,* 611–641.

Shen, N.; W. Fehr; P. White. High-temperature stabilities of oils from soybeans that lack lipoxygenases. *J. Am. Oil Chem. Soc.* **1997,** *74,* 381–385.

Sherwin, E.R. Antioxidants for vegetable oils. *J. Am. Oil Chem. Soc.* **1976,** *53,* 430–436.

Shimakata, T.; P.K. Stumpf. Isolation and function of spinach leaf β-ketoacyl-[acyl carrier protein] synthetases. *Proc. Natl. Acad. Sci. USA* **1982,** *79,* 5808–5812.

Shimakata, T.; P.K. Stumpf. Purification and characterization of β- ketoacyl-[acyl carrier protein] synthetase 1 from *Spinacea oleracea* Leaves. *Arch. Biochem. Biophys.* **1983,** *220,* 39–45.

Siedow, J.N. Plant lipoxygenase: structure and function. *Annu. Rev. Plant Physiol Plant Mol. Biol.* **1991,** *42,* 145–188.

Sims, R.J.; J.A. Fioriti; M.J. Kanuk. Sterol additives as polymerization inhibitors for frying oils. *J. Am. Oil Chem. Soc.* **1972,** *49,* 298–301.

Song, Y.; M.H. Love; P. Murphy. Subcellular localization of lipoxygenase-1 and -2 in germinating soybean seed and seedlings. *J. Am. Oil Chem. Soc.* **1990,** *67,* 961–965.

Sperling, P.; E. Heinz. Plant sphingolipids: structural diversity, biosynthesis, first genes and functions. *Biochim. Biophys. Acta* **2003,** *1632,* 1–15.

Su, C.; P.J. White. Flavor stability and quality of high-oleate and regular soybean oil blends during frying. *J. Am. Oil Chem. Soc.* **2004a,** *81,* 853–859.

Su, C; P.J. White. Frying stability of high-oleate and regular soybean oil blends during frying. *J. Am.*

Oil Chem. Soc. **2004b,** *81,* 783–788.

Sugawara, T.; and T. Miyazawa. Separation and detection of glycolipids from edible plant sources by high-performance liquid chromatography and evaporative light scattering detection. *Lipids* **1999,** *34,* 1231–1237.

Sullards, M.C.; D.V. Lynch; A.H. Merrill; J. Adams. Structure determination of soybean and wheat glucosylceramides by tandem mass spectrometry. *J. Mass Spectrm.* **2000,** *35,* 347–353.

Torres-Penaranda, A.V.; C.A. Reitmeier; L.A. Wilson; W.R. Fehr; J.M. Narvel. Sensory characteristics of soymilk and tofu made from lipoxygenase-free and normal soybeans. *J. Food Sci.* **1998,** *63,* 1084–1087.

Thijssen, M.A.; R.P. Mensink. Small differences in the effects of stearic acid, oleic acid, and linoleic acid on the serum lipoprotein profile of humans. *Am. J. Clin. Nutr.* **2005,** *85,* 510–516.

Trawatha, S.E.; D.M. TeKrony; D.F. Hildebrand. Soybean lipoxygenase mutants and seed longevity. *Crop. Sci.* **1995,** *35,* 862–868.

United Soybean Board. Evaluation and Analysis of Vegetable Oil Markets: The Implications of Increased Demand for Industrial Uses on Markets and USB Strategy, 2005. http://www.unitedsoybean.org/FileDownload.aspx?File=2005_USB_oil_strategy_report_final.pdf (accessed June, 2007).

Verleyen, T.; M. Forcades; R. Verhe; K. Dewettinck; A. Huyghebaert; and W. De Greyt. Analysis of Free and Esterified Sterols in Vegetable Oils. *J. Am. Oil Chem. Soc.* **2002a,** *79,* 117–122.

Verleyen, T.; U. Sosinska; S. Ioannidou; R. Verhe; K. Dewettinck; A. Huyghebaert; W. De Greyt. Influence of the vegetable oil refining process on free and esterified sterols. *J. Am. Oil Chem. Soc.* **2002b,** *79,* 947–953.

Vernooy-Gerritsen, M.; J.L.M. Leuinssen; G.A. Veldink; J.F.G. Vliegenthart. Intracellular localization of lipoxygenases-1 and -2 in germinating soybean seeds by indirect labeling with protein a-colloidal gold complexes. *Plant Physiol.* **1984,** *76,* 1070–1079.

Vlahakis, C.; J. Hazebroek. Phytosterol accumulation in canola, sunflower, and soybean oils: effects of genetics, planting location, and temperature. *J. Am. Oil Chem. Soc.* **2000,** *77,* 49–63.

Voelker, T.; A.T. Kinney. Variations in the biosynthesis of seed-storage lipids. *Annu. Rev. Plant Physiol. Plant Mol. Biol.* **2001,** *52,* 335–361.

Wang, C.; K.P.C. Croft; U. Järlfors; D. Hildebrand. Subcellular localization studies indicate that lipoxygenases 1 to 6 are not involved in lipid mobilization during soybean germination. *Plant Physiol.* **1999,** *120,* 227–235.

Wang, L.; T. Wang; W.R. Fehr. Effect of seed development stage on sphingolipid and phospholipid contents in soybean seeds. *J. Agric. Food Chem.* **2006a,** *54,* 7812–7816.

Wang, L.; T. Wang; W.R. Fehr. HPLC Quantification of sphingolipids in soybeans with modified palmitate content. *J. Agric. Food Chem.* **2006b,** *54,* 7422–7428.

Wang, T.; E.G. Hammond; W.R. Fehr. Phospholipid fatty acid composition and stereospecific distribution of soybeans with a wide range of fatty acid composition. *J. Am. Oil Chem. Soc.* **1997,** *74,* 1587–1594.

Warner, K. Chemical and Physical Reactions in Oil during Frying. In *Frying Technology and Practices.* Gupta, M.K.; K. Warner; P.J. White Eds.; AOCS Press: Champaign, IL, 2004; pp. 16–28.

Warner, K. Effects on the flavor and oxidative stability of stripped soybean and sunflower oils with added pure tocopherols. *J. Agric. Food Chem.* **2005,** *53,* 9906–9910.

Warner, K.; M. Gupta. Potato chip quality and frying oil stability of high oleic acid soybean oil. *J. Food Sci.* **2005,** *70,* S395–S400.

Weijrauch, J.L.; J. Gardner. Sterol content of foods of plant origin. *J. Am. Diet. Assoc.* **1978,** *73,* 39–47.

Wilson, L.A. Comparison of Lipoxygenase-Null and Lipoxygenase-Containing Soybeans for Foods. In *Lipoxygenase and Lipoxygenase Pathway Enzymes;* Piazza, G.J. Ed.; AOCS Press: Champaign, IL, 1996; pp. 209–225.

Wilson, R.F. Seed Composition. In *Soybeans: Improvement, Production, and Uses;* Boerma, H.; and J. Specht Eds.; American Society of Agronomy: Madison, WI, 2004; pp. 621–677.

Wilson, R.F.; H.H. Weissinger; J.A. Buck; G.D. Faulkner. Involvement of phospholipids in polyunsaturated fatty acid synthesis in developing soybean cotyledons. *Plant Physiol.* **1980,** *66,* 545–549.

Wojciechowski Z.A. Biochemistry of Phytosterol Conjugates. In *Physiology and Biochemistry of Sterols.* Patterson, G.W.; and D.W. Nes Eds.; AOCS Press: Champaign, IL, 1991; pp. 361–395.

Wolf R.B.; J.F. Cavins; R. Kleiman; L.T. Black. Effect of temperature on soybean seed constituents: oil, protein, moisture, fatty acids, amino acids and sugars. *J. Am. Oil Chem. Soc.* **1982,** *59,* 230–232.

Yadav, N.S; A. Wierzbicki; M. Aegerter; C.S. Caster; L. Perez-Grau; A.J. Kinney; W.D. Hitz; J.R. Booth; B. Schweiger; K.L. Stecca; et al. Cloning of higher plant ω3-desaturases. *Plant Physiol.* **1993,** *103,* 467–476.

Yamaya, A.; Y. Endo; K. Fujimoto; K. Kitamura. Effects of genetic variability and planting location on the phytosterol content and composition in soybean seeds. *Food Chem.* **2007,** *102,* 1071–1075.

Yan, P.S.; P.J. White. Linalyl acetate and other compounds with related structures as antioxidants in heated soybean oil. *J. Agric. Food Chem.* **1990,** *38,* 1904–1908.

Yoshida, H.; S. Kanrei; Y. Tomiyama; Y. Mizushina. Regional characterization of tocopherols and distribution of fatty acids within soybean seeds (*Glycine max* L.). *J. Food Lipids* **2006a,** *13,* 12–26.

Yoshida, H.; S. Takagi; H. Ienaga; C. Tsuchiya. Regional distribution of tocopherols and fatty acids within soybean seeds. *J. Am. Oil Chem. Soc.* **1998,** *75,* 767–773.

Yoshida, H.; Y. Tomiyama; S. Kanrei; Y. Mizushina. Tocopherol distribution and molecular species of triacylglycerols in soybean embryonic axes. *J. Am. Oil Chem. Soc.* **2006b,** *83,* 345–351.

8

Soybean Proteins

Patricia A. Murphy
University Professor, Department of Food Science and Human Nutrition, Iowa State University, Ames, IA 50011

Introduction

Soybeans contain a variety of proteins with unique properties for the germinating seed. The proteins were utilized as human foods earlier than 2800 BCE when the soybean was domesticated in China. The earliest records of soybean food products are about 1000 and 2000 years ago in Japan and China, respectively. The characteristics of these soybean foods are attributable in part to the proteins from the bean. The largest mass of the seed protein is the storage proteins, glycinin and β-conglycinin, which have no biological activity other than as amino nitrogen stores for the germinating seed. The structures of these two proteins were relatively conserved across many legume species and other related plants in major plant gene families of the legumins and vicilins. The other important soybean seed proteins that are discussed here have biological activity and include the lipoxygenases, the trypsin inhibitor family including Kunitz trypsin inhibitor and Bowman-Birk inhibitor, and the soy lectins. The seed proteins represent between 30 and 50% of seed mass with the storage proteins accounting for 65–80% of the seed protein. Soybean proteins have excellent amino acid profiles for humans although they are deficient in sulfur amino acids for livestock and rodents. In addition to soy's protein nutritional qualities, other biological activities are attributed to them including cholesterol-lowering abilities and anticancer activities. The storage proteins possess major flavor-binding ability that is a challenge in preparing bland soy protein products for Western tastes as well as formulation properties unique among food proteins.

Seed Subcellular Structures

Soybean proteins are packaged in discrete spherical subcellular structures called protein bodies in the palisade-like cells of the soybean cotyledons (Bair & Snyder, 1980). The soybean storage protein structures for glycinin and β-conglycinin are apparently highly conserved to maximize protein packaging in the protein bodies (Shewry et al.,

1995). The protein body diameters average 8 to 10 μm but can range from 2 to 20 μm (Snyder & Kwon, 1987a). In raw or minimally heat-treated soybean cotyledons, the soybean proteins are readily water-soluble at pH neutrality. The protein bodies can be isolated (Saio & Watanabe, 1966; Tombs, 1967; Wolf, 1970), but take great care not to fix the protein bodies by slight heat treatment, which will result in insoluble protein bodies (Tombs, 1967). Isolation of protein bodies is not a route of soybean protein purification due to this fixation problem. Excellent electron microscopic photography of soybean seed structures is found in Bair (1979). These photographs contribute to our understanding of the structural changes that occur in the protein bodies and their associated lipid bodies during soybean ingredient production.

Protein Levels, Crop and Cultivar Variations

Mahmoud et al. (2006) recently reviewed soybean protein levels in soybeans based on historical United States Department of Agriculture (USDA) records. While overall yield has steadily increased since the 1920s, the total percent protein has not. In fact, a noticeable 3% drop in total protein between ancestral and modern soybean was noted in the past 60 years with values ranging from 37 to 40% protein. Yaklich (2001) suggests that higher percentage protein soybean lines contain higher percentage of individual storage proteins. Krishnan's group (Mahmoud et al., 2006; Krishnan et al., 2007) suggest their observed higher seed protein levels are the result of increased expression of the storage proteins rather than other seed proteins. Soybeans intended for food use typically are higher percent protein varieties as well as larger seed size (Murphy et al., 1997). Glycinin and β-conglycinin content in soybeans varies with variety and environmental growing conditions (Murphy & Resurreccion, 1984; Murphy et al., 1997; Fehr et al., 2003). Fehr et al. (2003) reported no significant correlations between percent protein (or oil) and individual or total storage protein concentrations when growing different cultivars in different environments.

Storage Proteins

Soybean storage proteins are members of large families of related proteins, the vicilins and the legumins, with no unique biological activity other than to provide amino nitrogen for the germinating seed. Glycinin, the legumin family member, is the larger molecular weight hexamer. β-conglycinin, the vicilin family, is the smaller trimer. Early work by Wolfe and Cowen (1971) used ultracentrifugation sedimentation to discover the approximate molecular weights and assignment of Svedberg units to each protein, 11*S* and 7*S* as well as a 2*S* and a 15*S* fraction. These numerical names are unfortunately used widely as shorthand for the actual protein names. This shorthand nomenclature does not take into account that Wolf's 7*S* fraction also contains lipoxygenases, soy lectins, and β-amylase as well as β-conglycinin (Nielsen, 1985; Snyder & Kwon, 1987b). Early attempts to estimate β-conglycinin concentrations by analyzing the ultracentrifugal Schlieren optics pattern typically overestimated it. The 15*S* fraction reported for soy proteins apparently is a dimer of glycinin (Wolf & Nelson, 1996). The 2*S* fraction contains the trypsin inhibitors and other small molecular weight (MW) enzymes. Use of Svedberg units in describing many seed storage proteins is rather widely employed, although somewhat inaccurately.

According to Pernollet and Mosse (1983), the legumins are characterized with MW between 300 and 400 kilodaltons (kD), less solubility in neutral salts, higher temperature stability and higher amide nitrogen and sulfur amino acid content. Marcone (1999) reviewed the diversity of plant species with legumin-type proteins. These proteins are found in plant species ranging from legumes to sunflowers to buckwheat to pumpkins. The vicilins have smaller MW from 150 to 250 kD, higher salt solubility, lower temperature stability, and lower sulfur and nitrogen contents. Additionally, the legumins are typically hexamers with acidic and basic subunits produced from single genes covalently attached via a unique disulfide bond. The legumin hexamers associate via hydrogen and hydrophobic bonding. The vicilins are typically trimers with no covalent bonds between the individual polypeptide chains and are likely to be glycosylated. The two soy storage proteins are members of these two storage protein classifications.

Glycinin

Glycinin is composed of 12 unique polypeptides that were thought to be randomly associated in a mature dodecamer (Badley et al., 1975). In the late 1980s, the biosynthetic route for glycinin was established as we came to understand that plant storage proteins were produced with the same protein synthetic rules as animal proteins. The original 12 polypeptides were identified as acidic or basic depending on their migration in isoelectric focusing under reducing conditions. As the genes, mRNAs, and proglycinin peptides were identified, clearly the acidic and basic peptides of glycinin were not randomly associated, but rather unique acidic-basic pairs were produced from same gene and mRNA. The acidic-basic pairs were synthesized as a single polypeptide protomer with a signal peptide (Fig. 8.1). During post-translational processing, the signal peptide is hydrolyzed, a unique disulfide bond is formed that will serve as eventual single link between each acidic-basic peptide pair. A unique asparaginyl endopeptidase hydrolyzes the peptide to form the acidic and basic peptides now covalently linked by the disulfide bond (Scott et al., 1992). Following this proteolytic event, the individual acidic and basic peptide pairs are assembled into a trimer, consisting of 6 peptides, associated by hydrophobic and hydrogen bonding. Finally two trimers, which are in the donut form described from Badley et al. (1975)'s scanning electron microscope photographs, associate into the mature glycinin in a proscribed manner as reported by Adachi et al. (2001). Glycinin deposition in the seed occurs between 40 and 90 days after flowering (DAF). After synthesis, the mature glycinins are transported to the forming protein bodies with MW of 360 kD. Plietz et al.

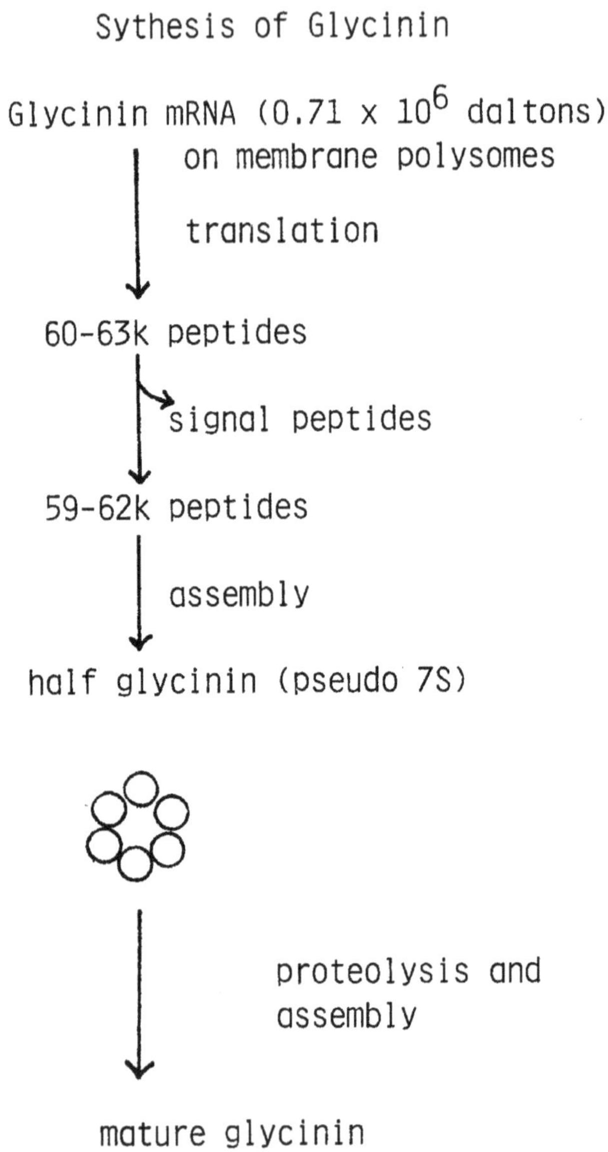

Fig. 8.1. Glycinin synthesis.

(1987) reported spectroscopic evidence to support the glycinin mature hexamer with the basic peptides buried in the internal volume of glycinin and the acidic peptides on the exterior of glycinin. Analysis of peptide fragments following proteolysis of mature glycinin supports the Plietz model (Shutov et al., 1996).

The acidic-basic peptides are the products of at least five identified gene families: Gy1 producing $A_{1a}B_{1b}$; Gy2 producing A_2B_{1a}; Gy3 producing $A_{1b}B_{1b}$; Gy4 producing $A_5A_4B_3$; and Gy5 producing A_3B_4 and the peptides sequenced (Nielsen, 1989). Some discrepancies exist in the literature in identification of acidic and basic glycinin pairs. The Ustumi group identifies the acidic-basic pairs as: Gly I = $A_{1a}B_{1b}$; Gly2 = A_2B_{1b}; Gly3 = $A_{1b}B_2$; Gly4 = A_4 A_5 B_3 ; and Gly5 = A_3B_4. The Utsumi nomenclature is used in this review (Utsumi et al., 1997). The unique glycinin pairs, according to Stastwick et al. (1981) and Utsumi et al. (1997), are shown in Table 8.1. Without a reducing agent, the disulfide linked dimers migrate at the MW equal to the sum of the acidic-basic pair. A reducing agent is required to separate the acidic and basic peptides for electrophoresis (Fig. 8.2). The basic peptides migrate as a single band on SDS-PAGE at 20 kD. The acidic peptides are more heterogeneous in MW. A_3 is the largest acidic peptide with apparent MW of 42–43 kD. Some but not all cultivars have A_4 which co-migrates with most of the acidic peptides at 40 kD unless urea is included in the SDS-PAGE. Then the A_4 appears above the main acidics' band but below A_3. One

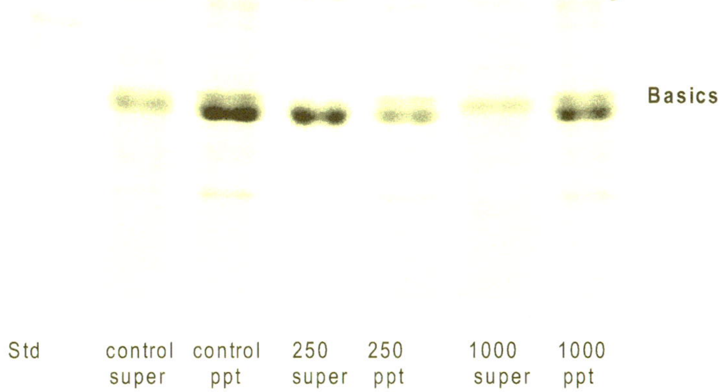

Fig. 8.2. Urea-SDS-PAGE of soy proteins with and without phytase treatment in supernatants (super) and precipitates (ppt) (Aldin, 2004).

acidic peptide, A_5, is smaller than all other glycinin peptides at 10 kD. The main acidic peptide band contains A1a, A_{1b} and A_2. Some cultivars have additional acidic peptides, such as A_6 in the Raiden variety, although its basic peptide pair is not known (Nielsen, 1985).

The amino acid sequences of each glycinin peptide are determined directly as well as deduced from the gene. In addition to the unique cysteine disulfide between each nonrandom acidic-basic peptide pair, the glycinin peptides contain cysteine residues that may form internal disulfide linkages, and methionine mainly within the acidic peptides. Cysteine residues may also remain reduced in mature glycinin. The distribution of the sulfur amino acids among the acidic-basic peptide pairs is not even. Plant breeders suggested that higher sulfur amino acid cultivars could be developed by selecting for the high-sulfur amino acid glycinin genes. This molecular approach certainly would have positive ramifications for animal agriculture because the sulfur amino acid content of soybeans is lower than the requirement. However, the sulfur amino acid content of soybean protein is adequate for human requirements. More importantly, the functional properties of the different acidic-basic peptide pairs in food products are quite different and are discussed below. Alterations of the peptide distribution in glycinin without appreciation of the functional properties of these unique peptides in foods will lead to varieties not useful in food systems.

Table 8.1. Sulfur Amino Acid Content of Glycinin Subunits (modified from Staswick et al., 1981 and Utsumi et al., 1997)

Acidic	# M	#C	Basic	# M	#C	AB-complex	Total C	Total S
Staswick et al., 1981								
A_{1a}	4	3	B_2	3	1	$A_{1a}B_2$	4	11
A_{1b}	4	6	B_{1b}	2	1	$A_{1b}B_{1b}$	7	13
A_2	6	6	B_{1a}	2	2	A_2B_{1a}	8	16
A_3	2	3	B_4	1	1	A_3B_4	4	7
A_4A_5	2	2	B_3	0	1	$A_4A_5B_3$	3	5
A_6	?							
Utsumi et al., 1997								
A_{1a}	3	6	B_{1b}	3	2	$A_{1a}B_2$	8	14
A_{1b}	3	6	B_2	2	2	$A_{1b}B_{1b}$	8	13
A_2	5	6	B_{1a}	2	2	A_2B_{1a}	8	15
A_3	2	4	B_4	2	2	A_3B_4	6	10
A_4A_5	2	4	B_3	0	2	$A_4A_5B_3$	6	8

M = methionine' C=cysteine

β-Conglycinin

β-Conglycinin is the vicilin storage protein of soybeans. It is composed of three unique peptides, α, α′ and β, that associate as trimers. The α and α′ peptides are synthesized about 5 days earlier in DAF than the β peptide (Gayler & Skyes, 1981). One can reasonably assume that the initial β-conglycinin trimers are α and α′ in a random manner. β trimers of β-conglycinin are generated later in DAF when this peptide is synthesized. β trimers are isolated from soybean seeds, usually as part of the glycinin fraction in most soybean storage fractionation protocols (Gayler & Skyes, 1981; Yamauchi et al., 1981). However, β-conglycinin trimers seem to be present in seeds in a nonrandom association of all three peptides according to Thanh and Shibasaki (1977) with trimers' MW ranging from 125 kD to 170 kD for α_3, $\alpha_2\beta$, $\alpha\alpha'\beta$, $\alpha\beta_2$, $\alpha_2\alpha'$ and β_3 (Thanh & Shibasaki, 1976a; Gayler & Skyes, 1981; Yamauchi et al., 1981). Recently, Muruyama (2002b) reported isolating native α_3, α'_3, $\alpha_2\beta$, $\alpha'_2\beta$, $\alpha\beta_2$, $\alpha'\beta_2$ and β_3.

All three β-conglycinin peptides are glycosylated via unique asparagine residues. α and α′ have two glycosylation sites while β has only one. The carbohydrate composition of the glycan units is: α (Asn 199 and Asn 455) and α′ (Asn 215 and Ash 471) have 2 carbohydrate chains with 2 moles of glucosamine and 3 moles of mannose per carbohydrate chain; and β (Asn 328) has 1 carbohydrate chain with the total accounting for 3–5% of the final MW (Thanh & Shibasaki, 1976a; Muruyama et al., 2002). The carbohydrate content of the β-conglycinin peptides may play a role in the confusion in the literature regarding the MW of the β-conglycinin subunits.

Thanh and Shibasaki published a number of papers characterizing the two proteins after reporting fractionation of the two proteins (Thanh & Shibasaki, 1976b). In their 1977 paper, they reported MW of α, α′ and β as determined by SDS-PAGE, urea-PAGE, urea-SDS-PAGE and gel filtration in presence of urea. Table 8.2 shows the MWs reported by Thanh and Shibasaki (1977), more recent MW by the Nielsen group and by the Utsumi groups and MW deduced from the amino acid sequence in the Protein Database (based on the gene) using UniProt (Berman et al., 2000). Until the mid-1980s, the lower MWs of Thanh and Shibasaki's three β-conglycinin peptides were taken as accurate. Many citations after 1985 reported higher MW of 72, 68, and 52 kD for α, α′, and β, respectively, from SDS-PAGE without recognition that glycopeptides migrate at higher apparent MW than deglycosylated peptides. The MW deduced from the DNA or amino acid sequences suggests MW between the lower Thanh and Shibasaki values and the overestimates reported later from SDS-PAGE. Our own recent experiments to estimate MW of the β-conglycinin peptides before and after chemical deglycosylation reveal that the deglycosylated β-conglycinin subunit migrate at lower MW than native peptides with the deglycosylated α, α′, and β at 58, 65.8, and 53 kD, respectively (Fig. 8.3; Table 8.2).

Apparently, no internal disulfide bonds nor interpeptide links in β-conglycinin exist, although two cysteines, one each in the α and the α′ peptides, do exist. The

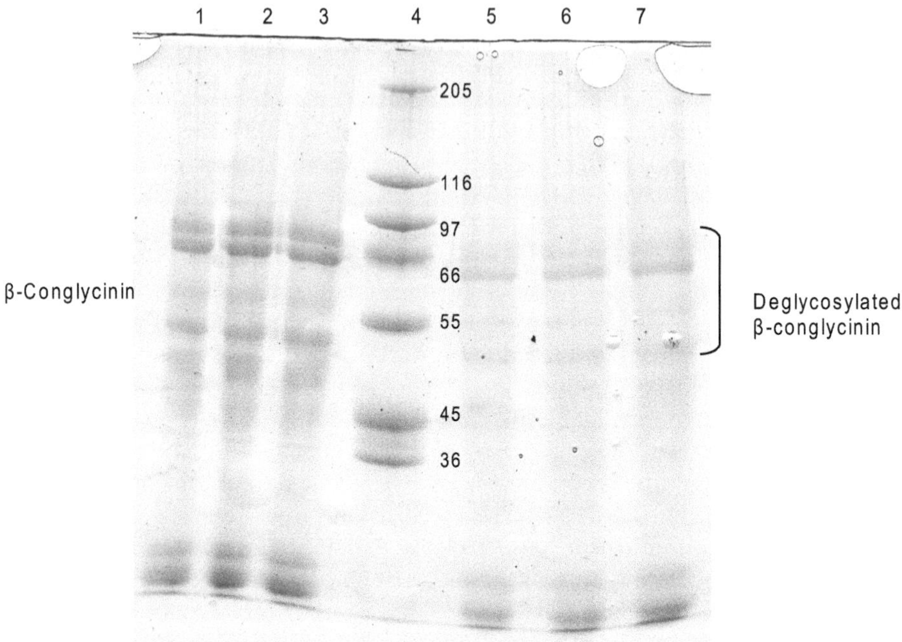

Fig. 8.3. Native and deglycosylated β-conglycinin in 9% urea-SDS PAGE gel.

Table 8.2. Molecular Weight of β-conglycinin Peptides (kD)

Method	α'	α	β	Reference
Urea/acetate/10% PAGE	68±4	68±4	42±3	Thanh & Shibasaki, 1977
SDS, 10% PAGE	59±3	59±3	44±2	Thanh & Shibasaki, 1977
Urea/SDS/9% PAGE	58±2	57±2	46±2	Thanh & Shibasaki, 1977
Guanidine gel filtration	57	57	42	Thanh & Shibasaki, 1977
SDS PAGE	71	67	50	Murayama et al., 1998
SDS PAGE	72	68	52	Medieros, 1982 cited by Nielsen, 1985
SDS PAGE	76	72	53	Sebastiani et al., 1990
Deduced from sequence	67	63	48	Utsumi et al., 1997
Deduced from sequence	65	63	48	UniProt www.pri.uniprot.org
9% SDS-urea PAGE	65.8	58	53	Figure 8.3

trimers associate through strong hydrophobic and hydrogen bonding. The trimers contain five methionines, one in α and four in α′ (Utsumi et al., 1997).

Structures

Only recently were some of the peptides of the two storage proteins crystallized to allow estimation of the three-dimensional molecular structures. Supposedly, the heterogeneous nature of the storage protein peptides inhibits crystallization (Adachi et al., 2001). Work with mutant soybean lines lacking certain subunit peptides did not lead to success in crystallization of either storage protein (Muruyama et al., 2002a; 2002b). Recently, crystallization work employed synthesis of glycinin homotrimers and hexamers using specific cDNAs for the individual acidic and basic glycinin pairs for $A_{1a}B_{1b}$ (Adachi et al., 2001) or from a soybean variety producing only A_3B_4 of glycinin (Adachi et al., 2003). β-Conglycinin was not successfully crystallized until recently as β trimers (Muruyama et al., 2001). The α′ and α trimers were not crystallized without construction of deletion mutants of these peptides (Muruyama et al., 2004). However, this work has led to a clear picture of the three-dimensional structure of the soybean storage proteins.

Proglycinin $A_{1a}B_{1b}$ was successfully expressed from its cDNA in *Escherichia coli* (Utsumi et al., 1988) and subsequently crystallized (Utsumi et al., 1993). Adachi et al. (2003) suggest that the X-ray crystallography data show glycinin protomers (acidic-basic peptide pairs) contain two jelly roll barrels and two α-helix domains that are similar to other vicilins (*7S* globulins), such as phaseolin from *Phaseolus vulgarus* and canavalin of castor beans. Adachi et al. (2001) show evidence that each $A_{1a}B_{1b}$ protomer has 25 strands and five α-helices that are able to fold into the two barrel domains in β-sheet confirmation and two extended helices. The $A_{1a}B_{1b}$ protomers are arranged as trimers with a three-dimensional size of 95 Å × 95 Å × 45 Å that are consistent with Badley et al. (1975) and other legumins (Adachi et al., 2001). Two highly conserved disulfide bonds are in legumins that are cysteine 12 and cysteine 45, the $A_{1a}B_{1b}$ interpeptide disulfide pair, and cysteine 88 and cysteine 298, an intra-acidic disulfide pair. These two disulfides appear on opposite faces of the trimers. Adachi et al. (2001) elegantly show evidence that the sum of hydrophobic, electrostatic, hydrogen, and salt bridge interactions favors the association of two trimers through the trimer side that contain the interpeptide disulfide pair in forming the mature hexamer. Adachi et al., (2003) provide similar evidence for A_3B_4 trimer association into the hexamer. Both of these three-dimensional structures are deposited in the Protein Data Bank (Adachi et al., 2000; Itoh et al., 2006). Adachi et al. (2003) suggest how pH changes allow dissociation of the hexamer during seed germination. These speculations support our understanding of association and dissociation of the glycinin hexamers and trimers in food processing.

The β-conglycinin three-dimensional structure was evaluated somewhat differently than glycinin. Native β-conglycinin was not crystallized probably due to its glycosylation and different permutations of α, α′, and β peptides in mature trimer (Morita et al., 1996). The β trimer was crystallized recently from a soybean line producing only β_3 type β-conglycinins and by expression of recombinant β_3 in *E. coli* systems (Muruyama et al., 2001). The size of the trimer was reported at 96 × 96 × 44 Å and almost identical to seed vicilins, canavalin, and phaseolin. The individual β monomers, in the trimer, appear to have the same β barrel (jelly roll) and α-helical structure as described above for glycinin protomers. These authors suggest no observable difference exists in the three-dimensional structure between the native (glycosylated) and recombinant β-conglycinin β_3. Five intramolecular salt bridges are reported and all in the core region of the peptide. One of the salt bridges was considered identical to a salt bridge identified in phaseolin and canavalin. Seventy percent of hydrophobic amino acid residues were buried in the monomer peptide suggesting a major role of hydrophobic bonding in monomer structure. Trimer association appears largely driven by hydrophobic interactions since 65% of surface hydrophobicity of the β monomer was buried upon formation of the β_3 although hydrogen bonding and one salt bridge were identified.

The α and α′ peptides and trimers were not crystallized. Muruyama et al. (2004) created deletion mutants of α′ (α'_c) that retained the core region and could be crystallized. The core regions of α, α′ and β of β-conglycinin are very homologous with each other with α and α′ 90%, α and β 76%, and α′ and β 76% (Muruyama et al., 2002a). The extension regions, not present in the deletion mutants described above, are not as homologous at 57% and have a low pI. The α'_c trimers contain the same β-barrel or jelly roll configuration with adjacent α-helical regions and were highly homologous to β_3 of β-conglycinin. Both of these three-dimensional structures are deposited in the Protein Data Bank (Muruyama et al., 2003a; 2003b).

Fractionation of Soybean Storage Proteins

Thanh and Shibasaki (1976b) reported the first generally reproducible method to isolate glycinin and β-conglycinin that is considered the gold standard in soy protein fractionation by taking advantage of the differential pH solubility of the two storage protein in tris or THAM (trihydroxyaminomethane) buffer with β-mercaptoethanol as the reductant. The full Thanh and Shibasaki method requires further clean-up using size-exclusion chromatography with Sepharose 6B and affinity chromatography with Con-A Sepharose 4B that preferentially binds the glycoprotein, β-conglycinin, which is later eluted with methyl-D-mannoside. The isoelectric fractionation followed by affinity and size exclusion chromatography result is quite pure glycinin and β-conglycinin; however, this is a lengthy process. Isoelectric precipitation alone results in a glycinin fraction of 79% purity with 6% β-conglycinin contamination and a β-conglycinin fraction of 52% purity with 3% glycinin contamination (Wu et al., 1999). O'Keefe et al. (1991a) added an intermediate pH step to the Thanh and Shibasaki Tris fractionation, resulting in higher purities of β-conglycinin but at the expense of yield. Clearly, attempts to improve glycinin and β-conglycinin purity were possible on a mg-laboratory-scale, but larger quantities needed for functionality evaluation and pilot-plant scale work needed a different approach than traditional protein chromatography.

Nagano et al. (1992) provided insight into a soy protein fractionation method that could be scaled up. These authors modified the isoelectric precipitation portion of the Thanh and Shibasaki method by extracting the soy proteins in pH 7.5 water, used sodium bisulfite as the reducing agent, and produced three isoelectric precipitation fractions at pH 6.5, 5.0, and 4.8. The pH 4.8 fraction was claimed to be >90% pure β-conglycinin. The intermediate fraction at pH 5.0 was a mixture of both storage proteins, and glycinin was relatively pure at pH 6.5. However, inspection of the SDS-PAGE shows the β peptide of β-conglycinin clearly in the glycinin purified fraction. However, in our hands on a laboratory-scale, although the glycinin purity was about 96%, the β-conglycinin purity was 78%, which was better than simple isoelectric fractionation of Thanh and Shibasaki (1976b) but still contained significant impurities. Scale-up of the process to the pilot-plant scale with 15 kg of starting material resulted in some decreases in glycinin and β-conglycinin purities, 84% and 72%, respectively. However, the possibilities of producing kg quantities of the two protein fractions appeared to be feasible. Further refinement of pilot-plant scale glycinin and β-conglycinin fractionation resulted in greater yields of protein products but at the expense of β-conglycinin purity by eliminating the intermediate protein fraction (Wu et al., 2000) or optimizing temperature of extract (Rickert et al., 2004a). Wu et al. (1999) and Rickert et al. (2004a) clearly show that the soy proteins need as little heat denaturation as possible in preparation of defatted flakes to obtain efficient fractionation of these two proteins. The intermediate fraction of these processes consists of denatured glycinin and β-conglycinin. Lower protein solubility of the starting materials results in larger intermediate fractions and lower yields of glycinin and β-conglycinin. Saito et al. (2001) proposed an enzymatic treatment to produce purified glycinin and β-conglycinin using phytase. The method involves creating a pH 7.5 water extract of defatted soy flour. The supernatant pH is adjusted to 6.0 at 40°C, and phytase is added. The precipitate is glycinin. The supernatant pH is dropped to 5.0, and the resulting precipitate is β-conglycinin. Saito et al. (2001) report 80% purity for β-conglycinin on a laboratory-scale. We replicated this work and found 69% purity for glycinin and 42% β-conglycinin without chilling the supernatant but 72% β-conglycinin with chilling (Table 8.3) (Aldin, 2004). Deak et al. (2006a) reported utilizing Ca^{+2} and sodium bisulfite in purification scheme with 86% purity glycinin and 81% purity β-conglycinin for a laboratory-scale procedure. Deak et al. (2006b) reported a simplified method utilizing Ca^{+2} in a purification scheme and reported 71% purity glycinin and 79% purity β-conglycinin at room temperature and 81% glycinin and 86% β-conglycinin when supernatants were chilled at 4°C for a laboratory-scale procedure. Each of the modifications gave us enriched fractions

Table 8.3. Protein Fraction Composition for Phytase Treatment by Urea-SDS-PAGE

Treatment and Fraction	*Glycinin (% of protein)*	*β-conglycinin (% of protein)*
Precipitate		
Control	77.6a	18.5a
250 FYT	70.6b	24.0b
1000FYT	68.8b	27.8b
250 FYT and chilling	65.7c	34.3c
1000 FYT and chilling	59.2d	35.3c
Supernatant		
Control	61.3a	34.4a
250 FYT	62.0a	34.3a
1000FYT	42.9b	49.7b
250 FYT and chilling	18.3c	71.2c
1000 FYT and chilling	15.0d	72.7c

[a]n = 3. Means in the same fraction and for each storage protein, with different letters are significantly different at $p < 0.05$
FTY = phytase activity units.

of glycinin and β-conglycinin but not pure. The effect of interactions among the two storage proteins' subunits in functionality evaluations is important in evaluating structure–function relationships.

Thermal Stability

The effect of heat processing on these proteins is an active research area. Thermal stability studies were conducted on the pure storage proteins and mixtures there of by evaluating the heat capacities, in gel forming capacity, in tofu gel formation (different from pure protein gel formation), and as parts of soy protein isolate and whole soy thermal processing. The individual subunits of each storage protein have unique thermal properties as well as the ability to interact with each other. Additionally, the effects of ionic strength, pH, and reducing agent profoundly affect thermal behavior of these proteins. The overall goal is to understand how these proteins would behave in food systems. But the first level of evaluation must be at the protein structure level. Considerable progress was made in the past 10 years due to improved methods to produce these storage proteins on a large scale as well as molecular techniques to produce native and mutant peptides from specific genes for these proteins.

According to Pernollet and Mosse (1983), the legumins have higher denaturation temperatures than vicilins, and the two soy storage proteins fit these models. Glycinin has an apparent denaturation temperature of 90°C while β-conglycinin denatures at 75°C (Hermansson, 1979a). The mechanism of denaturation is controlled by subunit composition, interactions between the subunits, ionic strength and reductant, and may lead to eventual aggregation and polymerization reactions (Wolf & Nielsen,

1996; Nakamura et al., 1984; Utsumi et al., 1997). Hermansson (1979b) showed β-conglycinin has different thermal transition temperatures, as measured by differential scanning calorimetry (DSC) of 67°C at 0 ionic strength (μ) up to 87°C at μ of 1.0 M. Glycinin showed a similar pattern with 80°C and 103°C, respectively. However, Hashizume and Watanabe (1979) showed that while glycinin was stabilized with increasing μ, β-conglycinin became more sensitive to temperature. We observed similar stability to Hashizume and Watanabe (1979) by measuring the loss of native structure of glycinin and β-conglycinin at μ between 0 and 0.5 M as measured by response to antibody recognition in the absence of reductants (Fig. 8.4). Glycinin denatured faster as temperature increased but slower at a given temperature as μ increased. β-Conglycinin denaturation rate slowed as temperature increased between μ of 0 and 0.2 M. β-Conglycinin denaturation rate increased as μ increased between 0 and 0.2 M. Only at μ of 0.5 M did β-conglycinin behave as in Hemansson (1979b). The behavior of β-conglycinin suggests hydrophobic (van der Waal) forces are involved with increasing salt concentration, resulting in faster denaturation while glycinin becomes more stable between μ of 0 and 0.2 M. Since most foods have a μ between 0.1 to 0.3 M, data derived in this salt range are of practical value.

Damodaran and Kinsella (1982) proposed a model for thermal-induced interactions between glycinin and β-conglycinin at 80°C, pH 8.0 in presence of a reducing agent. Glycinin acidic and basic groups dissociate from intact glycinin. The basic peptides aggregate. In the presence of β-conglycinin subunits, the glycinin subunits form soluble aggregates with β-conglycinin subunits due to electrostatic interactions. Addition of sodium chloride to μ of 0.5M causes the basic subunits to re-aggregate even with β-conglycinin subunits present. The aggregated particles were detected as turbidity. In a model system without reducing agent and more akin to a food system, Yamagishi et al. (1983) showed that precipitates were formed consisting of polymers of β subunits of β-conglycinin and basic subunits of glycinin. Soluble aggregates (oligomers) of acidic glycinin subunits and α and α′ subunits of β-conglycinin were formed in the supernatant. Clearly, interactions between glycinin and β-conglycinin subunits during thermal processing depend not only on chemical agents (pH, μ, reducing agent) but also on how much of each protein is present. Therefore, knowledge of the purity of soybean protein fractions is critical in understanding the interactions observed.

Glycinin Thermal Behavior

Nakamura et al. (1984a) were one of the first to present evidence that the type of acidic subunit, and now we know a specific basic subunit as well, had effects on gelling (or thermal denaturation rate) of glycinin gels. The content of the A_3 glycinin peptide correlated with increased gel strength. The lower sulfur amino acid peptides, A_3B_4 and $A_4A_5B_3$, apparently provide greater gelling strength compared to the other acidic-basic pairs (Nakamura et al., 1984a; Mori et al., 1982). Heat stability for the acidic-basic

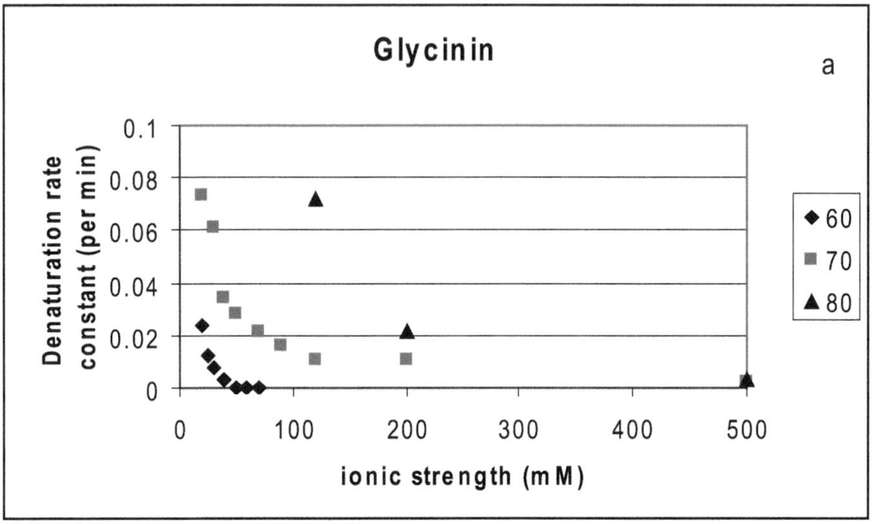

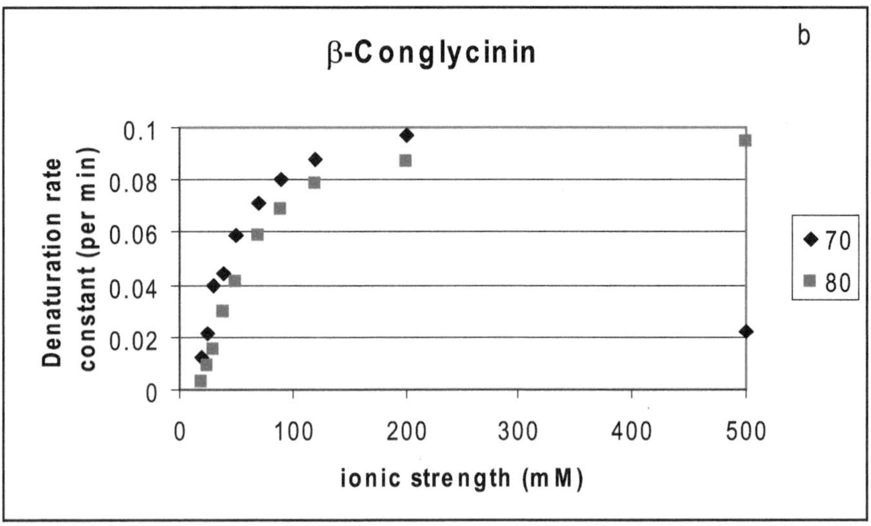

Fig. 8.4. Rate of native structure loss (measured as loss of antibody reactivity (Wu et al., 1999) of a) glycinin and b) β-conglycinin in pH 7.5 phosphate at 60, 70, and 80°C.

peptide pairs is in the order of A_3B_4 > $A_4A_5B_3$ > $A_{1a}B_{1b}$ ~ $A_{1b}B_2$ ~ A_2B_{1a} (Tezuka et al., 2000) in tofu gels. Lakemond et al. (2002) reported effects of isothermal heating of purified glycinin, resulting in increasing stabilities of A_4 A_5 B_3 > A_3B_4 > A_4 >> $A_{1a}B_{1b}$ ~ $A_{1b}B_{1b}$ ~ A_2B_2. Muruyama et al. (2004) report thermal stabilities (evaluated as transition temperatures) of isolated glycinin hexamers as native glycinin ~ group I ~ group II ~ A_4 A_5 B_3 > A_3B_4 where group I was a mixture of $A_{1a}B_2$, $A_{1b}B_{1b}$ and A_2B_{1b} and group II was a mixture of A_4 A_5 B_3 and A_3B_4. Group I glycinin actually showed two transition temperatures, one much lower than typical glycinin and a higher one attributed to a trimer and hexamer form by these authors. But Muruyama et al. (2007) concluded that subunit composition has no or a very small effect on denaturation rate. Tezuka et al. (2004) reported A_3B_4 ~ $A_4A_5B_3$ ~ $A_{1a}B_{1b}$ ~ $A_{1b}B_2$ ~ A_2B_{1a} for denaturation temperature (range 95–97.9°C). Prak et al. (2005) reported *E. coli* expressed glycinin protomer thermal stability in the order of $A_{1a}B_{1b}$ ≥ A_3B_4 > A_4 A_5 B_3 ≥ A_2B_{1b} >> $A_{1b}B_2$. Mori et al. (1982) suggested two protein concentration-dependent heat denaturation pathways for glycinin. Initial heating at 100°C resulted in soluble aggregates at 8000 kD. Further heating resulted in two concentration-dependent routes. Glycinin concentrations ≤0.5% produce large MW aggregates divided into acidic and basic subunits. At higher glycinin concentrations, the high MW aggregates form networks and gel. Nakamura et al. (1984b) extended these observations with electron microscopy and termed the process "a string of beads" model with the bead being the undissociated but slightly unfolded glycinin hexamer. The string of beads strands match the molecular dimensions predicted for glycinin. Initially, linear strands form upon heating at 100°C, followed by branching for gel formation. Hermanson (1985) suggested a similar model.

Disulfide bonds play a role in heat stability of glycinin but also in gel network formation. Nakamura et al. (1984a) was one of the first to show that blocking free sulfhydral groups resulted in no gelation of glycinin. These authors also correlated gel clarity to decreasing sulfhyral content. Utsumi et al. (1993), Adachi et al. (2003), and Adachi et al. (2004) produced a number of cysteine deletions or additions to glycinins by using altered genes for $A_{1a}B_{1b}$ by site-directed mutagenesis. The authors concluded that cross-linking via disulfide bonds is part of the thermal gelling mechanism. Adachi et al. (2003) produced cysteine to glycine at cysteine 12, the intrapeptide disulfide linkage site in $A_{1a}B_{1b}$, and in cysteine 88, the interpeptide disulfide linkage site, and wild-type glycinin in *E. coli* expression systems. They report little difference in the thermal stabilities of the different glycinins. However, the DSC scans show increasing thermal denaturation points, with wild-type > cysteine 88 mutant > cysteine 12 mutant, suggesting a modest contribution by disulfides to thermal denaturation. Adachi et al. (2004) introduced additional cysteine groups to form disulfide linkages between glycinin monomers (acidic-basic pair), free cysteines, and intrapeptide disulfide bonds. Compared to the wild-type glycinin $A_{1a}B_{1b}$, all mutants had greater gel hardness at pH 7.6, 10 mM phosphate. In terms of Tm, the model is more complicated because the effect of the amino acid replaced by a cysteine also impacts stability. Wild-type glycinin $A_{1a}B_{1b}$ at 0.1% had lowest Tm at 75.3°C at pH 7.6, μ of 0.435 M, but one interpeptide disulfide mutant and the free cysteine mutant had Tm only 1–2° higher, whereas the other interpeptide disulfide mutant Tm was 3.7° higher and the intrapeptide disulfide mutant was 4.8° higher. These Tm are lower than those of intact hexamer glycinin of 80 to 103°C for μ of 0 to 1.0 M, respectively.

β-Conglycinin Thermal Behavior

β-Conglycinin thermal denaturation was evaluated with native β-conglycinin isolated from soybeans and with peptides produced by recombinant technology. Since the *E. coli* recombinant systems do not glycosylate proteins, the effect of glycosylation can be evaluated. In contrast to glycinin gel formation, a time-dependent process, gelation of β-conglycinin apparently is independent of heating time at 100°C (Nakamura et al., 1986). These authors report β-conglycinin gels have no turbidity, whereas glycinin gels show greater turbidity at lower protein concentrations. Muruyama et al. (2002a) isolated homologous trimers from soybean varieties lacking other β-conglycinin peptides to compare with recombinant β-conglycinin trimers (Muruyama et al., 1999). Muruyama et al. (2002a) reported that the DSC Tm for native homotrimers was β_3 > α'_3 > α_3 at 87.0, 82.6, and 78.2°C. The Tm of α_3 and α'_3 were similar to recombinant trimers reported in Muruyama et al. (1999), but native β_3 was about 4°C lower than recombinant β_3. Muruyama et al. (2001) reported two amino acid differences between native β peptide and recombinant β peptide and small difference in crystalline structure of the two types of β_3. Muruyama et al. (2002a) suggested these structural differences may be the reason for the different Tm of the β_3. Apparently, glycosylation plays little role in Tm of β-conglycinin. Heat aggregation properties were similar between native and recombinant β-conglycinin peptides. α_3 and α'_3 aggregated into soluble aggregates while β_3 formed insoluble aggregates above their respective Tms. The differences were attributed to the extension regions of α and α′ discussed earlier. The soluble aggregates of native α_3 and α'_3 were about half the MW of recombinant α_3 and α'_3 suggesting glycosylation plays a role in limiting associations during thermal processing. Muruyama et al. (2002b) isolated heterologous trimers from soybean seeds to compare thermal properties. They isolated α_3, α'_3, $\alpha_2\beta$, $\alpha'_2\beta$, $\alpha\beta_2$, $\alpha'\beta_2$ and β_3. The order of thermal stabilities (Tm) of the trimers was β_3 at 87.0°C, α'_3 at 82.6°C, $\alpha\beta_2$ at 82.5°C, $\alpha'\beta_2$ at 82.1°C, $\alpha'_2\beta$ at 80.3°C, $\alpha_2\beta$ at 78.5°C, and α_3 at 78.2°C. These data suggest the subunits do not contribute equally to thermal stability, and the lower Tm subunit imparts greater effects on trimer stability. The subunits composition of the trimers contributed to aggregation behavior as well. Trimers with two or three α and α′ subunits formed soluble aggregates while those with two or three β subunits formed insoluble aggregates at pH 7.6, μ of 0.5 M. The soluble aggregates with a β peptide were much larger than α and α′ trimer aggregates. The differences in glycosylation and extension regions contributed to differences in thermal aggregation (Muruyama et al., 2002b).

Mixed Systems Thermal Behavior

Partially purified glycinin and β-conglycinin as well as soy protein foods as model systems should show interactions among the different subunits during thermal processing (Damodaran & Kinsella, 1982; Yamagishi et al., 1983). However, attempts to predict thermal processing effects for tofu production apparently do not follow a universal model when different varieties, soybean production location, and crop year are different (Murphy et al., 1997). Additionally commercially prepared soy protein isolates (SPI) may be prepared as proteins extracted between RT and 80°C, although, most processes are proprietary. However, model systems are still our best approach in attempting to explain thermal behavior in mixed systems. Many publications are available that evaluate the interactions and functionalities of the two soy proteins that are beyond the scope of this review. However, the major highlights include:

1. glycinin and β-conglycinin interact in formation of thermal gels via electrostatic bonds and hydrogen and van der Waals forces;

2. glycinin content is apparently related to gel hardness and unfracturability, whereas β-conglycinin contributes to elasticity;

3. the ratio of glycinin/β-conglycinin affects gel characteristics in both pure protein systems and food gels such as tofu;

4. the α and α′ subunits contain cysteines that will interact with glycinin subunits during gel formation via disulfide interchange;

5. the temperature used for gelation governs the contribution of which storage protein is contributor to observed thermal gels; and

6. the temperature treatment prior to gelation affects protein conformation and gel properties.

Wu et al. (1999) evaluated denaturation of the storage proteins isolated on a pilot-plant scale by measuring native structure recognition by polycolonal antibodies. Their data showed their intermediate fraction consisted of denatured glycinin and β-conglycinin while the native state of glycinin and β-conglycinin was retained in their respective fraction. Rickert et al. (2004b) reported enthalpies for Wu et al. (1999) method fractions that supported the immunological native structure recognition. Rickert et al. (2004b) also showed increasing extraction temperature decreased enthalpies of β-conglycinin but not glycinin in their respective fractions. SPI produced by Rickert et al. (2004b) at 60°C retained little native conformation of either protein based on enthalpies. Additionally, glycinin in the β-conglycinin fractions and β-conglycinin in the glycinin fractions retained little of their native structures. These changes in native structure may play a role in the differences in gelling characteristics reported by Rickert et al. (2004b). Nagano et al. (1996) evaluated gelling of SPIs made for low-β-conglycinin and low-glycinin soybeans at 80°C. These authors attributed larger effects due to preheated β-conglycinin compared to unheated, and both β-conglycinin treatments were more effective on gelation than glycinin, which is relatively unaffected at this temperature. Riblett et al. (2001) evaluated various soybean varieties with different compositions of glycinin and β-conglycinin and attributed differences in gel formation to differences in amounts of each storage protein as well as slight differences in enthalpies of denaturation for the two storage proteins compared across varieties. Khatib et al. (2002) reported on laboratory-scale glycinin and β-conglycinin fractionation and gel functionality although purity of protein fractions was not reported. These authors reported slight differences in gel modulus, G', for β-conglycinin among varieties but larger differences in glycinin G'. However, without knowing the purity of the fractions, to estimate the interactions between the proteins is difficult.

A different type of thermal gel, tofu, has served as a model system for examining interactions between glycinin and β-conglycinin interactions. Tofu gels are produced from soy milk, an emulsified soy protein–soy oil extract. After initial heating to approximately 80–95°C, either a metal ion coagulant, Ca or Mg, or a H^+ producer, δ-gluconolactone, is added to initiate gelation. Saio et al. (1969) may have first suggested that glycinin and β-conglycinin contribute differently to tofu gels. Glycinin contributed more to tofu gel hardness, cohesiveness, and elasticity than did β-conglycinin. The ratio of glycinin/β-conglycinin was reported by many to affect gel characteristics (Saio, 1979; Murphy et al., 1997; Guo & Ono, 2005; Mujoo et al., 2003). However, smaller-scale tofu or "test-tube" tofus were shown to not correlate well with production-scale tofus (Pesek & Wilson, 1982 cited in Wilson et al., 1992). Guo & Ono (2005) attribute tofu gelation to particle- size formation prior to coagulant addition. Kohyama et al. (1995) proposed a mechanism for tofu gelling as a two-step process: i) protein denaturation or unfolding by heat and ii) coagulation driven by hydrophobic interaction among the soy storage proteins promoted by H^+ of δ-gluconolactone or Ca^{+2} with the difference in the coagulant only a difference in ion-induced aggregation rate. Differences in storage protein subunit composition were evaluated to explain tofu gel differences. Murphy et al. (1997) reported differences in glycinin peptide composition were correlated with tofu texture characteristics but only within single varieties, not across different varieties. Poysa et al. (2005) evaluated 20 soybean varieties with different subunit compositions based on laboratory-scale tofu production. They concluded that the absences of α′ of β-conglycinin correlated with gel hardness. Glycinin A_3 correlated with tofu gel firmness while A_4 was reported to have negative effects. Tezuka et al. (2000) attributes tofu gelation to differences in glycinin composition. Tofu hardness and particle content were correlated with $A_4 A_5 B_3 > A_3B_4 > A_{1a}B_{1b} \sim A_2B_{1b} \sim A_{1b}B_2$. Liu et al. (2004) suggest thermal processing, by heating two cycles, selectively denatures β-conglycinin which forms a network with the first heating. Glycinin is unfolded in the second heating, and associates with β-conglycinin subunits in its network that leads to stronger tofu gels.

Flavor Binding

The predominance of beany and grassy off-flavors of soy protein foods, as well as other identified off-flavors, is well recognized by researchers and consumers and is probably the main reason for lower acceptance of soy protein foods in Western and Japanese cultures. Additionally, off-tastes including bitter and astringent are associated with soy proteins. In 1978, the National Science Foundation assigned the top priority in processing and utilization research to maximizing the acceptance of soy in human foods via the identification and removal of undesirable flavors (Milner et al., 1978). Almost 30 years later, we continue to try to solve the off-flavor problem. The ability of the soy storage proteins to form reversible, and probably irreversible, bonds with flavor chemicals is one of the remaining challenges to wider use of soy proteins. The old adage "if no one eats it, it has no nutritional value" is very true for soy protein foods in Western cultures and in Japan. Not only is soy protein well-known for binding off-flavors, but soy proteins have a major capacity to bind any flavor chemical that causes challenges in food formulations to provide proper flavorant profiles for soy ingredients (Malcolmson et al., 1987; Van den Ouweland & Schutte, 1978; Schutte & Van den Ouweland, 1979; MacLeod & Ames, 1988; Inouye et al., 2002).

Researchers have studied reversible flavor binding to soy protein in dry and aqueous model systems. Early work identified lipohydroperoxide breakdown products from the primary product of lipoxygenase action. However, lipid auto-oxidation products provide many of the same off-flavor constituents in lipoxygenase-null soybeans. Pentanol, hexanol, heptanol, hexanal, 3-*cis*-hexenal, 2-propanone, 2-pentylfuran, ethyl vinyl ketone, *trans-trans*-2,4-nonadienal, *trans-trans*-2,4-decadienal, *trans-cis*-2,4-decadienal, *trans*-2-nonenal, *trans*-2-octenal, 1-octen-3-one, 1-octen-3-ol, *trans-cis*-2,6-nonadienal, and 2-pentyl pyridine were identified as major lipid derived off-flavors of soy proteins (Hill & Hammond, 1965; Arai et al., 1967; Cowan et al., 1973; Maga, 1973; Sessa & Rackis, 1977; Rackis et al., 1979; Hsieh et al., 1982; Boatright & Crum, 1997, Lozano et al., 2007).

In addition to the products of lipid oxidation, methanethiol and dimethyl trisulfide were shown to contribute to the complex odor characteristic of soy protein products such as SPI and soy protein concentrates (Boatright & Lei, 2000; Lei & Boatright, 2001) and soymilk (Lozano et al., 2007) at concentrations comparable to hexanal. Since the threshold in water for methanethiol was reported at 0.02 ppb compared to hexanal at 4.5 ppb (MacLeod & Ames, 1988), these sulfur compounds are intense flavor notes in soy protein products. Lei and Boatright (2007) provided evidence that methanethiol is generated in aqueous slurries of SPI or defatted soy flake from methionine by a free radical mechanism involving manganese, sulfite, and oxygen. They showed that addition of cysteine or potassium iodate reduced free sulfite, thus leading to very low levels of methanethiol. Lozano et al. (2007) reported that neither methanethiol flavor dilution profiles nor absolute concentrations varied from that of the control with UHT treatments of soy milk except at the highest temperature treatment of 154°C for 29 sec. Lower temperatures and longer processing had no effect.

The mechanism of flavor constituents' interactions with dry soy proteins was evaluated by Aspelund and Wilson (1983) and Crowther et al. (1981) using dry soy protein isolate as gas chromatography packing material as a basis to evaluate equilibrium binding of off-flavors. Recently, Zhou and Cadwallader (2004) replicated this technique. These three groups used these techniques to estimate heats of adsorption (enthalpy) for a homologous series of alkanes, aldehydes, ketones, and methyl esters to obtain insight into the mechanism of flavor binding. Homologous series of flavor hydrocarbons have heats of adsorption for alkanes < ketones ~ methyl esters < aldehydes. Zhou and Cadwallader (2006) were able to refine the C_6 series in more detail because the unsaturated C_6 compounds are now commercially available. The heats of adsorptions (ΔH) at 0% relative humidity (RH) followed hexane ~ 1-hexene < ethyl butyrate ~ 2-hexanone ~ hexanal < *trans*-2-hexenal < 1-hexanol < *cis*-3-hexen-1-ol < *trans*-2-hexen-1-ol. When the effects of RH were examined over a range up to 50%, the order of adsorption remained the same; however, the heats of adsorption of the alcohols decreased bringing them closer to the aldehydes and ketones. At 50% RH the ranking was hexane ~ 1-hexene < ethyl butyrate ~ 2-hexanone ~ hexanal ≤ *trans*-2-hexenal < 1-hexanol ~ *cis*-3-hexen-1-ol ~ *trans*-2-hexen-1-ol. Aspelund and Wilson (1983) evaluated heats of absorption and free energies of binding for C_5 to C_{10} flavors and interpreted binding of alkanes to SPI as van der Waals interactions. Ketones, aldehydes, and methyl esters showed van der Waals and one hydrogen bond through the carbonyl oxygen while alcohols formed two hydrogen bonds with soy protein functional groups in this dry model system. The negative ΔG for binding indicates the equilibrium favors binding, thus making removal more difficult.

Aqueous flavor interactions with soy proteins were evaluated by a number of groups using a variety of techniques including equilibrium headspace sampling, equilibrium dialysis, and solvent extraction of the flavor ligand. However, only a few reported data with statistical evaluation of their binding data which make comparisons among studies difficult (O'Keefe et al., 1991a). Klotz (1982) warned about misinterpretation of binding site numbers and binding equilibrium constants if the inflection point on the binding saturation curve is not achieved.

The Kinsella group evaluated flavor binding for glycinin, β-conglycinin, and SPI for several authentic soy off-flavors and other carbonyl compounds (Damadaran & Kinsella, 1981a; b; O'Neill & Kinsella, 1987). They concluded aqueous flavor binding was by hydrophobic interactions, with β-conglycinin binding constants being larger but moles bound being greater for glycinin. In contrast, O'Keefe et al.

(1991a, b) showed binding affinities for either protein depended on aqueous conditions including ionic strength (μ). O'Keefe et al. (1981a) attributed the different results to extrapolation of data by the Kinsella group from well below Klotz' saturation curve inflection point. In pH 8.0 buffer, glycinin had 4× the binding site number of β-conglycinin for hexanal, although, the K_{eq} was similar for both proteins. Addition of 0.5 M NaCl decreased the numbers of binding sites for glycinin while increasing them for β-conglycinin and decreasing the K_{eq} for β-conglycinin. β-Mercaptoethanol did not change the binding parameters compared to the buffer control. O'Keefe et al. (1991b) reported binding affinities for aldehydes, ketones, and hexane all were greater for glycinin than β-conglycinin. Increasing aldehyde chain length increased binding affinity for glycinin but not β-conglycinin. Hexane affinity was only observed at 5°C. Klotz plots of these data were not linear, suggesting unfolding of protein as flavor titration proceeds (O'Keefe et al., 1991a,b). Maheshwari et al. (1997) data suggested Schiff base formation between lysine-ε-amino groups and off-flavor carbonyls that was controlled by pH. Boatright and Crum (1997) reported flavor thresholds for 2-pentyl-pyridine, a new off-flavor constituent, of 12 ppt compared to 7500 ppt for hexanal in SPI. Zhou et al. (2002) showed more 2-pentyl pyridine binding sites for glycinin than β-conglycinin and SPI. 2-Pentyl pyridine binding affinities increased for both proteins as pH increased but both decreased with increasing μ. These authors did not report thermodynamic values although they examined binding equilibria at three temperatures. Interactions of flavors in aqueous systems appear to be more complicated than simple binding kinetic models can explain. Neither soy storage protein is solely responsible for all flavor binding, apparently.

Major efforts made to improve soy protein food off-flavor problems are heating and other processing techniques such as supercritical CO_2 extraction (Maheshwari et al., 1995), solvent azeotrope extraction (Eldridge et al., 1971), grinding with ethanol (Borhan & Snyder, 1979) and removing the initiator of the lipid oxidation chain reaction by removing lipoxygenase isozymes from soybean lines (Hildebrand & Hymowitz, 1981). Lipoxygenase is discussed below, but agronomists were successful in creating soybean lines by traditional breeding and genetic engineering that are lipoxygenase triple-nulls. However, this approach is not entirely successful because many of the same off-flavors that are present due to lipid auto-oxidation are still present (Kitamura, 1984; Kobayashi et al., 1995). But hexanal, the main beany note, is not present in the lipoxygenase triple-null soy flours or extracts. Torres-Penaranda and Reitmeier (2001) reported decreased beany notes and increased grassy notes in lipoxygenase null soymilks. They concluded that lipoxygenase triple-null varieties create different flavor problems than do normal soybean varieties. The lack of lipoxygenase isozoymes does not solve the problems of soy protein's large capacity to reversibly bind flavor ligands.

Chiba et al. (1979) suggested using aldehyde dehydrogenase to convert aldehydes to higher detection threshold alcohols, but the expensive requirement for NAD^+ limited the practical application. Takahashi et al. (1979) suggested aldehyde oxidase as an enzymatic alternative with no cofactor requirement. Maheshwari et al. (1997) extensively characterized two porcine aldehyde oxidase isozymes and their practical ability to reduce aldehydic off-flavors in aqueous soy products. Sensory panelists could perceive differences in enzyme-treated soymilks with >90% hexanal and pentanal oxidized to the corresponding acids having higher flavor thresholds.

Bitter and astringent notes are associated with soy proteins (Torres-Penaranda & Reitmeier, 2001; Drewnowski & Gomez-Carneros, 2000; Matsuura et al., 1989; Okubo et al., 1992; Mahfuz et al., 2004; Robinson et al., 2004). A number of the constituents associated with health benefits of soy protein foods are in this bitter and/or astringent group. Torres-Penaranda and Reitmeier (2001) reported more intense bitter scores for lipoxygenase-null soymilks compared to the control soymilks. The lack of beany and grassy flavors in the lipoxygenase null products probably allows panelists to detect the bitter notes. Attempts to identify the components responsible for bitter and astringent notes in soy protein foods were mixed. The soy isoflavones, soyasaponins, and phenolic acids were suggested as the chemicals involved. Bitter peptides in hydrolyzed soybean products are associated with bitter taste (Cho et al., 2004). Some of the bitter peptides are also associated with the angiotensin-(I)-converting enzyme inhibition (Pripp & Ard, 2007). Tsukomoto et al. (1995) suggested isoflavone and saponins were bitter components of soy. Aldin et al. (2006) demonstrated that the isoflavone malonyl-β-glucosides and DDMP-soyasaponins, both in high concentration in raw soybeans but converted to other forms with heat processing, were the major bitter notes in soy extracts. However, Aldin et al. (2006) found aglucon isoflavones and DDMP-free saponins appear to have no bitter notes as judged by their panelists.

Lipoxygenases

Lipoxygenases (EC 1.13.11.12, linoleate: oxygen oxidoreductase) are found in many plants with soybeans having the highest activity. The characteristics of these enzymes were reviewed recently (Robinson et al., 1995; Gardner, 2003). Fujimaki et al. (1965) showed that lipoxygenase was associated with hexanal production in soybeans.

Soybean seeds contain at least three lipoxygenase isozymes. These isozymes are non-heme iron containing oxido-reductases with strict requirements for unsaturated lipid substrates possessing a *cis, cis*-1, 4-pentadiene configuration. The specificity is required for the antarafacial (or backside) addition of oxygen to form the hydroperoxide product. Soybean lipoxygenases (LOX) are single peptides with MW of 102 kD. LOX-1 is the most thoroughly studied. In terms of abundance, LOX-3 is the highest in terms of protein with LOX-1 almost as concentrated. LOX-1 has a pH optimum between 8 and 9 while LOX-2 and LOX 3 have optima of pH 6.5. The pI for the three isozymes are 5.68, 6.25, and 6.15 for LOX-1, LOX-2, and LOX-3, respectively. LOX-2 is apparently calcium-activated. LOX-1 is most active with linoleic acid at pH 9. LOX-2 and LOX-3 are more active with methyl esters of the fatty acids and triglycerides than the free fatty acids. LOX-1 has activity with water-soluble substrates such as linoleyl sulfate while LOX-2 and LOX-3 show little activity. The products of LOX-1 oxidation of linoleic acid result in a hydroperoxide at ω-6 or a 13-lipohydroperoxide and at ω-10 or a 9-lipohydroperoxide depending on the pH of the reaction. Four stereoisomers were identified on HPLC: 13-hydroperoxy-9.*3-cis*, 11-*trans*-; 13-hydroperoxy-9-*trans* 11-*trans*-; 9-hydroperoxy-10-*cis*, 12-*trans*-; and 9-hydroperoxy-10-*trans*, 12-*trans*-octadecenoic acid (Schwimmer, 1990). Linolenic acid only forms the 13-hydroperoxides (Whitaker, 1994). The iron in the native enzyme is Fe^{+2}. This form requires the hydroperoxide product to oxidize the iron to Fe^{+3}, the active enzyme form, which is similar to other oxidases. Lipoxygenase is an ordered Bi Uni sequential mechanism with the lipid substrate being added first, followed by oxygen (Chen & Whitaker, 1986). Few actual inhibitors of LOX are present although the lipohydroperoxide product is a suicide inhibitor by covalently binding around the active site and effectively blocking new substrate from entering the active site. Most "inhibitors" of LOX reported in the literature actually act on the secondary products resulting from the auto-oxidation of the lipohydroperoxide. Examples of these types of compounds include antioxidants, quenching the chain reaction with mannitol and ethanol, reducing agents such as ascorbic acid and Maillard reaction products. Enzymatic conversion of the secondary products can reduce their concentration and flavor profile. Interestingly, aspirin, which is an effective inhibitor of mammalian cyclo-oxygenases, has no activity with soybean lipoxygenases (Schwimmer, 1990). Of course, removal of oxygen, one of the substrates, is effective in stopping the reaction but is not usually feasible in food processing operations. The slightest damage to soybean seed cells allows LOX to begin its activity. The turnover or k_{cat} for LOX-1 is 280 to 350 sec^{-1} showing how exceedingly fast off-flavors are generated (Gardner, 2003).

Lipoxygenase may play a role in SPI production. When we examined SPI production on a laboratory scale using triple LOX-null soybeans compared to the parental variety, we observed no differences in yield (data not shown). However, when these varieties were compared using the Iowa State University Center for Crops Utilization Research pilot plant SPI procedure (Wu et al., 2000), it became apparent that the LOX-triple null SPI precipitates were much more difficult to centrifuge. The data in Table 8.4 compares the yield differences in SPI made from triple nulls from two crop years. Moisture, protein, and lipid compositions of the soybeans were not different between years or varieties.

Trypsin Inhibitors

The 2S fraction of soybean seeds contains the trypsin inhibitors as well as other smaller MW proteins (<25 kD) such as cytochrome c. The two best characterized trypsin inhibitors in soybeans are the Kunitz trypsin inhibitor (KTI) and the trypsin-chymotrypsin double-headed Bowman Birk inhibitor (BBI). Apparently several

Table 8.4. SPI Yields from Lipoxygenase (LOX) Triple-null Varieties

Yield	LOX triple-null	Parent
% SPI yield 1997	25.2	38.3
% SPI yield 1998	23.7	27.2
% protein yield 1997	93.7	92.7
% protein yield 1998	93.7	93.4

isoforms of each inhibitor exist (Friedman & Brandon, 2001). From observations in soy processing, application of moist heat treatment, but not dry, resulted in improved protein utilization by rodents concomitant with trypsin inhibitor activity reduction. However, usually 10–20% residual trypsin inhibitor activity remained. Because of the importance of their biological activity, both soy trypsin inhibitors were crystallized and their structure determined (Voss et al., 2001; Song & Suh, 1998; Koepke et al., 2000).

The presence of trypsin inhibitor activity in soy foods and feeds was a major concern based on early rodent feeding studies. Rats and mice fail to thrive on raw soybean foods, in part due to trypsin inhibitor activities. The role of sulfur amino acid requirements for rodents also played a role in low growth rates on minimally heat processed soy protein foods. A well-understood interaction occurs between soy trypsin inhibitor (STI) activity and sulfur amino acid requirements in rodents (Liener, 1981). Animals with a pancreas size of > ~0.3% (as % body weight) experience pancreatic hypertrophy when fed raw soybeans. Immature, but not adult, guinea pigs are the largest animals in which this effect was observed. Apparently in these smaller animals, as the pancreas attempts to compensate for the loss of gut trypsin and chymotrypsin, the pancreas hypertrophies, and an enhanced need arises for sulfur amino acids for enzyme synthesis. The cycle of trypsin/chymotrypsin loss through binding with the trypsin inhibitor coupled with increased synthesis of the proteases exacerbates the low sulfur amino acid balance from the soy diets for rodents. Thus, low PERs are also associated with raw soy protein fed to rats. In contrast, pancreatic hypertrophy is not observed in larger animals including humans. Of the two human forms of trypsin, the more abundant cationic trypsin is minimally affected by STI, whereas the anionic trypsin (~10–20% of human trypsin) is completely inhibited by STIs. Additionally, humans have a much lower sulfur amino acid requirement than rodents resulting in no sulfur amino acid deficiency for adult humans. Soy-based infant formulas are routinely assayed for residual STI activity, and most SIF commercial products are quite low in STI.

The KTI has a MW 21.5 kD with two disulfide bonds per mole protein (Koide & Ikenaka, 1973). KTI typically is 2–8× more abundant in soybeans than BBI. Kunitz is more heat-labile than the other trypsin inhibitor, Bowman Birk, mainly because of the low amount of disulfide linkages. Kunitz has little inhibitor activity toward chymotrypsin. Heat-processed soy foods, including soy-based infant formula, contain very low levels of KTI compared to raw soybeans measured by difference using BBI ELISA and enzymatic based trypsin inhibitor assay (Dipietro & Liener, 1989) or by KTI ELISA (Friedman & Brandon, 2001).

The Bowman Birk trypsin inhibitor is the smaller trypsin inhibitor with MW ~8.0kD and binding sites for trypsin and chymotrypsin (Birk, 1985). BBI is more heat-stable than the Kunitz inhibitor probably resulting from its great proportion of disulfide cross-linking at 7 per mole BBI. Although BBI is present in much lower amounts in raw soybeans, its relative heat stability may be the main reason for the residual STI activity in moist-heated soybean protein products. Friedman and Brandon (2001) reported about 7 μg/mL BBI in soy-based infant formula as measured by ELISA while Dipietro and Liener (1989) reported <0.1 μg/mg (their limit of detection by ELISA) for SIF.

In the 1980s, concern existed that STIs were carcinogens (Rackis, 1981). Extensive work was done by the Rackis group (Gummerman et al., 1985) with rats showing the pancreatic hypertrophy lead to cancer. However, mice did not respond in the same way (Gummermann et al., 1989). BBI was reported to act as an anticarcinogen after radiation treatment (Witschi & Kennedy, 1989). A number of different cancers were inhibited by BBI treatment including oral leukoplakia (Kennedy et al., 1993), lymphosarcoma (Evans et al., 1992), head and neck tumors (Meyskens et al., 2001) as well as prostatic hyperplasia (Malkowicz et al., 2001). Production of BBI was commercialized and patented (Malkowicz et al., 2001; Kennedy et al., 1993; Rostomi & Kennedy, 2004). Phase I and II clinical trials were completed with BBI or BBI concentrate (Armstrong et al., 2000; Armstrong et al., 2004). Recently, Kennedy's group reported BBI treatment of multiple sclerosis autoimmune effects (Gran et al., 2006). Clearly, numerous positive outcomes exist with BBI far beyond our initial concerns as an antinutrient. BBI falls into the beneficial and health-promoting arena of soy protein products.

Lectins are another class of potential antinutrient soy proteins (Sharon & Lis, 1972; Liener, 1974). The carbohydrate content and structure of soy lectins were determined (Lis & Sharon, 1978) and consist of mannose and *N*-acetyl-glucosamine at about 5% by weight. Soy lectin has MW 110 kD from four identical subunits. The soy lectins are partly responsible for weight loss in rodent feeding studies with raw soy protein. Soy lectins are relatively easy to heat denature compared to other legume lectins. Properly processed soy foods have little native lectin present. Soybean lectins are widely used in clinical studies because of their interaction with red blood cell surface features (Friedman & Brandon, 2001).

Bioactive Properties Associated with Soy Proteins

The biological effects of soy proteins and their digestive peptide products are a continuing area of interest for consumers and researchers. The range of biological effects evaluated or hypothesized to be related to soy proteins includes plasma cholesterol lowering, antimytotic activity, angiotensin or antihypertensive properties, phagocytosis stimulating activity, immunomodulatory activity (the Chapter: *Human Nutrition Value of Soybean Oil and Protein* of this monograph) as well as the anticancer activity attributed to BBI discussed above and to lunesin (Galvez, et al., 2001).

Numerous animal studies suggested a role for soy proteins or their individual storage proteins in serum lipid alteration (Sugano et al., 1988, 1990; Adams et al., 2004). Cell systems suggested a role for β-conglycinin or its peptides in cholesterol regulation (Lovati et al., 1992, 1996, 2000; Sirtori et al., 1993). However, not all the biological outcomes translated into the same benefits for humans. The FDA-approved health claim for soy protein and cardiovascular health as part of a heart-healthy diet led to increased numbers of soy protein-containing products in the marketplace. The health claim states that soy protein at 6 g per serving with four servings per day as part of a low saturated fat and low cholesterol diet will result in lowered levels of serum cholesterol. FDA evaluated 41 studies submitted by the health claim petitioner and concluded that sufficient evidence supported the soy protein health claim petition (Schultz, 1998). Since the soy protein health claim was approved, numerous well-designed studies with human subjects supported the hypothesis that soy protein, with their associated soy isoflavones, resulted in modest plasma cholesterol lowering (Crouse et al., 1999; Potter et al., 1998; Washburn et al., 1999; Baum et al., 1998; Merz-Demlow et al., 2000; Teixeira et al., 2000; Urban et al., 2001; Teede et al., 2001; Steinberg et al., 2003; Nagata et al., 1998; Wong et al., 1998). Crouse et al. (1999) estimated a dose response for isoflavones associated with soy proteins and suggested a minimum of 1.5 mg isoflavone aglucon/g is required for an observable plasma cholesterol reduction. Although a recent position statement by the American Heart Association (Sacks et al., 2006) stated soy protein had little effect on plasma cholesterol levels in humans, a more recent meta analysis of 41 studies still supports the modest LDL-cholesterol lowering in human subjects consuming soy protein (Reynolds et al., 2006). Studies examining soy isoflavones fed without soy protein resulted in no cholesterol lowering (Dewell et al., 2002; Nestel et al., 1997; Samman et al., 1999; Hodgeson et al., 1999). No published reports exist and only a few scientific meeting abstracts on feeding individual soy protein fractions to humans to detect plasma cholesterol alterations. To date, no one has compared glycinin and β-conglycinin side-by-side in a human feeding study. Clearly, dietary soy protein with its native isoflavone constituents has some effects on plasma cholesterol. The effects are larger with higher concentrations of LDL-cholesterol, as observed with other dietary interventions.

The cholesterol-lowering ability of soy storage proteins is associated with specific peptide sequences in glycinin (Makino et al., 1988; Kato & Iwami, 2002; Choi et al., 2002) and β-conglycinin (Tsuruki et al., 2003) via their bile acid binding ability. Maruyama et al. (2003) reported a phagocytosis stimulating peptide isolated from α′ subunit of β-conglyinin in a highly conserved region and called it soymetide. The α and β peptides contain similar peptide regions, but these peptide fragments did not stimulate the same activity. Angiotensin converting I enzyme (ACE) inhibitory peptides were evaluated from soy (Gouda et al., 2006; Gibbs et al., 2004; Lo et al., 2006; Kuba et al., 2005). In model systems, these peptides seem to have activities that would alter ACE in biological systems. Much work remains to be done to confirm the efficacy of this approach.

Conclusion

Over the past ten years, the three-dimensional structure of the soybean storage proteins, glycinin and β-conglycinin, was elucidated based on gene and protein sequences and X-ray crystallography of the native proteins or mutant constructs that will crystallize. However, the structure–functionality relationships will require much more research. Some inroads into understanding these relationship were made by producing large amounts of highly purified storage proteins by a variety of schemes and by utilizing genetically engineered soy storage proteins expressed in small amounts in microorganisms. However, these are very complex relationships requiring significant investments of funds and time to uncover the mechanisms of structure in functionality. The flavor issues with soy proteins are still a prominent problem more than 30 years after it was recognized as the major limiting reason in wider utilization of this excellent protein source. Traditional and molecular genetic approaches to remove lipoxygenase as the initiator of off-flavors was not universally successful in solving the flavor problems. Our understanding of the effects of thermal processing on individual soy proteins has improved; however, what is occurring during thermal processing in mixed protein systems requires more research. The formerly toxic constituents of soybean protein products are almost universally moved into the health protective constituents category that includes Bowman-Birk trypsin inhibitor, isoflavones, saponins, and soy protein peptides, to mention a few. Understanding soy proteins and their associated constituents is still a very vital field.

References

Adachi, M.; J. Kanamori; T. Masuda; K. Yagasaki; K. Kitamura; B. Mikami; S. Utsumi. Crystal structure of soybean 11S globulin: Glycinin A_3B_4 homohexamer. *Proc. Natl. Acad. Sci. U.S.A.* **2003,** *100,* 7395–7400.

Adachi, M.; Y. Takenaka; A.B. Gidamis; B. Mikami; S. Utsumi. Crystal structure of soybean proglycinin $A_{1A}B_{1B}$. Deposited 2000. http://www.rcsb.org/pdb/explore.do?structureId=1FXZ.

Adachi, M.; Y. Takenaka; A.B. Gidamis; B. Mikami; S. Utsumi. Crystal structure of soybean proglycinin $A_{1a}B_{1b}$ homotrimer. *J. Mol. Biol.* **2001,** *305,* 291–305.

Adams, M.R.; D.L. Golden; A.A. Franke; S.M. Potter; H.S. Smith; M.S. Anthony. Dietary soy β-conglycinin (7S globulin) inhibits atherosclerosis in mice. *J. Nutr.* **2004,** *134,* 511–516.

Aldin, E.D. Soy Protein: A Fractionation and Sensory Study. M.S. thesis, Iowa State University,

2004; 101 pp.

Aldin, E.A.; C.A. Reitmeier; P.A. Murphy. Bitterness threshold and descriptive analyses of soy isoflavones and saponins. *J. Food Sci.* **2006,** *71,* S211–215.

Arai, S.; O. Koyanagi; M. Fujimaki. Studies on flavor compounds in soybean. *Agric. Biol. Chem.* **1967,** *31,* 868–873.

Armstrong, W.B.; A.R. Kennedy; X.S. Wan; J. Atiba; C.E. McClaren; F.L. Meyskens, Jr. Single-dose administration of Bowman-Birk inhibitor concentrate in patients with oral leukoplakia. *Cancer Epidemiol. Biomark. Prev.* **2000,** *9,* 43–47.

Armstrong, W.B.; A.R. Kennedy; X.S. Wan; T.H. Taylor; Q.A. Nguyen; J. Jensen; W. Thompson; W. Lagerberg; F.L. Meyskens. Clinical modulation of oral leukoplakia and protease activity by bowman-birk inhibitor concentrate in a phase iia chemoprevention trial. *Clin. Cancer Res.,* **2000,** *6,* 4684–4691.

Aspelund, T.G.; L.A. Wilson. Adsorption of off-flavor compounds onto soy protein: a thermodynamic study. *J. Agric. Food Chem.* **1983,** *31,* 539–545.

Badley, R.A.; D. Atkinson; H. Hauser; D. Oldani; J.P. Green; J.M. Stubbs. The structure, physical and chemical properties of the soybean protein glycinin. *Biochim. Biophys. Acta* **1975,** *412,* 214–228.

Bair, C.W. Microscopy of Soybean Seeds: Cellular and Subcellular Structure During Germination, Development and Processing with Emphasis on Lipid Bodies. Ph.D. dissertation, Iowa State University.

Bair, C.W.; H.E. Snyder. Ultrastructural changes in soybeans during processing to yield desolventized-toasted meal. *J. Food Sci.* **1980,** *45,* 529–533.

Baum, J.A.; H. Teng; J.W. Erdman; R.M. Weigel; B.P. Klein; V.W. Persky; S. Freels; P. Surya; R.M. Bakhit; E. Ramos; N.F. Shay; S.M. Potter. Long-term intake of soy protein improves blood lipid profiles and increases mononuclear cell low-density-lipoproteinreceptor messenger RNA in hypercholesterolemic, postmenopausal women. *Am. J. Clin. Nutri.* **1998,** *68,* 545–551.

Berman, H.M.; J. Westbrook; Z. Feng; G. Gilliland; T.N. Bhat; H. Weissig; I.N. Shindyalov; P.E. Bourne. The Protein Data Bank. *Nucleic Acids Res.* **2000,** *28,* 235–242.

Birk, Y. The Bowman-Birk inhibitor. *Int. J. Pept. Protein Res.* **1985,** *25,* 113–131.

Boatright, W.L.; A.D. Crum. Odor and flavor contribution of 2-pentyl pyridine to soy isolate. *J. Am. Oil Chem. Soc.* **1997,** *74,* 1575–1581.

Boatright, W.L.; Q. Lei. Headspace evaluation of methanethiol and dimethyl trisulfide in aqueous solutions of soy protein isolates. *J. Food Sci.* **2000,** *65,* 819–821.

Borhan, M.; H.E. Snyder. Lipoxygenase destruction in whole soybeans by combinations of heating and soaking in ethanol. *J. Food Sci.* **1979,** *44,* 586.

Chen, A.O.; J.R. Whitaker. Purification and characterization of a lipoxygenase from immature English peas. *J. Agric. Food Chem.* **1986,** *34,* 203–211.

Chiba, H.; N. Takahashi; R. Sasaki. Enzymatic improvement of food flavor II. Removal of beany flavor from soybean products by aldehyde dehydrogenase. *Agric. Biol. Chem.* **1979,** *43,* 1883–1889.

Cho, M.J.; N. Unklesbay; F. Hsieh; A.D. Clarks. Hydrophobicity of bitter peptides from soy protein hydrolysates. *J. Agric. Food Chem.* **2004,** *52,* 5895–5901.

Choi, S.K.; M. Adachi; S. Utsumi. Identification of the bile acid-binding region in the soy glycinin A1aB1b subunit. *Biosci. Biotechnol. Biochem* **2002,** *66,* 2395–2401.

Cowan, J.C.; J.J. Rackis; W.J. Wolf. Soybean protein flavor components: A review. J. *Am. Oil Chem. Soc.* **1979,** *50,* 426A–4356.

Crouse, J.R.; T. Morgan; J.G. Terry; J. Ellis; M. Vitolins; G.L. Burke. A randomized trial comparing the effect of casein with that of soy protein containing varying amounts of isoflavones on plasma concentrations of lipids and lipoproteins, *Arch. Intern. Med.* **1999,** *159,* 2070–2076.

Crowther, A.; L.A. Wilson; C.E. Glatz. Effects of processing on adsorption of off-flavors onto soy protein. *J. Food Process Engineer.* **1981,** *4,* 99–115.

Damodaran, S.; J.E. Kinsella. Interaction of carbonyls with soy protein: thermodynamic effects. *J. Agric. Food Chem.* **1981a,** *29,* 1249–1253.

Damodaran, S.; J.E. Kinsella. Interaction of carbonyls with soy protein: conformational effects. *J. Agric. Food Chem.* **1981b,** *29,* 1253–1257.

Damodaran, S.; J.E. Kinsella. Effect of conglycinin on the thermal aggregation of glycinin. *J. Agric. Food Chem.* **1982,** *30,* 812.

Davies, C.S.; J.B. Coates; N.C. Nielsen. Inheritances and biochemical analysis of four electrophoretic varieants of β-conglycinin from soybean. *Theor. Appl. Genetics* **1985,** *71,* 351–308.

Deak, N.A.; P.A. Murphy; L.A. Johnson. Fractionation of soybean storage proteins using Ca^{+2} and $NaHSO_3$. *J. Food Sci.* **2006,** *71,* C413–424.

Deak, N.A.; P.A. Murphy; L.A. Johnson. Characterization of fractionated soy proteins Produced by a new simplified procedure. *J. Am. Oil Chem. Soc.* **2006b,** 10.1007/s11746-006-1017-1.

Dewell, A.; C.B. Hollenbeck; B. Bruce. The effects of soy-derived phytoestrogens on serum lipids and lipoproteins in moderately hypercholesterolemic postmenopausal women, *J. Clin. Endocrinol. Metab.* **2002,** *87,* 118–121.

Dipietro, C.M.; I.E. Liener. Soybean protease inhibitors in foods. *J. Food Sci.* **1989,** *54,* 606–617.

Drewnowski, A.; C. Gomez-Carneros. Bitter taste, phytonutrients, and the consumer: a review. *Am. J. Clin. Nutri.* **2000,** *72,* 1424–1435.

Eldridge, A.C.; J.E. Kalbrenner; H.A. Moser; D. Honig; J.J. Rakis; W.J. Wolf. Laboratory evaluation of hexane: alcohol azeotrope-extracted soybean flakes as a source for bland protein isolates. *Cereal Chem.* **1971,** *48,* 640–646.

Evans, S.M.; B.F. Szuhaj; T. Van Winkle; K. Michel; A.R. Kennedy. Protection against metastasis of radiation induced thymic lymphosarcoma and weight loss in C57B1/6NCr1BR mice by an autoclave resistant factor present in soybeans. *Radiat. Res.* **1992,** *132,* 259–262.

Fehr, W.R.; J.A. Hoeck; S.L. Johnson; P.A. Murphy; J.D. Nott; G.I. Padilla; G.A. Welke. Genotype and environment influence on protein components of soybean. *Crop Sci.* **2003** *43,* 511–514.

Friedman, M.; D.L. Brandon. Nutritional and health benefits of soy proteins. *J. Agric. Food Chem.* **2001,** *49,* 1069–1086.

Fujimaki, M.; S. Arai; N. Kirigaya; Y. Sakurai. Studies of flavour compounds in soybean. Part 1. Aliphatic carbonyl compounds. *Agric. Biol. Chem.* **1965,** *29,* 855–858.

Galvez, A.F.; N. Chen; J. Macasieb; B.O. de Lumen. Chemopreventive property of a soybean peptide (lunasin) that binds to deacetylated histones and inhibits acetylation. *Cancer Res.* **2001,** *61,* 7473–7478.

Galyer, K.R.; G.E. Sykes. Beta-conglycinins in developing soybean seeds. *Plant Physiol.* **1981,** *67(5),* 958–961.

Gardener, H.W. Lipoxygenase and associated enzymes. *Handbook of Food Enzymology;* J.R. Whitaker, A.G.J. Voragen, D.W.S. Wong. Eds.; Marcel Dekker: New York, NY, 2003; pp. 557–567.

Gibbs, B.F.; A. Zougman; R. Masse; C. Mulligan. Production and characterization of bioactive peptides from soy hydrolysate and soy-fermented food. *Food Res. Int.* **2004,** *37,* 123–131.

Gouda, K.G.M.; L.R. Gowda; A.G.A. Rao; V. Prakash. Angiotensin I-converting enzyme inhibitory peptide derived from glycinin, the 11S globulin of soybean (glycine max). *J. Agric. Food Chem.* **2006,** *54,* 4568–4573.

Gran, B.; N. Tabibzadeh; A. Martin; E.S. Ventura; J.H. Ware; G.-X. Zhang; J.L. Parr; A.R. Kennedy; A. M. Rostami. The protease inhibitor, Bowman-Birk Inhibitor, suppresses experimental autoimmune encephalomyelitis: a potential oral therapy for multiple sclerosis. *Mult. Scler.* **2006,** *12,* 688.

Gumermann, M.R.; W.L. Spangler; G.M. Dugan; E.C. Baker; J.J. Rakis. Pancreatic response in rats and mice to trypsin inhibitors from soy and potato after short- and long-term dietary exposure. *J. Nutr.* **1989,** *119,* 1598–1609.

Gumermann, M.R.; W.L. Spangler; G.M. Dugan; J.J. Rackis; I. Liener. The USDA trypsin inhibitor study. IV. The chronic effects of soy flour and soy protein isolate on the pancreas in rats after two years. *Qual. Plant. Plant Foods Human Nutr.* **1985,** *35,* 275–314.

Guo, S.T.; T. Ono. The role of composition and content of protein particles in soymilk on tofu curding by glucono-δ-lactone or calcium sulfate. *J. Food Sci.* **2005,** *70,* C258–C262.

Hashizume, K.; T. Watanabe. Influence of heating temperature on conformational changes of soybean proteins. *Agric. Biol. Chem.* **1979,** *43,* 683–690.

Hemansson, A.M. *Functionality and Protein Structure;* A. Pour-El, Ed.; American Chemical Society symposium series 92, 1979a; pp. 379.

Hemansson, A.M. Structure of soya glycinin and conglycinin gels. *J. Sci. Food Agric.* **1985,** *36,* 822–832.

Hermansson, A.M. *Food Texture and Rheology;* P. Sherman, Ed.; Academic Press: NY, 1979b; p. 265.

Hildebrand, D.F.; T. Hymowitz. Two soybean genotypes lacking lipixygease-1. *J. Am. Oil Chem. Soc.* **1981,** *58,* 249–253.

Hill, F.D.; E.G. Hammond. Studies on the flavor of autoxidized soybean oil. *J. Am. Oil Chem. Soc.* **1965,** *42,* 1148–1150.

Hodgson, J.M.; I.B. Puddey; L.J. Beilin; T.A. Mori; K.D. Croft. Supplementation with isoflavonoid phytoestrogens does not alter serum lipid concentrations: a randomized controlled trial in humans, *J. Nutri.* **1999,** *128,* 728–732.

Hsieh, O.A.L.; A.S. Huang; S.S. Chang, Isolation and identification of objectional volatile flavor compounds in defatted soybean flour. *J. Food Chem.* **1982,** *47,* 16–18.

Inouye, K.; M. Shihara; T. Uno; T. Takita. Deodorization of soybean proteins by enzymatic and physicochemical treatments. *J. Agric. Food Chem.* **2002,** *50,* 1652–1658.

Itoh, T.; M. Adachi; T. Masuda; B. Mikami; S. Utsumi. Crystal structure of recombinant soybean proglycinin A_3B_4 subunit, its comparison with mature glycinin A_3B_4 subunit, responsible for hexamer assembly. Deposited 2005. http://www.rcsb.org/pdb/explore.do?structureId=2D5F.

Kato, N.; K. Iwami. Resistant protein: its existence and function beneficial to health. *J. Nutr. Sci. Vitaminol.* **2002,** *48,* 1–5.

Kennedy, A.R.; P.C. Billings; P.A. Maki; P. Newberne. Effects of various protease inhibitor preparations on oral carcinogenesis in hamsters induced by 7,12 dimethylbenz(a)anthracene. *Nutr. Cancer,* **1993,** *19,* 191–200.

Kennedy , A.R.; B.F. Szuhaj. Method of making soybean Bowman-Birk inhibitor concentrate and use of same as a human cancer preventative and therapy. U.S. Patent 5,217,717 June 8, 1993.

Khatib, K.A.; T.J. Herald; F.M. Aramouni; F. MacRitchie; W.T. Schapaugh. Characterization and functional properties of soy β-conglycinin and glycinin of selected genotypes. *J. Food Sci.* **2002,** *67,* 2923–2929.

Kitamura, K. Biochemical characterization of lipoxygenase lacking mutants, L-1-less, L-2 less, and L-3-less soybeans. *Agric. Biol. Chem.* **1984,** *48,* 2339–2346.

Klotz, I.M. Number of receptor sites from Scatchard graphs: facts or fantasies. *Science* **1982,** *217,* 1247–1249.

Kobayashi, A.; Y. Tsuda; N. Hirata; K. Kubota; K. Kitamura. Aroma constituents of soybean [Glycine max (L.) Merrill] milk lacking lipoxygenase isozymes. *J. Agric. Food Chem.* **1995,** *43,* 2449–2452.

Koepke, J.; U. Ermler; E. Warkentin; G. Wenzl; P. Flecker. Crystal structure of cancer chemopreventive Bowman-Birk inhibitor in ternary complex with bovine trypsin at 2.3 Å resolution. Structural basis of Janus faced serine protease specificity. *J. Mol. Biol.* **2000,** *298,* 477–481.

Kohyama, K.; Y. Sano; E. Doi. Rheological characteristics and gelation mechanism of tofu (soybean curd). *J. Agric. Food Chem.* **1995,** *43,* 1808–1812.

Koide, T.; T. Ikenaka. Studies on soybean trypsin inhibitors 3. Amino acid sequence of the carboxyl terminal region and the complete amino acid sequence of soybean trypsin inhibitor (Kunitz). *Eur. J. Biochem.* **1973,** *32,* 417–431.

Krishnan, H.B.; S.S. Natarajan; A.A. Mahmoud; R. Nelson. Identification of glycinin and β-conglycinin subunits that contribute to the increased protein content of high-protein soybean lines. *J. Agric. Food Chem.* **2007,** *55,* 1839–1845.

Kuba, M.; C. Tana; S. Tawata; M. Yasuda. Production of angiotensin I-converting enzyme inhibitory peptides from soybean protein with Monascus purpureus acid proteinase. *Process Biochem.* **2005,** *40,* 2191–2196.

Lakemond, C.M.M.; H.H.J. de Jongh; H. Gruppen; A.G.J. Voragen. Differences in denaturation of genetic variants of soy glycinin. *J. Agric. Food Chem.* **2002,** *50,* 4275–4281.

Lei, Q.; W.L. Boatright. Compounds contributing to the odor of aqueous slurries of soy protein concentrates, *J. Food Sci.* **2001,** *66,* 1306–1310.

Lei, Q.; W.L. Boatright. Sulfite radical anions in isolated soy proteins, *J. Food Sci.* **2007,** *72,*

C302–307.

Liener, I.E. Factors affection the nutritional quality of soya products. *J. Am. Oil Chem. Soc.* **1981,** *58,* 406–415.

Liener, I.E. Phytohaemagglutinins: their nutritional significance. *J. Agric. Food Chem.* **1974,** *22,* 17–24.

Lis, H., N. Sharon. Soybean agglutinin-A plant glycoprotein, *J. Biol. Chem.* **1978,** *253,* 3468–3476.

Liu, Z.S.; S.C.K. Chang; L.T. Li; E. Tatsumi. Effect of selective thermal denaturation of soybean proteins on soymilk viscosity and tofu's physical properties. *Food Res. Int.* **2004,** *37,* 815–822.

Lo, W.M.Y.; E.R. Farnworth; E.C.Y. Li-Chan. Angiotensin I-converting enzyme inhibitory activity of soy protein digests in a dynamic model system simulating the upper gastrointestinal tract. *J. Food Sci.* **2006,** *71,* S231–S237.

Lovati, M.R.; C. Manzoni; A. Corsini; A. Granata; R. Frattini; R. Fumagalli; C.R. Sirtori. Low density lipoprotein receptor activity is modulated by soybean globulins in cell culture. *J. Nutr.* **1992,** *122,* 1971–1978.

Lovati, M.R.; C. Manzini; A. Corsini; A. Granata; R. Fumagalli; C.R. Sirtori. 7S globulin from soybean is metabolized in human cell cultures by a specific uptake and degradation system. *J. Nutr.* **1996,** *126,* 2831–2842.

Lovati, M.; C. Manzoni; E. Gianazza; A. Arnoldi; E. Kurowska; K.K. Carroll; C.R. Sirtori. Soy protein peptides regulate cholesterol homeostasis in Hep G2 cells. *J. Nutr.* **2000,** *130,* 2543–2549.

Lozano, P.R.; M. Drake; D. Benitez; K.R. Cadwallader. Instrumental and sensory characterization of heat-inducedodorants in aseptically packaged soy milk. *J. Agric. Food Chem.* **2007,** *55,* 3018–3026.

MacLeod, G.; J. Ames. Soy flavor and its improvement. *Crit. Rev. Food Sci. Nutr.* **1988,** *27,* 219–400.

Maga, J.A. A review of flavor investigations associated with the soy products raw soybeans, defatted flakes and flours, and isolates. *J. Agric. Food Chem.* **1973,** *21,* 864–868.

Maheshwari, P.; P.A. Murphy; Z.L. Nikolov. Characterization and application of porcine liver aldehyde oxidase in the off-flavor reduction of soy proteins. *J. Agric. Food Chem.* **1997**, *45,* 2488–2495.

Maheshwari, P.; E.T. Ooi; Z.L. Nikolov. Off-flavor removal from soy protein isolate by using liquid and supercritical carbon dioxide. *J. Am. Oil Chem. Soc.* **1995,** *72,* 1107–1115.

Mahfuz, A.A.; C. Tsukamoto; S. Kudou; T. Ono. Changes of astringent sensation of soy milk during tofu curd formation. *J. Agric. Food Chem.* **2004,** *52,* 7070–7074.

Mahmoud, A.A.; S.S. Natarajan; J.O. Bennett; T.P. Mawhinney; W.J. Wiebold; H.B. Krishnan. Effect of six decades of selective breeding on soybean protein composition and quality: A biochemical and molecular analysis. *J. Agric. Food Chem.* **2006,** *54,* 3916–3922.

Makino, S.; H. Nakashima; K. Minami; R. Moriyama; S. Takano. Bile acid-binding protein from soybean seed: isolation, partial characterization and insulin-stimulating activity. *Agric. Biol. Chem.* **1988,** *52,* 803–809.

Malcolmson, J.L.; M.R. McDaniel; E. Hoahn. Flavor protein interactions in a formulated soup containing flavored soy protein. *Can. Inst. Food Sci. Technol. J.* **1987,** *20,* 229.

Malkowicz, S.B.; W.G. McKenna; D.J. Vaughn; X.S. Wan; K.J. Propert; K. Rockwell; S.H.F. Marks; A.J. Wein; A.R. Kennedy. Effects of Bowman Birk inhibitor concentrate (BBIC) in patients with benign prostatic hyperplasia. *The Prostate* **2001,** *48,* 16–28.

Marcone, M.F. Biochemical and biophysical properties of plant storage proteins: a current understanding with emphasis on 11S seed globulins. *Food Res. Int.* **1999,** *32,* 79–92.

Maruyama, N.; T. Katsybe; Y. Wada; M.H. Oh; A.P. Barba de la Rosa; E. Okuda; S. Nakagawa; S. Utsumi. The roles of the N-linked glycans and extension regions of soybean β-conglycinin in folding, assembly and structural features. *Eur. J. Biochem.* **1998,** *258,* 854–862.

Maruyama, N.; M.R.M. Salleh; K.I. Takahashi; K. Yagasaki; H. Goto; N. Hontani; S. Nakagawa; S. Utsumi. The effect of the N-linked glycoans on structural features and physicochemical functions of soybean β-conglycinin homotrimers. *J. Am.. Oil Chem. Soc.* **2002a**, *79,* 134–144.

Maruyama, Y.; N. Maruyama; B. Mikami; S. Utsumi. Structure of the core region of the soybean β-conglycinin α′ subunit. *Acta Crystallogr.* **2004,** *D60,* 289–297.

Maruyama, N.; M. Adachi; K. Takahashi; K. Yagasaki; M. Kohno; Y. Takenaka; E. Okuda; S. Nakagawa; B. Mikami; S. Utsumi. Crystal structures of recombinant and native soybean β-conglycinin β homotrimers. *Eur. J. Biochem.* **2001,** *268,* 3595–3604.

Maruyama, Y.; N. Maruyama; B. Mikami; S. Utsumi. Crystal structure of soybean beta conglycinin alpha prime homotrimer. Deposited 2003a Top of Form http://pdbbeta.rcsb.org/pdb/explore.do?structureId=1UIK.

Maruyama, N.; Y. Maruyama; T. Tsuruki; E. Okuda; M. Yoshikawa; S. Utsumi. Crystal structure of soybean beta-conglycinin beta homotrimer (I122M/K124W) Deposited 2003b http://pdbbeta.rcsb.org/pdb/explore.do?structureId=1UIJ.

Maruyama, N.; E.M. Tecson-Mendoza; Y. Maruyama; M. Adachi; S. Utsumi. Molecular design of soybean proteins for enhanced food quality. *Functional Foods and Biotechnology;* K. Shetty, G. Paliyath, A.L. Pometto, R.E. Levin; CRC Taylor and Francis: Boca Raton, FL, 2007; pp. 26–50.

Maruyama, N.; R. Sato; Y. Wada; Y. Matsumura; H. Goto; E. Okuda; S. Nakagawa; S. Utsumi. Structure-physicochemical function relationships of soybean β-conglycinin constituent subunits. *J. Agric. Food Chem.* **1999,** *47,* 5278–5284.

Maruyama, N.; M.R.M. Salleh; K. Takahashi; K. Yagasaki; H. Goto; N. Oto, N. Hontani; S. Nakagawa; S. Utsumii. Structure-physicochemical function relationships of soybean β-conglycinin heterotrimers. *J. Agric. Food Chem.* **2002b,** *50,* 4323–4326.

Maruyama, N.; Y. Maruyama; T. Tsuruki; E. Okuda; M. Yoshikawa; S. Utsumi. Creation of soybean β-conglycinin β with strong phagocytosis-stimulating activity. *Biochim. Biophys. Acta* **2003,** *1648,* 99–104.

Matsuura M.; A. Obata; D. Fukushima. Objectionable flavor of soy milk developed during the soaking of soybeans and its control. *J Food Sci.* **1989,** *54,* 602–605.

Merz-Demlow, B.E.; A.M. Duncan; K.E. Wangen; X. Xu; T.P. Carr; W.R. Phipps; M.S. Kurzer. Soy isoflavones improve plasma lipids in normocholesterolemic, premenopausal women, *Am. J. Clin. Nutr.* **2000,** *71,* 1462–1469.

Meyskens, F.L. Development of Bowman-Birk inhibitor for chemoprevention of oral head and neck cancer. *Ann. N.Y. Acad. Sci.* **2001,** *952,* 116–123.

Milner, M.; N.S. Scrimshaw; D.I.C. Wang. *Protein Resources and Technology: Status and Research Needs*; Avi Publishing Co., Inc.: Westport, CT, 1978; pp. 288–290.

Mori, T.; T. Nakamura; S. Utsumi. Gelation mechanism of soybean 11S globulin: formation of soluble aggregates as transient intermediates. *J. Food Sci.* **1982,** *47,* 26–30.

Morita, S.; M. Fukase; Y. Morita. Thermal stability of individual molecular species of soybean β-conglycinin. *Biosci. Biotech. Biochem.* **1996,** *60,* 1870–1871.

Mujoo, R.; D.T. Trinh; P.K.W. Ng. Characterization of storage proteins in different soybean varieties and their relationship to tofu yield and texture. *Food Chem.* **2003,** *82,* 265–273.

Murphy, P.A.; H.P. Chen; C.C. Hauck; L.A. Wilson. Soybean storage protein composition and tofu quality. *Food Technol.* **1997,** *51(3),* 86–88, 111.

Murphy, P.A.; A.P. Resurreccion. Varietal and genetic differences in soybean glycinin and ß-conglycinin content. *J. Agric. Food Chem.* **1984,** *32,* 911–915.

Nagano, T.; Y. Fukuda; T. Akasaka. Dynamic viscoelastic study on the gelation properties of β-conglycinin-rich and glycinin-rich soybean protein isolates. *J. Agric. Food Chem.* **1996,** *44,* 3484–3488.

Nagano, T.; M. Hirotsuka; H. Mori; K. Kohyama; K. Nishinari. Dynamic viscoelastic study on the gelation of 75 globulin from soybeans. *J. Agrlc. Food Chem.* **1992,** *40,* 941–944.

Nagata, C.; N. Takatsuka; Y. Kurisu; H. Shimizu. Decreased serum total cholesterol concentration is associated with high intake of soy products in Japanese men and women. *J. Nutr.* **1998,** *128,* 209–213.

Nakamura, T.; S. Utsumi; K. Kitamura; K. Harada; T. Mori. Cultivar differences in gelling characteristics of soybean glycinin. *J. Agric. Food Chem.* **1984a,** *32,* 647–651.

Nakamura, T.; S. Utsumi; T. Mori. Network formation in thermally induced gelation of glycinin. *J. Agric. Food Chem.* **1984b,** *32,* 349–352.

Nakamura, T.; S. Utsumi; T. Mori. Mechanism of heat-induced gelation and gel properties of soybean 7S protein. *Agric. Biol. Chem.* **1986,** 1287–1293.

Nestel, P.J.; T. Yamashita; T. Sasahara; S. Pomeroy; A. Dart; P. Komesaroff; A. Owen; M. Abbey. Soy isoflavones improve systemic arterial compliances but not plasma lipids in menopausal and perimenopausal women. *Arterioscler. Thromb. Vasc. Biol.* **1997,** *17,* 3392–3398.

Nielsen, N.C. Structure of seed proteins. *New Protein Foods: Seed Storage Proteins;* A.M. Altschul, H.L. Wilcke, Eds.; Academic Press: New York, 1985; Vol. 5, pp. 31–34.

Nielsen, N.C. Soybean seed composition. *Soybean: Genetics, Molecular Biology and Biotechnology, Biotechnology in Agriculture*; D.P.S. Verma, R.C. Shoemaker, Eds., CAB International: Wallingford, UK, 1996; No. 14, pp. 127–163.

O'Keefe, S.F.; A.P. Resurreccion; L.A. Wilson; P.A. Murphy. Determination of the binding of hexanal to soy glycinin and ß-conglycinin in an aqueous model system using a headspace technique. *J. Agric. Food Chem.* **1991a,** *39,* 1022–1028.

O'Keefe, S.F.; A.P. Resurreccion; L.A. Wilson; P.A. Murphy. Temperature effect on binding of volatile flavor compounds to soy protein in aqueous model system. *J. Food Sci.* **1991b,** *56,* 802–806.

Okubo K.; M. Iijima; Y. Kobayashi, M. Yoshikoshi; T. Uchida; S. Kudou. Components responsible for the undesirable taste of soybean seeds. *Biosci. Biotechnol. Biochem.* **1992,** *56,* 99–103.

O'Neill, T.; J.E. Kinsella. Flavor protein interactions: Characteristics of 2-nonanone binding to isolated soy protein fractions. *J. Food Sci.* **1987,** *62,* 98–101.

Pernollet, J.C.; J. Mosse. Structure and location of legume and cereal seed storage proteins. *Seed Proteins;* J. Daussant, J. Mosse, J. Vaughan, Eds.; 1983; pp. 155–191.

Plietz, P.; B. Drescher; G. Damaschum. Relationship the amino acid sequence and the domain structure of the subunits of the 11S seed globulins, *Int. J. Biol. Macromol.* **1987,** *9,* 161–165.

Potter, S.M.; J.A. Baum; H. Teng; R.J. Stillman; N.F. Shay; J.W. Erdman. Soy protein and isoflavones: their effects on blood lipids and bone density in postmenopausal women, *Am. J. Clin. Nutr.* **1998,** *68(suppl),* 1375S–1379S.

Poysa, V.; L. Woodrow; K. Yu. Effect of soy protein subunit composition on tofu quality. *Food Res. Int.* **2006,** *39,* 309–317.

Prentice, R.D.M.; J.H. Bryce. A source of dimethyl disulfide and dimethyl trisulfide in grain spirit produced with a coffey still. *J. Am. Soc. Brew. Chem.* **1998,** *56,* 99–103.

Pripp, A.H.; Y. Ard. Modelling relationship between angiotensin-(I)-converting enzyme inhibition and the bitter taste of peptides. *Food Chem.* **2007,** *102,* 880–888.

Rackis, J.J. Significance of soya trypsin inhibitors in nutrition. *J. Am. Oil Chem. Soc.* **1981,** *58,* 495–503.

Rackis, J.J.; D.J. Sessa; D.H. Honig. Flavor problems of vegetable food proteins. *J. Am. Oil Chem. Soc.* **1979,** *56,* 262–267.

Reynolds, K.; A. Chin; K.A. Lees; A. Nguyen; D. Bujnowski; J. He. A meta-analysis of the effect of soy protein supplementation on serum lipids. *Am. J. Cardiol.* **2006,** *98,* 633–640.

Riblett, A.L.; T.J. Herald; K.A. Schmidt; K.A. Tilley. Characterization of β-conglycinin and glycinin soy protein fractions from four selected soybean genotypes. *J. Agric. Food Chem.* **2001,** *49,* 4983–4989.

Rickert, D.A.; L.A. Johnson; P.A. Murphy. Improved Fractionation of Glycinin and β-Conglycinin and Partitioning of Phytochemicals. *J. Agric. Food Chem.* **2004a,** *52,* 1726–1734.

Rickert, D.A.; L.A. Johnson; P.A. Murphy. Functional properties of improved glycinin and β-conglycinin fractions. *J. Food Sci.* **2004b,** *69,* 303–311.

Robinson, K.M.; B.P. Klein; S.Y. Lee. Utilizing the R-index measure for threshold testing of model soy isoflavone solutions. *J. Food Sci.* **2004,** *69,* SNQ1–4.

Robinson, D.S.; Z. Wu; C. Domoney; R. Casey. Lipoxygenases and the quality of foods. *Food Chem.* **1995,** *54,* 33–43.

Rostami, A.; A.R. Kennedy. Use of Bowman Birk inhibitor for treatment of multiple sclerosis and other autoimmune diseases. U.S. Patent 6,767,564, July 27, 2004.

Sacks, F.M.; A. Lichtenstein; L. Van Horn; W. Harris; P. Kris-Etherton; M. Winston. Soy protein, isoflavones, and cardiovascular health: an American Heart Association science advisory for professionals from the nutrition committee. *Circulation* **2006,** *113,* 1034–1044.

Saio, K.; M. Kamiya; T. Watanabe. Food processing characteristics of soybean 11S and 7S proteins. Part I. Effect of difference of protein component among soybean varieties on formation of tofu-curd. *Agric. Biol. Chem.* **1969,** *33,* 1301–1308.

Saio K. Tofu-relationships between texture and fine structure. *Cereal Foods World* **1979,** *24,* 342–354.

Saito, T.; M. Kohno; K. Tzumura; W. Kugimiya; M. Kito. Novel method using phytase for separating soybean β-conglycinin and glycinin. *Biosci. Biotechnol. Biochem.* **2001,** *65,* 884–887.

Samman, S.; P.M. Lyons-Wall; G.S. Chan; S.J. Smith; P. Petocz. The effect of supplementation with isoflavones on plasma lipids and oxidizability of low density lipoprotein in premenopausal women. *Atherosclerosis* **1999,** *147,* 277–283.

Schultz, W.B. Food labeling: Health claims; soy protein and coronary heart disease. *Fed. Reg.* **1998,** *63,* 62977–63015.

Schutte, L.; G.A.M. Van den Ouweland. Flavor problems in the application of soy protein materials. *J. Am. Oil Chem. Soc.* **1979,** *56,* 289.

Schwimmer, S. *Source Book of Food Enzymology,* AVI, 1990; pp. 421–439.

Scott, M.P.; R. Jung; K. Muntz; N.C. Nielsen. A protease responsible for post-translational cleavage of a conserved Asn-Gly linkage in glycinin, the major seed storage protein of soybean. *Proc. National Acad. Sci.* **1992,** *89,* 658–662.

Sebastiani, F.L.; L.B. Farrell; M.A. Schuler; R.N. Beachy. Complete sequence of a cDNA of the α subunit of soybean β-conglycinin. *Plant Mol. Biol.* **1990,** *15,* 197–201.

Sessa, D.J.; J.J. Rackis. Lipid-derived flavors of legume protein products. *J. Am. Oil Chem. Soc.* **1977,** *54,* 468–473.

Sharon, N.; H. Lis. Lectins: Cell-agglutinating and sugar-specific proteins. *Science* **1972,** *177,* 949–959.

Shewry, P.R.; J.A. Napier; A.S. Tatham. Seed storage proteins: Structures and biosynthesis. *The Plant Cell.* 1995; *7,* pp. 945–956.

Shutov, A.D.; I.A. Kakhovskaya; A.S. Bastrygina; V.P. Bulmaga; C. Horstmann; K. Muntz. Limited proteolysis of beta-conglycinin and glycinin, the 7S and 11S storage globulins from soybean [Glycine max (L.) Merr.]. Structural and evolutionary implications. *Euro. J. Biochem.* **1996,** *241,* 221–228.

Sirtori, C.R.; R. Even; M.R. Lovati. Soybean protein diet and plasma cholesterol: from therapy to molecular mechanisms. *Ann. N.Y. Acad. Sci.* **1993,** *676,* 188–201.

Snyder, H.E.; T.W. Kwon. *Soybean Utilization.* AVI: New York, 1987a; p. 25.

Snyder, H.E.; T.W. Kwon. *Soybean Utilization.* AVI: New York, 1987b; pp. 41–42.

Song, H.K.; S.W. Suh. Complex porcine pancreatic trypsin/soybean trypsin inhibitor, tetragonal crystal form. Deposited 1997. http://pdbbeta.rcsb.org/pdb/explore.do?structureId=1AVX.

Staswick, P.E.; M.A. Hermondson; N.C. Nielsen. Identification of the acidic and basic subunit complexes in glycinin. *J. Biol. Chem.* **1981,** *256,* 8752–8755.

Steinberg, F.M.; N.L. Guthrie; A.C. Villablanca; K. Kumar; M.J. Murray. Soy protein with isoflavones has favorable effects on endothelial function that are independent of lipid and antioxidant effects in healthy postmenopausal women. *Am. J. Clin. Nutr.* **2003,** *78,* 123–130.

Sugano, M.; S. Goto; Y. Yamada; K. Yoshida; Y. Hashimoto; T. Matsuo; M. Kimono. Cholesterol-lowering activity of various undigested fractions of soybean protein in rats. *J. Nutr.* **1990,** *120,*

977–985.

Sugano, M.; Y. Yamada; K. Yoshida; Y. Hashimoto; T. Matsuo; M. Kishimoto. The hypocholesteroletic action of the undigested fraction of soybean protein in rats. *Atherosclerosis* **1988,** *72,* 115–122.

Takahashi, N.; R. Sasaki; H. Chiba. Enzymatic improvement of food flavor IV. Oxidation of aldehydes in soybean extract by aldehyde oxidase. *Agric. Biol. Chem.* **1979,** *43,* 2557–2561.

Teede, H.; B.P. McGrath; L. DeSilva; M. Cehun; A. Fassoulakis; P.J. Nestel. Isoflavones reduce arterial stiffness: a placebo-controlled study in men and postmenopausal women. *Arterioscler. Thromb. Vasc. Biol.* **2003,** *23,* 1066–1071.

Teixeira, S.R.; S.M. Potter; R. Weigel; S. Hannum; J.W. Erdman; C.M. Hasler. Effects of feeding 4 levels of soy protein for 3 and 6 weeks on blood lipids and apolipoproteins in moderately hypercholesterolemic men, *Am. J. Clin. Nutr.* **2000,** *71,* 1077–1084.

Tezuka, M.; H. Taira; Y. Igarashi; K. Yagasaki; T. Ono. Properties of tofus and soymilks prepared from soybeans having different subunits of glycinin. *J. Agric. Food Chem.* **2000,** *48,* 1111–1117.

Tezuka, M.; K. Yagasaki; T. Ono. Changes in charactors of soybean glycinin groups I, IIa, and IIb caused by heating. *J. Agric. Food Chem.* **2004,** *52,* 1693–1699.

Thanh, V.H.; K. Shibasaki. Heterogeneity of β-conglycinin. *Biochim. Biophys. Acta* **1976a,** *439,* 339–348.

Thanh, V.H.; K. Shibasaki. Major proteins of soybean seeds. A straightforward fractionation and their characterization. *J. Agric. Food Chem.* **1976b,** *24,* 1117–1121.

Thanh, V.H.; K. Shibasaki. β-Conglycinin from soybean proteins: Isolation and immunological and physicochemical properties of the monomeric forms. *Biochim. Biophys. Acta* **1977,** *490,* 370–384.

Tombs, M.P. Protein bodies of the soybean. *Plant Physiol.* **1967,** *42,* 797

Torres-Penaranda, A.V.; C.A. Reitmeier. Sensory descriptive analysis of soymilk. *J. Food Sci.* **2001,** *66,* 352–356.

Tsukamoto, C.; S. Shimada; K. Igita; S. Kudou; M. Kokubun; K. Okubo; K. Kitamura. Factors affecting isoflavone content in soybean seeds: changes in isoflavones, saponins, and composition of fatty acids at different temperatures during seed development. *J. Agric. Food Chem.* **1995,** *43,* 1184–1192.

Tsuruki, T.; K. Kishi; M. Takahashi; M. Tanaka; T. Matsukawa; M. Yoshikawa. Soymetide, an immunostimulating peptide derived from soybean β-conglycinin, is an fMLP agonist. *FEBS Lett.* **2003,** *540,* 206–210.

Urban, D.; W. Irwin; M. Kirk; M.A. Markiewicz; R. Myers; M. Smith; H. Weiss; W.E. Grizzle; S. Barnes. The effect of isolated soy protein on plasma biomarkers in elderly men with elevated serum prostate specific antigen. *J. Urol.* **2001,** *165,* 294–300.

Utsumi, S.; A.B. Gidamis; B. Mikami; M. Kito. Crystallization and preliminary X-ray crystallagraphic analysis of the soyben proglycinin expressed in Escherichia coli. *J. Mol. Biol.* **1993,** *233,* 177–178.

Utsumi, S.; C.S. Kim; T. Sato; M. Kito. Signal sequence of preproglycinin affects production of the expressed protein in Escherichia coli. *Gene* **1988,** *71,* 349–358.

Utsumi, S.; Y. Matsumura; T. Mori. Structure-function relationships of soy proteins. *Food Pro-*

teins and Their Applications; S. Damodaran, A. Paraf, Eds.; Marcel Dekker: New York, 1997; pp. 257–291.

Van den Ouweland, G.A.M.; L. Schutte. Flavor problems in the application of soy protein materials as meat substitutes. *Flavor of Foods and Beverages;* G. Charlambous, G.E. Inglett, Eds.; Academic Press Inc.: New York, 1978; p. 33.

Voss, R.H.; U. Ermler; L.O. Essen; G. Wenzl; Y.M. Kim; P. Flecker. Crystal structure of the bifunctional soybean Bowman-Birk inhibitor at 0.28 nm resolution. Structural peculiarities in a folded protein conformation Deposited 200l. http://pdbbeta.rcsb.org/pdb/explore.do?structureId=1K9B

Washburn, S.; G.L. Burke; T. Morgan; M. Anthony. Effect of soy protein supplementation on serum lipoproteins, blood pressure, and menopausal symptoms in perimenopausal women, *Menopause* **1999,** *6,* 7–13.

Whitaker, J.R. *Principles of Enzymology for the Food Sciences,* Second ed.; Marcel Dekker: NY, 1994; pp. 579–593.

Wilson, L.A.; P.A. Murphy; P. Gallagher. *Soyfood Products Market in Japan: U.S. Export Opportunities.* MATRIC. 1992; 64 pp.

Witschi, H.; A.R. Kennedy. Modulation of lung tumor development in mice with the soybean-derived Bowman-Birk protease inhibitor. *Carcinogenesis* **1989,** *10,* 2275–2277.

Wolf, W.J. Scanning electron microscopy of soybean protein bodies. *J. Am.. Oil Chem. Soc.* **1970,** *47,* 107–108.

Wolf, W.J.; T.C. Nelsen. Partial purification and characterization of the 15S globulin of soybeans, a dimer of glycinin. *J. Agric. Food Chem.* **1996,** *44,* 785–791.

Wolfe, W.J.; J.C. Cowan. Soybeans as a food source. CRC Critical Reviews in *Food Technology,* CRC Press Inc.: Boca Raton, FL, 1971; Vol. 2, pp. 33–77.

Wong, W.W.; E.O. Smith; J.E. Stuff; D.L. Hachey; W.C. Heird; H.J. Pownell. Cholesterol lowering effect of soy protein in normocholesterolemic and hypercholesterolemic men. *Am. J. Clin. Nutr.* **1998,** *68,* 1385S–1389S.

Wu, S.; P.A. Murphy; L.A. Johnson; A.R. Fratzke; M.A. Rueber. Pilot-plant fractionation of soybean glycinin and β-conglycinin. *J. Am. Oil Chem. Soc.* **1999,** *76,* 285–293.

Wu, S.W.; P.A. Murphy; L.A. Johnson; M.A. Reuber; A.R. Fratzke. Simplified process for soybean glycinin and β-conglycinin fractionation. *J. Agric. Food Chem.* **2000,** *48,* 2702–2708.

Yaklich, R.W. β-Conglycinin and glycinin in high-protein soybean seeds. *J. Agric. Food Chem.* **2001,** *49,* 729–735.

Yamagishi, T.; A. Miyakawa; N. Noda; F. Yamauchi. Isolation and Electrophoretic Analysis of Heat-induced Products of Mixed Soybean 7S and 11S Globulins. *Agric. Biol. Chem.* **1983,** *47,* 1229–1237.

Yamauchi, F.; M. Sato; W. Sato; Y. Kamata; K. Shibasaki. Isolation and purification of new types of β-conglycinin in soybean globulins. *Agric. Biol. Chem.* **1981,** *45,* 2893.

Zhou, A.; W.L. Boatright; L.A. Johnson; M. Reuber. Binding properties of 2-pentyl pyridine to soy protein as measured by solid phase microextraction. *J. Food Sci.* **2002,** *67,* 142–145.

Zhou, Q.; K.R. Cadwallader. Inverse gas chromatographic method for measurement of interactions between soy protein isolate and selected flavor compounds under controlled relative humidity. *J. Agric. Food Chem.* **2004,** *52,* 6271–6277.

Zhou, Q.; K.R. Cadwallader. Effect of flavor compound chemical structure and environmental relative humidity on the binding of volatile flavor compounds to dehydrated soy protein isolates. *J. Agric. Food Chem.* **2006,** *54,* 1838–1843.

9

Soybean Carbohydrates

Ingomar S. Middelbos and George C. Fahey, Jr.
Department of Animal Sciences, University of Illinois, Urbana-Champaign, IL, 61801

Introduction

Traditionally, soybeans (SB) are used as food and feed ingredients for their protein content and protein composition. As a protein source, SB often are used in the form of soybean meal (SBM), which is the product of SB processing for SB oil extraction. The carbohydrate content and carbohydrate composition of SB and SBM have received limited attention compared to their protein and fat constituents. Nevertheless, carbohydrates make up a significant part of the SB seed. Typically, SB contain approximately 30–35% carbohydrates (Snyder & Kwon, 1987; NRC, 1998). In SBM, carbohydrate content may be as high as 40% (NRC, 1998).

Carbohydrates are usually divided into two main groups based on their physicochemical properties in plant material. The first group, the nonstructural carbohydrates, includes low molecular weight sugars, oligosaccharides, and storage polysaccharides (Karr-Lilienthal et al., 2005). The second group comprises the structural polysaccharides, and includes dietary fiber components (Bach Knudsen et al., 1987). Dietary fiber is a heterogeneous mixture of cell wall polysaccharides, noncellulose polysaccharides, and structural nonpolysaccharides (e.g., lignin, a phenolic compound). Components of dietary fiber have their own unique chemical, physical, and nutritional properties. The definition of dietary fiber is an ongoing topic of discussion (DeVries, 2003), although the recurring theme in all attempts at a definition is that dietary fiber components are indigestible by endogenous enzymes (Englyst, 1989; Eastwood, 1992; Theander et al., 1994; Institute of Medicine, 2001). Additionally, numerous methods of fiber analysis exist, all of which measure fiber in different ways or measure different components of dietary fiber. Selvendran et al. (1987) reviewed several fiber analysis methods, and Campbell et al. (1997) published a comparative study of some of the more common methodologies used in nutrition research.

In legume seeds such as SB, the main constituents of dietary fiber are cellulose, hemicelluloses, pectins, and glycoproteins or proteoglycans, found in the parenchymal cells of the cotyledons (Selvendran et al., 1987). The main fiber components of

SB hulls are galactomannans, xylan hemicelluloses, uronic acids, and cellulose (Aspinall & Whyte, 1964; Aspinall et al., 1966; Aspinall et al., 1967b, 1967d).

Research pertaining to the composition and structure of SB carbohydrates is ongoing, but is not yet as mature as research on proteins and lipids. Nevertheless, in this chapter, the available literature on SB and SBM carbohydrates is reviewed and categorized. Soybean and SBM carbohydrates are discussed based on their functionality in the plant (nonstructural or structural), and their value as feed or food ingredients.

Carbohydrates in Soybeans and Soybean Meal

The carbohydrate concentration in whole SB is lower than in SBM, due to the concentrating effect of lipid extraction on the latter. Whole SB contain approximately 27% nitrogen free extract (NFE) and 6% crude fiber (CF) on a dry matter basis (DMB), whereas SBM with or without hulls contains 36% NFE and 8% CF or 34% NFE and 4% CF, respectively (Potter & Potchanakorn, 1984). Although NFE (plus CF) is not commonly used today because of the limited information it provides on carbohydrates from a nutritional standpoint, it provides a reasonable indication of the gross carbohydrate content of a feed- or foodstuff. The NFE contains the nonstructural carbohydrates and part of the structural carbohydrates, and CF contains part of the structural carbohydrates.

Nonstructural Carbohydrates

Nonstructural carbohydrates often are divided into three groups: (i) low molecular weight sugars (mono- and disaccharides), (ii) oligosaccharides, and (iii) storage polysaccharides (e.g., starch). These groups of carbohydrates combined are referred to as total nonstructural carbohydrates (TNC), and make up about half of the total carbohydrates in SB and SBM. Grieshop et al. (2003) reported TNC concentrations (DMB) of 12.3–16.0% in whole SB samples taken from 10 different SB processing plants in the United States, while SBM samples taken from these plants contained 18.3–21.2% TNC. Similar concentrations of TNC in SBM (19.5–21.6% DMB) were reported in SBM samples from SB processing plants in four U.S. states and The Netherlands (Van Kempen et al., 2002).

Low Molecular Weight Sugars

In SB, approximately 40–45 % of total carbohydrate DM is represented by low molecular weight sugars, and this value increases to 50% when SB are processed to SBM (Grieshop et al., 2003). Not all sugars found in SB can be recovered in SBM, because they are either destroyed or removed by processing of SB. For example, glucose, galactose, and fructose are present in SB at 0.12–0.47, 0.07–0.40, and 0.11–0.47% DM, respectively, but are not present in SBM.

The primary sugars found in SB and SBM are sucrose (disaccharide) and the oligosaccharides stachyose, raffinose, and verbascose. Approximate concentrations of these sugars in SBM are presented in Table 9.1. Sucrose is the predominant sugar in SBM and can be as high as 9.5% DMB. In comparison with sucrose, oligosaccharides are found in lower concentrations, with stachyose as the main oligosaccharide (3.0–6.4% DMB), followed by raffinose (0.5–1.4% DMB), and verbascose which is present at low concentrations (<1% DM). The variability in sugar concentrations among studies is caused by use of different SB cultivars and also depends on environmental conditions during growth.

Oligosaccharides

The SB oligosaccharides are low molecular weight sugars, but they deserve separate attention because of their contribution to the nutritional attributes of SB and SBM. The soybean oligosaccharides (stachyose, raffinose, and verbascose) are galactooligosaccharides (GOS), and they consist of a terminal sucrose to which 1 (raffinose), 2 (stachyose), or 3 (verbascose) galactose monomers are linked (Fig. 9.1). The galactose units are α-1,6 linked, and the bond between galactose and the terminal sucrose is α-1,3 (Mul & Perry, 1994).

In SB, the oligosaccharides make up approximately 5% of DM. Ranges of stachyose, raffinose, and verbascose concentrations were 3.1–5.7, 0.50–0.74, and 0.12–0.20% DM, respectively, in SB samples taken from 10 U.S. SB processing plants (Grieshop et al., 2003). In SBM, oligosaccharides may make up as much as 7 or 8% DM (Grieshop et al., 2003; Van Kempen et al., 2006), because processing does not remove or destroy GOS. The proportions of oligosaccharides in SBM are approximately the same as in SB. Oligosaccharide concentrations in SB may depend on the cultivar

Table 9.1. Approximate Composition of Total Carbohydrates in Dehulled Soybean Meal[1,2]

Carbohydrate	% (DMB)		
Total oligo- and monosaccharides	11.7	–	17.9
Sucrose	3.1	–	9.5
Stachyose	3.0	–	6.4
Raffinose	0.5	–	1.4
Verbascose	0.12	–	0.30
Total nonstarch polysaccharides	16.4	–	22.2
Noncellulosic	13.4	–	16.4
Cellulose	2.9	–	5.8
Starch	0.26	–	1.2

[1]Sources: Honig & Rackis (1979), Daveby & Aman (1993), Irish & Balnave (1993), Bach Knudsen (1997), Gdala et al. (1997), Huisman et al. (1998), Parsons et al. (2000), Grieshop et al. (2003), Karr-Lilienthal et al. (2005), Van Kempen et al. (2006).
[2]Does not include low-oligosaccharide varieties.

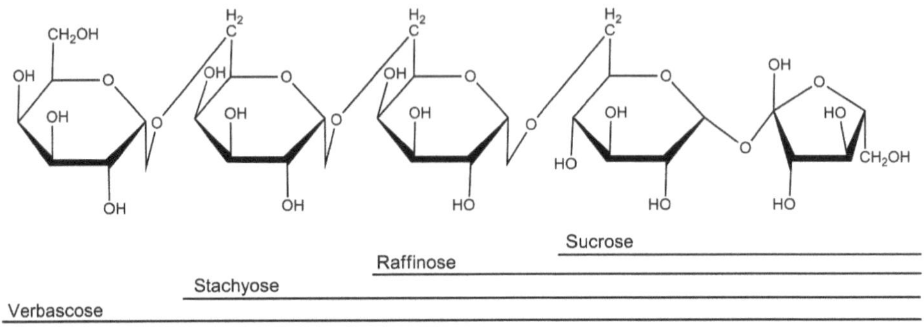

Fig. 9.1. Structure of the soy oligosaccharides, raffinose, stachyose, and verbascose.

used, but are very similar among cultivar groups selected for a specific purpose (e.g., high protein versus high oil) (Hartwig et al., 1997). Compared to conventional SB, oligosaccharide concentrations may be significantly lower in genetically modified SB. Low-oligosaccharide SB cultivars are of interest because of some of the antinutritional aspects of soy oligosaccharides, and may have up to 87% lower stachyose and raffinose concentrations (Parsons et al., 2000).

Nonstructural Polysaccharides

The nonstructural polysaccharides are primarily storage polysaccharides. The main storage polysaccharide in legumes is starch. In contrast to other members of the legume family, SB contain little starch. Where beans, lentils, and peas may contain as much as 40 –60% (DMB) starch (Bednar et al., 2001), SB and SBM tend to contain less than 5% (DMB) starch. Starch concentration in SB and SBM is dependent on the cultivar used and its range may be 0.2–2.7% (Wilson et al., 1978; Irish & Balnave, 1993; Bach Knudsen, 1997; Gdala et al., 1997; Thomas et al., 2003).

Starch is found in higher concentrations in immature SB. Stevenson et al. (2006) reported starch concentrations in SB 20 days preharvest of up to 11.7% (DMB). Soybean starch generally contains more amylopectin than amylose, and the ratio of amylopectin to amylose in immature SB may be as high as 7.5:1, depending on the specific type of cultivar (Stevenson et al., 2006). Other legumes contain lower ratios of amylopectin to amylose (e.g., peas [0.4:1–2.2:1] and common beans [~2:1]) in comparison to SB (Guillon & Champ, 2002).

Structural Carbohydrates

The remainder of carbohydrates in SB and SBM not accounted for in the nonstructural carbohydrates belongs to the structural carbohydrates. The structural carbohydrates are, by definition, polysaccharides because of their function in the plant and include those fractions that can be quantified by common fiber analysis assays. Concentrations of fiber components vary among whole SB, SBM, and SB hulls and are outlined in Table 9.2. Soybean hulls contribute the largest fiber fractions found in whole SB, while cotyledons contain only low concentrations of fiber fractions. Variability in the concentrations of the fiber fractions may depend on the geographical origin (Grieshop & Fahey, 2001) of SB, but also varies in SBM among processing plants (Grieshop et al., 2003).

Structural polysaccharides, also referred to as nonstarch polysaccharides (NSP), in SB are diverse, and some have complex structures. Total NSP concentration is the sum of water-insoluble and water-soluble NSP fractions. Nonstarch polysaccharides also can be divided into cellulosic and noncellulosic polysaccharides. The noncellulosic polysaccharides consist of a variety of monosaccharides (arabinose, galactose, glucose, mannose, xylose, and uronic acids; Table 9.3) that are arranged in complex combinations. As the difference in fiber fractions between SBM and SH (Table 9.2) indicates, the NSP composition is rather different in SB cotyledons compared to SH. Purified SB cotyledon cell walls contain approximately 73% NSP, and small amounts of noncarbohydrate matter consisting of protein, minerals, and phenolics (Brillouet & Carre, 1983).

Table 9.2. Approximate Composition of Nutritionally Important Carbohydrate-Containing Fractions of Soybeans, Soybean Meals, and Soybean Hulls[1]

Item, % DMB	Soybean	Soybean Meal (dehulled)	Soybean Hull
Crude fiber	40.9–41.6	2.6–6.2	32.1–37.3
Acid detergent fiber		2.4–10.5	35.1–49.5
Neutral detergent fiber	11.1–24.9	5.3–11.3	49.0–67.0
Total dietary fiber	34.4	15.6–24.1	63.8–88.0

[1]Sources: Erdman and Weingartner (1981), Jones (1984), Cole et al. (1991), Mansfield & Stern (1994), Parsons et al. (2000), Clapper et al. (2001), Grieshop & Fahey (2001), Cromwell et al. (2002), Van Kempen et al. (2002), Grieshop et al. (2003), Sessa (2003), Willis (2003), Coverdale et al. (2004), Dilger et al. (2004), Dust et al. (2004), Karr-Lilienthal et al. (2004), Bruce et al. (2006), Van Kempen et al. (2006).

Table 9.3. Approximate Monosaccharide Composition of Nonstarch, Noncellulosic Polysaccharides in Soybean Meal[1]

Monosaccharide	*% DMB*
Rhamnose	0.22–0.39
Fucose	0.32–0.40
Arabinose	2.6–3.1
Xylose	1.2–2.0
Mannose	0.67–1.4
Galactose	4.3–5.0
Glucose	0.26–0.52
Uronic acid	2.9–4.5

[1]Sources: Irish & Balnave (1993), Gdala et al. (1997).

Cotyledon Polysaccharides

The polysaccharide fraction in SB cotyledons consists of both neutral and acidic polysaccharides. In the 1960s, Aspinall and co-workers identified the main polysaccharides in SB cotyledons and SB hulls (see section: Soybean Hull Polysaccharides). Extraction of defatted SB cotyledon meal yields a mixture of arabinogalactan and an acidic polysaccharide complex. The arabinogalactan fraction in SB consists mainly of type I arabinogalactan that has a fairly simple structure with a linear 1→4 β-D-galactose chain that is branched at every fourth or fifth residue by (on average) two L-arabinose units (Aspinall et al., 1967c). The ratio of arabinose to galactose indicates the degree of branching in the arabinogalactan polysaccharide, and is approximately 1:2.8 (Aspinall & Cottrell, 1971), but lower ratios of 1:1.5–2.0 also are reported (Brillouet & Carre, 1983; Meng et al., 2005). Other legumes such as peas may have arabinose to galactose ratios opposite (as low as 1:0.3) to those of SB (Meng et al., 2005). Upon hydrolysis of the neutral polysaccharide, galactose and arabinose are the main monosaccharides found, but small amounts of xylose and galacturonic acid also may be present.

The complex acidic polysaccharides in SB cotyledons were found to contain approximately 30% uronic acids (Aspinall et al., 1967c). Structurally, they are similar to pectin, with high degrees of branching and a heterogeneous monosaccharide composition consisting mainly of galacturonic acid, galactose, arabinose, xylose, fucose, rhamnose, and traces of glucose (Aspinall et al., 1967a, 1967c, 1967d). Although the overall acidic polysaccharide is a pectinic acid, the interior structure contains high amounts of neutral polysaccharides, especially arabinose and galactose, when compared with citrus pectin (Yamaguchi et al., 1996). When studying the structure of SB pectic acid by degradation using cloned enzymes, Huisman et al. (1999) indicated that SB pectic polysaccharides have a different structure than other sources of pectic polysaccharides. Analysis of the structure of the acidic soluble SB polysaccharide fraction (SSPS) using enzymatic degradation indicates that purified SSPS contains three types of galacturonic backbones, G-1, G-2, and G-3 (Nakamura et al., 2000). The G-1 and G-2 galacturonate backbones of purified SSPS consist of short homogalacturonan regions interrupted by long rhamnogalacturonan regions. The G-3 backbone is more like that found in citrus pectin, with short rhamnogalacturonan regions interrupted by long homogalacturonan regions (Nakamura et al., 2000). Subsequent work by Nakamura et al. (2002) showed that the SSPS rhamnogalacturonan backbone comprises 15, 28, or 100 repeats of a (1→2)-α-L-Rha-(1→4)-α-GalA diglycosyl unit. Rhamnogalacturonan has long side chains of homogalactan and homoarabinan, but the galactans may be branched by neutral sugars (arabinose, xylose, fucose, or glucose). A similar branched rhamnogalacturonan structure was proposed by Huisman et al. (2001) using 1,2-diaminocyclohexane-*N,N,N',N'*-tetraacetic acid (CDTA) extracted SB pectin, but with a slightly different side chain composition. These differences may be due to the different fractions that are extracted using the SSPS method (hot water) and CDTA (chelating agent-containing buffer). Soybean hemicelluloses consist of mainly galactofucosylated xyloglucans and minor amounts of heteroxylans and heteromannans (Harris & Smith, 2006) and are structurally similar to those found in many plants (Huisman et al., 1999).

Soybean Hull Polysaccharides

Soybean hulls are a by-product of SB processing and consist of the SB seed coat. Soybean hulls make up approximately 8% of the whole seed and contain 86% complex carbohydrates (Gnanasambandam & Proctor, 1999). The insoluble carbohydrate fraction of SB hulls consists of approximately 30% pectin, 50% hemicelluloses, and 20% cellulose (Snyder & Kwon, 1987). These compositional characteristics make SB hulls a good source of dietary fiber. Summarized data on fiber fractions in SB hulls can be found in Table 9.2. Dust et al. (2004) presented a complete analysis of SB hull fiber fractions, and found unprocessed hulls to contain 67% neutral detergent fiber, 49.3% acid detergent fiber, 2.3% acid detergent lignin, 17.7% insoluble hemicelluloses, and 47% cellulose (DMB). The TDF fraction of 83.3% in SB hulls is 69.5% insoluble dietary fiber and 13.3% soluble dietary fiber (Dust et al., 2004).

The polysaccharides that make up the SB hull consist of galactomannans (Whistler & Saarnio, 1957; Aspinall & Whyte, 1964) and acidic polysaccharides (Aspinall et al., 1966; Aspinall et al., 1967b, 1967d), but also contain xylans and cellulose (Aspinall et al., 1966). The insoluble hemicelluloses contain xylan polymers in large quantities. This type of β-1→4 linked glycan does not have large numbers of side chains and is, therefore, not soluble in water. Upon hydrolysis, the xylan polymer yields xylose, small amounts (~4%) of glucuronic acid, higher oligosaccharides, traces of rhamnose, and a mixture of acidic sugars that are part of the SB acidic polysaccharide V. The xylan structure is essentially linear due to the β-1→4 linkage between the xylopyranose residues, and has the small proportion of glucuronic acid residues attached as monomeric side chains by 1→2 linkages (Aspinall et al., 1966). Xylans of this structure type often are found associated with cellulose in lignified tissues (Aspinall et al., 1967d).

The noncellulosic fraction of SB hull polysaccharides consists mainly of xylose and mannose. The acidic polysaccharides that can be extracted from SB hulls in sequential extractions steps (acidic polysaccharides I–V) are all very similar in structure with interior chains made up of D-galacturonic acid and L-rhamnopyranose residues. These interior acidic chains are associated with side chains consisting of neutral sugars, including L-arabinose and chains of 1→4 linked β-D-galactopyranose residues. In addition, 2-O-β-D-gal-D-xyl and 2-O-α-L-fuc-D-xyl residues also may be present (Aspinall et al., 1966). Although the acidic polysaccharide fractions I–V are similar structurally, upon hydrolysis they may yield different monosaccharide concentrations. For example, acidic polysaccharide III contains 76% uronic acids, whereas fractions IV and V contained approximately 45% uronic acids. Hydrolysis of the acidic polysaccharide fractions yields galacturonic acid, galactose, arabinose, xylose, fucose, rhamnose, and trace amounts of methylxylose and methylfucose (Aspinall et al., 1966).

Carbohydrates in Processed Soy Protein Products

Besides SBM, several processed soy protein products are available for use in human and animal nutrition. Soy protein concentrate (SPC) is produced by extracting soluble carbohydrates from defatted soy flour or flakes using alcohol, acid, or hot water. The resulting SPC contains at least 65% protein, and has decreased concentrations of soluble sugars, but total dietary fiber is usually increased (Riaz, 2006). Little quantitative information is available about the carbohydrate composition of SPC. Clapper et al. (2001) reported total dietary fiber concentrations in SPC products of up to 21.3% (DMB), approximately five percentage units higher than in soy flour or SBM. Total carbohydrate concentration in SPC ranges 15.8–17.0% in a study by Wang et al. (2004).

Soy protein isolate (SPI) contains at least 90% protein, and is produced by alkali solubilization of the protein, followed by centrifugation to separate the protein solution and solids. The resultant soluble protein is precipitated with acid and further concentrated by washing, and is eventually spray-dried (Riaz, 2006). The high protein concentration (>90%) of SPI leaves little room for other constituents, but SPI are not carbohydrate-free. Fernandez-Quintela et al. (1997) found 5.6% carbohydrates (DMB) in SPI, and Wang et al. (2004) reported up to 7.6% carbohydrates in SPI products derived from extruded-expelled SBM. Lower carbohydrate values in SPI were reported by Jung et al. (2006), who analyzed the sugar profile (stachyose, raffinose, sucrose, maltose, lactose, galactinol, glucose, and fructose) of SPI carbohydrates. These researchers reported total sugar concentrations of 2.3–3.0%, approximately fourfold lower values than in defatted soy flakes.

Texturized vegetable protein (TVP) is produced by extruding soy flour (Riaz, 2006). Although processing by extrusion may affect digestibility of carbohydrates (see section: Processing Conditions), the gross composition of TVP is similar to that of soy flour. Hill et al. (2001) reported TVP products to contain 31% NFE, 13–15% oligosaccharides, and 15–18% polysaccharides. The oligosaccharides were made up of 6–8% sucrose, 4–5% stachyose, and 1–2% raffinose. The polysaccharide fraction consisted of 8–10% acidic polysaccharide, approximately 5% arabinogalactan, 0.1% cellulose, and 0.5% starch.

Soybean Carbohydrates as Related to Nutrition

In nonruminant animal and human nutrition, SB (or SBM) often are added to a diet as a protein or amino acid source rather than a carbohydrate source. Soy fiber is an exception to this and may be added to a diet for its specific high fiber concentration, while whole SB hulls may be fed to ruminants as a source of carbohydrates (fiber).

Carbohydrates from SB may play a significant role in the nutrition of nonruminants in various ways.

Early in the twentieth century, Adolph and Kao (1934) established that rats utilize only 40% of the carbohydrates found in SBM. Similarly, the relative bioavailability of SB NFE was reported to be 38% in rats (Karimzadegan et al., 1979). Soybean soluble sugars and starch availability in vitro was reported to be only 24% (DMB), and availability of carbohydrates was even lower than that (14%) in chicks fed SBM (Lodhi et al., 1969).

Nonstructural Carbohydrates

The nonstructural carbohydrates in SB can be divided based on their chemical properties (See section: Carbohydrates in Soybeans and Soybean Meal), but also on their value as nutrients. The different categories of SB carbohydrates identified based on chemistry are profoundly different in their rate of digestion and their physiological behavior.

Low Molecular Weight Sugars and Storage Polysaccharides

The low molecular weight sugars and storage polysaccharides (starch) in SB and SBM are assumed to be readily accessible to nonruminants, and they provide rapidly available energy to the animal. In poultry, starch and sugars from SBM were reported to be 100% digestible (Bolton, 1957). Gdala et al. (1997) noted a starch digestibility of 98.8% in piglets fed a diet containing 36% SBM, while Marsman et al. (1997a) reported starch digestibilities of greater than 99% in poultry fed diets containing 38% SBM. In ruminants, diets often contain SB hulls or SBM. The low molecular weight sugars and storage polysaccharides are quickly fermented by ruminal microbiota, but no specific research is available on the nutritional value of these soy carbohydrates to ruminants.

Oligosaccharides

Soybean oligosaccharides are GOS and require α-galactosidase to be enzymatically digested. The intestinal tract of nonruminants does not produce this enzyme, and nonruminants, therefore, are dependent on microbiota in the lower gut for digestion of raffinose, stachyose, and verbascose. Soybean oligosaccharides in high concentrations are reported to depress small intestinal nutrient digestibility, and they, therefore, are considered antinutritional factors. Additionally, the microbial breakdown of these oligosaccharides produces large amounts of flatus in rats, humans, dogs, and swine (Steggerda, 1968; Smiricky-Tjardes et al., 2003a), and may increase incidence of diarrhea (Kuriyama & Mendel, 1917; Liying et al., 2003).

Despite the lack of α-galactosidase, animal models using ileal cannulated swine and cecectomized roosters show that a proportion of the SB GOS disappears prior to reaching the primary site of fermentation in the intestinal tract. Apparent ileal digestibility values of 58 to 80.3% for GOS are reported in growing swine (Smiricky et al., 2002; Smiricky-Tjardes et al., 2003b; Rubio et al., 2005), but even higher values (90.4% apparent raffinose digestibility) at the terminal ileum of piglets also are reported (Gdala et al., 1997). Cecectomized roosters are able to utilize 48% of stachyose and 25% of raffinose from SBM, compared to 79 and 74%, respectively, in intact roosters (Parsons et al., 2000). The reason for this "pre-fermentation" digestion of GOS is that the ileum of swine and the intestine of the cecectomized rooster are not devoid of microbiota. The innate concentrations of microbiota may be orders of magnitude below that in the colon or cecum, but GOS are easily and rapidly fermented. Smiricky-Tjardes et al. (2003a) investigated in vitro fermentation characteristics of several oligosaccharides, including raffinose and stachyose, using swine fecal inoculum. They reported that SB oligosaccharides, stachyose, and a mixture of raffinose and stachyose, had the highest rate of SCFA production as well as the shortest time to reach maximal SCFA production. Moreover, these oligosaccharides had the highest rate of gas production of all substrates tested, as well as the shortest time to reach maximal gas production (Smiricky-Tjardes et al., 2003a). Lan et al. (2005) noted similar results in an in vitro experiment using cecal contents from broilers. Purified raffinose and stachyose had the highest total gas production, as well as the highest rate of gas production compared to oligo- and polysaccharides extracted from SBM and alfalfa meal. Gas production (mainly H_2 and CO_2) from oligosaccharide fermentation in vivo may result in abdominal discomfort, flatulence, and nausea. Suarez et al. (1999) reported increased flatulence in humans consuming soy flour over white rice as the control, although no increased discomfort (e.g.,bloating, abdominal pain, nausea) was noted. Yamka et al. (2006) noted no difference in flatulence in dogs fed diets containing conventional SBM and low oligosaccharide SBM, although their analysis was limited to H_2S (as an odor component) that is not formed directly from oligosaccharide fermentation.

Soybean GOS also are reported to negatively affect nutrient digestibility and energy availability of SBM (Coon et al., 1990). Parsons et al. (2000) reported a 6.5% decrease in total net metabolizable energy in roosters when they were fed conventional SBM as compared to a low-oligosaccharide variety. In broilers, supplementation of diets containing Hamlet protein (enzymatically treated SB with decreased concentrations of oligosaccharides and trypsin inhibitor) with 0, 4, 8, 12, or 16 g/kg stachyose led to a linear decrease in average daily gain and a linear decrease in feed efficiency (gain:feed) (Jiang et al., 2006). Additionally, apparent digestibilities of DM and OM were depressed linearly with increasing stachyose supplementation, but apparent metabolizable energy and nitrogen digestion were unaffected. In ileal cannulated dogs, Yamka et al. (2003) reported a linear decrease in ileal nutrient digestibility and total tract DM digestibility with increasing SB oligosaccharide concentrations in the diet. However, Zuo et al. (1996) and Yamka et al. (2005) noted no negative effect of SB oligosaccharides concentration on apparent ileal nutrient digestibility. In swine, data are also conflicting. Smiricky-Tjardes et al. (2003b) noted a decrease of five percentage points in ileal digestibility of DM and nitrogen when they supplemented purified swine diets with 3.5 or 4.8% SB GOS. Using a semi-purified diet that contained 3.74% total GOS, Smiricky et al. (2002) noted no decrease in apparent ileal digestibility of DM and nitrogen in swine, although a diet containing 5.36% GOS decreased DM digestibility, but not nitrogen digestibility. Liying et al. (2003) noted that weanling pigs had lower (40–80 g/d) daily gain when fed diets supplemented with commercially available stachyose compared to a SBM-free control diet. However, no differences were noted between a SBM-containing diet and diets supplemented with stachyose. Although higher concentrations of SB GOS appear to affect nutrient digestibility, it is not likely that typical corn-SBM diets contain sufficient GOS to interfere with nutrient digestion.

In addition to flatus production, SB GOS also are linked to increased incidence of diarrhea. This effect was reported in rats (Kuriyama & Mendel, 1917) and weanling pigs (Liying et al., 2003). However, the fermentation of GOS in the intestine also may have beneficial properties. Liying et al. (2003) measured microbial populations in the intestinal tract of weanling pigs, and noted that 1% stachyose supplementation increased ileal lactobacilli and colonic bifidobacteria concentrations, while colonic enterobacteria concentrations decreased. Smiricky-Tjardes et al. (2003b) reported increased fecal lactobacilli and bifidobacteria in growing swine fed diets supplemented with GOS. Soybean GOS may, therefore, serve as prebiotics (see section: Functional Foods: Soy Oligosaccharides and Soy Fiber).

Structural Polysaccharides

The usefulness of SB structural polysaccharides in nutrition is directly associated with their fermentability. Both ruminants and nonruminants cannot utilize the energy that is present in structural polysaccharides without the assistance of microbiota that possess the necessary enzymes to break down polysaccharides. Limited information is available on fermentation of SB polysaccharides. In a 24-h in situ degradation experiment with SBM cell wall, Van Laar et al. (1999) reported different degradation rates of individual cell wall sugars. When the cell walls of the cotyledon and hulls were degraded separately in vitro (24 h, sheep ruminal fluid inoculum), cotyledon cell wall sugars were much more rapidly degraded than the hull cell wall sugars (Table 9.4; Van Laar et al., 1999). Although the ruminal environment is much more specialized in breaking down cell wall material, the large intestine of swine should be able to degrade significant amounts of cell wall material, given that intestinal retention time may be 20–38 h (Keys & DeBarthe, 1974). Indeed, Van Laar et al. (2000) reported a similar SB cell wall sugar disappearance profile in a swine in vitro experiment, but sugar disappearance was not as rapid as in their ruminant experiments. Dust et al.

Table 9.4. Disappearance of Sugars from Soybean Cell Walls in situ (dairy cow) and in vitro (sheep)[1]

Sugar	Sbm Cell Wall	Sb Cotyledon Cell Wall	Sb Hull Cell Wall	Combined Cotyledon And Hull
disappearance, %/h	(in situ)		(in vitro)	
Galactose	13.6	20.9	4.2	13.4
Arabinose	7.8	12.1	2.4	27.9
Uronic acid	5.1	7.0	4.3	7.0
Xylose	3.5	8.6	4.6	4.0
Glucose	3.2	7.6	3.9	2.6

[1]Source: Van Laar et al. (1999).

(2004) noted a 10% OM disappearance in SB hulls after 24 h of simulated hydrolytic digestion (attributed to a small fraction of starch and protein digestion). Subsequently, only a 9.7% OM disappearance and low to moderate SCFA production from SB hulls in a swine in vitro fermentation experiment after 8 h of incubation were observed (Dust et al., 2004). Rubio et al. (2005) found apparent ileal digestibility of SBM NSP in swine to be approximately 56% overall. Of the individual sugars present in SBM NSP, mannose had the highest digestibility (73%), followed by galactose (65%), arabinose (60%), xylose (59%), rhamnose (55%), and glucose (52%), while uronic acids (32%) were least digestible (Rubio et al., 2005). The presence of structural carbohydrates from SB hulls (6–10% supplementation) in swine diets may decrease DM, protein, and energy digestibility (Kornegay, 1978; Mitaru et al., 1984). Incremental inclusion of 0, 3, 6, or 9% SB hulls linearly decreased apparent ileal digestibility of DM and energy, but not total nitrogen. Nevertheless, apparent ileal digestibility of several individual essential amino acids was linearly decreased with increasing SB hull supplementation (Dilger et al., 2004). In vitro fermentation data on SB hulls using dog fecal inoculum showed that fermentation of SB hulls generates comparable OM disappearance and SCFA production to beet pulp from 6–24 h of incubation (Sunvold et al., 1995). In vivo, the dog appears to be able to utilize approximately 22% of SB hull polysaccharides when supplemented at intermediate levels (7.5 or 9.0% of the diet), based on TDF digestibility (Cole et al., 1991). In horses, SB hulls are reported to be a suitable replacement for hay in the diet, although a linear decrease in nitrogen digestibility was noted with increasing (up to 75% of the diet) SB hull supplementation (Coverdale et al., 2004).

In contrast to nonruminants, the structural carbohydrates from SB hulls may be used as a dietary supplement or an ingredient for ruminants. Limited data are available investigating the specific fermentability of SB structural carbohydrates in ruminants. Trater et al. (2001) found total tract digestibilities for sugars from SB polysaccharides of 97% for mannose, 87.2% for arabinose, 82.5% for galactose, 77.2% for rhamnose, 62.2% for glucose, and 36% for xylose in steers fed a diet consisting of 95.7% SB hulls, 3% molasses, 0.5% calcium phosphate, 0.5% urea, and 0.3% trace mineral salt. Feeding of SB hulls to ruminants is reported to have positive associative effects. In beef cows fed poor quality forage, supplementing SB hulls is reported to increase OM digestibility, and SB hulls are believed to be an effective energy supplement for beef cattle on poor quality forage (Martin & Hibberd, 1990). A similar observation was reported by Galloway et al. (1993), who supplemented Holstein steer calves fed moderate quality hay with SB hulls at 0.7% of bodyweight. Although the structural polysaccharides of SB hulls are well digested and utilized by ruminants based on performance measures, the specific kinetics involved in their fermentation and (or) digestion are unknown.

Variability in Nutritional Value of Soybean Carbohydrates

In the growth period of a plant, many environmental factors affect how its genetics will be expressed in its phenotype and, consequently, its composition from a nutritional point of view. Additionally, the genetics themselves affect the eventual composition of the plant. In food and feed, SB are usually processed in some way, and this processing may affect the nutritional aspects of SB or SB products. In recent years, the use of enzymes to treat SBM is on the rise to increase the utilization of its nutrients.

Environment and Genotype

The influence of genetics on nutritional value is illustrated by a point mutation in SB reported by Sebastian et al. (2000). In plants containing this mutation, the replacement of a single base in the genetic code resulted in several dramatic changes in phenotype. Compared to wild-type SB, the concentration of raffinose and stachyose in mutated plants is reduced by 85 and 75%, respectively (Hitz et al., 2002). Moreover, the mutation results in lower concentrations (~50%) of phytic acid in SB seeds. Both these aspects are considered beneficial, as SB oligosaccharides and phytic acid are regarded as anti-nutritional factors for nonruminant animals. Neus et al. (2005) concluded that the agronomic performance of these recessively mutated SB plants was similar to that of wild-type plants and that mutated cultivars could be of benefit to nonruminant animals. A combination of genetic and environmental effects can result in differences in the nutritional composition of SB. Mieth et al. (1988) reported a variability of up to 46% in NFE concentration in several SBM from cultivars grown under a range of different conditions, and a total cell wall variability of up to 37%. Thomas et al. (2003) reported a linear decrease in carbohydrate concentration in SB seeds grown under increasingly warmer temperature regimens. Similar results are reported by Wolf et al. (1982) for SB sucrose concentration which decreased approximately 50% when SB were grown at a 33/28°C temperature regimen versus a 18/13°C regimen. Concentrations of raffinose, stachyose, fructose, and glucose were not affected by growth temperature (Wolf et al., 1982). Grieshop et al. (2003) and

Karr-Lilienthal et al. (2005) reported differences in proximate and oligosaccharide composition, respectively, in SBM samples collected at 55 U.S. SB processing plants in different maturity zones. Unfortunately, no specific data are available evaluating the effects of SB genetics or the combination of genetics and growth conditions on carbohydrate digestibility.

Although SB genetics (cultivar) can affect the composition of SB seeds, growth conditions play a significant role in the composition of the mature SB seed. In a multi-year survey of SBM from 18 different feed companies, Jones (1984) concluded that SBM composition varied significantly from year to year. These differences could be due to growth conditions, but no cultivar information was provided. More research in this area is needed to delineate specific cultivar and environmental effects on nutritional value of SB and SBM.

Processing Conditions

Prior to using SB in feed, several processing steps are needed to extract or obtain a product that is useful in animal nutrition. The main product of SB processing is SB oil, and SBM is regarded as a by-product. Soybean oil can be extracted from SB by mechanical force (expelling) or by using solvents to extract the oil. In the latter process, SB are dehulled, flaked, and usually extruded prior to oil extraction (Proctor, 1997). These steps have profound effects on the carbohydrates in SB. Although SB are low in starch compared to other legumes, heating SB under moist conditions (such as extrusion) will lead to gelatinization of the starch molecules that are present, and make them irreversibly lose their native structure (Camire et al., 1990). Physical processing of SB such as flaking or grinding also affects carbohydrates. The reduction in particle size is accompanied with an increase in surface area, creating easier access for hydrolytic enzymes and, subsequently, an increased rate of hydrolysis (Würsch et al., 1986). Based on experiments with barley starch, it is suggested that extremely fine grinding or milling can physically break glycosidic bonds and lead to quick digestion or fermentation (Stark & Yin, 1986).

The use of thermal treatment to increase the nutritional value of SBM is common in the swine and poultry industries (Würsch et al., 1986; Bengala-Freire et al., 1991). The main goal of heat treatment is to eliminate trypsin inhibitors and other enzyme inhibitors that may affect protein utilization, and this can be done effectively (Trugo et al., 2000). However, the conditions under which processing takes place are not uniform. Factors that may affect the nutritional value of the product resulting from processing include temperature, moisture, pressure, and processing time. Grieshop et al. (2003) reported a significant effect of a processing plant when they compared the nutritional value of SB and SBM samples collected at 10 U.S. SB processing plants (Table 9.5). The SB samples collected at the plants were similar in proximate composition with the exception of OM and CP. The variability in the resultant SBM was much greater and significantly different for all components analyzed (except OM).

Table 9.5. Composition of Soybean Meals Collected at Ten U.S. Soybean Processing Plants[1]

	Plant no.										
Item	1	2	3	4	5	6	7	8	9	10	SEM
DM[2] (%)	89.0	88.9	88.3	88.9	88.2	89.4	90.2	88.2	88.3	90.2	0.64
					% of DM						
OM[3]	92.1	92.5	92.2	92.0	93.2	92.2	91.6	92.8	93.1	93.4	0.40
CP[4]	55.3^{ab}	55.1^{ab}	54.8^{ab}	54.9^{ab}	54.0^{b}	55.0^{ab}	53.5^{b}	56.2^{a}	55.1^{ab}	48.2^{c}	0.70
AHF[5]	3.6^{cd}	4.1^{bc}	3.5^{cd}	4.5^{b}	3.9^{bcd}	3.9^{bcd}	3.8^{bcd}	4.4^{bc}	3.3^{d}	9.2^{a}	0.27
TDF[6]	17.3^{b}	17.3^{b}	18.5^{b}	18.6^{b}	18.5^{b}	17.0^{b}	18.5^{b}	17.0^{b}	17.7^{b}	20.7^{a}	0.66
TNC[7]	19.4^{bcde}	19.8^{abcde}	20.7^{ab}	18.3^{e}	20.6^{abc}	20.1^{abcd}	20.5^{abc}	19.0^{cde}	21.2^{a}	18.8^{de}	0.57
Total sugars	17.7^{a}	17.0^{a}	17.0^{a}	13.6^{b}	17.9^{a}	16.9^{a}	17.5^{a}	16.9^{a}	17.5^{a}	14.4^{b}	0.43
	mg/g DM										
Sucrose	69.4^{ab}	72.2^{ab}	68.5^{b}	42.4^{c}	77.2^{a}	68.0^{b}	69.8^{ab}	69.3^{ab}	73.4^{ab}	48.2^{c}	2.92
Raffinose	13.3^{ab}	11.5^{cd}	10.8^{de}	13.4^{a}	12.7^{abc}	10.2^{de}	10.2^{de}	9.8^{e}	14.3^{a}	11.8^{cd}	0.54
Stachyose	57.2^{a}	49.3^{de}	51.6^{bcd}	41.5^{f}	50.6^{cde}	52.2^{bcd}	54.3^{ab}	53.0^{bc}	47.4^{e}	41.0^{f}	1.24
Verbascose	2.3^{bc}	2.0^{cd}	2.2^{bc}	2.9^{a}	1.6^{e}	2.3^{b}	2.3^{b}	2.4^{b}	1.9^{de}	2.0^{bcd}	0.10
Uronic acid	34.7^{b}	34.8^{b}	36.6^{b}	35.8^{b}	36.5^{b}	36.3^{b}	38.0^{b}	34.8^{b}	37.9^{b}	41.5^{a}	1.15

[1]Source: Grieshop et al. (2003).
[2]Dry matter.
[3]Organic matter.
[4]Crude protein.
[5]Acid-hydrolyzed fat.
[6]Total dietary fiber.
[7]Total non-structural carbohydrates.
[abcdef]Means within rows lacking common superscript letters are different ($P<0.05$).

The carbohydrate composition of the SBM samples also varied for each processing plant, and especially raffinose (9.8–14.3 g/kg DM) and stachyose (41.0–57.2 g/kg DM) showed large variations (Grieshop et al., 2003). Among the 10 plants evaluated by these authors, one plant (no.10, Table 9.5) used only mechanical oil extraction methods (expeller) as compared to the other plants using solvent extraction (with or without extrusion). The difference in the oil extraction method does not only affect fat and CP concentration in the SBM, but also affects TDF and total sugars. The exact effects of certain processing conditions on the carbohydrate quality of the resultant SBM are not well documented and deserve further attention.

Some of the processes used in oil extraction may increase the availability of SB carbohydrates. During extrusion, soluble dietary fiber increases at the expense of the insoluble fraction. In addition, the expansion that occurs during extrusion makes the resultant product fluffy and porous, making it easier for enzymes and bacteria to gain access and digest or ferment the product (Camire et al., 1990; Marsman, 1997a). Compared to toasting, extrusion of SBM increases ileal NSP digestibility in broilers, and significantly increases water-holding capacity, and increases gain:feed (0.62 versus 0.64). These effects were attributed to improved fermentation of cell wall as a result of extrusion (Marsman et al., 1997a). Varying the extrusion conditions by increasing shear force significantly increased the concentration of soluble NSP up to threefold, water- holding capacity by up to 14%, and broiler chyme viscosity by up to 60% (Marsman et al. 1997a). Marsman et al. (1997b) studied the effects of toasting and extrusion of SBM on the composition and in vitro accessibility by enzymes. They isolated the water-unextractable solids (WUS) to eliminate the contamination of soluble sugars (mono- and oligosaccharides) in their compositional analysis. The yield of WUS from raw SBM was much lower than that from toasted SBM and extruded SBM (Table 9.6). Processing of SBM not only lowered total NSP concentrations, but the individual sugar concentrations also were lower in toasted SBM and extruded SBM. Additionally, incubating the SBM WUS fraction with a carbohydrase resulted in a higher concentration of solubilized sugars after 24 h for both processed SBMs compared with the rSBM (Table 9.6, Marsman et al., 1997b). The extrusion of soy flour yields TVP that is useful in canned pet foods because of its meaty texture. Hill et al. (2001) demonstrated that high inclusion rates of TVP in canned foods for dogs are best avoided, as digestibility of DM, CP, and NFE may become depressed.

Enzyme Treatment

In recent years, treating SBM with exogenous enzymes prior to feeding has received attention. These enzymes could potentially break down some of the carbohydrate fraction, thereby making the carbohydrates more available to the animal. A difficulty exists with SB structural polysaccharides in that the SB polysaccharides are more complex than those found in cereal grains (Annison & Choct, 1993). This makes it potentially difficult to target cell wall components as a whole. Huisman et al. (1999)

Table 9.6. Yield and Carbohydrate Composition of Water-Unextractable Solids from Raw, Toasted, and Extruded Soybean Meal, and Molar Percentage of Solubilized Sugars after Incubation with a Carbohydrase for 24 h[1]

	Composition, % DMB			Soluble Sugars After Carbohydrase Incubation for 24 h, mol%		
Item	Raw SBM	Toasted SBM	Extruded SBM	Raw SBM	Toasted SBM	Extruded SBM
Yield2	40.0	67.4	70.7			
NSP3	43.0	22.8	18.5			
Rhamnose	1.3	0.6	0.5	5.5	5.5	5.3
Fucose	0.6	0.3	0.2	0.9	1.2	1.1
Arabinose	5.0	2.6	2.1	17.8	23.1	23.4
Xylose	1.2	0.7	0.6	<0.1	<0.1	<0.1
Mannose	1.5	1.0	0.9	1.4	<0.1	<0.1
Galactose	12.1	6.0	5.1	38.5	49.6	49.5
Glucose	11.2	6.5	5.1	10.1	2.2	<0.1
Uronic acid	10.1	5.1	4.0	25.7	18.3	20.6

[1]Source: Marsman et al. (1997b).
[2]As-is weight percentage of the respective SBMs.
[3]Nonstarch polysaccharides.

noted increased enzymatic degradability of SB cell wall components in sequentially extracted polysaccharide fractions (from SBM, WUS, and chelating agent soluble solids). Ouhida et al. (2002) reported similar observations. These authors also extracted cellulose by using increasingly strong alkaline agents and noted that the progressive fractionation increased the degradation of pectin- and hemicellulose-rich fractions. Nevertheless, cellulose degradation occurred only after all protein, pectin, and hemicelluloses were extracted (Ouhida et al., 2002). These results indicate that enzyme treatments yield the best results when specific parts of the cell wall are isolated and targeted. The combination of SBM processing and enzyme treatments also may be effective, as Marsman et al. (1997b) reported increased release of sugars from SB cell wall polysaccharides in processed SBM when incubated with a carbohydrase (Table 9.6). Meng et al. (2005) evaluated the effects of several enzyme combinations on SBM NSP degradation. They concluded that a multi-enzyme cocktail is more effective in breaking down SBM cell wall components as their most complex combination of enzymes (cellulase, pectinase, galactanase, xylanase, glucanase, and mannanase activity) was able to degrade approximately 26% of total NSP.

In animal experiments, the effect of enzyme supplementation is not clear-cut. Irish and Balnave (1993) supplemented SBM-based broiler diets with enzyme cocktails (pectinase, galactanase, arabinase, β-glucanase, and polygalacturonase) targeting SB NSP. They were unable to demonstrate an improvement in animal performance with enzyme supplementation. Similarly, Marsman et al. (1997a) noted no difference in performance of broiler chicks fed SBM-containing diets supplemented with protease and carbohydrase, despite higher apparent ileal NSP and CP digestibilities. Graham et al. (2002) reported an approximate 12% increase in TME of SBM for broilers when it was treated with α-galactosidase, and enzyme treatment eliminated raffinose and stachyose from the feces, but no effect on performance was noted. Supplementation of broiler diets with 80,000 units/kg of β-mannanase increased 42-day weight gains and gain:feed ratios (Jackson et al., 2004). Besides significantly affecting degradation of NSP of SB in vitro, Meng et al. (2005) noted increased performance characteristics of broilers fed diets treated with enzyme cocktails (Table 9.7). Both weight gain and gain:feed were increased, while all enzyme combinations increased nitrogen-corrected apparent metabolizable energy (AME_n) compared with the control diet. Additionally, enzymatic treatment of the diets more than doubled the total tract NSP digestion. Increased NSP digestion and AME_n concentration in the diet also were reported by Meng and Slominski (2005) in broilers fed a corn-SBM diet treated with an enzyme cocktail (cellulase, pectinase, xylanase, glucanase, mannanase, and galactanase). Performance characteristics of the birds in this experiment showed tendencies for improvement, but results were not significant. Kocher et al. (2002) noted increased AME_n but not feed efficiency, while ileal NSP disappearance was increased by enzyme supplementation. In leghorns, the use of β-mannanase is reported to improve energy utilization, and to potentially lower feeding costs of practical layer diets (Wu et al., 2005).

Supplementation of swine diets containing SBM with β-glucanase increased ileal

Table 9.7. Performance Characteristics, Total Tract Nonstarch Polysaccharide Digestibility, and Dietary Nitrogen-Corrected Apparent Metabolizable Energy Concentration of Enzyme-Treated Diets Fed to Broilers (5–18 d)[1]

Enzyme	Weight gain, g	Gain:feed	NSP digestion %	AME_n, kcal/kg[2]
None	436[b]	0.65[c]	6.3[b]	2,902[b]
C[3]+P[4]	459[a]	0.67[b]	14.0[a]	2,997[a]
C+XG[5]	470[a]	0.68[b]	14.0[a]	3,004[a]
C+P+XG	456[a]	0.67[b]	12.8[a]	3,001[a]
C+P+XG+MC[6]	466[a]	0.69[c]	14.9[a]	3,046[a]
SEM	6.5	0.004	1.37	21.0

[1]Source: Meng et al. (2005).
[2]Nitrogen-corrected apparent metabolizable energy.
[3]Cellulase.
[4]Pectinase.
[5]Xylanase and glucanase.
[6]Mannanase and cellulase.
[abc]Means within columns lacking common superscript letters are different (P<0.05).

but not total tract digestibilities of β-glucans. Additionally, ileal and total tract digestibility of DM, OM, CP, and GE were increased with supplementation of β-glucanase (Li et al., 1996). Gdala et al. (1997) fed 8- to 12-week-old piglets diets containing SBM supplemented with either single enzymes (galactosidase, xylanase, glucanase, amylase, or protease) or a mixture of several enzymes (galactosidase, xylanase, and proteases). The ileal digestibility of NSP in the diet was not affected by enzyme supplementation, although xylanase supplementation led to increased digestibility of xylose. Baucells et al. (2000) reported improved feed efficiency in grower (~8%) and finisher (~10%) gilts when a cereal-SBM-pea diet was supplemented with α-galactosidase. Supplementation of corn-SBM diets for nursery pigs with 0.1% carbohydrase (galactosidase, mannanase, and mannosidase) did not affect daily gain over the first five weeks post-weaning, but improved feed efficiency by 9%. In a similar experiment that lasted three weeks, daily gain was increased by 50 g/d in the third week for animals fed a diet supplemented with 0.1% carbohydrase. Over the entire three-week experimental period, carbohydrase supplementation increased feed efficiency by 7% (Kim et al., 2003).

Although not all evidence pertaining to exogenous enzyme supplementation to animal feed is conclusive, results appear to be affected strongly by the choice of enzymes used for a particular feed. This area of research is still developing, and as more results become available, more optimal enzyme cocktails can be formulated for use in typical corn-SBM diets for poultry and swine.

Functional Foods: Soy Oligosaccharides and Soy Fiber

Over the past decade, an increasing interest in the development of "functional foods" has grown. More and more research is geared toward testing and developing feed ingredients or nutrients that provide benefits beyond standard nutrition. Of the functional ingredients, oligosaccharides are among the most researched. Many types of oligosaccharides have been evaluated in a range of species, often for their assumed prebiotic effects. Additionally, despite the discussion on what exactly comprises dietary fiber, the consensus is that dietary fiber (both soluble and insoluble) is essential in maintaining a healthy gut. Soybeans contain oligosaccharides and large amounts of dietary fiber (hulls), and both these fractions could be used as functional ingredients.

The SB oligosaccharides tend to be a double-edged sword when it comes to their classification as a functional ingredient. In animal nutrition, the GOS found in SB often are considered to be anti-nutritional factors. In poultry, GOS tend to lower ME, and for that reason, low-oligosaccharide SB varieties have been evaluated in poultry (Parsons et al., 2000). In swine, rats, and humans, flatus production in combination with potential intestinal discomfort often results in a negative view of SB oligosaccharides (Liener, 1994). Nevertheless, SB oligosaccharides are able to stimulate growth of beneficial bacteria (lactobacilli and bifidobacteria) in swine (Smiricky-Tjardes et al., 2003b), and raffinose supplementation in humans downregulates detrimental bacteria such as clostridia (Benno et al., 1987). The negative side effects of consuming SB oligosaccharides, plus the fact that they are not easily purified (Espinosa-Martos & Ruperez, 2006), probably make SB oligosaccharides a less interesting functional food ingredient (prebiotic) than, for example, fructooligosaccharides.

Fiber from SB has been investigated as a functional food since the 1980s. Shorely et al. (1985) reported that humans on a nonrestricted diet that were supplemented with 25 g/d of SB polysaccharide decreased their total plasma cholesterol by up to 11%. Lo et al. (1986) reported similar results with an additional 13 mg/dL decrease in total cholesterol and 12 mg/dL decrease in LDL cholesterol in humans already on a low-cholesterol diet. In subsequent work, Lo and Cole (1990) noted a 7.7% decrease in total cholesterol and a 7.4% decrease in LDL cholesterol in individuals consuming 14 g/d of SB cotyledon fiber. Water-soluble SB fiber also prevented induced hypercholesterolemia by ovariectomy in rats (Mitamura et al., 2003). Additionally, SB polysaccharides may reduce insulin responses to oral glucose challenges by as much as 20% (Lo et al., 1986), which is potentially useful in pre-diabetic individuals. Vanderhoof et al. (1997) reported a significant shortening (from 23 to 10 h) of episodes of acute diarrhea in infants (>6 months of age) fed formulas containing soy fiber. More recently, Ostrom et al. (2006) reported decreased regurgitation in infants fed formulas containing soy fiber compared to cow milk-based formula. Most applications for SB fiber are geared toward humans. Soybean fiber application in food is an area of research (in human nutrition) that can be greatly expanded upon, as many effects and processes are yet to be elucidated.

Conclusion

Carbohydrates make up a significant part of the SB seed and the resultant SBM. The carbohydrate fraction of soybeans and SBM contains large amounts of sugars that are readily digested by humans and animals. Also a considerable amount of oligosaccharides is present in SB that are not digested but can be fermented by microbiota in the rumen or lower gut. Starch concentration is low in mature SB, and starch is readily digested. The structural carbohydrates found in SB and SBM are complex and have characteristic structures and monosaccharide composition compared to cereal grains, although the hemicelluloses are similar to those found in other plants. The structural carbohydrates are readily fermented by ruminants, but they can also be (partly) fermented in the lower gut of swine. Little information is available on the carbohydrate composition of SPC and SPI, and no feeding studies have investigated the effect of carbohydrate fractions in these products on the animal or human. The environment and genome can play a significant role in the nutritional value of SB carbohydrates. Low oligosaccharide varieties tend to increase ME in poultry, whereas flatulence in humans and swine may be reduced. Processing of SB and SBM generally does not alter carbohydrates directly (except starch), but may make carbohydrates more available to the animal and, therefore, have increased nutritional value. A similar philosophy is used in the enzyme treatment of SBM for animal feeds, but the effect is not always clear-cut. There is potential for use of SB carbohydrates as functional food ingredients, but more research is needed to elucidate potential uses and dosage.

In conclusion, SB carbohydrates are important nutritive components of SB and SBM. The utilization of these carbohydrates by humans and animals depends on several factors, including specific cultivar composition, growth conditions, and processing of SB and SBM. The research effort to improve utilization of SB carbohydrates still is ongoing, and should include carbohydrate fractions in purified soy protein products such as SPC and SPI. Additionally, it is important to investigate the possibilities of specific fractions of SB carbohydrates for use as functional food ingredients in human diets, as there is limited information in this area.

References

Adolph, W.H.; H.C. Kao. The biological availability of soybean carbohydrates. *J. Nutr.* **1934,** *7,* 395–406.

Annison, G.; M. Choct. Enzymes in poultry diets. *Enzymes in Animal Nutrition;* C. Wenk; M. Boessinger, Eds.; Zurich, ETH-Zurich, 1993; pp. 61–67.

Aspinall, G.O.; I.W. Cottrell; S.V. Egan; I.M. Morrison; J.N.C. Whyte. Polysaccharides of soybeans. Part IV. Partial hydrolysis of the acidic polysaccharide complex from cotyledon meal. *J. Chem. Soc.* **1967a,** 1071–1080.

Aspinall, G.O.; I.W. Cottrell. Polysaccharides of soybeans. VI. Neutral polysaccharides from cotyledon meal. *Can. J. Chem.* **1971,** *49,* 1019–1022.

Aspinall, G.O.; J.N.C. Whyte. Polysaccharides of soy-beans. Part I. Galactomannans from the hulls. *J. Chem. Soc.* **1964,** 5058–5063.

Aspinall, G.O.; K. Hunt; I.M. Morrison. Polysaccharides of soy-beans. Part V. Acidic polysaccharides from the hulls. *J. Chem. Soc.* **1967b,** 1080–1086.

Aspinall, G.O.; K. Hunt; I.M. Morrison. Polysaccharides of soy-beans. Part II. Fractionation of hull cell-wall polysaccharides and the structure of xylan. *J. Chem. Soc.* **1966,** 1945–1949.

Aspinall, G.O.; R. Begbie; A. Hamilton; J.N.C. Whyte. Polysaccharides of soy-beans. Part III. Extraction and fractionation of polysaccharides from cotyledon meal. *J. Chem. Soc.* **1967c,** 1065–1070.

Aspinall, G.O.; R. Begbie; J.E. McKay. Polysaccharide components of soybeans. *Cer. Sci. Today.* **1967d,** *12,* 223–261.

Bach Knudsen, K.E. Carbohydrate and lignin contents of plant materials used in animal feeding. *Anim. Feed Sci. Technol.* **1997,** *67,* 319–338.

Bach Knudsen, K.E.; P. Aman; B.O. Eggum. Nutritive value of Danish-grown barley varieties, I, carbohydrates and other major constituents. *J. Cereal Sci.* **1987,** *6,* 173–186.

Baucells, F.; J.F. Perez; J. Morales; J. Gasa. Effect of α-galactosidase supplementation of cereal-soya-bean-pea diets on the productive performances, digestibility and lower gut fermentation in growing and finishing pigs. *Anim. Sci.* **2000,** *71,* 157–164.

Bednar, G.E.; A.R. Patil; S.M. Murray; C.M. Grieshop; N.R. Merchen; G.C. Fahey, Jr. Starch and fiber fraction in selected food and feed ingredients affect their small intestinal digestibility and fermentability and their large bowel fermentability in vitro in a canine model. *J. Nutr.* **2001,** *131,* 276–286.

Bengala-Freire, J.; A. Aumaitre; J. Peiniau. Effects of feeding raw and extruded peas on ileal digestibility, pancreatic enzymes and plasma glucose and insulin in early weaned pigs. *J. Anim. Physiol. Anim. Nutr.* **1991,** *65,* 154–164.

Benno, Y.; K. Endo; N. Shiragami; K. Sayama; T. Mitsuoka. Effects of raffinose intake on human fecal microflora. *Bifido. Microflora* **1987,** *6,* 59–63.

Bolton, W. The determination of digestible carbohydrates in poultry foods. *Analyst* **1957,** *85,* 189–193.

Brillouet, J-M.; B. Carre. Composition of cell walls from cotyledons of *Pisum savitum, Vicia faba* and *Glycine max. Phytochem.* **1983,** *22,* 841–847.

Bruce, K.J.; L.K. Karr-Lilienthal; K.E. Zinn; L.L. Pope; D.C. Mahan; N.D. Fastinger; M. Watts; P.L. Utterback; C.M. Parsons; E.O. Castaneda; et al. Evaluation of the inclusion of soybean oil and soybean processing by-products to soybean meal on nutrient composition and digestibility in swine and poultry. *J. Anim. Sci.* **2006,** *84,* 1403–1414.

Camire, M.E.; A. Camire; K. Krumhar. Chemical and nutritional changes in foods during extrusion. *Crit. Rev. Food Sci. Nutr.* **1990,** *29,* 35–37.

Campbell, J.M.; E.A. Flickinger; G.C. Fahey, Jr. A comparative study of dietary fiber methodologies using pulsed electrochemical detection of monosaccharide constituents. *Sem. Food Anal.* **1997,** *2,* 43–53.

Clapper, G.M.; C.M. Grieshop; N.R. Merchen; J.C. Russet; J.L. Brent, Jr; G.C. Fahey, Jr. Ileal and total tract nutrient digestibilities and fecal characteristics of dogs as affected by soybean protein inclusion in dry, extruded diets. *J. Anim. Sci.* **2001,** *79,* 1523–1532.

Cole, J.T.; G.C. Fahey, Jr; N.R. Merchen; A.R. Patil; S.M. Murray; H.S. Hussein; J.L. Brent, Jr. Soybean hulls as a dietary fiber source for dogs. *J. Anim. Sci.* **1991,** *77,* 917–924.

Coon, C.N.; K.L. Leske; O. Akavanichan; T.K. Cheng. Effect of oligosaccharide-free soybean meal on true metabolizable *energy* and fiber digestion in adult roosters. *Poultry Sci.* **1990,** *69,* 787–793.

Coverdale, J.A.; J.A. Moore; H.D. Tyler; P.A. Miller-Auwerda. Soybean hulls as an alternative feed for horses. *J. Anim. Sci.* **2004,** *82,* 1663–1668.

Cromwell, G.L.; M.D. Lindemann; J.H. Randolph; G.R. Parker; R.D. Coffey; K.M. Laurent; C.L. Armstrong; W.B. Mikel; E.P. Stanisiewski; G.F. Hartnell. Soybean meal from roundup ready or conventional soybeans in diets for growing-finishing swine. *J. Anim. Sci.* **2002,** *80,* 708–715.

Daveby, Y.D.; P. Aman. Chemical composition of certain dehulled legume seeds and their hulls with special reference to carbohydrates. *Swed. J. Agric. Res.* **1993,** *23,* 133–139.

Devries, J.W. On defining dietary fibre. *Proc. Nutr. Soc.* **2003,** *62,* 37–43.

Dilger, R.N.; J.S. Sands; D. Ragland; O. Adeola. Digestibility of nitrogen and amino acids in soybean meal with added soyhulls. *J. Anim. Sci.* **2004,** *82,* 715–724.

Dust, J.M.; A.M. Gajda; E.A. Flickinger; T.M. Burkhalter; N.R. Merchen; G.C. Fahey, Jr. Extrusion conditions affect chemical composition and in vitro digestion of select food ingredients. *J. Agric. Food. Chem.* **2004,** *52,* 2989–2996.

Eastwood, M.A. The physiological effect of dietary fiber: An update. *Annu. Rev. Nutr.* **1992,** *12,* 19–35.

Englyst, H. Classification and measurement of polysaccharides. *Anim. Feed Sci. Technol.* **1989,** *23,* 27–42.

Erdman Jr, J.W.; K.E. Weingartner. Nutrition aspects of fiber in soya products. *J. Am. Oil Chem. Soc.* **1981,** *58,* 511–514.

Espinosa-Martos, I.; P. Ruperez. Soybean oligosaccharides. Potential as new ingredients in functional food. *Nutr. Hosp.* **2006,** *21,* 92–96.

Fernandez-Quintela, A.; M.T. Macarulla; A.S. Del Barrio; J.A. Martinez. Composition and functional properties of protein isolates obtained from commercial legumes grown in northern Spain. *Plant Foods Human Nutr.* **1997,** *51,* 331–342.

Galloway, Sr., D.L.; A.L.Goetsch; L.A. Forster, Jr.; A.R. Patil; W. Sun; Z.B. Johnson. Feed intake and digestibility by cattle consuming bermudagrass or orchardgrass hay supplemented with soybean hulls and (or) corn. *J. Anim. Sci.* **1993,** *71,* 3087–3095.

Gdala, J.; N.E. Johanson; K.E. Bach Knudsen; I.H. Knap; P. Wagner; O.B. Jorgensen. The digestibility of carbohydrates, protein and fat in the small and large intestine of piglets fed non-supplemented and enzyme supplemented diets. *Anim. Feed Sci. Technol.* **1997,** *65,* 15–33.

Gnanasambandam, R.; A. Proctor. Preparation of soy hull pectin. *Food. Chem.* **1999,** *65,* 461–467.

Graham, K.K.; M.S. Kerley; J.D. Firman; G.L. Allee. The effect of enzyme treatment of soybean meal on oligosaccharide disappearance and chick growth performance. *Poultry Sci.* **2002,** *81,* 1014–1019.

Grieshop, C.M.; G.C. Fahey, Jr. Comparison of quality characteristics of soybeans from Brazil, China, and the United States. *J. Agric. Food Chem.* **2001,** *49,* 2669–2673.

Grieshop, C.M.; C.T. Kadzere; G.M. Clapper; E.A. Flickinger; L.L. Bauer; R.L. Frazier; G.C. Fahey, Jr. Chemical and nutritional characteristics of United States soybeans and soybean meals. *J. Agric. Food Chem.* **2003,** *51,* 7684–7691.

Guillon, F.; M. M-J. Champ. Carbohydrate fractions of legumes: Uses in human nutrition and potential for health. *Brit. J. Nutr.* **2002,** *88,* S293–S306.

Harris, P.J.; B.G. Smith. Plant cell walls and cell-wall polysaccharides: Structures, properties and uses in food products. *Int. J. Food Sci. Technol.* **2006,** *41(Suppl. 2),* 129–143.

Hartwig, E.E.; T. Min Kuo; M.M. Kenty. Seed protein and its relationship to soluble sugars in soybeans. *Crop Sci.* **1997,** *37,* 770–774.

Hill, R.C.; C.F. Burrows; G.W. Ellison; J.E. Bauer. The effect of texturized vegetable protein from soy on nutrient digestibility compared to beef in cannulated dogs. *J. Anim. Sci.* **2001,** *79,* 2162–2171.

Hitz, W.D.; T.J. Carlson; P.S. Kerr; S.A. Sebastian. Biochemical and molecular characterization of a mutation that confers a decreased raffinosaccharide and phytic acid phenotype on soybean seeds. *Plant Physiol.* **2002,** *128,* 650–660.

Honig, D.H.; J.J. Rackis. Determination of the total pepsin-pancreatin indigestible content (dietary fiber) of soybean products, wheat bran, and corn bran. *J. Agric. Food Chem.* **1979,** *27.* 1262–1266.

Huisman, M.M.H.; C.T.M. Fransen; J.P. Kamerling; J.F.G. Vliegenthart; H.A. Schols; A.G.J. Voragen. The CDTA-soluble pectic substances from soybean meal are composed of rhamnogalacturonan and xylogalacturonan but not homogalacturonan. *Biopoly.* **2001,** *58,* 279–294.

Huisman, M.M.H.; H.A. Schols; A.G.J. Voragen. Cell wall polysaccharides from soybean (*Glycine max.*) meal. Isolation and characterisation. *Carbohydr. Poly.* **1998,** *37,* 87–95.

Huisman, M.M.H.; H.A. Schols; A.G.J. Voragen. Enzymatic degradation of cell wall polysaccharides from soybean meal. *Carbohydr. Poly.* **1999,** *38,* 299–307.

Institute of Medicine. Proposed definition of dietary fiber. A report of the panel on the definition of dietary fiber and the standing committee on the scientific evaluation of dietary reference intakes, food and nutrition board; National Academies Press: Washington, DC, 2001.

Irish, G.G.; D. Balnave. Non-starch polysaccharides and broiler performance on diets containing soyabean meal as the sole protein concentrate. *Aust. J. Agric. Res.* **1993,** *44,* 1483–1499.

Jackson, M.E.; K. Geronian; A. Knox; J. McNab; E. McCartney. A dose-response study with the feed enzyme β-mannanase in broilers provided with corn-soybean meal based diets in the absence of antibiotic growth promoters. *Poultry Sci.* **2004,** *83,* 1992–1996.

Jiang, H.Q.; L.M. Gong; Y.X. Ma; Y.H. He; D.F. Li; H.X. Zhai. Effect of stachyose supplementation on growth performance, nutrient digestibility and caecal fermentation characteristics in broilers. *Brit. Poultry Sci.* **2006,** *47,* 516–522.

Jones, F.T. A survey of soybean meal used in poultry feeds 1976 to 1982. *Poultry Sci.* **1984,** *63,* 1462–1463.

Jung, S.; B.P. Lamsal, V. Stepien, L.A. Johnson; P.A. Murphy. Functionality of soy protein produced by enzyme-assisted extraction. *J. Am. Oil Chem. Soc.* **2006,** *83,* 71–78.

Karimzadegan, E.; A.J. Clifford; F.W. Hill. A rat bioassay for measuring the comparative availability of carbohydrates and its application to legume foods, pure carbohydrates and polyols. *J. Nutr.* **1979,** *102,* 2247–2259.

Karr-Lilienthal, L.K.; C.M. Grieshop; N.R. Merchen; D.C. Mahan; G.C. Fahey, Jr. Chemical composition and protein quality comparisons of soybeans and soybean meals from five leading soybean-producing countries. *J. Agric. Food Chem.* **2004,** *52,* 6193–6199.

Karr-Lilienthal, L.K.; C.M. Grieshop; J.K. Spears; G.C. Fahey, Jr. Amino acid, carbohydrate, and fat composition of soybean meals prepared at 55 commercial U.S. soybean processing plants. *J. Agric. Food Chem.* **2005,** *53,* 2146–2150.

Keys, Jr., J.E.; J.V. DeBarthe. Cellulose and hemicellulose digestibility in the stomach, small intestine and large intestine of swine. *J. Anim. Sci.* **1974,** *39,* 53–56.

Kim, S.W.; D.A. Knabe; K.J. Hong; R.A. Easter. Use of carbohydrases in corn-soybean meal-based nursery diets. *J. Anim. Sci.* **2003,** *81,* 2496–2504.

Kocher, A.; M. Choct; M.D. Porter; J. Broz. Effects of feed enzymes on nutritive value of soyabean meal fed to broilers. *Brit. Poultry Sci.* **2002,** *43,* 53–63.

Kornegay, E.T. Feeding value and digestibility of soybean hulls for swine. *J. Anim. Sci.* **1978,** *47,* 1272–1280.

Kuriyama, S.; L.B. Mendel. The physiological behavior of raffinose. *J. Biol. Chem.* **1917,** *31,* 125–147.

Lan, Y.; B.A. Williams; S. Tamminga; H. Boer; A. Akkermans; G. Erdi; M.W.A. Verstegen. In vitro fermentation kinetics of some non-digestible carbohydrates by the caecal microbial community of broilers. *Anim. Feed Sci. Technol.* **2005,** *123–124,* 687–702.

Li, S.; W.C. Sauer; S.X. Huang; V.M. Gabert. Effect of β-glucanase supplementation to hulless barley-or wheat-soybean meal diets on the digestibilities of energy, protein, β-glucans, and amino acids in young pigs. *J. Anim. Sci.* **1996,** *74,* 1649–1656.

Liener, I.E. Implications of antinutritional components in soybean foods. *Crit. Rev. Food Sci. Nutr.* **1994,** *24,* 31–67.

Liying, Z.; D. Li; S. Qiao; E.W. Johnson; B. Li; P.A. Thacker; N.K. Han. Effects of stachyose on performance, diarrhoea incidence and intestinal bacteria in weanling pigs. *Arch. Anim. Nutr.* **2003,** *57,* 1–10.

Lo, G.S.; T.G. Cole. Soy cotyledon fiber products reduce plasma lipids. *Atherosclerosis* **1990,** *82,* 59–67.

Lo, G.S.; A.P. Goldberg; A. Lim; J.J. Grundhauser; C. Anderson; G. Schonfeld. Soy fiber improves lipid and carbohydrate metabolism in primary hyperlipidemic subjects. *Atherosclerosis* **1986,** *62,* 239–248.

Lodhi, G.N.; R. Renner; D.R. Clandinin. Available carbohydrate in rapeseed meal and soybean meal as determined by chemical method and chick bioassay. *J. Nutr.* **1969,** *99,* 413–418.

Mansfield, H.R.; M.D. Stern. Effects of soybean hulls and lignosulfonate-treated soybean meal on ruminal fermentation in lactating dairy cows. *J. Dairy Sci.* **1994,** *77,* 1070–1083.

Marsman, G.J.P.; H. Gruppen; A.J. Mul; A.G.J. Voragen. In vitro accessibility of untreated, toasted, and extruded soybean meals for proteases and carbohydrases. *J. Agric. Food Chem.* **1997b,** *45,* 4088–4095.

Marsman, G.J.P.; H. Gruppen; A.F.B. van der Poel; R.P. Kwakkel; M.W.A. Verstegen; A.G.J. Voragen. The effect of thermal processing and enzyme treatments of soybean meal on growth performance, ileal nutrient digestibilities, and chyme characteristics in broiler chicks. *Poultry Sci.* **1997a,** *76,* 864–872.

Martin, S.K.; C.A. Hibberd. Intake and digestibility of low-quality native grass hay by beef cows supplemented with graded levels of soybean hulls. *J. Anim. Sci.* **1990,** *68,* 4319–4325.

Meng, X.; B.A. Slominski. Nutritive values of corn, soybean meal, canola meal, and peas for broiler chickens as affected by a multicarbohydrase preparation of cell wall degrading enzymes. *Poultry Sci.* **2005,** *84,* 1242–1251.

Meng, X.; B.A. Slominski; C.M. Nyachoti; L.C. Campbell; W. Guenter. Degradation of cell wall polysaccharides by combinations of carbohydrase enzymes and their effect on nutrient utilization and broiler chicken performance. *Poultry Sci.* **2005,** *84,* 37–47.

Mieth, G.; G.W. Krausse; V. Erhardt; K. Marzilger. Zum Einfluss genetischer Merkmale und umweltspezifischer Faktoren auf die Zusammensetzung von Sojabohnen. *Die Nahrung.* **1988,** *32,* 911–921.

Mitamura, R.; H. Hara; Y. Aoyama; T. Takahashi; H. Furuta. Ingestion of water-soluble soybean fiber prevents osteopenia and hypercholesterolemia induced by ovariectomy in rats. *J. Agric. Food Chem.* **2003,** *51,* 1085–1089.

Mitaru, B.N.; R. Blair; R.D. Reichert; W.E. Roe. Dark and yellow rapeseed hulls, soybean hulls and a purified fiber source: their effects on dry matter, energy, protein and amino acid digestibilities in cannulated pigs. *J. Anim. Sci.* **1984,** *59,* 1510–1518.

Mul, A.J.; F.G. Perry. The role of fructo-oligosaccharides in animal nutrition. *Recent Advances in Animal Nutrition*; P.C. Garnsworthy; D.J.A. Cole, Eds.; Nottingham University Press, Nottingham, UK, 1994; pp. 57–79.

Nakamura, A.; H. Furuta; H. Maeda; Y. Nagamatsu; A. Yoshimoto. The structure of soluble soybean polysaccharide. *Hydrocolloids–Part 1;* K. Nishinari, Ed.; Elsevier Science: New York, NY, 2000, pp. 235–241.

Nakamura, A.; H. Furuta; H. Maeda; T. Takao; Y. Nagamatsu. Structural studies by stepwise enzymatic degradation of the main backbone of soybean soluble polysaccharides consisting of galacturonan and rhamnogalacturonan. *Biosci. Biotechnol. Biochem.* **2002,** *66,* 1301–1313.

National Research Council. *Nutrient Requirements of Swine,* Tenth rev. ed.; National Academies Press: Washington, DC, 1998.

Neus, J.D.; W.R. Fehr; S.R. Schnebly. Agronomic and seed characteristics of soybean with reduced raffinose and stachyose. *Crop Sci.* **2005,** *45,* 589–592.

Ostrom, K.M.; J.R. Jacobs; R.J. Merritt; R.D. Murray. Decreased regurgitation with a soy formula containing added soy fiber. *Clin. Pediatr.* **2006,** *45,* 29–36.

Ouhida, I.; J. F. Perez; J. Gasa. Soybean (*Glycine max*) cell wall composition and availability to feed enzymes. *J. Agric. Food Chem.* **2002,** *50,* 1933–1938.

Parsons, C.M.; Y. Zhang; M. Araba. Nutritional evaluation of soybean meals varying in oligosaccharide content. *Poultry Sci.* **2000,** *79*:1127-1131.

Potter, L.M.; M. Potchanakorn. Digestibility of carbohydrate fraction of soybean by poultry. *Proc. World Soybean Res. Conf. III;* R. Shibles, Ed.;Westview Press: Boulder, CO, 1984; pp. 218–223.

Proctor, A. Soybean oil extraction and processing. *Soy Applications in Food. Soybeans: Chemistry, Technology and Utilization;* K. Liu, Ed.; Chapman and Hall: New York, NY, 1997, pp. 297–346.

Riaz, M.N. *Processing of Soybeans into Ingredients.* M.N. Riaz, Ed.; CRC Press: Boca Raton, FL, 2006, pp. 39–62.

Rubio, L.A.; M.M. Pedrosa; A. Perez; C. Cuadrado; C. Burbano; M. Muzquiz. Ileal digestibility of defatted soybean, lupin and chickpea seed meals in cannulated Iberian pigs: II. Fatty acids and carbohydrates. *J. Sci. Food Agric.* **2005,** *85,* 1322–1328.

Sebastian, S.A.; P.S. Kerr; R.W. Pearlstein; W.D. Hitz. Soybean germplasm with novel genes for improved digestibility. *Soy in Animal Nutrition;* J.K. Drackely, Ed.; Federation of Animal Sciences Societies: Savoy, IL, 2000, pp. 56–74.

Selvendran, R.R.; B.J.H. Stevens; M.S. Du Pont. Dietary fiber: chemistry, analysis, and properties.

Adv. Food Res. **1987,** *31,* 117–209.

Sessa, D.J. Processing of soybean hulls to enhance the distribution and extraction of value-added proteins. *J. Sci. Food Agric.* **2003,** *84,* 75–82.

Shorely, R.L.; P.L. Day; R.A. Willis; G.S. Lo; F.H. Steinke. Effects of soybean polysaccharide on plasma lipids. *J. Am. Diet. Assoc.* **1985,** *85,* 1461–1465.

Smiricky-Tjardes, M.R.; E.A. Flickinger; C.M. Grieshop; L.L. Bauer; M.R. Murphy; G.C. Fahey, Jr. In vitro fermentation characteristics of selected oligosaccharides by swine fecal microflora. *J. Anim. Sci.* **2003a,** *81,* 2505–2514.

Smiricky, M.R.; C.M. Grieshop; D.M. Albin; J.E. Wubben; V.M. Gabert; G.C. Fahey, Jr. The influence of soy oligosaccharides on apparent and true ileal amino acid digestibilities and fecal consistency in growing pigs. *J. Anim. Sci.* **2002,** *80,* 2433–2441.

Smiricky-Tjardes, M.R.; C.M. Grieshop; E.A. Flickinger; L.L. Bauer; G.C. Fahey, Jr. Dietary galactooligosaccharides affect ileal and total-tract nutrient digestibility, ileal and fecal bacterial concentrations, and ileal fermentative characteristics of growing pigs. *J. Anim. Sci.* **2003b,** *81,* 2535–2545.

Snyder, H.E.; T.W. Kwon. *Soybean Utilization.* Van Nostrand Reinhold Co.: New York, NY, 1987.

Stark, J.R.; X.S. Yin. The effect of physical damage on large and small barley starch granules. *Starch* **1986,** *38,* 369–374.

Steggerda F.R. Gas production following food consumption. *Ann. NY Acad. Sci.* **1968,** *150,* 57–66.

Stevenson, D.G.; R.K. Doorenbos; J-L. Jane; G.E. Inglett. Structures and functional properties of starch from seeds of three soybean (*Glycine max* (L.) Merr.) varieties. *Starch* **2006,** *58,* 509–519.

Suarez, F.L.; J. Springfield; J.K. Furne; T.T. Lohrmann; P.S. Kerr; M.D. Levitt. Gas production in humans ingesting a soybean flour derived from beans naturally low in oligosaccharides. *Am. J. Clin. Nutr.* **1999,** *69,* 135–139.

Sunvold, G.D.; G.C. Fahey, Jr.; N.R. Merchen; G.A. Reinhart. In vitro fermentation of selected fibrous substrates by dog and cat fecal inoculum: Influence of diet composition on substrate organic matter disappearance and short-chain fatty acid production. *J. Anim. Sci.* **1995,** *73,* 1110–1122.

Theander, O.; P. Aman; E. Westerlund; H. Graham. Enzymatic/chemical analysis of dietary fiber. *J. Assoc. Off. Anal. Chem. Int.* **1994,** *77,* 703–709.

Thomas, J.M.G.; K.J. Boote; L.H. Allen, Jr.; M. Gallo-Meagher; J.M. Davis. Elevated temperature and carbon dioxide effects on soybean seed composition and transcript abundance. *Crop Sci.* **2003,** *43,* 1548–1557.

Trater, A.M.; E.C. Titgemeyer; C.A. Loest; B.D. Lambert. Effects of supplemental alfalfa hay on the digestion of soybean hull-based diets by cattle. *J. Anim. Sci.* **2001,** *79,* 1346–1351.

Trugo, L.C.; C.M. Donangelo; N.M.F. Trugo; K.E. Bach Knudsen. Effect of heat treatment on nutritional quality of germinated legume seeds. *J. Agric. Food Chem.* **2000,** *48,* 2082–2086.

Van Kempen, T.A.T.G.; E. van Heugten; A.J. Moeser; N.S. Muley; V.J.H. Sewalt. Selecting soybean meal characteristics preferred for swine nutrition. *J. Anim. Sci.* **2006,** *84,* 1387–1395.

Van Kempen, T.A.T.G.; I.B. Kim; A.J.M. Jansman; M.W.A. Verstegen; J.D. Hancock; D.J. Lee; V.M. Gabert; D.M. Albin; G.C. Fahey, Jr.; C.M. Grieshop; et al. *J. Anim. Sci.* **2002,** *80,* 429–439.

Van Laar, H.; S. Tamminga; B.A. Williams; M.W.A. Verstegen; F.M. Engels. Fermentation characteristics of cell-wall sugars from soya bean meal, and from separated endosperm and hulls of soya beans. *Anim. Feed Sci. Technol.* **1999,** *79,* 179–193.

Van Laar, H.; S. Tamminga; B.A. Williams; M.W.A. Verstegen; H.A. Schols. Fermentation characteristics of polysaccharide fractions extracted from the cell walls of soya bean cotyledons. *J. Sci. Food Agric.* **2000,** *80,* 1477–1485.

Vanderhoof, J.; N.D. Murray; C.L. Paule; K.M. Ostrom. Use of soy fiber in acute diarrhea in infants and toddlers. *Clin. Pediatr.* **1997,** *36,* 135–139.

Wang, H.; L.A. Johnson; T. Wang. Preparation of soy protein concentrate and isolate from extruded-expelled soybean meals. *J. Am. Oil Chem. Soc.* **2004,** *81,* 713–717.

Whistler, R.L.; J. Saarnio. Galactomannan from soy bean hulls. *J. Chem. Soc.* **1957,** 6055–6057.

Willis, S. The use of soybean meal and full fat soybean meal by the animal feed industry. *12th Australian Soybean Conf. Proc.;* Toowomba, Queensland: North Australian Soybean Industry Association, 2003, pp. 1–8.

Wilson, L.A.; V.A. Birmingham; D.P. Moon; H.E. Snyder. Isolation and characterization of starch from mature soybeans. *Cereal Chem.* **1978,** *55,* 661–670.

Wolf, R.B.; J.F. Cavins; R. Kleiman; L.T. Black. Effect of temperature on soybean seed constituents: Oil, protein, moisture, fatty acids, amino acids and sugars. *J. Am. Oil Chem. Soc.* **1982,** *59,* 230–232.

Wu, G.; M.M. Bryant; R.A. Votile; D.A. Roland, Sr. Effects of β-mannanase in corn-soy diets on commercial leghorns in second-cycle hens. *Poultry Sci.* **2005,** *84,* 894–897.

Würsch, P.; S. Del Vedovo; B. Koellreutter. Cell structure and starch nature as key determinants of the digestion rate of starch in legume. *Am. J. Clin. Nutr.* **1986,** *43,* 25–29

Yamaguchi, F.; H. Kojima; M. Muramoto; Y. Ota; C. Hatanaka. Effects of hexametaphosphate on soybean pectic polysaccharide extraction. *Biosci. Biotechnol. Biochem.* **1996,** *60,* 2028–2031.

Yamka, R.M.; D.L. Harmon; W.D. Schoenherr; C. Khoo; K.L. Gross; S.J. Davidson; D.K. Joshi. In vivo measurement of flatulence and nutrient digestibility in dogs fed poultry by-product meal, conventional soybean meal, and low-oligosaccharide low-phytate soybean meal. *Am. J. Vet. Res.* **2006,** *67,* 88–94.

Yamka, R.M.; B.M. Hetzler; D.L. Harmon. Evaluation of low-oligosaccharide, low-phytate whole soybeans and soybean meal in canine foods. *J. Anim. Sci.* **2005,** *83,* 393–399.

Yamka, R.M., U. Jamikorn, A.D. True, and D.L. Harmon. Evaluation of soyabean meal as a protein source in canine foods. *Anim. Feed. Sci. Technol.* **2003,** *109,* 121-132.

Zuo, Y.; G.C. Fahey, Jr.; N.R. Merchen; N.L. Bajjalieh. Digestion responses to low oligosaccharide soybean meal by ileally-cannulated dogs. *J. Anim. Sci.* **1996,** *74,* 2441–2449.

10

Minor Constituents and Phytochemicals of Soybeans

Tong Wang
Department of Food Science and Human Nutrition, Iowa State University, Ames, IA 50011

Soybeans are recognized as a storehouse of nutrients. The focus of this chapter is on composition of minor compounds or phytochemicals (Table 10.1), while the major components of the seeds (i.e., proteins and oil) are discussed in other chapters of the book.

Lipid Components

Tocopherols

Tocopherols are important antioxidants present in relatively high concentration in soybeans. The content of tocopherols in raw soybean is shown in Table 10.1 while the compositional comparison of tocopherols in crude soybean oil and wheat germ oil is shown in Table 10.2.

Vitamin E is a mixture of four different forms of tocopherols (Fig. 10.1) and four different forms of tocotrienols (having three double bonds on the side chain), with α-tocopherol being the most effective form of Vitamin E. Soybean only contains tocopherols. Vitamin E protects against the oxidation of polyunsaturated fatty acids in biological membranes and in plasma lipoproteins. The antioxidation mechanism is the termination of the free radical autooxidation of lipids by the reaction of the phenolic ring with the free radical, forming a stable phenoxyl radical. Some tocotrienols may have greater antioxidant activity than their counterpart tocopherols in certain model systems (Serbinova et al., 1991). A good review by White and Xing (1997) describes various investigations on comparisons of relative effectiveness of various forms of tocopherols.

Phytosterols

Phytosterols are sterols in plants, and they are structurally similar to sterols from animal sources, as illustrated in Fig. 10.2. Phytosterols are present at about 300–600 ppm concentrations based on the dry weight of the soybean. The primary soybean

Table 10.1. General Composition (dry seed weight basis) of Soybean Minor Constituents[1]

Constituent	Range	Typical Value	Reference
Lipids			
Tocopherols, ppm			
α-tocopherol	11–28		Guzman & Murphy, 1986
γ-tocopherol	150–90		
δ-tocopherol	25–73		
Phytosterols, ppm	300–600		Rao & Janezic, 1992
Phospholipids, %	0.3–0.6	0.74	Wang et al., 1997
Sphingolipids, ppm	193		Guiterrez & Wang, 2004a
Carotenoids, ppm	0.8–3.7		Kanamaru et al., 2006
Proteins			
Trypsin inhibitors, mg/g	16.7–27.2	22.3	Anderson & Wolf, 1995
Lectin (hemagglutinin) unit/mg protein	1.2–6.0	3.0	Padgette et al., 1996
Lunasin, % in defatted flour	0.33–0.95	0.65	De Mejia et al., 2004
Carbohydrates, %	26–38	34	Liu et al., 1995
Sucrose	2.5–8.2	5.5	Hymowitz et al., 1972
Raffinose	0.1–0.9	0.9	
Stachyose	1.4–4.1	3.5	
Other phytochemicals			
Isoflavones, %	0.1–0.4	2.5	Wang & Murphy, 1994
Saponins, % base on protein	0.1–0.3		Arditi et al., 2000
Phytate, %	1.0–1.5	1.1	Lolas et al., 1976
Water-soluble vitamins, ppm			
Thiamine	6.3–6.9		Fernando & Murphy, 1990
Riboflavin	0.9–1.1		

[1] Adopted from Liu, 2004a.

phytosterols are β-sitosterol, campesterol, and stigmasterol, and their compositions are shown in Table 10.3.

Both phytosterols and tocopherols are co-extracted with oil and partially removed during soybean oil refining. The refining by-product is one of the important sources for commercial phytosterol and tocopherol production.

Health benefits of phytosterols have been a topic of intense research in recent years. The main physiological effect of consuming phytosterols (2–3 g/day for 21–30 days) is their reported lowering of low density lipoprotein (LDL) cholesterol by 10–15%. The Food and Drug Administration (FDA) allows a health claim for food containing phytosterols because of the association with reduced risk of coronary heart disease (Federal Register, 2000).

Table 10.2. Tocopherol Content and Composition of Crude Soybean and Wheat Germ Oils[1]

Tocopherol	Mechanically Pressed Soybean Oil	Solvent-extracted Soybean Oil	Solvent-extracted Wheat Germ Oil
Total tocopherol, ppm	1257	1370	2682
α-Tocopherol, %	9.3	10.5	67.8
β-Tocopherol, %	1.2	1.2	32.2
γ-Tocopherol, %	62.8	63.5	-
δ-Tocopherol, %	26.7	25.0	-

[1]Wang, 2002.

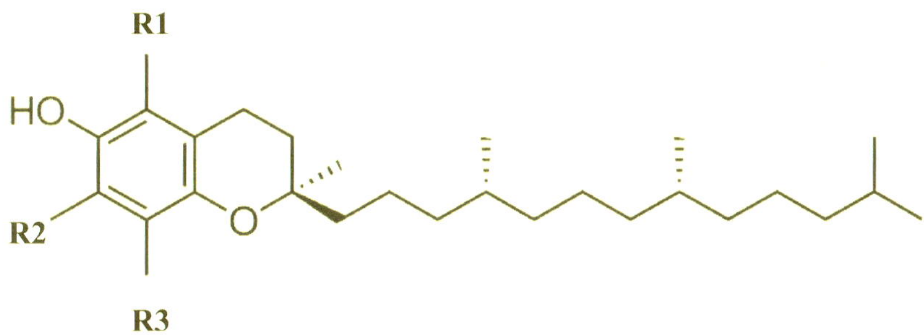

Name	R1	R2	R3
α-tocopherol	$-CH_3$	$-CH_3$	$-CH_3$
β-tocopherol	$-CH_3$	-H	$-CH_3$
γ-tocopherol	-H	$-CH_3$	$-CH_3$
δ-tocopherol	-H	-H	$-CH_3$

Fig. 10.1. Molecular structure of tocopherols present in soybean.

Some phytosterols also are shown to have antioxidant activities. The mechanism of antioxidation is different from the traditional phenolic compounds. They seem to be effective in preventing polymerization reaction in heated oils (Tian & White, 1994), and this effect is due to the structure on the side chain of some specific phytosterols.

Phospholipids (PLs)

Phospholipids are polar membrane lipids that are present in relatively high concentration (about 3% of total lipids) in soybeans, compared to their levels in other oilseeds

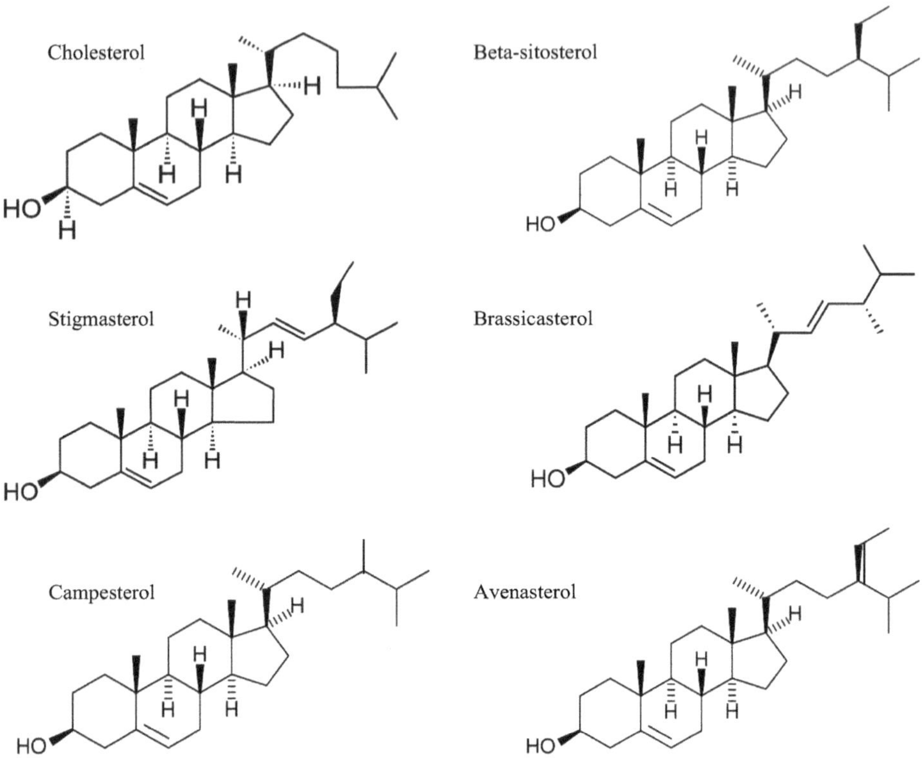

Fig. 10.2. Structure of cholesterol and soybean phytosterols.

Table 10.3. Phytosterol Content (mg/100 g) of Soybean Oils[1]

Study	Sterol	Crude	Refined
Study 1			
	β-Sitosterol	183	123
	Campesterol	68	47
	Stigmasterol	64	47
	Δ^5-Avenasterol	5	1
	Δ^7-Stigmasterol	5	1
	Δ^7-Avenasterol	2	<0.5
	Total	327	221
Study 2			
	β-Sitosterol	125–236	
	Campesterol	62–131	
	Stigmasterol	47–77	
	Total	235–405	

[1] Wang, 2002.

(Wang et al., 1997). On a dry seed-weight basis, soybean contains about 0.74% total PLs (Wang et al., 1997). The three major classes of soybean PLs are phosphatidylcholine (PC), phosphatidylethanolamine (PE), and phosphatidylinositol (PI), present in relative proportions of 55.3, 26.3 and 18.4%, respectively (Wang et al., 1997). The molecular structures of soybean PLs are shown in Fig. 10.3.

Class composition, fatty acyl stereospecific distribution, and molecular species composition of PLs in normal soybeans and seeds with modified fatty acid composition were studied by Wang et al. (1997) and Wang and Hammond (1999). The PI

Phosphatidylcholine (PC)

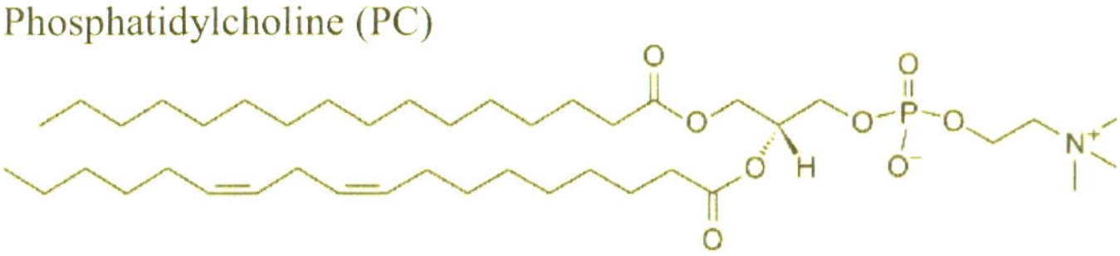

Phosphatidylethanolamine (PE)

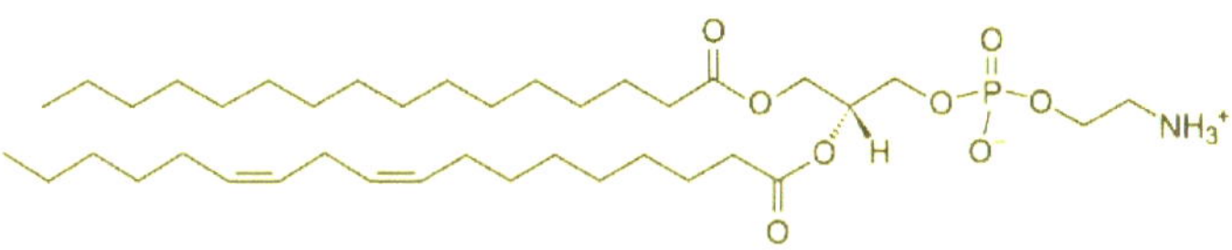

Phosphatidylserine (PS)

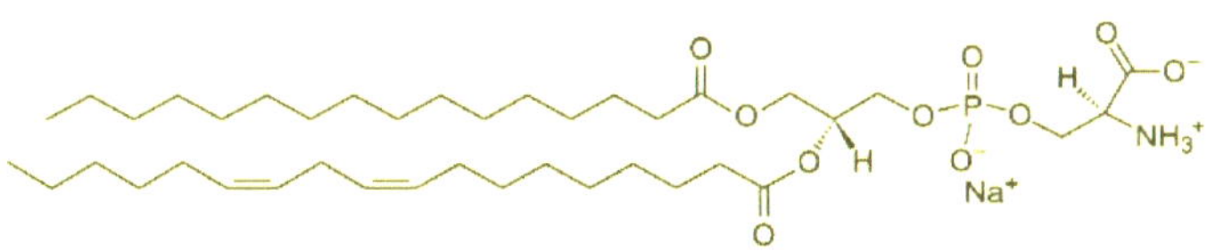

Phosphatidylinositol (PI)

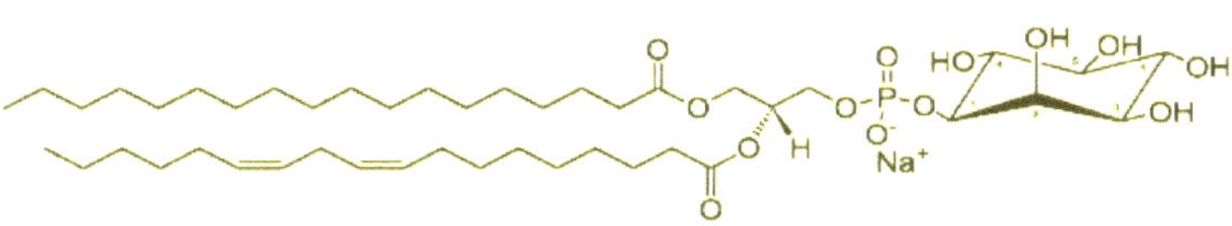

Phosphatidic acid (PA), sodium salt

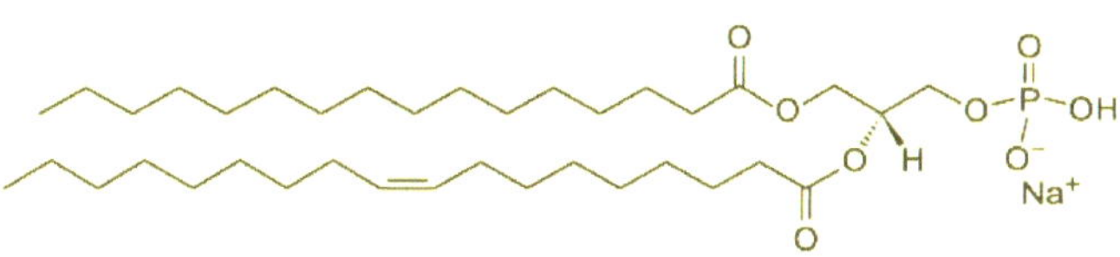

Fig. 10.3. Molecular structure of soybean major (PC, PE, and PI) and minor (PS and PA) phospholipids.

had greater palmitate (~25%) and stearate (~10%) than did PC (~12 and 5%) and PE (~17 and 3%), and PC had the lowest palmitate percentage, whereas PE had the lowest stearate percentage. Stereospecific analysis indicated that saturated fatty acids were concentrated on the *sn*-1 position, and the unsaturated fatty acids preferred the *sn*-2 position on the glycerol backbone.

The mixture of PLs in their commercial form (not highly pure) and applications is referred to as lecithin. Lecithin is an effective antioxidant. The antioxidant property of crude soybean lecithin was studied in various storage tests with sunflower oil and lard (Nasner, 1985). The addition of lecithin after refining improved oxidative stability, and the antioxidant activity depended on the composition of the PLs, and the tocopherol content of the oil. The synergistic effect of tocopherol, ascorbate, and lecithin was clearly shown in fish oil oxidation reduction (Hamilton et al., 1998; Segawa et al., 1995). A better understanding of the effective concentration level (i.e., dose–response relationship) effect of PL polar head group, fatty acid composition of the base oil is still needed. Some studies were performed under very high temperature, such as 100°C (Nwosu et al., 1997) and 180°C (King et al., 1992), where the oxidation mechanism is different than at lower temperatures. The effective concentration of PLs could be as low as 100 ppm in salmon oil, and even at 1% concentration, there was no prooxidant effect observed (King et al., 1992). However, at 0.5% concentration, PLs did not have antioxidant activity in menhaden oil (Nwosu et al., 1997). The need for tocopherols for PLs' antioxidant effect was also shown by Kashima et al. (1991) in a perilla oil model system, but the best synergism conditions for various oil types still need to be determined.

Lecithin's nutritional properties were reviewed by Orthoefer and List (2006). Lecithin is a good source of choline, an essential nutrient that acts as a precursor in the synthesis of the neurotransmitter acetylcholine. Choline is a major source of methyl groups that are involved in the formation of methionine from homocysteine, a sulfur-containing amino acid implicated in cardiovascular disease risk. High homocysteine levels increase the risk of cardiovascular disease, and medical therapy with choline can reduce homocysteine levels (Da Costa et al., 2005; Innis et al., 2007; Zeisel, 2005). Lecithin lowers serum cholesterol levels, and it is a component of lipoproteins, which transport fat and cholesterol (Jimenez et al., 1990). Choline prevents fat accumulation in the liver, and choline deficiency disturbs lecithin synthesis that is needed to export triacylglycerols from the liver as part of lipoproteins. Choline-deficient diets promote liver carcinogenesis because of the disturbance of the protein kinase C (PKC) transmembrane signaling system (Orthoefer & List, 2006). Lecithin and choline are also essential in brain and mental development in fetus and infant. Choline in mother's milk is at a much higher level than that in the maternal bloodstream (Zeisel, 2005), and it is a required nutrient during pregnancy and lactation (Zeisel, 1998). Many vital organs, such as the central nervous system, kidney, and liver, contain high levels of PLs. In several animal and human studies, lecithin was shown to improve memory and learning. For example, when aged male rats were given lecithin, they showed markedly higher spatial memory as demonstrated in a water maze than untreated rats, and the acetylcholine content in the brains of the treated rats was higher than in those of the control (Masuda et al., 1992). Lecithin supplementation (0.2 g lecithin/kg body mass) prevents the rapid decrease of choline during long-duration intense exercise and improves physical performance (von Allworden et al., 1993).

Sphingolipids (SLs)

SLs are also polar cell membrane lipids, but they are typically present in much lower concentration than PLs. Soybeans are a relatively rich source of SLs (Vesper et al., 1999), and ceramides and cerebrosides are the primary SL classes in soybeans. SLs contain a sphingoid long-chain (C18) dihydroxy base and an α-hydroxy fatty acyl chain that is linked to the base by an amide bond. The main soybean ceramide molecular species is a trihydroxy base (4-hydroxy-trans 8-sphingenine) *N*-acylated with α-hydroxy lignoceric acid (C24:0). The main soybean cerebroside molecular species is a dihydroxy base (*trans* 4-*trans* 8-sphingediene) *N*-acylated with α-hydroxy palmitic acid. The general molecular structures of ceramide and cerebroside are shown in Fig. 10.4.

Significant differences exist in cerebroside concentrations among soybean genotypes, with a range of 142 to 389 nmol/g seed (dry weight basis, equivalent to 102 to 286 ppm) (Table 10.4). The changes in PLs and SLs with seed development are presented in the following section of this chapter.

SLs are highly bioactive in animal cells. They can act as mediators of cell growth, differentiation, and programmed cell death (apoptosis). In vitro studies showed that ceramide and sphingosine were toxic for a variety of transformed cell lines and even inhibited cell transformation during the early events of carcinogenesis (Merill & Schmelz, 2001).

SLs reduced the risk of colon cancer as shown by Dillehay et al. (1994). The occurrence of aberrant colonic crypt foci (ACF, the early biomarker of tumor development) of mice fed sphingomyelin at 0.05% of their diets was reduced by 50% in comparison to the control group. In a longer term study, Schmelz et al. (1996) showed ACF were reduced in the mice fed sphingomyelin (0.1% of diet) by up to 70%. Although the incidence of colonic tumors was not reduced in the sphingomyelin-fed mice, the proportion of benign adenomas versus adenocarcinomas was higher in the sphingomyelin-fed mice than in the control group. It was suggested that sphingomyelin may prevent adenomas from progressing into adenocarcinomas. SLs, such as glucosylceramide and lactosylceramide, also were shown to reduce ACF by 50–80%, indicating that SLs suppress colon carcinogenesis through the release of their metabolites by hydrolysis (Schmelz, 2000).

SLs reduced hepatic cholesterol content in a short-term rat feeding study (Imai-

Table 10.4. Mean Composition and Cerebroside (GlcCer) Content of 10 Soybean Genotypes Grown Near Ames, IA (Guiterrez et al., 2004a)[1]

Genotype and Selectively Modified Trait(s)	GlcCer (nmol/g dry wt basis)
IA1008, Conventional	142
IA2021, low protein (36%)	283
IA2041, high protein (41%)	201
A00-815004, high palmitate (41%)	389
A97-877006, mid palmitate (27%)	221
FA22, high oleate (52%)	306
B0147B013, low palmitate (3.4%)	168
AX7019-12, mid palmitate (21%)/stearate (24%)	246
A97-552013, low linolenate (1.3%)	229
A99-144085, high stearate (28%)	197
MSD	122

[1]MSD = minimum significant difference determined by Tukey Kramer's mean comparison test ($P = 0.05$).

zumi et al, 1997) by decreasing cholesterol absorption and/or increasing fecal excretion. In a long-term feeding study of two generations, rats fed SLs at 1% of their diet had their total plasma cholesterol reduced by 30% (Kobayashi et al., 1997). Dietary SLs may influence plasma and liver lipid levels in humans, but more research is needed to better understand the mechanism of cholesterol reduction.

SLs offer protection against pathogenic microorganisms (bacteria and viruses) and toxins. Synthetic SLs were successfully used to prevent bacterial and viral infections by binding to pathogens and removing them from the intestine (Vesper et al., 1999). The primary compound in human milk that protects against pathogens is assumed to be glycosphingolipids (Vesper et al., 1999). SLs also reduced skin carcinoma development (Birt et al., 1998).

Carotenoids (pro-vitamin A)

Carotenoids are present in soybeans in a very low concentration (0.8–3.7 ppm), and the main forms are lutein and β-carotene. They are co-extracted with oil but are often removed or degraded by oil refining steps designed to remove the undesirable minor components that contribute to physical and chemical instability and undesirable color, such as degumming to remove PLs, neutralization to remove free fatty acids, bleaching to decompose lipid hydroperoxides, and deodorization to remove volatile oxidation products.

Lutein is the major carotenoid in common soybeans with a yellow seed coat, whereas soybeans with a green seed coat contain xanthophylls in addition to lutein.

Ceramide

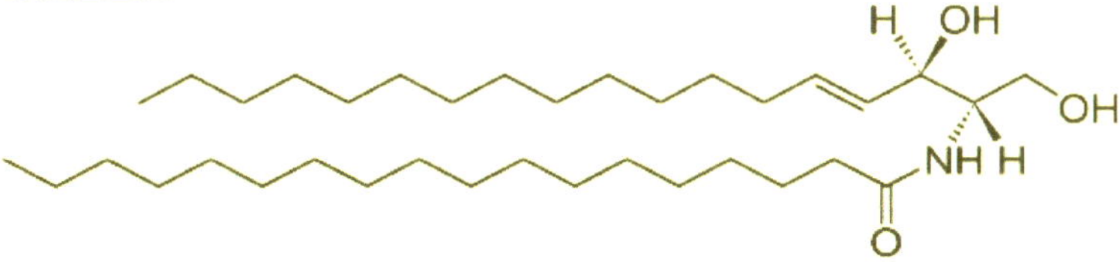

Cerebroside (Glucosylceramide)

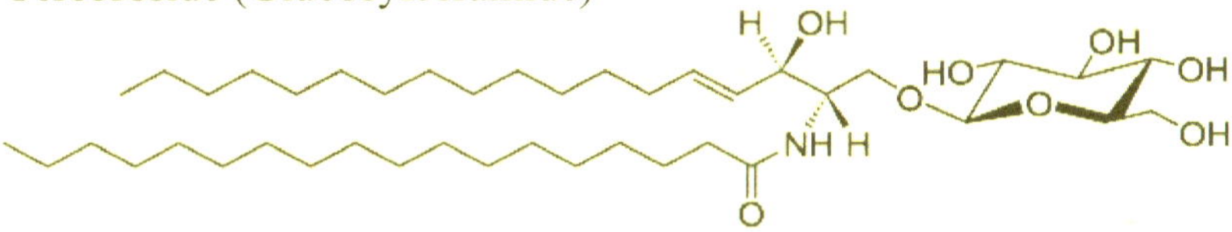

Cerebroside with hydroxyl fatty acid

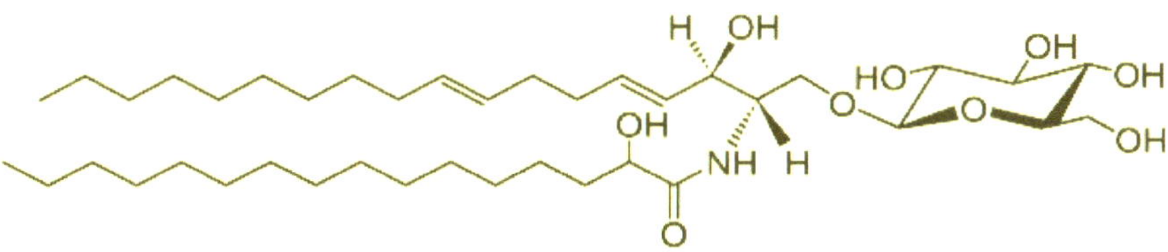

Fig. 10.4. General molecular structure of ceramide and cerebroside (glucosylceramide).

Carotenoid content in immature soybean was affected by genotype, with mean lutein contents ranging from 8.9 to 21.2 ppm and β-carotene from 2.9 to 4.9 ppm based on dried weight (Simonne et al., 2000). The amount of β-carotene decreased more rapidly than that of lutein and chlorophylls during seeds maturation. Mature soybean seeds contained little β-carotene. In commercial mature soybeans, lutein content range was 0.8–3.7 ppm in seed, and no detectable β-carotene was present (Kanamaru et al., 2006).

Protein Components

Trypsin inhibitors (TI)

TIs are protease-inhibiting factors that bind the protease enzymes to decrease their catalytic power. The two predominant protease inhibitors are Kunitz trypsin inhibitor and Bowman-Birk inhibitor (BBI), and they are both protein in nature.

Kunitz trypsin inhibitor has a molecular weight between 20 and 25 kDa. It consists of 181 amino acid residues and two disulfide bonds (Liu, 1999a). BBI is a protein with a molecular weight of 8 kDa, and with a single polypeptide chain of 71 amino acids with seven disulfide bonds. BBI is well-characterized for its ability to inhibit trypsin and chymotrypsin.

These protease inhibitors, if not inactivated, are responsible for growth depression for both animals and human by reducing the digestibility of dietary proteins and causing pancreatic hypertrophy (Chernick et al., 1948). They can be denatured, thus deactivated, by proper heating of the proteins or the seeds, such as with live steam, boiling in water, dry roasting, microwave radiation, and extrusion cooking. Chemicals, such as the reducing agents, cysteine, glutathione, and sodium sulfite, can inactivate TIs at relatively low temperatures (Liener, 1994). For example, treatment of raw soy flour at 75°C with 0.03M sodium sulfite for one hour can completely inactivate TIs, leaving no disulfide bonds in the protein. Feeding trials showed the mild treatment is more advantageous than heating treatment regarding nutritional improvement (Liu, 1999a).

The physiological roles of protease inhibitor are controversial, because medical research shows they have anticarcinogenic activities, and they are effective at extremely low concentrations. The soybean-derived BBI was particularly effective in suppressing carcinogenesis (Kennedy, 1998). Purified BBI and BBIC (BBI concentrate) have comparable suppressive effects on the carcinogenic process in a variety of in vivo and in vitro systems, and BBI appeared to be a universal cancer preventive agent (Kennedy, 1998). Purified BBI and BBIC suppressed carcinogenesis in three different animal species (mice, rats, and hamsters), and in organ systems, tissues, and cells of various types (Kennedy, 1998). They had no observed in vivo toxicity. BBIC now has the Investigational New Drug status from the FDA (in April 1992, IND no. 34671; sponsor, A.R. Kennedy).

Soybean BBI showed a significant and dose-dependent growth decrease of human colorectal adenocarcinoma HT29 cells in vitro (Lemente et al., 2005). The mechanism by which BBI suppresses carcinogenesis was studied by Chen et al. (2005). BBI specifically and potently inhibits the proteasomal chymotrypsin-like activity in vitro and in vivo in MCF7 breast cancer cells, leading to accumulation of ubiquitinated proteins. In addition, BBI suppressed cell growth and decreased the activities of phosphorylated extracellular signal-related kinases. Chen's results support a new mechanism of proteasome inhibition by BBI that prevents cancer development. For the first time, soybean TIs induced human leukemia Jurkat cell death, as measured by flow cytometry (Troncoso et al., 2007).

The effect of purified soybean Kunitz and BBI trypsin inhibitors as dietary supplements (5, 15, or 50 g/kg) on spontaneous pulmonary metastasis of lung carcinoma 3LL cells as well as human ovarian cancer HRA cells was investigated in mice (Kobayashi et al., 2004a). Only Kunitz inhibitor inhibited the formation of lung metastasis in a dose-dependent manner. These results suggest that dietary supplementation of Kunitz inhibitor could more efficiently regulate cell metastatic processes than BBI, and Kunitz inhibitor may also be beneficial for ovarian cancer patients by inhibiting phosphorylation of kinases. Kunitz but not BBI suppressed ovarian cancer cell invasion by blocking urokinase upregulation (Kobayashi et al., 2004b). Inagaki et al. (2005) also showed that Soybean Kunitz inhibitor inhibited signaling pathways in ovarian cancer cell growth.

Lectins

The occurrence of cell-agglutinating and sugar-specific proteins has been known for a long time. Very few lectins had been isolated, and they had attracted little attention until the early 1970s, when it was demonstrated that lectins were extremely useful tools for the study of cell malignant changes. More information is contained in two excellent historical reviews on lectins (Sharon & Lis, 2004; Sharon, 2007).

Lectins, also known as hemagglutinins, are proteins in nature and are composed of a 120 kDa tetrameric glycoprotein, possessing a single oligomannose chain per monomer. Lectins have a strong ability to agglutinate the red blood cells and intestinal mucosa cells by their strong affinity for cell surface carbohydrates. The destruction of the intestinal cell organization has a significant impact on nutrient absorption and utilization; therefore, they are considered as anti-nutritional factors. These proteins can be denatured by moist heat, as can TIs; thus, properly processed soy protein should have a low level of this anti-nutritional factor. However, lectins were reported to be resistant to digestion if not heat-denatured, could survive gut passage but bind to gastrointestinal cells and enter the circulation intact with full biological activity (De Mejia and Prisecaru, 2005). They are useful as cancer therapeutic agents because of their binding to cancer cell membranes, causing cytotoxicity, apoptosis, and inhibition of tumor growth.

Lectins can be mitogenic, non-mitogenic (potato lectin), and anti-mitogenic (tomato lectin). Mitogenic stimulation or mitogenecity is the ability to induce division in a mature quiescent cell that does not normally divide. Prolonged mitogenic activity results in cell proliferation as in cancer. The concentration of lectin seems to be one of the factors that determines proliferation (low concentration) or inhibition (high concentration). A recent review of the literature data concerning the biological activity of plant lectins is given by Abdullaev and de Mejia (1997), where a discussion on toxic, cytotoxic, antitumor, and anticarcinogenic properties of lectins is presented. A brief description of the biological properties of plant lectins, as well as the effect of plant lectins on normal and malignant cells and the antitumor properties of these lectins in vivo and in vitro, are included. These findings are interpreted, and possible mechanisms of the antitumor effect of plant lectins are discussed.

The study of effect of soybean lectin on intestinal morphology and lymphoid organ weight of poultry-fed diets containing soy lectin indicates that lectin up to 0.048% enhanced intestinal development by increasing villus crypt, but might alter the structural integrity of lymphoid organs (Fasina et al., 2006).

Lectins also affect the immune system by altering the production of various interleukins or by activating certain protein kinases. Lectins can bind to ribosomes and inhibit protein synthesis; can modify the cell cycle by inducing non-apoptotic mechanisms, cell cycle arrest and apoptosis; can activate the caspase cascade; and can also downregulate telomerase activity and inhibit angiogenesis (De Mejia & Prisecaru, 2005). The effect of purified soybean lectin on growth and immune function in rats fed diets containing 0, 0.05, 0.10, 0.15, or 0.20% lectin showed growth decline, and decline of the concentrations of interleukin-2, interferon-γ and tumor necrosis factor-α in plasma, spleen, and mesenteric lymph nodes, as well as plasma concentrations of IgA, IgG, and IgM; therefore, dietary soybean lectin has a negative effect on growth and immune function of rats. Although lectins seem to have great potential as anticancer agents, further research is still needed and should include a genomic and proteomic approach.

Lunasin

Lunasin, a naturally occurring peptide in soybeans classified as a 2*S* albumin, has 43 amino acids with a molecular weight of 4.7 kDa and contains nine Asp residues at its carboxyl end, an Arg-Gly-Asp cell adhesion motif, and a predicted helix with structural homology to a conserved region of chromatin-binding proteins.

The reduction in cancer risk resulting from soy protein consumption is attributed to the Bowman-Birk protease inhibitor and isoflavones, as well as naturally occurring lunasin peptide, which is cancer preventive (Jeong et al., 2007). Exogenous application of the lunasin peptide inhibited chemical carcinogen-induced transformation of fibroblast cells to cancerous foci in vitro by binding to the deacetylated histones and inhibiting its acetylation. In a mouse skin cancer model, dermal application of lunasin (250 μg/wk) decreased skin tumor incidence by approximately 70%, and delayed the appearance of tumors by 2 weeks relative to a control (Galvez et al., 2001). The results suggest lunasin can be a new chemopreventive agent that functions via a chromatin modification mechanism (Galvez et al., 2001). Lunasin was also shown to upregulate the genes involved in the control of tumor suppression, cell division, DNA repair, and cell death as measured with gene microarray analysis (Magbanua et al., 2004).

The anticancer potential of lunasin and soy hydrolyzates studied on leukemia cells (Wang & de Mejia, 2007) showed that lunasin-enriched soy flour caused cytotoxicity of leukemia cells. Simulated gastrointestinal hydrolysis of soy protein increased topoisomerase inhibitory activities and cytotoxicity. Such hydrolysates contain hydrophilic, small bioactive peptides (<3 kDa), and three novel topoisomerase inhibitory soy peptides that can be isolated.

Carbohydrate Components

Sucrose is present in a quantity of about 5.5% in soybean seeds. Oligosaccharides, i.e., raffinose and stachyose, are present at about 0.9 and 3.5% in the seed, respectively. They are soluble sugars with one or two galactose units linked by a α1-6 glycosidic bond to sucrose (Fig. 10.5). Soluble sugars are important for the flavor of certain soy foods, such as tofu. At physiological maturity, soybean has about 12% nonstructural carbohydrate on a dry-seed weight basis, of which, starch accounts for about 2%, and the other 10% is di- or oligosaccharides (sucrose, 41–68%; stachyose, 12–35%; and raffinose, 5–16%). The sugar content tends to be negatively correlated with protein content (Cui et al., 2004).

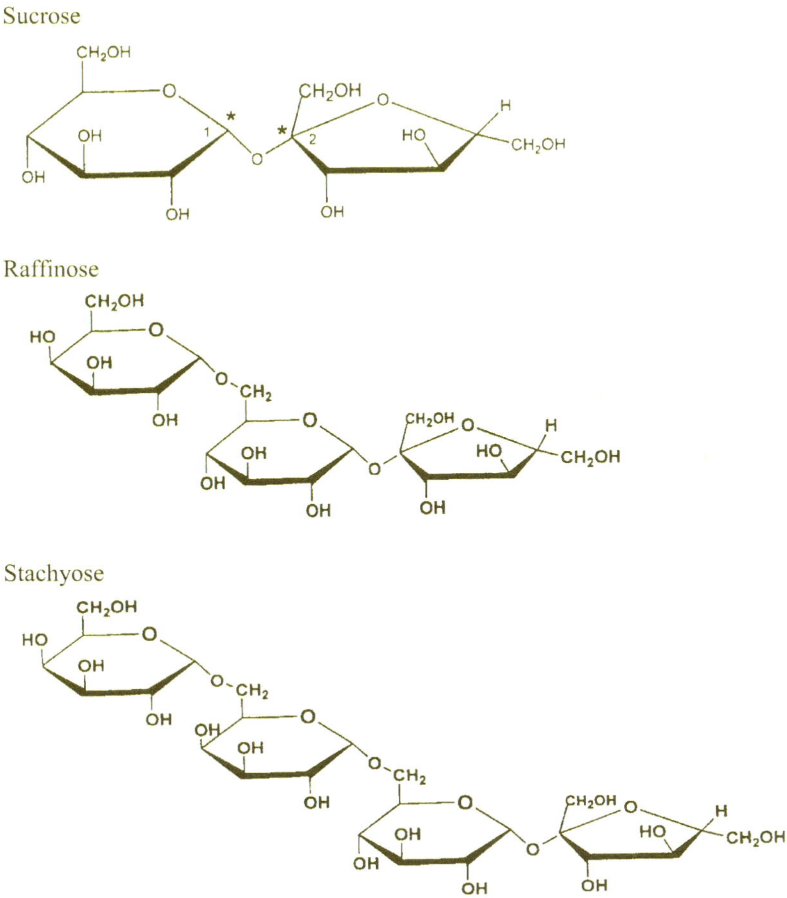

Fig. 10.5. Molecular structures of soluble sugars in soybean.

The raffinose and stachyose oligosaccharides are referred to as flatulence sugars because humans lack the enzyme (α-galactosidase) necessary to break down the molecule for metabolism. The intact sugars travel to the large intestine where the microflora is capable of utilizing them. Gases and acids are produced from microbiological action or fermentation to lead to bloating and diarrhea symptoms. This type of sugar is now used as a prebiotic to encourage the growth of the health-promoting microorganisms in the colon such as bifidobacteria (Bouhnik et al., 2007; Woodmansey, 2007).

Modification of soybeans to produce reduced levels of oligosaccharides for improved feed metabolizable energy is described in a later section.

Other Phytochemicals

Isoflavones and Total Phenolic Compounds

Flavonoids are a group of plant phenolic compounds having a carbon skeleton of C_6-C_3-C_6, with two aromatic rings linked together by a three carbon aliphatic chains, which normally is condensed to form a pyran. Isoflavones differ from the flavones in that the second aromatic ring is attached to position 3 instead of position 2 on the pyran ring. Isoflavones are one type of flavonoid, and the structures of the main soybean isoflavones are shown in Fig. 10.6.

Soybean is one of a few plants that contains high concentrations of isoflavones (Liu, 2004b). The three main types of isoflavones are daidzein, genistein, and glycitein. These are the free aglucone forms, and they can conjugate with glucose and its derivatives to form glycosides. The free and β-glucoside forms of daidzein and genistein, which are referred to as daizin and genistin, are the major soybean isoflavones. The structures of acetylated glucosides are also illustrated in Fig. 10.6. In addition, another glucoside conjugate form exists, which is malonylglucoside. Therefore, a total of 12 isoflavone isomers are in soybeans.

The concentration of isoflavones varies with variety and growing conditions and is reported as 1.2–2.5 mg/g in United States beans, 0.5–2.3 mg/g in Korean beans, and 0.2–3.5 mg/g in Japanese beans (Hammond et al., 2005). Wang and Murphy (1994) reported 12 isoflavones in eight American and three Japanese varieties, with total isoflavone concentrations ranging from 1.2 to 4.2 mg/g in American cultivars, and 1.3 to 2.3 mg/g in Japanese cultivars. Glucosides of genistein and daidzein account for about 90% of the total soybean isoflavones.

Many of the health benefits of consuming soybeans are attributed to its isoflavones. These compounds have estrogen-like activities and are believed to be beneficial for menopausal women (Messina & Hughes, 2003). Isoflavones also reportedly reduce the risks of coronary heart disease by reducing the degree of oxidation of cholesterol and reducing LDL cholesterol accumulation on the wall of blood vessels, thus enhancing arterial relaxation (Nestel, 2003). Isoflavones play a role in preventing cer-

Aglycones

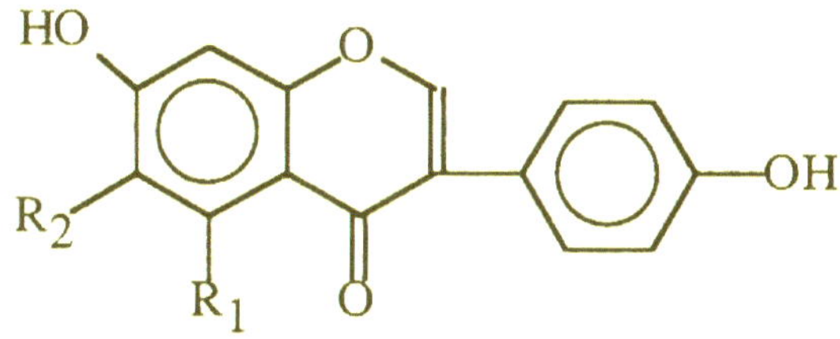

	R_1	R_2
daidzein	H	H
genistein	OH	H
glycitein	H	OCH_3

Glucosides

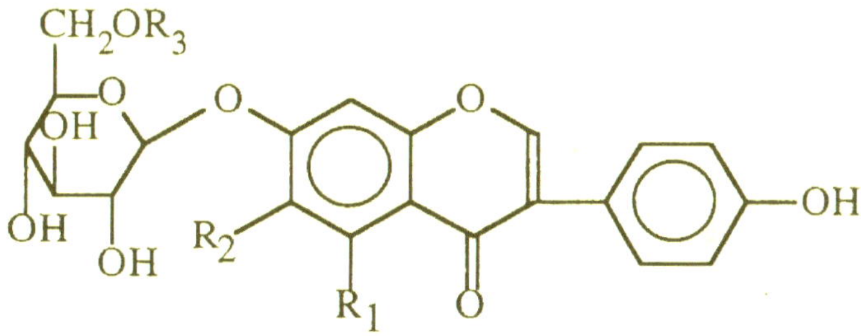

	R_1	R_2	R_3
daidzin	H	H	H
genistin	OH	H	H
glycitin	H	OCH_3	H
6″-*O*-Acetyldaidzin	H	H	$COCH_3$
6″-*O*-Acetylgenistin	OH	H	$COCH_3$
6″-*O*-Acetylglycitin	H	OCH_3	$COCH_3$

Fig. 10.6. Structures of soybean isoflavones (White & Xing, 1997).

tain cancers, such as breast cancer (Yan & Spitznagel, 2004; Hirose et al., 2005), and may reduce mortality of prostate cancer by preventing the latent cancer to progress into larger tumors (Griffiths, 2000). The American Cancer Society recommends men eat soyfoods to reduce their risk of prostate cancer. The bone loss following the onset of menopause may be alleviated by isoflavone intake (Cotter & Cashman, 2003).

Other phenolic compounds are in soybeans, mainly acids, such as chlorogenic, isochlorogenic, caffeic, ferulic, *p*-coumaric, syringic, vanillic, *p*-hydroxybenoic, salicylic, and sinapic acids. Some of these acids have strong antioxidant activity (White & Xing, 1997).

Saponins

Saponins are glycosylated alkaloids, steroid, or triterpenes. They are in low concentrations in soybeans, 0.1–0.3% based on protein content, and legumes are the major source of saponins in the human diet (Lin & Wang, 2004).

Saponins are amphiphilic in nature because of their hydrophilic sugar groups (galactose, arabinose, rhamnose, glucose, glucuronic acid, and fructose) and the hydrophobic aglycones, referred to as sapogenin. Therefore, they are excellent foaming and emulsifying agents. Five sapogenins were identified in soybeans, and their structures are illustrated in Fig. 10.7. Three groups of soyasaponins are classified: group A, B, and E. Group A saponins have soyasapogenol A as the aglycone and have two sugar chains attached on carbon 3 and 22 (Fig. 10.8). The hydroxyl groups on the sugar may be acetylated. Group B and E saponins have soyasapogenol B and E as the aglycones. Group B saponins, the main group of soy saponins, contain one sugar chain on carbon 3. Group E saponins have a different structural attachment on carbon 22 than that of Group B (Lin & Wang, 2004).

Studies on total content and composition of soy saponins are comprehensively reviewed by Lin and Wang (2004). Saponin composition was not affected by year of cultivation but was affected by soybean seed variety. Seed maturity stage was the most influential factor on content and composition of saponins. In general, saponin content decreases with maturity.

Hu et al. (2002) developed HPLC methods for quantitative determination of the group B soyasaponins. Saponin concentrations in 46 soybean varieties ranged from 2.5 to 5.9 µmol/g. In soy ingredients (soybean flour, toasted soy hypocotyls, soy protein isolates, textured vegetable protein, soy protein concentrates, and Novasoy) and soy foods (commercial soy milk, tofu, and tempeh), the group B soyasaponins ranged from 0.2 to 114 µmol/g (Table 10.5). Soy milk, tempeh, and tofu were low in soyasaponin content compared to the raw soybeans on an "as is" weight basis. However, the soyasaponin concentrations on a dry weight basis were 3.8 µmol/g in tempeh, 4.5 µmol/g in tofu, and 5.1 µmol/g in soy milk, which were greater than that in soy flour (3.3 µmol/g, as-is basis). No apparent correlation existed between isoflavone and soyasaponin concentrations in the soy products examined.

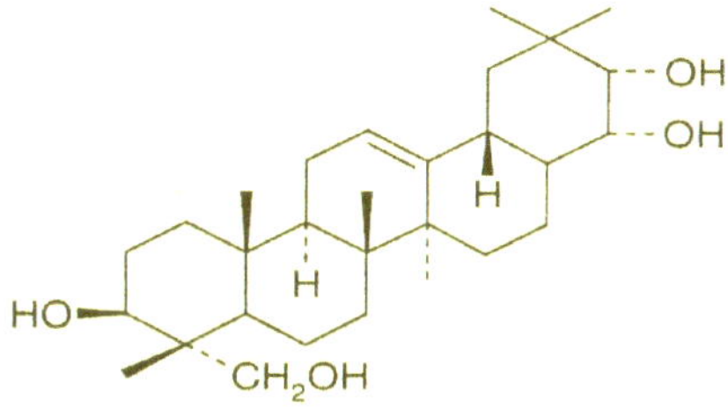

Soyasapogenol A

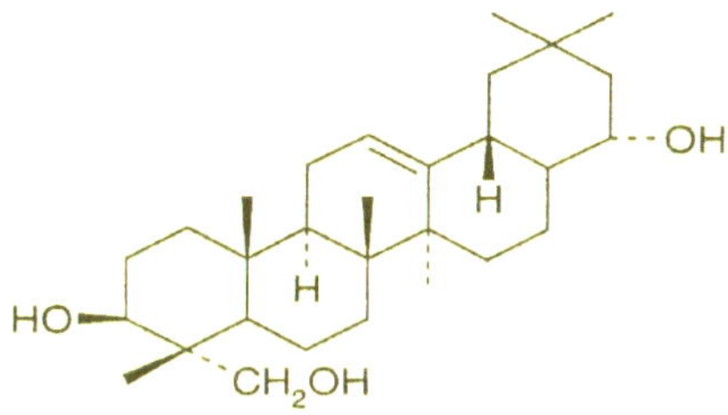

Soyasapogenol B

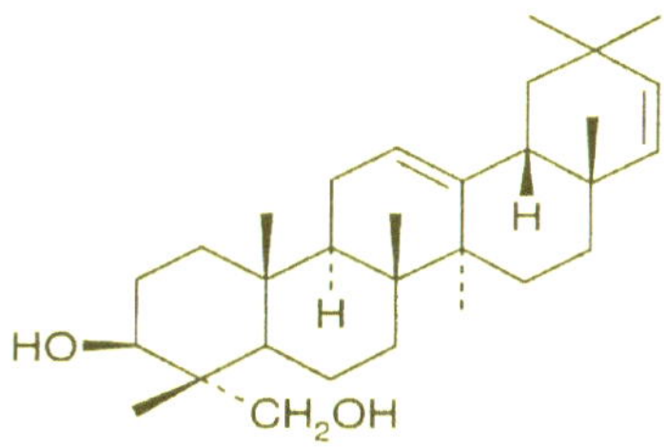

Soyasapogenol C

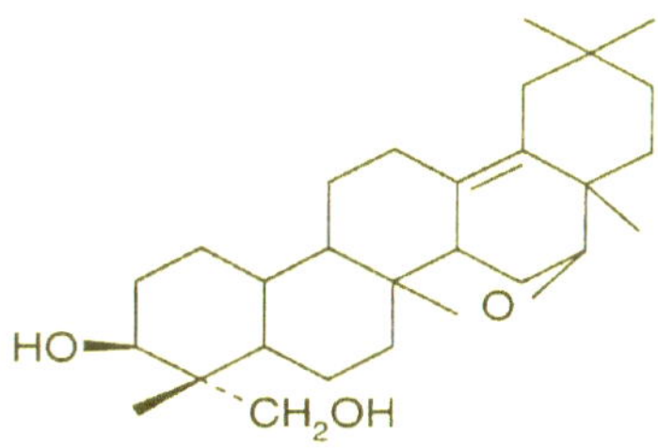

Soyasapogenol D

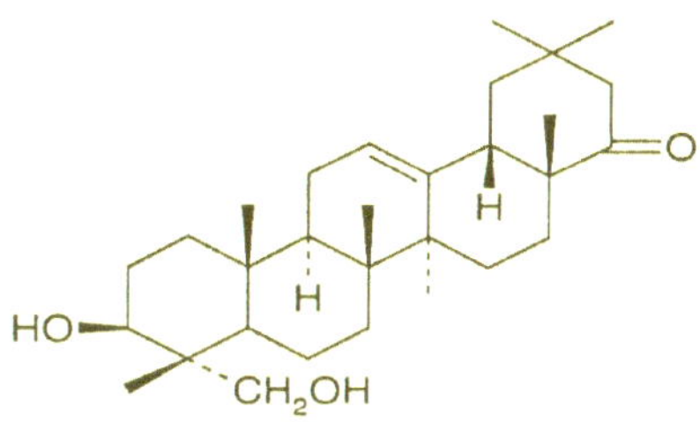

Soyasapogenol E

Fig. 10.7. Structures of soybean saponin aglycones (sapogenin) (Lin & Wang, 2004).

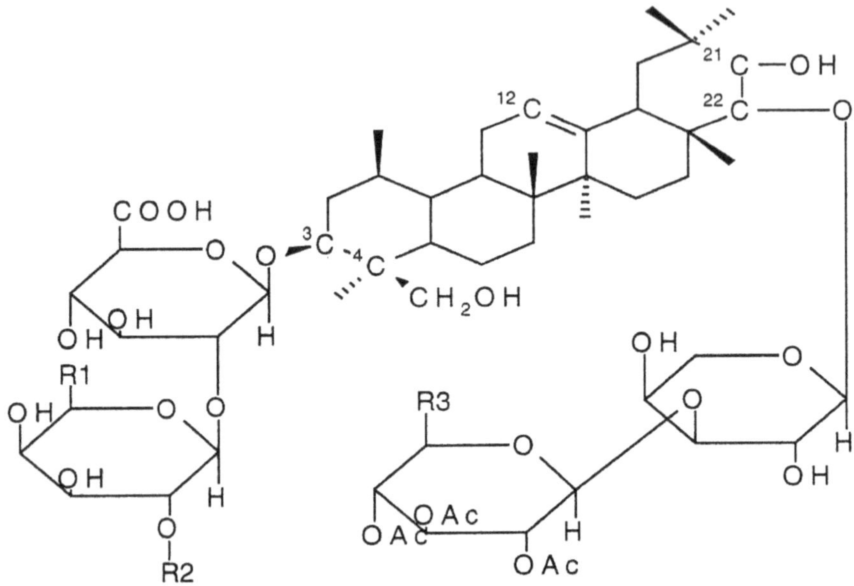

	R_1	R_2	R_3
Soyasaponin Aa (A4)	CH_2OH	β-D-Glc	H
Soyasaponin Ab (A1)	CH_2OH	β-D-Glc	CH_2Oac
Soyasaponin Ac	CH_2OH	α-L-Rha	CH_2Oac
Soyasaponin Ad	H	β-D-Glc	CH_2Oac
Soyasaponin Ae (A5)	CH_2OH	H	H
Soyasaponin Af (A2)	CH_2OH	H	CH_2Oac
Soyasaponin Ag (A6)	H	H	H
Soyasaponin Ah (A3)	H	H	CH_2Oac

Fig. 10.8. Structures of group A saponins (Lin & Wang, 2004).

Table 10.5. Molecular Weight of Soybean Minor and Bioactive Proteins

Protein	Molecular Weight, kDa
Kunitz trypsin inhibitor	20–25
Bowman-Birk inhibitor	8
Lectins	120
Lunasin	4.7

Saponins are historically considered as anti-nutritional factors, but they are recently regarded as functional food ingredients because of their cholesterol-lowering (Shibayama, 2003; Ueda et al., 1996), cancer-preventative (Jun, 2002; Oh & Sung, 2001), immune-modulating (Kang et al., 2005), and antioxidative properties (Rodrigues et al., 2005). The anti-nutritional properties are because of the hemolytic activity of these compounds, causing lysis of erythrocytes in vitro. Soy saponins did not impair the growth of chicks, rats, and mice; however, they caused slight growth retardation of *Tribolium castaneum* larvae, and were harmful to tadpoles and guppies (Lin & Wang, 2004).

Saponins and foods rich in saponins reduced plasma cholesterols in animal models by increasing the excretion of bile acids and neutral sterols in the feces, as reviewed by Lin and Wang (2004). Saponins are cytotoxic and inhibitory on the growth of tumor cells in cell culture studies. Dietary soy saponins reduced the incidence of aberrant crypt foci (ACF) in the colon of mice and, at 150–600 ppm concentration, there was a dose-dependent growth inhibition effect on human carcinoma cells (Koratkar & Rao, 1997). Soy saponins have antiviral activities for many types of viruses (Lin & Wang, 2004), and they also have inhibitory effect on HIV infection (Nakashima et al., 1989). Soybean saponins, in a dose-dependent manner, suppressed the release of proinflammatory mediators, which play a critical role in tumor development (Kang et al., 2005). Therefore, soy saponins may be useful for ameliorating inflammatory diseases as well as suppressing tumor progression.

Phytate

Phytate is the calcium, magnesium or potassium salt of phytic acid, which is inositol hexaphosphoric acid (Fig. 10.9). More than half of the total phosphorus in soybeans is in the form of phytic acid (Liu, 2004a). Because of its chelating power, phytic acid makes many essential minerals in soybeans or in diets unavailable for absorption and utilization for both human and domestic animals; thus phytic acid is known as an anti-nutritional factor.

Phytic acid or phytate can also bind with protein at the extreme pH ranges. At acidic conditions (pH below pI of soy protein), the positively charged protein binds strongly with phytic acid, and under alkaline conditions the phytate's positively charged mineral binds the negatively charged protein. These complexes make the protein less accessible by the digestive enzymes, thus affecting protein quality.

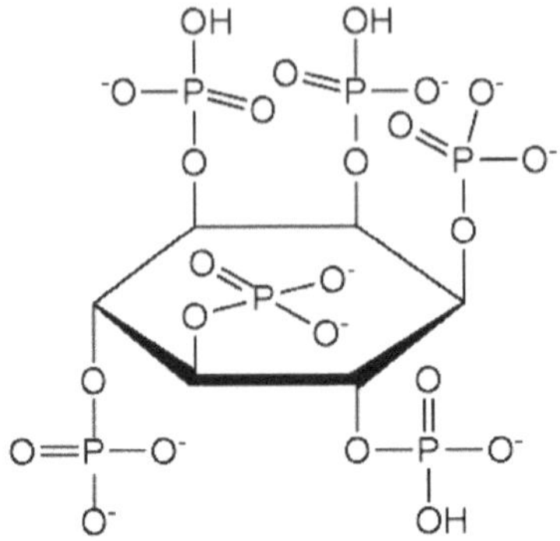

Fig. 10.9. Structure of phytic acid.

The influence of dietary phytate on hepatic activities of lipogenic and drug-metabolizing enzymes was examined in male Wistar rats, and it was found that phytate diminished the increases in hepatic lipids and activities of lipogenic enzymes induced by 1,1,1-trichloro-2,2-bis(*p*-chlorophenyl) ethane (DDT) (Okazaki et al., 2003). Therefore, dietary phytate may protect humans from accumulating hepatic lipids by depressing hepatic lipogenesis. It may also improve the function of the liver drug-metabolizing enzyme system. Dietary phytate can also protect sucrose-fed animals against an accumulation of hepatic lipids, as shown in a study on growing rats to compare the effects of dietary phytate on metabolism of hepatic lipids (Katayama, 1997). This study showed that dietary phytate significantly depressed the rises in liver weight and hepatic concentration of total lipids and triacylglycerols.

Phytic acid can chelate transition metal ions and inactivate its pro-oxidant effect. Consumption of diets rich in phytic acid may protect intestinal epithelial cells against iron-induced oxidative damage, as shown in some rat models (Miyamoto et al., 2002). Oral administration of phytic acid also protected large intestinal mucosa against iron-induced lipid peroxidation in male rats. Dietary phytic acid lowers the incidence of colonic cancer and protects against inflammatory bowel diseases (Graf & Eaton, 1990). During digestion, phytic acid is partially dephosphorylated; therefore, its antioxidant properties may decrease as a result of the reduction in chelating power.

Water-soluble Vitamins and Minerals

Soybean contains thiamine, riboflavin, niacin, pantothenic acid, and folic acid. An HPLC method developed to determine thiamin and riboflavin in soy products (Fernando & Murphy, 1990) showed vitamin contents of soy products were less than those reported in the literature for which AOAC methods had been employed. The thiamin and riboflavin contents ranged from 6.3 to 6.9 and 0.9 to 1.1 μg/g in three soybean varieties (Fernando & Murphy, 1990). Processing soybeans into tofu leads to

Table 10.6. Soyasaponin Contents and Compositions an Commercial Soy Products (Hu et al., 2002)

Product	Total Group B Soyasaponin Content[a] (μmol/g)
Soybean flour[b]	3.31
Tofu[c]	0.59
Tempeh[d]	1.53
Soy milk[f]	0.47
Acid-washed soy concentrates[g]	9.41
Ethanol-washed soy concentrates[g]	0.20
Isolated soy protein 500E[h]	10.60
Isolated soy protein Supro 670[h]	9.51
Textured vegetable protein[g]	4.51
Soy hypocotyl[i]	27.46
Novasoy[g]	114.02

[a] Mean value of duplicated analyses. Saponin contents are reported on an "as is" weight basis.
[b] Vinton 81, 1994 crop.
[c] Mori-nu, firm.
[d] Quong Hop and Co.
[f] White Wave, Inc.
[g] Archer Daniels Midland Co.
[h] Protein Technologies International.
[i] Schouten USA Inc., toasted.

retention of thiamin and riboflavin of 7.6–15.7% and 11.7–21.1%, respectively.

Soybeans have an ash content of about 5%, and potassium, phosphorus, magnesium, sulfur, calcium, chloride, and sodium are present in 0.2–2.1% (Liu, 1999a). There are other trace minerals as well.

Compositional Changes during Seed Maturation and Processing

Effect of Seed Development on Content of Minor Components in Soybeans

Soybean seeds of three cultivars (IA1008, IA1010, and IA1014) were harvested at 5-day intervals from 28 days after flowering (DAF) to 68 DAF (mature seed; Wang et al., 2006a). Sphingolipid (SL) and phospholipid (PL) concentrations decreased significantly during seed development (Table 10.7). Averaged across cultivars, ceramide content on a dry-weight basis decreased from 51.4 nmol/g at 28 DAF to 22.2 nmol/g at 68 DAF, whereas cerebroside content decreased from 522.8 nmol/g at 28 DAF to 135.8 nmol/g at 68 DAF. PL percentage of the total lipid decreased from 9.1% at 28 DAF to 3.5% at 68 DAF.

Table 10.7. Lipid Compositions of Developing Soybean Seeds (Wang et al., 2006a)

	Sphingolipid (SL)				Phospholipid (PL)		
DAF	Cer nmol/g	*GlcCer nmol/g*	*Cer mol% of GlcCer*	*SL,mol% in the polar lipid*[a]	%[b]	*μmol/g*	Neutral oil[b,c] %
28	51.4	522.8	9.8	2.8	9.1	18.7	81.5
33	43.4	511.5	8.5	2.7	7.8	18.1	86.0
38	44.5	428.0	10.4	2.5	6.3	16.1	87.2
43	33.3	330.0	10.1	1.9	6.4	16.7	88.3
48	32.6	286.9	11.4	1.9	5.8	14.6	89.8
53	27.0	199.5	13.5	1.7	5.1	12.1	90.2
58	20.7	163.1	12.7	1.7	4.0	10.2	91.3
63	22.1	146.7	15.1	1.7	3.8	9.0	92.3
68	22.2	135.8	16.3	1.4	3.5	10.4	92.9
$LSD_{0.05}$	14.0	86.2	5.1	0.6	1.8	4.9	2.2

[a] Polar lipids is the sum of GlcCer (glucosylceramide), Cer (ceramide), SG (steryl glucoside), ESG (esterified steryl glucoside), and PL.
[b] Percentage of PL and neutral oil calculated based on total lipid extract.
[c] Neutral oil was obtained by silica column fractionation and may include triacylglycerols, free fatty acids, mono- and diacylglycerols, tocopherols, and pigments.

Kim et al. (2006) reported changes in soybean composition, such as protein, lipid, free sugars, isoflavones, and saponins during soybean development and maturation in two Korean soybean cultivars. As soybean seed matured, total soy saponin concentration constantly decreased. The ratio of total isoflavone to total soyasaponin in the developing soybean increased from 0.06 to 1.31. Total soy saponin content was negatively correlated with isoflavone content.

During maturation, β-carotene content decreased, reaching its lowest level at maturity. The immature seeds contained 0.46 mg/100 g fresh weight β-carotene, the mature and soaked seeds had 0.12 mg/100 g soaked weight (Bates & Matthews, 1975).

Phytate concentration increased during seed development. Changes in trypsin inhibitors were somewhat controversial, and a general slight increase with seed maturation was demonstrated. Total isoflavone content also generally increased with seed development (Liu, 1999b).

Effect of Processing on Content of Minor Components in Soybeans

The effect of processing on minor lipid components is shown in Table 10.8. Guiterrez and Wang (2004) reported the effect of processing on the sphingolipid content of

Table 10.8. Effect of Processing on Contents of Tocopherols, Sterols, and Squalene in Soybean Oil[1]

	Tocopherols		Sterols		Squalene	
Processing Step	ppm	% Loss	ppm	% Loss	ppm	% Loss
Crude	1132	–	3870	–	143	–
Degummed	1116	1.4	3730	3.6	142	0.7
Neutralized	997	11.9	3010	22.2	140	2.1
Bleached	863	23.8	3050	21.2	137	4.2
Deodorized	726	35.9	2620	32.3	89	37.8

[1] Ramamurthi et al., 1998.

various soybean products. Glucosylceramide (GlcCer), the major sphingolipid type in soybeans, was measured in several processed soybean products to determine partitioning and loss during processing. Whole soybean was processed into full-fat flakes, from which crude oil was extracted. Crude oil was refined by conventional methods, and defatted soy flakes were further processed into alcohol-washed and acid-washed soy protein concentrates (SPC) and soy protein isolate (SPI) by laboratory-scale methods that simulate industrial practices. GlcCer was isolated from the samples by solvent extraction, solvent partition, and TLC, and quantified by HPLC. As shown in Table 10.9, GlcCer mostly remained with the defatted soy flakes (91%) rather than with the oil (9%) after oil extraction. Only 52, 42, and 26% of GlcCer from defatted soy flakes was recovered in the acid-washed SPC, alcohol-washed SPC, and SPI products, respectively. The minor quantity of GlcCer in the crude oil was almost completely removed by water degumming.

Table 10.9. Cerebroside (Glccer) Contents in Soybean Products (Guiterrez & Wang, 2004b)

Soy Product	GlcCer, nmol/g (dry wt basis)	GlcCer, ppm (dry wt basis)
Full-fat soy flakes	268.2	192.5
Defatted soy flakes	311.2	223.3
Soy protein concentrate (SPC, acid washed)	264.4	189.1
SPC (alcohol washed)	216.5	155.3
Soy protein isolate (SPI)	296.9	213.9
Crude oil	Not detected (ND)	ND
Gum	1678.9	1202.8
Soapstock	ND	ND
Alkaline refined oil	ND	ND
MSD[a]	113.4	78.4

[a] MSD = minimum significant differences between means in each column determined by Tukey Kramer's mean comparison ($P \leq 0.05$).

The effect of soybean processing on the distribution of isoflavones was reported by Wang and Murphy (1996). Soybeans (600 g) were used for tofu processing (Table 10.10), and finely ground soybean flour (50 g) was used for soy isolate production (Table 10.11). Isoflavone distribution measured in pilot-plant soymilk and tofu preparation showed no significant loss of isoflavones in soymilk production. However, tofu contained only 33% (based on dry matter) of the isoflavones in the starting raw soybeans. Isoflavone distribution during soy protein isolate preparation indicated a significant loss of 53% of total isoflavone contents in the processing steps between the raw material and the protein isolate (Table 10.11). The alkaline extraction step was the major step for isoflavone loss. In protein isolate processing, alkaline extraction causes the generation of daidzein and genistein, which is attributed to alkaline hydrolysis of the glucosides.

Composition Modification through Plant Breeding and Genetic Engineering

Genetic modification has shown to be an effective means to alter tocopherol contents of soybeans (Mounts et al, 1996; Almonor et al., 1998). The most abundant γ-tocopherol is positively correlated with the most unsaturated linolenic acid content. Therefore, soybean lines with reduced linolenate currently developed to replace some hydrogenated oils containing *trans*-fatty acids will have lower γ-tocopherol contents. However, these lines tend to have higher α-tocopherol contents.

A study by McCord et al. (2004) showed a relationship between reduced palmitate or reduced linolenate and tocopherol content in soybeans. A total of 41 soybean cultivars and lines with palmitate contents ranging from 3.7 to 12.4% and linolenate contents ranging from 1.2 to 8.3% were compared for tocopherol content. Lines with reduced palmitate had significantly greater mean total tocopherol contents than did lines with normal fatty ester contents. No significant difference in total tocopherol was observed between normal and low-linolenate lines containing 1.0 or 2.5% linolenate. In a further study, lines with 1% linolenate and lines with 7% linolenate were compared for tocopherol content. The mean total tocopherol of the 7% linolenate lines was significantly greater than that of the 1% linolenate lines in the three populations tested. However, an overlap in ranges of total tocopherol between the 1% and 7% linolenate lines indicates that it should be possible to develop 1% linolenate cultivars with acceptable contents of individual and total tocopherols compared with normal cultivars.

The relationship between reduced palmitate and high tocopherols was further studied in soybeans having similar genetic backgrounds (Scherder et al., 2006). The mean total tocopherols of the reduced palmitate lines was 15% greater than the normal palmitate lines, and the line with the greatest total tocopherols in each population had reduced palmitate.

Table 10.10. Fraction Weight, Moisture and Isoflavone Contents in Soymilk and Tofu Processing (Wang & Murphy, 1996)[1]

Step	Weight (g)	Moisture (%)	Total Daidzein (mg)	Total Genistein (mg)	*Total Glycitein (mg)*	*Total*
Raw soybeans	600	11.03 ± 0.07	59.3 ± 20.0a	124.0 ± 35.0a	33.9 ± 7.7ab	217.2 ± 63.0a
Soaked soybeans	1297	60.95 ± 0.48	54.2 ± 31.0ab	114.7 ± 47.0a	27.8 ± 4.1b	196.8 ± 82.0a
Soaking water	1063	99.67 ± 0.08	0.5 ± 0.3d	0.3 ± 0.0c	0.2 ± 0.0d	1.0 ± 0.3c
Cooked slurry	5976	91.85 ± 0.09	67.9 ± 17.0a	121.6 ± 22.0a	37.7 ± 2.7a	227.2 ± 41.0a
Soymilk	5581	93.93 ± 0.09	63.6 ± 12.0a	103.7 ± 16.0a	27.6 ± 1.9b	194.8 ± 30.0a
Okara	717	79.08 ± 0.67	4.1 ± 0.8d	14.0 ± 0.6bc	7.8 ± 2.1c	25.9 ± 0.7bc
Tofu	1390	82.11 ± 2.21	16.0 ± 1.5cd	40.2 ± 3.4b	14.6 ± 2.9c	70.8 ± 4.7b
Whey	5140	98.37 ± 0.13	35.9 ± 1.3bc	50.4 ± 1.8b	9.1 ± 0.1c	95.4 ± 3.2b

[1]Values represent the mean ± standard deviation; $n = 3$. Values in a column with different letters were significantly different ($P < 0.05$).

Table 10.11. Fraction Weights and Isoflavone Contents in Soy Protein Isolate Processing (Wang & Murphy, 1996)[1]

Step	Weight (g)	Total Daidzein (mg)	Total Genistein (mg)	Total Glycitein (mg)	Total (mg)
Soybean flour	50	9.1 ± 0.3a	18.4 ± 1.2a	2.4 ± 0.5a	30.0 ± 1.7a
Defatted flour	43	10.0 ± 0.6a	20.0 ± 2.4a	1.7 ± 0.4b	31.7 ± 3.2a
Oil	8	0.7 ± 0.6c	0.0 ± 0.0c	0.0 ± 0.0d	0.7 ± 0.6c
Alkaline soluble	949	7.8 ± 2.6c	8.4 ± 2.6b	1.7 ± 0.1b	17.8 ± 5.3b
Alkaline insoluble	76	6.0 ± 1.8b	9.0 ± 2.9b	0.8 ± 0.2c	15.8 ± 4.9b
Protein isolate	9	6.2 ± 0.6b	7.8 ± 1.0b	0.5 ± 0.1c	14.5 ± 1.5b
Whey	865	1.5 ± 0.1c	0.9 ± 0.1c	0.9 ± 0.1c	3.3 ± 0.1c

[1]Values represent the mean ± standard deviation; $n = 3$. Values in a column with different letters were significantly different ($P < 0.05$).

Isoflavone content in response to genetic modification also was reported. Although little is known about the genetic regulation of isoflavone synthesis, several pathways were studied and the relationship between protein and fatty acid composition and isoflavone content was observed (Tsukamoto et al., 1995). For example, isoflavone content was negatively correlated with linolenic acid content, and also with protein content.

Phospholipid (PL) fatty acid composition and stereospecific distribution of 25 genetically modified soybean lines having a wide range of compositions were determined by GC and phospholipase A_2 hydrolysis (Wang et al., 1997). PL class proportions were affected by changes in overall fatty acid composition. PL fatty acid composition was changed with oil fatty acid modification, especially for palmitate, stearate, and linolenate.

A review by Liu (1999b) summarizes the efforts of plant breeding to eliminate or reduce trypsin inhibitors, oligosaccharides (Table 10.12), and phytate. There is also interest in increasing isoflavone content in soybeans.

Raffinose saccharides are a group of D-galactose-containing oligosaccharides of sucrose that are widely distributed in plants, particularly in legumes. The number of galactose units ranges from 0 to 4, and they are known as *sucrose, raffinose, stachyose, verbascose,* and *ajugose*, respectively. Synthesis of the raffinose saccharide family from sucrose is thought to be catalyzed by distinct galactosyltransferases (Kerr & Sebastian, 2000). Reportedly, removal of raffinose saccharides from soybean meal results in an increase in the metabolizable energy for broilers. Low oligosaccharide soybean meals with raffinose and stachyose levels of 0.08 and 0.42%, compared to 0.58 and 3.23% in regular meal, had 7% higher energy value (Parsons et al., 2000).

Efforts have been made to identify soybean germplasm that may contain genes giving a low seed oligosaccharide content. Low oligosaccharide content is typically related to high sucrose content in soybeans, and a sucrose content range of 6.1% to 12.4% and genomic regions associated with sucrose synthesis were identified with

Table 10.12. Modification of Soybean Soluble Sugars by Traditional Plant Breeding, % Dry Weight Basis[1]

Lines	Sucrose	Raffinose	Stachyose
Normal	5.1	1.0	4.7
Modified 1	6.0	0.4	1.3
Modified 2	7.0	0.1	0.5
Modified 3	11.5	0.1	0.0

[1]Adopted from Liu, 1999c.

molecular markers (Maughan et al., 2000). A cDNA coding for a variant raffinose synthase from a Japanese soybean was found, and low raffinose content plants were obtained by Watanabe and Oeda (2001). However, surveys of the soybean germplasm suggested instability of the low oligosaccharide phenotype. Therefore, the practical means to achieve a desirable oligosaccharide content was to physically and/or chemically treat the soybean to cause mutation, and patents by Kerr and Sebastian (1998, 2000, and 2003) described soybeans with stachyose content as low as 1% in the seeds.

In summary, soybeans contain many bioactive minor components that are lipids, proteins, or carbohydrate in nature, or that are low-molecular weight phenolics or saponins, etcetera. Although much research has been conducted on isolated components, and many health benefits have been demonstrated, the interactions, synergism, or antagonism among various minor compounds and interactions between the minor and major components in soybeans are not yet fully understood. Extracting and concentrating these minor compounds may provide a convenient source for expected beneficial effects; however, the cost associated with the processing may be prohibitive. Increasing the consumption of whole soy foods may allow us to gain not only the health, but also the economic, benefits of this relatively inexpensive food.

References

Abdullaev, F. I.; E.G..de Mejia. Antitumor effect of plant lectins. *Nat. Toxins* **1997,** *5,* 157–163.

Almonor, G.O.; G.P. Fenner; R.F. Wilson. Temperature effect on tocopherol composition in soybeans with genetically improved oil quality. *J. Am. Oil Chem. Soc.* **1998,** *75,* 591–596.

Anderson, R.L.; W.J. Wolf. Compositional changes in trypsin inhibitors, phytic acid, saponins, and isoflavones related to soybean processing, *J. Nutr.* **1995,** *125,* 581S–588S.

Arditi, T.; T. Meredith; P. Flowerman. Renewed interest in soy isoflavones and saponins, *Cereal Foods World* **2000,** *45,* 414–417.

Bates, R.P.; R.F. Matthews. Ascorbic acid and β-carotene in soybeans as influenced by maturity, sprouting, processing and storage. *Proc. Fl. State Hort. Soc.* **1975,** *88,* 266–271.

Birt, D.A.; A.H. Merrill; T. Barnett; B. Enkvetchakul; P.M. Pour; D.C. Liotta; V. Geisler; D.S. Menaldino; J. Schwartzbauer. Inhibition of skin carcinomas but not papillomas by sphingosine, N-methylsphingosine, and N-acetylsphingosine, *Nutr. Cancer* **1998,** *31,* 119–126.

Bouhnik, Y.; L. Raskine; K. Champion; C. Andrieux; S. Penven; H. Jacobs; G. Simoneau. Prolonged administration of low-dose inulin stimulates the growth of bifidobacteria in humans. *Nutr. Res.*, **2007,** *27,* 187–193.

Chen, Y.; S. Huang; S. Lin-Shiau; J. Lin. Bowman-Birk inhibitor abates proteasome function and suppresses the proliferation of MCF7 breast cancer cells through accumulation of MAP kinase phosphatase-1. *Carcinogenesis,* **2005,** *26,* 1296–1306.

Chernick, S.S.; S.S. Lepkovsky; I.L. Chaikoff. A dietary factor regulating the enzyme content of the pancreas: Changes induced in size and proteolytic activity of the chick pancreas by the ingestion of raw soybean meal. *Am. J. Physiol.* **1948,** *155,* 33–41.

Cotter, A.; K.D. Cashman. Genistein appears to prevent early postmenopausal bone loss as effectively as hormone replacement therapy. *Nutr. Rev.* **2003,** *61,* 346–351.

Cui, Z.; A.T. James; S. Miyazaki; R.F. Wilson; T.E. Carter, Jr. Breeding Specialty Soybeans for Traditional and New Soyfoods, *Soybeans as Functional Foods and Ingredients.* K. Liu, Ed.; AOCS Press: Champaign, IL, 2004: pp. 264–322.

Da Costa, K.; C.E. Gaffney; L.M. Fischer; S.H. Zeisel. Choline deficiency in mice and humans is associated with increased plasma homocysteine concentration after a methionine load. *Am. J. Clin. Nutr.*, **2005,** *81,* 440–444.

De Mejia, E.G.; V.I. Prisecaru. Lectins as Bioactive plant proteins: A potential in cancer treatment. *Crit. Rev. Food Sci. Nutr.*, **2005,** *45,* 425–445.

De Mejia, E.G.; W. Wang; M. Vasconez-Costa; R. Nelson; B.O. de Lumen. Physiologically active peptides in soybean and soy products, *Proceedings of the VII World Soybean Research Conference and IV International Soybean Processing and Utilization Conference*, Foz do Iguassu, Brazil, Feb. 29–Mar. 5, 2004; pp. 775–779.

Dillehay, D.L; S.K. Webb; E.M. Schmelz; A.H. Merrill. Dietary sphingomyelin ***inhibits*** 1,2-dimethylhydrazine-induced colon cancer in CF1 mice, *J. Nutr.* **1994,** *124,* 614–620.

Fasina, Y.O.; H.L. Classen; J.D. Garlich; B.L. Black; P.R. Ferket; Z. Uni; A.A. Olkowski. Response of Turkey poults to soybean lectin levels typically encountered in commercial diets. 2. Effect on intestinal development and lymphoid organs. *Poultry Sci.* **2006,** *85,* 870–877.

Federal Register, published September 8, 2000, http://www.fda.gov/ohrms/dockets/98fr/090800d.pdf.

Fernando, S.M.; P.A. Murphy. HPLC determination of thiamin and riboflavin in soybeans and tofu *J. Agric. Food Chem.* **1990,** *38,*163–167.

Galvez, A.F.; N. Chen; J. Macasieb; B.O. De Lumen. Chemopreventive property of a soybean peptide (Lunasin) that binds to deacetylated histones and inhibits acetylation. *Cancer Res.*, **2001,** *61,* 7473–7478.

Graf, E.; J.W. Eaton. Antioxidant function of phytic acid, *Free Radical Biol. Med.* **1990,** *8,* 61–69.

Griffiths, K. Estrogens and prostatic disease, *Prostate* **2000,** *45,* 87–100.

Guiterrez, E.; T. Wang. Effect of processing on sphingolipid content in soybean products, *J. Am. Oil Chem. Soc.* **2004,** *81,* 971–977.

Guiterrez, E.; T. Wang; W.R. Fehr. Quantification of sphingolipids in soybeans, *J. Am. Oil Chem. Soc.* **2004,** *81,* 737–742.

Guzman, G.J.; P.J. Murphy. Tocopherols of soybean seeds ans soybean curd (Tofu), *J. Agric. Food Chem.* **1986,** *34,* 791–795.

Hamilton, R.J.; C. Kalu; G.P. McNeil; F.B. Padley; J.H. Pierce. Effects of tocopherols, ascorbyl palmitate, and lecithin on autoxidation of fish oil. *J. Am. Oil Chem. Soc.* **1998,** *75,* 813–822.

Hammond, E.G.; L.A. Johnson; C. Su; T. Wang; P.J. White. Soybean oil, *Bailey's Industrial Oil and Fat Products, Sixth ed.;* F. Shahidi, Ed.; John Wiley & Sons, Inc. Publication: New York, 2005; pp. 577–653.

Hirose, K.; N.Imaeda; Y.Tokudome; C.Goto; K.Wakai; K.Matsuo; H. Ito; T. Toyama; H. Iwata; S. Tokudome; K. Tajima. Soybean products and reduction of breast cancer risk: a case-control study in Japan. *Br. J. Cancer,* **2005,** *93,* 15–22.

Hu, J.; S. Lee; S. Hendrich; P.A. Murphy. Quantification of the group B soyasaponins by high-performance liquid chromatography. *J. Agric. Food Chem.* **2002,** *50,* 2587–2594.

Hymowitz, T.; F.I. Collins; J. Panczner; W.M. Walker. Relationship between the content of oil, protein, and sugar in soybean seed, *Agron. J.* **1972,** *64,* 613–616,

Imaizumi, K.; A. Tominaga; M. Sato; M. Sugano. Effects of dietary sphingolipids on levels of serum and liver lipids in rats, *Nutr. Res.* **1997,** *12,* 111–114.

Inagaki, K.; H. Kobayashi; R. Yoshida.; Y. Kanada; Y. Fukuda; T. Yagyu; T. Kondo; N. Kurita; T. Kitanaka; Y. Yamada; Y. Sakamoto; M. Suzuki; N. Kanayama; T. Terao. Suppression of urokinase expression and invasion by a soybean Kunitz trypsin inhibitor are mediated through inhibition of Src-dependent signaling pathways. *J. Biol. Chem.* **2005,** *280,* 31428–31437.

Innis, S.M.; A. Davidson; F. George; S. Melynk; S.J. James. Choline-related supplements improve abnormal plasma methionine- homocysteine metabolites and glutathione status in children with cystic fibrosis. *Am. J. Clin. Nutr.* **2007,** *85,* 702–708.

Jeong, H.; J. Jeong; D. Kim; B. de Lumen. Inhibition of core histone acetylation by the cancer preventive peptide lunasin. *J. Agric. Food Chem.* **2007,** *55,* 632–637.

Jimenez, M.A.; M.L. Scarino; F. Vignolini; E. Mengheri. Evidence that polyunsaturated lecithin induces a reduction in plasma cholesterol level and favorable changes in lipoprotein composition in hypercholesterolemic rats. *J. Nutr.* **1990,** *120,* 659–667.

Jun, H.; S. Kim; M. Sung. Protective effect of soybean saponins and major antioxidants against aflatoxin B1-induced mutagenicity and DNA-adduct formation. *J. Medicinal Food,* **2002,** *5,* 235–240.

Kanamaru, K.; S. Wang; J. Abe; T. Yamada; K. Kitamura. Identification and characterization of wild soybean (*Glycine soja* Sieb. et Zecc.) strains with high lutein content, *Breeding Sci.,* **2006,** *56,* 231–234.

Kang, J.; M. Sung; T. Kawada; H. Yoo; Y. Kim; J. Kim; R. Yu. Soybean saponins suppress the release of proinflammatory mediators by LPS-stimulated peritoneal macrophages. *Cancer Lett.* **2005,** *230,* 219–227.

Kashima, M.; G.S. Cha; Y. Isoda; J. Hirano; T. Miyazawa. The antioxidant effects of phospholipids on perilla oil. *J. Am. Oil Chem. Soc.* **1991,** *68,* 119–122.

Katayama, T. Effects of dietary myo-inositol or phytic acid on hepatic concentrations of lipids and hepatic activities of lipogenic enzymes in rats fed on corn starch or sucrose. *Nutr. Res.,* **1997,** *17,*

721–728.

Kennedy, A.R. The Bowman-Birk inhibitor from soybeans as an anticarcinogenic agent, *Am. J. Clin. Nutr.* **1998,** *68,* 1406–1412S.

Kerr, P.S.; S.A. Sebastian. Soybean products with improved carbohydrate composition and soybean plants, 2003, US patent 6,653,451.

Kerr, P.S.; S..A. Sebastian. Soybean products with improved carbohydrate composition and soybean plants, 2000, US patent 6,147,193.

Kerr, P.S.; S.A. Sebastian. Soybean products with improved carbohydrate composition and soybean plants, 1998, US Patent 5,710,365.

Kim, S.; M.A. Berhow; J. Kim; H. Chi; S. Lee; I. Chung. Evaluation of soyasaponin, isoflavone, protein, lipid, and free sugar accumulation in developing soybean seeds. *J. Agric. Food Chem.* **2006,** *54,* 10003–10010.

King, M. F.; L.C. Boyd; B.W. Sheldon. Antioxidant properties of individual phospholipids in a salmon oil model system. *J. Am. Oil Chem. Soc.* **1992**, *69,* 545–551.

Kobayashi, H.; Y. Fukuda; R. Yoshida; Y. Kanada; S. Nishiyama; M. Suzuki; N. Kanayama; T. Terao. Suppressing effects of dietary supplementation of soybean trypsin inhibitor on spontaneous, experimental and peritoneal disseminated metastasis in mouse model. *Int. J. Cancer*, **2004a,** *112,* 519–524.

Kobayashi, T.; T. Shimizugawa; T. Osakabe; S. Watanabe; H. Okuyama. A long-term feeding of sphingolipids affected the levels of plasma cholesterol and hepatic triacylglycerol but not tissue phospholipids and sphingolipids, *Nutr. Res.* **1997,** *17,* 111–114.

Kobayashi, H.; M. Suzuki; N. Kanayama; T.A. Terao. Soybean Kunitz trypsin inhibitor suppresses ovarian cancer cell invasion by blocking urokinase upregulation. *Clin. Exp. Metastasis,* **2004b**, *21* 159–166.

Koratkar, R.; A.V. Rao. Effect of soya bean saponins on azoxymethane-induced preneoplastic lesions in the colon of mice, *Nutr. Cancer* **1997,** *27,* 206–209.

Liener, I.E. Implications of antinutritional components in soybean foods. *CRC Crit. Rev. Food Sci. Nutr.* **1994,** *34,* 31–67.

Lin, J.; C. Wang. Soybean saponins: chemistry, analysis, and potential health effects. *Soybeans as Functional Foods and Ingredients;* K. Liu, Ed.; AOCS Press: Champaign, IL, 2004; pp. 73–100.

Liu, K. Chemistry and nutritional value of soybean components. *Soybeans: Chemistry, Technology, and Utilization;* K. Liu, Ed.; Aspen Publishers, Inc.: Gaithersburg, MD, 1999a; pp. 25–113.

Liu, K. Soy isoflavones: chemistry, processing effect, health benefits, and commercial production. *Soybeans as Functional Foods and Ingredients;* K. Liu, Ed., AOCS Press: Champaign, IL, 2004b; pp. 52–72.

Liu, K. Soybeans as a powerhouse of nutrients and phytochemicals. *Soybeans as Functional Foods and Ingredients;* K. Liu, Ed.; AOCS Press: Champaign, IL, 2004a; pp. 1–22.

Liu, K., Soybean mprovements through plant breeding and genetic engineering. *Soybeans: Chemistry, Technology, and Utilization;* K. Liu, Ed.; Aspen Publishers, Inc.: Gaithersburg, MD, 1999b; pp. 478–524.

Liu, K.; F.T. Orthoefer; E.A. Brown. Association of seed size with genotypic variation in the chemical constituents of soybeans. *J. Am. Oil Chem Soc.* **1995,** *72,* 189–192.

Lolas, G.M.; N. Palamidas; P. Markakis. The phytic acid— Total phosphorus relationship in barley, oats, soybeans and wheats. *Cereal Chem.* **1976,** *53,* 867–871.

Magbanua, M. J. M.; L. Huang; R.L. Rodriguez. Soy peptide up-regulates tumor suppressor and other chemopreventive genes in human cells. Abstracts, 39th Western Regional Meeting of the American Chemical Society: Sacramento, CA, October 27–30, 2004.

Masuda, Y.; T. Yamagata; Y. Shigematsu. Brain function improving compositions containing lecithin and nicergoline. Japanese patent. 1992. Application: JP 90-116436 19900501.

Maughan, P.J.; M.A. Maroof Saghai; G.R. Buss. Identification of quantitative trait loci controlling sucrose content in soybean (Glycine max). *Molecular Breeding* **2000,** *6,* 105–111.

McCord, K.L.; W.R. Fehr; T. Wang; G.A. Welke; S.R. Cianzio; S.R. Schnebly. Tocopherol content of soybean lines with reduced linolenate in the seed oil, *Crop Sci.* **2004**, *44,* 772–776.

Merrill, A.H.; E.M. Schmelz.. Sphingolipids: mechanism-based inhibitors of carcinogenesis produced by animals, plants, and other organisms. *Handbook of Nutraceuticals and Functional Foods;* R.C. Wildman, Ed.; CRC Press: New York, NY, 2001; pp. 377–392.

Messina, M.; C.L. Hughes. The efficacy of soyfoods and soybean isoflavone supplements for alleviating menopausal symptoms is positively related to initial hot flash frequency, *J. Medicinal Foods* **2003,** *6,* 1–11.

Miyamoto, S.; K. Murota; G. Kuwataz; M. Imai; A. Nagao; J. Terao. Antioxidant activity of phytic acid hydrolysis products on iron ion-induced oxidative damage in biological system. *ACS Symposium Series* **2002,** *807,* 241–250.

Mounts, T.L.; S.L. Abidi; K.A. Rennick. Effect of genetic modification on the content and composition of bioactive constituents in soybean oil, *J. Am. Oil Chem. Soc.* **1996**, *73,* 581–586.

Nakashima, H.; K. Okubo; Y. Honda; T. Tamura; S. Matsuda; N. Yamamoto. Inhibitory effect of glycosides like saponins from soybean on the infectivity of HIV in vitro, *AIDS,* **1989,** *3,* 655–658.

Nasner, A. Antioxidative properties of lecithin. *Fette, Seifen, Anstrichmittel,* **1985,** *87,* 477–481.

Nestel, P. Isoflavones: their effect on cardiovascular risk and functions. *Curr. Opin. Lipidol.* **2003,** *14,* 3–8.

Nwosu, C.V.; L.C. Boyd; B. Sheldon. Effect of fatty acid composition of phospholipids on their antioxidant properties and activity index. *J. Am. Oil Chem. Soc.* **1997,** *74,* 293–297.

Oh, Y.; M. Sung. Soybean saponins inhibit cell proliferation by suppressing PKC activation and induce differentiation of HT-29 human colon adenocarcinoma cells. *Nutr. Cancer,* **2001,** *39,* 132–138.

Okazaki, Y.; T. Kayashima; T. Katayama. Effect of dietary phytic acid on hepatic activities of lipogenic and drug-metabolizing enzymes in rats fed 1,1,1-trichloro-2,2-bis(p-chlorophenyl) ethane (DDT). *Nutr. Res.,* **2003,** *23,* 1089–1096.

Orthoefer, F.T.; G.R. List. Phospholipids/lecithin: a class of neutraceutical lipids. *Nutraceutical and Specialty Lipids and Their Co-Products.* F. Shahidi, Ed.; Taylor and Francis: St. John's, Newfoundland, 2006; 5, pp. 509–530.

Padgette, S.R.; N.B. Taylor; D.L. Nida; M.R. Bailey; J. MacDonald; L.R. Holden; R.L. Fuchs. The composition of glyphosate-tolerant soybean seeds is equivalent to that of conventional soybeans, *J. Nutr.* **1996,** *126,* 702–716.

Parsons, C.M.; Y. Zhang; M. Araba. Nutritional evaluation of soybean meals varying in oligosaccharide content. *Poultry Sci.* **2000,** *79,* 1127–1131.

Ramamurthi, S.; A.R. McCurdy; R.T. Tyler. Deodorizer distillate: A valuable byproduct. *Proc. World Conf. Oilseed Edible Oils Process,* S.S. Koseoglu, K.C. Rhee, R.F. Wilson, Eds.; AOCS Press: Champaign, IL, 1998; 1, pp. 130–134.

Rao, A.V.; S.A. Janezic. The role of dietary phytosterols in colon carcinogenesis, *Nutr. Cancer,* **1992,** *18,* 43–52.

Rodrigues, H.G.; Y.S. Diniz; L.A. Faine; C.M. Galhardi; R.C. Burneiko; J.A. Almeida; B.O. Ribas; E.L.B. Novelli. Antioxidant effect of saponin: potential action of a soybean flavonoid on glucose tolerance and risk factors for atherosclerosis. *Int. J. Food Sci. Nutr.* **2005,** *56,* 79–85.

Scherder, C.W.; W.R. Fehr; G.A. Welke; T. Wang. Tocopherol content and agronomic performance of soybean lines with reduced palmitate. *Crop Sci.* **2006,** *46,* 1286–1290.

Schmelz, EM. Dietary sphingomyelin and other sphingolipids in health and disease, *Nutr. Bull.* **2000,** *25,* 135–139.

Schmelz, E.M.; D.L. Dillehay; S.K. Webb; A. Reiter; J. Adams; A.H. Merrill. Sphingmyelin consumption suppresses aberrant colonic crypt foci and increases the proportion of adenomas versus adenocarcinomas in CF1 mice treated with 1,2-dimethylhydrazine: implications for dietary sphingolipids and colon carcinogenesis. *Cancer Res.* **1996,** *66,* 4936–4941.

Segawa, T.; M. Kamata; S. Hara; Y. Totani. Antioxidant activity of phospholipids for polyunsaturated fatty acids of fish oil. III. Synergism of nitrogen-containing phospholipids with tocopherol. *Yukagaku* **1995,** *44,* 36–42.

Serbinova, E.; V. Kagan; D. Han; L. Parker. Free radical recycling and intramembrane mobility in the antioxidant properties of α-tocopherol and α-tocotrienol, *Free Radical Biol. Med.* **1991,** *10,* 263–275.

Sharon, N. Lectins: carbohydrate-specific reagents and biological recognition molecules. *J. Biol. Chem.* **2007,** *282,* 2753–2764.

Sharon, N.; H. Lis. Lectins: from hemagglutinins to biological recognition molecules. A historical overview. *Glycobiol.* **2004,** *14,* 53R–62R.

Shibayama, N. Reduction of blood cholesterol level by soybean saponins. *Food Style 21,* **2003,** *7,* 73–77.

Simonne, H.; M. Smith; D.B. Weaver; T. Vail; S. Barnes; C.I. Wei. Retention and changes of soy isoflavones and carotenoids in immature soybean seeds (Edamame) during processing *J. Agric. Food Chem.* **2000,** *48,* 6061–6069.

Tang, S.; D. Li; S. Qiao; X. Piao; J. Zang. Effects of purified soybean agglutinin on growth and immune function in rats. *Arch. Anim. Nutr.* **2006,** *60,* 418–426.

Tian, L.L.; P.J. White. Antipolymerization activity of oat extract in soybean and cottonseed oils under frying conditions. *J. Am. Oil Chem. Soc.* **1994,** *71,* 1087–1094.

Troncoso, M.F.; V.A. Biron; S.A. Longhi; L.A. Retegui; C. Wolfenstein-Todel. Peltophorum dubium and soybean Kunitz-type trypsin inhibitors induce human Jurkat cell apoptosis. *Int.*

Immunopharmacol. **2007,** *7,* 625–636.

Tsukamoto, C.; S. Shimada; K. Igita; S. Kudou; M. Kokubun; K. Okubo; K. Kitamura. Factors affecting isoflavone content in soybean seeds: Changes in isoflavones, saponins, and composition of fatty acids at different temperatures during seed development, *J. Agric. Food Chem.* **1995,** *43,* 1184–1192.

Ueda, H.; A. Matsumoto; S. Goutani. Effects of soybean saponin and soybean protein on serum cholesterol concentration in cholesterol-fed chicks. *Anim. Sci. Technol.* **1996,** *67,* 415–422.

Vesper, H.; E.M. Schmelz; M.N. Nikolova-Karakashian; D.L. Dillehay; D.V. Lynch; A.H. Merrill. Sphingolipids in food and the emerging importance of sphingolipids to nutrition, *J. Nutr.* **1999,** *129,* 1239–1250.

von Allworden H.N.; S. Horn; J. Kahl; W. Feldheim. The influence of lecithin on plasma choline concentrations in triathletes and adolescent runners during exercise. *Eur. J. Aappl. Physiol. Occup.l Physiol.* **1993,** *67,* 87–91.

Wang, T. Soybean oil. *Vegetable Oils in Food Technology;* F. Gunstone, Ed.; CRC Press: Boca Raton, FL, 2002; pp. 18–58.

Wang, W.; E. de Mejia. Anticancer potential and mechanisms of lunasin and soy protein hydrolysates. *Abstracts, 233rd ACS National Meeting,* Chicago, IL, March 25–29, 2007.

Wang, T.; E.G. Hammond. Fractionation of soybean phospholipids by high-performance liquid chromatography with an evaporative light scattering detector. *J. Am. Oil Chem. Soc.* **1999,** *76,* 1313–1321.

Wang, T.; E.G. Hammond; W.R. Fehr. Phospholipid fatty acid composition and stereospecific distribution of soybeans with a wide range of fatty acid compositions. *J. Am. Oil Chem. Soc.* **1997,** *74,* 1587–1594.

Wang, H.; P.A. Murphy. Mass balance study of isoflavones during soybean processing. *J. Agric. Food Chem.* **1996,** *44,* 2377–2383.

Wang, H.; P.J. Murphy. Isoflavone composition of American and Japanese soybeans in Iowa: effects of variety, crop year, and location, *J. Agric. Food Chem.* **1994,** *42,* 1674–1677.

Wang, L.; T. Wang; W.R. Fehr. Effect of seed development stage on sphingolipid and phospholipid contents in soybean seeds, *J. Agric. Food Chem.* **2006a,** *54,* 7812–7816.

Watanabe, E.; K. Oeda. Variant soybean raffinose synthase cDNA and uses in obtaining low raffinose plants and genotyping. Japanese patent, 2001, JP 2001078783.

White, P.J.; Y. Xing. Antioxidants from cereals and legumes. *Natural Antioxidants, Chemistry, Health Effects, and Applications;* F. Shahidi, Ed.; Champaign, IL: AOCS Press, 1997; pp. 25–63.

Woodmansey, E.J. Intestinal bacteria and ageing. *J. Appl. Microbiol.* **2007,** *102,* 1178–1186.

Yan, L.; E. Spitznagel. A meta-analysis of soyfoods and risk of breast cancer in women. Nutrition Sciences, *Inter. J. Cancer Prevention.* **2004,** *1* 281–293.

Zeisel, S.H. Choline, homocysteine, and pregnancy. *Am. J. Clin. Nutr.* **2005,** *82,* 719–720.

Zeisel, S.H. Choline and choline esters as required nutrients during pregnancy and lactation. *Choline, Phospholipids, Health, and Disease;* S.H. Zeisel, B.F. Szuhaj, Eds.; AOCS Press: Champaign, IL, 1998; pp.131–142.

11

Oil Recovery from Soybeans

Lawrence A. Johnson
Department of Food Science and Human Nutrition; Director, Center for Crops Utilization Research, Iowa State University, Ames, IA 50011

Introduction

Fats and oils have been recovered for thousands of years from oil-bearing seeds and fruits, and fatty animal tissues. Soybeans in Asia and sesame seed and olives in the Middle East were the earliest sources of vegetable oils for food, cosmetics, lubricants, and chemicals. Soybeans is a very important crop to the economies of the United States and Brazil as a source of edible oil and high-protein meal for supplementing animal diets. No other crop rivals soybeans for high-quality feed protein, and only palms rival soybeans as a world source of edible oil.

Despite the wide diversity of today's sources of vegetable oils, all recovery processes, whether for soybeans or other oilseeds, are designed to obtain triglycerides in high yield and purity, and to produce co-products (usually high in protein content) of maximum value. Oilseeds, including soybeans, are processed by one or more of three types of processes—hard screw pressing, prepress solvent extraction, or direct solvent extraction. The preferred seed extraction process for different oilseeds depends on the oil content of the material to be processed, how much unextracted oil can remain in the meal without affecting its market value, how much protein denaturation is allowed in the meal, how much investment capital is available, and how restrictive local environmental laws are concerning emissions of volatile organic compounds (VOCs). In the case of soybeans, the vast majority (>97%) are processed today by direct solvent extraction and only modest amounts by various forms of hard screw pressing (~3%). Only in one instance has the author observed soybeans being crushed by prepress solvent extraction, and that plant normally crushed sunflower seed—the amount of oil in soybeans simply does not warrant this type of processing that is used for higher oil-containing seeds such as sunflower, peanuts, safflower, and canola.

Modern Soybean Processing

General Methods

Hard screw pressing is the oldest method used to process soybeans, is completely mechanical, and requires the least amount of capital investment and attention to safety. Although the Central Soya Company and Procter & Gamble installed solvent-extraction plants during 1937–1939, improved solvent extraction technology was brought to the United States from Germany following WWII. Until then, hydraulic pressing and screw pressing were the predominant methods used to process soybeans. Direct solvent extraction became the preferred processing method in the 1950s.

In hard screw pressing, the liquid oil is squeezed or pressed from the soybean solids known as *cake*, usually after cooking and/or drying the beans to enhance the efficiency of continuous screw presses in removing oil (Fig. 11.1). According to ancient history, batch systems incorporating lever presses and screw-operated presses powered by work animals were used. During the nineteenth century, batch hydraulic presses (similar to today's cider press) were used. By the turn of the twentieth century, continuous screw presses connected to line shafts driven by steam engines became prevalent. Today, electric-motor-powered, continuously fed screw presses, sometimes referred to in the industry as *expellers*, are used (the word *expeller* is a trademarked name for a continuous screw press manufactured by Anderson International, Cleveland, OH). The term *crushing* to denote oil recovery originates from the days when pressing was the predominant method used to process soybeans.

Normally, seeds containing >30% oil require pressing, either hard pressing or prepressing, prior to solvent extraction (so that the material is sufficiently dense to withstand solvent extraction and not to produce fine meal particles, which adversely affect downstream oil-processing operations). Hard pressing involves squeezing out as much oil as possible, whereas, in prepressing only the easily recoverable oil is removed before subjecting the partially de-oiled material to more complete extraction with solvent (over pressing impedes solvent diffusion and thus oil extraction). Sunflower seed, safflower, peanuts, wet-milled corn germ, rapeseed, and sesame are usually prepressed followed by solvent extraction, however, the recent adoption of the expander (an extruder-like device) has allowed some high-oil-content seeds to be solvent extracted without prepressing (Fig. 11.2). Although prior to WWII hard pressing soybeans was common, direct solvent extraction (without prepressing) is now the most widely practiced method for processing soybeans. Oil is more completely recovered when employing solvent extraction (typically <1% oil content in meal for solvent extraction versus 5–8% oil for hard screw pressing resulting in nearly 95% of the oil being recovered by solvent extraction compared with 70% for hard screw pressing).

For simplicity, solvent extracting soybeans, or for that matter any oilseed, has been likened to cleaning paint from a brush. To get the brush clean, the solvent must have good contact and penetration into the brush's bristles, sufficient heat to quickly do the cleaning, and enough clean solvent and time to remove all the paint.

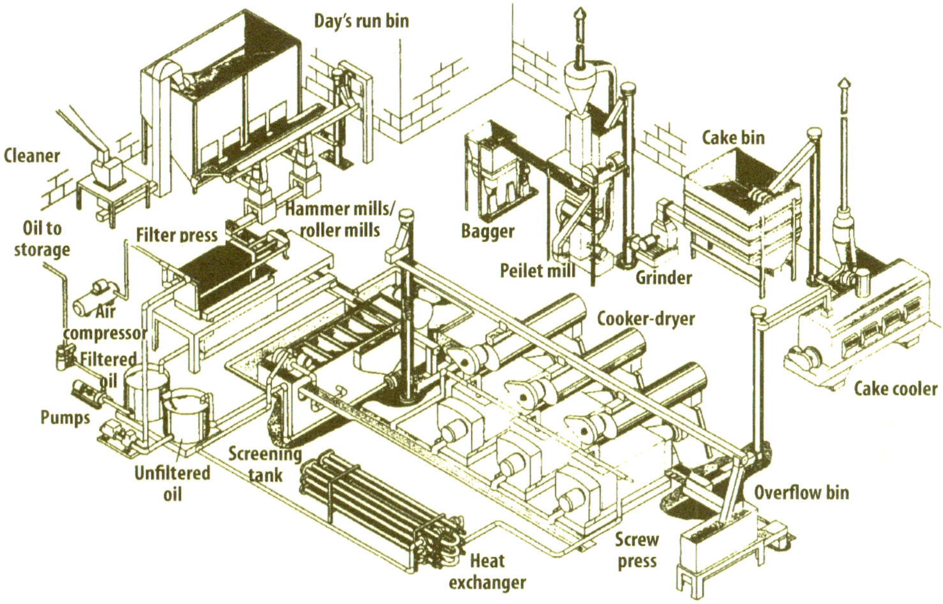

Fig. 11.1. Depiction of hard screw pressing (provided by Anderson International, Cleveland, OH).

This is somewhat an oversimplification because this form of leaching assumes all oil is freed from cells, which is not the case and is discussed further later in this chapter. Although many solvents were tried over the years, mixtures of hexanes are mostly used today. Most hexane mixtures used contain about two-thirds *n*-hexane, the remainder being mostly other isomers of hexane. Very little detectable benzene is allowed. Direct solvent extraction is usually preferred for soybeans; both prepress solvent extraction and direct solvent extraction are depicted in Fig. 11.3.

Protein Versus Oil

Soybeans are an important source of edible oil, but many have argued that soybean is actually a protein crop because 60–70% of the returns in processing soybeans is due to the sale of meal (Table 11.1). No other oilseed contains as much protein. Thus, processes used to extract soybean oil are designed to maximize meal quality for use in livestock feeds.

About 20 kg (44 lb) of meal (48% protein, dehulled), 5.2 kg (11.5 lb) of oil, and 1.4 kg (3 lb) of hulls are obtained per bushel (27 kg, 60 lb) of soybeans processed; the remaining 0.7 kg (1.5 lb) is shrinkage (includes moisture loss). The value of soybeans to processors, however, is dependent on the composition of the soybeans (protein and

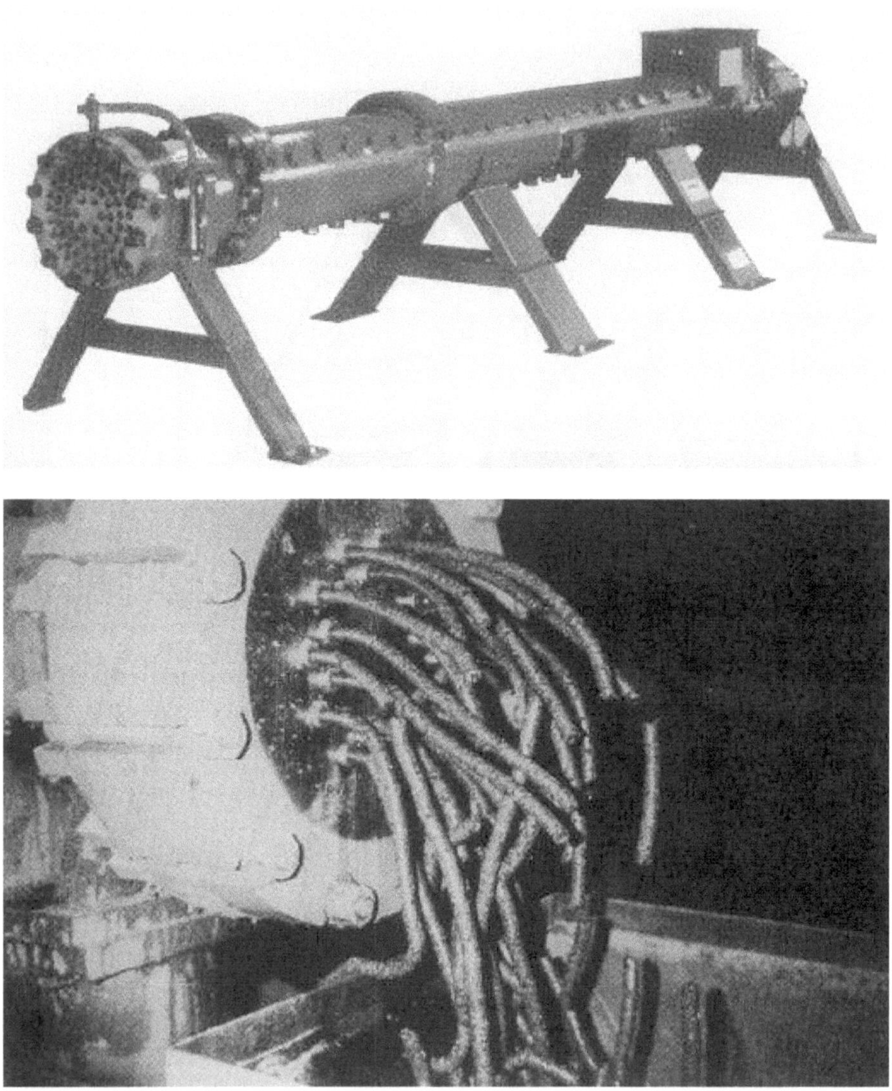

Fig. 11.2. Photographs of an expander (A) and collets exiting the die (B) (provided by Anderson International, Cleveland, OH).

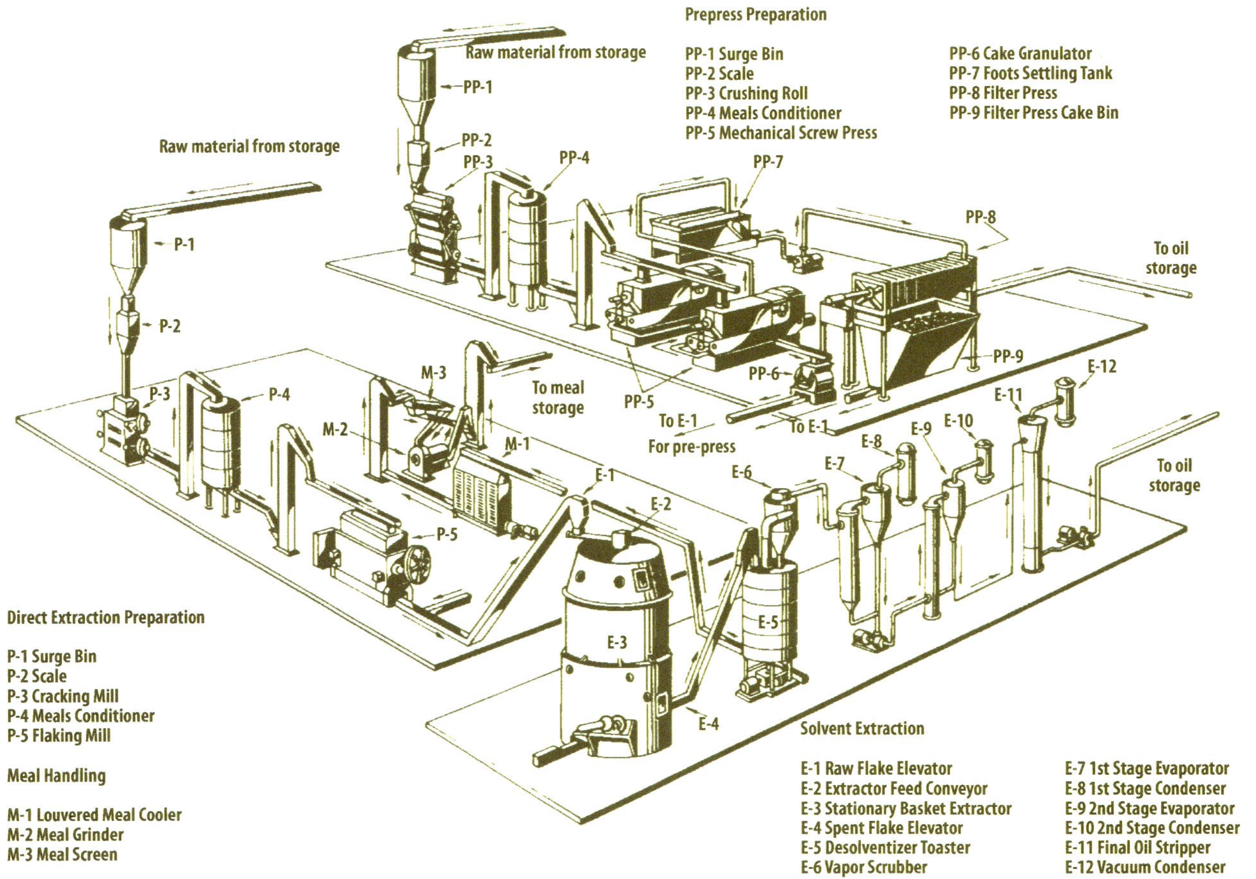

Fig. 11.3. Depiction of prepress solvent extraction and direct solvent extraction (provided by French Oil Mill Machinery Co., Piqua, OH).

Table 11.1. Source of Returns from Crushing Soybeans

	Value of Products per Bushel										Soybean Price		Margin Between Value of Products and Soybean Prices	
	Soy Oil			Soybean Meal			Soybean Hulls[‡]			Total	Rec'd by	No. 1 Y	Rec'd by	No. 1 Y
Year beg. Sept.	Yield* (lbs)	Price[†] (cents)	Value (dollars)	Yield* (lbs)	Price** (cents)	Value (dollars)	Yield* (lbs)	Price (cents)	Value (dollars)	Value (dollars)	Farmers (dollars)	IL pts. (dollars)	Farmers (dollars)	IL pts. (dollars)
1955-56	11.10	12.50	1.39	46.20	2.68	1.24				2.63	2.22	2.51	.41	.12
1960-61	11.00	11.20	1.23	47.00	3.00	1.41				2.64	2.13	2.53	.51	.11
1965-66	10.70	11.80	1.26	47.50	4.02	1.91				3.17	2.54	2.91	.63	.26
1970-71	10.80	12.80	1.38	47.40	3.96	1.88				3.26	2.85	3.00	.41	.26
1975-76	10.90	18.50	2.02	47.30	7.20	3.40				5.42	4.92	5.26	.50	.16
1980-81	11.10	23.30	2.58	47.90	11.10	5.32				7.90	7.57	7.67	.33	.23
1985-86	11.00	18.70	2.06	47.20	8.26	3.84				5.90	5.05	5.30	.90	.60
1990-91	11.23	21.31	2.39	47.47	9.01	4.28				6.67	5.74	5.90	.93	.77
1991-92	11.42	19.31	2.20	47.51	9.53	4.53				6.73	5.58	5.84	1.15	.89
1992-93	10.84	21.01	2.28	47.54	9.63	4.58				6.86	5.56	5.95	1.30	.91
1993-94	10.87	26.78	2.91	47.62	9.75	4.64				7.55	6.40	6.59	1.15	.96
1994-95	11.08	27.70	3.07	47.33	8.10	3.83				6.90	5.48	5.73	1.42	1.17
1995-96	11.15	24.89	2.78	47.69	11.40	5.44				8.21	6.72	7.39	1.49	.82
1996-97	10.91	22.60	2.47	47.36	13.54	6.41				8.88	7.35	7.80	1.53	1.08
1997-98	11.25	25.65	2.88	47.41	9.87	4.68				7.56	6.47	6.64	1.09	.92
1998-99	11.30	20.49	2.31	47.25	6.87	3.24				5.56	4.93	5.00	.63	.56
1999-00	11.34	15.81	1.79	47.76	7.90	3.77				5.57	4.65	4.90	.92	.67
2000-01	11.24	13.99	1.57	48.06	8.28	3.98				5.55	4.54	4.77	1.01	.78
2001-02	11.14	16.05	1.79	44.27	8.33	3.69	3.33	3.07	0.10	5.58	4.38	4.79	1.20	.79
2002-03	11.39	21.80	2.48	43.90	8.94	3.93	3.27	3.30	0.11	6.52	5.53	5.90	.99	.62
2003-04	11.20	29.74	3.33	44.32	12.98	5.75	3.37	3.87	0.13	9.21	7.34	8.22	1.87	.99
2004-05	11.33	23.24	2.63	44.26	9.15	4.05	3.41	2.83	0.10	6.78	5.74	5.98	1.04	.80
2005-06[1]	11.63	22.86	2.66	43.85	8.84	3.88	3.35	3.53	0.12	6.65	5.65	5.72	1.00	.93

*Because Census reports crush and production quarterly, yields of meal and oil will be adjusted quarterly. [†]Crude, tanks, f.o.b. Decatur. **Beg '85/'86 meal prices reported at 48% (solvent), Decatur. [1]Through May 2006. [‡]Not reported before Sept. 2001. Source: USDA ERS, June 2006.

oil contents), and some advocate that soybean prices should reflect their processing value; protein and oil contents are not factors in the United States Department of Agriculture (USDA) Federal Grain Inspection Service soybean grades and standards. Iowa State University researchers developed a computer model and program that estimates the values of soybeans with different compositions to processors based on current prices for soybean oil and meal (Brumm & Hurburgh, 1990). The program is accessible at the Internet site maintained by the University of Illinois http://www.stratsoy.uiuc.edu/epv/. The program calculates the amounts of products and the values of soybeans with different oil and protein contents. At least one soybean processor used a pricing scale to purchase soybeans based on protein and oil contents. This offers an incentive to seed providers and farmers to improve soybean quality.

Soybean meal may be sold at a variety of protein contents. If soybeans are not dehulled, the meal (low-pro) usually meets a minimum specification of 44% protein and is primarily used in cattle feeds. Usually, soybeans are dehulled to produce high-protein meal (hi-pro) to better compete for more lucrative swine and poultry feed markets. The Chicago Board of Trade contract specification is for hi-pro soybean meal containing 48% protein; however, in actual practice the protein contents of hi-pro meal range from 45 to 50% (with vast majority in the range of 47.0 to 48.5% protein), and hi-pro meal is sold on specification between processor and end-user. The meal protein range largely results from the range of protein contents of soybeans themselves due to environmental factors in which they are grown such as the lower protein content of soybeans grown in the northern part of the soybean belt. The United Soybean Board estimates that approximately 46% of the soybean meal produced is used in poultry diets (broilers and layers, and turkeys); 25% in swine diets; 21% ruminant diets (beef and dairy cattle, and sheep); and the remaining 8% is used in pet food, fish feeds, food protein ingredients, and industrial uses (adhesives, paper coatings, etc.).

Recovery of Oil from Soybeans

Today's new soybean processing plants using direct solvent extraction need to be capable of processing at least 2,700 metric tons per day (3,000 tons/day) to be economically viable in the United States, representing an investment of over $75 million. In recent years, only one plant was built in the United States having less capacity (725 metric tons/day, 800 tons/day), and it has had difficulties operating profitably. Solvent extraction plants as large as 5,450 metric tons per day (6,000 tons/day) capacity have been built in Europe and as large as 14,000 metric tons per day (15,400 tons/day) in Argentina. Under some conditions, plants as small as 23 metric tons per day (25 tons/day) capacity are profitable when using screw pressing, especially when processing identity-preserved beans to produce specialty oils and meals (e.g., low linolenic, ultra-low linolenic, organic, etc.). Process flow diagrams for direct solvent extraction of soybeans are shown in Figures 11.4 and 11.5, and aerial and interior photographs

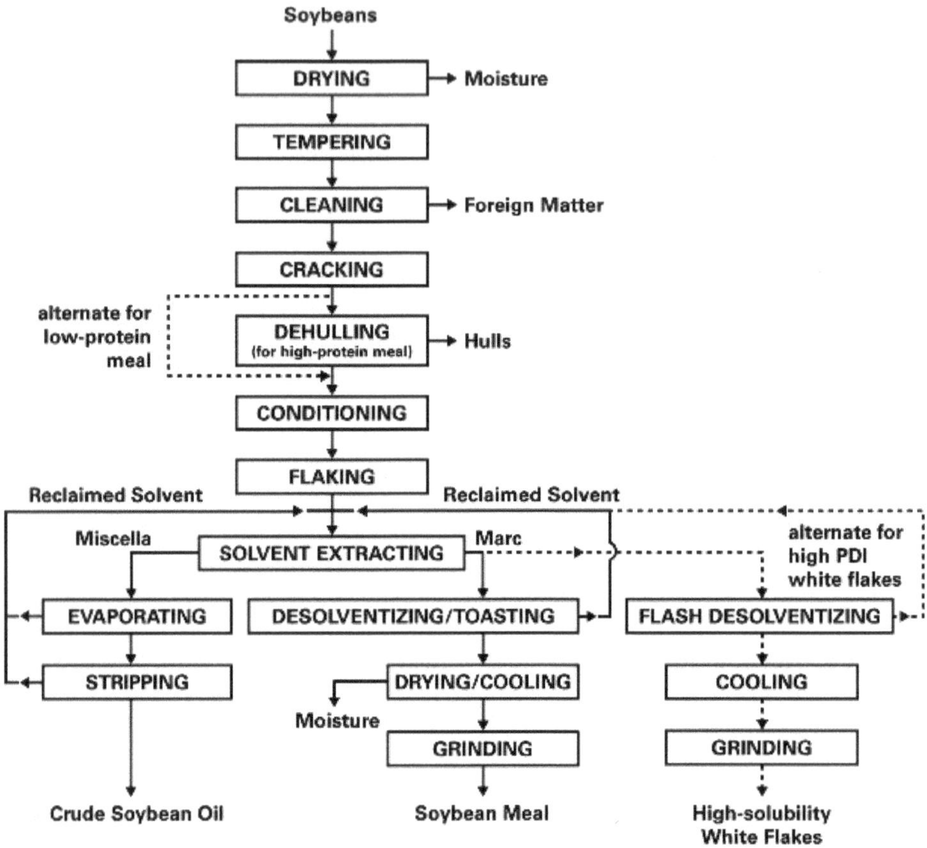

Fig. 11.4. Flow diagram for direct solvent extracting soybeans.

of a soybean plant are shown in Fig. 11.6. Soybean processing can be broken down into two (if screw pressing) or three (if solvent extracting) major areas: preparation (cleaning, dehulling, flaking, expanding), extraction (screw pressing, solvent extraction), and oil/meal finishing (oil and meal desolventizing, oil cooling and filtering, and meal toasting, drying, cooling and grinding).

Seed Handling

Soybean tissue is composed of many cells containing oil, protein and metabolites, which supply energy, nitrogen storage reserves, and other important compounds, respectively, to support the germination of new plants. The triglycerides are stored in discrete bodies called *oil bodies* or *spherozomes*. The preponderance of the protein is storage protein, which is concentrated in other discrete bodies known as *protein bodies*. Most of the phospholipids are associated with membranes around the protein bod-

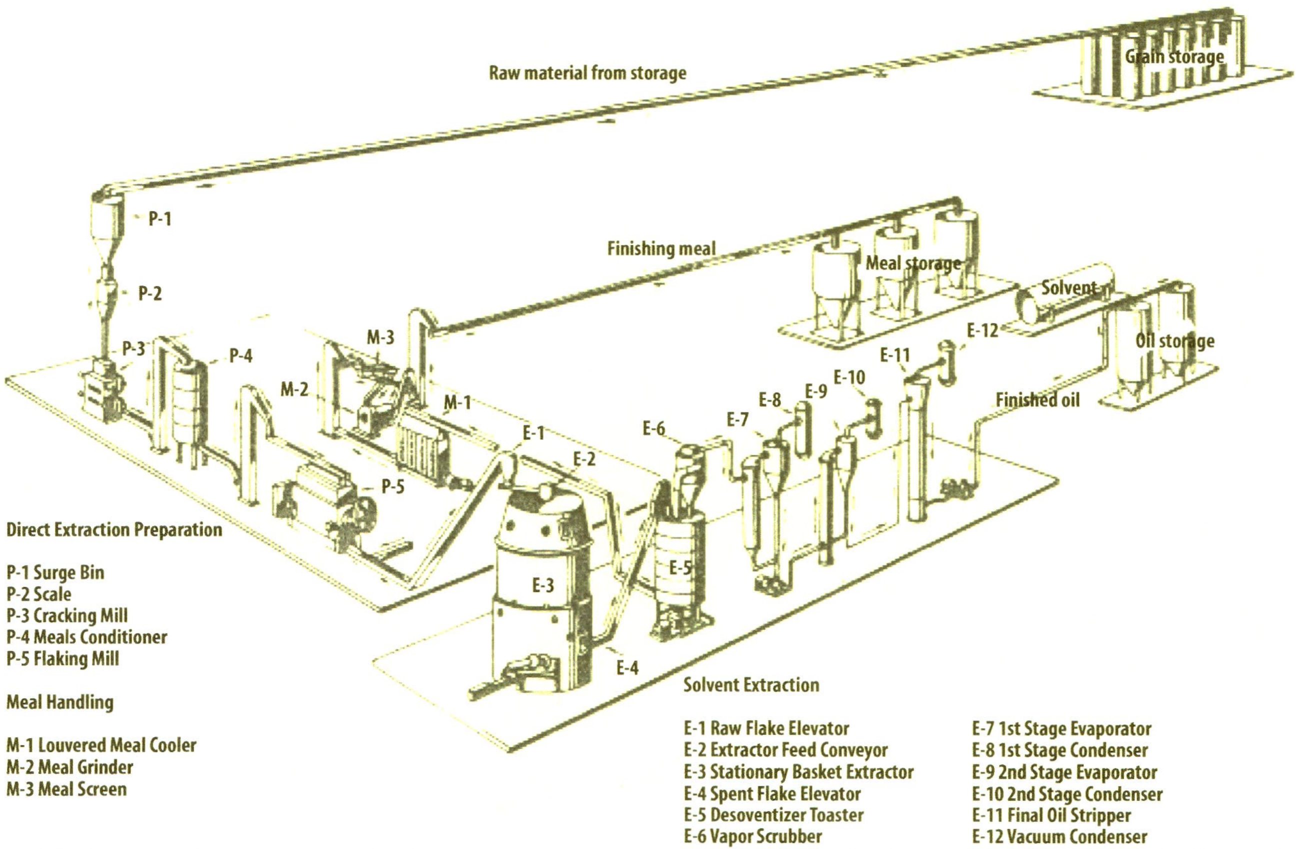

Fig. 11.5. Depiction of equipment used in direct solvent extracting soybeans (provided by French Oil Mill Machinery Co., Piqua, OH).

Fig. 11.6. Aerial (A) and interior (B) photographs of direct-solvent-extraction soybean plants (aerial photograph provided by Bunge Corporation, St. Louis, MO; interior photograph provided by Crown Iron Works, Minneapolis, MN).

ies and spherozomes. Catabolic enzymes (enzymes responsible for hydroysis of lipids and proteins) and cellular metabolites are present in the cytoplasm. This separation of enzymes and substrates is important to long-term storage because its breakdown in separation, such as caused by bruising (rupturing cell walls and organelle membranes) when hitting a hard surface or other forms of mechanical damage (e.g., in harvesting machinery), leads to enzymatic deterioration of oil quality (e.g., dark colors and high free fatty acid and phosphorus contents). In intact undamaged seeds, catabolic enzymes are present in the cytoplasm, and oil is protected within spherozomes. This natural compartmentalization separates the oil from oil-damaging enzymes.

Lipase is an enzyme that catalyzes triglyceride hydrolysis, increasing the free fatty acid content that must be removed by alkali refining (lipase action is not usually a problem in soybeans, but is quite important in storing and processing cottonseed and palm fruit). The class of enzymes known as *phospholipases* may catalyze the conversion of phospholipids to a nonhydratable form that is very difficult to remove from the oil by water degumming and can be a challenge in some soybeans. Nonhydratable phosphatides are believed to be Ca^{++} and Mg^{++} salts of phosphatidic acid. Still another class of enzymes in soybeans, known as the *lipoxygenases* (three known iso-enzymes), oxidizes the prevalent polyunsaturated fatty acids linoleic and linolenic acids causing beany flavors. Anyone who has chewed a raw soybean knows the consequences of lipoxygenase activity—rapid production of obnoxious paint-like and grassy/beany flavors after a short tasteless period prior to the enzyme becoming active.

The quality of soybeans is highest just before harvesting in a year with ideal weather. If cell walls and membranes become ruptured, oil and deteriorative enzymes come into contact with one another. As temperature, moisture content, and extent of damage increase, so do the rates of enzyme-catalyzed reactions that reduce oil quality. Some physical damage to seed is inevitable during mechanical harvesting, shipping, and storing, but it must be minimized (partly reflected in grades and standards as splits). For the same reasons, the oil must be extracted quickly to produce oil of maximum quality once seed processing begins and cells are crushed to liberate oil. In one alternative extraction process, known as the *ALCON Process*, soybeans are cooked in stack cookers prior to solvent extracting, and better quality oil may be produced. Cooking inactivates enzymes harmful to oil quality (i.e., phospholipase, lipase, lipoxygenase). This process of cooking soybeans is not frequently used (not at all in the United States), but expanding also accomplishes the same effect.

Soybean Drying

Usually soybeans are allowed to dry in the field to about 13% moisture before harvesting, which is considered safe for long-term storage, but sometimes soybeans are harvested with as much as 20% moisture. In the unfortunate event when the moisture content exceeds the 13% critical value, the beans must be dried incurring cost, which is usually done with open-flame grain dryers. Seed temperature should not exceed 76°C during drying to prevent seed and oil discoloration and protein denaturation. Normally the highest temperature the seed attains is 55–60°C. The dried seed should then be cooled to <10°C above the ambient temperature.

Soybean Storage

Farmers often store soybeans on the farm in aerated metal bins or at local grain elevators in concrete silos (100,000 to 700,000 bu/tank capacity). Higher prices are usually paid later in the crop year, and storing the grain for later sale can increase returns to farmers. Alternatively, some soybean processing plants may contract with farmers to deliver their grain later in the year or "on call" when needed.

Storing soybeans for extended periods at moisture contents exceeding 13% also reduces the recoveries of oil and protein in addition to adversely affecting the quality of the oil. In the spring and through the early summer, moisture may migrate from the cool core to outer parts of storage bins, and forced air may be used to control moisture migration and local accumulation when the relative humidity is <70%; otherwise, moisture may be added instead of removed (Barger, 1981). When aeration is called for, an aeration rate of 0.003 m^3/min × bu (0.1 ft^3/min × bu) is needed. Blowing air up through the seed and out the top dries and cools soybeans; thus, the seed at the bottom is dried first and the drying front moves from the bottom to the top.

Soybeans are alive at harvest and respire, converting seed mass to CO_2 and metabolites. The respiration rate is low when either the temperature is low (e.g., 5°C) or the moisture content is <13%. Oilseeds with higher oil contents than soybeans (20% oil) have lower critical moisture values for safe storage. When soybeans contain >13% moisture, respiration increases; soybeans may even germinate or be attacked by fungi during storage under less than ideal conditions. As soybeans respire, heat is produced, and the increased temperature further accelerates respiration. In the worst conditions, the soybeans may become heat-damaged and scorched, and such damage reduces prices that the market will pay because oil and meal qualities will be reduced. Modern storage facilities employ temperature-monitoring systems to alert operators when seed temperatures exceed set points. Soybeans that are heating are immediately processed or, if not possible, are immediately moved to another bin to disperse hot spots, and/or ambient air is blown through the beans to provide cooling. The amount of heat-damaged beans is a factor in the U.S. grades and standards for soybeans. Heat-damaged beans contain high free fatty acid levels and dark-colored oil, thereby increasing refining loss.

Over-drying soybeans also has consequences. The soybean cotyledon is prone to breaking into halves, called *splits*, when the hull becomes separated from the cotyledon during conveying and transporting. Seed breakage becomes worse when the beans are over-dried. Splits are difficult to separate from foreign matter, and their oil deteriorates at a faster rate. Catabolic enzymes are activated in splits, increasing the contents of free fatty acids, phosphatides, iron, and peroxides in the oil. Oils from field- and storage-damaged soybeans usually also have poor flavors. For these reasons, damaged kernels and splits are also factors in the U.S. grades and standards for soybeans. Over-dried beans may also fracture into excessive fines during dehulling, thereby increasing loss of meats (oil and protein) in the hulls.

When soybeans are received at the crushing plant, they are sampled to analyze moisture, foreign matter, splits, and damaged-seed contents. Sometimes oil and protein contents are also determined (only test weight, total damaged kernels, heat-damaged kernels, foreign matter, splits, and soybeans of other colors are considered in U.S. grading standards), and prices paid are usually based upon the values and/or grades obtained. Soybeans with excessive foreign matter are transferred to scalping operations for removing the foreign matter. Foreign matter usually contains more moisture than the beans. The foreign matter tends to become concentrated under discharge spouts during placement into storage bins and, consequently, this area of the bin is prone to heating. Removing foreign matter reduces the moisture content, improves storability of the beans, and reduces degradation of oil during storage. As mentioned earlier, sometimes farmers are paid on additional factors such as protein and oil contents. The advent of near-infrared reflectance/transmittance technology enables rapid estimation of soybean composition.

Inside storage systems, primarily concrete silos and metal bins, are used for storing soybeans at elevators and plant sites. Sometimes a shortage of available protected storage at peak harvest time exists. Unlike storing cottonseed in dry regions (e.g., west Texas) or corn in the Midwest, soybeans are rarely and for only very short periods stored on concrete pads without protection against the weather.

Cleaning

Stems, pods, leaves, broken grain, dirt, sand, small stones, and extraneous seeds are typical components of foreign matter in harvested soybeans. Cleaning is usually the first step in processing soybeans. Foreign matter (stems, pods, leaves, etc.) reduces oil and protein contents, adversely affects oil quality (especially color), increases wear and damage to expensive processing equipment, and increases the mass to be processed. Vibrating and/or shaker screens called a *scalper*, sometimes with aspiration, remove foreign materials. Shaker screens are used to separate particles on the basis of size, whereas aspiration separates on the basis of density and buoyancy in a stream of air (removing light materials). The cleanings, often known as *mill run*, may be added back to the meal to control protein content. Tramp iron, extraneous metal acquired during harvesting, transporting, storing or upstream processing steps, must be removed to prevent damage to expensive processing equipment by placing magnets in chutes just ahead of vulnerable machines.

Dehulling

It is desirable to remove the hull (seed coat) that covers the soybean cotyledon or *meat*. Soybean hulls contain much less oil and protein than do the meats. Soybean hulls account for about 8% of the bean dry matter but contain less than 1% lipid. Removing the hull reduces the amount of material that must be further processed, thus increasing downstream plant capacity and reducing energy consumption per unit processed. However, removing the hull is usually done to raise the protein level of the meal. The protein content of soybean meal increases by about 4 percentage points when removing the hull (generally, 44% protein meal is produced when not employing dehulling versus about 48% with dehulling). Hulls, however, can be helpful at times as they allow easier hard pressing and enhance solvent drainage. The majority of soybeans today are dehulled because the feed industry prefers and pays a premium for high-protein soybean meal.

Decortication, the freeing of the hull (pericarp) from the cotyledon, becomes easier after further drying the beans after storage to ~10% moisture (from 13% for safe storage). This practice shrinks the meat away from the hull. In traditional cold dehulling, the beans are sometimes tempered for 24 h to allow the moisture content to equilibrate throughout the bean before decorticating. Decorticating is usually accomplished by cracking the bean into 6–8 pieces with corrugated roller mills; although sometimes the bean is cracked into only 2–4 pieces for more efficient hull removal and to minimize producing fine meats that are easily lost during aspiration, and then the beans are recracked for proper flaking. Do cracking carefully to avoid breaking the cotyledon into fine particles that are difficult to separate from the hulls. Also avoid crushing the meats during dehulling because this causes oil cells to rupture freeing the oil. Minimize this damage to prevent absorption of liberated oil by hulls. Both problems increase the oil content of the hulls, which are removed, thereby reducing oil yield. Properly removed hulls should contain <2.5% oil (Serrato, 1981). Barger (1981) indicates that properly cracked soybeans will have a particle size distribution of 10–15% on a 6-mesh screen, 60–70% on 10-mesh, 5–15% on 20-mesh, and 0–3% through 20-mesh.

Soybeans are cracked using a cracking rollermill comprised of two corrugated rolls turning counter to each other with one roll turning slightly faster than the other, referred to as differential, to shear the seed. The sharp sides of the corrugations on the slow roll face up while the sharp side of the fast roll faces down. To achieve proper cracking, the rolls must have the right number of corrugations per unit of circumference and correct spiral, and the rolls must be parallel (same distance between rolls along the gap) and tram (horizontally parallel).

Shriveled, wrinkled, and misshapen beans caused by droughty growing conditions affect cracking and dehulling operations, and such beans are more difficult to dehull and control meal protein content (Brumm et al., 1990b). Equipment adjustments may be needed to minimize loss of meats in hulls and other problems and to maximize meal quality.

Shaker screens, aspirators, and/or gravity tables may be used to separate the hulls from the meats. Hulls should be larger in size, lower in density, and thus are more buoyant in an air stream than the oil-rich meat and are easily removed by cascade aspirators. The hulls-rich stream may then be sent to shaker screens and/or gravity tables to remove any small meat particles that were aspirated with the hulls. The recovered meat particles are recycled back to the dehulled meats stream to minimize oil and protein losses. The hulls may be heat-treated to inactivate anti-nutritional factors before being sold for feed. Part of the hulls may be blended back with the meal to control protein level, but the majority is sold as a separate co-product for cattle roughage.

Recently developed hot-dehulling systems eliminate the need for tempering dried beans. Cleaned beans are heated to 60°C over 20–30 min to allow moisture to migrate to the surface, then quickly heated to 85°C to lower the bean moisture content by 1–3 percentage points to help decortication. The beans are split into halves, and the hulls are further loosened by friction or impact and removed by aspiration. Hot dehulling is more efficient than conventional dehulling and is capable of more complete hull removal. Hot dehulling is becoming more important to make higher meal protein levels from lower protein content beans. Hot dehulling can handle up to 14% moisture soybeans, often eliminating the need for grain drying except where longer-term storage conditions dictate. In recent years, hot-dehulling systems have largely replaced drying/tempering/cold dehulling as the industry standard practice.

To minimize oil loss, complete hull removal is not desired when marketing the protein as meal for livestock feeding, only about one-half are removed. On the other hand, when making edible flours or *white flakes* (untoasted soybean meal) for manufacturing protein concentrates and isolates, near complete removal of hulls is required. Over 90% of the hulls from soybeans must be removed to assure that the minimum specification of 50% protein in the defatted meal is met. Hulls may also adversely affect the appearance, functionality, and performance characteristics of soy protein ingredients, especially flour. Livestock feeders are also moving toward higher-protein-containing meals, yet it is becoming increasingly difficult to produce soybean meal with the desired high protein content because increasing soybean yields on the farm are depressing protein content, especially in the northern soybean-growing areas. These trends have led to increased acceptance of hot-dehulling methods.

Hard Screw Pressing

In the United States, hard screw pressing is largely limited to the minor oilseeds (e.g., peanuts, rapeseed, and some cottonseed and sunflower seed) or locations where soybean supplies are not sufficient for large-scale solvent-extraction plants. Interest, however, is growing in hard screw pressing soybeans to produce organic, specialty (such as unique fatty acid composition), and nongenetically-modified vegetable oils and meals. In some locations, local laws prevent construction of new solvent plants, such as in California, and hard pressing is the only option for local processing. A small amount of soybeans is hard-pressed where the meal has particularly high value (e.g., high rumen by-pass protein and high metabolizable energy for dairy cattle feed) or where identity-preserved processing is desired (Fig. 11.7). The few remaining hard-screw-press plants crushing soybeans are quite small with the exception of one Ralston, Iowa, plant that processes over 910 metric tons per day (1,000 tons/day), where most of the oil is used to manufacture biodiesel and the meal captures a premium in dairy

Fig. 11.7. Photograph of the interior of a hard-screw-press plant operating on soybeans.

cattle feeds. Today, an estimated 3% of the U.S. soybean production is hard screw pressed including extruding-expelling and gas-supported screw pressing (both discussed later). This amount is up from an estimated 1% in 1995 (Erickson, 1995).

Prior to hard screw pressing, soybeans are usually cracked using corrugated roller-mills and heated in rotary-tube steam dryers or stack cookers (Fig. 11.1). The heated and dried beans are then conveyed to continuously fed screw presses. By employing proper drying and heating methods (2–3% moisture, 104–115°C and using a well-maintained, modern screw press, as low as 3–4% residual oil contents in cake can reportedly be achieved; but this is rarely, if ever, the case, and residual oil contents are more like 5–8%. Using more and more pressure becomes self-defeating, because pressure and heat cause capillaries to be reduced in volume, sheared, and eventually sealed by coagulation of protein.

A screw press is a continuous screw auger designed to subject the oil-rich material to increasing pressure as it is conveyed through the barrel. The barrel forms a cage of bars surrounding the screw in a parallel fashion. The bars of the cage are separated by spacers, which allow the oil to drain from the cage while the solids are conveyed down the barrel toward higher pressure. The discharge opening for the reduced-oil solids and, thus, the back pressure are controlled by a choking device. A plug of compressed oil-lean solids, termed *cake*, forms at the discharge. The pressure increases down the length of the barrel by increasing the root diameter of the screw, decreasing the pitch of the screw flights, and controlling the cake discharge opening by means of a choking device. The feed is rammed against the choking device causing the solids to be compressed and the oil to be squeezed out through the bars of the cage. In hard screw pressing, the presses are choked to achieve maximum pressure within the barrel so as to release as much oil as possible while maintaining desired cake discharge flow and an acceptable amount of *foots* (cellular debris) in the oil.

Considerable frictional heat is generated that must be removed to achieve low residual oil and prevent damage to the cake and oil. Some screw presses recycle cooled pressed oil over the cage to remove the heat, while others use water-cooled shafts and bar cages.

The oil is pumped or flows by gravity to a sedimentation basin to facilitate settling of the foots. The settled foots are recycled to the press to reclaim entrained oil. The oil is filtered and placed into storage tanks. The cake is then ground into meal.

In recent years, extruding and then hard screw pressing soybeans (Figs. 11.8, 11.9), often referred to as *extruding-expelling*, has become popular, especially where a local dairy cattle feeding industry is present to utilize the higher fat-containing meals. Unlike traditional hard pressing, however, extruding-expelling produces meal with low rumen by-pass protein, and beef cattle producers are unwilling to pay premium prices. Poultry feeders do not like the higher fat content compared to solvent-extracted meal. The result is that the extruder-expeller meal may be difficult to market. Because these plants are quite small by today's standards, few markets exist for the oil—a few major food companies and a few other companies producing biolubricants and other industrial products. This is changing, however, as more and more biodiesel plants come on line, and the marketability of locally extruded-expelled oil improves.

Interest is growing among these small extruding-expelling plants to process specialty soybean oils where the oil market does not justify larger plants. Some of these extrusion-expelling plants seek to develop low-cost refining techniques so that they also can bottle edible oils for niche markets. Some of the niche markets for specialty soybean oils include: oil high in oleic acid for better oxidative stability; oil low in saturated fat for health advantages; oil high in saturated fat to eliminate the need for hydrogenation; oil low in linolenic acid for better shelf life; organically produced oil; and oil certified to be free of transgenic modification (such as those tolerant to herbicides).

Prepress Solvent Extraction

Only when soybeans are processed in plants designed for other oilseeds (e.g., cottonseed, sunflower seed, etc.), which must be prepressed, are soybeans subjected to prepress solvent extraction. Soybeans do not contain sufficient oil to require or justify prepressing and are rarely, if ever, prepress solvent extracted. In prepress solvent extraction, part of the oil, that which is easily removed, is pressed out, generally as de-

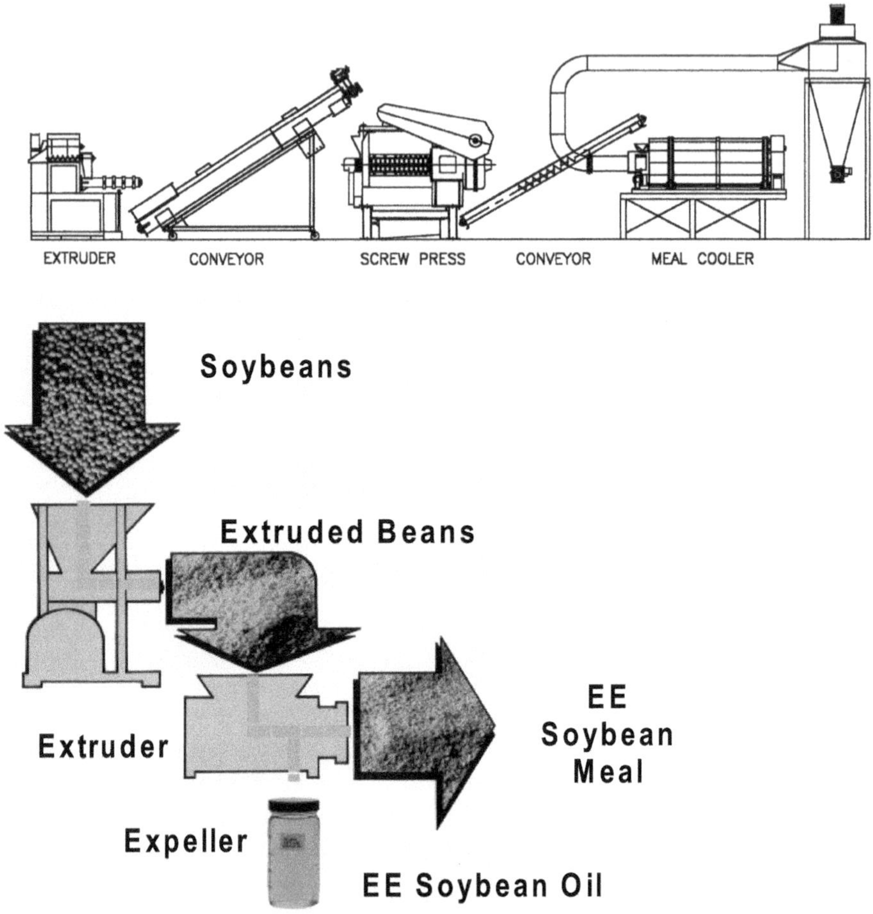

Fig. 11.8. Diagrams of the extruding-expelling process (adapted from Weijratne et al., 2004).

scribed for hard screw pressing. It may be helpful, however, to understand the process for comparing soybean processing with other oilseeds.

In prepressing, the screw press is choked so that less pressure is developed than in hard screw pressing; therefore, less press oil is recovered, and the capacity of the screw press is increased. The oil content of prepress cake is typically 14–16%, and the partially de-oiled cake is then extracted with solvent. The cake may be broken into pieces and even flaked to decrease particle size, to increase bulk density and extractor capacity, and to speed solvent extraction. The remaining steps are the same as for

Fig. 11.9. Photograph of the interior of an extruder-expeller plant operating on soybeans.

direct solvent extraction, which is described in the next section. Usually screw-pressed and solvent-extracted oils are mixed before transporting to a refinery.

Direct Solvent Extraction

Solvent Selection

The solvent of choice today is hexane, but other solvents have been tried (ethanol, isopropanol, acetone, isohexane, heptane, trichlorethylene). Alternative extraction solvents were extensively reviewed by Johnson and Lusas (1983), Johnson (1997), and Hron (1982). The disadvantages of hexane are its flammability, its price structure that is tied to petroleum prices, and increasing regulatory pressures on emissions; whereas, its advantages include high oil solubility, ease of evaporation, abundant availability, and a historically low price (but now hexane prices are escalating due to increases in petroleum prices). Because of the high flammability of industrial hexane, considerable effort and capital are invested in safety. The lower explosive limits are 1.2 to 6.9 vol% hexane in air, which is the concentration range at which hexane is flammable if exposed to an ignition source (spark). The National Fire Protection Agency (NFPA 36) provides guidelines for the design of oilseed extraction plants and the use of hexane.

Extraction Mechanism

Solvent extraction of soybeans is a combination of classic mass transfer unit operations to engineers; however, it does not conform to simple leaching theory because there are multiple barriers to mass transfer of oil in flaked soybeans into the bulk solvent. Leaching theory applies well only when all cell walls and intracellular membranes are ruptured and the material holds oil much like our example of a paint brush or a sponge. The walls of soybean cells and organelle membranes are largely impermeable to oil and nearly so to extraction solvents; consequently, they must be distorted (ruptured) prior to extraction. This requires reducing the size of soybean meats and flaking to crush cell walls by compressing the meat particle. In addition to cell distortion, the rate of oil extraction also depends upon flake thickness.

With the traditional cold-dehullung system, cracked soybeans are conditioned (heated without drying) prior to flaking by using vertical-stack cookers or rotary-tube conditioners to heat the seed to 60–70°C over a 10–15 min period while maintaining ~10% moisture (slight injection of steam is required). Small pieces transfer heat and moisture more readily during conditioning, but, excessive size reduction reduces mechanical distortion during flaking. With the new hot-dehulling systems, conditioning is performed in vertical seed conditioners or similar devices, eliminating traditional cookers and conditioners. These treatments make the meats pliable so that they can be flaked to about 0.25 mm (0.010–0.012 inches) thickness without producing excessive fines, which adversely affect other operations. Slightly thicker flakes (0.4–0.5 mm, 0.015–0.020 inches) are sometimes used in deep-bed extractors in order to better withstand the more rigorous handling. Proper plastic texture is necessary to produce thin, nonfragile flakes with minimum fines and maximum cell rupture.

Flaking is carried out by passing the cracked conditioned meats through a set of two smooth-surfaced roller mills. Just as in cracking, the rolls must be turning at proper speed and differential and must be in parallel and in tram. The feeder must evenly distribute the cracked meats along the gap or nip so that the rolls wear evenly and the proper gap is maintained across the entire roll. Flaking larger particles causes more cell rupture than flaking smaller particles. The flaked material must have tenacious, thin structure with porosity that allows transport of the oil or *miscella* (solvent–oil mixture).

The moisture content of the flakes is still another factor affecting the rate of solvent extraction. In most cases, 9.5–10.5% moisture is ideal. Hexane and water are immiscible, and higher moisture contents interfere with the penetration of hexane. Lower moisture levels reduce the structural strength of the flakes and produce more meal fines.

Solvent temperature also greatly impacts extraction rate, so the solvent is heated close to the boiling point (bp of industrial hexane is 65–70°C). Avoid solvent boiling, however, because extraction vessels are not designed for pressurization, and solvent losses increase and safety is compromised.

In actual practice, soybean extraction does not follow the single mechanism of leaching as described in the overly simplified brush-cleaning analogy. Instead, soybean extraction involves a combination of leaching, diffusion, and dialysis. This combination of mechanisms leads to an ever-decreasing rate of extraction as the relative importance of each mechanism changes during the course of extraction. For flakes, the larger proportion of readily extractable oil originates with ruptured cells, especially those near the flake surface. The transfer of oil from ruptured interior cells is governed by capillary flow, and the rate of oil transfer is partly dependent on the viscosity of the miscella. A portion of the slowly extracted oil is contained within intact unruptured cells and must be transferred by osmosis, which is very slow.

In recent years, some soybean plants, especially those using deep-bed Rotocel extractors, have adopted expanding soybean flakes prior to solvent extracting by using an extruder-like device called an *expander* (Williams, 1990; Lusas & Watkins, 1988; Pedrotti & Boling, 1996). Expanding flakes increases capacity of an extractor by producing collets with greater packing and bulk densities and reducing extraction time. Presumably, expanding shifts the relative importance toward the leaching mechanism of oil extraction because nearly all of the cells are ruptured, and the collet (extrudate) is quite porous. More complete rupture of cell walls is achieved by the additional work done to the soybeans due to the high shear and heat of expanding or extruding. The rapid pressure release creates internal moisture evaporation, which destroys cell wall integrity. This enables faster extraction and shorter extraction times to the same residual oil content. Collets also drain more completely than do flakes reducing the amount of energy required to desolventize the meal, which is becoming increasingly important as energy prices escalate. Expanders are most often seen in plants using deep-bed extractors (>1 m bed depth) because the benefits of higher density and increased percolation are not as significant with shallow-bed extractors.

Expanders also inactivate catabolic enzymes due to heat, just as cooking does in the ALCON process. Both the expanding and the cooking of flakes increase the levels and types of phosphatides present in the crude oil, but the phosphatides are more hydratable and more easily removed during water degumming. Expanders can affect the quality of edible lecithin, especially in the formation of white haze. Cooked or expanded soybean flakes need to be dried to <10% before solvent extracting to avoid color reversion (Erickson, 1995a).

The best quality oil, low in phosphatides, free fatty acids, nonsaponifiable matter and pigments, is extracted first, while poorer quality lipids are extracted with more exhaustive extraction. The industry, however, strives for the most complete extraction possible. Residual oil contents of finished solvent-extracted soybean meal range 0.5–1.5% (toasting defatted soybean meal liberates oil, and toasted meal will test higher in oil content than do drained untoasted extracted flakes). Some have advocated doing less complete oil extraction to reduce refining costs, but not all extraction plants have capabilities to alkali refine and, thus, there is little incentive to extraction plants to minimize refining loss.

Extractor Design
In all of today's extractors, the flow of solvent relative to the flakes is countercurrent to reduce the amount of solvent used (Fig. 11.10) and thus the amount of solvent that must be evaporated. In countercurrent systems, the freshest flakes contact the solvent richest in oil and progress through the extractor until the oil-free flakes contact fresh solvent. The extractor is a vessel enclosed to contain solvent vapors and designed to wash, extract, drain, and transport flakes relative to the solvent. Because hexane is highly flammable, controlling solvent vapor loss to an absolute minimum is critical for safety. Fire and explosions have caused loss of life and property damage. Insurance costs as well are significant to processors. Therefore, every effort and cost are expended to provide safe working conditions for operators.

Two principle types of extractors have been employed—*immersion extractors* and *percolation extractors* (Fig. 11.11). An immersion extractor immerses and soaks the material to be extracted in solvent (a laboratory example is the Soxhlet extractor). Immersion extractors require more solvent usage and have material conveyance problems; therefore, percolation extractors almost exclusively predominate in the industry today. Early soybean extractors were immersion extractors, but the industry quickly moved to percolation extractors. In a percolation extractor, the solvent percolates by

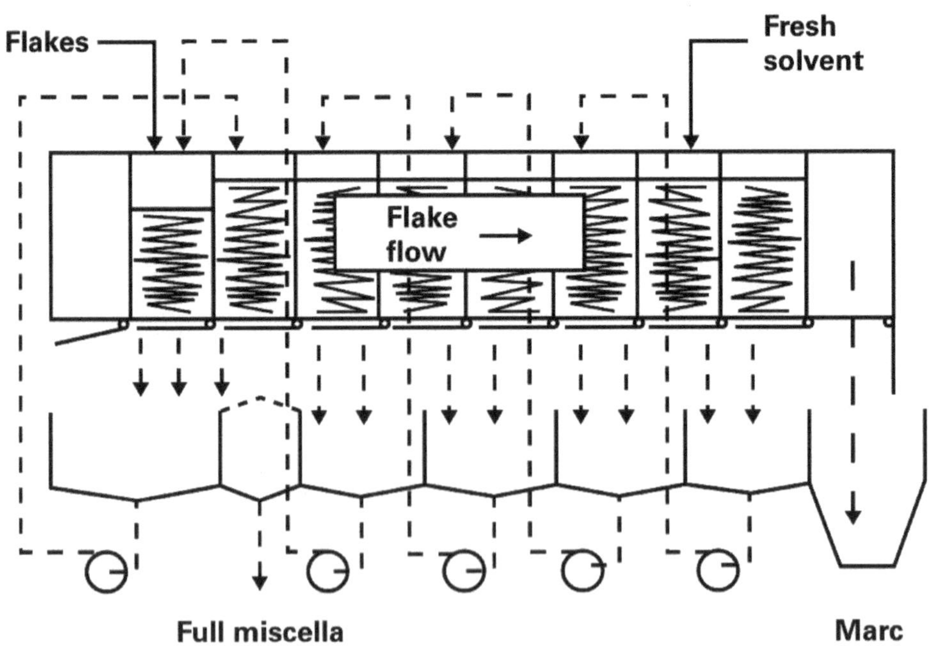

Fig. 11.10. Depiction of countercurrent extraction principles (redrawn from Milligan, 1976).

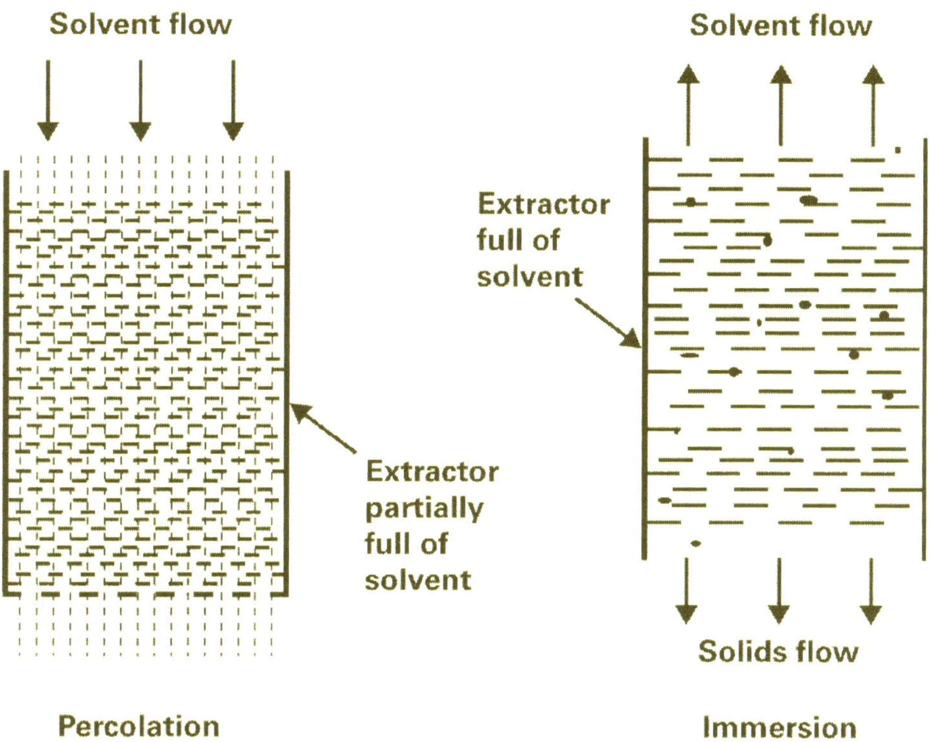

Fig. 11.11. Depiction of percolation-extraction and immersion-extraction principles (redrawn from Milligan, 1976).

gravity through a bed of material (a laboratory example of a percolation extractor is the Goldfisch apparatus). As the solvent percolates downward through the flake bed, the solvent flows over the flakes and diffuses through them dissolving the oil. The mixture of solvent and oil is termed *miscella.* Miscella flows in successive passes through the bed while the flake bed moves in opposite direction to the solvent flow. The oil concentration increases as the number of passes increases (usually 5–6 passes).

The extraction principles employed by most percolation extractors are the same, but the method by which each achieves countercurrent flow of solvent to flakes is different. The shallow-bed, chain extractor (Fig. 11.12), which resembles a full-loop conveyor, is one of today's widely used extractors. Crown Iron Works (Minneapolis, MN) manufactures this type of extractor. In early versions that still exist in the industry, flakes are fed into an inlet hopper and are conveyed down the first leg of the loop where they are washed with moderately dilute miscella to extract surface oil and penetrate the cells. As the flake bed moves into the bottom horizontal section, full miscella is recycled through the bed for filtering, and then to a liquid cyclone for final

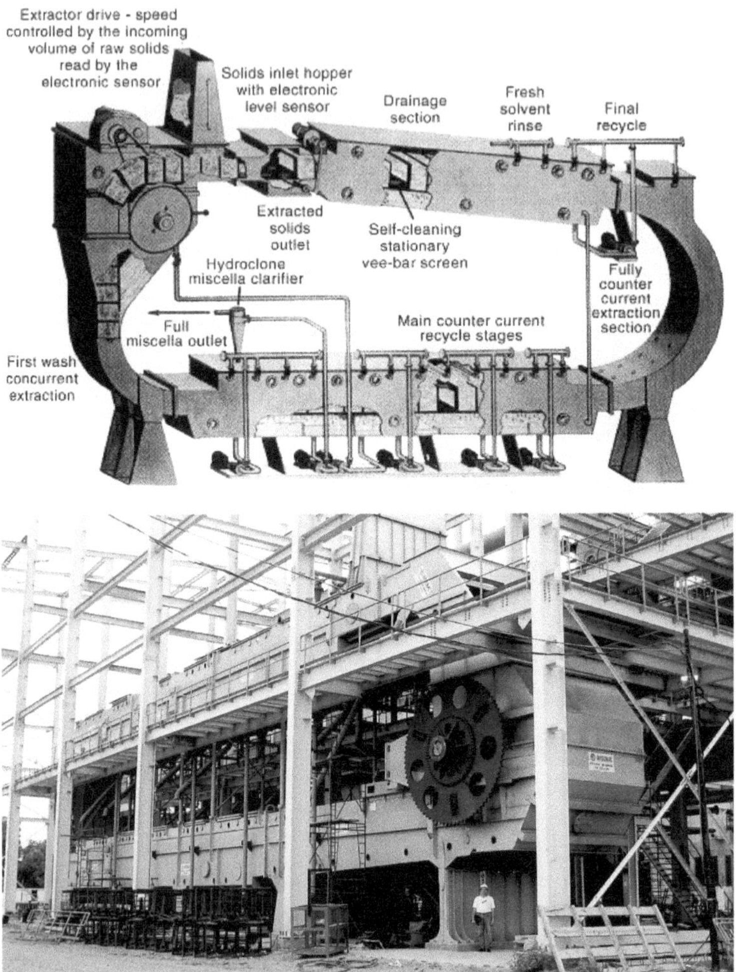

Fig. 11.12. Depiction of a shallow-bed, chain-type extractor; early design (A) and more recent design (B) (provided by Crown Iron Works, Minneapolis, MN).

removal of fines. The clarified full miscella goes to the evaporation system. Flakes are conveyed counterclockwise, through progressively more dilute miscella washes. A final wash with fresh solvent is used in the top horizontal section of the loop. The latter half of the top loop is for drainage, after which the *marc* (solvent-laden spent flakes) is conveyed to the meal desolventizer. The chain extractor uses relatively shallow bed depths (generally <1 m), which promote drainage, and thus has low solvent hold-up or carryover to the meal desolventizer. In recent designs to achieve greater capacity, the flakes are fed on the top, pass by one elbow and back discharging on the bottom, while one elbow is empty.

A second type of extractor widely used on soybeans is the deep-bed (34 m deep), rotary-basket extractor (Fig. 11.13), such as the "Reflex" extractor marketed by Desmet Ballestra. Countercurrent solvent-to-flake flow is accomplished by rotating the baskets of flaked material while the solvent and miscella sprays, the miscella collection cells, and the marc discharge remain stationary. The bed is divided into cells or baskets to prevent back mixing of oil-lean miscella with oil-rich miscella. The drained marc discharges when the basket rotates to the position above the discharge hopper.

Fig. 11.13. Depiction of a deep-bed, rotary basket extractor ("Reflex" extractor, Desmet Ballestra North America, Marietta, GA).

A modification of this type of extractor is the stationary basket extractor (Fig. 11.14), such as was manufactured by the French Oil Mill Machinery Co. (Piqua, OH). Unlike other extractors, the solids do not move. Countercurrent solvent-to-flake flow is accomplished by rotating the flake inlet, the solvent and miscella sprays, the miscella collection cells, and the marc discharge. The bed is divided into cells or baskets to prevent back mixing of oil-lean miscella with oil-rich miscella. Typically the bed depths used in this extractor are much deeper than those of the chain extractor. The drained marc discharges when an opening in the bottom screen and the discharge hopper rotate into appropriate positions. To the author's knowledge none of this type of extractor has been installed for over 12 years, but they do exist in the industry.

A third type of extractor is the belt-type (Fig. 11.15), such as one manufactured by Desmet Ballestra, but this extractor is more commonly used on oilseeds other than soybeans that are prepressed. In this type of extractor, the flakes are conveyed through a series of solvent sprays by means of a belt. Fresh solvent is introduced at the discharge end and is circulated countercurrent to the flow of flakes by a series of stage pumps. No dividers are in the belt, but the belt is inclined to assure countercurrent flow of solvent to solids. The drained marc discharges the extraction belt by means of rotary paddles.

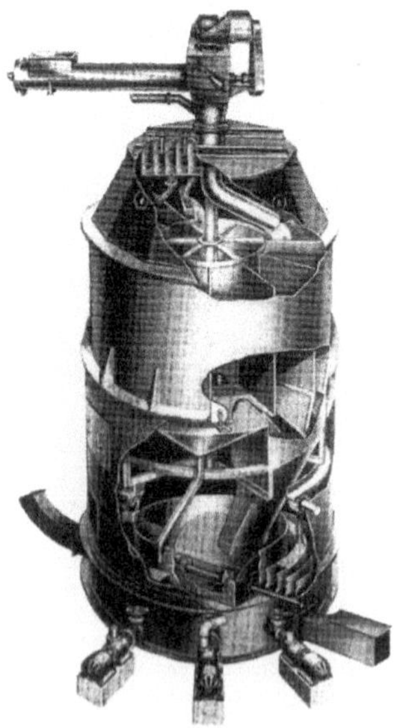

Fig. 11.14. Depiction of a deep-bed, stationary-basket extractor (provided by French Oil Mill Machinery Co., Piqua, OH, rights now owned by Desmet Process and Technology, Marietta, GA).

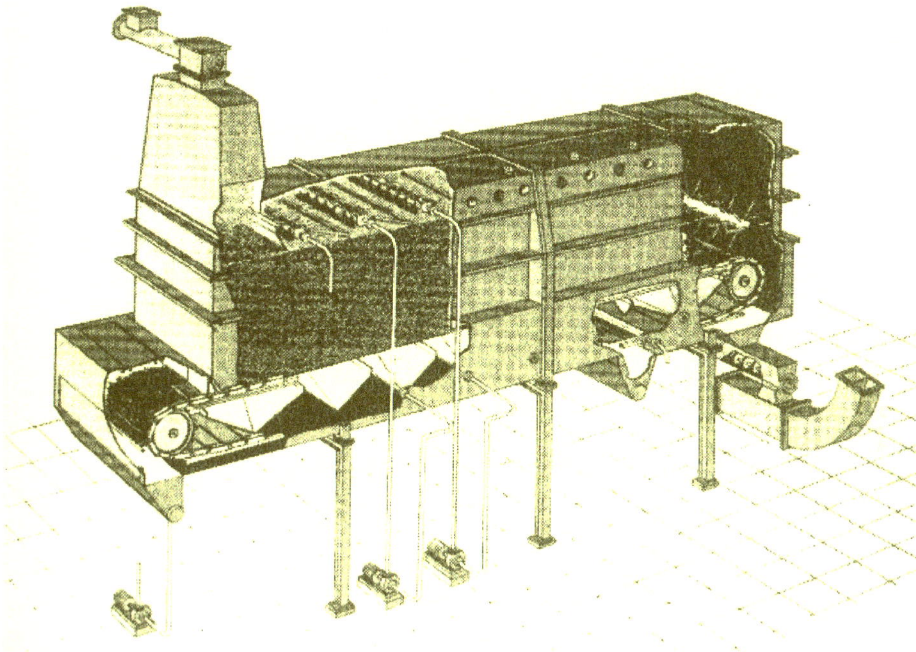

Fig. 11.15. Depiction of a deep-bed, belt-type extractor (provided by Desmet Process and Technology, Marietta, GA).

Extraction Operations

Soybean flakes enter the extractor by passing through a vapor seal that prevents flammable vapors from escaping. The flakes are extracted for 30–60 min, depending on type of extractor and method used to prepare the beans, to achieve <1% residual oil in the extracted material. In soybean extraction, less than one part of hexane is used for each part of soybean flake extracted, and the industry continually strives to reduce the amount of hexane used and the amount of energy consumed in evaporating.

The full miscella exiting the extractor contains 22–30% oil (the higher the better to reduce miscella evaporation costs) and is sent to evaporation and stripping columns to separate the oil from hexane. The full miscella is heated under vacuum to evaporate the solvent in two stages of evaporators. The first-stage evaporator concentrates the oil to 70–85% using reclaimed heat from hot solvent vapors from the meal desolventizer. The second-stage evaporator uses steam to concentrate the oil on up to 93–98%. This concentrated oil is sent to a stripping column where heat, vacuum, and live steam are used to remove most of the remaining hexane. The hexane is recycled to the extractor to be used again. After the stripping column, the crude oil contains <0.15% moisture and hexane. The stripped oil then goes to an oil dryer to remove any condensed moisture. Normally, the oil exiting the oil dryer contains <100 ppm of hexane. The oil must have a flash point greater than 121°C (250°F), which is equivalent to <800 ppm of hexane (the lower the flash point, the more residual hexane in the oil). The remaining traces of hexane are removed during deodorization at the oil refinery. Excessively high temperatures (>115°C, >240°F) in desolventizing miscella can increase nonhydratable phosphatide content (Erickson, 1995a) and cause dark-colored oil.

Unlike screw-pressing and extruding-expelling plants, solvent-extraction plants often partially degum crude soybean oil before transporting to centralized oil refineries. About one-third of the gums that can be produced are used to make edible lecithin, which is used as a common emulsifier in food products. The lecithin market is not sufficiently large to take all the lecithin that can be produced, and the remainder is added back onto the meal just ahead of the desolventizer/toaster to provide additional metabolizable energy in livestock rations. That way, at least, meal prices are obtained for gums, and the value of the oil is increased because degummed oil with lower refining loss is sold. Many U.S. soybean plants remove gums with soap and put both back onto the meal as a single stream. These differences in practice between extraction plants account for some of the differences in nutrient contents of soybean meals.

Meal Desolventizing

The marc generally contains 30–32% hexane (known as *solvent hold-up*), which also must be recovered and recycled to the extractor. Heat must be used to evaporate the hexane from the meal. Live steam is injected as the primary heat source and to provide moisture vapor as a stripping gas to transport solvent vapors to condensers. The extracted flakes, known as *spent flakes*, must be drained of solvent as much as possible to reduce the amount of energy required to desolventize the meal.

More than 97% of the available soybean meal is used for feed where extensive heat treatment is necessary to maximize feed conversion efficiency by livestock. Toasting inactivates protease inhibitors (especially trypsin and chymotrypsin inhibitors) and the enzyme urease, and improves protein digestibility. None of these objectives can be obtained without protein being denatured and loss in water solubility; however, depending on the method used, meals with great differences in protein solubilities or dispersibilities can be produced. The optimum amount of heat treatment in toasting soybean meal is still debated among animal nutritionists.

Very few tray desolventizer/toasters are left in the soybean industry but are described in the older literature and provide a benchmark for describing more current technology. A tray desolventizer/toaster (DT) is a vessel composed of about six stacked trays, all with indirect heating. The first two also have provisions for injecting live steam through nozzles in the sweep arms. The meal advances down through the trays evaporating more and more solvent. The lower four trays are essentially toasting/drying sections where the meal is held at a minimum temperature of 100°C, and the meal is partially dried before going to the meal dryer.

Soybean meal should be dried to about 13–14% moisture and cooled for safe storage. A hammer mill is used to grind the meal to uniform particle size before it is sold and shipped.

The Schumacher-type desolventizer/toaster/dryer/cooler was designed to reduce energy use and has become widely accepted (Fig. 11.16). This device consists of multiple trays where the top one or two trays are for pre-desolventizing; the second set of one or two trays is for desolventizing-toasting with injection of steam through perforated bottoms (achieving countercurrent use of steam relative to solvent evaporation); the third set of one tray is for sparging steam; the fourth set of one or more trays is for drying with hot air blown through perforated bottoms; and the fifth set of one or more trays is for cooling by blowing cold air through perforated bottoms.

Despite much research to demonstrate new uses, only about 3% of the available soybean meal is processed into edible flours and protein concentrates (>65% protein) and isolates (>90% protein). The *flash desolventizer* was developed to reduce protein denaturation and produce highly soluble protein food ingredients from soybeans (Fig. 11.17). Integrating these desolventizing systems with subsequent cooking systems produces edible protein flours with a broad spectrum of protein dispersibility characteristics. The system includes a desolventizing tube, a flake separator, a circulating

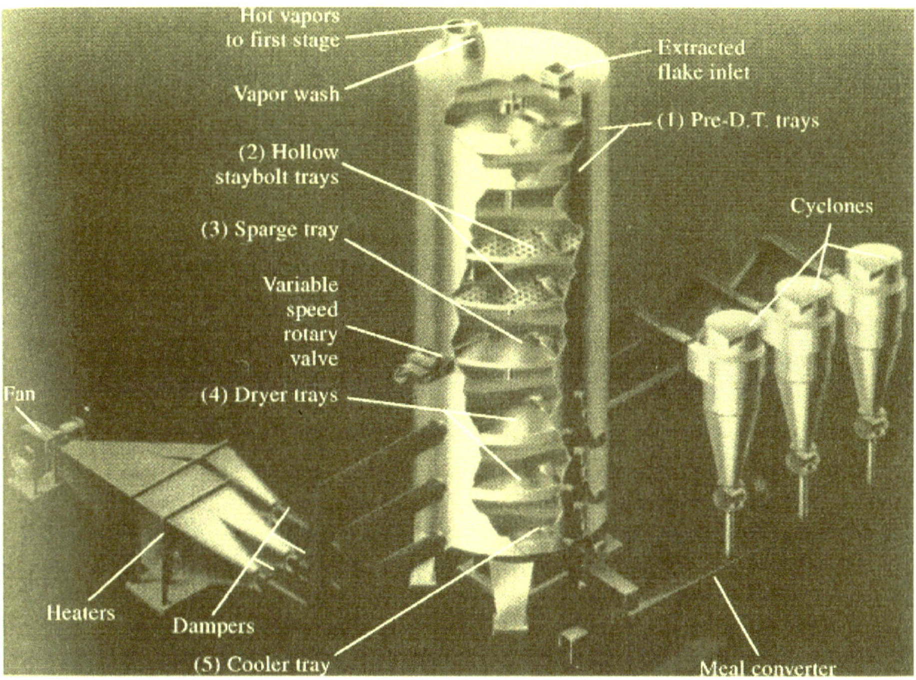

Fig. 11.16. Depiction of the Schumacher-type desolventizer/toaster/dryer/cooler (provided by Crown Iron Works, Minneapolis, MN).

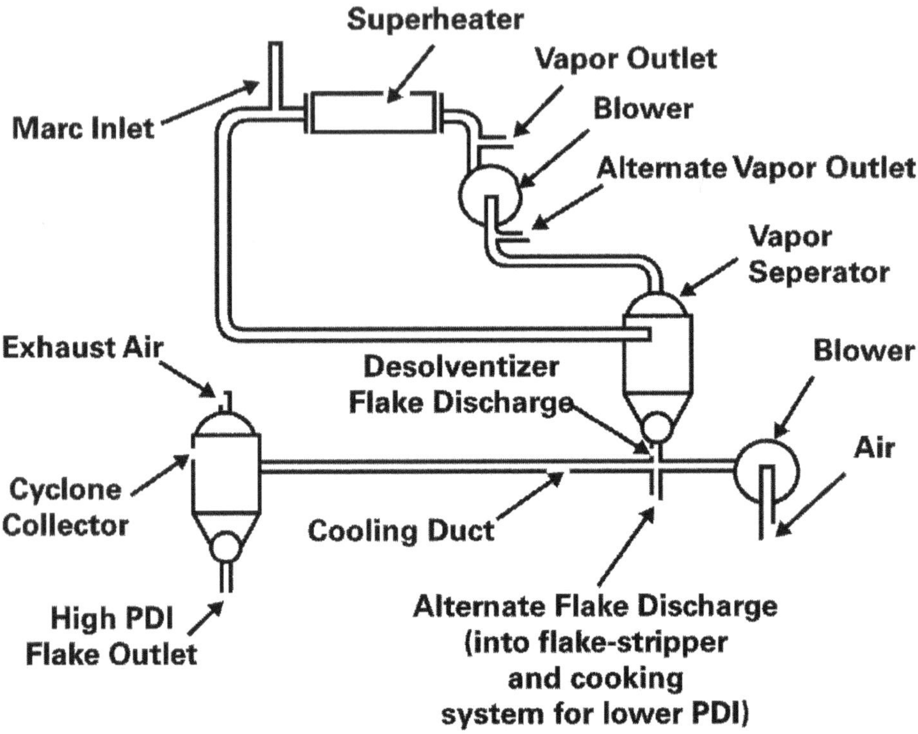

Fig. 11.17. Depiction of a flash-desolventizing system.

blower, and a vapor heater. These units are arranged in a closed loop in which hexane vapor is superheated under pressure and continuously circulated. Solvent-laden flakes, usually from dehulled soybeans, are fed into the system and conveyed by the high-velocity circulating vapor stream. The turbulent superheated vapor flow (160°C) elevates the temperature of the flakes to 77–88°C over a period of <3 sec, well above the boiling point of hexane (65°C). As the flakes travel through the tube to the cyclone separator, the greatest portion of the entrained hexane is evaporated. Because the flakes enter the flash desolventizer at low moisture for a very short period and no steam is injected into the vapor stream, little protein denaturation occurs. At the exit point, the protein dispersibility index (PDI) of the protein will be as high as 90, but normally >80. The substantially desolventized flakes, known as *white flakes*, are removed from the system through a cyclone with a vapor-tight, rotary airlock and go to deodorizers and then to cookers if moderate protein solubilities are desired.

A relatively recent alternative to traditional flash-desolventizing systems is the down-draft desolventizer (DDD system). The DDD system incorporates a low-temperature, low-pressure desolventizer to reduce the hexane content to 2500 ppm in the first stage, followed by a high-vacuum stripper to produce white flakes with a residual hexane content <500 ppm. The claimed major advantages of DDD systems are lower capital and operating costs, consistently producing white flakes with low residual hexane content, and maintaining flake integrity. White flake integrity is important for manufacturing soy protein concentrate where sugars are extracted with alcohol in a second extractor.

Condensed solvent and water from the meal desolventizer/toaster and the oil stripper must be separated in gravity water separators before recycling solvent to the extractor. The water phase is then heated to remove residual hexane. Hexane losses must be minimized, and much engineering has gone into reducing hexane loss, which is now typically ~1 L/metric ton of soybeans processed.

Meal Grinding

Desolventized meal is ground with a hammer mill so that 95% passes through a U.S. 10-mesh screen, and a maximum of 3–6% passes through a U.S. 80-mesh screen. Excessive meal grinding causes dust problems during feed handling. Meal for edible purposes is ground, sized and sold as grits in a wide variety of sizes and flour (<U.S. 100 mesh). Edible flours may have added refined oil or lecithin or undergone texturization by using extruders. When white flakes are used for manufacturing soy protein concentrates or isolates, the meal is not usually ground because protein purification is easier with flaked material.

Oil and Meal Storage

Both oil and meal must be cooled before placing in storage because high temperatures accelerate degradation reactions. Store neither oil nor meal any longer than necessary. Oil degrades through oxidation, and reducing contact with air is important. Crude oil is more stable to oxidation than refined oil because crude oil contains natural antioxidants that are removed in refining steps. Definitely prevent oil from contacting water so that hydrolysis does not occur, which increases free fatty acid content and refining losses. Thus, protections against water, heat, and air are important for maintaining oil quality.

Laboratory Simulation of Commercial Extraction

Oftentimes, commercial continuous extraction is simulated in the laboratory for research purposes, but a high frequency of faulty experiments occurs. The most frequent error is using *n*-hexane as solvent. Industry does not use *n*-hexane, but rather a narrow distillation cut, containing 40–70% *n*-hexane (Johnson, 1997; Lusas & Gregory, 1996). The remainder is various isomers of hexane and therefore is often termed *hexanes*. Table 11.2 shows a typical specification for industrial hexanes used in oilseeds extraction. The older literature, of which current researchers often seem to be unaware, documents well that these isomers act as extraction enhancers and improve oil extraction (Ayres & Dooley, 1948; Arnold & Choudhury, 1960).

Another common fault is the use of a Soxhlet extractor or other common lab extractors to simulate commercial oilseed extraction. A Soxhlet extractor is an immersion extractor with almost infinite number of fresh solvent stages and very long extraction times (>3 h), while today's commercial extractors are percolation extractors using only about six stages of solvent with declining oil concentrations within <1 h. It is far better to percolate heated solvent through the bed of flakes in a jacketed glass vessel in which the material to be extracted is held on a screen above the outlet. Restrict the solvent/miscella flow rate to give 6–10 min residence time per stage.

Further improve simulation of continuous solvent extraction by removing only the full miscella from the first extraction stage and recycling the other miscellas to the next extraction sequence by advancing by one stage. Use fresh solvent only at

Table 11.2. Typical Purchase Specifications for Oilseed Extraction Hexane

Property	Value	Test
Specific gravity @ 25 °C (g/cc)	0.6705-0.6805	ASTM D 1963-61
Distillation range (760 mm)		
Minimum initial boiling point (°C)	65.0	ASTM 1078-63
Typical 10% distillation (°C)	67.1	
Typical 50% distillation (°C)	67.7	
Typical 90% distillation (°C)	68.2	
Maximum dry point (°C)	70.0	
Maximum nonvolatile residue (g/100 mL)	0.001	
Acidity of distillation residue	Neutral	
Closed-cup flash point (°C)	-32 to -58	ASTM D 56-61
Maximum sulfur	10	ASTM D 1266-62T
Maximum vapor pressure (psia @ 35 °C)	6.0	ASTM D 323-58
Composition (GLC, % area)		
n-Hexane	45-70	
Methyl cyclopentane	10-25	
Total n-hexane and methyl cyclopentane	60-80	
Total 2-methyl pentane; 2,3 dimethylbutane; and 3-methyl pentane	18-36	
Maximum cyclohexane	2.5	
Maximum benzene	0.1	
Maximum APHA color	15	ASTM D 1209-62
General appearance	Free of foreign material	

the last stage. After 6–8 extraction trials, steady-state oil concentrations are achieved and experimental replications can start. This way, proper solvent usage and amount are achieved. The author has used this method successfully many times. Other than discarding six to seven runs to achieve steady-state oil concentrations in miscella, the method works extremely well.

A third frequent error is to extract ground material instead of flaked material. As discussed earlier, cell distortion is critical to any material being as much like a sponge, where the cell and membranes are all ruptured, for easy extraction as possible. Grinding does not achieve as extensive cell distortion as does flaking or flaking/expanding. Additionally, finely ground material does not allow the solvent to percolate efficiently through the extraction bed.

Product Qualities

Oil

As previously noted, oil quality can be affected by a number of soybean processing operations as well as post-processing operations (crude oil storage, handling and transport) (Boring, 1995). The trading rules for prime crude soybean oil and crude degummed soybean oil (NOPA, 2006a) as established by the National Oilseed Processing Association specify sampling and testing procedures, analytical requirements without penalty discounts, and a schedule of discounts (Table 11.3). These trading rules specify that "The Official Methods" of the American Oil Chemists' Society be used. Wang and Johnson (2001) and Deak et al. (2007) surveyed the quality of soybean oil produced by different extraction methods, and Table 11.4 shows their findings.

Meal

The trading rules for prime crude soybean oil and crude degummed soybean oil (NOPA, 2006b) as established by the National Oilseed Processing Association specify sampling and testing procedures (Table 11.5), analytical requirements without penalty discounts and a schedule of discounts. Wang and Johnson (2001) and Deak et al. (2007) surveyed the quality of soybean meal produced by different extraction methods, and Table 11.6 shows their findings. Typical dehulled, high-protein, solvent-extracted meal (the preponderance of soybean meal produced) will contain about 48% protein, 12% moisture, and <1.5% fat. Sometimes, soybean hulls or soybean mill-run are added back to the meal before grinding to adjust and/or precisely control protein levels to meet end-user specifications. Soybean meal is one of the most consistent and high-quality protein sources available to the livestock industry. Numerous studies show the amino acid digestibilities of soybean meal to be near 90%, which is significantly higher than alternative protein sources available to feed formulators. The

Table 11.3. NOPA Rules for Soybean Oil Qualities

Quality Trait	Prime Crude Soybean Oil	Crude Degummed Soybean Oil	AOCS Method
Moisture and volatile matter (%)	0.5 max	NA	Ca 2c-25
Moisture, volatiles and impurities (%)	NA	0.3 max	Ca 2c-25, Ca 3a-46
Free fatty acid content, as oleic (%)	NA	0.75 max	Ca 5a-40
Phosphorus (%)		0.02% max	Ca 12-55
Refined and bleached color (Lovibond red)	6.0 max	NA	
Neutral oil loss (%)	7.5 max	NA	
Unsaponifiable matter, exclusive of moisture and insoluble impurities (%)	1.5 max	NA	
Flash point (°F)	250 min	250 min	Cc 9c-95

Table 11.4. Soybean Oil Qualities Produced by Different Oil-extraction Methods

Quality Trait	Hard Screw Pressing[a]	Extruding-expelling[a]	Gas-assisted Screw Pressing[b]	Solvent Extracting[a]
Peroxide value (meq/kg)	1.76	1.73	0.35	0.96
Free fatty acid (%)	0.33	0.21	0.12	0.31
Phosphorus (ppm)	463	75	65	277[c]
AOM stability (h)	36.2	23.9	NA	39.8
Moisture (%)	0.05	0.08	0.33	0.08
Tocopherols (ppm)	1,217	1,257	NA	1,365
Color (Lovibond red)	17.5	10.2	4.1	11.1

[a]Data from Wang and Johnson (2001).
[b]Data from Deak et al. (2007).
[c]Oil is often degummed oil.

Table 11.5. NOPA Rules for Soybean Meal Qualities

Quality Trait	44% Protein Meal	High-protein Meal	AOCS Method
Moisture (%)	12.0% max	12.0% max	Ba 2a-38
Protein	44.0% min	47.5-49.0% min	Ba 4e-93
Crude fiber (%)	7.0% max	3.3-3.5% max	Ba 6-84
Oil (%)	0.5% min	0.5% min	Ba 3-38

Table 11.6. Soybean Meal Qualities Produced by Different Oil-extraction Methods

Quality Trait	Hard Screw Pressing[a]	Extruding-expelling[a]	Gas-assisted Screw Pressing[b]	Solvent Extracting[a]
Moisture (%)	11.0	6.9	7.1	11.7
Oil (%)	6.3	7.2	4.5	1.2
Protein (%)	43.2	42.5	47.3	48.8
Urease (ΔpH)	0.03	0.07	2.19	0.04
KOH solubility (%)	61.6	88.1	99.8	89.1
PDI	10.6	18.1	70.2	61.6
Rumen bypass (%)	48.1	37.6	NA	36.0
Color				
Hunter L	51.5	65.8	88.0	69.1
Hunter a	4.8	0.4	-1.61	2.0
Hunter b	14.8	16.6	15.1	4.8
Trypsin inhibitor				
(mg/g)	0.3	5.52	NA	5.46
(TIU/g)	2,000	12,254	53,400	5,275

[a]Data from Wang and Johnson (2001).
[b]Data from Deak et al. (2007).

livestock feed trade magazine *Feedstuffs* (Table 11.6) published more detailed analyses of meal compositional properties.

Animal nutritionists have long sought a rapid and easy chemical analysis that would predict performance of soybean meal in various species of livestock. Soybean meal must be toasted to inactivate antinutritional factors, however, to obtain maximum feed conversion and animal performance. Both quantity and quality or protein are important and, while simple crude protein and amino acid assays are useful measures of protein quantity, they do not provide information on protein quality. Trypsin inhibitor content, urease activity, protein solubility in potassium hydroxide or water, and dye-binding methods are used as measures of protein quality, especially heat damage. High urease activity indicates under-toasting, while low solubility in potassium hydroxide indicates over-toasting. More information on feed uses for soybean meal is presented in the Chapter *Nutritional Properties and Feeding Values of Soybeans and its Co-products.*

Alternative Solvents

A long-standing interest is shown in alternative extraction methods that use solvents other than hexane to extract soybeans due to safety and emission issues over hexane use as well as periodic hexane shortages when petroleum supplies are tight (Johnson & Lusas, 1983; Hron, 1982; Wan & Wakelyn, 1997). Hexane is highly flammable, and periodic accidents result in explosion and fire with loss of life or injury. Hexane is

Table 11.7. Nutritional Characteristics of Soybean Meal for Livestock Feed [Source: *Feedstuffs, 75*(38):16 (Sept. 17, 2003)]

Nutritional Characteristic	Units	Solvent-extracted Meal	Dehulled Solvent-extracted Meal	Hard Screw Pressed Meal	Full-fat Soybeans	Soybean Hulls	Soybean Millrun
Proximate analyses							
Dry matter	%	90	88	89	90	91.0	88.0
Crude protein	%	44.0	47.8	42.0	38.0	12.1	13.6
Ether extract	%	0.5	1.0	3.5	18.0	2.1	1.4
Crude fiber	%	7.0	3.0	6.5	5.0	40.1	40.7
Ash	%	6.0	6.0	6.0	4.6	5.1	5.1
Ruminant values							
Ruminant dig. protein	%	37.5	46.6	35.5	34.1		
Ruminant TDN	%	78	79	78	85	77	44.0
Energy values							
Poultry ME	Kcal/lb Kcal/lb	1020 1405	1125 1425	1100 1360	1520 1610		
Amino acids							
Methionine	%	0.65	0.70	0.6	0.54		
Cystine	%	0.67	0.71	0.62	0.55		
Lysine	%	2.9	3.02	2.7	2.4		
Tryptophan	%	0.60	0.70	0.58	0.52		
Threonine	%	1.7	2.0	1.7	1.69		
Isoluecine	%	2.5	2.6	2.8	2.18		
Histidine	%	1.1	1.3	1.1	1.01		
Valine	%	2.4	2.7	2.2	2.02		
Leucine	%	3.4	3.8	3.8	2.8		
Arginine	%	3.4	3.6	3.2	2.8		
Phenylalanine	%	2.2	2.7	2.1	2.1		
Minerals							
Calcium	%	0.25	0.20	0.20	0.3	0.69	0.42

Table 11.7., cont. Nutritional Characteristics of Soybean Meal for Livestock Feed [Source: *Feedstuffs, 75*(38):16 (Sept. 17, 2003)]

Total phosphorus	%	0.60	0.65	0.60	0.6	0.21	0.21
Available phosphorus	%	0.20	0.21	0.20	0.2		
Sodium	%	0.04	0.04	0.04	0.04		
Potassium	%	1.97	1.90	1.71	1.70	1.27	
Chloride	%	0.02	0.02	0.02	0.03		
Magnesium	%	0.27	0.27	0.25	0.21		
Sulfur	%	0.43	0.43	0.33	0.30		
Manganese	ppm	27.5	27.5	32.3	30.0		
Iron	ppm	120	120	160	75		
Copper	ppm	28	28	18	15		
Zinc	ppm	60	60	59	35		
Selenium	ppm	0.1	0.1	0.1	0.1		

a regulated pollutant, and pressure is increasing to reduce emissions posing engineering and financial challenges. Nevertheless, hexane remains the solvent of choice.

Alternate Hydrocarbons

Unlike hexane, heptane, and isohexane emissions are not regulated, and they perform similarly to hexane (Wan, 1994; Wan & Wakelyn, 1997). Some soybean processors have recently chosen to use isohexane instead of hexane, which requires little retrofitting but has less restrictive emissions requirements (Lusas & Gregory, 1996).

Propane

Propane is proposed as a solvent for extracting soybeans (Wan & Wakelyn, 1997), and reportedly several extraction plants were built in China using propane to extract soybeans. Propane must be pressurized to maintain its liquid form to be useful as an extraction solvent (Johnson & Lusas, 1983). Propane is highly flammable, and preventing its escape and, thus, preventing catastrophic explosions and fires in a commercial plant poses a major problem and no plants employing propane have been constructed in the United States or Brazil. Developers of the technology cite abundant, inexpensive supply and easy desolventization of both oil and meal with low energy input as the major advantages in using propane.

Acetone

Acetone also has some desirable attributes as an extraction solvent (Wan & Wakelyn, 1997). Vegetable oil is highly soluble in dry acetone but only sparingly when water is added. Thus, water can be added to the miscella to phase separate the oil by gravity settling, and then the oil-lean phase can be distilled to remove the water and restore solvating capacity. Reportedly, acetone was used for a short period in one soybean plant operating in Europe.

Alcohols

Considerable research was carried out in the 1980s using ethanol and isopropanol as oil extraction solvents. Ethanol is unusual because its oil solvating capacity is temperature and moisture-dependent. Oil solubility is relatively low at room temperature and moisture contents above the water:alcohol azeotrope. Thus, the moisture content of the flakes must be in equilibrium with the alcohol (e.g., 2% for 95% ethanol and 7% for 91% isopropanol) (Wan & Wakelyn, 1997), otherwise the solvency changes. Differences in oil solubility afford inexpensive means of oil separation from the solvent by merely cooling the miscella to separate an oil-rich phase without evaporating the bulk solvent.

New Soybean Processing Alternatives

Extrusion-expelling (EE)

Extrusion-expelling (Figs. 11.8, 11.9) has already been mentioned, but is one of the new soybean processing technologies being adopted for local processing. Extruding-expelling was developed by InstaPro, Des Moines, Iowa, to process soybeans in local communities and developing countries with low capital investment (Weijratne et al., 2004). An electric-powered dry extruder is used to generate heat by friction and mechanically ruptures cell walls replacing steam-heated dryers and flaking mills. This eliminates the need for steam-generating boilers, rotary dryers and roller mills, which represent significant capital investments, although need for electrical power is increased. In at least one instance, one such plant dehulls soybeans and produces partially defatted soy flour and extruder-texturized soy protein for use as food ingredients (Iowa Soy Specialties, Vinton, IA).

Supercritical Fluid Extraction (SFE)

Supercritical fluid extraction using CO_2 has long been an appealing technology to extract soybeans. At elevated temperatures and pressures (31°C and 73 bar, 88°F and 1,060 psi), a supercritical state is achieved where CO_2 has the diffusivity of a gas and the density of a liquid and has significant oil solvency. A simplified flow diagram is shown in Fig. 11.18. In practice, CO_2 is compressed to >350 bar (>5,100 psi) to increase oil solvency. When the pressure is reduced in two stages (<50 bar), oil solvency declines, and the free oil can be trapped and removed, while the oil-lean CO_2 can then be recompressed and oil solvency restored for recycling back to the extraction vessel. This technology has been commercially used to decaffeinate coffee and extract hops for beer production and other flavorings. Unfortunately, no one has yet devised a practical means of getting large masses of feed solids into and out of the high-pressure vessel in a continuous manner, and consequently all SFE processes thus far have been batch systems that are too expensive to apply to an inexpensive commodity like soybeans.

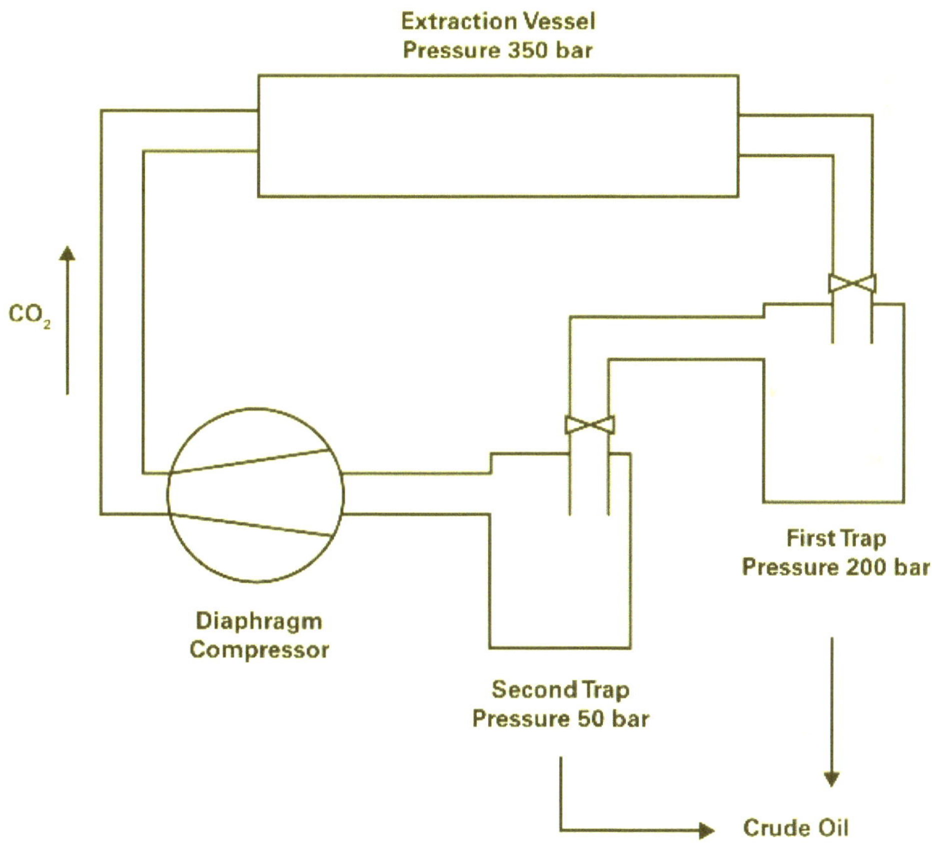

Fig. 11.18. Depiction of a supercritical-fluid-extraction process (redrawn from Stahl et al., 1980).

Gas-supported Screw Pressing (GSSP)

Recently, the companies Harburg-Freudenberger (Hamburg, Germany) and Crown Iron Works developed gas-supported screw pressing (HIPLEX® extraction; High Pressure Liquid Extraction) in which carbon dioxide is injected into the barrel of a screw press. A photograph of this new screw press and an interior photograph of the first plant operated by SafeSoy Technologies and constructed in Elsworth, Iowa, are shown in Fig. 11.19. The pressures achieved are not believed to be sufficient to achieve supercritical state to enable true oil solubilization. Extraction is attributed to action as

Fig. 11.19. Photographs of a gas-supported screw press (A) and the interior of the processing plant in Elsworth, IA (B).

a displacement fluid where CO_2 displaces the oil. Residual oil values of 3.5–4.5% oil are achieved on a dry weight basis (Deak et al., 2007). These low levels of residual oil are achieved with little heat generation and, thus, protein denaturation. Protein dispersibility indexes (PDI) of 70–80 can be achieved. This makes the process ideal for producing identity-preserved meal for preparing protein ingredients, especially functional soy flours, protein isolates, and fractionated soy proteins (discussed more in the Chapter: *Soy Protein Products, Processing, and Utilization*). Because the process is mechanical and uses an inert gas, it complies with organic processing and, when organic production practices are employed, enables efficient production of organic oil and protein ingredients. The oil is low in phosphorus content, and the residual oil in the meal is unusually high in lecithin content enhancing emulsification properties of the soy flour and protein ingredients that are produced (Tables 11.4, 11.6).

Aqueous Processing

Soybean oil is not appreciably soluble in water, but water is used as an extraction aid or medium for physical separation of oil from other soybean components in a process known as *aqueous extraction processing* (AEP) (Cater et al., 1974; Lawhon et al., 1981; Lusas et al., 1982). In this process, extraction of oil from other seed components is based on insolubility of oil rather than dissolution. The original process involved grinding soybeans, dispersing the ground solids in water, centrifuging to separate an oil-rich oil-in-water cream phase, a fiber-rich residue phase, a protein- and sugar-rich solubles phase; and breaking the cream phase into free oil and a water phase. Unlike other oilseeds, breaking the cream phase is difficult with soybeans because of the high levels of soluble protein and lecithin, which are efficient surfactants stabilizing the emulsion. Edible protein products, such as protein isolates and concentrates, may be simultaneously produced.

A recent resurgence of interest in AEP has occurred because it is regarded as "green" processing with little environmental impact. Most of the current work is focused on using enzymes to enhance oil and protein extraction in AEP (Rosenthal et al., 1996, 1998, 2001). Critical steps in improving oil extraction are those operations used to rupture cell walls and release the oil so that it can be recovered as an emulsified cream, or even more preferably, as free oil. Enzymes are helpful in such separations, and interest in enzyme-assisted AEP is increasing as enzyme costs decline.

Most of the work on AEP, with or without enzymes, uses ground material (full-fat flour) prior to extraction. Oil recovery is improved by reducing full-fat soybean flour to smaller particles (Rosenthal et al., 1996). Very fine grinding, however, produces smaller oil globules, smears oil over protein and fiber particles, and makes a more stable emulsion cream phase. Grinding alone also does not completely rupture cell walls, which is a key barrier to recovering oil by AEP. Oil extraction recoveries typically range around 60% of the total available oil. This comparatively low oil extraction recovery (>95% for hexane extraction) has discouraged commercial adoption of AEP.

Frietas et al. (1997) improved oil recovery by extruding dehulled soybean cotyledons prior to enzyme-assisted AEP, but extruding dehulled soybeans without flaking does not achieve as extensive cell distortion as does extrusion of flaked soybeans. AEP of extruded full-fat soy flakes gave 68% extraction of the total available oil without using enzymes but, with a protease enzyme treatment, oil extraction increased to >90% (Lamsal et al., 2006). Treating with cellulase did not enhance oil extraction either alone or in combination with protease. Low levels of proteolysis do not seem to affect protein precipitation as SPI.

Opportunities may be available to use the method as the front-end to a soybean biorefinery to produce oil for biodiesel, ethanol from soy fiber (cell walls), and value-added protein products, and to integrate into small-scale, "organic" and identity-preserved processing strategies. In this approach (Fig. 11.20), several new steps were incorporated including step of flaking and twin-screw extruding to achieve more

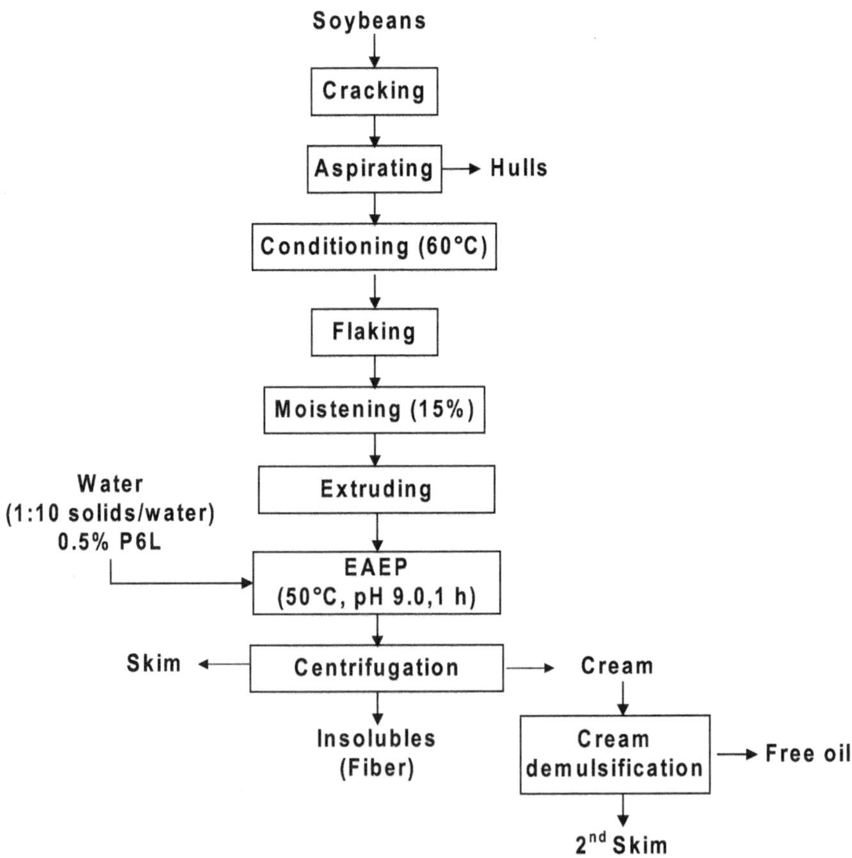

Fig. 11.20. Flow diagram for enzyme- and extrusion-assisted aqueous processing of soybeans.

complete cell wall disruption and freeing of oil for washing out of the solids, a step of using protease enzymes to assist separation of oil and solubilizing protein, and using enzymes to destroy the surfactants stabilizing the cream (Freitas et al, 1997; Lamsal et al, 2006; Lamsal & Johnson, 2007). When optimum conditions are used, over 82% of the oil can be separated as a cream, and all of the oil can be recovered from the cream (15% of the oil is retained in the high-protein skim fraction [Lamsal & Johnson, 2007]). Advances are quickly being made that lead this author to conclude >97% oil extraction, nearly equivalent to hexane extraction, is possible. The remaining issue is to devise systems to capture added value and to achieve economic use of wet protein in feeding swine.

Future Challenges

The challenges facing the soybean crushing industry in the future are considerable. Energy consumption in processing soybeans is high, and as energy prices increase, alternative processes that consume less energy will become attractive. Increasing pressures will continue to reduce emissions that pollute the air and contribute to greenhouse gases. The advent of soybean biorefineries will drive researchers and engineers to devise improved processing technologies that deliver biofuels, industrial chemicals, and biobased products as well as food and feed.

References

Arnold, L.K.; R.B.R. Choudhury. Extraction of soybeans and cottonseed oil by four solvents, *J. Am. Oil Chem. Soc.* **1960,** *37,* 458–459.

Ayres, A.L.; J.J. Dooley. Laboratory extraction of cottonseed with various petroleum hydrocarbons, *J. Am. Oil Chem. Soc.* **1948,** *25,* 372–379.

Barger, W.M. Handling, transport and preparation of soybeans, *J. Am. Oil Chem. Soc.* **1981,** *58,* 154–156.

Basiron, Y. Palm oil. *Bailey's Industrial Oil and Fat Products,* Fifth ed.; Y.H. Hui, Ed.; Wiley-Interscience: New York, NY, 1996; Vol. 5.

Bockisch, M. *Fats and Oils Handbook;* AOCS Press: Champaign, IL, 1993.

Boring, S. Soybean processing quality control. *Practical Handbook of Soybean Processing and Utilization,* Second ed.; D.R. Erickson, Ed.; AOCS Press: Champaign, IL, 1995, pp. 483–503.

Brumm, T.J.; C.R. Hurburgh. Estimating the processed value of soybeans. *J. Am. Oil Chem. Soc.* **1990a,** *67,* 302–307.

Brumm, T.J.; C.R. Hurburgh; L.Al Johnson. Cracking and dehulling shriveled and wrinkled soybeans. *J. Am. Oil Chem. Soc.* **1990b,** *67,* 750–756.

Cater, C.M.; K.C. Rhee.; R.D. Hagenmaier; K.F. Mattil. Aqueous extraction alternative oilseed milling process, *J. Am. Oil Chem. Soc.* **1974,** *51,* 178–181.

Deak, N.A.; Z.M. Nazareth.; L.A. Johnson. Compositions and properties of gas-supported screw-pressed meal. *98th AOCS Annual Meeting and Exposition.* Quebec City, Canada, May 14–16,

2007.

Erickson, D.R. Overview of modern soybean processing and links between processes. *Practical Handbook of Soybean Processing and Utilization,* Second ed.; D.R. Erickson, Ed.; AOCS Press: Champaign, IL, 1995a; pp. 65–92.

Erickson, D.R., editor, *Practical Handbook of Soybean Processing and Utilization,* Second ed.; AOCS Press: Champaign, IL, 1995b.

Freitas, S.P.; L. Hartman; S. Couri; F.H. Jablonka; C.W.P. de Carvalho. The combined application of extrusion and enzymatic technology for extraction of soybean oil, *Fett/Lipid* **1997,** *99,* 333–337.

Hron, R.J. Renewable solvents for vegetable oil extraction, *J. Am. Oil Chem. Soc.* **1982,** *59,* 674A–684A.

Johnson, L.A. Recovery, refining, converting, and stabilizing edible fats and oils. *Food Lipids;* C. Akoh, D. Min, Eds.; Marcel Dekker, Inc.: New York, NY, 1998.

Johnson, L.A. Theoretical, comparative and historical analyses of alternative technologies for oilseeds extraction. *Technology and Solvents for Extracting Non-Petroleum Oils;* P.J. Wan, P.J. Wakelyn, Eds.; AOCS Press: Champaign, IL, 1997.

Johnson, L.A.; E.W. Lusas. Comparison of alternative solvents for oils extraction. *J. Am. Oil Chem. Soc.* **1983,** *60,* 181A–193A.

Laisney, J. Processes for obtaining oils and fats. *Oils and Fats Manual;* A. Karleskind, Ed.; Intercept Limited: Andover, United Kingdom, 1996; Vol. 1.

Lamsal, B.P.; L.A. Johnson. Separating oil from aqueous extraction fraction of soybeans. *J. Am. Oil Chem. Soc.* **2007,** *85,* 785–792.

Lamsal, B.P.; P.A. Murphy; L.A. Johnson. Flaking and extrusion as a mechanical treatment for enzyme-assisted aqueous extraction of oil from soybeans, *J. Am. Oil Chem. Soc.* **2006,** *83,* 973–979.

Lawhon, J.T.; L.J. Manak; K.C. Rhee; E.W. Lusas. Combining aqueous extraction and membrane isolation techniques to recover protein and oil from soybeans, *J. Food Sci.* **1981,** *46,* 912–916, 919.

Lusas, E.W.; S.R. Gregory. New solvents and extractors. *Emerging Technologies, Current Practices, Quality Control, Technology Transfer and Environmental Issues;* S. Koseoglu, K. Rhee, R. Wilson, Eds.; *Proceedings of the World Conference on Oilseed and Edible Oils Processing;* AOCS Press: Champaign, IL, 1996; Vol. 1, pp. 208–217.

Lusas, E.W.; J.T. Lawhon; K.C. Rhee. Producing edible oil and protein from oilseeds by aqueous processing. *Oil Mill Gaz.* **1982,** *86(11),* 28–34.

Lusas, E.W.; L.R. Watkins. Extrusion for solvent extraction. *J. Am. Oil Chem. Soc.* **1988,** *65,* 1109–1114.

Milligan, E.D. Survey of current solvent extraction equipment. *J. Am. Oil Chem. Soc.* **1976,** *53,* 286–290.

Niranjan, K.; P. Hanmoungjai. Enzyme-added aqueous extraction. *Nutritionally Enhanced Edible Oil and Oilseed Processing;* N.T. Dunford, H.B. Dunford, Eds.: AOCS Press: Champaign, IL, 2004.

NOPA, *Trading Rules for the Purchase and Sale of Soybean Meal,* National Oilseed Processors Association; Washington, DC, 2006a.

NOPA, *Trading Rules for the Purchase and Sale of Soybean Oil,* National Oilseed Processors Association: Washington, DC, 2006b.

Pedrotti, S.; F. Boling. Expander technology in the modern soybean prep room. *Emerging Technologies, Current Practices, Quality Control, Technology Transfer and Environmental Issues;* S. Koseoglu, K. Rhee, R. Wilson, Eds.; Proceedings of the World Conference on Oilseed and Edible Oils Processing. AOCS Press: Champaign, IL, 1996; Vol. 1, pp. 201–203.

Rosenthal, A.; D.L. Pyle; K. Niranjan. Aqueous and enzymatic processes for edible oil extraction, *Enz. Microb. Technol.* **1996,** *19,* 402–420.

Rosenthal, A.; D.L. Pyle; K. Niranjan. Simultaneous aqueous extraction of oil and protein from soybean: mechanisms for process design, *Trans. IchemE* **1998,** *76,* 224–230.

Rosenthal, A.; D.L. Pyle; K. Niranjan; S. Gilmour; L. Trinca. Combined effect of operational variables and enzyme activity on aqueous enzymatic extraction of oil and protein from soybean, *Enz. Microb. Technol.* **2001,** *28,* 499–509.

Serrato, A.G. Extraction of oil from soybeans, *J. Am. Oil Chem. Soc.* **1981,** *58,* 157–159.

Stahl, E.; T.M. Schultz; H.K. Mangold. Extraction of seed oils with liquid and supercritical carbon dioxide, *J. Agric. Food Chem.* **1980,** *28,* 1153–1157.

Wan, P.J. Alternate hydrocarbon solvents for cottonseed extraction, *Oil Mill Gaz.* **1994,** *100(2),* 32–36.

Wan, P.J.; P.J. Wakelyn, (Eds.), *Technology and Solvents for Extracting Oilseeds and Nonpetroleum Oils.* AOCS Press: Champaign, IL, 1997.

Wang, T.; L.A. Johnson. Survey of soybean oil and meal qualities produced by different processes, *J. Am. Oil Chem. Soc.* **2001,** *78,* 311–318.

Weijratne, W.; T. Wang; L.A. Johnson. Extrusion-based oilseed processing methods. *Nutritionally Enhanced Edible Oil and Oilseed Processing.* N.T. Dunford, H.B. Dunford, Eds.; AOCS Press: Champaign, IL, 2004.

Williams, M. Using expanders to improve extractability, *Inform (AOCS)* **1990,** *1,* 959–963.

Williams, M.A.; R.J. Hron. Obtaining oils and fats from source materials. *Bailey's Industrial Oil and Fat Products,* Fifth ed.; Y.H. Hui, Ed.; Wiley-Interscience: New York, NY, 1996; Vol. 4.

Witte, N.H. Soybean meal processing and utilization. *Practical Handbook of Soybean Processing and Utilization, Second ed.;* D.R. Erickson, Ed.: AOCS Press: Champaign, IL, 1995; pp. 93–116.

Woerfel, J.B. Harvest, storage, handling, and trading of soybeans. *Practical Handbook of Soybean Processing and Utilization, Second ed.;* D.R. Erickson, Ed., AOCS Press: Champaign, IL, 1995; pp.161–173.

Woerfel, J.B. Extraction. *Practical Handbook of Soybean Processing and Utilization,* Second ed.; D.R. Erickson, Ed.; AOCS Press: Champaign, IL, 1995; pp. 65–92.

12

Soybean Oil Purification

Richard D. O'Brien
Consultant, Schulensburg, TX 78956

Introduction

Crude soybean oil is composed of triglycerides with oil-soluble and suspended non-glyceride materials (fatty acids, phosphatides, sterols, tocopherols, metals, hydrocarbons, pigments, and protein fragments). Of these, the triglycerides, tocopherols, and sterols each have commercial value. The remaining compounds are considered undesirable because they contribute offensive flavors and other undesirable reactions in the presence of oxygen and/or heat. The objective of the purification processes is to remove the undesirable materials with the least possible damage to either the triglycerides or the beneficial nonglycerides with a low loss of oil (Norris, 1982). Soybean oil purification, as used here, refers to the purification processes individually referred to as degumming, refining, bleaching, and deodorization. Each process is designed to remove certain minor components. The product of the purification processes is identified as RBD (refined, bleached, deodorized) soybean oil or soybean salad oil. Some of the principal uses for RBD soybean oil are consumer bottled oil, salad dressings, mayonnaise, sauces, and other products that require a liquid oil.

Degumming

Degumming is a process for removal of phosphatides from crude soybean and other vegetable oils to improve physical stability and facilitate further processing. The phosphatides are also called gums and lecithin. Lecithin is the common name for phosphatidylcholine, but common usage refers to all of the phosphatides present in vegetable oils. Soybean oil is the major source of commercial lecithin because it contains the highest level of gums and is the world's leading vegetable oil (Erickson, 1995a). It is preferable to degum soybean oil before caustic refining and mandatory with physical refining due the high levels of impurities, including phosphatides, proteinaceous and mucilaginous materials. These substances can be removed simultaneously with the free fatty acids (FFA) during chemical refining, but the operating efficiency, yield, and quality are enhanced when degumming is performed as a separate process. The decision to perform the degumming process or forego it to remove the gums during chemical refining is based on energy conservation and capital savings. However, separate degumming offers several potential advantages (O'Brien, 2004):

- It is necessary for lecithin production—the hydrated gums are the raw materials for lecithin production.
- It is necessary for export oil requirements that the oil be free of impurities that could settle out during shipment—sludges form when the hygroscopic phosphatides become hydrated by moisture from the air.
- It reduces chemical refining oil loss—phosphates can act as emulsifiers to increase neutral oil retention in soapstock.
- It reduces refinery wastewater load due to the lower oil losses and the reduction of gums discharged.
- It improves acidulation performance—soapstock from degummed oil has a lower emulsifier content, and the lower acid level required has less impact on the wastewater treatment system.
- It prepares the oil for physical or steam refining—degumming significantly reduces the nonvolatile phosphatides and metallic prooxidants.

Water Degumming

Degumming of oils intended for use in edible products is traditionally accomplished by hydrating the gums and similar materials to make them insoluble in oil. Hydrated phosphatides become more dense than the triglycerides and precipitate, or settle out of the oil. Water degumming reduces the phosphorus content to less than 50 ppm with good quality crude soybean oil. The usual relation between phosphorus and phosphatide content is that phosphatides are 30 times phosphorus. The water degumming process is simple, but the crude soybean oil quality has a significant influence on the efficiency of the process. The phosphatides in crude soybean oil exist in either hydratable or nonhydratable forms. The hydratable form is readily removed by the addition of water, but the nonhydratable phosphatides are unaffected and remain in the oil phase. The nonhydratable phosphatides are generally identified as the calcium and magnesium salts of phosphatidic acids that are produced by an enzymatic action of phospholipases released by damage to the soybean cellular structure (Hvolby, 1971). This damage may occur with handling, extraction practices, or both. Johnson reviewed these problems and remedies in the Chapter: *Oil Recovery from Soybeans*. The nonhydratable phosphatides in the oil can be reduced significantly by inactivating the phospholipases early in the extraction process with the use of the expander or the ALCON process, used in Europe, which cooks the soybeans to inactivate the enzymes harmful to oil quality. Soybean oil extracted with the use of an expander has high phosphatide levels of 2.5–4.0%, and the ALCON process levels are 4.0–6.0% (Erickson, 1995d), but the phosphatides are more hydratable and more easily removed by water degumming. Normally, soybean oil from conventional solvent extraction has about 90% hydratable phosphatides and 10% nonhydratable phosphatides, and the total phosphatide content ranges from 1.1 to 3.2%. The FFA of good quality crude soybean oil ranges from 0.5 to 1.0%, which is reduced by 20–40% in water-degummed oil. Poor quality soybean oil is identified by a high FFA (>1.0%), also indicating a higher than normal nonhydratable phosphatide content. Recognition of the function of calcium and magnesium led to the use of demineralized water for degumming and the use of citric or phosphoric acids to transform the nonhydratable to hydratable gums. The use of acids is not recommended for gums intended for lecithin production because they cause darkening of the lecithin (Erickson, 1995).

Batch and continuous water degumming systems are similar; the major differences are continuous flow versus batching. For the batch systems, the oil to be degummed is heated to 150 ± 10°F (65 ± 5°C), water added, and mixed for 30 min. The amount of water added should be 75% of the phosphate content of the oil. Too little water produces dark viscous gums and a hazy oil, while too much water causes excess oil losses through hydrolysis. The hydration temperature is important because degumming is temperature- sensitive; it is less complete at higher temperatures due to the increased solubility of the phosphatides, and the increased viscosity at lower temperatures makes separation of the phosphatides more difficult. After hydration, the oil and gums are separated with a centrifuge (Carr, 1978). For continuous systems, oil preheated to 65°C (150°F) is treated with water and mixed in a hydration vessel sized for a 45-min retention time. This retention time can be reduced to 1 min with the use of in-line agitators. Treat the hydrated oil very gently to avoid developing an emulsion. After hydration, centrifuge the oil to separate the gums from the neutral oil. The gums pass to a wiped-film evaporator to become soybean oil lecithin or may be added to animal feed. The degummed oil is vacuum-dried, or if close-coupled to the refining process, drying is not necessary (Farr, 2000).

Acid-degumming

The acid-degumming process is a variant of the water-degumming process in that it uses a combination of acid and water. Acid degumming leads to a lower residual phosphorus content than water degumming. The nonhydratable gums can be conditioned into hydratable forms with a degumming acid. Phosphoric and citric acids are used because they are food-grade and sufficiently strong, and they chelate divalent metal ions. Citric acid is usually preferable because it does not increase the phosphoric content of the oil. Dispersion of the acids is critical for maximal contact with the nonhydratable phosphatide complexes. The gums isolated with acid degumming are not suitable for standard lecithin because the phosphatides have a higher phosphatidic acid, and the degumming acid is present. Several acid-degumming processes can be developed to attain a phosphorus value lower than 5 ppm that is required for good quality physically refined oils. Unilever's super-degummed process uses mild temperatures with a complicated multiple-holding-steps process. The oil is heated to 70°C (158°F), modified lecithin is optionally mixed into the oil, and then a strong solution of citric acid is added as a degumming acid to decompose the nonhydratable phosphatides. After the reaction, the mixture is cooled to below 40°C (104°F), and water is added to promote the dissociation of the liberated free phosphatidic acid and phosphatidylethanolamine. A further 3-h holding time is provided to form liquid phosphatide crystals at this reduced temperature, which are removed by a centrifuge (Dijkstra, 1992).

Modified Acid-degumming

Modified acid-degumming is a physical refining pretreatment that incorporates the benefits of caustic soda neutralization. This physical refining preparatory process treats the oil with a degumming acid and then partially neutralizes it with NaOH in solution. The amount of NaOH used is limited to prevent soap formation. The metal–phospholipid complexes are dissociated by the acid into insoluble metal salts and phospholipids in their acid form, which are still soluble in oil. The NaOH addition raises the pH and converts the phospholipids into sodium salts that are hydratable. The hydrated salts can be centrifuged for separation or dried to form agglomerates for adsorption on silica for removal with filtration.

The acid-degumming treatment generally reduces phosphorus to between 25 and 35 ppm. Neutralization with NaOH after the acid treatment should reduce the phosphorus content to 15–25 ppm. These treatments, followed by either a water wash or the use of a silica adsorbent, further reduce the phosphorus to the 5-ppm maximum required for physical refining. Silica adsorbents are added with a separate mixing step before bleaching, and can be removed with a separate filtration or with the spent bleaching earth. Some consider modified acid-degumming an intermediate between acid degumming and chemical refining. This process should be applicable to all types of oils, either crude or previously degummed (Carlson, 1993; Dijkstra, 1992; Grace).

Enzymatic Degumming

Enzymatic degumming is a relatively new process. An enzyme, phospholipase, converts phospholipids into lysophospholipids that can be removed by centrifugation. Crude oil, pretreated with a combination of sodium hydroxide and citric acid, is mixed with water and enzymes by a high shear mixer, creating a stable emulsion. The emulsion allows the enzyme to react with the phospholipids, transforming them into water-soluble lysophospholipids. The emulsion is broken by centrifugation, separating the gums and phospholipids from the oil (ISEO, 2006). Enzymatic degumming advantages are (O'Brien, 2004):

- Generates a better oil yield than traditional degumming
- Reactions usually carried out under mild conditions
- Enzymes are highly specific
- Acceptable reaction rates
- Small quantities of enzyme required for the reaction
- Degummed oils with low phosphorus and iron contents produced even with poor-quality oils

Membrane Filter Degumming

Membrane processing has been applied to remove phospholipids from crude oil/hexane mixtures as well as from crude oil itself without the addition of an organic solvent. Pagliero and co-workers (2001) showed that membranes were suitable for removing phospholipids from the miscella of crude oil and hexane. Subramanian and co-workers (1999) reduced phospholipids in soybean oil in the range of 85.8 to 92.8% with surfactant-aided membrane degumming. The phosphorus content of the degummed oil was 20–58 ppm. The high membrane reduction level indicates that hydratable and nonhydratable phospholipids were removed from the soybean oil. Commercialization of a membrane filtering process to simultaneous degum and refine in a single step is reportedly in progress (Carlson, 2006). This system is expected to produce a membrane-separated oil with a phosphorous level of <2.0 ppm and reduction of the chlorophyll content by more than half (Farr, 2000).

Refining

Refining of soybean oil is practiced as a purifying treatment designed principally to remove FFA while reducing the phosphatides or gums, coloring matter, insoluble matter, settlings, and miscellaneous unsaponifiable materials. Failure to remove these impurities causes the oil to foam, darken, smoke, and become cloudy when heated. The time-honored or conventional method for removal of the impurities from the oils is by the use of a solution of sodium hydroxide to react with the FFA to produce a soap solution that phase separation can remove. The caustic or chemical refining process produces good quality oil and is flexible in its ability to treat different oils and different qualities of individual oils. However, it does have certain disadvantages such as the neutral oil emulsified in the soapstock and the soapstock itself. Neutral oil losses are high when treating oils with a high FFA content, and disposal of the soapstock can be a problem. Physical refining, another deacidification process, eliminates the soapstock problem; it removes fatty acids from the oil by steam distillation under vacuum with a minimal neutral oil loss. However, phospholipids must be reduced to less than 50 ppm and metal contamination minimized before physical refining to produce soybean oil with acceptable quality.

Chemical (Caustic) Refining

Several variations and equipment configurations are used for caustic refining by companies around the world, but the basic processing sequence necessary to produce optimal quality soybean oil remains fairly constant. Alkali refining practices are a result of the gradual application of science to the basic art of batch refining originally performed in open-top, cone-shaped kettles. For example, the efficiency of the separation of soapstock from neutralized oil was significantly improved with the replacement of gravity with centrifugal separators beginning in 1932, reduced losses by 20–30% (Norris, 1982). The chemical refining system continues to evolve to provide the flexibility to efficiently refine crude soybean and other oils. Nevertheless, the basic system used in the United States, also known as the long-mix system, is outlined as follows:

- Crude oil conditioning—a phosphoric or citric acid pretreatment is recommended even for degummed oil because nonhydratable phosphatides remain in the oil after water degumming. The addition of the preferred acid treatment to the crude oil in a day tank is a convenient method for pretreatment. Crude soybean oils with a high level of phosphatides are usually treated with 0.05–0.2% of 75% food-grade phosphoric acid, depending on the calcium/magnesium content of the oil, for a minimum of 4–8 h before refining. The phosphoric acid dosage is estimated by the formula: (Ca + Mg) divided by 2 times 10 equals ppm phosphoric acid (Farr, 2000). The purposes of the acid treatment are: to precipitate phosphatidic materials; to precipitate natural calcium and magnesium as insoluble phosphate salt; to inactivate trace metals; to reduce the neutral oil losses; to destabilize and improve the removal of chlorophyll in bleaching; and to improve the color and flavor stability of the deodorized oil (O'Brien, 2004). The acid pretreatment temperature should be controlled at 20– 30°C (68–86°F); treatment at 60°C (140°F) for longer than 20 min causes the phosphatidic acid to revert to the nonhydratable form (Erickson, 1995b).

- Alkali treatment—the compounds used for refining include sodium hydroxide

(NaOH), potassium carbonate (KOH), sodium bicarbonate ($NaHCO_3$), and sodium carbonate (Na_2CO_3). Sodium hydroxide is preferred by most soybean oil refiners, but some use the more expensive potassium hydroxide to provide an outlet for the soapstock as fertilizer. Typically, degummed or acid-conditioned crude soybean oil is continuously mixed with a proportioned stream of dilute sodium hydroxide solution and heated to break the emulsion. The selection of the amount and concentration of NaOH used for refining is critical. The strength of the caustic solution is measured in terms of specific gravity, expressed in degrees Baumé (°Bé). Typically, 14–18°Bé is used with long-mix refining system for soybean oil. A low concentration of the caustic solution and adequate hydration time limit the possibility of saponification (Erickson, 1995b). The caustic treat selected for the crude soybean oil varies with the FFA content, the amount of acid pretreatment, and the level of caustic *excess* over *theoretical*. Any excess NaOH over that required to neutralize FFA reacts with the phosphatidic materials and some of the color bodies as well as saponification of neutral oil. The refiner's charge is to determine a % excess and caustic strength that produce a quality refined oil with a low loss of oil (Hendrix, 1990). Typically, the % excess used for soybean oil is 0.01–0.05 for degummed oil and 0.15–0.25 for nondegummed oil (Farr, 2000). The theoretical quantity of NaOH based on the ratio of molecular weights of sodium hydroxide to oleic fatty acid is 0.142. Thus, the *treat* or the amount of NaOH solution to add is calculated as follows:

$$\%\ \text{Treat} = \frac{(\%\ \text{FFA} \times 0.142) + \text{Excess}}{\%\ \text{NaOH}} \times 100$$

- Caustic-oil Mixing—After the caustic reagent is proportioned into the crude soybean oil, it must be adequately blended to ensure sufficient contact with the FFA, phosphatides, and color pigments. The gums are hydrolyzed by the water in the caustic solution and become oil-insoluble. The caustic and oil are mixed at 30–35°C (86–95°F) in a dwell mixer with a 5- to 15-min residence time. After mixing with caustic solution, the oil is heated to 70–75°C (158–167°F) to provide the thermal shock necessary to break the emulsion.

- Soap-oil Separation—The primary centrifuge is a critical component of the refining process because it determines the refining yield efficiency. The soap-in-oil suspension is fed to high-speed centrifuges for separation into light- and heavy-density phases. The light phase is the neutral oil still containing traces of moisture and soap. The heavy phase, or soapstock, is primarily insoluble soap, meal, free caustic, phosphatides, and small quantities of neutral oil. Refined oil yield and quality depend on a uniform feedstock and separation of the soapstock with the least amount of entrained oil; however, even under the most optimal conditions, complete separation of the two phases cannot be achieved. Therefore, the primary separation is accomplished by allowing a small amount of the soapstock to remain in the refined oil for removal by the water-wash centrifuge (Sullivan, 1968). The soybean oil discharged from the primary centrifuge typically contains 200 to 400 ppm soap and 5 to 10 ppm phosphorus (Erickson, 1995b).

- Water Washing—Refined oil from the primary centrifuge is washed with hot softened water proportioned into the oil at a rate of 10–20% of the oil flow. Softened water must be used to avoid the formation of insoluble soaps. Sodium soaps remaining from the primary separation phase are readily washable and easily removed from the oil with either a single or double wash. A single wash is usually sufficient; however, two washes may provide savings in bleaching earth, as well as a reduction in wash-water volume. One water wash achieves ~95% reduction in both soap and phosphorus resulting in an oil with 10–20 ppm soap and <1.0 ppm phosphorus (Erickson, 2001).

  Wash-water temperature is important for efficient separation in the centrifuge. The water temperature should be 185–195°F (85–90°C), preferably 10–15°F (5–8°C) warmer than the oil temperature. The wash-water flow rate controls soap removal and affects the oil losses in the wash water. As with the primary centrifuge, a pulsating flow of water must be avoided.

  Two things that water washing will not do are remove phosphatides left in the oil after the primary centrifuge and remove unwashable soaps related to the calcium and magnesium contents of the crude oil. These metal complexes should have been removed in either the degumming or refining steps. Iron soaps are prooxidants, while calcium and magnesium result in nonwashable soaps (Weidermann, 1981). Some processors add up to 400 ppm citric acid to wash waters to effect the removal of these residuals (Erickson, 2001). The water-washing step may be eliminated with the use of soap-adsorbing silica in bleaching (Carlson, 2006).

- Vacuum Drying—Water-washed oil must be vacuum-dried if it is to be stored for any period of time. This is usually accomplished with a continuous spray drier, operating at 50 mm Hg absolute pressure. The vacuum-drying step is not required when the washed soybean oil is coupled directly to bleaching or silica purification (Farr, 2000).

Short-mix Caustic Refining

Refining practices vary between countries and plants due to the number, quality, and kind of source oils processed. The refining practices in Europe differ from those used in the United States mainly because of the need to process all types of oils and typically poorer-quality oils. The European oilseeds or crude oils must be imported, and a typical refinery must be equipped to handle all kinds of oils, depending on availability and price. The quality of the oil seeds and imported crude oils is variable, but normally a higher FFA oil is processed. The short-mix process was adopted in Europe after World War II because the relatively high FFA oils made it necessary to avoid the long contact time and the larger excess of caustic used with the long-mix caustic-soda refining system utilized in the United States (Hendrix, 1990; Braae, 1976). Figure 12.1 compares the differences between the long- and short-mix refining processes.

For the short-mix process, the oil temperature is raised to 80–90°C (175–195°F) before the addition of the caustic soda. A break between the neutral oil and soapstock takes place immediately, reducing the losses due to emulsification. The contact time between the caustic and oil is reduced to a 30-sec maximum, which helps to reduce the saponification losses. Because it is standard in Europe to degum solvent-extracted oils and to condition the oils with phosphoric acid before refining, the excess caustic treatment can be eliminated or reduced substantially. The oil is finally washed with demineralized water to help remove the traces of soap remaining in the oil and dried with processes similar to the systems used for the long-mix caustic refining process.

Physical Refining

Physical refining was utilized as early as 1930 as a process for the preneutralization of products with a high initial FFA content. In this case, it was followed by caustic refining. Later, it was found possible to physically refine lauric oils and tallow if the proper pretreatment was applied before steam distillation. Physical refining became a reality in the 1950s for processing palm oil, which typically contains high FFA and low-gum contents. The palm oil process subjected the crude feedstock first to pretreatment and then to deacidification. The pretreatment consisted of a degumming step and an earth bleaching step, which together remove certain nonvolatile impurities by filtration. Volatile and thermally labile components are removed during the conditions of steam distillation under vacuum, which originally gave the process its name of steam refining (Swoboda, 1985). For vegetable oils, such as soybean, that contain relatively low levels of FFAs and higher amounts of phosphatides, physical refining became a possibility only recently.

The traditional edible-oil processing system consists of caustic refining, bleaching, and deodorization. Caustic refining of vegetable oils with high phosphatide contents delivers a soapstock that is a mixture of sodium salts of fatty acids, neutral oil, water, unused caustic, and other compounds resulting from the reactions of the caustic with various impurities in the oil. Disposal of this soapstock or the waste streams from soapstock processing systems has become increasingly more expensive. A second problem associated with chemical neutralization is the loss of neutral oil, which re-

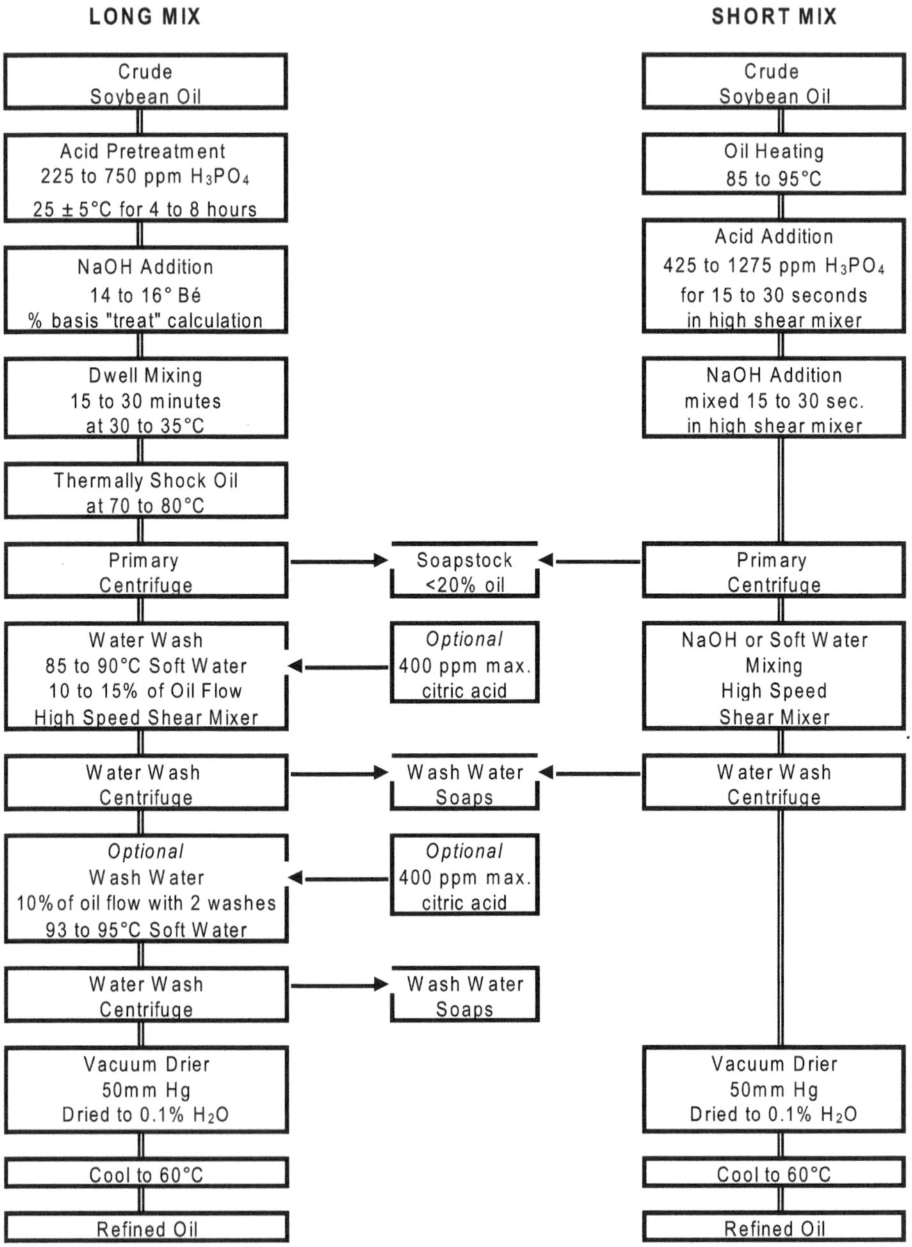

Fig. 12.1. Comparison of long- and short-mix chemical refining systems.

duces the overall yield from the crude oil. Elimination of the caustic refining step is economically attractive, but it means that degumming or some other pretreatment process or system must assume all the functions of the alkali-refining process, except for FFA removal.

Physical refining can remove the FFA, as well as the unsaponifiable and other impurities, by steam stripping, thus eliminating the production of soapstock and keeping neutral oil loss to a minimum. However, degumming and pretreatment of the oil are still required to remove those impurities that darken or otherwise cause a poor-quality product when heated to the temperatures required for steam distillation. Crude oil pretreatment is normally a two-step operation—the addition of a chemical is required to remove any trace quantities of gums remaining after water degumming and bleaching. Following pretreatment, all the FFA and any remaining trace impurities are removed by steam distillation in a single unit. Soapstock acidulation is eliminated with physical refining, and a higher grade distilled fatty acid is recovered directly from the oil without major pollution problems.

Vegetable-oil refining has to cope with many minor components. After water degumming, a number of impurities must still be removed or neutralized: carotenoids, chlorophyll, brown pigments, phosphatides, metals, free sugars, FFA, and oxidizing lipids. Steam stripping can convert the carotenoids and remove FFA, most off-flavors, and pesticides, but the other impurities must be handled before the distillation step (Segers, 1983); therefore, the pretreatment step is critical to the success of the physical refining process. The major process variables in pretreatment are (i) pretreatment chemical, concentration, and level; (ii) bleaching clay and level; and (iii) operating conditions. Normally, for a single-source oil with a history of consistent quality, the pretreatment process variables can be expected to remain fairly constant, but when more than one source oil is processed, varying conditions and chemical treatments must be considered (Tandy & McPherson, 1984).

Pretreated and bleached oil can be modified by hydrogenation, interesterification, or fractionation before steam refining to strip the fatty acids, flavors, and odors. The operating temperatures of the steam-refining deodorizer are the same as those used for deodorization of chemically refined oils, usually >240°C (>464°F), to reach the desired FFA level within the specifications for RBD soybean oil. Advantages for elevating the deodorization temperature are decreased residence time in the deodorizer due to an increase in efficiency and enhanced removal of health-risk components, such as pesticides and polycyclic aromatic hydrocarbons. Unfortunately, raised temperatures during deodorization also greatly increase the removal of nutritional components, including tocopherols, phenolic acids, and phytosterols. Furthermore, application of temperatures exceeding 200°C (392°F) rapidly induces the formation of the undesirable *trans* fatty acids. From a nutritional standpoint, the deodorization temperature and time must be adjusted for physical refining. The influence of deodorization temperature on the removal of minor constituents is shown in Table 12.1 (ten Brink & van Duijn, 2003).

The principal advantage for steam refining a low FFA oil, such as soybean, corn, peanut, sunflower, safflower, or canola, is the reduction of plant pollutants commonly caused by the acidulation of soapstock produced with conventional caustic refining. The economics for steam refining are usually favorable for high-FFA products such as palm and the lauric oils. Analyses indicate that no operating cost advantages can be gained by physically refining low-acidity oils (Carlson, 1993), which make up the majority of the oils processed in the United States. Additionally, flavor stability and potential unsatisfactory bleached color concerns still exist. Flavor evaluation work at the United States Department of Agriculture (USDA) Northern Regional Research Center (now known as the National Center for Agricultural Utilization Research) indicated that steam-refined soybean oil was equivalent to caustic-refined product; however, some of the test results indicated a potential problem with oxidative stability, which was not duplicated with further test work (List et al., 1978).

Miscella Refining

Facilities with an existing oilseed solvent extraction system may find miscella refining to be advantageous because the same solvent recovery unit can be used for both purposes. Miscella is the solution or mixture that contains the extracted oil. Both continuous and batch miscella refining processes are suitable for most fats and oils. This type of refining should be done at a solvent extraction plant as soon as possible, preferably within 6 h after the oil is extracted from the oilseed or animal. The advantages for miscella refining, as compared to conventional continuous caustic refining, are: higher oil yield; lighter color oil without bleaching; elimination of the water-wash step; and extraction of the color pigments before solvent stripping has set the color (Crauer, 1964).

Table 12.1 Deodorization Temperature Effect on Nonglycerides and *Trans* Development

	Reduction, %				Developed, %
Temperature	FFA	Tocopherols	Light PAH	Methidathion	*trans* fatty acids
160°C - 320°F	11	none	none	35	none
180°C - 356°F	25	none	50	82	none
200°C - 392°F	48	1	60	96	0.1
220°C - 428°F	75	4	75	All	0.2
240°C - 464°F	92	24	90	All	0.8
260°C - 500°F	99	51	All	All	2.5

PAH - polycyclic aromatic hydrocarbons; Methidathion - an organophosphorus insecticide; FFA - free fatty acid

For this purification process, the crude miscella source may be from: the pre-evaporator of a direct-solvent extraction plant; a blend of prepressed crude oil and solvent extracted miscella from the press cake; or a reconstituted blend of crude oil with solvent. In the process, a mixture of approximately 40 to 58% oil in solvent is heated or cooled to 104°F (40°C) and filtered to remove meal, scale, and other insoluble impurities. Two solvents used commercially for miscella refining are hexane and acetone.

Hydrolysis of phosphatides and pigments in the crude oil miscella requires an acid pretreatment, which usually varies between 100 and 500 ppm by weight of the oil, depending on the quality of the crude oil. An acid such as phosphoric or glacial acetic has been found effective in improving oil quality and reducing refining losses. Phosphoric acid is more commonly used due to its less corrosive properties and availability. The acid is mixed with the miscella in a static mixer to provide an intimately dispersed acid phase, which immediately reacts with the crude miscella.

The pretreated crude miscella is then alkali-refined using dilute caustic soda with a 16 to 24 °Bé and a 0.2–0.5% NaOH excess over the theoretical required to neutralize the FFA. The reaction of the caustic soda with the FFA proceeds rapidly at 130–135°F (54–57°C), using homogenizers with a shear mixing intensity capable of homogenizing milk and hydrolyzing the phosphatides and pigments with the caustic soda to produce a two-phase mixture. The miscella temperature is adjusted to 135°F (57°C) to obtain the best separation of the heavy phase or soapstock from the oil or the light phase with the centrifuge. The neutral oil is then filtered through a pressure leaf filter precoated with diatomaceous earth. At this point, the refined and filtered miscella can be stripped of the solvent to produce a neutral yellow oil, or it can be further processed as miscella to dewax, fractionate, or hydrogenate the oil (Cavanagh, 1976; Hendrix, 1984).

Obvious disadvantages for the miscella refining process that may have discouraged many processors from adopting this processing system include (Norris, 1982):

- Equipment—All equipment and facilities must be explosion-proof for solvent handling
- Maintenance—The equipment and facilities must be well-maintained to avoid excessive solvent losses and accidents

Silica Refining and Bleaching Modifications

Silica processing utilizes a chemically inert synthetic amorphous silica adsorbent with an affinity for polar contaminants. The surface area, porosity, and moisture content of the silica adsorbents provide them with the capability of removing soaps, phospholipids, sulfur compounds, and trace metals from edible oils. The function of the moisture is to hold the pores open and aid in the attraction of polar contaminants. Several different options are offered for the use of silica adsorbents. The simplest option adds silica adsorbent with the bleaching earth prior to bleaching to reduce the clay usage. A 40% reduction in bleaching-earth usage, less neutral-oil loss, and longer filter cycles are claimed by W.R. Grace for their TrySyl® product (Grace). A second bleaching option adds the silica separate from the bleaching clay, which is precoated on the filter press. The silica collapses in the bleaching vessel, trapping the contaminants to preserve the clay for color pigment removal. An 80% reduction in bleaching earth is claimed for this enhanced silica option. The modified caustic-refining procedure eliminates the need for a water-wash centrifuge. The high soap and gum adsorption capacity of the silica replaces the water-wash procedure. The usual washing process produces a high biological oxygen demand effluent as well as a loss of oil. About 0.05% of the oil washed is lost in the wastewater. Also, disposal of wastewater from conventional washing of alkali-refined oils presents a problem to processors because of increasingly stringent laws regarding oil-refinery effluents. The water-washing step may be eliminated by using 0.05 to 0.4% silica hydrogels to absorb residual soap and trace metals from the refined oil. The silica material has a higher ability to absorb soap, secondary oxidation products, and phospholipids than do traditional bleaching clays. The spent silica is removed by filtration before the oil is bleached. Bleaching with clay is still required to remove the color pigments and other impurities; however, the bleaching-earth usage has been reduced 40–80% in some operations (Grace).

Soapstock Processing

Soapstock from alkali refining is a source of fatty acids, but it also presents a handling, storage, and disposal problem. It is generated at a rate of ~6% of the volume of crude soybean oil caustic refined (Wang, 2002). Originally, many years ago, the caustic refining by-products were merely discarded. Then, it became a valuable source of fatty acids for the soapmaker and the fatty acid distiller. Soapstock was shipped from the refiner in the raw form as it was separated from the neutral oil. The growth of synthetic detergents over soaps reduced this market for soapstock considerably, and in the fatty acid field soapstock utilization was replaced with tall oil, a by-product of the paper industry. These changes turned edible-oil refiners to soapstock acidulation to produce acid oil, which is used as a high-energy ingredient in feeds or provides a more refined product for chemical use (Carr, 1978). Soybean oil can be refined using potassium hydroxide (KOH) rather than NaOH. This allows acidulation of the soapstock with sulfuric acid followed with neutralization with ammonia. The acid oil is still used for animal feed or chemical use, and the potassium and ammonium salts in the waste product are used as fertilizer (Carlson, 2006). Soybean oil methyl esters are also produced from soapstock for biodiesel application (Wang, 2002).

Bleaching

Color reduction occurs with each process in soybean oil neutralization processing: degumming, refining, bleaching, and deodorization. In fact, the usual color limits for other source oils, Lovibond color of less than 20 yellow and 1 red, are readily achievable without the bleaching process (Erickson, 1995c). The bleaching process is not just a color-reducing process, it is also instrumental in removing residual soaps, phosphatides, and oxidizing bodies. The soybean oil bleaching process must be assigned to remove chlorophyll, which is involved in photosensitized oxidation, and to break down peroxides into lower molecular weight carbonyl compounds that are removed with deodorization. Acid-activated bleaching clays are the most effective in adsorbing chlorophyll and in decomposing peroxides. Low levels of phosphorus, 5–10 ppm, and 10–30 ppm soap in the refined oil are required to maximize the bleaching effect (Wang, 2002). Soybean oil should be bleached to <50 ppb maximal chlorophyll content and a zero peroxide value to maintain the optimal oxidative stability of the finished oil (Hastert, 1991). The key parameters for the bleaching process are (i) procedure, (ii) adsorbent type and dosage, (iii) temperature, (iv) time, (v) moisture, and (vi) filtration.

Procedure

The bleaching process is relatively straightforward—refined and/or degummed soybean oil is mixed with the appropriate dosage of earth, heated to a bleaching temperature, and then filtered. The three most common types of contact bleaching methods are batch atmospheric, batch vacuum, and continuous vacuum. This sequence is also the chronological order in which the methods were developed:

- Atmospheric Batch Bleaching—This oldest bleaching process simply mixes the bleaching earth with oil at ~71°C (160°F) in an open vessel, followed by agitating and heating to the bleaching temperature of 100–104°C (212–220°F), holding for 15–20 min and then cooling to 49–54°C (120–130°F) and filtering. Precautions are necessary with atmospheric bleaching to avoid air incorporation, such as, the design of agitators and baffles to avoid vortexes and limiting the time at bleaching temperature to drive off the moisture from the bleaching earth. The oil must be cooled and filtered as quickly as possible and then further cooled and protected from exposure to the air to minimize oxidation and peroxide development. The low equipment cost advantages of atmospheric bleaching are probably outweighed by the risk of oxidation and the operator skill required for process control.

- Vacuum Batch Bleaching—The same operating conditions are applicable to vacuum bleaching as for atmospheric bleaching, except the temperatures may be decreased in proportion to the vacuum. The advantages of vacuum over atmospheric bleaching are: (i) better protection of the oil against oxidation, (ii) reduced bleaching time, (iii) lower bleaching earth levels, which result in lower oil loss in spent earth, (iv) lighter-colored oils, and (v) better soap removal (Sipos & Szuhaj, 1996).

- Continuous Vacuum Bleaching—Refined oil is pumped in a proportioned stream at ~54°C (130°F) into a slurry tank. Bleaching earth and filter aid are continuously fed into the slurry tank at a rate predetermined for the oil being bleached. After thorough mixing, the slurry is sprayed into the deaerating and dehydrating section of the vacuum bleacher that is maintained at 381 mm Hg absolute pressure. After 7 min retention time, the oil is pumped into the bleaching section through a heat exchanger to raise the oil to a bleaching temperature of 104–116°C (220–240°F). The oil is retained in the bleaching section for ~10 min before filtering in closed press and cooled before the vacuum is broken (Sipos & Szuhaj, 1996).

Bleaching Procedure Modifications

Recognition of the importance of moisture in acid-activated bleaching earth treatment has led to the practice of close coupling the refining and bleaching processes. Water-washed refined soybean oil normally contains about 0.2–0.4% moisture. This water is easily removed in the vacuum bleaching system, and theoretically, the extra moisture increases the acidity effect of the bleaching earth. One disadvantage may be a slight increase in FFA (Erickson, 1995c).

The used bleaching earth discharged from the filters is still active. Several methods for the use of this spent earth have been devised. One procedure utilizes two filtration steps. First, the oil is passed through a filter that has already been used and filled to capacity with spent bleaching earth. The prefiltered oil is then bleached and refiltered. The prefiltering step with spent earth can reduce the new bleaching earth usage by as much as 50% (Carlson, 2006; Erickson, 1995c).

The use of special silica adsorbents that specifically target soaps, phosphatides, and trace metals is reported to perform two functions: (i) elimination of the water-washing step in refining, and (ii) reduction of the bleaching earth requirement by up to 50%. Refined oil is treated with the silica adsorbent prior to the addition of the bleaching earth (Carlson, 2006). For the best effect, the silica should be added to the oil at ~70°C (158°F). Preferably, the oil contains ~0.15% moisture and is under deaeration but not drying conditions. The mixture is vigorously agitated for 15 min before the addition of bleaching earth or separation by filtration. The silica with adsorbed impurities is removed by filtering through a cake of spent bleaching earth as discussed above (Young, 1990).

Bleaching Agents and Dosage

Edible oils are not bleached chemically because the color reduction occurs because of oxidizing reactions that have an undesirable effect on the flavor and oxidative stability of the oil (Sipos & Szuhaj, 1996). The effective agents for edible-oil bleaching are natural clays, activated earths, carbon, and synthetic silicates (see detailed descriptions below). Bleaching earths consist of natural clays, which have *bleaching activity*, and others that become active only after a specific treatment. Bleaching earths are made from naturally occurring minerals such as palygorskite, which is also known as attapulgite, sepiolite, bentonite, and other minerals that all belong to the aluminum silicate family. Anhydrous silica gel and activated carbon are also used as bleaching adsorbents to a limited extent.

- Natural Bleaching Earths—Originally known as *Fuller's earth*, they are basically hydrated aluminum silicates, naturally varying in their ability to adsorb pigments. The better natural earths can adsorb 15% of their own weight in pigments and other impurities but also retain about 30% neutral oil. Natural bleaching earths perform best with atmospheric bleaching and are usually only employed for easily bleached oils such as coconut, lard, and tallow. The natural earths do not elevate FFA content nor isomerize unsaturated fatty acid groups; however, for dark or difficult-to-absorb pigments or impurities, prohibitive level of the earth is required, which makes the activated bleaching materials more attractive.

- Activated Bleaching Earths—These bleaching agents are made from bentonite clays that contain a high proportion of montmorillonite. This hydrous aluminum silicate has considerable capacity for exchanging part of the aluminum for magnesium, alkalies, and other bases. Treatment, to varying degrees, with sulfuric or hydrochloric acid, washing, drying, and milling alters the bleaching media's degree of acidity, adsorption capabilities, and particle size distribution (Richardson, 1978). Activated bleaching earths normally contain 10–18% moisture, which supports the montmorillonite layers in the clays. The layers collapse to decrease the surface area available to adsorb pigments and other impurities when the bleaching earth is dry. The bulk density of the bleaching earth is dependent on the amount of void space in the earth—the more void space, the lower the density. Acid activation decreases the bulk density to increase its oil retention activity. The lower activated earth usage level required results in a lower overall bleached oil loss with a lower bleach color and increased impurity removal (Wiedermann, 1981). In addition to removing color pigments, the activated bleaching agents split soap residues to elevate FFA, destroy peroxides and secondary oxidation products, and promote isomerization. The latter effect is more pronounced at temperatures above 160°C (320°F), which is well above the optimal bleaching conditions (Brekke, 1980). The amount of activated bleaching earth to be used should be the minimal amount needed to effect removal of impurities as measured by peroxide reduction with zero as the goal. Normally, this requires 0.3 to 0.5% earth on a weight basis and is dependent on the quality of oil to bleaching and press effect opportunities. The earth usage should be guided by an equal-performance philosophy rather than an established dosage practice (Wiedermann, 1981).

- Activated Carbon—A wide variety of materials can be used to form activated carbon by carbonization at high temperatures, combined with the use of activating materials such a phosphoric acid, metal salts, etcetera. The treated material is washed, dried, and ground to produce activated carbons of various pore sizes, internal specific surface areas, and alkalinity or acidity. Activity is determined by the chemical state and a large specific surface area. Carbon is used sparingly by most processors due to problems with filtration, relatively high costs, and high oil retention; carbon can retain up to 150% of its weight of oil (Rini, 1960). Activated carbons are effective in removing soaps, and pigments, especially chlorophyll (Erickson, 1995c), as well as some aromatic materials that are not volatilized by deodorization (Rini, 1960).

- Silicates—These chemically inert synthetic amorphous silica adsorbents have an affinity for polar contaminants. The surface area, porosity, and moisture content of the silica adsorbents provide them the capability of adsorbing secondary oxidation products (aldehydes, ketones), phosphatidic compounds, sulfur compounds, trace metals, and soap. Moisture functions to hold the pores open and aid in the attraction of the polar contaminants. Most of the synthetic silicas do not have significant direct adsorption capabilities for carotenoid or chlorophyll compounds, but the removal of the other impurities enhances the efficiency of the bleaching earths (Young, 1990).

Bleaching Temperature Influence

The activity of an absorbent in bleaching an edible fat or oil is at a maximum at some particular temperature that varies with oil type and process. Low temperatures favor the retention of the adsorbed pigment on the bleaching media surface, while higher temperatures favor movement into the pores where chemisorption is most likely, which promotes structural changes in the unsaturated fatty acid groups. Splitting of soap to fatty acid is most obvious with acid-activated earths above 95°C (203°F). Extremely high-temperature processing must be avoided to prevent oxidation and isomerization of the unsaturated fatty acid groups and excessive FFA development. The optimal bleaching temperatures of nearly all edible oils range between 70–110°C (160–230°F) (Patterson, 1976).

Both the synthetic silicas and bleaching earths should be slurried with the oil at relatively low temperatures (70°C or 158°F), and then the complete mixture is increased to the final bleach temperature (90–100°C or 194–212°F). Experience shows that final bleach colors are darker when the adsorbents are added to hot oil. Evidently, this effect is due to one or both of the following two factors (O'Brien, 2004).

i. Adding the adsorbent to hot oil reduces its adsorptive capacity because the moisture is driven off too rapidly, causing a collapse of the lattice structure, which reduces the effective surface area to adsorb impurities and pigments (Wiedermann, 1981).

ii. The oil is unprotected against oxidation when heated before the adsorbent is added, which can cause some color fixation or set (Rich, 1967).

If bleaching earths were simply adsorbents, the best color reduction would be expected at low temperatures. The adsorption equilibrium would be expected to shift toward desorption with high temperatures, and some of the adsorbed molecules would dissolve back into the oil. However, this is not observed. Color removal improves as the temperature is increased, indicating that bleaching earth is more than a simple adsorbent. Chemical reactions also take place on the surface of the bleaching earth. According to the rule of van't Hoff, the speed of reaction, wanted or unwanted, doubles with each 10°C temperature increase. Consequently, there must be an optimal bleaching temperature. This optimal temperature depends on the type of oil, the bleaching system type, the by-products, and the impurities (Zschau, 2000). Temperature requirements for vacuum bleaching systems are normally lower than those for atmospheric bleaching to reach optimal color removal. Since temperature also affects other properties of the oil, it must be kept as low as possible to minimize product damage but high enough for adequate adsorbance of the impurities and color pigments (Rich, 1967).

Oxidative stability is one of the major properties, other than color removal, that requires careful control of the process temperatures. Few problems are encountered when the bleaching temperatures remain below 110°C (230°F) and steps are taken to control air oxidation. Anisidine values begin to rise with bleaching temperatures above 110°C (230°F), indicating damage to the oxidative stability. Hydroperoxides, present in the refined oil or formed during bleaching, are decomposed during bleaching to form aldehydes and ketones, which are adsorbable because of their polar structure. A measure of the concentration of secondary oxidation products is commonly obtained with the anisidine value. Adsorption of the secondary products is relatively inefficient. An oil with a peroxide value of 5–10 meq/kg requires earth dosages of 2–3% to achieve an anisidine value of 2 to 3. Anisidine values of 2 to 3 after deodorization, which can remove ~50% of the secondary oxidation products, are requirements for good oil stability (Mag, 1990).

Bleaching Time Influence

In theory, adsorption should be practically instantaneous; however, in practice this is not the case. The rate of color decrease is very rapid during the first few minutes that the adsorbent is in contact with the oil and then decreases to a point where equilibrium is reached and no more color is removed. Time is required for the adsorbent to release all of the bound moisture and take up the color pigments and impurities to maximal capacity. Usually, a contact time of 15–20 min is adequate at a bleaching temperature above the boiling point of water (Patterson, 1976). The usual error is to extend bleaching time beyond the optimum (Rini, 1960). Side reactions are especially important to oil quality and oil losses, making it important to limit adsorbent/oil contact time and temperature. Longer oil contact time with a bleaching earth can cause color reversion and damage to the oxidative stability of the oil (Zschau, 2000).

Contact time for bleaching is made up of two time periods: the time in the bleaching vessel or continuous stream and the contact time in the filter during recirculation or final filtering. A significant portion of total adsorption in bleaching occurs during filtration (Mag, 1990). Continued or progressive reduction in peroxides and the other impurities during filtering is caused by *press effect*, a benefit provided by the earth buildup in the filter with continued use. Some processors take advantage of this effect by decreasing the level of earth used when the filters are partially filled (Brekke, 1980).

When silica is used, it is recommended that it be added to the oil first, with strong agitation and vacuum for 15 min, before the bleaching earth is introduced. During this time, the silica should adsorb soaps, secondary oxidation products, phosphatides, and trace metals, which normally compete with the color pigments for space on the bleaching earth surface. The absence of these impurities increases the efficiency of the bleaching earths to adsorb the chlorophylloid and carotenoid pigments. Experience has also indicated that the press effect may be more effective than the normal bleaching earth and oil mixing with the use of the synthetic silicas. W.R. Grace advocated the use of packed-bed filtration to take advantage of the press effect (Grace).

Adsorbent and/or Oil Moisture Influence

The presence of some moisture seems to be essential for good adsorbance and bleaching action. Bleaching earths completely dried before use were found to be inactive. The adsorbents normally contain from 10–18% moisture, which acts as a structural support to keep the montmorillonite layers apart. During bleaching, it is necessary to remove the moisture in the adsorbent to obtain optimal adsorption capacity; the color bodies and other impurities cannot be adsorbed to maximal capacity until all the water is removed. The bound moisture is not released until the elevated bleaching temperatures are attained (Wiederman, 1981). Refined oil can contain moisture levels from less than 0.1 to as high as 1.0%, which must also be removed for effective adsorption of the traces of soap remaining after refining. Experience indicates that a slightly wet oil may be beneficial for the removal of color pigments and flavor precursors to provide a lighter, more stable oil. Maximal adsorption is achieved when the silicas and bleaching earth are slurried with the oil below the boiling point of water, then gradually increasing the mix to bleaching temperature. Adding the bleaching earth before heating the oil inhibits heat darkening.

Filtration Influence

Good filtration is paramount in ensuring good quality oil products. After an adsorbent has selectively captured the impurities, it must be removed from the oil before it becomes a catalyst for color development or other undesirable reactions (Erickson, 1995c). Filtration, the separation method most often used for spent bleaching-earth removal, is the process of passing a fluid through a permeable filter material to separate particles from the fluid. Examples of the filtration materials utilized include filter paper, filter cloth, filter screen, and membranes. Filter aids such as diatomite, perlite, or cellulose are usually used in conjunction with the permeable filters for surface protection.

The three steps of filtration are precoating, filtering, and cleaning. The purpose of the precoat is to protect the filter screens, provide immediate clarity, improve the flow rate, and aid in filter cake removal during cleaning. It also helps to prevent blinding, which stops the product flow. Precoating is accomplished by slurring filter aid with previously filtered oil and allowing the oil to carry the filter aid to the filter, deposit it on the filter screen, and return to the precoat slurry tank to pick up more filter aid. The amount of precoat is determined by the filter area, usually 5–11 kg/m^2. The flow rate during precoating should be the same as during filtration to obtain an even coating on the filter. Uneven coatings result in blinded filters and short filtration cycles.

During filtration, *body feed*, or the continuous addition of filter aid, can help prevent blinding of the suspended solids on the precoat. The body feed surrounds the suspended solids to provide flow around them. The body feed slurry of filter aid and oil is injected into the system prior to the filter. The suspended solids are ridged or deformable and can elongate under pressure to extrude through the filter cake and slow or block the product flow. Body feed coats the deformed solids, allowing them to be retained on the filter cake.

Several indicators are utilized to determine the point at which the filter space is filled with solids from the bleached oil: when the pressure drop across the leaves reaches a predetermined level, when a predetermined decrease in flow rate occurs, or when a calculated load level is reached. Short cycles or premature filter stoppages are usually the result of (i) inadequate body feed; (ii) too high flow rate, which can cause the solids to pack; (iii) too low flow rate, which can allow the solids to settle and block the flow rate; (iv) blinded screens, which reduce the filter surface area; or (v) solid load exceeding filter capacity. The perfect circumstance is when the differential pressure is reached, and the flow rate is severely reduced at the same time that the calculated filter capacity is exhausted. Once the filter cycle is complete, the filter cake must be removed and the process repeated all over again (Butterworth, 1978).

Traditionally, either plate and frame filters or pressure leaf filters have been used for bleach-clay removal. The sequence of change in usage was approximately as follows: plate and frame filters, pressure leaf filters, self-cleaning closed filters, and automated filters. Pressure leaf filters began to replace plate and frame presses for several reasons. One of the major reasons was that the leaf filters were easier to clean than the plate and frame presses, and labor costs were less. Labor costs were the impetus for more complete automation of the bleaching operations and all the other processes. Currently, completely self-cleaning closed filters that operate on an automated cycle are available (Latondress, 1983). The general rule for determining the needed capacity ranges from 250–484 kg oil flow/h/m^2 of filter surface (50–100 lb/h/ft^2), with the lower flow rate preferred (Erickson, 1995c).

Bleaching By-product

The spent bleaching earth removed from the bleached oil with filters represents a substantial amount of waste material. The most common handling procedure is to discard spent bleaching earth directly from the filters to a landfill. The spent bleaching earth oxidizes rapidly when exposed to the air to develop a strong odor, and spontaneous combustion easily occurs, especially with oils high in polyunsaturates. Therefore, the spent bleaching earth must be covered with soil or sand soon after dumping.

The oil content of the spent bleaching earth may range from 25–75% of the weight of the earth. Oil retention is affected by the type of filters, the type of refined oil bleached, and the degree of color reduction. It is important to recover as much of this oil as possible, but methods that are too efficient may cause desorption of the impurities adsorbed by the bleaching earth from the refined oil. Because it is possible to remove a substantial portion of the oil from the spent earth, it may become a regulatory requirement in the future. Oil can be recovered by several methods, some performed on the cake while it is still in the filter and others after it has been removed from the filter. Some of the procedures for oil recovery include (Patterson, 1976; Ong, 1983; Hong, 1983; Svensson, 1976):

- Cake Steaming—Blowing steam through the cake in the filter can reduce the oil content to as low as 20%; however, the oil content should not be reduced below 25% because the steam wetting may cause desorption of the impurities below this point to lower the quality of the recovered oil. Also, spent earth with a low oil content oxidizes more rapidly when exposed to the atmosphere.

- Hot-water Extraction—Circulation of hot water at 200°F (95°C) through the filter cake while maintaining a pressure of 5 atm at a rapid flow rate can displace as much as 55–70% of the oil for collection and separation. Washing time may be extended to 30 min, but 90% of the recoverable oil is obtained in the first 10 min. After water washing, the filter cake may be partially dried with steam. Drying with air can cause the filter to catch on fire, especially when oils high in unsaturates are processed.

- Solvent Extraction—Organic solvents can be used to extract the oil from the filter cake in certain enclosed filters as a separate process. Hexane, a nonpolar solvent, performed well, but strong polar solvents such as acetone or trichlorethylene may also recover the impurities separated from the refined oil. Solvent extraction provides oil yields of over 95% with a quality comparable to the originally filtered oil. Explosion-proof environments, buildings, and equipment, which are quite expensive, are required for solvent extraction. In most cases, the less efficient hot-water extraction is more practical than solvent extraction, and it may be feasible only for very large processing facilities that generate large quantities of spent earth.

- Solvent Extraction with Oilseeds—Extraction of the bleaching earths in a mixture with oilseeds is practiced by some extraction plants with processing capabilities, but the potential problems for this type of recovery may outweigh the savings; for example, the mineral content of the meal may be increased beyond the acceptable limits, and the recovered oil may decrease the quality of the new oil extracted. The oxidation products and polymers from the recovered oil could contaminate the fresh oil.

- Water/lye Extraction—Oil can be extracted from the spent bleaching earth by suspending it in double the amount of water and boiling with concentrated lye. The oil accumulates on the surface of the slurry for recovery. The remaining slurry can be centrifuged with separations as high as 85% efficiency. The separated bleaching clay has a light gray color, is almost odorless, and does not ignite spontaneously. It can be used as a landfill material to cover other refuse, instead of requiring soil or sand to cover it. The procedure is simple and relatively inexpensive, but a dark-colored, low-quality oil suitable only for technical purposes or possibly cattle feed is obtained.

Deodorization

Deodorization of soybean oil and other edible fats and oils is the last step in preparing them as an ingredient for food preparation. It establishes the oil characteristics of *flavor* and *odor*, which are those most readily recognized by consumers. Deodorization is a steam distillation process in which good quality steam is injected into soybean oil under a high temperature and a high vacuum to remove FFA and volatile odiferous components to obtain a bland and odorless oil. Undeodorized soybean oil and other vegetable oils have characteristic undesirable flavors and odors and obtain others during processing; for example, bleaching imparts an *earthy* flavor. Steam deodorization is feasible because of the great differences in the volatility between the triglycerides and the substances that give oils and fats their flavors and odors. Most of the undesirable materials in undeodorized oil have about the same vapor pressure as fatty acids to allow them to be one measure of the completeness of deodorization. When the FFA content is reduced to 0.01–0.03% and the peroxide value to zero, most flavors and odors are eliminated. The known compounds removed by deodorization are aldehydes, ketones, alcohols, hydrocarbons, and various others formed by the heat decomposition of peroxides and pigments. The concentration of these minor compounds in well degummed, refined, and bleached oil is ≤200 ppm and usually no more than 1000 ppm. However, the flavor threshold for these materials is quite low; for example one component, decadienal, has a detectable flavor at 0.5 ppb (Norris, 1985).

Nutritional studies in the 1990s identifying the deleterious effect of *trans* fatty acids in the diet mandated closer examination of the isomerization of fatty acids during high-temperature regimes in all stages of fats and oils processing. RBD soybean oils should contain only a limited amount of *trans* fatty acids, mainly *trans* C-18:2 and C-18:3 isomers formed during deodorization: typically 1.0% maximum. Isomerization during deodorization is a function of time and temperature only with no influence from steam or pressure. Depending on deodorization conditions, 3–24% of the α-linolenic fatty acid is converted to the *trans* isomer, and a maximum of 2% of linoleic fatty acid is isomerized. In general the rate of *cis* to *trans* isomerization of α-linolenic fatty acid is 10 times higher than linoleic fatty acid and 100 times higher than oleic fatty acid (Harper, 2001). This means that soybean oil has a higher susceptibly to *trans* fatty acid development during deodorization than vegetable oils with lower levels or the absence of linolenic fatty acids.

Principles of Deodorization

Soybean oil that has been properly degummed, caustic refined, and bleached still contains some substances that are responsible for its unpleasant odor and flavor. These components are present in small amounts, ~200–300 ppm excluding FFA, in the oil both as natural components and some that formed during the preliminary treatments. The undesirable minor components consist of hydrocarbons, low-molecular-weight fatty acids, aldehydes, ketones, and alcohols. Removal of these substances and the high- molecular-weight FFA, while preserving the tocopherol, sterols, and other desirable components, influence the deodorization process conditions.

Deodorization conditions depend on the particular oil type, the oil quality, and the refining system used. A shift in refining technology toward physical refining where the FFA is removed exclusively by steam distillation requires more severe conditions than chemically refined oils. With chemical refining, most of the FFA content is removed before deodorization. Degummed and bleached soybean oils before physical refining have ≤1.5% FFA contents compared to 0.05–0.1% for chemically refined oils before deodorization. The steam distillation requirements for both physical and chemically refined oils are achieved by altering one or more of the operating variables. The four interrelated operating variables that influence deodorized oil quality are vacuum, temperature, stripping-steam rate, temperature, and retention time at deodorization temperatures.

- Vacuum—Vacuum systems affect the low absolute pressures necessary for low-temperature distillation of the odoriferous substances. The boiling points of the principal fatty acids and the vapor pressure of the odoriferous materials decrease as the absolute pressure decreases (Bernardini, 1993). The required low absolute pressure, usually 2 to 3 mbar, is commonly generated by vacuum systems consisting of a combination of steam-jet ejectors (boosters), vapor condensers, and mechanical (liquid ring) vacuum pumps. Special vacuum systems were developed to reach lower pressures and operating costs and, at the same time, reduce emissions by a more efficient condensing of the volatiles. In the dry condensing system, the sparge steam is condensed on surface condensers working alternatively. The remaining noncondensables are removed by mechanical pumps or by a vacuum ejector system (Kellens & De Greyt, 2000).

- Stripping Steam—Stripping steam is the primary motive force in the deodorization process as it is the carrier medium for moving the vaporized FFA, ketones, aldehydes, and other volatiles from the feedstock to the vacuum ejector and distillate recovery system. Raoult's law shows that steam volume rather than weight is important for the steam stripping process. Therefore, the quantity or percentage of stripping steam requirement is related to system absolute pressure; for example, 1.5% steam quantity at 3 mbar absolute pressure has the same volume as 3% steam quantity at 6 mbar. Another variable is the particular deodorizer design and efficiency. Current systems require stripping steam in the range of 0.6–1.2% of the oil throughput at an operating pressure of 3 mbar (Zehnder, 1995).

- Temperature—Temperature is the operating condition that can be classified as a variable. It directly affects the vapor pressure of the volatile constituents to be removed; therefore, increasing or decreasing the oil temperature produces corresponding removal rates for the odoriferous compounds and thermal decomposition of the carotenoid pigments. An excessive increase in temperature can result in polymerization, isomerization, thermal cracking with formation of odoriferous and low boiling point products, color reversion, and excessive removal of tocopherols and sterols (Zehnder, 1995). Thermal degradation of the tocopherols becomes significant at deodorization temperatures above 260°C (500°F). Twice as many tocopherols and sterols are stripped out at 275°C (525°F) than at 240°C (465°F), and pressure variations of 2–6 mbar had only a slight effect on tocopherol/sterol stripping. Higher deodorization temperatures can be avoided by lowering the pressure and increasing the stripping steam at an increased cost of deodorization. Another apparent alternative, reduction of the residence time, is harmful to the oxidative stability and color reduction of the oil. Most deodorizers used in the processing of soft oils, like soybean oil, operate at temperatures 230–260°C (446–500°F), a pressure of 3 mbar or lower, and 0.6–0.8% stripping steam (Kellens, 2000).

- Retention Time— Retention time is the period during which the oil is at deodorizing temperature and subjected to stripping steam. Stripping time for efficient deodorization must be long enough to reduce the odoriferous components to the required level. The holding times are affected by the deodorizing temperature, vacuum, depth of the layer of oil, and the source oil. For example, the holding time for a batch deodorizer, with 8–10 ft oil depth, is 3–8 h, whereas the holding time for continuous and semicontinuous systems, with shallow oil depths, varies from 15–60 min for oils caustic refined with an FFA level of <0.1%. Higher FFA levels, such as ≤1.5 for physical refining, can require as much as twice that of caustic refined and bleached oils. Normally, the holding time for soybean oil is 15–120 min. Additionally, certain reactions with the oils deodorized are not related to the removal of FFA but instead help provide a stable oil after deodorization. These reactions and the heat bleaching are time- and temperature-dependent; therefore, deodorization systems provide a retention period at deodorizing temperatures to allow these reactions and heat bleaching to occur (Zehnder, 1995). Formation of *trans* isomers depends on the deodorization temperature and time held at that temperature. At 220°C (428°F), few *trans* isomers are formed, but at 260°C (500°F) *trans* isomers are formed over 10 times the rate at 220°C; a typical soybean oil reaches 1.0% *trans* fatty acids in about 20 min at 260°C (Paine, 2002).

- Physical Refining— Physical refining imposes more severe conditions than caustic refined oils, since the FFA levels are considerably higher, ≤1.5% for most degummed and bleached soybean oils compared to ≤0.1% for caustic refined and bleached oils. To modify the operating conditions for physically refined oils to obtain a final FFA of 0.03–0.05% is necessary. The easiest remedy is to increase the deodorization temperature to ~250°C (482°F), but this results in a higher *trans* fatty acid content and a greater loss of tocopherols. Avoid higher temperatures by lowering the deodorizer pressure and increasing the amount of stripping steam, but this raises the overall cost of deodorizing. Reducing the deodorizing time is another alternative, but experience shows that certain time- and temperature-dependent reactions within the oil itself, unrelated to FFA removal, are necessary to provide a stable oil. Therefore, most deodorizers used for steam refining of oils operate at temperatures 230–260°C (446– 500°F), a pressure of 3 mbar or lower, with a stripping steam consumption of ~10kg/t of processed oil. All three of these parameters increase the demand on the vacuum unit. Consequently, the deodorization stage in physical refining is more expensive than with caustic refined and bleached oil (Kellens & De Greyt, 2000). The construction material of choice for commercial deodorizers operating above 150°C (300°F) is stainless steel in contact with the product. Type 304 stainless has proven satisfactory for caustic refined oils: however, type 316 stainless is recommended for physical refining because of its greater acid resistance (Zehnder, 1995).

Deodorization Systems

The system choice depends on several factors: such as the number of feedstock changes, heat recovery, investment, operating costs, ecology requirements, physical or chemical refining, construction materials, etcetera. Deodorizer types in use are classified into three principal groups: batch, continuous, and semi-continuous.

- Batch Deodorization— Batch deodorization is the simplest system that can be installed. It is suitable for small capacities, irregular production, or processing of small batches of different oils that demand minimal intermixing. The principal components consist of a vertical cylinder form vessel with dished or cone heads fabricated from type 304 or 316 stainless steel to avoid the deleterious catalytic activity of iron on oils, welded to prevent air leaks, and well-insulated to minimize heat loss. The usual capacity range is 10,000–40,000 lb with vessel diameter determined for an oil depth of 8–10 ft with a like headspace above the oil surface. The headspace is necessary to avoid excessive entrainment losses from the rolling and splashing of the oil caused by the stripping steam injected into the bottom of the vessel through a distributor. In addition to the steam ejector system, means for heating, cooling, pumping, and filtering the oil are required. The batch system controls include a device for indicating oil temperature and a pressure gauge designed to indicate accurately low pressure within the deodorizer. Operating at a high temperature and 6–12 mbar pressure batch deodorization requires 8 h for a complete cycle of charging, heating, deodorizing, cooling, and discharging the oil. Batch systems operating at higher pressures or lower deodorization temperatures may require as much as 10–12 h for a deodorization cycle. Stripping steam is ordinarily injected at 3 lb per 100 lb of oil per hour at 6 mbar pressure. The oil must be cooled before discharging to the atmosphere, preferably to 38–49°C (100–120°F) for RBD soybean oil (Gavin, 1978).

- Semi-continuous Deodorization—These systems operate on the basis of handling finite batches of oil in a timed sequence of deaeration-heating, holding-steam stripping, and cooling such that each quantum of oil is completely subjected to each condition before proceeding to the next step. The semi-continuous deodorizer consists principally of a tall cylindrical shell of carbon steel construction with five or more stainless steel trays stacked inside of, but not quite contacting, the outer shell. Each tray is fitted with a steam sparge and is capable of holding a measured batch of oil. By means of a measuring tank, oil is charged to the top tray, where it is deaerated while being heated with steam to about 320–330°F (160–166°C). At the end of the heating period, the charge is automatically dropped to the second tray, and the top tray is refilled. In the second tray the oil is heated to the operating temperature and again, after a timed period, is automatically dropped to the tray below. When the oil reaches the bottom tray, it is cooled to 300°F (149°C) with stripping steam and 50–100 ppm citric acid added to chelate metals. The deodorized oil is discharged to a drop tank, where it is quickly cooled to 100–130°F (38–54°C), after which it is pumped through a polishing filter to storage. Semi-continuous deodorizers are usually automated and controlled from a central panel with a time-cycle controller and interlocks such that the sequence steps are interrupted in the event of insufficient batch size, improper drop-valve opening or closing, or the oil not reaching the preset heating or cooling temperatures in the allotted time (Gavin, 1981).

One of the principal advantages of the semi-continuous deodorization system is that all of the trays are under the same relatively high vacuum. All oil receives substantially identical treatment, and the annular space between the trays and the shell provides some insurance against oxidation due to inward leakage of air. The deodorizer arrangement avoids refluxing of once-distilled undesirable materials back into the oil. This reflux, plus any mechanical carryover, is permitted to drain from the bottom of the deodorizer shell. The ability to accommodate frequent stock changes with a minimum of lost production and practically no intermixing is an important advantage for the semi-continuous systems over the continuous deodorization systems; however, heat recovery is less efficient than a continuous operation, and 10 to 20% more sparge steam is required (Zehnder, 1976, 1995).

- Continuous Deodorization— Continuous deodorization provides uniform utility consumption by not being subject to the peak loads attendant with batch-type heating and cooling of semi-continuous operations. This permits smaller heating and cooling auxiliaries and optimal heat recovery through interchange between incoming and outgoing oils. Continuous systems are suitable for high-capacity operations with fewer than three stock changes per 24-h day. However, continuous deodorization benefits are lost with multiple stock changes due to the lost production of 30 to 60 min for each stock change and the likelihood of commingling oil types.

Several configurations of continuous deodorizers exist. The interest in edible-oil processing with physical refining initiated the development of more efficient deodorization equipment designs. The industrial designs can involve tray or thin-film deodorizers. Tray deodorizers are based on a series of steam-agitated trays or compartments stacked vertically or horizontally in cylindrical vessels. Stripping of FFA and other volatile compounds are carried out simultaneously. The retention time per tray is usually 10–30 min with oil levels of typically 0.3–0.8 meters maintained by overflow pipes or weirs and drained by separate discharge valves. Thin-film deodorizers typically employ a structured packing at the top of the column to create a maximal surface-to-volume ratio. The oil flows by gravity over the packing and meets sparging steam concurrently for FFA stripping. The residence time of the oil in the structured packing is ~5 min. The steam passes upward through the deodorizer at minimal pressure drop. A three-tray retention section at the bottom of the column provides flexible retention times for heat bleaching of 0–90 min. The retention time at high temperatures is reduced to a minimum to retain a high level of tocopherols and limit *trans* fatty acid development (Ahrens, 1998). Due to the high oil–metal contact surface in packed columns, the risk of fouling is higher. The frequency of cleaning as well as the efficiency of the structured packing material is determined by the type of oil processed, the frequency of shutdowns, feedstock changes, and the purity of the feedstock. Packed-column processing physically refined oils requires more frequent cleaning, each 6–10 mo, and replacement of the structured packing every 2–3 yr. Deodorization of chemically refined oils with this system extends the cleaning frequency to once a year and the structured packing lifetime to 3–4 yr (Kellens, 2000).

Deodorizer Distillate

Deodorizer distillate is the material collected from the steam distillation of edible oils. As a general rule, 0.5% of the deodorizer feedstock approximates the amount of distillate produced by a typical processor of edible fats and oils. The composition of the distillate depends on the source oil, the refining technique utilized, and the deodorizer operating conditions. The distillate from physically refined oils consists mainly of FFA with low levels of unsaponifiable components. Distillates from chemically refined oils have higher economic value due to higher sterol and tocopherol contents that are sources for natural vitamin E, natural antioxidants, and other pharmaceuticals (Kellens, 2000). The use of deodorizer distillates in animal feeds is forbidden by U.S. regulations because any insecticides in the source oils are co-distilled with the other organic compounds. The distillate not used as a source of tocopherols, sterols, or industrial fatty acids can be mixed with the fuel oil used to fire the steam boilers; up to 10% of the distillate has been successfully used in fuel oil. The remaining disposal alternatives are depositing it in a refuse dump, if permitted, or combustion in a fluidized bed incinerator (Watson & Meiehoefer, 1976; Svensson, 1976).

Finished Oil Handling

The last process that can change the flavor, odor, color, nutritional attributes, and the oxidative stability of soybean oil is deodorization. If the conditions for extraction and crude oil processing are satisfactory, the result is a tasteless, odorless, light-colored oil, free from peroxides and other contaminants. Storage, handling, and packaging of the purified oil must protect the achieved quality prior to use by the consumer or use in a prepared food product. Deodorized oils must be protected from the five factors that contribute to the oxidative deterioration of oils: oxygen or air, heat, light, prooxidant metals, and time.

References

Ahrens, D. Comparison of tray, thin-film deodorization. *Inform* **1998,** *9,* 566–576.

Bernardini, M. Deodorization. *Proceedings of the World Conference on Oilseed Technology and Utilization;* T.H. Applewhite, Ed.; AOCS Press: Champaign, IL, 1993; pp. 186–193.

Braae, B. Degumming and refining practices in Europe. *J. Am. Oil Chem. Soc.* **1976,** *53,* 353–357.

Brekke, O.L. Bleaching. *Handbook of Soy Oil Processing and Utilization;* D.R. Erickson et al., Eds.; American Oil Chemists' Society: Champaign, IL, and American Soybean Association: St. Louis, MO, 1980; pp. 120, 122–123.

Butterworth, E.R. Separation methods in processing edible oils. *J. Am. Oil Chem. Soc.* **1978,** *55,* 781–782.

Carlson, K. Acid and alkali refining of canola oil. *Inform* **1993,** *4,* 273–281.

Carlson, K. Recent developments and trends in processing of fats and oils. *Inform* **2006,** *17,* 671–674.

Carr, R.A. Refining and degumming systems for edible fats and oils. *J. Am. Oil Chem. Soc.* **1978,** *55,* 765–771.

Crauer, L.S. Continuous refining of crude cottonseed miscella. *J. Am. Oil Chem. Soc.* **1964,** *41,* 656–661.

Cavanagh, G.C. Miscella refining. *J. Am. Oil Chem. Soc.* **1976,** *53,* 361–363.

Dijkstra, A.J. Degumming, refining, washing and drying fats and oils. *Proceedings of the World Conference on Oilseed Technology and Utilization;* T.H. Applewhite, Ed.; AOCS Press: Champaign, IL, 1992; pp. 140–145.

Erickson, D.R. Degumming and lecithin processing and utilization. *Practical Handbook of Soybean Processing and Utilization;* AOCS Press: Champaign, IL, and United Soybean Board: St. Louis, MO, 1995a; pp. 174–179.

Erickson, D.R. Neutralization. *Practical Handbook of Soybean Processing and Utilization;* AOCS Press: Champaign, IL, and United Soybean Board: St. Louis, MO, 1995b; pp. 184–201.

Erickson, D.R. Bleaching/adsorption treatment. *Practical Handbook of Soybean Processing and Utilization;* AOCS Press: Champaign, IL, and United Soybean Board: St. Louis, MO, 1995c; pp. 203, 211–216.

Erickson, D.R. Overview of modern soybean oil processing. *Practical Handbook of Soybean Processing and Utilization;* AOCS Press: Champaign, IL, and United Soybean Board: St. Louis, MO, 1995d; p. 60.

Erickson, D.R. Optimum and adjusted refining practices for soybean oil. *Proceedings of the World Conference on Oilseed Processing and Utilization;* R.F. Wilson, Ed.; AOCS Press: Champaign, IL, 2001; pp. 74–79.

Farr, W.E. Refining of fats and oils. *Introduction to Fats and Oils Technology,* Second ed.; R.D.O'Brien, W.E. Farr, P.J. Wan, Eds.; AOCS Press: Champaign, IL, 2000; pp. 136–157.

Gavin, A.M. Edible oil deodorization. *J. Am. Oil Chem. Soc.* **1978,** *55,* 785.

Gavin, A.M. Deodorization and finished oil handling. *Proceedings of the World Conference on Soya Processing and Utilization;* AOCS Press: Champaign, IL, 1981; pp. 175–184.

Grace, W.R. Co. *The Amazing Characteristics of Trisyl Silica* [brochure], Columbia, MD.

Harper, T. Chemical and physical refining. *Proceedings of the World Conference on Oilseed Processing and Utilization;* R.F. Wilson, Ed.; AOCS Press: Champaign, IL, 2001; pp. 21–26.

Hastert, R.C. Adsorptive treatment of edible oils. *Introduction to Fats and Oils Technology;* P.J. Wan, Ed.; AOCS Press: Champaign, IL, 1991; p. 96.

Hendrix, B. Neutralization 1. Theory and practice of conventional caustic (NaOH) refining. *Proceedings of the World Conference on Edible Fats and Oils Processing;* D.R. Erickson, Ed.; AOCS Press: Champaign, IL, 1990; pp. 94–100.

Hendrix, W.B. Current practices in continuous cottonseed miscella refining. *J. Am. Oil Chem. Soc.* **1984,** *61,* 1369–1372.

Hong, W.M. Quality of byproducts from chemical and physical refining of palm oil and other oils. *J. Am. Oil Chem. Soc.* **1983,** *60,* 316–318.

Hvolby, A. Removal of nonhydratable phospholipids from soybean oil. *J. Am. Oil Chem. Soc.* **1971,** *48,* 503.

ISEO Technical Committee. Degumming. *Food Fats and Oils,* Ninth ed.; 2006.

Kellens, M.; W. De Greyt. Deodorization. *Introduction to Fats and Oils Technology,* Second ed.; R.D. O'Brien, W.E. Farr, P.J. Wan, Eds.; AOCS Press: Champaign, IL, 2000; pp. 235–268.

Latondress, E.G. Oil-soilds separation in edible oil processing. *J. Am. Oil Chem. Soc.* **1983,** *60,* 259–260.

List, G.R.; T.L. Mounts; K. Warner; A.J. Heakin. Steam refined soybean oil: effect of refining and degumming methods on oil quality. *J. Am. Oil Chem. Soc.* **1978,** *55,* 277–279.

Mag, T.K. Bleaching—theory and practice. *Edible Fats and Oils Processing: Basic Principles and Modern Practices: World Conference Proceedings;* D.R. Erickson, Ed.; 1990; pp. 107–116.

Norris, F.A. Refining and bleaching. *Bailey's Industrial Oil and Fat Products,* Fourth ed.; D. Swern, Ed.; John Wiley & Sons, Inc.: New York, NY, 1982; Vol. 2, pp. 253, 259, 268–288, 290.

Norris, F.A. Deodorization. *Bailey's Industrial Oil and Fat Products,* Fourth ed.; T.H. Applewhite, Ed.; John Wiley & Sons: New York, NY, 1985; Vol. 3, pp. 128–129.

O'Brien, R.D. Fats and oils processing. *Fats and Oils: Formulating and Processing for Applications,* Second ed.; CRC Press: Boca Raton, FL, 2004; pp. 67–83, 83–92.

Ong, J.T. Oil recovery from spent bleaching earth and disposal of the extracted material. *J. Am. Oil Chem. Soc.* **1983,** *60,* 314–315.

Pagliero, C.; N. Ochoa; J. Marchese; M. Mattea. Degumming of crude soybean oil by ultrafiltration using polymeric membranes. *J. Am. Oil Chem. Soc.* **2001,** *78,* 793–796.

Paine, A.R. Latest developments in continuous vegetable oil deodorization. *Practical Short Course on Processing of Vegetable Oils;* Food Protein R&D Center, Texas A&M University System: College

Station, TX, 2002; p. 8.

Patterson, H.B.W. Bleaching practices in Europe. *J. Am. Oil Chem. Soc.* **1976,** *53,* 339–341.

Rich, A.D. Major factors that affect bleaching performance. *J. Am. Oil Chem. Soc.* **1967,** *44,* 298A–324A.

Richardson, L.L. Use of bleaching clays in processing edible oils. *J. Am. Oil Chem. Soc.* **1978,** *55,* 777–780.

Rini, S.J. Refining, bleaching, stabilization, deodorization, and plasticization of fats, oils, and shortening. *J. Am. Oil Chem. Soc.* **1960,** *37,* 515–520.

Segers, J.C. Pretreatment of edible oils for physical refining. *J. Am. Oil Chem. Soc.* **1983,** *60,* 262–264.

Sipos, E.F.; B.F. Szuhaj. Soybean oil. *Bailey's Industrial Oil and Fat Products;* Fifth ed.; Y.H. Hui, Ed.; John Wiley & Sons, Inc.: New York, NY, 1996; Vol. 2, pp. 522–527.

Subramanian, S.; M. Nakajima; A. Yasui1; H. Nabetani; T. Kimura; T. Maekawa. Evaluation of surfactant-aided degumming of vegetable oils by membrane technology. *J. Am. Oil Chem. Soc.* **1999,** *76,* 1247–1253.

Sullivan, F.E. Refining of oils and fats. *J. Am. Oil Chem. Soc.* **1968,** *45,* 564A–582A.

Svensson, C. Use or disposal of byproducts and spent material from vegetable oil processing industry in Europe. *J. Am. Oil Chem. Soc.* **1976,** *53,* 443–444.

Swoboda, P.A. Chemistry of refining. *J. Am. Oil Chem. Soc.* **1985,** *62,* 287–291.

Tandy, D.C.; W.J. McPherson. Physical refining of edible oil. *J. Am. Oil Chem. Soc.* **1984,** *62,* 1253–1261.

ten Brink, H.B.; G. van Duijn. Refined seed oil production: A milder future? *Inform* **2003,** *14,* 274–275.

Wang, T. Soybean oil. *Vegetable Oils in Food Technology Composition, Properties and Uses;* F.D. Gunstone, Ed.; CRC Press LLC, Boca Raton, FL, 2002; pp. 18–38.

Watson, K.S.; C.H. Meiehoefer. Use or disposal of by-products and spent material from the vegetable oil processing industry in the U.S. *J. Am. Oil Chem. Soc.* **1976,** *53,* 437–442.

Wiedermann, L.H. Degumming, refining and bleaching soybean oil. *J. Am. Oil Chem. Soc.* **1981,** *58,* 159–165.

Young, F.V.K. Physical refining. *Edible Fats and Oils Processing: Basic Principles and Modern Practices, World Conference Proceedings;* D.R. Erickson, Ed.; AOCS Press: Champaign, IL, 1990; pp. 124–135.

Zehnder, C.T. Deodorization 1975. *J. Am. Oil Chem. Soc.* **1976,** *53,* 364–369.

Zehnder, C.T. Deodorization. *Practical Handbook of Soybean Processing and Utilization;* D.R. Erickson, Ed.; AOCS Press: Champaign IL, and United Soybean Board: St. Louis, MO, 1995; pp. 239–257.

Zschau, W. Bleaching. *Introduction to Fats and Oils Technology.* Second ed.; R.D. O'Brien, W.E. Farr, P.J. Wan, Eds.; AOCS Press: Champaign, IL, 2000; pp. 169–171.

13

Soybean Oil Modification

Richard D. O'Brien
Consultant, Schulensburg, TX 78956

Introduction

Slightly less than one-half of the 19.2 billion pounds of crude soybean oil produced in 2005 for domestic use in the United States was utilized for food in its purified, natural state-refined, bleached, and deodorized RBD soybean oil (Ash & Dohlman, 2007). To extend its food applications, soybean oil is modified by chemical or physical manipulation. Modification processes to produce soybean oil products with suitable physical characteristics include: hydrogenation, interesterification, fractionation, and blending. These modification processes are performed after refining and bleaching but before deodorization to allow distillation of certain impurities developed during the modification processes. One objective of these processes is to produce higher melting fat forms suitable for the production of margarine, shortening, and other specialized fats and oils products. The finished product performance of the three modification processes is achieved differently for each process: hydrogenation changes the fatty acid composition; interesterification redistributes the fatty acids; and fractionation physically separates hard and soft fractions. Fats and oils products with higher melting characteristics require regimented crystallization processing to stabilize the crystal form; crystal types define the textural and functional characteristics of most fat-based products (O'Brien, 2004). The crystallization processes are performed after deodorization.

Modification processes have been instrumental in the growth of edible soybean oil utilization. Initially, soybean oil was not considered a good paint or food oil. Flavor and odor deficiencies were the major problems restricting its use as an edible oil (Dutton, 1981). Modification by hydrogenation provided a practical but not complete solution to the flavor reversion problem of soybean oil (Frankel, 1980). Reduction of the linolenic fatty acid content from ~9.0 to <3.0% increased the oxidative stability of soybean oil to an acceptable level (Dutton et al., 1953). This solution made hydrogenation crucial for maintaining flavor stability for shortenings, margarines, and salad oils produced with soybean oil. Recent nutritional research connected the *trans* isomers, primarily produced by partial hydrogenation, with unfavorable effects on both LDL and HDL cholesterol. As a result, health organizations recommended that consumption of *trans* fatty acids be reduced, and the U.S. FDA revised the labeling regulations to include a listing of the amount of *trans* on the Nutritional Facts Panel for food products beginning January 2006 (O'Brien & Wakelyn, 2005). Elimination or reduction of the *trans* fatty acids in edible oil products necessitated the reformulation of most shortening and margarine products to include naturally saturated oils, fully hydrogenated oils (*trans*-free), modified hydrogenation procedures, and other modification processes to produce suitable basestocks. Reformulation of existing products without the benefit of *trans* isomers is a challenge to fats and oils product developers. The replacement products must provide the functionality of flakiness, texture, appearance, and, above all, the oxidative stability of the product being replaced (List & Reeves, 2005).

Hydrogenation

In North America and northern Europe, animal fats in the form of butter, lard, and tallow were the major sources of edible fats until development of the hydrogenation process. This process made it possible for vegetable oils to be converted into plastic fat forms that people were accustomed to with greater flavor stability. From the time that the British patent on liquid-phase hydrogenation was issued to Norman in 1903 and its introduction in the United States in 1911, few chemical processes have made as great an impact on any industry. Liquid-phase, catalytic hydrogenation is the most complex chemical reaction carried out in the processing of edible fats and oils. Most chemistry textbooks describe hydrogenation of oils as a simple saturation of double bonds in an unsaturated fatty acid with hydrogen, using a catalyst. Actually, that is only one of several very complex reactions during hydrogenation. The products of hydrogenation are a very complex mixture because of the simultaneous reactions that occur: saturation of double bonds; *cis/trans* isomerization of double bonds; and shifts of double-bond locations, usually to the lower energy conjugated state (Allen, 1981, 1987).

The growth of the soybean oil industry has been paralleled by an increase in hydrogenation (Hastert, 1981). The instability of the flavor of soybean oil was instrumental in promoting the development of partial hydrogenation (Dutton, 1981). One reason for this connection was that it was widely accepted that the major precursor of soybean oil's flavor reversion was the linolenic fatty acid content (Frankel, 1980). Reduction of the linolenic fatty acid content from ~9.0 to <3.0% was found to increase the oxidative stability of soybean oil to an acceptable level (Dutton et al., 1953). Soybean salad oil brush hydrogenated to a 110 iodine value and then winterized to regain its salad oil characteristics was introduced in the late 1950s. The RBHWD (refined, bleached, hydrogenated, winterized, deodorized) soybean salad oil was accepted by the retail salad oil customers and also industrially as a component of salad dressings, mayonnaise, sauces, etcetera (Erickson, 1983). In the late 1970s, improvements in soybean seed handling, oil extraction, and oil processing produced an RBD soybean oil that was first accepted by industrial users and then introduced to the retail market. The processing techniques which helped achieve this improved flavor and oxidative stability and others that followed were: (i) inactivation of the lipoxygenase enzymes that oxidize the polyunsaturated fatty acids and the phospholipase enzymes that catalyze the conversion of phospholipids to a nonhydratable form during oil extraction (Johnson, 2008); (ii) degumming to remove phospholipids, both the hydratable and nonhydratable forms; (iii) bleaching to remove oxidation products (Erickson, 1995) and the chlorophyll pigment that has a role in photosynthesis (Hastert, 1991); (iv) milder processing temperatures; (v) use of stainless-steel transfer lines and tanks for storage and process vessels (Carlson, 2006); (vi) metal chelating (Dutton et al., 1948); and (vii) nitrogen protection for the oil throughout processing.

Oils are hydrogenated for two reasons: to alter the melting properties, and to increase oxidative stability (Okonek, 1987). A wide range of basestocks can be produced with the hydrogenation process depending on the conditions used and the degree of saturation and/or isomerization. Achievement of the desired hydrogenated oil product is usually measured with the solids fat index (SFI) or solids fat content (SFC), either of which measures the amount of solids present in a fat at different temperatures from below room temperature to above body temperature. Natural fats are not single compounds, and the hydrogenated products are even more complex mixtures due to the simultaneous reactions. Not only are double bonds saturated with hydrogen, but some of the remaining bonds are isomerized: geometric isomerization changes the low-melting *cis* form to a higher melting *trans* form, and positional isomers shift the double bond away from its natural position in the carbon chain. Extensive geometrical or *trans*-isomerization tends to provide products that are hard at low temperatures but soft at high temperatures, which results in steep SFI/SFC curves. A lesser but significant effect on melting points is contributed by the positional isomerization, as the shift of a double bond in a carbon chain affects the melting point of the hydrogenated oil. Additionally, the bonds that are shifted can be in either the *cis* or *trans* form, which further substantiates the complexity of the hydrogenated oil process.

Selective hydrogenation is the tool by which partial hydrogenation is accomplished in a controlled manner. Selectivity is the saturation with hydrogen of the double bonds in the most unsaturated fatty acid before that of a less unsaturated fatty acid. In a theoretical sense, an oil hardened with perfect preferential selectivity would first have all of its linolenic fatty acids (C-18:3) reduced to linoleic fatty acids (C-18:2) before any linoleic was reduced to oleic (C-18:1); then, all linoleic fatty acids would be reduced to oleic before any oleic was saturated to stearic (C-18:0). Unfortunately, this does not happen in actual practice, but it is possible to vary the hydrogenation rate of linoleic to that of oleic from the very selective conditions of 50 to 1 to the less selective conditions of 4 linoleic to 1 oleic. The latter is generally described as nonselective.

Formation of the high-melting unsaturated fats or isomerization accompanies hydrogenation and appears to be in proportion to the selectivity of the reaction. Therefore, compromises must be made between selectivity and isomer formation when determining the best hydrogenation conditions for the various basestocks. Control of the operating variables that affect the hydrogenation of fats and oils is necessary to produce the desired product functionality (Calsicat, 1992).

Hydrogenation Process Condition Variables

Hydrogenation is a reaction of three components: oil, hydrogen, and catalyst. The reaction takes place on the surface of the catalyst where the oil and gas molecules are adsorbed and brought into close contact. Therefore, any condition which affects the catalyst surface or controls the supply of gas to the catalyst surface will, in turn, affect the course and rate of the reaction. The variables that can affect the results of the hydrogenation are temperature, degree of agitation, hydrogen pressure in the reactor, catalyst amount, type and poisons, hydrogen gas purity, feedstock source, and feedstock quality. The effects of the variables include the following (Allen, 1960, 1967, 1978, 1981, 1982):

- Temperature—Hydrogenation, like most chemical reactions, proceeds at a faster rate with increased temperatures. An increase in temperature decreases the solubility of the hydrogen gas in the liquid oil while increasing the reaction rate. This causes quicker hydrogen removal from the catalyst to reduce the quantity of hydrogen on the catalyst surface, resulting in a high selectivity and isomer formation; therefore, increased temperature increases selectivity, *trans*-isomer development, and the reaction rate that results in a steep SFI curve.

  Because hydrogenation is an exothermic reaction, it will create heat as long as the reaction is active; a decrease of one iodine value increases the reaction temperature by 1.6 to 1.7°C (2.9 to 3.1°F). Temperature increases will increase the reaction rate until an optimum is reached. At this point, cooling of the reaction mixture is required to continue hydrogenation. The optimum temperature varies for different products, but most oils probably reach their maximum temperature at 230 to 260°C (450 to 500°F).

- Pressure—Most edible fats and oils hydrogenations are performed at hydrogen pressures ranging from 0.7–4.0 bar (10–60 psig). At low pressures, the hydrogen gas dissolved in the oil does not cover the catalyst surface, while at high pressures hydrogen is readily available for saturation of the double bonds. The increased saturation rate results in a decrease in *trans*-isomer development and selectivity to produce a flatter SFI curve.

- Agitation—The main function of agitation is to supply dissolved hydrogen to the catalyst surface, but the reaction mass must also be agitated for the distribution of heat or cooling for temperature control and suspension of the catalyst throughout the oil mixture for uniformity of reaction. Agitation has a significant effect upon selectivity and isomerization. Both are decreased because the catalyst is supplied with sufficient hydrogen to increase the reaction rate.

- Catalyst Level—The hydrogenation reaction rate increases as the catalyst concentration is increased up to a point and then levels off. The increase in rate is caused by an increase in active catalyst surface; however, a maximum is reached because at very high levels hydrogen will not dissolve fast enough to adequately supply the higher catalyst levels. Both selectivity and *trans*-isomer formation are increased with catalyst concentration increases, but only slightly.

- Catalyst Type—The choice of catalysts has a strong influence on the reaction rate, preferential selectivity, and geometric isomerization. Nickel catalysts are used almost exclusively for edible fats and oils hydrogenation. Catalysts are prepared by a variety of techniques, some propriety to the catalyst supplier; however, nickel catalyst is usually prepared by the reduction of a nickel salt and supported on an inert solid or flaked in hardfat or a combination of the two. The activity of a catalyst depends on the number of active sites available for hydrogenation. These active sites may be located on the surface of the catalyst or deep inside the pores. High-selectivity catalysts allow the processor to reduce the linolenic fatty acid without producing excessive amounts of stearic fatty acid, thus producing a product with good oxidative stability and a low melting point. The selectivity characteristics of a catalyst are unrelated to the ability of the catalyst to form *trans*-fatty acids because the catalyst may have a very low or very high selectivity, but all common nickel catalysts appear to produce the same level of *trans*-fatty acids at the same conditions. However, catalysts may be treated with other materials such as sulfur, which increases the amount of *trans*-fatty acids unsaturation.

Sulfur-poisoned catalysts produce larger quantities of *trans*-isomers in hydrogenated oils. Reaction with sulfur inhibits the capacity of nickel to adsorb and dissociate hydrogen, reducing the total activity of the catalyst. As the ability of the nickel to hydrogenate is reduced, its tendency to promote isomerization is enhanced. Hydrogenated oils with a relatively high melting point at a high iodine value, which results in a very steep SFI slopes, are the result of the high *trans*-isomer content. Commercially, sulfur-treated catalysts have been found to provide more uniform performance than products that are sulfur-poisoned during processing.

Copper-chromite catalyst has been used for selective hydrogenation of linolenic fatty acid to linoleic fatty acid in soybean oil for a more flavorful, stable salad oil with higher winterization yields. The selectivity offered by these catalysts is excellent, but the activity is poor and they are more sensitive to catalyst poisons.

Precious metals have been investigated and found effective as hydrogenation catalysts. Evaluations have shown that basestocks hydrogenated with 0.0005% palladium modified with silver and bismuth were exceedingly more active and slightly more selective with more *trans*-fatty acids development than were equivalent stocks prepared with nickel catalyst. Subsequent evaluations have shown that the precious metals are more active at lower temperatures than nickel. Oils have been hydrogenated at 60°C (140°F) with precious metals, while temperatures above 130 to 140°C (265 to 285°F) are required with nickel catalyst. *Trans*-isomer development is increased as the hydrogenation temperature is increased; therefore, less *trans*-isomer development should be obtained with precious metal utilization at low temperatures.(Haumann, 1994) Palladium has been found to be some 30 times as active as nickel, as only 6 ppm is required to replace 200 ppm nickel. The principal deterrent to the use of palladium has been economics, both in the initial costs and recovery problems associated with the minute quantities required. However, with an adequate recovery system the precious metal catalysts can be more cost effective than nickel catalyst due to lower utility costs, a long life-cycle and spent catalyst recycling directly into fresh catalyst (Beers, 2004).

- Catalyst Poisons—Refined oils and hydrogen gas can contain impurities that modify or poison the catalyst. Catalyst poisons are a factor that can have a significant effect upon the product. The poisons effectively reduce catalyst concentration with a consequent change in the selectivity, isomerization, and rate of reaction. Impurities present in both the feedstock oil and hydrogen gas are known to have a deleterious effect upon nickel catalyst. Hydrogen gas may contain carbon monoxide, hydrogen sulfide, or ammonia. Refined oil can contain soaps, sulfur compounds, phosphatides, moisture, free fatty acids, mineral acids, and a host of other materials that can change the catalyst. Studies determined that 1 ppm sulfur poisons 0.004% nickel, 1 ppm phosphorus poisons 0.0008% nickel, 1 ppm bromine poisons 0.00125% nickel, and 1 ppm nitrogen poisons 0.0014% nickel (Hastert, 1988). Sulfur primarily affects the activity to promote isomerization by inhibiting the capacity of the nickel catalyst to absorb and dissociate hydrogen. Phosphorus in the form of phosphatides and soaps affects selectivity by residing at the catalyst pore entrance to hinder the triglyceride exit for a higher degree of saturation (Beckmann, 1983)· Water or moisture and free fatty acids are deactivators that decrease the hydrogenation rate by reacting chemically with the catalyst to form nickel soaps.

Hydrogenation Systems

Batch hydrogenation is most commonly used in the edible-oil industry, primarily because of its simplicity and flexibility for use with different source oils. Essentially all that is required is a reaction vessel, usually referred to as a *converter*, that can withstand 7 to 10 bar (105 to psig) pressure, with an agitator, heating and cooling coils, a hydrogen gas inlet, piping and pumps to move the oil in and out, and a sample port for process control of the reaction. The converter must also be provided with the means to control two of the three reaction variables, pressure and temperature, the rate of agitation is usually fixed.

Two different batch converter designs utilized for the partial hydrogenation of edible fats and oils are recirculation and dead-end. In the recirculation system, hydrogen gas is introduced at the bottom of the vessel, and nonreacted hydrogen gas is withdrawn from the headspace, purified, and returned to the converter. The converter is almost always filled with hydrogen under pressure in the operation of the recirculation system. Hydrogenation begins immediately when the catalyst is added with the oil charge during the heating period and thereafter until the endpoint is attained when recirculation is discontinued. Reaction temperature is controlled by circulating water through the cooling coils to carry away the heat of reaction. The hydrogenated oil is pumped out of the converter through an external cooler to a filter for catalyst removal.

A dead-end hydrogenation system converter is loaded with oil from a scale tank or metering devise. Converter vacuum is utilized to deaerate, dry, and prevent any hydrogenation while heating with steam to reaction temperature. Catalyst, slurried in a portion of the feedstock, is added during the heating period. When the oil reaches reaction temperature, the vacuum is discontinued and hydrogen is added until the specified pressure is attained. This pressure is maintained during the hydrogenation. An agitator designed to provide efficient hydrogen dispersion is necessary to create a vortex to draw hydrogen from the headspace back into the oil. When the exothermic reaction has raised the oil temperature close to the maximum specified temperature, cooling water is introduced into the coils. Samples are drawn from the converter via the sample port as the reaction proceeds to measure the hydrogenation progress. Agitation is suspended whenever awaiting a laboratory analysis to confirm that the endpoint has been reached. When the endpoint has been attained, the hydrogen is vented to the atmosphere through the vacuum system, and the oil is cooled in the converter, in a drop tank, or with a heat exchanger. After cooling to 65°C (150°F), the oil is filtered through a black press to separate the catalyst from the oil. Hydrogenation black presses traditionally have been of the plate and frame or pressure leaf variety.

From an operations standpoint, the two types of converters do not differ very much. In general, the dead-end type is preferred by many processors because it (i) requires less energy, (ii) offers more versatility, (iii) requires less capital and operating costs, and (iv) is safer than the recirculation system. Quality and performance-wise, the advantages for the dead-end system are (i) oxidation and hydrolysis prevention through deaeration and dehydration provided by the vacuum during heat up and cooling, (ii) more positive control of the reaction for product uniformity, and (iii) the ability to vary the hydrogen pressure as well as temperature.

Most hydrogenations of edible fats and oils are performed both in the United States and in the rest of the world in batch converters. Continuous hydrogenation systems have been available for quite some time, but their commercial usage has been limited for several reasons. The maximum value for any continuous operation is realized when it is used to produce large quantities of the same product. Considerable out-of-specification product can be produced during a change from one product to another. Because most fats and oils processors produce a variety of products, several different basestocks are routinely required that can be produced more uniformly with batch hydrogenation systems (O'Brien, 2004).

Hydrogenated Basestock System

Most prepared foods are formulated with ingredients designed for their application or, in many cases, specifically for the particular product or processing technique employed by the producer. These customer-tailored products have expanded the product base for fats and oils processors from a few basic products to literally hundreds. Each of these products could be formulated to require a different hydrogenated product for each different product. This practice with the ever-increasing number of finished products would result in a scheduling nightmare with a large number of product heels tying-up tank space and inventory. Basestock systems with a limited number of hydrogenated stock products for blending to meet the finished product requirements are utilized by most fats and oils processors. The advantages provided by a well-designed basestock system are basically control and efficiency (Lantondress, 1981; O'Brien, 1987; 2004). The control advantages include:

- Hydrogenated oil batch blending to average minor variations
- Increased uniformity by the production of the same product more often
- Reduced contamination afforded by the ability to schedule compatible products together
- Elimination of product deviations generated from attempts to use product heels
- Elimination of rework generated by heel deterioration before use

The efficiency advantages contributed by a basestock system include:

- Hydrogenation scheduling to maintain basestock inventories rather than reacting to customer orders

- Hydrogenation of full batches instead of producing some partial batches to meet demands

- Better reaction time to meet customer requirements

Basestock requirements will vary with each processor, depending upon the customer requirements, which dictate the finished products produced. The basestock systems can include several source oils or can be limited to almost a single oil type. In either case, the basestock inventories usually consist of a few hydrogenated products that cover a wide range for blending to the desired consistencies (O'Brien, 2004):

- Brush Hydrogenated Basestocks—For many edible-fat ingredient specifications, a liquid oil is required. To guarantee an acceptable shelf life, the level of polyunsaturates should be low, with an absence or severely reduced level of linolenic fatty acids (C-18:3). This can be achieved by a light and highly selective hydrogenation of an oil within the oleic/linoleic fatty acid group such as soybean, sunflower, or canola. During hydrogenation, the iodine value drop is kept to a minimum to reduce the formation of saturated fatty acids, and the *trans*-isomers formation is largely suppressed. The hydrogenation should be performed at a low temperature to reduce the formation of *trans*-isomers. A high pressure of 3–4 bar (45–60 psig), in combination with new catalyst with high activity, selectivity, and poison resistance, should be used. Optimal conditions will vary considerably, depending upon the geometry of the converter, agitator, hydrogen gas purity, and the other hydrogenation variables. After hydrogenation, this basestock can be winterized or fractionated to produce a flavor-stable salad oil or a high-stability liquid oil depending on the extent of the hydrogenation. This basestock class is also very useful in margarine oil blends, snack-frying oil, and in specialty product formulations.

- Flat Partially Hydrogenated Basestocks—Many food products require fats and oils products that have an extended plastic range with good oxidative stability. These products must be relatively soft with a plastic consistency at room temperature and still possess some body at temperatures of 37.8°C (100°F) with a melting point only slightly above body temperature. Stability is important because of the probable exposure to baking or frying temperatures and long shelf-life expectancy. Usually, moderately selective conditions are utilized to produce these flat SFI basestocks, for example, relatively low temperatures of 300–350°F (150–175°C) with high pressures of 20–30 psig (1.3–2 bar) with a selective catalyst that has *trans*-isomer suppressant qualities. Catalyst reuse, should be avoided as they enhance *trans*-isomer formation.

- Steep Partially Hydrogenated Basestocks—The physical properties of these basestocks are characterized by steep SFI curves or high solids contents at the lower measuring temperatures with an absence of solids at temperatures higher than body temperature. Hydrogenation of these basestocks should be effected with highly selective conditions or high temperature and low pressure. These high-*trans* basestocks are beneficial in blends for margarine oils, high-stability frying shortenings, nondairy products, fillings, and other products requiring a sharp melting point with good flavor stability while providing the required firmness at room temperature.

- Low-IV Hardfats—These basestocks are often referred to as *fully hydrogenated hardfats*, or *stearines*; however, technically, that designation should require a zero iodine value (IV). Because catalyst activity is the only criterion with these hydrogenations, used catalyst can be utilized. In general, high-pressure (4 bar or 60 psig, or higher) and high-temperature (450°F or 230°C) conditions are used for these basestocks to make the reaction progress as rapidly as possible. Hydrogenation conditions to produce hardfats is the least critical of all hydrogenation operations since neither selectivity nor isomerization is a factor, and unsaturation is very minor as the oil is almost totally saturated.

Table 13.1 outlines a soybean oil basestock system with seven hydrogenated stock oils ranging from a lightly hydrogenated 109 IV to a saturated hardfat with a maximum IV of 8. Utilization of a similar basestock system designed for the required product mix should enable fats and oils processors to meet most shortening specifications by blending two or more basestocks, except for some specialty products that can be made only with special hydrogenation conditions (O'Brien, 2004).

SFI/SFC is one of the most important consistency measurements, and it also indicates the selectivity of the conditions used to prepare the individual basestocks. It measures the amount of solids present in a fat at different temperatures from below room temperature to above normal body temperature. A fat can appear to be a solid but really exist as a semisolid and does not have a distinct melting point. Natural and hydrogenated fats and oils melt over a wide range of temperatures. SFI/SFC analysis determines the solid or unmelted portion of a fat over a measured temperature range. These results relate to the consistency of the fats and oils product in terms of its softness, plasticity, organolepic, and other physical properties important for its use as an ingredient in prepared foods. The slope of the SFI/SFC curve shows the effects of hydrogenation selectivity as it affects consistency. The SFI/SFC curve slope becomes steeper as the hydrogenation conditions are made more selective — that is, the highest temperature, lowest pressure, and highest level of a selective catalyst. The slope of the SFI/SFC curve becomes flatter as the hydrogenation reaction conditions are made less selective with lower temperatures, higher pressure, and low catalyst levels. These effects are illustrated for the soybean oil basestocks graphically on Fig. 13.1. (O'Brien, 2004).

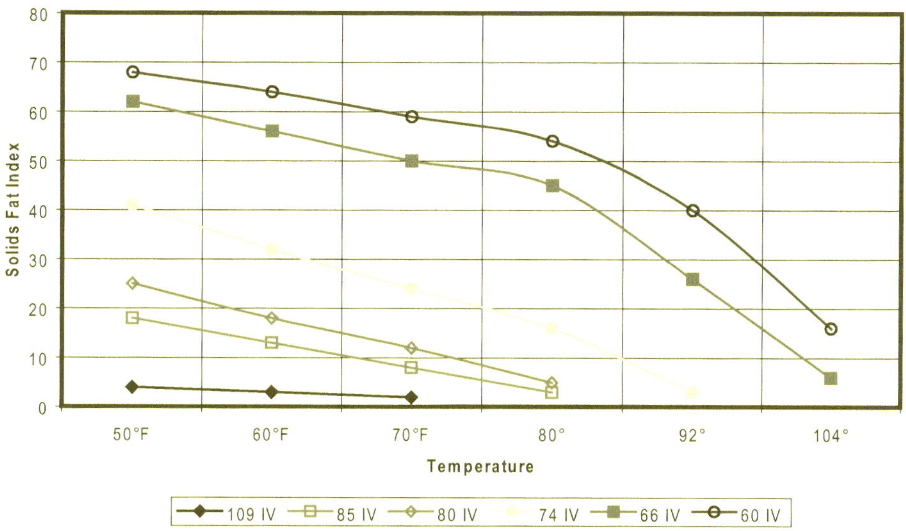

Fig. 13.1 Soybean oil basestocks.

Interesterification

Nutritional findings linking consumption of *trans* fatty acids to the risk of coronary heart disease and the required disclosure of the *trans* content on product labels have generated interest in alternate processes to replace hydrogenation for the modification of oils for the production of shortenings and margarines. One alternative for achieving *trans*-free comparable performance is random chemical or enzymatic interesterification of blends of liquid soybean oil with fully hydrogenated soybean oil. Random interesterification alters the SFI/SFC profile from a flat slope to a more desirable steeper one better suited for margarine and shortening formulations, and it slows down the transition to the stable β crystal form to stabilize the modified oil in the β′crystal while maintaining a high level of polyunsaturates. The directed interesterification variation of random rearrangement, can produce a high level of trisaturates with temperature manipulation during processing. The result is the formation of solids without the addition of a hardstock (Rozendaal, 1992).

Interesterification is a process used in the fats and oils industry to modify the properties of triglyceride mixtures. Interesterification can be visualized as a breakup of a specific triglyceride, removal of a fatty acid at random, shuffling it among the rest of the fatty acid pool, and replacement at random by another fatty acid. The reaction ultimately attains equilibrium when all possible combinations of glycerol and fatty acids have taken place. This process is commonly referred to as *randomization* because

Table 13.1 Hydrogenated Soybean Oil Basestock System

Basestock Type	Brush	Flat		Steep			Hard-fat
Iodine Value	109	85	80	74	66	60	<8
Mettler Dropping Point, °C	a	30±2	33±2	35±1	43±3	46.5±1.5	b
Solids Fat Index:							
10.0°C - 50°F	4 max	18±3	25±2	41±3	62±3	68±3	b
21.1°C - 70°F	2 max	8±2	12±3	24±3	50±3	59±3	b
26.7°C - 80°F	–	–	5±1	16±3	45±3	54±3	b
33.3°C - 92°F	–	–	–	<3.5	26±3	40±3	b
40.0°C - 104°F	–	–	–	–	6±3	16±2	b
Titer, °C	–	–	–	–	–	–	63±1
Fatty Acid Composition							
C-14:0 Myristic	–	–	–	–	–	–	
C-16:0 Palmitic	10.8	10.2	10.1	10.9	10.7	10.7	10.5
C-16:1 Palmitoleic	–	–	–	–	–	–	–
C-18:0 Stearic	4.4	6.4	7	8.2	15.2	20.8	86.9
C-18:1 Oleic [c]	44.3	68.2	72	75	70.7	66.3	1.7
C-18:2 Linoleic [c]	37.7	15.2	10.9	5.9	3.1	2.2	–
C-18:3 Linolenic [c]	2.8	–	–	–	–	–	–
trans isomers, %	14.8	22.7	25.3	44.7	45.5	45.0	nil

Hydrogenation Conditions:	Brush	Flat	Steep	Hard-fat
Gassing Temp., °F	300	300	300	300
Hydrogenation Temp., °F	325	350	440	450
Pressure, bar	3.0 to 4.0	1.3 to 1.5	0.7 to 1.0	4
Catalyst, % nickel	0.01 to 0.02	0.02	0.04 to 0.08	0.04 to 0.08
Agitation	fixed	fixed	fixed	fixed

a - too soft to analyze; b - too hard to analyze; c - including *trans* fatty acids; d - optimum conditions will vary dependent upon the converter, agitation, hydrogen gas purity, etc.

of the random rearrangement of the fatty acids. As the fatty acids rearrange, they reach an equilibrium that is based on the composition of the starting material and is predictable from the laws of probability (Sreenivasan, 1978).

Three general types of interesterification processes are practiced: random, enzymatic, and directed. In random interesterification all positions take part in the ester interchange in the liquid phase. This eventually results in a random or statistical distribution of the fatty acids. The resulting equilibrium triglyceride composition is therefore predictable from, and fully determined by, the overall fatty acid composition of the mixture. Enzymatic interesterification is possible with a number of lipases as biocatalysts. Most of these are 1,3 specific, which means that only the fatty acids at the external positions of the triglycerides take part in the interesterification process, while the fatty acids at the 2- position are not affected. Directed interesterification is carried out at such a low temperature that part of the triglycerides crystallize and no longer participate in the reaction. The random composition in the liquid phase is thus continuously disturbed. The net effect is that the saturated fatty acids tend to become concentrated into high-melting components to form solids without the addition of hardstock (Rozendaal, 1992).

Random Chemical Interesterification

Random rearrangement is accomplished with either batch or continuous processes. A typical batch reaction vessel is fitted with an agitator, heating and cooling coils, nitrogen sparging, and vacuum. Initially, in a batch process, the oil is dried by heating to 120–150°C (248–302°F) under vacuum to prevent moisture deactivation of the catalyst. After drying, the oil is cooled to reaction temperature of 70–100°C (158–212°F). Sodium methylate powder is then drawn into the vessel under the surface of the oil with the vacuum and agitated to form a white slurry. Only 0.1% of the sodium methylate is necessary to promote the interesterification reaction; however, 0.2–0.4% is the normal usage due to free fatty acid and moisture poisoning of the catalyst. The mixture is well agitated at reaction temperature (usually for 30–60 min) until formation of a distinctive brown color denoting randomization occurs. After laboratory analysis confirms the reaction, the catalyst is neutralized by washing with water or the addition of phosphoric or citric acid.

Continuous interesterification processes follow the same cycle as the batch process but utilize different equipment. Most continuous processes heat the oil with a heat exchanger and flash dry it with a vacuum oil dryer to bring the moisture level to 0.01% or less. The catalyst is continuously introduced into the hot oil stream and homogenized for dispersion. The homogenized mixture is then passed through a tubular reactor with variable residence time determined by length and flow rate. The catalyst is neutralized with water and passes on to centrifugal separation, drying and post bleaching (Laning, 1985).

Random interesterification is most commonly used in the food industry to prepare functional fats and oils by modifying the triglyceride structure of blends of a hardfat with a liquid oil. A mixture of two oils with radically different properties responds to random interesterification because the distribution of fatty acids is always far from random. Redistribution of the fatty acids to a random arrangement produces a steeper solids content and a reduced melting point when a high melting hardfat is interesterified with a liquid oil. It also slows down the transition of the crystal to the most stable β-form to stabilize the modified oil in the β′-form. Higher concentrations of a saturated soybean hardstock are identified as interesterification basestocks that are blended with refined bleached (RB) soybean oil to produce margarine oil and shortening fat solids and melting characteristics desired for a variety of products. The random interesterification of 80:20 through 50:50 blends of RB soybean oil and soybean oil hardfat results are shown in Table 13.2 (Petrauskaite et al., 1998). These results indicate that random interesterification can be used to produce margarine oils and shortenings directly or as basestocks for blending.

Directed Chemical Interesterification

In directed rearrangement processes, one or more of the triglyceride products of the interesterification reaction is selectively removed from the ongoing reaction. Some procedures involve distillation of low molecular weight fatty acids, but usually directed rearrangements involve the selective crystallization of the higher melting, saturated triglycerides. As these triglycerides are removed, the reaction balance, is disturbed and the remaining fatty acids continue to re-randomize to form more trisaturates. Theoretically, this process could continue until all the saturated fatty acids are converted into trisaturated triglycerides and separated from the reaction (Laning, 1985).

Continuous processes are normally used for directed interesterification because the batch process is difficult to control and would require a number of extra tanks. The process flow begins with the vacuum drying of the oil to 0.01% moisture. After drying, the oil is cooled to just above the melting point with a heat exchanger. The catalyst is metered into the product stream and mixed or homogenized to suspend the catalyst throughout the product. Sodium potassium (NaK) is a typical catalyst used at 0.2% for directed interesterification because of its activity at relatively low reaction temperatures. The homogeneous mixture is quick-chilled with a scraped-wall heat exchanger to a predetermined point to initiate crystallization of the trisaturated triglycerides. The cooled mixture is held in an agitated vessel, where interesterification proceeds under controlled agitation. At this stage, the trisaturated glycerides are crystallizing while the liquid phase is continuing to form more trisaturated triglycerides. Heat liberated by crystallization can cause the temperature of the mixture to rise above the desired temperature to require a second chilling with a scraped-wall heat exchanger. After rechilling, the product is transferred to another vessel where the directed reaction continues to the desired endpoint. Crystallization slows as the trisaturates diminish, so this stage requires more time to complete the reaction. The

Table 13.2. Properties of Random Interesterified RB Soybean Oil and Soybean Oil Hardfat Blends

RB Soybean Oil, %	80		70		60		50	
Soybean Oil Hardfat, %	20		30		40		50	
	Interesterification		Interesterification		Interesterification		Interesterification	
Properties:	Before	After	Before	After	Before	After	Before	After
Triglycerides, %								
U_3	45.3	37.0	40.5	26.0	37.0	17.1	31.0	11.1
U_2S	27.4	42.4	21.6	51.0	17.4	40.0	15.1	28.4
S_2U	4.0	17.3	5.4	14.5	4.9	31.9	3.5	47.4
S_3	23.3	3.3	32.5	8.5	40.7	11.0	50.4	13.1
Solids Fat Content:								
10.0 °C - 50 °F	12.3		22.4		38.3		55.0	
21.1 °C - 70 °F	9.3		19.5		21.5		47.0	
33.3 °C - 92 °F	3.5		8.5		17.5		29.0	
CMP, °C	31.0		47.0		53.0		56.0 [c]	
Iodine Value	107.4		93.8		81.5		67.6 [c]	
Laboratory Chemical	Dry fat blend at 110 °C and ~50 mbar for 40 minutes							
Interesterification	Add 0.2% sodium methoxide catalyst to fat blend at 90 °C under vacuum							
Procedure:	After 90 minutes of stirring, cool the mixture to 80 °C Break vacuum and add 20% citric acid solution to inactivate the catalyst, while mechanically stirring for 15 minutes Post bleach with 0.5% bleaching earth under vacuum for 30 minutes at 90 °C							

[c] - calculated; U - unsaturated; S - saturated; CMP - Capillary Melting Point; RB - refined and bleached

level of trisaturated triglycerides can be controlled by varying the time in the crystallizer, the crystallization temperature, or a combination of the two. After the desired endpoint is reached, the catalyst is inactivated by adding water. The amount of water is calculated to provide the desired fluidity for centrifuging to remove the soap phase. Saponification of the fat can be minimized somewhat by the addition of carbon dioxide with the water to buffer mixture to a lower pH. After neutralization of the catalyst, the mixture is heated to melt the trisaturated glyceride crystals for centrifugation followed by vacuum drying (Laning, 1985; Going, 1967; Hawley & Holman, 1956).

Directed interesterification can convert a liquid oil into a semisolid capable of the production of margarine oil or a shortening base. In this reaction, trisaturates are formed *in situ* to produce products with solids profiles that are soft at low temperatures and firm at high temperatures for improved plastic ranges (Going, 1967). Although soft zero-*trans* margarine oils can be prepared with some natural liquid oils containing a high level of saturated fatty acids, natural soybean oil interesterified with the directed procedure would probably be too soft. However, compositionally modified soybean oils with higher saturate levels have more potential to produce the firmer margarine oils and shortenings with directed interesterification.

Enzymatic Interesterification

Enzymes have been used for many years to modify the structure and composition of foods but only recently available on a large enough scale for industrial applications. Lipases have distinct advantages compared to classical chemical catalysts: (i) enzymes function under mild reaction conditions over a range of temperatures and pressures that minimize the production of side products, (ii) enzyme-catalyzed reactions are more efficient and easier to control, and (iii) unique specificities of lipases allow the selection of a particular lipase for the desired application. Enzymatic interesterification is now used to produce products free of *trans* fatty acids and high-value-added structured fats and oils (Macrae, 1983; Quinlan & Moore, 1993). Useful glyceride mixtures that cannot be obtained by chemical interesterification processes are possible by exploiting the specificity of lipases. In all glyceride reactions, lipases catalyze either the removal or the exchange of fatty acid groups on the glycerol backbone. Different lipases can show preferences for both the position of the fatty acid group on the triglyceride and the nature of the fatty acid. The types of lipase catalyst identified by application specificity are (Foglia & Villeneuve, 1997):

- Random lipases, which catalyze reactions at all three positions on the glyceride randomly

- *sn*-1,-3 specific lipases, which catalyze reactions only at the outer positions of the triglyceride

- A lipase with *sn*-2 selectivity has been reported: *C. antartica*
- Lipases specific for a particular class of fatty acids; *Geotrichum candidum* has been found specific for omega-9 fatty acids; others are specific for short-chain fatty acids, and still others for long-chain fatty acids

Again, the use of lipase catalyst for interesterification of edible fats and oils has advantages over the classical chemical catalysts. One of the most attractive features is the unique specificities possible with their use. Nonspecific lipases provide reactions like the random chemical catalyzed interesterification. Specific lipases make it possible to produce fats and oils with a customized triglyceride structure. The enzymatic process can be selective with the use of a positional specificity lipase. These processes are usually much slower and more sensitive to the reaction conditions to provide a better control over the reaction results. Also, the lipases can operate under milder reaction conditions, temperature and pressure, that minimize the formation of side products.

Some of the disadvantages for the enzymatic interesterification process for edible fats and oils are that it generates some free fatty acids and partial glycerides, and it is more expensive than the chemical interesterification. A FFA content of 4–6% has been reported, which must be removed for edible oil uses. However, the levels of FFA decrease with repeated use of the lipase after the initial stabilization period. Free fatty acid formation can also be reduced by operating at lower temperatures and by drying the enzyme substrate feedstock before use. Diglycerides are formed with both the enzyme and chemical interesterification processes; they are intermediates in the interesterification reaction. A high content of diglycerides can delay crystallization and lower the solid fat content of the interesterified fat, but diglycerides were used to stabilize the β'-beta prime crystal in margarine oils (Haumann, 1994).

Lipases are manufactured by fermentation of selected microorganisms followed by a purification process. The enzymatic interesterification catalysts are prepared by the addition of a solvent such as acetone, ethanol, or methanol to a slurry of an inorganic particulate material in buffered lipase solution. The precipitated enzyme coats the inorganic material, and the lipase-coated particles are recovered by filtration and dried. Various support materials have been used to immobilize lipases. Generally, porous particulate materials with high surface areas are preferred. Typical examples of the support materials are ion-exchange resins, silicas, macroporous polymers, clays, etcetera. Effective support functionality requirements include (i) the lipase must adsorb irreversibly with a suitable structure for functionality, (ii) pore sizes must not restrict reaction rates, (iii) the lipase must not contaminate the finished product, (iv) the lipase must be thermally stable, and (v) the lipase must be economical. The dried particles are almost inactive as interesterification catalyst until hydrated with up to 10% water prior to use.

Lipase-catalyzed interesterification of fats and oils can be accomplished either by using a stirred batch reactor or with continuous processing using a fixed-bed reactor.

The latter is the preferred process, offering the advantage of minimized reaction times due to the high catalyst substrate ratio along with the other advantages of (i) catalyst recovery, (ii) reduced catalyst damage, and (iii) improved operability. The continuous fixed-bed interesterification process begins by dissolving the feedstock in a solvent followed by treatment to remove enzyme catalyst inhibitors, poisons, and particulate materials. This solution is then partially saturated with water prior to pumping through a bed of hydrated catalyst particles. The reaction products are a mixture of triglycerides and free fatty acids. After the reaction, the FFAs are removed by evaporation and processed for recovery. The FFA-free oil is then solvent fractionated to yield the desired triglyceride composition.

Fractionation

Fractionation of edible oils consists of the separation of oils into two or more fractions with different melting characteristics. Separation of the oil fractions is based on a distribution of the triglycerides between different phases. The efficiency of the separation of the liquid and solid fractions depends particularly on the method of cooling, which determines the form and size of the crystals. Rapid cooling causes heavy supersaturation and gives a great number of small crystals, resulting in the formation of a shapeless, small, soft precipate with poor filtration properties. This form slowly transforms into the alpha form with a tendency to develop mixed crystals. Gradual cooling results in stable β and β′ crystal development which facilitate separation from the liquid form by filtration. Fractionation is a separation of triglycerides not individual fatty acids; it is a modification of the texture, crystallization, and melting behavior, defined by the triglyceride composition. (Kreulen, 1976)

Three distinct processes for the fractionation of triglycerides that couple crystallization and separation processes are practiced commercially to produce value-added fractionated oils (Kellens, 2000):

- Dry Fractionation—is the simplest fractionation technique because no additives or posttreatment of the end product is involved. Fractionation is basically a two-stage process. First the oil is crystallized by cooling the oil in a controlled manner to the required temperature in a crystallizer. The oil is then filtered to separate the liquid from the solid fraction by means of a vacuum filter or membrane filter press. Recent developments of new and more efficient crystallizers and reliable high-pressure membrane filters moved dry fractionation from a third choice to a good alternative for solvent fractionation in many cases.

- Detergent Fractionation—a wetting agent, usually sodium lauryl sulfate, in combination with an electrolyte, usually magnesium sulfate, is added to the crystallizing oil or fat to allow the crystals to be suspended easily in the aqueous phase. The water phase with the crystals is separated from the liquid phase by means of a centrifuge. After separation, the stearin and olein fractions are heated, washed, and dried to remove the additives.

- Solvent Fractionation—is the most efficient fractionation process. Crystallization is performed in the presence of a solvent, usually acetone, at a ratio of between 3 and 5 to l (solvent to oil). Separation is usually performed on a vacuum belt filter. The high separation efficiency and the purity and yield of the solid fraction are the main advantages of solvent fractionation. The high investment and operating costs prevented the adoption of this process for commodity products such as margarines and shortenings.

Winterization, a narrow form of dry fractionation from a technological point of view, became a staple for soybean oil processing when it was widely accepted that the major precursor of soybean oil's flavor reversion was the linolenic fatty acid content (Frankel, 1980). Reduction of the linolenic fatty acid content from ~9.0 to <3.0% was found to increase the oxidative stability of soybean oil to an acceptable level (Dutton et al., 1953). Soybean oil brush hydrogenated to a 110 IV and then winterized to regain its salad oil characteristics was introduced in the late 1950s. The RBHWD soybean salad oil was accepted by the retail salad oil customers and also industrially as a component of salad dressings, mayonnaise, sauces, etcetera to replace cottonseed salad oil (Erickson, 1983). In the late 1970s, improvements in soybean oil handling, oil extraction, and oil processing produced an RBD soybean oil that was accepted by industrial users and then introduced to the retail market. These soybean oil purification process improvements negated the requirement for brush hydrogenation and winterization of soybean oil in a very short time.

The objective in the application of fractionation technology for soybean oil is the commercial production of substances with unique properties. Two product categories utilizing a fractionation process to produce products with soybean oil as a portion of the feedstock are (O'Brien, 1995):

- *Domestic Hard Butters*—Hydrogenation and fractionation technology are combined to produce cocoa butter substitutes that are compatible with cocoa butter and do not require tempering of the coating.

- *High-Stability Liquid Oils*—Modification of oils by utilizing hydrogenation with fractionation permits the development of liquid oils with high-resistance oxidative degradation. Liquid oils with a 350-h AOM stability and others with exception frying stabilities are available commercially.

Postbleaching

A separate bleaching operation, immediately following the modification processes, has three general purposes: (i) insurance that all traces of the prooxidant hydrogenation catalyst and interesterification soaps have been captured; (ii) to remove undesirable colors which have been developed or accentuated during a modification process; and (iii) to remove peroxides and secondary oxidation products. Post bleach systems can be exact duplicates of the prebleach process. However, batch systems are usually preferred over continuous systems and this bleaching process generally employs a bleaching earth and a metal-chelating acid, in the case of hydrogenated oils, to reduce the residual nickel content to the lowest possible level. As much as 50 ppm nickel, mostly in colloidal form, can remain in hydrogenated oils after the black press filtration. Trace amounts of nickel remaining in the oil adversely affect the stability of the oil by accelerating the oxidation process. After postbleaching, the trace metal levels in the oils should be reduced to <0.1 ppm nickel and <0.02 ppm copper for oxidative stability. (O'Brien, 2004) Interesterified oils have traces of soap and a reddish brown color remaining after water washing as well as secondary oxidation products which must be removed by bleaching (Brinkmann, 2001)

Green colors can emerge in the hydrogenated oils because of the heat bleaching of yellow and red masking pigments during hydrogenation. In the course of hydrogenation, the carotenoid pigments can be reduced to a colorless form while the chlorophyllic pigments merely have their absorption maxima shifted from 660 to 640 μm. After the removal of the masking reddish pigments, the greenish pigments predominate, resulting in oils that appear green. The green pigments can be removed by adsorption on acid-activated clays with some difficulty, depending on the severity. Green colors are more easily removed in the pre-bleach process before the color is set by heating during hydrogenation.

Crystallization

Considerably more is involved in preparing a shortening or margarine for packaging and eventual use as an ingredient or spread than simply lowering the temperature to cause solidification. For example, a grainy, pasty, nonuniform mass is produced when shortenings and margarines are allowed to cool slowly. The more saturated triglycerides crystallize first and grow in size to produce an unsightly, difficult to handle, mushy product that lacks many of the basic qualities necessary for shortening or margarine performance. Development of the desired edibility, appearance, stability, texture, functionality, uniformity, and reproducibility in solidified fat and oil products is a function of controlled crystallization or plasticization (O'Brien, 2004).

Shortenings and margarine products appear to be soft homogeneous solids; however, microscopic examination shows a mass of very small interlocked crystals that trap and hold by surface tension a high percentage of liquid oil. The crystals are separate, discrete particles capable of moving independently of each other when a sufficient shearing force is applied to the mass. Therefore, shortening, margarine, and other solidified fat and oil products possess the characteristic structure of a plastic solid. The distinguishing feature of a plastic substance is the property of behaving as a solid by completely resisting small stresses but yielding at once and flowing like a liquid when subjected to deforming stress above a minimal value. A firm, plastic material will not flow or deform from its own weight; however, it may easily mold into any desired form with a slight pressure. Plastic solids derive their functionality from their unique nature. Three conditions are essential for plasticity (Mattil, 1964):

1. It must consist of two phases, one a solid and the other a liquid.

2. The solid phase must be dispersed finely enough to hold the mass together by internal cohesive forces.

3. The two phases must be in proper proportions. The solid portion must be capable of holding the liquid while enough liquid must be available to allow flow when stress is applied.

Plasticity and consistency of an edible fats and oils product depend on the amount, size, shape, and distribution of the solid material, as well as the development of crystal nuclei capable of surviving high-temperature abuse to serve as starting points for new desirable crystal growth. The factors that influence these characteristics are (Joyner, 1953; McMichael, 1956; O'Brien, 2004; Hoerr & Ziemba, 1965):

- Product Composition—The solid phase influences the consistency of a plastic shortening or margarine; the product becomes firmer as the solids fat contents increase. The solids contents are determined by the source oil, type, and degree of modification, and/or blends of naturally solid oils or fats such as palm, coconut, palm kernel, lard, or tallow. Solidified fats and oils products begin to have enough body to hold their shape well at a solids content as low as 5% and become rigid, losing elasticity as the solids contents reach 40–50%. A typical all-purpose bakery shortening, formulated for creaming properties and spreadability or workability, attempts to maintain a solids fat index of 15–25% over the widest temperature range possible; however, each product has a satisfactory plastic range, which is the temperature range in which the particular fat and oil product may be used with the intended results (O'Brien, 1987).

- Crystal Size—At elevated temperatures, fats retain enough molecular motion to preclude organization into stable crystal structures; however, edible fats and oils go through a series of increasingly organized crystal phases with cooling until a final stable crystal form is achieved. All fats crystallize from the liquid phase in the α form and transform rapidly in β' and subsequently to the intermediate or the β modification if they are likely to exhibit these higher polymorphs. This sequence is irreversible; once transformation to the more stable forms has occurred, lower polymorphs can be obtained only by melting the product and repeating the process. This process can occur in fractions of a second or in months. The crystal types define the texture and functional properties of most fat-based products. Each crystal form possesses its own specific physical properties: different melting point, solubility, specific heat, and dielectric constants. The crystal lattice formed when the molecules solidify is a relatively loose arrangement. As crystallization proceeds, the molecules tend to pack more closely together. With time the molecules pack together as closely as their structure permits; therefore, the molecules in the most stable crystal lattice require the least space. Fats and oils products become progressively firmer as the average size of the crystals decrease and become softer as the size increases. A fat that has been melted and allowed to crystallize slowly under static conditions will contain many large crystals plainly visible to the eye. Crystals formed in the same fat by rapid chilling are microscopic in size. Quick-chilled product with very small crystals will be firmer and have a consistency range much wider than that of a fat slowly crystallized. The slowly crystallized product will also be softer than the quick-chilled fat.

- Supercooling—The supercooling properties of the triglycerides are critical and complicating factors in plasticization of fats and oils products. The solidification and plasticization process requires careful attention because fats can remain liquid when chilled below their melting point, and they are polymorphic allowing them to crystallize in two or more crystal forms. The degree of supercooling and the temperature at which the supercooled product is allowed to reach crystal equilibrium are directly related to the temperature range over which the product is workable. In practice, the temperature to which the product is supercooled, worked, and packaged is controlled to produce the widest plastic range for the individual product formulation. The extent to which a product is supercooled can affect not only the consistency, but also the melting point of the solidified product.

- Mechanical Working—Optimal plasticity is produced when the supercooled product is mechanically worked during the crystal formation period until substantially all of the latent heat of crystallization is dissipated. Product allowed to solidify without agitation forms a strong crystal lattice and exhibits a narrow plastic range. This consistency is desirable for stick margarine or spread but is harmful for shortening and margarine products requiring a plastic-like consistency. Therefore, mechanical working must be adapted to the product consistency desired.

- Gas Incorporation—An inert gas, usually nitrogen, is incorporated into most standard shortenings at 13±1%, regular soft tub margarines or spreads at 4–8%, whipped margarines or spreads at 30–35%, and precreamed household shortenings at 18–25%. Creaming gas is added to these products to provide: a white, creamy appearance; a bright surface sheen; a less dense, easier handling product; texture improvement; homogeneity; increased volume; reduced calories per serving; and reduced saturated and *trans* fatty acid grams per serving. Stick, liquid, and most industrial margarines do not have creaming gas added during crystallization. The aqueous phase of a margarine emulsion has the same effect on appearance as gas incorporation.

Plasticized Shortening Crystallization Process

The ultimate polymorphic form of plastic shortening is determined by the triglyceride composition, but mechanical and thermal energy influence the rate at which the most stable form is reached. Thus, plastic shortenings are processed through various heat-exchanger working configurations to remove heat of crystallization and transformation. A typical shortening plasticization process begins when the deodorized shortening blend is transferred to the packaging department, had all of the specified additive materials incorporated, and met all of the process quality-control requirements. The sequence of operations for the plasticization of a shortening is described below (Rini, 1960; Joyner, 1953; McMichael, 1956; Slaughter, 1948; Brown, 1949; Fincher, 1953; Hoerr & Ziemba, 1965; Weiss, 1983):

> The deodorized shortening is transferred to a small float controlled supply tank adjacent to the scraped wall chilling unit. It is pumped by a gear pump at pressures 20–67 bar, depending on the system, to the chilling system; this pressure is maintained throughout the system. Creaming gas is introduced into the suction side of the gear pump. The oil and nitrogen mixture is precooled to 10–15°F (5.5–8.3°C) above the melting point of the product to reduce the load, to ensure a constant temperature, and to ensure a large number of crystal nuclei in the product as it is chilled in the scraped wall heat exchanger. The precooled shortening is rapidly chilled, usually in less than 30 sec, to temperatures ranging from 60–75°F (15.6–25.6°C) depending on the product type and desired firmness. The supercooled product is then subjected to a shearing action, while the heat of crystallization is dissipated. The shaft in the worker unit revolves at ~125 rpm, and the residence time of the chilled product is usually about 3 min. The temperature of the product rises ~10–15°F (5.6–8.3°C) during this time due to the heat of crystallization. The worked product is then forced through an extrusion valve to aid in making the product homogeneous by breaking up any remaining crystal aggregates with an intense shearing action. The solidified product is delivered by a rotary pump at pressures of 20–27 bar to a second extrusion valve located near the filling station. Packaged shortening temperature rise should not exceed 1 or 2°F (0.6–1.1°C); a temperature rise above this level is indicative of crystallization under static conditions and causes the consistency to be firmer than desired.

Liquid Shortening Crystallization

The major attribute of liquid shortenings is fluidity at room temperature. Liquid shortenings are easily poured, pumped, and metered under normal atmospheric conditions, which reduce handling problems for the consumer. Properly processed liquid shortenings do not require agitation to ensure uniformity. Also, oxidative stability is prolonged because no heat is required for fluidity at temperatures as low as 50°F (10°C) for most liquid shortenings. The products are milky white in appearance due to the dispersion of hardfats in the form of microcrystalline particles, which do not settle out because of the crystallization process. Fluid shortenings are composed of components that are stable in the β crystal form. Low-IV, β crystal-forming hardfats, usually hydrogenated soybean oil, seed crystallization for liquid shortenings. The hardfat level is limited by the desired fluidity of the shortening and the eating quality requirements of the finished food product. A stable fluid system in the β crystalline form does not increase in viscosity or gel once it is properly processed. Hardfats with β′ crystal habits are unacceptable for liquid shortenings because the tight-knit crystal lattice structure initiates a viscosity change with crystallization to a nonfluid product. Aeration properties normally associated with β′ small crystals are achieved by the addition of appropriate emulsifiers. Emulsifiers are also included in some formulations to retard staling of yeast-raised breads and rolls to increase shelf life.

The rate at which a liquid shortening transforms into its stable crystal form is important because it must be completed before packaging to avoid solidification after packaging. Therefore, liquid shortening crystallization must be accomplished in a few hours. The quickest transformation of a liquid shortening to its stable β-crystal form can be attained by heating the fat mixture until completely melted; then rapidly cooling to just above the α-crystal melting point (theoretically, the melting point of the α-crystal is the lower limit to which fatty materials can be cooled without forming any crystals); and then heating to just above the β′-crystal melting point followed by a crystallization period with gentle agitation to allow the heat of crystallization to dissipate and the crystal to stabilize (O'Brien, 2004).

Margarine Crystallization

Several different types of margarine and spread are produced for the retail, foodservice, and food processor markets. The processing parameters are tailored to the appropriate finished product. Both batch and continuous processing techniques, utilized to produce margarines and spreads, follow the same process flow. Processing for all of the margarine types begins with the preparation of a water-in-oil emulsion. The fat phase is prepared by heating the margarine oil to at least 10°F (5.6°C) above its melting point and adding all the oil-soluble ingredients. The aqueous phase is prepared by dissolving the water-soluble ingredients in the water. The water phase is added to oil phase and blended together with high-shear agitation to form a water-in-oil emulsion. The emulsion is continuously agitated until transfer to the crystallization system. The solidification process for the various margarine and spread types all employ a scraped-surface heat exchanger for rapid chilling, but the other steps are different than for shortening or other margarine and spread types. The solidification process conditions for the basic margarine or spread types are (O'Brien, 2004):

- Stick Margarine or Spread—The temperature of the emulsion is adjusted and maintained at 10°F (5.6°C) above the melting point of the margarine oil before pumping to the scraped-surface heat exchanger. Precrystallization structures can be formed when the emulsion temperature to the chiller is below the melting point of the oil. The emulsion is rapidly chilled to 40–45°F (4.4–7.2°C) in less than 30 sec. Stick tablespreads require a stiffer consistency than shortening, which is accomplished with the use of a quiescent tube immediately after the chilling unit. The supercooled mixture passes directly to the quiescent resting or aging tube for molded print-forming equipment. For filled print equipment, a small blender may be utilized before the resting tube to achieve the proper consistency for packaging and a slightly softer product. A remelt line is necessary because all closed filler systems must have some overfeeding to maintain a uniform supply of product to the filler for weight control. The excess is pumped to a remelt tank and then reintroduced into the product line.

- Soft Tub Margarine or Spread—The oil blends for these tablespread products are formulated with lower solids-oil-ratios than the stick-type products to produce a spreadable product directly out of the consumer's refrigerator or freezer. Crystallization techniques contribute to the desirable consistency as well, but the products are too soft to print into sticks; therefore, packaging in plastic tubs or cups with snap-on lids is utilized. To fill the container properly, the soft margarine or spread consistency must be semifluid like shortening. The temperature of a typical soft tub margarine is adjusted and maintained at 95–105°F (35.0–40.6°C) before transferring with a high-pressure pump to the scraped-surface heat exchanger. Creaming gas, added to further improve spreadability, is injected at the suction side of the pump at 8% for the most spreadable product; lower levels are used for a firmer product. The product is rapidly chilled to an exit temperature of 48–52°F (8.9– 11.1°C). Spread fill temperatures are higher than margarine products because the emulsion is more viscous. If the fill temperature is too low, the product mounds in the bowl for excessive lid coverage, and the product may become dry and crumbly. The supercooled product then passes through a worker unit to dissipate the heat of crystallization. The product is then delivered to the filler where it is forced through an extrusion valve to create pressures in the range of 20–26.7 bar. The excess product is remelted and eventually reenters the solidification system.

- Whipped Tub Margarine or Spread—The same equipment used to prepare, crystallize, and package regular soft-type tablespreads can be utilized for whipped tub products. The difference during crystallization is the addition of 33% nitrogen gas by volume for a 50% overrun. The nitrogen is injected inline through a flow meter into the suction side of the pump before the scraped-surface heat exchanger. The larger tubs required for the increased volume necessitate change parts for the filling, lidding, and packaging equipment.

- Industrial Margarines or Spreads—Foodservice and food processor margarines and spreads may either be duplicates or the retail products in larger packages or designed for a specific use, either product- or process-related. Specific-use products, puff paste, and Danish roll-in products are the most difficult regarding crystallization. The characteristic features of these products are plasticity and firmness. Plasticity is necessary, as the roll-in must remain in unbroken layers during repeated folding and rolling operations. Firmness is equally important, as soft and oily product is partly absorbed by the dough, thus destroying its role as a barrier between the dough layers. The ultimate polymorphic form for roll-in products is determined by the triglyceride composition, but the rate at which the most stable form is reached is influenced by mechanical and thermal energy. The crystallization process for roll-in and/or bakers margarines is the same as for the shortening plasticization process except that margarines containing water or milk in emulsion form are normally not aerated. The aqueous phase of a margarine emulsion has the same effect as gas incorporation on appearance and performance.

Edible-oil Flake Crystallization

Flakes describe the higher melting edible-oil products solidified into a thin flake form for ease of handling, quicker remelting, or for a specific function in a food product. Flaking rolls, initially used for chilling margarine and shortening prior to the introduction of scraped wall heat exchangers, are still used for the production of edible-oil flakes. Chill rolls have been adapted to produce several different flaked products used to provide distinctive performance characteristics in specialty formulated foods. The usual edible-oil products flaked are low-IV hardfats, high-melting mono- and diglycerides and other fat-based emulsifiers, icing stabilizers, shortening chips, and confectioner's fats.

Chill rolls are available in different sizes, configurations, surface treatments, feeding mechanisms, etcetera, but most are a hollow metal cylinder with a surface machined and ground smooth to true cylindrical form. Rolls, internally refrigerated with either flooded or spray systems, turn slowly on longitudinal and horizontal axes. Several options exist for feeding the melted oil product to the chill roll: (i) a trough arrangement positioned midway between the bottom and top of the roll; (ii) a dip pan at the bottom of the roll; (iii) overhead feeding between the chill roll and a smaller applicator roll; and (iv) a double or twin drum arrangement operating together with a very narrow space between them where the fat product is sprayed for application to both rolls. A coating of fat is carried over the roll to solidify and is removed by a doctor blade, positioned ahead of the feed mechanism with all of the designs.

During the flaking operation, a portion of the edible oil is supercooled sufficiently to cause very rapid crystallization. The latent heat released by fat crystallization is absorbed by the cooling medium inside the chill roll. In the crystallization of edible oil products, the sensible heat of the liquid is removed until the temperature of the product is equal to the melting point. At the melting point, heat must be removed to allow crystallization of the product to continue. The quantity of heat associated with this phenomenon is called *heat of crystallization.* The sensible heat of specific heat of soybean oil hardfat is equal to 27.8 calories per gram (50 BTU/lb). The amount of heat that must be removed to crystallize low IV hardfat is 100 times the amount of heat that must be removed to lower the product temperature. The flake product dictates the chill roll operating conditions and additional treatment necessary before and after packaging. Some generalizations relative to chill roll operations and product quality are (O'Brien, 1996):

- Crystal Structure—Each flaked product has crystallization requirements dependent on the source oil, melting point, degree of saturation, and the physical characteristics desired.

- Flake Thickness—Four controllable variables help determine flake thickness: oil temperature to the roll, chill roll temperature, speed of the chill roll rotation, and the feed mechanism.

- In Package Temperature—Heat of crystallization causes a product temperature rise after packaging unless it is dissipated prior to packaging. The product temperature can increase to the point where partial melting coupled with pressure from stacked containers causes the product to fuse together into a large lump.

- Flake Condition—Glossy or wet flakes are caused by a film of liquid oil on the flake surface due to incomplete solidification. Either too warm or too cold chill roll temperatures can cause this condition. High roll temperatures may not provide sufficient cooling to completely solidify the flake. Low roll temperatures may shock the oil film causing the flake to pull away from the surface before completely solidified. Wet flakes from either cause will lump in the container.

Tempering

Tempering is performed after packaging to control the consistency and plasticity of fats and oils products. Tempering conditions depend on the product and type. A plasticized product is tempered when the crystal structure of the hard fraction reaches equilibrium by forming a stable crystal matrix. The crystal structure entraps the liquid portion of the shortening or margarine. The mixture of low- and high-melting components of the solids undergoes a transformation in which the low-melting fractions remelt and then recrystallize into a higher melting, more stable form. This process can take 1–10 days, depending on the product formulation and package size. Small packages temper quicker than the larger packages, which have less surface area exposed to the conditioning temperature. After a plasticized product takes an initial set, some α-crystals are still present. These crystals remelt and slowly recrystallize into the β'-crystal form during tempering. The β'-crystal form is preferred for most plastic shortenings and margarines, especially those designed for creaming or roll-in applications (Hoerr & Ziemba, 1965). Soybean-oil-based shortenings requiring a wide plastic range are usually formulated with a cottonseed oil or palm oil hardfat to force the product to the β'-crystal form. The β' hardfat must have a higher melting point than the soybean oil basestock in order for the entire product to crystallize in the stable β'form. Random interesterification can also stabilize soybean oil in the β'-crystal form.

Shortenings and margarines requiring a wide plastic range for performance are usually tempered for 40 h minimum in a quiescent state at 85°F (29.4°C), which is slightly above the fill temperatures for this class of shortening and margarine. Frying and other shortenings, which are melted when used, are not normally tempered. These plasticized products are usually held for 24 h in a quiescent state to allow the crystal to stabilize somewhat before shipment. Opaque liquid shortenings also do not require any tempering after packaging, but the storage temperature is critical. Holding liquid shortening below 18°C (65°F) causes the liquid shortening to solidify with a loss of fluidity; storage above 35°C (95°F) results in partial melting of the suspended solids. Table and kitchen use margarines do not require any heat treatment after packaging; they are tempered at refrigerated temperatures to achieve a loose, structured, and brittle consistency (O'Brien, 1996).

References

Allen, R.R. Hydrogenation. *J. Am. Oil Chem. Soc.* **1960,** *37,* 521–523.

Allen, R.R. Hydrogenation, principals and catalyst. *J. Am. Oil Chem. Soc.* **1967,** *45,* 312A–314A, 340A–341A.

Allen, R.R. Principals and catalyst for hydrogenation of fats and oils. *J. Am. Oil Chem. Soc.* **1978,** *55,* 792–795.

Allen, R.R. Hydrogenation. *Proceedings of the World Conference on Soya Processing and Utilization;* AOCS Press: Champaign, IL, 1981; pp. 166–168.

Allen, R.R. Theory of hydrogenation and isomerization. *Hydrogenation: Proceedings of an AOCS Colloquium;* R. Hastert, Ed.; AOCS Press: Champaign, IL, 1987; pp. 1–10.

Allen, R.R. Hydrogenation. *Bailey's Industrial Oil and Fat Products;* D. Swern, Ed.; John Wiley & Sons: New York, NY, 1982; pp. 14–27.

Ash, M.; E. Dohlman. *Oil Crops Situation and Outlook Yearbook,* USDA, May, 2007, pp. 61, 71.

Beckmann, H.J. Hydrogenation process. *J. Am. Oil Chem. Soc.* **1983,** *60,* 286–288.

Beers, A.; G. Mangnus. Hydrogenation of edible oils for reducing *trans*-fatty acid content, *Inform* **2004,** *15,* 404–405.

Brinkmann, B. Interesterification as a tool to improve the nutritional value of fat blends for various applications. *Proceedings of the World Conference on Oilseed Processing and Utilization;* R.F. Wilson, Ed.; AOCS Press: Champaign, IL, 2001; pp. 31–33.

Brown, L.C. Emulsion food products. *J. Am. Oil Chem. Soc.* **1949,** *26,* 632–636.

Calsicat. *Fats and Oils Hydrogenation Manual;* Mallinckrodt Specialty Chemicals Co.: Erie, PA, 1992; pp. 26–40.

Carlson, K. Recent developments and trends in processing of fats and oils. *Inform* **2006,** *17,* 671–674.

Dutton, H.J. History of the development of soy oil for edible uses. *Proceedings of the World Conference on Soya Processing and Utilization,* AOCS Press: Champaign, IL, 1981; pp. 234–236.

Dutton, H.J., et al. The flavor problem of soybean oil. iv. Structure of compounds counteracting the effect of prooxidant metals. *J. Am. Oil Chem. Soc.* **1948,** *25,* 385–388.

Dutton, H.J.; C.D. Evans; J.C. Cowan. Status of research on the flavor problem of soybean oil at the northern research laboratory. *Am. Assoc. Cereal Chem.* **1953,** *11,* 116–135.

Erickson, D.R. Soybean oil: an update on number one. *J. Am. Oil Chem. Soc.* **1983,** *60,* 356.

Erickson, D.R. Bleaching/adsorption treatment. *Practical Handbook of Soybean Processing and Utilization;* AOCS Press: Champaign, IL, and United Soybean Board: St. Louis, MO, 1995; pp. 203–206.

Fincher, H.D. General discussion of processing edible oil seeds and edible oils. *J. Am. Oil Chem. Soc.* **1953,** *30,* 479–481.

Foglia, T.A.; P. Villeneuve. Lipase specifications: Potential applications in lipid bioconversions. *Inform* **1997,** *8,* 640–650.

Frankel, E.N. Soybean oil flavor stability. *Handbook of Soy Oil Processing and Utilization;* American Oil Chemists' Society: Champaign, IL, and American Soybean Association: St, Louis, MO, 1980; pp. 230, 240.

Going, L.H. Interesterification products and processes. *J. Am. Oil Chem. Soc.* **1967,** *44,* 414A–422A, 454A–456A.

Hastert, R.C. Practical aspects of hydrogenation and soybean salad oil manufacture. *Proceedings of the World Conference on Soya Processing and Utilization;* AOCS Press: Champaign, IL, 1981; pp. 169–174.

Hastert, R.C. The Partial Hydrogenation of Edible Oils, American Oil Chemists' Society Short Course, Phoenix, AZ, 1988.

Hastert, R.C. Adsorptive treatment of edible oils. *Introduction to Fats and Oils Technology;* P.J. Wan, Ed.; AOCS Press: Champaign, IL, 1991; pp. 96–100.

Haumann, B.F. Tools, hydrogenation, interesterification. *Inform* **1994,** *5,* 688–678.

Hawley, H.K.; G.W. Holman. Directed interesterification as a new processing tool for lard, *J. Am. Oil Chem. Soc.* **1956,** *33,* 29–35.

Hoerr, C.W.; J.V. Ziemba. Fat crystallography points way to quality. *Food Engineering* **1965,** *37(5),* 90–95.

Johnson, L. Oil recovery from soybeans. *Soybeans, Chemistry, Production, Processing and Utilization;* AOCS Press: Champaign, IL, 2008 (in press).

Joyner, N.T. The plasticization of edible fats. *J. Am. Oil Chem. Soc.* **1953,** *30,* 526–538.

Kellens, M.; M. Hendrix. Fractionation. *Introduction to Fats and Oils Technology;* R.D. O'Brien, W.E. Farr, P. J. Wan, Eds.; AOCS Press: Champaign, IL, 2000; pp. 196–207.

Kreulen, H.P. Fractionation and winterization of edible fats and oils. *J. Am. Oil Chem. Soc.* **1976,** *53,* 393–396.

Laning, S.J. Chemical interesterification of palm, palm kernel and coconut oils. *J. Am. Oil Chem. Soc.* **1985,** *62,* 400–404.

Lantondress, E.G. Formulation of products from soybean oil. *J. Am. Oil Chem. Soc.* **1981,** *58,* 185.

List, G.R.; R. Reeves. *Trans* reduction. *Trans Fats Alternatives;* D.R. Kodali, G.R. List, Eds.; AOCS Press: Champaign, IL, 2005; pp. 72–73.

Macrae, A.R. Lipase-catalyzed interesterification of oils and fats. *J. Am. Oil Chem Soc.* **1983,** *60,* 291–294.

Mattil, K.F. Plastic shortening agents; *Bailey's Industrial Oil and Fat Products,* Third ed.; D. Swern, Ed.; Wiley-Interscience: New York, NY, 1964: pp. 272–281.

McMichael, C.E. Finishing and packaging of edible fats. *J. Am. Oil Chem. Soc.* **1956,** *33,* 512–516.

O'Brien, R.D. Formulation—single feedstock situation. *Hydrogenation: Proceedings Of An AOCS Colloquium;* R. Hastert, Ed.; AOCS Press: Champaign, IL, 1987; pp. 155.

O'Brien, R.D. Soybean oil crystallization and fractionation. *Practical Handbook of Soybean Processing and Utilization;* D.R. Erickson, Ed.; AOCS Press: Champaign, IL, and United Soybean Oil Board: St. Louis, MO, 1995; pp. 268–275.

O'Brien, R.D. Shortening types and formulations. *Bailey's Industrial Oil & Fat Products,* Fifth ed.; Y.H. Hui, Ed., John Wiley & Sons, Inc.: New York, NY, 1996; Vol. 3, pp. 181–191.

O'Brien, R.D. Fats and oils processing. *Fats and Oils Formulating and Processing for Applications;* CRC Press: Boca Raton, FL, 2004; pp. 92, 101–105, 148–165, 236, 396–399.

O'Brien, R.D.; P.J. Wakelyn. Cottonseed oil: An oil for *trans*-free options. *Inform* **2005,** *16,* 677–679.

O'Brien, R.D.; P.J. Wan. Cottonseed oil: Processing and utilization. *Proceedings of the World Conference on Oilseed Processing and Utilization;* R.F. Wilson, Ed.; AOCS Press: Champaign, IL, 2001; pp. 113–114.

Okonek, D.V. Nickel-sulfur catalysts for edible oil hydrogenation. *Hydrogenation: Proceedings of An AOCS Colloquium;* R. Hastert, Ed.; 1987; pp. 65,

Petrauskaite, V. et al. Physical and chemical properties of *trans*-free fats produced by chemical interesterification of vegetable oil blends. *J. Am. Oil Chem. Soc.* **1998,** *75,* 489–493.

Quinlan, P.; S. Moore. Modification of triglycerides by lipases, process technology and its application to the production of nutritionally improved fats. *Inform* **1993,** *4,* 580–585.

Rini, S.J. Refining, bleaching, stabilization, deodorization, and plasticization of fats, oils, and shortening, *J. Am. Oil Chem. Soc.* **1960,** *37,* 515–520.

Rozendaal, A. Interesterification and fractionation. *Proceedings of the World Conference on Oilseed Technology and Utilization.* T.H. Applewhite, Ed.; AOCS Press: Champaign, IL, 1992; pp. 180–185.

Slaughter, J.E., Jr. Plasticization and packaging. *Processing of a Six-Day Short Course in Vegetable Oils;* The American Oil Chemists' Society, conducted at The University of Illinois, Urbana, IL, 1948; pp. 119–131.

Sreenivasan, B. Interesterification of fats. *J. Am. Oil Chem. Soc.* **1978,** *55,* 796–805.

Weiss, T.J. Basic processing of fats and oils. *Food Oils and Their Uses,* Second ed.; AVI Publishing: Westport, CT, 1983; pp. 96–97.

14

Food Use of Whole Soybeans

KeShun Liu
Grain Chemistry & Utilization Laboratory, USDA-ARS, Aberdeen, ID, 83210

Introduction

Based on historical and geographical evidence, the soybean [*Glycine max* (L. Merrill)] first emerged as a domesticated crop in China thousands of years ago, and was considered one of the five sacred grains (or *wu gu* in Chinese) along with rice, wheat, barley, and millet. During the course of soybean domestication, the Chinese gradually transformed soybeans into various forms of soyfoods. Tofu, soy sauce, soy paste, and soy sprouts are among the popular forms. This transformation makes soybeans as a food more versatile, more tasteful, and more digestible. Because of their high contents of protein and oil, with a fairly balanced amino acid profile and abundant essential fatty acids, traditional soyfoods have long nourished the Chinese people.

Along with the method of soybean cultivation, the ways of preparing soyfoods and eating soybeans were gradually introduced to Japan, Korea, and other nearby countries and regions. People in these regions not only accepted soyfoods, but also modified them and even created their own types to suit local taste. Japanese natto and Indonesia tempeh are just two examples. Since the beginning of the last century, the art of preparing soyfoods has now spread to the rest of the world, thanks to agricultural and processing innovation, cultural exchanges, and the influence of Chinese and other Asian immigrants. More importantly, many modern processing technologies have found applications in making traditional soyfoods. This has led to their large-scale production in many regions of the world, along with improved and consistent quality. In recent years, medical research has unveiled new insights on the role of soyfoods in preventing and treating chronic diseases (Qin et al., 2006; Messina, 2007). Soyfoods are gaining popularity throughout the world (Golbitz, 2002; Tripathi & Misra, 2005).

In general, traditional soyfoods, also known as Oriental soyfoods, are classified as non-fermented and fermented. Non-fermented soyfoods include soymilk, tofu, soy sprouts, yuba (soymilk film), okara (soy pulp), vegetable soybeans, soynuts and toasted soy flour, whereas fermented soyfoods include soy sauce, miso (fermented soy paste), natto, tempeh, soy yogurt (fermented soymilk), sufu (fermented tofu), and soy nuggets (fermented whole soybeans). Traditional soyfoods that are commonly marketed in North America include soy sauce, tofu, soymilk, tempeh, green vegetable soybeans, soynuts, and soy yogurt.

Most non-fermented soyfoods are consumed mainly for nourishment. In contrast, many fermented soyfoods are generally used as seasonings or condiments in cooking or making soups. They contribute more flavor than nutrition. The two exceptions for fermented products are tempeh and natto, which are consumed as a part of the main meal and contribute nutrients to the diet in addition to their characteristic flavor.

Historically, all traditional soyfoods were made from whole soybeans. However, with use of modern processing technologies, some traditional foods, such as soy sauce, soymilk, and tofu, can now be made from defatted soy meal or its derivative products, such as soy protein isolate. In this chapter, various traditional soyfoods are discussed with respect to their variety, preparation methods and principles. Because of the diversity of ethnic soyfoods and their preparation methods, it is impossible to cover each soyfood in detail. Additional information on the subject can be found in Shurtleff and Aoyagi (1976, 1979, 2001), Watanabe and Kishi (1984), Shi and Ren (1993), Imram (2003), and Liu (1999, 2005). With regards to processing and utilization of soy ingredients as food, such as soy oil and soy protein ingredients, discussion is covered elsewhere in this book.

Non-fermented Soyfoods

Soymilk

Based on the method of preparation, soymilk is generally divided into traditional soymilk and modern soymilk. Traditional soymilk, known as *dou jiang* in Chinese, is made by a traditional method in the home or at the village level. Being an intermediate product during tofu production, *dou jiang* is generally served fresh and hot during breakfast. The product not only has limited shelf-life, but also possesses a characteristic beany flavor and bitter or astringent taste, with all nutrients coming solely from original soybeans.

In contrast, modern soymilk, sometimes referred to as soy beverage or soy drink, is produced by using modern technology and equipment. Known as *dou ru* or *dou nai* in Chinese, these products have a relatively bland taste with their own commercial identity and standards. In most cases, they are flavored, sweetened and/or fortified for better taste and better nutrition, and packed for longer shelf-life, when compared with traditional soymilk. Sold as a milk substitute or a healthful soft drink, soymilk is particularly important to infants who suffer from malnutrition due to absence of dairy milk supply in certain regions of the world or who suffer from allergies and diseases associated with dairy milk consumption. Thus, for the past several decades, soymilk has been produced in large commercial scales throughout the world (Golbitz, 2002).

Traditional Soymilk

In China, traditional soymilk is made by soaking, rinsing, and grinding soybeans into a slurry. This is followed by filtering the slurry to separate the residue, and cooking the soy extract to become edible (Fig. 14.1). The method is basically the same as the one used originally for making tofu and its intermediate product—soymilk during the second century BCE Technically, it can be defined as the "cold grinding" method and involves the following steps:

- Soaking. Dry whole soybeans, preferably beans with large seed size and light hilum, are cleaned, measured (or weighed) and then soaked in water overnight. The volume of water used is normally 2-3 times the bean volume.

- Draining and rinsing. The soaked beans are drained and rinsed with fresh water 2–3 times.

- Grinding. The wet, clean soybeans are then ground in a stone mill or hammer-mill with additional fresh water. The water:bean ratio is normally in the range of 6:1 to 10:1. The slurry is collected in a large container.

- Filtering. The bean slurry is filtered through a screen, cloth or pressing sack, with or without a wooden level press. The residue, known as soy pulp in English, *dou zha* in Chinese, and okara in Japanese, is removed, and is normally washed once or twice with water (cold or hot), stirred and re-pressed to maximize milk yield. The total volume of the combined filtrate (raw soymilk) is about 6-10 times the original bean volume.

- Cooking. The raw milk is transferred to a large wok or pot, and then heated to boiling. To avoid burning at the bottom of the cooking pot, slow heating with frequent stirring is necessary. After boiling for about 10 min, the hot soymilk is ready to serve, or is transferred to another container for later consumption. The hot milk can be further processed into tofu by adding a coagulant. Alternatively, the soy slurry may be heated before filtering into soymilk. This procedure is particularly popular in Japan (Watanabe & Kishi, 1984; Kwok & Niranjan, 1995).

Modern Soymilk

The basic principle for making modern soymilk is very similar to that of the traditional Chinese method. The procedure also includes selecting and cleaning raw soybeans, incorporating water, grinding, separating the residue, and heating. However, in modern methods, several key steps are modified or added to reduce the beany flavor, increase production yield, and improve overall product quality and consumer acceptance (Liu, 1999; Imram, 2003).

Soymilks in North American markets nowadays are mostly chilled products, made by modern food-processing techniques. The starting material can be whole soybeans, soy flour, soy protein isolates, or combinations of the former.

Techniques to Reduce Beany Flavors

Many techniques have been used to reduce or eliminate beany flavors. Four strategies are used to reduce beany flavors: selecting the right beans, preventing beany flavor formation, striping off the responsible volatiles once they are formed, and masking the residual off-flavor with flavorings.

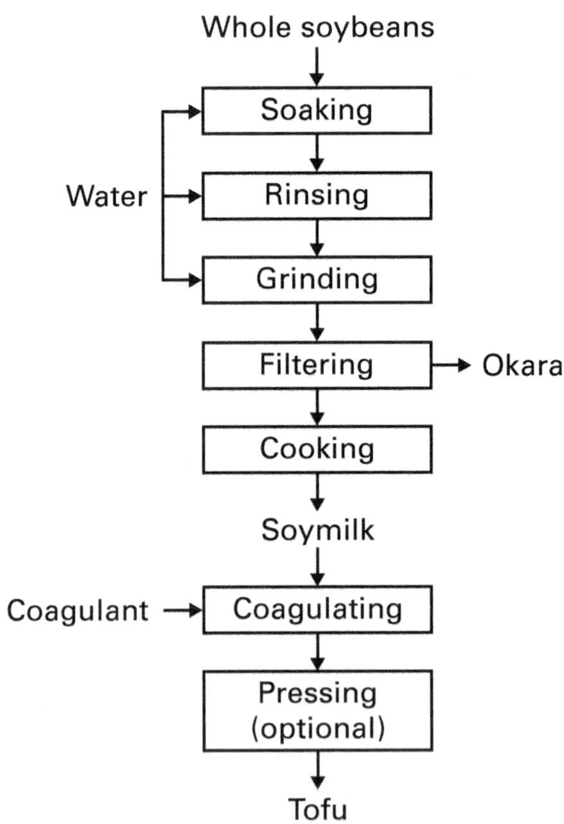

Fig. 14.1. A traditional Chinese method for making soymilk and tofu.

During the 1960s and 1970s, several studies were conducted to elucidate the chemistry and mechanism of beany flavor formation during preparation of soymilk (Wilkens et al., 1967; Nelson et al., 1976). The beany flavors of soymilk in particular and soy products in general result mainly from peroxidation of polyunsaturated fatty acids or esters catalyzed by an enzyme known as lipoxygenase. The reaction produces many volatile compounds, including ketones, aldehydes, and alcohols; most of which impart undesirable flavors. The traditional cold-grinding method promotes lipoxygenase activity during grinding and thus produces abundant beany flavor compounds.

Several procedures have been developed to prevent beany flavor formation by inactivating the enzyme. One is the hot-grinding method and the other is the pre-blanching method. These methods are contrasted to the cold-grinding method. In the hot-grinding method, also known as the *Cornell method*, unsoaked, dehulled soybeans are ground with hot water in a preheated grinder. The slurry is maintained between 80–100°C in the grinder to completely inactivate lipoxygenase, and then is boiled in a steam-jacketed kettle with constant stirring for 10 min. The insoluble solids of the heated slurry are removed by using a centrifuge or a filter press. The resulting soymilk is formulated, bottled, sealed, and finally sterilized at 121°C for 12 min (Wilkens et al., 1967). Recently, Prabhakaran and Perera (2006) showed that the hot grinding method extracted more of the isoflavones into the soymilk than does the cold-grinding process, thereby improving the phytochemical profile of the finished products in addition to controlling the beany flavor.

The pre-blanch method, known also as the Illinois method, starts with blanching pre-soaked soybeans in boiling water for 10 min or with placing dry beans directly into hot water for 20 min. Either of the procedures hydrates the soybeans and inactivates enzymes. The beans are then drained and ground with sufficient cold water to make a slurry containing 12% bean solids. The slurry is heated to about 93°C and then homogenized. The product may be formulated, pasteurized, homogenized again, and bottled. Soymilk thus produced has high yield since all original materials end in the final product (Nelson et al., 1976).

A steam-infusion cooking process, known as hydrothermal cooking (HTC), was developed to produce soymilk continuously from ground full-fat soy flour (Johnson et al., 1981). It was claimed that soymilk processed by the HTC process had less beany flavors because of the much shorter time for lipoxygenase to be active and because steam flashing stripped volatiles. The process also increased recovery of dry matter and protein in the soymilk.

Recently, a high-pressure procedure was found to inactivate lipoxygenase in soymilk. The beany flavor of the final product was not evaluated in the study (Wang et al., 2008).

Since lipoxygenase activity is a major cause of beany flavor formation during processing of soy products, an alternative strategy to control beany flavor in soymilk and other soy products is to remove lipoxygenases through genetic approaches. A soybean line lacking all three lipoxygenase isomers was successfully induced through γ-ray irradiation by a group of Japanese scientists (Hajika et al., 1991). These lines were later introduced into North America and elsewhere. Breeding was effective in reducing lipoxygenase activity in soybean seeds (Kitamura, 1995); soymilk and other products made from soybeans lacking one, two, or all three lipoxygenase isozymes had less beany flavor and less astringency (Davies et al., 1987; Torres-Penaranda et al., 1998). However, a disadvantage to using lipoxygenase-free beans is that they produce darker and more yellow colored soymilk than do soybeans with normal lipoxygenase activities (Torres-Penaranda et al., 1998).

Once off-flavors are formed, the only way to eliminate them is to strip off the responsible volatile compounds. To accomplish this task, a deodorization process is available for soymilk production. The process involves passing cooked soymilk through a vacuum pan at high temperature and high vacuum. The method is fairly complex and expensive, and is used in conjunction with other techniques by a number of large soymilk manufacturers.

Formulation and Fortification

One of the key factors to widespread acceptance of modern soymilk is the formulation of soymilk with sweeteners, flavoring agents, and other materials. Formulation not only masks the characteristic beany flavor and bitter taste associated with soy products, but also imparts different types of flavors and tastes to suit local customers. Formulation also improves nutritional value of soymilk. Common ingredients used in formulating include honey or maple syrup, sugars, vanilla extract, locust bean, cocoa powder, orange juice, and salt. Other occasionally used ingredients include malt or malt flavor, coffee, and almond extract. Mixing soymilk with non-fat cow's milk solids or coconut milk is also popular. In addition, fat and/or lecithin are often added to increase richness and creaminess of the final product, although such practice generally requires homogenization.

Typically, soymilk contains 8-12% total solids, depending on the water:bean ratio used during processing. Protein content is about 3.6%; fat, 2.0–3.2%; carbohydrates, 2.9–3.9%; and ash, ~0.5%. The composition of soymilk compares favorably with those of cow's milk and human milk (Chen, 1989). In addition, soymilk is lactose-free and contains higher levels of protein, iron, unsaturated fatty acids, and niacin; however, it contains lower amounts of fat, carbohydrates, calcium, riboflavin, thiamine, methionine, and lysine (Kosikowski, 1971). Therefore, many commercial soymilks are fortified with vitamins, minerals, and in some cases, amino acids. The most widely used nutrients for fortifying soymilk are vitamin B_{12}, calcium, and methionine.

Homogenization, Thermal Processing, and Packaging

Final processing steps of preparing modern soymilk may include homogenization, pasteurization or sterilization, and packaging. Homogenization is normally carried out after formulating and/or fortifying as a means to uniformly mix all ingredients. Homogenization also further stabilizes soymilk emulsions and reduces the tendency for chalkiness. Use of ultra-high-pressure homogenization or ultra-high-temperature processing of soymilk is more effective in stabilizing particles in soymilk (Lakshmanan et al., 2006; Cruz et al., 2007).

Pasteurization extends the shelf-life of the final product by destroying the vegetative bacteria. It is a critical control point for producing safe, nutritious, and flavorful soymilk. After the thermal treatment, the product is typically packaged in glass bottles, plastic pouches, or gable-top cartons and must be stored under refrigeration. The shelf-life of pasteurized products is usually 7 days. Because the product undergoes a low-temperature and short-duration treatment, the essential nutrients and the original flavor are generally preserved. Alternatively, soymilk may be sterilized under higher and more prolonged heat treatment. The resulting product has longer shelf-life even without refrigeration.

Aseptic packaging is a modern technology that not only extends product shelf-life but also maintains maximum original quality. In this process, the soymilk and packaging material are sterilized separately. Soymilk is first heated to 140–150°C for 2–8 sec to inactivate microbial spores, and then is sent to a vacuum system for flash-cooling to 60–75°C and stripping off undesirable flavor compounds. The product is finally filled into a sterilized package in a sterile environment and is hermetically sealed to prevent recontamination. The technique has a number of advantages compared with conventional packaging methods, including lower cost, lighter weight packaging, easy handling and stocking, longer shelf life, and favorable consumer acceptance. Thus, aseptic packaging is widely used in modern soymilk production (Imram, 2003).

Tofu

Tofu is a water-extracted and salt- or acid-precipitated soybase in the form of a curd, resembling a soft white cheese or a very firm yogurt. On a wet basis, typical pressed tofu contains about 85% moisture, 7.8% protein, 4.2% lipid, and 2 mg/g calcium. On a dry basis, tofu contains about 50% protein and 27% fat. The remaining components are carbohydrates and minerals (Wang et al., 1983).

Tofu is inexpensive, nutritious, versatile, and can be served as a meat or cheese substitute. Yet, compared with meat or cheese, tofu contains fewer calories because of its higher protein/fat ratio. Tofu is also cholesterol-free, lactose-free, and lower in saturated fat. Because of its bland taste and porous texture, tofu can be prepared with virtually any other food. It is frequently served in soups or separate stir-fried dishes with meat and/or vegetables. Tofu can also be further processed into secondary products such as deep-fried tofu, savory tofu, and fermented tofu (sufu).

Historians believe that the methods for preparing both soymilk and tofu were discovered by Liu An of the Han Dynasty in China, in about 164 BCE About 900 years later, it had spread to Japan and then to other Far East countries (Shurtleff & Aoyagi, 1979). Since then, tofu has been the most popular way to serve soybeans as a food in the Far East. Even today, there are thousands of tofu shops throughout China, Japan, and other Southeast Asian countries or regions, where many types of tofu are produced daily for local consumption. In recent years, tofu has become increasingly popular throughout the world; increasingly more consumers are seeking healthful food of plant origin. This has led to development of an infrastructure for large-scale commercial tofu production and distribution.

Preparation Methods

Many methods are available for making tofu today and all are derived from the traditional Chinese method developed over 2000 years ago. The procedure starts with preparing soymilk (Fig. 14.1). After the milk is boiled for about 10 min, it is transferred to another container, usually a wooden barrel or a pottery vat, and allowed to cool. At the same time, a coagulant suspension is prepared by mixing a powdered coagulant with hot water. Traditionally, either powdered gypsum or nigari is used. When soymilk cools to ~78°C, the coagulant solution is added with rigorous stirring. When tiny curds appear (normally within <30 sec), the container is covered and coagulation is allowed to go to completion over ~30 min.

The soy curd is ready for molding while it is still hot. The curd is first broken up by stirring, and then is transferred to a shallow forming box lined with cloth at each edge. The four ends of the forming box cloth are pulled up and folded over the curds in the box. The box is covered with a wooden lid, which is smaller than the box size. For weight, bricks or stones are placed on the top lid for about 30 min, and the whey is pressed out and the tofu becomes firm. The cooled tofu is finally cut into cakes, which are ready to be served or immersed in cold water for short storage or sale at local markets.

Today, the traditional method is still popular at home and village levels. Based on the same principle of the traditional method, many new methods have been developed to make different types of tofu products and to use different types of equipment for varying scales of production. These variations in tofu making will be discussed in the following sections.

Factors Involved in Tofu-making

In the Orient, tofu making has long been considered as an art. Without undertaking apprenticeships, not everyone can make high quality tofu. Even today, with our current extensive understanding of protein chemistry, it is a difficult task to make tofu with consistent quality and yield even under well-controlled processing conditions.

The major reason for this difficulty arises from the fact that there are complex interactions of many factors that are involved in the making of tofu (Saio, 1979; Skurray et al., 1980; deMan et al., 1986; Beddows & Wong, 1987a, 1987b; Ohara et al., 1992; Shi & Ren, 1993; Shih et al., 1997; Liu, 1999; Kim & Wicker, 2005; Mine et al., 2005; Poysa et al., 2006; Yoon & Kim, 2007). These factors (or variants) mostly center around three key areas: (i.) the way soymilk is prepared, (ii.) the way soy protein is coagulated, and (iii.) the way tofu is pressed and packaged.

In general, factors affecting soymilk preparation include soybean varieties, bean storage history and pretreatment, grinding temperature (hot or cold grinding), heat exposure of the soy slurry before or after filtration, the water-to-bean ratio, and the extent of heat applied to soymilk. Factors affecting coagulation include the temperature at which a coagulant is added, the type and concentration of coagulants, the mode of adding coagulants, and the duration of coagulation. Factors involving in the molding step include whether curds are broken and then pressed to separate whey, the temperature of curds, and the pressure and time used to press curds. To master tofu making, one needs to constantly make choices and to maintain a balance among these factors to maximize both tofu yield and quality. A few major factors are briefly discussed below; for detailed discussion refer to Liu (1999).

Soybean Varieties

Traditionally, soymilk and tofu manufacturers prefer large-seeded soybeans with clear hilum and high protein content. These beans are now specially bred in China, United States, and elsewhere (Liu et al., 1995). Such beans, known as tofu beans, are believed to produce tofu with whiter color, higher yield, and better overall quality, compared with regular commodity field beans known as oil beans. The effects of soybean varieties on soymilk and tofu qualities have been documented by many researchers (Skurray et al., 1980; Wang et al., 1983; Kim & Wicker, 2005; Poysa et al., 2006; Toda et al., 2006). Soybeans lacking lipoxygenases have also been tested for tofu making (Torres-Penaranda et al., 1998).

Storage and Pretreatment

Prolonged storage at relatively high temperature and humidity affects tofu quality (Hou & Chang, 2004). In addition, mild heat treatment of soybeans produces firmer and more elastic tofu (Yoon & Kim, 2007).

Solids Concentration

The concentration of soymilk, or total solids in a soymilk, is closely related to water:bean ratio. Solids content can be measured with a refractometer and expressed as °Brix. In general, higher solids content in soymilk correlates with harder texture and lower yield of tofu (Beddows & Wong, 1987a; Ohara et al., 1992; Shih et al., 1997). In general, tofu manufacturers use soymilk with solids contents ranging between 5 and 12%. Kim and Wicker (2005) showed that differences in protein content, 7S/11S protein ratio, viscosity, textural properties, and color of soymilk can be used as indicators of quality and functionality of its derivative product, tofu.

Heating

Heating soymilk after filtering (or in some processes, soy slurry before filtering) is essential not only for improving nutritional quality and reducing beany flavor but also for denaturing proteins so that they can coagulate into curds in the presence of a coagulant for tofu making. However, extended heat treatment should also be avoided because it leads to destruction of such nutrients as essential amino acids and vitamins, Maillard browning, and development of cooked flavor but also produces tofu with reduced yield and poor quality. Regardless of whether soymilk or soy slurry is heated, heating should be for 10 min at ~100°C (Saio, 1979).

Wang et al. (2007) proposed 2-stage ohmic heating of soymilk to make tofu and found that the treatment increased firmness, reduced syneresis, and increased yield or solids recovery of final tofu. Toda et al. (2007) compared heating soymilk with okara removed and heating soy slurry before okara removal and found that the components extracted from okara affected physicochemical properties of soymilk and texture of the resulting tofu.

Type of Coagulants

Coagulants, widely used for tofu making, are calcium sulfate, nigari, and glucono-δ-lactone (GDL, or simply known as lactone). For a comparison of various coagulants in tofu-forming properties refer to deMan et al. (1986) and Mine et al. (2005).

Calcium sulfate is the most widely used tofu coagulant. It is also the oldest one used in China, with over 2000 years of history. It comes from a translucent crystalline white stone named gypsum, a dihydrate form of calcium sulfate in mountainous areas. The stone is baked and crushed before being used as a coagulant.

Nigari is a by-product of producing table salt from sea water. It is a mixture of mineral compounds naturally found in and recovered from sea water, from which table salt has been mostly removed. It consists primarily of magnesium chloride plus all of the other salts and trace minerals in sea water.

GDL is a fine, white, odorless crystalline powder with a sour taste. An oxidation product of glucose, it is made industrially from corn starch using a fermentation process. Upon dissolving GDL in water, it is slowly hydrolyzed to gluconic acid by water. First used in Japan as a tofu coagulant during the 1960s, GDL is fundamentally different from nigari- and gypsum-types of coagulants since coagulation by GDL results from action of an acid rather than a salt. The great advantage of GDL as a tofu coagulant over alternatives is that by controlling temperature it allows completion of mixing and packaging before coagulating. Thus, it is particularly suitable for automation and aseptic packaging. Since each type of coagulant has its advantages and disadvantages a mixture of coagulants, such as that of GDL and calcium sulfate or magnesium chloride, is sometimes used in commercial processes.

Beside the acid and salt types of coagulants discussed above, the microbial enzyme transglutaminase has also been studied as a tofu coagulant (Tang 2007). Since it forms a structure with strong chemical bonds, the resulting tofu has a very smooth texture.

Coagulant Concentration

When the proper amount of coagulant is used, the whey becomes transparent with an amber or pale yellow color and sweet taste. No uncoagulated soymilk remains. If too much coagulant is added, however, the whey has a slightly bitter taste and yellowish color, and the curds have a coarse and hard texture. In contrast, if too little coagulant is used, the whey is cloudy and some uncoagulated soymilk remains. Within a certain concentration range, as the coagulant concentration increases, tofu bulk yield and protein recovery decrease while tofu hardness, fracturability, and elasticity increase. Furthermore, this changing pattern varies with soybean varieties (Shih et al., 1997; Liu & Chang, 2004).

Coagulation Temperature

The temperature of soymilk at the time of adding the coagulant affects coagulation rate as well as tofu texture and yield. At high temperature, the proteins possess high active energy, leading to fast coagulation. The resulting tofu tends to have low water-holding capacity, hard and coarse texture, and therefore low bulk yield. When the coagulation temperature is low, the effect is just opposite; however, if the temperature is too low (<60°C), the coagulation becomes incomplete, the tofu contains too much water and the tofu is too soft to retain its shape. Generally, the temperature falls in the range of 70 to 80°C (Beddows & Wong, 1987b).

Coagulation Time

It is desirable to let the soymilk-coagulant system stand still for a while after the coagulant is added, since completion of coagulation requires a certain period of time. If the time is too short, coagulation is incomplete. If too long, the temperature of the system decreases to such an extent that the subsequent molding step becomes difficult. In general, for silken tofu, the standing time should be about 30 min; for regular tofu, 20–25 min; and for firm tofu, 10–15 min (Shi & Ren, 1993).

Process Automation

During commercial tofu production, coagulation has been the most difficult step to automate. Since the 1980s, however, different types of tofu coagulation machines have been developed in Japan, China, and South Asia. These machines automatically add the calculated amount of coagulant to the heated soymilk with a stirring device in both batch and continuous operation modes. They also allow the coagulation system to stand for certain period of time before molding. Therefore, the finished product has uniform quality. In some large commercial processing plants, tofu is now made continuously and automatically from raw bean cleaning to the final stage of packaging.

Packaging

In rural areas, tofu is sold fresh in the local market in cakes covered with water in a wooden bail or pan. For commercial distribution, however, most tofu products are packaged in polyethylene containers, filled with water, sealed, pasteurized, and finally chilled before distributing. These products need to be refrigerated until consumption and have shelf-life of about 3–4 wk. In addition, there are special tofu products under the brand name of Mori-Nu. These products are processed under a patented technology and are aseptically packaged in a tetrahedral box made of laminated carton paper. It has a shelf-life up to a year without refrigeration.

Varieties of Tofu

There are many different types of tofu in the market. Based on water content and textural properties, tofu is generally classified into soft (silken), regular, and firm tofu. All tofu products are made in a similar fashion except for variations in the water:bean ratio, the type and concentration of coagulants, the way a coagulant is added, and the amount of whey pressed out.

Silken Tofu

Soft or silken tofu has a soft cheese-like texture but is sufficiently firm to retain its shape after slicing. Silken tofu is normally made from rich soymilk containing 10-12% solids. After fine filtering, the soymilk is allowed to cool to 65–70°C and is mixed with a relatively low concentration of coagulant. Over a period of 30–60 min, a fine, smooth yet firm curd forms. The curd is neither broken nor pressed. The entire box-sized block of silken tofu is then cut into cakes, removed from the box under water and cooled. In China, soft tofu is known as *shui doufu*, meaning watery tofu; *neng doufu*, meaning tender tofu, or *doufu hua*, meaning tofu flower. In Japan, tofu made without removing the whey is known as *kinugoshi*-tofu.

Regular and Firm Tofus

Regular and firm tofus are mostly pressed tofu, which is known as *momen* tofu in Japan. The difference between the two products is that firm tofu is harder than regular tofu. Sometimes, the term extra firm is used by some manufacturers. Whatever the terms, the textural differences among silken, regular, firm, or extra firm are relative. There are no standards in absolute texture properties for them, and the texture of the same type of tofu may vary with manufacturers, seasons, and even batches.

In making pressed tofu, the curds are broken and then pressed while they are still warm. The smaller the particles of broken curds are, the firmer the tofu. Also, the heavier the weight or the higher the pressure applied during pressing, the firmer the tofu. As the texture becomes firmer, the water content is reduced. Firmer texture and reduced water content make tofu easier to handle and more similar to meat or cheese. These factors also help tofu keep its shape during cooking. Therefore, pressed tofu is ideal for use in pan-frying, deep-frying, grilling, freezing-drying, and dicing into an ingredient for other foods or soups.

Varieties of Tofu Products

Tofu is very versatile as a food; it can be served fresh or cooked with vegetables and/or meat in thousands of different dishes and soups. It can also be further processed into various secondary tofu products, including deep-fried tofu, savory tofu (Fig. 14.2), grilled tofu, frozen tofu, dried-frozen tofu, fermented tofu, etcetera. In most cases, these processed tofu products have different characteristics, end uses, and commercial identities from the original plain tofu discussed previously.

In recent years, some tofu products, such as deep-fried tofu and savory tofu, have been observed in the Western market and are gaining popularity. In general, these products are ready-to-eat and have much less beany flavor. Thus, they have received higher acceptance levels than regular plain tofu, with which Westerners are still not familiar in terms of preparation and texture.

Green Vegetable Soybeans

With green or green-yellow color, soft texture, and large seed size (due to high moisture content and specially-selected varieties), green vegetable soybeans are normally harvested at about 80% maturity in the green-yellow pod from the field. Therefore, they are also known as immature soybeans or fresh green soybeans. Direct consumption of green vegetable soybeans is very popular in China, Japan, and some other Far East countries or regions. Steamed or boiled in water (normally for <20 min) before or after shelling and lightly salted or spiced, these immature beans can be served either as a delicious green vegetable with a main meal or as a tasty hors d'oeuvre, often with beer or other alcoholic drinks. In Japan, immature soybeans are known as *edamame*, and are sold fresh or frozen. They may also be made into roasted beans, which have a crunchy texture and greenish-beige color, and sold as *Irori mame.*

In North America, frozen vegetable soybeans have long appeared in the marketplace. Some of the products may be marketed under a brand name of Sweet Beans. They are gaining popularity due to their tender texture, low beany flavor, and resemblance in preparation to frozen corns and peas. In making frozen vegetable soybeans, freshly harvested beans are de-shelled, blanched, and frozen.

Depending on the stage of maturity, green vegetable soybeans have protein contents in the range of 11–16% and oil in the range of 8–11%, whereas on moisture-free

Fig. 14.2. Savory tofu dices.

basis, they have protein and oil composition very close to that of mature soybeans. However, compared with field-dried mature soybeans, fresh immature soybeans offer several distinct features, including less beany flavor, higher contents of vitamins, lower levels of antinutrients, tender texture, and less cooking time requirements (Liu, 1996). In addition, green color and soft texture enhance their appeal as a vegetable. The quality of green vegetable soybeans depends on variety, storage history, and blanching method used (Yong et al., 2000; Akazawa et al., 2002). For a recent review on green soybeans, refer to Mohamed and Mentreddy (2005).

Soybean Sprouts

Bean sprouts have been used as food in the Orient since ancient times. They are made either from soybeans or mung beans (*Phaseolus aureus*) by germinating in the dark. Soybean sprouts (Fig. 14.3) are more popular in Korea and Southern China, but less popular than mung bean sprouts than in most other parts of the world. Serving as a vegetable throughout the year, soybean sprouts are used in soups, salads, and side dishes.

To produce soybean sprouts, soybeans, preferably freshly harvested, small to medium-seeded beans with good vigor, are first soaked in warm water (40–50°C) for 3–4 h, washed well, and then spread into thin layers in a deep container (or bucket) with holes at the bottom for draining the water. A cloth is normally placed in the bottom of the container to prevent passage of the beans. The container is covered with hay, rice straw, or other material to screen out the light but allow air exchange, and then is placed where the temperature can be maintained ~23°C.

During germination, heat builds up due to active seed metabolism. Thus, it is necessary to sprinkle fresh water over the beans in the container 3–4 times a day. Addition of water not only provides moisture for the seeds to germinate and for new seedlings to grow but also helps to reduce metabolic heat. Excessive moisture, however, is unfavorable for rapid sprouting as it tends to limit oxygen supply and leads to decay by mold infection. Also, light should always be avoided during the process because it causes sprouts to develop roots and turn green, both of which are undesirable.

In less than a week the majority of sprouts reach a length of about 8 cm, they are washed, dehulled, and ready for serving or transporting to market. The finished product is crispy and has a distinct taste, and is composed of yellowish cotyledons and long, bright white sprouts (Fig. 14.3). In a typical germination process, 1 kg of dry soybeans can produce 7–9 kg of fresh bean sprouts.

Compared with original dry soybeans, soy sprouts offer several nutritional advantages. These include increased contents of vitamin C and ß-carotene (Bates &

Fig. 14.3. Soy sprouts.

Matthews, 1975; Xu et al., 2005) and free amino acids (Martinez-Villaluenga et al., 2006), and reduced levels of flatulence-causing oligosaccharides and mineral-binding phytic acid. During cooking, it is desirable to minimize heating to maintain the inherent crisp texture and distinct taste, and to minimize destruction of vitamins.

Yuba

Yuba is another soyfood derived from soymilk (Fig. 14.4). Named after a Japanese word for soymilk film, yuba is also known as dried bean curd in English, *doufupi* or *fuzhu* in Chinese, *kong kook* in Korean, and *fuchok* in Malaysian.

Yuba has very chewy texture and is one of the first texturized protein foods (Fig. 14.4). On average, yuba contains 55% protein, 26% neutral lipids, 2% phospholipids, 12% carbohydrate, 2% ash, and 9% moisture (Wu & Bate, 1972). Among several Oriental soyfoods, soy protein in yuba has the highest percentage of digestibility, ~100% (Ikeda et al., 1995). Due to limited production and high cost, yuba is considered as a delicacy.

To make yuba, one needs to make a rich soymilk. The soymilk is then heated in a flat, open pan to near boiling (~80–90°C); a film gradually forms on the liquid surface due to surface dehydration. After the film becomes toughened it can be lifted with two sticks or by passing a rod underneath. The film is hung on a line or spread

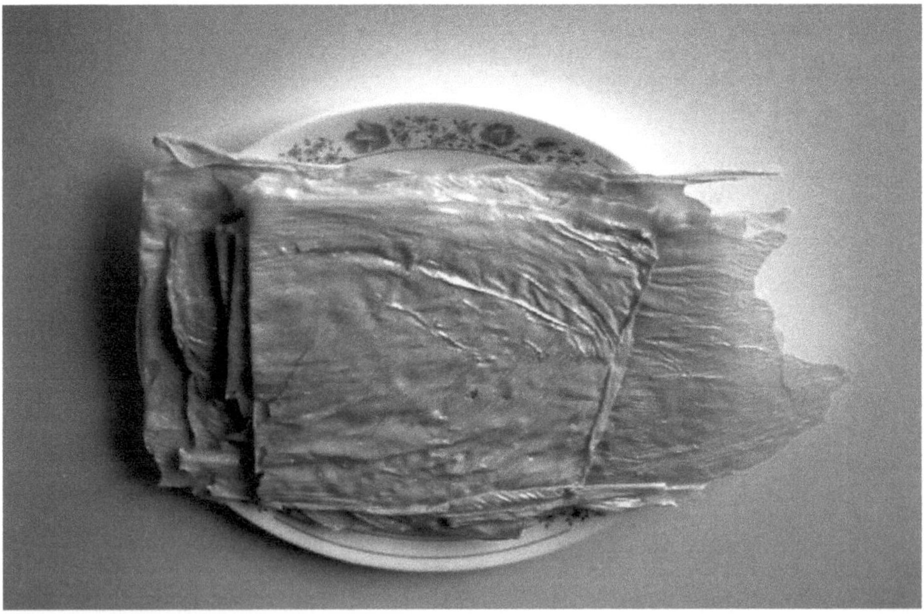

Fig. 14.4. Yuba (soymilk film).

on a galvanized wire mesh for drying. In this manner, films are continuously formed and removed from the surface of the soymilk until no further film formation occurs. Generally, 10–20 such sheets can be made before it is necessary to refill the pan with fresh soy milk.

During yuba production, protein and lipid contents in the film successively formed decrease, while the carbohydrate and ash contents gradually increase. Therefore, the first several pieces of yuba to be lifted off the heated soymilk are considered premium products. They have creamy white color, mild flavor, and less sweet taste. They stay relatively soft and flexible even when dried. The later formed products are regarded as second rate, because they become sweeter with a faintly reddish tinge, lack internal cohesiveness, tear more easily, and become brittle upon drying.

Yuba is commonly sold in three different states: fresh, semi-dried, and dried. Fresh yuba needs to be served as soon as possible after manufacture because it is highly perishable. It is usually regarded as the most delicious type. Semi-dry yuba has longer shelf-life than fresh yuba but not nearly as long as dried yuba. Dry yuba is quite brittle and has a relative long shelf-life. It is the most common of the three forms sold in the market. There are different shapes of yuba: flat sheets, long rolls, small rolls, U-shaped rolls, large spirals, etcetera. Immediately before cooking, semi-dry or dry yuba needs to be soaked until fully hydrated.

In spite of its nutritional excellence, yuba is appreciated primarily for its unique flavor and texture. There are various ways of using yuba. It can be used as a wrap for other foods or used in soups or cooked with other food materials. Fresh yuba can also be made into meat analogs by pressing into hinged molds and then steaming to create a number of forms resembling meat such as chicken, fish, and duck. The product is widely served in vegetarian restaurants under such names as vegetarian chicken, vegetarian fish, or Buddha's chicken. It may be packed or canned, and sold. Because yuba is a delicacy, a food containing yuba is commonly considered special.

Okara

Okara, also known as soy pulp in English, and *doufu zha* or *dou zha* in Chinese, is the insoluble residue after filtering soy slurry into soymilk. Therefore, it is considered to be a by-product of soymilk and tofu preparation. Yet, for every kg of dry soybeans made into soymilk or tofu, about 1 kg of okara is generated. More specifically, on average, 53% of the initial soybean dry mass is recovered in tofu, 34% in okara, and 16% in whey. About 72% of the protein is recovered in tofu, 23% in okara, and 8% in whey; the respective average soybean oil recoveries are 82, 16, and <1% (van der Riet et al., 1989).

Although the actual composition depends on the specific process as well as the soybean variety, fresh okara contains 76–80% moisture, 2.6–4.0% protein, and the remaining percentages for other solids. When dried, okara contains 25.4–28.4% protein, 9.3–10.9% oil, 40.2–43.6% insoluble fiber, 12.6–14.6% soluble fiber, and

3.8–5.3% soluble carbohydrates (van der Riet et al., 1989). Therefore, okara contains high fiber content and appreciable amounts of protein and oil.

The major use of okara is livestock feed, however, there are various ways of using okara as food. For examples, in some parts of China, okara is salted and spiced and served as a pickle, or simply made into a dish with meat or vegetable. In other parts of China, okara is pressed into cakes and allowed to ferment for 10-15 days until covered with the white mycelium of *Rhizopus* mold. The cakes are dried in the sun and then deep-fried or cooked with vegetables. A similar product, known as *tempeh gembus*, is also popular in Indonesia. Sometimes, okara may be mixed with soybeans before fermentation.

With growing awareness of the role of dietary fiber in human health, interest is increasing in using okara as a food ingredient. For example, plain or flavored okara can easily be dehydrated on a drum dryer to make a dry staple, which might be further milled into flour before being used as a high-fiber food ingredient. Furthermore, there is a growing research interest in okara due to potential health benefits, such as preventing obesity (Matsumoto et al., 2007) and suppressing plasma cholesterol levels (Fukuda et al., 2006), to drying and fractionation (Surel & Couplet, 2005), and to exploring utilization in food (Rinaldi et al., 2000).

Roasted or Cooked Soybeans

When clean, whole soybeans are roasted for about 30 min, they become brown and acquire a characteristic toasted flavor. Upon cooling, the roasted beans, known as soynuts, can be eaten like roasted peanuts as a snack or used as an ingredient to add crunchy texture and nutlike flavor to a wide variety of salads, sauces, casseroles, and miso preparations. In addition to dry-roasting, whole soybeans may be oil-roasted.

Roasted beans may be covered with various flavorings and coating materials, including sugar, chocolate, onion, garlic, etcetera. Such coated soynuts are now seen in the Western market. Compared with roasted peanuts, soynuts provide higher protein content and less fat.

When roasted soybeans are ground into powder, they become roasted soy powder (flour), which is similar to modern full-fat soy flour except that it contains the seed coat and has a nutty flavor. The product is known as *doufen* in Chinese, and *kinako* in Japanese. Like roasted soynuts, roasted soy flour is an inexpensive source of good quality protein for supplementing various types of food. In China, roasted soy flour may be mixed with lard and sugar, and used as a filling or coating material for pastry. In Japan, one favorite way of using roasted soy flour is to spread it on rice or rice cakes. In Indonesia, the flour is mixed with spices, such as garlic and chili powders, and served with *longtong*, which is boiled rice wrapped in banana leaves. In the West, soy butter is present in the marketplace, which is made of roasted soybean powder.

In the Far East, the whole soybeans are sometimes consumed directly after soaking and cooking (steaming or boiling) until their texture becomes tender. Salt, oil, soy sauce and other spices, and seasonings may be added during cooking. When meat is also added, the dish becomes a tasteful and Oriental version of pork and beans.

Fermented Soyfoods

Terms

Before describing methods of preparing fermented soy products, it is necessary to discuss the terms below.

Koji

The Chinese counterpart for the word *koji* is *qu*, meaning bloom of mold. Made by growing molds on rice, barley, wheat, soybeans, or a combination, koji contains a great variety of enzymes that digest the starch, protein, and lipid components in raw materials. It is an intermediate product for making various fermented products such as fermented soy paste, soy sauce, soy nuggets, and Japanese sake.

Fermentation

The microorganisms found in koji almost always belong to the fungi species, *Aspergillus oryzae* and/or *A. sojae*. *A. oryzae* molds reproduce only asexually and have ability to utilize starch, oligosaccharides, simple sugars, organic acids, alcohols, etcetera, as carbon sources, and protein, amino acids, urea, etcetera, as nitrogen sources. The mold is aerobic, with growth optima generally at pH 6.0, 37°C temperature, and 50% water content. When air is limited or water content of the medium is <30%, its growth slows down. When the temperature is <28°C, its growth slows but enzymatic activities remain high. A key secondary metabolite of koji production is kojic acid. A recent review article for this product is available (Bentley, 2006).

Koji Starter

Koji starter, also known as seed koji, koji seeds, or *tane-koji*, provides spores of microorganisms to make koji. Preparation of koji starter is essentially the same as making regular koji for soy paste and soy sauce except that in making koji starter, pure culture and different raw materials are used and longer fermentation is needed to produce abundant spores. In addition, a sterile condition is needed to avoid contamination.

Since many molds, including *A. oryzae*, are ubiquitous, up until several decades ago, wild spores of the species were used as the starters for jiang, miso, or soy sauce preparation. The modern process for making koji starter, however, begins with growing a selected *A. oryzae* strain on an agar slant in pure culture. The strain is selected for its special abilities by natural selection or by induced mutation to give a desirable koji for a particular fermentation. Therefore, there are many varieties of commercial tane-koji, each having different capacity to break down proteins, carbohydrates, and lipids in raw materials. It is very important to select a suitable variety for making a particular product. For example, for salty rice miso rich in protein, a tane-koji of high proteolytic activity is suitable whereas for sweet rice miso rich in starch, a koji-starter with high amylolytic activity is preferable.

Inoculum
In addition to the koji mold, halophilic yeasts and lactic acid producing bacteria play an important role in developing flavors during fermentation of other fermented soy products. This is particularly true in the later stage of the fermentation. In making traditional jiang, miso, or soy sauce, fermented products from a previous batch were used as inoculi to be mixed with salted koji and cooked soybeans. The inoculum generally contains a selected flora of salt-tolerant yeasts and bacteria capable of growing under anaerobic conditions. The dominant organisms are yeast *Zygosaccharomyces rouxii* and *Torulopsis* sp. and certain lactic acid producing bacteria, such as *Pediococcus halophilus* and *Streptococcus faecalis.* Since pure cultures of these organisms speed up fermentation and reduce wild yeasts and bacteria, its use in commercial preparation of certain fermented soy products has been popular in recent years.

Fermented Soy Paste

Soy paste is an important fermented soyfood in the Far East. It has a color varying from a light, bright yellow to very blackish brown, a distinctively pleasant aroma, and a salty taste (Fig. 14.5). Soy paste is commonly known as *jiang* (Mandarin) or *chiang* (Cantonese) in China, *miso* in Japan, *jang* in Korea, *taucho* in Indonesia, and *taotsi* in the Philippines.

Developed in China some 2500 years ago, *jiang* was the progenitor of the many varieties of soy paste and soy sauce that are now used throughout the world (Shurtleff & Aoyagi, 1983). At present, Chinese jiang and Japanese miso are the two most popular types of soy paste. Although sharing the same progenitor, the two differ in many aspects. Chinese jiang is made from soybeans and wheat flour. The finished product may be unground so that individual particles of soybeans are present. It is used mainly as an all-purpose seasoning for dishes and soups. Japanese miso, however, is made from soybeans mixed with rice or barley or from soybeans alone. The finished product is a paste resembling peanut butter in consistency and may have sweet taste. It is mainly dissolved in water as a base for various types of soups in Japan.

There is growing interest in fermented soy paste for its many health benefits, including antimutagenic and anticlastogenic effects (Jung et al., 2007), preventing hypertension (Watanabe et al., 2006), and fibrinolytic activity (Choi et al., 2007). Research on identifying chemical changes and flavor compounds formed in soy paste is also an active area (Park et al., 2003; Lertsiri et al., 2003; Ogasawara et al., 2006).

Fig. 14.5. Chinese jiang (bottom) and Japanese white (top) and red miso (middle).

Preparation Method

The method of making miso may vary slightly with type of product, but the basic processes are essentially the same (Shurtleff & Aoyagi, 1976; Watanabe & Kishi, 1984). For example, rice miso involves five distinct processing steps: preparing rice koji, treating soybeans, mixing of ingredients, fermenting, and pasteurizing and packaging (Fig. 14.6).

Preparing Rice Koji

Non-glutinous, polished rice is cleaned, washed, and soaked overnight in water at about 15°C or until the moisture content increases to ~35%. After draining the water, the rice is steamed for ~40 min. When cooled to 35°C, the cooked rice is inoculated with koji starter containing *A. oryzae* spores, at a concentration of about 0.1% of the rice. In modern plants, the cooking and inoculating process are carried out continuously.

After 15 h of incubation at 30–35°C and >90% relative humidity in a box, the incubated rice is transferred to the koji room where it is spread out to a 4 cm depth on koji-trays and allowed to ferment. When the temperature climbs to >35°C, the young koji is turned over and stirred for good aeration for further fermentation. After about

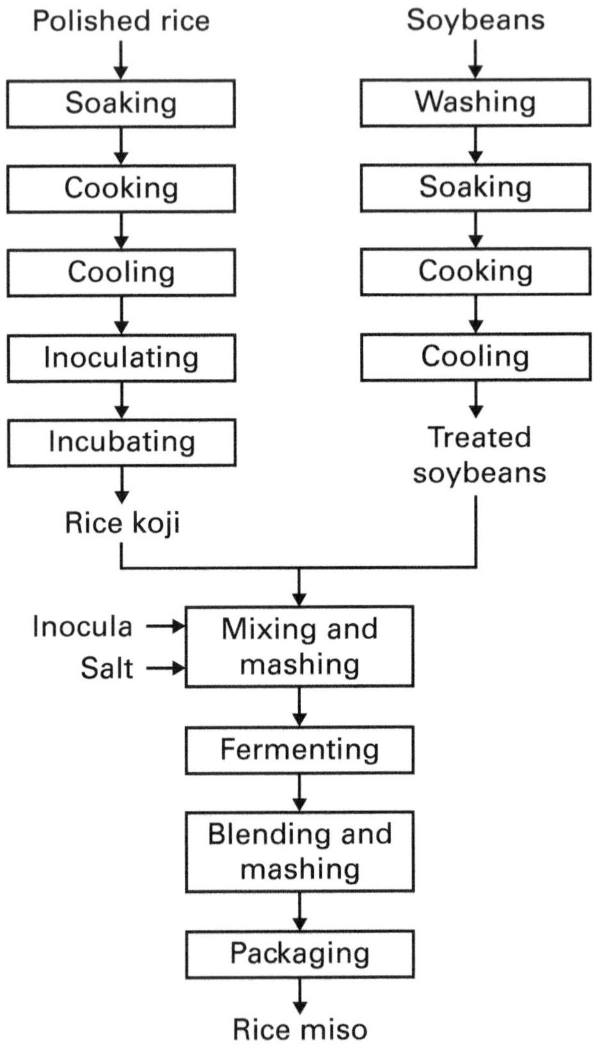

Fig. 14.6. A common method for making Japanese rice miso (adapted from Watanabe & Kishi, 1984).

40 h following inoculation, when the cooked rice is completely covered with white mycelium, koji becomes mature and ready to be harvested.

Good koji has a pleasant smell, lacks any musty or moldy odor, and is quite sweet in taste. The mature koji is taken out from the koji room to be mixed with salt to halt further mold development. Recently, a mechanical koji fermentor has been developed, which is equipped with an air conditioner and a mechanical stirrer to break the sponge-like lump during cultivation.

Treating Soybeans

Concurrent with the koji preparation, the whole soybeans are cleaned, washed, and soaked in water overnight. It is possible to reduce the soaking time by increasing the soaking temperature. The soaked beans are steamed under 0.7–1.0 kg/cm^2 pressure for 20–30 min or until they are sufficiently tender to be mashed between the fingers. Soybeans, preferably with light yellow seed coats and clear hilums to produce white or light-colored miso types, are generally cooked in boiling water with a bean:water ratio of 1:4 to prevent browning from steaming. Although batch-type cookers are widely used, continuous cookers have been recently adopted by large factories. Cooked beans are cooled using a belt-conveyor-type cooler to room temperature. Sometimes, they may be mashed with a chopper while hot.

Mixing and Mashing

After cooling to room temperature, the cooked soybeans are mixed with salted rice koji and water containing inoculum, which may come from a previous batch or pure culture. The mixed materials are mashed by passing them through a motor-driven chopper with 5-mm perforations. Homogeneous mixing is important to maintain normal fermentation.

The proportions among ingredients at the time of mixing determine the type, flavor, and appearance of the final product. These include water content, the proportion of rice koji to soybeans, and the amount of salt. Water content affects not only the rate of fermentation but also the consistency of the final product. For miso, the final mixture should contain 48–52% moisture. In addition, darker types of miso require higher portions of soybeans (50–90%) than rice compared with whiter miso. Red or brown miso contains 11–13% salt whereas white miso contains 4–8% salt. The lower salt content of white miso permits more rapid fermentation but gives the product a shorter shelf-life.

Fermenting

After mixing and mashing, the mixture is packed tightly into open tanks or vats. The containers are traditionally made of wood or concrete; however, some modern plants use steel vats coated with epoxy resin, stainless-steel vats, or glass-lined vats. Weight equivalent to 5–10% of the total miso is placed on the sheet to force liquid to the surface and ensure anaerobic conditions. The young miso is allowed to ferment in a controlled temperature (normally 30–38°C) for a period up to 6 mo, depending on the type of miso. To retain the homogeneity, miso is transferred from the original vat to another during fermentation.

Pasteurizing and Packaging
After ripening, miso is blended if necessary, and mashed again through a chopper with a plate cutter having 1–2 mm perforations. The mashed miso is then packaged in a resin bag or cubic container after being pasteurized with a steam jacket or mixing with preservatives such as 2% ethyl alcohol or 0.1% sorbic acid.

Processing Principles

Regardless of variation in preparing Chinese jiang and Japanese miso, the fermentation principles to produce these products are very similar to each other. In general, all preparations involve treating raw materials, preparing koji, mixing ingredients, and fermenting.

Treating the raw materials includes soaking and heating. Soaking hydrates proteins and other grain components, and thus promotes effects of subsequent heating. Insufficient hydration leads to insufficient denaturation of soy protein as well as softening of soybean texture during the subsequent heat treatment. The effects of heating include denaturing proteins so that they can be utilized by either koji mold or hydrolyzed by koji enzymes, inactivating trypsin inhibitors and lectins, softening soybeans, sterilizing soybeans, and removing the unpleasant bean odor (Nikkuni et al., 1988).

During koji making, various enzymes are produced as the spores of the koji mold germinate and grow into mycelia and eventually sporulate. Major enzymes include protease, amylase, glutaminase, lipase, hemicellulase, pectinase, and esterase. These enzymes become active during fermentation. As a result, components from the original materials are degraded. For example, starch from rice, barley, or wheat is converted by amylase into dextrin, maltose, and glucose, which contribute a sweet taste to the miso. Protein from soybeans is converted by proteases and peptidases into water-soluble nitrogen compounds containing mainly oligo-peptides and amino acids. The amino acids, particularly glutamic acid, contribute the delicious taste to miso (Hondo & Mochizuki, 1968). At the same time, oil from soybeans is partly hydrolyzed by lipase to free fatty acids and glycerol.

As enzymatic digestion progresses, it generates fermentable substances, such as simple sugars, for growth of yeasts and lactic bacteria, which come from inoculum. Under anaerobic conditions, these organisms further break down sugars to acids, alcohols, and other substances. As fermentation and aging continue, there are complex chemical and biological interactions among various components in miso (Park et al., 2003; Ogasawara et al., 2006). In general, acids react with alcohols to produce esters, which contribute a distinct aroma to miso. Amino acids and sugars interact to produce browning substances, which contribute in part to the color of miso. Since amino acids play the dual role of enhancing flavor and darking color, miso with the darkest color is often considered to be richest in flavor. Lertsiri et al. (2003) provided evidence of enzymatic browning during mash fermentation.

Soy Sauce

Soy sauce is a dark-brown liquid extracted from the fermented mixture of soybeans and wheat. Due to its salty taste and sharp flavor, it is served as an all-purpose seasoning for thousands of years. The product is known as *jiangyou* (Mandarin) or *chiangyu* (Cantonese) in China, meaning oil from jiang; *shoyu* in Japan; *tao-yu* in Indonesia; and *tayo* in the Philippines. Among all fermented soyfoods, soy sauce is now the widest-accepted product, not only in Far East but also in Western countries.

There are many types of soy sauce. Based on preparation principles, soy sauce is divided into three groups; fermented soy sauce, chemical soy sauce, and semi-chemical soy sauce. There are Chinese and Japanese soy sauces. Under each ethnic group, soy sauce is further divided based on differences in raw ingredients, methods of preparation, and duration of aging. More Information about soy sauce can be found in Yokotsuka (1986) and Liu (2005b). Recent research on soy sauce has focused on its health benefits (Kataoka, 2005; Kobayashi, 2005).

Preparation Method

Although there are some variations in making different types of soy sauce, their basic steps are the same, including treatment of raw materials, koji making, brine fermenting, pressing, and refining (Watanabe & Kishi, 1984; Yokotsuka, 1986; Liu, 2005b). A typical process for soy sauce is outlined below.

Treating Raw Materials

The initial step is to simultaneously treat soybeans and wheat. Whole soybeans are soaked overnight at ambient temperature. The soaked soybeans are cooked under steam pressure or in an open pan until very soft. If defatted soy products are used, which has become popular, they are first moistened by spraying with water in an amount of up to 30% of the soy product weight. This is followed by pressure-steaming for 45 min. The heated soybeans or soy grits are allowed to cool to <40°C within as short a time as possible.

Concurrent with treating soybeans, whole kernel wheat is roasted and cracked with roller mills into 4–5 pieces. When wheat flour and wheat bran are used, they are steamed after being moisturized.

Koji Making

The above two materials are mixed in proportions depending on what type of end products is desired. For *koikuchi* shoyu, the ratio of soybeans (or defatted soy meal) to water is about 1:1, whereas for *tamari* shoyu, the ratio is 9:1. The mixture is inoculated with 0.1–0.2% seed koji or a pure culture containing *Aspergillus oryzae* and/or *A. sojae*.

In traditional koji-making, the inoculated mixture is put into small wooden trays and kept for 3 or 4 days in a koji-making room. During mold growth, the temperature and moisture are controlled by manual stirring. In modern koji-making, however, the cultured mixture is put into a shallow, perforated vat and kept in a koji room where forced air is circulated and temperature and humidity are controlled (as in the case with an automatic koji-making system). After about 3–4 days, when the mixture turns green-yellow as a result of sporulation of the inoculated mold, it becomes mature koji.

In the early stage of koji making, temperatures as high as 30–35°C are preferable for mycelium growth and prevention of Bacillus as a contaminant. In the later stage, just before spore formation or after the second cooling, a lower temperature (20–25°C) is necessary to allow maximum enzyme production. Alternatively, koji may be prepared at a constant low temperature of 23–25°C for relatively longer times (66 h). In any cases, when temperature rises to >35°C due to active mold growth, it is advisable to cool the koji material twice either by hand mixing or a mechanical device.

Brine Fermentation

Mature koji is mixed with an equal amount or more (up to 120% by volume) of a salt solution to form the liquid mash known as *moromi* in Japan. The final salt concentration of the mash should be 17–19%. Lower salt concentration allows growth of undesirable putrefactive bacteria during subsequent fermentation and aging; however, higher salt concentration (>23%) may retard the growth of desirable halophilic bacteria and osmophilic yeasts. In the home, the mash is put in an earthen crock and fermented at ambient temperatures. In this case, a period of 10–12 mo may be necessary to complete the brine fermentation stage. On an industrial level, however, the mash is kept in large wooden containers or concrete vats with aeration devices. The temperature of their surroundings can be mechanically controlled. Thus, fermentation time can be shortened.

Temperature is also an important factor during brine fermentation. In general, the higher the temperature, the shorter the fermentation time; however, fermenting at lower temperatures gives better products because the rate of enzyme inactivation is slow. A good quality soy sauce can be made by 6-mo fermentation when the mash temperature is controlled to start at 15°C for 1 mo, followed by 28°C for 4 mo, and finishing at 15°C again for 1 mo (Watanabe & Kishi, 1984).

Pressing

After months of fermenting and aging, the mash becomes mature. In the case of home processing, raw sauce may be removed from the mash simply by siphoning it off from the top, or filtering through cloth under a simple mechanical press. In commercial operations, a batch-type hydraulic press is commonly used. Recently, automatic loading the mash into filter cloth or continuous pressing by a diaphragm-type machine has been developed for effective filtration. The filtrate is stored in a tank to separate the sediment at the bottom and the oil floating on the top.

Refining

Raw soy sauce may be adjusted to standard salt and nitrogen concentrations. It is then pasteurized at 70–80°C to inactivate enzymes and microorganisms, enhance the unique product aroma, darken color, and induce formation of flocs, which facilitates clarification. After heating, the soy sauce is clarified by either sedimentation or filtration. Kaolin, diatomite, or alum may be added before filtration to enhance clarification. The clear supernatant is packed immediately into cans or bottles. In some cases, preservatives, such as sodium benzoate and paraoxy-benzoate, may be used.

Processing Principles

Just like making jiang or miso, two stages of fermentation occur in soy sauce preparation. The first fermentation is solid-state and occurs during koji making where various enzymes are produced under aerobic conditions. The second fermentation occurs after the brine addition and is known as brine fermentation. It is mainly anaerobic. At the earlier stage of brine fermentation, enzymes from koji hydrolyze proteins to yield peptides and free amino acids. Starch is converted to simple sugars, which in turn serve as substrates for the growth of various types of salt-resistant bacteria and yeasts. These organisms become dominant in sequence as fermentation progresses. All these enzymatic and biological reactions, together with concurrent chemical reactions, lead to the formation of many new volatile and non-volatile substances, which contribute to the characteristic color, flavor, and taste of soy sauce (Yokotsuka, 1986).

Chemical Soy Sauce

Traditionally, soy sauce is made by fermentation as described above; however, soy sauce can also be made by acid hydrolysis. The resulting product is known as chemical soy sauce, or protein chemical hydrolysate. In this process, defatted soy products or other proteinous materials are first hydrolyzed by heating with 18% HCl for 8–12 h. After hydrolysis, the hydrolysate is neutralized with sodium carbonate and filtered to remove the insoluble materials. The resulting product (chemical soy sauce) is a clear dark-brown liquid. However, chemical soy sauce does not possess the flavor and odor of fermented shoyu. Therefore, to improve its quality, chemical soy sauce is often blended with fermented shoyu to become a semi-chemical product before being sold.

Japanese Natto

Originating in the northern part of Japan about 1000 years ago, natto is one of the few products in which bacteria predominate during fermentation. When properly prepared, it has a slimy appearance, sweet taste, and a characteristic aroma (Fig. 14.7). In Japan, natto is often eaten with soy sauce or mustard and served for breakfast and dinner along with rice. Similar products are also found in Indonesia and Thailand.

Recent research has showed natto has health benefits. In particular, natto has been shown to contain significant amount of vitamin K_2, which is derived from the microorganism, *Bacillus subtilis* (Yanogisawa & Sumi, 2005). Vitamin K_2 is the co-factor that converts nonactivated osteocalcin into activated osteocalcin by carboxylation. In rat as well as in in vitro studies, natto promotes formation of osteocalcin, a bone protein, and participates in bone formation (Yamaguchi et al., 2001).

Preparation Method

To make natto, soybeans, preferably small seeded, are washed and soaked overnight. The soaked beans are then steamed for ~30 min, drained, and cooled to ~40°C. Traditionally, the treated soybeans are wrapped with rice straw and set in a warm place for 1–2 days. Rice straw is credited for not only supplying the fermenting microorganism, *Bacillus natto*, but also absorbing the unpleasant odor of ammonia released from natto and imparting the aroma of straw to the product.

Fig. 14.7. Japanese natto.

However, since there is a great chance of contamination with unwanted microorganism from the rice straw, the quality of the product is rather difficult to control. Ever since the responsible microorganism *Bacillus natto* was isolated, the old straw method has been largely abandoned in favor of pure culture fermentation. Instead of wrapping with rice straw, the treated beans are inoculated with a pure culture suspension of *B. natto* at ~15 mL/100 kg raw beans, and thoroughly mixed before being packed in wooden boxes or perforated polyethylene bags. The packages are put into shallow sliced-wood or polystyrene trays and set in a warm, temperature-controlled chamber at 38°C. After 16–20 h of fermentation, the bacteria will have covered the beans with a white sticky coating, indicating the time for harvesting. For better quality, the package may be kept at a refrigerated temperature for 1-2 days to allow maturation and then removed for consumption or retail as needed. The practice is most common in the large plants.

Good quality natto should have a characteristic flavor, intact beans with viscous appearance and soft texture. There are many factors affecting the quality of natto. One key factor is selecting the right soybeans. Unlike for other fermented soyfoods, round, small seeded soybeans, with high soluble sugar content, yellowish seed coat, and a clear hilum, are preferred for producing natto. Small beans have larger surface area to volume than large- or medium-sized beans, and thus absorb water faster, require shorter steaming time, and allow faster growth of the natto organism. Because soluble sugars serve as an initial carbon and energy source for natto organisms, a high sugar content promotes microbial growth and makes the finished product taste sweeter. Since *B. natto* is aerobic, good aeration with a sufficient supply of oxygen is required during the entire fermentation. The temperature should be controlled at ~40°C, and over fermentation should be avoided since it leads to the release of ammonia, which not only spoils the natto flavor but also destroys the *B. natto* and promotes spoilage by other organisms.

Processing Principles

Unlike preparation of many other fermented soyfoods, which are complex and require actions of multiple microorganisms with a mold dominating, preparing natto is relatively simple and requires action of only one type of microorganism--bacteria, *Bacillus natto*. During fermentation, *B. natto* bacteria grow, multiply, and sporulate. One of the most remarkable features of the genus *Bacillus* is the secretion of various extracellular enzymes, including protease, amylase, γ-glutamyltranspeptidase (GTP), levansucrase, and phytase. As natto bacilli grow, the enzymes that they secrete catalyze many reactions that lead to production of the characteristic sticky material as well as formation of the characteristic aroma and flavor. The viscous material consists of polysaccharide (a levan-form fructan) and γ-polyglutamic acid (Hara et al., 1982). The latter contains D- and L-glutamate in varying proportions, depending on the amount of manganese ion and the type of amino acids in the media.

During natto fermentation, there are no significant changes in fat and fiber contents and in fatty acid composition of soy lipids; however, the soluble carbohydrates, such as sucrose, raffinose, and stachyose, almost completely disappear. Citric acid, the major organic acid in steamed soybeans also disappears. At the same time, many volatile components, which contribute the characteristic aroma and flavor of natto, are produced by the natto bacteria (Kanno & Takamatsu, 1987).

Indonesia Tempeh

Tempeh, or tempe in some literature, is made by fermenting dehulled and cooked soybeans with mold, *Rhizopus* sp. Freshly prepared tempeh is a cake-like product, covered and penetrated completely with white mycelium, and has a clean, yeasty odor. When sliced and deep-fat fried, it has a nutty flavor, pleasant aroma, and crunchy texture, serving as a main dish or meat substitute. In recent years, tempeh has been found to provide some health benefits, including antimicrobial and antioxidant effects and protection against diarrhea (Hachmeister & Fung, 1993; Nout & Kiers, 2005).

Tempeh is widely believed to have originated in Indonesia many centuries ago. Although relatively unknown in the surrounding countries, such as Thailand, China, and Japan, where soybeans form an important part of the diet, tempeh continues to be one of the most popular fermented foods in Indonesia. Because of its meat-like texture and mushroom flavor, tempeh is well suited to Western tastes. It is becoming a popular food for a number of vegetarians in the United States and other parts of the world (Nout & Kiers, 2005).

Processing Method

Traditionally, making tempeh is a household art in Indonesia. The method of preparation varies from one household to another (Hachmeister & Fung, 1993), but the principle steps are basically same (Fig. 14.8). Soybeans are cleaned and then boiled in water for 30 min before hand dehulling. The dehulled beans are soaked overnight to allow full hydration and lactic acid fermentation. The soaked, dehulled beans are cooked again for 60 min, drained using woven-bamboo baskets, and spread on a flat surface for cooling to room temperature. In some cultures, soybeans are soaked in water until the hulls can be easily removed by hand or feet and washed away with water, and then boiled until soft, normally for at least 30 min. This avoids twice-cooking procedures.

The treated beans are inoculated with a traditional starter known as *usar* or an inoculum from a previous batch, both containing *R. oligosporus* spores. The mixture is wrapped in banana leaves or perforated plastic bags, approximately 100 g per package. Fermentation is allowed to occur at room temperature for up to 18 h, or until the beans are bound together by white mycelium. Alternatively, inoculated beans are spread on shallow aluminum foil or mental trays with perforated bottoms and covered with layers of banana leaves, waxed paper or plastic films, which are also perforated.

Processing Principles

There are many critical aspects of tempeh fermentation, including temperature, pH, and chemical composition of the soybean substrate. All of these are brought about by microbial growth and enzyme actions.

During the initial stage of fermentation, the mold spores germinate, and the temperature of the mass rises gradually. Subsequently, the mold grows rapidly, reaches a

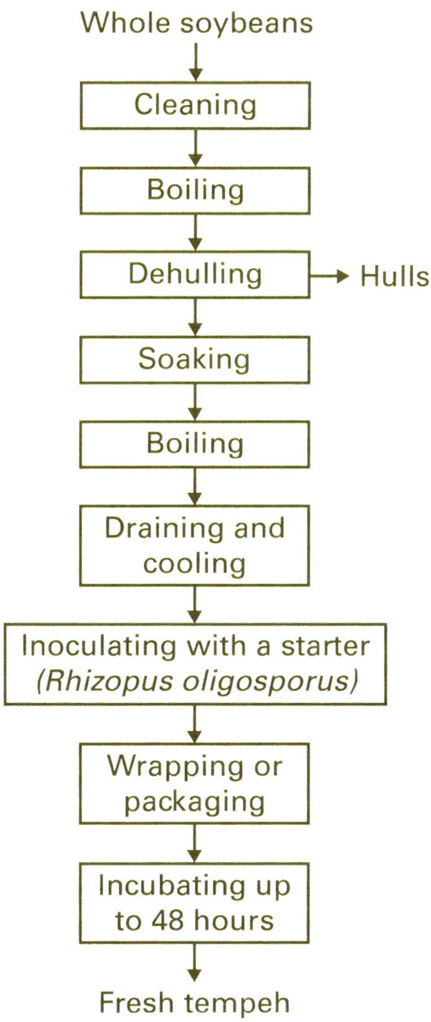

Fig. 14.8. A traditional Indonesian method for making tempeh.

peak, and then gradually subsides. Concomitant with mold growth is the rise and fall of the temperature. At the peak, the temperature may be as high as 45°C. By this time, the beans are already knitted into a compact mass by mold mycelia and the tempeh is ready to be harvested. Beyond this stage, the mold sporulates and NH_3 is produced due to protein breakdown (Hachmeister & Fung, 1993).

As microorganisms grow, they produce various enzymes, which breakdown soybean components. This leads to compositional changes. Compared with miso and soy sauce, these changes are much less vigorous due to limited production of enzymes by the tempeh mold. In comparison between tempeh and unfermented dehulled soybeans, there are increased levels of free amino acids and free fatty acids, a slight decrease in oil content, and no significant changes in protein and ash contents. In addition, there are significant increases in contents of several vitamins, including riboflavin, vitamin B_6, nicotinic acid, pantothenic acid, biotin, and folacin. Some of these increases are several-fold, although thiamin was found to change little (Murata et al., 1967). Furthermore, in some tempeh, vitamin B_{12} forms, which is widely attributed to the presence of certain contaminating bacteria, mainly *Klebsiella* (Liem et al., 1977).

Fermented Soymilk

Certain lactic acid producing bacteria have the ability to grow on dairy milk to produce various types of fermented dairy products, including acidophilus milk (sour milk), cultured buttermilk, yogurt, cheese, and other cultured milk products. The microbial action not only increases the shelf-life and nutritional value of these products, but also makes them more pleasant to eat or drink.

Soymilk resembles dairy milk in composition, so it can also be fermented by lactic acid producing bacteria to produce such products as sour soymilk and soy yogurt. Fermentation of soymilk offers not only a means of preserving soymilk but also a possibility for modifying or improving its flavor and texture so that it becomes more acceptable to Westerners. It also leads to new types of soy products, which resemble cultured dairy products but are at a low cost.

Certain lactic acid producing bacteria. such as *L. acidophilus, S. thermophilus, L. cellobiosis, L. plantarum,* and *L. lactis* have been shown to grow well in soymilk but produce less acid in soymilk than in cow's milk (Mital & Steinkraus, 1979). The major reason is that soymilk lacks monosaccharides and the disaccharide lactose. Instead, it contains such sugars as sucrose, raffinose, and stachyose, which are not readily digestible by many lactic starters due to lack of α-galactosidase in these organisms. Therefore, numerous efforts have been made to increase acid production during lactic fermentation of soymilk. These efforts include selecting culture strains, altering processing conditions during soymilk preparation, adding fermentable sugars, and/or enriching with dairy ingredients (Mital & Steinkraus, 1979; Chopra & Prasad, 1990; Omogbai et al., 2005). During fermentation of soymilk with lactic acid producing bacteria, isoflavone glucosides are converted to aglycones (Chien et al., 2006).

Fermented Tofu

When fresh tofu is fermented with a strain of certain fungi, such as *Mucor hiemalis* or *Actinomucor elegans*, it becomes a new product known as *sufu* or Chinese cheese. The product, known as *doufu ru* or *furu* in mandarin Chinese, and *toufu ju* or *fuju* in Cantonese, consists of tofu cubes covered with white or yellowish-white fungous mycelia, having firm texture, salty taste, and characteristic flavor. Although relatively unknown in some adjacent countries, such as Japan and Korea, sufu has been produced in China long before the *Ching* Dynasty and consumed mainly as an appetizer or relish by all segments of the Chinese people, including those living oversea.

Preparation Method

Preparation methods vary with type of sufu and region but all involve three basic steps; preparing tofu, molding (first fermentation), and brining (second fermentation) (Shi & Ren, 1993). Firm tofu is prepared and then cut into uniform dices with a size ranging from 2.5 to 7.5 cm (1 to 3 in.). In natural fermentation, the dices are arranged on woven bamboo trays or on rice straw in some places. The trays are placed in direct sunlight for at least several hours to let solar radiation naturally kill many unwanted microorganisms in tofu dices, and then they are stacked on shelves in an incubation room at 20–35°C. The molds already inhabiting the trays or rice straw begin to inoculate the tofu naturally. In pure culture fermentation, freshly cut tofu cubes are first immersed in an acid-saline solution (6% NaCl plus 2.5% citric acid) for 1 h and then subjected to sterilization at 100°C for 15 min. The cubes are separated from one another in a tray with small openings in the bottom and top to facilitate the circulation of air. This helps mycelium development on all sides of the cubes. After cooling, the cubes are then inoculated over their surfaces with a suspension or dried powder of spores from a selected, pure cultured microorganism.

The inoculated tofu dices, either naturally or with a pure culture, are now transferred to an incubation room. The recommended temperature and relative humidity are 25–30°C and ~97%, respectively. After several days of incubation, depending on the temperature and the type of culture, each cube is covered with a fragrant cotton-like mycelium. This intermediate product is known as *pehtze*. To ensure good formation of dense and thick texture of a mycelial mat, the mycelium on the pehtze is normally rubbed flat before pehtze are salted, seasoned, and aged in earthware crocks with brine. The crocks are tightly sealed, and put in a cool, dark place for several weeks or even several months.

The brine may contain different types of flavorings and colorings. This results in different types of sufu products. A typical brine is one containing 12% NaCl and rice wine (~10% ethanol). The end product has a characteristic flavor and color, and very salty taste. Some varieties may have putrid flavor and be objectionable to some individuals. In commercial production, the product is finally bottled with brine, sterilized, and marketed.

Processing Principles

Making sufu requires two stages of fermentation. The first stage allows molds to grow as much mycelia as possible. At the same time, various enzymes are produced, including lipases and proteases. The second stage of fermentation, known as brine aging, promotes major biological changes since many enzymes produced during the first fermentation are now released into the brine and become active. This process results in increased total soluble nitrogen, decreased total insoluble nitrogen, and increased free fatty acids, although total lipids and nitrogen remain unchanged. Many volatile compounds are produced (Chung et al., 2005). There are also changes in total amino acid and free amino acid profiles (Han et al., 2004).

Soy Nuggets

Soy nuggets, or *douchi* in mandarin Chinese, *toushih* in Cantonese, and *hamanatto* in Japanese, is made by fermenting whole soybeans with strains of *Aspergillus oryzae*, although some other strains of fungi or bacteria may also be responsible. The finished product (Fig. 14.9) has flavor similar to *jiang* or soy sauce. Because of its black color, it is also known as salted black beans in the West. Soy nuggets are commonly used as an appetizer with bland food, or as a flavoring agent cooked with vegetables, meats, and seafood.

Fig. 14.9. Chinese douchi (soy nugguts or fermented whole soybeans).

Developed in China before the Han dynasty (206 BCE), the soy nugget is considered to be the progenitor of many types of fermented soy pastes and soy sauces. It was the first soyfood to be described in written records. The product continues to be popular in China and certain regions of Japan. Similar products are also produced in Philippines (known as *tao-si*) and Eastern India (known as *tao-tjo*).

Due to variations in preparation methods, soy nuggets vary in texture, taste, salt, and moisture contents from country to country. For example, Japanese hamanatto is softer in texture and higher in moisture content compared with the Chinese counterpart, *douchi*. *Tao-tjo* in India has a sweet taste because sugar is often added to the brine. Even among the Chinese soy nuggets, there are many varieties. Based on raw material, soy nuggets are classified as those made from yellow soybeans and those from black beans. Based on taste, there are plain (less salty), salty, and wine types. Based on microorganisms involved, there are *Mucor*-type, *Aspergillus*-type, and bacterial-type (Shi & Ren, 1993).

Chinese Douchi

The methods of preparing soy nuggets may vary with region, but the essential features are similar to a traditional method that has handed down from generation to generation (Shi & Ren, 1993). Whole soybeans are soaked for 5–6 h and then steamed or boiled in water until soft. The cooked beans are inoculated either naturally or with a koji starter. When using natural inoculation, the predominance of a specific microorganism depends on incubation conditions. When air exchange is sufficient and incubated at 5–10°C for 15–20 days, *Mucor* sp. thrives, whereas incubating at 26–30°C for 5–6 days suits the growth of *Aspergillus* sp., and incubating at 20°C for 3–4 days with a covering of rice stock or pumpkin leaves promotes growth of *Bacillus* sp.

After the koji is mature, as evidenced by the appearance of abundant mycelium and spores, it is washed with water. Washing helps remove extra mycelium, spores, as well as contaminants, and ensures that the finished product is glossy and free of mold odor and bitter taste. After washing, the koji is mixed with 40–47% water, salt, and spices before being put into a jar for fermentation. Fermentation is normally carried out under natural conditions for several months. Sometimes, soy sauce is used in place of brine. In other regions, spices, wine, or sugar may be added at this stage. The aged beans become wet soy nuggets, which may be further dried to make dry soy nuggets. The finished product consists of intact beans with blackish color and has a salty taste and soy sauce flavor. Because of the relatively high salt and low water contents, the product can be kept for a long time.

Japanese Hamanatto

In Japan, a similar product known as *hamanatto* is produced, especially in the vicinity of *Hamanatsu*, Shizuoka Prefecture, from which the name of the product was perhaps derived (Watanabe & Kishi, 1984). Soybeans are soaked and steamed until soft. They are drained and cooled before being mixed with parched wheat flour in a soybean: wheat ratio of 2:1. The mixture is inoculated with a strain of *A. oryzae* (koji starter) and then distributed among shallow wooden boxes. After fermentation at 30–35°C for up to 50 h, the beans are dried in the sun until the moisture content of the mixture decreases from the original 30–35% to 20–25%. They are then covered with brine (15° Baume) and allowed to age under pressure in a tank for several weeks or months. The aged beans are dried in the sun. Sometimes, ginger pickled in soy sauce may be added after drying. The finished product normally contains about 10% salt and 38–38% water and can be stored for long time. An apparent difference in making Chinese douchi and Japanese hamanatto is that the koji for the former is made from only cooked soybeans while the koji for the latter is made from a mixture of cooked soybeans and parched wheat flour.

Conclusion

This chapter has covered various types of traditional soyfoods, nonfermented or fermented, made from whole soybeans, with respect to their variety, preparation methods, and principles. Additional Information on the subject can be found in Shurtleff and Aoyagi (1976, 1979, 2001), Watanabe and Kishi (1984), Shi and Ren (1993), Imram (2003), and Liu (1999, 2005).

In making any type of soyfoods, as we are moving from traditional arts to modern science and it is important to follow Good Manufacturing Practices (such as the ones outlined in FDA's 21 CFR 110 regulations). Because recalls of modern prepared soyfoods still occur due to microbial, chemical, and physical hazards, improving the sanitary conditions during soyfood preparation is just as important as learning the science behind it.

References

Akazawa, T.; E. Shiraiwa; N. Satoh; T. Sasahara. Variations in eating quality, flavor, and pod color of green soybeans from vegetable-type and grain-type plants stored at various temperatures. *Jap. J. Crop Sci.* **2002,** *71,* 62–67.

Bates, R.P.; R.F. Matthews. Ascorbic acid and ß-Carotene in soybeans as influenced by maturity, sprouting, processing and storage. *Proc. Fla. State Hort. Soc.* **1975,** *88,* 266–271.

Beddows, C.G.; J. Wong. Optimization of yield and properties of silken tofu from soybeans. I. The water:bean ratio. *Int'l. J. Food Sci. Technol.* **1987a,** *22,* 15–21.

Beddows, C.G.; J. Wong. Optimization of yield and properties of silken tofu from soybeans. Ii. Heating processing. *Int'l. J. Food Sci. Technol.* **1987b,** *22,* 23–27.

Bentley, R. From miso, sake and shoyu to cosmetics: a century of science for kojic acid. *Natural Product Rep.* **2006,** *23,* 1046–1062.

Chen, S. Preparation of Fluid Soymilk. In *Proceedings of the World Congress on Vegetable Protein Utilization in Human Foods and Animal Feedstuffs.* Applewhite, T.H. Ed.; AOCS Press: Champaign,

IL, 1989; pp. 341–351.

Chien, H.L.; H.Y. Huang; C.C. Chou. Transformation of isoflavone phytoestrogens during the fermentation of soymilk with lactic acid bacteria and bifidobacteria. *Food Microbiol.* **2006,** *23,* 772–778.

Choi, Y.M.; D.O. Noh; K.W. Yu; J.H. Koh; H.J. Suh. Physiological activities of chungkookjang (Korean traditional soybean paste) pill as a functional food. *J. Food Biochem.* **2007,** *31,* 108–120.

Chopra, R.; D.N. Prasad. Soymilk and lactic fermentation products–a reveiw. *Micro. Ali. Nutr.* **1990,** *8,* 1–13.

Chung, H.Y.; P.K. Fung; J.S. Kim. Aroma impact components in commercial plant sufu. *J. Agric. Food Chem.* **2005,** *53,* 1684–1691.

Cruz, N.; M. Capellas; M. Hernandex; A.J. Trujillo; B. Guamis; V. Ferragut. Ultra high pressure homogenization of soymilk: microbiological, physicochemical, and microstructural characteristics. *Food Res. Int.* **2007,** *40,* 725–732.

Davies, C.S.; S.S. Nielsen; N.C. Nielsen. Flavor improvement of soybean preparations by genetic removal of lipoxygenase-2. *J. Am. Oil Chem. Soc.* **1987,** *64,* 1428–1433.

deMan, J.M.; L. deMan; S. Gupta. Texture and microstructure of soybean curd (Tofu) as affected by different coagulants. *Food Microstruct.* **1986,** *5,* 83–89.

Fukuda, M.; Y. Sugihara; M. Itou; R. Horiuchi; H. Asao. Effects of feeding with okara on plasma and liver lipid levels in rats. *J. Jap Soc. Food Sci. Technol.* **2006,** *53,* 195–199.

Golbitz, P. Demand, Availability drive new growth in soyfoods market. *J. Am. Dietetic Assoc.* **2002,** *102,* 1–5.

Hachmeister, K.A.; D.Y.C. Fung. Tempeh: a mold–modified indigenous fermented food made from soybeans and/or cereal grains. *Crit. Rev. Microbiol.* **1993,** *19,* 137–188.

Han, B.Z.; F.M. Rombouts; M.J.R. Nout. Amino acid profiles of sufu, a Chinese fermented soybean food. *J. Food Comp. Anal.* **2004,** *17,* 689–698.

Hara, T.; Y. Fujio; S. Ueda. Polyglutamate [roduction by *Bacillus subtilis* (Natto). *J. Appl. Biochem.* **1982,** *2,* 112–120.

Hondo, S.; T. Mochizuki. Studies on the degradation process of soybean protein during miso making. Part II. Changes in peptides during miso making. *Nippon Shokuhin Kogyo Gakkaishi* **1968,** *15,* 414.

Hou, H.J.; K.C. Chang. Storage conditions affect soybean color, chemical composition and tofu qualities. *J. Food Proc. Preserv.* **2004,** *28,* 473–488

Ikeda, K.; Y. Matsuda; A. Katsumaru; M. Teranishi; T. Yamamoto; M. Kishida. Factors affecting protein digestibility in soybean foods. *Cereal Chem.* **1995,** *72,* 401.

Imram, N. *Soya Handbook.* Tetra Pak: Singapore. 2003.

Johnson, L.A.; C.W. Deyoe; W.J. Hoover. Yield and quality of soymik processed by steam-infusion cooking. *J. Food Sci.* **1981,** *46,* 239–243, 248.

Jung, K.O.; K.M. Kwang; J.H. Kil; Y.E. Lee; K.Y. Park. Increased antimutagenic and anticlastogenic effect of doenjang (Korean fermented soybean paste) prepared with bamboo salt. *FASEB J.* **2007,** *21,* A1078–A1078.

Kanno, A.; H. Takamatsu. Changes in the volatile components of "natto" during manufacturing and storage. *Nippon Shokunhin Kogyo Gakkaishi* **1987,** *34,* 330–335.

Kataoka, S. Functional effects of japanese style fermented soy sauce (shoyu) and its components. *J. Biosci. Bioeng.* **2005,** *100,* 227–234.

Kim, Y.; L. Wicker. Soybean cultivars impact quality and function of soymilk and tofu. *J. Sci. Food. Agric.* **2005,** *85,* 2514–2518.

Kitamura, K. Genetic improvement of nutritional and food process quality in soybean. *Jap. Agric. Res. Quant.* **1995,** *29,* 1–8.

Kobayashi, M. Immunological functions of soy sauce: hypoallergenicity and antiallergic activity of soy sauce. *J. Biosci. Bioeng.* **2005,** *100,* 144–151.

Kosikowski, F.V. Nutritive and organoleptic characteristics of non–dairy imitation milks. *J. Food Sci.* **1971,** *36,* 1021.

Kwok, K.C.; K. Niranjan. Effect of thermal processing on soymilk. *Intl. J. Food Sci. Technol.* **1995,** *30,* 263–265.

Lakshmanan, R.; M. de Lamballerie; S. Jung. Effect of soybean-to-water ratio and pH on pressurized soymilk properties. *J. Food Sci.* **2006,** *71,* E384–E391.

Lertsiri, S.; I.K. Phontree; W. Thepsingha; A. Bhumiratana. Evidence of enzymatic browning due to laccase–like enzyme during mash fermentation in thai soybean paste. *Food Chem.* **2003,** *80,* 171–176.

Liem, I.T.H.; K.H. Steinkraus; T.C. Cronk. Production of vitamin B_{12} in tempeh, a fermented soybean food. *Appl. Environ. Microbiol.* **1977,** *34,* 773.

Liu, Z.S.; S.K.C. Chang. Effect of soy milk characteristics and cooking conditions on coagulant requirements for making filled tofu. *J. Agric. Food Chem.* **2004,** *52,* 3405–3411.

Liu, K.; F. Orthoefer; K. Thompson. The case for food–grade soybean varieties. *INFORM* **1995,** *6,* 593.

Liu, K. Immature soybeans, nutritional quality and organoleptic feature. *INFORM* **1996,** *7,* 1217–1223.

Liu, K. *Soybeans: Chemistry, Technology, and Utilization.* Aspen Publisher: Gaithersburg, MD. 1999.

Liu, K. *Soybeans as Functional Foods and Ingredients,* AOCS Press: Champaign, IL, 2005a.

Liu, K. Soy Sauce as Natural Seasoning. In *Soybeans as Functional Foods and Ingredients.* Liu, K. Ed.; AOCS Press: Champaign, Illinois, 2005b; pp. 248–263.

Martinez–Villaluenga, C.; Y.H. Kuo; F. Lambein; J. Frias; C. Vidal–Valverde. Kinetics of free protein amino acids, free non–protein amino acids, and trigonelline in soybean (*Glycine max* L.) and lupin (*Lupinus angustifolius* L.) sprouts. *Euro. Food Res. Technol.* **2006,** *224,* 177–186.

Matsumoto, K.; Y. Watanabe; S. Yokoyama. Okara, soybean residue, prevents obesity in a diet–induced murine obesity model. *Bosci. Biotech. Bochem.* **2007,** *71,* 720–727.

Messina, M. Soyfoods and cholesterol reduction. *J. Am. Dietetic Assoc.* **2007,** *107,* 1–5.

Mine, Y.; K. Murakami; K. Azuma; S. Yoshihara; K. Fukunage; T. Saeki; E. Sawano. A comparison of various coagulants in tofu–forming properties. *J. Jap. Soc. Food Sci. Techol.* **2005,** *52,* 114–119.

Mital, B.K.; K.H. Steinkraus. Fermentation of soymilk by lactic acid bacteria, a review. *J. Food Prot.* **1979,** *42,* 895–899.

Mohamed, A.; R.S. Mentreddy. Vegetable Soybeans as Functional Food. In *Soybeans as Functional Foods and Ingredients;* Liu, K. Ed.; AOCS Press: Champaign, Illinois, 2005; pp. 209–238.

Murata, K.; H. Ikehata; T. Miyamoto. Studies on the nutritional value of tempeh. *J. Food Sci.* **1967,** *32,* 580.

Nelson, A.I.; M.P. Steinberg; L.S. Wei. Illinois process for separation of soymilk. *J. Food Sci.* **1976,** *41,* 57–61.

Nikkuni, S.; N. Okada; H. Itoh. Effect of soybean cooking temperature on the texture and protein digestibility of miso. *J. Food Sci.* **1988,** *53,* 445.

Nout, M.J.R.; J.L. Kiers. Tempeh as a Functional Food. In *Soybeans as Functional Foods and Ingredients.* Liu, K. Ed.; AOCS Press: Champaign, Illinois, 2005; pp. 239–247.

Ogasawara, M.; Y. Yamada; M. Egi. Taste enhancer from the long-term ripening of miso (Soybean Paste). *Food Chem.* **2006,** *99,* 736–741.

Ohara, T.; H. Karasawa; T. Matsuhashi. Relationships of coagulation characteristics and properties of "kori-tofu" in a controlled soymilk coagulation system. *J. Jap. Soc. Food Sci. Technol.* **1992,** *39,* 543–554.

Omogbai, B.A.; M.J. Ikenebomeh; S.I. Ojeaburu. Microbial utilization of starchyose in soymilk yogurt production. *African J. Biotech.* **2005,** *4,* 905–908.

Park, H.K.; B. Gill; J.K. Kim. Characteristic flavor compounds of commercial soybean paste. *Food Sci. Biotechnol.* **2003,** *12,* 607–611.

Poysa, V.; L. Woodrow; K. Yu. Effect of soy protein subunit composition on tofu quality. *Food Res. Intl.* **2006,** *39,* 309–317.

Prabhakaran, M.P.; C.O. Perera. Effect of extraction methods and UHT treatment conditions on the level of isoflavones during soymilk manufacture. *Food Chem.* **2006,** *99,* 231–237.

Qin, L.A.; L.Y. Xu; P.Y. Wang; K. Hoshi. Soyfood intake in the prevention of breast cancer risk in women: a meta–analysis of observational epidemiological studies. *J. Nutri. Sci. Vitaminology* **2006,** *52,* 428–436.

Ranaldi, V.E.A.; P.K.W. Ng; M.R. Bennik. Effects of extrusion on dietary fiber and isoflavone contents of wheat extrudates enriched with wet okara. *Cereal Chem.* **2000,** *77,* 237–240.

Saio, K. Tofu–relationships between texture and fine structure. *Cereal Foods World.* **1979,** *24,* 343.

Shi, Y.G.; L. Ren. *Soybean Products Technology.* China's Light Industry Publisher: Beijing, China. 1993.

Shih, M.C.; H.J. Hou; K.C. Chang. Process optimization of soft tofu. *J. Food Sci.* **1997,** *62,* 833–837.

Skurray, G.; J. Cunich; O. Carter. The effect of different varieties of soybean and calcium ion concentration on the quality of tofu. *Food Chem.* **1980,** *6,* 89–95.

Surel, O.; B. Couplet. Influence of the dehydration process on active compounds of okara during its fractionation. *J. Sci. Food Agric.* **2005,** *85,* 1343–1349.

Shurtleff, W.; A. Aoyagi. *The Book of Tempeh.* 2nd Edn. Ten Speed Press: Berkeley, California, 2001.

Shurtleff, W.; A. Aoyagi. *The Book of Miso.* Ten speed press: Berkeley, California, 1976.

Shurtleff, W.; A. Aoyagi. *Tofu and Soymilk Production.* The Soyfoods Center: Lafayette, California, 1979.

Tang, C.H. Effect of thermal pretreatment of raw soymilk on the gel strength and microstructure of tofu induced by microbial transglutaminase. *LWT–Food Sci. & Technol.* **2007,** *40,* 1403–1409.

Toda, K.; K. Chiba; T. One. Effect of components extracted from okara on the physicochemical properties of soymilk and tofu texture. *J. Food Sci.* **2007,** *72,* C108–C113.

Toda, K.; K. Takahashi; T. One; K. Kitamura; Y. Nakamura. Variation in the phytic acid content of soybeans and its effect on consistency of tofu made from soybean varieties with high protein content. *J. Sci. Food Agric.* **2006,** *86,* 212–219.

Torres–Penaranda, A.V.; C.A. Reitmeier; L.A. Wilson; W.R. Fehr; J.M. Narvel. Sensory characteristics of soymilk and tofu made from lipoxygenase–free and normal soybeans. *J. Food Sci.* **1998,** *63,* 1084–1087.

Tripathi, A.K.; A.K. Misra. Soybean–a consummate functional food: a review. *J. Food Sci. Technol. Mysore* **2005,** *42,* 111–119.

van der Riet, W.B.; A.W. Wight; J.J. Cilliers; J.M. Datel. Food chemical investigation of tofu and its byproduct okara. *Food Chem.* **1989,** *34,* 193–202.

Wang, H.L.; E.W. Swain; W.F. Kwolek. Effect of soybean varieties on the yield and quality of tofu. *Cereal Chem.* **1983,** *60,* 245.

Wang, L.J.; D. Li; E. Tatsumi; Z.S. Liu; X.D. Chen; L.T. Li. Application of two stage ohmic heating to tofu processing. *Chem. Eng. Proc.* **2007,** *46,* 486–490.

Wang, R.; X. Zhou; Z.X. Chen. High pressure inactivation of lipoxygenase in soy milk and crude soybean extract. *Food Chem.* **2008** *106,* 603–611.

Watanabe, H.; N. Kashimoto; J. Kajimura; K. Kamiya. A miso (Japanese soybean paste) diet conferred greater protection against hypertension than a sodium chloride diet in dahl salt–sensitive rats. *Hypertension Res.* **2006,** *29,* 731–738.

Watanabe, T.; A. Kishi. *Nature's Miracle Protein, The Book of Soybeans.* Japan Publications, Inc.: Tokyo and New York. 1984.

Wilkens, W.F.; L.R. Mattick; D.B. Hand. Effect of processing method on oxidative off-flavor of soybean milk. *Food Technol.* **1967,** *21,* 86.

Wu, L.C.; R.P. Bates. Soy protein-lipid films. 1. Studies on the film formation phenomenon. *J. Food Sci.* **1972,** *37,* 36.

Xu, M.J.; J.F. Dong; M.Y. Zhu. Effects of germination conditions on ascorbic acid level and yield of soybean sprouts. *J. Sci. Food Agric.* **2005,** *85,* 943–947.

Yamiguechi, M.; E. Sugimoto; S. Hachiya; Y. Tsukamoto. Stimulatory effect of menaquinone-7 (vitamin K_2) on osteoclastic bone formation in vitro. *Mol. Cell Biochem.* **2001,** *233,* 131–137.

Yanagisawa, Y.; H. Sumi. Natto bacillus contains a large amount of water-soluble vitamin K (Menaquinone-7). *J. Food Biochem.* **2005,** *29,* 267–277.

Yokotsuka, T. Soy Sauce Biochemistry. *Adv. Food Res.* **1986,** *30,* 196–329.

Yoon, H.H.; M. Kim. Physicochemical and sensory properties of tofu prepared with heat-treated soybeans. *J. Text. Studies* **2007,** *38,* 393–403.

Young, G.; T. Mebrahtu; J. Johnson. Acceptability of green soybeans as a vegetable entity. *Plant Foods Human Nutri.* **2000,** *55,* 323–333.

15

Food Uses for Soybean Oil and Alternatives to *Trans* Fatty Acids in Foods

Kathleen A. Warner
USDA, ARS, NCAUR, Food and Industrial Oil Research Unit, Peoria, IL 61604

The approximately 20% oil in the soybean seed is considered a co-product of the protein fraction; however, that oil fraction is a valuable commodity itself. In the United States, soybean oil products account for approximately 80% of the 18 billion lb of edible oils used every year. Home consumers buy liquid soybean salad oil sold in grocery stores for cooking, baking, and salad dressings. They also consume liquid oil incorporated into processed foods such as packaged mixes and prepared salad dressings. In contrast, restaurants and food manufacturers have traditionally used hydrogenated soybean oil (solid or pourable) for frying, baking, and processing into margarines and shortenings. The uses and amounts of soybean oil consumed in the United States. changed significantly over the past 45 yr. Fig. 15.1 shows that approximately 3 billion lb of soybean oil were utilized for food in 1960 with one-third each for salad/cooking oils, shortenings, and margarines (Anonymous, 2007). By 1987, about 10 billion lb of soybean oil for food applications were utilized per year (Bagby & Carlson, 1989). However, by 2005, the total soybean oil utilization had increased to 17 billion lb (Anonymous, 2007) with most of it divided equally between shortenings and salad/ cooking oils but with only a small amount for margarines that had not increased

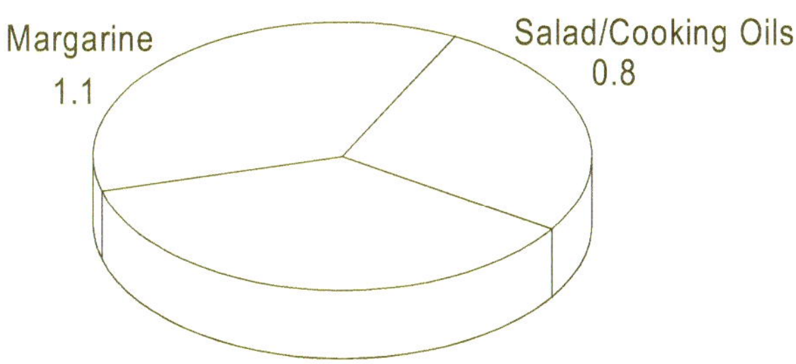

Fig. 15.1. Utilization of soybean oil in 1960.

from the 1960 level (Fig. 15.2). This chapter outlines the various food uses of this important oil including liquid soybean salad oils and solid/semi-solid fats. In addition, information is included on the soybean oils with modified fatty acid compositions that were developed over the past 20 yrs to improve oxidative stability, functionality, and/or nutritional properties.

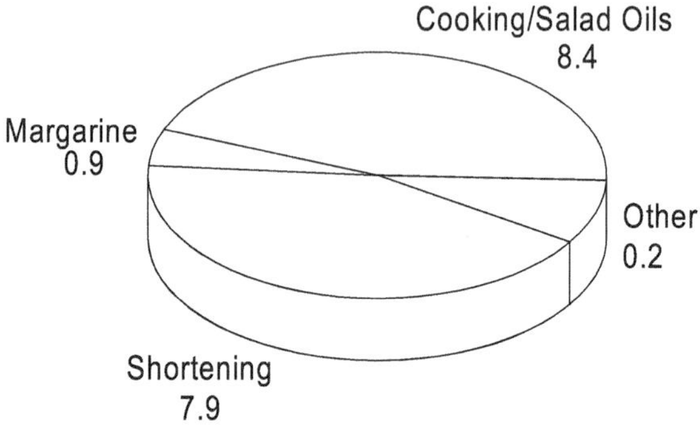

Fig. 15.2. Utilization of soybean oil in 2005.

Types of Soybean Oils

Liquid Soybean Oil

Soybean salad oil is liquid at room temperature (25°C) and at refrigeration temperature (4°C) because of its high linoleic acid fatty acid composition (Table 15.1). It is processed by refining, bleaching, and deodorizing (RBD) crude/degummed oil to remove color and off-flavor comlb. The RBD oil is a light-colored, bland product that is used for many food applications such as salad dressings, baking, and pan frying by the home consumer. Many years of research on the flavor problems of soybean oil (Dutton et al., 1951) resulted in an excellent quality soybean oil product when it is processed under standard procedures (Brekke, 1980). In fact, stabilizing options such as antioxidants and hydrogenation are not used for RBD soybean oil sold in grocery stores in the United States.

Table 15.1. Examples of Fatty Acid Compositions (%) of Regular and Hydrogenated Soybean Oils

Fatty Acid	Soybean	Hydrogenated
C16:0	9.7	13.1
C18:0	3.5	7.5
C18:1	25.2	67.9
C18:2	55.2	11.6
C18:3	6.4	0.8

In addition to the soybean salad oil that is bottled for consumer use, this oil is used by food manufacturers in a wide variety of products. Pourable and spoonable salad dressings routinely contain soybean salad oil rather than hydrogenated oil; however, some applications may require a higher level of oxidative stability than provided by soybean salad oil. Warner et al. (1986) found that substituting hydrogenated soybean oil for unhydrogenated soybean oil significantly improved the storage stability of a starch-based salad dressing (35% oil) at 21°C but not at 32°C. If the use of hydrogenated oil is not suitable, then other options are available. Warner et al. (1986) also reported that the use of BHA, ethylenediaminetetraacetate (EDTA) as a metal chelator, and packaging under nitrogen, was effective in extending the shelf life of the dressings made with unhydrogenated soybean oil. Moustafa (1995) thoroughly reviewed the preparation and stability of salad dressings. In addition to salad dressings, soy oil is used in cereals and packaged mixes sold on the retail market.

Solid and Semi-solid Soybean Oils

Although good quality soybean salad oil is achieved by oil processors because of proper processing and packaging, this oil does not have the functionality properties for texture and mouthfeel needed for margarines and shortenings. The solution to this problem in the past was to hydrogenate the oil, which changes its melting point for margarines and shortening applications. However, recent changes in the labeling of foods regarding declaration of the amount of trans fatty acids in the product caused the food industry to look for alternatives to hydrogenation to create oils for margarines and shortenings. The Chapters *Lipids* and *Nutritional Properties and Feeding Values of Soybeans and their Coproducts* discuss the nutritional implications of *trans* fatty acids. Liquid soybean oil is the starting point for most of the other edible forms of hydrogenated oil. Processes such as hydrogenation and interesterification can change the liquid oil into a product with a solid or semi-solid (pourable) consistency. These processes are discussed in detail in the Chapter: *Soybean Oil Modification.*

Changing the Fatty Acid Composition of Oils

The oil products from this major oilseed can range widely in fatty acid composition because of processing, such as hydrogenation, or from plant breeding or genetic modification. In the United States, hydrogenated oils are gradually being replaced by oils with fatty acid compositions modified by traditional plant breeding techniques and by transgenic means. However, hydrogenation is still necessary at this time because few alternatives exist that provide fats/oils with the right functional properties needed to create margarines and shortenings. In addition, the supply of oilseeds modified by plant breeding is not high enough to currently meet the needs of the restaurant and food industries for frying oils.

Hydrogenation

The hydrogenation process was developed in the early twentieth century and has been used extensively in the United States for soybean oil because the traditional fatty acid profile in this oil is too oxidatively unstable for high-stability food uses such as frying. For the past 50 yr, hydrogenation was also used to convert liquid soybean oil into margarines and shortenings as substitutes for butter and lard. Hydrogenated soybean oil provides functional characteristics contributing to the proper texture and mouth-feel of margarines and shortenings (Table 15.1). Hydrogenated oils were used for over 50 yr in the United States to make many frying oils. By 2000, food manufacturers were looking for alternatives to hydrogenated oils because the U.S. Food and Drug Administration (FDA) announced mandatory labeling of trans fatty acid content in foods as of January 2006 (List et al., 2007). Oil processors continue to prepare and market hydrogenated oils especially for those applications such as margarines and shortenings for which few alternatives are available in the United States at this time.

Hydrogenated oils also have characteristic off-flavors that are described as waxy, plastic, and fruity (Neff et al., 2000). Although many consumers are familiar with these characteristic flavors in their french-fried potatoes, some manufacturers want oils that do not impart such flavors and prefer that the natural flavor of the food or the deep-fried flavor of the oil be the primary flavors. Because of these problems with hydrogenated oils, oil processors and food manufacturers are currently looking for alternatives to hydrogenation to provide oxidatively stable, healthful, and functional oils for frying and margarines.

Plant Breeding

As an alternative to modifying the fatty acid composition of oil chemically by hydrogenation, oilseeds can be genetically modified to produce oils of desired functionality, oxidative stability, and nutritional attributes. Many of the oilseeds, including soybean, sunflower, corn, peanut, and canola, have their compositions modified to increase desirable or decrease undesirable fatty acids. In the 1980s, plant geneticists began to apply results of years of research on edible oils during 1940s–1970s (Dutton, 1981), showing that decreasing linolenic acid in soybean oil improved its flavor and oxidative stability. Therefore, decreasing linolenic acid became the first target in soybeans for geneticists because this fatty acid is the most easily oxidizable and is considered to be a major cause of flavor problems in soybean oil (Dutton et al., 1951). In the 1950s, breeding soybeans to decrease linolenic acid was not considered possible because of the genetic characteristics of soy. Therefore, researchers chose to reduce the amount of linolenic acid by hydrogenation (Dutton, 1981). Later, geneticists found that they could indeed reduce linolenic acid from the 7–9% levels regularly occurring in soybean oil to less than 4% by using traditional plant-breeding methods (Mounts et al., 1988).

Applications for Oils with Modified Fatty Acid Compositions

Genetic modification and breeding of soybeans is used to not only decrease linolenic acid but also to increase oleic acid that causes a corresponding decrease in linoleic acid. Plant-breeding techniques such as natural selection and mutation breeding are used to modify fatty acid composition in other oilseeds including sunflower, peanut, corn, and canola with the primary emphasis on increasing oleic acid to levels greater than 70%. All these modifications help to increase the oxidative stability of the oils. Other types of specialty soybean oils include high-palmitic acid and high-stearic acid for margarines and shortenings. In addition, low saturated fatty acid soybean oil would reduce the saturated fat content of soybean oil from approximately 13–7% (Table 15.2).

In the 1990s, the United Soybean Board's Better Bean Initiative proposed that a high-oleic acid soybean be developed. However, they subsequently recommended that a mid-oleic acid /low linolenic acid soybean oil be developed as a new commodity oilseed.

Researchers at Iowa State University developed a soybean variety with this profile that is currently being evaluated for its oxidative stability in foods. The intended food use of an oil is a primary factor in selecting the fatty acid composition of a modified oil. Therefore, oxidative stability in frying oils and functionality in margarine or shortening products are the high priorities. An oil's susceptibility to oxidation can be approximated by using the amounts of linoleic and linolenic acids to calculate an oxidizability rating (Frankel, 2005). For example, a soybean salad oil that has 55.2% linoleic acid and 6.4% linolenic acid (Table 15.1) has an oxidizability rating of 6.8 (Table 15.3). Decreasing the linolenic acid to 3% decreases the oxidizability to only 6.2. On the other hand, the mid- and high-oleic acid soybean oils have oxidizability values of 3.2 and 0.3, respectively. These two oils should be better frying oils than the low-linolenic acid oils. Although fatty acid composition is only one of several factors affecting oxidative stability, this calculation might give some idea of the stability to expect in a food application where oxidation is a problem.

Table 15.2. Fatty Acid Compositions (%) of Modified Soybean Oils

Fatty Acids	High-Saturates	Low-Linolenic	Ultra-Low-Linolenic	Mid-oleic	High-Oleic	High-Stearic
C16:0	23.3	10.8	11.5	8.9	7.3	9.3
C18:0	20.0	4.5	4.6	5.9	3.4	27.2
C18:1	10.5	26.1	24.8	52.0	85.1	16.7
C18:2	39.7	55.4	58.2	31.2	1.3	38.6
C18:3	6.5	3.0	0.8	1.0	2.0	6.5

Table 15.3. Oxidizability of Soybean Oil Types Listed in Tables 15.1 and Table 15.2

Oil Type	Oxidizability
Soybean	6.8
Hydrogenated	1.3
Low linolenic	6.1
Mid-oleic	3.2
High oleic	0.3
Ultra low linolenic	5.9

Oxidizability = [(% linoleic acid) + 2(% linolenic acid)]/100

Salad Oils

As indicated previously in this chapter, regular soybean salad oil, if processed and stored properly, has very good oxidative stability as a salad oil. In some instances, however, extra oxidative stability may be needed. In those cases, soybean oil with a modified fatty acid profile may be appropriate, such as one with reduced linolenic acid. Several studies on soybean oil modified to reduce the linolenate content showed that this approach improved the flavor quality and oxidative stability of the soybean salad oils (Mounts et al., 1988; Liu & White, 1992a; Mounts et al., 1994a; Su et al., 2003). Decreasing the linolenic acid helps to inhibit oxidation and the development of painty flavors derived from the oxidation of linolenic acid.

Deep-fat Frying

Deep-fat frying imparts desired sensory characteristics of fried-food flavor, golden-brown color, and crisp texture in foods. During frying, oils thermally and oxidatively decompose and form volatile and nonvolatile products that alter functional, sensory, and nutritional qualities of oils. During the past 30 yr, scientists reported extensively on the physical and chemical changes that occur during frying and on the wide variety of decomposition products formed in frying oils. A small amount of oxidation in frying oils is important to develop the delicious deep-fried flavor characteristics of most fried foods (Perkins, 1996; Warner et al.,1997). However, as oils break down further because of the processes of oxidation, hydrolysis, and polymerization, comlb are formed that can cause off-flavors (Fig. 15.3). Soybean salad oil is oxidatively unstable for the high stability required in applications such as frying (Frankel, 1995). Therefore, food manufacturers need a soybean oil with lower polyunsaturated fatty acids, especially linolenic acid.

Refined, bleached, and deodorized (RBD) soybean oil used for commercial frying has been routinely hydrogenated since the 1950s to increase stability during commercial frying and subsequent storage of fried food. However, nonhydrogenated oils

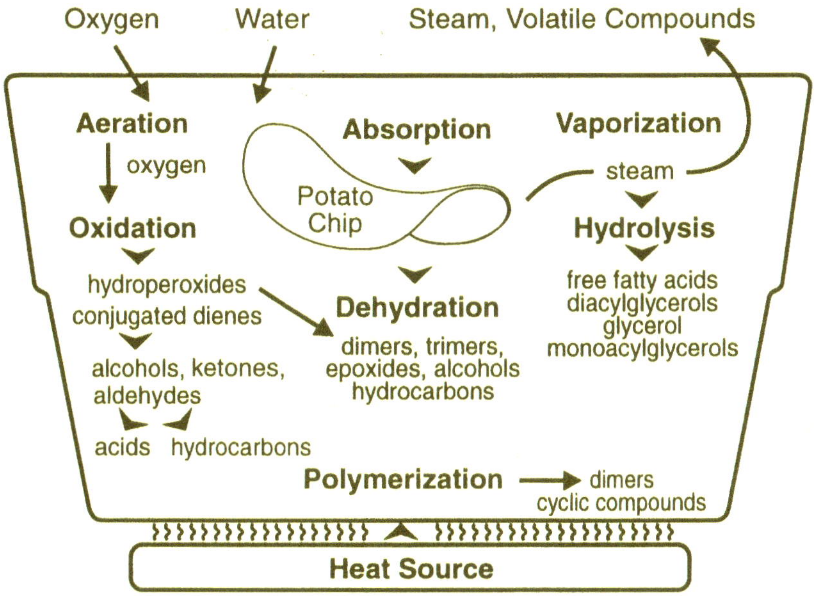

Fig. 15.3. Deterioration processes during frying.

are now available as alternatives to hydrogenated oils for frying. Because reducing the amount of linolenic acid significantly increases the oxidative stability of soybean salad oils, researchers tested this effect in frying oils. Liu and White (1992b) and Miller and White (1988) found that food fried in soybean oils with reduced linolenic content, plus increased concentration of saturated fatty acids, had better oxidative stability than regular soybean oil. Mounts et al. (1994b) reported that decreasing the linolenic content of soybean oil to 1.9% significantly decreased the fishy flavor in fried food compared to regular soybean oil. Oil with linolenic acid reduced to 3.7% produced significantly better quality french-fried potatoes and significantly lower polar comlb than did soybean oil with 6.2% 18:3 (Warner & Mounts, 1993).

Generally, the lower the linolenic acid content, the more oxidatively stable the oil. However, Warner and Gupta (2003) found that reducing the linolenic acid level in modified soybean oil from 2–0.8% only slightly improved frying oils and fried-food stability.

Although decreasing linolenic acid is important in enhancing the stability of frying oil and fried food, the oxidizability of the oil may still be too high for frying applications. Plant breeding is used to modify oleic acid in sunflower, corn, and canola oils to amounts greater than 70%. Increasing the oleic acid content is appropriate, because linoleic acid, which is easily oxidized, exists in an opposite ratio to oleic acid, that is,. as one increases, the other decreases. High-oleic oils had good stability during high-temperature heating and frying when measured by using instrumental and chemical tests, such as free fatty acids and total polar comlb (Warner et al., 1994; Warner & Knowlton,1997; Warner et al., 1997).

The oleic acid content of soybean oil was transgenically increased to 85% by DuPont. In a frying test with potato chips, high-oleic soybean oil had the best fry life as measured by total polar comlb compared to cottonseed oil and low linolenic acid soybean oil (Warner & Gupta, 2005). However, in flavor evaluations, fried foods including potato chips, tortilla chips or french-fried potatoes prepared in high-oleic oils—soybean, sunflower, and canola—had less deep-fried flavor than when prepared in oils with moderate or low-oleic acid content (Warner et al., 1994; Warner & Knowlton, 1997; Warner et al., 1997; Warner & Gupta, 2005). Frying tests with model systems of triolein and trilinolein heated to 180°C showed that triolein had predominant waxy and plastic odors (Neff et al., 2000). However, trilinolein had strong deep-fried odors. Trilinolein develops high amounts of 2, 4-decadienal, a compound that imparts the deep-fried odor; however, triolein does not develop much of this aldehyde (Warner et al., 2001). If a deep-fried flavor is important in a fried food, the oils should have at least 20–30% linoleic acid but not more than 70% oleic acid. As oleic acid content is increased in an oil, the level of linoleic acid decreases, so limiting the upper level of oleic acid keeps the linoleic acid at amounts high enough to produce a deep-fried flavor.

Margarines and Shortenings

Oils intended for use in manufacturing margarines and shortenings need specific fatty acid compositions to ensure proper functionality and mouthfeel. Although some oils have these compositions naturally, processing such as hydrogenation or interesterification is usually necessary to change the fatty acid composition of the liquid oil (List et al., 1995). List et al. (1996, 2001) used solid fat index (AOCS, 2005) to show that some soybean oils modified by plant breeding to have increased amounts of stearic acid could be used for margarines. Kok et al. (1999) used a soybean oil with high saturates (23.3% palmitic acid and 20% stearic acid) to produce margarines by interesterification.

Standards for Oils

The U.S. Department of Agriculture (USDA) quality standards for soybean oil state that soybean oil shall be clear and brilliant when held at 70–85°F and shall be free from sediment, such as metal, wood, dirt, glass, paint, insects, insect parts, or any other foreign material (Anonymous, 2005). In addition, the oil shall have a bland odor and flavor and shall be free of beany, rancid, painty, musty, metallic, fishy, putrid, or any other undesirable odor and/or flavor and have a light viscosity and no heavy oily mouthfeel. Table 15.4 lists the analytical requirements for a RBD and winterized soybean oil (Anonymous, 2005). If an oil does not have more than the maximum allowable values nor less than the minimum for the AOM, then the oil is probably of good quality.

In addition to the USDA standards, The Named Vegetable Oil Standard from the Codex Alimentarius Committee on Fats and Oils (Codex Alimentarius, 2006) identifies quality characteristics that oils must meet for international trade. For example, the standard states that an oil should be characteristic of the designated product and be free of foreign and rancid odor and taste. Maximum levels of matter volatile at 105°C (ISO 662:1998) (ISO, 2005) should be 0.2% m/m, insoluble purities (ISO 663:1998) (ISO, 2005) 0.05% m/m, and 0.005% m/m soap content [AOCS Ce17-95 (97)] (AOCS, 2005). Limits for metals are 1.5 mg/kg iron (Fe) in refined oils, whereas only 0.1 mg/kg of copper (Cu) is allowed in refined oils. For methods to measure Fe and Cu, use ISO 8294:1994 (ISO, 2005) or AOCS Ca 18b-91 (97) (AOCS, 2005). Limits for oil deterioration include peroxide value and acid value. Codex allows up to 10 meq/kg oil for refined oils. Acid value limits range from a low of 0.6 mg/KOH/g oil for refined oils.

The Named Vegetable Oil Standard also presents the compositional ranges for the 21 major oils of the world including soybeans. Compositions of crude oils extracted from known samples of seeds are given for fatty acids, tocopherols, sterols, and chemical and physical characteristics such as refractive index (Codex Alimentarius, 2006). The fatty acid composition ranges for regular soybean oil from Codex with minimum and maximum limits are presented in Table 15.5. The ranges of fatty acids for these oils for the Codex standard provide valuable information; however, they result from hundreds of authentic seed samples from throughout the world, which accounts for the wide ranges.

Table 15.4. Analytical Requirements for Refined, Bleached, Deodorized, and Winterized Soybean Oil (Brekke, 1980)

Item	Minimum	Maximum
Color Lovibond Scale		2.0 red 20.0 yellow
Moisture and Volatile Matter (percent by weight)		0.06
Insoluble Impurities		NONE
Free Fatty Acids (as % oleic, by weight)[1]		0.05
Peroxide Value (meg/kg) [1]		1.0
Stability, Active Oxygen Method (AOM) [1]	12 hours	

[1] Determination shall be made seven days after packaging.

Two modified soybean oils (low-linolenic acid and mid-oleic/low linolenic) were proposed for inclusion in the Codex standard at the February 2005 meeting of the Fats and Oils Committee; however, the decision to include these oils is still pending and will probably be based on the volume of these oils in future international trade.

Table 15.5. Fatty Acid Compositions of Crude Regular and Modified Soybean Oils

Fatty Acid	Soybean	Soybean (low-linolenic acid)	Soybean (mid-oleic acid)
6:0	ND	ND	ND
8:0	ND	ND	ND
10:0	ND	ND	ND
12:0	ND–0.1	ND–0.1	ND–0.1
14:0	ND–0.2	ND–0.2	ND–0.2
16:0	8.0–13.5	8.0–13.5	5.0–13.5
16:1	ND–0.2	ND–0.1	ND–0.2
17:0	ND–0.1	ND–0.1	ND–0.1
17:1	ND–0.1	ND–0.1	ND–0.1
18:0	2.0–5.4	2.0–5.4	2.0–5.4
18:1	17.7–28.0	22.0–33.0	45.0–70.0
18:2	49.8–59.0	48.0–60.0	15.0–40.0
18:3	5.0–11.0	0.5–4.5	0.5–4.5
20:0	0.1–0.6	ND	ND
20:1	ND–0.5	ND–0.5	ND–0.5
20:2	ND–0.1	ND–0.1	ND–0.1
22:0	ND–0.7	ND–0.7	ND–0.7
22:1	ND–0.3	ND–0.3	ND–0.3
22:2	ND	ND	ND
24:0	ND–0.5	ND–0.5	ND–0.5
24:1	ND	ND	ND

Source: http://www.codexalimentarius.net Official Standards; CXS_210_2003e [1].pdf

Analyses for Oils

Fatty Acid Composition

Determining fatty acid composition of oils is especially important today, because of the broad variety of soybean types available. Fatty acids can be measured by gas–liquid chromatography (GLC) using ISO 6321:1991 (ISO, 2005) plus Amendment 1:1998 (ISO, 2005) and AOCS Ce 1e-91(97)(AOCS, 2005). Method Ce 1e-91 (AOCS, 2005) can be used to measure the fatty acid composition of vegetable oils by capillary GC. If the oil is hydrogenated, then AOCS Method Ce 1f-96 (97) (AOCS, 2005) can be used to determine *cis-* and *trans-* fatty acids by capillary GLC. This method can identify and quantify the trans fatty acid isomers in vegetable oils by using a capillary column with a highly polar stationary phase, according to their chain length, degree of unsaturation, and geometry and position of the double bonds. Data on all other fatty acids can be obtained at the same time. A spectrophotometric method to determine trans fatty acids is available as AOCS Method Cd 14-95(97) (AOCS, 2005) and is appropriate for natural or processed long-chain acids, esters, and triglycerides with trans levels greater than 0.5%.

Chemical and Physical Characteristics

Some chemical and physical characteristics that may vary with fatty acid modification include relative density, refractive index, saponification value, iodine value, and unsaponifiable matter. Appropriate methods for these procedures are available from ISO or AOCS. Refractive index can be conducted according to ISO 6320:2000 (ISO< 2005) or AOCS Cc 7-25 (02) (AOCS, 2005). Saponification value procedures are available as ISO 3657:2000 (ISO, 2005) or AOCS Cd 3-25 (02) (AOCS, 2005). Unsaponifiable matter can be conducted according to ISO 3596:2000 (ISO, 2005) or ISO 18609:2000 (ISO, 2005) or AOCS Ca 6b-53(01) (AOCS, 2005).

Oxidative Stability

AOCS has a recommended practice (Cg 3-91) for assessing oil quality and stability (AOCS, 2005) for measuring primary and secondary oxidation products either directly or indirectly. For example, peroxide value analysis (AOCS method Cd 8-53) (AOCS, 2005) determines the hydroperoxide content and is a good analysis of primary oxidation products. To determine secondary oxidation products, the procedure recommends p-anisidine value (AOCS Method Cd 18-90, 2005) volatile comlb by gas chromatography (AOCS Method Cg 4-94, 2005) and flavor evaluation. (AOCS Method Cg 2-83, 2005). The anisidine value method determines the amounts of aldehydes, principally 2-alkenals and 2, 4-dienals, in oils. The volatile compound analysis method measures secondary oxidation products formed during the decomposition of fatty acids. These comlb can be primarily responsible for the flavors in oils. The flavor evaluation of oils by tasting is the best method for determining oil quality and stability although the procedure must be done by using trained sensory judges.

The Oil Stability Index (OSI) is another method to measure oil stability that can be conducted using AOCS Method Cd 12b-92 (AOCS, 2005) with a Rancimat instrument or an Oxidative Stability Instrument. The OSI may be run at temperatures of 100, 100, 120, 130, and 140°C. Although oil processors and food manufacturers are interested in rapid measurements of oxidation, the high temperature at which the procedure is conducted may not be relevant to ambient temperatures used for most oil storage. Frankel (1993) suggested that the variation in results at 110°C with the rapid analysis and ambient temperature storage may be because of differences in the oxidation mechanisms at the two temperatures.

Frying Oils and Fried-food Stability

Commercial and industrial frying-oil operators need to know when to discard frying oil. Some of the methods used to measure degradation products in frying oil are listed in Table 15.6. Nonspecific methods for measuring nonvolatile comlb in deteriorated frying oil include color, viscosity, smoke point, and foam height; however, none of these methods has proved to be a good measure of oil deterioration. Nonvolatile decomposition products such as polar comlb are a better measure of degradation of a frying oil than are volatile products, but more research is needed to determine the total polar component levels at which different frying oils should be discarded and to relate those levels to fried-food quality for each oil type. A total polar compound level of 24–26% is used in Europe to determine the endpoint of a frying oil's use in restaurants (Anonymous, 2000). A much lower level is required when fried food needs to be stable in shelf-life storage. Measuring the stability of fried food is necessary for food manufacturers. After the food is aging in shelf-life tests, the food can be analyzed for

Table 15.6. Methods to Measure Frying Oil and Fried-Food Deterioration

Frying-oil	Method Reference
Total polar comlb	AOCS Cd 20-91
Free fatty acids	AOCS Ca 5a-40/93
Color	AOCS Td 3a-64/93
Smoke point	AOCS Cc 9a-48/93
Fried food	
Peroxide value	AOCS Cd 8-53
Comjugated dienes	AOCS Ti 1a-64
Volatile comlb	AOCS Cg 4-94
Sensory analysis of odor and flavor	AOCS Cg 2-83

deterioration by sensory or gas chromatographic volatiles (Table 15.6). If necessary, the oil can be extracted and analyzed for peroxide level or conjugated dienes.

Margarines and Shortenings

Slip point (ISO 6321, 2005) and solid fat index (AOCS method Cd 10-57, 2005) can provide information as to the suitability of an oil for use in manufacturing margarines and shortenings. Triacylglycerol (TAG) composition is an additional compositional analysis that can provide information on the potential functionality of an oil as well as its potential oxidative stability. Reversed-phase HPLC with various detection methods such as flame ionization, refractive index, evaporative light scattering, or atmospheric chemical ionization (coupled with mass spectrometry) can be used to determine TAG composition (Neff et al., 1994; Neff et al., 2001).

Minor Oil Constituents

What is the next step in providing good quality, healthful oils? Past research on the factors that influence oil quality and stability shows that fatty acid composition is not the only determinant of oil quality. Oils contain a variety of natural antioxidants and stabilizers, such as tocopherols and phytosterols, that can also inhibit lipid degradation. Just as some vegetable oils have better fatty acid compositions than others to enhance oxidative stability, certain oils have a better profile of these minor constituents than other oils. For example, soybean oil is more stable than sunflower oil even though soy contains 6–8% of the highly unstable linolenic acid, whereas sunflower oil has no linolenic acid (Warner, 2005). However, soy has high levels of gamma and delta tocopherols that are much better antioxidants in vitro than alpha tocopherol, which makes up about 95% of sunflower oil's tocopherol profile (Warner, 2005). On the other hand, soy has low amounts of alpha tocopherol, which is a good antioxidant in vivo.

Plant geneticists revolutionized the vegetable oil industry with modified fatty acid composition oils, and we can expect that the amounts and types of minor oil constituents can be modified as well. The difference in the effects of tocopherols in vivo and in vitro highlights the conflict between choosing an oil based on nutritional benefits versus selecting an oil for its oxidative stability.

Some types of fatty acids and minor constituents are more beneficial in inhibiting oxidation in oils than others. As mentioned previously, polyunsaturated fatty acids such as linolenic acid oxidize easily and are undesirable in edible oils because of the off-flavors and potentially harmful comlb formed. However, linolenic acid is an omega-3 fatty acid that is shown to have positive health effects (Frankel, 2005). Food scientists recommend lower levels of 2–3% linolenic acid, but not complete removal. Tocopherols are another example of the paradox of food constituents. Although alpha tocopherol is the primary precursor to vitamin E in the body and is a good antioxidant in vivo, gamma and delta tocopherols are much better in vitro antioxidants than is alpha. Food researchers recommend that oils contain a balance of tocopherols with some alpha tocopherol for nutritional benefits and some gamma and delta tocopherols for oxidative stability.

References

American Oil Chemists' Society, *Official Methods and Recommended Practices of the American Oil Chemists' Society*, Fifth ed.; AOCS Press: Champaign, IL, 2005.

Anonymous. Recommendations of the 3rd international symposium on deep fat frying. *Eur. J. Lipid Sci. Technol.* **2000,** *102*, 594.

Anonymous, 2005, http://www.fas.usda.gov/excredits/FoodAid/commodities/soybean_oil.htm (3/2007).

Anonymous, 2007, http://www.soyatech.com/oilseed_statistics.htm (3/2007).

Bagby, M.O.; K.D. Carlson. Chemical and biological conversion of soybean oil for industrial products. Fats for the Future; R.C. Cambie, Ed.; Chichester: Ellis Horwood Limited: 1989; pp. 301–302.

Brekke, O.L. Edible Oil Processing, *Handbook of Soy Oil Processing and Utilization*, AOCS: Champaign, IL, 1980.

Codex Alimentarius, 2006, http://www.codexalimentarius.net/download/standards/336/CXS_210e.pdf (3/2007).

Dutton, H.J.; C.R. Lancaster; C.D. Evans; J.C. Cowan. The flavor problem of soybean oil. VIII. Linolenic acid. *J. Am. Oil Chem. Soc.* **1951,** *28,* 115–118.

Dutton, H.J. History of the development of soy oil for edible uses. *J. Am. Oil Chem. Soc.* **1981,** *58,* 234–236.

Frankel, E.N. In search of better methods to evaluate natural antioxidants and oxidative stability in food lipids. Trends in Food Sci. Tech. 1993, *4,* 220–225.

Frankel, E.N. Lipid Oxidation, The Oily Press: Bridgewater, England, 2005.

International Organization for Standardization (ISO) (2005); http://www.iso.org (3/2007).

Kok, L.L.; W.R. Fehr; E.G. Hammond; P.J. White. Trans-free margarine from highly saturated soybean oil. *J. Am. Oil Chem. Soc.* **1999,** *76,* 1175–1181.

List, G.R.; T.L. Mounts; F. Orthoefer; W.E. Neff. Margarine and shortening oils by interesterification of liquid and trisaturated triglycerides. *J. Am. Oil Chem. Soc.* **1995,** *72,* 379–382.

List, G.R.; T.L. Mounts; F. Orthoefer; W.E. Neff. Potential margarine oils from genetically modified soybeans. *J. Am. Oil Chem. Soc.* **1996,** *73,* 729–732.

List, G.R.; T. Pelloso; F. Orthoefer; K. Warner; W.E. Neff. Soft margarines from high stearic acid soybean oils. *J. Am. Oil Chem. Soc.* **2001,** *78,* 103–104.

List, G.R.; M. Jackson; F. Eller; R.O. Adlof. Low trans spreads and shortening oils by hydrogenation of soybean oil. *J. Am. Oil Chem. Soc.* **2007,** *84,* 609–612.

Liu, H.R.; P.J. White. Oxidative stability of soybean oils with altered fatty acid compositions. *J. Am.*

Oil Chem. Soc. **1992a,** *69,* 528–532.

Liu, H.R.; P.J. White. High temperature stability of soybean oils with altered fatty acid compositions. *J. Am. Oil Chem. Soc.* **1992b,** *69,* 533–537.

Miller, L.A.; P.J. White. High-temperature stabilities of low-linolenate, high-stearate and common soybean oils. *J. Am. Oil Chem. Soc.* **1988,** *65,* 1324–1327.

Mounts, T.L.; K. Warner; G.R. List. Performance evaluation of hexane-extracted oils from genetically modified soybeans. *J. Am. Oil Chem. Soc.* **1994a,** *71,* 157–161.

Mounts, T.L.; K. Warner.; G.R. List; W.E. Neff; R.F. Wilson. Low-linolenic acid soybean oils-alternatives to frying oils. *J. Am. Oil Chem. Soc.* **1994b,** *71,* 495–499.

Mounts, T.L.; K. Warner; G.R. List; R. Kleiman; W. Fehr; E.G. Hammond; J.R. Wilcox. Effect of altered fatty acid composition on soybean oil stability. *J. Am. Oil Chem. Soc.* **1988,** *65,* 624–628.

Moustafa, A. Practical Handbook of Soybean Processing and Utilization; D.R. Erickson, Ed.; AOCS Press: Champaign, IL, 1995; pp. 314.

Neff, W.E.; W.C. Byrdwell; G.R. List. A new method to analyze triacylglycerol composition of vegetable oils. *Cereal Foods World* **2001,** *46,* 6–10.

Neff, W.E.; T.L. Mounts; W. Rinsch; H. Konishi; M.A. El-Agaimy. Oxidative stability of purified canola oil triacylglycerols with altered fatty acid compositions as affected by triacylglycerol composition and structure. *J. Am. Oil Chem. Soc.* **1994,** *71,* 1101–1109.

Neff, W.E.; K. Warner; W.C. Byrdwell. Odor significance of undesirable degradation comlb in heated triolein and trilinolein. *J. Am. Oil Chem. Soc.* **2000,** *77,* 1303–1313.

Perkins, E.G. Lipid oxidation of deep fat frying. Food Lipids and Health; R.E. McDonald, D.B. Min, Eds.; Dekker: New York, 1996; p. 139.

Su, C.; M. Gupta; P.J. White. Oxidative and flavor stabilities of soybean oils with low- and ultra-low linolenic acid compositions. *J. Am. Oil Chem. Soc.* **2003,** *80,* 171–176.

Warner, K. Effects of the flavor and oxidative stability of stripped soybean and sunfloweroils with added pure tocopherols. *J. Agric. Food Chem.* **2005,** *53,* 9906–9910.

Warner, K.; E.N. Frankel; J.M. Snyder; W.L. Porter. Storage stability of soybean oil-based salad dressings: Effects of antioxidants and hydrogenation. *J. Food Sci.* **1986,** *51,* 703–708.

Warner, K.; M. Gupta. Frying quality and stability of low- and ultra-low-linolenic acid soybean oils. *J. Am. Oil Chem. Soc.* **2003,** *80,* 275–280.

Warner, K.; M. Gupta. Potato chip quality and frying oil stability of high oleic acid soybean oil. *J. Food Sci.* **2005,** *70,* 395–400.

Warner, K.; S. Knowlton. Frying quality and oxidative stability of high-oleic corn oils. *J. Am. Oil Chem. Soc.* **1997,** *74,* 1317–1322.

Warner, K.; T.L. Mounts. Frying stability of soybean and canola oils with modified fatty acid compositions. *J. Am. Oil Chem. Soc.* **1993,** *70,* 983–988.

Warner, K.; W.E. Neff; W.C. Byrdwell; H.W. Gardner. Effect of oleic and linoleic acids on the production of deep-fried odor in heated triolein and trilinolein. *J. Agric. Food Chem.* **2001,** *49,* 899–905.

Warner, K.; P. Orr; M. Glynn. Effect of fatty acid composition of oils on flavor and stability of fried foods. *J. Am. Oil Chem. Soc.* **1997,** *74,* 347–356.

Warner, K.; P. Orr; L. Parrott; M. Glynn. Effects of frying oil composition on potato chip stability. *J. Am. Oil Chem. Soc.* **1994,** *71,* 1117–1121.

16

Bioenergy and Biofuels from Soybeans

Jon Van Gerpen[1] and Gerhard Knothe[2]
[1]University of Idaho, Department of Biological and Agricultural Engineering, Moscow, ID 83844; [2] USDA ARS NCAUR, Peoria, IL 61604

Introduction

Vegetable oils and animal fats provide some of nature's most concentrated sources of energy. Plants and animals utilize this energy through metabolic processes but non-food uses for these materials currently focus on combustion to produce heat or work. Historically food uses have kept the price of soybean oil high enough that it was not economical for use as a fuel even when rendered fats and greases were viable alternatives to petroleum. Higher petroleum prices and government incentives have produced conditions where soybean oil can be used as fuel, usually in the form of biodiesel.

A general equation for the gross heat of combustion of vegetable oils developed by Bertram (1946) is:

$$-\Delta H_C \text{ (cal/g)} = 11{,}380 - (\text{IV}) - 9.15 \text{ (SV)}$$

Where IV is the iodine value and SV is the saponification value. Using an iodine value of 131 and a saponification value of 193 gives a value of 9,483 cal/g or 39.7 MJ/kg for soybean oil. Compared with petroleum-based No. 2 diesel fuel at about 45.3 MJ/kg, soybean oil has about 12.4% less energy. This is still one of the most concentrated sources of energy found in nature. Soybean oil can be burned directly for heat, although it is rare to do so because it is more expensive than other traditional heating fuels such as natural gas, heavy fuel oil or coal.

Transportation fuels tend to be in shortest supply and command high prices. Soybean oil and its derivatives do not generally have the volatility demanded by spark-ignited engines so most attempts to use these fuels have focused on diesel engines. Diesel engines can be run on straight vegetable oil, but the results have not been satisfactory, as noted below. The greatest problem is the high viscosity of vegetable oil, which is 10 to 15 times greater than that of the No. 2 diesel fuel that most diesel engines are designed to use. Emulsions of soybean oil have been tried but have not gained widespread acceptance due to concerns for the stability of the emulsion and the durability of the fuel injection system (Goering, 1982).

A new development in the area of alternative diesel fuels is a fuel produced from vegetable oils and animal fats using specially modified hydrogenation processes in a conventional petroleum processing facility (Rantanen et al., 2005). This fuel retains the low sulfur and low aromatic character of biodiesel but contains no oxygen and has a heating value that is similar to petroleum diesel fuel. Recent U.S. interest in this approach has expanded due to governmental announcements that the fuel qualifies for federal excise tax credits.

The best success and the greatest amount of experience with using vegetable oil in diesel engines has been with transesterification of the oils with simple alcohols to produce mono-alkyl esters, which have viscosities close to No. 2 diesel fuel. Most of the discussion in this chapter will focus on trandesterified fuels, known as *biodiesel.*

Biodiesel History

The first demonstration of a vegetable oil, peanut oil, as fuel for a diesel engine occurred at the 1900 World Exposition in Paris. One of five diesel engines shown at that event ran on peanut oil (Knothe, 2005). The use of peanut oil as fuel apparently occurred at the request of the French government, which was interested in a local energy source for its African colonies, as Rudolf Diesel (1858–1913), the inventor of the engine that bears his name, himself states (Diesel, 1912a; 1912b). The common assertion that Diesel invented "his" engine to specifically use vegetable oils as fuel is therefore incorrect. Rather, Diesel's objective was to develop a more efficient engine as he states in the first chapter of his book *Die Entstehung des Dieselmotors* (The Development (or Creation or Rise or Coming) of the Diesel Engine) (Diesel, 1913). However, Diesel conducted later experiments with vegetable oils as fuels.

Considerable interest existed in some European countries from the 1920s through the 1940s to use vegetable oils as diesel fuel, especially in countries with African colonies (Knothe, 2001; 2005). The objective was similar to the original demonstration in 1900 and to goals existing in a modified fashion today, namely to provide these colonies with a local and renewable source of energy. However, there was interest in countries such as Brazil, China and India also. Especially in China, some interest in using pyrolyzed vegetable oils existed.

This early work documented results that are still valid today. The high viscosities of vegetable oils were identified as major problems causing engine deposits (Mathot, 1921; Schmidt, 1932; Schmidt, 1933; Schmidt & Gaupp, 1934; Gaupp, 1937; Boiscorjon d'Ollivier, 1939). That exhaust emissions of diesel engines are "cleaner" when running on vegetable oils than with petroleum-based diesel fuel was observed visually (Knothe, 2001; 2005), although no quantitative exhaust emissions studies were performed.

Walton (1938) also recognized that the glycerol moiety has no fuel value and suggested splitting it off and running the engine on the residual acids. However, the first documentation of esterified vegetable oil, biodiesel, as a fuel is the Belgian patent 422,877 issued August 31, 1937, to Chavanne (Chavanne, 1937). Several other publications discuss the use of these esters as fuel (Chavanne, 1943; van den Abeele 1942). The fuel was ethyl esters of palm oil. A commercial passenger bus apparently used this fuel on the route from Brussels to Louvain (Leuven). In this work, the first cetane number of testing biodiesel, in form of ethyl esters of palm oil, was described (van den Abeele, 1942). The biodiesel fuel possessed a higher cetane number than the petroleum-based reference fuels.

In the United States, a *Dual Fuel* project utilizing vegetable oils and petrodiesel was carried out in the late 1940's and early 1950s at The Ohio State University (Huguenard, 1951; Lem, 1952) and other work was conducted at the Georgia Institute of Technology (Baker & Sweigert, 1947).

The energy crises of the 1970s and early 1980s sparked renewed interest in renewable and domestic sources of energy around the world. Vegetable oils were remembered as potential feedstocks for alternative diesel fuels. In 1980 and 1981, Bruwer et al. (1980a, 1980b, 1981) reported that diesel engines running on sunflower oil methyl esters were less prone to engine deposits build-up. Together with work in other countries, this research eventually led to the now existing interest in biodiesel. Later developments included the development of standards, and legislation and regulations around the world promoting the use of biodiesel.

Fats and Oils

The major components of fats and oils are compounds called *triacylglycerols* (triglycerides). They are esters of glycerol with long-chain fatty acids.

Besides these major components, vegetable oils usually contain a variety of other materials including phospholipids, sterols, and tocopherols. Some components are specific to a vegetable oil, such as gossypol in cottonseed oil or glucosinolates in rapeseed and mustard oils. Significant amounts of these materials are usually removed by the refining process. The remaining amounts of these materials are usually limited in biodiesel standards by a variety of specifications.

With triacylglycerols being the major components of vegetable oils and animal fats, the properties of biodiesel are significantly influenced by the fatty acids found in these triglycerides. The fatty acid profile of biodiesel corresponds to that of the feedstock used for its production. Table 16.1 gives the fatty acid profiles for several common vegetable oils.

The fatty acid profile determines the properties of the oil and the resulting biodiesel. Oils with higher levels of saturated fatty acids have more resistance to oxidation and higher cetane numbers but tend to gel at temperatures that limit their usefulness in cold climates (Klopfenstein, 1985; Knothe et al., 2003; 1997; Dunn, 2005a). Oils with high levels of polyunsaturates are prone to oxidation but frequently contain natural antioxidants that protect the oil.

Table 16.1. Major Fatty Acids (wt. %) in Some Oils and Fats Used as Biodiesel Feedstocks[a]

	Fatty Acid Composition (Wt. %)									
Oil or Fat	8:0	10:0	12:0	14:0	16:0	18:0	18:1	18:2	18:3	22:1
Canola					1.5-6	1-2.5	52-66.9	16.1-31	6.4-14.1	1-2
Coconut	4.6-9.5	4.5-9.7	44-51	13-20.6	7.5-10.5	1-3.5	5-8.2	1.0-2.6	0-0.2	
Corn				0-0.3	7-16.5	1-3.3	20-43	39-62.5	0.5-1.5	
Cottonseed				0.6-1.5	21.4-26.4	2.1-5	14.7-21.7	46.7-58.2		
Linseed					6-7	3.2-5	13-37	5-23	26-60	
Olive				0-1.3	7-20	0.5-5.0	55-84.5	3.5-21		
Palm			0-0.4	0.5-2.4	32-47.5	3.5-6.3	36-53	6-12		
Peanut				0-0.5	6-14	1.9-6	36.4-67.1	13-43		0-0.3
Rapeseed				0-1.5	1-6	0.5-3.5	8-60	9.5-23	1-13	5-64
Safflower					5.3-8.0	1.9-2.9	8.4-23.1	67.8-83.2		
Safflower, high-oleic					4-8	2.3-8	73.6-79	11-19		
Sesame					7.2-9.2	5.8-7.7	35-46	35-48		
Soybean					2.3-13.3	2.4-6	17.7-30.8	49-57.1	2-10.5	0-0.3
Sunflower					3.5-7.6	1.3-6.5	14-43	44-74		
Tallow (beef)				2.1-6.9	25-37	9.5-34.2	14-50	26-50		

[a] These oils and fats may contain small amounts of other fatty acids not listed here.
Source: Knothe et al., 2005.

Because of the dominance of soybean oil in the U.S. market, the majority of the biodiesel produced in the United States is made from this feedstock. A few plants use more costly feedstocks, such as canola, and others use low-cost materials such as yellow grease, tallow and chicken fat.

The lower cost materials tend to be high in free fatty acids and thus require pretreatment before they can be transesterified with alkaline catalysts. The pretreatment typically involves sulfuric acid-catalyzed esterification of the free fatty acids to methyl esters (Canakci & Van Gerpen, 2001).

The search for low-cost feedstocks has tended to focus on palm oil in the near term and algae oil in the long term. Palm oil is frequently winterized to produce palm olein but the cold-flow properties restrict use of the pure fuel to warm climates. Algae has been proposed as an oil source with very high yield but challenges associated with high water consumption, invasive species, and high cost have yet to be overcome (Sheehan et al., 1998a).

The Transesterification Reaction

Numerous reviews concerned with the transesterification reaction are available (Bondioli, 2004; Demirbas, 2003; Demirbas and Karslioğlu, 2007; Dimmig et al., 1999; Fukuda et al., 2001; Gutsche, 1997; Haas et al., 2002; Hoydonckx et al., 2004; Kulkarni et al., 2006; Lotero et al., 2005; Lotero et al., 2006; Ma & Hanna, 1999; Marchetti et al., 2007b; Mbaraka & Shanks, 2006; Meher et al., 2006; Nakazono, 2003; Shah et al., 2003; Schuchardt et al., 1998).

The transesterification reaction proceeds according to the general equation:

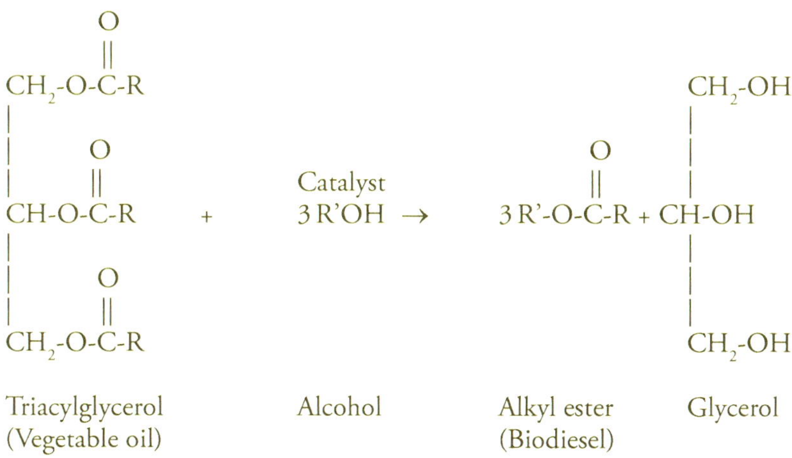

Scheme 16.1. Transesterification of triacylglycerol.

The most commonly prepared esters are methyl esters, which is largely the result of methanol being the least expensive alcohol in most countries. When using methanol (R' = CH_3 in the above equation), approximately 100 kg of vegetable oil are reacted with 10 kg of methanol to give approximately 100 kg of methyl esters (biodiesel) and 10 kg of glycerol. Glycerol and its uses are discussed later in this chapter.

The transesterification reaction is usually conducted with alkali catalysts (sodium or potassium hydroxide or methoxide). Alkali catalysis is much more rapid than acid catalysis in the transesterification reaction (Canakci & Van Gerpen, 1999; Freedman and Pryde 1982, Freedman et al. 1984). For transesterification to give maximum yield, the alcohol should be free of moisture and the free fatty acid (FFA) content of the vegetable oil should be <0.5% (Freedman et al., 1984).

When using an alkaline catalyst (NaOH or $NaOCH_3$) at 32°C, transesterification was 99% complete in 4 h (Freedman et al., 1984). At 60°C, with an alcohol-to-oil molar ratio of at least 6:1 and with fully refined oils, the reaction was complete in 1 h to give methyl, ethyl, or butyl esters. The reaction parameters investigated were the molar ratio of alcohol to vegetable oil, type of catalyst (alkaline vs. acidic), temperature, reaction time, degree of refinement of the vegetable oil, and effect of the presence of moisture and free fatty acid. Although the crude oils could be transesterified, ester yields were reduced because of gums and extraneous material.

The transesterification of beef tallow was studied with regard to the effects of mixing, catalyst, free fatty acids and water as well as the solubilities of different alcohols in the fat (Ma et al., 1998a; 1998b; 1999). Water had the greatest undesirable effect. As a result of the transesterification reaction, biodiesel contains small amounts of glycerol, free fatty acids, partially reacted acylglycerols (mono- and diacylglycerols), as well as residual starting material (triacylglycerols). These contaminating trace materials are limited by biodiesel standards such as ASTM (American Society for Testing and Materials) standard D6751 (ASTM 2007), and the European standard EN 14214 (CEN 2003), as well as other standards under development around the world.

As mentioned above, feedstocks that contain more than 4 to 5% of FFA (i.e. used cooking oils, grease, crude palm oil) require pre-treatment before the triacylglycerols can be subjected to alkaline transesterification. This pre-treatment usually consists of carrying out sulfuric acid-catalyzed esterification of the free fatty acids (Canakci & Van Gerpen, 2001). A pilot plant based on this process was described by Canakci and Van Gerpen (2003a). Once the amount of FFA has been reduced to a level making the material suitable for alkaline transesterification (<1%), this process can be carried out. The amount of FFA is monitored by determination of the acid value in this process.

Mechanism and Kinetics

Transesterification is a reversible reaction. The transesterification of soybean oil with methanol or 1-butanol was reported to proceed with pseudo-first-order or second-order kinetics, depending on the molar ratio of alcohol to soybean oil (30:1 pseudo-first order, 6:1 second order; NaOBu used as catalyst), while the reverse reaction was second order (Freedman et al., 1986). The methanolysis of sunflower oil at a molar ratio of methanol to sunflower oil equal to 3:1 was reported to begin with second-order kinetics and the rate decreased due to formation of glycerol (Mittelbach & Trathnigg, 1990). The originally reported kinetics (Freedman et al., 1986a) were reinvestigated (Boocock et al., 1996; 1998; Mittelbach & Trathnigg, 1990; Noureddini et al., 1997) and differences were found. A shunt reaction originally proposed (Freedman et al.,

1986a) as part of the forward reaction was shown to be unlikely, that second-order kinetics are not followed and that miscibility phenomena play a significant role (Boocock et al., 1996; 1998; Mittelbach & Trathnigg, 1990; Noureddini et al., 1997). The kinetics of non-catalyzed alcoholysis of soybean oil were also investigated (Dasari et al., 2003).

The distribution of methanol and catalyst between the ester and glycerol phases has been modeled using the Wilson equation and vapor-liquid-equilibrium data (Chiu et al., 2005). Using a polar dye, the phase behaviors of methanolysis, ethanolysis and butanolysis were visualized (Zhou & Boocock, 2006a). Ethanolysis was more easily initiated by mixing than methanolysis and showed a longer emulsion period and slower phase separation. Butanolysis remained one phase throughout the process (Zhou & Boocock, 2006a). In the case of methanolysis at 23°C, 42% of the methanol, 2.3% of glycerol and 5.9% of the catalyst were in the ester phase at steady state (Zhou & Boocock, 2006b). When carrying out ethanolysis, 75.4% of the ethanol, 19.3% of the glycerol and 7.5% of the catalyst were in the ester phase (Zhou & Boocock, 2006b). The glycerol phase dissolved most of the catalyst, thus transesterification became limited by mass transfer and conversion to meet biodiesel standards is not achieved in one reaction (Zhou & Boocock, 2006b). The use of membrane reactors may be a method for enabling separation of the reaction products and increasing product purity (Dubé et al., 2007).

The addition of co-solvents, such as tetrahydrofuran (THF) or methyl *tert.*-butyl ether (MTBE), to the methanolysis reaction was reported to accelerate the methanolysis of vegetable oils as a result of solubilizing the methanol in the oil and to a rate comparable to that of the faster butanolysis (Boocock et al., 1996; 1998). The production of ethyl esters, also in presence of THF, has been described (Zhou et al., 2003). Other possibilities for accelerating the transesterification appear to be microwave (Breccia et al., 1999) and ultrasonic treatment (Stavarache et al., 2003). Factorial experiment design and surface response methodology have been applied to different production systems (Vicente et al., 1998). A continuous pilot-plant-scale process for producing methyl esters with conversion rates >98% was reported (Noureddini & Zhu, 1997) as well as a discontinuous two-stage process with a total methanol to acyl ratio of 4:3 (Cvengroš & Povalanec, 1996).

Generally, alkoxides are preferable to hydroxides as catalysts for the transesterification reaction. The reason for this is that when using hydroxides, the following reaction, in which water is formed, is possible.

$$ROH + XOH \rightarrow ROX + H_2O \text{ (R = CH}_3\text{ or other alkyl rest; X = Na or K)}$$

Scheme 16.2. Water formation by hydroxide catalysts.

When using alkoxides at the beginning of the process, this reaction is not possible, leading to easier work-up and better quality of the product in terms of reduced potential minor components (contaminants).

Side reactions leading to undesirable minor components in the biodiesel fuel are largely due to the presence of water or other contaminants at the beginning of the reaction. For example, when using sodium or potassium hydroxide, OH^- can lead to the formation of free fatty acids, or rather, to the corresponding sodium or potassium salts. These salts are also known as *soap*.

Besides the methods discussed here, other catalysts have been applied in transesterification reactions. Many of the aspects related to such catalysts are summarized in the review articles cited above.

In situ transesterifications consist of directly exposing the plant material containing the oil to transesterification conditions. The alcohol acts as extraction solvent for the oil-containing material and as the esterifying reagent. An overall transesterification efficiency of 80% was achieved by subjecting soybean flakes to such a process (Haas et al., 2004). However, large amounts of alcohol are required. Drying the flakes prior to the process improves efficiency (Haas & Scott, 2007). Sunflower seed oils were transesterified *in situ* using macerated seeds with methanol in the presence of H_2SO_4 with yields reported to be higher than from transesterification of the extracted oils (Harrington & D'Arcy-Evans, 1985). Again, seed moisture reduced the yield of methyl esters. The cloud points of the *in situ* prepared esters appeared to be slightly lower than those prepared by conventional methods. In a related study, best yields were achieved with a 300:1 molar ratio of methanol to oil (Siler-Marinkovic & Tomasevic, 1998). Similarly, macerated soybeans were treated with methanol, ethanol, *n*-propanol and *n*-butanol to give the corresponding esters (Kıldıran et al., 1996), although due to the insolubility of soybean oil in methanol, conversion was low in that case.

The transesterification reaction with sodium or potassium hydroxide or alkoxide catalysts occurs in a homogeneous phase. Heterogeneous catalysis, in which the catalyst exists in a solid phase while the rest of the reaction mixture is in liquid phase, offers the desirable potential feature of catalyst retrieval and recycling. Because of their nature and the possibility of recycling, heterogeneous catalysts often cause less disposal and environmental concerns. However, cost and product yield can be concerns with such catalysts, as well as the sometimes extreme reaction conditions. Numerous reports have been published concerning heterogeneous processes, using a variety of Ca, Mg, zeolite-based and other catalysts (Abreu et al., 2005; Cantrell et al., 2005; Di Serio al., 2006; Dossin et al., 2006; Gryglewicz, 1999; Kim et al., 2004; Kiss et al., 2006; Leadbeater & Stencel, 2006; Leclercq et al., 2001; Marchetto et al., 2007a; Mazzocchia et al., 2004; Mbaraka & Shanks, 2005; Peter et al., 2002; Peterson & Scarrah, 1984; Reddy et al., 2006; Secheli et al., 1999; Shah et al., 2004; Srinivas et al., 2004; Shibasaki-Kitakawa et al., 2007; Suppes et al., 2001; 2004; Watkins et al., 2004; Xie & Huang, 2006; Xie & Li, 2006; Xie et al., 2006; 2007). An example is the use of alkylguanidines attached to modified polystyrene or siliceous MCM-41, encapsulated in the supercages of zeolite Y or entrapped in SiO_2 sol-gel matrices (Sercheli et al., 1999).

Catalyst-free reactions at supercritical conditions have also been employed for transesterification (Bunyakiat et al., 2006; Demirbas, 2002; 2003; Cao et al., 2005; He et al., 2007a; 2007b; Kusdiana & Saka, 2001; Minami et al., 2006; Varma & Madras, 2007; Wang & Yang, 2007; Warabi et al., 2004).

Enzymatic transesterification methods are receiving attention for producing esters suitable as biodiesel. Advantages of enzymatic reactions can be specificity, mild reaction conditions, reduced product isolation problems, water tolerance, and reduced waste (Posorske 1984), although they are more expensive and, as yet, have not been used for commercial biodiesel production. Lipases from *Pseudomonas fluorescens* with petroleum ether as solvent yielded methyl and ethyl esters of sunflower oil (Mittelbach, 1990). The lipase from *Mucor miehei* was the most efficient in yielding esters of primary alcohols while the lipase from *Candida antarctica* was most efficient for yielding branched esters from secondary alcohols (Nelson et al., 1996). Other reports on enzymatic production of esters mainly for fuel purposes include ethanolysis of sunflower oil with a solvent-free, immobilized 1,3-specific *Mucor miehei* lipase (Selmi & Thomas, 1998), a variety of enzymes used for producing different materials (Linko et al., 1998) with dependence on the presence of solvent (Soumanou & Bornscheuer, 2003) as well as stepwise addition of methanol (Soumanou & Bornscheuer, 2003; Bélafi-Bakó et al., 2002), the synthesis of esters of restaurant greases (Wu et al., 1999; Hsu et al., 2002; 2003) stepwise use of immobilized *Candida antarctica* lipase (Shimada et al., 1999) modified later for continuous use (Watanabe et al., 2000), methyl acetate as an acyl acceptor (Xu et al., 2003) use of *Rhizopus oryzae* lipase in a water-containing system without an organic solvent (Kaieda et al., 1999) and in the methanolysis of vegetable oils contained in waste activated bleaching earth (Pizarro & Park, 2003). An evaluation of several lipases showed dramatic differences towards alcoholysis (Deng et al., 2005), with increasing presence of water being beneficial in most cases and 96% ethanol being the preferred alcohol in most reactions.

Transesterification of Other Sources of Biodiesel

Potential low-cost sources of biodiesel, such as restaurant greases and soapstock, are of lower quality than refined vegetable oils. A major problem associated with them is the high content of free fatty acids, which, as indicated above, deactivate the catalyst by forming soap. Thus, the processing of feedstocks high in free fatty acid require some changes to the overall production process. For the production of biodiesel from soapstock, a byproduct of vegetable oil processing that contains a mixture of glycerides, phosphoglycerides and free fatty acids emulsified in a substantial amount of water, all ester bonds were first hydrolyzed by alkali catalysis and the resulting fatty acid sodium salts were converted to methyl esters by acid catalysis (Haas et al., 2000; 2003). The resulting ester preparation met the ASTM quality specifications and performed comparably to biodiesel produced from refined soybean oil when tested in a heavy duty engine (Haas et al., 2001).

Oils produced by algae are also being considered as possible feedstocks for biodiesel (Aresta et al., 2005; Chisti, 2007; Miao & Wu, 2006; Nagle & Lemke, 1990; Xu et al., 2006). The challenges associated with high-volume production of algae and the high content of polyunsaturated fatty acids in many algal oils are problems associated with use of algae for biodiesel production.

Analysis of the Transesterification Reaction Products

An important aspect of the final biodiesel product is verification that it satisfies the requirements of the accepted quality specification. Detailed overviews of the analytical methods used for biodiesel are available in the literature (Knothe, 2006). Some analytical methods used for biodiesel analysis are prescribed in standards. Of great significance is gas chromatography, the use of which is prescribed in most standards for analyzing the amount of glycerol and of unreacted triacylglycerols in biodiesel. The gas chromatographic method used for analyzing glycerol and the various acylglycerols is based on a literature method (Plank & Lorbeer, 1995) extended from another report (Freedman et al., 1986b). Various chromatographic, spectroscopic and other methods have been reported, an example being one describing simultaneous analysis of methanol and glycerol (Mittelbach et al., 1996).

Commercial Biodiesel Production

Oil Extraction

While most soybean oil is extracted in solvent extraction plants using hexanes as the solvent, there has been considerable interest from biodiesel producers in using expeller-extracted oils, sometimes known as *mechanical extraction* or *screw pressing*. While biodiesel producers have recognized the financial advantage of being closely associated with a source of oil, their size is generally too small to have dedicated solvent extraction plants. The mechanical extraction plants are much more suitable for smaller operations.

The primary soybean oil quality issues for biodiesel production are free fatty acids, moisture, and phosphorus. These correspond well to the requirements for edible oil but bleaching and deodorization are generally not required for biodiesel. It is desirable to have levels of free fatty acid and moisture as low as possible. If both contaminants are <0.1%, there will be no effect on the biodiesel process (Van Gerpen, 2005). In fact, levels of free fatty acids and moisture of 0.5%, as is common for crude as-pressed oil, can be used directly for biodiesel production. However, there will be some additional soap formation during processing that will need to be removed.

The specification for biodiesel, ASTM D 6751 (ASTM, 2007), requires that the phosphorus level in finished biodiesel must be <10 ppm. For this reason, most soy-

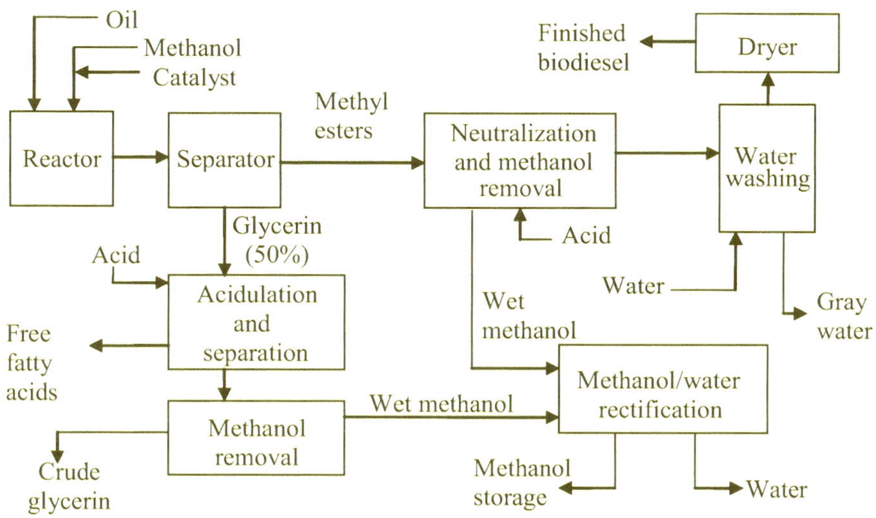

Fig. 16.1. Biodiesel production schematic diagram.

bean oil used for biodiesel production is degummed to <10 ppm phosphorus. It appears that there is not a direct transfer of phosphorus from the oil to biodiesel with a large portion of the phospholipids going into the glycerin during processing (Van Gerpen & Dvorak, 2002). In any case, most producers use oil that has been at least water-degummed to minimize the deposition of gum deposits in their equipment due to natural degumming.

Reaction Systems

Figure 16.1 shows a schematic diagram of the processes in a biodiesel production facility. It is assumed that the oil has been extracted, degummed and neutralized before entering the biodiesel process.

The transesterification reaction between soybean oil and methanol is initially a mass transfer limited reaction. The solubility of methanol in soybean oil is low enough that the reaction occurs at the interface of the two phases with the catalyst tending to concentrate in the alcohol phase. During the intermediate portion of the reaction, the presence of mono- and diglycerides produces a single-phase mixture where the reaction is limited by the chemical reaction rate. Then, as the reaction proceeds and significant amounts of product glycerin are produced, the reaction is limited by product inhibition and the tendency of the catalyst is to concentrate in the small droplets of insoluble glycerin. At least during the initial and final portions of the reaction, agitation is critical for acceptable reaction rate and complete reaction. Other reactions are occurring in parallel with the transesterification reaction. Free fatty acids will react with the alkali catalyst to produce soap. Water in the oil greatly increases the production of soap by facilitating the hydrolysis of glyceride-ester bonds to release free fatty acids.

A common reactor scheme is to use a series of two continuously stirred tank reactors (CSTR) with glycerin separation between and after the two reactors. Typically, 70 to 90% of the total alcohol and catalyst are added in the first reactor and the balance is added to the second reactor to compensate for the alcohol and catalyst lost with the glycerin separated after the first reaction (Van Gerpen, 2005).

Other reactor designs include batch reactors, plug-flow reactors, and packed-bed reactors. Batch reactors are usually found in small plants but have been used for up to 8–9 million gal/yr. Most plants above 2–3 million gal/yr use continuous flow processes. Packed-bed reactors using heterogeneous catalysts are becoming available (Stern et al., 1999) but still require higher temperatures and longer reaction times than conventional reactors utilizing homogenous catalysts such as sodium or potassium methylate. Detailed process models have been developed for biodiesel production to allow the advantages of different process strategies to be quantified and the economics to be determined (Haas et al., 2006).

Separation

Glycerin has very low solubility in methyl esters so it is relatively easy to separate by taking advantage of its higher density. Separation is accomplished with either centrifuges or decanters with both widely used in the industry. Glycerin separation is complicated by the presence of alcohol and soap. Methanol is completely soluble in glycerin and partially soluble in methyl esters. It tends to act as a co-solvent to increase the solubility of glycerin in the methyl esters and also decreases the density difference between the glycerin and methyl esters. With excessive amounts of methanol, there will be no glycerin separation.

Soap acts as an emulsifier and also inhibits glycerin separation. Oils and fats with high free fatty acid contents produce more soap and if the free fatty acid level is >5%, the glycerin may not be separable as a distinct phase. Feedstocks high in free fatty acid usually require pretreatment to lower the free fatty acid content before they can be converted to biodiesel.

Methanol Recovery

Most producers operate with a molar ratio of alcohol-to-oil of at least 6:1. This is 100% more than is consumed in the transesterification reaction so the excess must be removed and recycled. The excess methanol splits 60%/40% between the methyl esters and glycerin, so methanol must be removed from both streams (Ma et al., 1998a;1998b; 1999). Methanol recovery is frequently accomplished by flash vaporization, which yields the methanol plus any water that may have been present in the reaction mixture. Excessive water is removed by a distillation column.

Methanol recovery may be accompanied by soap precipitation if the soap level in the biodiesel is excessive resulting in plugged filters and pumps. Acid addition to lower the pH to 4.5 will split the soap and eliminate the soap precipitation issue.

Washing

After glycerin separation and removal of methanol, biodiesel still contains some contaminants such as soap, residual free glycerin, and a small amount of methanol. These contaminants must be removed before the fuel will meet the ASTM specification (ASTM 2007). Soap will be limited by the specifications for sodium or potassium and by the sulfated ash. Methanol is limited by the flash point and a limiting value for free glycerin is directly specified. Traditionally, these compounds have been removed by relying on their preferential solubility in water so that water can be used as a solvent in liquid-liquid extraction. An amount of water comparable to the amount of biodiesel is mixed with the biodiesel and gently agitated in a batch mixer, a rotary extractor, or a counterflow column. The column may contain packing to enhance mixing but excessive mixing can encourage emulsion formation, which is possible when soap or monoglyceride levels are high. The water should be deionized to prevent transfer of dissolved metals to the biodiesel and should be heated (60°C) to encourage maximum removal of free glycerin. It is common to expose the biodiesel to water in several stages in counterflow fashion to ensure removal of the contaminants. Even with this approach the total water consumption may be 1-2 L of water per L of biodiesel produced (Van Gerpen, 2005).

One approach to reducing water consumption is to lower the pH of the biodiesel, either by direct acid addition or by adding acid to the wash water. Below pH 4.5, the soap dissolved in the biodiesel will be split into free fatty acids and salts. The free fatty acids stay with the biodiesel and as long as they do not exceed the Acid Value specification in ASTM D 6751 (0.5 mg KOH/g), they do not cause a problem. The salts are removed with a small amount of water (3 to 10%).

Some biodiesel producers have sought to eliminate water from their process through the use of adsorbents. One type of adsorbent is applied similar to bleaching clay where approximately 1% of the adsorbent in the form of a fine powder is mixed with the biodiesel after the methanol has been removed and stirred for 20 min at an elevated temperature. Then, the biodiesel is filtered through a cake of the adsorbent and soap is removed along with glycerin, methanol, and some mono- and diglycerides. A second class of adsorbents are ion-exchange resins. These can be used in a packed bed and do not require methanol removal prior to use. They will remove soap and free glycerin, but generally do not reduce levels of mono- or diglycerides.

Additives

Petroleum-based diesel fuel is commonly treated with a large number of additives to enhance cetane number, improve cold flow and oxidative stability, lessen corrosiveness, etcetera. Additive technology for biodiesel is much less well developed. Most biodiesel in the United States is sold without additives. It may be treated with antioxidants such as TBHQ or BHT if customers require or if the additive is required for the fuel to meet the stability requirement in ASTM D 6751. Cold-flow additives may be added if the fuel is blended to lower the pour point or filter plugging point.

Effect of Alcohol Type

Virtually all biodiesel in the United States is made using methanol due to its cost advantage over other alcohols (Van Gerpen, 2005). In Europe, methanol is used because European Union specifications only recognize methyl esters as biodiesel. In Brazil and other parts of the world, ethanol is frequently used (Bikou et al., 1999; Encinar et al., 2002). Considerable work has also been done with isopropanol, which provides considerable cold flow advantages (Johnson & Hammond, 1996; Lee et al., 1995; Wang et al., 2005; Wu et al., 1998).

Other Uses of Methyl Esters

Vegetable oil alkyl esters (biodiesel) are used in a variety of other applications. These can be distinguished as fuel or non-fuel related uses.

Fuel-related Uses

Cetane improvers based on fatty compounds have been reported. The use of nitrate esters of fatty acids in diesel fuel was reported in a patent (Poirier et al., 1995). Multifunctional additives consisting of nitrated fatty esters for improving combustion and lubricity have been reported (Suppes et al., 2001; Suppes and Dasari, 2003). Glycol nitrates of acids of chain lengths C_6, C_8, C_{14}, C_{16} and C_{18} (oleic acid) were also prepared and tested as cetane improvers with C_6-C_{14} glycol nitrates showing better cetane-improving performance due to their balance of carbon numbers and nitrate groups (Suppes et al., 1999). These compounds are more stable and less volatile than ethylhexyl nitrate (EHN), the most common commercial cetane improver and their cetane-enhancing capability is up to 60% of that of EHN (Suppes et al., 1999; 2001; Suppes and Dasari, 2003).

Biodiesel can be used as heating oil (Mushrush et al., 2001). In Italy, the esters of vegetable oils serve as heating oil instead of diesel fuel (Staat & Vallet, 1994). A European standard, EN 14213, has been established for this purpose. A salient project in this regard has been the use of biodiesel as heating oil for the Reichstag building in Berlin, Germany (Anon., 1999). Other work reports the use of biodiesel derived from used cooking oils (Çetinkaya & Karaosmanoğlu, 2005) or soybean oil ethyl esters (Ferrari et al., 2005) as generator fuel.

Another suggested use of biodiesel as fuel has been in aviation (Dunn, 2001; Wardle, 2003). A major problem connected with this use is the low-temperature properties of biodiesel, thus making it more feasible only in lower-flying aircraft (Dunn, 2001). Biodiesel can be used to generate hydrogen for synthesis gas production (Czernichowski et al., 2006).

Non-fuel Related Uses

The classic use of methyl esters of vegetable oils has been as intermediates in the production of fatty alcohols from vegetable oils (Peters, 1996; Ahmad et al., 2007) or esterquats and methyl ester sulfonates (Ahmad et al., 2007). Fatty alcohols and the other products are used in surfactants and cleaning supplies. Intermediates were produced from polyisobutylene (PIB) maleic anhydride and rapeseed oil methyl esters which were used to acylate polyethylene polyamines (Hancsók et al., 2006). These additives showed corrosion-inhibiting and lubricity-improving effects.

Branched esters of fatty acids are used as lubricants because their improved biodegradability relative to petroleum lubricants makes them attractive from an environmental aspect (Willing, 1999). Vegetable oil esters also possess good solvent properties. This is expressed in their use as a medium for cleaning beaches contaminated with crude oil (petroleum) (Miller & Mudge, 1997; Mudge & Pereira, 1999; Pereira & Mudge, 2004; Fernández-Álvarez et al., 2006). Related results were obtained for crude palm oil and fatty acids (Obbard et al., 2004).

The high flash point, low volatile organic compounds and benign environmental properties of methyl soyate make it attractive as a cleaning agent (Wildes, 2002). The solvent strength of methyl soyate is also demonstrated by its high Kauri-Butanol value (relating to the solvent power of hydrocarbon solvents), which makes it similar or superior to many conventional organic solvents. In this connection, a variety of fatty esters were examined (Hu et al., 2004), with shorter fatty acid chains and straight-chain esters enhancing the Kauri-Butanol value. Methyl esters of rapeseed oil have been suggested as plasticizers in the production of plastics (Wehlmann, 1999) and as high-boiling absorbents for cleaning of industrial gases/emissions (Bay et al., 2004). Hydrogenated fatty alkyl esters mixed with paraffin-based or glyceride-based waxes provide improved combustion performance in candles (Schroeder et al., 2004).

Specifications and Standards

Biodiesel standards have been or are being developed in many countries or regions around the world. The American biodiesel standard ASTM D6751 and the European biodiesel standard EN 14214 are standards that have been utilized by other countries when developing their own standards. The current versions of ASTM D6751 and EN 14214 are summarized in Table 16.2.

Quality Concerns

While alkyl esters are the major components of biodiesel, a variety of minor components or contaminants can be found in biodiesel. Several sources can be identified for these compounds. The first source is the feedstock, i.e., naturally occurring materials such as sterols, or carryover materials containing phosphorus, sulfur, calcium and other elements in animal fats. The second source is biodiesel production, leading to unreacted triacylglycerols and partially converted mono- and diacylglycerols as well as residual glycerol, alcohol and catalyst. The third source is biodiesel storage, when it may come in contact with extraneous materials. During storage, biodiesel can also begin to slowly degrade, leading to a variety of compounds briefly discussed below under the header oxidative stability. Most of these issues are also addressed by various specifications in biodiesel standards.

Test Methods

Generally, test methods that should be observed are prescribed in standards such as ASTM D6751 or EN14214. However, in many cases, other test methods, usually developed by professional organizations, may be simpler, less expensive, and more suitable for process development. For example, since biodiesel is an oleochemical product, test methods developed by organizations, such as the American Oil Chemists' Society (AOCS), are often well-suited, while ASTM methods were often developed specifically for petrodiesel.

Oxidative Stability

Oxidative stability of biodiesel has been the subject of considerable research. This issue affects biodiesel primarily during extended storage. The effects of parameters, such as the presence of air, heat, traces of metals, antioxidants, peroxides as well as the nature of the storage container, were investigated in the aforementioned studies. Generally, factors, such as presence of air, elevated temperatures or the presence of metals, facilitate oxidation.

Oxidation, which can occur by autoxidation or photo-oxidation (initiation by light), is a complex chain reaction. The initial step of oxidation is the formation of hydroperoxides. This initial step is followed by secondary reactions in which species, such as aldehydes, acids, alcohols and hydrocarbons, are formed. Since it is a mechanism based on the formation of radicals, dimerization of some intermediates can occur, leading to formation of higher molecular weight products. Oxidative polymerization can also occur. In addition to these mechanisms, fuel deterioration can also occur hydrolytically through the presence of water. A detailed book on oxidation has been published by Frankel (2005).

The reason for autoxidation is the presence of double bonds in the chains of many fatty compounds. The autoxidation of unsaturated fatty compounds proceeds at different rates depending on the number and position of double bonds (Frankel 2005). The positions allylic to double bonds are especially susceptible to oxidation. The *bis*-allylic positions in common polyunsaturated fatty acids, such as linoleic acid (double bonds at 9 and 12, giving one *bis*-allylic position at C-11) and linolenic acid (double bonds at 9, 12, and 15, giving two *bis*-allylic positions at C-11 and C-14), are even more prone to autoxidation than allylic positions. The relative rates of oxidation given in the literature (Frankel, 2005) are 1 for oleates (methyl, ethyl esters), 41 for linoleates, and 98 for linolenates. This is essential because most biodiesel fuels contain significant amounts of esters of oleic, linoleic, or linolenic acids, which influence the oxidative stability of the fuels.

An overview of analytical methods used to study oxidation was edited by Kamal-Eldin and Pokorný (2005). Numerous methods, including wet-chemical methods, such as acid value and peroxide value, various oxidation tests, pressurized and conventional differential scanning calorimetry (P-DSC; DSC; see Dunn 2000, 2006), nuclear magnetic resonance (NMR) and others, have been applied in oxidation studies of biodiesel. NMR can be used to assess the fatty acid profile of oxidized biodiesel (Knothe, 2006a).

A European standard (EN 14112) assessing oxidative stability using the Rancimat method, which is very similar to the oil stability index (OSI) method (AOCS Cd 12b-93), is included in both the European biodiesel standard EN 14214 and the American standard ASTM D6751-07. In EN 14214, the requirement for the minimum Rancimat induction time is 6 h and in D6751 the prescribed minimum time is 3 h. Both standards prescribe a temperature of 110°C. The use of the Rancimat or OSI, as well as their limitations, have been described in several publications (Bondioli et al., 2004; Dittmar et al., 2004a; Knothe & Dunn, 2003; Lacoste & Lagardere, 2003). The induction period in such tests decreases with increasing oxidation (Mittelbach & Gangl, 2001).

In addition to the Rancimat-based oxidative stability specification, the European biodiesel standard EN 14214 contains other parameters dealing with oxidative stability (Knothe, 2006b). The most prominent is iodine value (IV), which is not without significant problems (Knothe, 2002). The IV is a method for determining the number of carbon-carbon double bonds of a lipid based on its tendency to add iodine. The IV depends on the molecular weight of the compound, thus using a higher ester, such as ethyl or propyl instead of methyl, reduces the IV without affecting the reactivity of the double bond towards oxidation. For mixtures, an infinite number of mixtures of fatty esters can yield the same the IV. Also, double bond(s) in polyunsaturated fatty esters separated by several CH_2 groups tend to behave more like the lone double bonds in monounsaturated fatty acid chains. In practical tests, the IV could not be used to predict storage stability and no correlation with polymerization or viscosity could be established (Bondioli & Folegatti, 1996).

The standard EN 14214 also limits oxidation-prone fatty acids with 3 double bonds to <12%, which permits rapeseed/canola oil to be used as feedstock, and fatty acids with >3 double bonds are limited to <1%. In the aforementioned tests (Bondioli & Folegatti 1996), oxidative stability could not be correlated with linoleic or linolenic acid content, although higher amounts of oleic acid increased oxidative stability. Accordingly, hydrogenation improved oxidative stability (Falk and Meyer-Pittroff 2004).

Stability tests developed for petrodiesel fuels reportedly are not suitable for biodiesel or biodiesel blends with petrodiesel (Canakci et al., 1999; Stavinoha & Howell, 1999; Westbrook & Stavinoha, 2003), however, appropriate modification may render them useful. Another study (Bondioli et al., 2003) states that the petrodiesel method ASTM D4625 (Standard Test Method for Distillate Fuel Storage Stability at 43°C (110°F) is suitable but not fast.

Long-term storage tests on biodiesel have been conducted. Viscosity, peroxide value, acid value and density increased in biodiesel stored for two years while heat of combustion decreased (Thompson et al., 1998). Viscosity and acid value, which can be strongly correlated (Monyem et al., 2000), changed dramatically over one year with changes in Rancimat induction period depending on the feedstock (Bondioli et al., 2003) but even during 90-day storage tests significant increases in viscosity, peroxide value, free fatty acid, anisidine value and UV absorption were found (Du Plessis, 1985). Biodiesel from different sources stored for 170–200d at 20–22°C did not exceed viscosity and acid value specifications but induction time decreased, with exposure to light and air having the most effect (Mittelbach and Gangl 2001). In another 52-week storage study, water enhanced degradation due to hydrolysis, but the effect was reportedly less than the combination of temperature and air (Leung et al., 2006). Acid value and viscosity are two facile methods that can be used to evaluate the oxidation status of aged biodiesel since they increase continuously with increasing fuel deterioration (Canakci et al., 1999; Dunn, 2002).

In addition to the composition of the fuel, the degradation of biodiesel is affected by factors such as elevated temperature, presence of air, presence of light, presence of extraneous materials such as metals, including the nature of the storage container (Bondioli et al., 1995; Du Plessis et al., 1985), and antioxidants. Studies performed with the automated oil stability index (OSI) method, confirmed the catalyzing effect of metals on oxidation, with copper being especially effective, however, the influence of compound structure of the fatty esters was even greater (Knothe & Dunn, 2003). Additives to encourage the regeneration of diesel particulate filters negatively affected biodiesel oxidative stability, possibly necessitating higher levels of antioxidants (Schober & Mittelbach, 2005). Some additives designed for improving the cold filter plugging point also had a small effect on oxidative stability. Antioxidants have been studied for prolonging the storage of biodiesel. In these studies, there is no agreement on which antioxidant performs best, although often propyl gallate is mentioned as an effective antioxidant. The varying results are underlined by studies where soy methyl esters from different producers or from different feedstocks responded differently to

Table 16.2. Biodiesel Standards ASTM D6751 and EN 14214

Property	ASTM D6751 (United States)			EN 14214 (Europe)		
	Test method	Units	Limits	Test method	Units	Limits
Flash point (closed cup)	D 93	°C	93 min	EN ISO 3679	°C	120 min
Methanol content	—	—	—	EN 14110	% (m/m)	0.2 max
Water and sediment	D 2709	% vol.	0.05 max	EN ISO 12937	mg/kg	500 max
Kinematic viscosity	D445	mm^2/s	1.9–6.0	EN ISO	mm^2/s	3.5–5.0
Density	—	—	—	EN ISO 3675; 12185	kg/m^3	860–900
Sulfated ash	D874	% mass	0.02 max	ISO 3987	% (m/m)	0.02
Sulfur	D5453	% mass (ppm)	0.0015 max (S15); 0.05 max (S500)[a]	EN ISO 20846; 20884	mg/kg	10.0
Copper strip corrosion	D 130	—	No. 3 max	EN ISO 2160	—	max 1
Cetane number	D 613	—	47 min	EN ISO 5165	—	51 min
Cloud point	D 2500	°C	Report	—	—	—
Carbon residue	D 4530	0.05 max	% mass	EN ISO 10370	% (m/m)	0.30
Total contamination	—	—	—	EN12662	mg/kg	24
Acid number	D 664	mg KOH/g	0.50 max	EN14104	mg KOH/g	0.50
Ester content	—	—	—	EN 14103	% (m/m)	96.5
Linolenic acid content	—	—	—	EN 14103	% (m/m)	12
FAME 4 double bonds	—	—	—	—	% (m/m)	1
Iodine value	—	—	—	EN 14111	g iodine /100g	120
Free glycerol	D 6584	% mass	0.02	EN 14105 / EN 14106	% (m/m)	0.02
Total glycerol	D 6584	% mass	0.24	EN 14105	% (m/m)	0.25

Cont. on p. 518.

Table 16.2., cont. Biodiesel Standards ASTM D6751 and EN 14214

Property	ASTM D6751 (United States)			EN 14214 (Europe)		
	Test method	Units	Limits	Test method	Units	Limits
Monoglyceride content	—	—	—	EN 14105	% (m/m)	0.80
Diglyceride content	—	—	—	EN 14105	% (m/m)	0.20
Triglyceride content	—	—	—	EN 14105	% (m/m)	0.20
Ca and Mg, combined	EN 14538	ppm (μg/g)	5 max	EN 14538	mg/kg	5.0
Phosphorus content	D 4951	% mass	0.001 max	EN 14107	mg/kg	10.0
Na and K, combined	EN 14538	ppm (μg/g)	5 max	EN ISO		
Oxidation stability	EN 14112	h	3 min	EN ISO	h	6 min
Distillation temp., atmospheric equivalent temp., 90% recovered	D1160	°C	360 max	—	—	—
Alcohol control: One of the following must be met:						
Methanol content	EN14110	% volume	0.2 max	—	—	—
Flash point	D93	°C	130 min	—	—	—

[a] Different specifications for 15 and 500 ppm sulfur.

different antioxidants (Dunn, 2005b; Schober & Mittelbach, 2004) and the nature of the production process affecting antioxidants (Dittmar et al., 2004b). Butyl hydroxytoluene (BHT) may have solubility problems because it may not be readily soluble in blends with larger methyl soyate ratios (Dunn, 2005a). A general consensus appears to be that synthetic antioxidants are more effective than naturally occurring tocopherols. However, crude esters or feedstocks are more stable than distilled or refined counterparts due to their content of natural antioxidants (Du Plessis et al., 1982; Liang et al., 2006). Synthetic antioxidants that have been studied include propyl gallate (PG), pyrogallol (PY), *tert.*-butylhydroquinone (TBHQ), BHT and butylated hydroxyanisole (BHA). Antioxidants in biodiesel can be analyzed by high-performance liquid chromatography (HPLC) (Tagliabue et al., 2004). Other studies on the use of antioxidants include Dunn (2006), Liang et al. (2006), Loh et al. (2006), Mittelbach and Schober (2003), Polavka et al. (2005), Simkovsky and Ecker (1998; 1999).

The effect of oxidation on exhaust emissions of biodiesel has been studied (Monyem & Van Gerpen, 2001; Monyem et al., 2001). A shorter ignition delay by 0.9° crank angle for oxidized biodiesel was observed and oxidized biodiesel produced 15 and 16% lower CO and unburned hydrocarbons (HC) exhaust emissions than unoxidized biodiesel. No statistically significant difference was found between NO_x and smoke emissions. Blends at 5% level using highly oxidized biodiesel were found to be compatible with fuel system components but 20% highly oxidized biodiesel showed that significant problems may occur with oxidized fuels (Terry et al., 2006).

Biodiesel blends can present some storage problems due to the degradation of biodiesel. It was reported that polymers formed during storage of biodiesel are soluble in oxidized biodiesel but are insoluble when biodiesel is mixed with petrodiesel (Bondioli et al., 2002). Sediments and gums formed can cause fuel filter plugging (Monyem et al., 2000). In one study, deposit formation was determined gravimetrically (Fang & McCormick, 2006).

Emissions

In 2006, the United States Environmental Protection Agency (EPA) regulations mandated that diesel fuel used for on-highway applications could contain no more than 15 ppm sulfur. This was a large decrease from the previous level of 500 ppm. The primary motivation for this reduction was to allow the introduction of exhaust catalysts to reduce oxides of nitrogen to the level required by the EPA in 2007. A consequence of the hydrotreating needed to reduce the sulfur to <15 ppm was that the fuel lost much of its lubricity. While biodiesel from virtually all vegetable oils is naturally low in sulfur, so compliance is not difficult, the need for lubricity in diesel fuel resulting from desulfurization provided an opportunity for biodiesel to be used at the 0.5 to 2.0 % level as a lubricity additive (Knothe & Steidley, 2005; Drown, 2001). The introduction of exhaust aftertreatment is also responsible for the tight limits on phosphorous, calcium, magnesium, sodium and potassium in the biodiesel as they can produce ash with the potential for deactivating or plugging particulate traps.

The pollutants of primary concern for diesel engines are carbon monoxide (CO), unburned hydrocarbons (HC), oxides of nitrogen (NO_x), and particulate matter (PM). Compared with spark-ignited, gasoline-fueled engines, CO and HC emissions from diesel engines are low and the regulations for these pollutants are relatively easy to satisfy. NO_x and PM are much more difficult and tend to respond in an opposing manner when engine design and operating parameters are changed. This is the termed the *NO_x – particulate tradeoff* (Heywood, 1988).

NO_x emissions are formed in virtually all combustion processes and originate from three fundamental mechanisms: fuel nitrogen, prompt NO_x and thermal NO_x. Fuel nitrogen can contribute to NO_x formation but since most transportation fuels contain little or no nitrogen, this is not a significant source of NO_x from engines. Similarly, prompt NO_x, which is NO_x formed by complex chemical reactions during the combustion itself, is also not believed to be a significant source of NO_x from engines (Heywood, 1988). Thermal NO_x is NO_x formed by elementary reactions between radicals, such as H, O, N, and OH, with stable species such as N_2 and O_2. It occurs in high-temperature post-flame gases because at the conditions found in engines the time and temperature conditions favor NO_x emissions by this mode (Heywood, 1988). One result has been the difficulty in treating NOx using additives. Since NO_x is formed in the post-flame gases, an additive must modify the time-temperature relationship of the post-flame gases. To have a direct effect on the chemical reactions that form NO_x, the additive would need to survive the flame and still be chemically active in the post-flame gases. Additives that change the combustion timing, such as cetane-enhancing additives, can influence NO_x production by shifting the timing of the combustion event or decreasing the initial rate of combustion of the fuel, and this can reduce NO_x (McCormick, 2002). Additives that can reduce NO_x by other means have not been found.

Biodiesel-fueled engines generally have higher levels of NO_x and lower emissions of PM than conventional diesel-fueled engines (Sharp et al., 2000a; 2000b). Figure 16.2 from an EPA report that surveyed biodiesel emissions data shows that the effects of blending biodiesel with diesel affect the emissions in an approximately linear manner (EPA, 2002). Increases in NO_x are expected to be 10–15% for B100 but only 2–3% for B20. The increases for B20 are difficult to measure consistently and differences between engines and testing protocols may be greater than the differences between diesel fuel and B20 (McCormick et al., 2006).

Energy Balance

Energy balance calculations are one element of the larger field of life-cycle analysis. These calculations usually focus on the ratio of the energy in the fuel to the energy required to produce the fuel. The most commonly cited analysis of the energy balance for soybean-based biodiesel was performed by Sheehan et al. (1998) who determined that the biodiesel contained 3.2 times more energy than was required to produce it. The reason that a fuel can contain more than 100% of its input energy is that the solar energy input is not included in the calculations.

Table 16.2 shows the fossil energy requirements for soybean-based biodiesel production from Sheehan et al. (1999). Most of the energy required can be divided into three main categories: soybean agriculture, extraction of the oil, and conversion to biodiesel. Approximately one-half of the input fossil energy is associated with the conversion of soybean oil to biodiesel. As crop yields increase and production technology improves, the energy inputs should decrease and the energy ratio should increase over time.

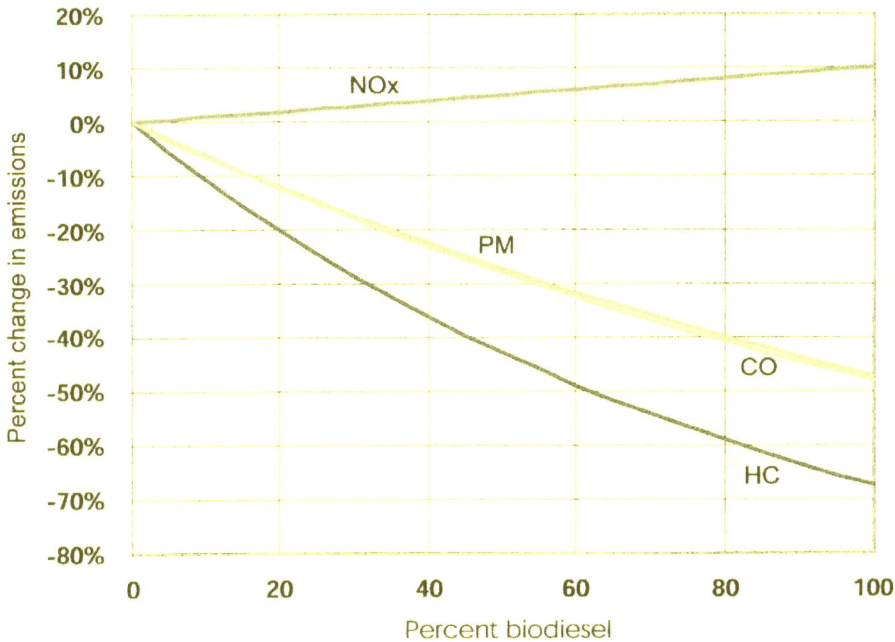

Fig. 16.2. Comprehensive analysis of biodiesel impacts on exhaust emissions (EPA 2002).

Table 16.2. Fossil Energy Requirements for the Biodiesel Life Cycle

Stage	Fossil Energy (MJ per MJ of Fuel)	Percent
Soybean Agriculture	0.0656	21.08%
Soybean Transport	0.0034	1.09%
Soybean Crushing	0.0796	25.61%
Soy Oil Transport	0.0072	2.31%
Soy Oil Conversion	0.1508	48.49%
Biodiesel Transport	0.0044	1.41%
Total	0.3110	100.00%

Other researchers have also studied biodiesel production from soybean oil (Ahmed, 1994; Hill et al, 2006; Pimentel & Patzek, 2005). Differences between these studies tend to be associated with the allocation of input energy between different byproduct streams and then how the byproduct energy is incorporated into the energy ratio calculations. For example, soybeans are crushed to separate the seeds into two primary products, high protein meal for livestock feed and oil for food use or biodiesel production. Approximately 18% of the seed weight is extracted as oil and 82% is meal. Researchers have used different approaches to dividing the input energy between these two products. Sheehan et al. (1998) chose to divide the input energy between the two products in proportion to their fraction of the incoming weight. Pimentel and Patzek (2005) estimated the energy value of the meal and then subtracted this as a credit from the input energy to leave the energy assumed to be needed to produce the oil. The energy content assigned to the meal by Pimentel and Patzek (2005) appears to be in error. They used 2 MJ/kg when the true value is about 19.95 MJ/kg (Beyer et al., 2003). With their assumption, the oil is associated with over 80% of the input energy, and the energy ratio is very close to 1.0. The problem with this approach is demonstrated when the actual energy content of the meal is used. In this case, the energy in the meal is greater than the input energy, so, when the meal energy is credited, the processing appears to produce rather than consume energy.

The energy balance for the production of an alternative fuel can be enhanced if the byproducts are used to produce some of the energy required to produce the fuel. For example, if the soybean meal or the glycerin were burned to provide process heat for the conversion of soybean oil to biodiesel, this would improve the overall energy balance. Currently, the economic value of these byproducts is higher for use as animal feed and other products than for use as fuel.

Glycerol Utilization

Besides alkyl esters, glycerol is the other product of the transesterification reaction (see Scheme 16.1). It may be noted that the term *glycerol* often denotes the pure compound while the term *glycerin* refers to the purified commercial products containing >95% glycerol (Appleby, 2005). However, in the literature these terms are generally used interchangeably.

Glycerol ($C_3H_8O_3$; 1,2,3-propanetriol) is a non-toxic, sweet-tasting, odorless, colorless, highly viscous and hygroscopic liquid, soluble in water and ethanol but insoluble in hydrocarbons. It boils at 290°C under decomposition. It has countless applications, which have been summarized previously (Jakobson et al., 1989; Morrison, 1994; Appleby, 2005). These applications include drugs, oral care products such as toothpaste and mouthwash, which are the two most significant uses of glycerol, cosmetics, urethane foams, lubricants, synthetic resins and ester gums as well as foods and tobacco. A more recent application for glycerol is in the formulation of aircraft anti- or deicing fluids (for example, see Samuels et al., 2006).

The increased production of biodiesel and thus glycerol has had a major effect on glycerol markets in the last few years, with glycerol prices dropping significantly. The increased production of glycerol from oleochemical sources has been accompanied by a sharp decline in synthetic glycerol, including even the closing of production facilities. In addition to the effect on traditional glycerol markets, the decrease in glycerol prices potentially can negatively affect the economics of biodiesel production. Therefore, it has become necessary to develop new uses for glycerol or expand existing markets in order to achieve some price stabilization.

Recent research concerning products from glycerol include its use as a feedstock and nutrient binder for the production of 1,2- and 1,3-propanediol, dihydroxyacetone, succinic acid, polyglycerols, polyesters, polyhydroxyalkanoates (Ashby et al., 2005a; 2005b; Koller et al., 2005), hyperbranched poly(glycerol-diacid) oligomers from by the acid-catalyzed condensation of glycerol with iminodiacetic, azelaic, or succinic acid (Wyatt et al., 2006), hydrogen and other materials (Claude, 1999; Pachauri & He, 2006). These materials themselves have numerous applications.

Some additional recent reports on hydrogen production from glycerol include a Ni catalyst on doped alumina improving selectivity in hydrogen production (Iriondo et al., 2006) and that the combination of Ru/C with Amberlyst was effective under mild conditions (Miyazawa et al., 2006). Steam reforming with a ruthenium catalyst was also used in hydrogen production from glycerol (Hirai et al., 2005).

Generally, 1,3-propanediol is currently one of the most "popular" products from syntheses with glycerol as starting material, since 1,2-propanediol is historically the more mature product and market. Within the project BIODIOL funded by the European Union, bioconversion processes of crude glycerol to 1,3-propanediol are being developed and evaluated (Hirschmann et al., 2005). The selective chemical dehydroxylation of glycerol to 1,3-propanediol from glycerol has been reported (Wang et al., 2003) and the mechanism of glycol formation from glycerol studied (Lahr & Shanks, 2003). Another recent report on the synthesis of 1,3-propanediol is Šmídová et al. (2006).

Fermentation of glycerol from biodiesel production gave citric acid (Imadi et al., 2007), while in other cases ethanol and succinate were obtained (Dharmadi et al., 2006). A biodiesel co-product stream consisting of 40% glycerol, 34% hexane-solubles (92% fatty acid soaps and methyl esters and 6% mono- and diacylglycerols) as well as 26% water was used as fermentation feedstock for the microbial synthesis of sophorolipids (Ashby et al., 2005a). Other syntheses using glycerol include carboxylation to glycerol carbonate with carbon dioxide in the presence of Sn catalysts (Aresta et al., 2006) and oxidation to ketomalonic acid using the catalyst TEMPO (2,2,6,6-tetramethylpiperidine-1-oxyl) with NaOCl as primary oxidant (Ciriminna & Pagliaro, 2003). One-pot electrocatalytic oxidation of glycerol to 1,3-dihydroxyacetone with a longer reaction under the applied conditions led to comparable amounts of hydroxypyruvic acid (Ciriminna et al., 2006). Instead of serving as a reactant, glycerol has also been used as a green solvent in the reduction of prochiral carbonyl compounds with baker's yeast (Wolfson et al., 2006).

Glycerol ethers have also been evaluated as fuel components. Ethers derived from glycerol and isobutene have been suggested as octane enhancers for automotive fuel, constituting an alternative to methyl *tert.*-butyl ether (MTBE) (Behr & Obendorf, 2003). The etherification of glycerol with *iso*-butene gives five ether products, two mono-ethers, two di-ethers and the tri-ether (Karinen & Krause, 2006). The products of the etherification of glycerol with *tert.*-butanol have been suggested as oxygenated additives for diesel fuel (Klepáová et al., 2003). A report on di-butoxy glycerol discusses this compound as a candidate material for blending with diesel fuel (Spooner-Wyman et al., 2003).

The transesterification of rapeseed oil with triacetin yielded a product in which all products could be used as fuel components, including triacetylglycerol, which was formed instead of glycerol (Lipkowski et al., 2005).

The crude glycerol phase from biodiesel production has several other potential applications. One use is to treat acid mine drainage. The crude glycerol served as carbon source for the sulfate-reducing bacteria in bioreactors used for this purpose (Zamzow et al., 2006). Other applications include the utilization as a feed component for swine (Kijora & Kupsch, 1996).

Conclusion

The decline in the supply of inexpensive petroleum-based fuels and the rise of concern for global climate change are driving the search for sustainable energy sources. Carbohydrate energy sources are attractive because of their wide spread availability but with the exception of starch-based ethanol, the technology to produce the liquid fuels needed for transportation from these materials is not commercially viable. In contrast, the high energy density and minimal processing requirements of fats and oils have made them an attractive option. Although the interaction of fuel and food production involves social and political issues that have not been resolved, it is likely that lipid-based fuels will play an increasing role in the world's energy future. In the United States, the source of these fuels will be predominately soybean oil for the foreseeable future.

References

Abreu, F.R.; M.B. Alves; C.C.S. Macêdo; L.F. Zara; P.A.Z. Suarez. New multi-phase catalytic systems based on tin compounds active for vegetable oil transesterification reaction. *J. Mol. Catal. A; Chemical* **2005,** *227,* 263–267.

Ahmad, S.; P. Siwayanan; Z.A. Murad; H.A. Aziz; H.S. Soi. Beyond biodiesel, methyl esters as the route for the production of surfactants feedstock. *INFORM* **2007,** *18,* 216–220.

Ahmed, I.; J. Decker; D. Morris. *How Much Energy Does It Take to Make a Gallon of Soydiesel?* National Soy Diesel Development Board: Jefferson City, MO, 1994.

Anon. Vegetable oil to heat new reichstag building. *INFORM* **1999,** *10,* 886.

Appleby, D.B. Glycerol, in The Biodiesel Handbook, G. Knothe, J. Van Gerpen, J. Krahl, Eds.; AOCS Press: Champaign, IL, 2005; pp. 264–276.

Aresta, M.; A. Dibenedetto; M. Carone; T. Colonna; C. Fragale. Production of biodiesel from macroalgae by supercritical CO_2 extraction and thermochemical liquefaction. *Environ. Chem. Lett.* **2005,** *3,* 136–139.

Aresta, M.; A. Dibenedetto; F. Nocito; C. Pastore. A study on the carboxylation of glycerol to glycerol carbonate with carbon dioxide: the role of the catalyst, solvent and reaction conditions. *J. Molec. Catal. A: Chemical* **2006,** *257,* 149–153.

Ashby, R.D.; A. Nunez; D.K.Y. Solaiman; T.A. Foglia. Sophorolipid biosynthesis from a biodiesel coproduct stream. *J. Am. Oil Chem. Soc.* **2005b,** *82,* 625–630.

Ashby, R.D.; D.K.Y. Solaiman; T.A. Foglia. Biomacromolecules **2005a,** *6,* 2106–2112.

ASTM. American Society for Testing and Materials. D 6751– Standard specification for biodiesel fuel (b100) blend stock for distillate fuels. *Annual Book of ASTM Standards;* ASTM International: West Conshohocken, PA, 2007; Vol. 05.04.

Baker, A.W.; R.L. Sweigert. *A Comparison of Various Vegetable Oils as Fuels for Compression-Ignition Engines.* Proc. Oil & Gas Power Meeting of the ASME; 1947, pp. 40–48.

Bay, K.; H. Wanko; J. Ulrich. Biodiesel—High-boiling absorbent for gas purification (Biodiesel: Hoch siedendes Absorbens für die Gasreinigung). *Chemie Ingenieur Technik* **2004,** *76,* 328–333.

Behr, A.; L. Obendorf, L. Development of a process for the acid-catalyzed etherification of glycerine and isobutene forming tertiary butyl ethers. *Eng. Life Sci.* **2003,** *2,* 185–189.

Bélafi-Bakó, K.; F. Kovács; L. Gubicza; J. Hancsök. Enzymatic biodiesel production from sunflower oil by candida antarctica lipase in a solvent-free system. *Biocatal. Biotransform.* **2002,** *20,* 437–439

Bertram, S.H. *Chem. Technol. (Dordrecht),* **1946,** *1.* 101.

Beyer, M.; A. Chudy; L. Hoffman; W. Jentsch; W. Laube; K. Nehring; R. Schiemann. *Rostock Feed Evaluation System.* Plexus Verlag: Miltenberg-Frankfurt, Germany, 2003.

Bikou, E.; A. Louloudi; N. Papayannakos. The effect of water on the transesterification kinetics of cotton seed oil with ethanol. *Chem. Eng. Technol.* **1999,** *22,* 70–75.

Boiscorjon d'Ollivier, A. French production of soybean oil. (La Production métropolitaine des Oléagineux: 'Le Soja'). *Rev. Combust. Liq.* **1939,** *17,* 225–235. *Chem. Abstr.* **1940,** *34,* 39377.

Bondioli, P. The preparation of fatty acid esters by means of catalytic reactions. *Topics in Catalysis* **2004,** *27,* 77–82.

Bondioli. P.; L. Folegatti. Evaluating the oxidation stability of biodiesel. An experimental contribution. *Riv. Ital. Sostanze Grasse* **1996,** *73,* 349–353.

Bondioli, P.; A. Gasparoli; L. Della Bella; S. Tagliabue. Evaluation of biodiesel storage stability using reference methods. *Eur. J. Lipid Sci. Technol.* **2002,** *104,* 777–784.

Bondioli, P.; A. Gasparoli; L. Della Bella; S. Tagliabue; F. Lacoste; L. Lagardere. The Prediction of biodiesel storage stability. proposal for a quick test. *Eur. J. Lipid Sci. Technol.* **2004,** *106,* 822–830.

Bondioli, P.;A. Gasparoli; L. Della Bella; S. Tagliabue; G. Toso. Biodiesel stability under commercial storage conditions over one year. *Eur. J. Lipid Sci. Technol.* **2003,** *105,* 735–741.

Bondioli, P.; A. Gasparoli; A. Lanzani; E. Fedeli; S. Veronese; M. Sala. Storage stability of biodiesel. *J. Am. Oil Chem. Soc.* **1995,** *72,* 699.

Boocock, D.G.B.; S.K. Konar; V. Mao; C. Lee; S. Buligan. Fast formation of high-purity methyl esters from vegetable oils. *J. Am. Oil Chem. Soc.* **1998,** *75,* 1167–1172.

Boocock, D.G.B., S.K. Konar; V. Mao; H. Sidi. Fast one-phase oil-rich processes for the preparation of vegetable oil methyl esters. *Biomass Bioenergy* **1996,** *11,* 43–50.

Breccia, A.; B. Esposito; G. Breccia Fratadocchi; A. Fini. Reaction between methanol and commercial seed oils under microwave irradiation. *J. Microwave Power Electromagn. Energy.* **1999,** *34,* 3–8.

Bruwer, J.J.; B.v.D. Boshoff; F.J.C. Hugo; L.M. du Plessis; J. Fuls; C. Hawkins; A.N. van der Walt; A. Engelbrecht. Sunflower Seed *Oil as an Extender for Diesel Fuel in Agricultural Tractors.* Symposium of the South African Institute of Agricultural Engineers, 1980a.

Bruwer, J.J.; B.v.D. Boshoff; F.J.C. Hugo; J. Fuls; C. Hawkins; A.N. van der Walt; A. Engelbrecht; L.M. du Plessis. *The Utilization of Sunflower Seed Oil as a Renewable Fuel for Diesel Engines.* National Energy Symposium, American Society of Agricultural Engineers, 1980b.

Bruwer, J.J.; B.v.D. Boshoff; F.J.C. Hugo; J. Fuls; C. Hawkins; A.N. van der Walt; A. Engelbrecht; L.M. du Plessis. *The Utilization of Sunflower Seed Oil as a Renewable Fuel for Diesel Engines Comm. Eur. Communities, [Rep.] EUR 1981, EUR 7091,* Energy Biomass Conf., 1981; 1st, pp. 934–940.

Bunyakiat, K.; S. Makmee; R. Sawangkeaw; S. Ngamprasertsith. Continuous production of biodiesel via transesterification from vegetable oils in supercritical methanol. *Energy & Fuels* **2006,** *20,* 812–817.

Canakci, M.; A. Monyem; J. Van Gerpen. Accelerated oxidation processes in biodiesel. *Trans. ASAE* **1999,** *42,* 1565.

Canakci, M.; J. Van Gerpen. A pilot plant to produce biodiesel from high free fatty acid feedstocks. *Trans. ASAE* **2003a,** *46,* 945–954.

Canakci, M.; J. Van Gerpen. Comparison of engine performance and emissions for petroleum diesel fuel, yellow grease biodiesel, and soybean oil biodiesel. *Trans. ASAE* **2003b,** *46,* 937–944.

Canakci, M.; J. Van Gerpen. Biodiesel production from oils and fats with high free fatty acids. *Trans. ASAE* **2001**, *44,* 1429–1436.

Canakci, M.; J. Van Gerpen. Biodiesel production via acid catalysis. *Trans. ASAE* **1999,** *42,* 1203–1210.

Cantrell, D.G.; L.J. Gillie; A.F. Lee; K. Wilson. Structure-reactivity correlations in mgal hydrotalcite catalysts for biodiesel synthesis. *Appl. Catal. A* **2005,** 287, 183–190.

Cao, W.; H. Han; J. Zhang. Preparation of biodiesel from soybean oil using supercritical methanol and co-solvent. *Fuel* **2005,** *84,* 347–351.

CEN. EN 14214 *Automotive Fuels: FAME for Diesel Engines; Requirements and Test Methods.* European Committee for Standardization, 2003.

Çetinkaya, M.; F. Karaosmanoglu. A new application area for used cooking oil originated biodiesel: generators. *Energy Fuels* **2005,** *19,* 645–652.

Chavanne, C.G. Procédé de transformation d'huiles végétales en vue de leur utilisation comme carburants (Procedure for the transformation of vegetable oils for their uses as fuels). Belgian Patent 422,877 (August 31, 1937). *Chem. Abstr.* **1938,** *32,* 43132.

Chavanne, C.G. Sur un Mode d'Utilization Possible de l'Huile de Palme à la Fabrication d'un Carburant Lourd (A method of possible utilization of palm oil for the manufacture of a heavy fuel). *Bull. Soc. Chim.* **1943,** *10,* 52–58. *Chem. Abst.* **1944,** *38,* 21839.

Chisti, Y. Biodiesel from microalgae. *Biotechnol. Advances* **2007,** *25,* 294–306.

Chiu, C.W.; M.J. Goff; G.J. Suppes. Distribution of methanol and catalysts between biodiesel and glycerin phases. *AIChE J.* **2005,** *51,* 1274–1278.

Ciriminna, R.; M. Pagliaro. One-pot homogeneous and heterogeneous oxidation of glycerol to ketomalonic acid mediated by tempo. *Adv. Synth. Catal.* **2003,** *345,* 383–388.

Ciriminna, R.; G. Palmisano; C. Della Pina; M. Rossi; M. Pagliaro. One-pot electrocatalytic oxidation of glycerol to dha. *Tetrahedron Lett.* **2006,** *47,* 6993–6995.

Claude, S. Research of new outlets for glycerol—Recent developments in France. *Fett/Lipid* **1999,** *101,* 101–104.

Cvengroš, J.; F. Povalanec. Production and treatment of rapeseed oil methyl esters as alternative fuels for diesel engines. *Bioresour. Technol.* **1996,** *55,* 145–152.

Czernichowski, A.; M. Czernichowski; K. Wesolowska. Generation of 1 kg/h of hydrogen from soybean biodiesel. Prepr. Pap. *Am. Chem. Soc., Div. Fuel Chem.* **2006,** *51,* 619–620.

Dasari, M.A.; M.J. Goff; G.J. Suppes. Noncatalytic alcoholysis kinetics of soybean oil. *J. Am. Oil Chem. Soc.* **2003,** *80,* 189–192.

Demirbas, A. Biodiesel from vegetable oils via transesterification in supercritical methanol. *Energy Convers. Managem.* **2002,** *43,* 2349–2356.

Demirbas, A. biodiesel fuels from vegetable oils via catalytic and non-catalytic supercritical alcohol transesterifications and other methods: a survey. *Energy Conv. Managem.* **2003**, *44,* 2093–2109.

Demirbas, A.; S. Karslioglu. Biodiesel production facilities from vegetable oils and animal fats. *Energy Sources, Part A* **2007,** *29,* 133–141.

Deng, L.; X. Xu; G.G. Haraldsson; T. Tan; F. Wang. Enzymatic production of alkyl esters through alcoholysis: a critical evaluation of lipases and alcohols. *J. Am. Oil Chem. Soc.* **2005,** *82,* 341–347.

Di Serio, M.; M. Ledda; M. Cozzolino; G. Minutillo; R. Tesser; E. Santacesaria. Transesterification of soybean oil to biodiesel by using heterogeneous basic catalysts. *Ind. Eng. Chem. Res.* **2006,** *45,* 3009–3014.

Diesel, R. *Die Entstehung des Dieselmotors.* Berlin, Verlag von Julius Springer: Berlin, 1913.

Diesel, R. The diesel oil-engine. *Engineering* **1912a,** *93,* 395–406. *Chem. Abstr.* **1912a,** *6,* 1984.

Diesel, R. The diesel oil-engine and its industrial importance particularly for great britain. *Proc. Inst. Mech. Eng.* **1912b,** 179–280. *Chem. Abstr.* **1913,** *7,* 1605.

Dimmig, T.; J. Haupt; W. Radig. *Biodiesel: Diesel Fuel Substitute from Vegetable Oils and Recycling Products.* (Biodiesel aus Pflanzenölen und nativen Recyclingprodukten.) Freiberger Foschungshefte 1999, A852 (Neue Erkenntnisse zur stofflichen Nutzung von nachwachsenden Rohstoffen) pp. 34–60.

Dittmar, T.; B. Ondruschka; J. Haupt; M. Lauterbach. determination of the influence of raw material and process technology on the oxidative stability of fatty acid methyl esters. (Ermittlung des Einflusses von Fettrohstoff und Prozesstechnologie auf die Oxidationsstabilität von Fettsäuremethylester.) *Chemie Ing. Techn.* **2004a,** *76,* 1170–1179.

Dittmar, T.; B. Ondruschka; J. Haupt; M. Lauterbach. Improvement of the oxidative stability of fatty acid methyl esters with antioxidants—limitations of the rancimat test. *Chemie Ing. Techn.*

2004b, *76,* 1167–1170. (Verbesserung der Oxidationsstabilität von Fettsäuremethylestern mit Antioxidantien—Grenzen des Rancimat-Tests.)

Dossin, T.F.; M.-F. Reyniers; R.J. Berger; G.B. Marin. Simulation of heterogeneously mgo-catalyzed transesterification for fine-chemical and biodiesel industrial production. *Appl. Catal. B: Environm.* **2006,** *67,* 136–148.

Drown, D.C.; K. Harper; E. Frame. Screening vegetable oil alcohol esters as fuel lubricity enhancers. *J. Am. Oil Chem. Soc.* **2001,** *78,* 579–584.

Du Plessis, L.M. Plant oils as diesel fuel extenders: stability tests and specifications on different grades of sunflower seed and soyabean oils. *CHEMSA* **1982**, *8*, 150–154.

Du Plessis, L.M.; J.B.M. de Villiers; W.H. van der Walt. Stability studies on methyl and ethyl fatty acid esters of sunflowerseed oil. *J. Am. Oil Chem. Soc.* **1985,** *62,* 748.

Dubé, M.; A.Y. Tremblay; J. Liu. Biodiesel production using a membrane reactor. *Bioresour. Technol.* **2007,** *98,* 639–647.

Dunn, R.O. Effect of oxidation under accelerated conditions on fuel properties of methyl soyate (biodiesel). *J. Am. Oil Chem. Soc.* **2002,** *79,* 915–920.

Dunn, R.O. Oxidative stability of soybean oil fatty acid methyl esters by oil stability index (osi). *J. Am. Oil Chem. Soc.* **2005b,** *82,* 381–387.

Dunn, R.O. Alternative jet fuels from vegetable oils. *Trans. ASAE* **2001,** *44,* 1751–1757.

Dunn, R.O. Analysis of oxidative stability of methyl soyate by pressurized-differential scanning calorimetry. *Trans. ASAE* **2000,** *43,* 1203–1208.

Dunn, R.O. Cold weather properties and performance of biodiesel. *The Biodiesel Handbook;* G. Knothe, J. Krahl, J. Van Gerpen, Eds.; AOCS Press: Champaign, IL, 2005a; pp. 83–121.

Dunn, R.O. Oxidative stability of biodiesel by dynamic mode pressurized-differential scanning calorimetry (p-dsc). *Trans. ASABE* **2006,** *49,* 1633–1641.

Dunn, R.O. Effect of antioxidants on the oxidative stability of methyl soyate (biodiesel). *Fuel Proc. Technol.* **2005a,** *86,* 1071–1085.

Encinar, J.M.; J.F. Gonzalez; J.J. Rodriguez; A. Tejedor. Biodiesel fuels from vegetable oils: transesterification of cynara cardunculus l. oils with ethanol. *Energy & Fuels* **2002**, *16,* 443–450.

EPA. Environmental Protection Agency. *A Comprehensive Analysis of Biodiesel Impacts on Exhaust Emissions.* Draft Technical Report. EPA 420-P-02-001. October 2002.

Falk, O.; R. Meyer-Pittroff. The effect of fatty acid composition on biodiesel oxidative stability. *Eur. J. Lipid Sci. Technol.* **2004**, *106,* 837–843.

Fang, H.L.; R.L. McCormick. Spectroscopic study of biodiesel degradation pathways. *SAE Techn. Pap. Ser.* 2006-01-3300; 2006.

Fernández-Álvarez, P.; J. Vila; J.M. Garrido-Fernández; M. Grifoll; J.M. Lema. Trials of bioremediation on a beach affected by the heavy oil spill of the prestige. *J. Hazardous Materials* **2006,** *137,* 1523–1531.

Ferrari, R.A.; V.d.S. Oliveira; A. Scabio. Biodiesel from soybean: characterization and consumption in an energy generator. (Biodiesel de Soja—Taxa de Conversão em Ésteres Etílicos, Caracterização Físico-Química e Consumo em Gerador de Energia.) *Quim. Nova* **2005,** *28,* 19–23.

Frankel, E.N. *Lipid Oxidation,* Second ed.; PJ Barnes and Associates: Bridgwater, England, 2005.

Freedman, B.; R.O. Butterfield; E.H. Pryde. Transesterification kinetics of soybean oil, *J. Am. Oil Chem. Soc.* **1986a,** *63,* 1375–1380.

Freedman, B.; W.F. Kwolek; E.H. Pryde. Quantitation in the analysis of transesterified soybean oil by capillary gas chromatography. *J. Am. Oil Chem. Soc.* **1986b,** *63,* 1370–1375.

Freedman, B.; E.H. Pryde. Fatty esters from vegetable oils for use as a diesel fuel. *Proceedings of the International Conference on Plant and Vegetable Oils as Fuels.* ASAE Publ. 4-82 (Veg. Oil Fuels). 1982; pp. 117–122.

Freedman, B., E.H. Pryde; T.L. Mounts. Variables affecting the yields of fatty esters from transesterified vegetable oils. *J. Am. Oil Chem. Soc.* **1984,** *61,* 1638–1643.

Fukuda, H.; A. Kondo; H. Noda. Biodiesel fuel production by transesterification of oils. *J. Biosci. Bioeng.* **2001,** *92,* 405–416.

Gaupp, K. Pflanzenöle als Dieselkraftstoffe (Plant oils as diesel fuel), *Automobiltech. Z.* **1937,** *40,* 203–207. *Chem. Abstr.* **1937,** *31,* 88765.

Goering, C.E.; R.M. Campion; A.W. Schwab; E.H. Pryde. Evaluation of soybean oil-aqueous ethanol microemulsions for diesel engines. Proceedings of the international conference on plant and vegetable oils as fuels, ASAE Publ. 4-82 (*Veg. Oil Fuels*) **1982,** 279–286.

Gryglewicz, S. Rapeseed oil methyl esters preparation using heterogeneous catalysts. *Bioresour. Technol.* **1999,** *70,* 249–253.

Gutsche, B. Technology of methyl ester production and its application to biofuels. (Technologie der Methylesterherstellung - Anwendung für die Biodieselproduktion.) *Fett/Lipid* **1997,** *99,* 418–427.

Haas, M.J.; S. Bloomer; K. Scott. Simple, high-efficiency synthesis of fatty acid methyl esters from soapstock. *J. Am. Oil Chem. Soc.* **2000,** *77,* 373–379.

Haas, M.J.; A. McAloon; W.C. Yee; T.A. Foglia. A process model to estimate biodiesel production costs. *Bioresour. Technol.* **2006,** *97,* 671–678.

Haas, M.J.; P.J. Michalski; S. Runyon; A. Nunez; K.M. Scott. Production of fame from acid oil, a by-product of vegetable oil refining. *J. Am. Oil Chem. Soc.* **2003,** *80,* 97–102.

Haas, M.J.; G.J. Piazza; T.A. Foglia. Enzymatic approaches to the production of biodiesel fuels. *Lipid Biotechnology;* T.M. Kuo, H.W. Gardner, Eds.; Marcel Dekker: New York, Basel, 2002; pp. 587–598.

Haas, M.J.; K.M. Scott. Moisture removal substantially improves the efficiency of in situ biodiesel production from soybeans. *J. Am. Oil Chem. Soc.* **2007,** *84,* 197–204.

Haas, M.J., K. M. Scott; T.L. Alleman; R.L. McCormick. Engine performance of biodiesel fuel prepared from soybean soapstock: a high quality renewable fuel produced from a waste feedstock. *Energy Fuels* **2001,** *15,* 1207–1212.

Haas, M.J.; K.M. Scott; W.N. Marmer; T.A. Foglia. In situ alkaline transesterification: an effective method for the production of fatty acid esters from vegetable oils. *J. Am. Oil Chem. Soc.* **2004,** *81,* 83–89.

Hancsók, J.; M. Bubálik; M. Tör; J. Baladincz. Synthesis of fuel additives on vegetable oil basis at laboratory scale. *Eur. J. Lipid Sci. Technol.* **2006,** *108,* 644–651.

Harrington, K.J.; C. D'Arcy-Evans. A comparison of conventional and in situ methods of transesterification of seed oil from a series of sunflower cultivars. *J. Am. Oil Chem. Soc.* **1985,** *62,* 1009–1013.

He, H.; S. Sun; T. Wang; S. Zhu. Transesterification kinetics of soybean oil for production of biodiesel in supercritical methanol. *J. Am. Oil Chem. Soc.* **2007a,** *84,* 399–404.

He, H.; T. Wang; S. Zhu. Continuous production of biodiesel fuel from vegetable oil using supercritical methanol process. *Fuel* **2007b,** *86,* 442–447.

Heywood, J.B. *Internal Combustion Engine Fundamentals.* McGraw Hill Book Co.: New York, NY, 1988

Hill, J.; E. Nelson; D. Tilman; S. Polasky; D. Tifanny. Environmental, economic and energetic costs and benefits of biodiesel and ethanol biofuels. *PNSS* **2006,** *103,* 11206–11210.

Hillion, S.R;, J.-J. Rouxel; S. Leporq. Process for the Production of Esters from Vegetable Oils or Animal Oils Alcohols. U.S. Patent 5,908,946. June 1, 1999.

Hirai, T.; N. Ikenaga; T. Miyake; T. Suzuki. Production of hydrogen by steam reforming of glycerin on ruthenium catalyst. *Energy & Fuels* **2005,** *19,* 1761–1762.

Hirschmann, S.; K. Baganz; I. Koschik; K.-D. Vorlop. Development of an integrated bioconversion process for the production of 1,3-propanediol from raw glycerol esters. *Landbauforschung Völkenrode* **2005,** *55,* 261–267.

Hoydonckx, H.E.; D.E. De Vos; S.A. Chavan; P.A. Jacobs. Esterification and transesterification of renewable chemicals. *Top. Catal.* **2004**, *27,* 83–96.

Hsu, A.-F.; K. Jones; T.A. Foglia; W.N. Marmer. Immobilized lipase-catalyzed production of alkyl esters of restaurant grease as biodiesel. *Biotechnol. Appl. Biochem.* **2002,** *36,* 181–186.

Hsu, A.-F.; K.C. Jones; T.A. Foglia; W.N. Marmer. Optimization of alkyl ester production from grease using a phyllosilicate sol-gel immobilized lipase. *Biotechnol. Lett.* **2003,** *25,* 1713–1716.

Hu, J.; Z. Du; Z. Tang; E. Min. Study on the solvent power of a new green solvent. *Biodiesel. Ind. Eng. Chem. Res.* **2004,** *43,* 7928–7931.

Huguenard, C.M. Dual Fuel for Diesel Engines Using Cottonseed Oil, M.S. Thesis, The Ohio State University, Columbus, OH, 1951.

Imandi, S.B.; V.V.R. Nabdaru; S.R. Somalanak; H.R. Garapati. Optimization of medium constituents for the production of citric acid from byproduct glycerol using doehlert experimental design. *Enzyme Microb. Technol.* **2007**, *40,* 1367–1372.

Iriondo, A.; V.L. Barrio; J.F. Cabra; P.L. Arias; M.B. Güemez; R.M. Navarro; M.C. Sánchez-Sanchez; J.L.G. Fierro. H2 Production from residual glycerol obtained from biomass transesterification. Prepr. Pap—*Am. Chem. Soc., Div. Fuel Chem.* **2006,** *51,* 557–558.

Jakobson, G.; F.W. Kathagen; M. Klatt; U. Steinberner. *Glycerol in Ullmann's Encyclopedia of Industrial Chemistry,* Fifth ed.;, VCH-Verlag: Weinheim, Germany, 1989; 477–489.

Johnson, L.A.; E.G. Hammond. Soybean Oil Ester Fuel Blends. U.S. Patent 5,520,708. 1996.

Kaieda, M.; T. Samukawa; T. Matsumoto; K. Ban; A. Kondo; Y. Shimada; H. Noda; F. Nomoto; K. Ohtsuka; E. Izumoto; H. Fukuda. Biodiesel fuel production from plant oil catalyzed by rhizopus oryzae lipase in a water-containing system without an organic solvent. *J. Biosci. Bioeng.* **1999,** *88,* 627–631.

Kamal-Eldin A.; J. Pokorný, Eds. *Analysis of Lipid Oxidation,* AOCS Press: Champaign, IL, 2005.

Karinen, R.S.; A.O.I. Krause. New biocomponents from glycerol. *Appl. Catal. A* **2006**, *306*, 128–133.

Kijora, C.; R.-D. Kupsch. Evaluation of technical glycerols from "biodiesel" production as a feed component in fattening of pigs. *Fett/Lipid* **1996,** *98 (7–8)*, 240–245.

Kildiran, G.; S. Ö. Yücel; S. Türkay. In-situ alcoholysis of soybean oil. *J. Am. Oil Chem. Soc.* **1996,** *73*, 225–228.

Kim, H.J.; B.-S. Kang; M.-J. Kim; Y.M. Park; D.K. Kim; J.-S. Lee; K.-Y. Lee. Transesterification of vegetable oil to biodiesel using heterogeneous base catalyst. *Catal. Today* **2004,** *93–95*, 315–320.

Kiss, A.A.; A.C. Dimian; G. Rothenberg. Solid acid catalysts for biodiesel production—towards sustainable energy *Adv. Synth. Catal.* **2006,** *348*, 75–81.

Klepáová, K.; D. Mravec; E. Hájeková; M. Bajus; Etherification of glycerol. *Petroleum & Coal* **2003,** *45*, 54–57.

Klopfenstein, N.W. Effect of molecular weights of fatty acid esters on cetane numbers as diesel fuels. *J. Am. Oil Chem. Soc.* **1985,** *62*, 1029–1031.

Knothe, G. Analysis of oxidized biodiesel by 1h-nmr and effect of contact area with air. *Eur. J. Lipid Sci. Technol.* **2006a,** *108*, 493–500.

Knothe, G. Analyzing biodiesel: standards and other methods. *J. Am. Oil Chem. Soc.* **2006b,** *83*, 823–833.

Knothe, G. Historical perspectives on vegetable oil-based diesel fuels. *INFORM* **2001,** *12*, 1103–1107.

Knothe, G. Structure indices in FA chemistry. How relevant is the iodine value? *J. Am. Oil Chem. Soc.* **2002,** *79*, 847–854.

Knothe, G. The history of vegetable oil-based diesel fuels. *The Biodiesel Handbook,* G. Knothe, J. Krahl, J. Van Gerpen, Eds; AOCS Press: Champaign, IL, 2005; pp. 4–16.

Knothe, G.; M.O. Bagby; T.W. Ryan III. *Cetane Numbers of Fatty Compounds: Influence of Compound Structure and of Various Potential Cetane Improvers.* SAE Technical Paper Series No. 971681 Society of Automotive Engineers, Warrendale, PA, 1997.

Knothe, G.; R.O. Dunn. Dependence of oil stability index of fatty compounds on their structure and concentration and presence of metals. *J. Am. Oil Chem. Soc.* **2003,** *80*, 1021–1026.

Knothe,G.; A.C. Matheaus; T.W. Ryan III. Cetane numbers of branched and straight-chain fatty esters determined in an ignition quality test. *Fuel* **2003,** *82*, 971–975.

Knothe, G.; R. Steidley. Lubricity of components of biodiesel and petrodiesel. the origin of biodiesel lubricity. *Energy & Fuels* **2005,** *19*, 1192–1200.

Knothe, G.; J. Van Gerpen; J. Krahl, (Eds.) *The Biodiesel Handbook 2005*, AOC Press: Champaign, IL.

Koller, M.; R. Bona; G. Braunegg; C. Hermann; P. Horvat; M. Kroutil; J. Martinz; J. Neto; L. Pereira; P. Varila. Production of polyhydroxyalkanoates from agricultural waste and surplus materials. *Biomacromolecules* **2005**, *6*, 561–565.

Kulkarni, M.G.; A.K. Dalai. Waste cooking oil—An economical source for biodiesel: A review. *Ind.*

Eng. Chem. Res. **2006,** *45,* 2901–2913.

Lacoste, F.; L. Lagardere. quality parameters evolution during biodiesel oxidation using rancimat test. *Eur. J. Lipid Sci. Technol.* **2003,** *105,* 149–155.

Lahr, D.G.; B.H. Shanks. Kinetic analysis of the hydrogenolysis of lower polyhydric alcohols: Glycerol to glycols. *Ind. Eng. Chem.* **2003,** *42,* 5467–5472.

Leadbeater, N.E.; L.M. Stencel. Fast, easy preparation of biodiesel using microwave heating. *Energy Fuels* **2006,** *20,* 2281–2283.

Leclercq, E.; A. Finiels; C. Moreau. Transesterification of rapeseed oil in the presence of basic zeolites and related solid catalysts. *J. Am. Oil Chem. Soc.* **2001,** *78,* 1161–1165.

Lee, I.; L.A. Johnson; E.G. Hammond. Use of branched-chain esters to reduce the crystallization temperature of biodiesel. *J. Am. Oil Chem. Soc.* **1995,** *72,* 1155–1160.

Lem, R.F.-A. Dual Fuel for Diesel Engines Using Corn Oil with Variable Injection Timing, M.S. Thesis, The Ohio State University, Columbus, OH, 1952.

Leung, D.Y.C.; B.C.P. Koo; Y. Guo. Degradation of biodiesel under different storage conditions. *Bioresour. Technol.* **2006,** *97,* 250–256.

Liang, Y.C.; C.Y. May; C.S. Foon; M.A. Ngan; C.C. Hock; Y. Basiron. The effect of natural and synthetic antioxidants on the oxidative stability of palm diesel. *Fuel* **2006,** *85,* 867–870.

Linko, Y.-Y.; M. Lämsä; X. Wu; E. Uosukainen; J. Seppälä; P. Linko. Biodegradable products by lipase biocatalysis. *J. Biotechnol.* **1998,** *66,* 41–50.

Lipkowski, A.W.; J. Kijenski; W. Walisiewicz-Niedbalska. Gliperol—New biofuel for diesel engines. *Chemik* **2005,** *58,* 238–240.

Loh, S.-K.; S.-M. Chew; Y.-M. Choo. Oxidative stability and storage behavior of fatty acid methyl esters derived from used palm oil. *J. Am. Oil Chem. Soc.* **2006,** *83,* 947–952.

Lotero, E.; J.G. Goodwin, Jr.; D.A. Bruce; K. Suwannakarn; Y. Liu; D.E. Lopez. The catalysis of biodiesel synthesis. *Catalysis* **2006,** *19,* 41–83.

Lotero, E.; Y. Liu; D.E. Lopez; K. Suwannakarn; D.A. Bruce; J.G. Goodwin, Jr. Synthesis of biodiesel via acid catalysis. *Ind. Eng. Chem. Res.* **2005.** *44,* 5353–5363.

Ma, F.; L.D. Clements; M.A. Hanna. biodiesel fuel from animal fat. ancillary studies on transesterification of beef tallow. *Ind. Eng. Chem. Res.* **1998a,** *37,* 3768–3771.

Ma, F.; L.D. Clements; M.A. Hanna. the effects of catalyst, free fatty acids, and water on transesterification of beef tallow. *Trans. ASAE.* **1998b,** *41,* 1261–1264.

Ma, F.; L.D. Clements; M.A. Hanna. The effect of mixing on transesterification of beef tallow, *Bioresour. Technol.* **1999,** *69,* 289–293.

Ma, F.; M.A. Hanna. Biodiesel production: a review. *Bioresour. Technol.* **1999,** *70,* 1–15.

Marchetti, J.M.; V.U. Miguel; A.F. Errazu. Heterogeneous esterification of oil with high amount of free fatty acids. *Fuel* **2007a,** *86,* 906–910.

Marchetti, J.M.; V.U. Miguel; A.F. Errazu. Possible methods for biodiesel production. *Renew. Sustain. Energy Rev.* **2007b,** *11,* 1300–1311.

Mathot, R.E. Vegetable oils for internal combustion engines. *Engineer* **1921**, *132,* 138–139. *Chem. Abstr.* **1921**, *15,* 3735.

Mazzocchia, C.; G. Modica; A. Kaddouri; R. Nannicini. Fatty acid methyl esters synthesis from triglycerides over heterogeneous catalysts in the presence of microwaves. *C.R. Chimie* **2004,** *7,* 601–605.

Mbaraka, I.K.; B.H. Shanks. Conversion of oils and fats using advanced mesoporous heterogeneous catalysts. *J. Am. Oil Chem. Soc.* **2006,** *83,* 79–91.

Mbaraka, I.K.; B.H. Shanks. Design of multifunctionalized mesoporous silicas for esterification of fatty acid. *J. Catalysis* **2005,** *229,* 365–373.

McCormick, R.L.:.J.R. Alvarez; M.S. Graboski; K.S. Tyson; K. Vertin. *Fuel Additive and Blending Approaches to Reducing No_x Emissions from Biodiesel,* Society of Automotive Engineers Technical paper Series. SAE 2002-01-1658. 2002. SAE: Warrendale, PA.

McCormick, R.L.; A. Williams; J. Ireland; M. Brimhall; R.R. Hayes. *Effects of Biodiesel Blends on Vehicle Emissions. Fiscal Year 2006 Annual Operating Plan—Milestone 10.4.* National Renewable Energy Laboratory. NREL/MP-540-40554, October 2006.

Meher, L.C.; D.V. Sagar; S.N. Nail. Technical aspects of biodiesel production by transesterification—A review. *Renew. Sustain. Energy Rev.* **2006,** *10,* 248–268.

Miao, X.; Q. Wu. Biodiesel production from heterotrophic microalgal oil. *Bioresour. Technol.* **2006,** *97,* 841–846.

Miller, N.J.; S.M. Mudge. The effect of biodiesel on the rate of removal and weathering characteristics of crude oil within artificial sand columns. *Spill Sci. Technol. Bull.* **1997,** *4,* 17–33.

Minami, E.; S. Saka. Kinetics of hydrolysis and methyl esterification for biodiesel production in two-step supercritical methanol process. *Fuel* **2006,** *85,* 2479–2483.

Mittelbach, M. Lipase catalyzed alcoholysis of sunflower oil. *J. Am. Oil Chem. Soc.* **1990,** *67,* 168–170.

Mittelbach, M.; G. Roth; A. Bergmann. Simultaneous gas chromatographic determination of methanol and free glycerol in biodiesel. *Chromatographia* **1996,** *Vol. 42,* 431–434.

Mittelbach, M.; S. Schober. The influence of antioxidants on the oxidation stability of biodiesel. *J. Am. Oil Chem. Soc.* **2003,** *80,* 817–823.

Mittelbach, M.; B. Trathnigg. Kinetics of alkaline catalyzed methanolysis of sunflower oil. *J. Am. Oil Chem. Soc.* **1990,** *92,* 145–148.

Miyazawa. T.; K. Kunimori; K. Tomishige. Glycerol hydrogenolysis in aqueous solution over ru/c+ an ion exchange resin and its reaction mechanism. Prepr. Pap.—*Am. Chem. Soc., Div. Fuel Chem.* **2006,** *51,* 530–531.

Monyem, A.; M. Canakci; J.H. Van Gerpen. Investigation of biodiesel thermal stability under simulated in-use conditions. *Appl. Eng. Agric.* **2000,** *16,* 373–378.

Monyem, A.; J.H. Van Gerpen. The effect of biodiesel oxidation on engine performance and emissions. *Biomass Bioenergy* **2001,** *20,* 317–325.

Monyem, A.; J.H. Van Gerpen; M. Canakci. The effect of timing and oxidation on emissions from biodiesel-fueled engines. *Trans. ASAE* **2001,** *44,* 35–42.

Morrison, L.R. Glycerol, *Kirk-Othmer Encyclopedia of Chemical Technology*, Fourth ed.; John Wiley & Sons: New York, 1994; Vol. 12, pp. 681–694.

Mudge, S.M.; G. Pereira. Stimulating the biodegradation of crude oil with biodiesel. preliminary results. *Spill Sci. Technol. Bull.* **1999,** *5,* 353–355.

Mushrush, G.; E.J. Beal; G. Spencer; J.H. Wynne; C.L. Lloyd; J.M. Hughes; C.L. Walls; D.R. Hardy. An environmentally benign soybean derived fuel as a blending stock or replacement for home heating oil. *J. Environ. Sci. Health* **2001**, *A36,* 613–622.

Nagle, N.; P. Lemke. Production of methyl ester fuel from microalgae. *Appl. Biochem. Biotechnol.* **1990,** 355–361.

Nakazono, Y. Production technology for biodiesel fuels. *Eco Industry* **2003,** *8,* 43–53. (Article in Japanese).

Nelson, L.A.; T.A. Foglia; W.N. Marmer. Lipase-catalyzed production of biodiesel, *J. Am. Oil Chem. Soc.* **1996,** *73,* 1191–1195.

Noureddini, H.; D. Zhu. Kinetics of transesterification of soybean oil. *J. Am. Oil Chem. Soc.* **1997,** *74,* 1457–1463.

Obbard, J.P.; K.L. Ng; R. Xu. Bioremediation of petroleum contaminated beach sediments: use of crude palm oil and fatty acids to enhance indigenous biodegradation. *Water, Air & Soil Pollut.* **2004,** *157,* 149–161.

Pachauri, N.; B. He. *Value-added Utilization of Crude Glycerol from Biodiesel Production: A Survey of Current Research Activities.* ASABE Paper No. 066223, 2006 ASABE Annual International Meeting.

Pereira, M.G.; S.M. Mudge. Cleaning oiled shores: laboratory experiments testing the potential use of vegetable oil biodiesels. *Chemosphere* **2004,** *54,* 297–304.

Peter, S.K.F.; R. Ganswindt; H.-P. Neuner; E. Weidner. Alcoholysis of triacylglycerols by heterogeneous catalysis. *Eur. J. Lipid Sci. Technol.* **2002,** *104,* 324–330.

Peters, R.A. Fatty alcohol production and use. INFORM **1996,** *7,* 502–504.

Peterson, G.R.; W.P. Scarrah. Rapeseed oil transesterification by heterogeneous catalysis. *J. Am. Oil Chem. Soc.* **1984**, *61,* 1593–1597.

Pimentel, D.; T.W. Patzek. Ethanol production using corn, switchgrass and wood; biodiesel production using soybean and sunflower. *Natural Resour. Res.* **2005,** *14,* 65–76.

Pizarro, A.V.L.; E.Y. Park. Lipase-catalyzed production of biodiesel fuel from vegetable oils contained in waste activated bleaching earth. *Process Biochem.* **2003,** *38,* 1077–1082.

Plank, C.; E. Lorbeer. Simultaneous determination of glycerol, and mono-, di-, and triglycerides in vegetable oil methyl esters by capillary gas chromatography. *J. Chromatogr., A* **1995,** *697,* 461–468.

Poirier, M.-A.; D.E. Steere; J.A. Krogh. Cetane Improver Compositions Comprising Nitrated Fatty Acid Derivatives. U.S. Patent 5,454,842, October 3, 1995.

Polavka, J.; J. Paligová; J. Cvengroš; P. Šimon. Oxidation stability of methyl esters studied by differential thermal analysis and rancimat. *J. Am. Oil Chem. Soc.* **2005,** *82,* 519–524.

Posorske, L.H. Industrial-scale application of enzymes to the fats and oil industry. *J. Am. Oil Chem. Soc.* **1984,** *61,* 1758–1760.

Rantanen, L., R. Linnaila; P. Aakko; T. Harju. *NExBTL— Biodiesel Fuel of the Second Genera-*

tion. Society of Automotive Engineers Technical paper Series. 2005-01-3771. SAE: Warrendale, PA, 2005.

Reddy, C.R.V.; R. Oshel; J.G. Verkade. Room-temperature conversion of soybean oil and poultry fat to biodiesel catalyzed by nanocrystalline calcium oxides. *Energy & Fuels* **2006,** *20,* 1310–1314.

Samuels, W.D.; H.N. Conkle; B.F. Monzyk; K.L. Simmons; J.G. Frye; T.A. Werpy; S.F. Kuczek; S.P. Chauhan. Deicing/anti-icing fluids, U.S. Patent 7105105, September 12, 2006.

Schmidt, A.W. Pflanzenöle als dieselkraftstoffe. *Tropenpflanzer* **1932,** *35,* 386–389. *Chem. Abstr.* **1933,** *27,* 1735.

Schmidt, A.W. Engine studies with diesel fuel (motorische untersuchungen mit dieselkraftstoffen), *Automobiltechn. Z.* **1933,** *36,* 212–214. *Chem. Abstr.* **1933,** *27,* 4055.

Schmidt, A.W.; K. Gaupp. Pflanzenöle als dieselkraftstoffe. *Tropenpflanzer* **1934,** *37,* 51–59. *Chem. Abstr.* **1934,** *28,* 69747.

Schober, S.; M. Mittelbach. Influence of diesel particulate filter additives on biodiesel quality. *Eur. J. Lipid Sci. Technol.* **2005,** *107,* 268–271.

Schober, S.; M. Mittelbach. The impact of antioxidants on biodiesel oxidation stability. *Eur. J. Lipid Sci. Technol.* **2004,** *106,* 382–389.

Schroeder, J.; I. Shapiro; J. Nelson. Candle mixtures comprising naturally derived alkyl esters, PCT Int. Appl. WO 2004 46,286; *Chem. Abstr. 141,* 8856j.

Schuchardt, U.; R. Serchelt; R.M. Vargas. Transesterification of vegetable oils: A review. *J. Braz. Chem. Soc.* **1998,** *9,* 199–210.

Selmi, B.; D. Thomas. Immobilized lipase-catalyzed ethanolysis of sunflower oil in a solvent-free medium. *J. Am. Oil Chem. Soc.* **1998,** *75,* 691–695.

Sercheli, R.; R.M. Vargas; U. Schuchardt. Alkylguanidine-catalyzed heterogeneous transesterification of soybean oil. *J. Am. Oil Chem. Soc.* **1999,** *76,* 1207–1210.

Shah, P.; A.V. Ramaswamy; K. Lazar; V. Ramaswamy. Synthesis and characterization of tin oxide-modified mesoporous sba-15 molecular sieves and catalytic activity in trans-esterification reaction. *Appl. Catal.* **2004,** *A 273,* 239–248.

Shah, S.; S. Sharma; M.N. Gupta. Enzymatic transesterification for biodiesel production. *Indian J. Biochem. Biophys.* **2003,** *40,* 392–399.

Sharp, C.A.; S.A. Howell; J. Jobe. *The Effect of Biodiesel Fuels on Transient Emissions from Modern Diesel Engines, Part I Regulated Emissions and Performance.* Society of Automotive Engineers Technical Paper Series SAE 2000-01-1967. SAE: Warrendale, PA, 2000a.

Sharp, C.A.; S.A. Howell; J. Jobe. *The Effect of Biodiesel Fuels on Transient Emissions from Modern Diesel Engines, Part II Unregulated Emissions and Chemical Characterization.* Society of Automotive Engineers Technical Paper Series SAE 2000-01-1968. SAE: Warrendale, PA, 2000b.

Sheehan, J.; T. Dunahry; J. Benemann; P. Roessler. *A Look Back at the U.S. Department of Energy's Aquatic Species Program Biodiesel from Algae.* National Renewable Energy Lab. NREL/TP -580-24190 July 1998a.

Sheehan, J.; V. Camobreco; J. Duffield; M. Graboski; H. Shapouri. *Life Cycle Inventory of Biodiesel and Petroleum Diesel for Use in and Urban Bus.* NREL / SR-580-24089; National Renewable Energy Laboratory. U.S. Department of Energy: Golden, CO, 1998b.

Shibasaki-Kitakawa, N.; H. Honda; H. Kuribayashi; T. Toda; T. Fukumura; T. Yonemoto. Biodiesel production using anionic ion-exchange resin as heterogeneous catalyst. *Bioresour. Technol.* **2007,** *98,* 416–421.

Shimada, Y.; Y. Watanabe; T. Samukawa; A. Sugihara; H. Noda; H. Fukuda; Y. Tominaga. Conversion of vegetable oil to biodiesel using immobilized candida antarctica lipase. *J. Am. Oil Chem. Soc.* **1999,** *76,* 789–793.

Siler-Marinkovic, S.; A. Tomasevic. Transesterification of sunflower oil in situ. *Fuel* **1998,** *77,* 1389–1391.

Simkovsky, N.M.; A. Ecker. Effect of antioxidants on the oxidative stability of rapeseed oil methyl esters. *Erdoel, Erdgas, Kohle* **1999,** *115,* 317–318.

Simkovsky, N.M.; A. Ecker. Influence of light and contents of tocopherol on the oxidative stability of fatty acid methyl esters. (Einfluß von licht und tocopherolgehalt auf die oxidationsstabilität von fettsäuremethylestern.) *Fett/Lipid* **1998,** *100,* 534–538.

Šmídová, D.; P. Mikulášek; R. Stloukal; S. Hirschmann. The biodiol production: Separation of model biomass by microfiltration. *Desalination* **2006,** *200,* 466–467.

Soumanou, M.M.; U.T. Bornscheuer; Lipase-catalyzed alcoholysis of vegetable oils. *Eur. J. Lipid Sci. Technol.* **2003,** *105,* 656–660.

Spooner-Wyman, J.K.; D.B. Appleby; D.M. Yost; *Evaluation of Di-Butoxy Glycerol (Dbg) for Use As a Diesel Fuel Blend Component.* Society of Automotive Engineers Technical Paper Series SAE 2003-01-2281. SAE: Warrendale, PA, 2003.

Srinivas, D.; R. Srivastava; P. Ratnasamy. Transesterifications over titanosilicate molecular sieves. *Catalysis Today* **2004,** *96,* 127–133.

Staat, F.; E. Vallet. Vegetable oil methyl ester as a diesel substitute. *Chem. Ind. (London)* **1994,** 863–865.

Stavarache, C.; M. Vinatoru; R. Nishimura; Y. Maeda. Conversion of vegetable oil to biodiesel using ultrasonic irradiation. *Chem. Lett.* **2003,** *32,* 716–717.

Stavinoha, L.; S. Howell. *Potential Analytical Methods for Stability Testing of Biodiesel and Biodiesel Blends.* Society of Automotive Engineers Technical Paper Series 1999-01-3520 SAE: Warrendale, PA, 1999.

Stern, R.; G. Hillion; J.J. Rouxel; S. Leporq. Process for the Production of Esters from Vegetable Oils or Animal Oils Alcohols. U.S. Patent No. 5,908,946. June 1, 1999.

Suppes, G.J.; K. Bockwinkel; S. Lucas; J.B. Botts; M.H. Mason; J.A. Heppert. Calcium carbonate catalyzed alcoholysis of fats and oils. *J. Am. Oil Chem. Soc.* **2001**, *78,* 139–145.

Suppes, G.J.; Z. Chen; Y. Rui; M. Mason; J.A. Heppert. Synthesis and cetane improver performance of fatty acid glycol nitrates. *Fuel* **1999,** *78,* 73–81.

Suppes, G.J.; M.A. Dasari. Synthesis and evaluation of alkyl nitrates from triglycerides as cetane improvers. *Ind. Eng. Chem.* **2003,** *42,* 5042–5053.

Suppes, G.J.; M.A. Dasari; E.J. Doskocil; P.J. Mankidy; M.J. Goff. Transesterification of soybean oil with zeolite and metal catalysts. *Appl. Catalysis A* **2004,** *257,* 213–223.

Suppes, G.J.; M. Goff; M.L. Burkhart; K. Bockwinkel; M.H. Mason; J.B. Botts; J.A. Heppert. Multifunctional diesel fuel additives from triglycerides. *Energy Fuels* **2001,** *15,* 151–157.

Tat, M.E.; J.H. Van Gerpen. Biodiesel blend detection using a fuel composition sensor. *Appl. Engin. Agric.* **2003,** *19,* 125–131.

Tat, M.E.; J.H. Van Gerpen. Effect of temperature and pressure on the speed of sound and isentropic bulk modules of mixtures of biodiesel and diesel fuel. *J. Am. Oil Chem. Soc.* **2003,** *80,* 1127–1130.

Tat, M.E.; J.H. Van Gerpen. Speed of sound and isentropic bulk modulus of alkyl monoesters at elevated temperatures and pressures. *J. Am. Oil Chem. Soc.* **2003,** *80,* 1249–1256.

Terry, B.; R.L. McCormick; M. Natarajanm. *Impact of Biodiesel Blends on Fuel System Component Durability.* SAE Techn. Pap. Ser. 2006-01-3279; 2006.

Thompson, J.C.; C.L. Peterson; D.L. Reece; S.M. Beck. Two-year storage study with methyl and ethyl esters of rapeseed. *Trans. ASAE* **1998,** *41,* 931–939.

van den Abeele, M. L'huile de palme: matière première pour la préparation d'un carburant lourd utilisable dans les moteurs à combustion interne (Palm oil as raw material for the production of a heavy motor fuel*). Bull. Agr. Congo Belge* **1942,** *33,* 3–90. *Chem. Abstr.* **1944,** *38,* 28051.

Van Gerpen, J. Biodiesel processing and production. *Fuel Process. Technol.* **2005**, *86,* 1097–1107.

Van Gerpen, J.; B. Dvorak. *The Effect of Phosphorus Level on the Total Glycerol and Reation Yield of Biodiesel.* Presented at Bioenergy 2002, the 10th Biennial Bioenergy Conference, Boise, ID. September 22–26, 2002.

Varma, M.N.; G. Madras. Synthesis of biodiesel from castor oil and linseed oil in supercritical fluids. *Ind. Eng. Chem. Res.* **2007,** *46,* 1–6.

Walton, J. The fuel possibilities of vegetable oils. *Gas Oil Power* **1938**, *33,* 167–168. *Chem. Abstr.* **1939,** *33,* 8336.

Wang, K.; M.C. Hawley; S.J. DeAthos. Conversion of glycerol to 1,3-propanediol via selective dehydroxylation. *Ind. Eng. Chem. Res.* **2003,** *42,* 2913–2923.

Wang, P.S.; M.E. Tat; J. Van Gerpen. The production of isopropyl esters and their use in a diesel engine. *J. Am. Oil Chem. Soc.* **2005,** *82,* 845–849.

Wang, L.; J. Yang. Transesterification of soybean oil with nano-mgo or not in supercritical and subcritical methanol. *Fuel* **2007,** *86,* 328–333.

Warabi, Y.; D. Kusdiana; S. Saka. Reactivity of triglycerides and fatty acids of rapeseed oil in supercritical alcohols. *Bioresour. Technol.* **2004,** *91,* 283–297.

Wardle, D.A. Global sale of green air travel supported using biodiesel. *Renew. Sustain. Energy Rev.* **2003,** *7,* 1–64.

Watanabe, Y.; Y. Shimada; A. Sugihara; H. Noda; H. Fukuda; Y. Tominaga. Continuous production of biodiesel fuel from vegetable oil using immobilized candida antarctica lipase. *J. Am. Oil Chem. Soc.* **2000,** *77,* 355–360.

Watkins, R.S.; A.F. Lee; K. Wilson. Li-CaO catalysed tri-glyceride transesterification for biodiesel applications. *Green Chem.* **2004,** *6,* 335–340.

Wehlmann, J. Use of esterified rapeseed oil as plasticizer in plastics processing. *Fett/Lipid* **1999,** *101,* 249–256.

Westbrook, S.R.; L.L. Stavinoha. *Biodiesel and B20 Blends: Stability Test Methods and Stability Char-*

acteristics. Proc. Int. Conf. on Stability and Handling of Liquid Fuels, 8th, 2003. 271–298.

Wildes, S. Methyl soyate: a new green alternative solvent. *Chemical Health & Safety* **2002,** 9 24–26.

Willing, A. Oleochemical esters—environmentally compatible raw materials for oils and lubricants from renewable resources. *Fett/Lipid* **1999**, *101,* 192–198.

Wolfson, A.; C. Dlugy; D. Tavor; J. Blumenfeld; Y. Shotland. Baker's yeast catalyzed asymmetric reduction in glycerol. *Tetrahedron: Asymmetry* **2006,** *17,* 2043–2045.

Wu, W.-H.; T.A. Foglia; W.N. Marmer; R.O. Dunn; C.E. Goering; T.E. Briggs. Low-temperature property and engine performance evaluation of ethyl and isopropyl esters of tallow and grease. *J. Am. Oil Chem. Soc.* **1998,** *75,* 1173–1178.

Wu, W.H.; T.A. Foglia; W.N. Marmer; J.G. Phillips. Optimizing production of ethyl esters of grease using 95% ethanol by response surface methodology. *J. Am. Oil Chem. Soc.* **1999,** *76,* 517–521.

Wyatt, V.T.; A. Nuñez; T.A. Foglia; W.N. Marmer. Synthesis of hyperbranched poly(glycerol-diacid) oligomers. *J. Am. Oil Chem. Soc.* **2006,** *83,* 1033–1039.

Xie, W.; X. Huang. Synthesis of biodiesel from soybean oil using heterogeneous kf/zno catalyst. *Catal. Lett.* **2006,** *107,* 53–59.

Xie, W.; X. Huang; H. Li. Soybean oil methyl esters preparation using nax zeolites loaded with koh as a heterogeneous catalyst. *Bioresour. Technol.* **2007,** *98,* 936–939.

Xie, W.; H. Li. Alumina-supported potassium iodide as a heterogeneous catalyst for biodiesel production from soybean oil. *J. Molec. Catal. A: Chemical* **2006,** *255,* 1–9.

Xie, W.; H. Peng; L. Chen. Transesterification of soybean oil catalyzed by potassium loaded on alumina as a solid-phase catalyst. *Appl. Catal. A: Gen.* **2006,** 300, 67–74.

Xu, Y.; W. Du; D. Liu; J. Zeng. A novel enzymatic route for biodiesel production from renewable oils in a solvent-free medium. *Biotechnol. Lett.* **2003,** *25,* 1239–1241.

Xu, H.; X. Miao; Q. Wu. High quality biodiesel production from a microalga chlorella protothecoides by heterotrophic growth in fermenters. *J. Biotechnol.* **2006,** *126,* 499–507.

Zamzow, K.L.; T.K. Tsukamoto; G.C. Miller. Waste from biodiesel manufacturing as an inexpensive carbon source for bioreactors treating acid mine drainage. *Mine Water Environ.* **2006,** *25,* 163–170.

Zhou, W.; D.G.B. Boocock. Phase behavior of the base-catalyzed transesterification of soybean oil. *J. Am. Oil Chem. Soc.* **2006a,** *83,* 1041–1045.

Zhou, W.; D.G.B. Boocock. Phase distributions of alcohol, glycerol, and catalyst in the transesterification of soybean oil. *J. Am. Oil Chem. Soc.* **2006b,** *83,* 1047–1052.

Zhou, W.; S.K. Konar; D.G.B. Boocock. Ethyl esters from the single-phase base-catalyzed ethanolysis of vegetable oils. *J. Am. Oil Chem. Soc.* **2003,** *80,* 367–371.

17

Biobased Products from Soybeans

John F. Schmitz[1], Sevim Z. Erhan[2], Brajendra K. Sharma[2,3], Lawrence A. Johnson,[1] and Deland J. Myers[4]

[1]Iowa State University, Center for Crops Utilization Research, Food Science & Human Nutrition, Ames, IA 50011; [2]USDA, ARS, NCAUR, Food and Industrial Oil Research Unit, Peoria, IL 61604; [3]The Pennsylvania State University, Chemical Engineering Department, University Park, PA 16802; [4]North Dakota State University, School of Food Systems, Fargo, ND 58105

Introduction

With recent rises in petroleum crude oil prices to as high as $100 per barrel in January 2008 and additional anticipated future price increases as petroleum becomes less available, many industrial applications and consumer goods that depend on petroleum need economically viable alternatives. Soybean oil and meal are two such resources capable of meeting the need for fuels and biobased products. U.S. market prices for crude soybean oil ranged from $311 – 661/MT, and U.S. market prices for high-protein soybean meal ranged from $153 – 289/MT over the period 1996–2006. Because of an especially strong recent demand for soybean oil by biodiesel producers, U.S. crude soybean oil prices rose to $1,012/MT ($0.46/lb) by November 2007. These unprecedented high market prices for soybean oil due to biodiesel demand and reduced supply as farmers shifted from soybeans to corn to meet demand for corn-based ethanol are not regarded as sustainable, and soybean oil is expected to be an attractive feedstock over the long term. Whereas similar increases in soybean meal stocks were observed before, prices of the two products have markedly different price trends. In the United States, the price of soybean oil increased by 50% over the last couple of years, whereas the price for meal decreased by 35% over the same time period (Golbitz, 2008).

Growing consumer and political interests are intent on achieving high contents of biobased materials in consumer goods to meet increasing expectations for sustainability and use of renewable resources. Along with these consumer and legislative trends, more stringent environmental standards, increased ability to deliver improved performance properties, and more cost-effective chemical conversion processes are driving increased usage of soybean products as feedstocks and materials for industrial products. From 1999–2007, worldwide production of soybean meal increased from approximately 108–150 million metric tons (MMT) [120 – 165 million tons (t)], and soybean oil increased from 25–35 MMT (28–39 million t). Ending stocks for soybean oils fluctuated around 2.5–3 MMT (2.8–3.3 million t), but meal stocks rose to as high as 6.8 MMT (7.5 million t) from ending stocks of 4.1 MMT (4.5 million t) in 1999 (Golbitz, 2008).

These factors provide incentive to develop new materials from soybeans, which can create new markets for this important crop. Additional initiatives to promote developing biobased products to replace petroleum-derived products were implemented by the United States through the Biomass Research and Development Act of 2000 and Executive Order 13134, "Developing and Promoting Bio-based Products and Bio-Energy," which set a national goal of "tripling the use of biobased products and bioenergy by 2010." During the last decade, public awareness on environmental issues rose considerably, and strict Federal Directives (i.e., Farm Bill 2002, Federal Executive orders 12873-1993, 13101-1998, and 13134-1999) will result in new regulations on the development and application of environmentally friendly base stocks. These Federal Directives have set goals for 25% of all government purchases to be biobased and for tripling the use of biobased products and bioenergy by 2010. This will create $15–20 billion in new income for farmers and will reduce fossil fuel emissions by up to 100 MMT of carbon.

Early Soybean Uses in Industrial Biobased Products

No current industrial uses for whole soybeans in biobased products are in place, but historical accounts do exist of soybeans being used as ship ballast (Anon, 1935) and of powdered soybeans being patented as a flooring cover. The first industrial uses of soybeans in the United States helped the country emerge from the Great Depression by providing various consumer goods. In current markets, the most prominent industrial soybean use is soybean oil for manufacturing biodiesel (a subject discussed in detail in the Chapter: *Bioenergy and Biofuels from Soybeans*), but the uses of soybean oil and meal are also growing in other markets.

Soybeans were first domesticated in northeastern China around the eleventh century BCE. The Chinese were the first to crush soybeans into oil and cake by using mechanical presses. Oil was primarily used for cooking, but records indicate it was also used for lubricating fluids, lamp oils, coatings, and marine caulking materials (Spon & Spon, 1980). The oldest known industrial use of soybeans dates from 980 CE., when soybean oil was first used in caulking compounds for boats (Shurtleff & Aoyagi, 1989). Up until the twentieth century, the Chinese used the deoiled cake for another important industrial purpose, fertilizer and soil amendment, often referred to in the early literature as *green manure* or *bean cake manure* (Montgaudry, 1855; Hance, 1980). Reportedly, soybeans were first introduced into North America in 1804 as the ballast of a Yankee sailing ship involved in the China trade (Anon, 1935); thus, ship ballast became the first industrial use for soybeans in the West. It was not until the late nineteenth century, however, that soybeans began to attract the serious attention of Western scientists, farmers, and businesses. The first commercial use for soybeans in the United States was industrial, because soybean oil was regarded as inferior to alternative food oils because of its flavor instability. In 1910, flax (linseed), the primary industrial oil crop at that time, escalated in price, and soybean oil began to be used as a substitute or extender for linseed oil, which was widely used in paints and varnishes.

Early Biobased Product Uses for Soybean Protein

The crudest form of soy protein, ground deoiled cake or meal, has had limited industrial uses in adhesives and paper sizings (Davidson et al., 1927), for binding charcoal briquettes (Rippey et al., 1929), as an additive in wall plaster, and as a spray emulsifier for dormant fruit trees (Johnson et al., 1992).

Alternatively, industrial-grade soy casein was prepared from oil-free meal by grinding in cold water, filtering, and treating with powdered gypsum (Anon, 1935). The mixture was boiled to precipitate the protein, which was collected and washed on filters. The soy casein was dissolved in dilute soda, filtered, precipitated with acetic acid, washed, collected by filtering, and dried. The isolated soy casein was white, relatively pure protein, and used to replace milk casein for paper sizing. Other early uses for soy casein were to prepare silk and artificial textiles, rubber, leathers, plastic materials, films, photographic emulsions, and paints (Johnson & Myers, 1995).

Early Biobased Product Uses for Soybean Oil

In 1908, some European countries, particularly Great Britain, began to import soybeans from Manchuria to supplement short supplies of cottonseed and flax to satisfy growing demand for edible oils, soaps, and glycerin. At the time, glycerin was in great demand for manufacturing explosives used in mining and in constructing the Panama Canal. Soybeans were pressed into oil and meal; oil was chiefly used in soaps, and meal was fed to dairy cattle. Soybean oil was used as a partial substitute for cottonseed oil in hard soap. For soft soaps, soybean oil could completely replace cottonseed oil and partially replace linseed oil. Glycerin, a by-product of soap- and candle-making, was distilled for explosives (dynamite, blasting gelatin, and cordite) and for manufacturing printing inks and printers' rollers. In 1909, the first of several patents was issued for the use of soybean oil in rubber substitutes (Goessel & Sauer, 1909), and in 1911, the first use of soybean oil in linoleum flooring was reported (Goessel, 1911).

In the United States during World War I (1914–1918), the largest industrial products market for soybean oil was the soap industry. Lesser amounts were used in paint, varnish, enamel, linoleum, oilcloth, asphalt, and other waterproofing materials.

As early as 1855, the potential for using soybean oil to replace linseed oil as a drying oil in wood finishes was recognized (Montgaudry, 1955). Replacing linseed oil was particularly important when the flax crop failed during 1910 in the United States and during World War I when shortages of linseed oil were severe due to military demand. This contributed immensely to the demand for soybeans, which in turn led to rapid growth in the number of acres that U.S. farmers devoted to soybeans.

Although soybeans typically contain only 5–11% linolenic acid compared with 35–60% for linseed, the amount in soybeans is sufficient for soybean oil to be classed as *semi-drying oil.* Soybean oil does not have drying properties as good as those of the *drying oil* linseed oil, which in those early formulations would dry to a nontacky film within several days (4.5 days with a linseed oil base but 6 days for a soybean oil base). Incorporating up to 25% soybean oil reportedly extended drying time by only several hours and produced marginally acceptable paint films. By 1919, it was learned that heating and blowing air through soybean oil, which gave rise to the term *blown oil,* would increase viscosity by initiating oxidation and partially polymerizing the oil. Blowing soybean oil also improved its properties in printing inks. Soybean oil also was used to replace linseed oil as a binder in foundry cores (Sefing & Surls, 1933). As early as 1926, R. Ditmar used soybean oil as an agent for plasticizing and increasing elongation of rubber (Johnson & Myers, 1995).

Soybean oil was not widely used for food until the 1930s, when oil-processing technologies advanced sufficiently to produce hydrogenated soybean oils with acceptable stability and flavor. Major industrial product uses for soybean oil today include paint and varnish, resins and plastics, and a source of fatty acids. The latter is often derived from refinery by-products (sometimes known as *foots*).

S. Satow (1917) obtained a U.S. patent for immersing coagulated (acid-treated) or *glutenized* soy proteins in a formaldehyde solution to produce a moldable plastic. This rigid, semi-transparent plastic was a good electric insulator and substitute for ebonite, celluloid, Bakelite, ivory, and marble. In 1923, the first of many patents was issued for a glue based on soybean meal (Johnson, 1923), and in 1926, the I.F. Laucks Co. began to market a soybean-oil-meal glue to the plywood industry.

Chemurgy Movement

The early successes of plastics and other industrial products made from soybeans and the economic stresses in the agricultural sector during the Great Depression of the 1930s promoted the establishment of various organizations fostering new industrial uses for agricultural products. This was the goal of the National Farm Chemurgic Council, formed during the early 1930s with the noted industrialist Francis P. Garvan as its first president. Many now-famous persons were involved in chemurgy, such as the scientists George Washington Carver, Leo M. Christiansen, Thomas A. Edison, and Percy Julian; the noted industrialist Henry Ford; and the journalist Wheeler McMillen. H. Ford founded the Edison Institute in Dearborn, Michigan, in honor of his close friend Thomas A. Edison for the purpose, among others, of developing uses for soybeans in manufacturing automobiles. Nearly 200 industrial product uses for soybeans were attributed to this movement (Myers, 1993).

H. Ford not only was a major force in the Chemurgy Movement but also is credited with being one of the few successful adopters of many chemurgy innovations. His interest in soybean-derived industrial products was based not on altruism but on raising farm-sector income so that farmers could purchase his automobiles, trucks, and farm tractors. In 1935 Ford automobiles, as many as 3.8 million L (1 million gal) of soybean oil were used in enamel paint; 2 million L (540,000 gal) were made into glycerin for use in shock absorbers; and 0.75 million L (200,000 gal) were used at engine foundries as core sand bond (Lougee, 1936). In 1935, H. Ford constructed his own soybean extraction and processing mill at his River Rouge, Michigan, automobile plant to process soybeans into industrial products. At one time, H. Ford had three such plants. In 1936, the Ford Motor Co. planted over 4,900 ha (12,000 A) of soybeans to support its industrial soybean interests. The 1937 automobiles reportedly used 180,000 kg (400,000 lb) of soybean meal in plastic parts, and by this time, development of soy fiber for automobile upholstery was underway.

On the national level, the Chemurgy Movement largely ended after World War II (1939–1945), and by the mid-1950s, programs continued at significant levels only at the U.S. Department of Agriculture's Northern Regional Research Center (USDA, NRRC, now known as the National Center for Agricultural Utilization Research). In 1949, Arthur D. Little, Inc. estimated that 23.4 million kg (51.5 million lb) of soy flour and 12.25 million kg (27 million lb) of soy protein isolate (SPI) were annually used in industrial products in the United States (Little, 1951). Based on the total meal produced, industrial uses of soy protein amounted to approximately 3.6% of the 1949 domestic supply, a much greater proportion than is used today.

Chemurgy interests resurfaced and are impacting federal purchasing policies. The New Uses Council, Memphis, Tennessee, was established in 1990 to deal with occasional surpluses in U.S. agricultural commodities and rising petroleum prices. In 1991, Congress passed legislation creating the Alternative Agricultural Research and Commercialization Center (AARCC), a program designed to foster commercialization of biobased product uses for agricultural materials. Today, the U.S. Departments of Energy, Agriculture, Defense, and Commerce are strongly encouraging biobased products to replace those derived from nonrenewable resources. The Farm Security and Rural Investment Act of 2002 included two provisions that launched the BioPreferred program (FB4P), a mandate to the U.S. Department of Agriculture to develop and implement a program for designating biobased products and a directive to all federal agencies to increase their procurement and use of products qualifying as *bioperferred.*

Soybean Protein in Biobased Products

Wood Adhesives

Background

Plywood adhesive was one of the major industrial uses for soybean products. The patents of O. Johnson (1923) and by I. Laucks and G. Davidson (1928) formed the basis for using soybean meal and protein in adhesives for the plywood industry in the late 1920s. Soy protein adhesives remained competitive in plywood formulations after World War II and into the 1960s. The first soy adhesives were used in cold-press, clamping applications (Lambuth, 1977; Johnson, 1928). The cited advantages of soy glues were soy flour's low cost and plentiful supply compared with casein; its relatively strong and water-resistant (even though not waterproof) bond; its lack of tackiness, making the glue-coated surfaces and materials easier to handle and improving manufacturing efficiencies; its ability to be spread or sprayed in either hot or cold application; and its compatibility with high-moisture-containing veneer without surface splitting (Burnett, 1951). A complete historical overview of soybean adhesives is provided by Johnson and Myers (1995) as well as an introductory view at the resurgence of soybean adhesives.

Markets

Adhesives are one of the most diverse polymer markets globally. From structural and decorative wood uses to packaging, adhesives are used in countless consumer goods. Soy protein adhesive research has been focused on wood products, but once adhesive formulations are developed, spreading into other markets involves little more than additional product testing. More than 9 billion kg (20 billion lb) of adhesives are used annually in the United States to manufacture plywood, oriented strandboard, particleboard, and packaging. The diversity of wood adhesive systems is among the largest for any type of adhesive in the world. Table 17.1 is a product summary of three current types of soy protein adhesives including general formulations, potential uses, perceived benefit(s), and relative cost structures.

Table 17.1. Comparison of Protein Adhesive Characteristics

Property	SPI-Kymene	Foamed Adhesive	Enzyme-modified Soybean Meal
General formulation	SPI (57%), Kymene (43%)	PF resin (86%), wheat flour (9%), soybean meal(5%), foaming additive	Soybean meal, protease (≤0.5% flour solids), cellulase (≤0.5% flour solids), antifoam
Potential uses	Plywood, particle board	Plywood	Exterior-grade OSB, particle board, plywood (blended with up to 70% PF)
Performance Benefit	Formaldehyde-free	Less expensive than animal blood (industry standard)	Controlled modification, tailor-made for multiple uses
Relative cost	High	Low	Low

Performance Properties

The strongest glue joints are formed by the chemical bonding between the adherand and the adhesive. In wood products, adhesives must be highly polar to bond to polar cellulose wood fibers. To enhance strength and water-resistance properties, it is also beneficial for adhesive systems to form high-molecular-weight, crosslinked networks. The adhesive performance of soybean proteins is dependent upon particle size, nature of the bonding surface, protein structure, viscosity, and pH (Lambuth, 1977). Other factors that can affect adhesive performance are processing parameters such as press temperature, pressure, and time (Lambuth, 1977). The particle size of soybean meal or SPIs used in adhesives has a significant effect on its suitability and performance. The smaller the particle size, the easier and more complete soy protein ingredients are dispersed and modified with chemicals or enzymes.

Factors important to making soy protein ingredients good adhesive components are not completely understood, in part, because of the diversity of product applications and the wide variety of adhesive functional needs. Protein is believed to be the primary wood-bonding component, although when soy flour is used, the 35% carbohydrate content may also provide some additional surface adhesion properties. In theory, one would expect that a higher-protein-content ingredient than soy flour (44 to 52% protein), such as a soy protein concentrate (SPC, >65% protein) or a soy protein isolate (SPI, >90% protein) (discussed in detail in the Chapter: *Soy Protein Products, Processing, and Utilization*), would give much greater bonding; however, research shows that using SPI does not sufficiently improve the adhesive to offset its higher cost.

Adhesive properties are also dependent upon the nature of the surface to be bonded. If the bonding surface is too rough, cohesive failure results; surfaces that are too smooth cause adhesive failure. Rough surfaces produce random micro *finger joint* structures under pressure, whereas, smooth surfaces may produce less micro random finger-joint effects, which may be responsible for the low bond strength. The major components in wood vary little from species to species, so variation in bond strength with the type of wood may be due to variation in physical properties such as surface roughness, grain, and porosity.

The bond strength of a protein glue depends on its ability to disperse in water and on the interaction of nonpolar and polar groups of the protein with wood material. In native protein, the majority of functional groups are unavailable for bonding and adhesion due to protein folding caused by van der Waals forces, hydrogen bonds, and hydrophobic interactions. As a result, unmodified soy flour is highly viscous and a poor adhesive material. Modifications change internal bonds and uncoil the protein molecules. Hydrolyzing the protein into smaller protein peptides can further enhance denatured proteins (Lambuth, 1977).

Viscosity is an important property, which largely governs adhesive behavior and performance (Lambuth, 1997). The operating viscosity limits of wood adhesives are very diverse, ranging from 500 to 75,000 cP, depending on the application. A working viscosity of 500 to 5000 cP is commonly needed for gluing materials that are highly absorbent like soft board, dried wood aggregates, or atomized/spray adhesive systems; 5,000 to 25,000 cP for either cold- or hot-press wood laminating applications; and >50,000 cP for wood laminating procedures. A viscosity range of 8,000 to 20,000 cP is specified for no-clamp, cold-press adhesive applications (Barth, 1977). Unmodified soybean adhesive viscosities are dependent on the solids content, but less water is preferable to shorten drying/curing times (Lambuth, 1977). In high-concentration adhesives, high viscosity results from intermolecular interactions due to unfolded protein molecules. Electrostatic interactions and disulfide bonds between peptide chains are the major viscosity-forming forces in soybean meal or SPI dispersions. Most wood adhesives fit in the low viscosity range, and therefore soy protein requires modification for use in adhesives.

The effects of wood product manufacturing conditions, such as press time, temperature and soy protein concentration, on gluing strength and water-resistance of soybean protein adhesives in fiberboard applications were reported by Zhong and Sun (2001a). Shear strength increases with increasing press time as well as press pressure at 25°C (77°F). Shear strength increases were observed for increased temperature, as well, primarily because curing and drying rates increase with temperature. Temperature effects were more pronounced at a higher temperature. Shear strength of soaked samples decreased by 12–25%. A maximum protein content of 12% was observed.

The major advantage of soy glue is that it can be cured either hot or cold. Hot-curing typically occurs at temperatures between 230 and 270°C (446–518°F), pressure of 1.21 MPa, and fast curing times (90–180 s) to prepare plywood panels (Lambuth, 1994). Another advantage of soy glue is that it can be used to bond green lumber without kiln drying. Using dry wood, cold-curing of soybean glues is recommended at 1.03–1.21 MPa pressure for 15 min. During clamping, soybean glues form films having sufficient gel strengths via dehydration to hold plywood sheets tightly even after pressure release. Complete adhesive cure is obtained at room temperature over several days, but machining can be done after 6 h.

Current Interests

Numerous adhesive models were developed over the years. However, most of the adhesion strength comes from three primary mechanisms: (i) chemical bonding, (ii) physical adsorption, and (iii) mechanical bonding (Schultz & Nardin, 1994). Improved protein functionality and performance are two reasons for modifying soy protein ingredients used in adhesive applications. Functional modifications for use in adhesives are achieved by altering molecular conformations through physical, chemical or enzymatic agents at the secondary, tertiary, and quaternary levels (Feeney & Whitaker, 1985). Denaturing and cleaving disulfide bonds enhance adhesion and water-resistance by unfolding the proteins and increasing their interaction with the wood (Kalapathy et al., 1997). Protein or flour modification is also utilized to increase water-resistance (Lambuth, 1994). Using salts or reducing agents can vary the viscosity of soy protein dispersions without negatively affecting bond strength (Hettiarachchy et al., 1995).

Alkali Modification
Alkaline hydrolysis is also an effective means of reducing viscosity. Alkali helps to: (i) unfold the protein structure, thus exposing all functional sites for interaction with wood, and (ii) enhance the hydrolysis reaction, which, in turn, affects viscosity as well as adhesive efficiency (Hettiarachchy et al., 1995). Higher pH increases the rate of hydrolysis and leads to better bond strength and water-resistance, but decreases storage life. At higher pH, viscosity decreases with storage time, which adversely affects the adhesive properties. Optimal treatment conditions for alkali-modified soy protein (AMSP) that resulted in the highest bond strengths were 9.0/70 (pH/temperature), 10.0/50, 11.0/50, and 12.0/40°C. Discoloration of wood products made with AMSP occurred with adhesives made at pH >11 because alkali salts react with wood to form brown color. Similar discoloration was noted in early adhesives in strong alkaline conditions (Lambuth, 1977), therefore limiting the potential use of adhesives from harsh treatments. Mild alkaline treatments, including calcium hydroxide, borax, disodium phosphate and ammonia hydroxide, were tested, but are not suitable for wood product applications due to poor bond strength (Lambuth, 1977).

Soy protein ingredients used in adhesives are typically modified using high sodium hydroxide concentrations and pressure. AMSP adhesive is stronger and more water-resistant compared with adhesives containing unmodified soy protein (Hettiarachchy et al., 1995).

Enzymatic Modification
Proteases, such as trypsin, pepsin, papain, and alcalase, have been examined as modifiers (Kalapathy et al., 1995; Sun & Bian, 1999a; Shera et al., 2007). The advantages of enzymatic modification include high reaction rates, mild conditions using low-cost processing equipment, and most importantly, the possibility of capitalizing on hydrolytic specificity to produce enhanced performance properties. Proteases hydrolyze peptide bonds, thereby modifying proteins but leaving carbohydrates untouched. Modification of SPI with papain affected hydrophobicity, solubility, and emulsifying properties (Wu et al., 1998). Papain-modified SPI has significantly higher solubility and better emulsifying properties. Trypsin-modified SPI (TMSPI) has lower viscosity than unmodified SPI, enabling adhesives with greater solids contents to be formulated (Shera et al., 2007). TMSPI and trypsin-modified soybean flour have much higher bond strengths with soft maple compared to unmodified SPI. Initially, bond strength increases with increased heating time (at 120°C), but strength decreases with treatments over 1 h (Kalapathy et al., 1995). Urea-formaldehyde (UF) can be partially

substituted with TMSPI adhesive. The highest shear strength is reached when 30% UF adhesive is replaced by trypsin-modified soy components.

Carbohydrases are also useful in preparing soy flour for adhesives. Depending on adhesive application requirements, protease treatment alone may not reduce the viscosity to a workable range in high-solids-content materials. In soy flour dispersions at 35% solids contents, the use of an endopeptidase and a mixture of cellulases reduces the viscosity from 4500 to 750 cP at 25°C (77°F), whereas protease treatment alone only reduces the viscosity to 2000 cP (Schmitz, 2006).

Chemical Modification

Certain reagents, such as urea, guanidine hydrochloride (GH), sodium dodecyl sulfate (SDS) and sodium dodecyl benzene sulfonate (SDBS), denature protein and improve bond strength and water-resistance (Burnett, 1951; Huang & Sun, 2000a,b; Bian & Sun, 1998). Chemical modifications with urea, GH, SDS and SDBS at low concentrations (<3 M) all increase the adhesive functionality of SPI (Sun & Bian, 1999a; Huang & Sun, 2000a,b). Urea and GH concentrations significantly affect the extent of protein unfolding and adhesive properties, but SDS- and SDBS-modified SPI give better water-resistance as well as improved bond strength (Huang & Sun, 2000b). Wet and dry heating, grinding, freezing, pressure, irradiating, and exposing to high-frequency sound waves can also be used to denature proteins, but adhesive functionality is diminished when soy protein is subjected to these treatments (Lambuth, 1977).

The effects of ionic strength on the functional properties of soy proteins are well documented (Kinsella, 1979; Klemazewski & Kinsella, 1991; Kella et al., 1998). Ionic surroundings weaken electrostatic interaction between protein molecules. Soy protein adhesives also were modified to reduce viscosity by using ionic solutions (i.e., sodium chloride or sodium sulfate). Concentrations of 0.1 M NaCl, Na_2SO_4 or Na_2SO_3 reduce the viscosity of soy protein with no significant adverse effects on bond strength and water-resistance. Viscosity was reduced from 30,000 to 6000 cP with 0.1 M NaCl and 1050 cP with 0.1 M Na_2SO_4. A similar Na_2SO_3 treatment results in modified SPI with 110 cP viscosity and 28% decrease in disufide linkages. Treating with >0.1 M of any ionic solution decreases viscosity further, but bond strength is also diminished (Kalapathy et al., 1996).

Chemical modification with dopamine was also used as strength and water-resistance aids for SPI adhesives (Liu et al., 2002). Dopamine is an amino acid with two adjacent phenolic hydroxyl groups, and is the primary component responsible for marine adhesive properties. The Liu modification scheme creates an SPI that is similar to mussel proteins used for surface adhesion. Increased water-resistance compared to other stand-alone SPI adhesives was achieved. Bond strength depends on the phenolic functionality in the synthesized compounds (Liu, 2002). Much interest in this adhesive has developed because it is a strong and resilient adhesive, which is formaldehyde-free, making it suitable for interior wood products.

Blended Adhesives

Blending soy protein adhesives with other protein or synthetic adhesives can produce adhesives with enhanced performance properties. Blends of soy flour with blood, casein, phenol formaldehyde (PF), and phenol-resorcinol formaldehyde (PRF) were used to develop wood glues with unique properties (Lambuth, 2001). Blended adhesives for biodegradable plant containers were obtained by blending SPI with varying amounts of poly-(vinyl alcohol) or poly-(vinyl acetate) (Brown, 1987; Zhao et al., 2000). Blends of soy protein and PRF resins are useful in finger-jointing green lumber with the *Honeymoon System* (Fig. 17.1) (Kreibich, 1995; Karcher, 1997; Steele et al., 1998; Clay et al., 1999). Soy protein and PRF blends cure rapidly at room temperature and have excellent water-resistance and reduced formaldehyde emissions. Soy protein is also much less expensive than PRF adhesives.

Soy flour dispersed in sodium bisulfite solution was blended with PF in a soy:PF ratio of up to 7:3. Particleboards made with the sprayable adhesive have acceptable strength attributes, and decreasing the mixture to as low as 20% soy flour produces boards with comparable strength and water-resistance as those made with only PF (Kuo et al., 2001; Kuo et al., 2003). PF is described as the primary cross-linking agent in the previously cited patents, and similar work was conducted by others. Wescott et al. (2005) reported that soy-based adhesives containing PF and 50–66%

Fig. 17.1. Finger-jointed lumber made with the honeymoon system (AMSP and PRF).

soybean flour were stable at room temperature for 100 days with no separation and maintained 300–600 cP viscosity. Petroleum-derived phenol amounts in these resins are ~75% lower than PF-only resins. Strandboard produced from the soy-based resin performed similarly to PF only, and when 10% methylene diphenyl-isocyanate (MDI) was added to the soy-based adhesive, thickness swell after 24 h soaking was 50% lower than that observed with PF only (Wescott et al., 2005). Adhesive viscosity can be an issue in blended adhesives as well. Gel permeation chromatography was used to determine optimal conditions for alkali treatment for soy flour blended with PF. Treating for <1 h at <100°C and pH 9–12 produced a modified soy flour with degraded components that are stable in adhesive formulations and do not lead to increased viscosity after blending (Lorenz et al., 2007).

Soy flour in a 30% dispersion treated with endopeptidases and a mixture of cellulases was blended with PF resins. Soy flour hydrolysate from a combination of enzyme treatments has viscosity similar to those produced by using alkali modification, whereas the endopeptidase or cellulase treatment alone has reduced viscosity, but is still twice as high as the combination of treatments or the alkaline hydrolysate (Schmitz, 2006). An ammonia-like odor is typical of alkali-treated soy flour, but enzyme treatments eliminate the noxious aroma. Medium-density fiberboards produced with resin blend amounts from 5 to 40% soy flour hydrolysate had internal bond strengths decreasing with increasing amounts of soy flour. After 24 h of soaking or 2 h of boiling, thickness swells do increase in blends with higher soy protein contents, but at lower amounts (5 and 10%), single treatments of cellulase or endopeptidases have more water resilience (less thickness swell) than the combination of treatments (Schmitz, 2006). Overall, similar properties to alkali-hydrolyzed soy flour can be achieved, but enzyme treatments can be tailored to enable the protein to co-react and polymerize with PF.

Another blended resin system is comprised of SPI and Kymene. Kymene is a commercial wet-strength agent for paper also known as a polyamidoamine-epichlorohydrin (PAE) adhesive resin. Kymene is multi-functional and undergoes a series of reactions depending on processing conditions. Lap-shear tests made with cherry wood veneer indicate SPI blended with Kymene produces adhesive bonds similar in strength to PF-only resin, but are less water-resistant (Li et al., 2004). One additional benefit of high importance to the industry using SPI–Kymene is being free of formaldehyde. The elimination of all formaldehyde is a key factor in commercial viability for future adhesive products because of growing concerns over carcinogenic compounds. Despite the benefits, the challenges to using SPI–Kymene systems are low solids content, relatively long press times, and use of expensive SPI.

Foaming Adhesives

Soy protein adhesives can also be used in foam adhesive applications for plywood manufacture. Plywood adhesives are typically extruded and foamed at the time of application. Spray-dried blood protein is a standard part of plywood adhesive formulations. Hojilla-Evangelista (2002) replaced blood protein with soy flour, SPI, and SPC in foamed PF formulations. All soy products produced adhesives with foaming and strength properties at least equal to those of blood protein. Further study into the workings of foamed adhesive formulations demonstrated soy protein with poor foaming had extensive modifications (Hojilla-Evangelista, 2001). Concerns with using animal blood in products are eliminated with extruded, foamed soy flour as well as lower product costs for soy flour relative to spray-dried animal blood.

Building Materials

Phoenix Composite (Mankato, MN) developed *Environ*, a composite material composed of used newspapers and soybean meal, having a wide range of uses as a construction material. Soy flour was used in a monolithic resin system to bind together recycled newspapers. *Environ* possessed easy-working properties of wood and the appearance and sales appeal of polished granite. It was introduced for use as decorative surfaces in furniture, cabinets, and recognition plaques. Although the cost of *Environ* is relatively high (\$130–150/m^2; \$12–14/ft^2), its excellent fire-resistance is a major factor in considering its use, in addition to the unique appearance and excellent working properties it offers.

Miscellaneous Adhesives

Soy protein is used in adhesive applications other than plywood, although at significantly smaller usage levels (Burnett, 1951). SPI is used in tacky and remoistening adhesives. SPI was used in formulating glue for shotgun shell casings in the late 1940s and early 1950s due to its initial tack and superior water-resistance compared to soy flour (Little, 1951). A process was also patented for using an alkaline dispersion of soy flour as an adhesive for charcoal briquettes (Burnett, 1951). Briquettes formulated with a soy flour dispersion along with other chemicals were resistant to weathering and breakage during handling.

Plastics

Background

Patents for preparing semi-plastic materials from soy protein were first issued in France (Contant & Perrot, 1913) and Great Britain (Dodd & Humphries, 1913), although S. Satow's U.S. patent of 1919 is credited with first stimulating widespread interest in soy protein plastics. Johnson and Myers (1995) provide a complete historical review on soybean plastics. No discussion of soy protein plastics is complete without at least mentioning the early research and commercialization efforts of H. Ford and the Edison Institute (Fig. 17.2). The Ford Motor Co. and the Edison Institute, headed by R. Boyer, were responsible for extensive protein plastic research, patents, and commercial applications in the late 1930s and 1940s.

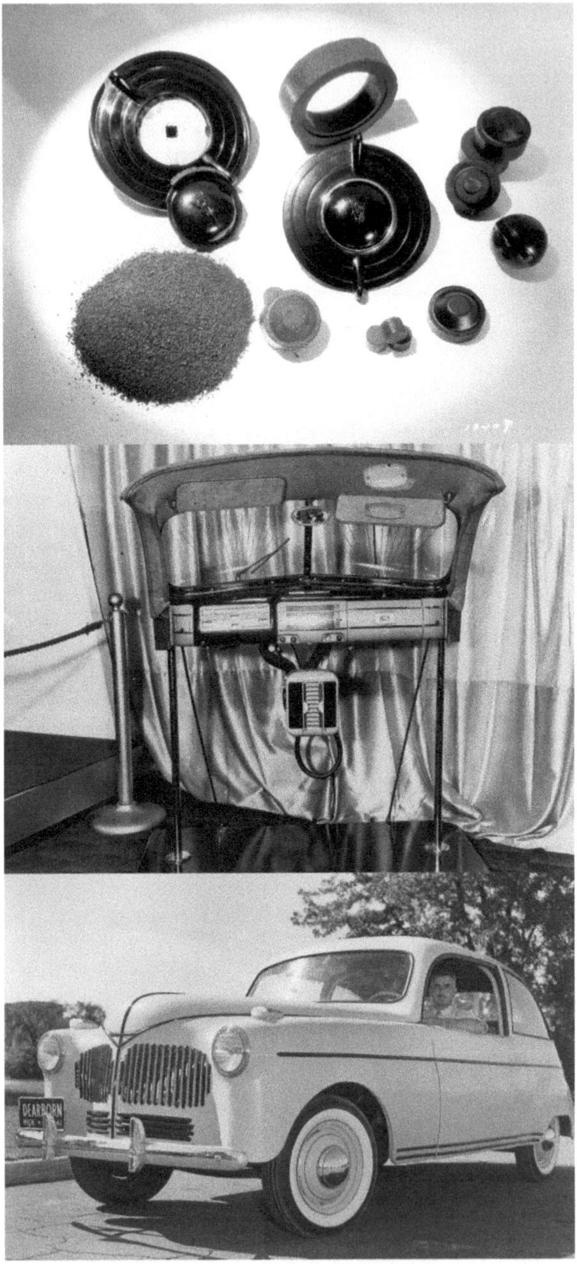

Fig. 17.2. Photos showing automotive parts molded from soy plastic (1939), an exhibit dash incorporating soy plastics (1940), and a plastic bodied car made from soy plastics. Photos from the Collections of the Henry Ford Museum.

Performance Properties

Compression-molded soy protein plastics are rigid and brittle without plasticizers present in the formulation (Paetau et al., 1994). Water absorption of early protein plastics could not be reduced to the very low levels that were obtained with synthetic, petroleum-derived resin products; however, color, dyeing, and strength properties were good and production costs were relatively low (Johnson & Myers, 1995).

SPI hardened with an aldehyde at or near the protein's isoelectric point (~pH 4.5 for soy protein) produces a material that is thermoplastic and absorbs <10% moisture during 24 h of soaking. An aqueous suspension of the formaldehyde-hardened SPI can be used as a laminating material for Kraft paper. Laminating material made from soy plastics has similar properties to the more expensive phenol and urea laminating materials, but is less water-resistant. SPI was discontinued during the early 1950s in favor of modified soybean meals used as modifiers and extenders in PF plastics.

Thermosetting properties of formaldehyde-hardened proteins were improved by combining them with phenolic resins. When combined with phenolic resin in a 50:50 mixture, long curing times and high water-absorption rates were observed. Blending formaldehyde-hardened protein with 25% wood filler results in a plastic with excellent strength, good curing properties, and acceptable finish.

Markets

Despite the early enthusiasm and fanfare, soy plastics have not measured up to alternative materials. As potential began to grow for protein-based plastics after World War II, inexpensive and adequate supplies of petroleum and better-performing synthetic materials were realized. Widespread availability of synthetic polymers discouraged attempts to make improvements. For a short period (1935–1943), H. Ford used soy plastics in his popular, inexpensive automobiles, but his plan to use over 1.4 million kg (4 million lb) of soybean meal for every million cars produced never materialized.

Current Interests

Environmental concerns and increased polymer prices relative to agricultural products are renewing interest in soy plastics. More than 70% of today's plastics are dominated by five petroleum-derived polymers: high-density polyethylene, low-density polyethylene, polyvinyl chloride, polypropylene, and polystyrene. The remaining plastics are comprised of more than 20 other polymer subunits. Despite the recent emphasis on recycling, <10% of our plastic wastes are recycled because recycling has not proven to be economically feasible. Mounting environmental concerns focused attention on the concept of designing materials from *cradle to grave*, especially in Europe.

The potential uses of soy plastics as commodity products (e.g., eating utensils, bags, toys) were successfully demonstrated (Paetau et al., 1994a,b; Wang et al., 1995;

Wang et al., 1996). However, the main problems of these plastics are pricing and performance. Numerous studies examined various mechanical and thermal properties of soy protein plastics for different types of systems and products. Early studies showed soy plastics can be engineered and manufactured to possess significantly higher tensile strength than petrochemical plastics if the moisture content is <5%. Dimensional stability and strength after storage are heavily dependent on plasticizers and compatibilizers formulated with soy protein (Sue et al., 1997).

Paetau et al. (1994a) studied the effects of preparation and processing on the mechanical properties and water absorption of biodegradable plastics prepared from SPI and SPC. Water acts as a plasticizer and aids processing of soy proteins. Compression-molded specimens prepared from SPC and SPI at different molding temperatures ranging from 80 to 160°C (176–320°F) were tested for mechanical performance. These plastics were rigid and brittle having tensile strengths of 10–40 MPa, yield strengths of 1–5.9 MPa, and elongations of 1.3–4.8%. Percentage elongation increased, and Young's modulus decreased with increasing moisture content in the molding material. Yield strength was maximum (4.9 MPa) at 10.0% moisture. Moisture absorption by molded protein plastic over 26 h ranged from 30 – 170%. Plastic samples from SPC and SPI had similar properties—increasing molding temperature increased strength but reduced flexibility and extensibility. Plastics molded at 140 and 160°C (284–320°F) were strongest having yield strengths of 49 MPa (Paetau et al., 1994a).

Morphology and mechanical behavior of high soy protein content plastics prepared by compression molding at 150°C (302°F) and 19.6 MPa for 6 min were determined by Sue et al. (1997). Keeping the moisture content <5% (w/w) could obtain a Young's modulus of 4.4 GPa. In a similar study, high tensile strengths were observed at as low as 2.6% moisture, but the materials were brittle. Maximal toughness was achieved at 3.6% moisture content. The ductility and dimensional stability strongly depended on moisture content. The glass transition temperature (Tg) of dry and moistened soy plastic was around 150°C (302°F), but upon adding 25% glycerol as plasticizer, Tg reduced to -50°C (-58°F) (Zhang et al., 1998).

The most effective plasticizers for soy protein processing have similar polarities to soy protein (Wang et al., 1996). The most common plasticizing agents for soy protein are water and polyols. Polyols are important to the mechanical strength of soy protein plastic (Wang et al., 1996, 1995; Zhang et al., 1998). Propylene glycol, triethylene glycol, ethylene glycol, and butane-diols all exhibited plasticizing compatibility with soy protein. Ethylene glycol, glycerol, and propylene glycol have the greatest effects on the percentage elongation at break of the plastic specimens. At 30% concentration, the plastic specimens containing ethylene glycol had ~400% elongation at break, while those containing glycerol had 330% elongation, and propylene glycol had 120%. This was attributed to low molecular weight and greater polarity of ethylene glycol. 1,3-Propanediol is less polar and less effective than ethylene glycol, glycerol, and propylene glycol in modifying tensile properties of soy protein plastics. In a comparative study between sorbitol and glycerol as a plasticizer, sorbitol gave protein plastics with higher tensile strengths and thermal stabilities, but impact strengths were greatly diminished relative to those plasticized with glycerol. Scanning electron microscopy indicated similar findings with sorbitol samples showing brittle fractures and glycerol showing local ductile fracture features (Tummala et al., 2006).

In examining the performance of glycinin and β-conglycinin fractions of soybean protein in plastics, materials made with glycinin were stronger and had lower water absorption than those made with β-conglycinin fractions (Sun et al., 1999b). Plastics made with mixtures of glycinin and β-conglycinin proteins exhibited the highest tensile strength and intermediate water-absorption capacity. When molding, the temperature was varied from 120 to 175°C (248–347°F); tensile strength and percentage elongation were highest at 145°C (293°F) for the β-conglycinin and the glycinin/β-conglycinin mixture plastics. Maximal tensile strength and elongation were achieved at 162.5°C (324.5°F) for plastics based on glycinin.

Blending soy protein with polyesters, such as polycaprolactone (PCL) and Biomax (a commercially available biodegradeable polyester), can produce molded soy-protein-based plastics (Fig. 17.3) (Zhong et al., 2001a,b; Mungara et al., 2002). Zhong et al. (2001a,b) examined the effects of SDS/guanidine hydrochloride (GH) modified glycinin soy protein and SPI blends with PCL for thermal, mechanical, and water absorption properties. Differential scanning calorimetry (DSC) and dynamic mechanical analysis (DMA) showed that SDS or guanidine hydrochloride acts as a plasticiser for glycinin and decreased the Tg with increasing SDS or GH concentration. All protein plastic blends made with both soy flour and SPI had high tensile strengths, but were brittle. In SPI/PCL blends, MDI acts as a compatibilizer, and the mechanical properties of 50/50 SPI/PCL blends increased with increasing MDI concentration (Zhong et al., 2001b). Likewise, plastic blends of SPI and polylactide increased tensile strength, elongation, and stability of extruded plastics (Zhang et al., 2006). Plastics made from soy protein/polyester blends had low water absorption and good stability at ambient conditions relative to the plastics made from soy protein alone. Plastics from soy flour had the lowest water absorption (Mungara et al., 2002; Zhong et al., 2001a,b).

Thermogravimetric analysis in nitrogen gas of molded protein plastics showed they were stable to 300°C, indicating good thermal stability. In the presence of oxygen, however, the plastics decomposed at 180°C (Wang et al., 1995). Biodegradation of molded soy protein in both soil and simulated marine environments is rapid, converting into carbon dioxide and water (Spence et al., 1996). Molded soy protein plastics degrade more quickly than the raw material. Soy protein/starch-blended plastics degrade very rapidly. The differences in degradation rates were attributed to heat denaturation of protein and the balanced carbon and nitrogen sources of the soy protein and starch blends. Soy protein plastic composites with enhanced stiffness,

Fig. 17.3. Plastic materials molded from soybean protein.

strength, and water-resistance were obtained using polyphosphate filler (Otaigbe & Adams, 1997a,b).

Textiles Fibers

Background

Soy protein was one of a number of proteins that were used to produce *regenerated protein textile fibers* in the late 1930s and 1940s. The discovery of regenerated protein fibers from casein is attributed to Todtenhaupt in 1904, but it was Parretti, who in 1935 successfully developed, patented, and produced a textile fiber with wool-like properties from soy protein. In the United States, soy protein textiles and felt materials were explored as replacements for wool, felt, and fur (Hartsuch, 1950; Moncrieff, 1975). In 1939, the Japanese reportedly produced about 450,000 kg (1 million lb) of soy protein fiber (Conner, 1989). The first U.S. patents for soy fibers were granted to T. Kajita and R. Inoue in 1940 (Kajita & Inoue, 1940).

Robert Boyer of H. Ford's Edison Institute was awarded an important patent in 1945 (Boyer et al., 1945) for producing textile fibers from soybean meal for use in automobile upholstery. In addition to soy protein, casein, corn zein, and peanut protein were also used to produce regenerated protein fibers. H. Ford once wore a suit made from soy protein fibers, which was reportedly quite itchy when dry and odiferous when wet. Soy protein fiber technology never attained commercial textile production in the United States. Daily pilot-plant production of soy protein fiber by the Ford Motor Co. reached 2,000 kg (4,400 lb) in 1940 (Burnett, 1951).

Technology

The strongest known protein fiber is spider web fiber. Silk and wool are also naturally occurring fibrous proteins and are used as textile fibers, which provide the technical basis for producing soy protein fibers. Protein fibers can be regenerated to have similar or improved properties compared to wool and silk by using the principles of the viscose process common to making rayon from cellulose (Hartsuch, 1950). This was never realized because the molecular structure required to produce good fiber qualities was lost due to denaturation, the relatively long repeat distance and bulky side chains, which prohibit protein chain interaction. The most significant problem was that denatured soy protein was extremely flexible and allowed internal rotational motion, resulting in random coil configurations.

SPI was used to produce regenerated protein fibers. According to Boyer's process soluble SPI was obtained from defatted meal that was not heat-treated (Boyer, 1940). The meal was then treated with weak alkali (0.1% sodium sulfite) for 30 min, and the solution was clarified by filtering or centrifuging. Protein was precipitated with acid, the curd was washed and dried, and the protein was dissolved to form a viscous stringy solution with >12% solids (Boyer, 1940). To unfold the protein, the solution needed to be highly alkaline, pH 12.5. A 12% solids solution was difficult to obtain because of the tendency for soy protein to gel, but 20% solids solutions were reported (Burnett, 1951). To increase the viscosity and develop greater stringiness, the solution was filtered, deaerated, and aged (Burnett, 1951; Boyer, 1940). Unfolding the proteins with detergents rather than alkali was rejected because of the large amounts required and the lack of adequate recovery processes. Heat, water, and mechanical shear were moderately successful, but did not produce fiber of high quality as did highly alkaline solutions. Finally, textile fibers were formed by a wet-spinning process, which forced the solubilized protein through a spinneret into an acid bath to coagulate the protein. The bath typically consisted of a solution of sulfuric acid, formaldehyde to harden the fiber, and salt to accelerate drying. The fibers were collected on a reel and stretched to orient them, improving strength and elasticity. The fibers were then immersed in a formaldehyde bath to crosslink the protein (improving the resistance to attack by water and dilute acids), cut into desired lengths, and dried.

Performance Properties

Soy fibers were white to tan in color and had warm soft feel, natural crimp, and high resilience (Hartsuch, 1950). Soy protein fibers had more elongation and 80% of wool's dry strength (Boyer, 1940). These textile efforts, however, were not as successful as those made with casein fibers and failed, largely because soy protein fibers had poor wet strength and could not withstand repeated washing or dry cleaning. The wet fibers also had a characteristic *wet dog* odor.

Early soy protein fibers resembled casein fibers, and later soy protein fibers even resembled rayon. The original intent of producing soy protein fibers was to compete with wool and silk (Moncrieff, 1975). The advantage of SPI over other protein sources was its relatively low price and high protein content. Soy protein fiber had one serious problem that was never solved: its low tenacity, particularly when wet. Compared with wool, soy protein fiber was 45% weaker when dry and 75% weaker when wet. Because of its low tenacity, the best application for the fiber was in blends with other fibers such as wool, rayon, nylon, and cotton (Johnson & Myers, 1995).

Soy protein fibers wet-spun into acid baths were so weak that they were impossible to collect and evaluate. The pH of the acid bath was irrelevant, and adding salts did affect fiber properties. The osmotic pressure was believed to help coagulation by dehydrating the fiber immediately in the acid bath (Huang et al., 1995). Water activity was a critical factor to soy protein fiber properties. At low water activity, spun fibers were dry and brittle, but at intermediate water activity, the fibers absorbed more water, tenacity decreased, and flexibility increased. Property changes resulted from increased molecular motion caused by less hydrogen bonding, and more protein–water interactions (Huang et al., 1995; Zhang et al., 1997).

Blending with corn zein protein can strengthen soy protein fibers. Soy–zein blends exhibited increased tenacity and more flexibility. As with soy protein fibers, properties of blended fibers changed with water activity, but the properties were always better with blended fiber systems (Zhang et al., 1997).

Zhang also explored the possibility of dry extruding soy fibers. Similar properties were observed between extruded and wet-spun fibers at varying water activities. It was not possible, however, to produce small-diameter fibers by extruding because extrusion dies quickly plugged (Zhang et al., 1997). Performance properties are clearly an issue, but both soy protein fibers and soy–zein blends could be improved by post-spinning modifications with glutaraldehyde, glyoxal, dimethylformamide, and dimethylsulfoxide (Huang et al., 1995; Zhang et al., 1997).

Markets

The Ford Motor Co. produced soy protein fibers on pilot-plant scale in the 1940s to test their use in automobile upholstery, but soy fibers were not commercialized in the United States. Fibers made from soy protein and soy–zein blends had inferior tenacity to natural and synthetic fibers. Proteins have been more expensive than raw materials for common synthetic fibers, and wet-spinning is a more demanding processing procedure. Wool and silk prices are much lower today relative to soy protein, a multitude of synthetic fibers with desired properties were developed, and environmental regulations are much more restrictive than in the 1940s. For any future protein textile to be successful, a dry-spinning process will likely be required to eliminate excessive wastes from the wet-spinning processes.

Paper Coatings

Background

The largest industrial biobased product use for soy protein is paper coatings. Paper is coated for a variety of reasons, including upgrading the paper, making it more printable, and providing increased resistance to water or grease (Little, 1951). Coating paper involves applying a thin layer of pigments on the paper to provide an improved surface. A binder is used to bind pigment particles to each other and to the paper. The binder also affects the size and volume of the pores in the paper, ultimately influencing the printing and gluing properties of the paper or paperboard (Garey, 1989). The binder also influences the runability of the coating during application by controlling rheological properties (Coco et al., 1990).

SPI was first utilized with the advent of machine-coated paper in 1935 (Burnett, 1951; Little, 1951). Starches were originally used for paper coatings, but casein and soy protein give coatings with better rheological properties and water-resistance (Little, 1951). Soy protein became predominantly utilized during World War II because most of the available milk casein was needed as dry milk for the military.

In the early 1940s, approximately 3,600 MT (3,966 t) of SPI were used in paper coatings, compared with 10,200 MT (11,243 t) of casein (Conner, 1989). Because of lower prices and more consistent properties, SPI began to out-compete casein in this market. By 1949, the amount of SPI used rose to 6,600 MT (7,275 t), while the amount of casein used declined to 9,000 MT (9,920 t). Current annual consumption of SPI in this market is about 24,500 MT (27,000 t). Soy protein is used primarily as a co-binder in these systems with latex-protein and starch-latex binding systems (Coco et al., 1990). Co-binders are used in amounts from 1 to 5% of the binding formulation depending on the application. The blend of latex and SPI produces a coating that imparts more brightness.

Performance Properties

Soy protein has a number of competitive advantages when used in paper coatings compared to other protein sources. Products with greater viscosity range and ability to be used at higher solids contents are two major factors. Soy protein does not agglomerate into strings, leaving tracks in the coating, as milk casein did (Garey, 1989). Soy protein may be hydrolyzed to various levels to tailor viscosities and rheological properties for different products. Additionally, hydrolyzed soy protein produced paper coatings with higher solids and lower moisture contents, facilitating faster machine speeds and decreasing drying costs. High solids content along with soy protein's extensive chemical interactions with the paper and other coating compounds led to faster coating immobilization and more flexibility in processing (Hiscock & Merrifield, 2000).

Technology

SPI continues to be used in paper manufacturing. Under selective conditions of pH and chemical reactants, the protein is modified to improve its influence as a coating modifier and dispersant (Coco et al., 1990). Only modified proteins are used in the industry today, and these modifications have continually improved the functionality of the protein (e.g., viscosity, water-retention, and pigment-stabilizing properties) and kept them competitive in the industry (Coco et al., 1990; Krinski & Hou, 2000).

One major consideration in binder formulating is the choice of alkali used to disperse the protein. The alkali influences the final properties (color, wet-rub resistance, and viscosity) of the protein in the binder formulation. Additionally, polymer matrices, such as polyacrylate, can be used as a co-binder for paper coatings. Krinski and Hou (2000) showed polyacrylate could help increase the coating functional properties while using considerably less protein. Another key consideration is the mixing regimen (mixing time, temperature, speed, and order of ingredient addition). The choice of other additives used to control the properties of the dispersion (flow modifiers to get workable viscosities at higher solids content, preservatives, foam controllers, and lubricants to help the release of the coating colors from the rollers) is also important to the formulation (Krinski & Hou, 2000).

SPI proved to be a better coating agent and inclusion matrix than modified starch for antimicrobial purposes. Losses of antimicrobials were always higher with modified starch compared to soy protein after applying and drying. Likewise after 60 days of storage, soy protein retained a higher retention rate (Ben Arfa et al., 2007). Retention of antimicrobials decreased initially after paper drying when antimicrobial and soy protein solutions were heated, although solution homogeneity and viscosity increased. Regardless of heat treatment, after 50 days' storage, antimicrobial retention was similar for paper coatings that were either untreated or treated up to 90°C (194°F) (Ben Arfa et al., 2006). In addition to water retention being a key factor, recent research shows water- and grease-resistance are improved through the use of soy protein coatings (Druckrey et al., 2005; Ha et al., 2006; Park et al., 2000).

Current Interests

SPI as a co-binder in paper coating binder formulations remains a vital part of the paper industry. Today, soy protein coatings are used in a variety of coated paper and paperboard applications. There should be steady growth in soy protein used in this application, provided SPI remains price-competitive with synthetic latex binders, which is likely, due to projected high petroleum prices relative to agricultural products.

Modifying SPI with alkoxy silanes is a useful treatment prior to paper coating application. Silanated SPI is produced by mixing alkali extracts of soybean flakes with alkoxy silanes at pH 11 and 50°C (122°F) for 1 h. Modified protein is then separated by isoelectric precipitation using H_2SO_4 (Krinski & Steinmetz, 1987). Silanated SPI is mixed with ammonium nitrate, clay and ammonium hydroxide, and heated at 60°C (140°F) to give a paper-coating binder. Silanation using 3-(2-aminoethyl)-aminopropyltrimethoxy silane as a coupling agent enhances interfacial adhesion between the soy matrix and glass fiber in producing fiber-reinforced composites (Liang & Wang, 1999). Copolymers of SPI are prepared by heating an alkaline dispersion of SPI with cationic epoxide or acrylate monomer to give modified protein materials (Krinski & Steinmetz, 1987). The pH is maintained during the reaction of SPI with hydroxy alkyl acrylate (Steinmetz & Krinski, 1987). The modified protein can be used alone or in combination with styrene-butadiene rubber (SBR) latex for paper coating. It has a high affinity for pigments in paper coating compositions and gives paper having good ink receptivity and printability. Vinyl monomers, such as styrene or butyl acrylate, in emulsion polymerization using ammonium persulfate as an initiator with SPI, gave graft co-polymers, which are used in combination with other synthetic latexes for coating applications (Dykstra & Hollingsworth, 1975).

Paper and Textile Sizings

I. Laucks (1928) is also credited with developing paper sizings from soy flour. Paper is sized to make it resistant to the penetration of moisture and liquids (Burnett, 1951). SPI improves the process by making the resin more uniform and stable upon addition of borax and by binding free alkali. *Prosize*, an industrial SPI, was used to stabilize an alkaline resin size dispersion for internal sizing of cellulosic fibers before complete dewatering (Sprague, 1944).

Textile fibers were sized for the process and end use. SPI was used as a warp sizing to manufacture rayon (Burnett, 1951). The advantages of SPI sizings were their ability to penetrate the yarn and heat stability in the operating temperature ranges of the machines. However, manufacturers had concerns about low resistance to abrasion and off-color (Little, 1951). These factors, along with the inroads of synthetics, prevented SPI from capturing greater market share.

Soy flour, alone or in combination with SPI, was used for a number of years as a coating for wallpaper. At concentrations <16.5%, the alkaline protein dispersion was thixotropic and able to suspend pigments and clays. The pH and viscosity remained fairly stable if preservatives were added to prevent mold and bacterial growth. Water-resistance was imparted to the paper coating by applying a tanning solution containing formaldehyde. High-grade washable wallpaper could be made with soy flour, and several million kg were used annually up to 1947 (Burnett, 1951); however, soy flour was used in only low-grade wallpaper, and those industries using soy flour declined significantly by 1949 (Little, 1951).

Other Uses for Soy Protein in Industrial Biobased Products

Cosmetics

Soy protein derivatives were examined in cosmetic applications during the past decade. Functional uses have varied from carrier systems for rheological properties (Sanbe & Toshishige, 2005) to SPI being an active ingredient for skin tightening, cleansing, or conditioning (Russ et al., 2001; Schultz & Tran, 2004). Cosmetics for skin-lightening are high in vitamin C or its derivatives, and a simple mix of vitamin C components and fermented soy protein provides physical and functional properties for skin-lightening (Sanbe & Toshishige, 2005). Wu et al. (1998) created blends with three common detergents. Native and modified SPI was used to replace sodium dodecyl sulfate, sodium lauryl sulfate, and disodium lauryl sulfosuccinate in bath soap, shampoo, and cream hand cleansers. Blends up to 25% SPI had good foaming or emulsifying properties. Above 25% SPI, performance properties changed for different soy protein modifications and commercial detergents, but even with total replacement (100% protein) foaming capacity and emulsion stability could be maintained at levels similar to pure detergent levels.

Printing Ink

Currently, a high interest is expressed in using soybean oil in printing inks as is discussed later; however, soy protein also was used in printing ink formulations. SPI was used in dispersions that were set by dehydration. SPI dispersed in polyhydroxyl alcohols replaced corn zein protein, once a widely used biobased ink component. Complexing the dispersed SPI with guanidine carbonate improved water-resistance (Johnson & Myers, 1995). At one time, an estimated 45,000 kg (100,000 lb) of SPI was used annually in this application (Little, 1951). Today, no evidence exists of protein being used in printing inks.

Fire-fighting Foams

During World War II, soy foams were used by the U.S. Navy for fighting shipboard fires. The soy foams were lime-hydrolyzed SPI, and were known as *bean soup* to sailors (Adams, 1944). During the war, fire-fighting foam for the military was an important use for SPI. Protein was fed into a water stream, and the foam was produced by means of an aerating nozzle that formed a tight foam blanket to completely smother flames. Due to the natural adhesive characteristics, the foam would stick to walls enhancing fire extinguishing. As late as 1949, 113 MT (250,000 lb) of SPI were used in this application (Little, 1951). No report is made of SPI being used in this application today.

Powder and Paste Paints

SPI was used in paste and powdered cold-water paint formulations because of its adhesive, film-forming, and water-proofing properties. The paste paints were more durable than the powder paints. The advantages of soy protein in paint formulations were: good washability of the painted surface, hiding power of a single application, durability on porous surfaces, rapid drying, and absence of odor (because they contained no organic solvents) (Burnett, 1951). The disadvantages were low water-resistance and durability, poor initial adhesion, and low resistance to microbial degradation if not properly preserved. These paints used higher protein concentrations (50 g/L, 0.4 lb/gal) than did the resin-oil emulsion paints (Little, 1951). Due to the superior properties of the resin-oil paints, like many other products after World War II, the importance and market share of these paints declined rapidly.

Agricultural Uses

Before World War II, substantial tonnage of soybean meal was used for mixed fertilizers because of its high nitrogen content. Today, no soybean meal is used for this purpose. In the late 1940s, the Glidden Co. marketed soy protein as a *sticker and spreader* in agricultural sprays. The product was known as *Spraysoy*.

Soybean Oil in Industrial Biobased Products

Over the past few decades, the world's fats and oils production has grown far beyond the economic demand for human food uses, and production is still growing. In 2005/06, production of oils and fats increased to 146 MMT (161 million t) from 130 MMT (144 million t) in 2003/04 (Gunstone, 2006). The United States annually produces ~408,000 MT (450,000 t) of soybean oil in excess of current domestic commercial need. According to Foreign Agricultural Services (FAS) of the United States Department of Agriculture (USDA), the worldwide production of soybean oil increased by 65% from 20.4 MMT (22.5 million t) in 1995 to 33.8 MMT (37.3 million t) in 2005. The United States produced an average of 8.6 MMT (9.5 million t) of soybean oil in 2005, which was 79% of the total vegetable oils consumed (10.9 MMT, 12 million t) in the United States, and of this amount only about 4% (0.443 MMT, 0.977 billion lb) was consumed in biobased products.

Natural soybean oil is too viscous and reactive to atmospheric oxygen to be used in many biobased product applications. These limitations must be overcome for soybean oil to be used in fuels, cosmetics, and lubricants, but on the other hand, soybean oil is not sufficiently reactive to be used in most paints and coatings. Important end-use categories for which economic data exist include fatty acids, paints and varnishes, resins and plastics, drying-oil products, and "other industrial products." Coating vehicles (paints and varnishes) and epoxidized oils (resins and plastics) comprise 50% of the industrial market for soybean oil used in biobased products. The market share for coating vehicles is shrinking due to competition from synthetic latex coating vehicles, while the resin and plastic markets are expanding.

Lubricants

Background

Vegetable oils were once widely used as lubricating fluids, but in the last half century petroleum-derived mineral oils were mainly used as lubricant base fluids. Until recently, mineral base oils were economical and provided certain superior performance characteristics in various applications, but they also present potential dangers because they are not readily biodegradable and are environmentally toxic. The recent rise in petroleum prices along with relatively low soybean oil prices eliminated these price differences; hence, renewed interest has arisen in soybean-oil-based lubricants. The last decade also has seen a slow but steady move toward the use of environmentally friendly or more readily biodegradable lubricants. This demand is due to growing concerns over the impact that technology is having on the environment. One-half of the lubricants sold worldwide ends up in the environment via total loss applications, spills, and volatility (Horner, 2002). Overall, 32% (1.635 billion L or 432 million gal) of the 5.114 billion L (1.351 billion gal) of lubricating oil sold in the United States ends up in landfills or is dumped (Adamczewska & Wilson, 1997), and 13% (660 million L or 174 million gal) of the 4.94 billion L (1.31 billion gal) of lubricants used in the European community in 1990 disappeared into the environment (Naegely, 1992).

Generally, lubricants are made from base oils and suitable additive packages. Mineral base oils are usually obtained from crude petroleum by means of various refining and extracting steps before blending with specialty chemical additives to enhance performance characteristics. Although most lubricants currently used originate from petroleum base stocks, the use of vegetable oils over the last decade has experienced a promising increase, especially as biodegradable fluids. Environmental concerns, as well as economic and performance issues, will drive the market for these oils. In addition to lower cost, soybean-oil-based lubricants are environmentally friendly, and thus, are safely disposable in case of spills or leaks.

Performance Properties

The beneficial aspects of soybean oils as base stocks are mainly biodegradability and nontoxicity, properties that are usually not exhibited by conventional mineral base oils (Battersby et al., 1992; Randles & Wright, 1992). They have very low volatility due to the high molecular weight of the triglyceride structure and small viscosity change with temperature (Table 17.2). Low volatility decreases exhaust emission and reduces engine sludge, while a high viscosity index means that the oil naturally satisfies the

Table 17.2. Properties of Lubricant Base Oils

Property	Mineral Oils	Soybean and Other Vegetable Oils	Synthetic Esters
Biodegradability, % (CEC-L33-T82)	<40	>90	>75
Biodegradability (EPA method)	42–48	72–80	55–65
Viscosity index	90–100	>200	>120
Pour point	Good	Poor	Excellent
Oxidation stability	Good	Poor/Good	Poor/Good
Compatibility with mineral oils	—	Good	Good
Relative cost	1	1.5 to 3	5 to 20
NOACK volatility	Normal	Low	Low

multi-grade requirements of modern engines. A high viscosity index eliminates the need for a polymeric viscosity index improver and yields high-shear stability. The ester linkages deliver inherent lubricity and are able to adsorb to metal surfaces. Lower friction in the engine results in more power and better fuel economy. Furthermore, soybean oil has superior solubilizing power for contaminants, polar deposits, and additive molecules compared to mineral base fluids, so no detergent additive is needed in formulated soybean-oil-based lubricants. Lubricity, antiwear protection, load-carrying capacity, rust prevention, foaming, demulsibility, and other important lubricating properties depend on additives to improve or optimize lubricant performance. Performance limitations of soybean oil base stocks include poor oxidative stability, tendency for deposit formation, low-temperature solidification and low hydrolytic stability, which restrict extensive use of soybean oil for lubricant applications.

The fatty acid composition of the vegetable oil affects the chemical properties as well as the thermal, oxidative, and low-temperature stabilities of the oils. With increasing unsaturation in soybean oil, the oxidation rate increases, resulting in polymerization and increased viscosity (Brodnitz, 1968). On the other hand, high saturation increases the melting point of the oil (Hagemann et al., 1972). Therefore, make a compromise between low-temperature and oxidative stabilities when selecting a vegetable oil base stock for a particular biobased product application. The chemical and physical properties of base oils (both mineral and vegetable origin) are controlled by the relative distribution of various molecular structures (Adhvaryu et al., 2000), which can be used to build structure-property predictive models to improve properties and cost-effectiveness of vegetable oils (Adhvaryu et al., 2006a,b).

Low resistance to oxidative degradation (Asadauskas et al., 1996, 1997, 2000); and poor low-temperature behavior (Erhan & Asadauskas, 2000; Hagemann & Rothfus, 1988; Rhee, 1996) remain the major impediments to using soybean oil as a base stock in biobased lubricants. Soybean oil is the most widely available and least expensive vegetable oil in the U.S. market, and soybean oil could have a distinct advantage over other oils if it could be modified to improve its oxidative and low-temperature stabilities.

To expand commercialization of soybean-oil-based lubricants, one of three factors must occur to eliminate cost as a deterrent: (i) the increased cost of the product must provide increased performance; (ii) the cost of competing petroleum products must increase significantly; or (iii) the use of the biobased product must provide the best way to meet required local, state, or federal regulations. Of these three factors, the first, improving the properties of soybean oil, is the most desirable scenario. Improve the low-temperature properties of soybean oil by blending with diluents, such as polyalphaolefins, diisodecyl adipate, and oleates (Asadauskas & Erhan, 1999; Erhan et al., 2006); improve oxidative stability by adding commercially available antioxidants (Becker & Knorr, 1996; Erhan et al., 2006; Sharma et al., 2007b).

The use of suitable additives with soybean oil may provide nearly equivalent performance to petroleum-based lubricants. Diluents and additives are used with soybean oil to meet performance standards for industrial and automotive applications (Erhan & Adhvaryu, 2002). The poor oxidative stability of soybean oil is due to the polyunsaturation in its triacylglycerol structure; therefore, reducing polyunsaturation through chemical or genetic modifications results in improved oxidation stability. Chemical modification of soybean oil at the double-bond sites of the component fatty acids also improves cold-flow properties (Erhan et al., 2003, 2005; Sharma et al., 2006b). Epoxidized soybean oil (ESO) is an important intermediate in utilizing soybean oil as a lubricant base stock because it is relatively inexpensive, and the epoxy group can be readily functionalized (Erhan et al., 2003, 2005; Sharma et al., 2006b).

Lubricant base stocks with good low-temperature properties and oxidative stabilities were prepared by acid-catalyzed ring opening of ESO using alcohol followed by esterification with an acid anhydride (Adhvaryu et al., 2006b; Hwang et al., 2003; Hwang & Erhan, 2001, 2002, 2006). Sulfurized fatty products also have good thermo-oxidation stability, and are used in extreme-pressure and antiwear additives. Studies show that they form metal sulfides that permit sliding contact and sloughing-off of the metal salt rather than the scarring associated with metal-to-metal contact (Kammann & Phillips, 1985). Oleic acid is more thermally stable than are polyunsaturated fatty acids, and therefore oleic acid is a highly desirable component in vegetable oils. Genetically engineered soybeans and other seeds contain oils that are high in oleic acid and have less polyunsaturation, such as linoleic and linolenic acids (Schmidt et al., 2005), and this improves oxidative stability.

Numerous oxidation tests are available to screen vegetable oil oxidative stability including: thin film oxygen uptake test (TFOUT, ASTM D 4742), rotating bomb oxidation test (RBOT, ASTM D 2272), panel coker test, and pressurized differential scanning calorimetry (Biswas et al., 2007; Erhan et al., 2006; Sharma et al., 2005,

2006b,c); testing for oxidation characteristics (ASTM D 2893), corrosiveness and oxidation stability (ASTM D 4636), and deposition tendencies of liquids in thin films (ASTM D 3711); estimating peroxide content (peroxide value, PV); and active oxygen method (AOCS Method Cd-12-57), Rancimat method (Laubli & Bruttel, 1986), chemiluminescence reaction (Matthaus et al., 1994), and oil stability index (AOCS Method Cd-12b-92).

Markets

The lubricating fluid market in the United States was estimated to be $20 billion (9.5 billion L or 2.5 billion gal) in 2005, with 57% of this total for automotive, 2% for greases, and the remainder for industrial application (process oils, 17%; general industrial oils, 14%; metalworking oils, 3%; and industrial engine oils, 6%) (Padavich & Honary, 1995; Persaud, 2007). In 2002, the vegetable-oil-based lubricant market share was 1% (45,000 MT/yr or 50,000 t/yr) of the total Western European market for all lubricants (5.02 MMT/yr or 5.5 million t/yr) and only 0.3 % (22,700 MT/yr or 25,000 t/yr) of the total U.S. lubricant market (8.25 MMT/yr or 9.1 million t/yr) (Whitby, 2004). Soybean-oil-based lubricants are of great value in areas where the lubricant is either lost in the environment or is in close contact with the environment (Canter, 2001). The primary area of their application has been in hydraulic fluids, which have the greatest need for biodegradability. Total-loss lubricants, where the lubricant is lost completely, account for 7–8% of total global lubricant demand and include bar-chain lubricants, wire rope oil, rail-flange lubricants, mold-release oils for the construction industry, drilling muds and oils, drip oils (oil for water and underground pumps in irrigation wells), two-stroke engine oils, outboard marine engine oils, cutting and drive-chain oils, metal-cutting oils, tractor oils, dedusting oils, etcetera. Several of these applications are in farming, mining, and forestry, where the lubricants are in close contact with the environment. Off-highway equipment used in waterways, mining, farming, and forestry constitutes a particularly attractive use for soybean-oil-based lubricants because this equipment typically uses large volumes of fluids [hydraulic-sump capacity of backhoes, 23–91 L (5–20 gal); excavators, 136–182 L (30–40 gal); tractors, 91–182 L (20–40 gal); mining shovels, 682–910 L (150–200 gal)], and operates in direct contact with the environment.

Industrial hydraulic fluids represent a 1.01 billion L (222 million gal) market in the United States. Soybean oil was successfully used in hydraulic elevator fluids (Adamczewska & Wilson, 1997; Honary, 1996; Kassfeldt & Dave, 1997; Padavich & Honary, 1995; Rhee, 1996). The market for motor oils in the United States alone approaches 5 billion L (1.1 billion gal) according to the used-oil Web site published by the American Petroleum Institute (http://www.api.org/ehs/performance/recycling/morerecycling.cfm; Johnson, 1999). Although crankcase oils have limited contact with the environment, active development work is in progress on using vegetable oils as base stocks (e.g., canola, corn, soybean oil, and their high-oleic varieties) in air-cooled engines (lawnmowers, chain saws, etc.) and automobile gasoline engines (Rhodes & Johnson, 2002). Environmentally friendly lubricants can meet engine oil specifications by using high-oleic varieties of vegetable oils and can have equivalent or better performance than petroleum-based oils in laboratory bench tests (Perez & Boehman, 2002). These vegetable-oil-based engine oils also provide the advantage of significantly reduced particulate emissions and higher lubricity. Nearly $700 million is annually spent in the United States on metalworking fluids (Asadauskas & Perez, 2002). Numerous oils including soybean oil and oleochemicals are primarily used to impart lubricity, rust prevention and emulsification, especially in aqueous metalworking fluids. Soybean oil also has advantages in formulating metalworking fluids, due to hazardous mist formation when mineral oils are used (Erhan, 2006; Herdan, 1999; John et al., 2004).

Greases comprise 8% of the industrial lubricant market, which corresponds to 68 million L (18 million gal) of annual use in the United States. Efforts to develop biodegradable soybean-oil-based greases resulted in alternatives to nonbiodegradable petroleum-based greases (Adhvaryu et al., 2004, 2005; Gangule & Dwivedi, 2001; Sharma et al., 2005, 2006c). These soybean-oil-based greases were used by railroads, where they reduce wheel-flange and rail-gauge face wear that occur when trains go around curves (Honary, 2000, 2003, 2004; Neuzil & Honary, 2003). Soybean-oil-based greases may have a competitive edge in lubricating rail tracks, because these greases can greatly impact the environment: for example, over 45,000 MT (50,000 t) of railroad grease were lost to the environment in Canada over the past century.

Current Interests

In all these areas, commercial acceptance of biodegradable soybean-oil-based products is good, and excellent opportunities are available to penetrate the existing mineral-oil market. The widespread use of soybean oils as lubricating basestocks depends mainly on how well they perform in high-temperature oxidation and low-temperature applications. Genetically modified high-oleic soybean oil performs better than normal soybean oils in terms of thermal and oxidative stabilities. Diluents and pour-point depressants lower the pour points and increase the cold-storage stabilities of soybean oils.

With improved properties of soybean-oil-based lubricants, consumer awareness and regulations that mandate the use of biodegradable lubricants will dictate the future outlook for these products. Currently, no regulations mandate the usage of such products in the United States. Several regulations require that preference be given in government procurement to biodegradable lubricants if the price is acceptable; however, the most encouraging development is original equipment manufacturer (OEM) interest in these types of fluids. Therefore, good opportunities are present for growth of this sector.

Printing Inks

Background

Ink is an organic or inorganic pigment or dye dissolved in a solvent or suspended in a carrier, similar to paint. The first inks were flowers, vegetable or fruit juices; protective secretions from cephalopods, such as cuttlefish, squid, and octopus; blood from some types of shellfish; and tannins from nuts, galls, or tree bark. The first synthetic ink appeared in Egypt about 4,500 yr ago and was made from animal or vegetable charcoal (lampblack) mixed with glue. Today's inks are divided into two classes, printing inks and writing inks.

Printing inks are further broken down into two subclasses, ink for conventional printing, in which a mechanical plate comes in contact with the paper or object being printed and ink for digital nonimpact printing, which includes ink-jet and electro-photographic technologies. The major printing processes used by the industry are lithography (offset), letterpress, flexography, gravure, and screen printing (Leach, 1988). Letterpress and lithographic inks are viscous and paste-like. Flexographic and gravure inks are extremely fluid and are called *liquid inks*. Screen inks are intermediate in viscosity between the above two categories. Vegetable oils, including soybean oil, are mainly used in paste inks (lithographic and letterpress).

Paste inks are made primarily with linseed oil, soybean oil, or a heavy petroleum distillate as the solvent (termed *vehicle*) combined with organic pigments, and other minor proprietary ingredients. The pigments are various organic compounds for colored inks, carbon black for black inks, and titanium dioxide for white pigments and to adjust the characteristics of color inks. Inks also contain additives, such as waxes, lubricants, surfactants, and drying agents, to aid printing and to impart any desired special characteristics.

The ink vehicle functions as the carrier and transport system for the pigment, which helps anchor the pigment to the substrate and becomes solid upon reaching the substrate. Historically, the vehicles were composed of a resin and a petroleum-derived hydrocarbon solvent such as mineral oil. The resin binds the pigment to the paper and provides appropriate viscosity and tackiness to the ink. The solvent is used to dissolve and disperse the resin and to achieve the desired flow properties of the ink for proper absorption by the paper. The solid ingredients are dispersed in the vehicle to form a paste ink. Typical paste ink consists of 50–70% mineral oil solvent, 15–25% hydrocarbon or alkyd resins, and 15–20% pigments. The petroleum shortage in the 1970s stimulated research into vegetable-oil-based inks as substitutes for petroleum-based products. At the same time, the ink industry was forced to use more expensive, highly refined mineral oils, because the less-refined oils contained polynuclear aromatics, shown to be carcinogenic. Additionally, waste and cleanout inks made with mineral oil are regarded as hazardous wastes. These wastes are becoming increasingly difficult to dispose, and disposal of waste ink can cost as much as $300 per barrel.

In the early 1980s, the American Newspaper Publishers Association (ANPA; renamed the Newspaper Association of America, NAA) developed a series of ink formulations using blends of gilsonite (resionous asphalt), tall-oil fatty acids, and carbon black pigment (Moynihan, 1983, 1985b). This worked well, but the cost and availability of tall oil, and the difficulty of equipment cleanup created by gilsonite, limited the acceptance of these inks by the industry. ANPA then focused on inks based on vegetable oils.

Technology

In 1985, ANPA came out with the first-generation soybean-oil-based lithographic news ink, comprised of alkali-refined soybean oil, hydrocarbon resin, carbon black pigment, and an antioxidant (Moynihan, 1985a). This black ink had similar printing properties to mineral-oil-based inks, but cost 30–50% more. The color inks were formulated in a similar way, had good print quality, but cost about 5–10% more than the petroleum-based commercial inks. Thus, industry continued to search for 100% vegetable-oil-based inks to replace petroleum-based inks.

Second-generation soybean-oil-based inks were developed at the USDA, NCAUR, Agricultural Research Service (ARS) in Peoria, Illinois (Erhan et al., 1992a, 1997a, b, 1999; Erhan & Bagby, 1991, 1992b, 1994, 1998; Erhan & Nelsen, 2001). The ink vehicles were totally derived from vegetable oils, thus eliminating the need for the petroleum-based resins (Erhan & Bagby, 1991). In one process, the vegetable oil was heat-polymerized at a constant temperature in a nitrogen atmosphere to achieve the desired viscosity. In a second process, the oil was heat-polymerized to a gel point, and then the gel was diluted with vegetable oil to obtain the desired viscosity (Erhan & Bagby, 1991, 1992b). The reaction time necessary to reach the desired viscosity depends on the mass and structures of the reactants, and the rate of heat transfer and agitation. As expected, vegetable oils with more unsaturation polymerize more rapidly than those with less unsaturation. Both methods facilitated tailoring the properties of the ink vehicle for a variety of news inks.

Blending these heat-bodied soybean oils of various viscosities will produce any desired viscosity. Blending different proportions of gels and unmodified soybean oils gives different vehicle viscosities. These vehicles typically have viscosities in the range of G-Y on the Gardner-Holdt viscometer scale, or about 1.6–18 P (Erhan & Bagby, 1992a, 1994). These viscosities correspond to apparent average molecular weights of 2600–8900 Daltons. Consider the thickening effect of the pigment on the base vehicle in selecting a vehicle viscosity (Erhan & Bagby, 1997, 1998). The oxidative polymerization tendency of polyunsaturated fatty acids present in triacylglycerols of vegetable oils was used in these ink vehicles. In fatty acids, such as linoleic and linolenic with two and three double bonds, respectively, oxidation occurs across the double bonds to form oxides, which are unstable, break down, and crosslink. The fatty acid chains of one triglyceride molecule attach to those in other triglyceride molecules and form a more rigid structure, giving rise to dry ink. Alkali-refined soybean oils are used in ink formulations, because refining removes the gums, waxes, and free fatty acids, and the presence of any of these compounds interferes with the desirable hydrophobic characteristics of the vehicle and ink formulations (Erhan, 1997).

Performance Properties

Important properties of the ink vehicle as a pigment carrier and transport system include pigment wetting and dispersion, ink rheology (ink transfer, ink misting, and press stability), and ink/water balance for lithography. Important properties of the ink vehicle as the pigment binder or anchor to the substrate area include ink setting, ink drying, and gloss- and rub-resistance of the dried ink. Setting has to occur very quickly. Drying occurs thereafter and is considered complete when the viscosity reaches 10^6 cP. The substrate also plays an important role in the setting and drying process. Setting and drying mechanisms include one or more of the following: solvent/oil penetration, solvent evaporation, oxidation polymerization, catalytic polymerization, and resin precipitation (*Printing Ink Handbook,* 1999). Printers are highly concerned about rub-resistance, print quality, and readability.

Properly formulated soybean-oil-based inks are biodegradable by microorganisms that are commonly found in soil (Erhan & Bagby, 1993; Erhan et al., 1997b) and also when using the modified-Sturm test (Erhan et al., 1995). Soybean-oil-derived components are almost completely degraded within 25 days. The 100% soybean oil-based vehicle degraded 82– 92% compared with 58–68% for the ANPA vehicle and 20% for the petroleum-based vehicle.

Soybean-oil-petroleum hybrid inks provide superior print qualities, brighter colors, higher rub-resistance, better mileage (spread quality), cleaner press runs, and environmental benefits. Replacing the solvent and resin with materials derived from soybean oil reduced the cost of ARS soy news inks, so that they are competitive with petroleum-based lithographic newspaper ink. The light color of ARS vehicle allows less pigment to be used, further reducing the cost of the colored inks. These inks meet or exceed industry standards for print quality, ease of cleanup, rub-resistance, viscosity, and tack.

Markets

Printing ink is a $14.5 billion global industry. According to the market research firm Freedonia Group, nearly $5 billion worth of printing inks (1.13 MMT or 1.25 million t) was sold in the United States in 2001, and is expected to grow to $9 billion by 2010 (Thayer, 2002). The European printing ink market accounted for another $2.7 billion. In the United States alone, more than 200 ink producers, ranging in size from local *mom-and-pop* firms to large international companies, are established. Publishing accounts for only 23% of ink demand, whereas the largest end-use markets are packaging at 36% and commercial printing at 33% (Thayer, 2002). The most widely used processes in publishing and commercial printing are gravure inks and lithographic inks. Lithographic ink represents about 43%, while gravure inks represent 17% of all ink shipments as measured in dollars in 2001 (Thayer, 2002). The lithographic inks are losing ground to flexographic inks (22%), particularly in printing packaging materials. Newspaper accounts for 40% of the total publication and commercial printing.

In ten short years, soy ink's U.S. market share quadrupled, from <5% in 1989 to 22.5% in 1999. In the United States alone, annual soybean oil usage in printing ink is 47 million kg (103 million lb) with room to grow. When soy ink reaches its full potential, it will annually consume 215 million kg (457 million lb) of soybean oil. Soy inks captured 27% of the color newsprint ink market because of their superior print qualities, despite high costs for these inks, but these inks captured only 6% of the black newsprint ink market (National Printing Ink Research Institute, Lehigh University, Bethlehem, PA). They have not yet achieved price-quality parity with conventional inks, and the price of the oil vehicle is a far more important factor in the pricing of black newsprint inks. The market shares captured by soy ink in sheet-fed and heatset ink are 9.3 and 7.3%, respectively. Today, soy ink (primarily color) is used by one-quarter of the nation's 50,000 commercial printers and one-third of the nation's nearly 10,000 newspapers (dailies, weeklies, and monthlies), including 90–95% of the 1,500 daily newspapers. Nearly 100 different U.S. ink manufacturing companies produce at least one soy ink product, representing 45% of the nation's ink manufacturers.

Current Interests

Complete soybean-oil-based technology is available to manufacture printing ink vehicles with the desired commercial characteristics. This technology allows manufacturers to increase the vegetable oil contents in ink formulations and, in turn, improve environmental properties by increasing biodegradability, reducing volatile organic compounds, and improving deinking properties. Soybean-oil-petroleum hybrid inks meet some of the properties achieved by the ARS soy news ink. With the adoption of either of these technologies, market share of soy ink should continue to rise. New government regulations, such as the 2002 Farm Bill, Executive order 13134-1999, may help the marketing of soybean-oil-based news inks.

Paints, Coatings, and Varnishes

Background

Paint is a historical use for soybean oil and other vegetable oils. Triacylglycerols of vegetable oils and their derivatives have been used as binders or additives in paints and coatings for 30,000 yr, going back to the days of cave paintings. Ancient Egyptians used linseed oil in decorative coatings. The early vegetable-oil-based paints and coatings were formulated using the drying oils selected from plant and fish oils, and naturally occurring pigments such as red iron oxide or carbon black. The curing or oxidation process was slow and yielded a soft coating since no catalyst was used. Typical drying oils include linseed oil with drying indexes of 123 and tung oil at >172, while soybean oil at 66, sunflower oil at 66, and corn oil at 54 are semi-drying oils because of insufficient polyunsaturation. Thus, soybean oil alone is a very poor paint, and usage was largely limited to extending linseed oil during periods of linseed oil shortages.

In the 1950s, with the advent of synthetic resins, particularly alkyd resins, it became possible to use semi-drying oils like soybean, sunflower, and safflower oils (Lin, 2004). Nondrying oils, such as coconut oil, were also used in coating materials, but they functioned as a plasticizer rather than as an active component. Soybean oil alkyd resins improved the drying, adherence, endurance, and color properties of soybean oil. Soybean oil alkyd resins were often blended with linseed and tung oil alkyd resins to improve performance at acceptable costs. H. Ford used soybean oil and soybean oil derivatives in enamel paints for his automobiles, and DuPont's development of the *four-hour enamel*, based on soybean alkyds, is generally considered to be the most important events in furthering the use of soybean oil paints. After World War II, vegetable-oil-based paints lost market share to less expensive and easier-to-use latexes (rubber-based) using water as solvent and lower levels of soybean-oil-derived ingredients.

Technology and Performance

An excellent description of technologies used to produce paints, coatings, and resins was discussed by Johnson and Myers (1995), Van De Mark and Sandefur (2005b), and Lin (2004). Before discussing paints, coatings and varnishes, as well as the functions of soybean oil in these products, it is desirable to first review and define some terms. A paint is a pigmented system applied to hide, protect, and decorate a surface, in which the pigment is bonded to the surface by a polymer (the resin) (Lin, 2004). Enamel is paint based on a vehicle that dries to a considerably harder film than paints derived from unmodified drying oils. Coatings are similar to paints in protecting and decorating surfaces, but are used in consumer goods such as automobile bodies and furniture. Protective coatings are materials that form durable films adhered to the surface to provide protection, and protective coatings must dry quickly. While the paints are applied with a brush or roller, coatings usually are applied by spraying, dipping, roller coating, air knifing, or other high-speed production application method, and are force-dried by baking. Varnishes are similar, except that no pigment is used to hide the surface, so they produce clear films. A varnish is a solvent-thinned combination of a drying oil and a hard resin. Vehicles used for clear films are called *varnishes*.

All three products contain vehicles consisting of combinations of oils, resins (an alkyd resin solution or a urethane-modified resin), polymers, and solvents; the nonvolatile portion is commonly called the *binder*. Paints and coatings also contain pigments. Apart from the vehicle, all three contain driers and solvents. Driers are metal salts that accelerate the conversion of a liquid film to a solid. Solvents or thinners control paint consistency and application properties, and can be either aqueous or an aliphatic or aromatic petrochemical. The solvent either dissolves or suspends the resin and the pigment during application. Once applied, the solvent evaporates slowly or rapidly depending on the type of solvent, leaving a thin film of pigment and resin on the surface. Then the liquid film polymerizes through oxidation of double bonds to form a solid.

The oxidation or polymerization is often accelerated by heating, by using oils having more polyunsaturated fatty acids, or by air-blowing the oil. Oxygen diffuses into the film and reacts with diallylic hydrogens to produce hydroperoxides. The hydroperoxide is formed at relative rates of 1:120:330 for trioleate:trilinoleate:trilinolenate (Wicks et al., 1998). The curing mechanism then involves the decomposition of the hydroperoxides by a redox catalyst, also called a *drier*. Common driers are cobalt, manganese naphthenate, zirconium, calcium, and iron salts. These salts add color to the coatings and thus are kept to minimum concentrations.

Alkyd resins have been the "workhorse" for the paints and coatings industry over the last half century. Alkyd resins are also used in core resins, adhesives, lithographic inks, and plasticizers (Pryde, 1980). The term *alkyd* was coined to define a reaction product of polyhydric alcohols and polybasic acids resulting in polyesters, but became narrowed to include only the polyesters containing monobasic acids such as long-chain fatty acids. The proliferation of literature on alkyd resins peaked during 1940–1960 and tapered off over the last two decades. The alkyd resins sustained popularity because they can be made at relatively low cost and are extremely versatile due to a large variety of reaction ingredients. The main reactions in synthesizing alkyd resins are polycondensation by esterification and ester interchange. The fatty acids from soybean triacylglycerol molecules are transferred to form esters with the polyols [$R(OH)_3$]. After alcoholysis, a dibasic acid [$R'(COOH)_2$] is added, and esterification takes place until the proper end point is achieved. The fatty acids, dibasic acids, and polyols are reacted in one step to give a resin molecule, if vegetable-oil-derived fatty acids are used in place of oils (Fig. 17.4).

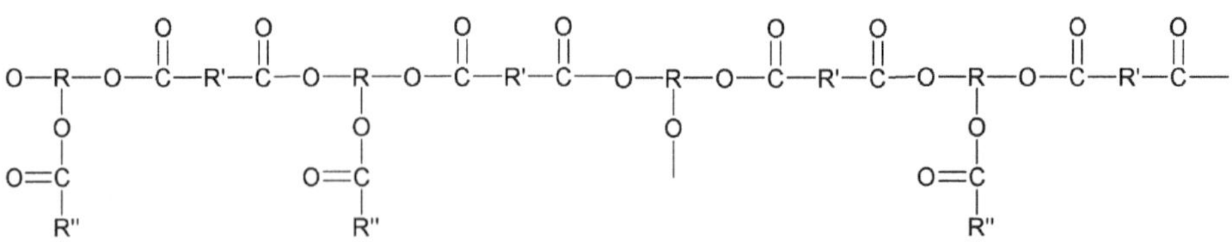

Fig. 17.4. Schematic representation of an alkyd resin molecule (where $R(O)_3$, a polyol molecule; -C(O)-R′-C(O)- is a polybasic acid molecule or radical; and -C(O)-R″, a mono-basic acid or fatty acid molecule or radical).

Based upon the amount of unsaturated oil used in the manufacture, the alkyd resins are classified as *very long oil*, *long oil*, *medium oil*, and *short oil*. Very-long-oil alkyds have >70% oil and <20% phthalic anhydride. Long-oil alkyds contain 56–70% oil and 20–30% phthalic anhydride, are soluble in mineral spirits, are rapid drying, produce good gloss and weathering, and produce little yellowing and soft resins. These alkyds are typically used in architectural paints, glossy and semigloss enamels in both interior and exterior applications, latex paints, stipple paints, spar varnishes, and fire-retardant paints (Pryde, 1980).

Medium-oil alkyds are typically used in architectural paints and as co-resin in some original equipment manufacturer (OEM) coatings. These alkyds are 46–55% oil and 30–35% phthalic anhydride. They are viscous, soluble in mineral spirits and aromatic solvents, and slightly harder, and have poorer application properties than long-oil alkyds.

Short-oil alkyds are used typically for OEM and do not dry well unless force-air dried or baked. These resins contain <45% oil and >35% phthalic anhydride. They are soluble in aromatics but not in mineral spirits, and are significantly harder than medium- or long-oil alkyds.

The oils used to make alkyds are selected on unsaturation number. In general, the higher the unsaturation number, the better drying. Excellent drying oils for alkyds include soybean, linseed, tung, and dehydrated castor oils.

The other classification for alkyd resins is based on drying properties of the fatty acids used in resin composition. Drying-type alkyd resins have polyunsaturated fatty acids derived from drying oils such as linseed oil. Nondrying alkyds have fatty acids derived from nondrying oils (coconut oil), and these resins have excellent color and gloss stability. More frequently, an alkyd resin is also classified by the source of the fatty acids (e.g., tung oil-modified soy alkyd, linseed alkyd, coconut alkyd). Soybean oil is the least expensive unsaturated oil, and its high linoleic acid content and low linolenic acid content compared to linseed oil impart good resistance to yellowing. High contents of linolenic acid present in linseed oils are responsible for high yellowing tendency. Therefore, alkyd resins intended for making white or light color enamels should avoid high contents of linolenic acids by choosing soybean, safflower, or dehydrated castor oil.

The drying rate of soybean oil can be improved using maleinized oils, so that soybean oil could be used in many of the same applications as linseed oil. By heating soybean oil, normal unconjugated fatty acids of soybean oil are converted to conjugated forms, which react with maleic anhydride in the Diels-Alder reaction (Fig. 17.5). The derivatives are then esterified to a polyol such as glycerol or pentaerythritol. This polymer is dissolved in a high-boiling, water-loving solvent such as ethylene glycol monobutyl ether. An amine, such as ammonia or triethylamine, is added to form the salt of the acid group. The pigments are dispersed in the resins, and water is finally added very slowly, which results in the formation of very small particles of

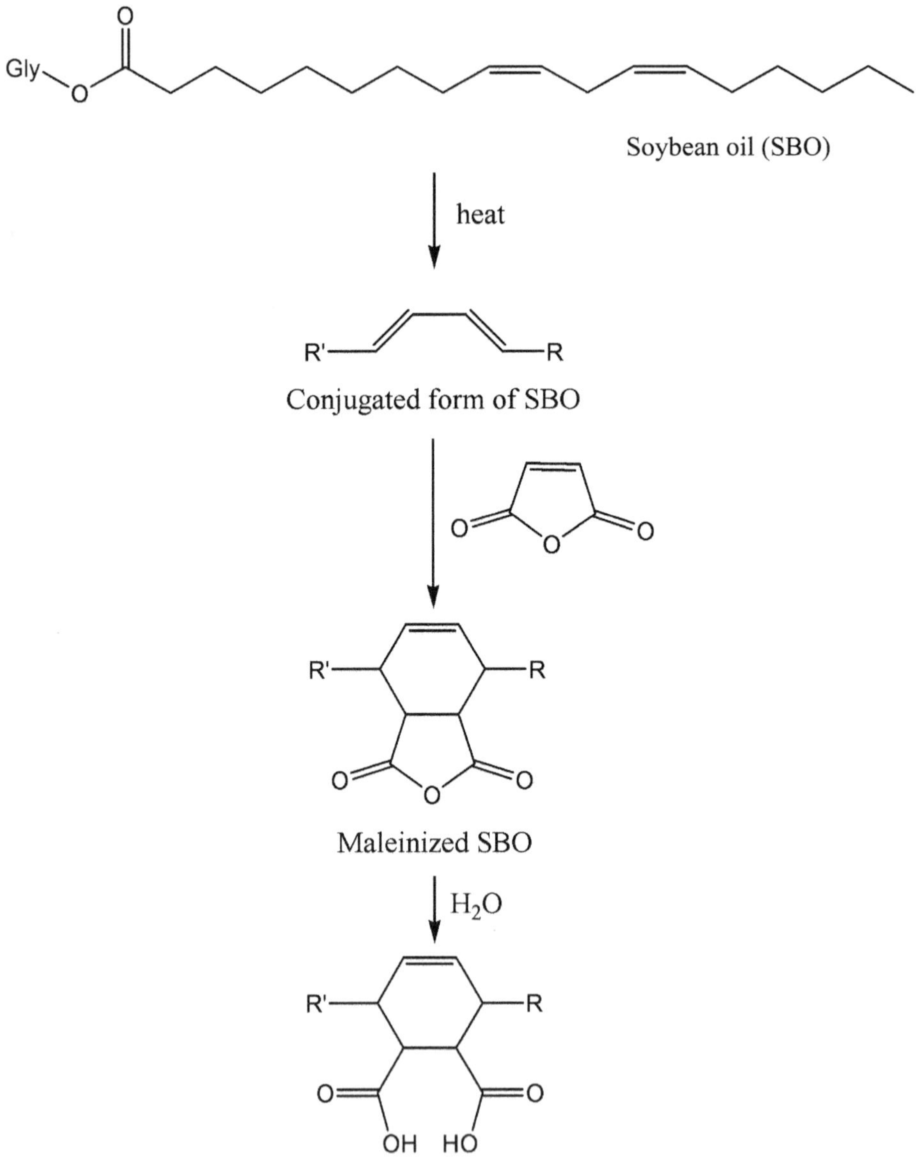

Fig. 17.5. Diels-Alder reaction forming maleinized soybean oil.

polymers, typically 5–100 nm. If formulated correctly, the polymer solution is clear, and the solution is thermodynamically stable due to Brownian motion, which keeps the particles suspended (Jiratumnukul & Van De Mark, 2000; Van De Mark, 2000; Van De Mark & Jones, 1998; Van De Mark & Loftin, 1999; Van De Mark & Sandefur, 2005a; Van De Mark & Schnelten, 1997). The polymer is not water-soluble but is water-dispersible, and many of the particles are single polymer chains. These water-reducible resin systems have been available for over 50 yr, but did not gain popularity until the mid 1980s when environmental regulations caused the industry to move to water-borne coatings. Water-reducible technology relies on the presence of water-loving groups, usually the salt of carboxylic acids, on the polymer chain to act as an internal emulsifying agent. These water-reducible coatings can be used for many OEM and maintenance applications.

Other water-borne coatings include water-soluble emulsions, dispersions, and latex resins. Water-soluble resins are rare because most resins derived from vegetable oils are insoluble in water. The true emulsions are based on the emulsification of the oil or alkyd through either the action of a surfactant or a resin that has a surfactant-like character. The alkyd emulsions are readily prepared and can be used for OEM coatings and architectural applications. The submicron size droplets are stabilized by the thickeners (El-Aasser & Sudol, 2004; Landfester, 2005; Landfester et al., 2004; Tsavalas et al., 2004; Weissenborn & Motiejauskaite, 2000a,b). In dispersions, the resin is a solid and is dispersed in water. The latex resin is usually vinyl acetate, styrene, acrylates, or methacrylates radically copolymerized in a micelle to form particles 0.1 µm in diameter (Bloom et al., 2005; Brister et al., 2000; Jiratumnukul & Van De Mark, 2000; Thames et al., 2005).

Epoxy alkyds typically have 5–45% resins based on epoxy chemistry. Bisphenol A along with other phenols is used with epichlorohydrin. The epoxy groups of the resin are esterified with unsaturated fatty acids of soybean and tall oils. These resins offer better performance than a simple alkyd with excellent chemical-resistance and strong adherence to a wide variety of materials.

Urethane-modified alkyds are similar to simple alkyds except that dibasic acid is replaced with a difunctional isocyanate such as toluene diisocyanate or hexamethylene diisocyanate. The process is also similar to simple alkyds. Coatings made with urethane-modified alkyds dry faster and harder than alkyds, yet retain flexibility. These systems have better water-, chemical- and abrasion-resistance than alkyd resins, and cost is also relatively low (Wicks et al., 1998). These are used in clear finishes for wood floors, cabinets, OEM, maintenance, and architectural coatings. The aliphatic-based systems are excellent for exterior use, or where UV exposure is possible, while aromatic-based systems usually have better abrasion-resistance. Chemo-enzymatic synthesis of urethane-based systems produces better control of stereochemistry and can impart unique properties (Athawale & Bhabhe, 1998; Athawale & Gaonkar, 1999; Athawale & Joshi, 2000, 2004; Bhabhe & Athawale, 1998).

Soybean oil reacts with various reactive monomers such as cyclopentadiene, styrene, acrylate, and vinyl toluene. Cyclopentadiene readily reacts with unsaturated fatty acids of oils to give products that dry hard and fast, and are soluble in aliphatic solvents. These are widely used in aluminum paints due to their good leafing properties. Acrylic modified alkyds are prepared using normal low-molecular-weight alkyd, acrylate monomer, and a radical initiator (benzoyl peroxide). The monomer polymerizes and is grafted onto the alkyd. These alkyds produce a fast-forming hard film and are used in OEM applications. Similarly, styrene and vinyl toluene can be grafted onto alkyds. The alkyds have improved gloss, cost, and hardness with rapid drying. Their disadvantages are poor UV stability with rapid yellowing and less adhesion to many substrates. Melamine-modified alkyds can be cured thermally, and are baking coatings. Melamine derivatives act as cross-linking agents to produce cross-linked films when baked. The resins are soluble in aromatics, ketones, esters, and hot alcohols. Silicone coatings are based mainly on dimethyl or diphenyl siloxane with significant cross-linking. These coatings have high thermal stability and are very expensive, relatively stain-resistant and nonsticking.

Durability is important to all paints, but many modern paints last longer than the customer desires the color. Therefore, for interior household paint, durability becomes less important than color, finish (gloss and texture), quick drying, and easy cleanup. Film durability is more important, where it is in direct contact with environment (sunlight, moisture, and thermal stress).

Markets

Paints, coatings, and varnishes are the largest industrial biobased product market for soybean oil. This market is relatively mature and growing only about as fast as the Gross National Product. Vegetable oils have declining shares of oil-based paints, and oil-based paints are a declining share of the overall paints, coatings, and varnishes market. Usage of vegetable oils in this market peaked around 1950, when 80% of paints, coatings, and varnishes was composed of vegetable oils. By 1987, market share had dropped to 30% with the remainder coming from petroleum-derived products. Of the 51,000 MT (56,000 t) of vegetable oil used in this sector during 1992, approximately 35% was supplied by soybeans. Soybean oil is the preferred vegetable oil in alkyd resins and some urethane and epoxy resins.

In the United States, the total consumption of alkyds increased from about 181,000 MT (200,000 t) in the mid 1950s to more than 272,000 MT (300,000 t) in the mid 1960s. It peaked in 1973 at about 313,000 MT (345,000 t), constituting about 33% of all synthetic coating resins. In 1980, alkyds still accounted for 30% of the 1 MMT (1.09 million t) of all resins consumed for coatings, while epoxy resin was 8.6% and urethane resin was 6.4%. From 1987 to 1989, although the consumption maintained at about 272,000 MT/yr (300,000 t/yr), its market share among all coating resins declined to 26% in 1987 and 25% in 1989. At present, 55–60%

of the alkyd resins consumed in the United States is used for architectural coatings (Lin, 2004). California and the northeastern United States are expected to adopt regulations that will severely restrict the use of solvent-borne coatings. The industry was hard hit in 2003–2004 with higher prices for raw materials such as linseed and soybean oils.

Current Interests

The environmental movement of the past 25 yr has driven interest in natural products and their potential use in paints and coatings (i.e., a drive to reduce solvent emissions), and a gradual decline in the market share of alkyd resins occurred. However, their versatility and low cost will undoubtedly continue to keep alkyds a major player in the coatings arena. Great strides in the development of water-borne types also were made in recent years; therefore, water-dispersible alkyds are playing an increasingly important role. Alkyds will continue to be used in the future, as a significant portion of their raw materials (fatty acids) is a renewable source. The next generation of paints will likely use nonvolatile and perhaps reactive solvents such as epoxy fatty acids. Redesigning soybeans using biotechnology to produce conjugated fatty acids (i.e., tung oil) and epoxy fatty acids (as in vernonia oil) could improve performance of soybean oil in paints, coatings, and varnishes (McKeon, 2005).

Polymeric Materials (Plastics and Plasticizers)

Background

Polymers are a group of materials comprised of long covalently bonded molecules, which are obtained either from natural or synthetic sources. Polymers in the form of plastics are used in making articles of daily use such as knobs, handles, switches, pipes, heart valves, and so on. Plasticizers are incorporated into plastics or elastomers to increase workability, flexibility, distensibility, and toughness. Narine and Kong (2004) published an excellent discussion of vegetable-oil-based polymers and plastics. Liu and Erhan (2005) discussed in detail the development of soybean-oil-based composites by direct deposition. Formo (1982) extensively discussed plasticizers. Without plasicizers, plastics, such as polyvinyl chloride (PVC), are hard, horn-like material with limited usefulness. Epoxidized soybean oil (ESO) is used extensively to plasticize PVC resins to improve stability to heat and light, and to reduce cost. Primary plasticizers, such as phthalate esters, can be used over a broad range. Secondary plasticizers, such as ESO, cannot be used alone over a wide range of plastic composition, because the plastic becomes opaque, sticky, and inflexible on aging.

Interest in polymeric materials prepared from renewable natural resources, such as soybean oil, has grown over the past decade. The advantages of these polymeric materials are their low cost, availability, and possible biodegradability (Kaplan, 1998). Among agricultural resources (starch, cellulose, fibers, polylactic acid, cashew nut shell liquid, vegetable oils), triacylglycerols of natural oils constitute significant raw materials for the production of biodegradable polymers.

The triacylglycerols of linseed, tung, lunaria, lesquerella, crambe, and soybean oils are used as sources of polymers. By virtue of their double bonds, they can undergo polymerization (Liu et al., 2007b; Nayak et al., 1999, 2004). The double bonds in these and other oils can also be epoxidized or converted into hydroxyls to increase their reactivity (Doll et al., 2007; Linne et al., 1984; Sharma et al., 2006a, 2007a). Swern (Findley et al., 1945) and others (Schmitz & Wallace, 1954) showed that the peroxy acid oxidation reaction could be stopped at the epoxy stage before forming dihydroxy compounds. The methods were refined and improved considerably over the years (Crocco et al., 1992; La Scala & Wool, 2002; Nowak et al., 2004; Petrovic et al., 2002c; Sinadinovic-Fiser et al., 2001; Vlcek & Petrovic, 2006). Only a few oils, however, contain naturally occurring special functional groups in addition to double bonds in their fatty acids (i.e., castor oil and lesquerella oil contain hydroxyl groups, and vernonia oil contains epoxy functional group). The double bonds, hydroxyl groups, and epoxy groups play important roles in polymer formation. Epoxies have good adhesion, mechanical properties, low moisture-absorption, chemical-resistance, little shrinkage, and ease of processing. These excellent properties make this family of compounds one of the best matrix materials for many composites.

Technology and Performance

The double bonds in soybean oil can be converted to reactive monomers (Khot et al., 2001; Wool et al., 2002a,b). These reactive monomers (maleates) are copolymerized with styrene through free radical mechanism to form rigid thermosetting resins. The maleates are obtained by glycerol transesterification of soybean oil followed by esterification with maleic anhydride. Several triacylglycerol-based polymers and composites were synthesized, and their properties compared.

The double bonds of soybean oil fatty acids can be converted into hydroxyls, making polyols (Erhan et al., 2003, 2005; Sharma et al., 2006b). Hydroxyl groups in vegetable oils can also be introduced at double-bond sites by hydroformylation with a rhodium-triphenylphosphine catalyst followed by hydrogenation (Frankel et al., 1974, 1973). These polyols are important for the production of cross-linked polymers, such as polyurethanes, by allowing reaction of hydroxyl groups with diisocyanates. The extent of cross-linking affects the stiffness of the polymer. The polymer structure must be highly cross-linked when rigid foam is required, while less cross-linking gives flexible foams. The degree of cross-linking is entirely dependent on the NCO/OH ratio. Branching occurs at the urethane linkage when NCO/OH ratio is low. A low degree of cross-linking allows the molecules freedom of movement, resulting in improved strength and creep-resistance, and slight loss in soft, flexible, rubbery behavior. With high NCO/OH ratio, the probability of forming urea linkages is greater, and therefore branching takes place at the urea linkage points. A high degree of cross-linking immobilizes the polymer molecule and results in a thermoset plastic. Less cross-linked polymers absorb a large amount of solvent and thus swell to form soft gels. Highly cross-linked polymers absorb less solvent as a result of less molecular mobility, and thus cannot move apart to accept solvent molecules.

Petrovic and coworkers (Djonlagic & Petrovic, 2004; Guo et al., 2000, 2002, 2006; Javni et al., 2000, 2004; Petrovic et al., 2002a,b, 2005a,b; Zlatanic et al., 2002, 2004) developed two technologies to prepare soybean-oil-based polyols for general polyurethane use. The first technology includes epoxidation of soybean oil followed by alcoholysis to form polyol (Fig. 17.6a and b). In the second technology, the double bonds are first converted to aldehydes by hydroformylation with either rhodium or cobalt catalysts, followed by hydrogenation to alcohols by nickel. These polyols are then reacted with diisocyanate to form polyurethanes. The resulting polyurethanes can behave as a hard rubber or a rigid plastic, depending on the methods used in the reaction process. By controlling the degrees of conversion, using different diisocyanates and varying the stoichiometry, a variety of products can be formed. John et al. (2002) found enhanced reactivity for soybean-oil-based polyols compared to synthetic polyols for making polyurethane foams.

Polyesters are formed when these polyols react with dibasic acids, such as sebacic acid, and the removal of water as a by-product (Fig. 17.6c). The hydroxyl groups can be exploited to make acrylates (Fig. 17.6d) and can be used as coating material, imparting excellent gloss to wood, aluminum, and steel and giving good adhesive properties. The hydroxyl group also reacts with propylene oxide, epichlorohydrin (Fig. 17.6e), and ethylene oxide, resulting in novel polyhydroxy compounds of improved reactivity.

The double bonds of soybean oil may also be converted into the more reactive oxirane moiety (Fig. 17.6a) by reaction with peracids, peroxides, or with hydrogen peroxide in the presence of a strong ion-exchange resin (Petrovic et al., 2002c; Pryde, 1980; Sinadinovic-Fiser et al., 2001). Most plastic formulations contain about 3% epoxidized oil (Pryde, 1980). ESO and other epoxidized vegetable oils are used in making adhesives, plasticizers, industrial coatings, varnishes, and paints. The ESO can be polymerized through a variety of reactions. A natural elastomer can be synthesized by reacting it with naturally occurring dibasic acids such as sebacic acid derived from castor oil (Sperling & Manson, 1983). The polyesters formed with high-epoxy oil, such as ESO, have higher glass-transition temperature than that with low-epoxy oil (vernonia oil) because of dense cross-linking in the former. The rubbery nature of polymerized oil may be used in toughening rigid epoxy materials because it phase separates into spherical domains when mixed and cured with bisphenol-A epoxy compounds (Dirlikov et al., 1996). The epoxide moieties also react with methacrylic acid in the presence of a tertiary amine to give acrylates useful in making UV-curing formulations. These acrylate esters easily polymerize through the acrylate vinyl moieties. Partially epoxidized soybean oils provide properties similar to vernonia oil and are

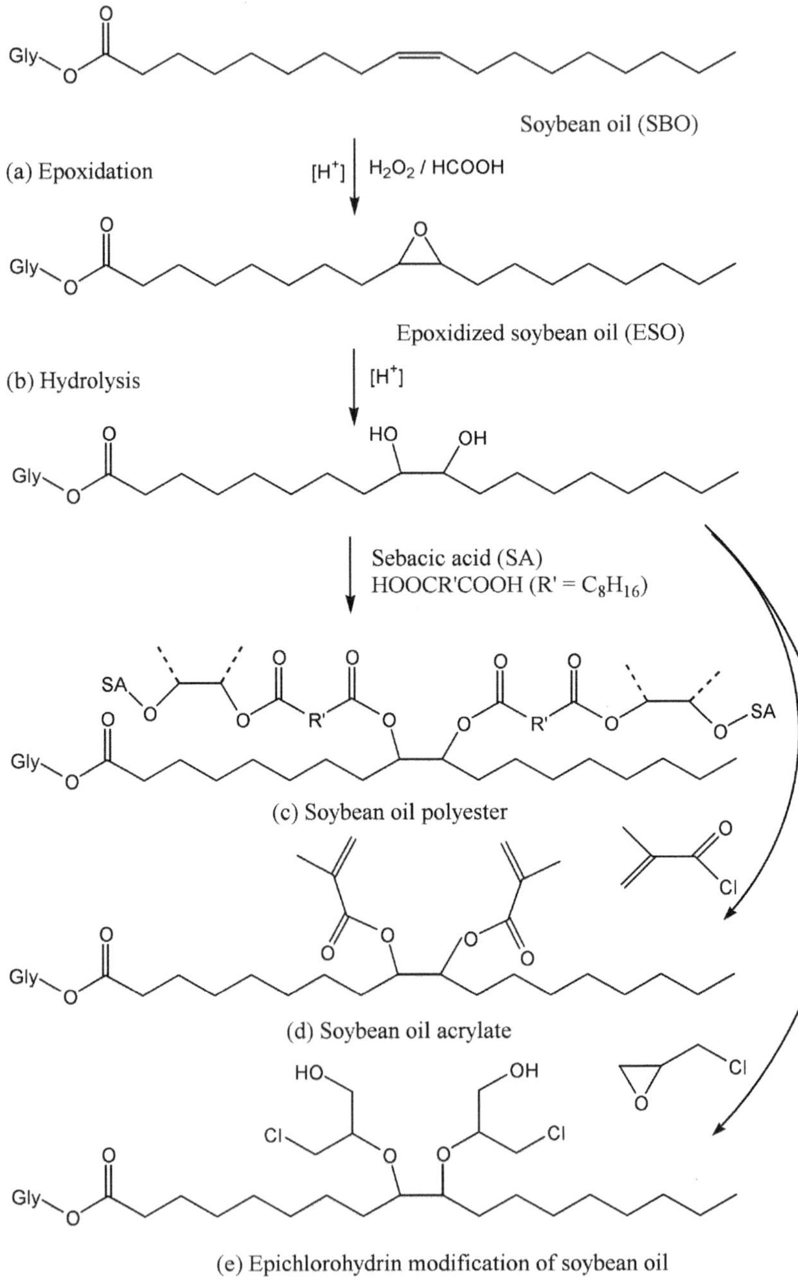

Fig. 17.6. Epoxidation of soybean oil (a), followed by hydrolysis to form polyols (b). Modification of polyols to form polyesters (c), soybean oil based acrylates (d), and epichlorohydrin modified soybean oil (e).

suitable for preparation of low-volatile-organic-compound alkyd and epoxy coating formulations.

To have plasticizing properties, the plasticizer must act as solvent for the polymeric materials and be a high-boiling polar liquid having proper balance of size and polarity of groups. ESO is widely applied as PVC additive to improve PVC processing, stability, and flexibility (Nayak et al., 1997). ESO has a stabilizing effect on PVC by extending its useful life. As PVC ages, hydrogen chloride is released, which accelerates deterioration. ESO minimizes this deterioration by scavenging the acid.

ESO was also used as a raw material for synthesizing new polymers suitable for liquid molding processes (Wool et al., 2002a,b). The preparation of structurally strong soybean-oil-based composites is attractive from both commercial and environmental prospectives. ESO has been used in preparing composites reinforced with a combination of organically modified clay and fibers (glass, carbon, mineral, and natural plant fibers). This process uses various curing agents, and extrusion solid freeform (Liu & Erhan, 2005) and compression-molding fabrication methods (Liu et al., 2002, 2004a,b, 2005, 2006, 2007a). Biopolymers made from ESO exhibit strong viscoelastic solid properties similar to synthetic rubber, and thus have the potential to replace some synthetic rubbers and/or plastics (Xu et al., 2002). Fiber-reinforced ESO-based composites of high strength and stiffness can be formed by the free-form fabrication method by using a combination of two fibers, and can be used in agricultural equipment, the automotive industry, civil engineering, marine infrastructure, and the construction industry. Liu and Erhan (2007) also synthesized hydrogels using ESO and demonstrated their use in drug delivery applications (Wong et al., 2006).

Markets

The major market for plasticizers is dominated by primary plasticizers, phthalate esters, while 10–15% of plasticizers are derived from vegetable oils. The annual plasticizer market is ~0.91 MMT (1 million t) and growing. Of the epoxidized vegetable oils, soybean oil is used in ~75% of production. Epoxy linseed oil is used in vinyl liners of bottle caps and medical tubing, where extraction of plasticizer must be avoided. Epoxy sunflower oil is more compatible with vinyl resin, while epoxy tall oils are used in some plastic products.

Current Interests

The double bonds in soybean oil fatty acids are good sites to add multiple functionalities such as hydroxyls and epoxies. The modified soybean oils containing multiple functionalities provide an alternative to petroleum as a chemical feedstock. Worldwide interest is expressed in using soybean oil, other vegetable oils, and their reactive modified versions as starting materials for polymer production because of the large quantities of nonrenewable petroleum being consumed (Barrett et al., 1993).

Drying Oil Products

Background

Historically, drying oils are the major film former of coatings, including paints, varnishes, and inks. The modest drying of soybean oil was used in numerous products up until the mid 1950s. These products include linoleum, oil cloth, sealing and caulking compounds, rubber-like materials, and core oils. A good treatise is available elsewhere on these applications (Formo, 1982; Rheineck & Austin, 1968). In the early 1950s, linoleum was the dominant flexible floor covering. Linoleum was manufactured as early as 1864 in England and 10 yr later in the United States; however, floor coverings based on vinyl and phenolic resins or asphalt have now captured this market.

Technology and Performance

Drying oils owe their value as raw material for various products to their ability to polymerize and cross-link, or dry. After application to a surface, they form tough, adherent, impervious, and abrasion-resistant films. Drying properties are closely related to degree of unsaturation, since polymerization and cross-linking take place through polyunsaturated centers. Preferred drying oils have high levels of polyunsaturation and conjugated unsaturation. Soybean oil is moderately polyunsaturated and naturally unconjugated. The drying of unconjugated oils is caused by the bis-allylic methylene group, which is more reactive than the allylic methylene group (Frankel, 2007). The average number of bis-allylic groups in triacylglycerol serves as a better indicator of the drying characteristics of the oil. Oils in which this number is >2.2 are considered to be *drying oils,* and ones with <2.2 are *semi-drying oils* (Wicks et al., 1998). Dryings oils were used directly as film formers, while the moderate drying oils like soybean oil were made more useful by altering their natural state. Aging in vats, heating, or blowing air through the oil oxidize the semi-drying oils and convert unconjugated to conjugated unsaturation (Cahoon et al., 1999; Larock et al., 2001). These processes change the viscosities and drying characteristics of semi-drying oils to improve properties for certain applications. The drying of oils can also be improved by using driers such as cobalt and manganese naphthenates (Oyman et al., 2005).

Linoleum typically consisted of one-third binder (oxidized soybean, linseed and tung oils, natural gums, and rosin), one-third inorganic fillers (pigments and ground limestone), and one-third organic fillers (ground cork and wood flour). The binder, known as *cement,* must be durable and resilient, yet thermoplastic. A highly oxidized and polymerizing drying oil works well when compounded with rosin and other natural resins.

Drying oils were used in two ways in oil cloth. One is impregnated cloth in which the entire cloth was impregnated with oil, and another was coated cloth in which only one side of the cloth was coated. The former was used in raincoats and machine-covers, while the latter was used as coverings for walls, tables, shelves, and other surfaces.

The polymerized oil on the fabric surface provided water-resistance with flexibility.

Sealing and caulking compounds use significant amounts of drying and semi-drying oils. Blown soybean oil is used in mastic products or glazing compounds, which have largely replaced putty. Other sealing materials have soybean oil bases to produce rubber-like properties.

Polyunsaturated fatty acids of triacylglycerol can polymerize to form various elastic, rubber-like materials. Factices are oils that were cross-polymerized using sulfur or disulfide bonds in the same manner as vulcanizing rubber. Factices are rubber-like, but don't have the combination of elasticity and tensile strength of vulcanized natural rubber because of the extensive cross-linking in the former, while the latter have long linear chains with occasional cross-linking. Factice is used in materials for gaskets, stoppers, bumpers, tubing, electrical insulation, and rubberized fabrics.

Soybean oil is also used in core oils for binding sand cores in manufacturing metal castings. Cores are formed by mixing about 2% oil in sand, molding in a wooden form, and baking until the core becomes hard through polymerization of the oil. The core must be sufficiently hard to retain its form during metal casting, but not so hard as to be difficult to break and remove the cooled metal casting.

Dimer Acids

Soybean fatty acids are conjugated thermally or catalytically to yield dimer and trimer polybasic acids (Erhan et al., 2005). Hydrolyzed dimer acids improve color and oxidative stability. These are used in polyamide resins, paints, plastics, and coatings, bodying/curing/flexibilizing agents, corrosion inhibitors, antiwear agents, lubricants, fuel, and lubricant additives (Antonucci et al., 1984; Bhowmick & Basu, 1988; Kale et al., 1991; Savastano, 2001; Watanabe et al., 1996). The annual dimer acid production is about 18,000 MT (20,000 t).

Polyamide Resins

Dimer acids can be converted into both reactive and nonreactive polyamide resins (Chen et al., 2001; Fan et al., 1998; Kale et al., 1994; Pryde, 1980). Nonreactive low-molecular-weight polyamides are formed by condensation of dimer acids with ethylenediamine. Dimer acids typically contain 10–30% trimer acids, which lead to three-dimensional polymers that cause stickiness, narrow melting ranges, cold-flow-resistance, and rheological properties useful in adhesives. Nonreactive polyamides are used without solvents as hot-melt adhesives in packaging and can seam solders, book bindings, and shoe soles. These are also used in flexographic printing inks for food packaging films and thixotropic alkyd paints. The annual market for nonreactive polyamides is about 10,000 MT (11,000 t).

Reactive polyamides resins are produced by reacting dimer acid with a slight excess of polyamines (diethyl triamine and triethyl tetramine) to produce low-molecular-weight polyamide resins having free amino groups in the chain (Kale et al., 1994).

These are used as curing agents in epoxy resins found in surface coatings, adhesives, potting and casting compounds, and patching and sealing compounds (Leonard, 1980). Epoxy adhesives are typically two-component systems including an epoxy-resin and a polyamide curing agent. The two-component epoxy systems are used in toppings for concrete floors, bridge decks, and airport landing strips. The market for reactive polyamide resins is about 4,900 MT (5,400 t).

Markets

Most drying-oil products no longer use significant amounts of soybean oil, so economic and market data are difficult to find. The market is dominated by petrochemicals because of better performance properties. Consumption of soybean oil in all other drying oil products peaked at 4,900 MT (5,400 t) in 1954 and declined to 1,600 MT (1,800 t) by 1978. Usage of soybean oil in linoleum and oil cloth peaked at 13,000 MT (14,000 t) in 1948, equivalent to 20% of total fats and oils in linoleum and oil cloth, and declined to 408 MT (450 t) by 1957. The direct use of drying oils accounted for only 4% of the total film formers consumed in the United States in 1990.

Oleochemicals

Background

Soybean oil can be converted into many different oleochemicals. Erhan et al. (2005) provides a good treatise on chemical modification of soybean oil. Possible reactions on soybean oil triacylglycerols involving hydrocarbon chains are epoxidation, hydrogenation, sulfation, olefin metathesis, carboxylation, oxidative scission, cyclization, acetylation, alkarylation, catalytic or thermal cracking, polymerization, etcetera, while the reactions at carboxyl group are hydrolysis (splitting), transesterification, amidation, etcetera as shown in Fig. 17.7. A large portion of soybean oil is currently transesterified with methanol to produce fatty acid methyl esters (FAME) and glycerol. A majority of FAME is used as fuel also called *biodiesel*, which is clean burning and produces no sulfur dioxide emissions. A detailed account on biodiesel is present in the Chapter: *Bioenergy and Biofuels from Soybeans*, and also elsewhere (Reaney et al., 2004).

FAME play a major role in the oleochemical industry by replacing fatty acids as starting materials. These are preferred over fatty acids as chemical intermediates for a number of oleochemicals, because of advantages such as lower energy consumption, less expensive equipment, a more concentrated glycerin by-product, and ease of distillation and transport. FAME can be converted to α-sulfo fatty acid esters on sulfonation, and fatty acid amides on amidation. Other fatty acid esters formed using primary and secondary alcohols, such as ethyl, isopropyl, butyl, octyl, decyl, myristyl, and cetyl, are commonly used as excipients, lubricants, plasticizers, softeners, solubilizers, and wetting agents in cosmetic and personal-care products (Meffert, 1983).

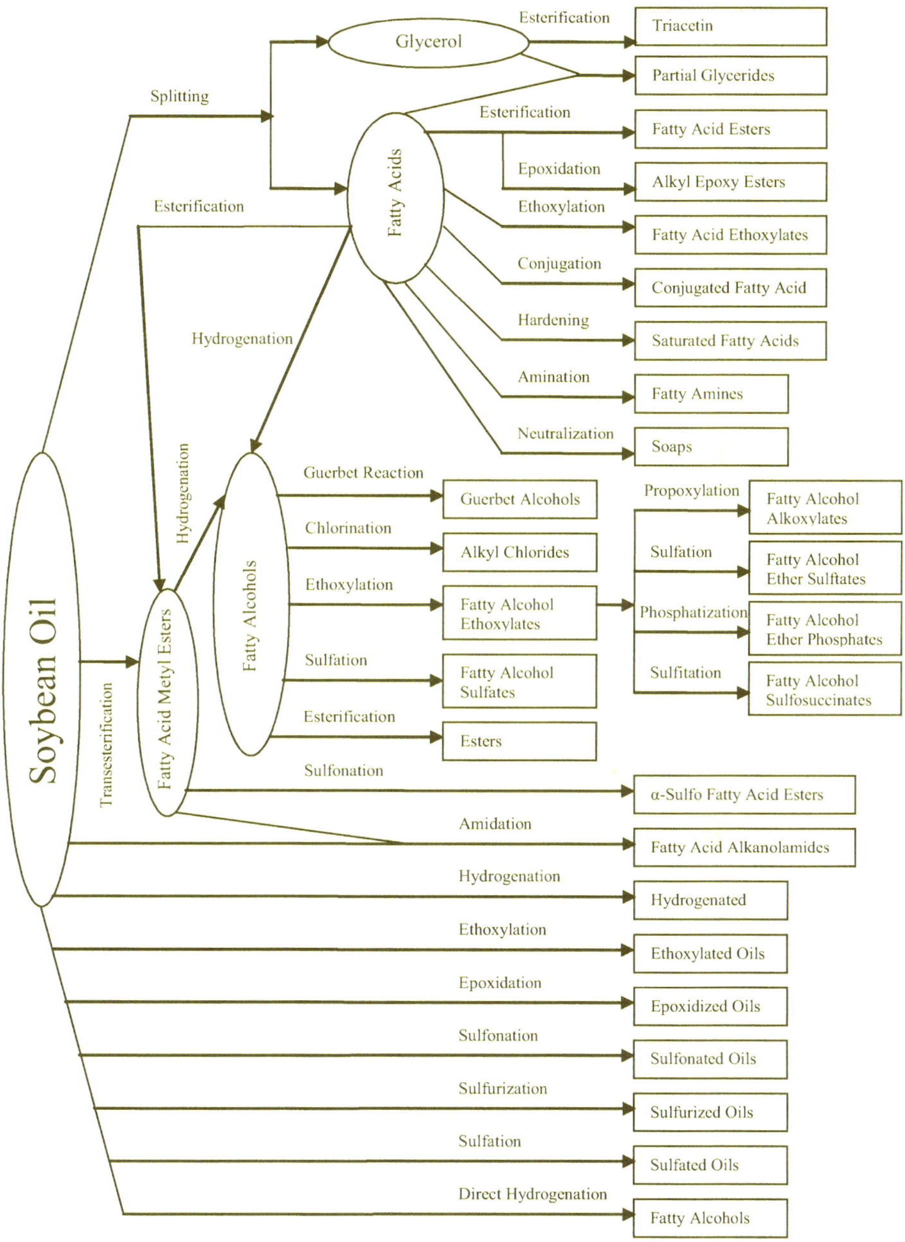

Fig. 17.7. Soy oleochemical raw materials and their derivatives.

Soybean oil may be hydrolyzed into glycerol and fatty acids, or soybean oil soapstocks (foots) may be acidified to produce fatty acids. Crude soybean fatty acids are used to make adhesive tape, shaving compounds, textile water repellents, carbon paper, and typewriter ribbons. Consumption of fatty acids in the United States, Western Europe, and Japan was 2.3 MMT (2.5 million t) in 2001. These soybean fatty acids can be separated into various fractions by distillation, and are used in candles, crayons, cosmetics, polishes, buffing compounds, and mold lubricants. These fatty acids can be converted to FAME by esterification, alkyl epoxy esters by epoxidation, fatty alcohols by hydrogenation (Kreutzer, 1983; Voeste & Buchold, 1983), and dimer and trimer acids by conjugation or amines and amides as shown in Fig. 17.7 (Maag, 1983).

Fatty alcohols make up one of the major basic oleochemicals having an increasing growth rate. As a primary raw material for surfactants, growth in fatty alcohol production parallels increasing economic prosperity and improved standards of living. Fatty alcohols are the raw materials of choice for surfactant manufacture because of their biodegradability and availability from renewable resources.

Glycerin is obtained as a by-product of splitting, saponification, or transesterification processes, and consists of glycerol and a small amount of water. Glycerin has a unique set of physical properties that allow it to be used in a variety of industries (1700 uses are identified). More prominent application areas for glycerin are alkyd resins (36%), cosmetics/pharmaceuticals (30%), tobacco products (16%), food/beverages (10%), urethane uses (6%), and explosives (2%). Glycerol is a tribasic alcohol occurring in nature in the form of triacylglycerols, which are glyceryl esters of fatty acids. The glycerol contents in fats and oils range from 9 to 12%. Although, it can be synthesized from petrochemicals, the increasing output of vegetable-oil-derived glycerol plays a significant role in the world supply of this material. In 1863, Alfred Nobel demonstrated nitroglycerin's (obtained from glycerol) explosive capabilities, and in 1866 he invented dynamite. Recently, the Dow Chemical Co. announced the introduction of monopropylene glycol derived from glycerin generated during manufacturing biodiesel from vegetable oil feedstocks (Rattray, 2007). This propylene glycol will be used in unsaturated polyester resins for boat hulls, bathroom fixtures, recreational vehicles, and marine units; aircraft deicers and antifreeze for automobiles; and heavy-duty laundry detergents. In 2003, market prices for natural USP glycerin in the United States were $1.21–1.65/kg ($0.55–0.75/lb), but have been as low as $0.55/kg ($0.25/lb) and as high as $2.20/kg ($1.00/lb) during the 1990s. The United States' annual production of glycerin is approximately 245,000 MT/yr (270,000 t/yr) and grows annually 2–5%. World glycerin production is well over 408,000 MT/yr (450,000 t/yr).

Surfactants, Soaps, and Detergents

Surfactants are chemical compounds that possess surface activity and act diversely because of their amphipathic molecular structures. A typical surfactant molecule has a head, which is hydrophilic (water-loving) and strongly polar, such as an anion, cation, or nonion; and a tail, that is hydrophobic (oil-loving) and nonpolar, such as a linear or branched hydrocarbon chain. These compounds find broad application in practically all industries (e.g., main ingredient of detergents and cleaners, foaming agents and emulsifiers in cosmetics and pharmaceuticals, emulsifiers for paints, scouring agents for textiles, floatation agents for mining industry, and emulsifiers and sanitizing agents for the food industry). Important vegetable-oil-based surfactants are fatty alcohol sulfate, fatty alcohol polyglycol ether, fatty alcohol ether sulfate, fatty acid methyl ester sulfonate, and alkyl polyglucoside. About 20% of the 103,000 MT/yr (113,000 t/yr) market is served by soybean fatty acids. Surfactants derived from soybean fatty acids are used in medium-grade liquid laundry detergent, dry-cleaning liquid, and building maintenance and dairy cleaners (Sonntag, 1984). For a detailed discussion of surfactants, detergents, and surfactant systems, the reader is referred elsewhere (Holser, 2005; Lynn, Jr., 2004).

Fatty alcohol sulfates are sodium salts of fatty alcohol sulfates of the C_{12}–C_{18} range and are common ingredients for laundry detergent products due to their good detergency, wetting and foaming properties, and biodegradability (Feitkenhauer & Meyer, 2002a,b). These compounds are also used extensively as foaming agents in personal-care products.

Fatty alcohol polyglycol ethers are nonionic surfactants, prepared by the reaction of fatty alcohols with ethylene oxide or propylene oxide using alkaline catalyst. The degree of ethoxylation can vary depending on the chain length of the fatty acid and the purpose for which it will be used. These are characterized by excellent wetting properties, low foam and high effectiveness even at low temperatures. They have replaced the alkylphenol polyglycol ethers in most applications (Knaebel et al., 1990; Tamura et al., 1995, 1998).

Fatty alcohol ether sulfates are produced by sulfation of the fatty alcohol ethoxylate (fatty alcohol containing 2–3 moles of ethylene oxide) with sulfur trioxide or chlorosulfonic acid, followed by neutralization with caustic soda, ammonia, or an alkanolamine (Weil et al., 1966). These ether sulfates possess superior properties compared to fatty alcohol sulfates, such as unlimited water solubility, tolerance to water hardness and superior skin compatibility, and are thus used in liquid shampoos and bath preparations.

The FAME sulfonate reaction mechanism involves two steps (Stirton et al., 1965; Weil et al., 1962). In the first step, the sulfur trioxide gas reacts quickly with FAME to form sulfoanhdyride. In the second step (40-90 min), the sulfoanhydride becomes the sulfonating agent, reacting with the still-unreacted ester. Finally, a neutralization step is carried out. These ester sulfonates exhibit favorable wetting, emulsifying and dispersing properties (Stirton et al., 1954, 1962), demonstrating yet another efficient method for producing anionic surfactants (Okano et al., 1996; Weil & Stirton, 1956).

FAME ethoxylates are prepared by catalytic ethoxylation of FAME, and are comparable to alcohol ethoxylates, but with increased water solubility (Ballun et al., 1954).

Fatty amines and their derivatives represent the most important nitrogen compounds of fatty acids, and are starting compounds for manufacturing quarternary ammonium compounds and various cationic and amphoteric substances. These products are prepared from catalytic hydrogenation of fatty nitriles derived from fatty acids and ammonia (Billenstein & Blaschke, 1983; Metzger & Bornscheuer, 2006; Zerkowski & Solaiman, 2006). These derivatives find wide application as biocides, sanitizing agents, algae control in water treatment, ore flotation agents in mining, effective corrosion inhibitors, and lubricants in drilling mud formulation.

Monoalkyl phosphate and phosphate esters are special types of phosphorus-containing anionic surfactants that are obtained from partial esterification fatty alcohols with phosphorus oxychloride followed by hydrolysis. Apart from use as surfactants, they are also used for flame-proofing, as antistatics for textiles, for foam inhibition, as extreme pressure lubricant additives, as acid cleaners, and for special cosmetic preparations. The composition of the phosphate ester greatly affects the functional properties of the product. The important properties they show are wetting, detergency, solubilizing, emulsification, surface tension reduction, foaming, dedusting, lubricity, antistatic, corrosion inhibition, chelating, dispersing, and antisoil redeposition (Basu & Dutta, 2003; Martinez-Palou et al., 2004).

Alkanolamides are condensation products of fatty acids with primary or secondary alkanolamines (Feairheller et al., 1994; Kopylov et al., 1980; Maltby & Read, 2000). These products are nonionic surface-active agents and find application in a multitude of uses including dry-cleaning soaps, fuel-oil additives, rust inhibitors, textile scouring, and dye-leveling agents.

Soaps are most commonly formed through either direct or indirect reaction of aqueous caustic soda (NaOH) with triacylglycerols from natural fats and oils. Soybean oil has been used in toilet soap bars and other solid soaps, but use was limited due to the high unsaturation content and high prices of soybean oil compared with alternative animal-derived fats. Soybean oil and fatty acids can be hydrogenated to make them more saturated and increase solids, but the cost can be higher than using naturally saturated animal fats. Most toilet soaps are 20% coconut oil and 80% inedible tallow. Most soybean soapstock is now used in animal feeds and pet food, and some as dust suppressants. Excellent description of technologies used to produce soaps is described by Burke (2004), which includes kettle boiling, continuous saponification, and hydrolysis/neutralization. The personal cleansing market in the United States represents nearly $2 billion in consumer sales divided among bar soaps (77%), liquid hand-soaps (13%), and body washes (10%).

Biofuels

Soybean oil is converted into soy FAME through transesterification with methanol. This product is an excellent substitute for diesel fuel with no engine adjustment required and no loss in efficiency (Knothe & Dunn, 2005). Besides transesterification to methyl esters, other approaches also were explored for utilizing soybean oil as fuel. These are diluted with conventional petroleum diesel fuel, microemulsions (co-solvent blending), and pyrolysis (Schwab et al., 1987). A detailed discussion on biodiesel is provided in the Chapter: *Bioenergy and Biofuels from Soybeans*, and also elsewhere (Knothe & Dunn, 2005; Reaney et al., 2004).

Other Industrial Uses for Soybean Oil

Home-heating Oils

Initially, wood and/or coal was used in fireplaces for interior home heating, but in the early 1900s, liquid fuels replaced solid fuels due to their ease of transportation and simplicity of use. In the mid-1900s, gaseous fuels, such as natural gas, became the fuel of choice due to ease of use and clean burning properties. Electric heating pumps have also become popular in the last four decades. Petroleum-based liquid home heating oil is used to heat over 8 million homes in the United States, comprising ~25 billion L/yr (6.6 billion gal/yr). Overall in the United States, fuel oil only currently accounts for 8% of the total residence heating fuel types, but with record high prices for natural gas, the heating-oil industry is eager to regain market share in the heating-fuel market. The Northeast region consumes 82% of the total fuel oil used in the United States. With higher petroleum prices and increasingly recognized environmental issues involving sulfur and nitrogen oxide emissions, alternatives to petroleum are needed to supply this market. Biofuels are particularly attractive due to their renewable nature, environmental benefits, and chemical similarity to petroleum hydrocarbons. Similar to automotive and industrial petroleum fuel use, environmental emissions and sulfur content are important for home-heating fuels. To simplify fuel distribution and to improve emissions, the National Oilheat Research Alliance (NORA) Board mandated that 80% of the heating oil consumed contain no more than 500 ppm sulfur by 2007.

Two kinds of plant-based fuels can be used as heating oil, degummed vegetable oil, and biodiesel. Degummed soybean or vegetable oil is an intermediate product in the production of food-grade refined vegetable oil and is approximately \$0.05–0.10/gal less expensive than the final product (\$0.44–0.55/kg, \$0.20–0.25/lb or about \$0.39/L and \$1.50/gal on average). Biodiesel or FAME is a product of transesterification of triacyglycerols of vegetable oils with methanol, and estimated current cost is \$0.66–0.79/L (\$2.50–3.00/gal). The #2 grade heating oil is currently selling at \$0.37–0.40/L (\$1.40–1.50/gal) wholesale, and ~\$0.53/L (~\$2.00/gal) retail. Higher blends of fuel oil can be an issue with respect to cold-temperature properties but are safer from a volatility perspective. Fuel oils contain ~10% more energy than soybean oil, while soybean oil has slightly higher energy content than biodiesel. A 20% blend of soybean oil in fuel oil has low cloud point (–27°C, –17°F), a flash point in the fuel oil range, energy content close to fuel oil, and viscosity of 3.46 cSt, which is within the ASTM recommendation for petroleum fuel oil (Tao, 2005). Recent results indicate lower emissions of SOx, NOx, and CO_2 based on a single burner/boiler using 20% soy-based biodiesel blended with low-sulfur diesel fuel (Batey, 2003). Field testing of 20% soybean oil in fuel oil conducted at two private residences for 2 yr demonstrated normal operations (Krishna, 2001).

The passage of the American JOBS Creation Act of 2004 created economic incentive for use of biodiesel fuels, including use as heating fuels. This incentive reduces the heating fuel cost approximately one cent per percentage point of biodiesel used in fuel blends. A similar type of incentive for soybean oil use as heating fuel would certainly be beneficial in opening new markets for soybean oil utilization.

Leather and Textiles

Vegetable oils are used as detergents and as leather-softening agents. Kronick and Kamath (2004) discussed the use of oils and fats in leather and textiles. Partially sulfated or sulfonated vegetable oils and fats are used, as self-emulsifying agents in aqueous emulsions, also called *fat liquors*, while the leather is still wet from tanning. The neutral portion of the fat liquor can be almost any vegetable oil, such as castor oil, soybean oil, or polymerizable oils (Kronick, 1998). Heavy grades of leather are softened by milling in the presence of warm molten fat, after the leather is dried. The fat dispersed among the fibers allows the leather to remain soft if it is dampened and dried again. This warm molten fat, also called *stuffing compound*, is made from wool grease or high melting mixtures of mineral waxes and fatty acids. The amount of these materials used in this industry is over 9,100 MT/yr (10,000 t/yr). A smaller amount of oils and fats is used as detergents in the preliminary cleaning of the raw skins and hides to suspend lime particles, buffer the alkaline solution that removes the epidermis and hair, suspend these materials when they come off, remove grease, tan certain type of leather, and control penetration of dyes.

A combination of oils, fats, and their derivatives is used extensively as lubricants and antistatic agents known as *spin finishes* in spinning, texturing of textile fibers. Oils and fats are also used as surface-active derivatives in the scouring, dyeing, and softening of textile yarns. Although synthetic oils replaced most of the natural oils in the processing of textiles, natural oils are still used mainly as their fatty acid derivatives (Proffitt & Patterson, 1988). The annual requirement of spin finishes is in the range of 63,500–127,000 MT (70,000–140,000 t).

Pharmaceuticals

A diverse number of applications of vegetable oils in pharmaceutical and cosmetic industry was published by Hernandez (2004). These applications include disease prevention and treatment, excipients and coadjuvants, transdermal carriers, and skin emolliency agents. The nontoxic nature of vegetable oils allows them to be used as reliable excipients or carriers to facilitate delivery of bioactive compounds, to act as fillers, binders, lubricants, solubilizers, emulsifiers, and emollients in a variety of delivery forms including tablets, capsules, suppositories, emulsions, ointments, creams, and lotions (Caliph et al., 2000; Charman, 2000; Porter et al., 2007). Other nondirect applications include artificial blood, gene delivery, diagnostic imaging, and medical devices (Porter et al., 1996a,b, 2004; Taillardat-Bertschinger et al., 2003). Use of controlled-release techniques of drugs and bioactive compounds is rapidly growing, and lipids are playing a major role in the formulation of new pharmaceutical products. Controlled-release delivery systems can improve drug efficacy and patient tolerance. They are used for cancer treatments (Wong et al., 2006), bacterial and fungal infections, and respiratory disease. A new soybean-oil-based polymer–lipid hybrid nanoparticle system was developed and evaluated. It can efficiently load and release the water-soluble anticancer drug doxorubicin hydrochloride (Dox) and enhance Dox toxicity against multi-drug-resistant cancer cells (Wong et al., 2006).

Cosmetics

A wide range of vegetable oils and fats is currently used as bases and bioactive ingredients in many cosmetic applications, such as emollients, specific ingredients for skin care and treatment, hair care, and makeup/decorative products. Commonly used lipids in cosmetics include triacylglycerols, emulsifiers, waxes, and structured lipids. The cosmetic market was more than $25 billion in 2000, and new products are continually introduced. The most common fatty acid used in cosmetics is oleic acid, which has high emolliency. Glycerol is a widely used humectant and moisturizer in cosmetics. As a result of the nature of most cosmetic oils and required specifications for finished products, stringent processing techniques are required to remove contaminants such as free fatty acids, peroxides, unpleasant odors, and other impurities. Refined vegetable oil for food usually contains <0.05% free fatty acids, while for cosmetics, values <0.03% are needed to avoid skin irritation.

Dust Suppressants

Airborne dust, if present in sufficient concentration in a confined space, can cause a disastrous explosion in the presence of oxygen and an ignition source. Every year about 30 explosions are attributed to grain dust, taking the lives of several workers and causing major losses of property. Barham and Barham (1980) were awarded a patent for using soybean oil to control dust in grain elevators and to reduce mold growth. Soybean oil reduces grain dust in elevators by 94%. In 1987, the United States Federal Grain Inspection Service ruled that soybean and other edible oils could be used to control grain dust in elevators. Spraying soybean oil is an inexpensive means of reducing the risk of dust explosion, while investing in equipment to use soybean oil is estimated to cost ~1% of that needed for dust collection equipment. Additionally, incorporating 1–2% degummed soybean oil in livestock feeds greatly reduces hog house dust and often gives a 5–10% increase in weight gains (Weigel, 1989). Controlling dust in hog houses leads to healthier pigs, improved weaning rates, and reduced odor, which is attributed to airborne dust. The market for using soybean oil to control dust in hog houses and grain elevators is estimated to be 544,000 MT (600,000 t). Recently, soybean oil soapstock also was used in similar dust control applications. Excellent control of road dust was achieved by spraying soybean refinery by-products onto gravel roads.

Herbicide and Insecticide Carriers

Herbicides and insecticides are usually diluted in a carrier so that these can be applied uniformly (Anon, 1985; Cox & Scherm, 2001; Kapusta, 1984; Woolwine & Reagan, 2001). Water almost exclusively was used for insecticides, but relatively large volumes (200 L/ha, 21 gal/A) are used requiring frequent time-consuming refilling of spray tanks. The introduction of rotary atomizing nozzles facilitated the application of very small volumes (2–3 L/ha or 0.2–0.3 gal/A), and has made it practical to use soybean oil as carrier. Additional benefits of this technology are reduced drift and evaporative losses and increased penetration that reduce effective application rates.

Growing widespread use of post-emergence herbicides on crops is happening, especially because of acceptance of no-till and other soil conservation tillage methods. Phytobland petroleum oil is used as the carrier in herbicide sprays applied at about 2.5 L/ha (0.3 gal/A) with rotary nozzle sprayers. Refined soybean oil can be used in this application as a substitute for petroleum oil because it has similar flow properties and the additional advantage of being biodegradable (Mack et al., 2003). Spray equipment may need to be cleaned more frequently because oil films build up on equipment, and those applying spray should wear protective clothing because soybean oil may increase absorption of active ingredients through the skin (Kapusta, 1984). Although soybean oil has proven to be a superior carrier for agricultural chemicals and have environmental benefits, it is not widely used because of its higher cost compared to petroleum products. Biodiesel from several oilseed sources (cuphea, lesquerella, meadowfoam, milkweed, and soybean) is useful as environmental friendly contact herbicides (Vaughn & Holser, 2007). These can control broadleaf weeds in turfgrasses, because few contact herbicide options are available for the homeowner market that will not cause turf injury. Their aqueous emulsions (1 and 2%, vol/vol) are more phytotoxic to sicklepod and velvetleaf than to perennial ryegrass.

Miscellaneous Uses

Soybean oil was used as an antifoam agent in aerated fermentations such as productions of penicillin, streptomycin, and tetracycline. It also markedly increases the yield of antibiotics, presumably by providing important nutrients (Smith & Thompson, 1989). Soybean oil also was observed to delay the onset of blooms on fruit trees, reducing susceptibility to frost damage.

References

Printing Ink Handbook; National Association of Printing Ink Manufacturers, Inc.: Harrison, NY, 1999.

Adamczewska, J.Z.; D. Wilson. Development of ecologically responsive lubricants. *J. Synth. Lubr.* **1997,** *14,* 129–142.

Adams, W.H. Now, it can be told: Soy saves ships. *Soybean Dig.* **1944,** *4,* 8.

Adhvaryu, A.; G. Biresaw; B.K. Sharma; S.Z. Erhan. Friction behavior of some seed oils: Biobased lubricant applications. *Ind. Eng. Chem. Res.* **2006a,** *45,* 3735–3740.

Adhvaryu, A.; S.Z. Erhan; Z.S. Liu; J.M. Perez. Oxidation kinetic studies of oils derived from unmodified and genetically modified vegetables using pressurized differential scanning calorimetry and nuclear magnetic resonance spectroscopy. *Thermochim. Acta.* **2000,** *364,* 87–97.

Adhvaryu, A.; S.Z. Erhan; J.M. Perez. Preparation of soybean oil-based greases: Effect of composition and structure on physical properties. *J. Agric. Food Chem.* **2004,** *52,* 6456–6459.

Adhvaryu, A.; B.K. Sharma; H.S. Hwang; S.Z. Erhan; J.M. Perez. Development of biobased synthetic fluids: Application of molecular modeling to structure-physical property relationship. *Ind. Eng. Chem. Res.* **2006b,** *45,* 928–933.

Adhvaryu, A.; C. Sung; S.Z. Erhan. Fatty acids and antioxidant effects on grease microstructures. *Ind. Crop Prod.* **2005,** *21,* 285–291.

Anon. Papers from the symposium on trends in industrial usage for vegetable oils. *J. Am. Oil Chem. Soc.* **1985,** *62,* 916–942.

Anon. Bean blast. *Time,* Oct. 21, 1935. pp. 34.

Antonucci, J.M.; S. Venz; D.J. Dudderar; M.C. Pham; J.W. Stansbury. Energy-absorbing, hydrophobic dental cements based on dimer and trimer acids. *Transactions of the Annual Meeting of the Society for Biomaterials in Conjunction with the International Biomaterials Symposium.* 1984, pp. 137.

Asadauskas, S.; S.Z. Erhan. Depression of pour points of vegetable oils by blending with diluents used for biodegradable lubricants. *J. Am. Oil Chem. Soc.* **1999,** *76,* 313–316.

Asadauskas, S.J.; J.M. Perez. Utilization of vegetable oils in metalworking fluids. *Biobased Industrial Fluids and Lubricants;* S.Z. Erhan, J.M. Perez, Eds.; AOCS Press: Champaign, IL, 2002; pp. 59–64.

Asadauskas, S.; J.M. Perez; J.L. Duda. Oxidative stability and antiwear properties of high oleic vegetable oils. *Lubr. Eng.* **1996,** *52,* 877–882.

Asadauskas, S.; J.M. Perez; J.L. Duda. Lubrication properties of castor oil—Potential basestock for biodegradable lubricants. *Lubr. Eng.* **1997,** *53,* 35–40.

Athawale, V.D.; M.D. Bhabhe. Chemoenzymatic synthesis and characterization of urethane oils for surface coatings. *J. Coating Technol.* **1998,** *70,* 43–48.

Athawale, V.D.; S.R. Gaonkar. Chemoenzymatic syntheses of optically active polyacrylates. *Macromolecules* **1999,** *32,* 6065–6068.

Athawale, V.D.; K.R. Joshi. Chemoenzymatic synthesis of PURs. *Eur. Coatings J.* **2000,** *6,* 42–46.

Athawale, V.D.; K.R. Joshi. The effect of HMMM crosslinker on the coating properties of chemoenzymatically synthesized aliphatic urethane oil. *Paint Coatings Ind.* **2004,** *20,* 108–116.

Ballun, A.T.; J.N. Schumacher; G.E. Kapella; J.V. Karabinos. Comparative detergency of several built polyethenoxy alkanoates. *J. Am. Oil Chem. Soc.* **1954,** *31,* 20–23.

Barham, H.N., Jr.; H.N. Barham. Method for the Adsorption of Solids by Whole Seeds. U.S. Patent 4,208,433 (1980).

Barrett, L.W.; L.H. Sperling; C.J. Murphy. Naturally functionalized triglyceride oils in interpenetrating polymer networks. *J. Am. Oil Chem. Soc.* **1993,** *70,* 523–534.

Barth, B.P. Phenolic Resin Adhesives. *Handbook of Adhesives,* Second ed.; I. Skeist, Ed.; Van Nostrand: New York, 1977; pp. 382–416.

Basu, A.K.; S. Dutta. Surfactant properties of monoalkyl ethanolamine phosphate esters. *J. Surface Sci. Technol.* **2003,** *19,* 79–84.

Batey, J. Combustion testing of biodiesel fuel oil blends in residential oil burning equipment. The 2003 National Oilheat Research Alliance Technology Symposium, Brookhaven National Laboratory: Boston, MA, 2003; pp. 101–116.

Battersby, N.S.; S.E. Pack; R.J. Watkinson. A correlation between the biodegradability of oil products in the CEC L-33-T-82 and modified strum tests. *Chemosphere* **1992,** *24,* 1989–2000.

Becker, R.; A. Knorr. An evaluation of antioxidants for vegetable oils at elevated temperatures. *Lubr. Sci.* **1996,** *8,* 95–117.

Ben Arfa, A.; Y. Chrakabandhu; L. Preziosi-Belloy; P. Chalier; N. Gontard. Coating papers with soy protein isolates as inclusion matrix of carvacrol. *Food Res. Int.* **2006,** *40,* 22–32.

Ben Arfa, A.; L. Preziosi-Belloy; P. Chalier; N. Gontard. Antimicrobial paper based on a soy protein isolate or modified starch coating including carvacrol and cinnamaldehyde. *J. Agric. Food Chem.* **2007,** *55,* 2155–2162.

Bhabhe, M.D.; V.D. Athawale. Chemoenzymatic synthesis of urethane oil based on special functional group oil. *J. Appl. Polym. Sci.* **1998,** *69,* 1451–1458.

Bhowmick, D.N.; S. Basu. Dimer acids—A chemical with immense potential in India. *Chem. Eng. World* **1988,** *23,* 53–55.

Bian, K.; S. Sun. Adhesive performance of modified soy proteins polymers. *Polym. Prepr.* **1998,** *39,* 72–73.

Billenstein, S.; G. Blaschke. Industrial production of fatty amines and their derivatives. *J. Am. Oil Chem. Soc.* **1983,** *61,* 353–357.

Biswas, A.; A. Adhvaryu; D.G. Stevenson; B.K. Sharma; J.L. Willet; S.Z. Erhan. Microwave irradiation effects on the structure, viscosity, thermal properties and lubricity of soybean oil. *Ind. Crop*

Prod. **2007,** *25,* 1–7.

Bloom, P.D.; G.B. Poppe; A.F. Rich. Polyunsaturated Fatty Acids as Part of Reactive Structures for Latex Paints: Thickeners, Surfactants, and Dispersants. U.S. Patent 6,924,333 (2005).

Boyer, R.A. Soybean protein fibers: Experimental production. *Ind. Eng. Chem.* **1940,** *32,* 154–155.

Boyer, R.A.; W.T. Atkinson; C.E. Robinette. Artificial Fibers and Manufacture Thereof. U.S. Patent 2,377,854 (1945).

Brister, E.H.; T. Johnston; C.L. King; S.F. Thames. New monomers from vegetable oils. *ACS Symp. Series* **2000,** *755,* 159–169.

Brodnitz, M.H. Autoxidation of saturated fatty acids. A review. *J. Agric. Food Chem.* **1968,** *16,* 994–999.

Brown, O.E. Labelling Adhesives. U.S. Patent 4,675,351 (1987).

Burke, M.R. Soaps. *Bailey's Industrial Oil & Fat Products,* Sixth ed.; F. Shahidi, Ed.; John Wiley & Sons, Inc.: Hoboken, New Jersey, 2004; Vol. 6, pp 103–136.

Burnett, R.S. Soybean protein industrial products. *Soybeans and Soybean Products;* K.S. Markley, Ed.; Interscience Publishers: New York, 1951; Vol. 2, pp. 1003–1054.

Cahoon, E.B.; T.J. Carlson; K.G. Ripp; B.J. Schweiger; G.A. Cook; S.E. Hall; A.J. Kinney. Biosynthetic origin of conjugated double bonds: Production of fatty acid components of high-value drying oils in transgenic soybean embryos. *P. Natl. Acad. Sci. USA* **1999,** *96,* 12935–12940.

Caliph, S.M.; W.N. Charman; C.J.H. Porter. Effect of short-, medium-, and long-chain fatty acid-based vehicles on the absolute oral bioavailability and intestinal lymphatic transport of halofantrine and assessment of mass balance in lymph-cannulated and non-cannulated rats. *J. Pharm. Sci.* **2000,** *89,* 1073–1084.48.

Canter, N. It isn't easy being green. *Lubricants World* **2001,** *11,* 16–21.

Charman, W.N. Lipids, lipophilic drugs, and oral drug delivery— Some emerging concepts. *J. Pharm. Sci.* **2000,** *89,* 967–978.

Chen, X.; H. Zhong; L. Jia; J. Ling; R. Tang; J. Qiao; Z. Zhang. Polyesteramides used for hot melt adhesives: Synthesis and effect of inherent viscosity on properties. *J. Appl. Polym. Sci.* **2001,** *81,* 2696–2701.

Clay, J.D.; B. Vijayendran; J. Moon. *Rheological Study of Soy-Protein-Based PRF Wood Adhesives.* Annu. Tech. Conf. Soc. Plast. Eng. 57th, pp. 1298–1301, 1999.

Coco, C.E.; D.R. Dill; T.L. Krinski. Short course proceedings, *Coatings Binders Short Course;* Technical Association of the Pulp and Paper Industry, Eds.; Coating Binders Committee: Boston, MA, 1990; pp. 71–80.

Conner, A.H.; L.A. Johnson (Eds.) *New Technologies for Value-Added Products from Protein and Co-Products;* Symposium Proceedings. 80th Annual Meeting and Exposition of the American Oil Chemists' Society, Cincinnati, OH, 1989, pp. 15–28.

Contant, P.J.; J.B.F. Perrot. [no title available]. French Patent 461,007 (1913).

Cox, K.D.; H. Scherm. Effect of desiccants and herbicides on germination of pseudosclerotia and development of apothecia of Monilinia vaccinii-corymbosi. *Plant Dis.* **2001,** *85,* 436–441.

Crocco, G.L.; W.F. Shum; J.G. Zajacek; H.S.J. Kesling. Epoxidation Process. U.S. Patent 5,166,372 (1992).

Davidson, G., et al.; H.F. Rippey; C.N. Cone; I.F. Laucks; H.P. Banks. Cellulose-Fiber Product Treated With a Size Embodying Soybean Flour and Process of Making the Same. U.S. Patent 1,622,496 (1927).

Dirlikov, S.; I. Frischinger; Z. Chen. Phase separation of two-phase epoxy thermosets that contain epoxidized triglyceride oils. *Adv. Chem. Ser.* **1996,** *252,* 95–108.

Djonlagic, J.; Z.S. Petrovic. Semi-interpenetrating polymer networks composed of poly(*N*-isopropyl acrylamide) and polyacrylamide hydrogels. *J. Polym. Sci. Pol. Phys.* **2004,** *42,* 3987–3999.

Dodd, R.; H.B.P. Humphries. Preparation of Plastic Substances and the Like From Protein Containing Materials. British Patent 15,316 (1913).

Doll, K.M.; B.K. Sharma; S.Z. Erhan. Synthesis of branched methyl hydroxy stearates including an ester from bio-based levulinic acid. *Ind. Eng. Chem. Res.* **2007,** *46,* 3513–3519.

Druckrey, A.K.; J.M. Lazar; M.H. Lang. Recyclable Moisture-Barrier Coated Paper Stock and Its Production. PCT Int. Appl. 2005.

Dykstra, G.M.; R.L. Hollingsworth. Method for Producing Paper Coating Binder Involving Grafting Unsaturated Acrylate Onto Proteinaceous Substrate in Water. U.S. Patent 3,920,592 (1975).

El-Aasser, M.S.; E.D. Sudol. Miniemulsions: Overview of research and applications. *J. Coating Technol.* **2004,** *1,* 20–31.

Erhan, S.Z. Vegetable oil-based printing inks and environmental advantages. *Recent Research Developments in Oil Chemistry I.* Transworld Research Network: Trivandrum, India, 1997; pp. 159–164.

Erhan, S.Z. Soybean oil metalworking lubricant. *Ind. Bioprocessing* **2006,** *28,* 3.

Erhan, S.Z.; A. Adhvaryu. Vegetable oil-based stocks. *Biobased Industrial Fluids and Lubricants;* S.Z. Erhan, J.M. Perez, Eds.; AOCS Press: Champaign, IL, 2002; pp. 1–19.

Erhan, S.Z.; A. Adhvaryu; Z. Liu. Chemically Modified Vegetable Oil-Based Industrial Fluid. U.S. Patent 6,583,302 (2003).

Erhan, S.Z.; A. Adhvaryu; B.K. Sharma. Chemically functionalized vegetable oils. Rudnick, L.R. (ed.), *Synthetics, Mineral Oils, and Bio-based Lubricants Chemistry and Technology;* L.R. Rudnick, Ed.; CRC Press: Boca Raton, FL, 2005; pp. 361–388.

Erhan, S.Z.; S. Asadauskas. Lubricant basestocks from vegetable oils. *Ind. Crop Prod.* **2000,** *11,* 277–282.

Erhan, S.Z.; S. Asadauskas; R.O. Dunn; G. Knothe. Vegetable oils for environmentally friendly applications. *Proceedings: 48th Oilseed Conference, Competing in World Markets in the New Millennium,* American Oil Chemists' Society: Champaign, IL, 1999.

Erhan, S.Z.; M.O. Bagby. Lithographic and letterpress ink vehicles from vegetable oils. *J. Am. Oil Chem. Soc.* **1991,** *68,* 635–638.

Erhan, S.Z.; M.O. Bagby. Gel permeation chromatography of vegetable-oil-based printing ink vehicles. *J. Appl. Polym. Sci.* **1992a,** *46,* 1859–1862.

Erhan, S.Z.; M.O. Bagby. Vegetable Oil-Based Printing Ink. U.S. Patent 5,122,188 (1992b).

Erhan, S.Z.; M.O. Bagby. Biodegradation of news ink vehicles. *TAGA Proceedings.* Technical Association of the Graphic Arts, TAGA Office. Book Crafters, Inc.: Chelsea, MI, 1993; pp. 314–326.

Erhan, S.Z.; M.O. Bagby. Polymerization of vegetable oils and their uses in printing inks. *J. Am. Oil Chem. Soc.* **1994,** *71,* 1223–1226.

Erhan, S.Z.; M.O. Bagby. Vegetable oil-based sheetfed and heatsink ink formulations and comparison of VOC test results with commercial inks. *TAGA Proceedings.* Technical Association of the Graphic Arts, TAGA Office. Book Crafters, Inc.: Chelsea, MI, 1997, pp. 807–814.

Erhan, S.Z.; M.O. Bagby. Vegetable Oil-Based Offset Printing Inks. U.S. Patent 5,713,990 (1998).

Erhan, S.Z.; M.O. Bagby; T.C. Nelsen. Biodegradation of news inks with modified strum test. *TAGA Proceedings.* Technical Association of the Graphic Arts, TAGA Office. Book Crafters, Inc.: Chelsea, MI, 1995; pp. 184–203.

Erhan, S.Z.; M.O. Bagby; T.C. Nelsen. Drying properties of metathesized soybean oil. *J. Am. Oil Chem. Soc.* **1997a,** *74,* 703–706.

Erhan, S.Z.; M.O. Bagby; T.C. Nelsen. Statistical evaluation of biodegradation of news ink vehicles and ink formulations. *J. Am. Oil Chem. Soc.* **1997b,** *74,* 707–712.

Erhan, S.Z.; T.C. Nelsen. Comparisons of volatile organic chemical content of news, sheetfed, and heatset ink formulations. *J. Am. Oil Chem. Soc.* **2001,** *78,* 419–422.

Erhan, S.Z.; B.K. Sharma; J.M. Perez. Oxidation and low temperature stability of vegetable oil-based lubricants. *Ind. Crop Prod.* **2006,** *24,* 292–299.

Fan, X.D.; Y. Deng; J. Waterhouse; P. Pfromm. Synthesis and characterization of polyamide resins from soy-based dimer acids and different amides. *J. Appl. Polym. Sci.* **1998,** *68,* 305–314.

Feairheller, S.H.; Bistline, R.G., Jr.; A. Bilyk; R.L. Dudley; M.F. Kozempel; M.J. Haas. Novel technique for the preparation of secondary fatty amides. III. Alkanolamides, diamides and aralkylamides. *J. Am. Oil Chem. Soc.* **1994,** *71,* 863–866.

Feeney, R.E., J.R. Whitaker. *Chemical and Enzymatic Modification of Plant Proteins in New Protein Foods;* A.M. Altschul, H.L. Wilcke, Eds.; Academic Press: New York, 1985; pp. 181–219.

Feitkenhauer, H.; U. Meyer. Anaerobic digestion of alcohol sulfate (anionic surfactant) rich wastewater—Batch experiments. Part I: Influence of the surfactant concentration. *Bioresour. Technol.* **2002a,** *82,* 115–121.

Feitkenhauer, H.; U. Meyer. Anaerobic digestion of alcohol sulfate (anionic surfactant) rich wastewater—Batch experiments. Part II: Influence of the hydrophobic chain length. *Bioresour. Technol.* **2002b,** *82,* 123–129.

Findley, T.W.; D. Swern; J.T. Scanlan. Epoxidation of unsaturated fatty materials with peracetic acid in glacial acetic acid solution. *J. Am. Oil Chem. Soc.* **1945,** *67,* 412–414.

Formo, M.W. *Bailey's Industrial Oil and Fat Products,* Fourth ed.; D. Swern, Ed.; John Wiley and Sons, Inc.: New York, NY, 1982; Vol. 2, pp. 343–405.

Frankel, E.N. The lipovore's challenge to food scientists. *INFORM* **2007,** *18,* 469–472.

Frankel, E.N.; F.L. Thomas; W.K. Rohwedder. Hydrocarboxylation of polyunsaturated fatty acids and esters with a palladium chloride-triphenylphosphine complex catalyst. *Adv. Chem. Ser.* **1974,** *13,* 145–165.

Frankel, E.N.; F.L. Thomas; W.K. Rohwedder; J.C. Cowan. Hydroformylation of methyl linoleate and linolenate with rhodium-triphenylphosphine catalyst. *Ind. Eng. Chem. Prod. RD.* **1973,** *12,* 47–53.

Gangule, N.G.; M.C. Dwivedi. Total vegetable o il greases. *J. Synth. Lubr.* **2001,** *17,* 333–349.

Garey, C.L. New technologies for value-added products from protein and co-products. *Symposium Proceedings;* L.A. Johnson, Ed.; 80th Annual Meeting and Exposition of the American Oil Chemists' Society, Cincinnati, OH, 1989, pp. 10–14.

Goessel, F.; A. Sauer. Artificial India rubber and an Improved Process of Producing It from Soya Oils. German Patent 228,887 (1909).

Golbitz, P. Statistics. *Soya & Oilseed Bluebook;* Soyatech LLC: Bar Harbor, ME, 2008; pp. 345–390.

Gunstone, F.D. Fats and oils forecast. *INFORM* **2006,** *17,* 667–669.

Guo, A.; Y. Cho; Z.S. Petrovic. Structure and properties of halogenated and nonhalogenated soy-based polyols. *J. Polym. Sci. A1.* **2000,** *38,* 3900–3910.

Guo, A.; D. Demydov; W. Zhang; Z.S. Petrovic. Polyols and polyurethanes from hydroformylation of soybean oil. *J. Polym. Environ.* **2002,** *10,* 49–52.

Guo, A.; W. Zhang; Z.S. Petrovic. Structure-property relationships in polyurethanes derived from soybean oil. *J. Mater. Sci.* **2006,** *41,* 4914–4920.

Ha, S.H.; C.S. Park; B.Y. Kim. Effect of water-resistance and physical properties of soy protein isolate-coated liner board. *Han'guk Sikp'um Yongyang Kwahak Hoechi.* **2006,** *35,* 1251–1255.

Hagemann, J.W.; J.A. Rothfus. Effects of chain length, conformation and α-form packing arrangement on theoretical monoacid triglyceride β'-forms. *J. Am. Oil Chem. Soc.* **1988,** *65,* 638.

Hagemann, J.W.; W.H. Tallent; K.E. Kolb. Differential scanning calorimetry of single acid tryglycerides: Effect of chain length and unsaturation. *J. Am. Oil Chem. Soc.* **1972,** *49,* 118–123.

Hance, H.F. *Gardeners' Chron.* **1980,** *13,* 209.

Hartsuch, B.E. *Introduction to Textile Chemistry.* John Wiley and Sons, Inc.: New York, NY, 1950.

Herdan, J.M. Rolling fluids based on vegetable oils. *J. Synth. Lubr.* **1999,** *16,* 201–210.

Hernandez, E. Pharmaceutical and cosmetic use of lipids. *Bailey's Industrial Oil & Fat Products, Sixth ed.;* F. Shahidi, Ed.; John Wiley & Sons, Inc.: Hoboken, NJ, 2004; Vol. 6, pp. 391–412.

Hettiarachchy, N.S.; U. Kalapathy; D.J. Meyer. Alkali modified soy protein with improved adhesive and hydrophobic properties. *J. Am. Oil Chem. Soc.* **1995,** *72,* 1461–1464.

Hinterwaldner, R. Plant proteins as resources for innovation in backbone binders. *Coatings* **1997,** *30,* 323–325.

Hiscock, D.F.; T.B. Merrifield. Coating immobilization using soy protein polymers: technical concepts and importance to quality. *Polpu, Chongi Gisul* **2000,** *32,* 60–66.

Hojilla-Evangelista, M.P. Adhesive qualities of soybean protein-based foamed plywood glues. *J. Am. Oil Chem. Soc.* **2002,** *79,* 1145–1149.

Hojillia-Evangelista, M.P.; L.B. Dunn. Foaming properties of soybean protein-based plywood adhesives. *J. Am. Oil Chem. Soc.* **2001,** *78,* 567–572.

Holser, R.A. Synthesis of surfactants from vegetable oil feedstocks. *Industrial Uses of Vegetable Oils;*

S.Z. Erhan, Ed.; AOCS Press: Champaign, IL, 2005; pp. 170–178.

Honary, L. Case study: Approval of soybean oil-based rail curve grease by a U.S. Class I railroad. *NLGI Spokesman.* **2003,** *67,* 19–20.

Honary, L.A.T. An investigation of the use of soybean oil in hydraulic systems. *Bioresour. Technol.* **1996,** *56,* 41–47.

Honary, L.A.T. Field test results of soybean-based greases developed by UNI-ABIL Research Program. *NLGI Spokesman.* **2000,** *64,* 22–28.

Honary, L.A.T. Case study: Approval of soybean oil-based rail curve grease by a U.S. Class I railroad. *NLGI Spokesman.* **2004,** *68,* 24–32.

Horner, D. Recent trends in environmentally friendly lubricants. *J. Syn. Lubr.* **2002,** *18,* 327–348.

Huang, H.C.; E.G. Hammond; C.A. Reitmeier; D.J. Myers. Properties of fibers produced from soy protein isolate by extrusion and wet-spinning. *J. Am. Oil Chem. Soc.* **1995,** *72,* 1453–1460.

Huang, W.N.; X. Sun. Adhesive properties of soy proteins modified by urea and guanidine hydrochloride. *J. Am. Oil Chem. Soc.* **2000a,** *77,* 101–104.

Huang, W.N.; X. Sun. Adhesive properties of soy proteins modified by sodium dodecyl sulfate and sodium dodecyl benzene sulfonate. *J. Am. Oil Chem. Soc.* **2000b,** *77,* 705–708.

Hwang, H.S.; A. Adhvaryu; S.Z. Erhan. Preparation and properties of lubricant basestocks from epoxidized soybean oil and 2-ethylhexanol. *J. Am. Oil Chem. Soc.* **2003,** *80,* 811–815.

Hwang, H.S.; S.Z. Erhan. Modification of epoxidized soybean oil for lubricant formulations with improved oxidative stability and low pour point. *J. Am. Oil Chem. Soc.* **2001,** *78,* 1179–1184.

Hwang, H.S.; S.Z. Erhan. Lubricant base stocks from modified soybean oil. *Biobased Industrial Fluids and Lubricants;* S.Z. Erhan; J.M. Perez, Eds.; AOCS Press: Champaign, IL, 2002; pp. 20–34.

Hwang, H.S.; S.Z. Erhan. Synthetic lubricant basestocks from epoxidized soybean oil and Guerbet alcohols. *Ind. Crop Prod.* **2006,** *23,* 311–317.

Javni, I.; Z.S. Petrovic; A. Guo; R. Fuller. Thermal stability of polyurethanes based on vegetable oils. *J. Appl. Polym. Sci.* **2000,** *77,* 1723–1734.

Javni, I.; W. Zhang; Z.S. Petrovic. Soybean-oil-based polyisocyanurate rigid foams. *J. Polym. Environ.* **2004,** *12,* 123–129.

Jiratumnukul, N.; M.R. Van De Mark. Preparation of glycol esters of soybean oil fatty acids and their potential as coalescent aids in paint formulations. *J. Am. Oil Chem. Soc.* **2000,** *77,* 691–697.

John, J.; M. Bhattacharya; P.C. Raynor. Emulsions containing vegetable oils for cutting fluid application. *Colloid Surface A.* **2004,** *237,* 141–150.

John, J.; M. Bhattacharya; R.B. Turner. Characterization of polyurethane foams from soybean oil. *J. Appl. Polym. Sci.* **2002,** *86,* 3097–3107.

Johnson, D.L. High performance 4-cycle lubricants from canola. *Perspectives on New Crops and New Uses;* J. Janick, Ed.; ASHS Press: Alexandria, VA, 1999; pp. 247–250.

Johnson, O. Adhesive. U.S. Patent 1,460,757 (1923).

Johnson, O. Process of Treating Soya Beans. U.S. Patent 1,680,264 (1928).

Johnson, L.A.; D.J. Myers. Industrial uses for soybeans. *Practical Handbook of Soybean Processing and Utilization;* D.R. Erickson, Ed.; AOCS Press: Champaign, IL, 1995; pp. 380–427.

Johnson, L.A.; D.J. Myers; D.J. Burden. Early uses of soy protein in Far East, US. *INFORM* **1992,** *3,* 282–290.

Kajita, T.; R. Inoue. Process for Manufacturing Artificial Fiber From Protein in Soya Bean. U.S. Patent 2,192,194 (1940).

Kalapathy, U.; N.S. Hettiarachchy; D. Myers; M.A. Hanna. Alkali modification of soy protein and their adhesive properties on woods. *J. Am. Oil Chem. Soc.* **1995,** *72,* 507–510.

Kalapathy, U.; N.S. Hettiarachchy; D.J. Myers; K.C. Rhee. Alkali modification of soy proteins: effects of salts and disulfide bond cleavage on adhesion and viscosity. *J. Am. Oil Chem. Soc.* **1996,** *73,* 1063–1066.

Kalapathy, U.; N.S. Hettiarachchy; K.C. Rhee. Effect on drying methods on molecular properties and functionalities of disulfide bond cleaved by soy proteins. *J. Am. Oil Chem. Soc.* **1997,** *74,* 195–199.

Kale, V.; R. Subbarao; G. Lakshminarayana; M.B. Rao. Studies on kinetics and preparation of C36 dimer acid esters. *J. Am. Oil Chem. Soc.* **1991,** *68,* 583–584.

Kale, V.; H.S. Vedanayagam; R. Subbarao; M.B. Rao. Kinetics of reaction of C36 dimer acids with diethylenetriamine and triethylenetetramine for reactive polyamides. *J. Appl. Polym. Sci.* **1994,** *53,* 807–814.

Kammann, K.P., Jr.; A.J. Phillips. Sulfurized vegetable oil products as lubricant additives. *J. Am. Oil Chem. Soc.* **1985,** *62,* 917–923.

Kaplan, D.L. (ed.), *Biopolymers from Renewable Resources.* Springer: New York, 1998, pp 249–280.

Kapusta, G. Uses of soybean oil in the application of herbicides. *J. Am. Oil Chem. Soc.* **1984,** *62,* 923–926.

Karcher, L.P. The incorporation of corn and soybean based materials into plywood adhesives. *Diss. Abstr. Int. B* **1997,** *57,* 7297.

Kassfeldt, E.; G. Dave. Environmentally adapted hydraulic oils. *Wear* **1997,** *207,* 41–45.

Kella, N.K.D.; Y.J. Kang; J.E. Kinsella. Effect of oxidative sulfitolysis of disulfide bonds of bovine serum albumin on its structural properties. *J. Protein Chem.* **1998,** *7,* 535–548.

Khot, S.N.; J.J. Lascala; E. Can; S.S. Morye; G.I. Williams; G.R. Palmese; S.H. Kusefoglu; R.P. Wool. Development and application of triglyceride-based polymers and composites. *J. Appl. Polym. Sci.* **2001,** *82,* 703–723.

Kinsella, J.E. Functional properties of soy proteins. *J. Am. Oil Chem. Soc.* **1979,** *56,* 242–258.

Klemazewski, J.L.; J.E. Kinsella. Sulfitolysis of whey proteins: Effect on emulsion properties. *J. Agric. Food Chem.* **1991,** 39, 1033–1036.

Knaebel, D.B.; T.W. Federle; J.R. Vestal. Mineralization of linear alkylbenzene sulfonate (LAS) and linear alcohol ethoxylate (LAE) in 11 contrasting soils. *Environ. Toxicol. Chem.* **1990,** *9,* 981–988.

Knothe, G.; R.O. Dunn. Biodiesel: An alternative diesel fuel from vegetable oils or animal fats. *Industrial Uses of Vegetable Oils;* S.Z. Erhan, Ed.; AOCS Press: Champaign, IL, 2005; pp. 42–89.

Kopylov, L.I.; Y.N. Shekhter; A.A. Gureev; M.B. Bakaleinikov. Physicochemical and functional properties of mixtures of akor-1 inhibitor and alkanolamides. *Chem. Tech. Fuels Oil.* **1980,** *16,*

62–64.

Kreibich, R.E. New adhesives based on soy protein. *Oils–Fats–Lipids 1995. Proc. 21st World Congr. Int. Soc. Fat Res;.* P.J. Barnes & Assoc.: Bridgwater, UK, 1995; pp. 503–509.

Kreutzer, U.R. Manufacture of fatty alcohols based on natural fats and oils. *J. Am. Oil Chem. Soc.* **1983,** *61,* 343–348.

Krinski, T.L.; K.C. Hou. Vegetable Protein-Polyacrylate Coprecipitate as Binder-Thickener for Paper Coating Compositions. Eur. Pat. Appl. (2000).

Krinski, T.L.; A.L. Steinmetz. Protein Modified With a Silanation Reagent as an Adhesive Binder and Process of Producing It. U.S. Patent 4,713,116 (1987).

Krishna, C.R. Biodiesel blends in space heating equipment. National Renewable Energy Laboratory: Upton, NY, NREL/SR-510-33579, 2001.

Kronick, P.L. Use of Polymerizable Oil for Leather Fat Liquor. U.S. Patent 5,853,427 (1998).

Kronick, P.; Y.K. Kamath. Leather and textile uses of fats and oils. *Bailey's Industrial Oil & Fat Products, Sixth ed.;* F. Shahidi, Ed.; John Wiley & Sons, Inc.: Hoboken, NJ, 2004; Vol. 6, pp. 353–370.

Kuo, M.L.; D.J. Myers; H. Heemstra; D. Curry; D.O. Adams; D.D. Stokke. Soybean-Based Adhesive Resins and Composite Products Utilizing Such Adhesives. U.S. Patent 6,518,387 (2001).

Kuo, M.L.; D.J. Myers; H. Heemstra; D. Curry; D.O. Adams; D.D. Stokke. Soybean-Based Adhesive Resins and Composite Products Utilizing Such Adhesives. U.S. Patent 6,306,997 (2003).

La Scala, J.; R.P. Wool. Effect of FA composition on epoxidation kinetics of TAG. *J. Am. Oil Chem. Soc.* **2002,** *79,* 373–378.

Lambuth, A.L. Soybean glues. *Handbook of Adhesives,* Second ed.; I. Skeist, Ed.; Van Nostrand: New York, 1977; pp. 172–180.

Lambuth, A.L. Protein adhesives for wood. *Advanced Wood Adhesive Technology;* A. Pizzi, K.L. Mittal, Eds.; Marcel Dekker Inc.: New York, 1994; pp. 259–281.

Lambuth, A.L. Blood and casein glues. *Coatings Technology Handbook;* D. Satas, A.A. Tracton, Eds.; Marcel Dekker Inc.: New York, 2001; pp. 519–530.

Landfester, K. Designing particles—Miniemulsion technology and its application in functional coating systems. *Eur. Coatings J.* **2005,** *12,* 20–25.

Landfester, K.; J. Eisenblatter; R. Rothe. Preparation of polymerizable miniemulsions by ultrasonication. *J. Coating Technol. Res.* **2004,** *1,* 65–68.

Larock, R.C.; X. Dong; S. Chung; C. Kishan Reddy; L.E. Ehlers. Preparation of conjugated soybean oil and other natural oils and fatty acids by homogeneous transition metal catalysis. *J. Am. Oil Chem. Soc.* **2001,** *78,* 447–453.

Laubli, M.W.; P.A. Bruttel. Determination of the oxidative stability of fats and oils: Comparison between the Active Oxygen Method (AOCS Cd 12-57) and the Rancimat Method. *J. Am. Oil Chem. Soc.* **1986,** *63,* 792–795.

Laucks, I.F.; G. Davidson. Vegetable Glue and Method of Making the Same. U.S. Patent 1,691,661 (1928).

Leach, R.H. (ed.). *The Printing Ink Manual.* Van Nostrand Reinhold (International) Co. Ltd.: London, 1988.

Leonard, E.C. The higher aliphatic di- and polycarboxylic acids. *Fatty Acids;* E.H. Pryde, Ed.; AOCS Press: Champaign, IL, 1980; pp. 504–526.

Li, K.; S. Peshkova; X. Geng. Investigation of soy protein-Kymene adhesive systems for wood composites. *J. Am. Oil Chem. Soc.* **2004,** *81,* 487–491.

Liang, F.; Y. Wang. Effect of silane coupling agents on the interface of soy protein and glass fibre. *Proc. Am. Soc. Compos., 14th Tech. Conf.,* pp. 511–520, 1999.

Lin, K.F. Paints, varnishes, and related products. *Bailey's Industrial Oil & Fat Products,* Sixth ed.; F. Shahidi, Ed.; John Wiley & Sons, Inc.: Hoboken, NJ,, 2004; Vol. 6, pp. 307–352.

Linne, M.A.; L.H. Sperling; A.M. Fernandez; S. Qureshi; J.A. Manson. Simultaneous interpenetrating networks prepared from special functional group triglyceride oils: lesquerella palmeri and other wild plant oils. *Adv. Chem. Ser.* **1984,** 37–49.

Little Inc., A.D. *Marketing Potential for Oilseed Proteins in Industrial Uses,* USDA Technical Bulletin 1043, 1951.

Liu, Z.S.; S.Z. Erhan. Development of soy composites by direct deposition. *Industrial Uses of Vegetable Oils;* S.Z. Erhan, Ed.; AOCS Press: Champaign, IL, 2005; pp. 131–142.

Liu, Z.S.; S.Z. Erhan. Soy-Based Thermosensitive Hydrogels for Controlled Release Systems. U.S. Patent Application 20070077298 (2007).

Liu, Z.; S.Z. Erhan; D.E. Akin; F.E. Barton. "Green" composites from renewable resources: Preparation of epoxidized soybean oil and flax fiber composites. *J. Agric. Food Chem.* **2006,** *54* 2134–2137.

Liu, Z.; S.Z. Erhan; P.D. Calvert. Solid freeform fabrication of soybean oil-based composites reinforced with clay and fibers. *J. Am. Oil Chem. Soc.* **2004a,** *81,* 605–610.

Liu, Z.S.; S.Z. Erhan; P.D. Calvert. Solid freeform fabrication of epoxidized soybean oil/epoxy composites with di-, tri-, and polyethylene amine curing agents. *J. Appl. Polym. Sci.* **2004b,** *93,* 356–363.

Liu, Z.S.; S.Z. Erhan; P.D. Calvert. Solid freeform fabrication of epoxidized soybean oil/epoxy composite with bis or polyalkyleneamine curing agents. *Compos. Part A-Appl. S.* **2007a,** *38,* 87–93.

Liu, Y.; K. Li. Chemical modification of soy protein for wood adhesives. *Macromol. Rapid Commun.* **2002,** *23,* 739–742.

Liu, Z.; S.Z. Erhan; J. Xu. Preparation, characterization and mechanical properties of epoxidized soybean oil/clay nanocomposites. *Polymer* **2005,** *46,* 10119–10127.

Liu, Z.S.; S.Z. Erhan; J. Xu; P.D. Calvert. Development of soybean oil-based composites by solid freeform fabrication method: Epoxidized soybean oil with bis or polyalkyleneamine curing agents system. *J. Appl. Polym. Sci.* **2002,** *85,* 2100–2107.

Liu, Z.S.; B.K. Sharma; S.Z. Erhan. From oligomers to molecular giants of soybean oil in supercritical carbon dioxide medium: 1. Preparation of polymers with lower molecular weight from soybean oil. *Biomacromolecules* **2007b,** *8,* 233–239.

Lorenz, L.; C.R. Frihart; J.M. Wescott. Chromatographic analysis of the reaction of soy flour with formaldehyde and phenol for wood adhesives. *J. Am. Oil Chem. Soc.* **2007,** *84,* 769–776.

Lougee, E.F. Industry and the soy bean. *Modern Plastics* **1936,** *13,* 13–15.

Lynn, J.L., Jr. Detergents and detergency. *Bailey's Industrial Oil & Fat Products,* Sixth ed.; F. Shahidi, Ed.; John Wiley & Sons, Inc.: Hoboken, NJ, 2004; Vol. 6, pp. 137–190.

Maag, H. Fatty acid derivatives: Important surfactants for household, cosmetic and industrial purposes. *J. Am. Oil Chem. Soc.* **1983,** *61,* 259–267.

Mack, R.E.; J.R. Roberts; G.C. Volgas. Comparisons of 2-ethylhexyl ester formulations of 2,4-d: chemistry and bioefficacy. *ASTM Special Technical Publication.* pp. 140–148, 2003.

Mackay, C.D. Good adhesive bonding starts with surface preparation. *Adhes. Age* **1998,** *41,* 30–32.

Maltby, A.J.; M. Read. Antistatic agents—The effectiveness of ethoxylated fatty acid amides in reducing the static charge decay time of LDPE film as compared with commercially available materials. *TAPPI—Polymers, Laminations and Coatings Conference.* pp. 1103–1114, Aug. 27–31, 2000, Chicago, IL, United States.

Mars-Proietti, L. *The Rauch Guide to the U.S. Ink Industry 2006 Edition;* Grey House Publishing: Millerton, NY, 2006.

Martinez-Palou, R.; J. Rivera; L.G. Zepeda; A.N. Rodriguez; M.A. Hernandez; J. Marin-Cruz; A. Estrada. Evaluation of corrosion inhibitors synthesized from fatty acids and fatty alcohols isolated from sugar cane wax. *Corrosion* **2004,** *60,* 465–470.

Matthaus, B.W.; C. Wiezorek; K. Eichner. Fast chemiluminescence method for detection of oxidized lipids. *Fat Sci. Technol.* **1994,** *96,* 95–99.

McKeon, T.A. Genetic modification of seed oils for industrial applications. *Industrial Uses of Vegetable Oils;* S.Z. Erhan, Ed.; AOCS Press: Champaign, IL, 2005; pp. 1–13.

Meffert, A. Technical uses of fatty acid esters. *J. Am. Oil Chem. Soc.* **1983,** *61,* 255–258.

Metzger, J.O.; U. Bornscheuer. Lipids as renewable resources: Current state of chemical and biotechnological conversion and diversification. *Appl. Microbiol. Biotechnol.* **2006,** *71,* 13–22.

Moncrieff, R.W. *Manmade Fibers,* Fifth ed.; Wiley Interscience Publishers: New York, NY, 1975.

Montgaudry, B. *Bulletin de la. Societe d'Acclimatation.* **1855,** *2,* 16.

Moynihan, J.T. Printing Ink. U.S. Patent 4,419,132 (1983).

Moynihan, J.T. Letterpress Printing Ink. U.S. Patent 4,554,019 (1985a).

Moynihan, J.T. Offset Printing Ink. U.S. Patent 4,519,841 (1985b).

Mungara, P.; T. Chang; J. Zhu; J.L. Jane. Processing and physical properties of plastics made from soy protein polyester blends. *J. Polym. Environ.* **2002,** *10,* 31–37.

Myers, D.J. Industrial applications for soy protein and potential for increased utilization. *Cereal Foods World* **1993,** *38,* 355–360.

Naegely, P.C. Environmentally acceptable lubricants. *Seed Oils for the Future.* S.L. MacKenzie, D.C. Taylor, Eds.; AOCS Press: Champaign, IL, 1992, pp 14–25.

Narine, S.S.; X. Kong. Vegetable oils in production of polymers and plastics. *Bailey's Industrial Oil & Fat Products,* Sixth ed.; F. Shahidi, Ed.; John Wiley & Sons, Inc.: Hoboken, NJ, 2004; Vol. 6, pp. 279–306.

Nayak, S.S.; S.K. Das; S. Lenka. Polymers from renewable resources: XIX: Synthesis and characterization of copolymers from cardanyl acrylate and vinyl monomers. *React. Func. Polym.* **1999,** *40,* 249–254.

Nayak, P.; D.K. Mishra; D. Parida; K.C. Sahoo; M. Nanda; S. Lenka; P.L. Nayak. Polymers from renewable resources. IX. Interpenetrating polymer networks based on castor oil polyurethane poly(hydroxyethyl methacrylate): Synthesis, chemical, thermal, and mechanical properties. *J. Appl. Polym. Sci.* **1997,** *63,* 671–679.

Nayak, R.R.; G. Ray; B. Guru; S. Lenka. Comparative studies of interpenetrating polymer networks derived from soybean oil-based polyurethane and cardanol m-aminophenol dye. *Polym-Plast. Technol.* **2004,** *43,* 261–272.

Neuzil, D.I.; L.A.T. Honary. Power Pods: Soy-based lubricants grease the way for a biobased future. *Resource* **2003,** *10,* 9–10.

Nowak, J.A.; T.A. Zillner; M.I. Leslie Patrick. Thin-film Epoxidation of an Unsaturated Oil or Alkyl Fatty Acid Ester. U.S. Patent 6,734,315 (2004).

Okano, T.; N. Egawa; M. Fujiwara; M. Fukuda. α-sulfonated fatty acid esters: II. Solution behavior of α-sulfonated fatty acid polyethylene glycol esters. *J. Am. Oil Chem. Soc.* **1996,** *73,* 31–37.

Otaigbe, U.J.; O.D. Adams. Novel bioabsorbable polyphosphate/soy protein plastic composites. *Int. SAMPE Tech. Conf. 29.;* 1997a; pp. 686–697.

Otaigbe, U.J.; O.D. Adams. Bioabsorbable soy protein plastic composites: Effect of polyphosphate fillers on water absorbtion and mechanical properties. *J. Environ. Polym. Degrad.* **1997b,** *5,* 199–208.

Oyman, Z.O.; W. Ming; R. Van Der Linde. Oxidation of drying oils containing non-conjugated and conjugated double bonds catalyzed by a cobalt catalyst. *Prog. Org. Coat.* **2005,** *54,* 198–204.

Padavich, R.A.; L. Honary. A market research and analysis report on vegetable-based industrial lubricants. *SAE Technical Publication* **1995,** *952077,* 13.

Paetau, I.; C.Z. Chen; J.L. Jane. Biodegradable plastics made from soybean products. *Ind. Eng. Chem. Res.* **1994a,** *33,* 1821–1827.

Paetau, I.; C.Z. Chen; J.L. Jane. Effects of crosslinking and cellulose incorporation on mechanical properties and water absorption. *J. Environ. Polym. Degrad.* **1994b,** *2,* 211–217.

Park, H.J.; S.H. Kim; S.T. Lim; D.H. Shin; S.Y. Choi; K.T. Hwang. Grease resistance and mechanical properties of isolated soy protein-coated paper. *J. Am. Oil Chem. Soc.* **2000,** *77,* 269–273.

Perez, J.M.; A.L. Boehman. Environmentally friendly fuels and lubricants. *Biobased Industrial Fluids and Lubricants;* S.Z. Erhan, J.M. Perez, Eds.; AOCS Press: Champaign, IL, 2002; pp. 65–77.

Persaud, M. (ed.), *2007–2008 Lubricant Industry Sourcebook, A Supplement to Lubes' n' Greases;* LNG Publishing Co. Inc.: Falls Church, VA, 2007, pp 4–13.

Petrovic, Z.S.; M.J. Cevallos; I. Javni; D.W. Schaefer; R. Justice. Soy-oil-based segmented polyurethanes. *J. Polym. Sci. Pol. Phys.* **2005a,** *43,* 3178–3190.

Petrovic, Z.S.; W. Zhang; I. Javni. Structure and properties of polyurethanes prepared from triglyceride polyols by ozonolysis. *Biomacromolecules* **2005b,** *6,* 713–719.

Petrovic, Z.S.; W. Zhang; I. Javni; A. Guo. Laminates from the soy-based polyurethanes and natural and synthetic fibers. *Materials Research Society Symposium—Proceedings*, 2002a. Boston, Massachusetts, November 26–30, 2001. pp 141–147.

Petrovic, Z.S.; W. Zhang; A. Zlatanic; C.C. Lava; M. Ilavsky. Effect of OH/NCO molar ratio on properties of soy-based polyurethane networks. *J. Polym. Environ.* **2002b,** *10*,5–12.

Petrovic, Z.S.; A. Zlatanic; C.C. Lava; S. Sinadinovic-Fiser. Epoxidation of soybean oil in toluene with peroxoacetic and peroxoformic acids—Kinetics and side reactions. *Eur. J. Lipid Sci. Tech.* **2002c,** *104,* 293–299.

Porter, C.J.H.; S.A. Charman; W.N. Charman. Lymphatic transport of halofantrine in the triple-cannulated anesthetized rat model: Effect of lipid vehicle dispersion. *J. Pharm. Sci.* **1996a,** *85,* 351–356.

Porter, C.J.H.; S.A. Charman; A.J. Humberstone; W.N. Charman. Lymphatic transport of halofantrine in the conscious rat when administered as either the free base or the hydrochloride salt: Effect of lipid class and lipid vehicle dispersion. *J. Pharm. Sci.* **1996b,** *85,* 357–361.

Porter, C.J.H.; A.M. Kaukonen; A. Taillardat-Bertschinger; B.J. Boyd; J.M. O'Connor; G.A. Edwards; W.N. Charman. Use of in vitro lipid digestion data to explain the in vivo performance of triglyceride-based oral lipid formulations of poorly water-soluble drugs: Studies with halofantrine. *J. Pharm. Sci.* **2004,** *93,* 1110–1121.

Porter, C.J.H.; N.L. Trevaskis; W.N. Charman. Lipids and lipid-based formulations: Optimizing the oral delivery of lipophilic drugs. *Nat. Rev. Drug Discov.* **2007,** *6,* 231–248.

Proffitt, T.J.; H.T. Patterson. Oleochemical surfactants and lubricants in the textile industry. *J. Am. Oil Chem. Soc.* **1988,** *65,* 1682–1694.

Pryde, E.H. *Handbook of Soy Oil Processing and Utilization;* D.R. Erickson, Ed.; AOCS Press: Champaign, IL, 1980; pp. 459–481.

Randles S.J.; M. Wright. Environmentally considerate ester lubricants for automotive and engineering industries. *J. Syn. Lubr.* **1992,** *9,* 145–161.

Rattray, J.B. Renewable propylene glycol introduced. *INFORM* **2007,** *18,* 318.

Reaney, M.J.T.; P.B. Hertz; W.W. McCalley. Vegetable oil as biodiesel. *Bailey's Industrial Oil & Fat Products,* Sixth ed.; F. Shahidi, Ed.; John Wiley & Sons, Inc.: Hoboken, NJ, 2004; Vol. 6, pp. 223–258.

Rhee, I.S. Evaluation of environmentally acceptable hydraulic fluids. *NLGI Spokesman.* **1996,** *60,* 28–34.

Rheineck, A.E.; R.O. Austin. Drying oils. Modifications and use. *Treatise on Coatings* **1968,** *1,* 181–248.

Rhodes, B.N.; D. Johnson. Vegetable oil-based motor oils. *Biobased Industrial Fluids and Lubricants;* S.Z. Erhan, J.M. Perez, Eds.; AOCS Press: Champaign, IL, 2002; pp. 85–109.

Rippey, H.F.; G. Davidson; C.N. Cone; I.F. Laucks; H.P. Banks. Carbonaceous Briquette and Process of Making the Same. U.S. Patent 1,735,506 (1929).

Russ, J.G.; I.M. Sandewicz; T. Zamyatin. Makeup Compositions Containing Soy Proteins. U.S. Patent Appl. 1999469825 (2001).

Sanbe, I.; Y. Toshishige. Skin-Lightening Cosmetics Containing Ascorbates. Japanese Pat. Appl. JP2003-436491 (2005).

Satow, S. Vegetable Proteid Product and Process of Making the Same. U.S. Patent 1,245,975 (1917).

Savastano, D. The resin report: Rising raw material costs, slowing economy are major concerns. *Ink World* **2001,** *7,* 36–38.

Schmidt, M.A.; C.R. Dietrich; E.B. Cahoon. Biotechnological enhancement of soybean oil for lubricant applications. *Synthetics, Mineral Oils and Bio-based Lubricants Chemistry and Technology;* L.R. Rudnick, Ed.; CRC Press: Boca Raton, FL, 2005; pp. 389–398.

Schmitz, J.F. Comparison of Soy-Flour Hydrolysates for Wood Adhesive Systems. M.S. Thesis; Iowa State University, Ames, IA. 2006.

Schmitz, W.R.; J.G. Wallace. Epoxidation of methyl oleate with hydrogen peroxide. *J. Am. Oil Chem. Soc.* **1954,** *31,* 363–365.

Schultz, J.; M. Nardin. Theories and mechanisms of adhesion. *Handbook of Adhesive Technology;* A. Pizzi, K.L. Mittal, Eds.; Marcel Dekker: New York, 1994.

Schultz, T.M.; H.T. Tran. Modified Soy Proteins in Personal Care Compositions. PCT Int. Appl. 2004112745 (2004).

Schwab, A.W.; M.O. Bagby; B. Freedman. Preparation and properties of diesel fuels from vegetable oils. *Fuel* **1987,** *66,* 1372–1378.

Sefing, F.G.; M.P. Surls. *Michigan Engineering Experiment Station Bulletin No. 54,* 1933, pp. 12.

Sharma, B.K.; A. Adhvaryu; S.Z. Erhan. Synthesis of hydroxy thio-ether derivatives of vegetable oil. *J. Agric. Food Chem.* **2006a,** *54,* 9866–9872.

Sharma, B.K.; A. Adhvaryu; Z. Liu; S.Z. Erhan. Chemical modification of vegetable oils for lubricant applications. *J. Am. Oil Chem. Soc.* **2006b,** *83,* 129–136.

Sharma, B.K.; A. Adhvaryu; J.M. Perez; S.Z. Erhan. Soybean oil-based greases: Influence of composition on thermo-oxidative and tribochemical behavior. *J. Agric. Food Chem.* **2005,** *53,* 2961–2968.

Sharma, B.K.; A. Adhvaryu; J.M. Perez; S.Z. Erhan. Biobased grease with improved oxidation performance for industrial application. *J. Agric. Food Chem.* **2006c,** *54,* 7594–7599.

Sharma, B.K.; K.M. Doll; S.Z. Erhan. Oxidation, friction reducing, and low temperature properties of epoxy fatty acid methyl esters. *Green Chem.* **2007a,** *9,* 469–474.

Sharma, B.K.; J.M. Perez; S.Z. Erhan. Soybean oil-based lubricants: A search for synergistic antioxidants. *Energy & Fuels* **2007b,** 10.1021/ef0605854: in press.

Shera, J.N.; J.W. Rawlins; S.F. Thames. Characterization of enzyme-modified soy protein isolate. *Abstracts of Papers, 233rd ACS National Meeting, Chicago, IL;* March 25–29, 2007.

Shurtleff, W.; A. Aoyagi. *Henry Ford and His Researchers' Work with Soybeans, Soyfoods, and Chemurgy–Bibliography and Sourcebook;* W. Shurtleff, Ed.; Soyfoods Center: Lafayette, CA, 1997.

Sinadinovic-Fiser, S.; M. Jankovic; Z.S. Petrovic. Kinetics of in situ epoxidation of soybean oil in bulk catalyzed by ion exchange resin. *J. Am. Oil Chem. Soc.* **2001,** *78,* 725–731.

Smith, K.; J. Thompson. *New Technologies for Value-added Products from Protein and Coproducts;* L.A. Johnson, Ed.; Symposium Proceedings, 80th Annual Meeting and Exposition of the American Oil Chemists' Society, Cincinnati, OH, AOCS Press: Champaign, IL, 1989; pp. 2–9.

Sonntag, N.O.V. Growth potential for soybean oil products as industrial materials. *J. Am. Oil Chem. Soc.* **1984,** *62,* 928–933.

Spence, K.E.; A.L. Allen; S. Wang; J.L. Jane. Soil and Marine. Biodegradation of Protein-Starch Plastics. In R. M. Ottenbrite, S. J. Huang and K. Park., editors Hydrogels and biodegradable polymers for bioapplications. ACS symposium series 627; American Chemical Society: Washington DC, 1996; pp. 149–158 (Chapter 12).

Sperling, L.H.; J.A. Manson. Interpenetrating polymer networks from triglyceride oils containing special functional groups: A brief review. *J. Am. Oil Chem. Soc.* **1983,** *60,* 1887–1892.

Spon, W.; F.N. Spon. *Span's Encyclopedia of the Industrial Arts, Manufacturers, and Commercial Products;* F.N. Spon: New York, 1980; pp. 1378, 1814.

Sprague, P.E. Utilization of soya proteins in industry. *Soybean Dig.* **1944,** *4,* 47–51.

Steele, P.H.; R.E. Kreibich; P.J. Steynberg; R.W. Hemingway. Finger jointing green southern yellow pine with a soy-based adhesives. *Adhes. Age* **1998,** *41,* 49–56.

Steinmetz, A.L.; T.L. Krinski. Modified Protein Adhesive Binder. U.S. Patent 4,554,337 (1985).

Steinmetz, A.L.; T.L. Krinski. Modified Protein Adhesive Binder and Process for Producing It. U.S. Patent 4,687,826 (1987).

Stirton, A.J.; R.G. Bistline, Jr.; E.A. Barr; M.V. Nunez-Ponzoa. Salts of alkyl esters of α-sulfopalmitic and α-sulfostearic acids. *J. Am. Oil Chem. Soc.* **1965,** *42,* 1078–1081.

Stirton, A.J.; R.G. Bistline, Jr.; J.K. Weil; W.C. Ault; E.W. Maurer. Sodium salts of alkyl esters of α-sulfo fatty acids. Wetting, lime soap dispersion, and related properties. *J. Am. Oil Chem. Soc.* **1962,** *39,* 128–131.

Stirton, A.J.; J.K. Weil; R.G. Bistline, Jr. Properties of salts of alpha-sulfonated acids and esters. *J. Am. Oil Chem. Soc.* **1954,** *31,* 13–16.

Sue, H.J.; S. Wang; J.L. Jane. Morphology and mechanical behavior of engineering soy plastics. *Polymer* **1997,** *38,* 5035–5040.

Sun, X.; K. Bian. Shear strength and water-resistance of modified soy protein adhesives. *J. Am. Oil Chem. Soc.* **1999a,** *76,* 977–980.

Sun, X.S.; H.R. Kim; X.Q. Mo. Plastic performance of soybean protein components. *J. Am. Oil Chem. Soc.* **1999b,** *76,* 117–123.

Taillardat-Bertschinger, A.; C.S. Perry; A. Galland; R.J. Prankerd; W.N. Charman. Partitioning of halofantrine hydrochloride between water, micellar solutions, and soybean oil: Effects on its apparent ionization constant. *J. Pharm. Sci.* **2003,** *92,* 2217–2228.

Tamura, T.; Y. Kaneko; M. Ohyama. Dynamic surface tension and foaming properties of aqueous polyoxyethylene n-dodecyl ether solutions. *J. Coll. Interf. Sci.* **1995,** *173,* 493–499.

Tamura, T.; Y. Takeuchi; Y. Kaneko. Influence of surfactant structure on the drainage of nonionic surfactant foam films. *J. Coll. Interf. Sci.* **1998,** *206,* 112–121.298.

Tao, B.Y. Biofuels for home heating oils. *Industrial Uses of Vegetable Oils;* S.Z. Erhan, Ed.; AOCS Press: Champaign, IL, 2005; pp. 90–109.

Thames, S.F.; K.G. Panjnani; O.S. Fruchey. Latex Compositions Containing Ethylenically Unsaturated Esters of Long-Chain Alkenols and Applications Thereof. U.S. Patent 6,897,255 (2005).

Thayer, A.M. Fit to print. *Chem. Eng. News* **2002,** *80,* 23–30.

Tsavalas, J.G.; F.J. Schork; K. Landfester. Particle morphology development in hybrid miniemulsion polymerization. *J. Coating Technol. Res.* **2004,** *1,* 53–63.

Tummala, P. W. Liu; L.T. Drzal; A.K. Mohanty; M. Misra. Influence of plasticizers on thermal and mechanical properties and morphology of soy-based bioplastics. *Ind. Eng. Chem. Res.* **2006,** *45,* 7491–7496.

Van De Mark, M. Solvents—What lies ahead? *Ind. Paint Powder Mag.* **2000,** *76,* 36–37.

Van De Mark, M.R.; J.A. Jones. Solvent selection and minimization for waterbornes. *Ind. Paint Powder Mag.* **1998,** *74,* 18.

Van De Mark, M.R.; K. Loftin. Advances in waterborne resin technology for OEM application. *Ind. Paint Powder Mag.* **1999,** *75,* 54–55.

Van De Mark, M.R.; K. Sandefur. Vegetable oils in paint and coatings. *INFORM* **2005a,** *16,* 478.

Van de Mark, M.R.; K. Sandefur. Vegetable oils in paints and coatings. *Industrial Uses of Vegetable Oils*; S.Z. Erhan, Ed.; AOCS Press: Champaign, IL, 2005b; pp. 143–162.

Van De Mark, M.R.; J.R. Schnelten. Water reducible acrylic resins. *Am. Paint Coatings J.* **1997,** *14.*

Vaughn, S.F.; R.A. Holser. Evaluation of biodiesels from several oilseed sources as environmental friendly contact herbicides. *Ind. Crop Prod.* **2007,** *26,* 63–68.

Vlcek, T.; Z.S. Petrovic. Optimization of the chemoenzymatic epoxidation of soybean oil. *J. Am. Oil Chem. Soc.* **2006,** *83,* 247–252.

Voeste, T.; H. Buchold. Production of fatty alcohols from fatty acids. *J. Am. Oil Chem. Soc.* **1983,** *61,* 350–352.

Wang, S.; H.J. Sue; J.L. Jane. Mechanical properties of soy protein plastics. *J. Macromol. Sci. Pure Appl. Chem.* **1996,** *A33,* 557–569.

Wang, S.; S. Zhang; J.L. Jane; H.J. Sue. Effect of polyols on mechanical properties of soy protein. *Polym. Mater. Sci. Eng.* **1995,** *72,* 88–89.

Watanabe, S.; H. Nakagawa; Y. Ohmori; T. Fujita; M. Sakamoto; T. Haga. Characteristic properties of cutting fluid additives derived from half esters of dimer acids. *J. Am. Oil Chem. Soc.* **1996,** *73,* 807–809.

Weigel, J.C. L.A. Johnson, Ed.; *New Technologies for Value-added Products from Protein and Coproducts;* Symposium Proceedings, 80th Annual Meeting and Exposition of the American Oil Chemists' Society, Cincinnati, OH, AOCS Press: Champaign, IL, 1989; pp. 48–51.

Weil, J.K.; F.D. Smith; A.J. Stirton. Alkanesulfonates and hydroxyallsanesulfonates from α-sulfo fatty acids. *J. Org. Chem.* **1962,** *27,* 2950–2952.

Weil, J.K.; A.J. Stirton. Critical micelle concentrations of α-sulfonated fatty acids and their esters. *J. Phys. Chem-US.* **1956,** *60,* 899–901.

Weil, J.K.; A.J. Stirton; M.V. Nunez-Ponzoa. Ether alcohol sulfates. The effect of oxypropylation and oxybutylation on surface active properties. *J. Am. Oil Chem. Soc.* **1966,** *43,* 603–606.

Weissenborn, P.K.; A. Motiejauskaite. Drying of alkyd emulsion paints. *J. Coating. Technol.* **2000a,** *72,* 65–74.

Weissenborn, P.K.; A. Motiejauskaite. Emulsification, drying and film formation of alkyd emulsions. *Prog. Org. Coat.* **2000b,** *40,* 253–266.

Wescott, J.M.; A. Traska; C.R. Frihart; L. Lorenz. Durable soy-based adhesive dispersions. *Wood Adhesives 2005*, [Proceedings]; C.R. Frihart, Ed.; San Diego, CA. Nov. 2–4, 2005.

Whitby, D. Market share of bio-lubricants in Europe and the USA. *Lipid Technol.* **2004,** *16,* 125–129.

Wicks, Z.W., Jr.; F.N. Jones; S.P. Pappas (ed.). *Organic Coatings Science and Technology;* Wiley-Interscience: New York, 1998.

Wong, H.L.; A.M. Rauth; R. Bendayan; J.L. Manias; M. Ramaswamy; Z. Liu; S.Z. Erhan; X.Y. Wu. A new polymer-lipid hybrid nanoparticle system increases cytotoxicity of doxorubicin against multidrug-resistant human breast cancer cells. *Pharm. Res.* **2006,** *23,* 1574–1585.

Wool, R.P.; S.N. Khot; J.J. LaScala; S.P. Bunker; J. Lu; W. Thielemans; E. Can; S.S. Morye; G.I. Williams. Affordable composites and plastics from renewable resources: Part I: Synthesis of monomers and polymers. *ACS Symposium Series.* **2002a,** *823,* 177–204.

Wool, R.P.; S.N. Khot; J.J. LaScala; S.P. Bunker; J. Lu; W. Thielemans; E. Can; S.S. Morye; G.I. Williams. Affordable composites and plastics from renewable resources: Part II: Manufacture of composites. *ACS Symposium Series.* **2002b,** *823,* 205–224.

Woolwine, A.E.; T.E. Reagan. Potential of winter cover crops to increase abundance of Solenopsis invicta (Hymenoptera: Formicidae) and other arthropods in sugarcane. *Environ. Entomol.* **2001,** *30,* 1017–1020.

Wu, W.U.; N.S. Hettiarachchy; M. Qi. Hydrophobicity, solubility and emulsifying of soy protein peptides prepared by papain modification and ultrafiltration. *J. Am. Oil Chem. Soc.* **1998,** *75,* 845–850.

Xu, J.; Z. Liu; S.Z. Erhan; C.J. Carriere. A potential biodegradable rubber—Viscoelastic properties of a soybean oil-based composite. *J. Am. Oil Chem. Soc.* **2002,** *79,* 593–596.

Zerkowski, J.A.; D.K.Y. Solaiman. Synthesis of polyfunctional fatty amines from sophorolipid-derived 17-hydroxy oleic acid. *J. Am. Oil Chem. Soc.* **2006,** *83,* 621–628.

Zhang, J.; J. Long; L. Zhu. Morphology and properties of soy protein and polylactide blends. *Biomacromolecules* **2006,** *7,* 1551–1561.

Zhang, J.; P. Mungara; J.L. Jane. Effect of plasticization and crosslinking on properties of soy protein based plastics. *Polymer* **1998,** *39,* 162–163.

Zhang, M.; C.A. Reitmeier; E.G. Hammond; D.J. Myers. The productions of textile fibers from zein and a soy protein-zein blend. *Cereal Chem.* **1997,** *74,* 594–598.

Zhao, K.; X.F. Hao; D.J. Liu. Soy protein isolate composite adhesive. *Zhengzhou Gongye Daxue Xuebao.* **2000,** *21,* 15–18.

Zhong, Z.; X.S. Sun. Properties of soy protein isolate/polycaprolactone blends compatalized by methylene diphenyl diisocyanate. *Polymer* **2001a,** *42,* 6961–6969.

Zhong, Z.K.; X.S. Sun. Thermal and mechanical properties and water absorption of sodium dodecyl sulfate modified soy protein (11*S*). *J. Appl. Polym. Sci.* **2001b,** *81,* 166–175.

Zhong, Z.K.; X.S. Sun; X. Fang; J.A. Ratto. Adhesion properties of soy protein with fiber cardboard. *J. Am. Oil Chem. Soc.* **2001,** *78,* 37–41.

Zlatanic, A.; C. Lava; W. Zhang; Z.S. Petrovic. Effect of structure on properties of polyols and polyurethanes based on different vegetable oils. *J. Polym. Sci. Pol. Phys.* **2004,** *42,* 809–816.

Zlatanic, A.; Z.S. Petrovic; K. Dusek. Structure and properties of triolein-based polyurethane networks. *Biomacromolecules* **2002,** *3,* 1048–1056.

18

Nutritional Properties and Feeding Values of Soybeans and Their Coproducts

Hans H. Stein, Larry L. Berger, James K. Drackley, George C. Fahey, Jr., David C. Hernot, and Carl M. Parsons
Department of Animal Sciences, University of Illinois, Urbana-Champaign, IL 61801

Introduction

Soybean meal (SBM) is the number-one protein source used in the poultry and livestock industries throughout the world. Of all the SBM that is sold in the United States, >50% is used in diets fed to poultry, and 26% is used in diets fed to swine. Ruminant animals, dogs, cats, and others account for the remaining portion of this usage (Fig. 18.1).

The main reason for the popularity of SBM is the unique composition of amino acids (AAs) that complements the AA compositions of many cereal grains. The excellent AA quality in SBM is also the reason why SBM is now increasingly being used in the pet-food industry. While SBM is by far the most popular soybean product in livestock diets, other products are also being used to a varying degree. These products include full-fat soybeans, soy protein concentrate (SPC), soy protein isolate (SPI) soy-

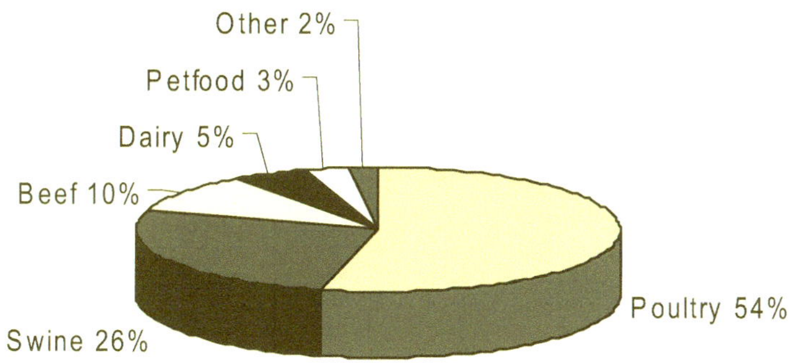

Fig. 18.1. Use of soybean meal in the United States by livestock, poultry, and companion animals.

bean oil, and soybean hulls. Each of these products have unique nutritional properties that make them appropriate for inclusion in diets fed to certain categories of animals. The objective of this chapter is to summarize current knowledge about the nutritional values of soybean products fed to poultry, livestock, and companion animals.

Soybean Products in Diets Fed to Poultry

SBM is an extensively used ingredient in poultry diets and is the largest source of protein in poultry diets in much of the world. Dehulled solvent-extracted meal is the most widely used SBM product because of its large production and higher protein and energy content than lower protein meals that contain hulls. Poultry derive very little, if any, energy from soybean hulls. SBM has advantages over most other oilseed meals with respect to digestible energy and protein/AA (Table 18.1). This is important because providing adequate quantities of energy and protein or amino acid (AA) accounts for >90% of the feed costs in most poultry diets. The concentration of metabolizable energy (ME) in SBM is 11 to 25% greater than that of other commonly used oilseed meals. This difference is largely due to the lower fiber concentration of SBM compared with most other meals. The digestibilities of AAs in SBM are generally greater than in other oilseed meals. This difference is usually greatest for lysine.

Poultry are by far the largest consumers of SBM in the United States. Poultry diets in the United States and much of the world are composed primarily of grain and SBM. Corn and sorghum are the two most common grain sources used in the United

Table 18.1. Metabolizable Energy and Protein Concentration and True Digestibilities of Amino Acids in Soybean Meal and Other Oilseeds Fed to Poultry[a]

Item	Dehulled Soybean Meal	Canola Meal	Dehulled Sunflower Meal	Cottonseed Meal[b]	Peanut Meal[c]
Energy, ME_n, kcal/kg	2,711	2,150	2,495	2,041	2,391
Protein, %	53.9	40.9	48.8	49.1	55.1
Digestibility of AA, %					
Arginine	92	90	93	87	84
Cysteine	82	75	78	73	78
Lysine	91	80	84	67	83
Methionine	92	90	93	73	88
Threonine	88	78	85	71	82
Valine	91	82	86	78	88

[a] Values for metabolizable energy (ME_n) and protein are on a dry matter basis. All values are from NRC (1994).
[b] Prepressed solvent-extracted, 44% protein.
[c] Solvent-extracted.

States. Corn or sorghum and SBM complement one another very well in meeting the protein and AA requirements of poultry. For example, the grains generally contain low concentrations of protein, lysine, and tryptophan, whereas SBM contains high concentrations of these nutrients. For many years, the main limiting factor for SBM use in poultry feeds was its deficiency in the sulfur AAs (methionine and cysteine). However, the commercial availability of inexpensive feed-grade sources of methionine resulted in the routine addition of this AA to grain-SBM diets. Also, for many years people believed that grain-SBM diets were deficient in certain "unidentified growth factors" and that ingredients, such as fish meal, were needed to obtain maximal growth performance. Subsequent research showed that most of the unexplained growth response often obtained from these ingredients, such as fish meal, was due to nutrients such as vitamin B_{12} and selenium. Consequently, the routine supplementation of poultry feeds with these and other nutrients today enables producers to obtain optimal performance using grain–SBM diets.

Soybean Products as Protein Sources for Poultry

Protein Quality of Soybean Products

The protein quality of SBM was reviewed by Baker (2000). The protein quality of SBM is high for poultry, and SBM is a particularly good source of both lysine and tryptophan. SBM is also an especially good source of lysine. When the digestible lysine concentration in SBM is compared to the required amount of lysine for chicks (per unit of protein), the amount of digestible lysine in SBM actually exceeds the requirement (Baker, 2000). No other oilseed comes close to being as good a source of lysine as SBM for poultry. SBM, however, is not a perfect protein source. When compared to the ideal AA contents needed by poultry, the protein in SBM is deficient in methionine plus cysteine, threonine, and valine. Consequently, virtually all poultry diets that contain large amounts of SBM are supplemented with a source of methionine (e.g., DL-methionine or the hydroxy analog of methionine). Soybean protein and SBM–grain combinations contain excesses of some AAs, particularly leucine, but these excesses are generally less than those for other oilseed meals and other oilseed–grain combinations. SBM is also a good source of arginine, which is beneficial for poultry because they cannot synthesize arginine, and thus, have much higher requirements for this AA than mammals.

When examining the protein quality of other soy products, such as SPC (approximately 64% protein) and SPI (approximately 85% protein), compared with SBM, both similarities and differences are found. SPC and SPI are first-limiting in methionine + cysteine and second-limiting in threonine, the same as for SBM. The overall protein quality of SPI, however, is lower than that of SBM (Emmert & Baker, 1995), which is due to the lower concentrations of total and digestible methionine + cysteine and threonine in the protein of the SPI than in the protein of SBM or SPC (Emmert

& Baker, 1995). The latter study also showed that the true digestibilities of AAs in SBM, SPC, and SPI were similar. More recent work by Batal and Parsons (2003), however, indicated that the apparent digestibilities of AAs in SPC and in SPI fed to chicks are greater than in SBM. When chicks were fed dextrose-based diets containing the various soy products, true digestibility of AAs increased with increasing age from 3 or 4 days to 21 days of age, and true digestibility coefficients for AAs were generally greater for SPC and SPI than for SBM.

Soybean Products as Protein Sources in Feeds for Broiler Chickens and Turkeys

This subject was reviewed for broiler chickens by Penz and Brugali (2000). The primary type of SBM used in broiler chicken diets is dehulled, solvent-extracted SBM, which contains ~48% protein. The lower protein SBM with the hulls, containing 44–45% protein, can also be used; however, growth performance, particularly feed efficiency, will be better for chicks fed dehulled SBM (Penz & Brugali, 2000). Full-fat soybeans, either toasted or extruded, are also an excellent protein source for broilers. The inclusion rates of full-fat soybeans may depend on the physical form in which they are fed. When high amounts of full-fat soybeans are fed, the diets may need to be pelleted to improve diet density or breakdown of plant cells to better release nutrients (Waldroup & Cotton, 1974); these latter researchers concluded that most diets should not contain >25% of full-fat soybeans. It is possible that greater concentrations may be used in pelleted diets; however, other studies indicated that full-fat soybeans can replace up to 100% of the SBM in broiler diets (Penz & Brugali, 2000). The principles for using SBM in turkey diets are similar to those for broiler chickens, but SBM is often used at higher concentrations in diets fed to young turkeys due to their higher AA requirement compared with broiler chickens.

Soybean Products as Protein Sources in Feed for Laying Hens

The above discussion for broilers and turkeys also applies to laying hens. Dehulled SBM is generally preferred over SBM with hulls due to its higher protein and metabolizable energy concentration. As reviewed by Penz and Brugali (2000), full-fat soybeans are an excellent ingredient for laying-hen diets if the soybeans are heated properly. Studies with laying hens reported adverse effects of feeding high levels of toasted or extruded soybeans; however, these results may be explained by the underheating of the soybeans. Thus, the effective utilization of full-fat soybeans in laying-hen diets depends greatly on the proper processing of the soybeans.

Assessment of Protein Quality of Soybean Products

Parsons (2000) reviewed this topic. Variation in protein quality among soybean products is due to the protein and AA concentrations of the product and the bioavailability of the AAs in the product. Variation in AA bioavailability among soybean products is primarily due to either insufficient or excessive heat processing. Several antinutritional factors (e.g., protease inhibitors, lectins) must be inactivated, and heating is the primary means of accomplishing this. Several different animal assays can be used to estimate protein quality of soy products. The three most commonly used procedures are protein efficiency ratio (PER) assays, slope–ratio growth assays, and digestibility or balance assays. In the PER assay for poultry, soy products are fed as the only source of protein (~10% protein in the diet) for 10 to 14 days, and PER is calculated by dividing weight gain (g) by protein intake (g). This type of assay was used to evaluate several different soy products (Emmert & Baker, 1995). The PER value of SBM is greater than the PER of SPC and SPI, and the PER values vary among different isolates (Emmert & Baker, 1995). Thus, the PER assay was shown to be sensitive for detecting differences in protein quality among soy products. The PER assay, however, has limited usefulness from a practical standpoint because it provides no direct information on bioavailability or digestibility of specific AAs, and it is not sensitive in detecting the reduction in protein quality or lysine digestibility due to excessive heating.

Slope–ratio growth assays are usually considered the best standard assay for measuring bioavailability of AAs in soy products. These assays, however, have several disadvantages, such as expense and time, and dietary factors other than the limiting AAs can affect growth, which was illustrated for SBM by Baker (1978). Due largely to the disadvantages of the slope–ratio assays, digestibility or balance assays are used more extensively to estimate bioavailability of AAs. The two most common assays for poultry are the precision-fed cecectomized rooster assay (Parsons, 1985) and the ileal digestibility assay using the slaughter method (Angkanaporn et al., 1996). The cecectomized rooster assay is faster and less expensive, but the ileal assay has the advantage that no surgery on the animals is needed. Both of these assays were used to evaluate SBM and other soy products, and results indicate that true digestibility coefficients for AAs in high-quality soy products are usually 90% or greater. The primary factors that cause reduced AA digestibility are insufficient or excessive heating. The effects of insufficient heating are not the same as those for excessive heating. The digestibilities of all AAs are reduced by underheating, whereas only the digestibility of lysine, and to some extent cysteine, is reduced by overheating (Parsons, 2000).

In addition to the in vivo or animal assays, several in vitro assays can be used to estimate protein quality of soy products. Analyzing for crude protein and lysine and then calculating lysine as a percentage of the protein may be a useful indicator of overprocessing or excessive heating. In addition to the digestibility of lysine being reduced by excessive heating, the analyzable lysine level may also be reduced due to total destruction during the formation of advanced Maillard reaction products (Hurrell & Carpenter, 1981). Consequently, overheating may reduce the analyzed lysine to crude protein ratio. For example, high-quality SBM usually has a lysine-to-protein ratio of 6.2 to 6.6. If the ratio is <6.0, then the SBM may be heat-damaged (Parsons, 2000).

The in vitro assay that is used most extensively for SBM in the poultry industry is the urease assay (AOCS, 1973). It is widely used because it is simple and is a reasonably good indirect indicator of the level of active trypsin and chymotrypsin inhibitors and lectins in soy products. The general or optimal desired range in urease pH change for poultry is 0.05–0.20. The exact critical lower and upper limits for the urease index, however, are controversial (Waldroup et al., 1985). For example, commercial SBM often has a urease pH change value of <0.05. This low value only indicates that the SBM perhaps was overheated and does not mean that the SBM or soy product was indeed overheated. In fact, the primary weakness of the urease assay is that, although it is a good indicator of insufficient heating, it is not a good indicator of overheating. Consequently, the KOH protein solubility assay (Araba & Dale, 1990; Parsons et al., 1991) was evaluated and shown to be a reasonably good method for determining overheating of SBM. Araba and Dale (1990) concluded that the critical limit for KOH protein solubility was approximately 70% and that values below this level are indicative of overprocessing. Protein solubility in water (e.g., protein dispersibility index), dye binding, and colorimetric assays were also shown to be useful assays, but these are not used to a great extent commercially.

Soybean Products as Energy Sources for Poultry

Dale (2000) reviewed the topic of soybean products as energy sources for poultry. Although soybean products are primarily considered as protein sources for poultry, they also contribute a large amount of energy to poultry diets. In typical grain–SBM diets, SBM furnishes ~25% of the metabolizable energy (ME) in the diet. The primary weakness of SBM as an energy source is the very poor digestibility of the carbohydrate fraction. As calculated by Dale (2000), SBM contains approximately 10% more gross energy (GE) than corn, but SBM contains only about 72% of the ME of corn because of the poor digestibility of the carbohydrates in SBM. This low digestibility results in a large amount of energy being lost in the excreta, a large amount of dry matter being excreted as manure, and a dilution of energy and other nutrients in the diet.

Carbohydrates make up 32–35% of SBM. The carbohydrate fraction is composed mainly of nonstarch polysaccharides and oligosaccharides such as sucrose, raffinose, and stachyose. The reason for the extremely low digestibility of the soy carbohydrates is unknown and somewhat controversial. The oligosaccharides, raffinose and stachyose, are often cited as the main culprits, but disagreement exists as to the extent of their role. Removing the oligosaccharides by ethanol extraction was reported to greatly increase the ME of SBM (Coon et al., 1990). However, Irish et al. (1995) later reported that removing the oligosaccharides had little or no effect on the ME of SBM. Parsons et al. (2000) reported that removing the oligosaccharides by genetics or plant breeding increased the ME of SBM by 7–9%. Thus, although the raffinose and stachyose in SBM are contributors to its low energy content, the poor digestibility of the nonstarch polysaccharides may be the main reason for the low ME of SBM.

The most effective way to increase the energy values of SBM and other soybean products is to use different processing procedures. One processing modification is to remove less or none of the oil. Properly heated full-fat soybeans (18% oil) have ME values of 3,300–3,350 kcal/kg compared with 2,440 kcal/kg for SBM (Dale, 2006; NRC, 1994). Mechanically processing soybeans by extruding and expelling (screw pressing) to yield a SBM with 6–8% oil also results in more energy than what is present in solvent-extracted SBM (Zhang et al., 1993). Another alternative is to remove the carbohydrates during processing. For example, the MEs of SPC and SPI are considerably greater than in SBM (Batal & Parsons, 2003). Not one of these methods increases the digestibility of carbohydrates in SBM and other soy products. This is an area of great importance and potential in poultry nutrition. It also, however, seems to be a very challenging and difficult area. Plant breeding and the use of exogenous feed enzymes demonstrated limited or no commercial success.

Genetically Modified Soybean Products

During the last 15 to 20 years, considerable interest and activity arose in developing new genetically modified or genetically enhanced crops and feed ingredients that have an increased nutritional value. For soybeans, these genetically modified plants include those with modified input traits for insect protection and herbicide tolerance and those with modified output traits for increased nutritional value. The insect-protected and herbicide-tolerant traits primarily include inserting genes from *Bacillus thuringensis* (*Bt*) and for glyphosate tolerance (Roundup family of herbicides). Although these genetic modifications were very successful commercially, they result in no change in nutritional composition or value of the soybeans (Hammond et al., 1996; Kan & Hartnell, 2004). Soybeans with genetic modifications for output traits that increase the nutritional value of soybeans or SBM for poultry include reduced trypsin inhibitor soybeans and low-lectin soybeans (Batal & Parsons, 2003; Douglas et al., 1999; Han et al., 1991), low-oligosaccharide SBM (Parsons et al., 2000), and high-protein soybeans (Edwards et al., 2000). Soybeans with reduced concentrations of phytate and increased digestibility of phosphorus were also developed. All of these modifications result in a substantial increase in nutritional value for poultry, but not one of these modified soybeans or SBM was successfully commercialized. This lack of success was mainly due to agronomic problems with the nutritionally enhanced crops and also to a market infrastructure that is primarily commodity-based, making identity preservation and marketing of the modified soybeans and/or SBM difficult.

Soybean Products in Diets Fed to Swine

Soybean products are included in most diets fed to swine in the United States and in most countries in the world because soybean protein is recognized as the premier protein source for pigs (Shelton et al., 2001). Intact soybeans may be used in swine feeding, but SBM is by far the most common protein source used for all categories of pigs (Cromwell, 2000). Processed products, such as SPC or SPI, are sometimes used in diets fed to weanling pigs. Newer enzymatically treated or fermented soybean products, such as Hamlet protein and PepSoyGen, were recently introduced to the feed industry and are mainly used in diets fed to weanling pigs (Pahm & Stein, 2007b).

The majority of soybean products included in diets fed to pigs are used to increase the dietary concentrations of AAs. However, soybean oil may also be used in diets fed to weanling, growing, and finishing pigs as an important source of energy (Mahan, 1991; Owen et al., 1996). Soybean hulls that are produced by dehulling soybeans may also be included in diets fed to sows and growing-finishing pigs in quantities of up to 15% without negatively affecting performance (Kornegay, 1981). However, because of the high concentration of fiber in soybean hulls, the digestibilities of energy and most dietary nutrients, including AAs, are reduced if soybean hulls are included in the diets (Dilger et al., 2004; Kornegay, 1981). Therefore, soybean hulls are usually not used in diets fed to swine.

Nutrients and Energy in Soybean Products

Nutrient and Energy Concentrations in Soy Products

Although all soybean products except soybean oil are included in diets fed to swine to increase the concentration of AAs in the diets, soybean products also contribute energy and other nutrients. It is beyond the scope of this contribution to discuss nutrients other than AAs and P along with energy provided by soybean products, but it is recognized that soybean products also supply significant quantities of vitamins and many minerals to swine diets.

The concentrations of energy, AAs, and P in full-fat soybeans as well as in SBM, SPC, and SPI have been published (NRC, 1998; Table 18.2). The concentration of GE in soybean meal is relatively constant across sources of soybean meal collected at different locations (van Kempen et al., 2006). However, the processing procedure used to produce the SBM has an impact on the total amount of energy in the meal. Screw-pressed meals usually contain more energy than solvent-extracted meals because screw-pressed meals have a greater concentration of fat (Woodworth et al., 2001). Likewise, dehulled SBM contains more energy than nondehulled SBM because of the lower concentration of fiber (Woodworth et al., 2001).

Compared with most other protein sources, soy protein has a relatively high concentration of lysine and tryptophan. The concentrations of these two AAs are relatively low in most cereal grains, and particularly so in corn. Therefore, lysine is

Table 18.2. Concentration of Energy, P, and Amino Acids in Soybean Products Fed to Swine [a]

Item	Soybean Meal		Full-fat Soybeans	Soy Protein Concentrate	Soy Protein Isolate
	Nondehulled	Dehulled			
Energy, kcal DE/kg	3,490	3,685	4,140	4,100	4,150
Energy, kcal ME/kg	3,180	3,380	3,690	3,500	3,560
Crude protein, %	43.80	47.50	35.20	64.00	85.80
Phosphorus, %	0.65	0.69	0.59	0.81	0.65
Calcium, %	0.32	0.34	0.25	0.35	0.15
Amino acids, %					
Arginine	3.23	3.48	2.60	5.79	6.87
Histidine	1.17	1.28	0.96	1.80	2.25
Isoleucine	1.99	2.16	1.61	3.30	4.25
Leucine	3.42	3.66	2.75	5.30	6.64
Lysine	2.83	3.02	2.22	4.20	5.26
Methionine	0.61	0.67	0.53	0.90	1.01
Cysteine	0.70	0.74	0.55	1.00	1.19
Phenylalanine	2.18	2.39	1.83	3.40	4.34
Tyrosine	1.69	1.82	1.32	2.50	3.10
Threonine	1.73	1.85	1.41	2.80	3.17
Tryptophan	0.61	0.65	0.48	0.90	1.08
Valine	2.06	2.27	1.68	3.40	4.21

[a] All data are from NRC (1998).

the first limiting AA in grain-based diets fed to pigs, and tryptophan is the second, third, or fourth limiting AA in diets based on corn and fed to monogastric animals (Sharda et al., 1976). The AA profile of soybean protein, however, complements the AA profile of cereal grains because of the relatively high concentrations of lysine and tryptophan.

Variabilities in the nutrient compositions of different sources of soybeans exist. This is true not only of soybeans and SBM obtained from different countries, but also of samples obtained from different locations within the United States (Grieshop et al., 2003; Karr-Lilienthal et al., 2004). However, the mean concentrations of nutrients in 10 soybean samples and in 10 SBM samples obtained from different locations in the United States are in good agreement with the values published by NRC (Grieshop et al., 2003). Likewise, when samples of nondehulled and dehulled SBM from 16 different sources in the United States were analyzed for AA composition, mean values that are close to NRC values were obtained (Cromwell et al., 1999). Therefore, good agreement exists on the average nutrient composition in soybean products, but variability among sources may exist. Within the United States, concentrations of protein and AAs in soybean products are reduced for soybeans grown in the northern part of the country as compared with the central or southern states (Cromwell et al., 1999; Grieshop et al., 2003). In addition, newer varieties of soybeans that were specifically selected for greater concentrations of proteins are now available; SBM obtained from these beans have greater concentrations of protein and AAs than meals produced from conventional soybeans (Pahm & Stein, 2007a).

Amino Acid Digestibility of Soybean Proteins by Pigs

The apparent and standardized ileal digestibilities of AAs in SBM (with or without hulls) and in other soybean products have been measured in numerous experiments, and results were summarized (NRC, 1998; Table 18.3). The digestibilities of AAs in SBM collected from different geographical locations in the United States are relatively constant (van Kempen et al., 2002). All sources of soy protein need to be heated prior to feeding to inactivate antinutritional factors, mainly protease inhibitors and lectins, present in raw soybeans (Qin et al., 1996). However, the form of heat applied to soy protein may influence the digestibilities of AAs in the product (Opapeju et al., 2006; Woodworth et al., 2001), and several other factors were shown to influence the digestibilities of AAs in soy protein. In general, the more processed the soy protein is, the greater are the digestibilities of AAs. Therefore, AAs in SPI and SPC are usually more digestible than AAs in SBM (Pahm & Stein, 2007b), and AAs in dehulled SBM have a greater digestibilities than AAs in nondehulled SBM (NRC, 1998). The latter observation is consistent with reports showing that soybean hulls reduce the digestibilities of AA in SBM (Dilger et al., 2004). Also, likely the reason for the increased

Table 18.3. Standardized Ileal Digestibilities of Amino Acids (%) in Soybean Products Fed to Swine [a]

Amino Acid	Soybean Meal		Full-fat Soybeans	Soy Protein Concentrate	Soy Protein Isolate
	Nondehulled	Dehulled			
Arginine	93	94	93	99	99
Histidine	90	91	88	97	91
Isoleucine	88	89	84	95	90
Leucine	88	89	86	95	89
Lysine	89-	90	86	95	91
Methionine	91	91	85	94	92
Cysteine	84	87	80	94	82
Phenylalanine	88	89	88	97	92
Tyrosine	90	90	87	96	91
Threonine	85	87	83	94	85
Tryptophan	87	90	82	93	88
Valine	86	88	83	94	89

[a] Data for soy protein isolate measured in weanling pigs (Pahm & Stein, 2007a). All other data measured in growing-finishing pigs (NRC, 1998).

digestibilities of AAs in SPI compared with SBM is that fiber and oligosaccharides are removed from the defatted meals, which may impact the digestibilities of AAs (Smirickey et al., 2002).

The digestibilities of AAs in full-fat soybeans are greater than in SBM (Pahm & Stein, 2007a). The reason for this observation is most likely that the addition of oil to SBM or SPC increases the digestibilities of AAs (Albin et al., 2001; Li and Sauer, 1994). In fact, Pahm and Stein (2007a) demonstrated that adding soybean oil to SBM increased the digestibilities of AA in SBM to levels that were not different from the digestibilities obtained in full-fat soybeans.

The particle size of SBM also influences the digestibilities of AAs, and the digestibilities are improved in SBM having a particle size of 600 microns compared with SBM with a particle size of 900 microns (Fastinger & Mahan, 2003). This observation concurs with the fact that the performance of pigs fed diets based on corn and SBM improves if the particle size is reduced (Lawrence et al., 2003). Microbial phytase does not influence ileal digestibility of AAs in SBM (Traylor et al., 2001), but measured values for the standardized ileal digestibility of AAs are reduced as feed intake is increased (Motor & Stein, 2004). This observation mainly has implications for pigs fed experimental diets used to measure AA digestibility of soybean protein because under commercial conditions, most pigs are allowed ad libitum access to feed.

Phosphorus Digestibility of Soybean Products by Pigs

Historically, values for relative availability of phosphorus rather than digestibility of phosphorus were measured (Cromwell et al., 1993), and relative availability values of 31 and 23% for nondehulled and dehulled SBM, respectively, were reported (NRC, 1998). However, values for the relative availability of phosphorus in other soybean products are not available.

The apparent total tract digestibility of phosphorus in dehulled SBM was reported to be 38% (Bohlke et al., 2005). Apparent total tract digestibility of phosphorus in nondehulled SBM was reported at 48.1 and 34.9% for diets containing approximately 41 and 55% SBM, respectively (Ajakaiye et al., 2003). However, Rodehutscord et al. (1996) measured a value of only 31% for apparent total tract digestibility of phosphorus in nondehulled SBM. Based on these results, the values for relative availability of phosphorus in SBM that are published by NRC (1998) probably are too low, because the apparent total tract digestibility of phosphorus in SBM seems to be 30–40% for both dehulled and nondehulled SBM.

The digestibility of phosphorus in SBM is improved by more than 100% if microbial phytase is added to the diet (Cromwell et al., 1993; Rodehutscord et al., 1996; Traylor et al., 2001). Dietary microbial phytase, therefore, is very effective in improving the digestibility of phosphorus in SBM.

Energy Digestibility of Soybean Products by Pigs

The digestibility of energy in soybean products has not been measured in many experiments, but values for digestible energy (DE) and ME in dehulled SBM of 3,685 and 3,380 kcal/kg, respectively, are published by NRC (1998). These values are in good agreement with recently measured values of 3,660 and 3,410 kcal/kg of DE and ME, respectively (Woodworth et al., 2001), but greater than an average DE value of 3,383 kcal of DE per kg calculated from van Kempen et al. (2006). SBM from soybeans that were not dehulled contains less digestible energy than dehulled SBM (3,490 and 3,180 kcal/kg of DE and ME, respectively), whereas SPC and SPI contain slightly more DE and ME than dehulled SBM (NRC, 1998). However, full-fat soybeans contain more energy (4,140 and 3,690 kcal of DE and ME, respectively) than any of the defatted soybean products. This observation is in agreement with the fact that soybean oil has a high concentration of energy (8,750 and 8,400 kcal of DE and ME, respectively).

Utilization of Soybean Products in Diets fed to Swine

Soybean Meal in Swine Diets

SBM is one of the best protein sources available for swine diets (Shelton et al., 2001), and both dehulled and nondehulled SBM are excellent sources of AAs for swine. However, new varieties of soybeans are constantly being developed. These new varieties have specific nutritional characteristics that influence the quality of the SBM being produced from the beans. Examples of such new varieties are soybeans with higher protein concentration or lower concentrations of oligosaccharides, but only limited information exists about the nutritional values of these varieties as compared with conventional varieties (Pahm & Stein, 2007a). Most of the soybeans that are grown were developed using genetically modified seeds that have specific agronomic traits. However, the nutritional composition and the feeding value of SBM produced from genetically modified beans are not different from the nutritional value of conventional soybeans (Cromwell et al., 2002).

In diets fed to growing–finishing and reproducing swine, SBM may provide all the AAs needed by the animals. However, newly weaned pigs do not tolerate soy protein as well as older pigs (Sohn et al., 1994), and they may develop allergenic reactions followed by immunological responses if they are fed large quantities of SBM (Li et al., 1990; 1991). Therefore the concentration of SBM in diets fed to pigs immediately after weaning should be limited and other protein sources need to be included in these diets. The inclusion rates of SBM can gradually be increased as the pigs grow older, and when they reach a weight of 20–25 kg, SBM can be used as the only protein source in the diet.

Soy Protein Concentrates and Soy Protein Isolates in Swine Diets

SPC and SPI may be used in diets fed to weanling pigs instead of SBM may be used because the ingredients are thought not to elicit antigenic responses in the pigs (Sohn et al., 1994). However, differences may exist among sources of SPCs, and extrusion of SPC may improve the nutritional value (Li et al., 1991). However, the cost of SPI is usually at a level that is prohibitive for use in diets fed to swine.

Soybean Oil in Swine Diets

Soybean oil is recognized as an excellent energy source in diets fed to all categories of swine. In addition, dietary soybean oil may reduce the dustiness of diets fed in meal form and the pelletability of pelleted diets. Addition of fat to diets fed to weanling pigs during the initial two weeks post-weaning usually does not increase performance. However, from approximately day-15 post-weaning and during the remaining nursery period, average daily gain may be improved if soybean oil is added to the diet (Howard et al., 1990; Owen et al., 1996) although that is not always the case (Hoffman et al., 1993; Tokach et al., 1995). Diets containing soybean oil usually have greater energy concentrations than diets containing no soybean oil, and feed utilization is, therefore, often improved if measured on a kg-per-kg basis (Owen et al., 1996). However, if measured on the basis of calories used per unit of gain, adding soybean oil to the diet has no effect (Hoffman et al., 1993).

In diets fed to growing pigs, fat addition often improves daily gain, but that is not always the case for finishing pigs (de la Llata et al., 2001; Overland et al., 1999). Feed utilization is usually not improved if measured on a caloric basis.

Dietary fat in diets fed to lactating sows increases milk fat yield and results in heavier pigs being weaned (Tilton et al., 1999; van den Brand et al., 2000). Soybean oil has been shown to be effective in promoting these improvements (Yen et al., 1991), but to our knowledge, no studies were conducted comparing the effects of soybean oil to the effects of feeding from other fat sources.

Full-fat Soybeans in Swine Diets

Full-fat soybeans may be used in diets fed to pigs provided that they are heat-treated prior to feeding. Because of the relatively high oil content in full-fat soybeans, the energy concentration of diets usually is improved if full-fat soybeans are included. The digestibilities of AAs in full-fat soybeans are greater than in SBM (Pahm & Stein, 2007a), and the concentrations of DE and ME in full-fat soybeans are also greater than in SBM (Woodworth et al., 2001).

Full-fat soybeans are often included in diets fed to nursery pigs and finishing pigs usually do not contain full-fat soybeans because the oil in full-fat soybeans may reduce the quality of the belly of the pigs. Diets containing full-fat soybeans may completely replace SBM in diets fed to growing–finishing pigs without any negative impact on pig performance (Leszczynski et al., 1992). If pigs are offered a diet containing no full-fat soybeans during the final three weeks prior to slaughter, belly quality is not impaired by feeding full-fat soybeans (Leszczynski et al., 1992). Full-fat soybeans may also be included in diets fed to sows and can potentially replace all SBM in gestating as well as lactating diets.

New Soybean Protein Sources in Diets Fed to Pigs

Enzymatic preparation of SBM was shown to remove the antigens in SBM that cause allergenic reactions in weanling pigs. It is, therefore, possible to produce enzyme-treated SBM that is tolerated by weanling pigs. A product produced using this technology is HP 300 (Hamlet Protein, Horsens, Denmark). This product has greater concentrations of protein and AAs than regular SBM because carbohydrates, antigens, and oligosaccharides were removed during the enzymatic treatment (Zhu et al., 1998). The digestibilities of AAs in HP 300 are greater than the digestibilities of AAs in conventional SBM, and most AAs have standardized ileal digestibility values that are similar to SPI and fish meal (Pahm & Stein, 2007b). Inclusion of HP 300 in diets fed to weanling pigs will, therefore, result in pig performance that is similar to that obtained in diets based on animal proteins.

Fermentation of SBM using *Apergillus oryzae* or *Bacillus subtillis* may also remove antigens, antinutritional factors, and oligosaccharides from soybeans or SBM (Hong et al., 2004; Yang et al., 2007). The fermentation partly hydrolyzes the soy proteins, which results in reduced peptide size in fermented SBM (Hong et al., 2004). The fermented SBM contains ~10% more protein than conventional SBM, but the AA sequence is similar to conventional SBM (Hong et al., 2004). The digestibilities of AAs in fermented SBM are similar to conventional SBM (Pahm & Stein, 2007b), but the inclusion of fermented SBM in diets fed to weanling pigs at the expense of conventional SBM was shown to improve pig performance (Feng et al., 2007). Therefore, possibly, fermented SBM may become an attractive ingredient in weanling pig diets in the future.

Conclusion on Soy Products in Diets Fed to Swine

Dehulled solvent-extracted SBM is a highly popular feed ingredient in diets fed to all categories of swine except for newly weaned pigs. The main reason for including SBM in diets fed to pigs is to provide AAs that are required by the animals. The nutrient composition of SBM is relatively constant although the protein and AA concentrations tend to be lower if the soybeans are grown in the Northern regions of the United States rather than in the Central or Southern regions. The digestibilities in pigs of the AAs in SBM have been measured in numerous experiments, and the results showed that the digestibilities are relatively high and constant in various sources of SBM. In contrast, only few reports exist on the digestibility of phosphorus in SBM, and the results are somewhat conflicting. However, the digestibility of phosphorus in SBM is relatively low because most of the phosphorus is bound in the phytate molecule, but it is possible to increase the digestibility of phosphorus in SBM by more than 100% by adding microbial phytase to the diet. Also, there is a need for more information on the concentration of digestible energy in SBM.

Other sources of soybean protein, such as nondehulled SBM and full-fat soybeans, are also valuable protein sources that may replace dehulled SBM. Excellent performance was demonstrated in pigs fed diets containing SPC because of the low concentration of antigens and the high AA digestibility in these products. However, due to the higher costs of these products compared with dehulled SBM, they are not used in diets fed to growing–finishing pigs or sows. Fermented SBM and enzyme-treated SBM recently became available to the industry. These products may potentially replace SPC in diets fed to weanling pigs.

Soybean oil is an excellent source of energy, and diets fed to weanling pigs after day-15 post-weaning and to growing–finishing pigs may be fortified with soybean oil. This usually results in an improvement in average daily gain and no change in the caloric utilization of the feed. If included in diets fed to sows, soybean oil will result in increased weaning weights of the pigs and reduced weight loss of the sows.

Soybean Products In Diets Fed To Companion Animals

A key component of diets fed to companion animals is the protein source that is used in the diets. As an economical source of protein, SBM is commonly included in companion animals diets. However, in 2006, the 1.0 million metric tons (1.1 million tons) used in commercial pet foods represented only 3% of total SBM used by animals in the United States (American Soybean Association, 2007; Fig. 18.1).

For companion animals, SBM is a readily available source of high-quality protein with a balanced AA profile (Grieshop et al., 2003; van Kempen et al., 2002). Moreover, soy protein ingredients have functional properties that make them desirable for use in manufacturing such as absorption, elasticity, and water- and fat-binding properties (Hill, 2003). Unfortunately, SBM also possesses undesirable characteristics such as a low concentration of methionine, high concentrations of trypsin inhibitors, and the presence of flatulence-causing oligosaccharides (Grieshop & Fahey, 2000). Dietary SBM may also cause poor stool consistency (Grieshop & Fahey, 2000). The cat was reported to have greater fecal losses of taurocholate when fed high soybean protein diets (Kim et al., 1995). Table 18.4 summarizes advantages and disadvantages of using soybean products in pet foods.

Soybean Products Used in Pet Foods

Examples of soybean products used in pet foods include SBM, soy flour, SPCs, SPI,

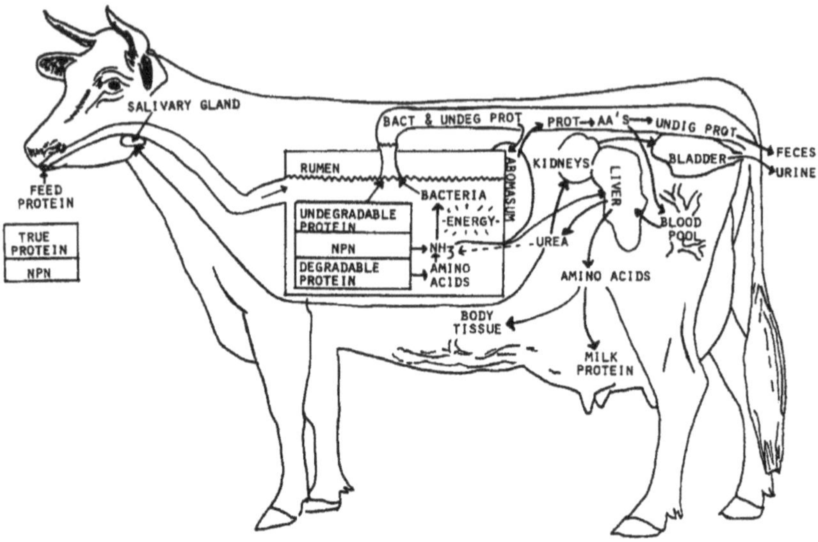

Fig. 18.2. Schematic of protein metabolism in the lactating cow.

and textured vegetable protein. Solvent-extracted SBM contains approximately 44.5% crude protein, 1.4% fat, and 7% crude fiber (NRC, 2006). Soy flour is finely ground SBM and can be used directly or subjected to further processing. Textured vegetable protein is produced by extrusion of defatted soy flour and contains approximately 51% crude protein, 1% fat, and 31% carbohydrates (Hill et al., 2001). Textured vegetable protein from soybeans is commonly included in canned pet foods because it retains the appearance of meat during the canning process. SPIs and concentrates are "purified SBM" made by separating the soy carbohydrates from the proteins (described in detail in the Chapter: *Soy Protein Products, Processing, and Utilization*). Soy protein fractions that contain a minimum of 65% protein are termed *soy protein concentrates*. Table 18.5 summarizes the chemical composition of selected soybean products used in pet foods.

Nutritional Characteristics of Soy Products Used in Pet Food

The soybean is an excellent source of protein and carbohydrates. It is rich in polyunsaturated fatty acids, very low in saturated fats, and contains no cholesterol (Zhang & Laflamme, 1999). However, environmental conditions under which soybeans are grown greatly impact chemical composition and nutrient quality(Grieshop et al., 2003; Grieshop & Fahey, 2001). Because variations in processing conditions can mask both the positive and negative effects of environment, it is critical know the chemical

Table 18.4. Advantages and Disadvantages of Using Soybean Products in Companion Animal Diets [a]

Advantages	Disadvantages
Economical source of protein Readily available and consistent quality	Contains antinutritional factors such as trypsin inhibitors and oligosaccharides
Balanced AA profile complements that of other ingredients, such as corn	Low methionine content; possible increased taurine requirement in cats
Textured vegetable protein retains appearance of meat during canning	Low digestibility Flatulence
Improved product texture Source of protein for vegetarian diets	Poor fecal quality (wet and loose feces) Source of excessive soluble fiber
Source of dietary fiber	Reduces trace mineral availability Allergic reaction to protein Negative connotation for pet owners

[a] From Grieshop and Fahey, 2000.

Table 18.5. Chemical Composition of Soy Protein Sources Used in Pet Foods.

Item	SBM[a]	Soybean, Full-fat[b]	Soybean, Defatted[a]	SPC1[a]	SPC2[a]	SPC3[a]
DM[c], %	87.4	96.2	92.7	94.9	94.3	94.5
			% of DM[c]			
OM[c]	92.6	–	93.0	93.9	93.0	95.9
CP[c]	56.6	39.6	55.3	72.2	70.4	70.5
Fat[c]	2.5	22.8	2.8	1.1	0.8	3.2
TDF[c]	15.7	–	16.2	21.3	17.5	21.1

[a] Data from Clapper et al. (2001). SBM = soybean meal; SPC1 = soy protein concentrate (traditional aqueous alcohol-extracted SPC); SPC2 = texturized soy protein (extruded SPC); SPC3 = modified molecular weight SPC (low antigen product).

[b] From NRC, 2006. Unit for CP and fat is percentage of DM.

[c] DM = dry matter; OM = organic matter; CP = crude protein; TDF = total dietary fiber.

characteristics of soybeans from different origins when diets are formulated.

Soybean Products as Protein Sources in Pet Diets

Soy protein is typically added to pet foods as a complementary protein in primarily grain-based diets (Hill, 2003). Although SBM contains high-quality protein (Dust et al., 2005), the AA profile of soy protein is not complete. Methionine and cysteine are the limiting AA in SBM in meeting the nutritional requirements of the dog and, especially, the cat (Table 18.6). Soy protein must be combined with a complementary protein to provide all the indispensable AAs that are needed by the animals. Wiernusz et al. (1995) compared five isonitrogenous (11% crude protein) canned dog foods containing 13.5% soybean grits, 13.3% soybean flour, 10.7% SPC, 8.1% SPI, or 8.9% wheat gluten. Diets containing SPC, SPI, and wheat gluten resulted in improved apparent digestibilities of dry matter (84.6, 86.4, and 88.4%, respectively), crude protein (89.8, 89.7, and 93.8%, respectively), and energy compared with diets containing soy grits or soy flour (81.1 and 82.6% for dry matter, and 86.7 and 87% for crude protein, respectively). Fecal output was reduced for all diets except the one containing soy grits, and stool quality was improved by the SPI and wheat gluten diets. These data suggest that the adverse effects reported with soybean product usage may be reduced and the nutritional value of soy improved by further processing the soybean products. Hullar et al. (1998) obtained similar results in castrated adult cats fed a diet containing 20% full-fat SBM, 40% meat meal, and 39% maize in the form of a raw mixture or an extruded pellet. Feeding the raw mixture decreased digestibilities of dry matter, organic matter, crude protein, ether extract, and N-free extract when compared with the extruded diet. However, in a more recent study in dogs, Clapper et al. (2001) showed that neither ileal nor total tract nutrient digestibility was affected by soy protein processing. The authors noted that total tract digestibilities of dry matter, organic matter, and GE were not different among treatments. Yamka et al. (2005) evaluated ileal and total tract nutrient digestibilities of isonitrogenous dry dog foods containing low-oligosaccharide, low-phytate SBM, conventional SBM; low-oligosaccharide, low-phytate whole soybeans; or conventional whole soybeans. They reported an intestinal dry matter digestibility ranging from 80.9 to 74%. Total tract dry matter digestibility was greater ($P = 0.02$) for low-oligosaccharide, low-phytate SBM (87%) than for conventional SBM (84.8%), but no difference existed between the two full-fat soybean diets (average = 83.3%).

Several studies evaluated soy products as pet-food ingredients compared with products of animal origin. Murray et al. (1997) conducted a study comparing the ileal and total tract nutrient digestibility of five isonitrogenous dry dog foods containing defatted soy flour, rendered beef meat and bone meal, fresh beef, poultry by-product meal, and fresh poultry. No differences were observed among treatments in ileal digestibilities of dry matter, organic matter, crude protein, fat, or GE. Bednar et al. (2000) confirmed these results in a study comparing SBM to beef meat and bone meal, poultry meal, and poultry by-product meal. Clapper et al. (2001) compared SBM, soy flour, three SPCs, and poultry meal, and found similar ileal digestibilities of dry matter, organic matter, fat, and GE, but greater ileal crude-protein digestibilities when soy protein-containing diets or SPC-containing diets were compared to the poultry meal diet. No differences in apparent total tract digestibility of dry matter were observed by Murray et al. (1997); however, they found increased organic matter and crude protein digestibilities with diets containing animal by-products (averages = 91.7 and 89.3%, respectively) when compared with the diet containing defatted

Table 18.6. Comparison of Protein and Amino Acid Profile of Soybean Meal and AAFCO Standard for Maintenance of Adult Dogs and Cats

Nutrient[b]	Soybean Meal[c]	AAFCO Food Nutrient Profile[a]	
		Dog	Cat
Crude protein	49.2	18	26
Arginine	7.37	2.83	4.00
Histidine	2.67	1.00	1.19
Isoleucine	4.54	2.06	2.00
Leucine	7.81	3.28	4.81
Lysine	6.46	3.50	3.19
Methionine + cysteine	3.08	2.39	4.23
Methionine	1.39	Not specified	2.38
Phenylalanine	4.98	Not specified	1.62
Threonine	3.95	2.67	2.81
Tryptophan	1.39	0.89	0.62
Valine	4.70	2.17	2.38
Taurine	0.00	Not specified	0.38[d]

[a] From AAFCO, 2007. Minimal requirement for maintenance for adult animals.
[b] Protein is in percentage of dry matter. Unit for AA is percentage of protein.
[c] From NRC, 1998.

soy flour (90.2 and 88.3%, respectively). Bednar et al. (2000) confirmed higher total tract organic matter digestibilities with diets containing animal by-products (average = 87.4%) compared with the SBM-containing diet (82.7%), but they found similar total tract digestibilities of crude protein with diets containing SBM, poultry by-product meal, and beef meat and bone meal (82.7, 81.6, and 82.4%, respectively). Clapper et al. (2001) found no differences in total tract digestibilities of dry matter, organic matter, fat, or GE between soy protein-containing diets and the poultry meal diet, but they observed greater total tract crude protein digestibilities when soy protein-containing diets or the SPC diets were compared to the poultry meal diet. In canned dog foods, the use of texturized vegetable protein from soy had a negative effect on nutrient digestibilities. Hill et al. (2001) showed that both ileal and total tract dry matter and crude protein digestibilities decreased linearly with increasing texturized vegetable protein concentration in canned diets. Moreover, texturized vegetable protein led to increased fecal output and fecal moisture (Hill et al., 2001; 2006) in the same way as fecal excretion was greater with SBM dry diets versus animal by-product diets (Bednar et al., 2000; Clapper et al., 2001). Nevertheless, soy flour and SPCs obtained through different processes seemed to improve fecal consistency and reduce fecal output (Clapper et al., 2001).

Several soy products may be used as protein sources in dog foods. In dry extruded diets, both ileal and total tract crude protein digestibilities of soy-containing diets appear to be equal or superior to diets containing animal protein by-products. On the other hand, in canned foods, texturized vegetable protein can reduce dry matter and crude protein digestibilities by dogs, and soy protein sources usually increase fecal output. A lack of information is apparent on the effects of including soy protein sources in diets fed to cats.

Soybean Products as Fiber Sources in Pet Diets

Although soy protein products are commonly added to companion animal diets as a source of protein, some soy products also can be used as sources of fiber or energy (Grieshop & Fahey, 2000). On a dry matter basis, soybean hulls may contain nearly 83% total dietary fiber when unprocessed and up to 85% total dietary fiber after extreme extrusion (Dust et al., 2004). Soybean carbohydrates make up approximately 35% of soybean seed and 40% of SBM dry matter (Karr-Lilienthal et al., 2005). Sunvold et al. (1995) conducted an in vitro fermentation of several fibrous substrates including soybean hulls. Soybean hull fermentation by dog fecal microflora resulted in relatively low organic matter disappearance (<20%) and low total short-chain fatty acid production (<15 mmol/g of substrate organic matter). Nevertheless, in this study, soy hull fermentation data were similar to those obtained with beet pulp fermentation, which is a commonly used fiber source in companion animal diets (Table 18.7). Cole et al. (1999) reported that inclusion of 3.0, 4.5, 6.0, 7.5, or 9.0% soybean hulls in diets fed to dogs had negative effects on apparent digestibilities of dry matter, organic matter, total dietary fiber, and GE, although crude protein and fat digestibilities were unaffected. Cole et al. (1999) also compared diets containing 0, 6.0, 7.5, or 9.0% soybean hulls with a diet containing 7.5% beet pulp. No differences appeared in nutrient digestibility among diets containing soybean hulls or beet pulp. These results are supported by those of Harmon et al. (1999), who also observed a decrease in dry matter, energy, crude protein, and neutral detergent fiber apparent digestibilities with soy fiber- and beet pulp-containing diets compared with diets containing corn or cellulose. Total fecal output was not affected by increased soy hull supplementation, but fecal quality was improved with increasing soy hull concentrations in the diet (Cole et al., 1999).

In dog foods, soy hulls seem to provide a relatively large fraction of readily fermentable fibers, and have a fermentation profile comparable to beet pulp. Moreover, soy hulls have a positive effect on fecal characteristics (consistency and output).

Table 18.7. In Vitro Soy Hull Fermentability by Dog Fecal Microflora[a]

Item	Organic Matter Disappearance, %			Total SCFA Production, mmol/g Substrate OM[b]		
	6 h	12 h	24 h	6 h	12 h	24 h
Beet pulp	17.2	17.7	24.5	0.66	0.71	1.96
Soy hulls	14.7	16.0	16.2	0.60	1.02	1.40
SEM	3.1			0.25		

[a] Data from Sunvold et al. (1995).
[b] OM = organic matter; SCFA = short-chain fatty acids.

Antinutritional Factors in Soy Products

Known factors with adverse effects on nutrient digestibility in whole soybeans are trypsin inhibitors and phytate (Zhang & Laflamme, 1999). Phytate has a high affinity for di- and trivalent metals, such as calcium, magnesium, iron, phosphorus, and zinc. The presence of phytate in the diet can decrease nutrient and mineral availability due to the lack of phytase in the gastrointestinal tract of dogs and other monogastric species (Schoenherr et al., 2000; Traylor et al., 2001). Phytate cannot be inactivated by heating, so mineral availability from soy-based diets must be addressed by other means (Zhang & Laflamme, 1999). Yamka et al. (2005) conducted a study evaluating the effect of low-oligosaccharide, low-phytate whole soybeans and SBM on nutrient digestibilities. The authors found no differences in ileal or total tract digestibilities of dry matter, nitrogen, and dispensable AA between treatments. Only histidine and tryptophan digestibilities were lower in low-oligosaccharide, low-phytate whole soybeans compared with conventional whole soybean-containing diets.

Raw soybeans contain numerous trypsin inhibitors. These compounds block the action of trypsin and other enzymes, such as chymotrypsin, elastase, and other serine proteases, which decreases protein digestibility and bioavailability (Liener, 1994). Trypsin inhibitors can be inactivated by moist heat (Osborne & Mendel, 1917), such as extrusion (Alomso et al., 1998). Romarheim et al. (2005) showed that extrusion sufficiently eliminated trypsin inhibitors in SBM-based diets fed to mink (2.7–0.2 mg of trypsin inhibitors/g diet, and 8.3–3.1 mg of trypsin inhibitors/g diet for defatted SBM diet). With similar heating conditions (>116°C) in both extrusion and canning processes, trypsin inhibitors should not be a problem in pet foods.

Physiological and Gastrointestinal Effects of Soy Products

The biggest limiting factor for increasing soybean usage in pet foods is the presence of flatulence-causing oligosaccharides. SBM contains 6–8% sucrose, 3–5% stachyose, and 1–2% raffinose. The colonic fermentation of these nondigestible oligosaccharides can lead to gas production in the gastrointestinal tract of dogs. Genetic manipulation of soybeans resulted in the creation of varieties that contain negligible quantities of raffinose and stachyose. Some studies investigated the digestive response of dogs to diets containing conventional and low-oligosaccharides SBM (Yamka et al., 2005; 2006; Zuo et al., 1996). Yamka et al. (2006) evaluated the effect of conventional SBM, low-oligosaccharide, low-phytate SBM, and poultry by-product meal containing up to 22.4 g/kg stachyose on nutrient digestibility and flatulence (H_2S production). The poultry by-product-based diet had higher dry matter digestibility and digestible energy concentration compared with soy-based diets. No differences were detected for any treatment regardless of protein source or addition of supplemental enzyme for any flatulence component analyses. The authors concluded that diets containing <2.4 g/kg of stachyose and <2 g/kg of raffinose did not alter digestibility or increase flatulence in dogs. Zentek (1995) showed that the in vitro fermentation of SBM and SPC decreased concentrations of H_2S compared with a meat meal-based diet (2.69, 2.57, and 3.28 volume percentage, respectively). These results show that processed or modified soybean products (SPC or low-oligosaccharide SBM) have a positive effect on gas production.

Hill et al. (2006) evaluated the effect of increased soy total dietary textured vegetable protein:beef protein ratio in canned high-fat diets on glucose and insulin responses in dogs. Adding total dietary textured vegetable protein reduced the insulin response during the first 2 h after a meal by 63%. According to the authors, this effect was due to the soy carbohydrates.

Feeding SBM-containing diets is associated with morphological or physiological changes in the gastrointestinal tract of dogs. Dogs fed a soy protein diet (23.5% crude protein) had greater net colonic fluxes of acetate, propionate, butyrate, and total short-chain fatty acids than did dogs fed a meat diet (Hallman et al., 1994). In contrast, dogs fed a casein-based diet had less damage in the colon than dogs fed diets containing SPC (Hallman et al., 1994). In this study, dogs fed the SPC diet had colons that contained more protein and fat than dogs fed diets containing freeze-dried beef or casein, although the colons of dogs fed the casein diet had a greater crypt depth. The effect of several dietary treatments including soy product-based diets on precancerous colon lesions (assessed as foci with aberrant crypts) in rats was evaluated. Among the soy treatments, defatted soy flour and full-fat soy flakes reduced the early stages of colon cancer by inhibiting the formation of foci with aberrant crypts (Thiagarajan et al., 1998).

Conclusion on Usage of Soy Products in Pet-food Diets

The volume of soybean products used in companion animal diets is significant, and the potential for higher rates of inclusion is great. Soy protein, when combined with other protein sources that contain complementary AA, can provide an economical source of highly available and high- quality protein to companion animals. In the past 10 years, chemical composition of relevant soy products was evaluated. These data indicated a significant variation in chemical and nutritional characteristics among SBM from different U.S. sources or from different countries. This revealed that the nutritive value of SBM was determined not only by the quantity and availability of AAs but also by origin and the processing conditions used in the preparation of the products. Fecal bulk and flatulence are the greatest concerns in promoting the benefits of soy product inclusion to pet owners. Research showed that, at the terminal ileum, soy protein fractions are equal to or superior to animal protein by-products in terms of dry matter and nutrient digestibility. When soybeans are processed beyond the meal and flour forms into SPC and SPI, nutrient digestibility may be increased and the problem with fecal bulk reduced, resulting in fecal characteristics comparable to those of dogs consuming animal protein by-products.

The knowledge about the usage of soybean products in companion animal diets is quite limited compared with the knowledge about using soybean products in diets for poultry and livestock. A limited database exists regarding the effects of soy products on growth performance, gastrointestinal tract characteristics, and physiological events in the dog and cat. Much is known about the protein and lipid components of soy products, but little published information is available on other major components such as carbohydrates. Some future research activity should center on genetically modified soybeans and the resultant products from them. Elimination or reduction of antinutritional factors in soy products (such as antitrypsin and flatulence- causing oligosaccharides) may improve the connotation of soy use and potentially increase demand for soy products by pet-food manufacturers.

Soybean Products in Diets Fed to Beef Cattle

Soybean products are excellent sources of protein and energy for beef cattle. Approximately 7.0% of the SBM utilized in the United States is fed to beef cattle. This is a much smaller portion than the quantities utilized by swine and poultry (Fig. 18.1). The purpose of this section is to describe the reason for this relationship and how recent research may increase the use of SBM in beef cattle diets in the future. With over 40 million growing and finishing beef cattle produced in the United States each year, this is a huge potential market for soybean products. To understand why only 7% of the SBM consumed by livestock and poultry is fed to beef cattle, one has to begin with a discussion of the digestive physiology of ruminants.

Protein Digestion in Ruminant Animals

Ruminants (cattle, sheep, and goats) are unique as food-producing animals because they possess a large fermentation vat called a *rumen*, which is inhabited by billions of bacteria and protozoa. The bacteria in the rumen produce an enzyme complex called *cellulase*, which allows ruminants to get most of their energy from forages and fibrous feedstuffs. As the bacteria digest fiber, they grow and divide, producing bacterial protein. This protein then passes out of the rumen and enters the abomasum (true stomach) and small intestine where it is digested and absorbed. Because rumen microbes produce urease and can combine free ammonia and carbon skeletons to form bacterial AAs, nonprotein nitrogen (NPN) like urea can be converted to animal protein. Consequently, the two major nutrients required for growth and milk production, energy and protein, can come from sources for which humans and nonruminant animals do not compete.

However, most high-producing ruminants require supplemental true protein in their diet for optimum production. Once consumed, protein can be metabolized in one of two ways in the rumen (Fig. 18.2). The undegradable (escape or bypass) protein escapes microbial degradation, passes out of the rumen, and enters the abomasum and small intestine, where the digestible portion is absorbed as AAs. Alternatively, true proteins may be degraded by bacteria to AAs that are deaminated to ammonia and short-chained carbon skeletons or incorporated intact into microbial protein. As is discussed later, SBM is a valuable source of degradable protein that allows bacteria to grow efficiently.

Ruminants are frequently fed diets that contain NPN because the bacteria can convert the ammonia and carbon skeletons produced in the rumen to indispensable and dispensable AAs. NPN sources usually cost 10–20% of the price of SBM on an equal crude protein basis. Consequently, economics encourages the maximal use of NPN that can be converted to bacterial protein. However, the ability of the bacteria to convert NPN to bacterial AAs is dependent on the amount of energy fermented in the rumen. In general, NPN is utilized more efficiently in high-grain diets than in high-forage diets. For the past 25 years, the approach to balancing feedlot diets was to maximize the utilization of urea or other NPN sources and to optimize the utilization of bypass or less-degradable protein sources.

The chemical nature of crude protein in feedstuffs is the primary factor determining how rapidly it is degraded to ammonia or escapes microbial degradation. To compare feedstuffs, feed nitrogen can be divided into NPN, true protein, and unavailable fractions, which Pichard and van Soest (1977) labeled as the *A*, *B*, and *C fractions*, respectively (Fig. 18.3). The A fraction is rapidly attacked by rumen bacteria and converted to ammonia. Approximately 20% of the crude protein in SBM is in the A fraction and is degraded in the rumen at a rate of 300%/h (NRC, 1996). In contrast, a more undegradable protein source like distillers grains has 6% of the crude protein in the A fraction.

The B fraction is composed of true protein and is usually degraded at a much slower rate than the A fraction (Fig. 18.3). With many feedstuffs, different types of proteins may reside in the B fraction, all with their own degradation rate (B_1, B_2, etc.). Feed proteins are composed of four major types of proteins: albumins, globulins, prolamines, and glutelins. In general, albumins and globulins are the most rapidly degraded proteins, and prolamines and glutelins are more slowly degraded (Sniffin, 1974). Soy protein is relatively high in albumins and globulins. Unfortunately, these fractions are the most rapidly degraded, because they contain the highest concentrations of essential AAs like lysine and arginine (Tamminga, 1979).

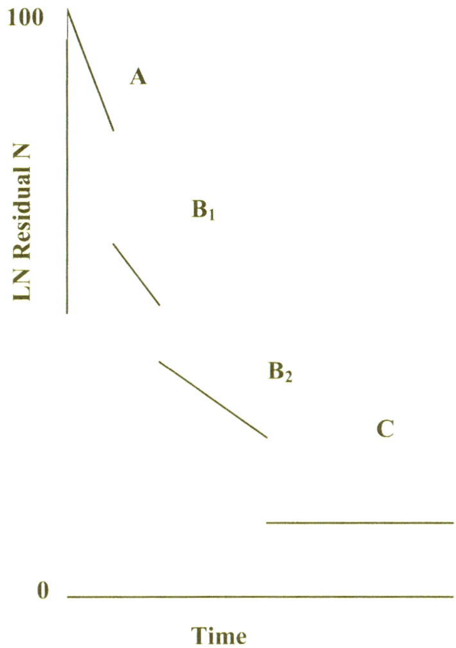

Fig. 18.3. Model illustrating the digestion in the rumen of the A, B_1, B_2, and C protein fractions in feedstuffs.

The unavailable or C fraction nitrogen is estimated by measuring the amount of acid detergent insoluble protein (van Soest, 1991). This fraction is assumed to have zero availability in the rumen and small intestine, and thus has no nutritional value. SBM is a highly digestible protein source with only 2% of the protein in the C fraction (NRC, 1996). In contrast, many slowly degraded protein sources have 10–20% of the protein in the C fraction.

Increasing the Bypass Proteins in Soybean Products

During the past 20 years, scientists sought to improve the nutritional value of SBM for ruminants by reducing the amount of protein in the A fraction and by slowing the degradation rate of the B fraction. Some of the methods explored include formaldehyde treatment, sodium hydroxide treatment, coating with lignin sulfonate, coating with blood proteins, Maillard reactions with xylose or other sugars, and binding with zinc or other metals. Two additional methods include heat and alcohol treatment.

Application of dry heat by baking SBM improved the utilization of SBM by sheep and cattle (Glimp et al., 1967; Sherrod & Tillman, 1962; Thomas et al., 1979). However, roasting has the potential to be a faster and more efficient heat treatment process for SBM. The purpose of this research was to evaluate the effect of roasting temperature on the nutritional value of SBM for beef cattle (Plegge et al., 1985). A commercial continuous flow grain roaster was used to roast SBM to 115, 130, or 145°C. After roasting, the SBM was steeped for 2 h and then cooled to ambient temperatures.

Effects of roasting temperature on nutrient composition, acid detergent insoluble nitrogen, and in situ rate of degradation of the B fraction are shown in Table 18.8. The percentage of the nitrogen recovered as acid detergent insoluble nitrogen increased from 4.1% for the unroasted to 4.9, 4.7, and 15.8%, respectively, for SBM that was roasted at 115, 130, and 145°C, respectively. These data showed that roasting to 130°C would not depress protein digestibility, but going above that temperature could cause a significant reduction in available protein.

In situ rate of protein degradation was estimated by placing samples of the different SBM in small dacron bags and incubating them in the rumen of a fistulated cow for 3, 6, 9, 12, 18, or 24 h. Roasting decreased the rate of in situ nitrogen degradation from 11.3%/h for the unroasted meal to around 4.0%/h for the two lowest roasting temperatures, to only 1.9%/h for the 145°C meal (Table 18.8). These data showed that roasting can dramatically decrease the degradation rate of the B fraction.

To estimate the effect of roasting temperature on the amount of SBM protein escaping ruminal degradation, five ruminal and duodenally cannulated steers were used. The dietary treatments compared were: a urea control, unroasted SBM, and the three roasted SBMs. A balanced 5 × 5 Latin-square design was used with 16-day periods, divided into 10 days for diet adaptation and 6 days for sampling. Lanthanum oxide and chromium-EDTA were used as markers to estimate digesta flow to the small intestine. Steers were fed a diet containing 45% ground corn cobs, 28% ground corn, 10% alfalfa hay, 9% SBM, and minerals and vitamins. All steers were fed 7.1 kg/day, corresponding to 1.8% of body weight.

Soybean meal nitrogen intake averaged 57 g/day, which was 40% of the total nitrogen intake (Table 18.9). Total nitrogen flowing to the duodenum increased linearly ($P < 0.05$) from 112.6 to 147.3 g/day for the unroasted and 145°C SBM, respectively. Bacterial nitrogen flows were not affected by treatment averaging 52 g/day. Soybean meal nitrogen escaping ruminal digestion increased ($P < 0.05$) from 8.3 g/day for the unroasted to 16.5, 20.4, and 27.3 g/day, respectively, for steers fed the 115, 130, and 145°C roasted meals. The SBM nitrogen escaping the rumen increased from 14.7% for the control to 47.3% for the 145°C roasted meal. Acid detergent insoluble N flow also increased with roasting temperature from 8.6 g/day for the unroasted meal to 14.4 g/day for the 145°C meal. These data showed that roasting could dramatically increase the amount of SBM escaping ruminal degradation, but it could also increase the unavailable soy protein.

Table 18.8. Nutrient Composition and In Situ Rate of Degradation of Soybean Meal (SBM) Roasted to Various Temperatures [a]

Item	Control SBM	Roasted SBM 115 °C	130 °C	145 °C
Dry matter, %	89.0	93.0	95.0	95.0
Dry matter composition, %				
Ash	7.3	7.6	7.1	7.3
N	8.8	8.9	8.8	8.9
N composition, %				
Acid detergent insoluble N	4.1	4.9	4.7	15.8
Acid pepsin insoluble N	6.8	7.2	7.3	21.0
In situ rate of degradation[b], %/h	11.3[x]	4.3[y]	4.1[y]	1.9[z]
SE[c]	0.0085	0.0058	0.0057	0.0032

[x-z] Values within a row lacking a common superscript letter are different ($P < 0.05$).
[a] Data from Pleege et al. (1985).
[b] Rate of in situ N (non-acid pepsin insoluble) disappearance from 3 to 12 h.
[c] Standard error of rate of degradation.

A reduction in ruminal degradation without increasing the acid detergent insoluble nitrogen can also be achieved using alcohol treatment. A mixture of alcohol and water can change the three-dimensional structure of the soybean proteins. The water penetrates the hydrophilic region of the protein, and the alcohol disrupts the hydrophobic region. This allows a change in the structure of the protein so that more of the hydrophobic AA side chains point to the exterior. This reduces the solubility of the protein in rumen fluid and slows bacterial attack. When combined with moderate heat (78°C), the effectiveness of the treatment process is improved and the alcohol can be recycled (Table 18.10) (Lynch et al., 1987).

Factors Affecting Degradability of Soybean Protein in the Rumen

The first factor that can affect the degradability of soybean protein is rumen pH. Loerch et al. (1983) reported that the disappearance of SBM protein from dacron bags decreased dramatically as the percentage of concentrate in the diet was increased from 20 to 80% (Table 18.11). Sodium hydroxide (3% concentration) was added to the corn to maintain the average rumen pH >6.6 as the corn was increased. When corn was included at 20 and 40% of the diet DM, rumen pHs were similar (above 6.2), and the 12-h disappearance of the SBM was 66.8–79.8%. However, when corn was included at 80% and the pH dropped to 5.5, only 45.2% of the SBM disappeared compared with 81.1% when the pH was 6.64. The changes in degradability are probably due to the isoelectric point of the SBM protein. The isoelectric point of a protein is the pH at which it has no net charge, which should also be the pH where

Table 18.9. Duodenal Nitrogen Flow and Ruminal Escape of Roasted Soybean Meal (SBM)[a]

Item	Urea	Control SBM	Roasted SBM 115 °C	Roasted SBM 130 °C	Roasted SBM 145 °C	MSE
Number of steers	5	4	5	5	5	–
Dry matter intake, g/day	7,110	7,123	7,112	7,117	7,121	–
Organic matter intake, g/day	6,517	6,493	6,480	6,488	6,491	–
N intake, g/day	141	140	141	142	142	–
SBM N intake, g/day	–	57	55	56	58	–
Duodenal organic matter flow, g/day	3,330	3,170	3,208	3,278	3,149	981.1
Total N flow[b], g/day	119.4	112.6	135.8	138.6	147.3	50.4
Particulate N flow[b], g/day	83.6	80.1	91.7	94.6	101.7	41.0
Liquid N flow, g/day	35.9	42.8	44.1	44	45.6	39.0
Liquid ammonia flow[c], g/day	5.4	5.4	6.4	5.6	5.4	0.3
Bacterial N flow, g/day	53.3	48.2	52.2	51.9	53.9	32.2
Plant N flow[b,d], g/day	60.7	69	77.2	81.1	87.9	33.1
Acid detergent insoluble N (ADIN) flow[b,c], g/day	9.5	8.6	9.6	9.8	14.4	0.4
SBM N flow[b,c,e], g/day	-	8.3	16.5	20.4	27.3	33.1
Percentage nitrogen escape[b,f]	-	14.7	29.7	36.2	47.3	121.5

[a] Data from Pleege et al. (1985).
[b] Linear effect of temperature ($P < 0.05$).
[c] Quadratic effect of temperature ($P < 0.05$).
[d] Flow of nonammonia, nonbacterial N.
[e] Flow of plant N minus flow of plant N in the urea basal.
[f] Flow of SBM N divided by SBM N consumed.

the protein is least soluble in rumen fluid, and thus, less susceptible to bacterial attack. The isoelectric point of SBM protein is reached at approximately pH 5.5, which was the average ruminal pH when the 80% rolled high-moisture corn diet was fed. In contrast, degradation of slowly degraded protein sources like blood meal and corn gluten meal (Table 18.11) is not affected by changes in rumen pH. These data help explain why SBM has supported feedlot performance equal or superior to those from cattle fed slowly degraded protein sources in high- concentrate diets. In general, SBM is considered a rapidly degraded protein, but when fed in a diet that produces a pH near its isoelectric point, it becomes a more slowly degraded protein.

The second factor that may influence the degradability of soybean protein in the rumen is if an ionophore, such as monensin, is fed to ruminant animals, it decreases the proteolytic activity of bacteria in the rumen. Poos et al. (1979) and Whetsone et al. (1981) show that the degradability of dietary protein is reduced with monensin

Table 18.10. Effects of Ethanol and Heat Treatment of Soybean Meal (SBM) on In Situ Nitrogen Disappearance and Nitrogen Solubility (%)[a]

Treatment	Time, h 0	3	6	9	12	18	Rate of N Loss Between 3 and 12 h, %/h	Soluble N, %
	% N remaining in rumen							
SBM[b]	75.5[x]	60.5[x]	50.6[x]	35.7[x]	13.6[x]	6.0[x]	16.9[z]	12.5[x]
ET-23[b]	85.5[y]	78.3[y]	68.9[y]	58.0[y]	29.7[y]	12.3[y]	11.4[y]	7.6[y]
ET-78[b]	92.9[z]	89.9[z]	84.3[z]	74.6[z]	55.4[v]	37.5[v]	4.9[x]	3.7[z]
SEM	0.9	1.4	1.4	1.2	1.5	2.0	0.7	0.9

[v-z] Values within a column lacking a common superscript letter are different ($P < 0.05$).
[a] Data from Lynch et al. (1987).
[b] SBM = untreated SBM; ET-23 = SBM treated in ethanol at 23 °C; ET-78 = SBM treated in ethanol at 78 °C.

Table 18.11. Effect of Concentration of Dietary Corn and Rumen Ph on Nitrogen Disappearance From Dacron Bags [a]

	% Corn							
	20		40		60		80	
Item	NaOH[b]	HMC[c]	NaOH[b]	HMC[c]	NaOH[b]	HMC[c]	NaOH[b]	HMC[c]
Ruminal pH	6.78	6.52	6.80	6.27	6.61	6.20	6.64	5.52
Ingredient	% disappearance after 12 h							
Soybean meal	76.2	66.2	79.8	66.8	75.6	47.1	81.1	45.2
Dehydrated alfalfa	69.3	72.3	72.5	62.9	62.8	55.1	71.4	35.6
Blood meal	14.0	15.4	11.6	11.5	10.7	11.6	14.7	15.2
Corn gluten meal	19.1	16.5	19.6	15.0	10.8	15.4	14.0	18.3

[a] Data from Loerch et al. (1983).
[b] Whole shelled corn was treated with 3% sodium hydroxide.
[c] High-moisture corn was rolled prior to feeding.

feeding. Over 95% of the feedlot cattle in the United States are fed an ionophore, with monensin being the most common. The average rumen undegradability value for SBM is 34% with a standard deviation of 12% (NRC, 1996). However, the undegradability value may be increased at least one standard deviation when soluble proteins like SBM are fed in diets that produce a low rumen pH and contain monensin (NRC, 1996). This increases the undegradability value to 46% for soybean protein fed in commercial diets.

The third factor that is believed to influence the degradability of soybean protein in the rumen is that bacterial growth in the rumen can be stimulated by the soluble proteins from SBM. Bacterial growth is rapid as long as AAs are readily available, but when the AA supply is depleted, bacteria must synthesize AAs from ammonia. This requires increased production by the bacteria and results in a slower growth rate (Owens, 1988), and a steady supply of AAs and peptides may be required to achieve optimal fermentation of high-grain finishing diets (Russell et al., 1992). When proteins like SBM are present in the rumen, a continuous supply of AAs is provided, which may result in greater bacterial protein synthesis.

Economic Implications of Using Soybean Meal in Diets Fed to Cattle

In the final analysis, economics dictate how much SBM is fed to growing and finishing beef cattle in the future. Because urea is a less expensive source of nitrogen than SBM, costs per ton of diet always favor diets containing high levels of urea. However, cattle producers need to look at the profit potential from feeding SBM and not just feed cost. They also need to recognize that current management strategies—such as feeding very high levels of concentrate, the use of monensin and other ionophores—and aggressive implant programs combine to make SBM supplementation more economically competitive.

The potential economic advantage of SBM supplementation compared with urea was demonstrated (Trenkle, 1995). Diets used in these experiments contained only 7% roughage (12% corn silage), steers were implanted aggressively, and the cattle had the genetic potential to gain 1.8 kg/day. The cattle were fed for a constant number of days and sold on a carcass basis when 60–80% of the cattle would grade "Choice." In all cases, feed cost per unit of dry matter was lowest for the diets containing 1.04% urea (Table 18.13).

Feeding Soybean Hulls to Beef Cattle

A second soybean product that has great potential for growing beef cattle is soy hulls, which is a highly digestible fiber source. The nutritional advantage of soy hulls is that they can increase the energy density of the diet without affecting the fiber-digesting bacteria in the rumen. When grains are added to high-forage diets, they can cause a shift in the bacterial population in the rumen or lower the pH so that less of the fiber is digested. Consequently, soy hulls are equal to corn as an energy source in grazing situations or when high-forage growing diets are fed. For example, supplementing steers grazing tall fescue with 1.8 kg/day of corn or soy hulls increased gains by 0.2 kg/day (Kerley & Williams, 1995). Although cost/kg of gain was increased with supplementation, total weight gains per steer were increased by 40 kg over the 160-day grazing period (Table 18.14). The extra weight gain is worth $60 per head, and soy hulls will likely become more popular as an energy source for growing beef cattle in the future.

Table 18.12. Effect of Dietary Nutrition Source and Concentration on Performance of Finishing Steers Fed Diets Based on Dry-Rolled Corn[a]

	Urea		SBM[b]		CSM[b]	
Item	1.93% N	2.24% N	1.93% N	2.24% N	2.24% N	SEM
Initial weight, kg	334	336	335	335	335	0.8
Final weight[c], kg	499	499	515	527	523	6.0
Day 0–132						
Dry matter intake, kg/day	9.820	9.330	9.870	10.020	10.340	0.290
Average daily gain[d], kg	1.250	1.240	0.136	1.460	1.420	0.050
Gain:feed[e]	0.128	0.133	0.139	0.145	0.137	0.003
Day 70–132						
Dry matter intake, kg/day	9.870	9.650	10.020	10.040	10.610	0.360
Average daily gain[d], kg	1.650	1.640	1.760	1.850	1.790	0.100
Gain:feed[e]	0.167	0.170	0.178	0.184	0.169	0.006
Day 70–132						
Dry matter intake, kg/day	9.770	8.960	9.730	10.000	10.090	0.370
Average daily gain[d], kg	0.800	0.790	0.910	1.020	1.000	0.080
Gain/feed[e]	0.083	0.088	0.093	0.102	0.100	0.006
Adjusted performance[f], day 0–132						
Average daily gain[g], kg	1.35[y]	1.24[x]	1.41d[yz]	1.510[z]	1.510[l]	0.040
Gain:feed[d]	0.137	0.130	0.144	0.151	0.146	0.004

[x-z] Data within a row lacking a common superscript letter are different ($P < 0.05$).
[a] Data from Milton et al. (1997).
[b] Supplemental N source: SBM = soybean meal; CSM = cottonseed meal.
[c] Final weight shrunk 4%.
[d] Urea vs. SBM ($P < 0.05$).
[e] Urea vs. SBM ($P < 0.10$).
[f] Average daily gain and gain:feed calculated using final weight = hot carcass weight divided by 0.63.
[g] Interaction between dietary N source and N concentration ($P < 0.05$).

Soybean Products In Diets Fed To Dairy Cattle

Soybean products are widely used in dairy cattle nutrition. The principal products used are SBM (Firkins & Fluharty, 2000), full-fat soybeans (Grummer & Rabelo, 2000), and soybean hulls (Titgemeyer, 2000). Readers are referred to those reviews for additional background and summation of older scientific literature. Little soybean oil is used directly in feeding dairy cattle, partly because of its highly unsaturated fatty acid profile (which can disrupt fermentation of fibrous carbohydrates in the rumen), but mainly because it is usually more expensive than other commodity fats. Soy protein products also are used as components of "alternate protein" milk replacers for young calves, the main products being soy flour, SPC, and SPI (Lallès, 2000).

Table 18.13. Economic Analysis of Feeding Soybean Meal (SBM) to Steers Implanted With Estradiol and Trenbolone Acetate[a,b]

Item	Diet			
	1.04 Urea	5% SBM	10% SBM	SBM-Urea[c]
Experiment 1—Implanted				
Feed DM, $/45.45 kg	4.81	5.08	5.29	4.99
Feed cost of gain, $/45.45 kg	31.01	30.53	32.21	30.01
Total cost, $/steer	809.00	816.52	827.80	816.32
Profit, $/steer	47.14	66.98	57.98	68.32
Experiment 2—Nonimplanted				
Feed DM, $/45.45 kg	4.81	5.08	5.29	
Feed cost of gain, $/45.45 kg	32.59	32.50	35.55	
Total cost, $/steer	862.91	866.49	883.44	
Profit, $/steer	26.27	26.13	22.86	
Experiment 2—Implanted				
Feed DM, $/45.45 kg	4.81	5.08	5.29	4.97
Feed cost of gain, $/45.45 kg	27.26	28.24	28.49	27.01
Total cost, $/steer	883.16	890.73	900.34	889.21
Profit, $/steer	90.40	97.65	107.42	101.45

[a] Data from Trenkle (1995).
[b] Based on feed prices of corn $98.21/metric ton; corn silage, $27.50/metric ton; molasses, $110/metric ton; SBM, $220/metric ton; urea, $0.286/kg; dicalcium phosphate, $0.352/kg; and all other additives, $0.44/kg. Other costs were implants, $3.30/dose; nonfeed, $0.35/day; and purchase price of steers, $1.72/kg. Selling price of carcasses was $2.51/kg.
[c] Diet changed from 10% SBM to 1.04% urea at 56 and 62 days in Experiments 1 and 2, respectively.

Table 18.14. Use of Soy Hulls as an Energy Supplement for Backgrounding Steers[a]

Treatment[b]	Average Daily Gain, kg/steer	Cost of Gain, $/kg	Total Gain, kg
Fescue	0.7	0.38	109[x]
Corn	0.9	0.49	141[y]
Soy hulls	0.9	0.48	149[y]

[x,y] Values within a column lacking a common superscript letter are different ($P < 0.05$).
[a] Data from Kerley and Williams (1995).
[b] Fescue = grazed tall fescue pastures for 160 days; Corn = grazed tall fescue pastures plus fed 1.8 kg of corn per steer/day; Soyhulls = grazed tall fescue pastures plus fed 1.8 kg of soybean hulls per steer/day.

Protein Utilization in Dairy Cattle

Protein is needed as a source of absorbable AAs in ruminant animals just as in monogastric animals, but the presence of the complex ruminant stomach complicates protein supply considerably. In the rumen, resident microorganisms (bacteria and protozoa) anaerobically ferment carbohydrates, such as cellulose and starch, producing volatile (or short-chain) fatty acids that are absorbed and used by the animal as energy sources. Simple NPN sources, such as ammonia, along with dietary true proteins that are degraded by the rumen microbes, are used as the nitrogen source for synthesis of microbial proteins needed by the rapidly growing microbial population. The microbial cells are washed out of the rumen with the digesta, and serve as a major protein source of digestible AAs for the host animal. The portion of dietary protein and NPN needed for the microbial population is defined as rumen-degradable protein (RDP; NRC, 2001).

Not all dietary proteins are degraded by the microbial population, and some pass unaltered or with minimal alteration into the small intestine where they are digested much as in a nonruminant animal. Dietary proteins that are not degraded by the ruminal microbes are referred to as rumen-undegradable protein (RUP; NRC, 2001). This fraction also is sometimes referred to as *escape* or *bypass* protein. The total amount of protein that reaches the intestine for digestion is the sum of microbial protein and RUP, and is called *metabolizable protein* (MP; NRC, 2001). This is the actual supply of digested proteins that furnishes AAs to the animal.

Ruminant animals that are essentially at maintenance, growing very slowly, or producing only small amounts of milk can meet virtually their entire AA requirements from microbial protein if moderately fermentable feeds are fed in properly balanced diets. However, as productivity increases, rumen microbial protein supply becomes inadequate to provide all the protein needed by the animal. This deficiency is particularly true for high-producing dairy cows. For example, a 650-kg Holstein cow producing 50 kg of milk with 3.6% fat daily requires a total of 3.25 kg of MP (NRC, 2001). A diet formulated for this cow would require about 2.8 kg of RDP daily and about 1.8 kg of RUP daily (NRC, 2001). Depending on the type and quality of forages available, as well as supply of various by-product feeds, such a cow might be fed 3–5 kg of soybean meal daily as part of the strategy to meet those large protein requirements.

Soybean Meal in Diets Fed to Dairy Cattle

SBM is the most widely used protein supplement for dairy cattle in North America and much of the world. Reasons for the dominance in the market include high nutritional quality, consistency, widespread availability, and cost-competitiveness. SBM usage by the dairy industry in the United States is estimated to be only about one-third of the total potential protein utilization if all supplemental protein was supplied by SBM (Clark & Bateman, 1999).

SBM added to diets for dairy cattle supplies both RDP and RUP. The proteins in typical solvent-extracted SBM are generally ~65% degradable in the rumen (NRC, 2001). Ruminal proteolysis of SBM results in formation of peptides, AAs, and ammonia, all of which may be utilized for growth by particular species of microorganisms in the rumen. This value is not a constant, however, and will be affected by several variables, including cow factors such as the amount of feed intake and rumen pH. In addition, various chemical- or heat-processing methodologies can be applied to alter the protein degradation characteristics of SBM.

As feed intake increases, rates of digesta passage through the digestive tract also increase (NRC, 2001), which decreases the time that SBM is available to the microbial population in the rumen, and therefore, increases the RUP value. Lower pH in the rumen (e.g., 5.8–6.0 instead of 6.0–6.2) also increases the RUP value because the proteins become less soluble as the ruminal pH approaches the isoelectric point of soy proteins; therefore, the proteins are less susceptible to microbial proteolysis. The practical significance of these principles is that RUP values for soybean meal are greater for high-producing cows consuming large amounts of feed than for animals near maintenance (Ipharraguerre & Clark, 2005).

A variety of chemical and heat treatments can be applied to SBM during processing that will increase the RUP values (Firkins & Fluharty, 2000). Treatment of SBM with aldehydes causes cross-linking between peptide chains, thereby decreasing protein solubility and microbial degradation. Formaldehyde has been used most commonly and is effective in decreasing protein degradability. However, toxicity concerns make this treatment less desirable in many countries. Lignosulfonate treatment involves a heat-induced chemical reaction between xylose in sulfite liquor (a waste product from paper manufacture) and AA residues in SBM. The result is the formation of early Maillard products (Friedman, 1996) that decrease solubility of the protein and hence increase resistance to microbial breakdown.

Additional heat applied during processing of SBM is a commonly used approach to manipulate the RUP value. Increasing heat causes more formation of Maillard-type cross-linkages between amino groups (especially lysine) and carbohydrates within the SBM, which decreases solubility and microbial access for proteolysis. As long as the protein is not heated too extensively so that Maillard reactions are irreversible, the protein still can be digested in the acidic abomasum (true stomach) and the enzymatic environment of the small intestine (Firkins & Fluharty, 2000). Several heating processes are used commercially to produce SBM with greater RUP, including screw pressing, extruding, roasting, and toasting. A variation is production of a heat-treated SBM–hulls combination. SBM is combined with soybean hulls in a 10:1 ratio, water is added to bring total moisture to 30–50%, and the mixture is cooked at 95°C until final moisture is between 12–16% (Borucki Castro et al., 2007).

A recent experiment compared solvent-extracted SBM with screw-pressed SBM, lignosulfonate-treated SBM, and SBM–hulls product (Borucki Castro et al., 2007).

The RUP estimates were 42, 68, 62, and 65%, respectively, indicating that all treatments were effective at decreasing the amount of soy protein degraded in the rumen relative to solvent-extracted soybean meal. Ipharraguerre et al. (2005) also showed that lignosulfonate treatment and screw press treatment resulted in increased passage of feed protein to the small intestine.

Heated SBMs are widely used for feeding high-producing dairy cows because of the challenge in providing adequate MP to these cows. Animal protein sources, such as meat and bone meal, were once widely used to supply RUP with good AA balance in diets for dairy cows. However, restrictions on the use of animal proteins in North American and European Union markets severely limited the use of these animal protein supplements. Use of heated or chemically processed SBM to increase dietary supply of RUP for high-producing dairy cows is likely to continue to increase to attempt to meet the AA requirements for copious milk production.

Unfortunately, substitution of more expensive altered proteins does not always lead to improvements in milk production, as summarized by Ipharraguerre and Clark (2005) on the basis of a large meta-analysis of experiments that tested dietary inclusion of high-RUP supplements. This lack of response was confirmed directly in more recent experiments. For example, Colmenero and Broderick (2006) found that solvent-extracted SBM resulted in similar milk yields as replacement with screw-pressed SBM. Possible reasons for the lack of expected improvement were summarized (Ipharraguerre & Clark, 2005) and may include reductions in microbial protein synthesis, decreased feed intake, or lack of improvement in limiting AAs in the intestine.

Similar to the general scheme for protein digestion, AA nutrition in ruminants is also complicated by the existence of the rumen. Among oilseed meals, SBM has the greatest content of indispensable AAs (NRC, 2001). However, the AA profile of proteins in RUP may be altered as a result of preferential degradation of certain AAs in the rumen (Borucki Castro et al., 2007). A fundamental challenge with use of proteins intended to increase RUP supply is that they are supplemented in small proportions relative to the total amount of microbial protein reaching the intestine (Ipharraguerre & Clark, 2005). The AA profile of microbial proteins is essentially constant and of very high nutritional value for milk production (NRC, 2001). In the RUP fraction of SBM, methionine may be limiting for optimal performance (NRC, 2001). Supplemental rumen-protected methionine products are often used in conjunction with SBM to balance the AA profile reaching the intestine.

Full-fat Soybeans in Diets Fed to Dairy Cows

While most soybeans are processed to extract the oil for food uses and to produce SBM for animal feeding, whole full-fat soybeans are an attractive and widely used feedstuff for dairy cows. The high protein concentration of the whole bean coupled with its high-fat concentration results in a feed rich in both protein and energy. While some soybeans are fed raw, most commonly the soybeans are heat-treated to inactivate urease and trypsin inhibitors and to decrease protein degradability (Aldrich et al., 1995). The most widely used treatment is roasting, in which the beans are passed through a forced-air oven or a flame in a rotating drum. Typically, soybeans are heated in a drum roaster to a 146°C exit temperature and then allowed to steep for 30 min before cooling. Optimal heating in forced-air ovens is 140°C for 120 min, 150°C for 60 min, or 160°C for 30 min (Clark & Bateman, 1999; Grummer & Rabelo, 2000). Because of the greater variability in heat penetration in whole beans relative to SBM, variation in the RUP content of whole beans is greater than that for SBM. Ipharraguerre et al. (2005) reported that whole roasted soybeans increased the intestinal supply of feed protein compared with a similar amount of protein only from solvent-extracted SBM.

Whole soybeans also can be heat-treated by extrusion, although this process results in greater rupture of oil-containing vesicles and increases the availability of oil in the rumen. More rapidly available unsaturated oil can result in decreased fiber digestion, lower feed intakes, and depression of milk fat synthesis at lower inclusion rates than with whole roasted soybeans (Grummer & Rabelo, 2000). Where available economically, roasted soybeans are widely used to provide protein and some additional fat to the dairy diet.

Soybean Hulls in Diets Fed to Dairy Cows

Soybean hulls are a by-product of soybean processing and are an excellent supplemental feed for dairy cattle. Because they are low in lignin, the cellulose is highly digestible in the rumen, and fermentation rates are rapid. Soybean hulls were used to replace forage and to replace cereals in concentrates (Titgemeyer, 2000). While soybean hulls ferment differently than starchy concentrate feeds, they cannot replace fibrous forages completely because they do not have sufficient structural fiber to stimulate rumination and maintain rumen pH. Conversely, excessive replacement of starch with soybean hulls may limit production of microbial protein and milk components (Titgemeyer, 2000).

Ipharraguerre and Clark (2003) reviewed the literature on use of soy hulls for dairy cows. They concluded that soybean hulls can replace corn grain to supply ~30% of the dry matter in high-grain diets without negatively affecting either the fermentation or digestion of nutrients in the gastrointestinal tract or the performance of dairy cows. They also concluded from the data that soybean hulls might successfully replace forage to supply up to 25% of the dry matter in the diets of dairy cows as long as the supply of effective fiber remains adequate after including the soybean hulls in the diets. In experiments designed to test these predictions, Ipharraguerre et al. (2002a,b) replaced corn grain with soy hulls to supply 10, 20, 30, or 40% of dietary dry matter. Inclusion of more than 30% soy hulls decreased milk yield despite similar passage of AAs to the small intestine.

Soy Proteins in Milk Replacers for Young Calves

As milk proteins continue to increase in price, manufacturers of milk replacers strive to find lower-cost proteins that will provide adequate growth and maintain health in young dairy calves. Because of the generally good AA profile, soybean proteins have long been considered as an alternate protein to replace milk proteins in young calf nutrition. However, replacement of milk proteins with soy proteins results in inferior growth performance and often impaired health. Similar to young nonruminants, the pre-ruminant calf is sensitive to trypsin inhibitors and the indigestible oligosaccharides found in raw soy. Like young pigs, calves are markedly susceptible to antigenic proteins (primarily β-conglycinin but also glycinin, α-conglycinin, Bowman-Birk inhibitor, and lectins; Lallès, 2000) present in soybean proteins. In contrast to young pigs, however, these allergenic responses persist in calves (Lallès, 2000). Indigestible carbohydrates and many of the offending allergens can be removed by hot aqueous ethanol processing during the production of SPC. However, even good quality feed-grade SPCs do not restore growth to levels comparable to milk proteins (Drackley et al., 2006). Although high-quality SPIs can result in satisfactory performance relative to milk proteins (Lallès, 2000), their higher cost, attributable to the additional processing steps, negates much of the potential advantage from replacing milk proteins. Moreover, many of the SPIs available to the feed industry are off-specification or low-quality materials not suited for the human market.

Reasons for the poor calf performance on soy proteins remain unidentified. Antinutritional factors and antigenic proteins present in raw soybeans are greatly decreased by the hot aqueous ethanol treatment involved in the production of SPC, yet adverse effects on growth and intestinal function still occur when SPC is fed to calves (Lallès, 2000). Changes in intestinal histomorphology occur when calves are fed SPC (Drackley et al., 2006) as well as soy flour (Kilshaw & Slade, 1982). The cellulose and hemicellulose present in SPCs may increase villus abrasion and cell desquamation and also increase mucus loss in the terminal small intestine (Leterme et al., 1998). In addition to alterations in villus size, a variety of other intestinal abnormalities was observed in calves fed low-antigenic soy protein products, including decreases in protein synthetic capacity (Grant et al., 1989), mucosal digestive enzyme activities (Grant et al., 1989; Montagne et al., 1999), and absorptive capacity (Grant et al., 1989) and increases in mucin secretion (Montagne et al., 2000), immune activation (Lallès, 2000), and specific endogenous protein loss (Montagne et al., 2001).

Plant-based proteins, such as soy, have high true digestibilities but lower apparent digestibilities because specific endogenous protein losses at the ileum are increased (Montagne et al., 2000, 2001). Montagne et al. (2003) suggested that resistant dietary oligopeptides may interact with intestinal mucosa to stimulate endogenous protein secretion. SPC increased ileal flow of diet-specific host protein (Montagne et al., 2001), which would increase dietary protein requirements by about 2 g/day for calves fed typical amounts of milk replacer (Drackley et al., 2006). Consequently, it is unlikely that intake of protein limits growth of calves when that occurs. Although dietary contents of lysine and methionine were equalized in most experiments, it is possible that another indispensable AA, such as threonine, might limit growth relative to whey proteins (Kanjanapruthipong, 1998).

Regardless of the mechanism, average daily gains and gain:feed ratios usually are lower than milk-fed controls when calves are fed milk replacers containing a substantial amount of protein from soybeans. Identification of factors limiting calf performance when soy-containing milk replacers are fed would be an enormous benefit to both the dairy and soybean industries.

Conclusion

Soybean products are primarily used in diets fed to poultry, livestock, and companion animals to supply indispensable AAs to the diets. The AA composition in SBM and other soybean products complements the AA profile in corn and other cereal grains. This is particularly important in diets fed to poultry and pigs, where lysine is often the first limiting AA. Soy protein also contains relatively high concentrations of arginine that is needed in poultry diets, and in tryptophan that is often limiting in diets fed to swine. Dehulled SBM may supply all AAs other than those supplied by the cereal grains in diets fed to poultry and pigs heavier than 20–25 kg. In diets fed to younger pigs, SPC may be used while SBM is slowly introduced during the post-weaning period. Full-fat soybeans are not used as frequently as SBM, but may be used as a total or partial replacement for SBM in diets fed to pigs and poultry. Soy protein may also supply a large proportion of the AAs needed in diets fed to cats, dogs, and ruminant animals. Soy oil, if available at a competitive price, may be used to increase the energy concentration in diets fed to pigs and poultry. In contrast, soybean hulls are usually not used in diets fed to monogastric animals, but they may be used as a valuable fiber source in diets fed to ruminants.

All soybean products, except soybean oil, fed to monogastric animals and preruminant calves need to be heat-treated prior to usage because of the presence of protease inhibitors and lectins in soybeans. Care must be taken to make sure that enough heat is applied for a complete inactivation of these factors, but the heating should not be excessive because that may reduce the digestibility of certain AAs, and lysine in particular. However, if properly heat-treated, the AAs in soybean proteins are usually well digested by all groups of animals. In contrast, the phosphorus in soybean products has low digestibility to most monogastric animals because most of it is bound in the phytate complex. The addition of microbial phytase, however, may improve the phosphorus digestibility by more than 100%.

The usage of soybean products may be increased in the future if modifications to the products can be made. A lower concentration of oligosaccharides will be beneficial to the poultry industry because it may increase the digestibility of dietary energy. A reduced concentration of oligosaccharides may also reduce the flatulence-enhancing properties in soy products, which may increase the usage of soy products in the pet-food industry. Likewise, the usage of soybean proteins in milk replacers to young dairy calves may be increased if varieties with lower concentrations of oligosaccharides and antigens are developed. In the swine industry, the main concern is also the presence of antigens in SBM that prevents the usage of large quantities of conventional SBM in diets fed to newly weaned pigs. Development of soybeans with a reduced concentration of antigens, therefore, will increase the utilization of SBM to this category of swine. The development of enzyme-treated SBM and fermented SBM also has the potential to increase the usage of soy products in diets fed to weanling pigs. The greatest potential for increased usage, however, is by the beef cattle industry that potentially could increase the usage of both SBM and soybean hulls. However, more education on the usage of soybean products to cattle producers may be needed for this to happen.

References

AAFCO (Association of American Feed Control Officials). *Official Publication;* 2007, pp. 131–137.

Ajakaiye, A.; M.Z. Fan; T. Archbold; R.R. Hacker; C.W. Forsberg; J.P. Phillips. Determination of true digestive utilization of phosphorus and the endogenous phosphorus outputs associated with soybean meal for growing pigs. *J. Anim. Sci.* **2003,** *81,* 2766–2775.

Albin, D.M.; M.R. Smiricky; J.E. Wubben; V.M. Gabert. The effect of dietary level of soybean oil and palm oil on apparent ileal AA digestibility and postprandial flow patterns of chromic oxide and AAs in pigs. *Can. J. Anim. Sci.* **2001,** *81,* 495–503.

Aldrich, C.G.; N.R. Merchen; D.R. Nelson; J.A. Barmore. The effects of roasting temperature applied to whole soybeans on site of digestion by steers: II. Protein and AA digestion. *J. Anim Sci.* **1995,** *73,* 2131–2140.

Aldrich, C.G.; N.R. Merchen; C.M. Parsons; H.S. Hussein; S. Ingram; J.R. Clodfelter. Assessment of postruminal AA digestibility of roasted and extruded whole soybeans with the precision-fed rooster assay. *J. Anim. Sci.* **1997,** *75,* 3046–3051.

Alonso, R.; E. Orue; F. Marzo. Effects of extrusion and conventional processing methods on protein and antinutritional factor contents in pea seeds. *Food Chem.* **1998,** *63,* 505–512.

American Soybean Association. U.S. soybean use by livestock. *Soy Stat: A Reference Guide to Important Soybean Facts and Figures;* The American Soybean Association, 2007.

Angkanaporn, K.; V. Ravindran; W.L. Bryden. Additivity of apparent and true ileal AA digestibilities in soybean meal, sunflower meal, and meat and bone meal for broilers. *Poult. Sci.* **1996,** *75,* 1098–1103.

AOCS. Official Method Ba 9-58, Urease Activity. American Oil Chemists' Society: Champaign, IL, 1973.

Araba, M.; N.M. Dale. Evaluation of protein solubility as an indicator of overprocessing of soybean meal. *Poult. Sci.* **1990,** *69,* 76–83.

Baker, D.H. Nutrient bioavailability in feedstuffs: Methodology for determining AA and B-vitamin availability in cereal grains and soybean meal. *Proceedings of the Georgia Nutrition Conference*, Athens, 1978, pp. 1–12.

Baker, D.H. Nutritional constraints to use of soy products by animals. *Soy in Animal Nutrition;* J.K. Drackley, Ed.; Federation of Animal Science Societies: Savoy, IL, 2000; pp. 1–12.

Batal, A.B.; C.M. Parsons. Utilization of different soy products as affected by age in chicks. *Poult. Sci.* **2003,** *82,* 454–462.

Bednar, G.E.; S.M. Murray; A.R. Patil; E.A. Flickinger; N.R. Merchen; G.C. Fahey. Selected animal and plant protein sources affect nutrient digestibility and fecal characteristics of ileally cannulated dogs. *Arch. Anim. Nutr.* **2000,** *53,* 127–140.

Bohlke, R.A.; R.C. Thaler; H.H. Stein. Calcium, phosphorus, and AA digestibility in low-phytate corn, normal corn, and soybean meal by growing pigs. *J. Anim. Sci.* **2005,** *83,* 2396–2403.

Borucki Castro, S.I.; L.E. Phillip; H. Lapierre; P.W. Jardon; R. Berthiaume. Ruminal degradability and intestinal digestibility of protein and AAs in treated soybean meal products. *J. Dairy Sci.* **2007,** *90,* 810–822.

Clapper, G.M.; C.M. Grieshop; N.R. Merchen; J.C. Russett; J.L. Brent; G.C. Fahey. Ileal and total tract nutrient digestibilities and fecal characteristics of dogs as affected by soybean protein inclusion in dry, extruded diets. *J. Anim. Sci.* **2001,** *79,* 1523–1532.

Clark, J.H.; H.G. Bateman, II. Use and future prospects for use of soy products in dairy cattle diets. *Opportunities for Soy Products in Animal Nutrition;* J.K. Drackley, Ed.; Federation of Animal Science Societies: Savoy, IL, 1999; pp. 87–105.

Cole, J.T.; G.C. Fahey; N.R. Merchen; A.R. Patil; S.M. Murray; H.S. Hussein; J.L. Brent. Soybean hulls as a dietary fiber source for dogs. *J. Anim. Sci.* **1999,** *77,* 917–924.

Colmenero, J.J.; G.A. Broderick. Effect of amount and ruminal degradability of soybean meal protein on performance of lactating dairy cows. *J. Dairy Sci.* **2006,** *89,* 1635–1643.

Coon, C.N.; K.L. Leske; O. Akavanichan; T.K. Cheng. Effect of oligosaccharide-free soybean meal on true metabolizable energy and fiber digestion in adult roosters. *Poult. Sci.* **1990,** *69,* 787–793.

Cromwell, G.L. Utilization of soy products in swine diets. *Soy in Animal Nutrition*; J.K. Drackley, Ed.; Federation of Animal Science Societies: Savoy, IL, 2000; pp. 274–280.

Cromwell, G.L.; C.C. Calvert; T.R. Cline; J.D. Crenshaw; T.D. Crenshaw; R.A. Easter; R.C. Ewan; C.R. Hamilton; G.M. Hill; A.J. Lewis et al. Variability among sources and laboratories in nutrient analyses of corn and soybean meal. *J. Anim. Sci.* **1999,** *77,* 3262–3273.

Cromwell, G.L.; M.D. Lindemann; J.H. Randolph; G.R. Parker; R.D. Coffey; K.M. Laurent; C.L. Armstrong; W.B. Mikel; E.P. Stanisiewski; G.F. Hartnell. Soybean meal from Roundup Ready or conventional soybeans in diets for growing-finishing swine. *J. Anim. Sci.* **2002,** *80,* 708–715.

Cromwell, G.L.; T.S. Stahly; R.D. Coffey; H.J. Monegue; J.H. Randolph. Efficacy of phytase in improving the bioavailability of phosphorus in soybean meal and corn-soybean meal diets for pigs. *J. Anim. Sci.* **1993,** *71,* 1831–1840.

Dale, N. Soy products as protein sources in poultry diets. *Soy in Animal Nutrition.* J.K. Drackley, Ed.; Federation of Animal Science Societies: Savoy, IL, 2000; pp. 283–288.

Dale, N.M. *Feedstuffs Ingredient Analysis Table.* Miller Publishing Co.: Minnetonka, MN, 2006.

de la Llata, M.; S.S. Dritz; M.D. Tokach; R.D. Goodband; J.L. Nelssen; T.M. Loughin. Effects of dietary fat on growth performance and carcass characteristics of growing-finishing pigs reared in a commercial environment. *J. Anim. Sci.* **2001,** *79,* 2643–2650.

Dilger, R.N.; J.S. Sands; D. Ragland; O. Adeola. Digestibility of nitrogen and AAs in soybean meal with added soyhulls. *J. Anim. Sci.* **2004,** *82,* 715–724.

Douglas, M.W.; C.M. Parsons; T. Hymowitz. Nutritional evaluation of lectin-free soybeans for poultry. *Poult. Sci.* **1999,** *78,* 91–95.

Drackley, J.K.; R.M. Blome; K.S. Bartlett; K.L. Bailey. Supplementation of 1% L-glutamine to milk replacer does not overcome the growth depression in calves caused by soy protein concentrate. *J. Dairy Sci.* **2006,** *89,* 1688–1693.

Dust, J.M.; A.M. Gadja; E.A. Flickinger; T.M. Burkhalter; N.R. Merchen; G.C. Fahey. Extrusion conditions affect chemical composition and in vitro digestion of select food ingredients. *J. Agric. Food Chem.* **2004,** *52,* 2989–2996.

Dust, J.M.; C.M. Grieshop; C.M. Parsons; L.K. Karr-Lilienthal; C.S. Shasteen; J.D. Quigley; N.R. Merchen; G.C. Fahey. Chemical composition, protein quality, palatability, and digestibility of alternative protein sources for dogs. *J. Anim. Sci.* **2005,** *83,* 2414–2422.

Edwards, III, H.M.; M.W. Douglas; C.M. Parsons; D.H. Baker Protein and energy evaluation of soybean meals processed from genetically modified high-protein soybeans. *Poult. Sci.* **2000,** *79,* 525–527.

Emmert, J.L.; D.H. Baker. Protein quality assessment of soy products. *Nutr. Res.* **1995,** *11,* 1647–1656.

Fastinger, N.D.; D.C. Mahan. Effect of soybean meal particle size on AA and energy digestibility in grower-finisher swine. *J. Anim. Sci.* **2003,** *81,* 697–704.

Feng, J.; X. Liu; Z.R. Xu; Y.P. Lu; Y.Y. Liu. The effect of *Aspergillus oryzae* fermented soybean meal on growth performance, digestibility of dietary components and activities if intestinal enzymes in weaned pigs. *Anim. Feed Sci. Technol.* **2007,** *134,* 295–303.

Firkins, J.L.; F.L. Fluharty. Soy products as protein sources for beef and dairy cattle. *Soy in Animal Nutrition;* J.K. Drackley, Ed.; Federation of Animal Science Societies: Savoy, IL, 2000; pp. 182–214.

Friedman, M. Food browning and its prevention: An overview. *J. Agric. Food Chem.* **1996,** *44,* 631–653.

Glimp, H.A.; M.R. Karr; C.O. Little; P.G. Woolfolk; G.E. Mitchell, Jr.; L.W. Hudson. Effect of reducing soybean protein solubility by dry heat on the protein utilization of young lambs. *J. Anim. Sci.* **1967,** *26,* 858–865.

Grant, A.L.; R.E. Holland; J.W. Thomas; K.J. King; J.S. Liesman. Effects of dietary amines on the small intestine in calves fed soybean protein. *J. Nutr.* **1989,** *119,* 1034–1041.

Grieshop, C.M.; G.C. Fahey. Comparison of quality characteristics of soybeans from Brazil, China, and the United States. *J. Agric. Food Chem.* **2001,** *49,* 2669–2673.

Grieshop, C.M.; G.C. Fahey. The role of soy in companion animal nutrition. *Soy in Animal Nutrition*; J.K. Drackley, Ed.; Federation of Animal Science Societies: Savoy, IL, 2000; pp. 171–181.

Grieshop, C.M.; C.T. Kadzere; G.M. Clapper; E.A. Flickinger; L.L. Bauer; R.L. Frazier; G.C. Fahey, Jr. Chemical and nutritional characteristics of United States soybeans and soybean meal. *J. Agric. Food Chem.* **2003,** *51,* 7684–7691.

Grummer, R.R.; E. Rabelo. Utilization of whole soybeans in dairy cattle diets. *Soy in Animal Nutrition;* J.K. Drackley, Ed.; Federation of Animal Science Societies: Savoy, IL, 2000; pp. 215–237.

Hallman, J.; E. Wallace; A. Milliken; E. Clemens. In vivo colonic electrolyte flux in dogs fed soyprotein, casein or meat as their primary protein source. *Nutr. Res.* **1994,** *14,* 1503–1511.

Hammond, B.G.; J.L. Vicini; G.F. Hartness; M.W. Naylor; C.D. Knight; E.H. Robinson; R.L. Fuchs; S.R. Padgette. The feeding value of soybeans fed to rats, chickens, catfish and dairy cattle is not altered by genetic incorporation of glyphosate tolerance. *J. Nutr.* **1996,** *126,* 717–727.

Han, Y.; C.M. Parsons; T. Hymowitz. Nutritional evaluation of soybeans varying in trypsin inhibitor content. *Poult. Sci.* **1991,** *70,* 896–906.

Harmon, D.L.; J.A. Walker; J.M. Silvio; A. Jamicorn; K.L. Gross. Nutrient digestibility in dogs fed fiber-containing diets. *Vet. Clin. Nutr.* **1999,** *6,* 6–10.

Hill, D. Fiber, texturized protein and extrusion. *Petfood Technology;* J.L. Kvamme, T.D. Phillips, Eds.; Watt Publishing Co.: Mt. Morris, 2003.

Hill, R.C.; C.F. Burrows; J.E. Bauer; G.W. Ellison; M.D. Finke; G.L. Jones. Texturized vegetable protein containing indigestible soy carbohydrate affects blood insulin concentrations in dogs fed high fat diets. *J. Nutr.* **2006,** *136,* 2024S–2027S.

Hill, R.C.; C.F. Burrows; G.W. Ellison; J.E. Bauer. The effect of texturized vegetable protein from soy on nutrient digestibility compared to beef in cannulated dogs. *J. Anim. Sci.* **2001,** *79,* 2162–2171.

Hoffman, L.A.; D.J. Ivers; M.R. Ellersieck; T.L. Veum. The effect of L-Carnitine and soybean oil on performance and nitrogen and energy utilization by neonatal and young pigs. *J. Anim. Sci.* **1993,** *71,* 132–138.

Hong, K.J.; C.H. Lee; S.W. Kim. Aspergillus *oryzae* GB-107 fermentation improves nutritional quality of food soybeans and feed soybean meal. *J. Med. Food.* **2004,** *7,* 430–436.

Howard, K.A.; D.M. Forsyth; T.R. Cline. The effect of an adaptation period to soybean oil additions in the diets of young pigs. *J. Anim. Sci.* **1990,** *68,* 678–683.

Hullar, I.; S. Fekete; Z. Szocs; E. Kienzle. Effect of extrusion on the quality of soybean-based catfood. *J. Anim. Physiol. Anim. Nutr.* **1998,** *80,* 201–206.

Hurrell, R.F.; K.J. Carpenter. The estimation of available lysine in foodstuffs after Maillard reactions. *Prog. Food Nutr. Sci.* **1981,** *5,* 159–176.

Ipharraguerre, I.R.; J.H. Clark. Impacts of the source and amount of crude protein on the intestinal supply of nitrogen fractions and performance of dairy cows. *J. Dairy Sci.* **2005,** *88 (Suppl. 1),* E22–E37.

Ipharraguerre, I.R.; J.H. Clark. Soyhulls as an alternative feed for lactating dairy cows: A review. *J. Dairy Sci.* **2003,** *86,* 1052–1073.

Ipharraguerre, I.R.; J.H. Clark; D.E. Freeman. Rumen fermentation and intestinal supply of nutrients in dairy cows fed rumen-protected soy products. *J. Dairy Sci.* **2005,** *88,* 2879–2892.

Ipharraguerre, I.R.; R.R. Ipharraguerre; J.H. Clark. Performance of lactating dairy cows fed varying amounts of soyhulls as a replacement for corn grain. *J. Dairy Sci.* **2002a,** *85,* 2905–2912.

Ipharraguerre, I.R.; Z. Shabi; J.H. Clark; D.E. Freeman. Ruminal fermentation and nutrient digestion by dairy cows fed varying amounts of soyhulls as a replacement for corn grain. *J. Dairy Sci.* **2002b,** *85,* 2890–2904.

Irish, G.G.; G.W. Barbour; H.L. Classen; R.T. Ryler; M.R. Bedford. Removal of the α-galactosides of sucrose from soybean meal using either ethanol extraction or exogenous α-galactosidase and broiler performance. *Poult. Sci.* **1995,** *74,* 1484–1494.

Kan, C.A.; G.F. Hartnell. Evaluation of broiler performance when fed insect-protected, control, or commercial varieties of dehulled soybean meal. *Poult. Sci.* **2004,** *83,* 2029–2038.

Kanjanapruthipong, J. Supplementation of milk replacers containing soy protein with threonine, methionine, and lysine in the diets of calves. *J. Dairy Sci.* **1998,** *81,* 2912–2915.

Karr-Lilienthal, L.K.; C.M. Grieshop; N.R. Merchen; D.C. Mahan; G.C. Fahey, Jr. Chemical composition and protein quality comparisons of soybeans and soybean meals from five leading soybean-producing countries. *J. Agric. Food Chem.* **2004,** *52,* 6193–6199.

Karr-Lilienthal, L.K.; C.T. Kadzere; C.M. Grieshop; G.C. Fahey. Chemical and nutritional properties of soybean carbohydrates as related to nonruminants: A review. *Livestock Prod. Sci.* **2005,** *97,* 1–12.

Kerley, M.S.; J.E. Williams. Alternative feeds for beef cattle on pasture. *Proceedings of 2nd National Alternative Feeds Symposium.* University Extension Conference Office: University of Missouri-Columbia, 1995; pp. 147–154

Kilshaw, P.J.; H. Slade. Villus atrophy and crypt elongation in the small intestine of preruminant calves fed with heated soyabean flour or wheat gluten. *Res. Vet. Sci.* **1982,** *33,* 305–308.

Kim, S.W.; J.G. Morris; Q.R. Rogers. Dietary soybean protein decreases plasma taurine in cats. *J. Nutr.* **1995,** *125,* 2831–2837.

Kornegay, E.T. Soybean hull digestibility by sows and feeding value for growing-finishing swine. *J. Anim. Sci.* **1981,** *53,* 138–145.

Lallès, J.P. Soy products as protein sources for preruminants and young pigs. *Soy in Animal Nutrition;* J.K. Drackley, Ed.; Federation of Animal Science Societies: Savoy, IL, 2000; pp. 106–126.

Lawrence, K.R.; C.W. Hastad; R.D. Goodband; M.D. Tokach; S.S. Dritz; J.L. Nelssen; J.M. DeRouchey; M.J. Webster. Effect of soybean meal particle size on growth performance of nursery pigs. *J. Anim. Sci.* **2003,** *81,* 2118–2122.

Leszczynski, D.E.; J. Pikul; F.K. McKeith; D.G. McLaren; J. Novakofski; P.J. Bechtel; D.E. Jewell. Effect of feeding finishing pigs extruded full-fat soybeans on performance and pork quality. *J. Anim. Sci.* **1992,** *70,* 2167–2174.

Leterme, P.; E. Froidmont; F. Rossi; A. Thewis. The high water-holding capacity of pea inner fibers affects the ileal flow of endogenous AAs in pigs. *J. Agric. Food Chem.* **1998,** *46,* 1927–1934.

Li, D.F.; J.L. Nelssen; P.G. Reddy; F. Blecha; J.D. Hancock; G.L. Allee; R.D. Goodband; R.D. Klemm. Transient hypersensitivity to soybean meal in the early-weaned pig. *J. Anim. Sci.* **1990,** *68,* 1790–1799.

Li, D.F.; J.L. Nelssen; P.G. Reddy; F. Blecha; R.D. Klemm; D.W. Giesting; J.D. Hancock; G.L. Allee; R.D. Goodband. Measuring suitability of soybean products for early-weaned pigs with immunological criteria. *J. Anim. Sci.* **1991,** *69,* 3299–3307.

Li, S., W.C. Sauer. The effect of dietary fat content on AA digestibility in young pigs. *J. Anim. Sci.* **1994,** *72,* 1737–1743.

Liener, I.E. Implications of antinutritional components in soybean foods. *Crit. Rev. Food Sci.* **1994,** *34,* 31.

Loerch, S.C.; L.L. Berger; D. Gianola; G.C. Fahey, Jr. Effects of dietary protein source and energy level on in situ nitrogen disappearance of various protein sources. *J. Anim. Sci.* **1983,** *56,* 206–216.

Lynch, G.L.; L.L. Berger; G.C. Fahey, Jr. Effects of ethanol, heat, and lipid treatment of soybean meal on nitrogen utilization by ruminants. *J. Dairy Sci.* **1986,** *70,* 91–97.

Mahan, D.C. Efficacy of initial postweaning diet and supplemental coconut oil or soybean oil for weanling swine. *J. Anim. Sci.* **1991,** *69,* 1397–1402.

Milton, C.T.; R.T. Brandt, Jr.; E.C. Titgemeyer. Effects of dietary nitrogen source and concentration in high-grain diets for finishing steers on performance and nutrient digestion. *J. Anim. Sci.* **1997,** *75,* 2813–2823.

Montagne, L.; I. Crévieu-Gabriel; R. Toullec; J.P. Lallès. Influence of dietary protein level and source on the course of protein digestion along the small intestine of the veal calf. *J. Dairy Sci.* **2003,** *86,* 934–943.

Montagne, L.; R. Toullec; M. Formal; J.P. Lallès. Influence of dietary protein level and origin on the flow of mucin along the small intestine of the preruminant calf. *J. Dairy Sci.* **2000,** *83,* 2820–2828.

Montagne, L.; R. Toullec; J.P. Lallès. Intestinal digestion of dietary and endogenous proteins along the small intestine of calves fed soybean or potato. *J. Anim. Sci.* **2001,** *79,* 2719–2730.

Montagne, L.; R. Toullec; T. Savidge; J.P. Lallès. Morphology and enzyme activities of the small intestine are modulated by dietary protein source in the preruminant calf. *Reprod. Nutr. Dev.* **1999,** *39,* 455–466.

Motor, V.; H.H. Stein. Effect of feed intake on endogenouslosses and AA and energy digestibility by growing pigs. *J. Anim. Sci.* **2004,** *82,* 3518–3525.

Murray, S.M.; A.R. Patil; G.C. Fahey; N.R. Merchen; D.M. Hughes. Raw and rendered animal by-products as ingredients in dog diets. *J. Anim. Sci.* **1997,** *75,* 2497–2505.

National Research Council. *Nutrient Requirements of Dairy Cattle,* Seventh rev. ed.; National Academy Press: Washington, DC, 2001.

NRC. *Nutrient Requirements of Dogs and Cats;* National Academy Press: Washington, DC, 2006.

NRC. *Nutrient Requirements of Poultry,* Ninth ed.; National Academy Press: Washington, DC, 1994.

NRC. *Nutrients Requirements of Beef Cattle.* Seventh ed.; National Academy Press: Washington, DC, 1996.

NRC. *Nutrient Requirements of Swine,* Tenth rev. ed.; National Academy Press: Washington, DC, 1998; pp. 110–142.

Opapeju, F.O.; A. Golian; C.M. Nyachoti; L.D. Campbell. AA digestibility in dry-expelled soybean meal fed to pigs and poultry. *J. Anim. Sci.* **2006,** *84,* 1130–1137.

Osborne, T.B.; L.B. Mendel. The use of soy bean as food. *J. Biol. Chem.* **1917,** *32,* 369–387.

Overland, M.; K.A. Rorvik; A. Skrede. High-fat diets improve the performance of growing-finishing pigs. *Acta Scand., Sect A. Anim. Sci.* **1999,** *49,* 83–88.

Owen, K.Q.; J.L. Nelssen; R.D. Goodband; T.L. Weeden; S.A. Blum. Effect of L-carnitine and soybean oil on growth performance and body composition of early weaned pigs. *J. Anim. Sci.* **1996,** *74,* 1612–1619.

Owen, F.N.; R. Zinn. Protein metabolism in ruminants. *The Ruminant Animal, Digestive Physiology and Nutrition;* D.C. Church, Ed.; Prentice Hall: NJ, 1988; pp. 227–249.

Pahm, S.C.; H.H. Stein. Ileal AA digestibility of a new high protein variety of soybeans fed to growing pigs. *J. Anim. Sci.* **2007a,** *85 (Suppl. 2)* 84 (Abstr.).

Pahm, S.C.; H.H. Stein. AA digestibility of protein sources fed to weanling pigs. *J. Anim. Sci.* **2007b,** *85 (Suppl. 2),* 105 (Abstr.).

Parsons, C.M. Assessment of nutritional quality of soy products. *Soy in Animal Nutrition;* J.K. Drackley, Ed.; Federation of Animal Science Societies: Savoy, IL, 2000; pp. 90–105.

Parsons, C.M. Influence of cecectomy on digestibility of AAs by roosters fed distillers dried grains with solubles. *J. Agric. Sci. Camb.* **1985,** *104,* 469–472.

Parsons, C.M; K. Hashimoto; K.J. Wedekind; D.H. Baker. Soybean protein solubility in potassium hydroxide: An in vitro test of in vivo protein quality. *J. Anim. Sci.* **1991,** *69,* 2918–2924.

Parsons, C.M.; Y. Zhang; M. Araba. Nutritional evaluation of soybean meals varying in oligosaccharide content. *Poult. Sci.* **2000,** *79,* 1127–1131.

Penz ,Jr., A.M.; I. Brugalli. Soybean products as protein sources in poultry diets. *Soy in Animal Nutrition;* J.K. Drackley, Ed.; Federation of Animal Science Societies: Savoy, IL, 2000; pp. 289–306.

Pichard, G.; P.J. van Soest. Protein solubility of ruminant feeds. *Proceedings of the Cornell Nutrition Conference,* 1997; pp. 91–97.

Plegge, S.D.; L.L. Berger; G.C. Fahey, Jr. Effect of roasting temperature on the proportion of soybean meal nitrogen escaping degradation in the rumen. *J. Anim. Sci.* **1985,** *61,* 1211–1218.

Poos, M.I.; T.L. Hanson; T.J. Klopfenstein. Monensin effect on diet digestibility, ruminal protein bypass and microbial protein synthesis. *J. Anim. Sci.* **1979,** *45,* 1516–1524.

Qin, G.; E.R. Elst; M.W. Bosch; A.F.B. van der Poel. Thermal processing of whole soy beans: Studies on the inactivation of antinutritional factors and the effects on ileal digestibility in piglets. *Anim. Feed Sci. Technol.* **1996,** *57,* 313–324.

Rodehutscord, M.; M. Faust; H. Lorenz. Digestibility of phosphorus contained in soybean meal, barley, and different varieties of wheat, without and with supplemental phytase fed to pigs and additivity of digestibility in a wheat-soybean meal diet. *J. Anim. Phys. Anim. Nutr.* **1996,** *75,* 40–48.

Romarheim, O.H.; M.A. Aslaksen; T. Storebakken; A. Krogdahl; A. Skrede. Effect of extrusion on trypsin inhibitor activity and nutrient digestibility of diets based on fish meal, soybean meal and white flakes. *Arch. Anim. Nutr.* **2005,** *59,* 365–375.

Russell, J.B.; J.D. O'Connor; D.G. Fox; P.J. van Soest; C.J. Sniffen. A net carbohydrate and protein system for evaluating cattle diets: I. Ruminal fermentation. *J. Anim. Sci.* **1992,** *70,* 3351–3361.

Schoenherr, W.; S. Davidson; L. Stephens; K. Wedekind; D. Harmon. Digestion and absorption of phytate in mature dogs. *FASEB J.* **2000,** *14,* A218 (Abstr.).

Sharda, D.P.; D.C. Mahan; R.F. Wilson. Limiting AAs in low-protein corn-soybean meal diets for growing-finishing swine. *J. Anim. Sci.* **1976,** *42,* 1175–1181.

Shelton, J.L.; M.D. Heman; R.M. Strode; G.L. Brashear; M. Ellis; F.K. McKeith; T.D. Bidner; L.L. Southern. Effect of different protein sources in growth and carcass traits in growing-finishing pigs. *J. Anim. Sci.* **2001,** *79,* 2428–2435.

Sherrod, L.B.; A.D. Tillman. Effects of varying the processing temperatures upon the nutritive value for sheep of solvent-extracted soybean and cottonseed meals. *J. Anim. Sci.* **1962,** *21,* 901–908.

Smiricky, M.R.; C.M. Grieshop; D.M. Albin; J.E. Wubben; V.M. Gabert; G.C. Fahey, Jr. The influence of soy oligosaccharides on apparent and true AA digestibilitties and fecal consistency in growing pigs. *J. Anim. Sci.* **2002,** *80,* 2433–2441.

Sniffen, C.J. Nitrogen utilization as related to the solubility of NPN and proteins in feeds. *Proceedings of the Cornell Nutrition Conference;* 1974; pp. 12–19.

Sohn, K.S.; C.V. Maxwell; D.S. Buchanan; L.L. Southern. Improved soybean protein sources for weanling pigs: I. Effects on performance and total tract AA digestibility. *J. Anim. Sci.* **1994,** *72,* 622–630.

Stern, M.D.; L.D. Satter. In vivo estimation of protein degradability in the rumen. *Protein Requirement for Cattle Symposium. Oklahoma State University;* 1982; pp. 57–71.

Sunvold, G.D.; G.C. Fahey; N.R. Merchen; G.A. Reinhart. In vitro fermentation of selected fibrous substrates by dog and cat fecal inoculum: Influence of diet composition on substrate organic matter disappearance and short-chain fatty acid production. *J. Anim. Sci.* **1995,** *73,* 1110–1122.

Tamminga, S. Protein degradation in the forestomachs of ruminants. *J. Anim. Sci.* **1979,** *49,* 1615–1624.

Thiagarajan, D.G.; M.R. Bennink; L.D. Bourquin; F.A. Kavas. Prevention of precancerous colonic lesions in rats by soy flakes, soy flour, genistein, and calcium. *Am. J. Clin. Nutr.* **1998,** *68,* 1394S–1399S.

Thomas, E.; A. Trenkle; W. Burroughs. Evaluation of protective agents applied to soybean meal and fed to cattle. II. Feedlot trials. *J. Anim. Sci.* **1979,** *49,* 1346–1355.

Tilton, S.L.; P.S. Miller; A.J. Lewis; D.E. Reese; P.M. Ermer. Addition of fat to the diets of lactating sows: I. Effects on milk production and composition and carcass composition of the litter at weaning. *J. Anim. Sci.* **1999,** *77,* 2491–2500.

Titgemeyer, E.C. Soy by-products as energy sources for beef and dairy cattle. *Soy in Animal Nutrition;* J.K. Drackley, Ed.; Federation of Animal Science Societies: Savoy, IL, 2000; pp. 238–257.

Titgemeyer, E.; N. Merchen; Y. Han; C. Parsons; D. Baker. Assessment of intestinal AA availability in cattle by use of the precision-fed cecectomized rooster assay. *J. Dairy Sci.* **1990,** *73,* 690–693.

Tokach, M.D.; J.E. Pettigrew; L.J. Johnston; M. Overland; J.W. Rust; S.G. Cornelius. Effect of adding fat and(or) milk products to the weanling pig diet on performance in the nursery and subsequent grow-finish stages. *J. Anim. Sci.* **1995,** *73,* 3358–3368.

Traylor, S.L.; G.L. Cromwell; M.D. Lindemann; D.A.Knabe. Effects of level of supplemental phytase on ileal digestibility of AAs, calcium, and phosphorus in dehulled soybean meal for growing pigs. *J. Anim. Sci.* **2001,** *79,* 2634–2642.

Trenkle, A. Response of finishing steers implanted with estradiol and trenbolone acetate to varying concentrations of dietary urea and soybean meal. *Iowa State Beef Report,* **1995,** *AS630,* 85–91.

van den Brand, H.; M.J.W. Heetkamp; N.M. Soede; J.W. Schrama; B. Kemp. Energy balance of lactating primiparous sows as affected by feeding level and dietary energy source. *J. Anim. Sci.* **2000,** *78,* 1520–1528.

van Kempen, T.A.T.G.; I.B. Kim; A.J.M. Jansman; M.W.A. Verstegen; J.D. Hancock; D.J. Lee; V.M. Gabert; D.M. Albin; G.C. Fahey, Jr.; C.M. Grieshop et al. Regional and processor variation in the ileal digestible AA content of soybean meals measured in growing swine. *J. Anim. Sci.* **2002,** *80,* 429–439.

van Kempen, T.A.T.G.; E. van Heughten; A.J. Moeser; N.S. Muley; V.J.H. Sewalt. Selecting soybean meal characteristics preferred for swine nutrition. *J. Anim. Sci.* **2006,** *84,* 1387–1395.

van Soest, P.J.; J.B. Robertson; B.A. Lewis. Methods for dietary fiber, neutral detergent fiber, and nonstarch polysaccharides in relation to animal nutrition. *J. Dairy Sci.* **1991,** *74,* 3583–3597.

Waldroup, P.W.; T.L. Cotton. Maximum usage levels of cooked full-fat soybean in all-mash broiler diets. *Poult. Sci.* **1974,** *53,* 677–680.

Waldroup, P.W.; B.E. Ramsey; H.M. Helwig; N.K. Smith. Optimum processing for soybean meal used in broiler diets. *Poult. Sci.* **1985,** *64,* 2314–2320.

Whetsone, H.D.; C.L. Davis; M.P. Bryant. Effect of monesin on breakdown of protein by ruminal microorganisms in vitro. *J. Anim. Sci.* **1981,** *53,* 803–809.

Wiernusz, C.J.; R.G. Shields; D.J. Van Vlierbergen; P.D. Kigin; R. Ballard. Canine nutrient digestibility and stool quality evaluation of canned diets containing various soy protein supplements. *Vet. Clin. Nutr.* **1995,** *2,* 49–56.

Woodworth, J.C.; M.D. Tokach; R.D. Goodband; J.L. Nelssen; P.R. O'Quinn; D.A. Knabe; N.W. Said. Apparent ileal digestibility of AAs and the digestible and metabolizable energy of dry extruded-expelled soybean meal and its effect on growth performance of pigs. *J. Anim. Sci.* **2001,** *79,* 1280–1287.

Yamka, R.M.; D.L. Harmon; W.D. Schoenherr; C. Khoo; K.L. Gross; S.J. Davidson; D.K. Joshi. In vivo measurement of flatulence and nutrient digestibility in dogs fed poultry by-product meal, conventional soybean meal, and low-oligosaccharide low-phytate soybean meal. *Am. J. Vet. Res.* **2006,** *67,* 88–94.

Yamka, R.M.; B.M. Hetzler; D.L. Harmon. Evaluation of low-oligosaccharide, low-phytate whole soybeans and soybean meal in canine foods. *J. Anim. Sci.* **2005,** *83,* 393–399.

Yang, Y.X.; Y.G. Kim; J.D. Lohakare; J.H. Yun; J.K. Lee; M.S. Kwon; J.I. Park; J.Y. Choi; B.J. Chae. Comparative efficacy of different soy protein sources on growth performance, nutrient digestibility and intestinal morphology in weaned pigs. *Asian-Aust. J. Anim. Sci.* **2007,** *20,* 775–783.

Yen, J.T.; G.L. Cromwell; G.L. Allee; C.C. Calvert; T.D. Crenshaw; E.R. Miller. Value of raw soybeans and soybean oil supplementation in sow gestation and lactation diets: A cooperative study. *J. Anim. Sci.* **1991,** *69,* 656–663.

Zentek, J. Influence of diet composition on the microbial activity in the gastro-intestinal tract of dogs. III. In vitro studies on the metabolic activities of the small-intestinal microflora. *J. Anim.*

Physiol. Anim. Nutr. **1995,** *74,* 62–73.

Zhang, P.; D.P. Laflamme. Use and future prospects for use of soy products in companion animal diets. *Opportunities for Soy Products in Animal Nutrition;* J.K. Drackley, Ed.; Federation of Animal Science Societies: Savoy, IL, 1999; pp. 45–55.

Zhang, Y.; C.M. Parsons; K.E. Weingartner; W. Wijeratne. Effects of extrusion and expelling on the nutritional quality of conventional and Kunitz trypsin inhibitor-free soybeans. *Poult. Sci.* **1993,** *72,* 2299–2308.

Zhu, X.; D. Li; S. Qiao; C. Xiao; Q. Qiao; C. Ji. Evaluation of HP 300 soybean protein in starter pig diets. *Asian-Aust. J. Anim. Sci.* **1998,** *11,* 201–207.

Zuo, Y.; G.C. Fahey; N.R. Merchen; N.L. Bajjalieh. Digestion responses to low oligosaccharide soybean meal by ileally-cannulated dogs. *J. Anim. Sci.* **1996,** *74,* 2441–2449.

19

Soy Protein Products, Processing, and Utilization

Nicolas A. Deak[1], Lawrence A. Johnson[2], Edmund W. Lusas[3], Khee Choon Rhee[3]

[1]Center for Crops Utilization Research, Iowa State University, Ames, IA 50011; [2]Department of Food Science & Human Nutrition; Director, Center for Crops Utilization Research, Iowa State University, Ames, IA 50011; [3]Retired, Food Protein Research and Development Center, Texas A&M University, College Station, TX 77843

Introduction and Definitions

Soybean Proteins

Probably no other plant protein, or for that matter any other crop, is studied more than soy protein and soybeans. Perhaps this is due to the fact that no other grain contains as much protein—on the order of 40% moisture-free basis (mfb). Soybeans deposit nitrogen and photosynthate as storage proteins to support the germinating plant. Storage proteins are stored in protein bodies and are thus separated from degrading enzymes until germination.

Soy Protein Structure

In the older literature, soy proteins were classified into three groups by their sedimentation coefficients using ultracentrifugation, namely 2*S*, 7*S*, and 11*S* proteins. Each group, however, is a heterogeneous mixture of proteins, not a single protein. The 2*S* proteins account for 8–22% of the extractable protein, are heat-sensitive, and predominantly consist of protease inhibitors and enzymes. The 7*S* proteins account for about 35% of the soluble protein and comprise enzymes, hemagglutenins, and storage proteins known as the 7*S* globulin or β-conglycinin. β-Conglycinin, a globular storage protein (Fig. 19.1), comprises about 85% of the 7*S* protein fraction. β-Conglycinin is a trimer or hexamer composed of two similar cystiene-containing peptides, α and α′, and a glycosylated, noncystiene-containing β peptide. These peptides are present in seven different combinations ($\alpha'\beta_2$, $\alpha\beta_2$, $\alpha\alpha'\beta$, $\alpha_2\beta$, $\alpha_2\alpha'$, α_3, and β_3) and undergo complex associations–disassociations depending on pH and ionic strength. The 11*S* proteins are mostly glycinin and comprise 31–52% of the soluble protein. Glycinin is a hexamer, although older literature indicates it is dodecamer, of nonrandomly paired acidic and basic peptides (Fig. 19.2). Seven acidic and eight basic peptides are identified. The Chapter *Soybean Proteins* addresses soy protein structure and chemistry in more detail.

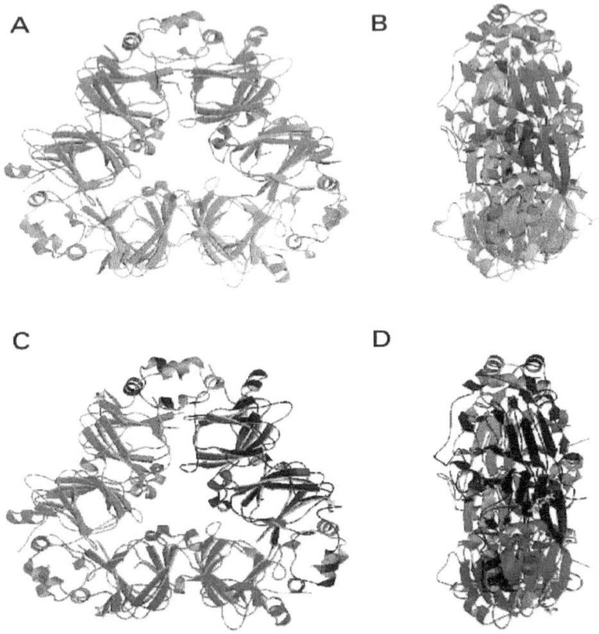

Fig. 19.1. β-Conglycinin structure (Source: Maruyama et al., 2001).

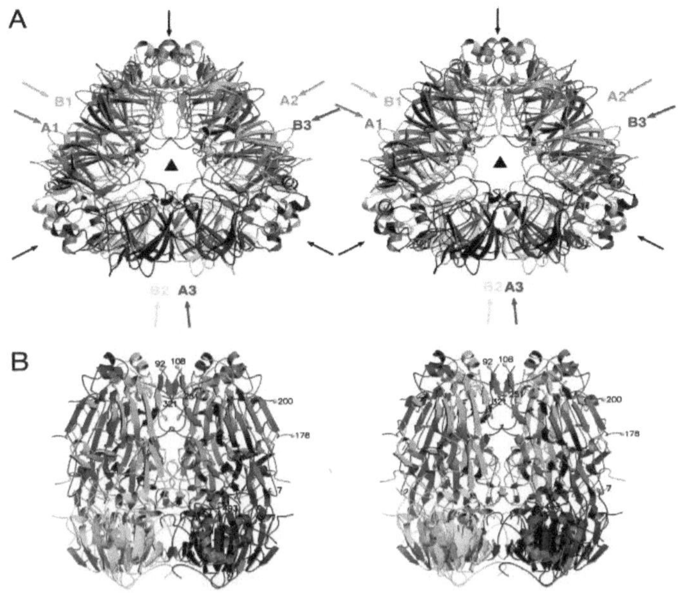

Fig. 19.2. Glycinin structure (Source: Adachi et al., 2003).

Glycinin and β-conglycinin comprise about 65–80% of the total protein in soybeans, and nearly all of the protein recovered in refined soy protein products are these two proteins. Recently, soybeans and soy protein products became recognized as having potential health benefits, and the U.S. Food and Drug Administration (FDA) approved a cholesterol-lowering health claim for soy proteins, indicating that daily consumption of 25 g of soy protein (6.25 g of soy protein per serving) may lower LDL cholesterol in individuals who have high cholesterol and who adhere to a low-fat diet (FDA, 1999). β-Conglycinin is believed by some experts to be responsible for those benefits, and this thinking has driven recent interest in producing fractionated soy protein products.

Protease Inhibitors

Protease inhibitors in soybeans, known as *trypsin inhibitors* (TIs), play important roles in nutritional properties of soybeans and soy protein products. Two types of TI are the Kunitz inhibitor and the Bowman-Birk inhibitor. The Kunitz inhibitor has a MW of 21,500 with two disulfide bonds, while the Bowman-Birk inhibitor has a MW of 7,900 with seven disulfide bonds (Wolf, 1977). The large ratio of disulfide bonds to MW in the Bowman-Birk inhibitor stabilizes protein conformation and makes the Bowman-Birk inhibitor highly resistant to heat denaturation and inactivation. The Kunitz inhibitor inhibits trypsin, while the Bowman-Birk inhibitor inhibits both trypsin and chymotrypsin. The kinetics of TI inactivation when heating at high water activity were determined by Johnson et al. (1980); they estimated that 83–91% of the TI activity in soybeans is due to the Kunitz inhibitor. Recently, the Bowman-Birk inhibitor was attributed cancer-protecting qualities; and also interest exists in using purified soy TI to treat AIDS patients (Kennedy, 1995, 1998; Kennedy & Szuhaj, 1994).

Soy Protein and Health

Considerable evidence establishes that soy protein products have important health benefits. Soy protein products (6.25 g of soy protein/serving and 25 g soy protein/day) may reduce the risk of heart disease when consumed in diets that are low in fat (<3 g/serving), saturated fat (<1 g/serving), cholesterol (<20 mg/serving), and sodium (<480 mg/serving for individual foods, <720 mg/serving for a main dish, <960 mg for a meal). Obesity is also a growing national epidemic in the United States, and soy protein products can assist in weight reduction and control. Food allergens are prevalent in infants; as many as 7% of children may be allergic to cow's milk. Soy protein ingredients are often used in food products recommended for those who are allergenic to cow's milk or are intolerant of lactose. Soy protein products may play a role in preventing osteoporosis. Intake of soy foods is also linked with reduced risk of certain cancers (Messina et al., 1994). Much research has recently focused on potential health-promoting isoflavones in soybean products. P. Murphy at Iowa State University established a database on isoflavone contents of various soy products that is found on the Internet at http://nal.usda.gov./fnic/foodcomp/Data/iosflav/html. Soy isoflavones may also reduce the discomfort experienced by menopausal women. Additional information on health benefits of soybean products is discussed in the Chapter: *Human Nutrition Value of Soybean Oil and Protein.*

Types of Products

The term *soy proteins* typically refers to processed, dry soybean products other than animal feed meals. Many types of protein products are produced for use in human and pet foods and in milk replacers and starter feeds for young animals. Some soy protein products are also used in biobased products (e.g., plastics, adhesives, paper coatings), and others are being aggressively researched by industry and public-sector research institutions, especially recently as petroleum prices rise (Johnson, 1992a; ISU, 1994). The many soy protein products and their uses are shown in Fig. 19.3.

Full-fat soy flours (FFSFs) and *full-fat grits* are the least processed soy protein products that are produced by grinding dehulled soy cotyledons to specific sizes and typically contain 40% protein (N × 6.25) on an "as is" basis (with moisture). *Defatted soy flours (SFs) and grits* are prepared by milling soybean meal from soybeans that have had the oil removed by solvent extracting dehulled and flaked soy cotyledons and contain 52 to 54% protein (as is). For flour, at least 97% of the material must pass through a U.S. Standard No. 100 sieve, whereas grits are milled to specific particle size ranges to pass through sieves between U.S. Nos. 8 and 80, depending on the manufacturer's or buyer's specifications. FFSFs and defatted SFs are available in enzyme-active forms or in various degrees of water solubility, expressed as Protein Dispersibility Index (PDI) or Nitrogen Solubility Index (NSI) with the differences being in the amounts of heat treatment and extents of protein denaturation (higher PDI and NSI indicate less protein denaturation because denatured soy protein, which forms insoluble protein, aggregates). *Enzyme-active* products are often used in bakery applications that are discussed later. *Re-fatted (0.5 to 30%) flours* are used in applications where crude soybean oil flavor is not acceptable, but dustiness must be minimized or fat must be provided in the formulation. *Lecithinated (0.5 to 30%) flours* are made for applications in which rapid dispersibility of powders facilitated by the emulsifying action of lecithin is desired.

Soy protein concentrates (SPCs) contain 65% minimum protein (mfb) and for the most part, are flours from which the water- or alcohol-soluble components, especially flatulence-promoting sugars and strong flavor compounds, are leached (or extracted). Accumulation of intestinal gas, flatulence, results from the presence of α-linked oligosaccharides, mainly raffinose and stachyose. These two sugars are nonreducing and composed of one or two galactose units linked to sucrose. Humans and other monogastric animals lack α-1,6-galactosidase in their intestinal mucosa. When ingested,

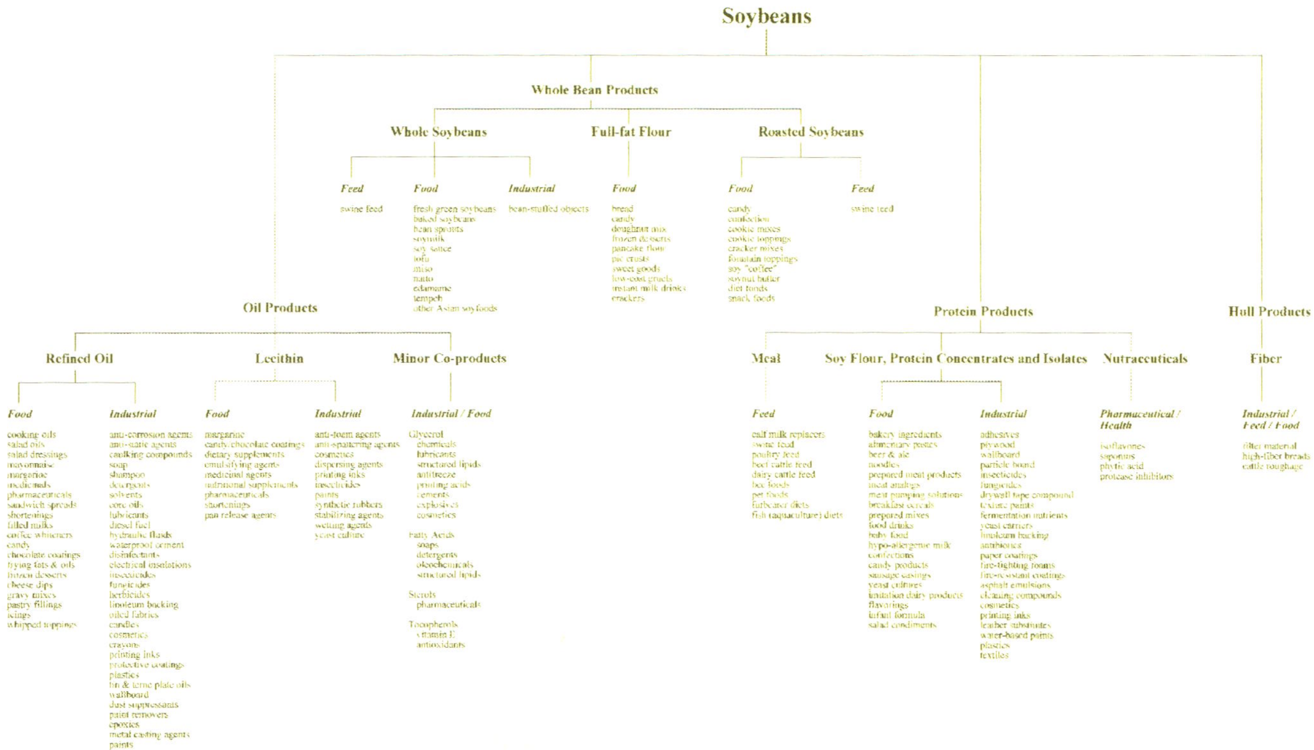

Fig. 19.3. Soybean uses (Source: printed with permission of L.A. Johnson, Center for Crops Utilization Research, Iowa State University, Ames, IA).

these soluble sugars remain unabsorbed in the small intestine and pass into the lower large intestine, where they are metabolized by intestinal microflora, which contain the enzyme, leading to gas production (Liener, 1994).

In making *soy protein isolates* (SPIs), which contain 90% protein minimum (mfb), the fiber (cell walls referred to as *soy cotyledon fiber*) is removed as well as the sugars from dehulled and defatted SF. Functionalities (performance properties in food and biobased products) of SPC and SPI may be modified by adjusting the pH with sodium or calcium bases, application of mechanical stress (i.e., homogenization), hydrolysis by proteolytic enzymes before drying (Soy Protein Council, 1987), and jet cooking (Wang & Johnson, 2001). Proximate composition ranges for soy protein products on as is basis and mfb, reported by the Soy Protein Council (Endres, 2001), are shown in Table 19.1.

Other soy protein products include dried soy milk and tofu (see the Chapter: *Food Use of Whole Soybeans*), as well as mixtures of SF, SPC, or SPI with milk or egg protein, gelatin, or other components for specific functional applications. *Extruder-texturized flours and concentrates* and *spun fiber isolates*, which resemble meat in appearance, may be made at the point of use but usually are supplied in bulk from strategically located production facilities. Edible co-products of soy protein ingredient manufacture include the hulls and the cotyledon fiber (cell walls) from SPI or soymilk production. The extract by-product from SPC may be used as the feedstock for isolating isoflavones.

Most soy protein products are sold in bulk as ingredients for meat processing, baking, and remanufacturing into retail grocery store, fast-food shop, institutional, and restaurant convenience foods. Except for the garnish *imitation bacon bits*, most forms of soy protein ingredients are seldom recognized by consumers. Limited quantities of SFs, SPCs, and SPIs, texturized chunks, and processed organically grown

Table 19.1. Typical Compositions (%) of Soy Protein Products

	Defatted Flours and Grits		Protein Concentrates		Protein Isolates	
Constituent	as is	mfb[a]	as is	mfb[a]	as is	mfb[a]
Crude protein (N × 6.25)	52–54	56–59	62–69	65–72	86–87	90–92
Crude free lipid (pet. ether)	0.5–1.0	0.5–1.0	0.5–1.0	0.5–1.0	0.5–1.0	0.5–1.0
Crude fiber	2.5–3.5	2.7–3.8	3.4–4.8	3.5–5.0	0.1–0.2	0.1–0.2
Ash	5.0–6.0	5.4–6.5	3.8–6.2	4.0–6.5	3.8–4.8	4.0–5.0
Moisture	6–8	0	4–6	0	4–6	0
Carbohydrates (by difference)	30–32	32–34	19–21	20–22	3–4	3–4

[a]mfb: moisture-free basis.
Source: Endres (2001).

soybean products are sold through health food stores. Products with *pareve* and *kosher* certification are available.

History of Soy Protein Products

Follow the development of soy protein processes and uses through publications including the summaries of Johnson et al. (1992a,b); handbooks by Circle (1950) and Smith and Circle (1972); patent abstracts (Hanson, 1974; Martz, 1981); and proceedings of the American Oil Chemists' Society World Conferences in Munich, 1973 (AOCS, 1974), Amsterdam, 1978 (Baldwin, 1976), Acapulco, 1980 (AOCS, 1981), Singapore, 1988 (Applewhite, 1989), and Budapest, 1992 (Applewhite, 1993). Notable bibliographies were prepared by Wolf and Cowan (1975) and Shurtleff and Aoyagi (1989abcd). The major utilization research efforts for soybeans focus on soy milk and tofu, soy protein products as aids in meat processing, extrusion processing, genetic modification of end-user traits, and functional foods and nutraceutical properties. Since the global enforcement of 200-mile offshore commercial fishing limits and the consequential interest in substitutes for fish products, Japanese scientists have become the most prominent national group in soy protein research publications; South Korea and the United States are also well-represented.

The preparation of soy protein products is described in this chapter, from the simpler to the more complex, but the technology did not develop in such an orderly fashion. Crude forms of SF were sold in the United States as early as 1926, as *health flours* (Smith & Circle, 1972). SFs were initially made from whole soybeans, then from batch hydraulic-pressed or continuous screw-pressed (expeller) cakes, and later from solvent-extracted meals.

Flavor

The strong beany flavor in early soy protein products limited market growth in food of Western cultures, and considerable research effort was invested in developing *debittering* processes to remove objectionable odor and flavor compounds. *Debittering* is a term erroneously used in early literature to refer to removing beany and grassy flavors. The source of these flavors is attributed to the enzyme lipoxygenase that catalyzes lipid oxidation of linoleic and linolenic fatty acids. Three lipoxygenase isoenzymes are known, and soybean varieties devoid of these isoenzymes were developed by traditional breeding, but, while products made from these beans have improved flavor, they are not totally bland. Some speculate that additional iso-lipoxygenases or other flavor-altering enzymes are also active in soybeans. In the presence of moisture, the lipoxygenases act rapidly in disrupted soy cotyledon tissue during processing to develop beany flavor and odor, and these enzymes continue to foster oxidative rancidity during storage unless deactivated. At high water activity, lipoxygenase can cause strong beany flavors within less than 1 min, possibly within a few seconds (Johnson,

1978). Recognition of this source of bad flavors led to improved processes and products. In 1966, Mustakas et al. (1966) reported on an extruder cooking process for making toasted FFSF for use in developing countries. Heating soybeans to deactivate lipoxygenases and other enzymes before milling, and rushing the meats (cotyledons) into heat treatment after cracking and dehulling resulted in modern SFs low in beany flavor.

Lipoxygenase- and other enzyme-catalyzed reactions, however, are not the sole source of beany flavors in soy products. Soybean oil is highly polyunsaturated, and these lipids are also prone to nonenzymatic oxidation, which can affect flavor characteristics of soy products.

Chemurgy

SPIs were developed next and introduced into the marketplace in 1937 by the Glidden Company (Chicago, IL) as replacements for bovine milk casein used for binding pigments in paper coatings and for producing blanketing foams for fire fighting (Johnson et al., 1992b). Henry Ford, the founder of the Ford Motor Company (Detroit, MI), and his research engineer R. Boyer were also early developers of biobased products from soybeans and used SPI in plastics. The 1930s and 1940s were periods of intense interest in using agricultural products in what is now referred to as *biobased products* but known at the time as *chemurgy*. In 1941, Ford promoted a prototype automobile in which the body was made with soy protein plastic; a few of these types of automobile parts were used in his automobiles for one to two years. The availability of SPI also led to research on spinning soy protein into textile fibers, which did not materialize as a U.S. industry; but soy protein textile fibers were reportedly produced in Japan during WWII. SPI fibers, however, were commercially spun as tows and worked, flavored, and colored to mimic meat products during the late 1960s and early 1970s (Smith & Circle, 1972; Johnson et al., 1992b). Short spun fibers were developed by Worthington Foods (Worthington, OH) under the guidance of R. Boyer and are produced today for use as texturizing agents in fabricated food products.

Early Food Uses

Research on debittering (or deflavoring) SF showed that extraction with aqueous ethanol was the preferred means for removing strong flavor components and flatulence sugars, despite causing protein denaturation and loss of solubility and native functionality. SPCs were first produced commercially in the early 1950s, also for industrial applications in biobased products (Campbell et al., 1985). A patent issued to E. Morse in 1943 for a process to remove sugars, salts, and other soluble materials by leaching SF at the isoelectric point (pH 4.5) was one of the first descriptions of commercial SPC (Morse, 1943).

In 1959, L. Sair (Sair, 1959) developed the first process specifically for making

SPCs for food use. Defatted SFs were acid-leached and then neutralized before drying. With this development, SPCs were rapidly accepted as food ingredients, intermediate (between SF and SPI) in cost, protein content, and flavor. Cooking-texturizing extruders also were introduced at about this time and enabled the production of texturized SF or SPC for use as meat analogs at appreciably lower costs than those produced by the SPI spinning process.

Water solubility is considered an index of a soy protein's functional and physical properties, and biological and enzymatic activities. Hexane extraction of soy flakes is essentially an anhydrous process, with very little water present to react with the protein. Approximately 40% of the weight of the *marc* (extracted drained flakes) is *hold-up* solvent, and energy must be supplied for its vaporization. Energy for evaporating hexane in desolventizer-toasters (DTs) is provided by contact heating and direct steam injection. Hexane and water form a low-boiling azeotrope at about 94.5:5.6 wt:wt ratio. Complete evaporation of hexane is favored by excess steam condensate in the extracted flakes. Both moisture and heat exposure are detrimental to protein solubility but inactivate TI that reduces feed conversion in livestock (especially swine and poultry). NSI values seldom reach >70 in meals processed by contact-heated DTs, nor >50 NSI when steam is injected. A recent survey by Wang and Johnson (2001) indicates the usual PDI of meal processed by a DT ranged from 27 to 62 with the majority in the range of 40 to 50.

The development of flash-desolventizing systems (FDSs) capable of desolventizing marc without the addition of steam enabled the production of *white flakes* (WFs) with high NSIs. In this process, the marc is transported by superheated solvent vapors through a desolventizing tube. As heat is surrendered, the adhering solvent is evaporated, and the vapors are swept away to a condenser. Little protein denaturation or loss in protein solubility occurs due to the absence of moisture or steam contact. Alternatively, the desolventized WFs may be passed to a heated tube to achieve any PDI desired. Additional information preparing WFs is provided in the Chapter: *Oil Recovery from Soybeans.*

Analysis

The soy protein industry uses the applicable *Official Methods and Recommended Practices of the American Oil Chemists' Society* (AOCS, 1999) along with annual updates for analyses in trading and litigation. Access these methods through the Internet at www.aocs.org/tech/onlinemethods/.

Protein Content

The vast majority, but not all, of the nitrogen in soy protein ingredients is of protein origin. The AOCS conversion factor for soybean protein is N × 6.25, consistent with industry practice of calculating protein content. Nitrogen may be determined by

Kjeldahl analysis or by the Dumas (combustion analysis) method using the Dumas-to-Kjeldahl conversion relationship of Jung et al. (2003), where:

$$\text{Kjeldahl Protein Content} = -0.00536 + 0.97188 \times \text{Dumas Protein Content}$$
$$R^2 = 0.9997$$

Protein Solubility

As soy protein denatures, it forms insoluble aggregates. Two methods, NSI and PDI, are broadly used to evaluate protein solubility/dispersibility in soy protein products. The PDI (AOCS Official Method Ba-10-65, 1993) *rapid stir* method uses a blender to disperse the sample, and the NSI (AOCS Official Method Ba 11-65, 1993) *slow stir* method uses a laboratory stirrer. In both methods, the protein or nitrogen leached into the liquid phase is compared with the total protein or nitrogen in the sample as determined by Kjeldahl analysis. The NSI method usually gives lower values and is related to PDI by the formula (Central Soya Company, 1988):

$$\text{PDI} = 1.07(\text{NSI}) + 1$$

One exception that does not respond as indicated by the formula above is meal produced by gas-supported screw pressing (Deak et al., 2007b).

Antinutritional Factors

One objective in heating soy protein products is to inactivate TI (primarily Kunitz-type), which acts as protease inhibitors and anti-growth factors by restricting protein digestion in monogastric animals. At least 80% reduction of the approximately 85 to 95 trypsin inhibitor units (TIU)/mg of solids, normally present in raw soy flour, is sought. The rationale for this value is that test animals consistently demonstrate tolerance to low levels of TI and the soy ingredient will receive additional heat treatment before consumption (Rackis, 1981). Over-toasting soybean meal will damage essential amino acids such as lysine.

A relationship between TI activity, protein efficiency ratio (PER), and steaming time of soybean meal is shown in Fig. 19.4 (Rackis, 1974). PER values, determined by rat-feeding tests, are no longer used to assess quality of food proteins (the balance of essential amino acids), because of extraordinarily high requirements for sulfur-containing amino acids by rats and fur-bearing animals. These values, however, are cited as the best indexes of protein quality in early research literature. The current technique for evaluating protein quality for adults and children over one year of age is the Protein Digestibility-Corrected Amino Acid Score (PDCAAS) (FAO, 1990; FDA, 1991; Madl, 1993).

Only 40% of growth inhibition in test animals, however, is related to TI activ-

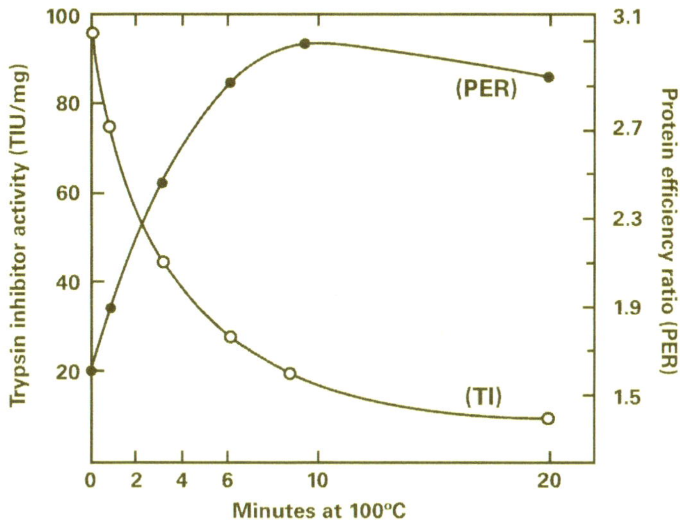

Fig. 19.4. Effect of atmospheric steaming on trypsin inhibitor activity and protein efficiency ratios of soybean meal fed to rats (Source: Rackis, 1974).

ity. Heating also partially inactivates heat-labile hemagglutinins (lectins), goitrogens, anti-vitamins, and phytates, but not the heat-stable saponins, estrogens, flatulence factors, and allergens. Furthermore, heat denaturation in itself increases digestibility of soy proteins (Liener, 1981). Remember that many of the antinutritional factors and enzymes that affect product quality in soybeans are rich in essential amino acids and beneficial to the diet once inactivated.

Urease

Urease activity has come to be used as an index for TI activity in some applications and is much easier to analyze. Urease catalyzes the hydrolysis of urea to ammonia and carbamate, and activity can be easily measured by changes in pH. Albrecht et al. (1966) showed that TI activity is destroyed at approximately the same rate as urease in soybean meal (Fig. 19.5); however, the authors' own experiences indicate this is not always the case, and one must be very cautious in using urease activity as a test for adequate heat treatment because the kinetics of urease and TI inactivation are very different under some conditions. High initial moisture promotes rapid decrease in both NSI and urease. Particle size influences the reduction of urease activity but has little effect on the rate of NSI reduction. By steaming soybean pieces of small particle size (<20 mesh) and low moisture (<8%), it is practical to destroy urease activity and retain high NSI. Figure 19.6 shows. the relationship between urease activity and NSI in a typical atmospheric steaming process (Wright, 1981), and Fig. 19.7 shows the relationship between PDI and TI for solvent-extracted meal.

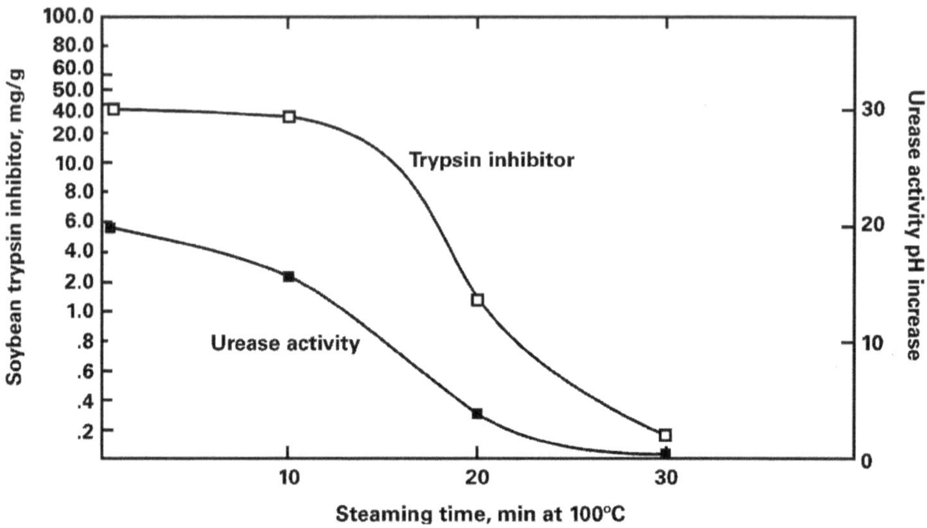

Fig. 19.5. Relationship of urease activity to trypsin inhibitor (Source: Wright, 1981).

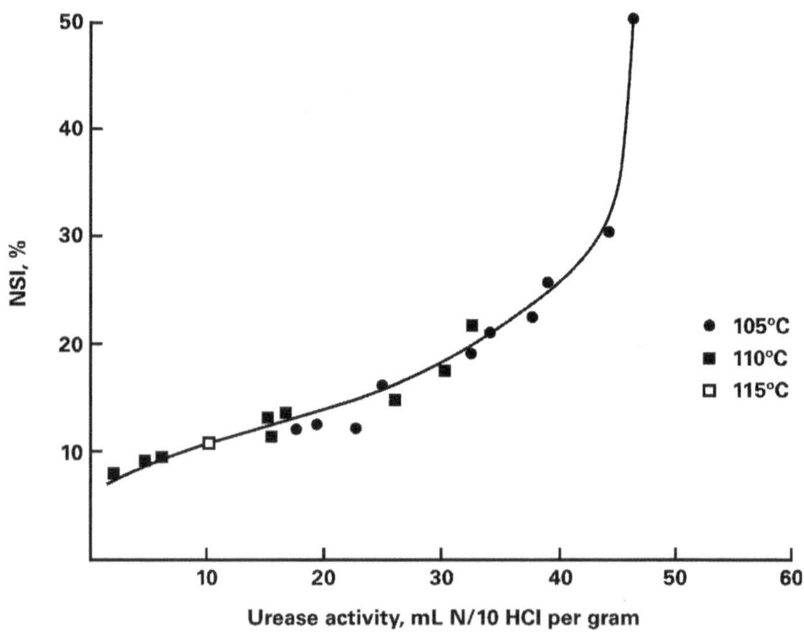

Fig. 19.6. Relationship of urease activity to Nitrogen Solubility Index (Source: Wright, 1981).

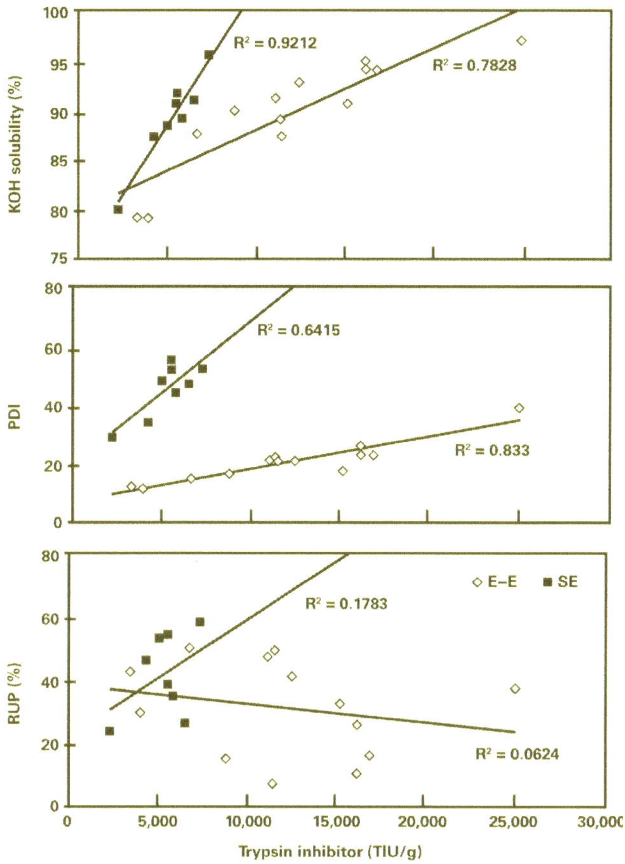

Fig. 19.7. Correlation between trypsin inhibitor activity and alkaline (HOH) solubility, PDI, and RUP of soybean meals; E-E designates extrusion-expelling and SE designates solvent extraction (Source: Wang & Johnson, 2001).

Processing Soybeans and Soy Protein Products

Considerable trade secrecy exists about the processing of commercial soy protein products, and the combinations of options for making different products are limitless. Even in making the same type of product, techniques and equipment differ between manufacturers, which result in different products.

For all soy protein products, it is essential to start with thoroughly cleaned, sound, mature, yellow soybeans sorted to uniform size. They should be dried and subsequently handled at low moisture (9–10%) and mild heat if high-PDI (85+) enzyme-active products are to be made. Typically, food-grade soy protein products are produced on different lines than are used for feed meals, with split and rejected soybeans diverted to animal feed meal extraction operations (Kanzamar et al., 1993; Johnson, 1989). Some manufacturers have washed soybeans to remove dirt and small stones (Pringle, 1974). Specially designed extractors (with self-cleaning, no-flake-breakage features) and use of a narrow-boiling-range (66 to 68°C, 151 to 154°F) hexane are recommended for producing high-PDI WFs because mild heat treatment reduces brown color formation (Kanzamar et al., 1993).

Nearly complete dehulling is desired to produce WFs for edible protein products. McDonald (1978) reported that dehulling of soybeans is enhanced by drying on receipt—first rapidly at 79°C (174°F) to lower the moisture content to the range of 12.5 to 13.0%, and then at the lower temperature of 65°C (150°F) to the range of 9.0 to 10.0% moisture. The soybeans are stored in this condition for 15 to 30 days before processing. Two-stage conditioning, however, is not normally practiced when making only feed meals, and hot-dehulling methods may be employed in making feed meals.

Loss of protein solubility during storage is well known in whole soybeans and in soy proteins (McDonald, 1978; Saio and Ariska, 1978; Chiba et al., 1981). Changes in protein solubility can be slowed by lowering storage temperature and relative humidity (RH). Thomas et al. (1989) reported a 14% decrease in protein extracted into soy milk from soybeans stored for 8 mo at 30°C (86°F) and 85% RH, compared to those stored at 20°C (68°F) at 65% RH. Tofu from soybeans stored at 85% RH became less uniform in microstructure toward the end of the storage period. Glycinin extractability declines more rapidly than β-conglycinin (Saio et al., 1982). Storage at higher RH is more deleterious than higher temperature. Furthermore, at high storage RH and temperature for soybeans, the color of extracted soy milk darkened and pH decreased, and phosphatidylcholine became more associated with the proteins in soy milk (Saio et al., 1980). Some soy protein manufacturers recommend storing their products in multi-wall bags at <24°C (75°F) and <60% RH.

Full-fat Soy Flours and Grits

FFSFs and grits are the least refined soy protein ingredients. Three types of FFSFs are produced: (i) enzyme-active, (ii) toasted, and (iii) extruder-processed. Re-fatted flours are sometimes referred to as *fully-fatted* or *full-fat*, but they are described in this chapter in the section on defatted SF because the original fat is removed and replaced with refined, bleached, and deodorized (RBD) oil.

Enzyme-active Soy Flours

Enzyme-active SFs are used for the action of their native lipoxygenases in bleaching wheat flours by oxidizing β-carotene and in conditioning doughs for Western-type breads (lipoxygenases give more mixing tolerance). Soy β-amylases are more heat-stable than those of wheat or barley and remain active longer in the early stages of baking, improving texture. Enzyme-active SFs are available in full-fat and defatted forms, with the former more popular in Europe and the latter in the United States (Johnson, 1989).

The cleaned soybeans are cracked into 6 to 8 pieces, and the hulls are removed by aspirating. The hulls are loosened by adjusting moisture content and mild heating before cracking. Hulls are separated from the meats (cotyledon pieces) with shaker screeners and/or aspirators (Kanzamar et al., 1993; Johnson, 1989). The dehulled pieces then are ground with a hammer mill or an impact pin mill into flours with desired particle sizes. Full-fat products are very difficult to pulverize and sieve. Customarily, they are not screened but are milled in two steps with separation of the coarse from the fine particles by air-classification between grindings (Smith and Circle, 1978; Kanzamar et al., 1993; Johnson, 1989). One U.S. manufacturer's specifications for a commercial enzyme-active FFSF are: 42.0 ± 1% protein (mfb), 21.0 ± 0.5% fat, 4.7 ± 0.2% ash, 10.0% moisture (max), and granulation of <1% on a U.S. No. 45 screen. In recent years, keen interest arose in grinding to extremely fine particle sizes (<1000 mesh).

In 1991, a processing plant was constructed in Jefferson, Iowa (MicroSoy Corp.), to make easily-extractable full-fat flakes from white hilum soybeans for export to make soy milk and tofu. Using these flakes eliminates the need for soaking the beans prior to wet grinding reducing soymilk preparation time.

Toasted Full-fat Soy Flours and Grits

Toasted FFSFs are also called *heat-treated full-fat soy flours*. To minimize development of beany flavor by lipoxygenase, the cleaned whole soybeans are often steamed under slight pressure for 20 or 30 min, then cooled, dried, cracked, passed over a shaker screen and aspirated to remove hulls, and milled with sieving to produce full-fat grits or FFSF. When properly processed, these products are yellow to slightly tan in color, with nutty flavors and aromas. The undesirable enzymes are destroyed, and the product is in the range of 20 to 35 PDI. Toasted SFs, ground to U.S. No. 100 or 200 mesh, are available with special granulations possible. Specifications for grits vary with manufacturer, but one manufacturer's definitions are: Coarse Grits, through No. 10 screen but on No. 20; Medium Grits, through No. 20 but on No. 40; and Fine Grits, through No. 40 but on No. 80 (Smith & Circle, 1978; Kanzamar et al., 1993; Johnson, 1989).

Alternatively, dehulled soybean chips may be heated in rotary steam tube dryers or tray cookers and screw pressed to partially remove the oil and produce press cake (6–10% residual oil content). The press cake is broken, ground, and screened to produce low-fat SF or grits (Smith & Circle, 1978).

Extruder-prepared Full-fat Soy Flours

Extruded FFSFs were pioneered by Mustakas et al. (1966, 1970) at the USDA Northern Regional Research Center (now known as the National Center for Agricultural Utilization Research), Peoria, Illinois. An extruder consists of a rotating screw in a barrel, both designed to compress relatively low-moisture (<18%) powders into a flowing mass, which can be sheared, cooked, cooled, and shaped into continuous extrudates (collets) that then can be cut into pieces. The extrudate expands (*puffs*) at the die and retains expanded volume if sufficient protein or starch is present to form a rigid matrix. If discrete shapes are not desired, the face plate, dies, and cutter are often replaced with one of several types of adjustable-cone discharges to produce granulated products. Screw and barrel designs vary with the process, desired product characteristics, and extruder manufacturer.

Depending on capital investment limitations, extruders may be heated by passing steam through the barrel jacket and/or by direct steam injection. Low-cost *autogenous* machines, which create heat by friction between the feed material and the surfaces of the barrel, screw and steam locks, and do not require steam-generating equipment, are also used (Wiejratne et al., 2004). This process is referred to as *dry extrusion*, and the extruder can be powered by a farm tractor and was originally developed to cook whole soybeans on-farm to prepare feeds.

Mustakas et al. (1970) obtained good results by cracking soybeans in corrugated roller mills and dehulling by using shaker screens and aspirators to obtain grits in the 12- to 30-mesh screen range. These were conditioned to inactivate lipoxygenase by dry heating for 6 to 8 min attaining 103 to 104°C (218 to 220°F) discharge temperature. Tempering to 20% moisture and extruding at 135°C (275°F) with 2-min retention resulted in 2.15 maximum dietary PER, 89% TI inactivation, 0.1 pH change urease activity, and 21 NSI.

Several types of extruders are used to prepare FFSFs, including the interrupted-cut flight extruder shown in Fig. 19.8. A flow diagram for making extruded FFSF is shown in Fig. 19.9.

Extruder-processed FFSFs were used to prepare high-protein content foods and beverage bases for worldwide child and infant feeding programs (Mustakas et al., 1971). Studies also show that lipids are quite stable in full-fat flours cooked to 30 NSI, but antioxidants are required to ensure stability in FFSFs processed to <20 NSI (Howard et al., 1980).

Considerable research in producing weaning foods for famine relief and improved nutrition in developing countries was conducted using low-cost extrusion cookers (LECs) (Jansen & Harper, 1980; Lorenz et al., 1980; Harper & Jansen, 1985). These are autogenous machines with simple screw and barrel configurations for dry extrusion (moisture <15%) and included the Brady Crop Cooker, the InstaPro Extruder (Fig. 19.10), the Anderson Grain Expander, and local designs. Approximately a dozen LEC plants were brought into production or designed for developing countries. FFS-

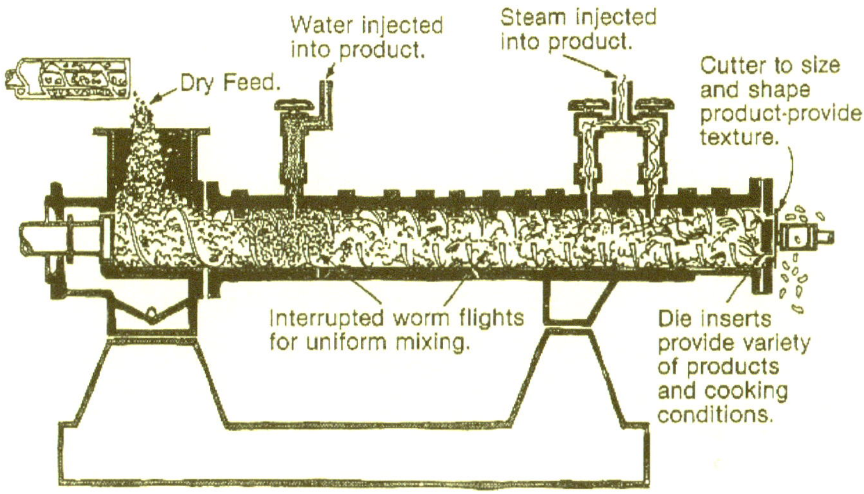

Fig. 19.8. Cross-section of an interrupted-flight extruder used for producing toasted full-fat soy flour (Source: Provided by the Anderson International Corporation, Cleveland, OH).

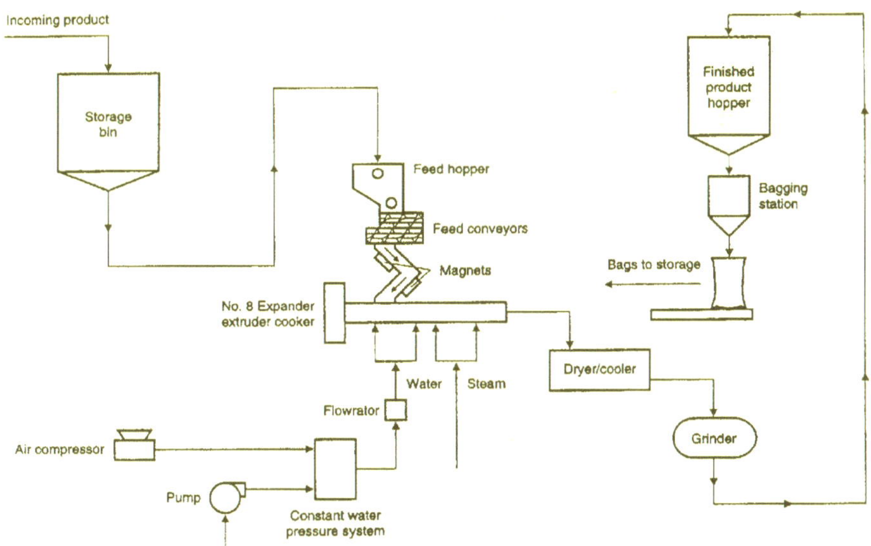

Fig. 19.9. Process diagram for making extrusion-cooked full-fat soy flour (Source: provided by the Anderson International Corporation, Cleveland, OH).

Fig. 19.10. Dry extruder used to prepare infant and child foods in developing countries (Source: provided by Insta-Pro International, Des Moines, IA).

Fs and co-processed mixtures of dehulled soy grits with corn, wheat, and rice were produced.

Nutritionists found that mixtures of approximately 50% oilseed or legume protein and 50% cereal protein are complementary in producing the essential-amino-acids profiles desired in infant and child foods. Because of differences in protein contents, this has meant the use of 30% soybeans with 70% cereal grains (Molina et al., 1983; Patil et al., 1990). Because of limitations in screw designs, coarsely ground ingredients worked better than flours in LEC machines. The protein efficiency ratio of FFSF was optimized by inactivating antigrowth factors by dry extrusion at 143°C (289°F) (Jansen & Harper, 1980). Others found that lipoxygenase is completely inactivated when extruding soybean and corn mixtures at 10% moisture in an InstaPro extruder at 127 to 160°C (260 to 320°F) die temperature. TI activity was reduced by 49–99% (Guzman et al., 1989).

Some LECs have the ability to grind whole soybeans and corn kernels into flours during processing. As an example, whole soybeans are pulverized and heated in InstaPro extruders to make TI-inactivated full-fat soybean meals for feeding poultry and swine. Doing so, however, bypasses the opportunity to reduce the fiber content through a separate dehulling step. In large installations, it is often more profitable to select specialized machines for grinding, preheating, extruding, and cooling when sufficient capital is available.

In some cases, soybean chips are passed through an autogenous extruder and then hard screw pressed to produce partially defatted cake, which is then ground and screened. Many of these operations, often known as extruding-expelling (EE), were constructed in the Midwest United States in recent years for preparing livestock feeds from undehulled soybeans and for recovering oil for food use or to convert to biodiesel. Only one EE plant, Iowa Soy Specialties, LLC (Vinton, IA; division of Kerry Food Ingredients), is manufacturing food-grade protein products from dehulled soybeans. This plant also re-extrudes EE press cake often from organic and/or non-GMO soybeans to produce partially defatted texturized products (http://www.uky.edu/Ag/AgEcon/pubs/ext_aec/ext2001-01.pdf).

Extracted Flake Products

By far, the majority of soy protein products are made from WFs (flash-desolventized, hexane-defatted flakes of dehulled soybeans graded for food use). The production of WFs is described in the Chapter: *Oil Recovery from Soybeans*. Johnson (1989), Fulmer (1989), and Kanzamar et al. (1993) also described the making of WFs. The Soy Protein Council (Endres, 2001) defined the following NSI categories for defatted flakes: White, 85+; Cooked, 20–60; and Toasted, <20. These terms, however, are used loosely in the industry. For example, *toasting* means steam cooking rather than dry heating, and *white flakes* means enzyme-active flakes or simply defatted flakes or meal used for making SF, SPC, or SPI.

The majority of soybean extraction plants produce oil and feed-grade meal by using a DT equipped for direct steam injection and sometimes an additional dryer to remove the condensate. Expanders, to shear soybean flakes and produce collets with enhanced oil extractability, also are common. The resulting extracted products have NSIs of 50 or less, because of protein denaturation by high temperatures and water activities.

In contrast, the production of flakes with high PDIs/NSIs typically utilizes a flash desolventizing system (FDS), also sometimes called a *white flake system*. Superheated gaseous hexane [boiling point 70°C (160°F)] under pressure at 116 to 138°C (240 to 280°F) pneumatically transports the extracted drained flakes in a desolventizing tube and evaporates the solvent hold-up within 2 to 5 sec. Moisture in the meal is also reduced by 3 to 5% during the process. The remaining hexane (about 0.3 to 0.5%) is then removed in a flake stripper using superheated steam under vacuum to obtain high-PDI flakes. Take care to avoid condensation of steam into water on flake surfaces (Smith & Circle, 1978; Kanzamar et al., 1993; Vavlitis & Mulligan, 1993). Once a FDS with solvent vapor circulation is installed, a person can operate it to produce flakes with a PDI range of 10 to 85%, depending on how much steam is applied.

White Flakes

WFs are an item of commerce in their own right. Some WFs are produced by soybean processors who supply soy protein to ingredient manufacturers. The major differences between WFs, grits, and SFs are granulations. White flours must be made carefully, since the properties and yields of subsequent products are critically important. One domestic supplier of high-PDI WFs offers a product with 86 to 88 PDI, 80% of the original lipoxygenase activity (bleaching activity), 2.2 minimum pH rise urease activity, and the granulation of 35% on U.S. No. 20, 45% through U.S. No. 20 but on U.S. No. 100, and 15% through U.S. No. 100.

Defatted Soy Flours and Grits

A general flowsheet for manufacturing FFSFs and defatted SF is shown in Fig. 19.11. Grinding WFs into grits is typically done by using hammer mills and sifters for sizing. Grinding into flours may be done by using hammer mills, pin mills, or classifier mills. Particle size distribution is controlled by air classifiers, with narrower particle size distributions possible when using sifters. Grinding capacity is dictated by flake PDI value and product mesh size. Commercial lines for grinding with a mill and air classifier are shown in Fig. 19.12, and for a classifier mill in Fig. 19.13 (Kanzamar et al., 1993).

During processing, protein content increases from 41% mfb for soybeans, to 49% for defatted, nondehulled soybean meal, and to 54% for dehulled soybean meal. Defatted SF contains about 38% total carbohydrates, including 15% soluble mono- and oligosaccharides, and 20% polysaccharides that are removed if SPIs are made (Table 19.2). Maximum total bacterial count specifications range up to 50,000/g depending on supplier and product. The presence of known disease-producing microorganisms, such as *Salmonella* and *E. coli,* must be negative (Fulmer, 1989a).

Relationships between heat processing and nutritional indicators of SFs are shown in Table 19.3. Increased heating reduces TI activity, which is reflected in lessened enlargement of the pancreas in test rats. Although minimum protein denaturation is desired to maximize SF solubility and functionality, the fabricated product must be adequately cooked before consumption (Fulmer, 1989a).

Recommended SF PDI values for various applications are shown in Table 19.4 (Fulmer, 1989ab). Domestic manufacturers generally offer SFs with PDIs of 90 (enzyme-active), 70, 65, and 20, with granulations of U.S. 100 and 200 mesh. Grits are offered in coarse, medium, and fine granulations. The bulk density of heat-treated

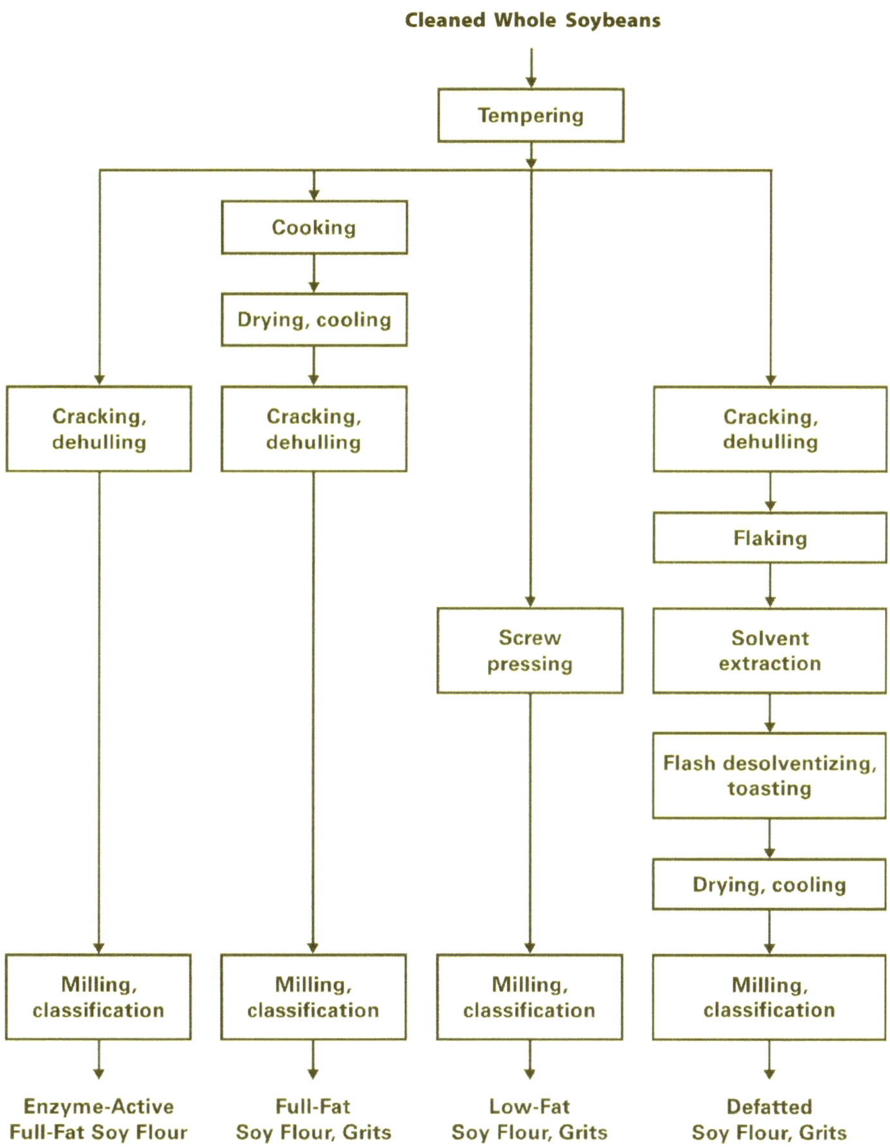

Fig. 19.11. Process flowsheet for manufacturing full-fat and defatted soy flours.

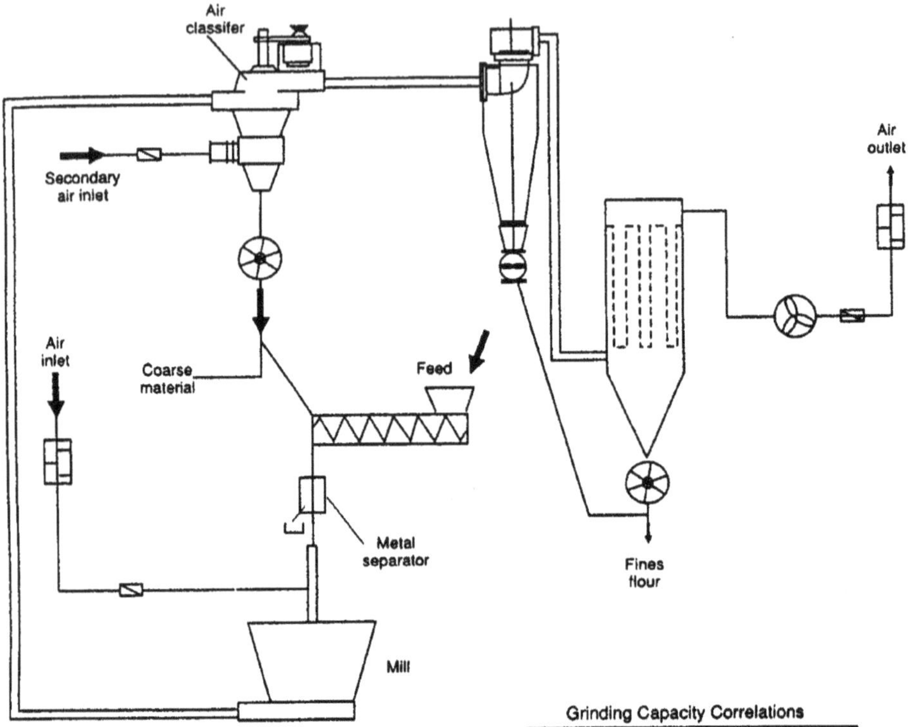

Grinding Capacity Correlations

WF \ Flour	Mesh 100	200	300
60 PDI WF	4	3	—
20 PDI WF	2.5 t/h	2	1.2
Full Fat	1.5	1.2	—

Fig. 19.12. Mill and air-classifying system for grinding defatted soy flour (Source: Kanzanar et al., 1993).

defatted SF is approximately 0.6 to 0.7/g/cm³ (37 to 44 lb/ft³).

Re-fatted or Lecithinated Soy Flours

SFs are re-fatted with 1 to 15% added fat to reduce dustiness and to provide fat for a product formula. Re-fatting extracted SF allows the use of bland RBD oil. Lecithinated SFs are offered with 3, 6, and 15% added lecithin. Lecithin improves dispersion of the SF and other admixed ingredients, especially in confection and cold beverage products. Generally, oil or lecithin is added to highly toasted flours. The compositions

Table 19.2. Carbohydrate Constituents of Dehulled Defatted Soybean Flakes

Carbohydrate	Source	Content, %
Monosaccharides		
Glucose	Cotyledons	0.3
Arabinose	Hulls	Trace to 0.1
Ribose	Nucleic acids	Trace to 0.1
Oligosaccharides		
Sucrose	Cotyledons	8.1
Maltose	Cotyledons	0.6
Raffinose	Cotyledons	1.1
Stachyose	Cotyledons	4.9
Verbascose	Cotyledons	Trace
Polysaccharides		
Arabinan	Cotyledons	15.0
Arabinogalactan	Cotyledons	5
Xylan (hemicellulose)	Hulls	3.5
Galactomannans	Hulls	Trace
Cellulose	Hulls	1-2

Source: Fulmer (1989a).

Table 19.3. Processing and Nutritional Parameters of Heat-treated Soy Flours

Heat[a], min	NSI[b]	TI[c], TIU/mg	PER[d]	Pancreas wt, g/100 g body wt
0	97.2	96.9	1.13	0.68
1	78.2	74.9	1.35	0.58
3	69.6	45.0	1.75	0.51
6	56.5	28.0	2.07	0.52
9	51.3	20.5	2.19	0.48
20	37.9	10.1	2.08	0.49
30	28.2	8.0		

[a]Live steam at 100 °C.
[b]NSI denotes nitrogen solubility index.
[c]TI denotes trypsin inhibitor; TIU, trypsin inhibitor units.
[d]PER denotes protein efficiency ratio, corrected on a basis of PER = 2.5 for casein.
Source: Fulmer (1989a).

Table 19.4. Applications of Defatted Soy Products in Foods

PDI[a]	Application
90+	White bread bleaching agent Fermentation Soy protein isolates, fibers
60-75	Controlled fat and water absorption Doughnut mixes Bakery mixes Pastas Baby foods Meat products Breakfast cereals Soy protein concentrates
30-45	Meat products Bakery mixes Nutrition, fat and water absorption, emulsification
10-25	Baby foods Protein beverages Comminuted meat products Soups, sauces and gravies Hydrolyzed vegetable proteins
Soy grits	Nutrition, meat extender Patties, meatballs and loaves, chili, sloppy Joes Soups, sauces and gravies

[a]Protein Dispersibility Index is a standard AOCS method (Ba 10-65) for measuring the amount of heat treatment used in the processing of soybean meal products.
Source: Fulmer (1989a).

of re-fatted and lecithinated SFs are primarily those of the carrier flour, diluted by the amount of oil or lecithin added.

Soy Protein Concentrates

SPCs contain at least 65% protein (N × 6.25) but < 90% mfb. The older definition of 70% minimum protein was replaced by the lower level in a definition promulgated by the U.S. Department of Agriculture's Food and Nutrition Service (USDA-FNS) in January 1983 (Campbell et al., 1985). Sucrose and total nondigestible oligosaccharides each account for about 8% weight of defatted soy flakes. SPC yield is about 75% of defatted flake weight.

In making SPCs, the objective is to insolubilize the protein while leaching away the solubles. Products are produced by three basic procedures (Fig. 19.14): (i) extraction of WFs with aqueous 20 to 80% ethyl alcohol; (ii) acid leaching of WFs or SF; and (iii) denaturing the protein with moist heat and extracting with water (Campbell

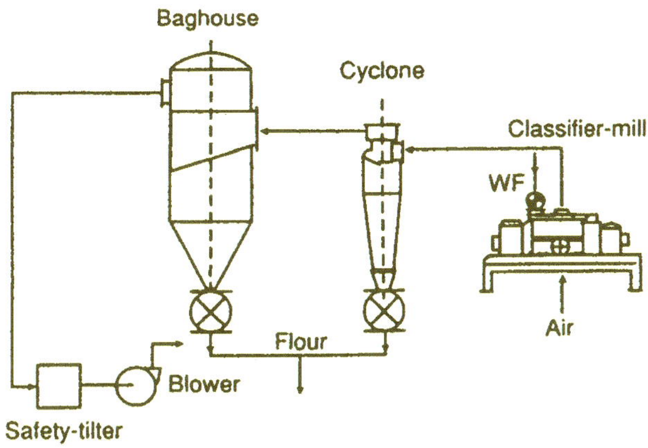

Fig. 19.13. Classifier mill system for grinding defatted soy flakes (Source: Kanzamar et al., 1993).

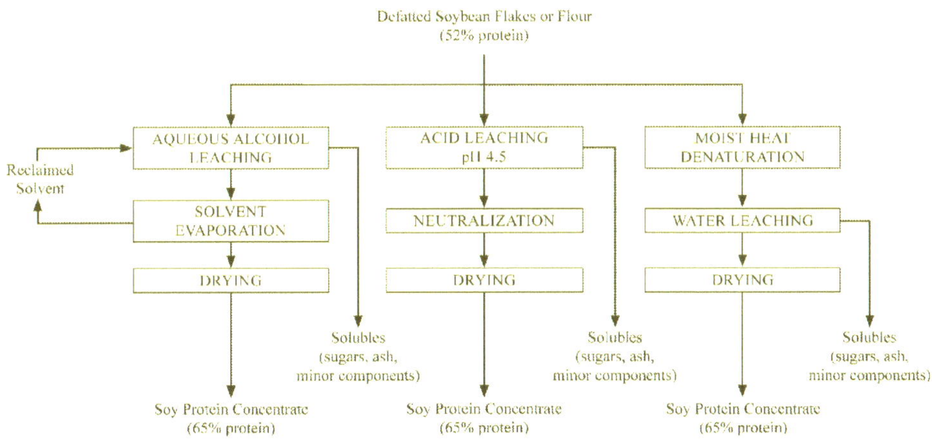

Fig. 19.14. Methods to prepare soy protein concentrates.

et al., 1985; Ohren, 1981). The preponderance of SPC today is made by the aqueous alcohol process because this process gives the least flavored product.

By using unground defatted flakes, SPC is produced with a countercurrent continuous chain or belt extractor, such as those made by Crown Iron Works (Minneapolis, MN) or Desmet Ballestra (Edegem, Belgium), respectively, and dried by a flash or vacuum drying system and ground. When WF flour is used, it is suspended at 1:10 or higher flour-to-solvent ratio, concentrated by decanter centrifuges, optionally extracted a second time, and spray-dried. For spray drying, domestic processors prefer nozzle sprayers, whereas rotary atomizers are more common in Europe.

Aqueous Alcohol Process

Mustakas et al. (1962) reported on a process for flash desolventizing soybean meals extracted with 50 to 70% alcohol. According to Campbell et al. (1985), the preferred ethanol concentration is 60% by weight, since soy protein solubility increases on either side of that concentration. The NSIs of SPCs made by aqueous alcohol extraction are low, sometimes in the range of 5 to 10, but are not necessarily related to poor functionality because the mechanism of denaturation is different from heating. Commercial versions can hold about 2.6 times their own weight in water of low-fat meat juice. They are used in meat patties, pizza toppings, and meat sauces and in conditions that stress the product, such as freeze-thaw cycles and extended holding times of precooked or cooked products. Alcohol-extracted SPCs are sometimes referred to as made by the *traditional process*.

Howard et al. (1980) patented a method in 1980 to regenerate high NSIs in ethanol-extracted SPCs. SPCs are subjected to successive pressure and cavitation, as by homogenization, at elevated temperatures and slightly alkaline conditions. The products, along with acid-leached SPCs, are sometimes called *functional SPC* and have very bland flavor. Recently, much of the lost solubility and functionality caused by ethanol extraction also was restored by jet cooking with or without alkali (Wang & Johnson, 2001). Presumably, disrupting the large aggregates of denatured protein by exposure to the combination of high temperature and shear is responsible for the improved properties (Wang et al., 2005).

Acid-leaching Process

The majority of soy proteins are globulins that are insoluble in water at their isoelectric point of pH 4.5. In the Sair process (1959), defatted soy flakes are leached with water at pH 4.5 to remove soluble sugars, neutralized, and spray-dried. Some loss of soluble proteins occurs, but the resulting SPC has a relatively high NSI, about 65 to 75 (Campbell et al., 1985).

A typical acid-leaching process uses 10:1 to 20:1 water to WF or flour ratio, hydrochloric acid for adjusting the pH to 4.5, and 30 to 45 min extraction at 40°C (104°F). A decanter centrifuge is used to separate the extract and concentrate the solids to about 20%. A second acid leach and centrifugation may be employed. The solids-rich slurry may be dried in acidic form, but is usually neutralized to pH 6.8 with sodium or calcium hydroxide before spray drying at 157°C (315°F) inlet air temperature and 86°C (187°F) outlet temperature.

Hot-water Leaching Process

A patent was issued to J. McAnelly (McAnelly, 1964) for a process in which a dough-like mass of defatted SF and water is developed, which then is heated under pressure to denature the protein and extruded to impart a porous structure that is leached with hot water (Campbell et al., 1985). This process is no longer commercially used.

Heat-denatured SF Process

An alternative approach to hot-water leaching in which heat denaturation is used to insolubilize soy protein for preparing SPC is to use EE or screw-pressed meal. Wang et al. (2004a) showed that SPC can be made with higher yield from EE soybean meal than from WFs, but the protein contents were lower.

Soy Protein Extracts

Considerable variation exists in raffinose (0.1-0.9%) and stachyose (1.4-4.1%) contents among soybean varieties (Hymowitz et al., 1972). Using molecular biology to produce genetically modified soybeans that are low in oligosaccharides and high in sucrose is possible (Crank & Sebastian, 2003). The utilization of high-sucrose/low-stachyose (HS/LS) soybean lines as starting SF offers the possibility for new soy protein ingredients. Two recent U.S. patents, Crank and Kerr (1999) and Johnson (1999), disclosed a new type of SPC based on removing the fiber while retaining the sucrose by merely extracting protein and sugars with alkali, centrifuging to separate the fiber-rich insoluble residue and spray drying the extract.

These new SPCs have somewhat lower solids and protein yields (~70 and ~81%, respectively) than conventional ethanol-leached SPC (~77 and ~93%, respectively). The new SPCs contained >65% protein (mfb) and high levels of isoflavones (Deak et al., 2006a). They also have higher sugar contents (~15%) than either traditional ethanol-leached SPC (~2.5%) or SPI (~1.5%); but the sums of stachyose and raffinose were only ~1% compared to ~1% for ethanol-leached SPC and 0.5% for SPI prepared from normal soybeans. The new SPCs have much higher water solubilities, better emulsification and foaming properties, and lower viscosities than ethanol-leached SPC (Deak & Johnson, 2006a). These new SPCs have a wide range of desirable functional properties that make them suitable as food additives and ingredients. Only small amounts of these new SPCs were produced for test marketing by one manufacturer (Solae, St. Louis, MO).

SPC Characteristics

Approximate compositions of SPCs made by the three processes are shown in Table 19.5. The most obvious difference is that ash contents are lower in SPCs prepared by acid or hot-water extraction, indicating more thorough removal of minerals. About 5 to 10% of the carbohydrates remaining in SPCs after leaching are soluble sugars, with the balance being insoluble polysaccharides. The amino acid compositions of SF, SPCs made by ethanol or acid extraction, and soy solubles from alcohol extraction are shown in Table 19.6. Of the essential amino acids, phenylalanine, tryptophan, methionine, and cystine, concentrate in the soy solubles fraction during alcohol extraction.

SPCs are offered in powder (95% through U.S. No. 100) or granular (90% retention on U.S. No. 60) forms as well as re-fatted or lecithinated forms. Typical bulk densities are: 0.40 to 0.45 g/cm^3 (25 to 28 lb/ft^3) for powders, 0.54 to 0.61 g/cm^3 (34 to 38 lb/ft^3) for granules, and 0.43 to 0.48 g/cm^3 (27 to 30 lb/ft^3) for 9% lecithinated products.

SPCs, SPIs, and texturized SFs and SPCs used for meeting a portion of the meat or meat alternative requirement in domestic school lunch and child nutrition programs must be fortified with vitamins and minerals, according to USDA-FNS requirements (Table 19.7). Separate fortification requirements exist for military ground beef applications (PP-B-2120B).

Soy Protein Isolates

Many processing options exist for making SPIs. A broad variety of commercial SPIs and SPCs already exists and can be readily put into use, including products designed for general use in processed meats, dry beverage and sauce mix formulations food bars, bakery, frozen desserts, and whipped products.

pH Extraction-precipitation

A water extraction/solubility curve for proteins from defatted soybean meal, in the pH range 0.5 to 12, is shown in Fig. 19.15 (Wolf & Cowan, 1975). Nondenatured soy protein is most soluble at pH values of 1.5 to 2.5 and 7 to 12 and least soluble at

Table 19.5. Approximate Composition of Soy Protein Concentrates Made by Three Extraction Processes[a]

Component	Alcohol Washing	Acid Washing	Hot-water Washing
Protein (N × 6.25)[b]	71.0	70.0	72.0
Protein	67.0	66.0	68.0
Moisture	6.0	6.0	5.0
Fat	0.3	0.3	0.1
Crude fiber	3.5	3.4	3.8
Ash	5.6	4.8	3.0
Carbohydrate[c]	17.6	19.5	20.1

[a]Data expressed as percentages.
[b]Dry solids basis; all other data expressed on an "as-is" basis.
[c]Percentage by difference.
Source: Campbell et al. (1985), Lusas & Rhee (1995).

Table 19.6. Amino Acid Composition of Soy Protein Concentrates, Soy Solubles, and Soy Flours[a]

		Soy Protein Concentrate		Soy Solubles from
Amino Acid	Soy Flour	Alcohol Washed	Acid Washed	Alcohol Washing
Alanine	4.00	4.86	4.03	3.94
Arginine	6.95	7.98	6.46	7.36
Aspartic acid	11.26	12.84	11.28	15.0
Half-cystine	1.45	1.40	1.36	4.14
Glutamic acid	17.18	20.20	18.52	20.7
Glycine	3.99	4.60	4.60	3.47
Histidine	2.60	2.64	2.59	2.50
Isoleucine	4.80	4.80	5.26	2.11
Leucine	6.50	7.90	8.13	3.17
Lysine	5.70	6.40	6.67	3.53
Methionine	1.34	1.40	1.40	3.60
Phenylalanine	4.72	5.20	5.61	5.65
Proline	4.72	6.00	5.32	3.48
Serine	5.00	5.70	5.97	3.38
Threonine	4.27	4.46	3.93	3.36
Tryptophan	1.80	1.60	1.35	7.00
Tyrosine	3.40	3.70	4.37	5.47
Valine	4.60	5.00	5.57	2.12

[a]Data expressed as g amino acid per 16 g nitrogen.

Table 19.7. Vitamin and Mineral Fortification Requirements for USDS-FNS Child Feeding Programs

Vitamins and Minerals	Min./g Protein
Vitamin A, I.U.	13.00
Thiamine, mg	0.02
Riboflavin, mg	0.01
Niacin, mg	0.30
Pantothenic acid, mg	0.04
Vitamin B6, mg	0.02
Vitamin B12, μg	0.10
Iron, mg	0.15
Magnesium, mg	1.15
Zinc, mg	0.50
Copper, μg	24.00
Potassium, mg	17.00

its isoelectric region of pH 4.2 to 4.6. These solubility properties are utilized to prepare SPI. A traditional SPI production process is shown in Fig. 19.16 (Wolf, 1983). The basic steps include: (i) solubilizing the protein in ground white flakes, at a 1:10 to 20 solids:solvent ratio, in 60°C (140°F) water adjusted to pH 9 to 11 with sodium hydroxide; (ii) removing the insoluble fiber by centrifuging; (iii) precipitating the protein at pH 4.2 to 4.5 by acidifying with hydrochloric acid; (iv) centrifuging to separate the protein curd from the whey containing the soluble sugars; (v) washing the curd with water and re-concentrating by centrifuging; (vi) neutralizing to pH 6.8 with sodium or calcium hydroxide; and (vii) spray-drying the washed, neutralized protein curd at 157°C (315°F) inlet air temperature and 86°C (187°F) outlet. Some processes wash the fiber a second time to improve protein yield. Use deionized process water when only naturally hard or alkaline water is available. Recently, ultrasonics is promoted as a means of increasing protein extraction. Supplement SPI with calcium, if intended for use in dairy product replacement applications; agglomerated to increase density; and lecithinated to improve dispersing properties.

High pH values and temperatures and prolonged processes favor production of lysinoalanine in many proteins. This reaction compound, formed mostly at the expense of lysine and cystine in soy protein (Savoie and Parent, 1983), causes nephrocytomegaly in rats (Karayianis et al., 1979). While the effects of lysinoalanine on humans are not known (Struthers, 1981), conditions favoring its production (Freidman, 1982) should be minimized. Lysinoalanine became a perceived problem when SPI produced for industrial applications in biobased products inadvertently got into food; lysinoalanine is not a current problem to manufacturers of soy protein food ingredients.

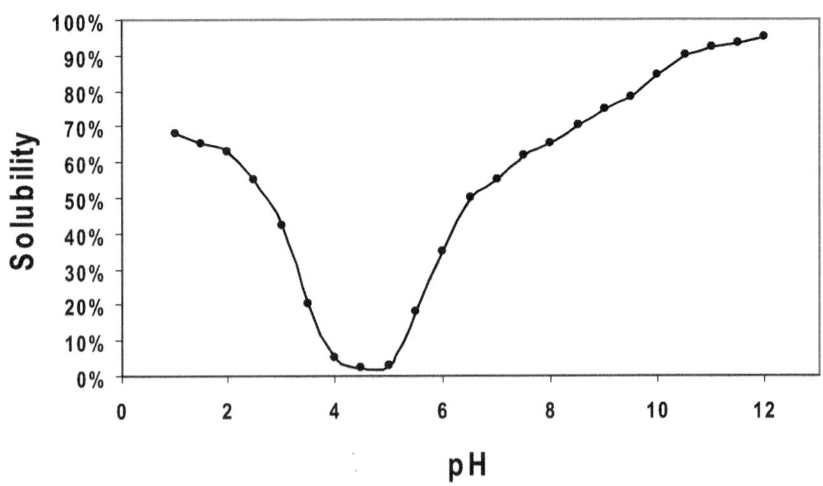

Fig. 19.15. pH-solubility profile of soy protein isolate in water (Source: adapted from Shen, 1976).

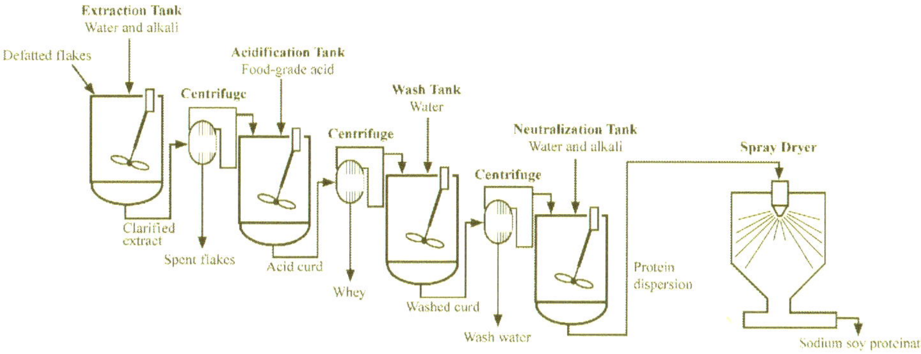

Fig. 19.16. Flow diagram for preparing soy protein isolate (Source: Wolf, 1983).

Usually, SPIs are prepared from WFs, but recent interest has arisen in using partially defatted soybean meals prepared by extruding-expelling (EE) and gas-supported screw pressing (GSSP). These two methods are strictly mechanical and comply with *organic* processing. The Chapter: *Oil Recovery from Soybeans* provides additional information on these two processes. The interest in using these partially defatted feedstocks (3–10% fat) is that the production plants have small capacity (<40 mt/day) making them ideal for producing specialty products using specialty beans modified by traditional breeding (e.g., low-linolenic acid soybeans), molecular biology (e.g., HS/LS soybeans or high β-conglycinin), or special production methods (e.g., certified non-GMO or organic). With EE, the lower the oil content achieved usually lowers the PDI and yield of SPI. With the widespread adoption of Roundup Ready® soybeans, some export markets pay premiums for non-GMO soy protein ingredients that compensate for lower product yields (Wang et al., 2004a). Kerry Food Ingredients (Beloit, WI) uses EE meal from which they produce SPI. GSSP, on the other hand, injects CO_2 into a screw press and achieves low oil content (3–5%) without denaturing protein. SPI with unique functionality is produced in high yields from GSSP meal. Table 19.8 compares the yields and compositions of SPIs from defatted meals prepared in different ways.

Fractionating Soy Proteins

Wolf et al. (1962) used ultracentrifuge sedimentation techniques to determine the protein components of soy protein and reported: 2*S* type, 22% of total, 8,000–21,500 MW; 7*S* type, 37% of total, 180,000–210,000 MW; 11*S* type, 31% of total, 350,000 MW; and 15*S* type, 11% of total, 600,000 MW. Much research focused on fractionating soy protein into fractions relatively pure in one of the specific proteins comprising bulk soy protein. β-Conglycinin (7*S*) and glycinin (11*S*) proteins are the major storage proteins of soybeans and comprise nearly 70% of the total protein in soy-

Table 19.8. Yields and Compositions of Isolated Soy Protein from Meals Produced by Various Oil Extraction Methods[a]

Starting Material	Yields, %		Compositions, %	
	Solids	Protein	Protein	Fat
White flakes	34.2	61.1	93.1	nd
GSSP	35.7	61.4	92.7	0.4
EE60	33.4	60.9	80.8	11.7
EE35	25.0	40.5	79.6	9.2

[a]White flakes denote flash desolventized hexane-defatted dehulled soybean meal; GSSP, gas-supported screw-pressed dehulled soybean cake; EE60, extruded-expelled, dehulled soybean cake having 60 PDI; EE35, extruded-expelled, dehulled soybean cake having 35 PDI.
Source: Wang and Johnson (2001), Deak et al. (2007b).

beans. While early research focused on obtaining pure glycinin and β-conglycinin to study structure–function relationships, recent interest in fractionating soy protein is focused on understanding the roles of each in health. Figure 19.17 shows that each fraction has a slightly different solubility curve, with the maximum for *7S* at about pH 5 and for *11S* at pH 5.8 in low-ionic-strength solution (0.03 M).

Many laboratory methods to fractionate soy proteins were reported (Wolf, 1956; Roberts & Briggs, 1965; Wolf & Sly, 1967; Eldridge & Wolf, 1967; Koshiyama 1965, 1968a, 1968b, 1972; Thanh & Shibasaki, 1976; Saio & Watanabe, 1973; Saio et al., 1974, 1975; Nagano et al., 1992; Wu et al., 1999; Rickert et al., 2004a; Deak & Johnson, 2005; Deak et al., 2006d). One of the first attempts to fractionate soy proteins used low temperatures to precipitate a glycinin-rich fraction and termed the fraction *cold-insoluble fraction* (Wolf, 1956). Others describe this method as *cryoprecipitation* and glycinin as *cryoprotein* (Wolf and Sly, 1967). These methods focused on recovering a glycinin-rich fraction and did not address β-conglycinin.

Probably the most widely used laboratory method to fractionate soy protein is one described by Thanh and Shibasaki (1976), in which soybean meal is extracted with Tris-buffer solution containing β-mercaptoethanol at pH 7.8, centrifuging to remove the insoluble material, then precipitating glycinin at pH 6.6 and β-conglycinin at pH 4.8, dialyzing, washing, and freeze-drying. To complete purification, column chromatography of the precipitated fractions was used.

Nagano et al. (1992) developed a fractionation method using three precipitation steps based on differences in solubilities of glycinin and β-conglycinin in the presence of NaCl. Sodium bisulfite was added to soy protein extract, the pH was adjusted to 6.4, and the solution was cooled in an ice bath to precipitate a glycinin-rich fraction. NaCl was added to the supernatant to salt-in β-conglycinin and the pH adjusted to 5 to precipitate an intermediate fraction comprising a mixture of glycinin and β-conglycinin. The supernatant was diluted with water to salt-out β-conglycinin and

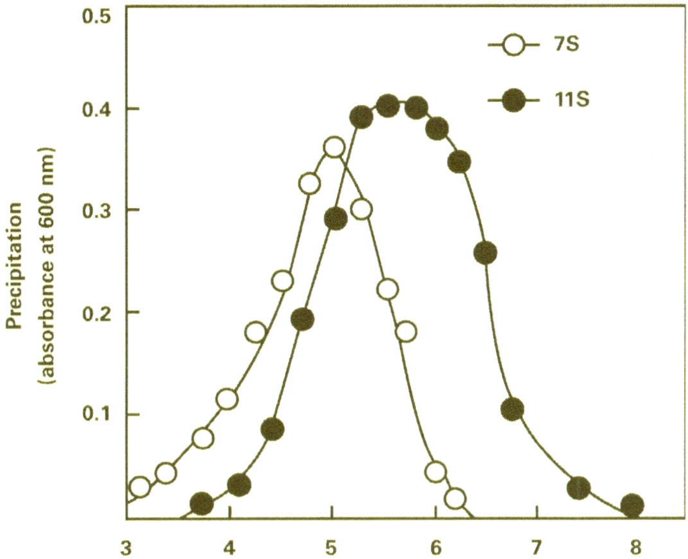

Fig. 19.17. Susceptibility of 7S and 11S soy protein fractions to precipitate from solutions of low ionic strength (0.03 M) (Source: Kinsella, 1979).

the pH adjusted to 4.8 to precipitate a β-conglycinin-rich fraction. The yields were 10% glycinin-rich fraction and 6% β-conglycinin-rich fraction, in >90% purities. This fractionation method differed from earlier methods in that it used simple precipitation steps, and no column purification nor use of β-mercaptoethanol.

Wu et al. (1999) scaled up the Nagano method to pilot-plant scale to obtain kg-quantities of the individual protein fractions. Relatively high yields of the individual protein fractions were obtained, 11% glycinin- and 11% β-conglycinin-rich fractions. The glycinin-rich fraction contained 84% glycinin and the β-conglycinin-rich fraction contained 72% β-conglycinin. The β-conglycinin contaminant in the glycinin-rich fraction was denatured, while only one-half of the glycinin contaminant in the β-conglycinin-rich fraction was denatured. The intermediate fraction had little native structures. This pilot-plant procedure was improved by decreasing the solvent to flake ratio from 15:1 to 10:1 and increasing the extraction temperature from 20 to 45°C (Rickert et al., 2004a). Phytochemical recovery (Rickert et al., 2004a) and functional properties of the fractions obtained (Rickert et al., 2004b) were improved, but purities were not enhanced.

Deak et al. (2006b) found that the reducing agent concentration significantly affected fraction yields, purities, and compositions, especially the purity of the glycinin-rich fraction. The optimal amount of reducing agent was 5 mM SO_2. The glycinin-rich fraction contained 23% of the total protein with 82% glycinin, and the β-conglycinin-rich fraction contained 17% of the total protein with 84% β-conglycinin. Increasing amounts of storage proteins were lost in the whey fraction as SO_2 concentration increased. Deak et al. (2006c) also evaluated different NaCl concentrations on yields and purities. NaCl concentration significantly affected the yields of the intermediate and β-conglycinin fractions. Maximum yield of the β-conglycinin-rich fraction was obtained at 500 mM NaCl, but at expense of purity; the ideal NaCl concentration was 250 mM. At higher NaCl concentrations, the intermediate fraction yields were significantly lower, and protein loss in the whey increased. Fractionation efficiency was also improved by using one-fold instead of two-fold dilution to salt-out the β-conglycinin-rich fraction.

Wu et al. (2000) also simplified soy protein fractionation obtaining two protein fractions, by using membrane filtration to obtain a β-conglycinin-rich fraction after precipitating the glycinin-rich fraction. The yield of the β-conglycinin-rich fraction was improved, but at the expense of purity.

Thiering et al. (2001) developed a fractionation method using pressurized CO_2 as a volatile electrolyte and pH adjustments to fractionate glycinin-rich, intermediate, and β-conglycinin-rich fractions. They reported 28% yield of glycinin-rich fraction with 95% purity and 21% yield of β-conglycinin-rich fraction with 80% purity.

Saito et al. (2001) reported on a method where soy protein extract was treated with phytase to hydrolyze phytic acid. Phytate hydrolysis disrupts phytate–protein complexes improving fractionation. Two fractions were obtained by adjusting the pH to 6 to precipitate a glycinin-rich fraction and to pH 5 to precipitate a β-conglycinin-rich fraction. About 22 and 36% protein yields of β-conglycinin-rich and glycinin-rich fractions were achieved, respectively. The purities for both fractions were about 80%. This process is used by Fuji Oil Co. (Osaka, Japan) to commercially produce β-conglycinin-rich food ingredients and products that command exceedingly high retail prices.

Using a different approach to eliminate phytate–protein interactions and to improve soy protein fractionation, Deak et al. (2006d) developed a process using calcium salts in combination with sulfites to achieve glycinin-rich and β-conglycinin-rich fractions in high yields and purities. By using 5 mM SO_2 in combination with 5 mM $CaCl_2$, this two-step fractionation procedure produced high purities in the glycinin-rich (85%) and β-conglycinin-rich (81%) fractions. It was possible to fractionate soy protein into two soy protein isolate fractions (>90% protein) enriched in either glycinin or β-conglycinin by using a new simplified procedure (referred to as the Deak procedure) employing $CaCl_2$ and $NaHSO_3$. The Deak procedure produced fractions with high yields of solids, protein, and isoflavones, with high purities and improved functional properties compared to previous methods. The glycinin-rich fraction comprised 16% of the solids, 24% of the protein, and 21% of the isoflavones in the starting soy flour, whereas the β-conglycinin-rich fraction comprised 23% of the solids, 37% of the protein, and 38% of the isoflavones.

The patent literature on fractionating soy protein is also abundant. A process patented by Davidson et al. (1979) involves extracting soy flakes with water at 55 to 70°C, cooling the extract slowly to precipitate 11*S* protein, and collecting the remaining proteins by precipitating at pH 4.5 or by ultrafiltering. Shemer (1980) patented a process for making a protein isolate rich in 7*S* globulins by extracting at pH 5.1 to 5.9 where the 11*S* protein is only slightly soluble and separating an intermediate fraction of 30% 7*S* and 70% 11*S* protein was necessary to obtain a relatively pure 7*S* fraction. Gibson and Yackel (1989) described fractionating 7*S* and 11*S* proteins using a combination of sodium chloride and sodium bisulfite. Howard et al. (1983) disclosed a method to fractionate soy storage proteins by means of pH adjustments in the presence of sulfite ions and water-soluble salts. Kolar et al. (1985) disclosed a method to separate 7*S* and 11*S* proteins on the basis of differences in their isoelectric points and the tendency of 11*S* protein to precipitate at low temperatures. Bringe and Charles (2006) patented a method to obtain food ingredients with increased proportions of glycinin or β-conglycinin by using genetically modified lines rich in glycinin or β-conglycinin. Deak and Johnson (2005) disclosed what they regard to be the only commercially viable processes to scalp β-conglycinin and to fractionate soybean storage proteins based on differences in protein solubility in the presence of calcium ions and reducing agents.

β-Conglycinin and glycinin are globular proteins and differ in functional properties; for example, glycinin plays an important role in crosslinking with divalent cations to form tofu-like curds. Solubility behaviors of the individual storage proteins were reported by Yuan et al. (2002), Bian et al. (2003), and Rickert et al. (2004b); glycinin is less soluble over a much wider pH range than β-conglycinin (Yuan et al., 2002) (Fig. 19.18), which was attributed to the greater Van der Waals and hydrophobic forces among glycinin molecules. The presence of salts slightly raises the isoelectric points of soy proteins. The glycinin-rich fraction has thermoplastic properties that may be useful in imitation cheese and comminuted luncheon meats.

Membrane Processing

Ultrafiltration (UF) and reverse osmosis (RO) are other forms of separation based on molecular size. UF retains or permeates (pass) molecules according to the size of membrane pores selected (molecular-weight cutoff, MWCO), and RO is used for dewatering and concentrating. Reduce energy costs by up to 90% by employing RO membranes for dewatering, compared to evaporative processes and up to 70% when a single-effect evaporator is used. The principles of membrane and adsorptive separations for vegetable proteins werereviewed by Koseoglu and Lusas (1989). Common practice is to incorporate diafiltration (maintaining a constant ratio of water- or solvent-to-solids ratio) to minimize problems of retentate-side concentration and surface fouling.

Lawhon et al. (1981) described a process for making SPI from defatted soybean

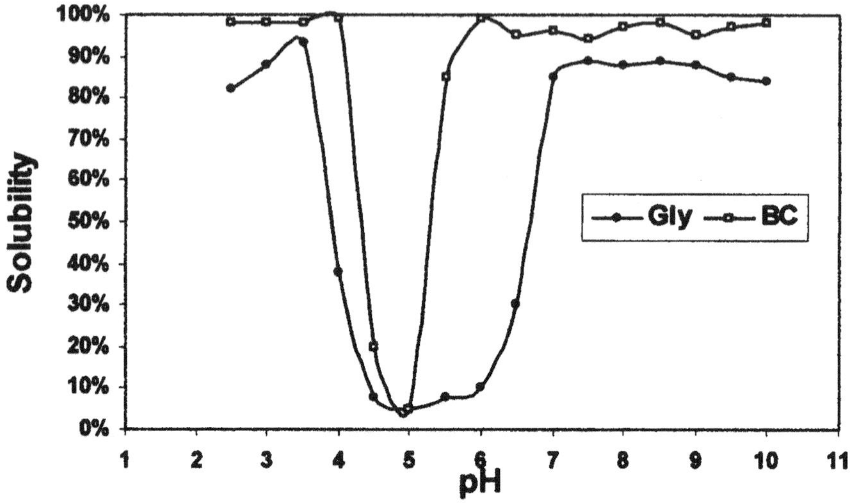

Fig. 19.18. Solubility of soy glycinin (Gly) and β-congylcinin (BC) in water (Source: adapted from Yuan et al., 2002).

flakes using UF-RO and diafiltration (Fig. 19.19). Ground SF was extracted (a single extraction of 30:1 water-to-flour ratio; or two extractions, 10:1 followed by 8:1) with water adjusted to pH 8 to 9 with calcium or sodium hydroxide. Calcium hydroxide was preferred as the base for solubilizing protein, because it produced higher SPI yield than sodium hydroxide. The extraction temperature was 43°C (110°F) for high-NSI flour or 55°C (132°F) for toasted flour. After 40 min of extraction, the material was centrifuged to remove the fiber and passed through a cross-flow 70,000-MWCO membrane. The retentate protein fraction was concentrated by RO and spray-dried. The permeate (soluble sugars, minerals, and small protein molecules) was concentrated by RO and spray-dried.

Advantages of membrane processing include: the ability to recover certain proteins without alkali solubilization and acid precipitation and accompanying protein damage; the potential for recovering small (12,000 to 20,000 MW) proteins if a membrane with sufficiently small pore size is selected; opportunities to remove small molecules, such as phytates; and opportunities to greatly reduce water consumption and processing discharge streams that contain significant BOD (biological oxygen demand). SPCs were also prepared by UF/RO processing.

Lawhon (1983) was granted a patent for preparing light-colored and bland SPI, using 70,000- to 100,000-MWCO membranes. In the process, flavor- and color-producing compounds apparently associate preferentially with the smaller (<20,000 MW) proteins and pass through the membrane, leaving the larger MW, bland, and light-colored fractions behind. Some industrially practiced membrane processes for

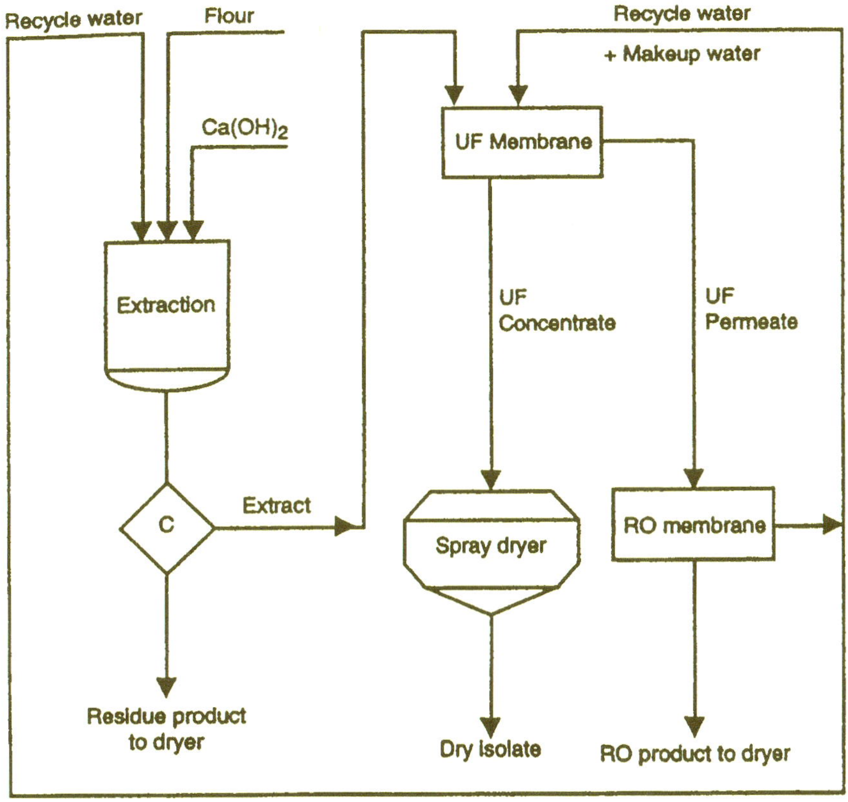

Fig. 19.19. Flowsheet for using UF and RO membranes in preparing soy protein isolates (Source: Lawhon et al., 1981).

SPC and SPI were not publicized and are kept as trade secrets but are known to be practiced in the United States, Japan, and Europe. Other industrial practices are known from the patent literature such as key patents of Singh (2003) and Muralidhara et al. (2003).

Aqueous Extraction Processing

Lawhon et al. (1981) described an aqueous extraction process (AEP) for removing oil and preparing SPCs and SPIs from soybeans without hexane and using water. In SPI preparation by AEP (Fig. 19.20), cleaned soybeans are dried at 70°C (158°F) to 6% moisture, dehulled by cracking and aspirating, and reduced in particle size (to 99% <70 mesh) by pin milling. Oil extraction is conducted at 1:12 solids-to-water ratio at 60°C, pH 9 and 0.01% hydrogen peroxide to inactivate lipoxygenase. After

30 min of extraction, the slurry is centrifuged to separate an aqueous phase, a solids phase, and an oil/emulsion phase, which later must be broken to obtain the oil. The aqueous phase is adjusted to pH 4.5 with hydrochloric acid to precipitate a protein curd, which is separated by centrifugation. Washing the curd before drying increased the protein content to almost 90%, mfb. The alkali-neutralized, spray-dried SPI contained as much as 8 to 10% residual oil.

Optionally, the solids phase from the first alkali extraction may be extracted a second time, with 1:5 solids-to-water ratio at pH 9, and re-centrifuged, with the resulting fractions combined with those from the initial centrifugation. Although the process is not yet sufficiently efficient to become the main commercial means for extracting soybean oil, the resulting SPCs and SPIs are extremely stable to oxidation and have properties that may be functionally useful in selected applications. UF and RO membrane techniques were also tried with AEP. Additional information on AEP is provided in the Chapter: *Oil Recovery from Soybeans.*

Salt Extraction

Glycinin and β-conglycinin have several subunits that can be dissociated by salt solution. Murray et al. (1980) patented a process that employs salt for extracting SPI at ionic strengths of 0.3 to 0.6 M, pH 5.0 to 6.8 and 15 to 25°C (60 to 78°F). The extract is then concentrated to one-fourth to one-third its volume and diluted to an ionic strength of <0.2 M to form protein micelles that precipitate into an amorphous mass and are dried or further processed.

Separation of Intact Protein Bodies

Storage proteins in soybean cotyledon cells are deposited in discrete protein bodies. Attempts were made to separate them from other cellular constituents by fine milling and density flotation using glycerin, other polyhydric alcohols, sodium chloride, sucrose, and metal salts of organic acids. A density of 1.2 to 1.5 g/mL is required to float protein bodies, and water activity must be maintained at less than 0.85 to prevent hydration from occurring. The separated bodies contain >80% protein mfb (Kolar et al., 1985).

Enzyme-modified SPI

The period after extracting SPIs and just before spray-drying provides an opportunity for treating with enzymes or chemicals. Although considerable research in succinylated and acetylated derivatives has occurred, these modifications are not allowed in food and are restricted to industrial applications. Likewise, the enzyme-catalyzed reaction plastein synthesis, the reassembling of proteins from peptides, is not practiced in domestic commercially prepared food proteins.

Various commercial proteolytic enzymes are available from plant, microbial, and animal intestinal sources. These enzymes differ in their affinities for various proteins, the location where they cleave the peptide bond between different amino acids, and conditions (pH, temperature, inhibitors) affecting their reaction rates. The effects of enzymatic hydrolysis on functional properties of soy protein depend on extent of prior protein denaturation, pH at which hydrolysis is carried out, enzyme specificity, and extent of hydrolysis (DH, degree of hydrolysis) (Jung et al., 2004). Numerous reports on enzymatic modification of soy proteins to improve functional properties are available, and key references include Kim et al. (1990), Achouri et al. (1998), Hrckova et al. (2002), Surowka et al. (2004), Calderon-de-la Barca et al. (2000), Jung et al., (2004, 2006), and Lamsal et al. (2006), which show that functional properties can be improved and enzyme-modified soy protein can be utilized in highly nutritional products such as beverages, confections, and infant formulas. Usually limited hydrolysis is preferred (i.e., 4% DH), which can be largely controlled by the enzyme-to-substrate ratio. Proteolysis can cause bitterness, which is a significant limitation in marketing enzyme-modified soy protein products.

Whipping Proteins

Enzyme-modified soy whipping proteins are an example of a special enzyme application that has led to a new industry (Gunther, 1979). Shortages of egg albumen, caused by World War II, led to a market for three types of whipping (aerating) ingredients: soy albumens, enzyme hydrolyzates made from wet SPI, and enzyme hydrolyzates made directly from SF. Enzymatic hydrolysis of whipping proteins typically is conducted in the pH 2.0–3.5 range, below the isoelectric point of soy protein, and results in MWs of less than 14,000. Figure 19.21 shows the change in nitrogen solubility of a 10% slurry of soybean flakes over time, during hydrolysis by 0.5% pepsin at pH 2 to 3.5 and 38°C (100°F).

A process for making whipping proteins via an intermediate SPI is shown in Fig. 19.22. The protein in WFs is first solubilized by alkali, and the fiber separated by a centrifuge. The protein in the extract is precipitated as an isoelectric curd, which is washed to remove solubles and then solubilized with additional acid. Enzyme hydrolysis is conducted at low pH for 12 to 24 h under controlled conditions, followed by centrifuging to remove the insoluble residue, concentrating the solubles, neutralizing to about pH 5.2, and spray-drying. In an alternative process (Fig. 19.23), WFs are first washed to partially remove the solubles and sugars, then directly treated with acid and enzyme to solubilize and hydrolyze the protein. The insolubles are separated by centrifuging, concentrated by evaporating, adjusting to about pH 6.6, and spray-drying.

Whipping preparations are used with different background formulas and sometimes contain sucrose, sodium hexametaphosphate, or a polysorbate emulsifier (Tween 60) to improve stability. Many of these preparations can whip to twice the volume

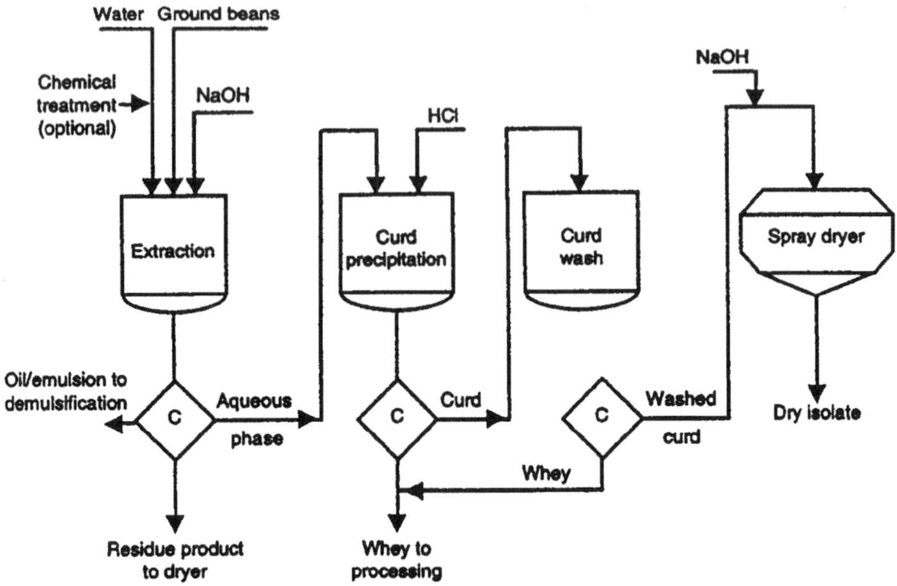

Fig. 19.20. Flowsheet for processing soybeans by aqueous extraction processing (Source; Lawhon et al., 1981).

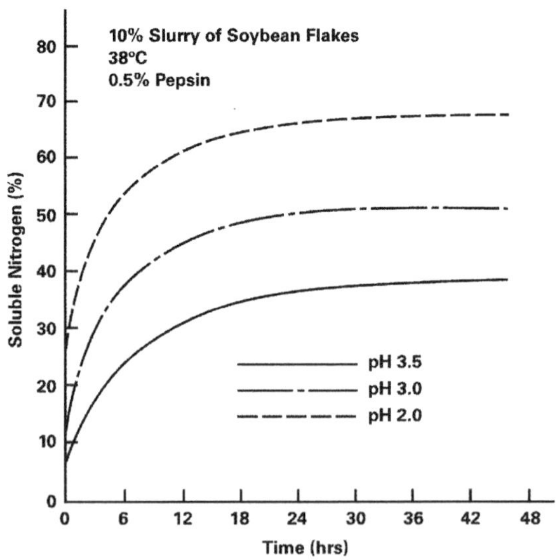

Fig. 19.21. Effects of pH on nitrogen solubility of a 10% soybean flake slurry after hydrolysis with 0.5% pepsin (Source: Gunther, 1979).

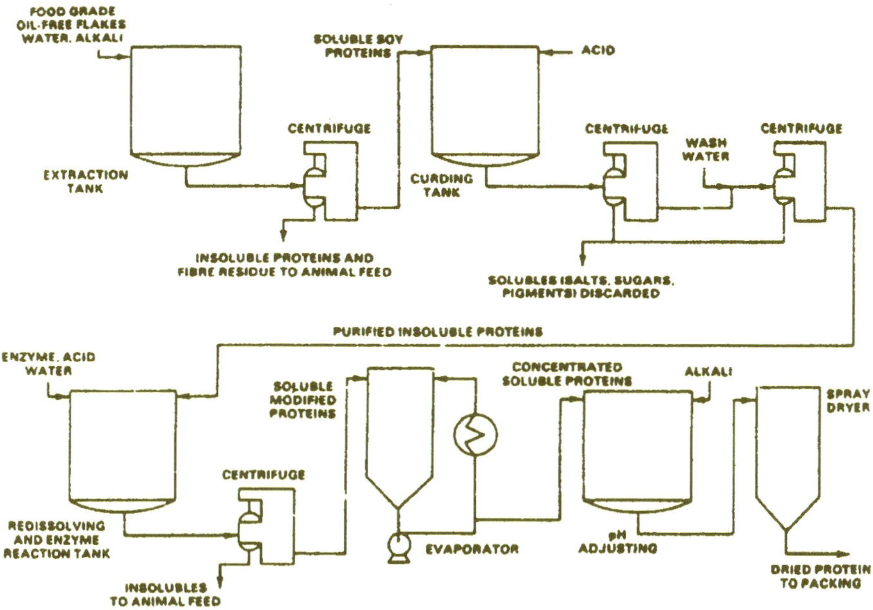

Fig. 19.22. Process diagram for preparing enzyme-modified whipping proteins from soy protein isolate (Source: Gunther, 1979).

of egg whites and are substituted at 25 to 100%. They differ from egg whites in not being heat-setting, but they will extend heat-set egg whites to lower usage levels.

Microbiological Stability

Microbiological safety for all food ingredients and food products is critical. Most soy protein ingredients are now heat-treated to reduce microbiological loads and to assure safety. Both indirect heat and direct injection of food-grade steam into the protein slurries just ahead of spray-driers are common. For controlling microbiological load with jet cooking typically involves heating for 2–3 sec at 105°C (Egbert, 2004). Jet cooking was also used to alter or restore functional properties (Wang & Johnson, 2001; Wang et al., 2004, 2005). All protein products except for SFs are heat-treated and contain denatured proteins with low solubilities unless enzyme hydrolyzed (Lee et al., 2003). Hydrogen peroxide can also be used to control microbial growth during processing, 0.1% treatment is effective (unpublished results of the authors).

Impact of Soybeans with Modified Compositions

In recent years, substantial changes were made to soybean traits having the potential

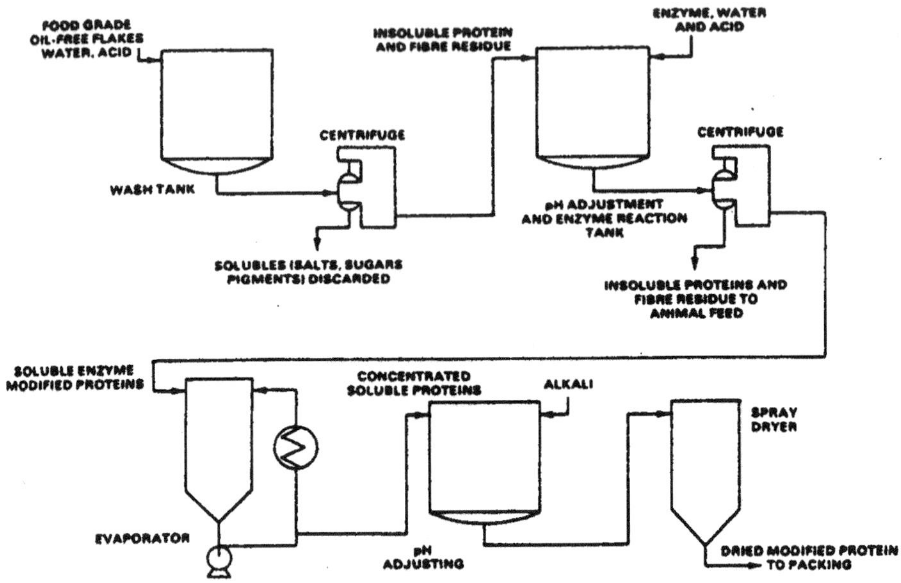

Fig. 19.23. Process diagram for preparing enzyme-modified whipping proteins by direct hydrolysis of soy flakes (Gunther, 1979).

to impact utilization of soy protein ingredients. These changes were accomplished through traditional breeding and genetic engineering. One of the earliest traits to be modified was to reduce TI activity and lipoxygenase contents, while recent changes in oligosaccharide contents, fatty acid composition, and protein composition were achieved.

Orf and Hymowitz (1979) identified soybean lines without Kunitz TI, which enable soybeans to be utilized without heating in feed and some food applications. The Kunitz inhibitor accounts for about 80-90% of the TI native activity. This modification was achieved through traditional breeding methods, but has not become a commercialized trait.

Lipoxygenase-null soybeans, devoid of the three known iso forms of lipoxygenase, were developed to improve the flavor and eating quality of soybeans and soy protein ingredients. As noted earlier, while beany flavors were substantially reduced, products made from lipoxygenase-null beans are not totally bland, probably due to the inherent oxidative instability of linolenic and linoleic acids comprising residual fat. Only small amounts of lipoxygenase-null soybeans are produced, and so far, lipoxygenase-null soybeans have had little commercial significance to soy protein ingredient manufacture.

High-sucrose-containing soybeans with much-reduced oligosaccharides (stachyose, raffinose) were recently developed by DuPont (Wilmington, DE) to eliminate flatulence (intestinal gas) when consumed by humans, pets, and livestock. Feed conversion can also be enhanced in some species, especially poultry, by eliminating oligosaccharides. This modification was accomplished by employing germplasm screening, chemical mutagenesis and conventional breeding, and enables new processing approaches to producing food protein ingredients as developed by Solae (St. Louis, MO).

Recently, Monsanto (St. Louis, MO) patented (Bringe & Charles, 2006) and announced that they are commercializing soybeans in which they have altered the glycinin/β-conglycinin ratio. These soybeans reportedly contain >40% of the protein as β-conglycinin, compared to the normal 30% and <10% of the protein as glycinin, compared to the normal 40%.

Many alterations in fatty acid composition were made. While these changes were driven by oil markets, they could impact full-fat and partially defatted soy protein products. These modifications include low-linolenic acid for enhanced oxidative stability, high-oleic acid, low-saturates, and high-saturates. Many of these traits are being marketed by Pioneer a DuPont Company (Wilmington, DE), Monsanto (St. Louis, MO), and Asoyia (Winfield, IA). Because many protein products contain residual oil in varying contents, differences in fatty acid compositions of that fat can affect properties of protein products, particularly flavor.

Dietary Fiber Products

Two types of edible fiber co-products are produced from soybean processing operations—soy hulls (seed pericarp) and soy cotyledon fiber (cotyledon cell walls). Dietary fiber is becoming increasingly recognized as important to good intestinal health.

Soy Hulls

Soy hulls are mainly used as animal feeds, but a small quantity is cleaned and sterilized for use as a dietary fiber source in breads. The natural grittiness of the product typically requires fine grinding. Product specifications for a domestic product include: 92% total dietary fiber, 3.5% moisture, 0.5% fat, 1.5% protein, 2.5% ash, 0.1 Kcal/g caloric content; 350–400% water absorption, 6.57–7.5 pH, and sieve analysis of 0% on U.S. No. 80, 0% on U.S. No. 100, 2% on U.S. No. 140, 7% on U.S. No. 200, and 91% through U.S. 200.

Soy Cotyledon Fiber

The insoluble fiber-rich portion separated by centrifugation during alkali extraction to produce SPI may be used as a source of dietary fiber ingredients. These solids are analogous to the okara co-product from making soy milk or tofu, but they differ in that they do not contain hulls and were defatted and treated with mild alkali. This material comes from the cell walls.

The cotyledon co-product is processed, dried, and sold as a dietary fiber in competition with other sources such as α-cellulose, psyllium seed, guar gum, locust bean gum, pectin, and wheat, corn, and oat brans. Manufacturer's specifications for a domestic product include: 75% (mfb) dietary fiber (65% noncellulosic polysaccharides and 10% cellulosic), 12% moisture, 0.2% fat, and 4.5% (as is) ash.

Texturized Products

Two general types of mechanical texturization are used, although some agglomerated soy proteins may also contribute a texturized-like appearance on hydration.

Spun and Fiber-like Products

Spun protein products are intriguing because of the many technical skills that were used for their production and early marketing. Peanut, casein, and corn zein proteins were processed into textile fibers and marketed during 1935 to 1945; soy protein textile fibers were also developed at that time but did not reach the commercial market (Smith & Circle, 1978). R. Boyer, who had worked on the protein spinning process in the mid-1940s, modified and patented processes in 1954 (Boyer, 1954, 1956) to produce a fibrous mass simulating meat in texture and appearance (Fig. 19.24). With later inputs from other scientists, techniques were developed to disperse 14 to 18% SPI in sodium hydroxide at pH 10 to 11 and age at 40 to 50°C (104 to 122°F) until the dispersion becomes a *spinnable dope*, which can be forced through a platinum spinneret with 15,000 or more holes 0.20 to 0.25 mm (0.008 to 0.01 inches) diameter into an acid coagulating bath. The parallel fibers form a tow, which goes through a second heated bath where the fibers are stretched. Egg albumin, fat, flavor, and coloring materials are also added at this point for eventual forming into meat analog products. The toughness of the fibers is controlled by the pH, salt concentration, and temperature of the bath. Although spun-fiber food products are no longer sold in the United States, research on improvement of dopes and spun soybean fibers continued in Japan into the late 1980s (Sogo et al., 1985).

A frozen SPI filament product, with fiber-like texture, was sold in the United States for improving textural characteristics of fabricated foods, including structuring mechanically deboned meat and poultry. The manufacturer's specifications for the product include: >93% protein (N × 6.25, mfb), <0.2% crude fiber, 65% moisture, 0.9% ash, and <0.1% fat.

Extruder-texturized Products

The meat-like appearance in spun protein isolates results from strands of parallel fibers, but in extruded SFs, SPCs, and SPI, the meat-like structure is created from multi-laminate palisade layers. Extruder-texturized proteins are readily acceptable to

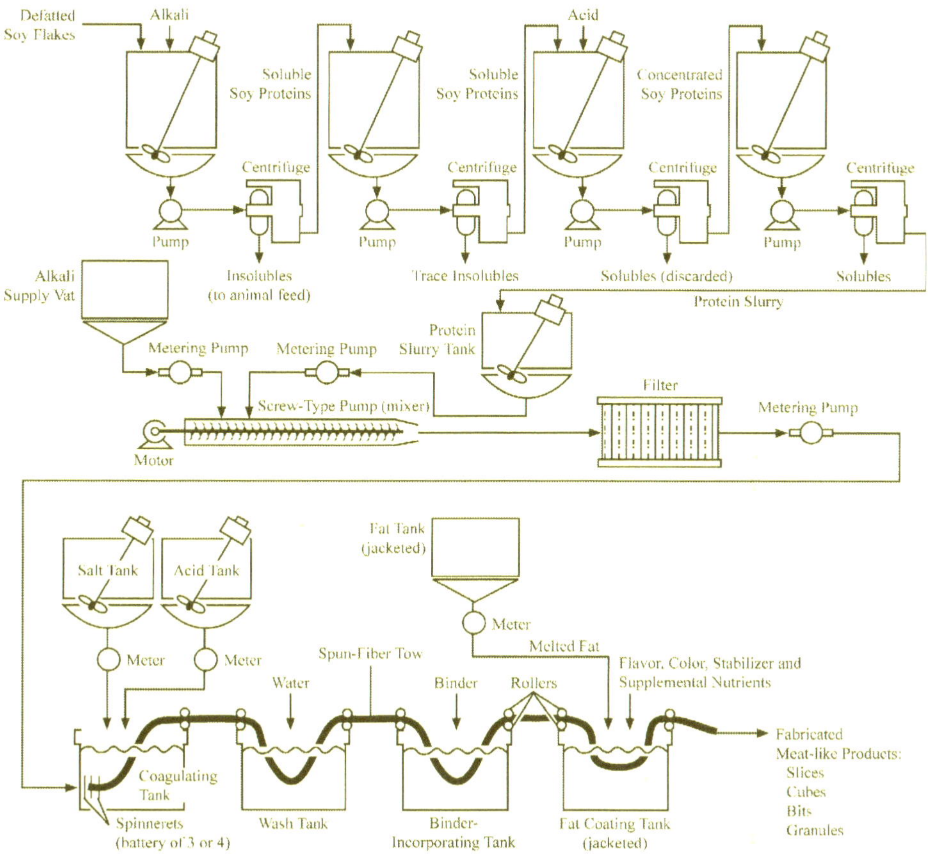

Fig. 19.24. Process diagram for spinning soy protein (Source: adapted from Ziemba, 1969).

the public, as demonstrated by their essentially complete replacement of spun products. Extrusion texturization also has the advantages of being a less complicated process and able to texturize lower-cost ingredients, including SFs and SPCs. A relatively small extruder-texturized SPI industry exists in the United States, which sells products in frozen form.

The principles of extrusion were described by Mercier et al. (1989) and the processing of proteins by Stanley (1989) and Rokey et al. (1993). *Texturized Vegetable Protein* and *TVP* are registered trademarks of the Archer Daniels Midland Company, Decatur, Illinois, and the generic terms *texturized soy protein*, *TSP*, or *texturized vegetable food protein* are used. Two types of products are made: (i) extrusion-cooked *meat extenders*, which are made from SF or flakes or SPC and are rehydrated to 60 to 65% moisture before blending with meats or meat emulsions at levels of 20 to 30%; and

(ii) extrusion-cooked *meat analogs*, which have similar appearances but are intended solely for use in meatless products.

The extrusion process restructures protein-based foodstuffs by applying mechanical and thermal energy, causing the macromolecules to lose their native structure and form a continuous viscoelastic mass (or melt). The extruder barrel, screw, and die align the molecules in the direction of flow, exposing bonding sites that cross-link into a reformed, expandable structure that creates a chewy texture in fabricated foods. In addition to restructuring vegetable food proteins, extrusion cooking does the following: (i) denatures proteins, lowers solubility, improves digestibility, and destroys biologically active enzymes and toxic proteins; (ii) inactivates residual heat-labile growth inhibitors native to many vegetable proteins in raw or partially processed states; (iii) prevents development of raw or bitter flavors commonly associated with many vegetable food sources; (iv) creates a homogeneous, irreversible, bonded dispersion of all micro-ingredients in a protein matrix; and (v) shapes and sizes the final product into desirable portions for packaging and sales (Smith & Circle, 1978).

An extrusion system consists of several important subsystems: (i) a feed delivery and proportioning system; (ii) a preconditioning area, which enables the raw materials to equilibrate in moisture content and heat; (iii) the cooking extruder itself; (iv) a laminar-flow area or die that allows aligning of molecules to occur; (v) a die and cutter to shape and cut the product into pieces; and (vi) a dryer/cooler to reduce moisture in the final product to a microbiologically stable level.

Barrels and screws evolved over the years into increasingly efficient designs. Twin-screw extruders cost more to acquire per unit of throughput capacity, but they provide nonpulsating discharge and steady operation (Sogo et al., 1985). A single-screw extruder, as used for making texturized soy protein, is shown in Fig. 19.25, and a flow sheet of a production line using a twin-screw extruder in Fig. 19.26.

The general parameters for raw ingredient specifications for texturized flours and concentrates include: 20 to 80 PDI, 0.5 to 6.5% fat, <7% fiber, and up to U.S. No. 8 mesh particle size. An exciting development in extruder operation is the ability to induce additional shear and to laminate low-NSI proteins that were once considered untexturizable (Sogo et al., 1985).

A variety of texturized soy food proteins is available from manufacturers, including products made from SF or SPC, colored and sized to different specifications. The volatile constituents are customarily added after extrusion by one of several enrobing processes. Specifically fortified products are available for use in school-lunch and child-feeding programs and in military-feeding applications.

Recently, Crowe and Johnson (2001) and Riaz (2001) showed that partially defatted screw-press cakes could be texturized. Previously it was believed that the fat content, which provides lubrication, was too high to allow proteins to texturize correctly. One manufacturer (Iowa Soy Specialties, Vinton, IA, a division of Kerry Food Ingredients) produces extruder-texturized products from EE press cake.

Fig. 19.25. Single-screw extruder used for preparing full-fat flours and texturized soy flours and concentrates (Source: provided by Wenger Manufacturing Company, Sebetha, KS).

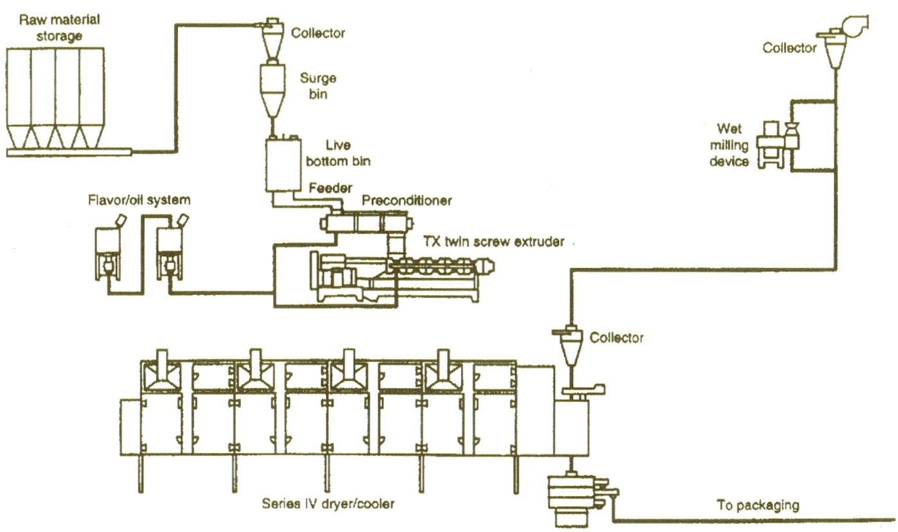

Fig. 19.26. Process diagram for preparing texturized soy protein (Source: provided by Wenger Manufacturing Company, Sebetha, KS).

Also, advances were made in co-extrusion (Simelunas et al., 1989; Chu et al., 2007). This is a process in which two or more materials are extruded through a die with two or more orifices, which enables encasing or enrobing one component within another. One application is to encase a food product within a soy protein dough. This technology is used as the basis for manufacturing protein nuggets that are a key component in certain food bars.

Applications of Soy Food Proteins

Functionality

Soy proteins are accepted in many applications because they provide desirable functionalities (performance properties) in fabricated foods at less cost than animal-source alternatives such as dried milk solids, casein, egg yolks, egg whites, or gelatin. Mimicking more expensive animal proteins has long been an objective of processing soy proteins. Reviews on soy protein functionality, modifications, and applications were prepared by Kinsella (1979), Kinsella and coworkers (Kinsella & Soucie, 1989; Kinsella et al., 1985), Cherry (l981), Rhee (1989), and Lusas & Rhee (1986).

The most sought-after functionalities in compounded foods, their modes of action, and the types of soy proteins used are shown in Table 19.9 (Fulmer, 1989a). Generally solubility is regarded as the most critical functional property, because without solubilization other properties are generally not possible. The soy ingredient is also expected to provide a concentrated source of protein as well as caloric density appropriate to the traditional or *light* product. In recent years, functional properties of fat were provided by protein ingredients with fewer calories. The soy ingredient also should not detract from the product in color or flavor, unless texturized and used to impart meat-like appearance. Important functionalities not included in Table 19.9 are thermoplasticity—the ability to solidify and re-melt repeatedly with temperature changes, as shown by bovine casein, and the ability to form edible films.

Selection of Soy Protein Preparations

If the food manufacturing or feeding institution is large enough, the setting of nutritional objectives usually is done by nutritionists or registered dieticians. Federally supported feeding programs require professional oversight of menus and approval of the overall diet. However, decisions of which types or forms of ingredients to use are typically left to the formulating technologist as guided by marketing objectives for the product.

Generally, one can predict about a soy protein's functional performance by examining chemical compositions in manufacturers' product specification sheets. Matrix tables, showing potential applications of a manufacturer's product line, also teach little. But for general guidance tables, such as Table 19.10, provide guidance about what types of soy protein ingredients should be considered for specific applications.

Table 19.9. Functional Properties Performed by Soy Protein Ingredients in Foods[a]

Functional Property	Mode of Action	Food System	Ingredient Used
Solubility	Protein salvation, pH dependent	Beverages	F, C, I, H
Water absorption and binding	Hydrogen-bonding of water, entrapment of water, no drip	Meats, sausages, breads, cakes	F, C
Viscosity control	Thickening, water binding	Soups, gravies	F, C, I
Gelation	Protein matrix formation and setting	Meats, curds, cheese	C, I
Cohesion-adhesion	Protein acts as adhesive	Meats, sausages, baked goods, pasta products	F, C, I
Elasticity	Disufide links in deformable gells	Meats, baked goods	I
Emulsification	Formation and stabilization of fat in emulsions, surfactant	Sausages, bologna, soup, cakes	F, C, I
Fat adsorption	Binding of free fat	Meats, sausages, donuts	F, C, I
Flavor binding	Adsorption, entrapment, release of flavor compounds	Simulated meats, baked goods	C, I, H
Foaming	Forms stable films to entrap gas	Whipped toppings, chiffon desserts, angel food cakes	I, W, H
Color control	Lypoxygenase bleaching of carotenoids	Breads	F

[a]F,denotes soy flour; C, soy protein concentrate; I, soy protein isolate; H, enzyme hydrolyzed soy protein; W, soy whey.
Source: Kinsella (1979), Fulmer (1989a).

Manufacturer's product specification sheets often tell only whether the ingredient is a SF, SPC, or SPI and its granulation. Inquiries to the supplier's technical service department, however, will usually yield additional information about the ingredient's production, functional properties, and limitations. The formulating technologist is advised to get several opinions about which ingredients to use from competing manufacturers. Processes differ between soy protein producers, sometimes resulting in subtle differences in performance, and the final selection of any ingredient should be based on its performance in the end product. Users very quickly develop proprietary expertise in soy protein applications exceeding that of the manufacturer.

It is important that formulators keep updated in soy protein developments. Over the years, the flavor of SFs has improved to where they might be substituted in former SPC applications, and enzyme modification of SPCs has made them contenders for applications formerly using only SPIs.

Table 19.10. Important Food Uses for Soy Protein Products[a]

Product	Soy Protein Isolate	Soy Protein Concentrate	Soy Flour or Grits	Textured Soy Protein
Bakery products				
Milk products	X	X	X	
Bread, rolls			X	
Breads (specialty)	X	X	X	
Cakes, cake mixtures	X	X	X	
Cookies, biscuits, crackers, pancakes, sweet pastry, snacks, etc.	X	X	X	
Doughnuts		X	X	
Pasta products	X	X	X	
Breakfast cereals	X	X	X	
Dairy-type products				
Beverage powders		X		
Cheeses		X		
Coffee whiteners		X		
Frozen desserts		X		
Whipped toppings		X		
Infant formulas	X	X	X	
Milk replacers for young animals	X	X	X	
Meat food products				
Emulsified meat products				
Bologna, frankfurters	X	X		
Miscellaneous sausage	X	X		
Luncheon loaves	X	X		
Luncheon loaves (canned)	X	X		
Seafoods	X	X		
Coarsely ground meat products				
Chili con carne, sloppy Joes	X	X	X	X
Meat balls	X	X	X	X
Patties	X	X	X	X
Pizza toppings	X	X	X	X
School lunch/military	X	X		X
Seafood				X
Whole muscle meat				
Analogs	X	X		X
Ham		X		
Meat bits (dried)				X

Table 19.10. cont. Important Food Uses for Soy Protein Products[a]

Product	Soy Protein Isolate	Soy Protein Concentrate	Soy Flour or Grits	Textured Soy Protein
Poultry breast		X		X
Seafood (surimi)	X	X		X
Stews	X	X		X
Miscellaneous applications				
Candies, confection, desserts	X	X	X	
Dietary items	X	X	X	
Asian foods		X		
Pet foods			X	X
Soup mixes, gravies	X	X		X

[a]Source: Rakes (1993), technical brochures.

Meat Applications

Meat products are expensive and attract cost-cutting technologies in all countries. The regulations for meat-type foods containing soy protein ingredients are shown in Table 19.11. In the United States, soy proteins are used: (i) as processing aids in the manufacture of frankfurters, sausages, and comminuted meat products; (ii) in marinades and tumbling solutions for restructured meats; (iii) in injection pumping proteins to increase the weight of intact muscles and cuts; and (iv) as extruder-texturized SFs and SPCs that are rehydrated and used at about the 20% level in hamburgers.

Processed Meats

The U.S. Department of Agriculture permits use of up to 3.5% SF or SPC in standard-of-identity frankfurters, up to 8% SF in scrapple and chili con carne, and up to 2% SPI in standard-of-identity frankfurters. SFs and SPCs can bind up to three times their weight in water, whereas nonfat dry milk solids bind only an equal weight of water. These ingredients reduce shrinkage due to moisture and fat losses during cooking. The use of SPIs globally, in making skin and fat emulsions for later inclusion in processed meats and other applications, is described in detail by Bonkowski (1989). Broad latitudes in formulation for processed meats exist outside of the United States, and also within the United States for nonstandard-of-identity meat products.

Restructured Meats

Principles of restructuring meats were reviewed by Pearson and Dutson (1987). Basically, red or poultry meats are flaked or chunked into small pieces, mixed or tumbled with salt and polyphosphates to extract heat-coagulable protein, shaped into loaf pans

Table 19.11. Regulations for Meat-type Foods Containing Soy Protein Products

Manufactured Product	Soy Product and Permitted Level	Comments
Cooked sausage	Soy flour, 3.5% Soy protein concentrate, 3.5% Soy protein isolate, 2%	Individually or collectively with other approved extenders. Where isolate is used, 2% is equivalent to 3.5% of others.
Fresh sausage	Same as for cooked sausage	Same as above.
Chili con carne	Soy flour, 8% Soy grits, 8% Soy protein concentrate, 8% Soy protein isolate, 8%	Individually or collectively with other approved extenders.
Spaghetti with meat balls, Salisbury steak	Soy four, 12% Soy grits, 12% Soy protein concentrate, 12% Soy protein isolate, 2%	Same as above.
Imitation sausage, soups, stews, nonspecific loaves, scrapple, tamales, meat pies, pork with barbecue sauce, beef with barbecue sauce, patties	All products: sufficient for the purpose	Provided meat and moisture requirements are met where such requirements may exist.

Source: Rakes (1993).

or other still-forming devices, heat-set at about 68°C (154°F), and cut into desirable shapes and thicknesses. SFs, SPCs, and SPIs are used at approximately the same levels as in processed meats to improve textural stability and minimize shrinkage.

Pumped Meats

Brines consisting of water, salt, polyphosphates, and SPIs or functional SPCs are prepared and pumped into muscle cuts using stitch pumps. Various domestic federal regulations apply; for example, hams and corned beef can be pumped to achieve cooked yields of 130%, provided a minimum protein content of 17% is maintained. Reviews on meat pumping technology were prepared by Bonkowski (1989) and Rakes (1993).

Extruder-texturized Soy Proteins

Texturized SFs or SPCs may be rehydrated to 18% protein content (60 to 65% moisture content) and used at levels up to 30% reconstituted soy protein in ground meat blends and hamburgers. In domestic practice, however, the reconstituted portion usually is used at about 20%, because of texture and flavor problems accompanying higher levels of meat substitution. Special vitamin- and mineral-fortified TSP products are required for school lunch and military feeding. TSPs are sometimes included in standard-of-identity canned meat products above the meat requirement to improve product attractiveness.

Baking Applications

The use of soy proteins in baked foods was reviewed by Hoover (1979), Dubois and Hoover (1981), and Fulmer (1989b). Many applications in this industry have long been served by SFs. Bakery applications of various soy protein products are shown in Table 19.12. The increased absorption of lightly toasted (PDI 60 to 80) SF requires an additional 0.75–1.0 lb of water for each pound of flour added. Examples of soy protein uses include:

- *Bread and buns.* 1 to 3% defatted SF (on flour basis) increases absorption of water by one pound for each pound of soy flour and improves crumb body, resilience, crust color (from sugars), and toasting characteristics.

- *Cakes.* 3 to 6% defatted SF improves batter smoothness and distribution of air cells and gives a more even texture and a softer, more tender crumb.

- *Sweet goods.* 2 to 4% defatted SF improves water-holding capacity and sheeting properties. This level should also be used for yeast-raised doughnuts.

- *Cake doughnuts. 2* to 4% defatted SF improves structure, gives an excellent star formation (hole), and reduces fat absorption during frying. The improved moisture retention improves product yield and shelf life.

- *Hard (snap) cookies. 2* to 5% defatted SF improves dough machining and imparts a crisp bite to cookies.

Toasted defatted SFs with about 20 PDI add color to the crumb, and nutty toasted flavor to whole-grain and specialty breads. Up to 15% of toasted defatted SFs can be added to leavened quick breads. Enzyme-active FFSF and defatted SFs, at 0.5% (flour basis), bleach carotenoid pigments in wheat flours and produce peroxides that strengthen gluten proteins and give the dough greater tolerance to mixing. Lecithinated SFs in cakes improve emulsification of fats, ingredient blending, pan release, and machinability, and they partially replace egg yolks.

Dairy and Beverage Applications

Regulatory principles have long required that a new food, intended and promoted as a

Table 19.12. Baking Applications for Various Soy Protein Ingredients

Soy Protein Ingredient	White Bread and Rolls	Specialty Breads and Rolls	Cakes	Cake Donuts	Yeast-raised Donuts	Sweet Goods	Cookies
Defatted soy flour	X	X	X	X	X	X	X
Enzyme-active soy flour	X						
Low-fat soy flour			X	X		X	
High-fat soy flour			X	X		X	
Full-fat soy flour			X	X		X	
Lecithinated soy flour			X	X			X
Soy grits		X					
Soy protein concentrates		X					
Soy protein isolates		X	X	X	X		
Soy fiber		X					

Source: Fulmer (1989a).

replacement for a major constituent of the current diet, be at least as nutritious as the product being replaced. The relative nutritional rating of soy protein has improved with abandonment of the PER system for evaluating protein quality. Although soy protein ranks high under the PDCAAS system, it normally contains less calcium than bovine milk and requires supplementation with this mineral. Calcium often renders soy protein poorly soluble and causes soy protein to precipitate from solution. Lin and Cho (1987) developed technology to produce mineral-fortified protein compositions, and this technology is the basis for the use of SPI in ready-to-drink, neutral beverages.

SPIs have long been used in coffee creamers, whipped toppings, dairy-type dips, and frozen desserts, including frozen tofus (Wilding, 1979; Kolar et al., 1979). Lecithinated SFs and SPCs are commonly used to assist dispersibility of powdered beverages. Another major application is in formulas for children allergic to milk proteins or showing lactose intolerance as well as in famine relief and nutritional improvement for infants and children at weaning in developing countries.

Future Considerations

The production of renewable fuels and biobased products are consuming more and more grains, farmland is becoming more costly, feed costs are escalating, and poultry, beef and pork will also soon become more expensive. The health benefits of increasing the proportion of plant proteins relative to animal proteins in the human diet are becoming increasingly recognized. These trends are causing food processors to seek new sources of cost-effective and functional proteins. Soy protein products are increasingly being used to manufacture bio-based products emanating from new soybean-based biorefineries and new uses as adhesives, plastics, fermentation ingredients, and construction materials. For these reasons, the future of soy protein ingredients is very bright indeed.

References

Achouri, A.; W. Zhang; X. Shiying. Enzymatic hydrolysis of soy protein and effect of succinylation on the functional properties of resulting protein hydrolysates, *Food Res. Int.* **1998,** *31,* 617–623.

Adachi, M.; J. Kanamori; T. Masuda; K. Yagasaki; K. Kitamura; B. Mikami; S. Utsumi. Crystal Structure of Soybean 11S Globulin: Glycinin A3B4 Homohexamer, *Proc. Natl. Acad. Sci. USA 100*:7395–7400, 2003.

Albrecht, W.J.; G.C. Mustakas; J.E. McGhee. Rate studies on atmosphere steaming and immersion cooking of soybeans, *Cereal Chem.* **1966,** *43,*400–408.

AOCS. Proceedings of the world conference on soya processing and utilization, Munich, Germany, Nov. 1114, 1973. *J. Am. Oil Chem. Soc.* **1974,** *51(1).*

AOCS. Proceedings of the world conference on vegetable food proteins, Amsterdam, The Netherlands, Oct. 29–Nov. 3, 1978. *J. Am. Oil Chem. Soc.* **1979,** *56(3).*

AOCS. Proceedings of the world conference on soya processing and utilization, Acapulco, Mexico, Nov. 9–14, 1980. *J. Am. Oil Chem. Soc.* **1981,** *58(3).*

AOCS. *Official Methods and Recommended Practices of the American Oil Chemists' Society,* Fifth ed.; AOCS: Champaign, IL, 1999.

Applewhite, T.H. (ed.) *Proceedings of the World Conference on Vegetable Protein Utilization in Human Foods and Animal Feedstuffs;* American Oil Chemists' Society: Champaign, IL, 1989.

Applewhite, T.H. (ed.) *Proceedings of the World Conference on Oilseed Technology and Utilization;* American Oil Chemists' Society: Champaign, IL, 1993.

Baldwin, A.R. (ed.) *Proceedings of the World Conference on Oilseed and Vegetable Oil Processing Technology;* American Oil Chemists' Society: Champaign, IL, 1976.

Bian, Y.; D.J. Myers; K. Dias; M.A. Lihono; S. Wu; P.A. Murphy. Functional properties of soy protein fractions produced using a pilot plant-scale process. *J. Am. Oil Chem. Soc.* 2003, *80,* 545–549.

Bonkowski, A.T. The utilization of soy proteins from hot dogs to haramaki. *Proceedings of the World Conference on Vegetable Protein Utilization in Human Foods and Animal Feedstuffs;* T.H. Applewhite, Ed.; American Oil Chemists' Society: Champaign, IL, 1989; pp. 430–438.

Bookwalter, G.N.; G.C. Mustakas; W.F. Kwoiek; J.E. McGhee; W.J. Albrecht. Full-fat soy flour extrusion cooked: properties and food uses. *J. Food Sci.* **1971,** *36,* 5–9.

Boyer, R.A. High Protein Food Product and Process for Its Preparation. U.S. Patent 2,682,466

(1954).

Boyer, R.A. Method of Manufacturing a High Protein Food Product and the Resulting Product. U.S. Patent 2,730,447 (1956).

Bringe, N.A.; S. Charles.. High Beta-Conglycinin Products and Their Use. U.S. Patent 7,084,751 (2006).

Calderon-de-la Barca, A.M.; R.A. Ruiz-Salazar; M.E. Jara-Maarini. Enzymatic hydrolysis and synthesis of soy protein to improve its amino acid composition and functional properties. *J. Food Sci.* **2000,** *65,* 246–253.

Campbell, M.F.; R.J. Fiala; J.D. Wideman; J.P. Rasche. Bland Vegetable Protein Product and Method of Manufacture, U.S. Patent 4,265,925 (1981).

Campbell, M.R.; C.W. Kraut; W.C. Yackel; H.S. Yang. Soy protein concentrate. *New Protein Foods;* A.M. Altschul, H.L. Wilcke, Eds.; Academic Press: New York, NY, 1985; Vol. 5, pp. 301–337.

Central Soya Company. *Soy Flour Product Line Summary.* Central Soya Co., Ft. Wayne, IN, 1988.

Cherry, J.P. *Protein Functionality in Foods,* ACS Symposium Series 147, 1981.

Chiba, H.; R. Sasaki; M. Yoshikawa; K. Ikura. Storage of soybean under high temperature and low humidity. *J. Japan. Soc. Food Nutr.* **1981,** *34,* 201–210.

Chu, H.; M.K. McMindes; S. Wu. Encased Food Product and Process for Producing the Same. U.S. Patent Application 20,070,031,542 (2007).

Circle, S.J. Proteins and other nitrogenous constituents. *Soybeans and Soybean Products;* K.S. Markley, Ed.; John Wiley & Sons: New York, NY, 1950; Vol. 1, pp. 275–370.

Crank, D.L.; P.S. Kerr. Isoflavone-Enriched Soy Protein Product and Method for Its Manufacture. U.S. Patent 5,858,449 (1999).

Crank, Sebastian. Soybean Products with Improved Carbohydrate Composition and Soybean Plants. U.S. Patent 6,653,451 (2003).

Crowe, T.W.; L.A. Johnson. Twin-screw texturization of extruded-expelled soybean flours. *J. Am. Oil Chem. Soc.* **2001,** *78,* 781–786.

Davidson, R.M.; R.E. Sand; R.E. Johnson. Method for Processing Soy Protein and Composition of Matter. U.S. Patent 4,172,828 (1979).

Deak, N.; L.A. Johnson. Novel Vegetable Protein Fractionization Process and Compositions. U.S. Provisional Patent Application (2005).

Deak, N.A.; L.A. Johnson. Functional properties of soy protein ingredients prepared from high-sucrose/low-stachyose soybeans. *J. Am. Oil Chem. Soc.* **2006,** *83,* 811–818.

Deak, N.A.; L.A. Johnson. Fractionation of glycinin and β-conglycinin from high-sucrose/low-stachyose soybeans. *J. Am. Oil Chem. Soc.* **2007a,** *84,* 137–149.

Deak, N.A.; L.A. Johnson. Effects of extraction temperature and preservation method on functionality of soy protein. *J. Am. Oil Chem. Soc.* **2007b,** *84,* 259–268.

Deak, N.A.; L.A. Johnson. Fate of phytic acid in producing soy protein ingredients. *J. Am. Oil Chem. Soc.* **2007c,** *84,* 369–376 2007.

Deak, N.A.; P.A. Murphy; L.A. Johnson. Compositional characteristics of protein ingredients prepared from high-sucrose/low-stachyose soybeans. *J. Am. Oil Chem. Soc.* **2006a,** *83,* 803–810.

Deak, N.A.; P.A. Murphy; L.A. Johnson. Effects of reducing agent concentration on soy protein fractionation and functionality. *J. Food Sci.* **2006b,** *71,* C200–208.

Deak, N.A.; P.A. Murphy; L.A. Johnson. Effects of NaCl concentration on salting-in and dilution during salting-out on soy protein fractionation. *J. Food Sci.* **2006c,** *71,* C247–254.

Deak, N.A.; P.A. Murphy; L.A. Johnson. Fractionating soybean storage proteins using calcium and sodium bisulfite. *J. Food Sci.* **2006d,** *71,* C413–424.

Deak, N.A.; P.A. Murphy; L.A. Johnson. Characterization of fractionated soy proteins produced by a simplified procedure. *J. Am. Oil Chem. Soc.* **2007a,** *84,* 137–149.

Deak, N.A.; Z.M. Nazareth; L.A. Johnson. *Compositions and Properties of Gas-supported Screw-pressed Meal.* 98th AOCS Annual Meeting and Exposition, Quebec City, Canada, May 14–16, 2007b.

Dubois, D.K.; W.J. Hoover. Soya protein products in cereal grain foods. *J. Am. Oil Chem. Soc.* **1981,** *58,* 343–346

Egbert, W.R. Isolated soy protein: technology, properties, and applications. *Soybeans as Functional Foods and Ingredients;* K. Liu, K., Ed.; AOCS Press: Champaign, IL, 2004; pp. 136–137.

Eldrige, A.C.; W.J. Wolf. Purification of the 11s component of soybean protein. *Cereal Chem.* **1967,** *44,* 645–652.

Endres, J.G. *Soy Protein Products: Characteristics, Nutritional Aspects, and Utilization.* AOCS Press: Champaign, IL, 2001.

FAO. *Protein Quality Evaluation,* Report of a Joint FAO/WHO Expert Consultation, Food and Nutrition Paper 51, Rome, 1990.

FDA. Proposed rules on food labeling, general provisions, nutritional labeling, nutrient content claims, health claims, ingredient labeling, state and local requirements, exemption (21 CFR 101, 104, 105), Food and Drug Administration: Washington, DC. *Fed. Reg.* **1991,** *56,* p. 603092.

FDA. Food labeling health claims: Soy protein and coronary heart disease, Final rule, Food and Drug Administration: Washington, DC. *Fed. Reg.* **1999,** *64, 57700–57733.*

Freidman, M. Lysinoalanine formation. *Soy Proteins: Kinetics and Mechanism, in Food Protein Deterioration Mechanisms and Functionality;* ACS *Symposium Series* **1982,** *206,* 231–273.

Fulmer, R.W. The preparation and properties of defatted soy flours and their products. *Proceedings of the World Conference on Vegetable Protein Utilization in Human Foods and Animal Feedstuffs;* T.H. Applewhite, Ed.; Singapore, Malaysia, Oct. 1988. American Oil Chemists' Society: Champaign, IL, 1989a; pp. 55–65.

Fulmer, R.W. Uses of soy proteins in bakery and cereal products. *Proceedings of the World Conference on Vegetable Protein Utilization in Human Foods and Animal Feedstuffs*; T.H. Applewhite, Ed.; Singapore, Malaysia, Oct. 1988. American Oil Chemists' Society: Champaign, IL, 1989b; pp. 424–429.

Gibson, P.W.; W.C. Yackel. Soy protein fractionation and applications. *Proceedings of the World Conference on Vegetable Protein Utilization in Human Foods and Animal Feedstuffs;* T.H. Applewhite, Ed.; American Oil Chemists' Society: Champaign, IL, 1989; pp. 507–509.

Gray, J.T.; T.R. Dutson. Rancidty and warmed-over flavor. *Advances in Meat Research;* A.M. Pearson, Ed.; *Restructured Meat and Poultry Products.* AVI-Van Nostrand Reinhold Co.: New York, NY, 1987; Vol. 3, pp. 221–269.

Gunther, R.C. Chemistry and characteristics of enzyme-modified whipping proteins, *J. Am. Oil Chem. Soc.* **1979,** *56,* 345–349.

Guzman, G.J.; P.A. Murphy; L.A. Johnson. Nutritional and organoleptic properties of soybean-corn mixtures processed by low-cost extrusion. *J. Food Sci.* **1989,** *54,* 1590–1593.

Hanson, L.P. Vegetable Protein Processing. *Food Technol. Rev.* No. 16, Noyes Data Corporation, 1974.

Harper, J.M.; G.R. Jansen. Indices of vitamin status in a man. *Food Rev. Int.* **1985,** *7,* 27–31.

Heywood, A.A.; D.J. Myers; T.B. Bailey; L.A. Johnson. Functional properties of extruded-expelled soybean flours originating from value-enhanced soybean varieties. *J. Am. Oil Chem. Soc.* **2002a,** *79,* 699–702.

Heywood, A.A.; D.J. Myers; T.B. Bailey; L.A. Johnson. Effect of value-enhanced texturized soy protein on the sensory and cooking properties of beef patties. *J. Am. Oil Chem. Soc.* **2002b,** *79,* 703–707.

Heywood, A.A.; D.J. Myers; T.B. Bailey; L.A. Johnson. Functional properties of low-fat soy flour produced by an extrusion-expelling system. *J. Am. Oil Chem. Soc.* **2002c,** *79,* 1249–1253.

Hoover, W. Use of soy proteins in baked foods. *J. Am. Oil Chem. Soc.* **1979,** *53,* 301–303.

Howard, P.A.; M.F. Campbell; D.T. Zollinger. Water Soluble Vegetable Protein Aggregates, U.S. Patent 4,234,620 (1980).

Howard, P.A.; W.F. Lehnhardt; F.T. Orthoefer. 7S and 11S Vegetable Protein Fractionation and Isolation, U.S. Patent 4,368,151 (1983).

Hrckova, M.; M. Rusnakova; J. Zemanovic. Enzymatic hydrolysis of defatted soy flour by three different proteases and their effect on the functional properties of resulting protein hydrolysates. *Czech. J. Food Sci.* **2002,** *20,* 7–14.

Hymowitz, T.; F.I. Collins; J. Panczner; J.M. Walker. Relationship between the content of oil, protein, and sugar in soybean seed. *Agron. J.* **1972,** *64,* 613–616.

ISU. *Identifying New Industrial Uses for Soy Protein*, Special Report 95, Iowa Agriculture and Home Economics Experiment Station, Iowa State University: Ames, IA, 1994.

Jansen, R.; J.M. Harper. Allocution á la conférence internationale sur la nutrition. *Food Nutr. Bull.* **1980,** *6,* 2–8.

Johnson, D.W. General uses of whole soybeans. *Food Uses of Whole Oil and Protein Seeds.* E.W. Lusas, D.R. Erickson, W.K. Nip, Eds.; American Oil Chemists' Society: Champaign, IL, 1989; pp. 12-–39.

Johnson, L.A. Processing Aqueous Extracts of Soybeans by Rapid-Hydration Hydrothermal Cooking, Ph.D. Dissertation, Kansas State University, Manhattan, KS, 1978.

Johnson, L.A. Process for Producing Improved Soy Protein Concentrate from Genetically Modified Soybeans, U.S. Patent 5,936,069 (1999).

Johnson, L.A.; C.W. Deyoe; W.J. Hoover; J.R. Schwenke. Inactivation of trypsin inhibitors in aqueous extracts of soybeans by direct steam infusion. *Cereal Chem.* **1980,** *57,* 376–379.

Johnson, L.A.; D.J. Myers; D.J. Burden. Historical perspectives on production and utilization of soy proteins. i. early oriental uses and industrial uses. *INFORM (AOCS)* **1992a,** *3,* 282–290.

Johnson, L.A.; D.J. Myers; D.J. Burden. Historical perspectives on production and utilization of soy proteins. i. early oriental uses and industrial uses. *INFORM (AOCS)* **1992b,** *3,* 282–290.

Johnson, L.A.; D.J. Myers; D.J. Burden. Historical perspectives on production and utilization of soy proteins. ii. feed and food uses and future prospects. *INFORM (AOCS)* **1992c,** *3,* 429–444.

Jung, S.; D. Rickert; N. Deak; E. Aldin; J. Recknor; L.A. Johnson; P.A. Murphy. Comparison of kjeldahl and dumas methods for protein content determination in soy products. *J. Am. Oil Chem. Soc.* **2003,** *80,* 1169–1173.

Jung, S.; B.P. Lamsal; V. Stepien; L.A. Johnson; P.A. Murphy. Functionality of soy proteins produced by enzyme-assisted extraction. *J. Am. Oil Chem. Soc.* **2006,** *83,* 71–78.

Jung, S.; P.A. Murphy; L.A. Johnson. Physicochemical and functional properties of soy protein substrates modified by low levels of protease hydrolysis. *J. Food Sci.* **2005,** *70,* C180–187.

Jung, S.; D. Rousel-Philippe; J. Briggs; P.A. Murphy; L.A. Johnson. Limited hydrolysis of soy proteins with endo- and exo-proteases. *J. Am. Oil Chem. Soc.* **2004,** *81,* 953–960.

Kanzamar, G.J.; S.J. Predin; D.A. Oreg; Z.M. Csehak. Processing of soy flours/grits and textured soy flours. *Proceedings of the World Conference on Oilseed Technology and Utilization.* T.H. Applewhite, Ed.; American Oil Chemists' Society: Champaign, IL, 1993; pp. 226–240.

Karayianis, N.I.; J.T. MacGregor; L.P. Bjelfanes. Lysinoalanine formation in alkali-treated proteins and model peptides. *Food Cosmetics Tox.* **1979,** *17,* 585–590.

Kennedy, A. The evidence for soybean products as cancer preventive agents. *J. Nutr.* **1995,** 733S–743S.

Kennedy, A. The bowman-birk inhibitor from soybeans as an anticarcinogenic agent. *Am. J. Clin. Nutr.* **1998,** *68 (suppl.),* 1406S–1412S.

Kennedy, A.; B. Szuhaj. Bowman-Birk Inhibitor Product for Use as an Anticarcinogenesis Agent, U.S. Patent 5,338,547 (1994).

Kim, S.K.; P.S. Park; K.C. Rhee. Functional properties of proteolytic enzyme modified soy protein isolate. *J. Agric. Food Chem.* **1990,** *38,* 651–656.

Kinsella, J.E. Functional properties of soy proteins. *J. Am. Oil Chem. Soc.* **1979,** *56,* 242–258.

Kinsella, J.E.; S. Damodaran; B. German. Physiochemistry and function of oilseed proteins. *New Protein Foods, 5, Seed Storage Proteins;* A.M. Altschul, H.L. Wilcke, Eds.; Academic Press, Inc.: New York, NY, 1985; pp. 109–179.

Kinsella, J.E.; W.G. Soucie. (ed.) *Food Proteins,* AOCS Press: Champaign, IL, 1989.

Kolar, C.W.; I.C. Cho; W.L. Watrous. Vegetable protein applications in yogurt, coffee creamers and whip toppings. *J. Am. Oil Chem. Soc.* **1979,** *56,* 389–391.

Kolar, C.W.; S.W. Richert; C.D. Decker; F.H. Steinke; R.J. Vander Zanden. *New Protein Foods;* A.M. Altschul, H.L. Wilcke Eds.; Academic Press: New York, NY, 1985, Vol. 5, pp. 259–299.

Koseoglu, S.S.; E.W. Lusas. Aqueous, membrane and adsorptive separations of vegetable proteins. *Proceedings of the World Conference on Vegetable Protein Utilization in Human Foods and Animal Feedstuffs;* T.H. Applewhite, Ed.; American Oil Chemists' Society: Champaign, IL, 1989; pp. 528–547.

Koshiyama, I. Purification of the 7s component of soybean proteins. *Agric. Biol. Chem.* **1965,** *29,* 885–887.

Koshiyama, I. Chemical and physical properties of a 7s protein in soybean globulins. *Cereal Chem.* **1968a,** *45,* 394–404.

Koshiyama, I. Chromatographic and sedimentation behavior of a purified 7s protein in soybean globulins. *Cereal Chem.* **1968b**, *45,* 405–412.

Koshiyama, I. A newer method for isolation of the 7s globulin in soybean seeds. *Agric. Biol. Chem.* **1972,** *36,* 2255–2257.

Lamsal, B.P.; C. Reitmeier; P.A. Murphy; L.A. Johnson. Enzymatic hydrolysis of extruded-expelled soy protein and resulting functional properties. *J. Am. Oil Chem. Soc.* **2006,** *83,* 732–737.

Lawhon, J.T. Method for Processing Protein From Nonbinding Oilseed by Ultrafiltration and Solubilization, U.S. Patent 4,420,425 (1983).

Lawhon, J.T.; K.C. Rhee; E.W. Lusas. Soy protein ingredients prepared by new processes—aqueous processing and industrial membrane isolation. *J. Am. Oil Chem. Soc.* **1981,** *58,* 377–384.

Lee, K.H.; H.S. Ryu; K.C. Rhee. Protein solubility characteristics of commercial soy protein products. *J. Am. Oil Chem. Soc.* **2003,** *80,* 85–90.

Lehnhardt, W.R.; P.W. Gibson; F.T. Orthoefer. Fractionation and Isolation of 7S and 11S from Isoelectrically Precipitated Vegetable Protein Mixtures, U.S. Patent 4,370,267 (1983).

Liener, I.E. Factors affecting the nutrition and qualities of soya products. *J. Am. Oil Chem. Soc.* **1981,** *58,* 406–416.

Liener, I.E. Implications of antinutritional components in soybean foods. *Crit. Rev. Food Sci. Nutr.* **1994,** *34,* 31–67.

Lin, S.H.A.; M. Cho. Process for the Production of a Mineral Fortified Protein Composition, U.S. Patent 4,642,238 (1987).

Lorenz, K.; G.R. Jansen; J. Harper. Nutrition and stability of full-fat soy flour and corn-soy blends produced by low-cost extrusion. *Cereal Foods World* **1980,** *25,* 161–162, 171.

Lusas, E.W.; K.C. Rhee. Soybean protein processing and utilization. *Practical Handbook of Soybean Processing and Utilization;* D. Erickson, Ed.; AOCS Press: Champaign, IL, 1995; pp. 117–160.

Lusas, E.W.; K.C. Rhee. Plant proteins. *ACS Symposium Series* ***1986,*** *312,* 32–44.

Ly, Y.; J. Jane; L.A. Johnson. Soy proteins as biopolymers. *Biopolymers from Renewable Resources;* D. Kaplan, Ed., Springer-Verlag: Heidelberg, Germany, 1998; pp. 144–176.

Madl, R.L. Evolution of protein quality determination. *Cereal Foods World* **1993,** *38,* 576–577.

Martz, M.A. Protein food supplements. *Food Technol. Rev.* No. 54, Noyes Data Corp., 1981.

Maruyama, M.; M. Adachi; K. Takahashi; K. Yagasaki; M. Kohno; Y. Takenaka; E. Okuda; S. Nakagawa; B. Mikami; S. Utsumi. Crystal structures of recombinant and native soybean β-conglycinin β homotrimers. *Eur. J. Biochem.* **2001,** *268,* 3595–3604.

McAnelly, J.K. Bland Textured Product from the Water-insoluble Portion of Defatted Cooked Soybean Flour, U.S. Patent 3,142,571 (1964).

McDonald, F.M. Problems in soybean processing. *Oil Mill Gaz.* **1978,** *85(3),* 8–9.

Mercier, C.; P. Linko; J.M. Harper. *Extrusion Cooking.* American Association of Cereal Chemists, Inc.: St. Paul, MN, 1989.

Messina, M.J.; V. Perskey; K. Setchell; S. Barnes. Soy isolate and cancer risk: a review on the in-vitro and in-vivo data. *Nutr. Cancer* **1994,** *21,* 113–131.

Molina, M.R.; J.E. Braham; R. Bressani. Some characteristics of whole corn:whole soybean (70:30) and rice:whole soybean (70:30) mixtures processed by simple extrusion cooking. *J. Food Sci.* **1983,** *48,* 434–437.

Morse, E.H. Process for Extraction of Vegetable Proteins, U.S. Patent 2,331,619 (1943).

Muralidhara, H.S.; M.A. Porter; J.V. Satyavolu; W.H. Sperber; I. Purtle. Process for Producing Oilseed Protein Products. U.S. Patent 6,630,195 (2003).

Murray, E.D.; T.J. Maurice; L.D. Barker; C.D. Davis. Process for Isolation pf Proteins Using Food-grade Salt Solutions at Specified pH and Ionic Strength, U.S. Patent 4,208,323 (1980).

Mustakas, G.C.; J.W. Albrecht; G.N. Bookwalter; J.E. McGhee; W.F. Kwoiek; E.L. Griffin, Jr. Extruder —processing to improve nutritional quality flavor, and keeping quality of full-fat soy flour. *J. Food Technol.* **1970,** *24,* 1290–1301.

Mustakas, G.C.; W.J. Albrecht; G.N. Bookwalter; V.E. Sohns; E.L. Griffin, Jr. New process for low-cost, high-protein beverage base. *Food Technol.* **1971,** 25, 534–546.

Mustakas, G.C.; E.L. Griffin, Jr.; V.E. Sohns. Full-fat soybean by continuous extrusion cooking. *Am.. Chem. Soc. Advan. Chem. Ser.* **1966,** *57,* 101–111.

Mustakas, G.C.; L.D. Kirk; E.L. Griffin. Flash desolventizing defatted soybean meals washed with aqueous alcohols to yield a high protein product. *J. Am. Oil Chem. Soc.* **1962,** *39,* 222–226.

Nagano, T.; M. Hirotsuka; H. Mori; K. Kohyana; K. Nishinari. Dynamic viscoelastic study on the gelation of 7s globulin from soybeans. *J. Agric. Food Chem.* **1992,** *40,* 941–944.

Ohren, J.A. Process and product characteristics for soya concentrates and isolates. *J. Am. Oil Chem. Soc.* **1981,** *58,* 333–335.

Orf, J.H.; T. Hymowitz. Genetics of the kunitz trypsin inhibitor: an antinutritional factor in soya beans. *J. Am. Oil Chem. Soc.* **1979,** *56,* 722–726.

Patil, R.T.; D.S. Singh; R.E. Tribelhorn. Effect of processing conditions on extrusion cooking of soy-rice blend with a dry extrusion cooker. *J. Food Sci. Technol. India* **1990,** 27, 376–378.

Pearson, A.M; T.R. Dutson. *Advances in Meat Research, 3, – Restructured Meat and Poultry Products.* AVI-VanNostrand Reinhold Company, New York, 1987.

Pringle, W., Full-fat soy flour, *J. Am. Oil Chem. Soc.* **1974,** *51,* 74A–76A.

Rackis, J.J. Biological and physiological factors in soybeans. *J. Am. Oil Chem. Soc.* **1974,** *51,* 161A–177A.

Rackis, J.J. Significance of soya trypsin inhibitors in nutrition. *J. Am. Oil Chem. Soc.* **1981,** *58,* 495501.

Rackis, J.J.; D.J. Sessa; D.H. Honig. Flavor problems of vegetable food proteins. *J. Am. Oil Chem. Soc.* **1979,** *56,* 262–271.

Rakes, G.A. Meat applications of soy protein concentrate. *Proceedings of the World Conference on Oilseed Technology and Utilization;* T.H. Applewhite, Ed.; American Oil Chemists' Society: Champaign, IL, 1993; pp. 311–319.

Rhee, K.C. Determining and modifying protein functionality. *Proceedings of the World Conference on Vegetable Protein Utilization in Human Foods and Animal Feedstuffs;* T.H. Applewhite, Ed.;

American Oil Chemists' Society: Champaign, IL, 1989; pp. 323–333.

Riaz, M.N. Extrusion-expelling of soybeans for texturized soy protein. *Proceedings of the World Conference on Oilseed Processing and Utilization; R. Wilson, Ed.;* AOCS Press: Champaign, IL, 2001; pp. 171–175.

Rickert, D.A.; L.A. Johnson; P.A. Murphy. Improved fractionation of glycinin and β-conglycinin and partitioning of phytochemicals. *J. Agric. Food Chem.* **2004a,** *52,* 1726-1734.

Rickert, D.A.; L.A. Johnson; P.A. Murphy. Functional properties of improved glycinin and β-conglycinin fractions. *J. Food Sci.* **2004b,** *69,* FCT303–311.

Roberts, R.C.; D.R. Briggs. Isolation and characterization of the 7s component of soybean globulins. *Cereal Chem.* **1965,** *42,* 71–85.

Rokey, G.J.; G.R. Huber; I. Ben-Gera. Extrusion-cooked and textured defatted soybean flours and protein concentrates. *Proceedings of the World Conference on Oilseed Technology and Utilization;* T.H. Applewhite, Ed.; American Oil Chemists' Society: Champaign, IL, 1993; pp. 290–298.

Saio, K.; M. Ariska. Deterioration of soybeans during storage under high moisture and temperature. *J. Japan. Soc. Food. Sci. Technol.* **1978,** *25,* 451-457.

Saio, K.; K. Kobayakawa; M. Kito. Protein denaturation during model storage studies of soybeans and meals. *Cereal Chem.* **1982,** *59,* 408-412.

Saio, K.; I. Nikkuni; Y. Ando; M. Ortsura; Y. Terauchi; M. Kito. Soybean quality. *Cereal Chem.* **1980,** *57,* 77–82.

Saio, K.; I. Sato; T. Watanabe. Food use of soybean 7s and 11s proteins—high temperature expansion characteristics of gels. *J. Food Sci.* **1974,** *39,* 777-781.

Saio, K.; M. Terashima; T. Watanabe. Food use of soybean 7s and 11s proteins—heat denaturation of soybean proteins at high temperature. *J. Food Sci.* **1975,** *40,* 537-540.

Saio, K.; T. Watanabe. food use of soybean 7s and 11s proteins/extraction and functional properties of their fractions. *J. Food Sci.* **1973,** *38,* 1139-1144.

Sair, L. Proteinaceous Soy Composition and Method of Preparing, U.S. Patent 2,881,076, (1959).

Saito, T.; M. Kohno; K. Tsumura; W. Kugimiya; M. Kito. Novel method using phytase for separating soybean β-conglycinin and glycinin. *Biosci. Biotechnol. Biochem.* **2001,** *65,* 884–887.

Savoie, L.; G. Parent. Susceptibility of protein fractions to lysinoalanine formation. *J. Food Sci.* **1983,** *48,* 1876–1877.

Shemer, M. Process for Preparing a Heat Coagulable Viscous Protein, U.S. Patent 4,188,399 (1980).

Shen, J.L. Solubility profile, intrinsic viscosity, and optical rotation studies of acid precipitated soy protein and of commercial soy isolate. *J. Agric. Food Chem.* **1976,** *24,* 784–795.

Shurtleff, W.; A. Aoyagi. *Bibliography of Soymilk and Soymilk Products: 2612 References from A.D. 1500 to 1989.* Soyfoods Center: Lafayette, CA, 1989a.

Shurtleff, W.; A. Aoyagi. *Bibliography of Soy Ice Cream, Yoghurt, and Cheese: 1071 References from 1910 to 1989.* Soyfoods Center: Lafayette, CA, 1989b.

Shurtleff, W.; A. Aoyagi. *Bibliography of Soy Flour and Cereal-Soy Blends: 308 References from 3rd Century B.C. to 1900.* Extensively Annotated, Soyfoods Center: Lafayette, CA, 1989c.

Shurtleff, W.; A. Aoyagi. *Bibliography of Tofu.* Soyfoods Center: Lafayette, CA, 1989d.

Simelunas, W.J.; N.R. Polifroni; H.N. Shoiket; M.M. Stefan. Method and Apparatus for Severing a Coextrusion for Making an Enrobed Food Piece, U.S. Patent 4,882,185 (1989).

Singh, N. Process for Manufacturing a Soy Protein Concentrate Having High Isoflavone Content, U.S. Patent 7,083,819 (2003).

Smith, A.K.; S.J. Circle. *Soybeans: Chemistry and Technology* Vol. 1, Proteins., Revised second ed.; AVI Publishing Co.: Westport, CT, 1972.

Sogo, Y.; S. Dosako; Y. Honda; K. Ahiko; S. Kawamura; T. Izutysu; S. Tanyea. *Reports of Research Laboratory;* Snow Brand Milk Products Co.: Kawagoe, Saitama, Japan, 1985.

Soy Protein Council. *Soy Protein Products,* Washington, DC, 1987.

Stanley, D.W. Protein reactions during extrusion processing. *Extrusion Cooking;* C. Mercier, P. Linko, J.M. Harper, Eds.; American Association of Cereal Chemists, Inc.: St. Paul, MN, 1989; pp. 321–341.

Struthers, B.J. Lysinoalanine: production, significance and control in preparation and use of soya and other food proteins. *J. Am. Oil Chem. Soc.* **1981,** *58,* 501–503.

Surowka, K.; D. Zmudzinski; M. Fik; R. Macura; W. Lasocha. New protein preparations from soy flour obtained by limited enzymic hydrolysis of extrudates. *Innov. Food Sci. Emer. Technol.* **2004,** *5,* 225–234.

Thanh, V.H.; K. Shibasaki. major proteins of soybean seeds—a straightforward fractionation and their characterization. *J. Agric. Food Chem.* **1976,** *24,* 1117–1121.

Thiering, R.; G. Hofland; N. Foster; G.J. Witkamp; L.V.D. Wielen. Fractionation of soybean proteins with pressurized carbon dioxide as a volatile electrolyte. *Biotechnol. Bioeng.* **2001,** *73,* 1–11.

Thomas, R.; J.M. deMan; L. deMan. Soymilk and tofu properties as influenced by soybean storage conditions. *J. Am. Oil Chem. Soc.* **1989,** *66,* 777–782.

Vavlitis, A.; E.D. Mulligan. Flash desolventizing. *Proceedings of the World Conference on Oilseed Technology and Utilization;* T.H. Applewhite, Ed.; American Oil Chemists' Society: Champaign, IL, 1993; pp. 286–289.

Wang, C.; L.A. Johnson. Functional properties of hydrothermally cooked protein products. *J. Am. Oil Chem. Soc.* **2001a,** *78,* 189–195.

Wang, T.; L.A. Johnson. Survey of soybean oil and meal qualities produced by different processes. *J. Am. Oil Chem. Soc.* **2001b,** *78,* 311–318.

Wang, T.; D.J. Myers; L.A. Johnson. Value-added products from extruding-expelling soybean. *Soybeans as a Functional Food;* K. Liu, C.Y. Wang, Eds.; AOCS Press: Champaign, IL, 2004.

Wang, H.; T. Wang; L.A. Johnson. Preparation of soy protein concentrate and isolate from extruded-expelled soybean meals. *J. Am. Oil Chem. Soc.* **2004a,** *81,* 713–717.

Wang, H.; T. Wang; L.A. Johnson. Refunctionalization of extruded-expelled soybean meals. *J. Am. Oil Chem. Soc.* **2004b,** *81,* 789–794.

Wang, H.; T. Wang; L.A. Johnson. Effect of alkali on the refunctionalization of soy protein by hydrothermal cooking. *J. Am. Oil Chem. Soc.* **2005,** *82,* 451–456.

Wang, H.; T. Wang; L.A. Johnson. Mechanism for refunctionalizing heat-denatured soy protein by alkaline hydrothermal cooking. *J. Am. Oil Chem. Soc.* **2006,** *83,* 39–45.

Wijeratne, W.B.; T. Wang; L.A. Johnson. Extrusion-based oilseed processing methods. *Nutritionally Enhanced Edible Oil Processing;* N.T. Dunford, H.B. Dunford, Eds., AOCS Press: Champaign, IL, 2004.

Wilding, M.D. Vegetable protein application in whey soy drink mix and ice cream. *J. Am. Oil Chem. Soc.* **1979,** *56,* 392–395.

Wilcke, H.L.; D.F.H. Waggle; C.K. Kolar. Textural contribution of vegetable protein products. *J. Am. Oil Chem. Soc.* **1979,** *56,* 259–261.

Wolf, W.J. Physical and Chemical Studies on Soybean Proteins, Thesis, University of Minnesota, Minneapolis, MN, 1956.

Wolf, W.J. Physical and chemical properties of soybean proteins. *J. Am. Oil Chem. Soc.* **1977,** *54,* 112A–117A.

Wolf, W.J. *Handbook of Processing and Utilization in Agriculture;* I.A. Wolf, Ed.; CRC Press, Inc.: Boca Raton, FL, 1983; Vol. 2, Part 2, pp. 23–55.

Wolf, W.J.; G.E. Babcock; A.K. Smith. Purification and stability studies of the 11s component of soybean proteins. *Arch. Biochem. Biophys.* **1962,** *99,* 265–274.

Wolf, W.J.; C.J. Cowan. *Soybeans as a Food Source.* CRC Press, Inc.: Cleveland, OH, 1975.

Wolf, W.J.; D.A. Sly. Cryoprecipitation of soybean 11s protein. *Cereal Chem.* **1967,** *44,* 653–668.

Wright, K.N. Soya meal. *J. Am. Oil Chem. Soc.* **1981,** *58,* 294–300.

Wu, S.; P.A. Murphy; L.A. Johnson; A.R. Fratzke; M.A. Reuber. Pilot-plant fractionation of soybean glycinin and β-conglycinin. *J. Am. Oil Chem. Soc.* **1999,** *76,* 285–293.

Wu, S.; P.A. Murphy; L.A. Johnson; A.R. Fratzke; M.A. Reuber. Simplified pilot-plant process for soybean glycinin and β-conglycinin fractionation. *J. Agric. Food Chem.* **2000,** *48,* 2702–2708.

Yuan, Y.J.; O.D. Velev; K. Chen; B.E. Campbell; E.W. Kaler; A.M. Lenhoff. Effect of pH and Ca^{2+}-induced associations of soybean proteins. *J. Agric. Food Chem.* **2002,** *50,* 4953–4958.

Ziemba, J.V. Simulated meats: how they are made. *Food Eng.* Nov. 1969; pp. 7275.

20

Human Nutrition Value of Soybean Oil and Soy Protein

Alison M. Hill[1], Heather I. Katcher[1], Brent D. Flickinger[2], and Penny M. Kris-Etherton[1]

[1]Department of Nutritional Sciences, S-126 Henderson Building, Penn State University, University Park, PA 16802 USA; [2]Archer Daniels Midland, Decatur, IL 62521

Introduction

The United States currently produces over 3.1 billion bushels of soybeans per year (The American Soybean Association, 2007). In 2006, over one-third (1.1 billion bushels) of the crop was exported, while the remainder was sold on the domestic market. Of the edible soybean products in the U.S. market, the consumption of soybean oil (SBO) is greatest; U.S. production of SBO has surpassed 20 billion pounds per year (The American Soybean Association, 2007). SBO is the major edible oil in the United States and represented 75% of the total edible oils and fats consumed in 2006. The predominant dietary sources of SBO are salad and cooking oil (48%) and baking and frying fats (34%). SBO has a unique fatty acid profile; it is comprised predominantly of unsaturated fatty acids including monounsaturated fatty acids (MUFA) and polyunsaturated fatty acids (PUFA), of which the predominant PUFA is linoleic acid (LA), and to a much lesser extent, α-linolenic acid (ALA).

Soybeans also are a source of protein. Consumption of soybean products containing soy protein has risen in recent years, although this remains significantly less than the consumption of SBO. This increase may be due, in part, to the U.S. Food and Drug Administration (FDA) approved soy protein health claim: "Diets low in saturated fat and cholesterol that include 25 grams of soy protein a day may reduce the risk of heart disease" (Food and Drug Administration, 1999). More recent research questioned the efficacy of this health claim on the basis of less than previously reported cholesterol-lowering effects (Balk et al., 2005). One component of soy protein that also gained interest is isoflavones (bioactive molecules contained in soy protein) because of their potential health effects.

Based on dietary assessment data for energy and fat (Food Surveys Research Group, 2005), SBO accounts for about 12% of calories in the average American diet. Given the contribution of SBO to the diet, this chapter discusses the health effects of SBO and its constituent fatty acid profile, as well as soy protein and its bioactive components.

Absorption of Fatty Acids

Edible oils and fats are well absorbed by humans. Absorption coefficients for commonly consumed vegetable oils and animal fats are above 91% (Deuel, 1955) and are a function of their melting point. A high melting point, as characteristic of saturated and *trans* fatty acids, is associated with lower absorption (Deuel, 1955). Hydrogenation of unsaturated fatty acids raises the melting point and, hence, decreases absorption. SBO, like other liquid vegetable oils, has an absorption coefficient greater than 97%. The absorption of individual fatty acids in SBO, specifically lauric, myristic, palmitic, stearic, oleic, and elaidic acids, and LA also is consistently high; absorption coefficients in excess of 93% were reported (Baer et al., 2003). Nonetheless, subtle differences exist among fatty acids. Palmitic acid (97.3%) is more readily absorbed than stearic acid (94.1%), while other fatty acids (lauric, myristic, oleic, elaidic, and linoleic acids) are nearly completely absorbed (>99.0%) (Baer et al., 2003). Other factors that affect absorption include positional distribution on the glycerol molecule, the presence of calcium and magnesium in the diet, and the mode of preparation and emulsification of the fat system. With respect to the position of fatty acids in the triglyceride (TG), those in the sn-2 position are better absorbed than in the sn-1 or sn-3 position (Mattson et al., 1979). High intakes of calcium and magnesium modestly impair stearic acid absorption via the formation of fatty acid soaps (37–98% absorbed depending on fatty acid esterification) (Mattson et al., 1979). Lastly, the absorption of fats, particularly those with high melting points, is influenced by emulsification. Emken and coworkers (1993) reported that tristearin is more readily absorbed (90%) when emulsified in a sugar–casein–water mixture prepared at 80–85°C than a mixture prepared at 70°C (34%). Despite these factors that affect fat and fatty acid absorption, the differences are not markedly dissimilar, and, therefore, all fats and fatty acids are highly absorbed, and, consequently, are a readily available fuel source.

Soybean Oil

Fatty Acid and Nutrient Profile of Soybean Oil, Mid-/High-Oleic Soybean Oil, Partially Hydrogenated Soybean Oil, and Fully Hydrogenated Soybean Oil

Liquid vegetable oils are low in saturated fatty acids (SFA) and high in unsaturated fatty acids, with varying amounts of MUFA and PUFA depending on the oil. Figure 20.1 presents the fatty acid composition of common fats in the food supply. Compared to other popular vegetable oils, SBO contains substantially less MUFA than canola or olive oil. SBO and olive oil contain similar amounts of SFA, whereas canola oil contains about half that found in SBO and olive oil. Compared to animal-derived fats and oils such as butter and tallow, SBO is substantially lower in MUFA and SFA. SBO is high in LA (89% of total PUFA), and also is a source of ALA (11% of total PUFA). Wheat germ and walnut oils provide similar amounts of ALA. Hempseed oil

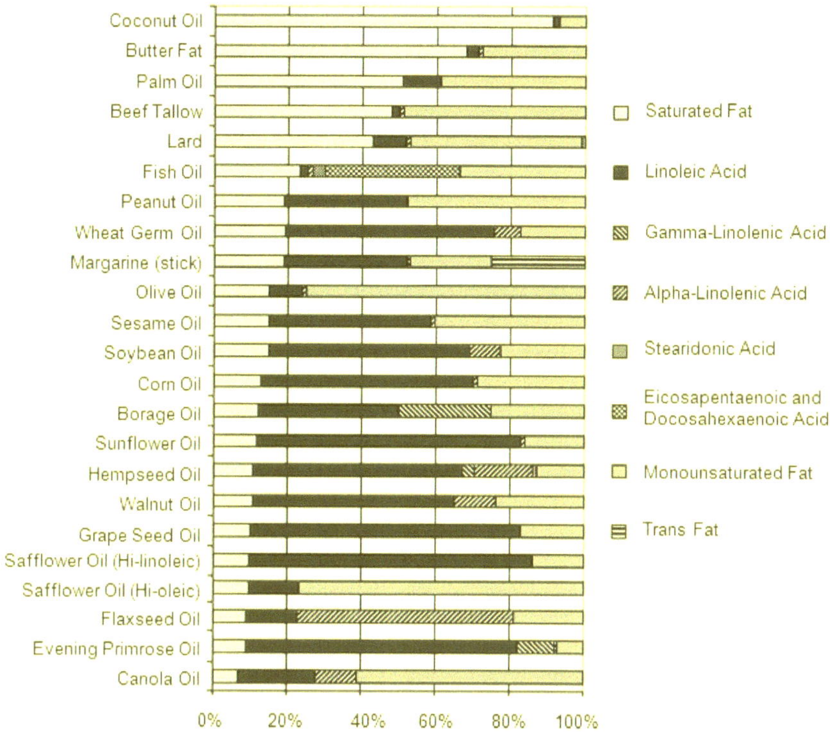

Fig. 20.1. Fatty acid contents of fats and oils (adapted from Pass & Pierce, 2002).

and canola oil have slightly higher levels of ALA, whereas flaxseed oil is appreciably higher. Corn oil, sunflower oil, and olive oil provide very small quantities of ALA. Since SBO is the predominant vegetable oil in the U.S. diet, it is a major contributor of dietary ALA.

In recent years, interest increased in the ratio of omega-6 (n-6) to omega-3 (n-3) PUFA, or LA:ALA, in part due to the link between inflammation and several lifestyle diseases, such as cardiovascular disease (CVD) and Type II diabetes. However, whether this ratio is directly associated with an increased risk of inflammatory diseases is unclear. Furthermore, the low conversion of dietary ALA to eicosapentaenoic acid (EPA) and docosahexaenoic acid (DHA) (Goyens et al., 2005; Hussein et al., 2005; Pawlosky et al., 2001) means that a lower n-6:n-3 PUFA ratio does not necessarily reflect physiologically important increases in EPA and DHA (Harris, 2006). Consequently, evaluating absolute dietary intakes of specific n-6 and n-3 PUFAs may be most appropriate, particularly when few human experimental and clinical trial data exist to support the use of an n-6:n-3 PUFA ratio. Nevertheless, when considering the composition of SBO, notably, SBO has a lower n-6:n-3 PUFA ratio than other commonly used vegetable oils, such as corn oil.

In the marketplace, many SBO varieties that vary in fatty acid composition are available (Table 20.1). The predominant fatty acids in conventional SBO are LA (51%) and oleic acid (23%), with smaller proportions of palmitic acid (10.3%), ALA (6.8%), and stearic acid (3.8%). Other SBO varieties with altered ALA and oleic acid contents also were developed (see Table 20.1). As their names suggest, the mid-oleic variety has over twice as much oleic acid as conventional SBO, whereas the high-oleic variety is largely oleic acid (84% of total fatty acids). The 1% linolenic variety has a reduced ALA content that is offset by a slight increase in oleic acid and LA. Hydrogenation markedly alters the fatty acid profile of SBO (Table 20.1). The fatty acid composition of partially hydrogenated SBO (PHSBO) can vary depending on the extent of hydrogenation, which affects the amount of *trans* fatty acids. Partial hydrogenation reduces both LA and ALA, with a concomitant increase in SFA and MUFA. The increase in MUFA is predominantly due to an increase in 18:1 *trans* isomers. Fully hydrogenated soybean oil is comprised of palmitic (11%) and stearic (89%) acids.

Similar to other vegetable oils (e.g., canola, corn, and olive oils), SBO is a source of fat-soluble vitamins including alpha-tocopherol (9.21 mg/100 g), and vitamin K (184 mcg/100 mg). The recommended daily intake of vitamin E for adults is 15 mg/day (Institute of Medicine, 2000), and consistent consumption of SBO can therefore play a role in meeting these dietary recommendations. One serving of SBO (14 g) provides 30% of the current daily value for vitamin K (as defined in the Code of Federal Regulations at 21CFR101.9(c)(8)(iv)) and 20% of the updated Adequate Intake from the Institute of Medicine (2001). SBO also provides a small amount of phytosterols (~250 mg/100 g SBO), which inhibit cholesterol absorption (reviewed by Berger et al., 2004), although in significantly larger quantities (0.8–1.0 g/day) than provided in a single serving of SBO. In comparison, the phytosterol concentration of corn oil is nearly four times that of SBO, but still not high enough to attain the dose recommended for cholesterol lowering. Crude SBO also is a source of lecithin, which contains the essential nutrient choline. Choline has several physiological functions and may have potential health and nutritive benefits. However, lecithin is removed from crude SBO during processing for functional reasons.

Fatty Acids and Risk of Chronic Disease

A robust database exists from observational studies, clinical studies/trials, animal studies, and in vitro studies demonstrating that fatty acids play a key role in chronic disease risk. Population studies show a positive correlation between coronary heart disease (CHD) risk and dietary SFA (Hu et al., 1997; Keys, 1970; Kromhout et al., 1995) and an inverse association with PUFA (Hu et al., 1997). The Nurses' Health Study (Hu et al., 1997) reported an inverse relationship between the PUFA:SFA ratio and CHD risk. By all estimates, replacing 5% of energy from SFA with PUFA or MUFA would reduce risk of CHD by ~48 and 36%, respectively. The relationship between dietary fat intake and risk of CHD is illustrated in Figure 20.2. *Trans* fatty

Table 20.1. Fatty Acid Compositions (% of total) of Soybean Oil (SBO), Partially Hydrogenated Soybean Oil (PHSBO), and Fully Hydrogenated Soybean Oil (HSBO)[a]

Fatty Acid	SBO (Conventional)	SBO (High-oleic)	SBO (Mid-oleic)	SBO (1% Linolenic)	PHSBO	HSBO
Total SFA	14.4	10.0	14.2	14.9	24.8	100.0
14:0	0.1				0.1	
15:0					0.0	
16:0	10.3	6.3	9.8	9.7	11.2	11.0
17:0					0.1	
18:0	3.8	3.7	4.4	5.2	12.6	89.0
20:0					0.4	
22:0					0.4	
Total MUFA	23.3	84.0	51.9	27.9	61.2	0.0
16:1	0.2				0.1	
18:1	22.8	84.0	51.9	27.9	61.1	
18:1 *cis*					31.5	
18:1 *trans*					29.7	
20:1	0.2				0.0	
Total PUFA	57.9	4.0	33.8	57.2	9.3	0.0
18:2	51.0	1.6	32.7	56.2	8.6	
18:2 *cis*					4.6	
18:2 i					4.0	
18:3	6.8	2.4	1.1	1.0	0.2	

[a]Data for mid-oleic and 1% linoleic SBO obtained from Iowa State University (http://www.notrans.iastate.edu/default.html).
Data for high-oleic SBO obtained from Oil Research Unit, U.S. Department of Agriculture (http://pbi-ibp.nrc-cnrc.ge.ca/en/bulletin/2002issue1/page6.html).
Data for conventional SBO and PHSBO obtained from the USDA Nutrient data laboratory.
Data for HSBO; personal communication.

acids also have an adverse effect on other chronic diseases including CHD (Ascherio et al., 1996b; Kromhout et al., 1995; Oh et al., 2005; Oomen et al., 2001; Pietinen et al., 1997). Replacing SFA and *trans* fats with unsaturated fats is associated with the greatest reduction in CHD risk (Hu et al., 1997). Consistent with the effects of individual fatty acids, by all suggestions, dietary interventions that incorporate several of these fatty acid substitutions can reduce the incidence of coronary artery disease (Fig. 20.3) (Sacks & Katan, 2002). More recently, both plant- and marine-derived n-3 PUFA were shown to have numerous health benefits. Several studies reported that

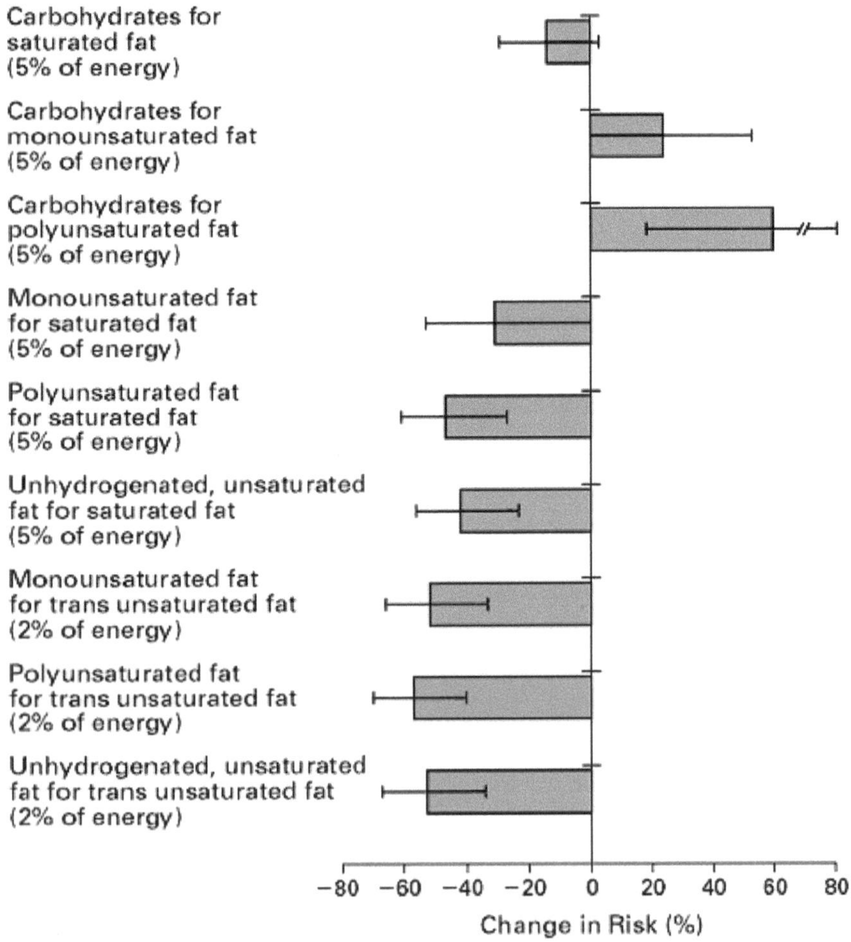

Fig. 20.2. Estimated changes (percent with 95% confidence intervals) in risk of coronary heart disease associated with isocaloric dietary substitutions. (Reprinted from Hu et al., 1997.

n-3 PUFA intake from fish is associated with reduced mortality from CHD (Mozaffarian & Rimm, 2006). The efficacy of PUFA from plant sources to reduce CVD risk/incidence was demonstrated in several early clinical trials (Dayton et al., 1969; Leren, 1970; Turpeinen et al., 1979). Total fat intake in these trials ranged from 34 to 46% of total energy intake, and differences between the intervention and control groups were achieved by replacing dietary SFA with PUFA (13–21% of energy) from several vegetable oil sources, including SBO. These trials reported significant reductions in total plasma cholesterol (TC; 13–15%), and the incidence of cardiovascular disease (25–43% decrease) following dietary interventions with PUFA compared to nonintervention controls (Dayton et al., 1969; Leren, 1970; Turpeinen et al., 1979).

Lipids and Lipoproteins

Fatty acids affect CHD risk, in part, via effects on plasma lipids and lipoproteins. A meta-analysis of 60 controlled trials (Mensink et al., 2003) reported that saturated and *trans* fatty acids increase low-density lipoprotein cholesterol (LDL-C), whereas unsaturated fatty acids decrease LDL-C. Saturated fatty acids, MUFA, and PUFA all increase high-density lipoprotein cholesterol (HDL-C), whereas *trans* fatty acids do not. Both MUFA and PUFA decrease the TC to HDL-C ratio, whereas *trans* fatty acids increase it, and SFA have little effect (Fig. 20.4).

On the basis of the low SFA and high PUFA profile of SBO, a blood cholesterol-lowering effect would be expected when SBO replaces mixed fats (constituting 10%

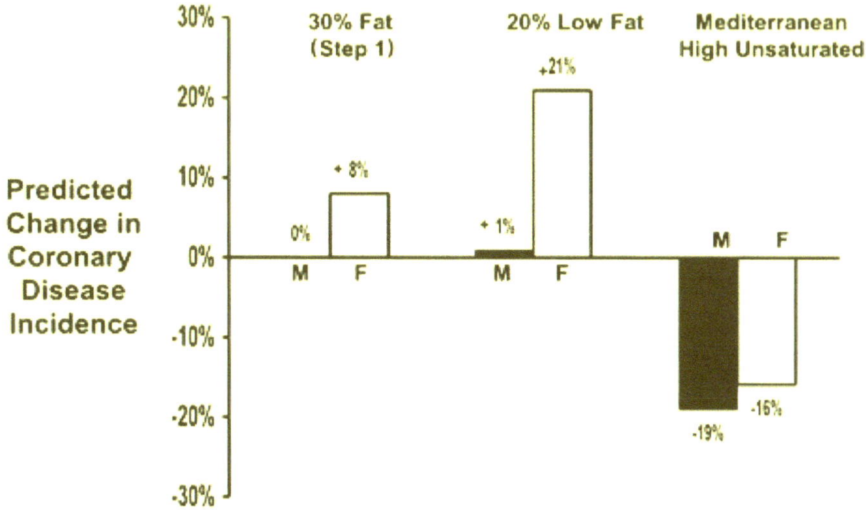

Fig. 20.3. Predicted effects on coronary artery disease of a 30% fat (step 1) diet, a 20% low-fat diet, and a mediterranean diet. (Reprinted from Sacks & Katan (2002). Copyright 2002, with permission from Elsevier.)

of total energy) isocalorically, thereby reducing the TC to HDL-C ratio (Mensink et al., 2003) (Fig. 20.5). In an early study, Kris-Etherton et al. (1993) compared the effects of commonly consumed fats in a whole-food diet and their constituent fatty acids on plasma lipids and lipoproteins. The diets provided 37% of total calories from fat, and the test fats, consisting of butter, cocoa butter, olive and SBO, contributed to 81% of total fat. The investigators reported that, compared to butter and cocoa butter, olive and SBO lowered TC and LDL-C, with the greatest hypocholesterolemic response achieved with SBO (–24.0 mg/dl). Both SBO and olive oil improved the LDL- to HDL-C ratio, a response that agrees with the predicted changes calculated by Mensink and colleagues (2003) (Fig. 20.5). The predicted change in TC and HDL-C shown in Figure 20.5 is consistent with other studies that used SBO to manipulate the fatty acid profiles of experimental diets to examine lipid and lipoprotein effects. These studies reported more favorable lipid profiles, i.e., reductions in TC and LDL-C, when a diet rich in SBO is compared to diets substituted with saturated fat from butter (Matthan et al., 2004) or palm oil (Vega-Lopez et al., 2006). Matthan and coworkers (2004) reported that TC and LDL-C increased by 21 mg/dl and 17 mg/dl, respectively, following a butter diet (20% of total energy) compared to an SBO diet. Similarly, a palm oil diet (20% of total energy) increased both TC and LDL-C by 20 mg/dl, compared to a SBO diet (Vega-Lopez et al., 2006).

The effects of *trans* fatty acids from PHSBO were evaluated extensively. Clinical trials that evaluated the effects of SFA and *trans* fatty acids on blood lipids and lipoproteins concluded that saturated and *trans* fatty acids in PHSBO elicit a similar TC and LDL-C raising effect; however, *trans* fats lower HDL-C (de Roos et al., 2001; Dyerberg et al., 2004), thereby creating an unfavorable LDL-C to HDL-C ratio and TC to HDL-C ratio. These effects of *trans* and SFA were first described by Mensink and Katan in 1990. This trial and several subsequent studies were later incorporated in a meta-analysis by Ascherio et al. (1999), which reported that the net effect of *trans* fatty acids from hydrogenated vegetable oils on the ratio of LDL- to HDL-C was approximately double that of SFA (Fig. 20.6). Diets high in PUFA reduce TC and LDL-C when compared to diets high in *trans* fatty acids (from PHSBO) or SFA (Lichtenstein et al., 1999; 2003; Matthan et al., 2004; Müller et al., 1998; Vega-Lopez et al., 2006). Diets substituting 20% of total calories with SBO consistently showed reductions of around 7–11% and 8–15% for TC and LDL-C, respectively, compared to diets substituting the same proportion of SFA or *trans* fatty acids (Lichtenstein et al., 1999; 2003; Matthan et al., 2004; Vega-Lopez et al., 2006).

Lichtenstein and colleagues (1999; 2003) conducted a comprehensive study to evaluate the effects of several SBO products differing in degrees of hydrogenation on the lipid and lipoprotein levels of individuals with elevated LDL-C. They compared six test diets that provided 30% of calories from fat; two-thirds of the fat (20% of calories) came from SBO, SBO-based margarines, SBO-based shortening, or butter. A positive, linear association existed between the extent of hydrogenation or *trans* fatty

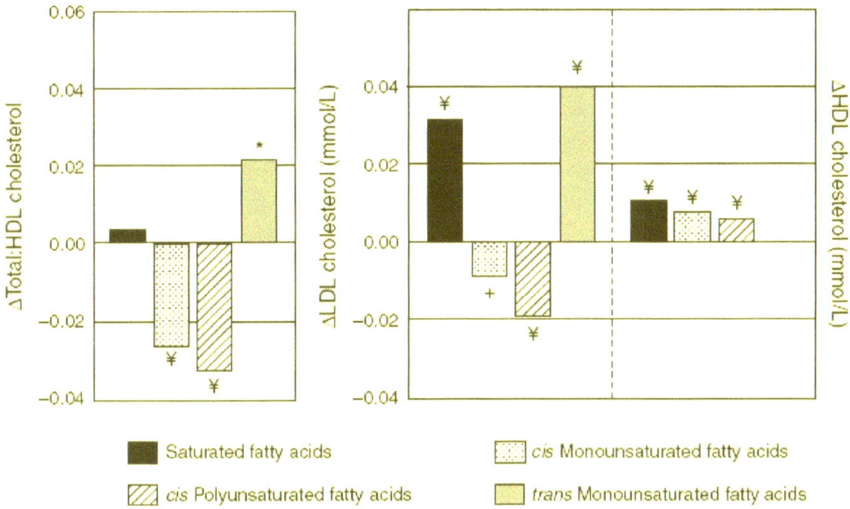

Fig. 20.4. Predicted changes (Δ) in the ratio of serum total cholesterol to HDL-C and in LDL- and HDL-C concentrations when carbohydrates constituting 1% of energy are replaced isoenergetically with saturated, *cis* monounsaturated, *cis* polyunsaturated, or *trans* monounsaturated fatty acids. *$P<0.05$; +$P<0.01$; ¥$P<0.001$.

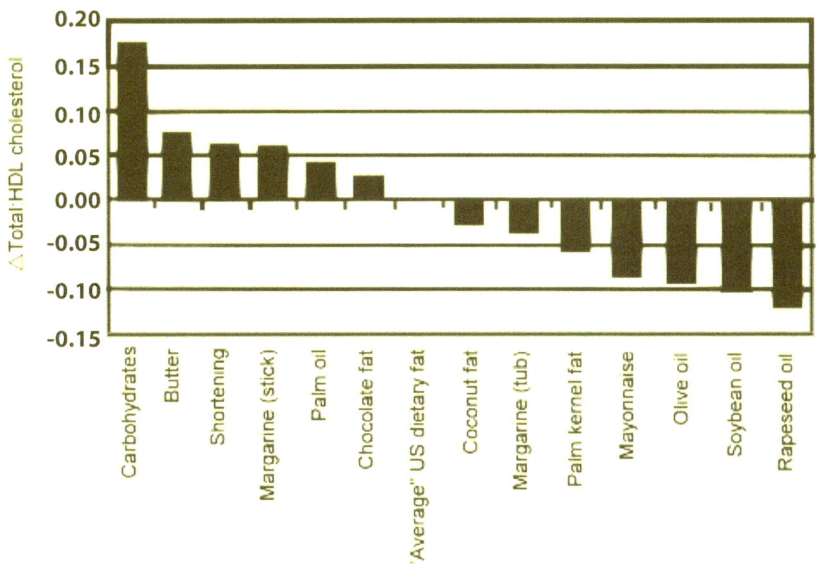

Fig. 20.5. Predicted change in the ratio of serum total to HDL-C when mixed fat constituting 10% of energy in the "average" American diet is replaced isoenergetically with a particular fat or with carbohydrates. Reprinted from Mensink et al., 2003, with permission from the American Society for Nutrition.

acid intake and TC and LDL-C (Fig. 20.7). The liquid SBO diet caused the greatest reduction in LDL-C and TC, whereas the moderate and high *trans* diets showed the least reduction. However, although stick margarine, which contained the most *trans* fatty acids, raised LDL-C; this effect was significantly less than butter. Furthermore, while the stick margarine diet reduced HDL-C, the butter diet resulted in a significant increase. The most unfavorable TC to HDL-C ratios therefore resulted from an elevated intake of *trans* fatty acids or butter. This effect of butter and high-*trans* margarines is consistent with that described in an early meta-analysis by Zock and Katan (1997).

The adverse effects of *trans* fatty acids on serum lipids and lipoproteins are thought to be mediated by alterations in lipid catabolism and metabolism. *Trans* fatty acids increase the catabolism rates of apolipoprotein A-I and decrease apolipoprotein B catabolism rates (Matthan et al., 2004), reduce LDL-C particle size (Mauger et al., 2003), and can increase cholesteryl ester transfer protein (CETP) activity (van Tol et al., 1995). CETP mediates the transfer of cholesterol esters from HDL- to LDL- and very-low-density lipoprotein (VLDL)-C, thereby offering a potential explanation for the LDL-C-raising and HDL-C-lowering effect of *trans* fatty acids.

Given the adverse physiological effects of saturated and *trans* fatty acids, the potential exists for vegetable oils with modified fatty acid profiles that have improved oxidative stability to serve as an alternative to hydrogenated oils. Outcomes from several clinical trials evaluating the effects of modified SBO on cardiovascular outcomes show beneficial effects on blood lipids and lipoproteins. Lu et al. (1997) reported similar reductions in serum TC, LDL-C, and TG concentrations from baseline when normolipidemic women were fed diets that provided either a standard SBO or a low linolenic acid SBO. In a recent study, Lichtenstein et al. (2006) showed that, compared to PHSBO, diets that were designed with novel SBO varieties (20% of total energy intake), including high-oleic acid SBO, low-SFA SBO and low-ALA SBO, improved the LDL- to HDL-C ratio in subjects with elevated LDL-C. Relative to the SBO diet, the PHSBO diet increased the ratio of TC to HDL-C by 4.0%, whereas the high- oleic acid SBO variety showed a modest improvement (–3%). However, the other varieties had a minimal effect (low-SFA SBO; –1.0%, low-ALA SBO; –0.01%) on the ratio of TC to HDL-C.

Solid fats have an important function in certain foods, and they differ markedly in their fatty acid profile. Effects on CVD risk status can vary appreciably depending on this profile, and health issues continue to surround the use of solid fats containing saturated and/or *trans* fatty acids in foods. Recent research by Mensink (2007) indicates that the total fatty acid profile is very important when considering the impact of a solid fat on cardiovascular risk profiles. Foods should therefore be reviewed on an individual basis with respect to amounts of saturated and/or *trans* fatty acids. As well, consider the specific individual SFA and all other fatty acids in the product.

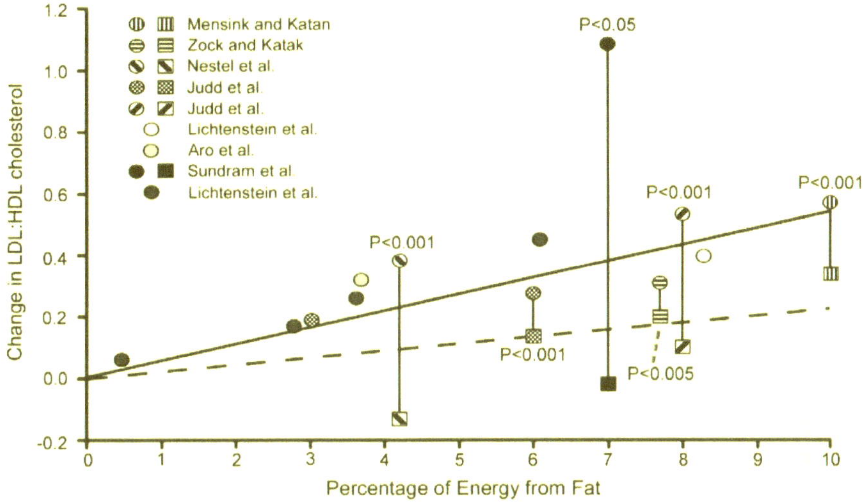

Fig. 20.6. Dose-dependent relationship between *trans* fatty acid intake ● and saturated fatty acid intake ■ and the LDL-C to HDL-C ratio. The solid line indicates the best-fit regression for *trans* fatty acids. The dashed line indicates the best-fit regression for saturated fatty acids. (Reprinted from Ascherio et al., 1999, copyright 1999 Massachusetts Medical Society. All rights reserved.)

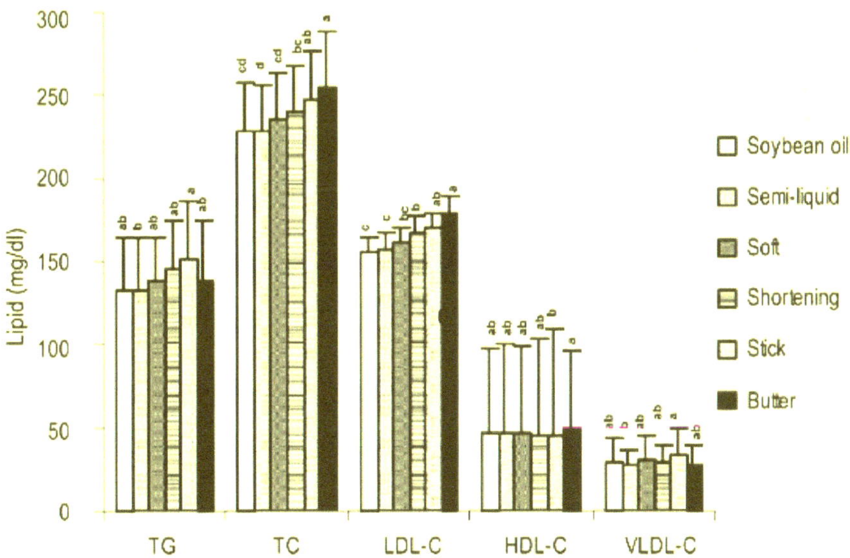

Fig. 20.7. Fasting plasma lipid and lipoprotein levels at the end of each diet phase. Numbers without common letters are significantly different at $P < 0.05$. (Reprinted from Lichtenstein et al., 2003, with permission from the American Society for Nutrition.)

Blood Pressure

The role of dietary fats in blood pressure (BP) regulation is inconclusive. Some epidemiologic studies reported a linear relationship between BP and SFA intake, while an inverse association was shown for PUFA intake and the PUFA to SFA ratio (Stamler et al., 1996). Although modest, a lower SFA intake (10.4% of total energy) was associated with a lower systolic BP (SBP; –0.5 mmHg) and diastolic BP (DBP; –1.2 mmHg) than that observed with a higher, less favorable SFA intake (13.8% of total energy). However, the Nurses' Health Study reported no association between intakes of total, saturated, or unsaturated fat and risk of hypertension (Ascherio et al., 1996a). Similarly, the U.S. Physicians' Health Study found no relationship between BP and intake of SFA, PUFA, or *trans* fatty acids (Ascherio et al., 1992). An early, although small, meta-analysis reported that MUFA was not effective in reducing BP in normotensive persons (Morris, 1994). However, some clinical trials reported a BP-lowering effect when MUFA are substituted for SFA (Lahoz et al., 1997) or carbohydrates (Rasmussen et al., 1993). A recent meta-analysis also reported that in comparison to carbohydrate-rich diets, MUFA-rich diets (19–30% of total energy) lower SBP (–2.6 mmHg) and DBP (–1.8 mmHg) (Shah et al., 2007). Furthermore, other recent studies show that healthy diets rich in MUFA (21% of total energy), including Mediterranean-type diets are effective in lowering blood pressure (Appel et al., 2005; Rasmussen et al., 2006). The OmniHeart Trial (Appel et al., 2005) demonstrated a substantial reduction in SBP and DBP (–9.5 and –5.2 mmHg, respectively) in borderline hypertensive subjects randomized to a MUFA-rich diet (37% of total energy from fat; 21% of total energy from MUFA), an effect that was significantly greater than that seen in subjects allocated to a carbohydrate-rich diet (58% of total energy). The KANWU study investigators (Rasmussen et al., 2006) reported that a MUFA-rich diet (37% of total energy from fat; 21% of total energy from MUFA) reduced both SBP and DBP by ~5 mmHg in healthy subjects, compared to nearly a 3 mmHg increase in SBP and DBP with a SFA-rich diet (18% of total energy). In comparison, evidence shows a modest hypotensive effect of n-3 PUFA from fish or fish oil, particularly in hypertensive individuals (Appel et al., 1993; Dickinson et al., 2006; Geleijnse et al., 2002; Morris et al., 1993). These meta-analyses suggest that >3 g/day of n-3 PUFA can lower BP by at least 3.0/2.0 mmHg (SBP/DBP) in hypertensive individuals. At the present time, the limited research on the effects of dietary fats on blood pressure is uncertain; only a few studies with SBO (non- and hydrogenated) evaluated this topic, and most (Dyerberg et al., 2004; Lichtenstein et al., 2003), although not all (Trifiletti et al., 2005), reported no effect.

Inflammation

Fatty acids play a key role in regulating inflammation and immune responses. A high intake of SFA and *trans* fatty acids is positively correlated with elevated levels of serum C-reactive protein (CRP) (King et al., 2003; Lopez-Garcia et al., 2004), which increases risk for CVD (Libby & Ridker, 2004). King and colleagues (2003) demonstrated that, compared with the lowest quartile, a higher intake of SFA (>24.0 g/day) was modestly correlated with an elevated level of CRP (>3.0 mg/L). Similarly, in the Nurses' Health study (Lopez-Garcia et al., 2004) a positive relationship exists between CRP and adherence to a Western dietary pattern (high intake of SFA and *trans* fatty acids), such that women in the highest quintile had the greatest concentrations of CRP. Positive associations also were demonstrated between several markers of endothelial dysfunction [plasma E-selectin, soluble vascular cell adhesion molecule-1 (sVCAM-1) and soluble intercellular adhesion molecule-1 (sICAM-1)] and a high intake of *trans* and SFA (>3.0 and 22.0 g/day, respectively) as is typical in a Western diet (Lopez-Garcia et al., 2004). Although relatively few intervention studies evaluated the effects of diets that provide different amounts of *trans* and SFA on markers of inflammation, some demonstrated that high SFA diets (>15% of total energy) and high intakes of *trans* fatty acids (>7% of total energy) increase several inflammatory markers, including CRP, E-selectin, tumor necrosis factor-alpha (TNF-α), and interleukin (IL)-6 (Baer et al., 2004; Han et al., 2002b). Han and colleagues (2002b) reported that the production of TNF-α and IL-6 from stimulated peripheral blood mononuclear cells was 58 and 36% higher, respectively, after subjects consumed a stick margarine compared with a SBO diet. Similarly, plasma levels of CRP, IL-6 and E-selectin were 79, 33, and 13% greater following consumption of a diet enriched with *trans* fatty acids compared to a diet enriched with oleic acid (Baer et al., 2004). In contrast, Lichtenstein et al. (2003) reported no effect of dietary fat types (SBO, PHSBO, or butter) on CRP. Similar inconsistencies exist for the effects of MUFA as part of a Mediterranean-type diet (Esposito et al., 2004; Michalsen et al., 2005). Interestingly, Yaqoob et al. (1998) reported a reduction in the expression of ICAM-1 by peripheral blood mononuclear cells following a MUFA diet enriched with olive oil (18.4% of total energy). However, whether this change was associated specifically with a reduction in SFA intake, which was 3% lower in the MUFA-rich diet compared to the control diet, is unclear. The control diet was designed to represent the average UK fatty acid intake, with SFA, MUFA, and PUFA providing 13.3, 11.3, and 6.9% of total energy, respectively.

A substantial body of literature exists reporting the inflammatory (or anti-inflammatory) effects of n-6 and n-3 PUFA. LA and ALA undergo a series of desaturation and elongation steps that yield arachidonic acid (AA, 20:4n-6) and EPA (20:5n-3), respectively. Additional metabolism of EPA produces the n-3 PUFA DHA (22:6n-3). However, the conversion of ALA to EPA, and especially DHA, is limited in humans (Goyens et al., 2005; Hussein et al., 2005; Pawlosky et al., 2001). Based on a metabolic model, Goyens et al. (2005) estimated that 7% of dietary ALA is converted to long-chain PUFA: 99% is converted to EPA and 1% to docosapentaenoic acid (DPA). DHA is subsequently produced via elongation and desaturation of DPA. Therefore, while dietary intake of ALA can increase cellular concentrations of EPA, it has a negligible impact on DHA. As shown in Figure 20.1, vegetable, nut, and seed oils are good sources of LA, although these oils contain significantly less ALA. The richest source of EPA and DHA is fish or fish oil. Omega-6 and n-3 PUFA exert their biological effects through alterations in the lipid membrane environment (i.e., an increase in membrane fluidity) or by acting directly as substrates for metabolism (Demaison & Moreau, 2002). A high concentration of AA results in increased synthesis of the pro-inflammatory and pro-aggregatory 2-series prostaglandins (PGE2) and thromboxanes (TXA2), and 4-series leukotrienes (LTB4), whereas a high concentration of EPA results in increased synthesis of the less inflammatory 3-series PG and TX (PGE3 and TXA3) and the 5-series LT (LTB5) (James et al., 2000).

Eicosanoids also influence the production of several inflammatory cytokines (Kinsella et al., 1990; Tilley et al., 2001). PGE2 inhibits production of interferon gamma (IFN-γ), TNF-α, IL-1, IL-2, and IL-6, while LTB4 has opposing effects and increases production of IFN-γ, TNF-α, IL-1, IL-2, and IL-6 (Calder & Grimble, 2002). In this respect, AA has some anti-inflammatory activity, as it is the precursor for PGE2. Therefore, increasing the availability of AA and EPA in the diet, as provided by SBO (a result of conversion of LA and ALA to longer chain fatty acids), may alter eicosanoid synthesis and cytokine production in a way that could elicit a beneficial shift in the ratio of inflammatory to anti-inflammatory cytokines. However, currently no studies exist reporting the effects of SBO on eicosanoid synthesis in humans.

Summary

The unique fatty acid composition of SBO undoubtedly contributes to its beneficial effect on some cardiovascular risk factors, particularly TC and LDL-C levels. Studies show that substituting a proportion of total calories (20%) with SBO can result in a 7–11% and 8–15% reduction in TC and LDL-C, respectively. However, only a few studies with SBO evaluated effects on BP, and most reported no benefit. Similarly, some evidence shows that SBO may reduce inflammation (specifically TNF-α and IL-6), but conclusions are limited by the small number of studies that investigated these effects. Nevertheless, the low-SFA and high-MUFA and -PUFA contents of SBO make it a healthy alternative to fats that are high in saturated and *trans* fatty acids.

Fatty Acids and Cancer

The rationale for an association between dietary fat intake and cancer comes from epidemiologic studies that observed ethnic differences in the incidence of cancer, which were linked to differences in fat intake (Kolonel et al., 1981). Initially, total fat intake was thought to be of primary importance, but now evidence supports a role of different types/classes of dietary fats in carcinogenesis. Experimental animal models suggest that LA upregulates tumor growth in vivo and in vitro, whereas n-3 PUFA and *trans* fatty acids inhibit growth (reviewed by Sauer et al., 2007). The effect of LA on tumor growth appears to be consistent in animals fed both essential fatty acid deficient and sufficient diets. MUFA appear to have a neutral effect on tumor growth, while SFA may be a weak stimulant (Fay et al., 1997). Omega-3 PUFA and *trans* fatty acids suppress LA-mediated tumor growth by binding their respective inhibitory G protein-coupled receptors, stimulating a cascade of cell-signaling events that results in the suppression of LA uptake and metabolism (Sauer et al., 2007). This finding suggests that the ratio of n-3 to n-6 PUFA in the diet may be important, particularly in relation to eicosanoid production. Eicosanoids synthesized from LA are linked to increased tumor growth and metastasis (Larsson et al., 2004). SBO, which has a lower n-6 to n-3 PUFA ratio than other commonly used vegetable oils, such as corn oil, may therefore be a preferable dietary fat substitute.

Evidence from human trials for an association between dietary fat intake and cancer is less clear. Hanf and Gonder (2005) reviewed the outcomes from several prospective cohort studies that evaluated the association between dietary fat intake and breast cancer. They failed to identify any clear associations for either total fat or individual fatty acid (SFA, MUFA, PUFA or *trans*) intakes. Some intervention studies that intervened with low-fat diets reported a significant reduction in circulating estradiol levels in pre- and postmenopausal women (reviewed by Wu et al., 1999). Recent reports show that a low-fat diet with modest weight loss may improve the relapse-free survival rate of postmenopausal women with breast cancer (Chlebowski et al., 2006). Although not statistically significant, the Women's Health Initiative Study (Prentice et al., 2006) reported a trend toward a reduced risk of invasive breast cancer with a low-fat dietary pattern (24.3% of total energy intake from fat), which may translate to a reduced risk in future years. However, the collective evaluation of several cohort studies suggests that it is still unclear whether manipulating the hormonal response through a low-fat diet will translate to a reduced risk of breast cancer (Hanf & Gonder, 2005).

Seemingly, no consistent relationship exists between intake of total fat or saturated fat and increased risk of prostate or colorectal cancer (Howe et al., 1997; Kolonel, 1996). This is in contrast to earlier epidemiologic studies, but is likely explained by adjustments for total energy intake in more recent studies (Kolonel, 1996; Kushi & Giovannucci, 2002). A higher dietary intake of ALA may increase risk for prostate cancer (Brouwer et al., 2004). However, the evidence for such a relationship is not conclusive (de Lorgeril & Salen, 2004), and a more recent study reported no association between ALA intake and prostate cancer (Koralek et al., 2006). In comparison, increased consumption of n-3 PUFA, specifically DHA and EPA, may reduce colorectal cancer risk, even after adjusting for total energy intake (Theodoratou et al., 2007). Compared with individuals in the lowest quartile of n-3 PUFA intake (0–1.85

g/day), those in the highest quartile (>2.82 g/day) had a 40% lower risk of colorectal cancer. The relationship between dietary fatty acids and risk of ovarian cancer also is unclear, although a weak association with SFA intake was suggested (Genkinger et al., 2006). Following recent evaluation of data from the Women's Health Initiative Study, Prentice et al. (2007) reported that a low-fat diet (desired intake ~20% total energy) may reduce the risk of ovarian cancer in postmenopausal women. Compared to the control group, women following a low-fat diet had a lower incidence of ovarian cancer; more pronounced differences were observed in the final 4 years of follow-up (incidence rate 0.38, versus 0.64 in the comparison group).

Fatty Acids, Diabetes, and Insulin Resistance

A plethora of observational studies exists that investigates the role of dietary fat in insulin resistance and the development of Type II diabetes (reviewed by Hu et al., 2001; Parillo & Riccardi, 2004). Although not conclusive, these studies generally support a positive association between elevated *trans* and saturated fatty acid intake and increased risk for insulin resistance and Type II diabetes, whereas an inverse association exists for PUFA. In the Nurses' Health Study (Salmeron et al., 2001), replacing 5% of SFA with PUFA was associated with a 35% lower incidence of Type II diabetes. Moreover, substituting 2% of *trans* fatty acids with PUFA lowered risk by 40%. The mechanism by which specific dietary fatty acids induce insulin resistance is thought to be related to changes in the lipid environment of the cell membrane. SFA increase the rigidity of cell membranes, thereby inducing changes in cell signaling and gene expression that can lead to insulin resistance. Increasing the supply of unsaturated fatty acids in the diet may therefore increase membrane fluidity, and alter intracellular signaling pathways (Hulbert et al., 2005). Borkman et al. (1993) demonstrated an inverse association between PUFA concentration in muscle and serum insulin levels in healthy men, suggesting that dietary interventions that aim to alter skeletal muscle composition by increasing PUFA and decreasing SFA intake may be beneficial. However, observational studies indicate that changes in dietary PUFA and SFA intake do not necessarily translate to an improvement in insulin sensitivity. Several human intervention studies evaluating changes in insulin sensitivity (assessed by fasting plasma glucose and insulin, and/or intravenous glucose tolerance tests) following a high SFA (13–20% of total energy) (Lichtenstein et al., 2003; Schwab et al., 1995) or *trans* fatty acid diet (5–7% of total energy) (Lichtenstein et al., 2003; Louheranta et al., 1999; Matthan et al., 2001) reported no adverse effects. However, this may be due to methodological issues including a relatively short duration of dietary intervention and a small number of study participants. More conclusive evidence for a detrimental effect of SFA is provided by the KANWU Study (Vessby et al., 2001). In this study, a 3-month dietary intervention with SFA (18% of total energy) reduced insulin sensitivity by 10% compared to a nonsignificant 2% improvement following a similar diet high in MUFA (21% of total energy). These effects were even more pronounced when stratified by median total fat intake. Individuals with a total fat intake <37% showed an 8.8% improvement in insulin sensitivity on the MUFA-rich diet in comparison to a 12.5% reduction in sensitivity on the SFA-rich diet.

Given the limited evidence from clinical trials, the Canadian Diabetes Association issued low-level evidence ratings for the beneficial effects of PUFA, particularly LA, on diabetes (Canadian Diabetes Association, 2003). The current nutrition recommendations from the American Diabetes Association (ADA) do not include guidance about LA intake (American Diabetes Association, 2008).

Soy Protein

Nutritional Aspects of Soy Protein

Soy protein is the major component of soybeans, typically comprising 40% of the total bean. Many types of edible soy proteins are produced for consumption by humans, including full-fat and defatted flours and grits, protein concentrates, and protein isolates. Isolated soy protein contains the greatest concentration of soy protein (typically 90% protein), whereas whole ground soybeans contain the least amount of soy protein (typically 40% protein). Nutritionally, based on amino acid composition, soy protein is one of the few plant proteins to be considered a complete protein in that it provides all the essential amino acids for human nutrition. Soybeans are similar in amino acid composition to legumes, although legumes tend to be higher in lysine and slightly deficient in methionine and tryptophan (Jansman, 1996). In comparison, proteins from cereal grains have higher levels of methionine but are low in lysine (Harper & Yoshimura, 1993).

The nutritional value of soy protein is dependent on its essential amino acid composition and digestibility. Plant proteins are incompletely digested by humans, and this is a key consideration for determining protein quality. The preferred method for determining protein quality is the Protein Digestibility-Corrected Amino Acid Score (PDCAAS) (FAO/WHO Expert Consultation, 1990). Depending on the level of processing, the overall protein quality for soy (as expressed by the PDCAAS) varies between 0.90 to 0.99, which is comparable to animal proteins and higher than most plant proteins (FAO/WHO Expert Consultation, 1990; Schaafsma, 2000).

Soy protein is present to varying degrees in traditional soy foods, foods formulated with soy protein ingredients, and supplements containing soy protein ingredients. The soy protein content of common soy foods is presented in Table 20.2. As expected, when standardized by weight (per 100 g of product), soy protein powder provides the greatest amount of soy protein (83%), followed by soy nuts (~40%), and full-fat soy flour (35%). Several nontraditional soy products, including protein bars and breakfast patties, also provide substantial amounts of protein, although the protein in these products is not always exclusively from soy.

Other components that may accompany soy protein are fiber, carbohydrates (including prebiotic sugars), and fat depending on the degree of protein isolation. Soybeans also contain isoflavones, which are the most prevalent bioactive compounds present with soy protein, regardless of the food or ingredient providing soy protein. The prominent isoflavones in soybeans are genistein, daidzein, glycitein, and their glucosides. Metabolism of the glucoside forms of isoflavones produces free forms known as aglycones, which the gut can absorb and may exert several biological effects. Isoflavones from soy (daidzein, genistein, and glycitein) are classified as phytoestrogens as a result of their chemical structure. Soy isoflavones bind to estrogen receptors with greater affinity for estrogen receptor beta versus estrogen receptor alpha, and soy isoflavones are not homogenous with regard to their ability to bind estrogen receptors (genistein>daidzein>glycitein) (Setchell, 1998). This distinction is not generally drawn when evaluating the impact of soy isoflavones. Additionally, in vivo bioconversion products such as equol have to be considered, as individuals with intestinal capacity to convert daidzein to equol may benefit from its greater biological activity and superior antioxidant activity.

Table 20.2. Soy Protein Content of Common Soy Foods [a]

Food	Standard Serving	Protein (g)/serving	Protein (g)/100 g
Raw tofu	3 ounces	6.87	8.08
Silken firm tofu	3 ounces	5.87	6.9
Plain soymilk (Vitasoy organic)	8 ounces	7.26	3.2
Vanilla soymilk	8 ounces	7.42	3.27
Soy nuts	1 ounce	11.08	39.58
Margarine; soybean oil/butter blend	2 tbsp	0.03	0.31
Protein powder (Genisoy)	3 heaping tbsp (30 g)	25	83.3
Soy ground crumbles (MorningStar Farms®)	2/3 cup (55 g)	11 (total protein)	20
Soy burger	1 burger	10.99	15.70
Breakfast links (MorningStar Farms®)	2 links (45 g)	9 (total protein)	20
Breakfast pattie (MorningStar Farms®)	1 pattie (38 g)	8 (total protein)	21.1
Tempeh	1/2 cup	15.39	18.54
Green soybeans, cooked	1/2 cup	11.12	12.35
Green soybeans, raw	1/2 cup	16.58	12.95
Soy flour (full-fat, roasted)	1/4 cup	7.39	34.8
Soy protein bar (Genisoy)	1 bar (61.5 g)	14	22.76

[a]Source: U.S.D.A. National nutrient database and specified product nutrition labels.

Bioavailability of isoflavones is a topic that was proposed to be involved in the efficacy of various soy preparations. Isoflavones in glycoside form are recognized as having greater bioavailability when compared to their aglycone counterparts. Although aglycones can be absorbed slightly faster and to a greater degree, glycosides provide greater bioavailability. To a certain extent, this difference may explain some observations between isoflavone bioavailability in nonfermented and fermented soy foods that contain glycoside or aglycone forms, respectively. The food matrix is another factor that also may explain some differences. Comprehensive review articles on the topic of isoflavone bioavailability were published recently (Nielsen & Williamson, 2007; Rowland et al., 2003).

Conversion of isoflavones by intestinal flora has become an area of considerable interest (Turner et al., 2003). The conversion of diadzein to equol has driven most of this interest (Atkinson et al., 2005; Yuan et al., 2007). Equol conversion is specific to individuals with certain intestinal flora, but the specific gut microorganism or combination of gut microorganisms responsible has yet to be fully elucidated. The prevalence of equol producers is believed to range between 30 and 50% (Atkinson et al., 2005; Yuan et al., 2007). Some observational studies suggest that fiber and/or prebiotics can induce equol production in individuals previously identified as non-producers (Lampe et al., 1998; Rowland et al., 2000). However, this initial suggestion was not supported by recent direct clinical intervention studies (Lampe et al., 2001; Larkin et al., 2007). Depending on the form of soy food or soy-containing food, compounds such as phenolic acids, phytic acid, protease inhibitors (i.e., Kunitz and Bowman-Birk inhibitors), peptides (i.e., lunasin), and tocopherols can be present also. The role of these bioactive compounds in human health continues to be investigated, although it is not discussed in this chapter.

Soy Protein and Risk of Chronic Disease

Observations of lower rates of certain chronic diseases in several Asian countries compared to the United States served as the impetus for studies that were conducted to identify the factors that were accountable (Tham et al., 1998). Consumption of soy is one factor that correlated with these lower rates. From this information, much research focused on soy protein and accompanying bioactive compounds with regard to elucidating mechanisms associated with risk markers for chronic disease, particularly CVD. The combination of macro- (protein, fiber, fat) and micro-components (isoflavones, saponins, tocopherols) as part of many traditional soy foods may underlie the epidemiological observations associated with soy intake. A body of cell culture and animal data shows potential health effects of a multitude of bioactive components in soy (i.e., saponins, phenolic acids, peptides). However, studies determining the independent effects of saponins, phenolic acids, and protease inhibitors in humans are yet to be conducted.

Lipids and Lipoproteins

The effects of soy protein on serum cholesterol levels have been an active area of investigation. An early meta-analysis of 38 controlled trials (Anderson et al., 1995) reported that soy protein intake (average 47 g/day) was associated with significant reductions in TC (9.3%), LDL-C (12.9%), and TG (10.5%). These changes in TC and LDL-C were directly related to initial serum cholesterol concentrations. In 1999, the U.S. FDA approved a soy protein health claim that advocated regular consumption of soy protein for protection against coronary heart disease. This health claim is based on scientific studies that show 25 g of soy protein daily can significantly lower cholesterol. For a food to qualify for the health claim, the FDA requires that a serving contain at least 6.25 g of soy protein, which is 25% of the required daily amount (25 g) (Food and Drug Administration, 1999). However, recent studies reported that soy protein may not be as effective at reducing LDL-C as previously thought. As part of an American Heart Association (AHA) Science Advisory, Sacks et al. (2006) reviewed 22 randomized trials conducted between 1998 and 2005 that evaluated the effects of isolated soy protein with isoflavones on LDL-C. In the 22 studies reviewed, soy protein (25–135 g/day) and isoflavone (40–318 mg/day) intake varied considerably; average soy protein intake was 50 g/day and was associated with a 3% reduction in LDL-C. A report published in 2005 by the Agency for Healthcare Research and Quality (AHRQ) (Balk et al., 2005) reported similar results: soy protein (median intake 38 g) has a modest TC (–5 mg/dL; 2.5%), LDL-C (–5mg/dL; 3%), and TG (–8 mg/dL; 6%) lowering effect. In a meta-analysis of 27 studies, Reynolds et al. (2006) assessed the effects of soy protein supplementation in the form of isolated soy protein on plasma lipids. They reported an improvement in TC (–5.26 mg/dl), LDL-C (–4.25 mg/dl), TG (–6.26 mg/dl), and HDL-C (+0.77 mg/dl) after 3–52 weeks of supplementation with 20->61 g/d (upper mean intake: 106.2 g/d) of soy protein. Isoflavone intake ranged from 2 to 192 mg/day among the studies within which isoflavones were present. These effects are consistent with those reported by the AHRQ (Balk et al., 2005). Figure 20.8 shows the effect of soy protein consumption on changes in LDL-C concentration. As shown in this figure, and as originally reported by Anderson et al. (1995), a key determinant of the effectiveness of soy protein is baseline LDL-C concentration; with increasing baseline LDL-C, a linear decrease in LDL-C occurs. Several other factors, including the amount of soy and isoflavones, the bioavailability of isoflavones, the type and preparation of the soy foods delivered and the population (e.g., hypercholesterolemic) to which they are administered, can influence the magnitude of changes in LDL-C (Fig. 20.9).

What remains less clear is the effect of isoflavones on LDL-C. As illustrated in Figure 20.9, the greatest variability in LDL-C effects is for "isoflavone only" studies (range: 0–185 mg/day) (Balk et al., 2005). The data supporting an independent cholesterol-lowering effect by soy isoflavones are inconsistent. Zhan and Ho (2005) reported that soy protein with isoflavones reduced TC (3.8%), LDL-C (5.3%), and

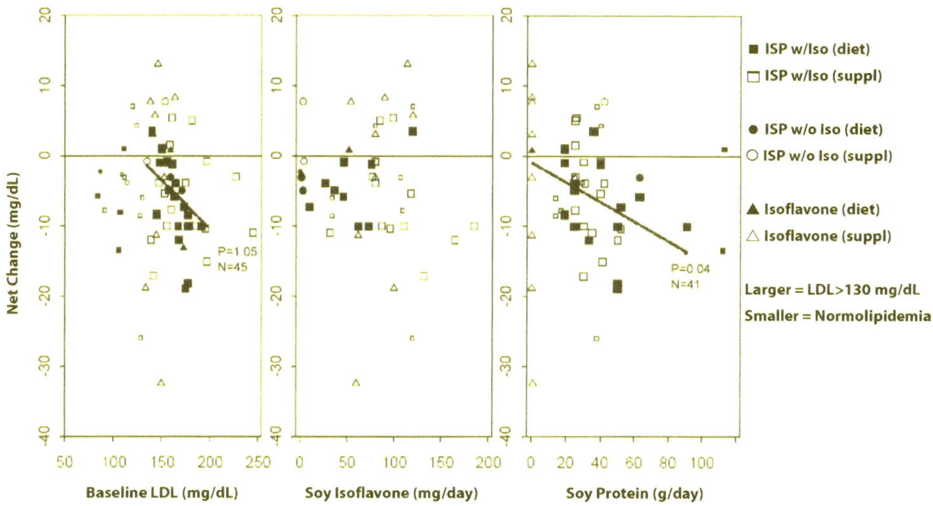

Fig. 20.8. Change in LDL-C concentration following consumption of soy products compared to control, by baseline level, isoflavone content, and soy protein content. Studies without a non-soy control are not included. Studies without data on isoflavone or protein content are omitted from relevant graphs. ISP w/Iso = soy protein with isoflavons; ISP w/o Iso = soy protein without isoflavones; suppl = supplement. Dashed lines represent adjusted regressions for studies with sufficient date fro regression. Regression lines are drawn only within the range of independent variable (x-axis) data examined. *P*-values and number of studies included in regressions are shown. Both regression lines drawn are for all studies with abnormal baseline LDL-C. Reprinted from Balk et al., 2005.

TG (7.3%), and increased HDL-C (3.0%), with higher intakes of soy isoflavones (>80 mg/d) producing more substantial effects on plasma lipids. However, both the AHRQ report (Balk et al., 2005) and the AHA Science Advisory on soy protein (Sacks et al., 2006) reported that isoflavone intake has no effect on LDL-C or other lipids. Nevertheless, more recent meta-analyses continue to suggest that soy isoflavones can independently lower TC (1.8%) and LDL-C (3.6%) (Taku et al., 2007), and, when combined with soy protein, may work synergistically or additively to lower LDL-C (soy protein + isoflavones, –4.98% versus soy protein without isoflavones, –2.77%). Additionally, oxidation of LDL is one area where isoflavones appear to exert potential protection (Balk et al., 2005), but large well-designed, long-term clinical studies are needed to confirm these observations.

Blood Pressure

Over 25 clinical trials evaluated the effects of soy products on SBP and DBP (Balk et al., 2005; Hermansen et al., 2005; Hutchins et al., 2005; Kreijkamp-Kaspers et al., 2005; Matthan et al., 2007; Teede et al., 2006; Tormala et al., 2007b). These

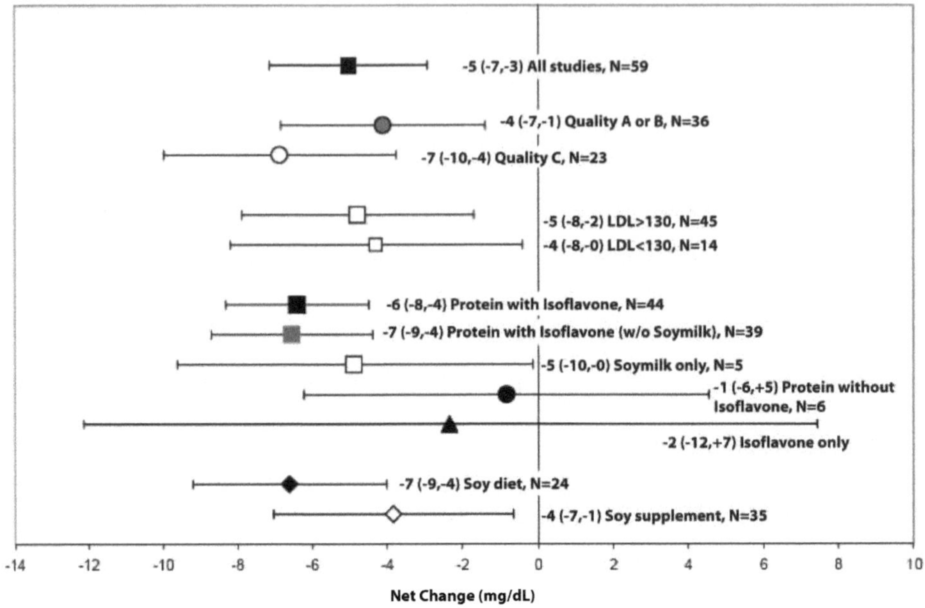

Fig. 20.9. Meta-analysis summary estimates of net change in low-density lipoprotein (LDL-C) for different subanalyses. Point estimate, 95% confidence interval, analysis group, and number of studies in each analysis group displayed. Reprinted from Balk et al., 2005.

studies range from 4 to 52 weeks and include normotensive, pre-hypertensive, and hypertensive men and women. All but two studies (Rivas et al., 2002; Welty et al., 2007) reported a change in SBP between –7 to +8 mm Hg and –5 to +4 mm Hg for DBP following soy supplementation. Rivas et al. (2002) investigated the effects of soy milk (500 mL 2×/d) versus cow's milk in a 3-month double-blind randomized trial of 40 hypertensive men and women. After 3 months of soy milk consumption, SBP decreased 17 mmHg, and DBP decreased 12 mmHg relative to the cow's milk group. The authors attributed the BP-lowering effect to twice daily consumption of soy milk high in soy protein and isoflavones. However, a similar study conducted over 6 weeks reported a substantially lower effect of soy protein (20 g, 2×/d) on BP (–1 mmHg SBP; –5 mm Hg DBP) compared with a complex carbohydrate supplement (Washburn et al., 1999).

A recent study reported that soy nuts, which are high in soy protein, reduced BP in postmenopausal women following a Therapeutic Lifestyle Changes (TLC) diet for 8 weeks (Welty et al., 2007). SBP and DBP decreased by 9.9 and 6.8%, respectively, in hypertensive women following the soy-nut diet compared with the TLC diet without soy nuts. Smaller reductions were observed in normotensive women on the soy-nut diet: –5.2% and –2.9% for SBP and DBP, respectively. However, other dietary changes possibly contributed to the BP- lowering effect observed in this study as participants were not following their habitual diets. Further research is needed, as this was the first clinical trial specifically designed to study the effects of soy nuts on CVD risk factors.

Although a few studies show a benefit of soy products on BP, the majority of evidence from clinical trials indicates that soy consumption does not significantly improve BP (Balk et al., 2005). A meta-analysis of 21 studies conducted through 2004 reported a net change of –1 mm Hg for both SBP and DBP (Balk et al., 2005). Changes in BP do not appear to be influenced by baseline BP, independent soy protein or isoflavone consumption, soy incorporated into the diet or taken as a supplement, gender, or menopausal status (Balk et al., 2005).

Vascular Function

Approximately half of randomized clinical trials demonstrated a beneficial effect of soy supplementation on endothelial function, shown either by increased brachial artery diameter or flow-mediated dilation, or decreased peak flow velocity. Of the studies reporting a positive effect, four reported an improvement in endothelial function after 4–6 weeks of supplementation with 20–25 g/d soy protein (Cuevas et al., 2003; Cupisti et al., 2007; Steinberg et al., 2003; Yildirir et al., 2001). The study populations in these four studies consisted of renal transplant patients and postmenopausal women. Three studies ranging from 6 weeks to 1 year also demonstrated an improvement in endothelial function in postmenopausal women following daily supplementation with soy isoflavones (54–90 mg/d) (Colacurci et al., 2005;Lissin et al., 2004;Squadrito et al., 2003). In addition, Azadbakht et al. (2007a) reported a decrease in E-selectin, a biomarker of endothelial function, after 8 weeks of soy-nut consumption in postmenopausal women. However, insufficient evidence is available to determine differential effects of soy protein dose or type of soy products on changes in endothelial function (Balk et al., 2005).

A similar number of clinical trials reported no effect of soy supplementation on endothelial function. Four studies reported no change in endothelial function after 4–8 weeks of supplementation with 25–52 g/d of soy protein in postmenopausal women and another showed no effect in hypercholesterolemic men and women (Blum et al., 2003; Evans et al., 2007; Katz et al., 2007; Matthan et al., 2007; Tormala et al., 2007a). In a large, long-term study, Kreijkamp-Kaspers et al. (2005) observed no effect of supplementation with 26 g/d soy protein for 12 months in 202 postmenopausal women. Likewise, no effect on endothelial function was observed following supplementation with 80 mg/d of soy isoflavones in postmenopausal women for 2–10 weeks (Hale et al., 2002; Nestel et al., 1997; Simons et al., 2000), nor with 55 mg of soy phytoestrogens for 6 weeks in menopausal women (Katz et al., 2007). Only Teede et al. (2001) found a statistically significant worsening of endothelial function, as indicated by a net decrease in flow-mediated dilation among men.

Inflammation

In vitro experiments demonstrated an anti-inflammatory effect of soy isoflavones, including inhibition of monocyte adhesion to vascular endothelial cells (Chacko et al., 2005), decreasing secretion of adhesion molecules (Gottstein et al., 2003), enhancing nitric oxide release (Walker et al., 2001), and reducing endothelin-1 concentrations (Altavilla et al., 2004; Minchenko & Caro, 2000). However, the majority of clinical trials in humans did not demonstrate an anti-inflammatory effect of soy supplementation.

Several short-term randomized, clinical trials ranging from 4 to 24 weeks reported no effect of soy supplementation on CRP (Balk et al., 2005; D'Anna et al., 2005; Fanti et al., 2006; Greany et al., 2007; Hanson et al., 2006; Hilpert et al., 2005; McVeigh et al., 2006; Ryan-Borchers et al., 2006; Teede et al., 2004; Yildiz et al., 2005). Hall et al. (2005) observed a significant decrease in CRP after 4 weeks of supplementation with isoflavone-enriched (50 mg/d) cereal bars in healthy postmenopausal women. However, CRP levels returned to baseline at 8 weeks. Fanti et al. (2006) also reported no effect of short-term soy supplementation (26–54 g/d) on plasma CRP concentrations in end-stage renal-disease patients on chronic hemodialysis. However, they observed a significant inverse relationship between changes in blood concentrations of isoflavones and changes in CRP from baseline (Fanti et al., 2006). One long-term clinical trial conducted over 4 years studied the anti-inflammatory effects of soy protein in Type II diabetic patients with nephropathy (Azadbakht et al., 2008). Patients randomized to a soy protein group consumed a diet containing 0.8 g protein/kg body weight (35% animal proteins, 35% textured soy protein, and 30% vegetable proteins), while those in the control group consumed a similar diet containing 70% animal protein and 30% vegetable protein. Serum CRP levels decreased significantly in the soy protein group compared with the control group (1.31 ± 0.6 vs. 0.33 ± 0.1 mg/L).

Several studies ranging from 4 to 24 weeks in duration showed no effect of soy supplementation on other inflammatory markers, including IL-6 (Fanti et al., 2006; Hilpert et al., 2005; Jenkins et al., 2002) and TNF-α (Fanti et al., 2006; Hermansen et al., 2005; Jenkins et al., 2002; Ryan-Borchers et al., 2006). One small study by Huang et al. (2005) reported a reduction in TNF-α in postmenopausal women following daily consumption of soymilk (1.065L) containing 112 mg isoflavones for 16 weeks. By 2 weeks, serum levels of TNF-α decreased by 25.1%, and by 66.7% after 10 wk of soy consumption. TNF-α concentration returned to pre-diet levels 4 weeks after soy consumption ended.

Taken together, the majority of short-term clinical trials showed no effect of short-term soy supplementation on markers of inflammation. However, the beneficial effect of soy supplementation on CRP in the longitudinal study by Azadbakht et al. (2008) suggests that a long-term anti-inflammatory benefit of soy products may exist. Further research is needed to determine if this long-term effect is consistent.

Summary

Conflicting evidence exists regarding the role of soy protein on blood lipids, and beneficial effects appear to be limited to modest reductions in LDL-C, particularly in individuals with elevated serum levels. Whether isoflavones can independently improve lipids is still a matter of discussion, although recent analyses suggest that small changes may result. The majority of evidence from clinical trials indicates that soy protein and/or isoflavone consumption does not significantly improve BP. While some studies reported favorable changes in endothelial function, an equal number have shown no effect. Similarly, short-term studies show that soy protein and/or isoflavones do not improve several markers of inflammation, including CRP, IL-6 and TNF-α, although a longer-term study shows promise. Future research is needed to identify the components of soy that are responsible for its modest lipid- lowering effects.

Soy Protein and Hormone-sensitive Conditions: Menopause, Cancer, and Bone Health

Epidemiologic observations consistently show lower rates of prostate and breast cancer, and osteoporosis in populations that consume soy (Anderson & Garner, 1997; Moyad, 1999; Nagata et al., 2001a; Wu et al., 1998). Intervention studies were conducted to investigate whether these observations are reproducible. These studies will help clarify whether the results reported in epidemiologic studies are causal, and due specifically to soy products (Reinwald & Weaver, 2006; Rice & Whitehead, 2006).

In their systematic review, the AHRQ described various health outcomes of a range of soy products, including both protein and isoflavones (Balk et al., 2005). The health outcomes relevant to this section include menopause, cancer and bone health. In general, far fewer studies are reported for each hormone-sensitive condition described in this section than for CVD risk factors. In addition, different soy products and endpoints were evaluated, and numerous experimental designs were employed. A summary of the AHRQ Report and accompanying update are presented.

Hormonally-sensitive conditions (such as menopause and cancer) are areas that were correlated with soy intake (Anderson & Garner, 1997; Moyad, 1999; Nagata et al., 2001a; Wu et al., 1998). From a mechanistic approach, these observations correlate with the phytoestrogen properties of soy isoflavones. Historically, synthetic estrogen is prescribed for women seeking relief of menopausal symptoms. Natural sources of phytoestrogens are sought based on this rational. Recent reviews of clinical studies concluded that isoflavones are not effective or are inconsistent in relieving hot flushes (Huntley & Ernst, 2004; Kronenberg & Fugh-Berman, 2002; The North American Menopause Society, 2000) which is inconsistent with epidemiologic observations. These reviews consistently treat isoflavones as a homogenous class of compounds, which is discouraged by leading researchers (Erdman et al., 2004). When the distinction between isoflavone composition and source is made, the amount of genistein and soy does appear to be an important determinant of whether hot flushes are significantly impacted (Cassidy et al., 2006; Williamson-Hughes et al., 2006).

Menopausal Symptoms

In the AHRQ Report, 21 trials were reviewed that examined the effects of soy and/or its isoflavones mainly on the frequency and severity of hot flashes and night sweats in post- and peri-menopausal women (Albert et al., 2002; Albertazzi et al., 1998; Balk et al., 2002; Burke et al., 2003; Crisafulli et al., 2004; Dalais et al., 1998; Faure et al., 2002; Han et al., 2002a; Knight et al., 2001; Kotsopoulos et al., 2000; Murkies et al., 1995; Nikander et al., 2003; Penotti et al., 2003; Quella et al., 2000; Russo & Corosu, 2003; Scambia et al., 2000; Secreto et al., 2004; St Germain et al., 2001; Upmalis et al., 2000; Van Patten et al., 2002; Washburn et al., 1999). The major conclusion was that soy isoflavone supplements may reduce hot flashes in symptomatic postmenopausal women, compared to a placebo. The net reduction in weekly hot flash frequency ranged from 7 to 40% in the six randomized trials reporting a beneficial effect. However, no significant effect was reported for soy and/or its isoflavone treatments in peri-menopausal women. No effect was seen in women who had breast cancer therapies.

The majority of trials that examined the effects of soy and/or its isoflavones on menstrual cycle length in pre-menopausal women reported no effect (Brown et al., 2002; Cassidy et al., 1995; Cassidy et al., 1994; Duncan et al., 1999; Kumar et al., 2002; Lu et al., 2000a; Lu et al., 1996; Martini et al., 1999; Maskarinec et al., 2002; Nagata et al., 1998; Wu et al., 2000). Only one randomized controlled trial showed that pre-menopausal women who took supplements of soy protein with isoflavones for 12 weeks had a significant net increase in their menstrual cycle lengths, compared with those who took the placebo (isocaloric milk protein) (Kumar et al., 2002).

Endocrine Function

Several studies evaluated endocrine measurements, specifically testosterone and follicle stimulating hormone (FSH), as primary or secondary endpoints. Testosterone is of clinical importance both as a risk factor for cancer and as part of the initial evaluation of male infertility. FSH is also measured as an initial evaluation of infertility in both men and women. Despite the substantial number of studies, results were conflicting and no significant effect of soy product consumption was found in men, pre- or postmenopausal women (Balk et al., 2005).

Cancer and Tumor-related Biomarkers

Numerous trials in cancer-free subjects evaluated the effects of soy consumption on risk factors or tumor markers related to the following types of cancer: breast (Bazzoli et al., 2002; Cassidy et al., 1994; Duncan et al., 1999; Hsu et al., 2001; Kumar et al., 2002; Lu et al., 2000a; Lu et al., 2001; Lu et al., 2000b; Lu et al., 1996; Maskarinec et al., 2003; Maskarinec et al., 2002; Petrakis et al., 1996; Xu et al., 2000), prostate (Davis et al., 2001; Gardner-Thorpe et al., 2003; Habito et al., 2000; Jenkins et al., 2003; Nagata et al., 2001b), endometrial (Murray et al., 2003), and colon (Adams et al., 2003). However, according to the National Cancer Institute (NCI), none of the markers evaluated are considered a risk factor relevant to the types of cancers studied (www.nci.nih.gov/cancertopics). Based on the data reported in the literature, the AHRQ Report concluded that it was unclear whether soy plays a beneficial role in preventing certain types of cancer.

Bone Endpoints

The AHRQ Report summarized numerous studies that evaluated the effects of soy products, including both protein and isoflavones, on various markers of bone health, such as bone mineral density (BMD) and biomarkers related to bone formation (bone-specific alkaline phosphatase and osteocalcin) and resorption (urinary hydroxyproline, urinary pyridinoline, and urinary deoxypyridinoline). In general, no effect of soy consumption on BMD or on biomarkers of bone formation resulted. Although a number of studies observed reductions in markers of bone resorption, these were restricted to only two biomarkers: urinary pyridinoline and deoxypyridinoline. Moreover, the effects were not consistent across studies. The AHRQ report found no consistent evidence of dose-response effects for either soy isoflavones or soy protein on markers of bone turnover (Balk et al., 2005).

However, a recent meta-analysis of ten studies with a total of 608 subjects reported that spine BMD increased significantly in subjects who consumed isoflavones; with higher intakes the effect became more significant (Ma et al., 2007). One earlier study reported a consistent effect on several markers of bone health in early postmenopausal women (Morabito et al., 2002). BMD in the femur and spine increased significantly following 6 months of supplementation with genistein (54 mg/d). This was accompanied by favorable effects on markers of bone metabolism, indicating a reduction in bone resorption and an increase in bone formation. Several factors might explain the more positive results from this study. The subjects in this study had a lower BMD than about 50% of the general population. In addition, the investigators used purified genistein tablets that may have different properties than other preparations of soy isoflavone extracts. Whether this suggests that genistein is the bioactive compound exerting endocrine-like activity, and accounting for the effects of soy products, remains to be determined. It also may be that these effects are life-stage specific (i.e., effective in perimenopausal and early-menopausal women) (Reinwald & Weaver, 2006).

Since the publication of the AHRQ report, the role of soy metabolite production by gut microflora as related to bone health continues to be examined in literature reviews (Vatanparast & Chilibeck, 2007). These reviews suggest that the production of equol from daidzein may protect against bone loss, and equol is therefore implicated as a potential determinant of beneficial bone-response to soy isoflavone-containing products in clinical studies (Frankenfeld et al., 2006; Wu et al., 2007; Wu et al., 2006).

Summary

The evidence base for soy products having effects on hormone-sensitive conditions is inconclusive. Soy products may reduce menopausal symptoms in postmenopausal women, although the current literature does not demonstrate effects of soy products on other hormone-sensitive conditions. However, importantly, clinical outcomes were not evaluated for most hormone-sensitive conditions. Conclusions are limited because of the relatively small numbers of studies and marked heterogeneity in design and in the profile of soy isoflavones across studies.

Given the inconclusive evidence base for soy protein and isoflavones, many questions abound about whether specific soy products in adequate doses may have a health benefit. As noted in the AHRQ Report, further well-conducted studies are needed to determine whether soy protein or isoflavones affect hormone-related conditions. Numerous clinical trials are underway (www.clinicaltrials.gov) to evaluate effects of soy products on multiple clinical conditions including menopause, cancer, and bone health, among others.

Soy Protein, Diabetes, and Insulin Resistance

In comparison to studies on lipids and lipoproteins, substantially fewer reports exist investigating the effects of soy protein and/or isoflavones on plasma glucose and insulin. The AHRQ report identified only six studies in individuals without diabetes (Han et al., 2002a; Huff et al., 1984; Nikander et al., 2004; Onning et al., 1998; Washburn et al., 1999; Yamashita et al., 1998): none of these studies reported any significant changes in fasting blood glucose following intervention. The AHA Scientific Advisory Committee for soy did not evaluate changes in blood glucose (Sacks et al., 2006).

Of the studies of soy protein including isoflavones published after the AHRQ report, three reported no effect on fasting glucose and/or insulin after 4–24 weeks of supplementation with 20–30 g/d soy protein providing 39–160 mg/d of isoflavones (Gardner et al., 2007; Hermansen et al., 2005; Sites et al., 2007). The populations in these three studies consisted of postmenopausal women and hypercholesterolemic adults. One study in postmenopausal women with metabolic syndrome reported that soy consumption improves glycemic control (Azadbakht et al., 2007b). In this study, women were randomly assigned to consume one of three isocaloric diets for 8 weeks: a control diet (Dietary Approaches to Stop Hypertension, DASH), a DASH plus soy-nut diet, or a DASH with soy-protein diet. Insulin resistance (determined by homeostasis model of assessment score; HOMA) decreased significantly at the end of the soy-nut diet compared with the soy-protein and control diets. The authors suggest that the PUFA in combination with the higher level of isoflavones in the soy-nut period (102 mg/d) as compared to the soy-protein period (84 mg/d) may be responsible for this effect. However, the results of studies evaluating the independent effects of soy isoflavones on glucose metabolism are conflicting. Ho et al. (2007) reported that over one year, changes in fasting glucose were significantly decreased in postmenopausal Chinese women taking 40 mg/d soy isoflavones, compared to a placebo. Improvements were also greater in women with elevated fasting glucose levels at baseline (>100 mg/dL). However, intention to treat analysis showed no differences between groups (placebo, 40 or 80 mg/d isoflavones) for mean fasting glucose concentrations at one year. In a longer-term intervention, daily supplementation, with genistein (54 mg/d) in conjunction with an isocaloric fat-reduced diet, significantly decreased fasting glucose and insulin in 389 postmenopausal women, compared with diet alone (Atteritano et al., 2007). Genistein may influence glucose metabolism by activating the transcription factor peroxisome proliferator-activated receptor-γ (PPAR-γ) (Dang et al., 2003). However, two studies in healthy postmenopausal women reported no effect of isoflavone supplementation (50 and 100 mg/d) on fasting glucose or insulin (Garrido et al., 2006; Hall et al., 2006), even with similar concentrations of genistein (50 mg/d) (Garrido et al., 2006).

Although limited, some recent studies suggest that the inclusion of soy as part of a weight-loss program may improve glucose control. In a 12-week trial, fasting insulin and hemoglobin A1c levels decreased significantly in healthy postmenopausal women following a low glycemic index diet with a functional food delivering 30 g of soy protein and 4 g of phytosterols per day, compared to women consuming an AHA Step 1 diet (Lukaczer et al., 2006). Li et al. (2005) reported a significant reduction in fasting blood glucose in Type II diabetics consuming a soy-based meal-replacement as compared to a diet based on ADA recommendations. However, these differences may be explained by changes in body weight, which was greater in subjects allocated to the soy intervention, or changes in other dietary components (Li et al., 2005; Lukaczer et al., 2006). In two studies reporting similar weight loss in the placebo and soy groups, changes in glycemic control did not differ between groups (Anderson et al., 2007; Liao et al., 2007).

When included as part of an isocaloric diet or a weight-loss intervention, the current literature does not show a consistent beneficial effect of soy protein and/or isoflavones on glycemic control. This is largely due to the small number of studies. Additional research is required to increase our understanding of the effects of soy protein on measures of glycemic control.

Dietary Recommendations for Soybean Oil and Soy-protein Foods

Food-based dietary recommendations in the United States were made for health promotion and chronic disease risk reduction (U.S. Department of Health and Human Services and the U.S. Department of Agriculture, 2005). In addition, food-based guidance meets current recommendations for micronutrients and macronutrients to achieve a nutritionally adequate diet, a core foundation for health and well-being. Table 20.3 shows the specific food-group recommendations for 13 calorie levels. Specific recommendations were made for liquid vegetable oils, and 27 g per day of soybean oil can be included in a 2000-calorie diet that meets current food-based dietary recommendations. The amount of vegetable oil that can be included in the diet varies as a function of calorie intake. For example, a 1600-calorie diet can include 22 g of liquid vegetable oil, while 44 g can be included in a 3000-calorie diet. Soy protein foods can be incorporated as meat alternatives and dairy alternatives. Thus, up to three servings of meat alternatives and three servings of soy milk can be included in a 2000-calorie diet that is consistent with current dietary guidelines.

Conclusion

SBO production has increased steadily since the early 1970s, and today accounts for 75% of the edible fats and oils consumed in the United States (The American Soybean Association, 2007). Current recommendations encourage a diet that provides 5 to 10% of calories from PUFA. The unique fatty acid profile of SBO is characterized as a good source of both LA and ALA, providing 89 and 11% of total PUFA, respectively. In addition to PUFA, SBO is an excellent source of MUFA, and is low in SFA. Likely, the beneficial effects of SBO on several cardiovascular risk factors, most notably blood lipids and lipoproteins, are attributable to this fatty acid profile. Unfortunately, the commercial use of SBO in the form of PHSBO contributes substantially to the total intake of *trans* fatty acids in the United States, which is associated with an increased risk of CVD. Strong evidence shows that PHSBO increases TC and LDL-C and reduces HDL-C, thereby creating an unfavorable LDL- to HDL-C ratio and TC to HDL-C ratio. In addition, *trans* fatty acids have many other adverse effects that adversely affect CVD risk, as well as other diseases/conditions and metabolic events. Thus, obviously, current dietary guidelines advocate a reduced intake of *trans* fatty acids, that is as low as possible and/or <1% of calories. Given these current recommendations to decrease *trans* fatty acids, intensive efforts are underway by the food industry to reduce/eliminate *trans* fatty acids in the food supply with notable progress having been made to date (Eckel et al., 2007).

Soy protein and the bioactive isoflavones may provide additional benefits of soy. When compared to animal protein, soy protein lowers serum LDL-C levels, although modestly, as demonstrated by more recent clinical studies (Balk et al., 2005; Sacks et

Table 20.3. USDA Food Guide[a]

Daily Amount of Food From Each Group (vegetable subgroup amounts are per week)												
Calorie Level	1000	1200	1400	1600	1800	2000	2200	2400	2600	2800	3000	3200
Food Group Food group amounts shown in cup (c) or ounce-equivalents (oz-eq), with number of servings (sv) in parentheses when it differs from the other units. See note for quantity equivalents for foods in each group. Oils are shown in grams (g).												
Fruits	1 c (2 srv)	1 c (2 srv)	1.5 c (3 srv)	1.5 c (3 srv)	1.5 c (3 srv)	2 c (4 srv)	2 c (4 srv)	2 c (4 srv)	2 c (4 srv)	2.5 c (5 srv)	2.5 c (5 srv)	2.5 c (5 srv)
Vegetables	1 c (2 srv)	1 c (2srv)	1.5 c (3 srv)	1.5 c (3 srv)	1.5 c (3 srv)	2 c (4 srv)	2 c (4 srv)	2 c (4 srv)	2 c (4 srv)	2.5 c (5 srv)	2.5 c (5 srv)	2.5 c (5 srv)
Dark green veg.	1 c/wk	1.5 c/wk	1.5 c/wk	2 c/wk	3 c/wk	3 c/wk	3 c/wk	3 c/wk	3 c/wk	3 c/wk	3 c/wk	3 c/wk
Orange veg.	.5 c/wk	1 c/wk	1 c/wk	1.5 c/wk	2 c/wk	2 c/wk	2 c/wk	2 c/wk	2.5 c/wk	2.5 c/wk	2.5 c/wk	2.5 c/wk
Legumes	.5 c/wk	1 c/wk	1 c/wk	2.5 c/wk	3 c/wk	3 c/wk	3 c/wk	3 c/wk	3.5 c/wk	3.5 c/wk	3.5 c/wk	3.5 c/wk
Starchy veg.	1.5 c/wk	2.5 c/wk	2.5 c/wk	2.5 c/wk	3 c/wk	3 c/wk	6 c/wk	6 c/wk	7 c/wk	7 c/wk	9 c/wk	9 c/wk
Other veg.	3.5 c/wk	4.5 c/wk	4.5 c/wk	5.5 c/wk	6.5 c/wk	6.5 c/wk	7 c/wk	7 c/wk	8.5 c/wk	8.5 c/wk	10 c/wk	10 c/wk
Grains	3 oz-eq	4 oz-eq	5 oz-eq	5 oz-eq	6 oz-eq	6 oz-eq	7 oz-eq	8 oz-eq	9 oz-eq	10 oz-eq	10 oz-eq	10 oz-eq
Whole grains	1.5	2	2.5	3	3	3	3.5	4	4.5	5	5	5
Other grains	1.5	2	2.5	2	3	3	3.5	4	4.5	5	5	5
Lean meat and beans	2 oz-eq	3 oz-eq	4 oz-eq	5 oz-eq	5 oz-eq	5.5 oz-eq	6 oz-eq	6.5 oz-eq	6.5 oz-eq	7 oz-eq	7 oz-eq	7 oz-eq
Milk	2 c	2 c	2 c	3 c	3 c	3 c	3 c	3 c	3 c	3 c	3 c	3 c
Oils	15 g	17 g	17 g	22 g	24 g	27 g	29 g	31 g	34 g	36 g	44 g	51 g
Discretionary calorie allowance	165	171	171	132	195	267	290	362	410	426	512	648

[a]The suggested amounts of food to consume from the basic food groups, subgroups, and oils to meet recommended nutrient intakes at 12 different calorie levels. Nutrient and energy contributions from each group are calculated according to the nutrient-dense forms of foods in each group (e.g., lean meats and fat-free milk). The table also shows the discretionary calorie allowance that can be accommodated within each calorie level, in addition to the suggested amounts of nutrient-dense forms of foods in each group.

From U.S. Department of Health and Human Services and the U.S. Department of Agriculture (2005).

al., 2006). Less consistent evidence exists for the independent cholesterol-lowering effect of soy isoflavones and their potential to influence hormonally-regulated conditions such as cancer and menopause, and osteoporosis. Future research that considers isoflavone composition and source may provide better insight, and will undoubtedly provide valuable information for future dietary recommendations. The widespread availability of foods that provide SBO and/or soy protein facilitates inclusion in the diet in a way that is consistent with food-based dietary guidelines.

References

Adams, K.F.; K.M. Newton; C. Chen; S.S. Emerson; J.D. Potter; E. White; J.W. Lampe. Soy isoflavones do not modulate circulating insulin-like growth factor concentrations in an older population in an intervention trial. *J. Nutr.* **2003**, *133*, 1316–1319.

Albert, A.; C. Altabre; F. Baró; E. Buendía; A. Cabero; M.J. Cancelo; C. Castelo-Branco; P. Chantre; M. Duran; J. Haya, et al. Efficacy and safety of a phytoestrogen preparation derived from Glycine max (L.) Merr in climacteric symptomatology: A multicentric, open, prospective and non-randomized trial. *Phytomedicine* **2002**, *9*, 85–92.

Albertazzi, P.; F. Pansini; G. Bonaccorsi; L. Zanotti; E. Forini; D. De Aloysio. The effect of dietary soy supplementation on hot flushes. *Obstet. Gynecol.* **1998**, *91*, 6–11.

Altavilla, D.; A. Crisafulli; H. Marini; M. Esposito; R. D'Anna; F. Corrado; A. Bitto; F. Squadrito. Cardiovascular effects of the phytoestrogen genistein. *Curr. Med. Chem. Cardiovasc. Hematol. Agents* **2004**, *2*, 179–186.

American Diabetes Association. J.P. Bantle; J. Wylie-Rosett; A.L. Albright; C.M. Apovian; N.G. Clark; M.J. Franz; B.J. Hoogwerf; A.H. Lichtenstein; E. Mayer-Davis, et al. Nutrition recommendations and interventions for diabetes: A position statement of the American Diabetes Association. *Diabetes Care* **2008**, *31*, S61–78.

Anderson, J.J.B.; S.C. Garner. The effects of phytoestrogens on bone. *Nutr. Res.* **1997**, *17*, 1617–1632.

Anderson, J.W.; J. Fuller; K. Patterson; R. Blair; A. Tabor. Soy compared to casein meal replacement shakes with energy-restricted diets for obese women: Randomized controlled trial. *Metabolism* **2007**, *56*, 280–288.

Anderson, J.W.; B.M. Johnstone; M.E. Cook-Newell. Meta-analysis of the effects of soy protein intake on serum lipids. *N. Engl. J. Med.* **1995**, *333*, 276–282.

Appel, L.J.; E.R.R. Miller; A.J. Seidler; P.K. Whelton. Does supplementation of diet with "fish oil" reduce blood pressure? A meta-analysis of controlled clinical trials. *Arch. Intern. Med.* **1993**, *153*, 1429–1438.

Appel, L.J.; F.M. Sacks; V.J. Carey; E. Obarzanek; J.F. Swain; E.R. Miller, III; P.R. Conlin; T.P. Erlinger; B.A. Rosner; N.M. Laranjo, et al. Effects of protein, monounsaturated fat, and carbohydrate intake on blood pressure and serum lipids: Results of the omniheart randomized trial. *J. Amer. Med. Assoc.* **2005**, *294*, 2455–2464.

Aro, A.; M. Jauhiainen; R. Partanen; I. Salminen; M. Mutanen. Stearic acid, *trans* fatty acids, and dairy fat: Effects on serum and lipoprotein lipids, apolipoproteins, lipoprotein(a), and lipid transfer proteins in healthy subjects. *Am. J. Clin. Nutr.* **1997**, *65*, 1419–1426.

Ascherio, A.; C. Hennekens; W.C. Willett; F. Sacks; B. Rosner; J. Manson; J. Witteman; M.J. Stampfer. Prospective study of nutritional factors, blood pressure, and hypertension among US women. *Hypertension* **1996a**, *27*, 1065–1072.

Ascherio, A.; M.B. Katan; P.L. Zock; M.J. Stampfer; W.C. Willett. *Trans* fatty acids and coronary heart disease. *N. Engl. J. Med.* **1999**, *340*, 1994–1998.

Ascherio, A.; E.B. Rimm; E.L. Giovannucci; G.A. Colditz; B. Rosner; W.C. Willett; F. Sacks; M.J. Stampfer. A prospective study of nutritional factors and hypertension among US men. *Circulation* **1992**, *86*, 1475–1484.

Ascherio, A.; E.B. Rimm; E.L. Giovannucci; D. Spiegelman; M. Stampfer; W.C. Willett. Dietary fat and risk of coronary heart disease in men: Cohort follow up study in the United States. *Br. Med. J.* **1996b**, *313*, 84–90.

Atkinson, C.; C.L. Frankenfeld; J.W. Lampe. Gut bacterial metabolism of the soy isoflavone daidzein: Exploring the relevance to human health. *Exp. Biol. Med. (Maywood)* **2005**, *230*, 155–170.

Atteritano, M.; H. Marini; L. Minutoli; F. Polito; A. Bitto; D. Altavilla; S. Mazzaferro; R. D'Anna; M.L. Cannata; A. Gaudio, et al. Effects of the phytoestrogen genistein on some predictors of cardiovascular risk in osteopenic, postmenopausal women: A two-year randomized, double-blind, placebo-controlled study. *J. Clin. Endocrinol. Metab.* **2007**, *92*, 3068–3075.

Azadbakht, L.; S. Atabak; A. Esmaillzadeh. Soy protein intake, cardio-renal indices and C-reactive protein in type 2 diabetes with nephropathy: A longitudinal randomized clinical trial. *Diabetes Care* **2008**, *Jan 9; [Epub ahead of print]*.

Azadbakht, L.; M. Kimiagar; Y. Mehrabi; A. Esmaillzadeh; F.B. Hu; W.C. Willett. Soy consumption, markers of inflammation, and endothelial function: A cross-over study in postmenopausal women with the metabolic syndrome. *Diabetes Care* **2007a**, *30*, 967–973.

Azadbakht, L.; M. Kimiagar; Y. Mehrabi; A. Esmaillzadeh; M. Padyab; F.B. Hu; W.C. Willett. Soy inclusion in the diet improves features of the metabolic syndrome: A randomized crossover study in postmenopausal women. *Am. J. Clin. Nutr.* **2007b**, *85*, 735–741.

Baer, D.J.; J.T. Judd; B.A. Clevidence; R.P. Tracy. Dietary fatty acids affect plasma markers of inflammation in healthy men fed controlled diets: A randomized crossover study. *Am. J. Clin. Nutr.* **2004**, *79*, 969–973.

Baer, D.J.; J.T. Judd; P.M. Kris-Etherton; G. Zhao; E.A. Emken. Stearic acid absorption and its metabolizable energy value are minimally lower than those of other fatty acids in healthy men fed mixed diets. *J. Nutr.* **2003**, *133*, 4129–4134.

Balk, E.; M. Chung; P. Chew; S. Ip; G. Raman; B. Kupelnick; A. Tatsioni; Y. Sun; B. Wolk; D. DeVine, et al. *Effects of Soy on Health Outcomes. Evidence Report/Technology Assessment No. 126.*; Agency for Healthcare Research and Quality: Rockville, MD, 2005.

Balk, J.L.; D.A. Whiteside; G. Naus; E. DeFerrari; J.M. Roberts. A pilot study of the effects of phytoestrogen supplementation on postmenopausal endometrium. *J. Soc. Gynecol. Investig.* **2002**, *9*, 238–242.

Bazzoli, D.L.; S. Hill; R.A. DiSilvestro. Soy protein antioxidant actions in active, young adult women. *Nutr. Res.* **2002**, *22*, 807–815.

Berger, A.; P.J. Jones; S.S. Abumweis. Plant sterols: Factors affecting their efficacy and safety as functional food ingredients. *Lipids Health Dis.* **2004**, *3*, 5.

Blum, A.; N. Lang; F. Vigder; P. Israeli; M. Gumanovsky; S. Lupovitz; A. Elgazi; A. Peleg; M. Ben-Ami. Effects of soy protein on endothelium-dependent vasodilatation and lipid profile in post-menopausal women with mild hypercholesterolemia. *Clin. Invest. Med.* **2003**, *26*, 20–26.

Borkman, M.; L.H. Storlien; D.A. Pan; A.B. Jenkins; D.J. Chisholm; L.V. Campbell. The relation between insulin sensitivity and the fatty-acid composition of skeletal-muscle phospholipids. *N. Engl. J. Med.* **1993**, *328*, 238–244.

Brouwer, I.A.; M.B. Katan; P.L. Zock. Dietary {alpha}-linolenic acid is associated with reduced risk of fatal coronary heart disease, but increased prostate cancer risk: A meta-analysis. *J. Nutr.* **2004**, *134*, 919–922.

Brown, B.D.; W. Thomas; A. Hutchins; M.C. Martini; J.L. Slavin. Types of dietary fat and soy minimally affect hormones and biomarkers associated with breast cancer risk in premenopausal women. *Nutr. Cancer* **2002**, *43*, 22–30.

Burke, G.L.; C. Legault; M. Anthony; D.R. Bland; T.M. Morgan; M.J. Naughton; K. Leggett; S.A. Washburn; M.Z. Vitolins. Soy protein and isoflavone effects on vasomotor symptoms in peri- and postmenopausal women: The soy estrogen alternative study. *Menopause* **2003**, *10*, 147–153.

Calder, P.C.; R.F. Grimble. Polyunsaturated fatty acids, inflammation and immunity. *Eur. J. Clin. Nutr.* **2002**, *56*, S14–19.

Canadian Diabetes Association Clinical Practice Guidelines Expert Committee. *Clinical practice guidelines for the prevention and management of diabetes in Canada. Can. J. Diabetes* **2003;** *Suppl. 27*, pp. S1–S152.

Cassidy, A.; P. Albertazzi; I. Lise Nielsen; W. Hall; G. Williamson; I. Tetens; S. Atkins; H. Cross; Y. Manios; A. Wolk, et al. Critical review of health effects of soyabean phyto-oestrogens in post-menopausal women. *Proc. Nutr. Soc.* **2006**, *65*, 76–92.

Cassidy, A.; S. Bingham; K. Setchell. Biological effects of isoflavones in young women: Importance of the chemical composition of soyabean products. *Br. J. Nutr.* **1995**, *74*, 587–601.

Cassidy, A.; S. Bingham; K.D. Setchell. Biological effects of a diet of soy protein rich in isoflavones on the menstrual cycle of premenopausal women. *Am. J. Clin. Nutr.* **1994**, *60*, 333–340.

Chacko, B.K.; R.T. Chandler; A. Mundhekar; N. Khoo; H.M. Pruitt; D.F. Kucik; D.A. Parks; C.G. Kevil; S. Barnes; R.P. Patel. Revealing anti-inflammatory mechanisms of soy isoflavones by flow: Modulation of leukocyte-endothelial cell interactions. *Am. J. Physiol. Heart Circ. Physiol.* **2005**, *289*, 908–915.

Chlebowski, R.T.; G.L. Blackburn; C.A. Thomson; D.W. Nixon; A. Shapiro; M.K. Hoy; M.T. Goodman; A.E. Giuliano; N. Karanja; P. McAndrew, et al. Dietary fat reduction and breast cancer outcome: Interim efficacy results from the women's intervention study. *J. Natl. Cancer Inst.* **2006,** *98,* 1767–1776.

Colacurci, N.; A. Chiantera; F. Fornaro; V. de Novellis; D. Manzella; A. Arciello; V. Chiantera; L. Improta; G. Paolisso. Effects of soy isoflavones on endothelial function in healthy postmenopausal women. *Menopause* **2005**, *12*, 299–307.

Crisafulli, A.; H. Marini; A. Bitto; D. Altavilla; G. Squadrito; A. Romeo; E.B. Adamo; R. Marini;

R. D'Anna; F. Corrado, et al. Effects of genistein on hot flushes in early postmenopausal women: A randomized, double-blind EPT- and placebo-controlled study. *Menopause* **2004**, *11*, 400–404.

Cuevas, A.M.; V.L. Irribarra; O.A. Castillo; M.D. Yanez; A.M. Germain. Isolated soy protein improves endothelial function in postmenopausal hypercholesterolemic women. *Eur. J. Clin. Nutr.* **2003**, *57*, 889–894.

Cupisti, A.; L. Ghiadoni; C. D'Alessandro; I. Kardasz; E. Morelli; V. Panichi; D. Locati; S. Morandi; A. Saba; G. Barsotti, et al. Soy protein diet improves endothelial dysfunction in renal transplant patients. *Nephrol. Dial. Transplant* **2007**, *22*, 229–234.

D'Anna, R.; G. Baviera; F. Corrado; F. Cancellieri; A. Crisafulli; F. Squadrito. The effect of the phytoestrogen genistein and hormone replacement therapy on homocysteine and C-reactive protein level in postmenopausal women. *Acta Obstet. Gynecol. Scand.* **2005**, *84*, 474–477.

Dalais, F.S.; G.E. Rice; M.L. Wahlqvist; M. Grehan; A.L. Murkies; G. Medley; R. Ayton; B.J. Strauss. Effects of dietary phytoestrogens in postmenopausal women. *Climacteric.* **1998**, *1*, 124–129.

Dang, Z.-C.; V. Audinot; S.E. Papapoulos; J.A. Boutin; C.W.G.M. Lowik. Peroxisome proliferator-activated receptor gamma (ppargamma) as a molecular target for the soy phytoestrogen genistein. *J. Biol. Chem.* **2003**, *278*, 962–967.

Davis, J.N.; O. Kucuk; Z. Djuric; F.H. Sarkar. Soy isoflavone supplementation in healthy men prevents NF-[kappa]B activation by TNF-[alpha] in blood lymphocytes. *Free Radic. Biol. Med.* **2001**, *30*, 1293–1302.

Dayton, S.; M.L. Pearce; S. Hashimoto; W.J. Dixon; U. Tomiyasu. A controlled clinical trial of a diet high in unsaturated fat in preventing complications of atherosclerosis. *Circulation* **1969**, *40*, 1–63.

de Lorgeril, M.; P. Salen. {alpha}-Linolenic acid, coronary heart disease, and prostate cancer. *J. Nutr.* **2004**, *134*, 3385.

de Roos, N.M.; E.G. Schouten; M.B. Katan. Consumption of a solid fat rich in lauric acid results in a more favorable serum lipid profile in healthy men and women than consumption of a solid fat rich in *trans*-fatty acids. *J. Nutr.* **2001**, *131*, 242–245.

Demaison, L.; D. Moreau. Dietary n-3 polyunsaturated fatty acids and coronary heart disease-related mortality: A possible mechanism of action. *Cell Mol. Life Sci.* **2002**, *59*, 463–477.

Deuel, H.J. *The Lipids, Their Chemistry and Biochemistry;* Interscience Publishers Inc.: New York, London, 1955; Vol. II, p. 919.

Dickinson, H.O.; J.M. Mason; D.J. Nicolson; F. Campbell; F.R. Beyer; J.V. Cook; B. Williams; G.A. Ford. Lifestyle interventions to reduce raised blood pressure: A systematic review of randomized controlled trials. *J. Hypertens.* **2006**, *24*, 215–233.

Duncan, A.M.; B.E. Merz; X. Xu; T.C. Nagel; W.R. Phipps; M.S. Kurzer. Soy isoflavones exert modest hormonal effects in premenopausal women. *J. Clin. Endocrinol. Metab.* **1999**, *84*, 192–197.

Dyerberg, J.; D.C. Eskesen; P.W. Andersen; A. Astrup; B. Buemann; J.H. Christensen; P. Clausen; B.F. Rasmussen; E.B. Schmidt; T. Tholstrup, et al. Effects of *trans*- and n-3 unsaturated fatty acids on cardiovascular risk markers in healthy males. An 8 weeks dietary intervention study. *Eur. J. Clin. Nutr.* **2004**, *58*, 1062–1070.

Eckel, R.H.; S. Borra; A.H. Lichtenstein; S.Y. Yin-Piazza. Understanding the complexity of *trans* fatty acid reduction in the American diet: American heart association *trans* fat conference 2006: Report of the *trans* fat conference planning group. *Circulation* **2007**, *115*, 2231–2246.

Emken, E.A.; R.O. Adlof; W.K. Rohwedder; R.M. Gulley. Influence of linoleic acid on desaturation and uptake of deuterium-labeled palmitic and stearic acids in humans. *Biochim. Biophys. Acta* **1993**, *1170*, 173–181.

Erdman, J.W., Jr.; T.M. Badger; J.W. Lampe; K.D.R. Setchell; M. Messina. Not all soy products are created equal: Caution needed in interpretation of research results. *J. Nutr.* **2004**, *134*, 1229–1233.

Esposito, K.; R. Marfella; M. Ciotola; C. Di Palo; F. Giugliano; G. Giugliano; M. D'Armiento; F. D'Andrea; D. Giugliano. Effect of a mediterranean-style diet on endothelial dysfunction and markers of vascular inflammation in the metabolic syndrome: A randomized trial. *J. Amer. Med. Assoc.* **2004**, *292*, 1440–1446.

Evans, M.; V.Y. Njike; M. Hoxley; M. Pearson; D.L. Katz. Effect of soy isoflavone protein and soy lecithin on endothelial function in healthy postmenopausal women. *Menopause* **2007**, *14*, 141–149.

Fanti, P.; R. Asmis; T.J. Stephenson; B.P. Sawaya; A.A. Franke. Positive effect of dietary soy in ESRD patients with systemic inflammation—correlation between blood levels of the soy isoflavones and the acute-phase reactants. *Nephrol. Dial. Transplant* **2006**, *21*, 2239–2246.

FAO/WHO Expert Consultation *Protein Quality Evaluation. Food and Agricultural Organization of the United Nations, FAO Food and Nutrition Paper 51*; Rome, 1990.

Faure, E.D.; P. Chantre; P. Mares. Effects of a standardized soy extract on hot flushes: A multicenter, double-blind, randomized, placebo-controlled study. *Menopause* **2002**, *9*, 329–334.

Fay, M.P.; L.S. Freedman; C.K. Clifford; D.N. Midthune. Effect of different types and amounts of fat on the development of mammary tumors in rodents: A review. *Cancer Res.* **1997**, *57*, 3979–3988.

Food and Drug Administration. Food labeling: Health claims; soy protein and coronary heart disease. *Fed. Regist.* **1999**, *64*, 57700–57733.

Food Surveys Research Group. *Intakes of 19 individual fatty acids: Results from the 1994–1996 Continuing Survey of Food Intakes by Individuals.* **2005** http://www.ars.usda.gov (20 August 2007).

Frankenfeld, C.L.; A. McTiernan; W.K. Thomas; K. LaCroix; L. McVarish; V.L. Holt; S.M. Schwartz; J.W. Lampe. Postmenopausal bone mineral density in relation to soy isoflavone-metabolizing phenotypes. *Maturitas* **2006**, *53*, 315–324.

Gardner, C.D.; M. Messina; A. Kiazand; J.L. Morris; A.A. Franke. Effect of two types of soy milk and dairy milk on plasma lipids in hypercholesterolemic ad ults: A randomized trial. *J. Am. Coll. Nutr.* **2007**, *26*, 669–677.

Gardner-Thorpe, D.; C. O'Hagen; I. Young; S.J. Lewis. Dietary supplements of soya flour lower serum testosterone concentrations and improve markers of oxidative stress in men. *Eur. J. Clin. Nutr.* **2003**, *57*, 100–106.

Garrido, A.; M.P. De la Maza; S. Hirsch; L. Valladares. Soy isoflavones affect platelet thromboxane A2 receptor density but not plasma lipids in menopausal women. *Maturitas* **2006**, *54*, 270–276.

Geleijnse, J.M.; E.J. Giltay; D.E. Grobbee; A.R. Donders; F.J. Kok. Blood pressure response to fish oil supplementation: Metaregression analysis of randomized trials. *J. Hypertens.* **2002**, *20*, 1493–1499.

Genkinger, J.; D. Hunter; D. Spiegelman; K. Anderson; W. Beeson; J. Buring; G. Colditz; G. Fraser; J. Freudenheim; R. Goldbohm, et al. A pooled analysis of 12 cohort studies of dietary fat, cholesterol and egg intake and ovarian cancer. *Cancer Causes and Control* **2006**, *17*, 273–285.

Gottstein, N.; B.A. Ewins; C. Eccleston; G.P. Hubbard; I.C. Kavanagh; A.M. Minihane; P.D. Weinberg; G. Rimbach. Effect of genistein and daidzein on platelet aggregation and monocyte and endothelial function. *Br. J. Nutr.* **2003**, *89*, 607–616.

Goyens, P.L.L.; M.E. Spilker; P.L. Zock; M.B. Katan; R.P. Mensink. Compartmental modeling to quantify {alpha}-linolenic acid conversion after longer term intake of multiple tracer boluses. *J. Lipid Res.* **2005**, *46*, 1474–1483.

Greany, K.A.; J.A. Nettleton; K.E. Wangen; W. Thomas; M.S. Kurzer. Consumption of isoflavone-rich soy protein does not alter homocysteine or markers of inflammation in postmenopausal women. *Eur. J. Clin. Nutr.* **2007**, *Sep 5 [Epub ahead of print]*.

Habito, R.C.; J. Montalto; E. Leslie; M.J. Ball. Effects of replacing meat with soyabean in the diet on sex hormone concentrations in healthy adult males. *Br. J. Nutr.* **2000**, *84*, 557–563.

Hale, G.; M. Paul-Labrador; J.H. Dwyer; C.N. Merz. Isoflavone supplementation and endothelial function in menopausal women. *Clin. Endocrinol. (Oxf)* **2002**, *56*, 693–701.

Hall, W.L.; K. Vafeiadou; J. Hallund; S. Bugel; C. Koebnick; M. Reimann; M. Ferrari; F. Branca; D. Talbot; T. Dadd, et al. Soy-isoflavone-enriched foods and inflammatory biomarkers of cardiovascular disease risk in postmenopausal women: Interactions with genotype and equol production. *Am. J. Clin. Nutr.* **2005**, *82*, 1260–1268.

Hall, W.L.; K. Vafeiadou; J. Hallund; S. Bugel; M. Reimann; C. Koebnick; H.J.F. Zunft; M. Ferrari; F. Branca; T. Dadd, et al. Soy-isoflavone-enriched foods and markers of lipid and glucose metabolism in postmenopausal women: Interactions with genotype and equol production. *Am. J. Clin. Nutr.* **2006**, *83*, 592–600.

Han, K.K.; J.M. Soares, Jr.; M.A. Haidar; G.R. de Lima; E.C. Baracat. Benefits of soy isoflavone therapeutic regimen on menopausal symptoms. *Obstet. Gynecol.* **2002a**, *99*, 389–394.

Han, S.N.; L.S. Leka; A.H. Lichtenstein; L.M. Ausman; E.J. Schaefer; S.N. Meydani. Effect of hydrogenated and saturated, relative to polyunsaturated, fat on immune and inflammatory responses of adults with moderate hypercholesterolemia. *J. Lipid Res.* **2002b**, *43*, 445–452.

Hanf, V.; U. Gonder. Nutrition and primary prevention of breast cancer: Foods, nutrients and breast cancer risk. *Eur. J. Obstet. Gynecol. Reprod. Bio.* **2005**, *123*, 139–149.

Hanson, L.N.; H.M. Engelman; D.L. Alekel; K.L. Schalinske; M.L. Kohut; M.B. Reddy. Effects of soy isoflavones and phytate on homocysteine, C-reactive protein, and iron status in postmenopausal women. *Am. J. Clin. Nutr.* **2006**, *84*, 774–780.

Harper, A.E.; N.N. Yoshimura. Protein quality, amino acid balance, utilization, and evaluation of diets containing amino acids as therapeutic agents. *Nutrition* **1993**, *9*, 460–469.

Harris, W.S. The omega-6/omega-3 ratio and cardiovascular disease risk: Uses and abuses. *Curr. Atheroscler. Rep.* **2006**, *8*, 453–459.

Hermansen, K.; B. Hansen; R. Jacobsen; P. Clausen; M. Dalgaard; B. Dinesen; J.J. Holst; E. Pedersen; A. Astrup. Effects of soy supplementation on blood lipids and arterial function in hypercholesterolaemic subjects. *Eur. J. Clin. Nutr.* **2005**, *59*, 843–850.

Hilpert, K.F.; P.M. Kris-Etherton; S.G. West. Lipid response to a low-fat diet with or without soy is modified by C-reactive protein status in moderately hypercholesterolemic adults. *J. Nutr.* **2005**, *135*, 1075–1079.

Ho, S.C.; Y.M. Chen; S.S. Ho; J.L. Woo. Soy isoflavone supplementation and fasting serum glucose and lipid profile among postmenopausal Chinese women: A double-blind, randomized, placebo-controlled trial. *Menopause* **2007**, *14*, 905–912.

Howe, G.R.; K.J. Aronson; E. Benito; R. Castelleto; J. Cornée; S. Duffy; R.P. Gallagher; J.M. Iscovich; J. Deng-ao; R. Kaaks, et al. The relationship between dietary fat intake and risk of colorectal cancer: Evidence from the combined analysis of 13 case-control studies. *Cancer Causes & Control* **1997**, *8*, 215–228.

Hsu, C.S.; W.W. Shen; Y.M. Hsueh; S.L. Yeh. Soy isoflavone supplementation in postmenopausal women. Effects on plasma lipids, antioxidant enzyme activities and bone density. *J. Reprod. Med.* **2001**, *46*, 221–226.

Hu, F.B.; M.J. Stampfer; J.E. Manson; E. Rimm; G.A. Colditz; B.A. Rosner; C.H. Hennekens; W.C. Willett. Dietary fat intake and the risk of coronary heart disease in women. *N. Engl. J. Med.* **1997**, *337*, 1491–1499.

Hu, F.B.; R.M. van Dam; S. Liu. Diet and risk of Type II diabetesl: The role of types of fat and carbohydrate. *Diabetologia* **2001**, *44*, 805–817.

Huang, Y.; S. Cao; M. Nagamani; K.E. Anderson; J.J. Grady; L.J. Lu. Decreased circulating levels of tumor necrosis factor-alpha in postmenopausal women during consumption of soy-containing isoflavones. *J. Clin. Endocrinol. Metab.* **2005**, *90*, 3956–3962.

Huff, M.W.; P.M. Giovannetti; B.M. Wolfe. Turnover of very low-density lipoprotein-apoprotein B is increased by substitution of soybean protein for meat and dairy protein in the diets of hypercholesterolemic men. *Am. J. Clin. Nutr.* **1984**, *39*, 888–897.

Hulbert, A.J.; N. Turner; L.H. Storlien; P.L. Else. Dietary fats and membrane function: Implications for metabolism and disease. *Biol. Rev. Camb. Philos. Soc.* **2005**, *80*, 155–169.

Huntley, A.L.; E. Ernst. Soy for the treatment of perimenopausal symptoms—A systematic review. *Maturitas* **2004**, *47*, 1–9.

Hussein, N.; E. Ah-Sing; P. Wilkinson; C. Leach; B.A. Griffin; D.J. Millward. Long-chain conversion of [13C]linoleic acid and {alpha}-linolenic acid in response to marked changes in their dietary intake in men. *J. Lipid Res.* **2005**, *46*, 269–280.

Hutchins, A.M.; I.E. McIver; C.S. Johnston. Soy isoflavone and ascorbic acid supplementation alone or in combination minimally affect plasma lipid peroxides in healthy postmenopausal women. *J. Am. Diet. Assoc.* **2005**, *105*, 1134–1137.

Institute of Medicine. *Dietary Reference Intakes for Vitamin C, Vitamin E, Selenium, and Carotenoids;* National Academy Press: Washington, DC, 2000.

Institute of Medicine. *Dietary Reference Intakes for Vitamin A, Vitamin K, Arsenic, Boron, Chromium, Copper, Iodine, Iron, Manganese, Molybdenum, Nickel, Silicon, Vanadium, and Zinc;* National

Academy Press: Washington, DC, 2001.

James, M.J.; R.A. Gibson; L.G. Cleland. Dietary polyunsaturated fatty acids and inflammatory mediator production. *Am. J. Clin. Nutr.* **2000**, *71*, 343–348.

Jansman, A.J.M. Bioavailability of proteins in legume seeds. *Grain Legumes (AEP)* **1996**, *11*, 19.

Jenkins, D.J.; C.W. Kendall; P.W. Connelly; C.J. Jackson; T. Parker; D. Faulkner; E. Vidgen. Effects of high- and low-isoflavone (phytoestrogen) soy foods on inflammatory biomarkers and proinflammatory cytokines in middle-aged men and women. *Metabolism* **2002**, *51*, 919–924.

Jenkins, D.J.A.; C.W.C. Kendall; M.A. D'Costa; C.-J. Jackson; E. Vidgen; W. Singer; J.A. Silverman; G. Koumbridis; J. Honey; A.V. Rao, et al. Soy consumption and phytoestrogens: Effect on serum prostate specific antigen when blood lipids and oxidized low-density lipoprotein are reduced in hyperlipidemic men. *J. Urol.* **2003**, *169*, 507–511.

Judd, J.; D. Baer; B. Clevidence; et al. Blood lipid and lipoprotein modifying effects of *trans* monounsaturated fatty acids compared to carbohydrate, oleic acid, stearic acid, and C 12:0-16:0 saturated fatty acids in men fed controlled diets. *FASEB J.* **1998**, *12*, A229-A229.abstract.

Judd, J.T.; B.A. Clevidence; R.A. Muesing; J. Wittes; M.E. Sunkin; J.J. Podczasy. Dietary *trans* fatty acids: effects on plasma lipids and lipoprotiens of healthy men and women. *Am. J. Clin. Nutr.* **1994,** *59*, 861–868.

Katz, D.L.; M.A. Evans; V.Y. Njike; M.L. Hoxley; H. Nawaz; B.P. Comerford; P.M. Sarrel. Raloxifene, soy phytoestrogens and endothelial function in postmenopausal women. *Climacteric* **2007**, *10*, 500–507.

Keys, A. Coronary heart disease in seven countries. *Circulation* **1970**, *41*, 11–211.

King, D.E.; B.M. Egan; M.E. Geesey. Relation of dietary fat and fiber to elevation of C-reactive protein. *Am. J. Cardiol.* **2003**, *92*, 1335–1339.

Kinsella, J.E.; B. Lokesh; S. Broughton; J. Whelan. Dietary polyunsaturated fatty acids and eicosanoids: potential effects on the modulation of inflammatory and immune cells: An overview. *Nutrition* **1990**, *6*, 24–44; 59–62.

Knight, D.C.; J.B. Howes; J.A. Eden; L.G. Howes. Effects on menopausal symptoms and acceptability of isoflavone-containing soy powder dietary supplementation. *Climacteric* **2001**, *4*, 13–18.

Kolonel, L.N. Nutrition and prostate cancer. *Cancer Causes and Control* **1996**, *7*, 83–94.

Kolonel, L.N.; J.H. Hankin; A.M. Nomura; S.Y. Chu. Dietary fat intake and cancer incidence among five ethnic groups in Hawaii. *Cancer Res.* **1981**, *41*, 3727–3728.

Koralek, D.; U. Peters; G. Andriole; D. Reding; V. Kirsh; A. Subar; A. Schatzkin; R. Hayes; M. Leitzmann. A prospective study of dietary alpha-linolenic acid and the risk of prostate cancer (United States). *Cancer Causes and Control* **2006**, *17*, 783–791.

Kotsopoulos, D.; F.S. Dalais; Y.L. Liang; B.P. McGrath; H.J. Teede. The effects of soy protein containing phytoestrogens on menopausal symptoms in postmenopausal women. *Climacteric* **2000**, *3*, 161–167.

Kreijkamp-Kaspers, S.; L. Kok; M.L. Bots; D.E. Grobbee; J.W. Lampe; Y.T. van der Schouw. Randomized controlled trial of the effects of soy protein containing isoflavones on vascular function in postmenopausal women. *Am. J. Clin. Nutr.* **2005**, *81*, 189–195.

Kris-Etherton, P.M.; J. Derr; D.C. Mitchell; V.A. Mustad; M.E. Russell; E.T. McDonnell; D. Salabsky; T.A. Pearson. The role of fatty acid saturation on plasma lipids, lipoproteins, and apolipoproteins: I. Effects of whole food diets high in cocoa butter, olive oil, soybean oil, dairy butter, and milk chocolate on the plasma lipids of young men. *Metabolism* **1993**, *42*, 121–129.

Kromhout, D.; A. Menotti; B. Bloemberg; C. Aravanis; H. Blackburn; R. Buzina; A.S. Dontas; F. Fidanza; S. Giaipaoli; A. Jansen, et al. Dietary saturated and transfatty acids and cholesterol and 25-year mortality from coronary heart disease: The seven countries study. *Prev. Med.* **1995**, *24*, 308–315.

Kronenberg, F.; A. Fugh-Berman. Complementary and alternative medicine for menopausal symptoms: A review of randomized, controlled trials. *Ann. Intern. Med.* **2002**, *137*, 805–813.

Kumar, N.B.; A. Cantor; K. Allen; D. Riccardi; C.E. Cox. The specific role of isoflavones on estrogen metabolism in premenopausal women. *Cancer* **2002**, *94*, 1166–1174.

Kushi, L.; E. Giovannucci. Dietary fat and cancer. *Am. J. Med.* **2002**, *113*, 63–70.

Lahoz, C.; R. Alonso; J.M. Ordovás; A. López-Farré; M. de Oya; P. Mata. Effects of dietary fat saturation on eicosanoid production, platelet aggregation and blood pressure. *Eur. J. Clin. Invest.* **1997**, *27*, 780–787.

Lampe, J.W.; S.C. Karr; A.M. Hutchins; J.L. Slavin. Urinary equol excretion with a soy challenge: Influence of habitual diet. *Proc. Soc. Exp. Biol. Med.* **1998**, *217*, 335–339.

Lampe, J.W.; H.E. Skor; S. Li; K. Wahala; W.N. Howald; C. Chen. Wheat bran and soy protein feeding do not alter urinary excretion of the isoflavan equol in premenopausal women. *J. Nutr.* **2001**, *131*, 740–744.

Larkin, T.A.; W.E. Price; L.B. Astheimer. Increased probiotic yogurt or resistant starch intake does not affect isoflavone bioavailability in subjects consuming a high soy diet. *Nutrition* **2007**, *23*, 709–718.

Larsson, S.C.; M. Kumlin; M. Ingelman-Sundberg; A. Wolk. Dietary long-chain n-3 fatty acids for the prevention of cancer: A review of potential mechanisms. *Am. J. Clin. Nutr.* **2004**, *79*, 935–945.

Leren, P. The Oslo diet-heart study. Eleven-year report. *Circulation* **1970**, *42*, 935–942.

Li, Z.; K. Hong; P. Saltsman; S. DeShields; M. Bellman; G. Thames; Y. Liu; H.J. Wang; R. Elashoff; D. Heber. Long-term efficacy of soy-based meal replacements vs an individualized diet plan in obese Type II DM patients: relative effects on weight loss, metabolic parameters, and C-reactive protein. *Eur. J. Clin. Nutr.* **2005**, *59*, 411–418.

Liao, F.-H.; M.-J. Shieh; S.-C. Yang; S.-H. Lin; Y.-W. Chien. Effectiveness of a soy-based compared with a traditional low-calorie diet on weight loss and lipid levels in overweight adults. *Nutrition* **2007**, *23*, 551–556.

Libby, P.; P.M. Ridker. Inflammation and atherosclerosis: Role of C-Reactive protein in risk assessment. *Am. J. Med.* **2004**, *116*, 9–16.

Lichtenstein, A.H.; L.M. Ausman; W. Carrasco; J.L. Jenner; J.M. Ordovas; E.J. Schaefer. Hydrogenation impairs the hypolipidemic effect of corn oil in humans. Hydrogenation, *trans* fatty acids, and plasma lipids. *Arterioscler. Thromb.* **1993**, *13*, 154–161.

Lichtenstein, A.H.; L.M. Ausman; S.M. Jalbert; E.J. Schaefer. Effects of different forms of dietary hydrogenated fats on serum lipoprotein cholesterol levels. *N. Engl. J. Med.* **1999**, *340*, 1933–1940.

Lichtenstein, A.H.; A.T. Erkkila; B. Lamarche; U.S. Schwab; S.M. Jalbert; L.M. Ausman. Influence of hydrogenated fat and butter on CVD risk factors: Remnant-like particles, glucose and insulin, blood pressure and C-reactive protein. *Atherosclerosis* **2003**, *171*, 97–107.

Lichtenstein, A.H.; N.R. Matthan; S.M. Jalbert; N.A. Resteghini; E.J. Schaefer; L.M. Ausman. Novel soybean oils with different fatty acid profiles alter cardiovascular disease risk factors in moderately hyperlipidemic subjects. *Am. J. Clin. Nutr.* **2006**, *84*, 497–504.

Lissin, L.W.; R. Oka; S. Lakshmi; J.P. Cooke. Isoflavones improve vascular reactivity in postmenopausal women with hypercholesterolemia. *Vasc. Med.* **2004**, *9*, 26–30.

Lopez-Garcia, E.; M.B. Schulze; T.T. Fung; J.B. Meigs; N. Rifai; J.E. Manson; F.B. Hu. Major dietary patterns are related to plasma concentrations of markers of inflammation and endothelial dysfunction. *Am. J. Clin. Nutr.* **2004**, *80*, 1029–1035.

Louheranta, A.M.; A.K. Turpeinen; H.M. Vidgren; U.S. Schwab; M.I.J. Uusitupa. A high-*trans* fatty acid diet and insulin sensitivity in young healthy women. *Metabolism* **1999**, *48*, 870–875.

Lu, L.-J.W.; K.E. Anderson; J.J. Grady; F. Kohen; M. Nagamani. Decreased ovarian hormones during a soya diet: Implications for breast cancer prevention. *Cancer Res.* **2000a**, *60*, 4112–4121.

Lu, L.-J.; K.E. Anderson; J.J. Grady; M. Nagamani. Effects of soya consumption for one month on steroid hormones in premenopausal women: Implications for breast cancer risk reduction. *Cancer Epidemiol. Biomarkers Prev.* **1996**, *5*, 63–70.

Lu, L.-J.W.; K.E. Anderson; J.J. Grady; M. Nagamani. Effects of an isoflavone-free soy diet on ovarian hormones in premenopausal women. *J. Clin. Endocrinol. Metab.* **2001**, *86*, 3045–3052.

Lu, L.-J.W.; M. Cree; S. Josyula; M. Nagamani; J.J. Grady; K.E. Anderson. Increased urinary excretion of 2-hydroxyestrone but not 16{{alpha}}-hydroxyestrone in premenopausal women during a soya diet containing isoflavones. *Cancer Res.* **2000b**, *60*, 1299–1305.

Lu, Z.; S. Hendrich; N. Shen; P.J. White; L.R. Cook. Low linolenate and commercial soybean oils diminish serum HDL cholesterol in young free-living adult females. *J. Am. Coll. Nutr.* **1997**, *16*, 562–969.

Lukaczer, D.; J.L. DeAnn; R.H. Lerman; G. Darland; B. Schiltz; M. Tripp; J.S. Bland. Effect of a low glycemic index diet with soy protein and phytosterols on CVD risk factors in postmenopausal women. *Nutrition* **2006**, *22*, 104–113.

Ma, D.-F.; L.-Q. Qin; P.-Y. Wang; R. Katoh. Soy isoflavone intake increases bone mineral density in the spine of menopausal women: Meta-analysis of randomized controlled trials. *Clin. Nutr.* **2007**, *Dec 4; [Epub ahead of print]*.

Martini, M.C.; B.B. Dancisak; C.J. Haggans; W. Thomas; J.L. Slavin. Effects of Soy Intake on Sex Hormone Metabolism in Premenopausal Women. *Nutr. Cancer* **1999**, *34*, 133–139.

Maskarinec, G.; A.E. Williams; L. Carlin. Mammographic densities in a one-year isoflavone intervention. *Eur. J. Cancer Prev.* **2003**, *12*, 165–169.

Maskarinec, G.; A.E. Williams; J.S. Inouye; F.Z. Stanczyk; A.A. Franke. A randomized isoflavone intervention among premenopausal women. *Cancer Epidemiol. Biomarkers Prev.* **2002**, *11*, 195–201.

Matthan, N.R.; K. Cianflone; A.H. Lichtenstein; L.M. Ausman; M. Jauhiainen; P.J.H. Jones. Hydrogenated fat consumption affects acylation- stimulating protein levels and cholesterol esterification rates in moderately hypercholesterolemic women. *J. Lipid Res.* **2001**, *42*, 1841–1848.

Matthan, N.R.; S.M. Jalbert; L.M. Ausman; J.T. Kuvin; R.H. Karas; A.H. Lichtenstein. Effect of soy protein from differently processed products on cardiovascular disease risk factors and vascular endothelial function in hypercholesterolemic subjects. *Am. J. Clin. Nutr.* **2007**, *85*, 960–966.

Matthan, N.R.; F.K. Welty; P.H.R. Barrett; C. Harausz; G.G. Dolnikowski; J.S. Parks; R.H. Eckel; E.J. Schaefer; A.H. Lichtenstein. Dietary hydrogenated fat increases high-density lipoprotein apoa-i catabolism and decreases low-density lipoprotein apob-100 catabolism in hypercholesterolemic women. *Arterioscler. Thromb. Vasc. Biol.* **2004**, *24*, 1092–1097.

Mattson, F.H.; G.A. Nolen; M.R. Webb. The absorbability by rats of various triglycerides of stearic and oleic acid and the effect of dietary calcium and magnesium. *J. Nutr.* **1979**, *109*, 1682–1687.

Mauger, J.-F.; A.H. Lichtenstein; L.M. Ausman; S.M. Jalbert; M. Jauhiainen; C. Ehnholm; B. Lamarche. Effect of different forms of dietary hydrogenated fats on LDL particle size. *Am. J. Clin. Nutr.* **2003**, *78*, 370–375.

McVeigh, B.L.; B.L. Dillingham; J.W. Lampe; A.M. Duncan. Effect of soy protein varying in isoflavone content on serum lipids in healthy young men. *Am. J. Clin. Nutr.* **2006**, *83*, 244–251.

Mensink, R.P. Effects of products made from a high-palmitic acid, *trans*-free semiliquid fat or a high-oleic acid, low-*trans* semiliquid fat on the serum lipoprotein profile and on C-reactive protein concentrations in humans. *Eur. J. Clin. Nutr.* **2007**, *Apr 18; [Epub ahead of print].*

Mensink, R.P.; M.B. Katan. Effect of dietary *trans* fatty acids on high-density and low-density lipoprotein cholesterol levels in healthy subjects. *N. Engl. J. Med.* **1990**, *323*, 439–445.

Mensink, R.P.; P.L. Zock; A.D.M. Kester; M.B. Katan. Effects of dietary fatty acids and carbohydrates on the ratio of serum total to HDL cholesterol and on serum lipids and apolipoproteins: A meta-analysis of 60 controlled trials. *Am. J. Clin. Nutr.* **2003**, *77*, 1146–1155.

Michalsen, A.; N. Lehmann; C. Pithan; N.T.M. Knoblauch; S. Moebus; F. Kannenberg; L. Binder; T. Budde; G.J. Dobos. Mediterranean diet has no effect on markers of inflammation and metabolic risk factors in patients with coronary artery disease. *Eur. J. Clin. Nutr.* **2005**, *60*, 478–485.

Minchenko, A.; J. Caro. Regulation of endothelin-1 gene expression in human microvascular endothelial cells by hypoxia and cobalt: Role of hypoxia responsive element. *Mol. Cell Biochem.* **2000**, *208*, 53–62.

Morabito, N.; A. Crisafulli; C. Vergara; A. Gaudio; A. Lasco; N. Frisina; R. D'Anna; F. Corrado; M.A. Pizzoleo; M. Cincotta, et al. Effects of genistein and hormone-replacement therapy on bone loss in early postmenopausal women: A randomized double-blind placebo-controlled study. *J. Bone Miner. Res.* **2002**, *17*, 1904–1912.

Morris, M.C. Dietary fats and blood pressure. *J. Cardiovasc. Risk* **1994**, *1*, 21–30.

Morris, M.C.; F. Sacks; B. Rosner. Does fish oil lower blood pressure? A meta-analysis of controlled trials. *Circulation* **1993**, *88*, 523–533.

Moyad, M.A. Soy, disease prevention, and prostate cancer. *Semin. Urol. Oncol.* **1999**, *17*, 97–102.

Mozaffarian, D.; E.B. Rimm. Fish intake, contaminants, and human health: Evaluating the risks and the benefits. *J. Amer. Med. Assoc.* **2006**, *296*, 1885–1899.

Müller, H.; O. Jordal; P. Kierulf; B. Kirkhus; J.I. Pedersen. Replacement of partially hydrogenated soybean oil by palm oil in margarine without unfavorable effects on serum lipoproteins. *Lipids*

1998, *33*, 879–887.

Murkies, A.L.; C. Lombard; B.J.G. Strauss; G. Wilcox; H.G. Burger; M.S. Morton. Dietary flour supplementation decreases postmenopausal hot flushes: Effect of soy and wheat. *Maturitas* **1995**, *21*, 189–195.

Murray, M.J.; W.R. Meyer; B.A. Lessey; R.H. Oi; R.E. DeWire; M.A. Fritz. Soy protein isolate with isoflavones does not prevent estradiol-induced endometrial hyperplasia in postmenopausal women: A pilot trial. *Menopause* **2003**, *10*, 456–464.

Nagata, C.; N. Takatsuka; S. Inaba; N. Kawakami; H. Shimizu. Effect of soymilk consumption on serum estrogen concentrations in premenopausal Japanese women. *J. Natl. Cancer Inst.* **1998**, *90*, 1830–1835.

Nagata, C.; N. Takatsuka; N. Kawakami; H. Shimizu. Soy product intake and hot flashes in Japanese women: Results from a community-based prospective study. *Am. J. Epidemiol.* **2001a**, *153*, 790–793.

Nagata, C.; N. Takatsuka; H. Shimizu; H. Hayashi; T. Akamatsu; K. Murase. Effect of soymilk consumption on serum estrogen and androgen concentrations in Japanese men. *Cancer Epidemiol. Biomarkers Prev.* **2001b**, *10*, 179–184.

Nestel, P.; M. Noakes; B. Belling; R. McArthur; P. Clifton; E. Janus; M. Abbey. Plasma lipoprotein lipid and Lp[a] changes with substitution of elaidic acid for oleic acid in the diet. *J. Lipid Res.* **1992**, *33*, 1029–1036.

Nestel, P.J.; T. Yamashita; T. Sasahara; S. Pomeroy; A. Dart; P. Komesaroff; A. Owen; M. Abbey. Soy isoflavones improve systemic arterial compliance but not plasma lipids in menopausal and perimenopausal women. *Arterioscler. Thromb. Vasc. Biol.* **1997**, *17*, 3392–3398.

Nielsen, I.L.; G. Williamson. Review of the factors affecting bioavailability of soy isoflavones in humans. *Nutr. Cancer* **2007**, *57*, 1–10.

Nikander, E.; A. Kilkkinen; M. Metsa-Heikkila; H. Adlercreutz; P. Pietinen; A. Tiitinen; O. Ylikorkala. A randomized placebo-controlled crossover trial with phytoestrogens in treatment of menopause in breast cancer patients. *Obstet. Gynecol.* **2003**, *101*, 1213–1220.

Nikander, E.; A. Tiitinen; K. Laitinen; M. Tikkanen; O. Ylikorkala. Effects of isolated isoflavonoids on lipids, lipoproteins, insulin sensitivity, and ghrelin in postmenopausal women. *Clin. Endocrinol. Metab.* **2004**, *89*, 3567–3572.

Oh, K.; F.B. Hu; J.E. Manson; M.J. Stampfer; W.C. Willett. Dietary fat intake and risk of coronary heart disease in women: 20 years of follow-up of the nurses' health study. *Am. J. Epidemiol.* **2005**, *161*, 672–679.

Onning, G.; B. Akesson; R. Oste; I. Lundquist. Effects of consumption of oat milk, soya milk, or cow's milk on plasma lipids and antioxidative capacity in healthy subjects. *Ann. Nutr. Metab.* **1998**, *42*, 211–220.

Oomen, C.M.; M.C. Ocke; E.J.M. Feskens; M.-A. J. v. Erp-Baart; F.J. Kok; D. Kromhout. Association between *trans* fatty acid intake and 10-year risk of coronary heart disease in the Zutphen Elderly Study: A prospective population-based study. *The Lancet* **2001**, *357*, 746–751.

Paas, E.; G. Pierce. *Canola Oil*; National Centre for Agrifood Research in Medicine: Winnipeg, Canada, 2002.

Parillo, M.; G. Riccardi. Diet composition and the risk of type 2 diabetes: Epidemiological and clinical evidence. *Br. J. Nutr.* **2004**, *92*, 7–19.

Pawlosky, R.J.; J.R. Hibbeln; J.A. Novotny; N. Salem, Jr. Physiological compartmental analysis of {alpha}-linolenic acid metabolism in adult humans. *J. Lipid Res.* **2001**, *42*, 1257–1265.

Penotti, M.; E. Fabio; A.B. Modena; M. Rinaldi; U. Omodei; P. Vigano. Effect of soy-derived isoflavones on hot flushes, endometrial thickness, and the pulsatility index of the uterine and cerebral arteries. *Fertil. Steril.* **2003**, *79*, 1112–1117.

Petrakis, N.L.; S. Barnes; E.B. King; J. Lowenstein; J. Wiencke; M.M. Lee; R. Miike; M. Kirk; L. Coward. Stimulatory influence of soy protein isolate on breast secretion in pre- and postmenopausal women. *Cancer Epidemiol. Biomarkers Prev.* **1996**, *5*, 785–794.

Pietinen, P.; A. Ascherio; P. Korhonen; A.M. Hartman; W.C. Willett; D. Albanes; J. Virtamo. Intake of fatty acids and risk of coronary heart disease in a cohort of Finnish men: The alpha-tocopherol, beta-carotene cancer prevention study. *Am. J. Epidemiol.* **1997**, *145*, 876–887.

Prentice, R.L.; B. Caan; R.T. Chlebowski; R. Patterson; L.H. Kuller; J.K. Ockene; K.L. Margolis; M.C. Limacher; J.E. Manson; L.M. Parker, et al. Low-fat dietary pattern and risk of invasive breast cancer: the women's health initiative randomized controlled dietary modification trial. *J. Amer. Med. Assoc.* **2006**, *295*, 629–642.

Prentice, R.L.; C.A. Thomson; B. Caan; F.A. Hubbell; G.L. Anderson; S.A.A. Beresford; M. Pettinger; D.S. Lane; L. Lessin; S. Yasmeen, et al. Low-fat dietary pattern and cancer incidence in the women's health initiative dietary modification randomized controlled trial. *J. Natl. Cancer Inst.* **2007**, *99*, 1534–1543.

Quella, S.K. ; C.L. Loprinzi; D.L. Barton; J.A. Knost; J.A. Sloan; B.I. LaVasseur; D. Swan; K.R. Krupp; K.D. Miller; P.J. Novotny. Evaluation of soy phytoestrogens for the treatment of hot flashes in breast cancer survivors: A north central cancer treatment group trial. *J. Clin. Oncol.* **2000**, *18*, 1068–1074.

Rasmussen, O.W.; C. Thomsen; K.W. Hansen; M. Vesterlund; E. Winther; K. Hermansen. Effects on blood pressure, glucose, and lipid levels of a high-monounsaturated fat diet compared with a high-carbohydrate diet in NIDDM subjects. *Diabetes Care* **1993**, *16*, 1565–1571.

Rasmussen, B.M.; B. Vessby; M. Uusitupa; L. Berglund; E. Pedersen; G. Riccardi; A.A. Rivellese; L. Tapsell; K. Hermansen; K. S. G. for The; Effects of dietary saturated, monounsaturated, and n-3 fatty acids on blood pressure in healthy subjects. *Am. J. Clin. Nutr.* **2006**, *83*, 221–226.

Reinwald, S.; C.M. Weaver. Soy isoflavones and bone health: A double-edged sword? *J. Nat. Prod.* **2006**, *69*, 450–459.

Reynolds, K.; A. Chin; K.A. Lees; A. Nguyen; D. Bujnowski; J. He. A meta-analysis of the effect of soy protein supplementation on serum lipids. *Am. J. Cardiol.* **2006**, *98*, 633–640.

Rice, S.; S.A. Whitehead. Phytoestrogens and breast cancer—promoters or protectors? *Endocr. Relat. Cancer* **2006**, *13*, 995–1015.

Rivas, M.; R.P. Garay; J.F. Escanero; P. Cia, Jr.; P. Cia; J.O. Alda. Soy milk lowers blood pressure in men and women with mild to moderate essential hypertension. *J. Nutr.* **2002**, *132*, 1900–1902.

Rowland, I.; M. Faughnan; L. Hoey; K. Wähälä; G. Williamson; A. Cassidy. Bioavailability of phyto-oestrogens. *Br. J. Nutr.* **2003**, *89*, S45–58.

Rowland, I.R.; H. Wiseman; T.A.B. Sanders; H. Adlercreutz; E.A. Bowey. Interindividual variation in metabolism of soy isoflavones and lignans: Influence of habitual diet on equol production by the gut microflora. *Nutr. Cancer* **2000**, *36*, 27–32.

Russo, R.; R. Corosu. The clinical use of a preparation based on phyto-oestrogens in the treatment of menopausal disorders. *Acta Biomed.* **2003**, *74*, 137–143.

Ryan-Borchers, T.A.; J.S. Park; B.P. Chew; M.K. McGuire; L.R. Fournier; K.A. Beerman. Soy isoflavones modulate immune function in healthy postmenopausal women. *Am. J. Clin. Nutr.* **2006**, *83*, 1118–1125.

Sacks, F.M.; M. Katan. Randomized clinical trials on the effects of dietary fat and carbohydrate on plasma lipoproteins and cardiovascular disease. *Am. J. Med.* **2002**, *113*, 13S–24S.

Sacks, F.M.; A. Lichtenstein; L. Van Horn; W. Harris; P. Kris-Etherton; M. Winston; Soy protein, isoflavones, and cardiovascular health: An American Heart Association science advisory for professionals from the nutrition committee. *Circulation* **2006**, *113*, 1034–1044.

Salmeron, J.; F.B. Hu; J.E. Manson; M.J. Stampfer; G.A. Colditz; E.B. Rimm; W.C. Willett. Dietary fat intake and risk of type 2 diabetes in women. *Am. J. Clin. Nutr.* **2001**, *73*, 1019–1026.

Sauer, L.A.; D.E. Blask; R.T. Dauchy. Dietary factors and growth and metabolism in experimental tumors. *J. Nutr. Biochem.* **2007**, *18*, 637–649.

Scambia, G.; D. Mango; P.G. Signorile; R.A. Anselmi Angeli; C. Palena; D. Gallo; E. Bombardelli; P. Morazzoni; A. Riva; S. Mancuso. Clinical effects of a standardized soy extract in postmenopausal women: A pilot study. *Menopause* **2000**, *7*, 105–111.

Schaafsma, G. The protein digestibility-corrected amino acid score. *J. Nutr.* **2000**, *130*, 1865–1867.

Schwab, U.S.; L.K. Niskanen; H.M. Maliranta; M.J. Savolainen; Y.A. Kesaniemi; M.I.J. Uusitupa. Lauric and palmitic acid-enriched diets have minimal impact on serum lipid and lipoprotein concentrations and glucose metabolism in healthy young women. *J. Nutr.* **1995**, *125*, 466–473.

Secreto, G.; L.M. Chiechi; A. Amadori; R. Miceli; E. Venturelli; T. Valerio; E. Marubini. Soy isoflavones and melatonin for the relief of climacteric symptoms: A multicenter, double-blind, randomized study. *Maturitas* **2004**, *47*, 11–20.

Setchell, K.D. Phytoestrogens: The biochemistry, physiology, and implications for human health of soy isoflavones. *Am. J. Clin. Nutr.* **1998**, *68*, 1333S–1346.

Shah, M.; B. Adams-Huet; A. Garg. Effect of high-carbohydrate or high-*cis*-monounsaturated fat diets on blood pressure: A meta-analysis of intervention trials. *Am. J. Clin. Nutr.* **2007**, *85*, 1251–1256.

Simons, L.A.; M. von Konigsmark; J. Simons; D.S. Celermajer. Phytoestrogens do not influence lipoprotein levels or endothelial function in healthy, postmenopausal women. *Am. J. Cardiol.* **2000**, *85*, 1297–1301.

Sites, C.K.; B.C. Cooper; M.J. Toth; A. Gastaldelli; A. Arabshahi; S. Barnes. Effect of a daily supplement of soy protein on body composition and insulin secretion in postmenopausal women. *Fertil. Steril.* **2007**, *88*, 1609–1617.

Squadrito, F.; D. Altavilla; A. Crisafulli; A. Saitta; D. Cucinotta; N. Morabito; R. D'Anna; F. Corrado; P. Ruggeri; N. Frisina, et al. Effect of genistein on endothelial function in postmenopausal women: A randomized, double-blind, controlled study. *Am. J. Med.* **2003**, *114*, 470–476.

St Germain, A.; C.T. Peterson; J.G. Robinson; D.L. Alekel. Isoflavone-rich or isoflavone-poor soy protein does not reduce menopausal symptoms during 24 weeks of treatment. *Menopause* **2001**, *8*, 17–26.

Stamler, J.; A. Caggiula; G.A. Grandits; M. Kjelsberg; J.A. Cutler. Relationship to blood pressure of combinations of dietary macronutrients: findings of the multiple risk factor intervention trial (MRFIT). *Circulation* **1996,** *94,* 2417–2423.

Steinberg, F.M.; N.L. Guthrie; A.C. Villablanca; K. Kumar; M.J. Murray. Soy protein with isoflavones has favorable effects on endothelial function that are independent of lipid and antioxidant effects in healthy postmenopausal women. *Am. J. Clin. Nutr.* **2003**, *78*, 123–130.

Sundram, K.; A. Ismail; K.C. Hayes; R. Jeyamalar; R. Pathmanathan. *Trans* (elaidic) fatty acids adversely affect the lipoprotein profile relative to specific saturated fatty acids in humans. *J. Nutr.* **1997**, *127*, 514S–520S.

Taku, K.; K. Umegaki; Y. Sato; Y. Taki; K. Endoh; S. Watanabe. Soy isoflavones lower serum total and LDL cholesterol in humans: A meta-analysis of 11 randomized controlled trials. *Am. J. Clin. Nutr.* **2007**, *85*, 1148–1156.

Teede, H.J.; F.S. Dalais; D. Kotsopoulos; Y.L. Liang; S. Davis; B.P. McGrath. Dietary soy has both beneficial and potentially adverse cardiovascular effects: A placebo-controlled study in men and postmenopausal women. *J. Clin. Endocrinol. Metab.* **2001**, *86*, 3053–3060.

Teede, H.J.; F.S. Dalais; B.P. McGrath. Dietary soy containing phytoestrogens does not have detectable estrogenic effects on hepatic protein synthesis in postmenopausal women. *Am. J. Clin. Nutr.* **2004**, *79*, 396–401.

Teede, H.J.; D. Giannopoulos; F.S. Dalais; J. Hodgson; B.P. McGrath. Randomised, controlled, cross-over trial of soy protein with isoflavones on blood pressure and arterial function in hypertensive subjects. *J. Am. Coll. Nutr.* **2006**, *25*, 533–540.

Tham, D.M.; C.D. Gardner; W.L. Haskell. Potential health benefits of dietary phytoestrogens: A review of the clinical, epidemiological, and mechanistic evidence. *J. Clin. Endocrinol. Metab.* **1998**, *83*, 2223–2235.

The American Soybean Association. *Soy Stats 2007.* http://www.soystats.com/2007/Default-frames.htm (accessed 7 October 2007).

The North American Menopause Society. The role of isoflavones in menopausal health: Consensus opinion of The North American Menopause Society. *Menopause* **2000**, *7*, 215–229.

Theodoratou, E.; G. McNeill; R. Cetnarskyj; S.M. Farrington; A. Tenesa; R. Barnetson; M. Porteous; M. Dunlop; H. Campbell. Dietary fatty acids and colorectal cancer: A case-control study. *Am. J. Epidemiol.* **2007**, *166*, 181–195.

Tilley, S.L.; T.M. Coffman; B.H. Koller. Mixed messages: Modulation of inflammation and immune responses by prostaglandins and thromboxanes. *J. Clin. Invest.* **2001**, *108*, 15–23.

Tormala, R.; S. Appt; T.B. Clarkson; P.H. Groop; M. Ronnback; O. Ylikorkala; T.S. Mikkola. Equol production capability is associated with favorable vascular function in postmenopausal women using tibolone; no effect with soy supplementation. *Atherosclerosis* **2007a**, *Oct 23; [Epub ahead of print].*

Tormala, R.M.; S. Appt; T.B. Clarkson; M.J. Tikkanen; O. Ylikorkala; T.S. Mikkola. Individual differences in equol production capability modulate blood pressure in tibolone-treated postmenopausal women: Lack of effect of soy supplementation. *Climacteric* **2007b**, *10*, 471–479.

Trifiletti, A.; R. Scamardi; A. Gaudio; A. Lasco; N. Frisina. Hemostatic effects of diets containing olive or soy oil in hypertensive patients. *J. Thromb. Haemost.* **2005**, *3*, 179–180.

Turner, N.J.; B.M. Thomson; I.C. Shaw. Bioactive isoflavones in functional foods: The importance of gut microflora on bioavailability. *Nutr. Rev.* **2003**, *61*, 204–213.

Turpeinen, O.; M.J.M. Pekkarinen; M. Miettinen; R. Elosuo; E. Paavilainen. Dietary prevention of coronary heart disease: The Finnish mental hospital study. *Int. J. Epidemiol.* **1979**, *8*, 99–118.

U.S. Department of Health and Human Services and the U.S. Department of Agriculture. *Dietary Guidelines for Americans. 6th Edition*; U.S. Government Printing Office: Washington, DC, January 2005.

Upmalis, D.H.; R. Lobo; L. Bradley; M. Warren; F.L. Cone; C.A. Lamia. Vasomotor symptom relief by soy isoflavone extract tablets in postmenopausal women: A multicenter, double-blind, randomized, placebo-controlled study. *Menopause* **2000**, *7*, 236–242.

Van Patten, C.L.; I.A. Olivotto; G.K. Chambers; K.A. Gelmon; T.G. Hislop; E. Templeton; A. Wattie; J.C. Prior. Effect of soy phytoestrogens on hot flashes in postmenopausal women with breast cancer: A randomized, controlled clinical trial. *J. Clin. Oncol.* **2002**, *20*, 1449–1455.

van Tol, A.; P.L. Zock; T. van Gent; L.M. Scheek; M.B. Katan. Dietary *trans* fatty acids increase serum cholesterylester transfer protein activity in man. *Atherosclerosis* **1995**, *115*, 129–134.

Vatanparast, H.; P.D. Chilibeck. Does the effect of soy phytoestrogens on bone in postmenopausal women depend on the equol-producing phenotype? *Nutr. Rev.* **2007**, *65*, 294–299.

Vega-Lopez, S.; L.M. Ausman; S.M. Jalbert; A.T. Erkkila; A.H. Lichtenstein. Palm and partially hydrogenated soybean oils adversely alter lipoprotein profiles compared with soybean and canola oils in moderately hyperlipidemic subjects. *Am. J. Clin. Nutr.* **2006**, *84*, 54–62.

Vessby, B.; M. Uusitupa; K. Hermansen; G. Riccardi; Rivellese, A.A.; L.C. Tapsell; C. Nälsén; L. Berglund; A. Louheranta; B.M. Rasmussen, et al. Substituting dietary saturated for monounsaturated fat impairs insulin sensitivity in healthy men and women: The KANWU study. *Diabetologia* **2001**, *44*, 312–319.

Walker, H.A.; T.S. Dean; T.A. Sanders; G. Jackson; Ritter, J.M.; P.J. Chowienczyk. The phytoestrogen genistein produces acute nitric oxide-dependent dilation of human forearm vasculature with similar potency to 17beta-estradiol. *Circulation* **2001**, *103*, 258–262.

Washburn, S.; G.L. Burke; T. Morgan; M. Anthony. Effect of soy protein supplementation on serum lipoproteins, blood pressure, and menopausal symptoms in perimenopausal women. *Menopause* **1999**, *6*, 7–13.

Welty, F.K.; K.S. Lee; N.S. Lew; J.-R. Zhou. Effect of soy nuts on blood pressure and lipid levels in hypertensive, prehypertensive, and normotensive postmenopausal women. *Arch. Intern. Med.* **2007**, *167*, 1060–1067.

Williamson-Hughes, P.S.; B.D. Flickinger; M.J. Messina; M.W. Empie. Isoflavone supplements containing predominantly genistein reduce hot flash symptoms: A critical review of published studies. *Menopause* **2006**, *13*, 831–839.

Wu, J.; J. Oka; J. Ezaki; T. Ohtomo; T. Ueno; S. Uchiyama; T. Toda; M. Uehara; Y. Ishimi. Possible role of equol status in the effects of isoflavone on bone and fat mass in postmenopausal Japanese women: A double-blind, randomized, controlled trial. *Menopause* **2007** ***Vol.*** *14*, 866–874.

Wu, J.; J. Oka; M. Higuchi; I. Tabata; T. Toda; M. Fujioka; N. Fuku; T. Teramoto; T. Okuhira; T. Ueno, et al. Cooperative effects of isoflavones and exercise on bone and lipid metabolism in postmenopausal Japanese women: A randomized placebo-controlled trial. *Metabolism* **2006**, *55*, 423–433.

Wu, A.H.; M.C. Pike; D.O. Stram. Meta-analysis: dietary fat intake, serum estrogen levels, and the risk of breast cancer. *J. Natl. Cancer Inst.* **1999**, *91*, 529–534.

Wu, A.H.; F.Z. Stanczyk; S. Hendrich; P.A. Murphy; C. Zhang; P. Wan; M.C. Pike. Effects of soy foods on ovarian function in premenopausal women. *Br. J. Cancer* **2000**, *82*, 1879–1886.

Wu, A.H.; R.G. Ziegler; A.M. Nomura; D.W. West; Kolonel, L.N.; P.L. Horn-Ross; R.N. Hoover; M.C. Pike. Soy intake and risk of breast cancer in Asians and Asian Americans. *Am. J. Clin. Nutr.* **1998**, *68*, 1437S–1443.

Xu, X.; A.M. Duncan; K.E. Wangen; M.S. Kurzer. Soy consumption alters endogenous estrogen metabolism in postmenopausal women. *Cancer Epidemiol. Biomarkers Prev.* **2000**, *9*, 781–786.

Yamashita, T.; T. Sasahara; S.E. Pomeroy; G. Collier; P.J. Nestel. Arterial compliance, blood pressure, plasma leptin, and plasma lipids in women are improved with weight reduction equally with a meat-based diet and a plant-based diet. *Metabolism* **1998**, *47*, 1308–1314.

Yaqoob, P.; J.A. Knapper; D.H. Webb; C.M. Williams; E.A. Newsholme; P.C. Calder. Effect of olive oil on immune function in middle-aged men. *Am. J. Clin. Nutr.* **1998**, *67*, 129–135.

Yildirir, A.; S.L. Tokgozoglu; T. Oduncu; A. Oto; I. Haznedaroglu; D. Akinci; G. Koksal; E. Sade; S. Kirazli; S. Kes. Soy protein diet significantly improves endothelial function and lipid parameters. *Clin. Cardiol.* **2001**, *24*, 711–716.

Yildiz, M.F.; S. Kumru; A. Godekmerdan; S. Kutlu. Effects of raloxifene, hormone therapy, and soy isoflavone on serum high-sensitive C-reactive protein in postmenopausal women. *Int. J. Gynaecol. Obstet.* **2005**, *90*, 128–133.

Yuan, J.P.; J.H. Wang; X. Liu. Metabolism of dietary soy isoflavones to equol by human intestinal microflora —Implications for health. *Mol. Nutr. Food Res.* **2007**, *51*, 765–781.

Zock, P.L.; M.B. Katan. Hydrogenation alternatives: effects of *trans* fatty acids and stearic acid versus linoleic acid on serum lipids and lipoproteins in humans. *J. Lipid Res.* **1992**, *33*, 399–410.

Zock, P.L.; M.B. Katan. Butter, margarine and serum lipoproteins. *Atherosclerosis* **1997**, *131*, 7–16.

21

Soybean Production and Processing in Brazil

Peter D. Goldsmith
National Soybean Research Laboratory, University of Illinois, Urbana-Champaign, IL 61801

Introduction

The success of soybean production and utilization in Brazil actually begins with the development of the poultry sector during the 1950s in the southern United States (Kiel, 2005). Researchers in the United States sought to adapt soybeans to lower latitudes in order to provide southern poultry farmers with a local high-quality protein meal. Researchers quickly developed varieties adapted to the longer growing season and warmer climates by focusing on the role of the nighttime photoperiod in soybeans' growth and development (Kiel, 2005).

These new varieties became the opening for the Brazilians. Researchers took the low-latitude technology and developed germ plasm that could be deployed in the Southern three states of Brazil (Rio Grande do Sul, Santa Catarina, and Parana) with a growing climate similar to the Southern United States (Schnepf et al., 2001). Brazil's soybean industry began in the South of the country in the late 1960s, supporting both soybean processing and poultry production.

By the 1980s, the federal agricultural research institute [Empresa Brasileira de Pesquisa Agropecuária (EMBRAPA)] had advanced the photoperiod line of research even further. EMBRAPA successfully adapted soybeans to grow in the tropics at even lower latitudes. Developing this technology opened up soybean production to the West and North regions of the country that lie between 15 degrees south latitude and 5 degrees north latitude. Of greatest potential was the Cerrado region encompassing over 200 million hectares (the equivalent of the combined land areas of the 12 Midwestern U.S. states stretching from Ohio to North Dakota) of low brush-like forest that was easy to clear and had predictable rainfall. The development of the lowest-latitude varieties begins the real story of the Brazilian soybean complex.

Compared to the South region of Brazil, Cerrado farming could take advantage of huge economies of scale. U.S. agricultural development and land privatization began before the age of mechanization. The U.S. Midwest was settled using the concept of a section, where 80 A (32.4 ha) was sufficient to support a homesteading family.

Brazil's Cerrado region has none of that social, political, or normative legacy as to what is an appropriate unit of production. The rapid expansion of soybean production in the 1980s arose because of the availability of large tracts of arable land, soybean technology that produced yields equal to those of the United States, mechanization that allowed operational efficiency, and the lowest operating costs per hectare in the world. Cerrado farming also has great challenges. The production, transportation, and processing infrastructure is underdeveloped; markets are distant; soils are relatively poor; and environmental concerns exist.

Brazil did not become a significant player in the world soybean scene until the mid 1970s when low-latitude varieties were commercialized, production topped 10 MMT (11 million t), and 10% of the world's product (Fig. 21.1). In the 30 yr since, Brazil has expanded its soybean production fivefold. South America—led by Brazil, Argentina, Paraguay, and Bolivia— as a region, recently surpassed the United States' output, and it now produces 48% of the world's needs (Fig. 21.2).

The United States still holds the most soybean-processing capacity, followed by China and Brazil (Fig. 21.3). Following the expansion of soybean production outside the United States though, a fundamental shift occurred in soybean-processing investment, away from the United States and Europe toward China, Argentina, and Brazil (Fig. 21.4). Capital for soybean processing is increasingly invested outside the United States because of superior procurement economies, lower costs of plant operation, and close proximity to high-growth livestock industries (Goldsmith et al., 2004). Soybean meal and oil demand growth is most active outside the United States, so many times foreign crush facilities are better able to supply these new customers. For example, two of the fastest growing poultry and pork sectors are in Brazil and China, which are able to utilize their domestically produced meal (Fig. 21.5).

Soybean Industry in Brazil

Production and Yield

Brazil produced 51 MMT (56 million t) of soybeans on 23 million ha (57 million A) in 2005 (Fig. 21.6). Since 1990, the size of Brazil's soybean crop increased 10.5% per year. The value of the national crop is $14B and has more than doubled over the last five years (Fig. 21.7). Yields over that period were flat, but significant expansion of soybean acreage combined with increasing prices is behind the growth in the industry.

The leading states producing soybeans are located in the Southeast and the Center West regions of the country (Fig. 21.8). Mato Grosso, in the Center–West part of the country, produces almost 16 MMT (17.6 million t), about 70% greater than the number-two state, Parana, and double the number-three state, Rio Grande Do Sul.

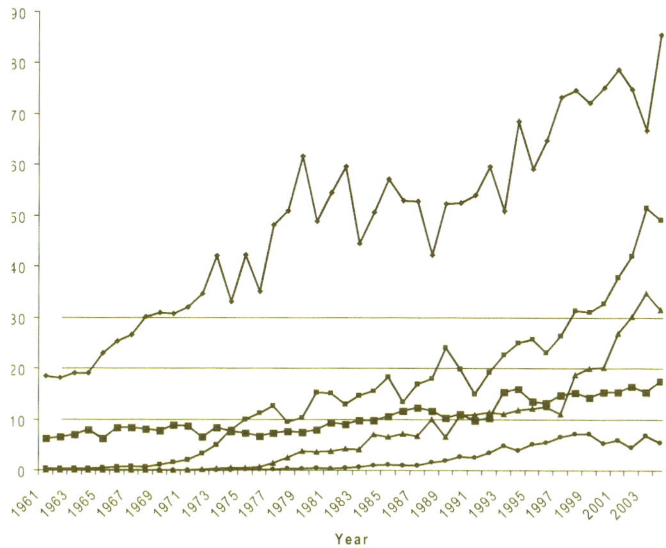

Fig. 21.1. World soybean production (Source: FAO, 2005; author's calculations).

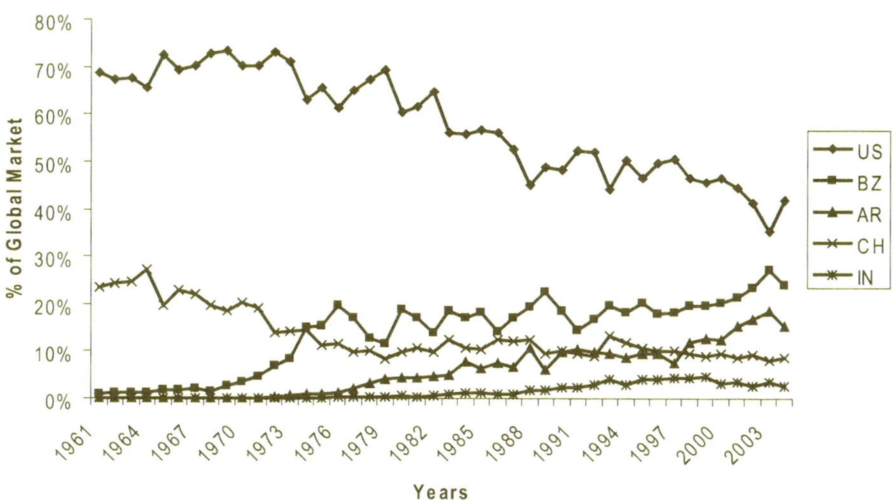

Fig. 21.2. Global soybean market shares (Source: FAO, 2005; author's calculations).

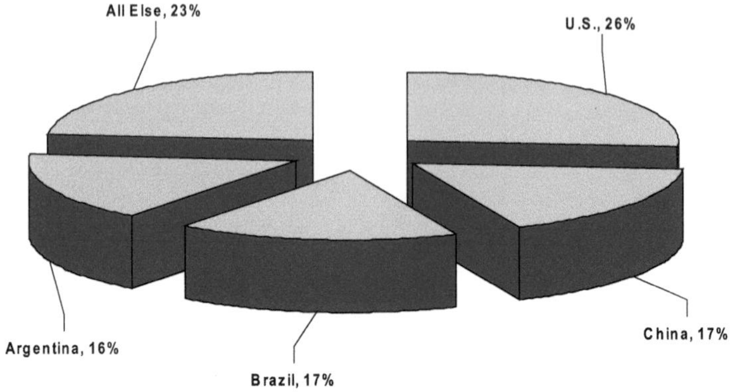

Fig. 21.3. World soybean crushing capacity shares in 2005 (Source: FAO, 2007; author's calculations).

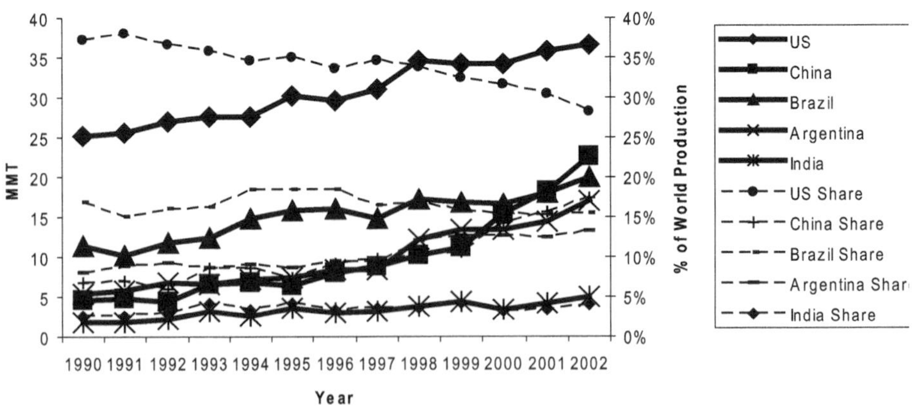

Fig. 21.4. Leading soybean meal producers (Source: FAO, 2005; author's calculations).

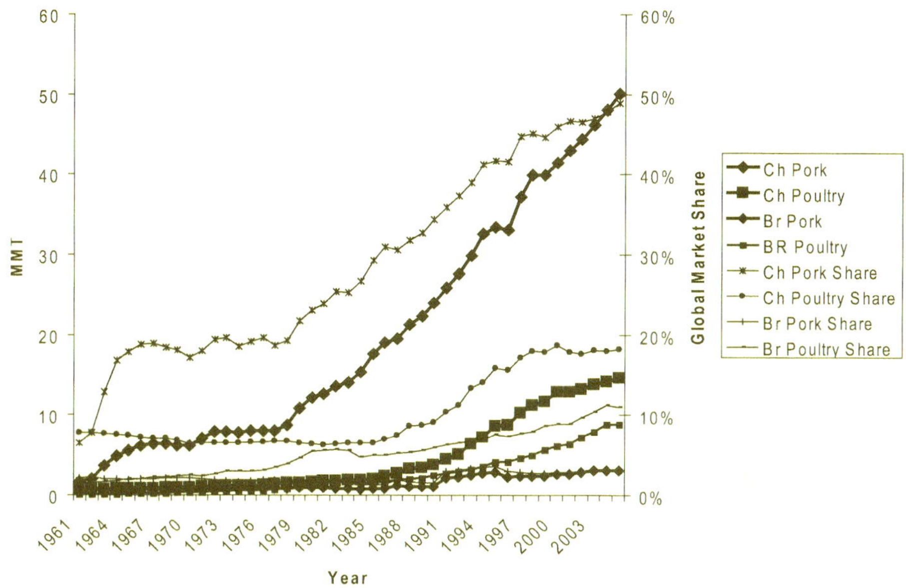

Fig. 21.5. Pork and poultry production in China and Brazil (Source: FAO, 2005; author's calculations).

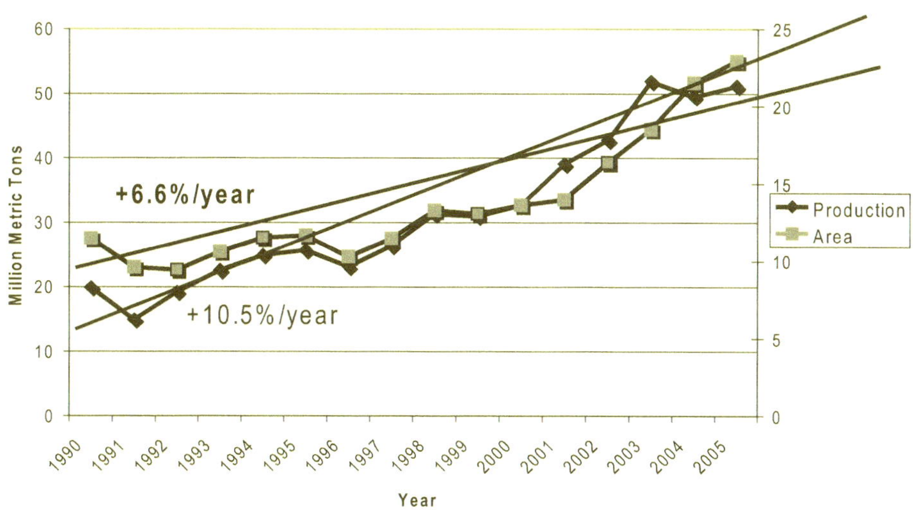

Fig. 21.6. Brazil soybean production and A harvested (1990–2005) (Source: FAO, 2007; author's calculations).

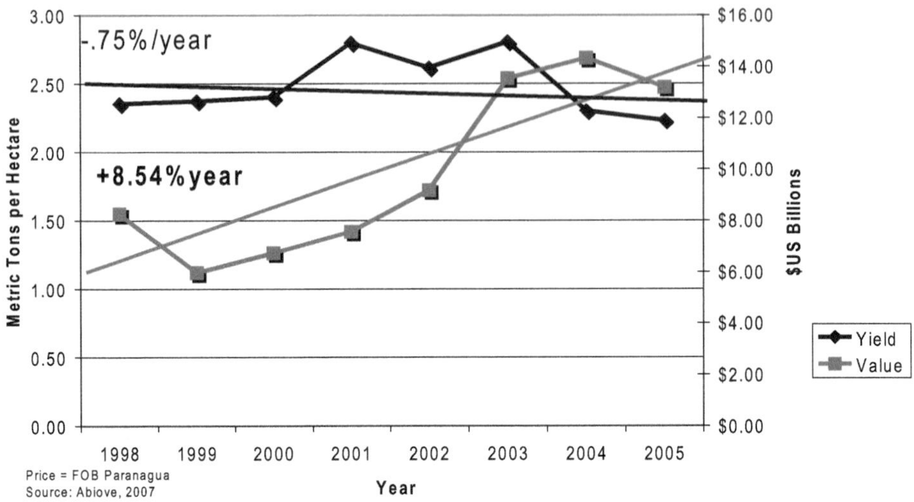

Fig. 21.7. Brazil soybean yield and value of the national crop (1998–2005).

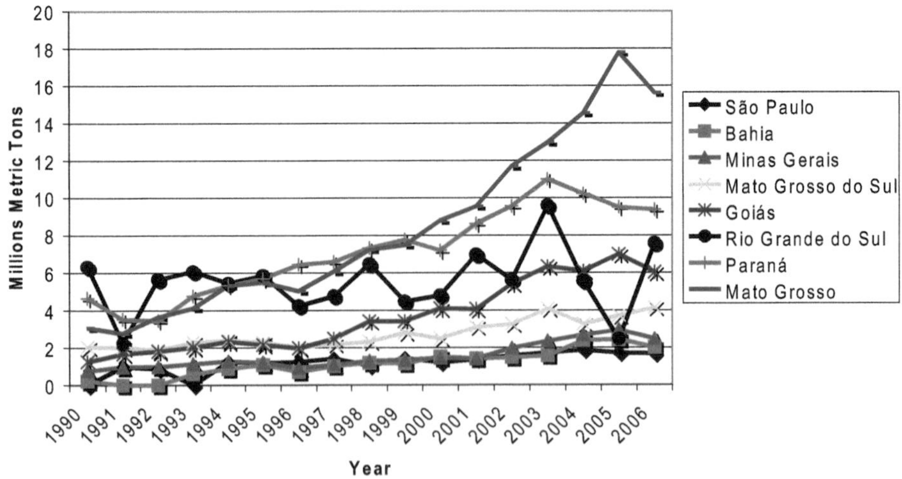

Fig. 21.8. Brazilian soybean production by state (1990–2006) (Source: IBGE, 2007).

Prices

The prices received by farmers in Brazil vary considerably across the country. Prices nationally averaged $5.39/bu over the 2003–2006 period, $1.10 or 20% less than the Chicago Cash price for the same period. The price in Ponta Grossa, Parana, located in a major soybean region in Southeastern Brazil had an average price of $6.14/ bu, only $0.34/bu (5%) less than the average Chicago Cash price (Fig. 21.9). On the other hand, Sorriso, Mato Grosso, in Central Mato Grosso had an average price of $4.76/ bu, $1.72/bu (27%) less than the Chicago Cash price.

Late in the year, Eastern Brazil shows a positive or neutral basis compared to the Chicago Cash price. These months correspond to the harvest period in the Northern Hemisphere, thus driving U.S. prices down combined with binding storage constraints that create late-season shortages in Brazil (Fig. 21.10).

Soybean prices are seasonal. The highest average daily prices (e.g., Parana) occur in November at $6.33/bu, as new crop supplies of soybeans are exhausted six months after harvest. The lowest average monthly prices ($5.71/bu) are seen in January as U.S. selling drives down world prices at the beginning of the tax year.

Prices fall, moving from the East to the West in Brazil. The difference between the Center–West with the coastal regions widened over the period to over $2.00/bu (32%) by late 2006. The interior price in central Mato Grosso was on average about 23% or $1.38/bu lower than in the coastal state of Parana (Fig. 21.11).

The price disparities are due to two factors: the decreasing quality of the infrastructure and a lack of local agro-industrial activity as one moves west. The agro-industrial complex is much larger in the historically more highly populated and developed Eastern states of Brazil. For example, 64% of all the soybeans produced in the state of Mato Grosso are exported internationally (51%) or domestically (13%) as whole grain (Goldsmith et al., 2007). Of the remaining soybeans, 34% are converted into meal and oil, of which 95% is sold outside of the state. So, in summary, 96% of the soybeans are not converted in-state to higher valued goods, such as meat, food, or energy, but are exported.

The soybean cluster in Mato Grosso was estimated in 2004 to be about $8 billion (Goldsmith et al., 2008). For example, Illinois, a U.S. state with a similar size soybean crop, has a soybean cluster over three times as large, at $25 billion. Soybean production comprises about 11% of the Illinois cluster that incorporates processing and meat production, while soybean production in Mato Grosso comprises close to 60%.

Cost of Production

Soybean costs of production are about 38% lower in the high-growth regions of the Center-West of Brazil compared to the Midwest United States (Hirsch, 2004). Fixed costs per acre in the Center-West are about one-fifth the costs in the Midwest in the

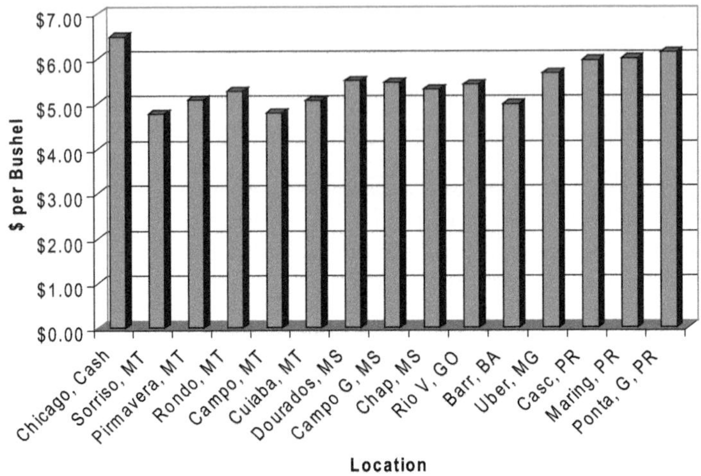

Fig. 21.9. Selected regional average soybean prices in Brazil (2003–2006) (Source: IBGE, 2007).

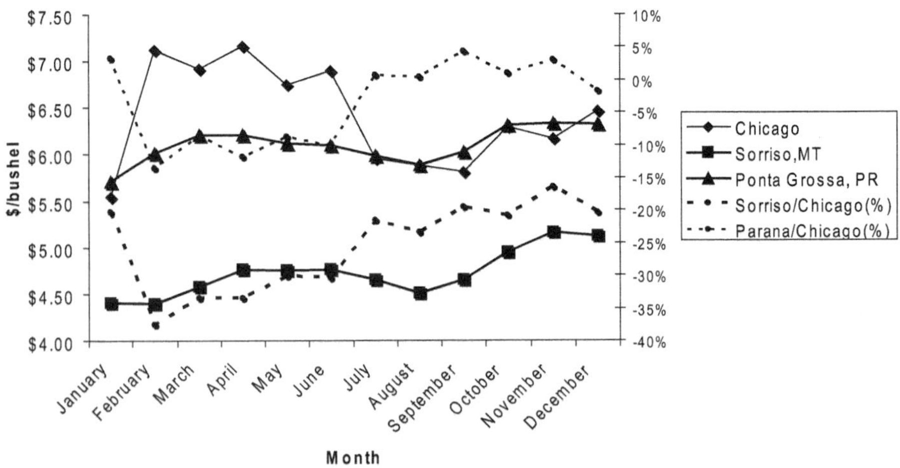

Fig. 21.10. Monthly average prices in Chicago, Mato Grosso, and Parana (2003–2006) (Source: Barchart.com, 2007; IBGE, 2007; author's calculations).

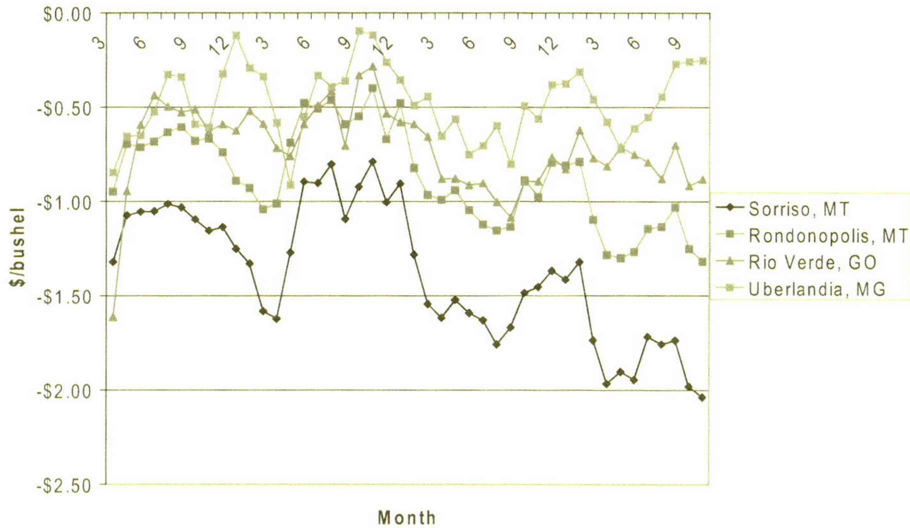

Fig. 21.11. Local soybean price differences ($) with Ponta Grossa and Parana (2003–2006) (Source: IBGE, 2007; author's calculations).

United States due to differences in land prices (Fig. 21.12). Operating costs and ocean freight (FOB Rotterdam) are quite comparable. Internal freight costs to the port are almost three times greater from the Center-West, though the distances are comparable. Transport from the interior of Brazil involves significant usage of trucks over a very poor highway system. The United States relies much more heavily on rail and water transport, which are much less expensive per kilometer per MT.

Soybean operating costs of production in the Center-West region of Brazil rose 17% per year between 2000 and 2006 while gross revenue rose only 4%/yr (Table 21.1). Soybean operating costs averaged $141/A over the 2000–2006 period. Operating cost variability was exceptionally high as costs have ranged from a low of $89/A in 2000 to a high of $202/A in 2004. Much of the cost increase was attributed to rising fertilizer (+$38), fungicide (+$24), and insecticide (+$19) costs per acre. In 2005 and 2006, gross margins approached zero as costs of production outpaced soybean price increases, and yields faltered due to soybean rust. Fertilizer costs average 34% of the costs of production and are the single largest cost item (Fig. 21.13). Fungicide costs quintupled as farmers were forced to combat the devastating disease Asian Rust. For example in 2004, fungicide costs per acre increased $20/A over the previous year while gross margins were only $13/A. During the same period, insecticides costs increased 475%. The increase in costs caused tremendous financial stress in the region. Debt repayment became difficult for highly leveraged producers who had little cash flow to use toward principal and interest payments.

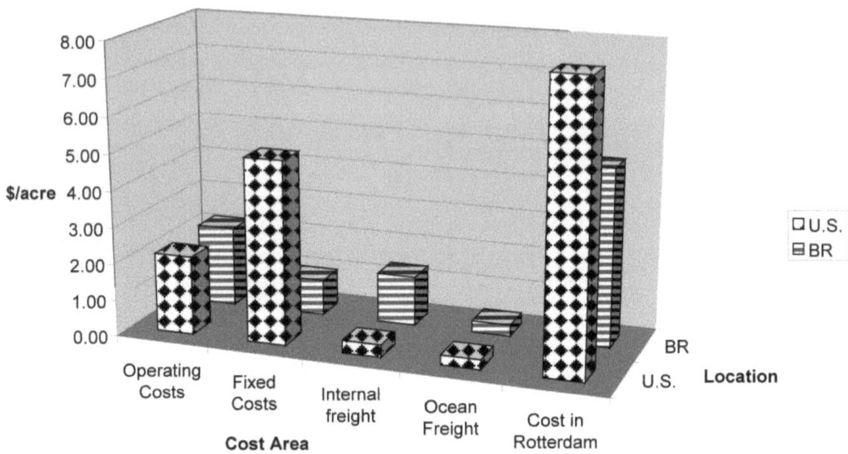

Fig. 21.12. Cost of soybean production comparison for Center-West Brazil with the Midwest in the United States (Source: adapted from Goldsmith & Hirsh, 2006).

Costs of Addressing Asian Soybean Rust

Asian Soybean Rust became an economic threat to the Brazilian soybean crop in 2002. Fungicide applications rose from less than $5.00/A (<1% of operating costs in 2001) to over $40/A (>15% of operating costs) by 2005 (Fig. 21.14). Costs rose for three reasons: (i) the disease spread and affected more regions; (ii) the Real strengthened and as a result increased the cost of the base products; and (iii) the intensity of the disease increased, causing farmers to spray multiple times (Fig. 21.15).

Since 2005, costs per acre have fallen as producers learned to manage rust more effectively, and a ban was imposed on second-crop soybeans. Eliminating the second crop or a mid-year crop for seed significantly reduced the quantity of host material for the fungus to reside. This, in part, broke the cycle of infection and re-infection. Starting in crop year 2009–2010, rust-resistant soybean varieties will be commercially available (Hirimoto, 2007). This will give farmers another tool to treat this devastating disease.

Costs of Transportation

The most limiting factor affecting agro-industrial development in the Center-West region of Brazil is the lack of transportation infrastructure (Hirsch, 2004). For example, the state of Mato Grosso comprises a land area almost 30% larger than the U.S. state of Texas and is ~1,600 km (1,000 miles) from an ocean port. It is Brazil's leading agricultural (soybean) state, but it has no expressways, no commercial waterways, and

Table 21.1. Costs of Production for Central-West Brazil (2000–2006)[a]

	2000	2001	2002	2003	2004	2005	2006	Average	Annual Change
Exchange Rate	2.38	2.97	3.12	2.93	2.43	2.18	1.87		
Fertilizer	$34.51	$50.40	$36.98	$56.34	$72.20	$64.31	$67.50	$54.60	15.94%
Fungicides	$3.12	$2.83	$3.23	$4.69	$26.70	$26.02	$19.42	$12.29	86.99%
Herbicides	$21.20	$13.58	$20.22	$23.58	$30.35	$28.16	$21.41	$22.64	0.17%
Insecticides	$2.94	$2.82	$3.27	$4.76	$22.27	$19.74	$16.95	$10.39	79.25%
Seeds	$4.96	$5.10	$7.29	$13.30	$18.71	$16.89	$9.26	$10.79	14.44%
Other Costs	$22.44	$19.39	$23.26	$31.47	$31.49	$35.54	$45.69	$29.90	17.27%
Effective Operational Costs	$89.17	$94.13	$94.25	$134.14	$201.72	$190.64	$180.24	$140.61	17.02%
Assumed Yield *	47.64	47.64	43.31	43.31	39.84	45.04		44.46	1.09%
Sale Price**	$3.25	$2.86	$4.81	$7.11	$5.40	$4.11		4.59	5.34%
Gross Revenue	$154.67	$136.48	$208.33	$307.84	$215.16	$185.28		201.29	3.96%
Return over Variable Costs	$65.50	$42.35	$114.08	$173.70	$13.44	-$5.36		67.28	-21.64%

[a]All prices in U.S. Dollars. All land units in A. All quantities in bushels. Source: EMBRAPA and author's calculations.

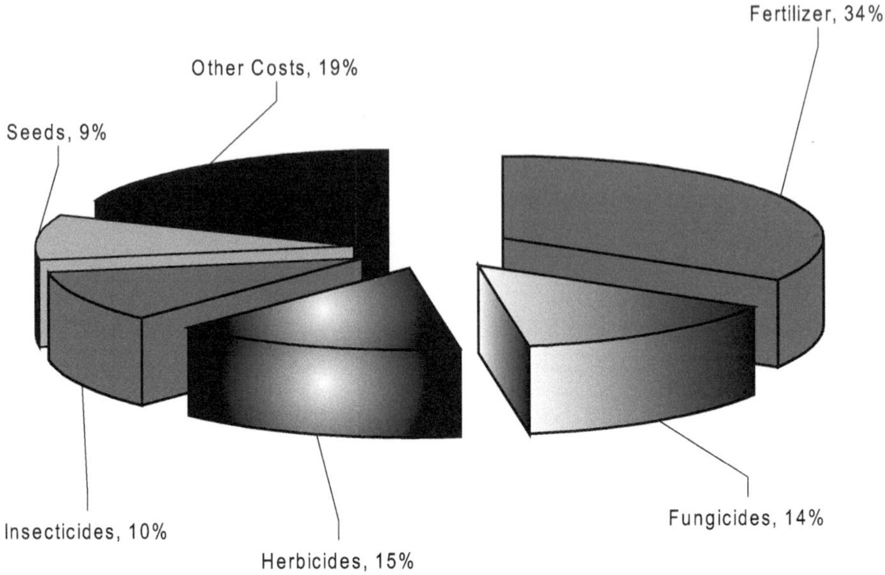

Fig. 21.13. Soybean operating costs in the Center-West Brazil in 2006 (Source: EMPRAPA and author's calculations).

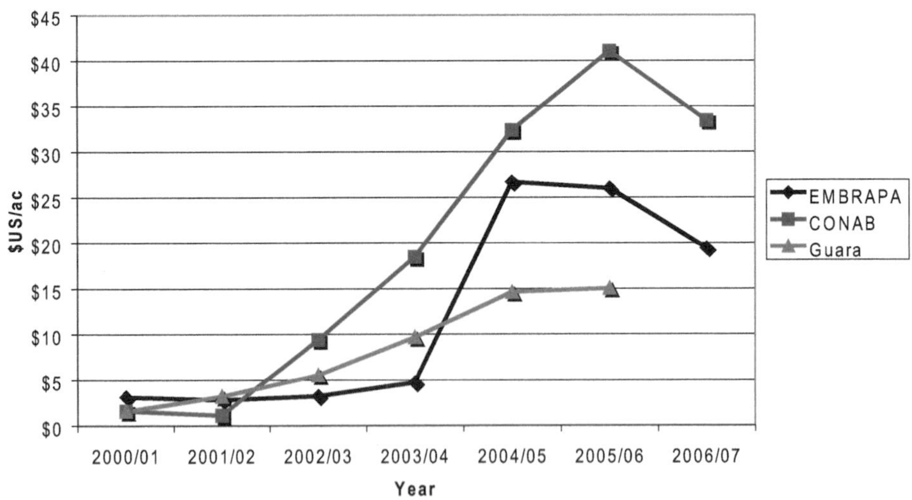

Fig. 21.14. Costs associated with Asian Soybean Rust in Brazil (2000–2006) in Mato Grosso, Brazil (Source: Ma, 2007).

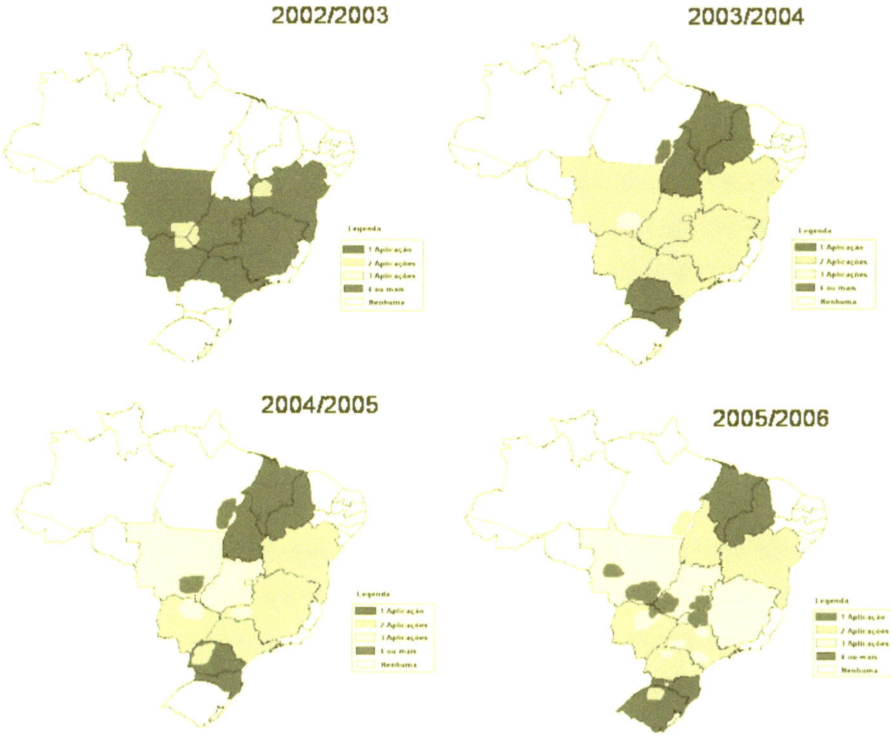

Fig. 21.15. Progression of rust impacts in Brazil (2002–2006) (Source: Ma, 2006).

<160 km (100 miles) of rail.

As a result, transportation costs in Mato Grosso are equal to 71% of the price paid to local farmers, and 41% of the landed price in Europe (Table 21.2). Freight costs add $1.58/bu in the more traditional eastern areas of soybean production as compared to $3.36/bu for soybeans sourced from the Center-West region.

The cost of the weak transportation system is borne by the producers in the prices they receive. Most of the difference between the prices received by farmers in the East compared to the farmers in the Center-West is attributed to the high freight costs. The higher costs are not just a function of distance. Goias, a state to the east of Mato Grosso, has superior infrastructure with access to both rail and water transport. Its cost per km is 22% lower than in Mato Grosso because more expensive truck transport is not as prevalent.

About 55% of Brazil's soybeans move out from two ports in the Southeast, Paranagua and Santos (SECEX, 2007). Both ports receive truck shipments, but Santos is connected directly by rail to Southeast Mato Grosso. The next most active area is in the Northeast, from the ports of Vitoria (13%), Sao Francisco (12%), and Sao Luiz

Table 21.2. Cost of Transportation in Brazil (Source: IBGE, 2007; USDA, 2006)

Soybean Source			$6.14	$5.39	$4.76
(Distance to the Port)			% of Landed Price (% of Soybean Price)		
Parana	Domestic Freight	$0.53	7% (9%)	8% (10%)	
(204 km)	Ocean Freight	$1.05	14% (17%)	15% (19%)	
	Total Freight	$1.58	20% (26%)	23% (29%)	
	Landed Price**		$7.72	$6.97	
Goias	Domestic Freight	$1.16	14% (19%)	15% (21%)	
(726 km)	Ocean Freight	$1.08	13% (18%)	14% (20%	
	Total Freight	$2.24	27% (36%)	29% (41%)	
	Landed Price**		$8.38	$7.63	
Mato Grosso	Domestic Freight	$2.31		26% (43%)	28% (49%)
(1190 km)	Ocean Freight	$1.05		12% (19%)	13% (22%)
	Total Freight	$3.36		38% (62%)	41% (71%)
	Landed Price**			$8.75	$8.12

* $U.S./bushel
** in Hamburg, Germany

(8%). The Amazon port of Manaus mostly depends on barge service via the river Madeira and accounts for 7% of soybean exports grown primarily in the State of Rondonia and western Mato Grosso.

Soybean Processing

Soybean processing or *crush* involves purchasing and processing raw soybeans into the principal products of soybean meal, used for animal feed, and soybean oil, used for human consumption or biodiesel. The crush margin is the difference between the cost of the soybeans and the value of the meal and oil. So, much of the value derived from the processing of soybeans is in the form of high-protein soybean meal, making soybeans principally a protein crop, not an oil crop. A *rule of thumb* holds that 78–80% of the soybean results in meal, 18–20% in oil, and the rest in the form of a low-value high-fiber mill feed.

Recently, the price of oil began to rise because of a strong demand for food oil in Asia and biodiesel feedstocks around the world (Fig. 21.16). Simultaneously, soybean meal faces increasing competition in the United States from dried distillers grains and solubles: a medium protein coproduct from the corn ethanol industry.

The average monthly ratio of the price of soybean oil to soybean meal in Brazil began to rise from a low of 1.5:1 in 2000 (Fig. 21.17). The ratio has averaged about 2.5:1 since 1998. If the ratio were to move above four, because oil was becoming more valuable and soy protein less, then the value of the oil from soybean processing would

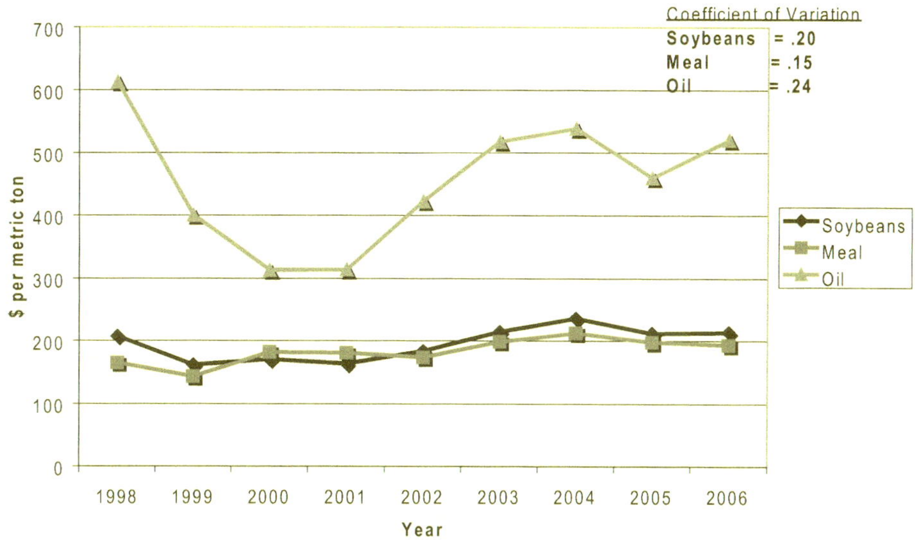

Fig. 21.16. Annual average prices (F.O.B) in Paranaqua, Brazil (1998–2006) (Source: IBGE, 2007).

surpass that of the meal. The shift in soybean's value from protein to oil, though, is unlikely in the long run because if oil were to become valuable, processors would seek out higher oil-yielding feedstocks, and farmers, in turn, would begin to switch to higher oil-yielding crops, such as canola and sunflower.

Month-to-month variability in prices and the level of correlation among prices are primary sources of uncertainty for processors. Soybean oil prices are 60% more variable from month to month than soybean meal, and 20% more variable than soybeans. Oil, though, impacts crush margin less than meal or soybeans because it is a small component of processing output.

Brazil's Soybean Crushing Plants

Brazil produces 17% of the world's soybean meal and oil (FAO, 2007). Currently, 96 plants operate in the country, representing 47 firms (Hinrichsen, 2006). The plants have the capacity to produce 141,000 MT/day (155,000 t/day) or 42 MMT/yr (46 million t/yr). The annual capacity based on 300 days of operation is about double the 22 MMT (24 million t) of meal produced in 2006 (ABIOVE, 2007). Thus, it appears that Brazil is over capacity.

The state of Parana has 21 crushing plants, representing 15 companies. It is the leading processing state in Brazil with a capacity to produce 28,700 MT/day (31,600 t/day) of soybean meal or 20% of the nation's output (Fig. 21.18). Mato Grosso is a

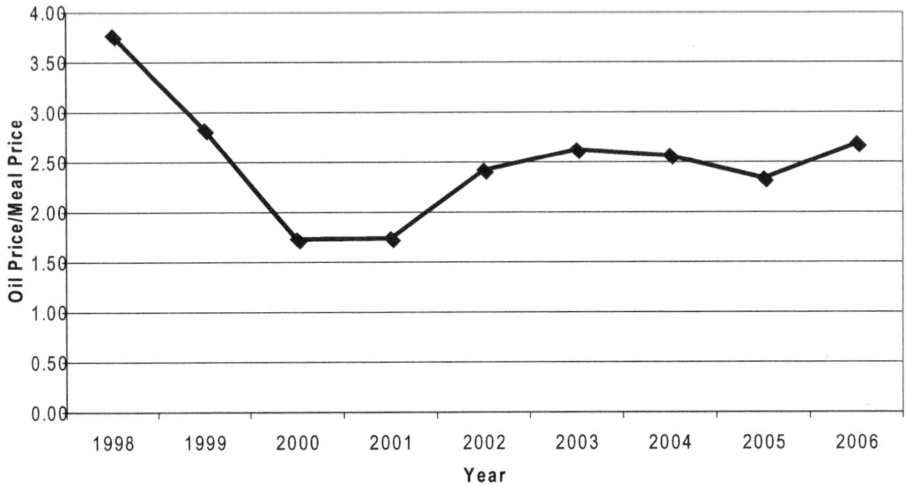

Fig. 21.17. Ratio of price of oil to the price of meal at Paranagua, Brazil (1998–2006) (Source: IBGE, 2007).

close second with 11 plants, representing 9 companies, 28,300 MT/day (31,200 t/day) of capacity, and also 20% of the nation's output.

Plant size increased in size as the industry moved from the Southeast to the Center-West. The two largest plants in the country are located in Mato Grosso and have a capacity of 6,500 MT/day (7,200 t/day) and 5,000 MT/day (5,500 t/day), respectively. The average plant size in the new growth states of the Center-West is about 1.5 times the size of the plants in the Southeast. Center-West plants average about 1,900 MT/day (2,100 t/day) of capacity, while the older plants in the Southeast average about 1,200 MY/day (1,300 t/day). The industry remains relatively unconcentrated: 20% of the crush plants produce 40% of the nation's soybean meal (Fig. 21.19), multiple firms operate any single state, and the average two-firm concentration level for the six leading processing states is 48%.

Crush Margin

Processors actively use risk- management tools to help manage the volatility of prices, and to hedge against shortages and unfavorable pricing as local supplies become scarce. Additionally, crush does not uniformly take place year-round because local production is seasonal, and, as a result, prices of soybeans, soybean meal, and oil are seasonal as well. Globally, two harvest seasons occur, one in the Northern Hemisphere and one in the Southern Hemisphere. The seasonality and the relative movement among the three products determine the decision whether or not to crush. Leveraging the two harvest seasons is a powerful incentive behind the globalization of soybean processing

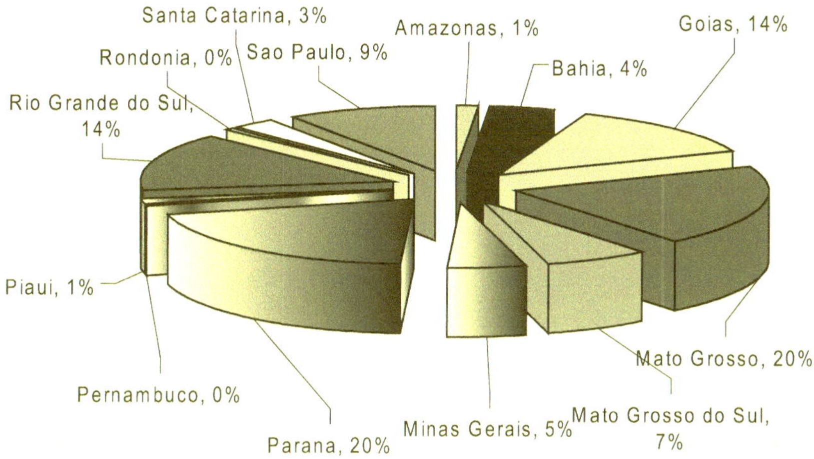

Fig. 21.18. Brazilian soybean crushing capacity (% of national total) in 2006 (Source: Hinrichsen, 2006).

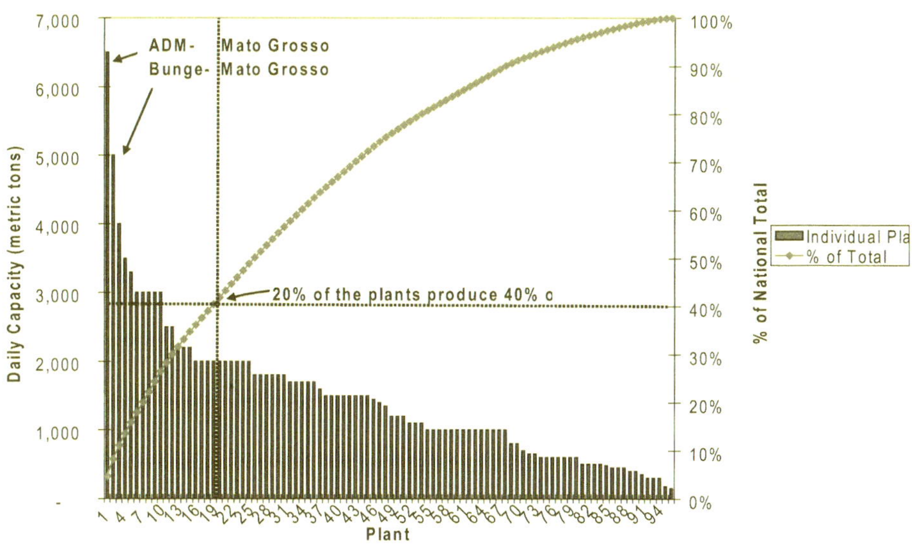

Fig. 21.19. Overview of Brazilian crushing plants in 2006 (N = 96) (Source: Hinrichsen, 2006).

plant investment. Strategically locating allows a processor's operating season to more closely match harvest, when raw material prices are at their lowest. Storing soybeans for processing later in the season can reduce the competitiveness of a plant. The cost of storage reduces already low margins and places the processor in an unfavorable competitive position with competition in the other hemisphere. For example, China, as the leading global importer of whole beans, switches its source of supply with the season to take advantage of hemispheric price differences.

The average monthly crush margin at the port of Paranagua, Brazil, was $31.55/MT or 14.55% (Fig. 21.20). The margin during the Southern harvest season of March–August was 42% higher (17% compared to 12%) than during the Northern harvest season. In 2001, margins averaged close to $38/MT (19%) as the ratio of the soybean meal price to soybean price exceeded 1.10 (Fig. 21.21). While, in 1999, the ratio of meal to soybeans sank to 0.89, reducing margins to $25/MT.

Soybean crush margin is principally driven by soybean prices with a correlation coefficient between the two of -.47 (Table 21.3). The relationship between meal and oil prices and crush margin is not statistically significant. So, operating a plant when meal and oil prices are high does not guarantee satisfactory margins, if soybean prices are not sufficiently low. Therefore, processors are most profitable when intensifying their operations close to harvest when soybean prices are at their lowest. As a result, Brazilian crushers produce the most meal, 10% their annual total, in May as local grain prices remain low, yet meal and oil prices rise as the Northern Hemisphere processing season comes to a close (Fig. 21.22).

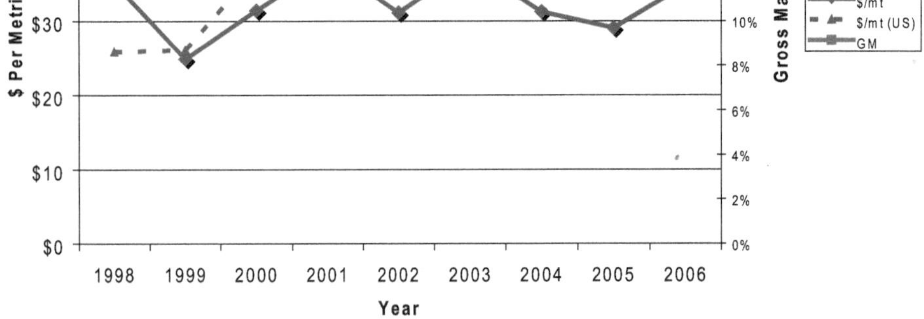

Fig. 21.20. Average crush margins in Brazil (1998–2006) (Source: ABIOVE, 2007; author's calculations).

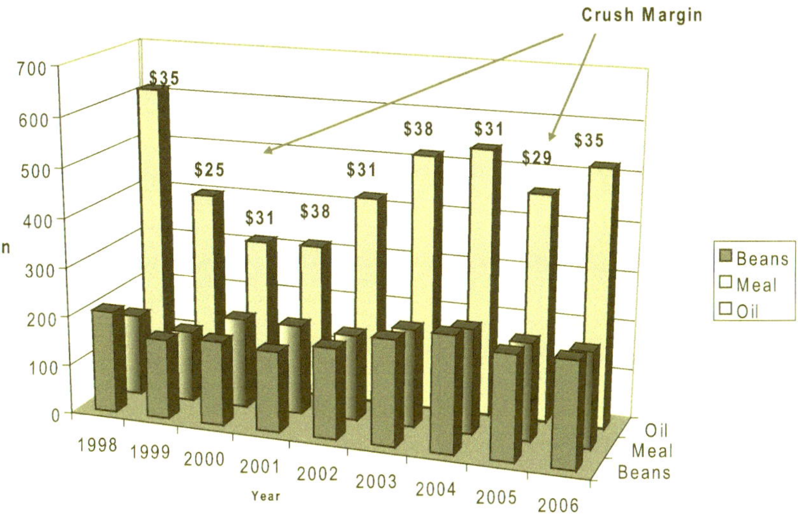

Fig. 21.21. Annual average soybean, oil and meal prices (1998–2006) (Source: IBGE, 2007).

Crush margin variability is greatest in the months leading up to the soybean harvest with several monthly periods having negative margins (Fig. 21.23). The preharvest months in Brazil have crush margin coefficients of variation over one with a range in prices more than double the average price (Fig. 21.24). Oftentimes, crushers choose to shut down during the preharvest period because of the variable of prices and competition from overseas (Ciappa et al., 2005).

Trade

Brazil is the second-leading soybean, meal, and oil exporter in the world by providing 35% of the world's trade in soybeans, 26% of the meal, and 26% of the oil (Fig. 21.25). Argentina is the leader in the export of meal and oil, while the United States is the number- one exporter of raw soybeans. The crushing sectors in Brazil and Argentina are still decidedly export- oriented because of a very small domestic agro-industrial capacity. The United States is the opposite with a domestic-oriented crushing sector that serves a large domestic agro-industrial complex involving livestock and food production.

China is the leading importer of soybeans and soybean oil, with 43 and 18% of world trade, respectively (FAO, 2007). Brazil provides 25% of China's soybean needs and 100% of the needs of The Netherlands, the world's second-largest importer (Table 21.4). Brazil also provides The Netherlands, the world's leading meal importer, with 71% of its imports, and France, the number-two importer, with 70%. Brazil

Table 21.3. Statistical Relationships* Between Soybeans, Meal, Oil, and Crush Margin in Brazil (1998–2006)**

	Correlation Coefficient			
	Soybeans	Meal	Oil	Crush Margin
Soybeans		.68	.71	-.47
Meal			.31	.11
Oil				.-04
Crush Margin				
	Regression t-Statistics			
Independent Variable				
Soybeans		9.6***	10.9***	5.4***
Meal				1.2
Oil			3.4***	-0.4
Crush Margin				

* Monthly averages
** Location (interior Pananagua soybean prices, port F.O.B. meal and oil prices)
*** Significant at the .01 level.

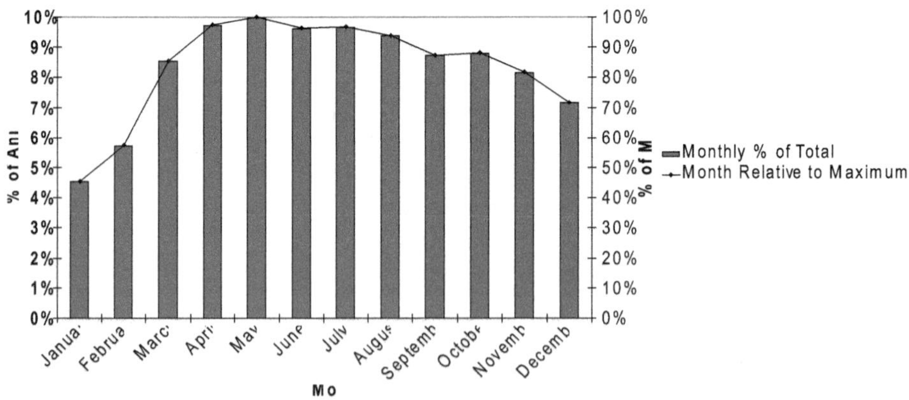

Fig. 21.22. Average monthly plant utilization in Brazil (2000–2006) (Source: ABIOVE, 2007).

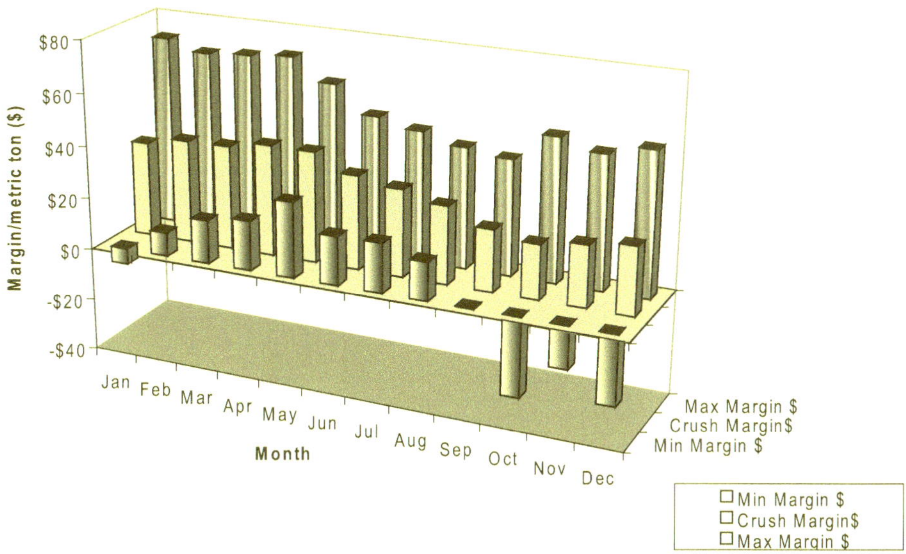

Fig. 21.23. Average monthly crush margin range (1998–2006)(Source: IBGE, 2007).

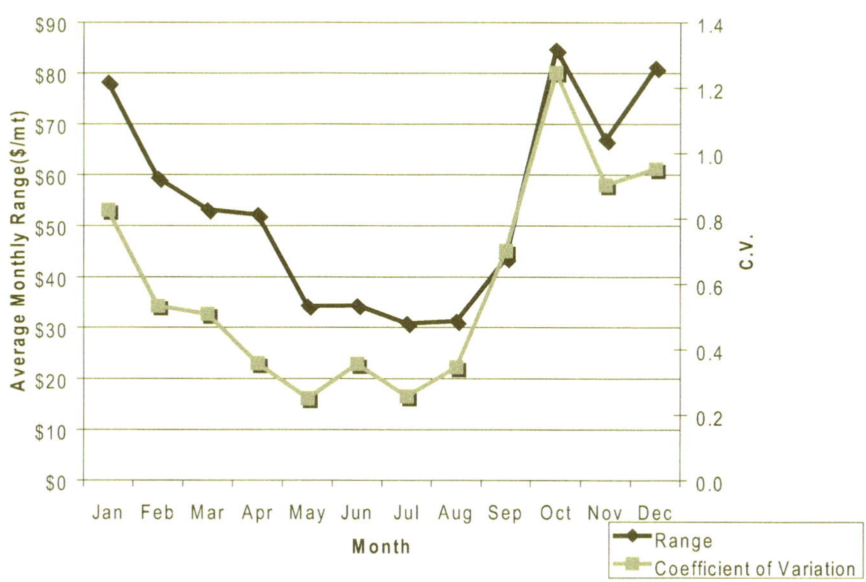

Fig. 21.24. Seasonal variability in Brazil crush margins (1998–2006) (Source: IBGE, 2007).

Fig. 21.25. The world's leading soybean, oil, and meal exporters in 2005 (Source: FAO, 2007).

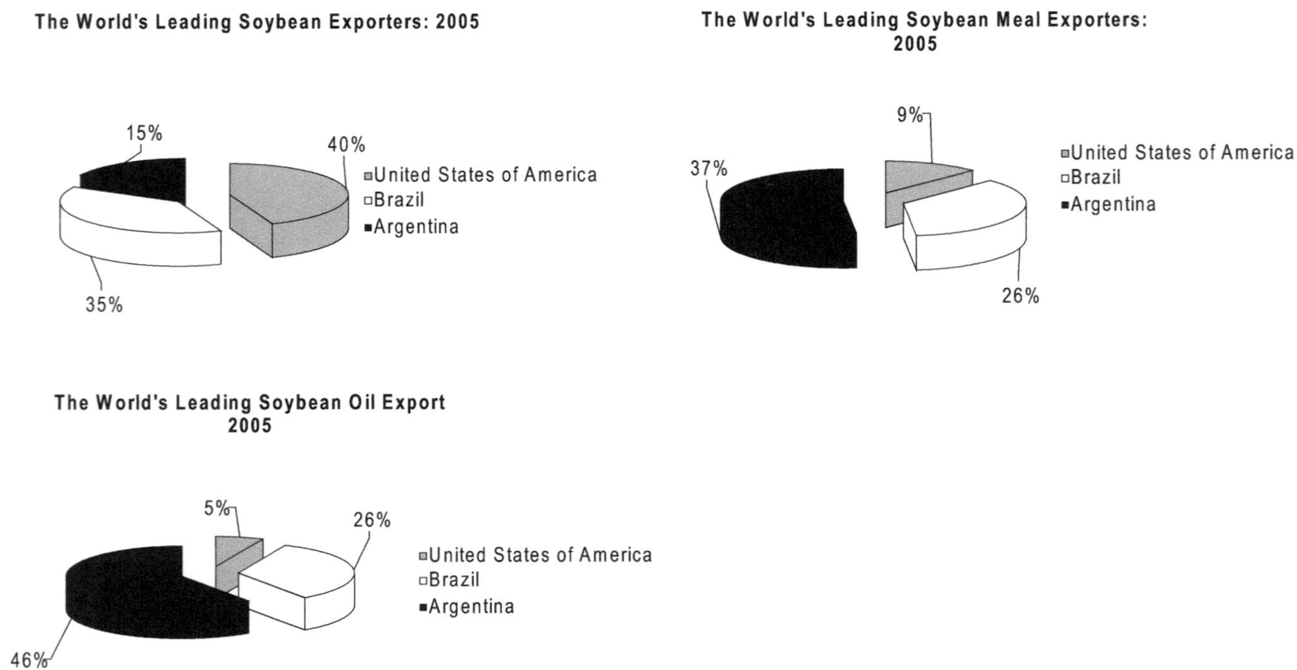

provides 100% of Iran's imported soybean oil needs, its number-one export customer. India and China are the number-two and number-three soybean oil customers for Brazil.

Since the mid-1990s, China has dedicated itself to increasing its processing capacity. It shifted domestic policy to favor soybean meal for livestock feed, and soybean oil for human consumption. This policy causes China to import large quantities of soybeans, mostly from Brazil and the United States to fuel its growing processing industry. China's demand combined with Brazil's relatively small animal industry results in Brazil exporting 73% of the soybeans it produces (production + a small amount of imports), 48% in the form of meal and 52% as raw soybeans.

Argentina also is a major exporter with superior logistics due to geography. The main soybean- growing region lies within 480 km (300 miles) of the deep water port at Rosario. Argentina also maintains tax policies favoring processing over direct grain exportation. Like Brazil, Argentina exports most of its soybeans (97%), but, in contrast, 74% is in the form of meal and only 26% are raw soybeans.

Alternatively, the United States is primarily a domestic user of its soybeans and soybean meal, producing six times the soybean meal it exports (Fig. 21.26). This, in part, is due to its large domestic agro-industrial complex that increasingly focuses on domestic demand, not exports. Correspondingly, Argentina and Brazil export most of what they produce due to a much smaller local agro-industrial/livestock complex.

Conclusion

The development of low-latitude soybean germ plasm by EMBRAPA, Brazil's national research agency, is one of the most important technological innovations in agriculture since the green revolution. The doubling of production in the last 10 yr is attributable to the development of soybean technology and practices adapted to the Cerrado region in the Center-West region of the country. The Cerrado is a low growth savannah that, prior to the 1970s, was not considered suitable for broad acre crop production.

Most of the world's reserves of arable land reside in the low-latitude region. Important as well, most of the world's malnutrition occurs in low-latitude regions. Low-latitude technologies like those developed by public and private researchers in Brazil will play an increasing role addressing the world's fast-growing food and bioenergy demand. The shift to the low latitudes is already well underway as 68% of the crops and 53% of the livestock products come from developing countries, an increase of 14% in the last 10 yr. (FAO, 2007).

Soybean farmers in the Center-West currently have a comparative advantage in soybean production compared with producers in the United States because of their low opportunity costs. U.S. farmers have increasingly chosen to grow corn, while farmers in the Center-West presently have few better alternatives than soybeans. As result, the tendency will be for the United States' role in the global soy complex to

Table 21.4. Brazil's Leading Soybean Trading Partners in 2007 (Source: SECEX, 2007)

Soybeans			Meal			Oil		
Leading Trading Partner	% of Brazilian Exports	% of Country Imports	Leading Trading Partner	% of Brazil-ian Exports	% of Country Imports	Leading Trading Partner	% of Brazilian Exports	% of Country Imports
(1) China	33%	25%	(1) Netherlands	26%	71%	(1) Iran	28%	100%
(2) Netherlands	24%	100%	(2) France	23%	70%	(2) India	16%	26%
(3) Spain	10%	81%	(3) Thailand	8%	54%	(3) China	14%	20%

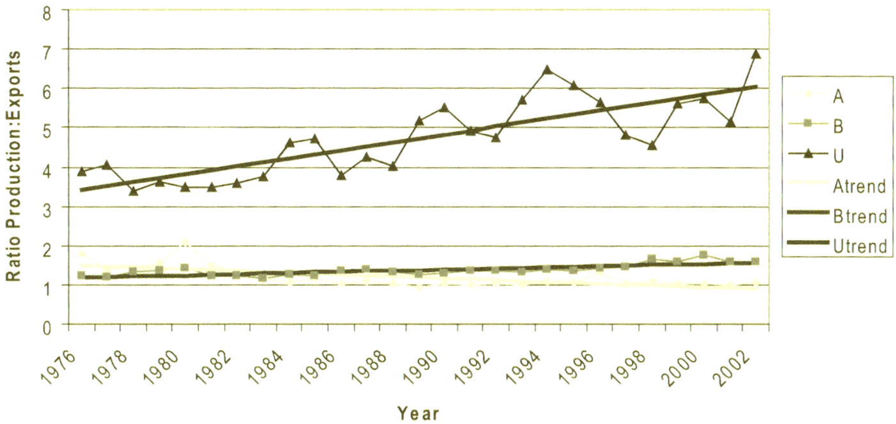

Fig. 21.26. Soybean meal domestic use rates (Source: FAO, 2005; author's calculations).

decline, while Brazil's will increase.

Poor infrastructure will continue to limit the industry's growth even though new low-latitude soybean technologies and global protein demand growth portend a bright future for the Brazilian soybean complex. Poor infrastructure indirectly affects soybean producers as the pace of agro-industrial investment is reduced, which, in turn, limits market opportunities in the region.

Balancing environmental stewardship, while meeting the world's increasing demand for soybeans, is probably the greatest challenge facing the industry. The state of Mato Grosso borders some of the most ecologically important rain forests in the world. One quarter of the land in Mato Grosso is classified as rain forest, yet most of the land is Cerrado or dry land forest, both of which are suitable for soybean production. A main north-south highway, the BR163 that connects Mato Grosso to the northern port city of Santarem, is highly controversial because it passes through major rain-forest regions and has the potential to contribute to ecosystem degradation. The road is critical for the economic development of the land-locked state of Mato Grosso. At the same time, demand for soybeans has never been greater. So while the market is signaling for Brazilians to expand soybean production in the Cerrado region, some policymakers and NGOs are concerned about the environmental impact.

References

ABIOVE. The Association of Oilseed Processors of Brazil. http://www.abiove.com.br. 2007.

Aliceweb, http://aliceweb.desenvolvimento.gov.br/default.asp. 2007.

Barchart.com. http://barchart.com/. 2007.

Ciappa, C.; P.D. Goldsmith; C.M. Acosta. Understanding the Crusher's Problem. Selected paper.

The Annual Meeting of the International Food and Agribusiness Management Association. Chicago. June 25, 2005.

FAO. FAOSTATS, Food and Agricultural Organization. http://faostat.fao.org/site/336/default.aspx. 2005.

FAO. FAOSTATS, Food and Agricultural Organization. http://faostat.fao.org/site/336/default.aspx. 2007.

Goldsmith, P.D.; H.L.G. Gastaldi; J. Martines, Agro-industrial Development in Mato Grosso: Cluster Economies, Social Responsibility, and the Case of the Soybean Complex. Presentation at the Bienal, Cuiaba, Mato Grosso, Brazil. August 2006.

Goldsmith, P.D.; H.L.G. Gastaldi; J. Martines; T. Masuda. The soybean complex in mato grosso. Int. Food & Agribus. Man. Rev. (IFAMR) Under Review. January 2008.

Goldsmith, P.D.; R. Hirsch. The Brazilian soybean complex. Choices. July 2006.

Goldsmith, P.D.; B. Li; J. Fruin; R. Hirsch. Global shifts in agro-industrial capital and the case of soybean crushing: Implications for managers and policy makers. Int. Food & Agribus. Man. Rev. (IFAMR) 2004, 7, 87–115.

Goldsmith, P.D.; G. Ramos; C. Steiger. Intellectual property piracy in a north-south context: Empirical evidence. Agric. Economics 2006, 35, 335–349.

Goldsmith, P.D.; G. Schnitkey. Soybean rust scenario model: Crop year 2005 decision-making requires planning. Feedstuffs 2005, 77 (10), March 7, 2005.

Hinrichsen, J.J. 2006 Yearbook # 41. J.J. Hinrichsen, ed: 136 pages.

Hirimoto, D. 2007. Executive Director, Mato Grosso Foundation. Personal Communication.

Hirsch, R. Regional Competitiveness Analysis of the Soybean Industry and Transportation Infrastructure in Brazil. Masters Thesis. The Department of Agricultural and Consumer Economics, University of Illinois. 2004.

IBGE, 2007 (Brazilian Institute of Geography and Statistics) http://www.ibge.gov.br/english/

Kiel, R. Personal Communication. 2005

Ma, E. Cost of Soybean Production in Mato Grosso. Undergraduate Thesis, University of Sao Paulo and University of Illinois. 2006.

Schnepf, R.D.; E. Dohlman; C. Bolling. Agriculture in Brazil and Argentina, WRS-01-3. Economic Research Service, USDA: Washington, DC, 2001; p. 77.

SECEX. http://www.desenvolvimento.gov.br/sitio/interna/index.php?area=5. 2007.

Reviewers

We gratefully acknowledge the following authorities, who graciously gave their time and expertise to review, clarify, and correct errors in the book contents.

Dan Anderson, Director of Asian Operations, Crown Iron Works Company, Minneapolis, MN

William L. Boatright, Associate Professor, Animal and Food Sciences, University of Kentucky

Kristjan Bregendahl, Assistant Professor, Animal Science, Iowa State University

Thomas J. Brumm, Associate Professor, Agricultural and Biosystems Engineering, Iowa State University

Gary L. Cromwell, Professor, Animal and Food Science, University of Kentucky

Joe Endres, Research Fellow, Central Soya, Ft. Wayne, IN (retired)

Jose Gerde, Graduate Research Assistant, Food Science & Human Nutrition, Iowa State University

William H. Johnson, Professor, Agricultural Engineering and Director, Engineering Experiment Station, Kansas State University (retired)

Michael J. Haas, Research Chemist, Eastern Regional Research Center, U.S. Department of Agriculture

Clifford A. Hall III, Assistant Professor, Cereal and Food Science, North Dakota State University

Earl Hammond, Professor Emeritus, Food Science & Human Nutrition, Iowa State University

Jules Janick, James Troop Distinguished Professor in Horticulture, Horticulture and Landscape Architecture, Purdue University

Tim Kemper, President and CEO, DeSmet Ballestra North America, Marietta, GA

Phil Kerr, The Solae Company, St. Louis, MO

Gary R. List, Research Chemist, National Center for Agricultural Utilization Research, U.S. Department of Agriculture

Keshun Liu, Research Chemist, Grain Chemistry and Utilization Laboratory, USDA-ARS, Aberdeen, ID
John C. McKinney, Illinois Crop Improvement Association, Champaign, IL
Gary Munkvold, Associate Professor and Seed Science Endowed Chair, Plant Pathology, Iowa State University
Patricia A. Murphy, University Professor, Food Science & Human Nutrition, Iowa State University
Andrew Proctor, Professor, Food Science, University of Arkansas
Rudy Pruszko, Center for Industrial Research and Service, Iowa State University Extension, Dubuque, IA
Graeme R. Quick, Professor, Agricultural and Biosystems Engineering, Iowa State University (retired)
R. Christopher Schroeder, Partner, Centrec Consulting Group, Savoy, IL
Brent H. Shanks, Professor, Chemical and Biological Engineering, Iowa State University
Randy C. Shoemaker, Research Geneticist, USDA-ARS, Collaborating Professor, Agronomy, Iowa State University
William Shurtleff, Soyinfo Center, Layfayette, CA
Steve Sonka, Ph.D., Interim Vice Chancellor for Public Engagement, University of Illinois
Greg L. Tylka, Professor, Plant Pathology, Iowa State University
C.Y. Wang, Department Head, Department of Nutrition, Food Science & Hospitality, South Dakota State University
Mark Westgate, Professor, Agronomy, Iowa State University
Maurice A. Williams, Anderson International Corporation, Cleveland, OH
Lester A. Wilson, Professor, Food Science & Human Nutrition, Iowa State University
Richard F. Wilson, USDA-ARS, Beltsville, MD (retired)

Contributors

Larry L. Berger, Professor, Department of Animal Sciences, University of Illinois, Urbana-Champaign, IL 61801

Carl J. Bern, University Professor, Department of Agricultural and Biosystems Engineering, Iowa State University, Ames, IA 50011

Nicolas A. Deak, Research Scientist, Center for Crops Utilization Research, Iowa State University, Ames, IA 50011

James K. Drackley, Professor, Department of Animal Sciences, University of Illinois, Urbana-Champaign, IL 61801

Sevim Z. Erhan, Supervisory Research Chemist, U.S. Department of Agriculture, Agricultural Research Service, National Center for Agricultural Utilization Research, Food and Industrial Oil Research Unit, Peoria, IL 61604

George C. Fahey, Jr., Professor, Department of Animal Sciences, University of Illinois, Urbana-Champaign, IL 61801

Brent D. Flickinger, Senior Research Manager, Nutritional Sciences, Archer Daniels Midland Co., Decatur, IL 62521

Jose A. Gerde, Research Assistant, Department of Food Science and Human Nutrition, Iowa State University, Ames, IA 50011

Peter D. Goldsmith, Soybean Industry Endowed Associate Professor in Agricultural Strategy, Department of Agricultural and Consumer Economics; Executive Director, National Soybean Research Laboratory, University of Illinois, Urbana-Champaign, IL 61801

H. Mark Hanna, Extension Agricultural Engineer, Department of Agricultural and Biosystems Engineering, Iowa State University, Ames, IA 50011

David C. Hernot, Research Fellow, Department of Animal Sciences, University of Illinois, Urbana-Champaign, IL 61801

Alison M. Hill, Research Fellow, Department of Nutritional Sciences, Penn State University, University Park 16802

Theodore Hymowitz, Professor Emeritus, Department of Crop Sciences, University of Illinois, Urbana-Champaign, IL 61801

Lawrence A. Johnson, Professor, Department of Food Science and Human Nutrition; Director, Center for Crops Utilization Research, Iowa State University, Ames, IA 50011

Heather I. Katcher, Research Assistant, Department of Nutritional Sciences, Pennsylvania State University, University Park 16802

Gerhard Knothe, Research Chemist, U.S. Department of Agriculture, Agricultural Research Service, National Center for Agricultural Utilization Research, Food and Industrial Oil Research Unit, Peoria, IL 61604

Penny M. Kris-Etherton, Distinguished Professor of Nutrition, Department of Nutritional Sciences, Pennsylvania State University, University Park 16802

KeShun Liu, Research Chemist, U.S. Department of Agriculture, Agricultural Research Service, Grain Chemistry and Utilization Laboratory, Aberdeen, ID 83210

Edmund W. Lusas, Professor Emeritus, Department of Soil and Crop Sciences; Director (retired), Food Protein Research and Development Center, Texas A&M University, College Station, TX 77843

Randall G. Luttrell, Professor, Department of Entomology, University of Arkansas, Fayetteville, AR 72701

Ingomar S. Middelbos, Research Associate, Department of Animal Sciences, University of Illinois, Urbana-Champaign, IL 61801

Patricia A. Murphy, University Professor, Department of Food Science and Human Nutrition, Iowa State University, Ames, IA 50011

Deland J. Myers, Professor, School of Food Systems, North Dakota State University, Fargo, ND 58105

Richard D. O'Brien, Consultant, Schulensburg, TX 78956

James H. Orf, Professor, Department of Agronomy and Plant Genetics, University of Minnesota, St. Paul, MN 55108

Carl M. Parsons, Professor, Department of Animal Sciences, University of Illinois, Urbana-Champaign, IL 61801

Marvin R. Paulsen, Professor Emeritus, Department of Agricultural & Biological Engineering, University of Illinois, Champaign-Urbana, IL 61801

Khee Choon Rhee, Professor Emeritus, Department of Soil and Crop Sciences; Director (retired), Food Protein Research and Development Center, Texas A&M University, College Station, TX 77843

John Rupe, Professor, Department of Plant Pathology, University of Arkansas, Fayetteville, AR 72701

John F. Schmitz, Research Assistant, Center for Crops Utilization Research, Iowa State University, Ames, IA 50011

Brajendra K. Sharma, Research Chemist, U.S. Department of Agriculture, Agricultural Research Service, National Center for Agricultural Utilization Research, Food and Industrial Oil Research Unit, Peoria, IL 61604; Senior Research Chemical Engineer, Department of Chemical Engineering, Pennsylvania State University, University Park 16802

Hans H. Stein, Associate Professor, Department of Animal Sciences, University of Illinois, Urbana-Champaign, IL 61801

Jon Van Gerpen, Professor and Head, Department of Biological and Agricultural Engineering, University of Idaho, Moscow, ID 83844

Tong Wang, Associate Professor, Department of Food Science and Human Nutrition, Iowa State University, Ames, IA 50011

Kathleen A. Warner, Research Chemist, U.S. Department of Agriculture, Agricultural Research Service, National Center for Agricultural Utilization Research, Food and Industrial Oil Research Unit, Peoria, IL 61604

Pamela J. White, University Professor, Department of Food Science and Human Nutrition, Iowa State University, Ames, IA 50011

William F. Wilcke, Professor, Department of Bioproducts and Biosystems Engineering, University of Minnesota, St. Paul, MN 55108

Index

A

C

D

G

H

M

N

O

P

Q

R

S

T

U

V

FSC
www.fsc.org
MIX
Papier | Fördert gute Waldnutzung
FSC® C083411

Kolibri 360 GmbH
Ferdinand-Jühlke-Straße 7
99095 Erfurt, Deutschland
produktsicherheit@kolibri360.de